Modern Mathematical Statistics with Applications

Jay L. Devore

California Polytechnic State University

Kenneth N. Berk

Illinois State University

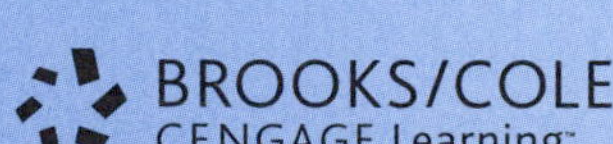

Australia • Brazil • Japan • Korea • Mexico • Singapore • Spain • United Kingdom • United States

Modern Mathematical Statistics with Applications
Jay L. Devore and Kenneth N. Berk

Acquisitions Editor: Carolyn Crockett

Editorial Assistant: Daniel Geller

Technology Project Manager: Fiona Chong

Senior Assistant Editor: Ann Day

Marketing Manager: Joseph Rogove

Marketing Assistant: Brian Smith

Marketing Communications Manager: Darlene Amidon-Brent

Manager, Editorial Production: Kelsey McGee

Creative Director: Rob Hugel

Art Director: Lee Friedman

Print Buyer: Rebecca Cross

Permissions Editor: Joohee Lee

Production Service and Composition: Newgen

Text Designer: Carolyn Deacy

Copy Editor: Anita Wagner

Cover Designer: Eric Adigard

Cover Image: Carl Russo

Library of Congress Control Number: 2005929405

ISBN-13: 978-0-534-40473-4

ISBN-10: 0-534-40473-1

Cengage Learning
10 Davis Drive
Belmont, CA 94002-3098
USA

Cengage Learning is a leading provider of customized learning solutions with office locations around the globe, including Singapore, the United Kingdom, Australia, Mexico, Brazil, and Japan. Locate your local office at: **www.cengage.com/global**

Cengage Learning products are represented in Canada by Nelson Education, Ltd.

To learn more aboutBrooks/Cole, visit **www.cengage.com/brookscole**

Purchase any of our products at your local college store or at our preferred online store **www.ichapters.com**

Printed in the United States of America
4 5 6 7 11 10

To my wife Carol

whose continuing support of my writing efforts over the years has made all the difference.

To my wife Laura

who, as a successful author, is my mentor and role model.

About the Authors

Jay L. Devore

Jay Devore received a B.S. in Engineering Science from the University of California, Berkeley, and a Ph.D. in Statistics from Stanford University. He previously taught at the University of Florida and Oberlin College, and has had visiting positions at Stanford, Harvard, the University of Washington, and New York University. He has been at California Polytechnic State University, San Luis Obispo, since 1977, where he is currently a professor and chair of the Department of Statistics.

Jay has previously authored five other books, including *Probability and Statistics for Engineering and the Sciences*, currently in its 6th edition. He is a Fellow of the American Statistical Association, an associate editor for the *Journal of the American Statistical Association*, and received the Distinguished Teaching Award from Cal Poly in 1991. His recreational interests include reading, playing tennis, traveling, and cooking and eating good food.

Kenneth N. Berk

Ken Berk has a B.S. in Physics from Carnegie Tech (now Carnegie Mellon) and a Ph.D. in Mathematics from the University of Minnesota. He is Professor Emeritus of Mathematics at Illinois State University and a Fellow of the American Statistical Association. He founded the Software Reviews section of *The American Statistician* and edited it for six years. He served as secretary/treasurer, program chair, and chair of the Statistical Computing Section of the American Statistical Association, and he twice co-chaired the Interface Symposium, the main annual meeting in statistical computing. His published work includes papers on time series, statistical computing, regression analysis, and statistical graphics and the book *Data Analysis with Microsoft Excel* (with Patrick Carey).

Brief Contents

Contents

Preface

Purpose

Our objective is to provide a postcalculus introduction to the discipline of statistics that

- Has mathematical integrity and contains *some* underlying theory.
- Shows students a broad range of applications involving real data.
- Is very current in its selection of topics.
- Illustrates the importance of statistical software.
- Is accessible to a wide audience, including mathematics and statistics majors (yes, there are a few of the latter), prospective engineers and scientists, and those business and social science majors interested in the quantitative aspects of their disciplines.

A number of currently available mathematical statistics texts are heavily oriented toward a rigorous mathematical development of probability and statistics, with much emphasis on theorems, proofs, and derivations. The emphasis is more on mathematics than on statistical practice. Even when applied material is included, the scenarios are often contrived (many examples and exercises involving dice, coins, cards, widgets, or a comparison of treatment A to treatment B).

So in our exposition we have tried to achieve a balance between mathematical foundations and statistical practice. Some may feel discomfort on grounds that because a mathematical statistics course has traditionally been a feeder into graduate programs in statistics, students coming out of such a course must be well prepared for that path. But that view presumes that the mathematics will provide the hook to get students interested in our discipline. That may happen for a few mathematics majors. However, our experience is that the application of statistics to real-world problems is far more persuasive in getting quantitatively oriented students to pursue a career or take further coursework in statistics. Let's first draw them in with intriguing problem scenarios and applications. Opportunities for exposing them to mathematical foundations will follow in due course. In our view it is more important for students coming out of this course to be able to carry out and interpret the results of a two-sample t test or simple regression analysis than to manipulate joint moment generating functions or discourse on various modes of convergence.

Content

The book certainly does include core material in probability (Chapter 2), random variables and their distributions (Chapters 3–5), and sampling theory (Chapter 6). But our desire to balance theory with application/data analysis is reflected in the way the book starts out, with a chapter on descriptive and exploratory statistical techniques rather than an immediate foray into the axioms of probability and their consequences. After the distributional infrastructure is in place, the remaining statistical chapters cover the basics of inference. In addition to introducing core ideas from estimation and hypothesis testing (Chapters 7–10), there is emphasis on checking assumptions and looking at the data prior to formal analysis. Modern topics such as bootstrapping, permutation tests, residual analysis, and logistic regression are included. Our treatment of regression, analysis of variance, and categorical data analysis (Chapters 11–13) is definitely more oriented to dealing with real data than with theoretical properties of models. We also show many examples of output from commonly used statistical software packages, something noticeably absent in most other books pitched at this audience and level. (Figures 10.1 and 11.14 have been reproduced here for illustrative purposes.) For example, the first section on multiple regression toward the end of Chapter 12 uses no matrix algebra but instead relies on output from software as a basis for making inferences.

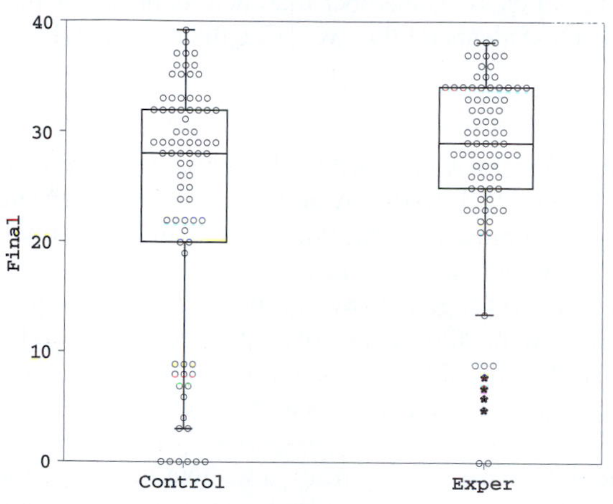

Figure 10.1

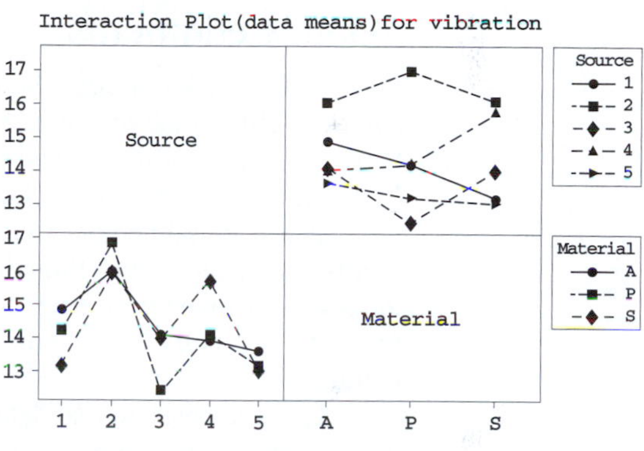

Figure 11.14

Mathematical Level

The challenge for students at this level should lie with mastery of statistical concepts as well as with mathematical wizardry. Consequently, the mathematical prerequisites and demands are reasonably modest. Mathematical sophistication and quantitative reasoning ability are, of course, crucial to the enterprise. Students with a solid grounding in univariate calculus and some exposure to multivariate calculus should feel comfortable with what we are asking of them. The several sections where matrix algebra appears (transformations in Chapter 5 and the matrix approach to regression in the last section of Chapter 12) can easily be deemphasized or skipped entirely.

Our goal is to redress the balance between mathematics and statistics by putting more emphasis on the latter. The concepts, arguments, and notation contained herein will certainly stretch the intellects of many students. And a solid mastery of the material will be required in order for them to solve many of the roughly 1300 exercises included in the book. Proofs and derivations are included where appropriate, but we think it likely that obtaining a conceptual understanding of the statistical enterprise will be the major challenge for readers.

Recommended Coverage

There should be more than enough material in our book for a year-long course. Those wanting to emphasize some of the more theoretical aspects of the subject (e.g., moment generating functions, conditional expectation, transformations, order statistics, sufficiency) should plan to spend correspondingly less time on inferential methodology in the latter part of the book. We have tried to help instructors by marking certain sections as optional (using an *). "Optional" is not synonymous with "unimportant"; an * is just an indication that what comes afterward makes at most minimal use of what is contained in a section so marked. Other than that, we prefer to rely on the experience and tastes of individual instructors in deciding what should be presented. We would also like to think that students could be asked to read an occasional subsection or even section on their own and then work exercises to demonstrate understanding, so that not everything would need to be presented in class. Remember that there is never enough time in a course of any duration to teach students all that we'd like them to know!

Acknowledgments

We gratefully acknowledge the plentiful feedback provided by the following reviewers: Bhaskar Bhattacharya, Southern Illinois University; Ann Gironella, Idaho State University; Tiefeng Jiang, University of Minnesota; Iwan Praton, Franklin & Marshall College; and Bruce Trumbo, California State University, East Bay.

A special salute goes to Bruce Trumbo for going way beyond his mandate in providing us an incredibly thoughtful review of 40+ pages containing many wonderful ideas and pertinent criticisms. Matt Carlton, a Cal Poly colleague of one of the authors, has provided stellar service as an accuracy checker, and has also prepared a solutions manual.

Our emphasis on real data would not have come to fruition without help from the many individuals who provided us with data in published sources or in personal communications; we greatly appreciate all their contributions.

We very much appreciate the production services provided by the folks at G&S Book Services. Our production editor, Gretchen Otto, did a first-rate job of moving the book through the production process, and was always prompt and considerate in her communications with us. Thanks to our copy editor, Anita Wagner, for employing a light touch and not taking us too much to task for our occasional grammatical and technical lapses. The staff at Brooks/Cole, Cengage Learning has been as supportive on this project as on ones with which we have previously been involved. Special kudos go to Carolyn Crockett, Ann Day, Dan Geller, and Kelsey McGee, and apologies to any whose names were inadvertently omitted from this list.

A Final Thought

It is our hope that students completing a course taught from this book will feel as passionately about the subject of statistics as we still do after so many years in the profession. Only teachers can really appreciate how gratifying it is to hear from a student after he or she has completed a course that the experience had a positive impact and maybe even affected a career choice.

Jay Devore
Ken Berk

CHAPTER ONE

Overview and Descriptive Statistics

Introduction

Statistical concepts and methods are not only useful but indeed often indispensable in understanding the world around us. They provide ways of gaining new insights into the behavior of many phenomena that you will encounter in your chosen field of specialization.

The discipline of statistics teaches us how to make intelligent judgments and informed decisions in the presence of uncertainty and variation. Without uncertainty or variation, there would be little need for statistical methods or statisticians. If the yield of a crop were the same in every field, if all individuals reacted the same way to a drug, if everyone gave the same response to an opinion survey, and so on, then a single observation would reveal all desired information.

An interesting example of variation arises in the course of performing emissions testing on motor vehicles. The expense and time requirements of the Federal Test Procedure (FTP) preclude its widespread use in vehicle inspection programs. As a result, many agencies have developed less costly and quicker tests, which it is hoped replicate FTP results. According to the journal article "Motor Vehicle Emissions Variability" (*J. Air Waste Manag. Assoc.*, 1996: 667–675), the acceptance of the FTP as a gold standard has led to the widespread belief that repeated measurements on the same vehicle would yield identical (or nearly identical) results. The authors of the article applied the FTP to seven vehicles characterized as "high emitters." Here are the results of four hydrocarbon and carbon dioxide tests on one such vehicle:

HC (*g/mile*)	13.8	18.3	32.2	32.5
CO (*g/mile*)	118	149	232	236

The substantial variation in both the HC and CO measurements casts considerable doubt on conventional wisdom and makes it much more difficult to make precise assessments about emissions levels.

How can statistical techniques be used to gather information and draw conclusions? Suppose, for example, that a biochemist has developed a medication for relieving headaches. If this medication is given to different individuals, variation in conditions and in the people themselves will result in more substantial relief for some individuals than others. Methods of statistical analysis could be used on data from such an experiment to determine on the average how much relief to expect.

Alternatively, suppose the biochemist has developed a headache medication in the belief that it will be superior to the currently best medication. A comparative experiment could be carried out to investigate this issue by giving the current medication to some headache sufferers and the new medication to others. This must be done with care lest the wrong conclusion emerge. For example, perhaps the average amount of improvement is identical for the two medications. However, the new medication may be applied to people who have less severe headaches and have less stressful lives. The investigator would then likely observe a difference between the two medications attributable not to the medications themselves, but just to extraneous variation. Statistics offers not only methods for analyzing the results of experiments once they have been carried out but also suggestions for how experiments can be performed in an efficient manner to lessen the effects of variation and have a better chance of producing correct conclusions.

1.1 Populations and Samples

We are constantly exposed to collections of facts, or **data**, both in our professional capacities and in everyday activities. The discipline of statistics provides methods for organizing and summarizing data and for drawing conclusions based on information contained in the data.

An investigation will typically focus on a well-defined collection of objects constituting a **population** of interest. In one study, the population might consist of all gelatin capsules of a particular type produced during a specified period. Another investigation might involve the population consisting of all individuals who received a B.S. in mathematics during the most recent academic year. When desired information is available for all objects in the population, we have what is called a **census**. Constraints on time, money, and other scarce resources usually make a census impractical or infeasible. Instead, a subset of the population—a **sample**—is selected in some prescribed manner. Thus we might obtain a sample of pills from a particular production run as a basis for investigating whether pills are conforming to manufacturing specifications, or we might select a sample of last year's graduates to obtain feedback about the quality of the curriculum.

We are usually interested only in certain characteristics of the objects in a population: the milligrams of vitamin C in the pill, the gender of a mathematics graduate, the age at which the individual graduated, and so on. A characteristic may be categorical, such as gender or year in college, or it may be numerical in nature. In the former case, the *value* of the characteristic is a category (e.g., female or sophomore), whereas in the latter case, the value is a number (e.g., age = 23 years or vitamin C content = 65 mg). A **variable** is any characteristic whose value may change from one object to another in the population. We shall initially denote variables by lowercase letters from the end of our alphabet. Examples include

x = brand of calculator owned by a student

y = number of major defects on a newly manufactured automobile

z = braking distance of an automobile under specified conditions

Data comes from making observations either on a single variable or simultaneously on two or more variables. A **univariate** data set consists of observations on a single variable. For example, we might determine the type of transmission, automatic (A) or manual (M), on each of ten automobiles recently purchased at a certain dealership, resulting in the categorical data set

M A A A M A A M A A

The following sample of lifetimes (hours) of brand D batteries put to a certain use is a numerical univariate data set:

5.6 5.1 6.2 6.0 5.8 6.5 5.8 5.5

We have **bivariate** data when observations are made on each of two variables. Our data set might consist of a (height, weight) pair for each basketball player on a team, with the first observation as (72, 168), the second as (75, 212), and so on. If a kinesiologist determines the values of x = recuperation time from an injury and y = type of injury, the resulting data set is bivariate with one variable numerical and the other categorical. **Multivariate** data arises when observations are made on more than two variables. For example, a research physician might determine the systolic blood pressure, diastolic blood pressure, and serum cholesterol level for each patient participating in a study. Each observation would be a triple of numbers, such as (120, 80, 146). In many multivariate data sets, some variables are numerical and others are categorical. Thus the annual automobile issue of *Consumer Reports* gives values of such variables as type of vehicle (small, sporty, compact, midsize, large), city fuel efficiency (mpg), highway fuel efficiency (mpg), drive train type (rear wheel, front wheel, four wheel), and so on.

Branches of Statistics

An investigator who has collected data may wish simply to summarize and describe important features of the data. This entails using methods from **descriptive statistics**. Some of these methods are graphical in nature; the construction of histograms, boxplots, and scatter plots are primary examples. Other descriptive methods involve calculation of numerical summary measures, such as means, standard deviations, and correlation coefficients. The wide availability of statistical computer software packages has made these tasks much easier to carry out than they used to be. Computers are much more efficient than human beings at calculation and the creation of pictures (once they have received appropriate instructions from the user!). This means that the investigator doesn't have to expend much effort on "grunt work" and will have more time to study the data and extract important messages. Throughout this book, we will present output from various packages such as MINITAB, SAS, and S-Plus.

Example 1.1 The tragedy that befell the space shuttle *Challenger* and its astronauts in 1986 led to a number of studies to investigate the reasons for mission failure. Attention quickly focused on the behavior of the rocket engine's O-rings. Data consisting of observations on x = O-ring temperature (°F) for each test firing or actual launch of the shuttle rocket engine appears on the following page (*Presidential Commission on the Space Shuttle Challenger Accident*, Vol. 1, 1986: 129–131).

```
Stem-and-leaf of temp N = 36
Leaf Unit = 1.0
  1   3   1
  1   3
  2   4   0
  4   4   59
  6   5   23
  9   5   788
 13   6   0113
 (7)  6   6777789
 16   7   000023
 10   7   556689
  4   8   0134
```

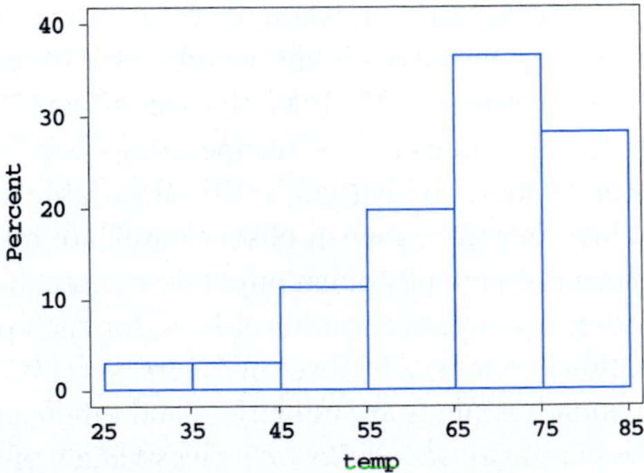

Figure 1.1 A MINITAB stem-and-leaf display and histogram of the O-ring temperature data

84 49 61 40 83 67 45 66 70 69 80 58
68 60 67 72 73 70 57 63 70 78 52 67
53 67 75 61 70 81 76 79 75 76 58 31

Without any organization, it is difficult to get a sense of what a typical or representative temperature might be, whether the values are highly concentrated about a typical value or quite spread out, whether there are any gaps in the data, what percentage of the values are in the 60's, and so on. Figure 1.1 shows what is called a *stem-and-leaf display* of the data, as well as a *histogram*. Shortly, we will discuss construction and interpretation of these pictorial summaries; for the moment, we hope you see how they begin to tell us how the values of temperature are distributed along the measurement scale. The lowest temperature is 31 degrees, much lower than the next-lowest temperature, and this is the observation for the *Challenger* disaster. The presidential investigation discovered that warm temperatures were needed for successful operation of the O-rings, and that 31 degrees was much too cold. In Chapter 12 we do a statistical analysis showing that the likelihood of failure increased as the temperature dropped. ■

Having obtained a sample from a population, an investigator would frequently like to use sample information to draw some type of conclusion (make an inference of some sort) about the population. That is, the sample is a means to an end rather than an end in itself. Techniques for generalizing from a sample to a population are gathered within the branch of our discipline called **inferential statistics**.

Example 1.2 Human measurements provide a rich area of application for statistical methods. The article "A Longitudinal Study of the Development of Elementary School Children's Private Speech" (*Merrill-Palmer Q.*, 1990: 443–463) reported on a study of children talking to themselves (private speech). It was thought that private speech would be related to IQ, because IQ is supposed to measure mental maturity, and it was known that private speech decreases as students progress through the primary grades. The study included 33 students whose first-grade IQ scores are given here:

82 96 99 102 103 103 106 107 108 108 108 108 109 110 110 111 113
113 113 113 115 115 118 118 119 121 122 122 127 132 136 140 146

Suppose we want an *estimate* of the average value of IQ for the first graders served by this school (if we conceptualize a population of all such IQs, we are trying to estimate the population mean). It can be shown that, with a high degree of confidence, the population mean IQ is between 109.2 and 118.2; we call this *a confidence interval* or *interval estimate*. The interval suggests that this is an above average class, because the nationwide IQ average is around 100. ■

The main focus of this book is on presenting and illustrating methods of inferential statistics that are useful in research. The most important types of inferential procedures—point estimation, hypothesis testing, and estimation by confidence intervals—are introduced in Chapters 7–9 and then used in more complicated settings in Chapters 10–14. The remainder of this chapter presents methods from descriptive statistics that are most used in the development of inference.

Chapters 2–6 present material from the discipline of probability. This material ultimately forms a bridge between the descriptive and inferential techniques. Mastery of probability leads to a better understanding of how inferential procedures are developed and used, how statistical conclusions can be translated into everyday language and interpreted, and when and where pitfalls can occur in applying the methods. Probability and statistics both deal with questions involving populations and samples, but do so in an "inverse manner" to one another.

In a probability problem, properties of the population under study are assumed known (e.g., in a numerical population, some specified distribution of the population values may be assumed), and questions regarding a sample taken from the population are posed and answered. In a statistics problem, characteristics of a sample are available to the experimenter, and this information enables the experimenter to draw conclusions about the population. The relationship between the two disciplines can be summarized by saying that probability reasons from the population to the sample (deductive reasoning), whereas inferential statistics reasons from the sample to the population (inductive reasoning). This is illustrated in Figure 1.2.

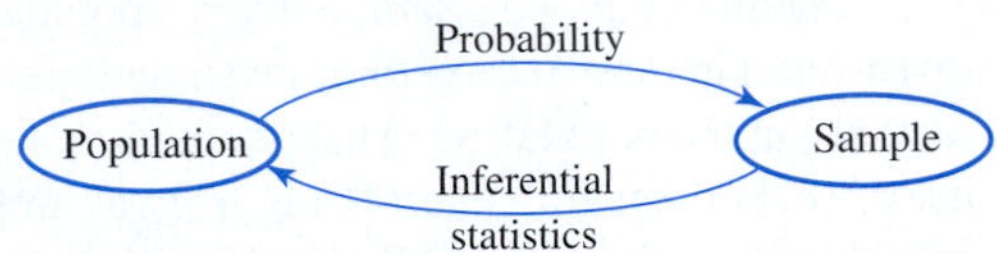

Figure 1.2 The relationship between probability and inferential statistics

Before we can understand what a particular sample can tell us about the population, we should first understand the uncertainty associated with taking a sample from a given population. This is why we study probability before statistics.

As an example of the contrasting focus of probability and inferential statistics, consider drivers' use of manual lap belts in cars equipped with automatic shoulder belt systems. (The article "Automobile Seat Belts: Usage Patterns in Automatic Belt Systems," *Human Factors*, 1998: 126–135, summarizes usage data.) In probability, we might assume that 50% of all drivers of cars equipped in this way in a certain metropolitan area regularly use their lap belt (an assumption about the population), so we might ask, "How likely is it that a sample of 100 such drivers will include at least 70 who regularly use their lap belt?" or "How many of the drivers in a sample of size 100 can we expect to regularly use their lap belt?" On the other hand, in inferential statistics we have sample information available; for example, a sample of 100 drivers of such cars revealed that 65 regularly use their lap belt. We might then ask, "Does this provide substantial evidence for concluding that more than 50% of all such drivers in this area regularly use their lap belt?" In this latter scenario, we are attempting to use sample information to answer a question about the structure of the entire population from which the sample was selected.

In the lap belt example, the population is well defined and concrete: all drivers of cars equipped in a certain way in a particular metropolitan area. In Example 1.1, however, a sample of O-ring temperatures is available, but it is from a population that does not actually exist. Instead, it is convenient to think of the population as consisting of all possible temperature measurements that might be made under similar experimental conditions. Such a population is referred to as a **conceptual** or **hypothetical population**.

There are a number of problem situations in which we fit questions into the framework of inferential statistics by conceptualizing a population.

Sometimes an investigator must be very cautious about generalizing from the circumstances under which data has been gathered. For example, a sample of five engines with a new design may be experimentally manufactured and tested to investigate efficiency. These five could be viewed as a sample from the conceptual population of all prototypes that could be manufactured under similar conditions, but *not* necessarily as representative of the population of units manufactured once regular production gets under way. Methods for using sample information to draw conclusions about future production units may be problematic. Similarly, a new drug may be tried on patients who arrive at a clinic, but there may be some question about how typical these patients are. They may not be representative of patients elsewhere or patients at the clinic next year. A good exposition of these issues is contained in the article "Assumptions for Statistical Inference" by Gerald Hahn and William Meeker (*Amer. Statist.*, 1993: 1–11).

Collecting Data

Statistics deals not only with the organization and analysis of data once it has been collected but also with the development of techniques for collecting the data. If data is not properly collected, an investigator may not be able to answer the questions under consideration with a reasonable degree of confidence. One common problem is that the target population—the one about which conclusions are to be drawn—may be different from the population actually sampled. For example, advertisers would like various kinds of information about the television-viewing habits of potential customers. The most systematic information of this sort comes from placing monitoring devices in a small number of homes across the United States. It has been conjectured that placement of such devices in and of itself alters viewing behavior, so that characteristics of the sample may be different from those of the target population.

When data collection entails selecting individuals or objects from a frame, the simplest method for ensuring a representative selection is to take a *simple random sample*. This is one for which any particular subset of the specified size (e.g., a sample of size 100) has the same chance of being selected. For example, if the frame consists of 1,000,000 serial numbers, the numbers 1, 2, . . . , up to 1,000,000 could be placed on identical slips of paper. After placing these slips in a box and thoroughly mixing, slips could be drawn one by one until the requisite sample size has been obtained. Alternatively (and much to be preferred), a table of random numbers or a computer's random number generator could be employed.

Sometimes alternative sampling methods can be used to make the selection process easier, to obtain extra information, or to increase the degree of confidence in conclusions. One such method, *stratified sampling*, entails separating the population units into nonoverlapping groups and taking a sample from each one. For example, a manufacturer of DVD players might want information about customer satisfaction for units produced during the previous year. If three different models were manufactured and sold, a separate sample could be selected from each of the three corresponding strata. This would result in information on all three models and ensure that no one model was over- or underrepresented in the entire sample.

Frequently a "convenience" sample is obtained by selecting individuals or objects without systematic randomization. As an example, a collection of bricks may be stacked in such a way that it is extremely difficult for those in the center to be selected. If the bricks on the top and sides of the stack were somehow different from the others, resulting sample data would not be representative of the population. Often an investigator will assume that such a convenience sample approximates a random sample, in which case a statistician's repertoire of inferential methods can be used; however, this is a judgment call. Most of the methods discussed herein are based on a variation of simple random sampling described in Chapter 6.

Researchers often collect data by carrying out some sort of designed experiment. This may involve deciding how to allocate several different treatments (such as fertilizers or drugs) to the various experimental units (plots of land or patients). Alternatively, an investigator may systematically vary the levels or categories of certain factors (e.g., amount of fertilizer or dose of a drug) and observe the effect on some response variable (such as corn yield or blood pressure).

Example 1.3 An article in the *New York Times* (Jan. 27, 1987) reported that heart attack risk could be reduced by taking aspirin. This conclusion was based on a designed experiment involving both a control group of individuals, who took a placebo having the appearance of aspirin but known to be inert, and a treatment group who took aspirin according to a specified regimen. Subjects were randomly assigned to the groups to protect against any biases and so that probability-based methods could be used to analyze the data. Of the 11,034 individuals in the control group, 189 subsequently experienced heart attacks, whereas only 104 of the 11,037 in the aspirin group had a heart attack. The incidence rate of heart attacks in the treatment group was only about half that in the control group. One possible explanation for this result is chance variation—that aspirin really doesn't have the desired effect and the observed difference is just typical variation in the same way that tossing two identical coins would usually produce different numbers of heads. However, in this case, inferential methods suggest that chance variation by itself cannot adequately explain the magnitude of the observed difference. ■

Exercises Section 1.1 (1–9)

1. Give one possible sample of size 4 from each of the following populations:

a. All daily newspapers published in the United States

b. All companies listed on the New York Stock Exchange

c. All students at your college or university

d. All grade point averages of students at your college or university

2. For each of the following hypothetical populations, give a plausible sample of size 4:

a. All distances that might result when you throw a football

b. Page lengths of books published 5 years from now

c. All possible earthquake-strength measurements (Richter scale) that might be recorded in California during the next year

d. All possible yields (in grams) from a certain chemical reaction carried out in a laboratory

3. Consider the population consisting of all DVD players of a certain brand and model, and focus on whether a DVD player needs service while under warranty.

a. Pose several probability questions based on selecting a sample of 100 such DVD players.

b. What inferential statistics question might be answered by determining the number of such DVD players in a sample of size 100 that need warranty service?

4. a. Give three different examples of concrete populations and three different examples of hypothetical populations.

b. For one each of your concrete and your hypothetical populations, give an example of a probability question and an example of an inferential statistics question.

5. Many universities and colleges have instituted supplemental instruction (SI) programs, in which a student facilitator meets regularly with a small group of students enrolled in the course to promote discussion of course material and enhance subject mastery. Suppose that students in a large statistics course (what else?) are randomly divided into a control group that will not participate in SI and a treatment group that will participate. At the end of the term, each student's total score in the course is determined.

a. Are the scores from the SI group a sample from an existing population? If so, what is it? If not, what is the relevant conceptual population?

b. What do you think is the advantage of randomly dividing the students into the two groups rather than letting each student choose which group to join?

c. Why didn't the investigators put all students in the treatment group? *Note*: The article "Supplemental Instruction: An Effective Component of Student Affairs Programming" (*J. College Student Dev*, 1997: 577–586) discusses the analysis of data from several SI programs.

6. The California State University (CSU) system consists of 23 campuses, from San Diego State in the south to Humboldt State near the Oregon border. A CSU administrator wishes to make an inference about the average distance between the hometowns of students and their campuses. Describe and discuss several different sampling methods that might be employed.

7. A certain city divides naturally into ten district neighborhoods. How might a real estate appraiser select a sample of single-family homes that could be used as a basis for developing an equation to predict appraised value from characteristics such as age, size, number of bathrooms, distance to the nearest school, and so on?

8. The amount of flow through a solenoid valve in an automobile's pollution-control system is an important characteristic. An experiment was carried out to study how flow rate depended on three factors: armature length, spring load, and bobbin depth. Two different levels (low and high) of each factor were chosen, and a single observation on flow was made for each combination of levels.

a. The resulting data set consisted of how many observations?

b. Does this study involve sampling an existing population or a conceptual population?

9. In a famous experiment carried out in 1882, Michelson and Newcomb obtained 66 observations on the time it took for light to travel between two locations in Washington, D.C. A few of the measurements (coded in a certain manner) were 31, 23, 32, 36, −2, 26, 27, and 31.

a. Why are these measurements not identical?

b. Does this study involve sampling an existing population or a conceptual population?

1.2 Pictorial and Tabular Methods in Descriptive Statistics

There are two general types of methods within descriptive statistics. In this section we will discuss the first of these types—representing a data set using visual techniques. In Sections 1.3 and 1.4, we will develop some numerical summary measures for data sets. Many visual techniques may already be familiar to you: frequency tables, tally sheets, histograms, pie charts, bar graphs, scatter diagrams, and the like. Here we focus on a selected few of these techniques that are most useful and relevant to probability and inferential statistics.

Notation

Some general notation will make it easier to apply our methods and formulas to a wide variety of practical problems. The number of observations in a single sample, that is, the *sample size*, will often be denoted by n, so that $n = 4$ for the sample of universities {Stanford, Iowa State, Wyoming, Rochester} and also for the sample of pH measurements {6.3, 6.2, 5.9, 6.5}. If two samples are simultaneously under consideration, either m and n or n_1 and n_2 can be used to denote the numbers of observations. Thus if {3.75, 2.60, 3.20, 3.79} and {2.75, 1.20, 2.45} are grade point averages for students on a mathematics floor and the rest of the dorm, respectively, then $m = 4$ and $n = 3$.

Given a data set consisting of n observations on some variable x, the individual observations will be denoted by $x_1, x_2, x_3, \ldots, x_n$. The subscript bears no relation to the magnitude of a particular observation. Thus x_1 will not in general be the smallest observation in the set, nor will x_n typically be the largest. In many applications, x_1 will be the first observation gathered by the experimenter, x_2 the second, and so on. The ith observation in the data set will be denoted by x_i.

Stem-and-Leaf Displays

Consider a numerical data set $x_1, x_2, \ldots, x_n$ for which each x_i consists of at least two digits. A quick way to obtain an informative visual representation of the data set is to construct a *stem-and-leaf display*.

STEPS FOR CONSTRUCTING A STEM-AND-LEAF DISPLAY

1. Select one or more leading digits for the stem values. The trailing digits become the leaves.
2. List possible stem values in a vertical column.
3. Record the leaf for every observation beside the corresponding stem value.
4. Indicate the units for stems and leaves someplace in the display.

If the data set consists of exam scores, each between 0 and 100, the score of 83 would have a stem of 8 and a leaf of 3. For a data set of automobile fuel efficiencies (mpg), all between 8.1 and 47.8, we could use the tens digit as the stem, so 32.6 would then have a leaf of 2.6. Usually, a display based on between 5 and 20 stems is appropriate.

Example 1.4 The use of alcohol by college students is of great concern not only to those in the academic community but also, because of potential health and safety consequences, to society at large. The article "Health and Behavioral Consequences of Binge Drinking in College" (*J. Amer. Med. Assoc.*, 1994: 1672–1677) reported on a comprehensive study of heavy drinking on campuses across the United States. A binge episode was defined as five or more drinks in a row for males and four or more for females. Figure 1.3 shows a stem-and-leaf display of 140 values of x = the percentage of undergraduate students who are binge drinkers. (These values were not given in the cited article, but our display agrees with a picture of the data that did appear.)

```
0 | 4
1 | 1345678889
2 | 1223456666777889999                       Stem: tens digit
3 | 0112233344555666677777888899999           Leaf: ones digit
4 | 111222223344445566666667778888899 9
5 | 00111222223345566666777788889 9
6 | 01111244455666778
```

Figure 1.3 Stem-and-leaf display for percentage binge drinkers at each of 140 colleges

The first leaf on the stem 2 row is 1, which tells us that 21% of the students at one of the colleges in the sample were binge drinkers. Without the identification of stem digits and leaf digits on the display, we wouldn't know whether the stem 2, leaf 1 observation should be read as 21%, 2.1%, or .21%.

When creating a display by hand, ordering the leaves from smallest to largest on each line can be time-consuming, and this ordering usually contributes little if any extra information. Suppose the observations had been listed in alphabetical order by school name, as

16% 33% 64% 37% 31% . . .

Then placing these values on the display in this order would result in the stem 1 row having 6 as its first leaf, and the beginning of the stem 3 row would be

3 | 371 . . .

The display suggests that a typical or representative value is in the stem 4 row, perhaps in the mid-40% range. The observations are not highly concentrated about this typical value, as would be the case if all values were between 20% and 49%. The display rises to a single peak as we move downward, and then declines; there are no gaps in the display. The shape of the display is not perfectly symmetric, but instead appears to stretch out a bit more in the direction of low leaves than in the direction of high leaves. Lastly, there are no observations that are unusually far from the bulk of the data (no *outliers*), as would be the case if one of the 26% values had instead been 86%. The most surprising feature of this data is that, at most colleges in the sample, at least one-quarter of the students are binge drinkers. The problem of heavy drinking on campuses is much more pervasive than many had suspected. ■

A stem-and-leaf display conveys information about the following aspects of the data:

- Identification of a typical or representative value
- Extent of spread about the typical value
- Presence of any gaps in the data
- Extent of symmetry in the distribution of values
- Number and location of peaks
- Presence of any outlying values

Example 1.5 Figure 1.4 presents stem-and-leaf displays for a random sample of lengths of golf courses (yards) that have been designated by *Golf Magazine* as among the most challenging in the United States. Among the sample of 40 courses, the shortest is 6433 yards long, and the longest is 7280 yards. The lengths appear to be distributed in a roughly uniform fashion over the range of values in the sample. Notice that a stem choice here of either a single digit (6 or 7) or three digits (643, . . . , 728) would yield an uninformative display, the first because of too few stems and the latter because of too many.

Stem: Thousands and hundreds digits
Leaf: Tens and ones digits

Stem	Leaves
64	35 64 33 70
65	26 27 06 83
66	05 94 14
67	90 70 00 98 70 45 13
68	90 70 73 50
69	00 27 36 04
70	51 05 11 40 50 22
71	31 69 68 05 13 65
72	80 09

(a)

Stem-and-leaf of yardage N = 40
Leaf Unit = 10

4	64	3367
8	65	0228
11	66	019
18	67	0147799
(4)	68	5779
18	69	0023
14	70	012455
8	71	013666
2	72	08

(b)

Figure 1.4 Stem-and-leaf displays of golf course yardages: (a) two-digit leaves; (b) display from MINITAB with truncated one-digit leaves ■

Dotplots

A dotplot is an attractive summary of numerical data when the data set is reasonably small or there are relatively few distinct data values. Each observation is represented by a dot above the corresponding location on a horizontal measurement scale. When a value occurs more than once, there is a dot for each occurrence, and these dots are stacked vertically. As with a stem-and-leaf display, a dotplot gives information about location, spread, extremes, and gaps.

Example 1.6 Figure 1.5 shows a dotplot for the O-ring temperature data introduced in Example 1.1 in the previous section. A representative temperature value is one in the mid-60's (°F), and there is quite a bit of spread about the center. The data stretches out more at the lower end than at the upper end, and the smallest observation, 31, can fairly be described as an outlier. This is the observation from the disastrous 1986 *Challenger* launch. An inquiry later concluded that 31 was much too cold for effective operation of the O-rings.

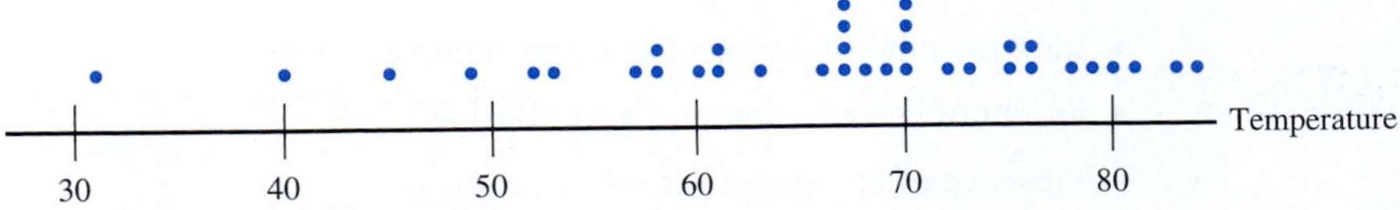

Figure 1.5 A dotplot of the O-ring temperature data (°F) ■

If the data set discussed in Example 1.6 had consisted of 50 or 100 temperature observations, each recorded to a tenth of a degree, it would have been much more cumbersome to construct a dotplot. Our next technique is well suited to such situations.

Histograms

Some numerical data is obtained by counting to determine the value of a variable (the number of traffic citations a person received during the last year, the number of persons arriving for service during a particular period), whereas other data is obtained by taking measurements (weight of an individual, reaction time to a particular stimulus). The prescription for drawing a histogram is generally different for these two cases.

Consider first data resulting from observations on a "counting variable" x. The **frequency** of any particular x value is the number of times that value occurs in the data set. The **relative frequency** of a value is the fraction or proportion of times the value occurs:

$$\text{relative frequency of a value} = \frac{\text{number of times the value occurs}}{\text{number of observations in the data set}}$$

Suppose, for example, that our data set consists of 200 observations on x = the number of major defects in a new car of a certain type. If 70 of these x values are 1, then

$$\text{frequency of the } x \text{ value 1:} \quad 70$$

$$\text{relative frequency of the } x \text{ value 1:} \quad \frac{70}{200} = .35$$

Multiplying a relative frequency by 100 gives a percentage; in the defect example, 35% of the cars in the sample had just one major defect. The relative frequencies, or percentages, are usually of more interest than the frequencies themselves. In theory, the relative frequencies should sum to 1, but in practice the sum may differ slightly from 1 because of rounding. A **frequency distribution** is a tabulation of the frequencies and/or relative frequencies.

A HISTOGRAM FOR COUNTING DATA

First, determine the frequency and relative frequency of each x value. Then mark possible x values on a horizontal scale. Above each value, draw a rectangle whose height is the relative frequency (or alternatively, the frequency) of that value.

This construction ensures that the *area* of each rectangle is proportional to the relative frequency of the value. Thus if the relative frequencies of $x = 1$ and $x = 5$ are .35 and .07, respectively, then the area of the rectangle above 1 is five times the area of the rectangle above 5.

Example 1.7

How unusual is a no-hitter or a one-hitter in a major league baseball game, and how frequently does a team get more than 10, 15, or even 20 hits? Table 1.1 is a frequency distribution for the number of hits per team per game for all nine-inning games that were played between 1989 and 1993. Notice that a no-hitter happens only about once in a thousand games, and 22 or more hits occurs with about the same frequency.

The corresponding histogram in Figure 1.6 rises rather smoothly to a single peak and then declines. The histogram extends a bit more on the right (toward large values) than it does on the left—a slight "positive skew."

Table 1.1 Frequency distribution for hits in nine-inning games

Hits/Game	Number of Games	Relative Frequency	Hits/Game	Number of Games	Relative Frequency
0	20	.0010	14	569	.0294
1	72	.0037	15	393	.0203
2	209	.0108	16	253	.0131
3	527	.0272	17	171	.0088
4	1048	.0541	18	97	.0050
5	1457	.0752	19	53	.0027
6	1988	.1026	20	31	.0016
7	2256	.1164	21	19	.0010
8	2403	.1240	22	13	.0007
9	2256	.1164	23	5	.0003
10	1967	.1015	24	1	.0001
11	1509	.0779	25	0	.0000
12	1230	.0635	26	1	.0001
13	834	.0430	27	1	.0001
				19,383	1.0005

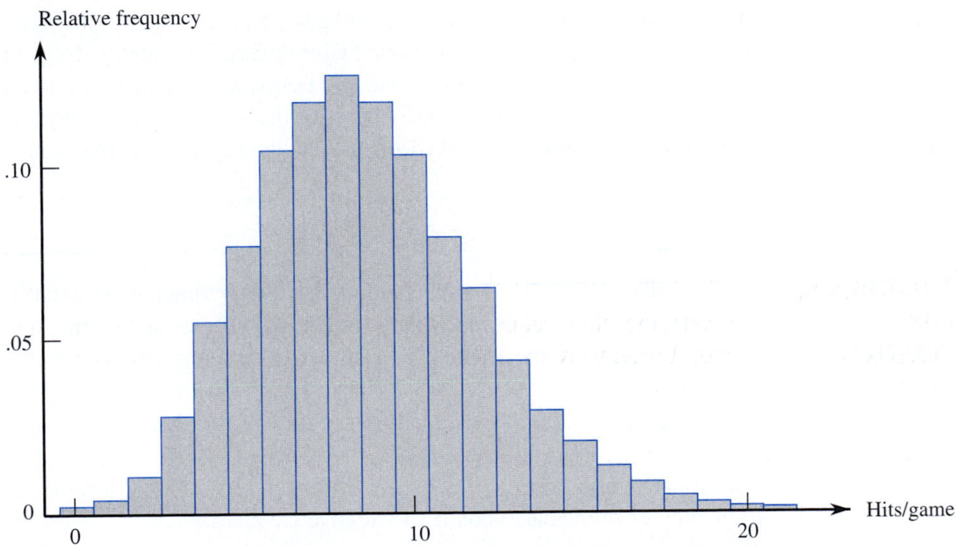

Figure 1.6 Histogram of number of hits per nine-inning game

Either from the tabulated information or from the histogram itself, we can determine the following:

$$\text{proportion of games with at most two hits} = \begin{matrix}\text{relative}\\ \text{frequency}\\ \text{for } x = 0\end{matrix} + \begin{matrix}\text{relative}\\ \text{frequency}\\ \text{for } x = 1\end{matrix} + \begin{matrix}\text{relative}\\ \text{frequency}\\ \text{for } x = 2\end{matrix}$$

$$= .0010 + .0037 + .0108 = .0155$$

Similarly,

$$\begin{array}{l}\text{proportion of games with} \\ \text{between 5 and 10 hits (inclusive)}\end{array} = .0752 + .1026 + \cdots + .1015 = .6361$$

That is, roughly 64% of all these games resulted in between 5 and 10 (inclusive) hits. ■

Constructing a histogram for measurement data (observations on a "measurement variable") entails subdividing the measurement axis into a suitable number of **class intervals** or **classes**, such that each observation is contained in exactly one class. Suppose, for example, that we have 50 observations on x = fuel efficiency of an automobile (mpg), the smallest of which is 27.8 and the largest of which is 31.4. Then we could use the class boundaries 27.5, 28.0, 28.5, . . . , and 31.5 as shown here:

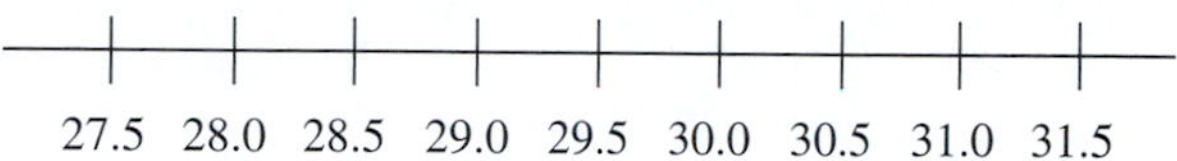

One potential difficulty is that occasionally an observation falls on a class boundary and therefore does not lie in exactly one interval, for example, 29.0. One way to deal with this problem is to use boundaries like 27.55, 28.05, . . . , 31.55. Adding a hundredths digit to the class boundaries prevents observations from falling on the resulting boundaries. The approach that we will follow is to write the class intervals as 27.5–28, 28–28.5, and so on and use the convention that *any observation falling on a class boundary will be included in the class to the right of the observation.* Thus 29.0 would go in the 29–29.5 class rather than the 28.5–29 class. This is how MINITAB constructs a histogram.

A HISTOGRAM FOR MEASUREMENT DATA: EQUAL CLASS WIDTHS

Determine the frequency and relative frequency for each class. Mark the class boundaries on a horizontal measurement axis. Above each class interval, draw a rectangle whose height is the corresponding relative frequency (or frequency).

Example 1.8 Power companies need information about customer usage to obtain accurate forecasts of demands. Investigators from Wisconsin Power and Light determined energy consumption (BTUs) during a particular period for a sample of 90 gas-heated homes. An adjusted consumption value was calculated as follows:

$$\text{adjusted consumption} = \frac{\text{consumption}}{(\text{weather, in degree days})(\text{house area})}$$

This resulted in the accompanying data (part of the stored data set FURNACE.MTW available in MINITAB), which we have ordered from smallest to largest.

2.97	4.00	5.20	5.56	5.94	5.98	6.35	6.62	6.72	6.78
6.80	6.85	6.94	7.15	7.16	7.23	7.29	7.62	7.62	7.69
7.73	7.87	7.93	8.00	8.26	8.29	8.37	8.47	8.54	8.58
8.61	8.67	8.69	8.81	9.07	9.27	9.37	9.43	9.52	9.58
9.60	9.76	9.82	9.83	9.83	9.84	9.96	10.04	10.21	10.28
10.28	10.30	10.35	10.36	10.40	10.49	10.50	10.64	10.95	11.09
11.12	11.21	11.29	11.43	11.62	11.70	11.70	12.16	12.19	12.28
12.31	12.62	12.69	12.71	12.91	12.92	13.11	13.38	13.42	13.43
13.47	13.60	13.96	14.24	14.35	15.12	15.24	16.06	16.90	18.26

We let MINITAB select the class intervals. The most striking feature of the histogram in Figure 1.7 is its resemblance to a bell-shaped (and therefore symmetric) curve, with the point of symmetry roughly at 10.

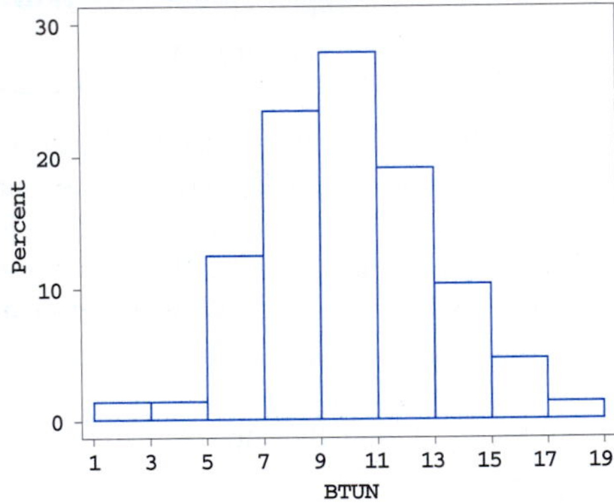

Figure 1.7 Histogram of the energy consumption data from Example 1.8

Class	1–3	3–5	5–7	7–9	9–11	11–13	13–15	15–17	17–19
Frequency	1	1	11	21	25	17	9	4	1
Relative frequency	.011	.011	.122	.233	.278	.189	.100	.044	.011

From the histogram,

$$\text{proportion of observations less than 9} \approx .01 + .01 + .12 + .23 = .37 \qquad \left(\text{exact value} = \frac{34}{90} = .378\right)$$

The relative frequency for the 9–11 class is about .27, so we estimate that roughly half of this, or .135, is between 9 and 10. Thus

$$\text{proportion of observations less than 10} \approx .37 + .135 = .505 \text{ (slightly more than 50\%)}$$

The exact value of this proportion is 47/90 = .522. ■

There are no hard-and-fast rules concerning either the number of classes or the choice of classes themselves. Between 5 and 20 classes will be satisfactory for most data sets. Generally, the larger the number of observations in a data set, the more classes should be used. A reasonable rule of thumb is

$$\text{number of classes} \approx \sqrt{\text{number of observations}}$$

Equal-width classes may not be a sensible choice if a data set "stretches out" to one side or the other. Figure 1.8 shows a dotplot of such a data set. Using a small number of equal-width classes results in almost all observations falling in just one or two of the classes. If a large number of equal-width classes are used, many classes will have zero frequency. A sound choice is to use a few wider intervals near extreme observations and narrower intervals in the region of high concentration.

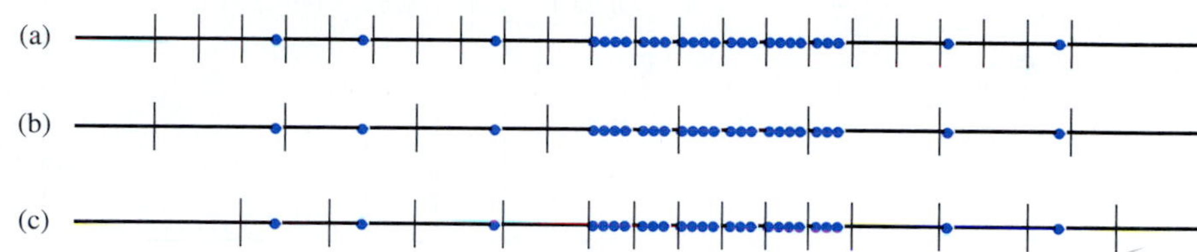

Figure 1.8 Selecting class intervals for "stretched-out" dots: (a) many short equal-width intervals; (b) a few wide equal-width intervals; (c) unequal-width intervals

A HISTOGRAM FOR MEASUREMENT DATA: UNEQUAL CLASS WIDTHS

After determining frequencies and relative frequencies, calculate the height of each rectangle using the formula

$$\text{rectangle height} = \frac{\text{relative frequency of the class}}{\text{class width}}$$

The resulting rectangle heights are usually called *densities*, and the vertical scale is the **density scale**. This prescription will also work when class widths are equal.

Example 1.9 Corrosion of reinforcing steel is a serious problem in concrete structures located in environments affected by severe weather conditions. For this reason, researchers have been investigating the use of reinforcing bars made of composite material. One study was carried out to develop guidelines for bonding glass-fiber-reinforced plastic rebars to concrete ("Design Recommendations for Bond of GFRP Rebars to Concrete," *J. Struct. Engrg.*, 1996: 247–254). Consider the following 48 observations on measured bond strength:

11.5	12.1	9.9	9.3	7.8	6.2	6.6	7.0	13.4	17.1	9.3	5.6
5.7	5.4	5.2	5.1	4.9	10.7	15.2	8.5	4.2	4.0	3.9	3.8
3.6	3.4	20.6	25.5	13.8	12.6	13.1	8.9	8.2	10.7	14.2	7.6
5.2	5.5	5.1	5.0	5.2	4.8	4.1	3.8	3.7	3.6	3.6	3.6

Class	2–4	4–6	6–8	8–12	12–20	20–30
Frequency	9	15	5	9	8	2
Relative frequency	.1875	.3125	.1042	.1875	.1667	.0417
Density	.094	.156	.052	.047	.021	.004

The resulting histogram appears in Figure 1.9. The right or upper tail stretches out much farther than does the left or lower tail—a substantial departure from symmetry.

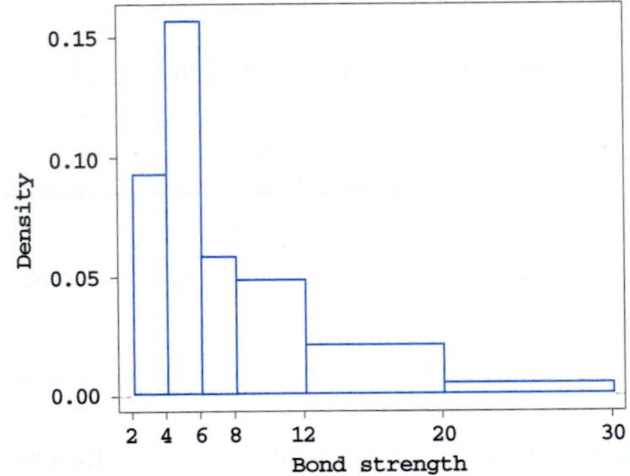

Figure 1.9 A MINITAB density histogram for the bond strength data of Example 1.9 ■

When class widths are unequal, not using a density scale will give a picture with distorted areas. For equal-class widths, the divisor is the same in each density calculation, and the extra arithmetic simply results in a rescaling of the vertical axis (i.e., the histogram using relative frequency and the one using density will have exactly the same appearance). A density histogram does have one interesting property. Multiplying both sides of the formula for density by the class width gives

$$\begin{aligned}\text{relative frequency} &= (\text{class width})(\text{density}) = (\text{rectangle width})(\text{rectangle height}) \\ &= \text{rectangle area}\end{aligned}$$

That is, *the area of each rectangle is the relative frequency of the corresponding class.* Furthermore, since the sum of relative frequencies must be 1.0 (except for roundoff), *the total area of all rectangles in a density histogram is* 1. It is always possible to draw a histogram so that the area equals the relative frequency (this is true also for a histogram of counting data)—just use the density scale. This property will play an important role in creating models for distributions in Chapter 4.

Histogram Shapes

Histograms come in a variety of shapes. A **unimodal** histogram is one that rises to a single peak and then declines. A **bimodal** histogram has two different peaks. Bimodality can occur when the data set consists of observations on two quite different kinds of individuals or objects. For example, consider a large data set consisting of driving times for automobiles traveling between San Luis Obispo and Monterey in California (exclusive of stopping time for sightseeing, eating, etc.). This histogram would show two peaks, one for those cars that took the inland route (roughly 2.5 hours) and another for those cars traveling up the coast (3.5–4 hours). However, bimodality does not automatically follow in such situations. Only if the two separate histograms are "far apart" relative to their spreads will bimodality occur in the histogram of combined data. Thus a large data set consisting of heights of college students should not result in a bimodal histogram because the typical male height of about 69 inches is not far enough above the typical female height of about 64–65 inches. A histogram with more than two peaks is said to be **multimodal**. Of course, the number of peaks may well depend on the choice of class intervals, particularly with a small number of observations. The larger the number of classes, the more likely it is that bimodality or multimodality will manifest itself.

A histogram is **symmetric** if the left half is a mirror image of the right half. A unimodal histogram is **positively skewed** if the right or upper tail is stretched out compared with the left or lower tail and **negatively skewed** if the stretching is to the left. Figure 1.10 shows "smoothed" histograms, obtained by superimposing a smooth curve on the rectangles, that illustrate the various possibilities.

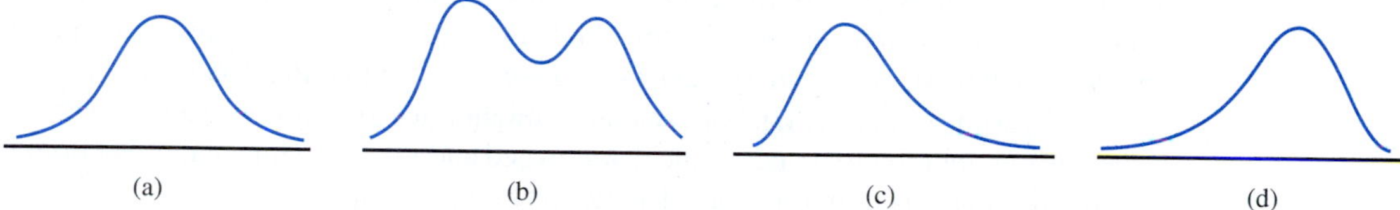

Figure 1.10 Smoothed histograms: (a) symmetric unimodal; (b) bimodal; (c) positively skewed; and (d) negatively skewed

Qualitative Data

Both a frequency distribution and a histogram can be constructed when the data set is qualitative (categorical) in nature; in this case, "bar graph" is synonymous with "histogram." Sometimes there will be a natural ordering of classes (for example, freshmen, sophomores, juniors, seniors, graduate students) whereas in other cases the order will be arbitrary (for example, Catholic, Jewish, Protestant, and the like). With such categorical data, the intervals above which rectangles are constructed should have equal width.

Example 1.10 Each member of a sample of 120 individuals owning motorcycles was asked for the name of the manufacturer of his or her bike. The frequency distribution for the resulting data is given in Table 1.2 and the histogram is shown in Figure 1.11.

Table 1.2 Frequency distribution for motorcycle data

Manufacturer	Frequency	Relative Frequency
1. Honda	41	.34
2. Yamaha	27	.23
3. Kawasaki	20	.17
4. Harley-Davidson	18	.15
5. BMW	3	.03
6. Other	11	.09
	120	1.01

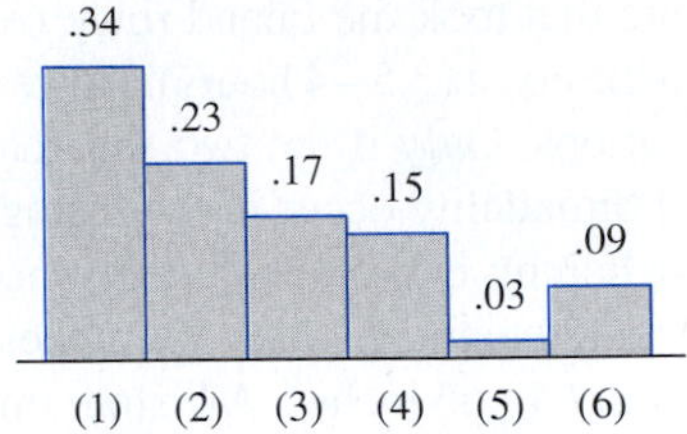

Figure 1.11 Histogram for motorcycle data ■

Multivariate Data

The techniques presented so far have been exclusively for situations in which each observation in a data set is either a single number or a single category. Often, however, the data is *multivariate* in nature. That is, if we obtain a sample of individuals or objects and on each one we make two or more measurements, then each "observation" would consist of several measurements on one individual or object. The sample is bivariate if each observation consists of two measurements or responses, so that the data set can be represented as $(x_1, y_1), \ldots, (x_n, y_n)$. For example, x might refer to engine size and y to horsepower, or x might refer to brand of calculator owned and y to academic major. We briefly consider the analysis of multivariate data in several later chapters.

Exercises Section 1.2 (10–29)

10. Consider the IQ data given in Example 1.2.

a. Construct a stem-and-leaf display of the data. What appears to be a representative IQ value? Do the observations appear to be highly concentrated about the representative value or rather spread out?

b. Does the display appear to be reasonably symmetric about a representative value, or would you describe its shape in some other way?

c. Do there appear to be any outlying IQ values?

d. What proportion of IQ values in this sample exceed 100?

11. Every score in the following batch of exam scores is in the 60's, 70's, 80's, or 90's. A stem-and-leaf display with only the four stems 6, 7, 8, and 9 would not give a very detailed description of the distribution of scores. In such situations, it is desirable to use repeated stems. Here we could repeat the stem 6 twice, using 6L for scores in the low 60's (leaves 0, 1, 2, 3, and 4) and 6H for scores in the high 60's (leaves 5, 6, 7, 8, and 9). Similarly, the other stems can be repeated twice to obtain a display consisting of eight rows. Construct such a display for the given scores. What feature of the data is highlighted by this display?

74 89 80 93 64 67 72 70 66 85 89 81 81
71 74 82 85 63 72 81 81 95 84 81 80 70
69 66 60 83 85 98 84 68 90 82 69 72 87
88

12. The accompanying specific gravity values for various wood types used in construction appeared in the article "Bolted Connection Design Values Based on European Yield Model" (*J. Struct. Engrg.*, 1993: 2169–2186):

.31 .35 .36 .36 .37 .38 .40 .40 .40
.41 .41 .42 .42 .42 .42 .42 .43 .44
.45 .46 .46 .47 .48 .48 .48 .51 .54
.54 .55 .58 .62 .66 .66 .67 .68 .75

Construct a stem-and-leaf display using repeated stems (see the previous exercise), and comment on any interesting features of the display.

13. The accompanying data set consists of observations on shower-flow rate (L/min) for a sample of $n = 129$

houses in Perth, Australia ("An Application of Bayes Methodology to the Analysis of Diary Records in a Water Use Study," *J. Amer. Statist. Assoc.*, 1987: 705–711):

4.6	12.3	7.1	7.0	4.0	9.2	6.7	6.9	11.5	5.1
11.2	10.5	14.3	8.0	8.8	6.4	5.1	5.6	9.6	7.5
7.5	6.2	5.8	2.3	3.4	10.4	9.8	6.6	3.7	6.4
8.3	6.5	7.6	9.3	9.2	7.3	5.0	6.3	13.8	6.2
5.4	4.8	7.5	6.0	6.9	10.8	7.5	6.6	5.0	3.3
7.6	3.9	11.9	2.2	15.0	7.2	6.1	15.3	18.9	7.2
5.4	5.5	4.3	9.0	12.7	11.3	7.4	5.0	3.5	8.2
8.4	7.3	10.3	11.9	6.0	5.6	9.5	9.3	10.4	9.7
5.1	6.7	10.2	6.2	8.4	7.0	4.8	5.6	10.5	14.6
10.8	15.5	7.5	6.4	3.4	5.5	6.6	5.9	15.0	9.6
7.8	7.0	6.9	4.1	3.6	11.9	3.7	5.7	6.8	11.3
9.3	9.6	10.4	9.3	6.9	9.8	9.1	10.6	4.5	6.2
8.3	3.2	4.9	5.0	6.0	8.2	6.3	3.8	6.0	

a. Construct a stem-and-leaf display of the data.
b. What is a typical, or representative, flow rate?
c. Does the display appear to be highly concentrated or spread out?
d. Does the distribution of values appear to be reasonably symmetric? If not, how would you describe the departure from symmetry?
e. Would you describe any observation as being far from the rest of the data (an outlier)?

14. A *Consumer Reports* article on peanut butter (Sept. 1990) reported the following scores for various brands:

Creamy 56 44 62 36 39 53 50 65 45 40
56 68 41 30 40 50 56 30 22

Crunchy 62 53 75 42 47 40 34 62 52
50 34 42 36 75 80 47 56 62

Construct a *comparative* stem-and-leaf display by listing stems in the middle of your page and then displaying the creamy leaves out to the right and the crunchy leaves out to the left. Describe similarities and differences for the two types.

15. Temperature transducers of a certain type are shipped in batches of 50. A sample of 60 batches was selected, and the number of transducers in each batch not conforming to design specifications was determined, resulting in the following data:

2 1 2 4 0 1 3 2 0 5 3 3 1 3 2 4 7 0 2 3
0 4 2 1 3 1 1 3 4 1 2 3 2 2 8 4 5 1 3 1
5 0 2 3 2 1 0 6 4 2 1 6 0 3 3 3 6 1 2 3

a. Determine frequencies and relative frequencies for the observed values of x = number of nonconforming transducers in a batch.
b. What proportion of batches in the sample have at most five nonconforming transducers? What proportion have fewer than five? What proportion have at least five nonconforming units?
c. Draw a histogram of the data using relative frequency on the vertical scale, and comment on its features.

16. In a study of author productivity ("Lotka's Test," *Collection Manag.*, 1982: 111–118), a large number of authors were classified according to the number of articles they had published during a certain period. The results were presented in the accompanying frequency distribution:

Number of papers	1	2	3	4	5	6	7	8
Frequency	784	204	127	50	33	28	19	19

Number of papers	9	10	11	12	13	14	15	16	17
Frequency	6	7	6	7	4	4	5	3	3

a. Construct a histogram corresponding to this frequency distribution. What is the most interesting feature of the shape of the distribution?
b. What proportion of these authors published at least five papers? At least ten papers? More than ten papers?
c. Suppose the five 15's, three 16's, and three 17's had been lumped into a single category displayed as "$\geq$15." Would you be able to draw a histogram? Explain.
d. Suppose that instead of the values 15, 16, and 17 being listed separately, they had been combined into a 15–17 category with frequency 11. Would you be able to draw a histogram? Explain.

17. The number of contaminating particles on a silicon wafer prior to a certain rinsing process was determined for each wafer in a sample of size 100, resulting in the following frequencies:

Number of particles	0	1	2	3	4	5	6	7
Frequency	1	2	3	12	11	15	18	10

Number of particles	8	9	10	11	12	13	14
Frequency	12	4	5	3	1	2	1

a. What proportion of the sampled wafers had at least one particle? At least five particles?

b. What proportion of the sampled wafers had between five and ten particles, inclusive? Strictly between five and ten particles?

c. Draw a histogram using relative frequency on the vertical axis. How would you describe the shape of the histogram?

18. The article "Determination of Most Representative Subdivision" (*J. Energy Engrg.*, 1993: 43–55) gave data on various characteristics of subdivisions that could be used in deciding whether to provide electrical power using overhead lines or underground lines. Here are the values of the variable x = total length of streets within a subdivision:

1280	5320	4390	2100	1240	3060	4770
1050	360	3330	3380	340	1000	960
1320	530	3350	540	3870	1250	2400
960	1120	2120	450	2250	2320	2400
3150	5700	5220	500	1850	2460	5850
2700	2730	1670	100	5770	3150	1890
510	240	396	1419	2109		

a. Construct a stem-and-leaf display using the thousands digit as the stem and the hundreds digit as the leaf, and comment on the various features of the display.

b. Construct a histogram using class boundaries 0, 1000, 2000, 3000, 4000, 5000, and 6000. What proportion of subdivisions have total length less than 2000? Between 2000 and 4000? How would you describe the shape of the histogram?

19. The article cited in Exercise 18 also gave the following values of the variables y = number of culs-de-sac and z = number of intersections:

y	1	0	1	0	0	2	0	1	1	1	2	1	0	0	1	1	0	1	1
z	1	8	6	1	1	5	3	0	0	4	4	0	0	1	2	1	4	0	4
y	1	1	0	0	0	1	1	2	0	1	2	2	1	1	0	2	1	1	0
z	0	3	0	1	1	0	1	3	2	4	6	6	0	1	1	8	3	3	5
y	1	5	0	3	0	1	1	0	0										
z	0	5	2	3	1	0	0	0	3										

a. Construct a histogram for the y data. What proportion of these subdivisions had no culs-de-sac? At least one cul-de-sac?

b. Construct a histogram for the z data. What proportion of these subdivisions had at most five intersections? Fewer than five intersections?

20. How does the speed of a runner vary over the course of a marathon (a distance of 42.195 km)? Consider determining both the time to run the first 5 km and the time to run between the 35-km and 40-km points, and then subtracting the former time from the latter time. A positive value of this difference corresponds to a runner slowing down toward the end of the race. The accompanying histogram is based on times of runners who participated in several different Japanese marathons ("Factors Affecting Runners' Marathon Performance," *Chance*, Fall 1993: 24–30).

What are some interesting features of this histogram? What is a typical difference value? Roughly what proportion of the runners ran the late distance more quickly than the early distance?

Histogram for Exercise 20

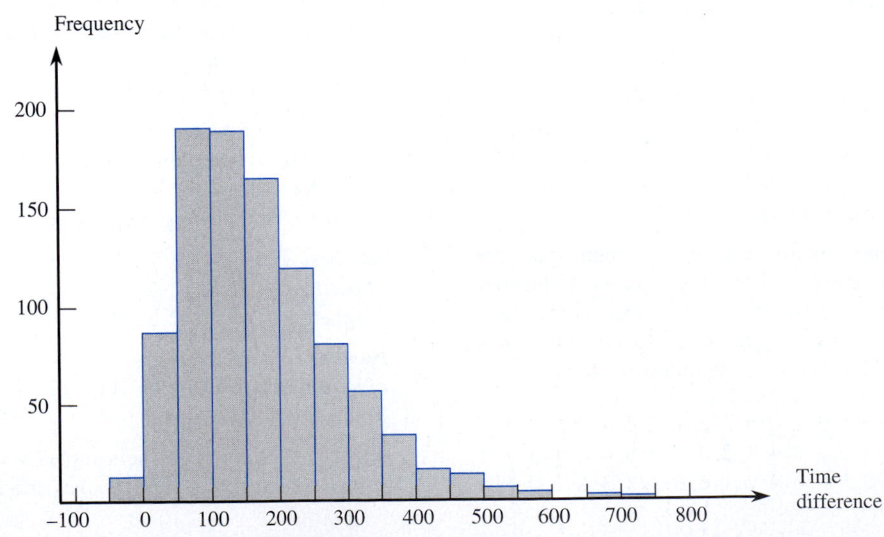

21. In a study of warp breakage during the weaving of fabric (*Technometrics*, 1982: 63), 100 specimens of yarn were tested. The number of cycles of strain to breakage was determined for each yarn specimen, resulting in the following data:

86	146	251	653	98	249	400	292	131	169
175	176	76	264	15	364	195	262	88	264
157	220	42	321	180	198	38	20	61	121
282	224	149	180	325	250	196	90	229	166
38	337	65	151	341	40	40	135	597	246
211	180	93	315	353	571	124	279	81	186
497	182	423	185	229	400	338	290	398	71
246	185	188	568	55	55	61	244	20	284
393	396	203	829	239	236	286	194	277	143
198	264	105	203	124	137	135	350	193	188

a. Construct a relative frequency histogram based on the class intervals 0–100, 100–200, . . . , and comment on features of the distribution.

b. Construct a histogram based on the following class intervals: 0–50, 50–100, 100–150, 150–200, 200–300, 300–400, 400–500, 500–600, 600–900.

c. If weaving specifications require a breaking strength of at least 100 cycles, what proportion of the yarn specimens in this sample would be considered satisfactory?

22. The accompanying data set consists of observations on shear strength (lb) of ultrasonic spot welds made on a certain type of alclad sheet. Construct a relative frequency histogram based on ten equal-width classes with boundaries 4000, 4200, ... [The histogram will agree with the one in "Comparison of Properties of Joints Prepared by Ultrasonic Welding and Other Means" (*J. Aircraft*, 1983: 552–556).] Comment on its features.

5434	4948	4521	4570	4990	5702	5241
5112	5015	4659	4806	4637	5670	4381
4820	5043	4886	4599	5288	5299	4848
5378	5260	5055	5828	5218	4859	4780
5027	5008	4609	4772	5133	5095	4618
4848	5089	5518	5333	5164	5342	5069
4755	4925	5001	4803	4951	5679	5256
5207	5621	4918	5138	4786	4500	5461
5049	4974	4592	4173	5296	4965	5170
4740	5173	4568	5653	5078	4900	4968
5248	5245	4723	5275	5419	5205	4452
5227	5555	5388	5498	4681	5076	4774
4931	4493	5309	5582	4308	4823	4417
5364	5640	5069	5188	5764	5273	5042
5189	4986					

23. A transformation of data values by means of some mathematical function, such as $\sqrt{x}$ or $1/x$, can often yield a set of numbers that has "nicer" statistical properties than the original data. In particular, it may be possible to find a function for which the histogram of transformed values is more symmetric (or, even better, more like a bell-shaped curve) than the original data. As an example, the article "Time Lapse Cinematographic Analysis of Beryllium-Lung Fibroblast Interactions" (*Environ. Res.*, 1983: 34–43) reported the results of experiments designed to study the behavior of certain individual cells that had been exposed to beryllium. An important characteristic of such an individual cell is its interdivision time (IDT). IDTs were determined for a large number of cells both in exposed (treatment) and unexposed (control) conditions. The authors of the article used a logarithmic transformation, that is, transformed value = log(original value). Consider the following representative IDT data:

IDT	$\log_{10}$ (IDT)	IDT	$\log_{10}$ (IDT)	IDT	$\log_{10}$ (IDT)
28.1	1.45	60.1	1.78	21.0	1.32
31.2	1.49	23.7	1.37	22.3	1.35
13.7	1.14	18.6	1.27	15.5	1.19
46.0	1.66	21.4	1.33	36.3	1.56
25.8	1.41	26.6	1.42	19.1	1.28
16.8	1.23	26.2	1.42	38.4	1.58
34.8	1.54	32.0	1.51	72.8	1.86
62.3	1.79	43.5	1.64	48.9	1.69
28.0	1.45	17.4	1.24	21.4	1.33
17.9	1.25	38.8	1.59	20.7	1.32
19.5	1.29	30.6	1.49	57.3	1.76
21.1	1.32	55.6	1.75	40.9	1.61
31.9	1.50	25.5	1.41		
28.9	1.46	52.1	1.72		

Use class intervals 10–20, 20–30, . . . to construct a histogram of the original data. Use intervals 1.1–1.2, 1.2–1.3, . . . to do the same for the transformed data. What is the effect of the transformation?

24. The clearness index was determined for the skies over Baghdad for each of the 365 days during a particular year ("Contribution to the Study of the Solar Radiation Climate of the Baghdad Environment," *Solar Energy*, 1990: 7–12). The accompanying table gives the results.

Class	Frequency
.15–.25	8
.25–.35	14
.35–.45	28
.45–.50	24
.50–.55	39
.55–.60	51
.60–.65	106
.65–.70	84
.70–.75	11

a. Determine relative frequencies and draw the corresponding histogram.

b. Cloudy days are those with a clearness index smaller than .35. What percentage of the days were cloudy?

c. Clear days are those for which the index is at least .65. What percentage of the days were clear?

25. The paper "Study on the Life Distribution of Microdrills" (*J. Engrg. Manufacture*, 2002: 301–305) reported the following observations, listed in increasing order, on drill lifetime (number of holes that a drill machines before it breaks) when holes were drilled in a certain brass alloy.

11	14	20	23	31	36	39	44	47	50
59	61	65	67	68	71	74	76	78	79
81	84	85	89	91	93	96	99	101	104
105	105	112	118	123	136	139	141	148	158
161	168	184	206	248	263	289	322	388	513

a. Construct a frequency distribution and histogram of the data using class boundaries 0, 50, 100, . . . , and then comment on interesting characteristics.

b. Construct a frequency distribution and histogram of the natural logarithms of the lifetime observations, and comment on interesting characteristics.

c. What proportion of the lifetime observations in this sample are less than 100? What proportion of the observations are at least 200?

26. Consider the following data on type of health complaint (J = joint swelling, F = fatigue, B = back pain, M = muscle weakness, C = coughing, N = nose running/irritation, O = other) made by tree planters. Obtain frequencies and relative frequencies for the various categories, and draw a histogram. (The data is consistent with percentages given in the article "Physiological Effects of Work Stress and Pesticide Exposure in Tree Planting by British Columbia Silviculture Workers," *Ergonomics*, 1993: 951–961.)

O O N J C F B B F O J O O M
O F F O O N O N J F J B O C
J O J J F N O B M O J M O B
O F J O O B N C O O O M B F
J O F N

27. A **Pareto diagram** is a variation of a histogram for categorical data resulting from a quality control study. Each category represents a different type of product nonconformity or production problem. The categories are ordered so that the one with the largest frequency appears on the far left, then the category with the second largest frequency, and so on. Suppose the following information on nonconformities in circuit packs is obtained: failed component, 126; incorrect component, 210; insufficient solder, 67; excess solder, 54; missing component, 131. Construct a Pareto diagram.

28. The **cumulative frequency** and cumulative relative frequency for a particular class interval are the sum of frequencies and relative frequencies, respectively, for that interval and all intervals lying below it. If, for example, there are four intervals with frequencies 9, 16, 13, and 12, then the cumulative frequencies are 9, 25, 38, and 50, and the cumulative relative frequencies are .18, .50, .76, and 1.00. Compute the cumulative frequencies and cumulative relative frequencies for the data of Exercise 22.

29. Fire load (MJ/m^2) is the heat energy that could be released per square meter of floor area by combustion of contents and the structure itself. The article "Fire Loads in Office Buildings" (*J. Struct. Engrg.*, 1997: 365–368) gave the following cumulative percentages (read from a graph) for fire loads in a sample of 388 rooms:

Value	0	150	300	450	600
Cumulative %	0	19.3	37.6	62.7	77.5
Value	750	900	1050	1200	1350
Cumulative %	87.2	93.8	95.7	98.6	99.1
Value	1500	1650	1800	1950	
Cumulative %	99.5	99.6	99.8	100.0	

a. Construct a relative frequency histogram and comment on interesting features.

b. What proportion of fire loads are less than 600? At least 1200?

c. What proportion of the loads are between 600 and 1200?

1.3 Measures of Location

Visual summaries of data are excellent tools for obtaining preliminary impressions and insights. More formal data analysis often requires the calculation and interpretation of numerical summary measures. That is, from the data we try to extract several summarizing numbers — numbers that might serve to characterize the data set and convey some of its salient features. Our primary concern will be with numerical data; some comments regarding categorical data appear at the end of the section.

Suppose, then, that our data set is of the form $x_1, x_2, \ldots, x_n$, where each x_i is a number. What features of such a set of numbers are of most interest and deserve emphasis? One important characteristic of a set of numbers is its location, and in particular its center. This section presents methods for describing the location of a data set; in Section 1.4 we will turn to methods for measuring variability in a set of numbers.

The Mean

For a given set of numbers $x_1, x_2, \ldots, x_n$, the most familiar and useful measure of the center is the *mean*, or arithmetic average of the set. Because we will almost always think of the x_i's as constituting a sample, we will often refer to the arithmetic average as the *sample mean* and denote it by $\bar{x}$.

DEFINITION

The **sample mean** $\bar{x}$ of observations $x_1, x_2, \ldots, x_n$ is given by

$$\bar{x} = \frac{x_1 + x_2 + \cdots + x_n}{n} = \frac{\sum_{i=1}^{n} x_i}{n}$$

The numerator of $\bar{x}$ can be written more informally as Σx_i where the summation is over all sample observations.

For reporting $\bar{x}$, we recommend using decimal accuracy of one digit more than the accuracy of the x_i's. Thus if observations are stopping distances with $x_1 = 125$, $x_2 = 131$, and so on, we might have $\bar{x} = 127.3$ ft.

Example 1.11 A class was assigned to make wingspan measurements at home. The wingspan is the horizontal measurement from fingertip to fingertip with outstretched arms. Here are the measurements given by 21 of the students.

$$\begin{array}{lllllll} x_1 = 60 & x_2 = 64 & x_3 = 72 & x_4 = 63 & x_5 = 66 & x_6 = 62 & x_7 = 75 \\ x_8 = 66 & x_9 = 59 & x_{10} = 75 & x_{11} = 69 & x_{12} = 62 & x_{13} = 63 & x_{14} = 61 \\ x_{15} = 65 & x_{16} = 67 & x_{17} = 65 & x_{18} = 69 & x_{19} = 95 & x_{20} = 60 & x_{21} = 70 \end{array}$$

Figure 1.12 shows a stem-and-leaf display of the data; a wingspan in the 60's appears to be "typical."

5H	9
6L	00122334
6H	5566799
7L	02
7H	55
8L	
8H	
9L	
9H	5

Figure 1.12 A stem-and-leaf display of the wingspan data

With $\Sigma x_i = 1408$, the sample mean is

$$\bar{x} = \frac{1408}{21} = 67.0$$

a value consistent with information conveyed by the stem-and-leaf display. ■

A physical interpretation of $\bar{x}$ demonstrates how it measures the location (center) of a sample. Think of drawing and scaling a horizontal measurement axis, and then representing each sample observation by a 1-lb weight placed at the corresponding point on the axis. The only point at which a fulcrum can be placed to balance the system of weights is the point corresponding to the value of $\bar{x}$ (see Figure 1.13). The system balances because, as shown in the next section, $\Sigma(x_i - \bar{x}) = 0$, so the net total tendency to turn about $\bar{x}$ is 0.

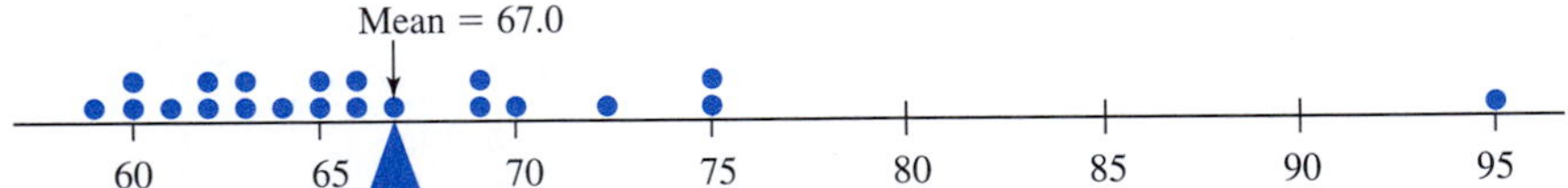

Figure 1.13 The mean as the balance point for a system of weights

Just as $\bar{x}$ represents the average value of the observations in a sample, the average of all values in the population can in principle be calculated. This average is called the **population mean** and is denoted by the Greek letter μ. When there are N values in the population (a finite population), then μ = (sum of the N population values)/N. In Chapters 3 and 4, we will give a more general definition for μ that applies to both finite and (conceptually) infinite populations. Just as $\bar{x}$ is an interesting and important measure of sample location, μ is an interesting and important (often the most important) characteristic of a population. In the chapters on statistical inference, we will present methods based on the sample mean for drawing conclusions about a population mean. For example, we might use the sample mean $\bar{x} = 67.0$ computed in Example 1.11 as a *point estimate* (a single number that is our "best" guess) of μ = the true average wingspan for all students in introductory statistics classes.

The mean suffers from one deficiency that makes it an inappropriate measure of center under some circumstances: Its value can be greatly affected by the presence of even a single outlier (unusually large or small observation). In Example 1.11, the value

$x_{19} = 95$ is obviously an outlier. Without this observation, $\bar{x} = 1313/20 = 65.7$; the outlier increases the mean by 1.4 inches. The value 95 is clearly an error—this student is only 70 inches tall, and there is no way such a student could have a wingspan of almost 8 feet. As Leonardo da Vinci noticed, wingspan is usually quite close to height.

Data on housing prices in various metropolitan areas often contains outliers (those lucky enough to live in palatial accommodations), in which case the use of average price as a measure of center will typically be misleading. We will momentarily propose an alternative to the mean, namely the median, that is insensitive to outliers (recent New York City data gave a median price of less than \$700,000 and a mean price exceeding \$1,000,000). However, the mean is still by far the most widely used measure of center, largely because there are many populations for which outliers are very scarce. When sampling from such a population (a normal or bell-shaped distribution being the most important example), outliers are highly unlikely to enter the sample. The sample mean will then tend to be stable and quite representative of the sample.

The Median

The word *median* is synonymous with "middle," and the sample median is indeed the middle value when the observations are ordered from smallest to largest. When the observations are denoted by $x_1, \ldots, x_n$, we will use the symbol $\tilde{x}$ to represent the sample median.

DEFINITION

The **sample median** is obtained by first ordering the n observations from smallest to largest (with any repeated values included so that every sample observation appears in the ordered list). Then,

$$\tilde{x} = \begin{cases} \text{The single middle value if } n \text{ is odd} & = \left(\dfrac{n+1}{2}\right)^{\text{th}} \text{ ordered value} \\ \text{The average of the two middle values if } n \text{ is even} & = \text{average of } \left(\dfrac{n}{2}\right)^{\text{th}} \text{ and } \left(\dfrac{n}{2}+1\right)^{\text{th}} \text{ ordered values} \end{cases}$$

Example 1.12

The risk of developing iron deficiency is especially high during pregnancy. The problem with detecting such deficiency is that some methods for determining iron status can be affected by the state of pregnancy itself. Consider the following data on transferrin receptor concentration for a sample of women with laboratory evidence of overt iron-deficiency anemia ("Serum Transferrin Receptor for the Detection of Iron Deficiency in Pregnancy," *Amer. J. Clin. Nutrit.*, 1991: 1077–1081):

$$x_1 = 15.2 \quad x_2 = 9.3 \quad x_3 = 7.6 \quad x_4 = 11.9 \quad x_5 = 10.4 \quad x_6 = 9.7$$
$$x_7 = 20.4 \quad x_8 = 9.4 \quad x_9 = 11.5 \quad x_{10} = 16.2 \quad x_{11} = 9.4 \quad x_{12} = 8.3$$

The list of ordered values is

$$7.6 \quad 8.3 \quad 9.3 \quad 9.4 \quad 9.4 \quad 9.7 \quad 10.4 \quad 11.5 \quad 11.9 \quad 15.2 \quad 16.2 \quad 20.4$$

Since $n = 12$ is even, we average the $n/2 =$ sixth- and seventh-ordered values:

$$\text{sample median} = \frac{9.7 + 10.4}{2} = 10.05$$

Notice that if the largest observation, 20.4, had not appeared in the sample, the resulting sample median for the $n = 11$ observations would have been the single middle value, 9.7 [the $(n + 1)/2 =$ sixth-ordered value]. The sample mean is $\bar{x} = \Sigma x_i/n = 139.3/12 = 11.61$, which is somewhat larger than the median because of the outliers, 15.2, 16.2, and 20.4. ■

The data in Example 1.12 illustrates an important property of $\tilde{x}$ in contrast to $\bar{x}$: The sample median is very insensitive to a number of extremely small or extremely large data values. If, for example, we increased the two largest x_i's from 16.2 and 20.4 to 26.2 and 30.4, respectively, $\tilde{x}$ would be unaffected. Thus, in the treatment of outlying data values, $\bar{x}$ and $\tilde{x}$ are at opposite ends of a spectrum: $\bar{x}$ is sensitive to even one such value, whereas $\tilde{x}$ is insensitive to a large number of outlying values.

Because the large values in the sample of Example 1.12 affect $\bar{x}$ more than $\tilde{x}$, $\tilde{x} < \bar{x}$ for that data. Although $\bar{x}$ and $\tilde{x}$ both provide a measure for the center of a data set, they will not in general be equal because they focus on different aspects of the sample.

Analogous to $\tilde{x}$ as the middle value in the sample is a middle value in the population, the **population median**, denoted by $\tilde{\mu}$. As with $\bar{x}$ and μ, we can think of using the sample median $\tilde{x}$ to make an inference about $\tilde{\mu}$. In Example 1.12, we might use $\tilde{x} = 10.05$ as an estimate of the median concentration in the entire population from which the sample was selected. A median is often used to describe income or salary data (because it is not greatly influenced by a few large salaries). If the median salary for a sample of statisticians were $\tilde{x} = \$66,416$, we might use this as a basis for concluding that the median salary for all statisticians exceeds \$60,000.

The population mean μ and median $\tilde{\mu}$ will not generally be identical. If the population distribution is positively or negatively skewed, as pictured in Figure 1.14, then $\mu \neq \tilde{\mu}$. When this is the case, in making inferences we must first decide which of the two population characteristics is of greater interest and then proceed accordingly.

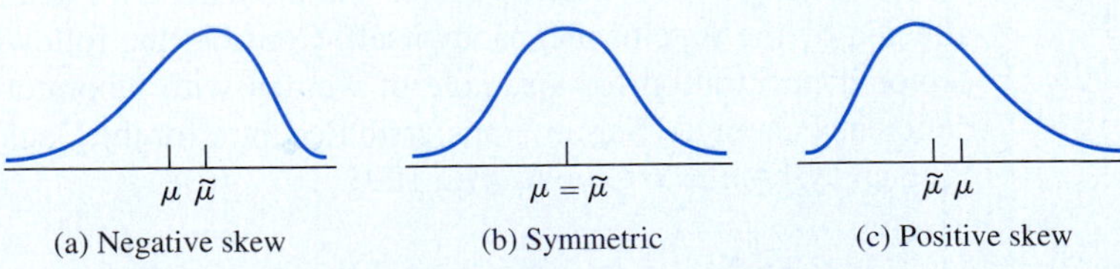

Figure 1.14 Three different shapes for a population distribution

Other Measures of Location: Quartiles, Percentiles, and Trimmed Means

The median (population or sample) divides the data set into two parts of equal size. To obtain finer measures of location, we could divide the data into more than two such parts. Roughly speaking, quartiles divide the data set into four equal parts, with the observations above the third quartile constituting the upper quarter of the data set, the second quartile being identical to the median, and the first quartile separating the lower quarter from the upper three-quarters. Similarly, a data set (sample or population) can be even more finely divided using percentiles; the 99th percentile separates the highest 1% from the bottom 99%, and so on. Unless the number of observations is a multiple of 100, care must be exercised in obtaining percentiles. We will use percentiles in Chapter 4 in connection with certain models for infinite populations and so postpone discussion until that point.

The sample mean and sample median are influenced by outlying values in a very different manner—the mean greatly and the median not at all. Since extreme behavior of either type might be undesirable, we briefly consider alternative measures that are neither as sensitive as $\bar{x}$ nor as insensitive as $\tilde{x}$. To motivate these alternatives, note that $\bar{x}$ and $\tilde{x}$ are at opposite extremes of the same "family" of measures. After the data set is ordered, $\tilde{x}$ is computed by throwing away as many values on each end as one can without eliminating everything (leaving just one or two middle values) and averaging what is left. On the other hand, to compute $\bar{x}$ one throws away nothing before averaging. To paraphrase, the mean involves trimming 0% from each end of the sample, whereas for the median the maximum possible amount is trimmed from each end. A **trimmed mean** is a compromise between $\bar{x}$ and $\tilde{x}$. A 10% trimmed mean, for example, would be computed by eliminating the smallest 10% and the largest 10% of the sample and then averaging what remains.

Example 1.13 Consider the following 20 observations, ordered from smallest to largest, each one representing the lifetime (in hours) of a certain type of incandescent lamp:

612	623	666	744	883	898	964	970	983	1003
1016	1022	1029	1058	1085	1088	1122	1135	1197	1201

The average of all 20 observations is $\bar{x} = 965.0$, and $\tilde{x} = 1009.5$. The 10% trimmed mean is obtained by deleting the smallest two observations (612 and 623) and the largest two (1197 and 1201) and then averaging the remaining 16 to obtain $\bar{x}_{\text{tr}(10)} = 979.1$. The effect of trimming here is to produce a "central value" that is somewhat above the mean ($\bar{x}$ is pulled down by a few small lifetimes) and yet considerably below the median. Similarly, the 20% trimmed mean averages the middle 12 values to obtain $\bar{x}_{\text{tr}(20)} = 999.9$, even closer to the median. (See Figure 1.15.)

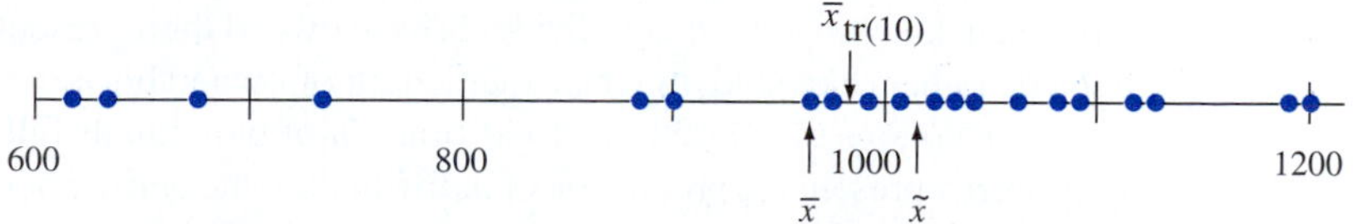

Figure 1.15 Dotplot of lifetimes (in hours) of incandescent lamps ■

Generally speaking, using a trimmed mean with a moderate trimming proportion (between 5 and 25%) will yield a measure that is neither as sensitive to outliers as the mean nor as insensitive as the median. For this reason, trimmed means have merited increasing attention from statisticians for both descriptive and inferential purposes. More will be said about trimmed means when point estimation is discussed in Chapter 7. As a final point, if the trimming proportion is denoted by α and $n\alpha$ is not an integer, then it is not obvious how the $100\alpha\%$ trimmed mean should be computed. For example, if $\alpha = .10$ (10%) and $n = 22$, then $n\alpha = (22)(.10) = 2.2$, and we cannot trim 2.2 observations from each end of the ordered sample. In this case, the 10% trimmed mean would be obtained by first trimming two observations from each end and calculating $\bar{x}_{tr}$, then trimming three and calculating $\bar{x}_{tr}$, and finally interpolating between the two values to obtain $\bar{x}_{tr(10)}$.

Categorical Data and Sample Proportions

When the data is categorical, a frequency distribution or relative frequency distribution provides an effective tabular summary of the data. The natural numerical summary quantities in this situation are the individual frequencies and the relative frequencies. For example, if a survey of individuals who own stereo receivers is undertaken to study brand preference, then each individual in the sample would identify the brand of receiver that he or she owned, from which we could count the number owning Sony, Marantz, Pioneer, and so on. Consider sampling a dichotomous population—one that consists of only two categories (such as voted or did not vote in the last election, does or does not own a stereo receiver, etc.). If we let x denote the number in the sample falling in category A, then the number in category B is $n - x$. The relative frequency or *sample proportion* in category A is x/n and the sample proportion in category B is $1 - x/n$. Let's denote a response that falls in category A by a 1 and a response that falls in category B by a 0. A sample size of $n = 10$ might then yield the responses 1, 1, 0, 1, 1, 1, 0, 0, 1, 1. The sample mean for this numerical sample is (since number of 1's $= x = 7$)

$$\frac{x_1 + \cdots + x_n}{n} = \frac{1 + 1 + 0 + \cdots + 1 + 1}{10} = \frac{7}{10} = \frac{x}{n} = \text{sample proportion}$$

This result can be generalized and summarized as follows: *If in a categorical data situation we focus attention on a particular category and code the sample results so that a* 1 *is recorded for an individual in the category and a* 0 *for an individual not in the category, then the sample proportion of individuals in the category is the sample mean of the sequence of* 1*'s and* 0*'s*. Thus a sample mean can be used to summarize the results of a categorical sample. These remarks also apply to situations in which categories are defined by grouping values in a numerical sample or population (e.g., we might be interested in knowing whether individuals have owned their present automobile for at least 5 years, rather than studying the exact length of ownership).

Analogous to the sample proportion x/n of individuals falling in a particular category, let p represent the proportion of individuals in the entire population falling in the category. As with x/n, p is a quantity between 0 and 1. While x/n is a sample characteristic,

p is a characteristic of the population. The relationship between the two parallels the relationship between $\tilde{x}$ and $\tilde{\mu}$ and between $\bar{x}$ and μ. In particular, we will subsequently use x/n to make inferences about p. If, for example, a sample of 100 car owners reveals that 22 owned their car at least 5 years, then we might use $22/100 = .22$ as a point estimate of the proportion of all owners who have owned their car at least 5 years. We will study the properties of x/n as an estimator of p and see how x/n can be used to answer other inferential questions. With k categories ($k > 2$), we can use the k sample proportions to answer questions about the population proportions $p_1, \ldots, p_k$.

Exercises Section 1.3 (30–40)

30. The article "The Pedaling Technique of Elite Endurance Cyclists" (*Int. J. Sport Biomechanics*, 1991: 29–53) reported the accompanying data on single-leg power at a high workload:

244	191	160	187	180	176	174
205	211	183	211	180	194	200

a. Calculate and interpret the sample mean and median.

b. Suppose that the first observation had been 204 rather than 244. How would the mean and median change?

c. Calculate a trimmed mean by eliminating the smallest and largest sample observations. What is the corresponding trimming percentage?

d. The article also reported values of single-leg power for a low workload. The sample mean for $n = 13$ observations was $\bar{x} = 119.8$ (actually 119.7692), and the 14th observation, somewhat of an outlier, was 159. What is the value of $\bar{x}$ for the entire sample?

31. In Superbowl XXXVII, Michael Pittman of Tampa Bay rushed (ran with the football) 17 times on first down, and the results were the following gains in yards:

23	1	4	1	6	5	9	6	2
−1	3	2	0	2	24	1	1	

a. Determine the value of the sample mean.

b. Determine the value of the sample median. Why is it so different from the mean?

c. Calculate a trimmed mean by deleting the smallest and largest observations. What is the corresponding trimming percentage? How does the value of this $\bar{x}_{tr}$ compare to the mean and median?

32. The minimum injection pressure (psi) for injection molding specimens of high amylose corn was determined for eight different specimens (higher pressure corresponds to greater processing difficulty), resulting in the following observations (from "Thermoplastic Starch Blends with a Polyethylene-Co-Vinyl Alcohol: Processability and Physical Properties," *Polymer Engrg. & Sci.*, 1994: 17–23):

15.0	13.0	18.0	14.5	12.0	11.0	8.9	8.0

a. Determine the values of the sample mean, sample median, and 12.5% trimmed mean, and compare these values.

b. By how much could the smallest sample observation, currently 8.0, be increased without affecting the value of the sample median?

c. Suppose we want the values of the sample mean and median when the observations are expressed in kilograms per square inch (ksi) rather than psi. Is it necessary to reexpress each observation in ksi, or can the values calculated in part (a) be used directly? *Hint*: 1 kg = 2.2 lb.

33. A sample of 26 offshore oil workers took part in a simulated escape exercise, resulting in the accompanying data on time (sec) to complete the escape ("Oxygen Consumption and Ventilation During Escape from an Offshore Platform," *Ergonomics*, 1997: 281–292):

389	356	359	363	375	424	325	394	402
373	373	370	364	366	364	325	339	393
392	369	374	359	356	403	334	397	

a. Construct a stem-and-leaf display of the data. How does it suggest that the sample mean and median will compare?

b. Calculate the values of the sample mean and median. *Hint*: $\Sigma x_i = 9638$.

c. By how much could the largest time, currently 424, be increased without affecting the value of the sample median? By how much could this value be decreased without affecting the value of the sample median?

d. What are the values of $\bar{x}$ and $\tilde{x}$ when the observations are reexpressed in minutes?

34. The article "Snow Cover and Temperature Relationships in North America and Eurasia" (*J. Climate Appl. Meteorol.*, 1983: 460–469) used statistical techniques to relate the amount of snow cover on each continent to average continental temperature. Data presented there included the following ten observations on October snow cover for Eurasia during the years 1970–1979 (in million km^2):

6.5 12.0 14.9 10.0 10.7 7.9 21.9 12.5 14.5 9.2

What would you report as a representative, or typical, value of October snow cover for this period, and what prompted your choice?

35. Blood pressure values are often reported to the nearest 5 mmHg (100, 105, 110, etc.). Suppose the actual blood pressure values for nine randomly selected individuals are

118.6 127.4 138.4 130.0 113.7 122.0 108.3
131.5 133.2

a. What is the median of the *reported* blood pressure values?

b. Suppose the blood pressure of the second individual is 127.6 rather than 127.4 (a small change in a single value). How does this affect the median of the reported values? What does this say about the sensitivity of the median to rounding or grouping in the data?

36. The propagation of fatigue cracks in various aircraft parts has been the subject of extensive study in recent years. The accompanying data consists of propagation lives (flight hours/10^4) to reach a given crack size in fastener holes intended for use in military aircraft ("Statistical Crack Propagation in Fastener Holes under Spectrum Loading," *J. Aircraft*, 1983: 1028–1032):

.736 .863 .865 .913 .915 .937 .983 1.007
1.011 1.064 1.109 1.132 1.140 1.153 1.253 1.394

a. Compute and compare the values of the sample mean and median.

b. By how much could the largest sample observation be decreased without affecting the value of the median?

37. Compute the sample median, 25% trimmed mean, 10% trimmed mean, and sample mean for the microdrill data given in Exercise 25, and compare these measures.

38. A sample of $n = 10$ automobiles was selected, and each was subjected to a 5-mph crash test. Denoting a car with no visible damage by S (for success) and a car with such damage by F, results were as follows:

S S F S S S F F S S

a. What is the value of the sample proportion of successes x/n?

b. Replace each S with a 1 and each F with a 0. Then calculate $\bar{x}$ for this numerically coded sample. How does $\bar{x}$ compare to x/n?

c. Suppose it is decided to include 15 more cars in the experiment. How many of these would have to be S's to give $x/n = .80$ for the entire sample of 25 cars?

39. **a.** If a constant c is added to each x_i in a sample, yielding $y_i = x_i + c$, how do the sample mean and median of the y_i's relate to the mean and median of the x_i's? Verify your conjectures.

b. If each x_i is multiplied by a constant c, yielding $y_i = cx_i$, answer the question of part (a). Again, verify your conjectures.

40. An experiment to study the lifetime (in hours) for a certain type of component involved putting ten components into operation and observing them for 100 hours. Eight of the components failed during that period, and those lifetimes were recorded. Denote the lifetimes of the two components still functioning after 100 hours by 100+. The resulting sample observations were

48 79 100+ 35 92 86 57 100+ 17 29

Which of the measures of center discussed in this section can be calculated, and what are the values of those measures? (*Note*: The data from this experiment is said to be "censored on the right.")

1.4 Measures of Variability

Reporting a measure of center gives only partial information about a data set or distribution. Different samples or populations may have identical measures of center yet differ from one another in other important ways. Figure 1.16 shows dotplots of three samples with the same mean and median, yet the extent of spread about the center is different for all three samples. The first sample has the largest amount of variability, the third has the smallest amount, and the second is intermediate to the other two in this respect.

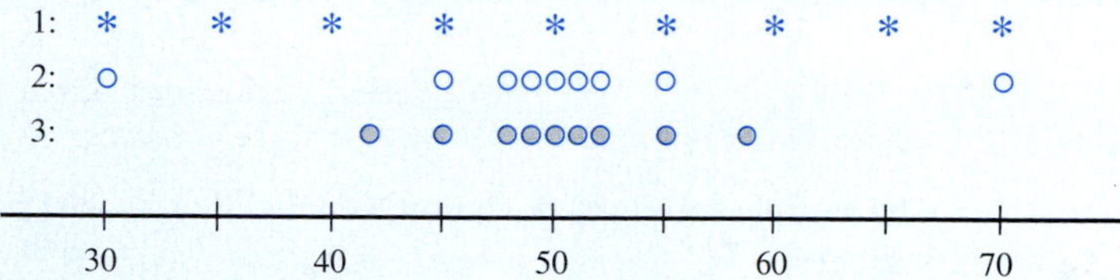

Figure 1.16 Samples with identical measures of center but different amounts of variability

Measures of Variability for Sample Data

The simplest measure of variability in a sample is the **range**, which is the difference between the largest and smallest sample values. Notice that the value of the range for sample 1 in Figure 1.16 is much larger than it is for sample 3, reflecting more variability in the first sample than in the third. A defect of the range, though, is that it depends on only the two most extreme observations and disregards the positions of the remaining $n - 2$ values. Samples 1 and 2 in Figure 1.16 have identical ranges, yet when we take into account the observations between the two extremes, there is much less variability or dispersion in the second sample than in the first.

Our primary measures of variability involve the **deviations from the mean**, $x_1 - \bar{x}, x_2 - \bar{x}, \ldots, x_n - \bar{x}$. That is, the deviations from the mean are obtained by subtracting $\bar{x}$ from each of the n sample observations. A deviation will be positive if the observation is larger than the mean (to the right of the mean on the measurement axis) and negative if the observation is smaller than the mean. If all the deviations are small in magnitude, then all x_i's are close to the mean and there is little variability. On the other hand, if some of the deviations are large in magnitude, then some x_i's lie far from $\bar{x}$, suggesting a greater amount of variability. A simple way to combine the deviations into a single quantity is to average them (sum them and divide by n). Unfortunately, there is a major problem with this suggestion:

$$\text{sum of deviations} = \sum_{i=1}^{n} (x_i - \bar{x}) = 0$$

so that the average deviation is always zero. The verification uses several standard rules of summation and the fact that $\Sigma\bar{x} = \bar{x} + \bar{x} + \cdots + \bar{x} = n\bar{x}$:

$$\sum (x_i - \bar{x}) = \sum x_i - \sum \bar{x} = \sum x_i - n\bar{x} = \sum x_i - n\left(\frac{1}{n}\sum x_i\right) = 0$$

How can we change the deviations to nonnegative quantities so the positive and negative deviations do not counteract one another when they are combined? One possibility is to work with the absolute values of the deviations and calculate the average absolute deviation $\Sigma|x_i - \bar{x}|/n$. Because the absolute value operation leads to a number of theoretical difficulties, consider instead the squared deviations $(x_1 - \bar{x})^2, (x_2 - \bar{x})^2, \ldots, (x_n - \bar{x})^2$. Rather than use the average squared deviation $\Sigma(x_i - \bar{x})^2/n$, for several reasons we will divide the sum of squared deviations by $n - 1$ rather than n.

DEFINITION

The **sample variance**, denoted by s^2, is given by

$$s^2 = \frac{\sum (x_i - \bar{x})^2}{n - 1} = \frac{S_{xx}}{n - 1}$$

The **sample standard deviation**, denoted by s, is the (positive) square root of the variance:

$$s = \sqrt{s^2}$$

The unit for s is the same as the unit for each of the x_i's. If, for example, the observations are fuel efficiencies in miles per gallon, then we might have $s = 2.0$ mpg. A rough interpretation of the sample standard deviation is that it is the size of a typical or representative deviation from the sample mean within the given sample. Thus if $s = 2.0$ mpg, then some x_i's in the sample are closer than 2.0 to $\bar{x}$, whereas others are farther away; 2.0 is a representative (or "standard") deviation from the mean fuel efficiency. If $s = 3.0$ for a second sample of cars of another type, a typical deviation in this sample is roughly 1.5 times what it is in the first sample, an indication of more variability in the second sample.

Example 1.14 Traumatic knee dislocation often requires surgery to repair ruptured ligaments. One measure of recovery is range of motion (measured as the angle formed when, starting with the leg straight, the knee is bent as far as possible). The given data on postsurgical range of motion (Table 1.3 on the next page) appeared in the article "Reconstruction of the Anterior and Posterior Cruciate Ligaments After Knee Dislocation" (*Amer. J. Sports Med.*, 1999: 189–197). Effects of rounding account for the sum of the deviations not being exactly zero. The numerator of s^2 is 1579.1; therefore $s^2 = 1579.1/(13 - 1) = 1579.1/12 = 131.59$ and $s = \sqrt{131.59} = 11.47$. ■

Motivation for s^2

To explain why s^2 rather than the average squared deviation is used to measure variability, note first that whereas s^2 measures sample variability, there is a measure of variability in the population called the *population variance*. We will use σ^2 (the square of the lowercase Greek letter sigma) to denote the population variance and σ to denote the population standard deviation (the square root of σ^2). When the population is finite and consists of N values,

Table 1.3 Data for Example 1.14

x_i	$x_i - \bar{x}$	$(x_i - \bar{x})^2$
154	23.62	557.904
142	11.62	135.024
137	6.62	43.824
133	2.62	6.864
122	−8.38	70.224
126	−4.38	19.184
135	4.62	21.344
135	4.62	21.344
108	−22.38	500.864
120	−10.38	107.744
127	−3.38	11.424
134	3.62	13.104
122	−8.38	70.224
$\sum x_i = 1695$	$\sum (x_i - \bar{x}) = .06$	$\sum (x_i - \bar{x})^2 = 1579.1$

$\bar{x} = \dfrac{1695}{13} = 130.38$

$$\sigma^2 = \sum_{i=1}^{N} (x_i - \mu)^2/N$$

which is the average of all squared deviations from the population mean (for the population, the divisor is N and not $N - 1$). More general definitions of σ^2 appear in Chapters 3 and 4.

Just as $\bar{x}$ will be used to make inferences about the population mean μ, we should define the sample variance so that it can be used to make inferences about σ^2. Now note that σ^2 involves squared deviations about the population mean μ. If we actually knew the value of μ, then we could define the sample variance as the average squared deviation of the sample x_i's about μ. However, the value of μ is almost never known, so the sum of squared deviations about $\bar{x}$ must be used. But *the x_i's tend to be closer to their average $\bar{x}$ than to the population average μ, so to compensate for this the divisor $n - 1$ is used rather than n.* In other words, if we used a divisor n in the sample variance, then the resulting quantity would tend to underestimate σ^2 (produce estimated values that are too small on the average), whereas dividing by the slightly smaller $n - 1$ corrects this underestimating.

It is customary to refer to s^2 as being based on $n - 1$ **degrees of freedom** (df). This terminology results from the fact that although s^2 is based on the n quantities $x_1 - \bar{x}, x_2 - \bar{x}, \ldots, x_n - \bar{x}$, these sum to 0, so specifying the values of any $n - 1$ of the quantities determines the remaining value. For example, if $n = 4$ and $x_1 - \bar{x} = 8, x_2 - \bar{x} = -6$, and $x_4 - \bar{x} = -4$, then automatically $x_3 - \bar{x} = 2$, so only three of the four values of $x_i - \bar{x}$ are freely determined (3 df).

A Computing Formula for s^2

Computing and squaring the deviations can be tedious, especially if enough decimal accuracy is being used in $\bar{x}$ to guard against the effects of rounding. An alternative formula for the numerator of s^2 circumvents the need for all the subtraction necessary to obtain the deviations. The formula involves both $(\sum x_i)^2$, summing and then squaring, and $\sum x_i^2$, squaring and then summing.

An alternative expression for the numerator of s^2 is

$$S_{xx} = \sum (x_i - \bar{x})^2 = \sum x_i^2 - \frac{\left(\sum x_i\right)^2}{n}$$

Proof Because $\bar{x} = \sum x_i/n$, $n\bar{x}^2 = (\sum x_i)^2/n$. Then,

$$\sum (x_i - \bar{x})^2 = \sum (x_i^2 - 2\bar{x} \cdot x_i + \bar{x}^2) = \sum x_i^2 - 2\bar{x} \sum x_i + \sum (\bar{x})^2$$

$$= \sum x_i^2 - 2\bar{x} \cdot n\bar{x} + n(\bar{x})^2 = \sum x_i^2 - n(\bar{x})^2$$ ■

Example 1.15 The amount of light reflectance by leaves has been used for various purposes, including evaluation of turf color, estimation of nitrogen status, and measurement of biomass. The article "Leaf Reflectance–Nitrogen–Chlorophyll Relations in Buffel-Grass" (*Photogrammetric Engrg. Remote Sensing*, 1985: 463–466) gave the following observations, obtained using spectrophotogrammetry, on leaf reflectance under specified experimental conditions.

Observation	x_i	x_i^2	Observation	x_i	x_i^2
1	15.2	231.04	9	12.7	161.29
2	16.8	282.24	10	15.8	249.64
3	12.6	158.76	11	19.2	368.64
4	13.2	174.24	12	12.7	161.29
5	12.8	163.84	13	15.6	243.36
6	13.8	190.44	14	13.5	182.25
7	16.3	265.69	15	12.9	166.41
8	13.0	169.00			
				$\sum x_i = 216.1$	$\sum x_i^2 = 3168.13$

The computational formula now gives

$$S_{xx} = \sum x_i^2 - \frac{\left(\sum x_i\right)^2}{n} = 3168.13 - \frac{(216.1)^2}{15}$$

$$= 3168.13 - 3113.28 = 54.85$$

from which $s^2 = S_{xx}/(n - 1) = 54.85/14 = 3.92$ and $s = 1.98$. ■

The shortcut method can yield values of s^2 and s that differ from the values computed using the definitions. These differences are due to effects of rounding and will not be important in most samples. To minimize the effects of rounding when using the shortcut formula, intermediate calculations should be done using several more significant digits than are to be retained in the final answer. Because the numerator of s^2 is the sum of nonnegative quantities (squared deviations), s^2 is guaranteed to be nonnegative. Yet if the shortcut method is used, particularly with data having little variability, a slight numerical error can result in a negative numerator $[\Sigma x_i^2$ smaller than $(\Sigma x_i)^2/n]$. If your value of s^2 is negative, you have made a computational error.

Several other properties of s^2 can facilitate its computation.

PROPOSITION Let $x_1, x_2, \ldots, x_n$ be a sample and c be a constant.

1. If $y_1 = x_1 + c, y_2 = x_2 + c, \ldots, y_n = x_n + c$, then $s_y^2 = s_x^2$, and

2. If $y_1 = cx_1, \ldots, y_n = cx_n$, then $s_y^2 = c^2 s_x^2$, $s_y = |c| s_x$,

where s_x^2 is the sample variance of the x's and s_y^2 is the sample variance of the y's.

In words, Result 1 says that if a constant c is added to (or subtracted from) each data value, the variance is unchanged. This is intuitive, since adding or subtracting c shifts the location of the data set but leaves distances between data values unchanged. According to Result 2, multiplication of each x_i by c results in s^2 being multiplied by a factor of c^2. These properties can be proved by noting in Result 1 that $\bar{y} = \bar{x} + c$ and in Result 2 that $\bar{y} = c\bar{x}$ (see Exercise 59).

Boxplots

Stem-and-leaf displays and histograms convey rather general impressions about a data set, whereas a single summary such as the mean or standard deviation focuses on just one aspect of the data. In recent years, a pictorial summary called a *boxplot* has been used successfully to describe several of a data set's most prominent features. These features include (1) center, (2) spread, (3) the extent and nature of any departure from symmetry, and (4) identification of "outliers," observations that lie unusually far from the main body of the data. Because even a single outlier can drastically affect the values of $\bar{x}$ and s, a boxplot is based on measures that are "resistant" to the presence of a few outliers—the median and a measure of spread called the *fourth spread*.

DEFINITION Order the n observations from smallest to largest and separate the smallest half from the largest half; the median $\tilde{x}$ is included in both halves if n is odd. Then the **lower fourth** is the median of the smallest half and the **upper fourth** is the median of the largest half. A measure of spread that is resistant to outliers is the **fourth spread** f_s, given by

$$f_s = \text{upper fourth} - \text{lower fourth}$$

Roughly speaking, the fourth spread is unaffected by the positions of those observations in the smallest 25% or the largest 25% of the data.

The simplest boxplot is based on the following five-number summary:

smallest x_i lower fourth median upper fourth largest x_i

First, draw a horizontal measurement scale. Then place a rectangle above this axis; the left edge of the rectangle is at the lower fourth, and the right edge is at the upper fourth (so box width $= f_s$). Place a vertical line segment or some other symbol inside the rectangle at the location of the median; the position of the median symbol relative to the two edges conveys information about skewness in the middle 50% of the data. Finally, draw "whiskers" out from either end of the rectangle to the smallest and largest observations. A boxplot with a vertical orientation can also be drawn by making obvious modifications in the construction process.

Example 1.16 Ultrasound was used to gather the accompanying corrosion data on the thickness of the floor plate of an aboveground tank used to store crude oil ("Statistical Analysis of UT Corrosion Data from Floor Plates of a Crude Oil Aboveground Storage Tank," *Materials Eval.*, 1994: 846–849); each observation is the largest pit depth in the plate, expressed in milli-in.

40 52 55 60 70 75 85 85 90 90 92 94 94 95 98 100 115 125 125

The five-number summary is as follows:

smallest $x_i = 40$ lower fourth $= 72.5$ $\tilde{x} = 90$ upper fourth $= 96.5$
largest $x_i = 125$

Figure 1.17 shows the resulting boxplot. The right edge of the box is much closer to the median than is the left edge, indicating a very substantial skew in the middle half of the data. The box width (f_s) is also reasonably large relative to the range of the data (distance between the tips of the whiskers).

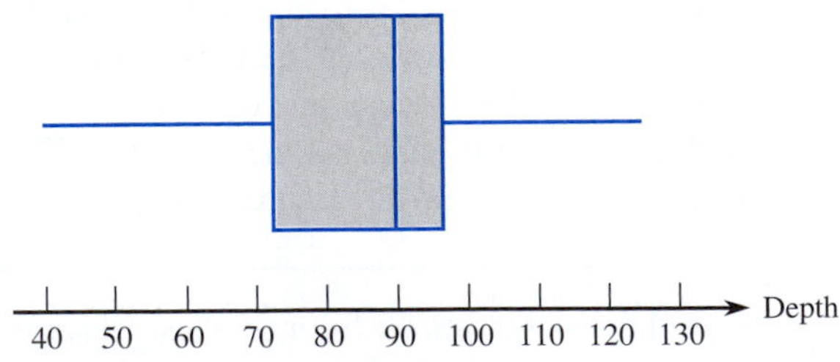

Figure 1.17 A boxplot of the corrosion data

Figure 1.18 shows MINITAB output from a request to describe the corrosion data. The trimmed mean is the average of the 17 observations that remain after the largest and smallest values are deleted (trimming percentage $\approx$ 5%). Q1 and Q3 are the lower and upper quartiles; these are similar to the fourths but are calculated in a slightly different manner. SE Mean is $s/\sqrt{n}$; this will be an important quantity in our subsequent work concerning inferences about μ.

```
Variable          N        Mean     Median     TrMean     StDev    SE Mean
depth            19       86.32      90.00      86.76     23.32       5.35
Variable    Minimum     Maximum         Q1         Q3
depth         40.00      125.00      70.00      98.00
```

Figure 1.18 MINITAB description of the pit-depth data ■

Boxplots That Show Outliers

A boxplot can be embellished to indicate explicitly the presence of outliers.

DEFINITION

Any observation farther than $1.5f_s$ from the closest fourth is an **outlier**. An outlier is **extreme** if it is more than $3f_s$ from the nearest fourth, and it is **mild** otherwise.

Many inferential procedures are based on the assumption that the sample came from a normal distribution. Even a single extreme outlier in the sample warns the investigator that such procedures should not be used, and the presence of several mild outliers conveys the same message.

Let's now modify our previous construction of a boxplot by drawing a whisker out from each end of the box to the smallest and largest observations that are *not* outliers. Each mild outlier is represented by a closed circle and each extreme outlier by an open circle. Some statistical computer packages do not distinguish between mild and extreme outliers.

Example 1.17 The effects of partial discharges on the degradation of insulation cavity material have important implications for the lifetimes of high-voltage components. Consider the following sample of $n = 25$ pulse widths from slow discharges in a cylindrical cavity made of polyethylene. (This data is consistent with a histogram of 250 observations in the article "Assessment of Dielectric Degradation by Ultrawide-band PD Detection," *IEEE Trans. Dielectrics Electr. Insul.*, 1995: 744–760.) The article's author notes the impact of a wide variety of statistical tools on the interpretation of discharge data.

5.3 8.2 13.8 74.1 85.3 88.0 90.2 91.5 92.4 92.9 93.6 94.3 94.8
94.9 95.5 95.8 95.9 96.6 96.7 98.1 99.0 101.4 103.7 106.0 113.5

Relevant quantities are

$$\tilde{x} = 94.8 \qquad \text{lower fourth} = 90.2 \qquad \text{upper fourth} = 96.7$$
$$f_s = 6.5 \qquad 1.5f_s = 9.75 \qquad 3f_s = 19.50$$

Thus any observation smaller than $90.2 - 9.75 = 80.45$ or larger than $96.7 + 9.75 = 106.45$ is an outlier. There is one outlier at the upper end of the sample, and four outliers are at the lower end. Because $90.2 - 19.5 = 70.7$, the three observations 5.3, 8.2, and 13.8 are extreme outliers; the other two outliers are mild. The whiskers extend out to

85.3 and 106.0, the most extreme observations that are not outliers. The resulting boxplot is in Figure 1.19. There is a great deal of negative skewness in the middle half of the sample as well as in the entire sample.

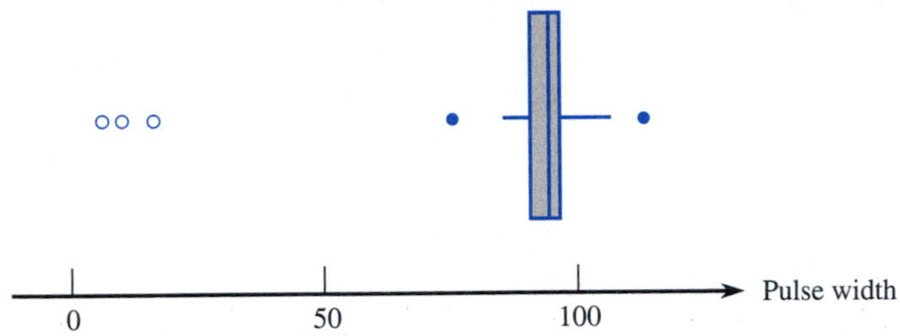

Figure 1.19 A boxplot of the pulse width data showing mild and extreme outliers ■

Comparative Boxplots

A comparative or side-by-side boxplot is a very effective way of revealing similarities and differences between two or more data sets consisting of observations on the same variable.

Example 1.18 In recent years, some evidence suggests that high indoor radon concentration may be linked to the development of childhood cancers, but many health professionals remain unconvinced. The article "Indoor Radon and Childhood Cancer" (*Lancet*, 1991: 1537–1538) presented the accompanying data on radon concentration (Bq/m^3) in two different samples of houses. The first sample consisted of houses in which a child diagnosed with cancer had been residing. Houses in the second sample had no recorded cases of childhood cancer. Figure 1.20 presents a stem-and-leaf display of the data.

1. Cancer		2. No cancer
9683795	0	95768397678993
86071815066815233150	1	12271713114
12302731	2	99494191
8349	3	839
5	4	
7	5	55
	6	
	7	
HI: 210	8	5

Stem: Tens digit
Leaf: Ones digit

Figure 1.20 Stem-and-leaf display for Example 1.18

Numerical summary quantities are as follows:

	$\bar{x}$	$\tilde{x}$	s	f_s
Cancer	22.8	16.0	31.7	11.0
No cancer	19.2	12.0	17.0	18.0

The values of both the mean and median suggest that the cancer sample is centered somewhat to the right of the no-cancer sample on the measurement scale. The mean, however, exaggerates the magnitude of this shift, largely because of the observation 210 in the cancer sample. The values of s suggest more variability in the cancer sample than in the no-cancer sample, but this impression is contradicted by the fourth spreads. Again, the observation 210, an extreme outlier, is the culprit. Figure 1.21 shows a comparative boxplot from the S-Plus computer package. The no-cancer box is stretched out compared with the cancer box ($f_s = 18$ vs. $f_s = 11$), and the positions of the median lines in the two boxes show much more skewness in the middle half of the no-cancer sample than the cancer sample. Outliers are represented by horizontal line segments, and there is no distinction between mild and extreme outliers.

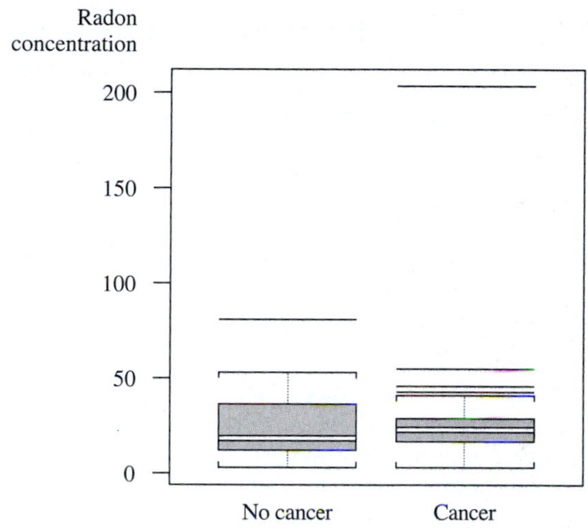

Figure 1.21 A boxplot of the data in Example 1.18, from S-Plus ■

Exercises Section 1.4 (41–59)

41. The article "Oxygen Consumption During Fire Suppression: Error of Heart Rate Estimation" (*Ergonomics*, 1991: 1469–1474) reported the following data on oxygen consumption (mL/kg/min) for a sample of ten firefighters performing a fire-suppression simulation:

29.5 49.3 30.6 28.2 28.0 26.3 33.9 29.4 23.5 31.6

Compute the following:

a. The sample range

b. The sample variance s^2 from the definition (i.e., by first computing deviations, then squaring them, etc.)

c. The sample standard deviation

d. s^2 using the shortcut method

42. The value of Young's modulus (GPa) was determined for cast plates consisting of certain intermetallic substrates, resulting in the following sample observations ("Strength and Modulus of a Molybdenum-Coated Ti-25Al-10Nb-3U-1Mo Intermetallic," *J. Mater. Engrg. Perform.*, 1997: 46–50):

116.4 115.9 114.6 115.2 115.8

a. Calculate $\bar{x}$ and the deviations from the mean.

b. Use the deviations calculated in part (a) to obtain the sample variance and the sample standard deviation.

c. Calculate s^2 by using the computational formula for the numerator S_{xx}.

d. Subtract 100 from each observation to obtain a sample of transformed values. Now calculate the sample variance of these transformed values, and compare it to s^2 for the original data.

43. The accompanying observations on stabilized viscosity (cP) for specimens of a certain grade of asphalt with 18% rubber added are from the article "Viscosity Characteristics of Rubber-Modified Asphalts" (*J. Mater. Civil Engrg.*, 1996: 153–156):

2781 2900 3013 2856 2888

a. What are the values of the sample mean and sample median?

b. Calculate the sample variance using the computational formula. (*Hint*: First subtract a convenient number from each observation.)

44. Calculate and interpret the values of the sample median, sample mean, and sample standard deviation for the following observations on fracture strength (MPa, read from a graph in "Heat-Resistant Active Brazing of Silicon Nitride: Mechanical Evaluation of Braze Joints," *Welding J.*, Aug. 1997):

87 93 96 98 105 114 128 131 142 168

45. Exercise 33 in Section 1.3 presented a sample of 26 escape times for oil workers in a simulated escape exercise. Calculate and interpret the sample standard deviation. (*Hint*: $\Sigma x_i = 9638$ and $\Sigma x_i^2 = 3{,}587{,}566$).

46. A study of the relationship between age and various visual functions (such as acuity and depth perception) reported the following observations on area of scleral lamina (mm^2) from human optic nerve heads ("Morphometry of Nerve Fiber Bundle Pores in the Optic Nerve Head of the Human," *Experiment. Eye Res.*, 1988: 559–568):

2.75 2.62 2.74 3.85 2.34 2.74 3.93 4.21 3.88
4.33 3.46 4.52 2.43 3.65 2.78 3.56 3.01

a. Calculate Σx_i and Σx_i^2.

b. Use the values calculated in part (a) to compute the sample variance s^2 and then the sample standard deviation s.

47. In 1997 a woman sued a computer keyboard manufacturer, charging that her repetitive stress injuries were caused by the keyboard (*Genessy v. Digital Equipment Corp.*). The injury awarded about \$3.5 million for pain and suffering, but the court then set aside that award as being unreasonable compensation. In making this determination, the court identified a "normative" group of 27 similar cases and specified a reasonable award as one within two standard deviations of the mean of the awards in the 27 cases. The 27 awards were (in \$1000s) 37, 60, 75, 115, 135, 140, 149, 150, 238, 290, 340, 410, 600, 750, 750, 750, 1050, 1100, 1139, 1150, 1200, 1200, 1250, 1576, 1700, 1825, and 2000, from which $\Sigma x_i = 20{,}179$, $\Sigma x_i^2 = 24{,}657{,}511$. What is the maximum possible amount that could be awarded under the two-standard-deviation rule?

48. The article "A Thin-Film Oxygen Uptake Test for the Evaluation of Automotive Crankcase Lubricants" (*Lubric. Engrg.*, 1984: 75–83) reported the following data on oxidation-induction time (min) for various commercial oils:

87 103 130 160 180 195 132 145 211 105 145
153 152 138 87 99 93 119 129

a. Calculate the sample variance and standard deviation.

b. If the observations were reexpressed in hours, what would be the resulting values of the sample variance and sample standard deviation? Answer without actually performing the reexpression.

49. The first four deviations from the mean in a sample of $n = 5$ reaction times were .3, .9, 1.0, and 1.3. What is the fifth deviation from the mean? Give a sample for which these are the five deviations from the mean.

50. Reconsider the data on area of scleral lamina given in Exercise 46.

a. Determine the lower and upper fourths.

b. Calculate the value of the fourth spread.

c. If the two largest sample values, 4.33 and 4.52, had instead been 5.33 and 5.52, how would this affect f_s? Explain.

d. By how much could the observation 2.34 be increased without affecting f_s? Explain.

e. If an 18th observation, $x_{18} = 4.60$, is added to the sample, what is f_s?

51. Reconsider these values of rushing yardage from Exercise 31 of this chapter:

23 1 4 1 6 5 9 6 2
−1 3 2 0 2 24 1 1

a. What are the values of the fourths, and what is the value of f_s?
b. Construct a boxplot based on the five-number summary, and comment on its features.
c. How large or small does an observation have to be to qualify as an outlier? As an extreme outlier?
d. By how much could the largest observation be decreased without affecting f_s?

52. Here is a stem-and-leaf display of the escape time data introduced in Exercise 33 of this chapter.

32	55
33	49
34	
35	6699
36	34469
37	03345
38	9
39	2347
40	23
41	
42	4

a. Determine the value of the fourth spread.
b. Are there any outliers in the sample? Any extreme outliers?
c. Construct a boxplot and comment on its features.
d. By how much could the largest observation, currently 424, be decreased without affecting the value of the fourth spread?

53. The amount of aluminum contamination (ppm) in plastic of a certain type was determined for a sample of 26 plastic specimens, resulting in the following data ("The Lognormal Distribution for Modeling Quality Data when the Mean Is Near Zero," *J. Qual. Tech.*, 1990: 105–110):

30	30	60	63	70	79	87	90	101
102	115	118	119	119	120	125	140	145
172	182	183	191	222	244	291	511	

Construct a boxplot that shows outliers, and comment on its features.

54. A sample of 20 glass bottles of a particular type was selected, and the internal pressure strength of each bottle was determined. Consider the following partial sample information:

median = 202.2 lower fourth = 196.0
upper fourth = 216.8

Three smallest observations 125.8 188.1 193.7
Three largest observations 221.3 230.5 250.2

a. Are there any outliers in the sample? Any extreme outliers?
b. Construct a boxplot that shows outliers, and comment on any interesting features.

55. A company utilizes two different machines to manufacture parts of a certain type. During a single shift, a sample of n = 20 parts produced by each machine is obtained, and the value of a particular critical dimension for each part is determined. The comparative boxplot below is constructed from the resulting data. Compare and contrast the two samples.

56. Blood cocaine concentration (mg/L) was determined both for a sample of individuals who had died from cocaine-induced excited delirium (ED) and for

Comparative boxplot for Exercise 55

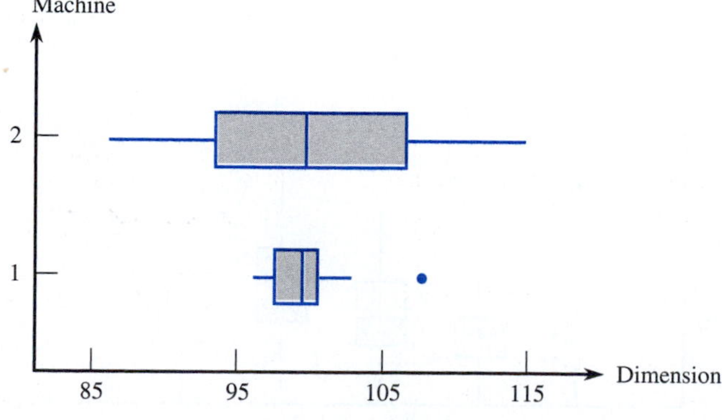

a sample of those who had died from a cocaine overdose without excited delirium; survival time for people in both groups was at most 6 hours. The accompanying data was read from a comparative boxplot in the article "Fatal Excited Delirium Following Cocaine Use" (*J. Forensic Sci.*, 1997: 25–31).

ED	0 0 0 0 .1 .1 .1 .1 .2 .2 .3 .3 .3 .4 .5 .7 .8 1.0 1.5 2.7 2.8 3.5 4.0 8.9 9.2 11.7 21.0
Non-ED	0 0 0 0 0 .1 .1 .1 .1 .2 .2 .2 .3 .3 .3 .4 .5 .5 .6 .8 .9 1.0 1.2 1.4 1.5 1.7 2.0 3.2 3.5 4.1 4.3 4.8 5.0 5.6 5.9 6.0 6.4 7.9 8.3 8.7 9.1 9.6 9.9 11.0 11.5 12.2 12.7 14.0 16.6 17.8

a. Determine the medians, fourths, and fourth spreads for the two samples.

b. Are there any outliers in either sample? Any extreme outliers?

c. Construct a comparative boxplot, and use it as a basis for comparing and contrasting the ED and non-ED samples.

57. Observations on burst strength (lb/in^2) were obtained both for test nozzle closure welds and for production cannister nozzle welds ("Proper Procedures Are the Key to Welding Radioactive Waste Cannisters," *Welding J.*, Aug. 1997: 61–67).

Test	7200 6100 7300 7300 8000 7400 7300 7300 8000 6700 8300
Cannister	5250 5625 5900 5900 5700 6050 5800 6000 5875 6100 5850 6600

Construct a comparative boxplot and comment on interesting features (the cited article did not include such a picture, but the authors commented that they had looked at one).

58. The comparative boxplot below of gasoline vapor coefficients for vehicles in Detroit appeared in the article "Receptor Modeling Approach to VOC Emission Inventory Validation" (*J. Environ. Engrg.*, 1995: 483–490). Discuss any interesting features.

59. Let $x_1, \ldots, x_n$ be a sample and let a and b be constants. If $y_i = ax_i + b$ for $i = 1, 2, \ldots, n$, how does f_s (the fourth spread) for the y_i's relate to f_s for the x_i's? Substantiate your assertion.

Comparative boxplot for Exercise 58

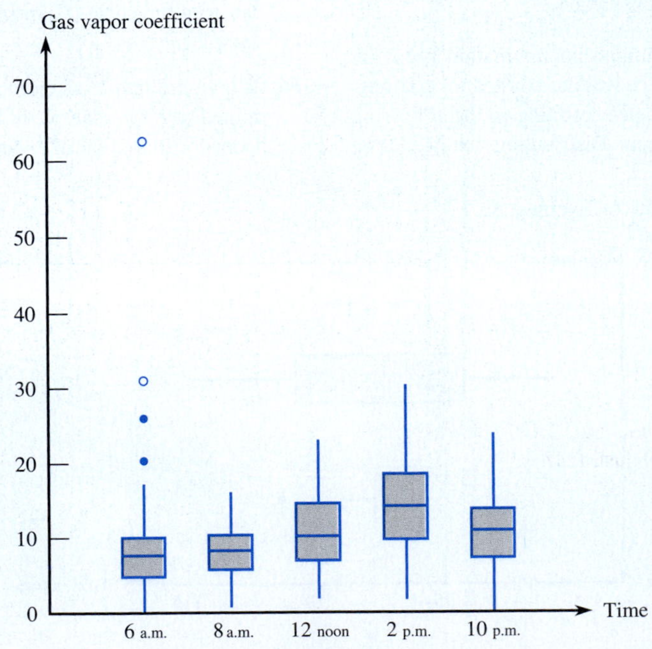

Supplementary Exercises (60–80)

60. Consider the following information from a sample of four Wolferman's cranberry citrus English muffins, which are said on the package label to weigh 116 g: $\bar{x} = 104.4$ g; $s = 4.1497$ g, smallest weighs 98.7 g, largest weighs 108.0 g. Determine the values of the two middle sample observations (and don't do it by successive guessing!).

61. Three different C_2F_6 flow rates (SCCM) were considered in an experiment to investigate the effect of flow rate on the uniformity (%) of the etch on a silicon wafer used in the manufacture of integrated circuits, resulting in the following data:

Flow rate						
125	2.6	2.7	3.0	3.2	3.8	4.6
160	3.6	4.2	4.2	4.6	4.9	5.0
200	2.9	3.4	3.5	4.1	4.6	5.1

Compare and contrast the uniformity observations resulting from these three different flow rates.

62. The amount of radiation received at a greenhouse plays an important role in determining the rate of photosynthesis. The accompanying observations on incoming solar radiation were read from a graph in the article "Radiation Components over Bare and Planted Soils in a Greenhouse" (*Solar Energy*, 1990: 1011–1016).

6.3	6.4	7.7	8.4	8.5	8.8	8.9
9.0	9.1	10.0	10.1	10.2	10.6	10.6
10.7	10.7	10.8	10.9	11.1	11.2	11.2
11.4	11.9	11.9	12.2	13.1		

Use some of the methods discussed in this chapter to describe and summarize this data.

63. The following data on HC and CO emissions for one particular vehicle was given in the chapter introduction.

HC (*g/mile*)	13.8	18.3	32.2	32.5
CO (*g/mile*)	118	149	232	236

a. Compute the sample standard deviations for the HC and CO observations. Does the widespread belief appear to be justified?

b. The *sample coefficient of variation* $s/\bar{x}$ (or $100 \cdot s/\bar{x}$) assesses the extent of variability relative to the mean. Values of this coefficient for several different data sets can be compared to determine which data sets exhibit more or less variation. Carry out such a comparison for the given data.

64. The accompanying frequency distribution of fracture strength (MPa) observations for ceramic bars fired in a particular kiln appeared in the article "Evaluating Tunnel Kiln Performance" (*Amer. Ceramic Soc. Bull.*, Aug. 1997: 59–63).

Class	81–83	83–85	85–87	87–89	89–91
Freq.	6	7	17	30	43
Class	91–93	93–95	95–97	97–99	
Freq.	28	22	13	3	

a. Construct a histogram based on relative frequencies, and comment on any interesting features.

b. What proportion of the strength observations are at least 85? Less than 95?

c. Roughly what proportion of the observations are less than 90?

65. Fifteen air samples from a certain region were obtained, and for each one the carbon monoxide concentration was determined. The results (in ppm) were

9.3	10.7	8.5	9.6	12.2	15.6	9.2	10.5
9.0	13.2	11.0	8.8	13.7	12.1	9.8	

Using the interpolation method suggested in Section 1.3, compute the 10% trimmed mean.

66. **a.** For what value of c is the quantity $\Sigma(x_i - c)^2$ minimized? (*Hint*: Take the derivative with respect to c, set equal to 0, and solve.)

b. Using the result of part (a), which of the two quantities $\Sigma(x_i - \bar{x})^2$ and $\Sigma(x_i - \mu)^2$ will be smaller than the other (assuming that $\bar{x} \neq \mu$)?

67. **a.** Let a and b be constants and let $y_i = ax_i + b$ for $i = 1, 2, \ldots, n$. What are the relationships between $\bar{x}$ and $\bar{y}$ and between s_x^2 and s_y^2?

b. The Australian army studied the effect of high temperatures and humidity on human body temperature (*Neural Network Training on Human Body Core Temperature Data*, Technical Report DSTO TN-0241, Combatant Protection Nutrition Branch, Aeronautical and Maritime Research Laboratory). They found that, at 30°C and 60% relative humidity, the sample average body temperature for nine soldiers was 38.21°C, with standard deviation .318°C. What are the sample average and the standard deviation in °F?

68. Elevated energy consumption during exercise continues after the workout ends. Because calories burned after exercise contribute to weight loss and have other consequences, it is important to understand this process. The paper "Effect of Weight Training Exercise and Treadmill Exercise on Post-Exercise Oxygen Consumption" (*Med. Sci. Sports Exercise*, 1998: 518–522) reported the accompanying data from a study in which oxygen consumption (liters) was measured continuously for 30 minutes for each of 15 subjects both after a weight training exercise and after a treadmill exercise.

Subject	1	2	3	4	5	6
Weight (x)	14.6	14.4	19.5	24.3	16.3	22.1
Treadmill (y)	11.3	5.3	9.1	15.2	10.1	19.6

Subject	7	8	9	10	11	12
Weight (x)	23.0	18.7	19.0	17.0	19.1	19.6
Treadmill (y)	20.8	10.3	10.3	2.6	16.6	22.4

Subject	13	14	15
Weight (x)	23.2	18.5	15.9
Treadmill (y)	23.6	12.6	4.4

a. Construct a comparative boxplot of the weight and treadmill observations, and comment on what you see.

b. Because the data is in the form of (x, y) pairs, with x and y measurements on the same variable under two different conditions, it is natural to focus on the differences within pairs: $d_1 = x_1 - y_1, \ldots, d_n = x_n - y_n$. Construct a boxplot of the sample differences. What does it suggest?

69. Anxiety disorders and symptoms can often be effectively treated with benzodiazepine medications. It is known that animals exposed to stress exhibit a decrease in benzodiazepine receptor binding in the frontal cortex. The paper "Decreased Benzodiazepine Receptor Binding in Prefrontal Cortex in Combat-Related Posttraumatic Stress Disorder" (*Amer. J. Psychiatry*, 2000: 1120–1126) described the first study of benzodiazepine receptor binding in individuals suffering from PTSD. The accompanying data on a receptor binding measure (adjusted distribution volume) was read from a graph in the paper.

PTSD: 10, 20, 25, 28, 31, 35, 37, 38, 38, 39, 39, 42, 46

Healthy: 23, 39, 40, 41, 43, 47, 51, 58, 63, 66, 67, 69, 72

Use various methods from this chapter to describe and summarize the data.

70. The article "Can We Really Walk Straight?" (*Amer. J. Phys. Anthropol.*, 1992: 19–27) reported on an experiment in which each of 20 healthy men was asked to walk as straight as possible to a target 60 m away at normal speed. Consider the following observations on cadence (number of strides per second):

.95 .85 .92 .95 .93 .86 1.00 .92 .85 .81
.78 .93 .93 1.05 .93 1.06 1.06 .96 .81 .96

Use the methods developed in this chapter to summarize the data; include an interpretation or discussion wherever appropriate. (*Note*: The author of the article used a rather sophisticated statistical analysis to conclude that people cannot walk in a straight line and suggested several explanations for this.)

71. The **mode** of a numerical data set is the value that occurs most frequently in the set.

a. Determine the mode for the cadence data given in Exercise 70.

b. For a categorical sample, how would you define the modal category?

72. Specimens of three different types of rope wire were selected, and the fatigue limit (MPa) was determined for each specimen, resulting in the accompanying data.

Type 1	350	350	350	358	370	370	370	371
	371	372	372	384	391	391	392	
Type 2	350	354	359	363	365	368	369	371
	373	374	376	380	383	388	392	
Type 3	350	361	362	364	364	365	366	371
	377	377	377	379	380	380	392	

a. Construct a comparative boxplot, and comment on similarities and differences.

b. Construct a comparative dotplot (a dotplot for each sample with a common scale). Comment on similarities and differences.

c. Does the comparative boxplot of part (a) give an informative assessment of similarities and differences? Explain your reasoning.

73. The three measures of center introduced in this chapter are the mean, median, and trimmed mean. Two additional measures of center that are occasionally used are the *midrange*, which is the average of the smallest and largest observations, and the *midfourth*, which is the average of the two fourths. Which of these five measures of center are resistant to the effects of outliers and which are not? Explain your reasoning.

74. Consider the following data on active repair time (hours) for a sample of $n = 46$ airborne communications receivers:

.2	.3	.5	.5	.5	.6	.6	.7	.7	.7	.8	.8
.8	1.0	1.0	1.0	1.0	1.1	1.3	1.5	1.5	1.5	1.5	2.0
2.0	2.2	2.5	2.7	3.0	3.0	3.3	3.3	4.0	4.0	4.5	4.7
5.0	5.4	5.4	7.0	7.5	8.8	9.0	10.3	22.0	24.5		

Construct the following:

a. A stem-and-leaf display in which the two largest values are displayed separately in a row labeled HI.

b. A histogram based on six class intervals with 0 as the lower limit of the first interval and interval widths of 2, 2, 2, 4, 10, and 10, respectively.

75. Consider a sample $x_1, x_2, \ldots, x_n$ and suppose that the values of $\bar{x}$, s^2, and s have been calculated.

a. Let $y_i = x_i - \bar{x}$ for $i = 1, \ldots, n$. How do the values of s^2 and s for the y_i's compare to the corresponding values for the x_i's? Explain.

b. Let $z_i = (x_i - \bar{x})/s$ for $i = 1, \ldots, n$. What are the values of the sample variance and sample standard deviation for the z_i's?

76. Let $\bar{x}_n$ and s_n^2 denote the sample mean and variance for the sample $x_1, \ldots, x_n$ and let $\bar{x}_{n+1}$ and s_{n+1}^2 denote these quantities when an additional observation x_{n+1} is added to the sample.

a. Show how $\bar{x}_{n+1}$ can be computed from $\bar{x}_n$ and x_{n+1}.

b. Show that

$$ns_{n+1}^2 = (n - 1)s_n^2 + \frac{n}{n + 1}(x_{n+1} - \bar{x}_n)^2$$

so that s_{n+1}^2 can be computed from x_{n+1}, $\bar{x}_n$, and s_n^2.

c. Suppose that a sample of 15 strands of drapery yarn has resulted in a sample mean thread elongation of 12.58 mm and a sample standard deviation of .512 mm. A 16th strand results in an elongation value of 11.8. What are the values of the sample mean and sample standard deviation for all 16 elongation observations?

77. Lengths of bus routes for any particular transit system will typically vary from one route to another. The article "Planning of City Bus Routes" (*J. Institut. Engrs.*, 1995: 211–215) gives the following information on lengths (km) for one particular system:

Length	6–8	8–10	10–12	12–14	14–16
Freq.	6	23	30	35	32

Length	16–18	18–20	20–22	22–24	24–26
Freq.	48	42	40	28	27

Length	26–28	28–30	30–35	35–40	40–45
Freq.	26	14	27	11	2

a. Draw a histogram corresponding to these frequencies.

b. What proportion of these route lengths are less than 20? What proportion of these routes have lengths of at least 30?

c. Roughly what is the value of the 90th percentile of the route length distribution?

d. Roughly what is the median route length?

78. A study carried out to investigate the distribution of total braking time (reaction time plus accelerator-to-brake movement time, in msec) during real driving conditions at 60 km/hr gave the following summary information on the distribution of times ("A Field Study on Braking Responses during Driving," *Ergonomics*, 1995: 1903–1910):

mean = 535 median = 500 mode = 500
sd = 96 minimum = 220 maximum = 925

5th percentile = 400 10th percentile = 430
90th percentile = 640 95th percentile = 720

What can you conclude about the shape of a histogram of this data? Explain your reasoning.

79. The sample data $x_1, x_2, \ldots, x_n$ sometimes represents a **time series**, where x_t = the observed value of a response variable x at time t. Often the observed series shows a great deal of random variation, which makes it difficult to study longer-term behavior. In such situations, it is desirable to produce a smoothed version of the series. One technique for doing so involves **exponential smoothing**. The value of a smoothing constant α is chosen ($0 < \alpha < 1$). Then with $\bar{x}_t$ = smoothed value at time t, we set $\bar{x}_1 = x_1$, and for $t = 2, 3, \ldots, n$, $\bar{x}_t = \alpha x_t + (1 - \alpha)\bar{x}_{t-1}$.

a. Consider the following time series in which x_t = temperature (°F) of effluent at a sewage treatment plant on day t: 47, 54, 53, 50, 46, 46, 47, 50, 51, 50, 46, 52, 50, 50. Plot each x_t against t on a two-dimensional coordinate system (a time-series plot). Does there appear to be any pattern?

b. Calculate the $\bar{x}_t$'s using $\alpha = .1$. Repeat using $\alpha = .5$. Which value of α gives a smoother $\bar{x}_t$ series?

c. Substitute $\bar{x}_{t-1} = \alpha x_{t-1} + (1 - \alpha)\bar{x}_{t-2}$ on the right-hand side of the expression for $\bar{x}_t$, then substitute $\bar{x}_{t-2}$ in terms of x_{t-2} and $\bar{x}_{t-3}$, and so on. On how many of the values $x_t, x_{t-1}, \ldots, x_1$ does $\bar{x}_t$ depend? What happens to the coefficient on x_{t-k} as k increases?

d. Refer to part (c). If t is large, how sensitive is $\bar{x}_t$ to the initialization $\bar{x}_1 = x_1$? Explain.

(*Note*: A relevant reference is the article "Simple Statistics for Interpreting Environmental Data," *Water Pollution Control Fed. J.*, 1981: 167–175.)

80. Consider numerical observations $x_1, \ldots, x_n$. It is frequently of interest to know whether the x_i's are (at least approximately) symmetrically distributed about some value. If n is at least moderately large, the extent of symmetry can be assessed from a stem-and-leaf display or histogram. However, if n is not very large, such pictures are not particularly informative. Consider the following alternative. Let y_1 denote the smallest x_i, y_2 the second smallest x_i, and so on. Then plot the following pairs as points on a two-dimensional coordinate system: $(y_n - \tilde{x}, \tilde{x} - y_1)$, $(y_{n-1} - \tilde{x}, \tilde{x} - y_2)$, $(y_{n-2} - \tilde{x}, \tilde{x} - y_3), \ldots$. There are $n/2$ points when n is even and $(n-1)/2$ when n is odd.

a. What does this plot look like when there is perfect symmetry in the data? What does it look like when observations stretch out more above the median than below it (a long upper tail)?

b. The accompanying data on rainfall (acre-feet) from 26 seeded clouds is taken from the article "A Bayesian Analysis of a Multiplicative Treatment Effect in Weather Modification" (*Technometrics*, 1975: 161–166). Construct the plot and comment on the extent of symmetry or nature of departure from symmetry.

4.1	7.7	17.5	31.4	32.7	40.6	92.4
115.3	118.3	119.0	129.6	198.6	200.7	242.5
255.0	274.7	274.7	302.8	334.1	430.0	489.1
703.4	978.0	1656.0	1697.8	2745.6		

Bibliography

Chambers, John, William Cleveland, Beat Kleiner, and Paul Tukey, *Graphical Methods for Data Analysis*, Brooks/Cole, Pacific Grove, CA, 1983. A highly recommended presentation of both older and more recent graphical and pictorial methodology in statistics.

Devore, Jay, and Roxy Peck, *Statistics: The Exploration and Analysis of Data* (5th ed.), Thomson-Brooks/Cole, Pacific Grove, CA, 2005. The first few chapters give a very nonmathematical survey of methods for describing and summarizing data.

Freedman, David, Robert Pisani, and Roger Purves, *Statistics* (3rd ed.), Norton, New York, 1998. An excellent, very nonmathematical survey of basic statistical reasoning and methodology.

Hoaglin, David, Frederick Mosteller, and John Tukey, *Understanding Robust and Exploratory Data Analysis*, Wiley, New York, 1983. Discusses why, as well as how, exploratory methods should be employed; it is good on details of stem-and-leaf displays and boxplots.

Hoaglin, David, and Paul Velleman, *Applications, Basics, and Computing of Exploratory Data Analysis*, Duxbury Press, Boston, 1980. A good discussion of some basic exploratory methods.

Moore, David, *Statistics: Concepts and Controversies* (5th ed.), Freeman, San Francisco, 2001. An extremely readable and entertaining paperback that contains an intuitive discussion of problems connected with sampling and designed experiments.

Peck, Roxy, et al. (eds.), *Statistics: A Guide to the Unknown* (4th ed.), Thomson-Brooks/Cole, Belmont, CA, 2006. Contains many short, nontechnical articles describing various applications of statistics.

CHAPTER TWO

Probability

Introduction

The term **probability** refers to the study of randomness and uncertainty. In any situation in which one of a number of possible outcomes may occur, the theory of probability provides methods for quantifying the chances, or likelihoods, associated with the various outcomes. The language of probability is constantly used in an informal manner in both written and spoken contexts. Examples include such statements as "It is likely that the Dow Jones Industrial Average will increase by the end of the year," "There is a 50–50 chance that the incumbent will seek reelection," "There will probably be at least one section of that course offered next year," "The odds favor a quick settlement of the strike," and "It is expected that at least 20,000 concert tickets will be sold." In this chapter, we introduce some elementary probability concepts, indicate how probabilities can be interpreted, and show how the rules of probability can be applied to compute the probabilities of many interesting events. The methodology of probability will then permit us to express in precise language such informal statements as those given above.

The study of probability as a branch of mathematics goes back over 300 years, where it had its genesis in connection with questions involving games of chance. Many books are devoted exclusively to probability and explore in great detail numerous interesting aspects and applications of this lovely branch of mathematics. Our objective here is more limited in scope: We will focus on those topics that are central to a basic understanding and also have the most direct bearing on problems of statistical inference.

2.1 Sample Spaces and Events

An **experiment** is any action or process whose outcome is subject to uncertainty. Although the word *experiment* generally suggests a planned or carefully controlled laboratory testing situation, we use it here in a much wider sense. Thus experiments that may be of interest include tossing a coin once or several times, selecting a card or cards from a deck, weighing a loaf of bread, ascertaining the commuting time from home to work on a particular morning, obtaining blood types from a group of individuals, or calling people to conduct a survey.

The Sample Space of an Experiment

DEFINITION The **sample space** of an experiment, denoted by $\mathcal{S}$, is the set of all possible outcomes of that experiment.

Example 2.1 The simplest experiment to which probability applies is one with two possible outcomes. One such experiment consists of examining a single fuse to see whether it is defective. The sample space for this experiment can be abbreviated as $\mathcal{S} = \{N, D\}$, where N represents not defective, D represents defective, and the braces are used to enclose the elements of a set. Another such experiment would involve tossing a thumbtack and noting whether it landed point up or point down, with sample space $\mathcal{S} = \{U, D\}$, and yet another would consist of observing the gender of the next child born at the local hospital, with $\mathcal{S} = \{M, F\}$. ■

Example 2.2 If we examine three fuses in sequence and note the result of each examination, then an outcome for the entire experiment is any sequence of N's and D's of length 3, so

$$\mathcal{S} = \{NNN, NND, NDN, NDD, DNN, DND, DDN, DDD\}$$

If we had tossed a thumbtack three times, the sample space would be obtained by replacing N by U in $\mathcal{S}$ above. A similar notational change would yield the sample space for the experiment in which the genders of three newborn children are observed. ■

Example 2.3 Two gas stations are located at a certain intersection. Each one has six gas pumps. Consider the experiment in which the number of pumps in use at a particular time of day is determined for each of the stations. An experimental outcome specifies how many pumps are in use at the first station and how many are in use at the second one. One possible outcome is (2, 2), another is (4, 1), and yet another is (1, 4). The 49 outcomes in $\mathcal{S}$ are displayed in the accompanying table. The sample space for the experiment in which a six-sided die is thrown twice results from deleting the 0 row and 0 column from the table, giving 36 outcomes.

	Second Station						
First Station	**0**	**1**	**2**	**3**	**4**	**5**	**6**
0	(0, 0)	(0, 1)	(0, 2)	(0, 3)	(0, 4)	(0, 5)	(0, 6)
1	(1, 0)	(1, 1)	(1, 2)	(1, 3)	(1, 4)	(1, 5)	(1, 6)
2	(2, 0)	(2, 1)	(2, 2)	(2, 3)	(2, 4)	(2, 5)	(2, 6)
3	(3, 0)	(3, 1)	(3, 2)	(3, 3)	(3, 4)	(3, 5)	(3, 6)
4	(4, 0)	(4, 1)	(4, 2)	(4, 3)	(4, 4)	(4, 5)	(4, 6)
5	(5, 0)	(5, 1)	(5, 2)	(5, 3)	(5, 4)	(5, 5)	(5, 6)
6	(6, 0)	(6, 1)	(6, 2)	(6, 3)	(6, 4)	(6, 5)	(6, 6)

■

Example 2.4 If a new type-D flashlight battery has a voltage that is outside certain limits, that battery is characterized as a failure (F); if the battery has a voltage within the prescribed limits, it is a success (S). Suppose an experiment consists of testing each battery as it comes off an assembly line until we first observe a success. Although it may not be very likely, a possible outcome of this experiment is that the first 10 (or 100 or 1000 or . . .) are F's and the next one is an S. That is, for any positive integer n, we may have to examine n batteries before seeing the first S. The sample space is $\mathcal{S} = \{S, FS, FFS, FFFS, \ldots\}$, which contains an infinite number of possible outcomes. The same abbreviated form of the sample space is appropriate for an experiment in which, starting at a specified time, the gender of each newborn infant is recorded until the birth of a male is observed. ■

Events

In our study of probability, we will be interested not only in the individual outcomes of $\mathcal{S}$ but also in any collection of outcomes from $\mathcal{S}$.

DEFINITION An **event** is any collection (subset) of outcomes contained in the sample space $\mathcal{S}$. An event is said to be **simple** if it consists of exactly one outcome and **compound** if it consists of more than one outcome.

When an experiment is performed, a particular event A is said to occur if the resulting experimental outcome is contained in A. In general, exactly one simple event will occur, but many compound events will occur simultaneously.

Example 2.5 Consider an experiment in which each of three vehicles taking a particular freeway exit turns left (L) or right (R) at the end of the exit ramp. The eight possible outcomes that comprise the sample space are LLL, RLL, LRL, LLR, LRR, RLR, RRL, and RRR. Thus there are eight simple events, among which are $E_1 = \{LLL\}$ and $E_5 = \{LRR\}$. Some compound events include

$A = \{RLL, LRL, LLR\}$ = the event that exactly one of the three vehicles turns right

$B = \{LLL, RLL, LRL, LLR\}$ = the event that at most one of the vehicles turns right

$C = \{LLL, RRR\}$ = the event that all three vehicles turn in the same direction

Suppose that when the experiment is performed, the outcome is *LLL*. Then the simple event E_1 has occurred and so also have the events B and C (but not A). ■

Example 2.6 (Example 2.3 continued)

When the number of pumps in use at each of two six-pump gas stations is observed, there are 49 possible outcomes, so there are 49 simple events: $E_1 = \{(0, 0)\}$, $E_2 = \{(0, 1)\}, \ldots, E_{49} = \{(6, 6)\}$. Examples of compound events are

$A = \{(0, 0), (1, 1), (2, 2), (3, 3), (4, 4), (5, 5), (6, 6)\}$ = the event that the number of pumps in use is the same for both stations

$B = \{(0, 4), (1, 3), (2, 2), (3, 1), (4, 0)\}$ = the event that the total number of pumps in use is four

$C = \{(0, 0), (0, 1), (1, 0), (1, 1)\}$ = the event that at most one pump is in use at each station ■

Example 2.7 (Example 2.4 continued)

The sample space for the battery examination experiment contains an infinite number of outcomes, so there are an infinite number of simple events. Compound events include

$A = \{S, FS, FFS\}$ = the event that at most three batteries are examined

$E = \{FS, FFFS, FFFFFS, \ldots\}$ = the event that an even number of batteries are examined ■

Some Relations from Set Theory

An event is nothing but a set, so relationships and results from elementary set theory can be used to study events. The following operations will be used to construct new events from given events.

DEFINITION

1. The **union** of two events A and B, denoted by $A \cup B$ and read "*A or B*," is the event consisting of all outcomes that are *either in A or in B or in both events* (so that the union includes outcomes for which both A and B occur as well as outcomes for which exactly one occurs)—that is, all outcomes in at least one of the events.
2. The **intersection** of two events A and B, denoted by $A \cap B$ and read "*A and B*," is the event consisting of all outcomes that are in *both A and B*.

3. The **complement** of an event A, denoted by A', is the set of all outcomes in $\mathcal{S}$ that are not contained in A.

Example 2.8
(Example 2.3 continued)

For the experiment in which the number of pumps in use at a single six-pump gas station is observed, let $A = \{0, 1, 2, 3, 4\}$, $B = \{3, 4, 5, 6\}$, and $C = \{1, 3, 5\}$. Then

$$A \cup B = \{0, 1, 2, 3, 4, 5, 6\} = \mathcal{S} \qquad A \cup C = \{0, 1, 2, 3, 4, 5\}$$

$$A \cap B = \{3, 4\} \qquad A \cap C = \{1, 3\} \qquad A' = \{5, 6\} \qquad \{A \cup C\}' = \{6\}$$ ■

Example 2.9
(Example 2.4 continued)

In the battery experiment, define A, B, and C by

$$A = \{S, FS, FFS\}$$

$$B = \{S, FFS, FFFFS\}$$

and

$$C = \{FS, FFFS, FFFFFS, \ldots\}$$

Then

$$A \cup B = \{S, FS, FFS, FFFFS\}$$

$$A \cap B = \{S, FFS\}$$

$$A' = \{FFFS, FFFFS, FFFFFS, \ldots\}$$

and

$$C' = \{S, FFS, FFFFS, \ldots\} = \{\text{an odd number of batteries are examined}\}$$ ■

Sometimes A and B have no outcomes in common, so that the intersection of A and B contains no outcomes.

DEFINITION

When A and B have no outcomes in common, they are said to be **disjoint** or **mutually exclusive** events. Mathematicians write this compactly as $A \cap B = \varnothing$, where $\varnothing$ denotes the event consisting of no outcomes whatsoever (the "null" or "empty" event).

Example 2.10

A small city has three automobile dealerships: a GM dealer selling Chevrolets, Pontiacs, and Buicks; a Ford dealer selling Fords and Mercurys; and a Chrysler dealer selling Plymouths and Chryslers. If an experiment consists of observing the brand of the next car sold, then the events $A = \{\text{Chevrolet, Pontiac, Buick}\}$ and $B = \{\text{Ford, Mercury}\}$ are mutually exclusive because the next car sold cannot be both a GM product and a Ford product. ■

The operations of union and intersection can be extended to more than two events. For any three events A, B, and C, the event $A \cup B \cup C$ is the set of outcomes contained in at least one of the three events, whereas $A \cap B \cap C$ is the set of outcomes contained in all three events. Given events $A_1, A_2, A_3, \ldots$, these events are said to be mutually exclusive (or pairwise disjoint) if no two events have any outcomes in common.

A pictorial representation of events and manipulations with events is obtained by using Venn diagrams. To construct a Venn diagram, draw a rectangle whose interior will represent the sample space $\mathcal{S}$. Then any event A is represented as the interior of a closed curve (often a circle) contained in $\mathcal{S}$. Figure 2.1 shows examples of Venn diagrams.

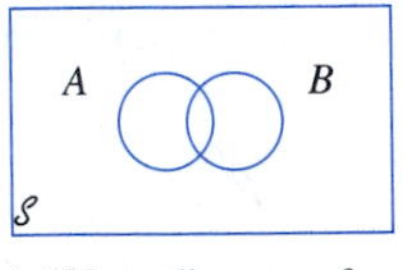

(a) Venn diagram of events A and B

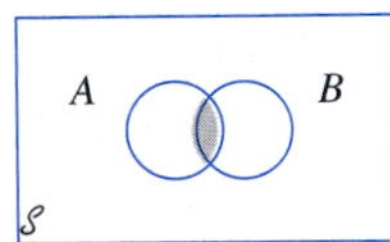

(b) Shaded region is $A \cap B$

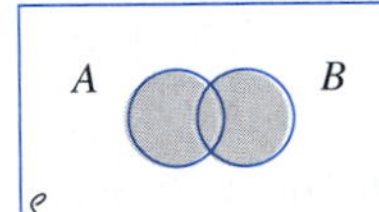

(c) Shaded region is $A \cup B$

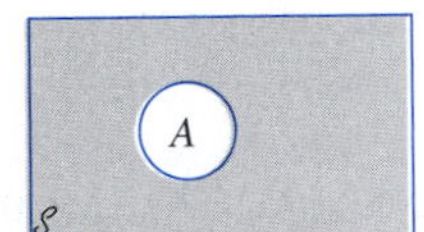

(d) Shaded region is A'

(e) Mutually exclusive events

Figure 2.1 Venn diagrams

Exercises Section 2.1 (1–12)

1. Several jobs are advertised and both Ann and Bev apply. Let A be the event that Ann is hired and let B be the event that Bev is hired. Express in terms of A and B the events
   a. Ann is hired but not Bev.
   b. At least one of them is hired.
   c. Exactly one of them is hired.

2. Each voter makes a choice from three candidates, 1, 2, and 3. We consider the votes of just two voters, Al and Bill, so the sample space $\mathcal{S}$ consists of all pairs, where the members of the pair are chosen from 1, 2, and 3.
   a. List all elements of $\mathcal{S}$.
   b. List all outcomes in the event A that Al and Bill make the same choice.
   c. List all outcomes in the event B that neither of them vote for candidate 2.

3. Four universities—1, 2, 3, and 4—are participating in a holiday basketball tournament. In the first round, 1 will play 2 and 3 will play 4. Then the two winners will play for the championship, and the two losers will also play. One possible outcome can be denoted by 1324 (1 beats 2 and 3 beats 4 in first-round games, and then 1 beats 3 and 2 beats 4).
   a. List all outcomes in $\mathcal{S}$.
   b. Let A denote the event that 1 wins the tournament. List outcomes in A.
   c. Let B denote the event that 2 gets into the championship game. List outcomes in B.
   d. What are the outcomes in $A \cup B$ and in $A \cap B$? What are the outcomes in A'?

4. Suppose that vehicles taking a particular freeway exit can turn right (R), turn left (L), or go straight (S). Consider observing the direction for each of three successive vehicles.
   a. List all outcomes in the event A that all three vehicles go in the same direction.
   b. List all outcomes in the event B that all three vehicles take different directions.
   c. List all outcomes in the event C that exactly two of the three vehicles turn right.
   d. List all outcomes in the event D that exactly two vehicles go in the same direction.
   e. List outcomes in D', $C \cup D$, and $C \cap D$.

5. Three components are connected to form a system as shown in the accompanying diagram. Because the components in the 2–3 subsystem are connected in parallel, that subsystem will function if at least one of the two individual components functions. For

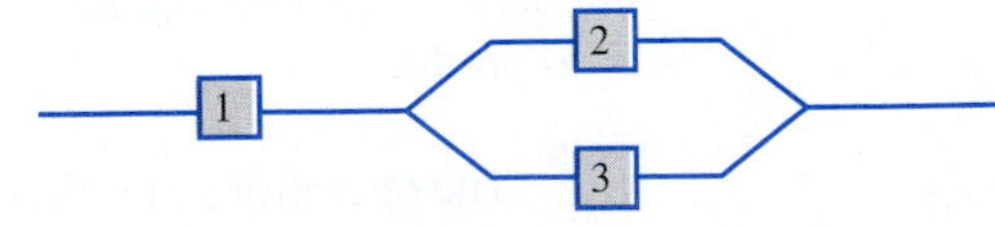

the entire system to function, component 1 must function and so must the 2–3 subsystem.

The experiment consists of determining the condition of each component [S (success) for a functioning component and F (failure) for a nonfunctioning component].

a. What outcomes are contained in the event A that exactly two out of the three components function?

b. What outcomes are contained in the event B that at least two of the components function?

c. What outcomes are contained in the event C that the system functions?

d. List outcomes in C', $A \cup C$, $A \cap C$, $B \cup C$, and $B \cap C$.

6. Each of a sample of four home mortgages is classified as fixed rate (F) or variable rate (V).

a. What are the 16 outcomes in $\mathcal{S}$?

b. Which outcomes are in the event that exactly three of the selected mortgages are fixed rate?

c. Which outcomes are in the event that all four mortgages are of the same type?

d. Which outcomes are in the event that at most one of the four is a variable-rate mortgage?

e. What is the union of the events in parts (c) and (d), and what is the intersection of these two events?

f. What are the union and intersection of the two events in parts (b) and (c)?

7. A family consisting of three persons—A, B, and C—belongs to a medical clinic that always has a doctor at each of stations 1, 2, and 3. During a certain week, each member of the family visits the clinic once and is assigned at random to a station. The experiment consists of recording the station number for each member. One outcome is (1, 2, 1) for A to station 1, B to station 2, and C to station 1.

a. List the 27 outcomes in the sample space.

b. List all outcomes in the event that all three members go to the same station.

c. List all outcomes in the event that all members go to different stations.

d. List all outcomes in the event that no one goes to station 2.

8. A college library has five copies of a certain text on reserve. Two copies (1 and 2) are first printings, and the other three (3, 4, and 5) are second printings. A student examines these books in random order, stopping only when a second printing has been selected. One possible outcome is 5, and another is 213.

a. List the outcomes in $\mathcal{S}$.

b. Let A denote the event that exactly one book must be examined. What outcomes are in A?

c. Let B be the event that book 5 is the one selected. What outcomes are in B?

d. Let C be the event that book 1 is not examined. What outcomes are in C?

9. An academic department has just completed voting by secret ballot for a department head. The ballot box contains four slips with votes for candidate A and three slips with votes for candidate B. Suppose these slips are removed from the box one by one.

a. List all possible outcomes.

b. Suppose a running tally is kept as slips are removed. For what outcomes does A remain ahead of B throughout the tally?

10. A construction firm is currently working on three different buildings. Let A_i denote the event that the ith building is completed by the contract date. Use the operations of union, intersection, and complementation to describe each of the following events in terms of A_1, A_2, and A_3, draw a Venn diagram, and shade the region corresponding to each one.

a. At least one building is completed by the contract date.

b. All buildings are completed by the contract date.

c. Only the first building is completed by the contract date.

d. Exactly one building is completed by the contract date.

e. Either the first building or both of the other two buildings are completed by the contract date.

11. Use Venn diagrams to verify the following two relationships for any events A and B (these are called De Morgan's laws):

a. $(A \cup B)' = A' \cap B'$

b. $(A \cap B)' = A' \cup B'$

12. **a.** In Example 2.10, identify three events that are mutually exclusive.

b. Suppose there is no outcome common to all three of the events A, B, and C. Are these three events necessarily mutually exclusive? If your answer is yes, explain why; if your answer is no, give a counterexample using the experiment of Example 2.10.

2.2 Axioms, Interpretations, and Properties of Probability

Given an experiment and a sample space $\mathcal{S}$, *the objective of probability is to assign to each event A a number P(A), called the probability of the event A, which will give a precise measure of the chance that A will occur*. To ensure that the probability assignments will be consistent with our intuitive notions of probability, all assignments should satisfy the following axioms (basic properties) of probability.

AXIOM 1 For any event A, $P(A) \geq 0$.

AXIOM 2 $P(\mathcal{S}) = 1$.

AXIOM 3 If $A_1, A_2, A_3, \ldots$ is an infinite collection of disjoint events, then

$$P(A_1 \cup A_2 \cup A_3 \cup \cdots) = \sum_{i=1}^{\infty} P(A_i)$$

You might wonder why the third axiom contains no reference to a *finite* collection of disjoint events. It is because the corresponding property for a finite collection can be derived from our three axioms. We want our axiom list to be as short as possible and not contain any property that can be derived from others on the list. Axiom 1 reflects the intuitive notion that the chance of A occurring should be nonnegative. The sample space is by definition the event that must occur when the experiment is performed ($\mathcal{S}$ contains all possible outcomes), so Axiom 2 says that the maximum possible probability of 1 is assigned to $\mathcal{S}$. The third axiom formalizes the idea that if we wish the probability that at least one of a number of events will occur and no two of the events can occur simultaneously, then the chance of at least one occurring is the sum of the chances of the individual events.

PROPOSITION $P(\emptyset) = 0$ where $\emptyset$ is the null event. This in turn implies that the property contained in Axiom 3 is valid for a *finite* collection of events.

Proof First consider the infinite collection $A_1 = \emptyset, A_2 = \emptyset, A_3 = \emptyset, \ldots$. Since $\emptyset \cap \emptyset = \emptyset$, the events in this collection are disjoint and $\cup A_i = \emptyset$. The third axiom then gives

$$P(\emptyset) = \sum P(\emptyset)$$

This can happen only if $P(\emptyset) = 0$.

Now suppose that $A_1, A_2, \ldots, A_k$ are disjoint events, and append to these the infinite collection $A_{k+1} = \emptyset, A_{k+2} = \emptyset, A_{k+3} = \emptyset, \ldots$. Again invoking the third axiom,

$$P\left(\bigcup_{i=1}^{k} A_i\right) = P\left(\bigcup_{i=1}^{\infty} A_i\right) = \sum_{i=1}^{\infty} P(A_i) = \sum_{i=1}^{k} P(A_i)$$

as desired. ■

Example 2.11 Consider tossing a thumbtack in the air. When it comes to rest on the ground, either its point will be up (the outcome U) or down (the outcome D). The sample space for this event is therefore $\mathcal{S} = \{U, D\}$. The axioms specify $P(\mathcal{S}) = 1$, so the probability assignment will be completed by determining $P(U)$ and $P(D)$. Since U and D are disjoint and their union is $\mathcal{S}$, the foregoing proposition implies that

$$1 = P(\mathcal{S}) = P(U) + P(D)$$

It follows that $P(D) = 1 - P(U)$. One possible assignment of probabilities is $P(U) = .5$, $P(D) = .5$, whereas another possible assignment is $P(U) = .75$, $P(D) = .25$. In fact, letting p represent any fixed number between 0 and 1, $P(U) = p$, $P(D) = 1 - p$ is an assignment consistent with the axioms. ■

Example 2.12 Consider the experiment in Example 2.4, in which batteries coming off an assembly line are tested one by one until one having a voltage within prescribed limits is found. The simple events are $E_1 = \{S\}$, $E_2 = \{FS\}$, $E_3 = \{FFS\}$, $E_4 = \{FFFS\}$, ... Suppose the probability of any particular battery being satisfactory is .99. Then it can be shown that $P(E_1) = .99$, $P(E_2) = (.01)(.99)$, $P(E_3) = (.01)^2(.99)$, . . . is an assignment of probabilities to the simple events that satisfies the axioms. In particular, because the E_i's are disjoint and $\mathcal{S} = E_1 \cup E_2 \cup E_3 \cup \ldots$, it must be the case that

$$\begin{aligned} 1 = P(\mathcal{S}) &= P(E_1) + P(E_2) + P(E_3) + \cdots \\ &= .99[1 + .01 + (.01)^2 + (.01)^3 + \cdots] \end{aligned}$$

Here we have used the formula for the sum of a geometric series:

$$a + ar + ar^2 + ar^3 + \cdots = \frac{a}{1 - r}$$

However, another legitimate (according to the axioms) probability assignment of the same "geometric" type is obtained by replacing .99 by any other number p between 0 and 1 (and .01 by $1 - p$). ■

Interpreting Probability

Examples 2.11 and 2.12 show that the axioms do not completely determine an assignment of probabilities to events. The axioms serve only to rule out assignments inconsistent with our intuitive notions of probability. In the tack-tossing experiment of Example 2.11, two particular assignments were suggested. The appropriate or correct assignment depends on the nature of the thumbtack and also on one's interpretation of probability. The interpretation that is most frequently used and most easily understood is based on the notion of relative frequencies.

Consider an experiment that can be repeatedly performed in an identical and independent fashion, and let A be an event consisting of a fixed set of outcomes of the experiment. Simple examples of such repeatable experiments include the tack-tossing and die-tossing experiments previously discussed. If the experiment is performed n times, on some of the replications the event A will occur (the outcome will be in the set A), and on others, A will not occur. Let $n(A)$ denote the number of replications on which A does occur. Then the ratio $n(A)/n$ is called the *relative frequency* of occurrence of the event A

in the sequence of n replications. Empirical evidence, based on the results of many of these sequences of repeatable experiments, indicates that as n grows large, the relative frequency $n(A)/n$ stabilizes, as pictured in Figure 2.2. That is, as n gets arbitrarily large, the relative frequency approaches a limiting value we refer to as the *limiting relative frequency* of the event A. The objective interpretation of probability identifies this limiting relative frequency with $P(A)$.

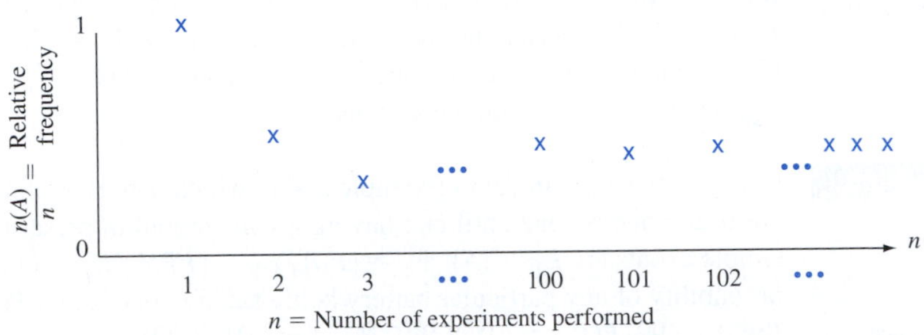

Figure 2.2 Stabilization of relative frequency

If probabilities are assigned to events in accordance with their limiting relative frequencies, then we can interpret a statement such as "The probability of that coin landing with the head facing up when it is tossed is .5" to mean that in a large number of such tosses, a head will appear on approximately half the tosses and a tail on the other half.

This relative frequency interpretation of probability is said to be objective because it rests on a property of the experiment rather than on any particular individual concerned with the experiment. For example, two different observers of a sequence of coin tosses should both use the same probability assignments since the observers have nothing to do with limiting relative frequency. In practice, this interpretation is not as objective as it might seem, since the limiting relative frequency of an event will not be known. Thus we will have to assign probabilities based on our beliefs about the limiting relative frequency of events under study. Fortunately, there are many experiments for which there will be a consensus with respect to probability assignments. When we speak of a fair coin, we shall mean $P(H) = P(T) = .5$, and a fair die is one for which limiting relative frequencies of the six outcomes are all $\frac{1}{6}$, suggesting probability assignments $P(\{1\}) = \cdots = P(\{6\}) = \frac{1}{6}$.

Because the objective interpretation of probability is based on the notion of limiting frequency, its applicability is limited to experimental situations that are repeatable. Yet the language of probability is often used in connection with situations that are inherently unrepeatable. Examples include: "The chances are good for a peace agreement"; "It is likely that our company will be awarded the contract"; and "Because their best quarterback is injured, I expect them to score no more than 10 points against us." In such situations we would like, as before, to assign numerical probabilities to various outcomes and events (e.g., the probability is .9 that we will get the contract). We must therefore adopt an alternative interpretation of these probabilities. Because different observers may have different prior information and opinions concerning such experimental situations, probability assignments may now differ from individual to individual.

Interpretations in such situations are thus referred to as *subjective*. The book by Robert Winkler listed in the chapter references gives a very readable survey of several subjective interpretations.

More Probability Properties

PROPOSITION For any event A, $P(A) = 1 - P(A')$.

Proof Since by definition of A', $A \cup A' = \mathcal{S}$ while A and A' are disjoint, $1 = P(\mathcal{S}) = P(A \cup A') = P(A) + P(A')$, from which the desired result follows.

This proposition is surprisingly useful because there are many situations in which $P(A')$ is more easily obtained by direct methods than is $P(A)$.

Example 2.13 Consider a system of five identical components connected in series, as illustrated in Figure 2.3.

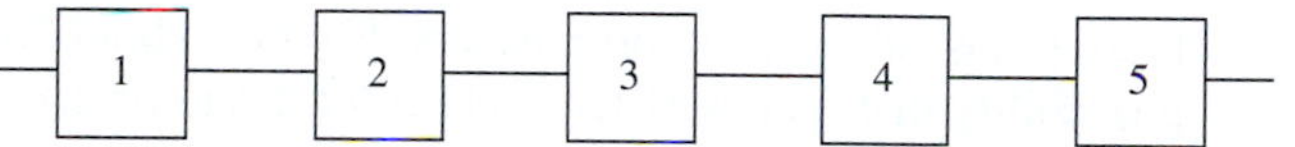

Figure 2.3 A system of five components connected in series

Denote a component that fails by F and one that doesn't fail by S (for success). Let A be the event that the *system* fails. For A to occur, at least one of the individual components must fail. Outcomes in A include $SSFSS$ (1, 2, 4, and 5 all work, but 3 does not), $FFSSS$, and so on. There are in fact 31 different outcomes in A. However, A', the event that the system works, consists of the single outcome $SSSSS$. We will see in Section 2.5 that if 90% of all these components do not fail and different components fail independently of one another, then $P(A') = P(SSSSS) = .9^5 = .59$. Thus $P(A) = 1 - .59 = .41$; so among a large number of such systems, roughly 41% will fail. ■

In general, the foregoing proposition is useful when the event of interest can be expressed as "at least . . . ," because the complement "less than . . ." may be easier to work with. (In some problems, "more than . . ." is easier to deal with than "at most. . . .") When you are having difficulty calculating $P(A)$ directly, think of determining $P(A')$.

When A and B are disjoint, we know that $P(A \cup B) = P(A) + P(B)$. How can this union probability be obtained when the events are not disjoint?

PROPOSITION For any events A and B,

$$P(A \cup B) = P(A) + P(B) - P(A \cap B)$$

Notice that the proposition is valid even if A and B are disjoint, since then $P(A \cap B) = 0$. The key idea is that, in adding $P(A)$ and $P(B)$, the probability of the intersection $A \cap B$ is actually counted twice, so $P(A \cap B)$ must be subtracted out.

Proof Note first that $A \cup B = A \cup (B \cap A')$, as illustrated in Figure 2.4. Since A and $(B \cap A')$ are disjoint, $P(A \cup B) = P(A) + P(B \cap A')$. But $B = (B \cap A) \cup (B \cap A')$ (the union of that part of B in A and that part of B not in A). Furthermore, $(B \cap A)$ and $(B \cap A')$ are disjoint, so that $P(B) = P(B \cap A) + P(B \cap A')$. Combining these results gives

$$\begin{aligned} P(A \cup B) = P(A) + P(B \cap A') &= P(A) + [P(B) - P(A \cap B)] \\ &= P(A) + P(B) - P(A \cap B) \end{aligned}$$

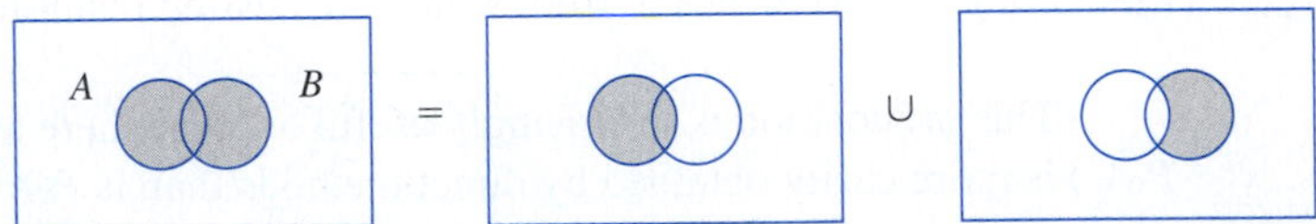

Figure 2.4 Representing $A \cup B$ as a union of disjoint events ■

Example 2.14 In a certain residential suburb, 60% of all households subscribe to the metropolitan newspaper published in a nearby city, 80% subscribe to the local paper, and 50% of all households subscribe to both papers. If a household is selected at random, what is the probability that it subscribes to (1) at least one of the two newspapers and (2) exactly one of the two newspapers?

With $A = \{\text{subscribes to the metropolitan paper}\}$ and $B = \{\text{subscribes to the local paper}\}$, the given information implies that $P(A) = .6$, $P(B) = .8$, and $P(A \cap B) = .5$. The previous proposition then applies to give

$$\begin{aligned} &P(\text{subscribes to at least one of the two newspapers}) \\ &\qquad = P(A \cup B) = P(A) + P(B) - P(A \cap B) = .6 + .8 - .5 = .9 \end{aligned}$$

The event that a household subscribes only to the local paper can be written as $A' \cap B$ [(not metropolitan) and local]. Now Figure 2.4 implies that

$$.9 = P(A \cup B) = P(A) + P(A' \cap B) = .6 + P(A' \cap B)$$

from which $P(A' \cap B) = .3$. Similarly, $P(A \cap B') = P(A \cup B) - P(B) = .1$. This is all illustrated in Figure 2.5, from which we see that

$$P(\text{exactly one}) = P(A \cap B') + P(A' \cap B) = .1 + .3 = .4$$

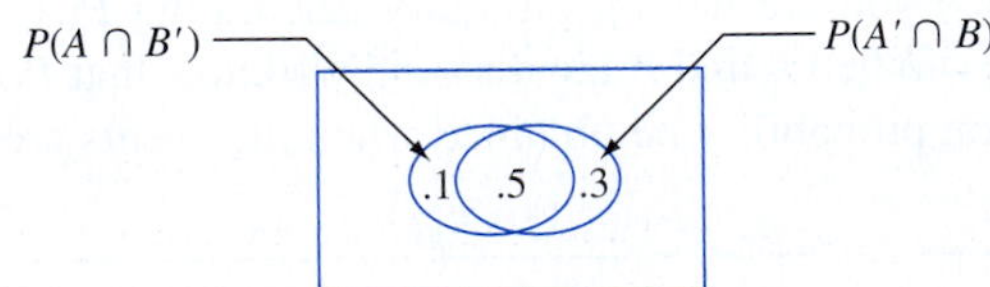

Figure 2.5 Probabilities for Example 2.14 ■

The probability of a union of more than two events can be computed analogously. For three events A, B, and C, the result is

$$P(A \cup B \cup C) = P(A) + P(B) + P(C) - P(A \cap B) - P(A \cap C) \\ - P(B \cap C) + P(A \cap B \cap C)$$

This can be seen by examining a Venn diagram of $A \cup B \cup C$, which is shown in Figure 2.6. When $P(A)$, $P(B)$, and $P(C)$ are added, outcomes in certain intersections are double counted and the corresponding probabilities must be subtracted. But this results in $P(A \cap B \cap C)$ being subtracted once too often, so it must be added back. One formal proof involves applying the previous proposition to $P((A \cup B) \cup C)$, the probability of the union of the two events $A \cup B$ and C. More generally, a result concerning $P(A_1 \cup \cdots \cup A_k)$ can be proved by induction or by other methods.

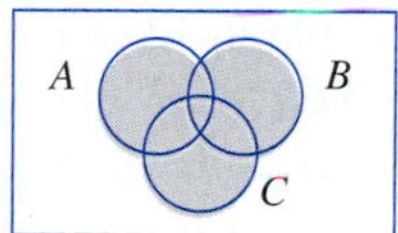

Figure 2.6 $A \cup B \cup C$

Determining Probabilities Systematically

When the number of possible outcomes (simple events) is large, there will be many compound events. A simple way to determine probabilities for these events that avoids violating the axioms and derived properties is to first determine probabilities $P(E_i)$ for all simple events. These should satisfy $P(E_i) \geq 0$ and $\Sigma_{\text{all } i} P(E_i) = 1$. Then the probability of any compound event A is computed by adding together the $P(E_i)$'s for all E_i's in A:

$$P(A) = \sum_{\text{all } E_i\text{'s in } A} P(E_i)$$

Example 2.15 Denote the six elementary events $\{1\}, \ldots, \{6\}$ associated with tossing a six-sided die once by $E_1, \ldots, E_6$. If the die is constructed so that any of the three even outcomes is twice as likely to occur as any of the three odd outcomes, then an appropriate assignment of probabilities to elementary events is $P(E_1) = P(E_3) = P(E_5) = \frac{1}{9}$, $P(E_2) = P(E_4) = P(E_6) = \frac{2}{9}$. Then for the event $A = \{\text{outcome is even}\} = E_2 \cup E_4 \cup E_6$, $P(A) = P(E_2) + P(E_4) + P(E_6) = \frac{6}{9} = \frac{2}{3}$; for $B = \{\text{outcome} \leq 3\} = E_1 \cup E_2 \cup E_3$, $P(B) = \frac{1}{9} + \frac{2}{9} + \frac{1}{9} = \frac{4}{9}$. ■

Equally Likely Outcomes

In many experiments consisting of N outcomes, it is reasonable to assign equal probabilities to all N simple events. These include such obvious examples as tossing a fair coin or fair die once or twice (or any fixed number of times), or selecting one or several cards from a well-shuffled deck of 52. With $p = P(E_i)$ for every i,

$$1 = \sum_{i=1}^{N} P(E_i) = \sum_{i=1}^{N} p = p \cdot N \quad \text{so} \quad p = \frac{1}{N}$$

That is, if there are N possible outcomes, then the probability assigned to each is $1/N$.

Now consider an event A, with $N(A)$ denoting the number of outcomes contained in A. Then

$$P(A) = \sum_{E_i \text{ in } A} P(E_i) = \sum_{E_i \text{ in } A} \frac{1}{N} = \frac{N(A)}{N}$$

Once we have counted the number N of outcomes in the sample space, to compute the probability of any event we must count the number of outcomes contained in that event and take the ratio of the two numbers. Thus when outcomes are equally likely, computing probabilities reduces to counting.

Example 2.16 When two dice are rolled separately, there are $N = 36$ outcomes (delete the first row and column from the table in Example 2.3). If both the dice are fair, all 36 outcomes are equally likely, so $P(E_i) = \frac{1}{36}$. Then the event $A = \{\text{sum of two numbers} = 7\}$ consists of the six outcomes (1, 6), (2, 5), (3, 4), (4, 3), (5, 2), and (6, 1), so

$$P(A) = \frac{N(A)}{N} = \frac{6}{36} = \frac{1}{6}$$ ■

Exercises Section 2.2 (13–30)

13. A mutual fund company offers its customers several different funds: a money-market fund, three different bond funds (short, intermediate, and long-term), two stock funds (moderate and high-risk), and a balanced fund. Among customers who own shares in just one fund, the percentages of customers in the different funds are as follows:

Money-market	20%	High-risk stock	18%
Short bond	15%	Moderate-risk stock	25%
Intermediate bond	10%	Balanced	7%
Long bond	5%		

A customer who owns shares in just one fund is randomly selected.

a. What is the probability that the selected individual owns shares in the balanced fund?

b. What is the probability that the individual owns shares in a bond fund?

c. What is the probability that the selected individual does not own shares in a stock fund?

14. Consider randomly selecting a student at a certain university, and let A denote the event that the selected individual has a Visa credit card and B be the analogous event for a MasterCard. Suppose that $P(A) = .5$, $P(B) = .4$, and $P(A \cap B) = .25$.

a. Compute the probability that the selected individual has at least one of the two types of cards (i.e., the probability of the event $A \cup B$).

b. What is the probability that the selected individual has neither type of card?

c. Describe, in terms of A and B, the event that the selected student has a Visa card but not a MasterCard, and then calculate the probability of this event.

15. A computer consulting firm presently has bids out on three projects. Let $A_i = \{\text{awarded project } i\}$, for $i = 1, 2, 3$, and suppose that $P(A_1) = .22$, $P(A_2) = .25$, $P(A_3) = .28$, $P(A_1 \cap A_2) = .11$, $P(A_1 \cap A_3) = .05$, $P(A_2 \cap A_3) = .07$, $P(A_1 \cap A_2 \cap A_3) = .01$. Express in words each of the following events, and compute the probability of each event:

a. $A_1 \cup A_2$

b. $A_1' \cap A_2'$ [*Hint*: $(A_1 \cup A_2)' = A_1' \cap A_2'$]

c. $A_1 \cup A_2 \cup A_3$

d. $A_1' \cap A_2' \cap A_3'$

e. $A_1' \cap A_2' \cap A_3$

f. $(A_1' \cap A_2') \cup A_3$

16. A utility company offers a lifeline rate to any household whose electricity usage falls below 240 kWh during a particular month. Let A denote the event that a randomly selected household in a certain community does not exceed the lifeline usage during January, and let B be the analogous event for the month of July (A and B refer to the same household). Suppose $P(A) = .8$, $P(B) = .7$, and $P(A \cup B) = .9$. Compute the following:

a. $P(A \cap B)$.

b. The probability that the lifeline usage amount is exceeded in exactly one of the two months. Describe this event in terms of A and B.

17. Consider the type of clothes dryer (gas or electric) purchased by each of five different customers at a certain store.

a. If the probability that at most one of these purchases an electric dryer is .428, what is the probability that at least two purchase an electric dryer?

b. If $P(\text{all five purchase gas}) = .116$ and $P(\text{all five purchase electric}) = .005$, what is the probability that at least one of each type is purchased?

18. An individual is presented with three different glasses of cola, labeled C, D, and P. He is asked to taste all three and then list them in order of preference. Suppose the same cola has actually been put into all three glasses.

a. What are the simple events in this ranking experiment, and what probability would you assign to each one?

b. What is the probability that C is ranked first?

c. What is the probability that C is ranked first and D is ranked last?

19. Let A denote the event that the next request for assistance from a statistical software consultant relates to the SPSS package, and let B be the event that the next request is for help with SAS. Suppose that $P(A) = .30$ and $P(B) = .50$.

a. Why is it not the case that $P(A) + P(B) = 1$?

b. Calculate $P(A')$.

c. Calculate $P(A \cup B)$.

d. Calculate $P(A' \cap B')$.

20. A box contains four 40-W bulbs, five 60-W bulbs, and six 75-W bulbs. If bulbs are selected one by one in random order, what is the probability that at least two bulbs must be selected to obtain one that is rated 75 W?

21. Human visual inspection of solder joints on printed circuit boards can be very subjective. Part of the problem stems from the numerous types of solder defects (e.g., pad nonwetting, knee visibility, voids) and even the degree to which a joint possesses one or more of these defects. Consequently, even highly trained inspectors can disagree on the disposition of a particular joint. In one batch of 10,000 joints, inspector A found 724 that were judged defective, inspector B found 751 such joints, and 1159 of the joints were judged defective by at least one of the inspectors. Suppose that one of the 10,000 joints is randomly selected.

a. What is the probability that the selected joint was judged to be defective by neither of the two inspectors?

b. What is the probability that the selected joint was judged to be defective by inspector B but not by inspector A?

22. A certain factory operates three different shifts. Over the last year, 200 accidents have occurred at the factory. Some of these can be attributed at least in part to unsafe working conditions, whereas the others are unrelated to working conditions. The accompanying table gives the percentage of accidents falling in each type of accident–shift category.

Shift	Unsafe Conditions	Unrelated to Conditions
Day	10%	35%
Swing	8%	20%
Night	5%	22%

Suppose one of the 200 accident reports is randomly selected from a file of reports, and the shift and type of accident are determined.

a. What are the simple events?

b. What is the probability that the selected accident was attributed to unsafe conditions?

c. What is the probability that the selected accident did not occur on the day shift?

23. An insurance company offers four different deductible levels—none, low, medium, and high—for its homeowner's policyholders and three different levels—low, medium, and high—for its automobile policyholders. The accompanying table gives proportions for the various categories of policyholders who have both types of insurance. For example, the proportion of individuals with both low homeowner's deductible and low auto deductible is .06 (6% of all such individuals).

	Homeowner's			
Auto	**N**	**L**	**M**	**H**
L	.04	.06	.05	.03
M	.07	.10	.20	.10
H	.02	.03	.15	.15

Suppose an individual having both types of policies is randomly selected.

a. What is the probability that the individual has a medium auto deductible and a high homeowner's deductible?

b. What is the probability that the individual has a low auto deductible? A low homeowner's deductible?

c. What is the probability that the individual is in the same category for both auto and homeowner's deductibles?

d. Based on your answer in part (c), what is the probability that the two categories are different?

e. What is the probability that the individual has at least one low deductible level?

f. Using the answer in part (e), what is the probability that neither deductible level is low?

24. The route used by a certain motorist in commuting to work contains two intersections with traffic signals. The probability that he must stop at the first signal is .4, the analogous probability for the second signal is .5, and the probability that he must stop at at least one of the two signals is .6. What is the probability that he must stop

a. At both signals?

b. At the first signal but not at the second one?

c. At exactly one signal?

25. The computers of six faculty members in a certain department are to be replaced. Two of the faculty members have selected laptop machines and the other four have chosen desktop machines. Suppose that only two of the setups can be done on a particular day, and the two computers to be set up are randomly selected from the six (implying 15 equally likely outcomes; if the computers are numbered 1, 2, . . . , 6, then one outcome consists of computers 1 and 2, another consists of computers 1 and 3, and so on).

a. What is the probability that both selected setups are for laptop computers?

b. What is the probability that both selected setups are desktop machines?

c. What is the probability that at least one selected setup is for a desktop computer?

d. What is the probability that at least one computer of each type is chosen for setup?

26. Use the axioms to show that if one event A is contained in another event B (i.e., A is a subset of B), then $P(A) \le P(B)$. [*Hint*: For such A and B, A and $B \cap A'$ are disjoint and $B = A \cup (B \cap A')$, as can be seen from a Venn diagram.] For general A and B, what does this imply about the relationship among $P(A \cap B)$, $P(A)$, and $P(A \cup B)$?

27. The three major options on a certain type of new car are an automatic transmission (A), a sunroof (B), and a stereo with compact disc player (C). If 70% of all purchasers request A, 80% request B, 75% request C, 85% request A or B, 90% request A or C, 95% request B or C, and 98% request A or B or C, compute the probabilities of the following events. [*Hint*: "A or B" is the event that at least one of the two options is requested; try drawing a Venn diagram and labeling all regions.]

a. The next purchaser will request at least one of the three options.

b. The next purchaser will select none of the three options.

c. The next purchaser will request only an automatic transmission and not either of the other two options.

d. The next purchaser will select exactly one of these three options.

28. A certain system can experience three different types of defects. Let A_i ($i = 1, 2, 3$) denote the event that the system has a defect of type i. Suppose that

$$P(A_1) = .12 \quad P(A_2) = .07 \quad P(A_3) = .05$$
$$P(A_1 \cup A_2) = .13 \quad P(A_1 \cup A_3) = .14$$
$$P(A_2 \cup A_3) = .10 \quad P(A_1 \cap A_2 \cap A_3) = .01$$

a. What is the probability that the system does not have a type 1 defect?

b. What is the probability that the system has both type 1 and type 2 defects?

c. What is the probability that the system has both type 1 and type 2 defects but not a type 3 defect?

d. What is the probability that the system has at most two of these defects?

29. In Exercise 7, suppose that any incoming individual is equally likely to be assigned to any of the three stations irrespective of where other individuals have been assigned. What is the probability that

a. All three family members are assigned to the same station?

b. At most two family members are assigned to the same station?

c. Every family member is assigned to a different station?

30. Apply the proposition involving the probability of $A \cup B$ to the union of the two events $(A \cup B)$ and C in order to verify the result for $P(A \cup B \cup C)$.

2.3 Counting Techniques

When the various outcomes of an experiment are equally likely (the same probability is assigned to each simple event), the task of computing probabilities reduces to counting. In particular, if N is the number of outcomes in a sample space and $N(A)$ is the number of outcomes contained in an event A, then

$$P(A) = \frac{N(A)}{N} \tag{2.1}$$

If a list of the outcomes is available or easy to construct and N is small, then the numerator and denominator of Equation (2.1) can be obtained without the benefit of any general counting principles.

There are, however, many experiments for which the effort involved in constructing such a list is prohibitive because N is quite large. By exploiting some general counting rules, it is possible to compute probabilities of the form (2.1) without a listing of outcomes. These rules are also useful in many problems involving outcomes that are not equally likely. Several of the rules developed here will be used in studying probability distributions in the next chapter.

The Product Rule for Ordered Pairs

Our first counting rule applies to any situation in which a set (event) consists of ordered pairs of objects and we wish to count the number of such pairs. By an ordered pair, we mean that, if O_1 and O_2 are objects, then the pair (O_1, O_2) is different from the pair (O_2, O_1). For example, if an individual selects one airline for a trip from Los Angeles to Chicago and (after transacting business in Chicago) a second one for continuing on to New York, one possibility is (American, United), another is (United, American), and still another is (United, United).

PROPOSITION If the first element or object of an ordered pair can be selected in n_1 ways, and for each of these n_1 ways the second element of the pair can be selected in n_2 ways, then the number of pairs is $n_1 n_2$.

Example 2.17 A homeowner doing some remodeling requires the services of both a plumbing contractor and an electrical contractor. If there are 12 plumbing contractors and 9 electrical contractors available in the area, in how many ways can the contractors be chosen? If we denote the plumbers by $P_1, \ldots, P_{12}$ and the electricians by $Q_1, \ldots, Q_9$, then we wish the number of pairs of the form (P_i, Q_j). With $n_1 = 12$ and $n_2 = 9$, the product rule yields $N = (12)(9) = 108$ possible ways of choosing the two types of contractors. ■

In Example 2.17, the choice of the second element of the pair did not depend on which first element was chosen or occurred. As long as there is the same number of choices of the second element for each first element, the product rule is valid even when the set of possible second elements depends on the first element.

Example 2.18 A family has just moved to a new city and requires the services of both an obstetrician and a pediatrician. There are two easily accessible medical clinics, each having two obstetricians and three pediatricians. The family will obtain maximum health insurance benefits by joining a clinic and selecting both doctors from that clinic. In how many ways can this be done? Denote the obstetricians by O_1, O_2, O_3, and O_4 and the pediatricians by $P_1, \ldots, P_6$. Then we wish the number of pairs (O_i, P_j) for which O_i and P_j are associated with the same clinic. Because there are four obstetricians, $n_1 = 4$, and for each there are three choices of pediatrician, so $n_2 = 3$. Applying the product rule gives $N = n_1 n_2 = 12$ possible choices. ■

Tree Diagrams

In many counting and probability problems, a configuration called a *tree diagram* can be used to represent pictorially all the possibilities. The tree diagram associated with Example 2.18 appears in Figure 2.7. Starting from a point on the left side of the diagram, for each possible first element of a pair a straight-line segment emanates rightward. Each of these lines is referred to as a first-generation branch. Now for any given first-generation branch we construct another line segment emanating from the tip of the branch for each possible choice of a second element of the pair. Each such line segment is a second-generation branch. Because there are four obstetricians, there are four first-generation branches, and three pediatricians for each obstetrician yields three second-generation branches emanating from each first-generation branch.

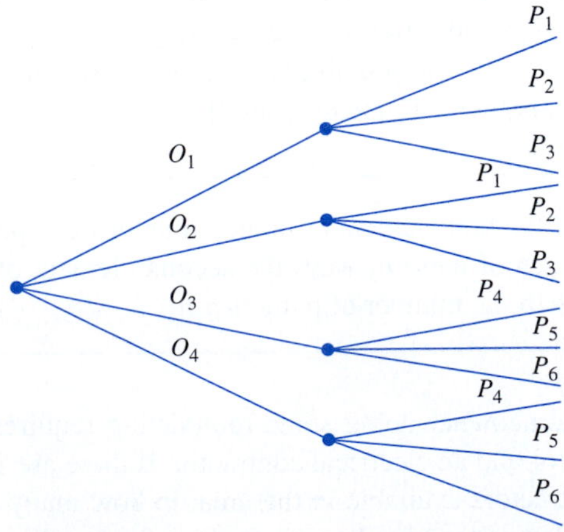

Figure 2.7 Tree diagram for Example 2.18

Generalizing, suppose there are n_1 first-generation branches, and for each first-generation branch there are n_2 second-generation branches. The total number of second-generation branches is then $n_1 n_2$. Since the end of each second-generation branch corresponds to exactly one possible pair (choosing a first element and then a second puts us at the end of exactly one second-generation branch), there are $n_1 n_2$ pairs, verifying the product rule.

The construction of a tree diagram does not depend on having the same number of second-generation branches emanating from each first-generation branch. If the second clinic had four pediatricians, then there would be only three branches emanating from two of the first-generation branches and four emanating from each of the other two first-generation branches. A tree diagram can thus be used to represent pictorially experiments other than those to which the product rule applies.

A More General Product Rule

If a six-sided die is tossed five times in succession rather than just twice, then each possible outcome is an ordered collection of five numbers such as (1, 3, 1, 2, 4) or (6, 5, 2, 2, 2). We will call an ordered collection of k objects a ***k*-tuple** (so a pair is a 2-tuple and a triple is a 3-tuple). Each outcome of the die-tossing experiment is then a 5-tuple.

PRODUCT RULE FOR *k*-TUPLES

Suppose a set consists of ordered collections of k elements (k-tuples) and that there are n_1 possible choices for the first element; for each choice of the first element, there are n_2 possible choices of the second element; . . . ; for each possible choice of the first $k - 1$ elements, there are n_k choices of the kth element. Then there are $n_1 n_2 \cdot \cdots \cdot n_k$ possible k-tuples.

This more general rule can also be illustrated by a tree diagram; simply construct a more elaborate diagram by adding third-generation branches emanating from the tip of each second generation, then fourth-generation branches, and so on, until finally kth-generation branches are added.

Example 2.19 (Example 2.17 continued)

Suppose the home remodeling job involves first purchasing several kitchen appliances. They will all be purchased from the same dealer, and there are five dealers in the area. With the dealers denoted by $D_1, \ldots, D_5$, there are $N = n_1 n_2 n_3 = (5)(12)(9) = 540$ 3-tuples of the form (D_i, P_j, Q_k), so there are 540 ways to choose first an appliance dealer, then a plumbing contractor, and finally an electrical contractor. ■

Example 2.20 (Example 2.18 continued)

If each clinic has both three specialists in internal medicine and two general surgeons, there are $n_1 n_2 n_3 n_4 = (4)(3)(3)(2) = 72$ ways to select one doctor of each type such that all doctors practice at the same clinic. ■

Permutations

So far the successive elements of a k-tuple were selected from entirely different sets (e.g., appliance dealers, then plumbers, and finally electricians). In several tosses of a die, the set from which successive elements are chosen is always {1, 2, 3, 4, 5, 6}, but the choices are made "with replacement" so that the same element can appear more than once. We now consider a fixed set consisting of n distinct elements and suppose that a k-tuple is formed by selecting successively from this set *without replacement* so that an element can appear in at most one of the k positions.

DEFINITION Any ordered sequence of k objects taken from a set of n distinct objects is called a **permutation** of size k of the objects. The number of permutations of size k that can be constructed from the n objects is denoted by $P_{k,n}$.

The number of permutations of size k is obtained immediately from the general product rule. The first element can be chosen in n ways, for each of these n ways the second element can be chosen in $n-1$ ways, and so on; finally, for each way of choosing the first $k-1$ elements, the kth element can be chosen in $n-(k-1)=n-k+1$ ways, so

$$P_{k,n} = n(n-1)(n-2) \cdot \cdots \cdot (n-k+2)(n-k+1)$$

Example 2.21 Ten teaching assistants are available for grading papers in a particular course. The first exam consists of four questions, and the professor wishes to select a different assistant to grade each question (only one assistant per question). In how many ways can assistants be chosen to grade the exam? Here n = the number of assistants = 10 and k = the number of questions = 4. The number of different grading assignments is then $P_{4,10} = (10)(9)(8)(7) = 5040$. ■

The use of factorial notation allows $P_{k,n}$ to be expressed more compactly.

DEFINITION For any positive integer m, $m!$ is read "m factorial" and is defined by $m! = m(m-1) \cdot \cdots \cdot (2)(1)$. Also, $0! = 1$.

Using factorial notation, $(10)(9)(8)(7) = (10)(9)(8)(7)(6!)/6! = 10!/6!$. More generally,

$$\begin{aligned} P_{k,n} &= n(n-1) \cdot \cdots \cdot (n-k+1) \\ &= \frac{n(n-1) \cdot \cdots \cdot (n-k+1)(n-k)(n-k-1) \cdot \cdots \cdot (2)(1)}{(n-k)(n-k-1) \cdot \cdots \cdot (2)(1)} \end{aligned}$$

which becomes

$$P_{k,n} = \frac{n!}{(n-k)!}$$

For example, $P_{3,9} = 9!/(9-3)! = 9!/6! = 9 \cdot 8 \cdot 7 \cdot 6!/6! = 9 \cdot 8 \cdot 7$. Note also that because $0! = 1$, $P_{n,n} = n!/(n-n)! = n!/0! = n!/1 = n!$, as it should.

Combinations

Often the objective is to count the number of *unordered* subsets of size k that can be formed from a set consisting of n distinct objects. For example, in bridge it is only the 13 cards in a hand and not the order in which they are dealt that is important; in the formation of a committee, the order in which committee members are listed is frequently unimportant.

DEFINITION

Given a set of n distinct objects, any unordered subset of size k of the objects is called a **combination**. The number of combinations of size k that can be formed from n distinct objects will be denoted by $\binom{n}{k}$. (This notation is more common in probability than $C_{k,n}$, which would be analogous to notation for permutations.)

The number of combinations of size k from a particular set is smaller than the number of permutations because, when order is disregarded, some of the permutations correspond to the same combination. Consider, for example, the set $\{A, B, C, D, E\}$ consisting of five elements. There are $5!/(5-3)! = 60$ permutations of size 3. There are six permutations of size 3 consisting of the elements A, B, and C since these three can be ordered $3 \cdot 2 \cdot 1 = 3! = 6$ ways: (A, B, C), (A, C, B), (B, A, C), (B, C, A), (C, A, B), and (C, B, A). These six permutations are equivalent to the single combination $\{A, B, C\}$. Similarly, for any other combination of size 3, there are 3! permutations, each obtained by ordering the three objects. Thus,

$$60 = P_{3,5} = \binom{5}{3} \cdot 3! \quad \text{so} \quad \binom{5}{3} = \frac{60}{3!} = 10$$

These ten combinations are

$$\{A, B, C\}\ \{A, B, D\}\ \{A, B, E\}\ \{A, C, D\}\ \{A, C, E\}\ \{A, D, E\},\ \{B, C, D\}$$
$$\{B, C, E\}\ \{B, D, E\}\ \{C, D, E\}$$

When there are n distinct objects, any permutation of size k is obtained by ordering the k unordered objects of a combination in one of $k!$ ways, so the number of permutations is the product of $k!$ and the number of combinations. This gives

$$\binom{n}{k} = \frac{P_{k,n}}{k!} = \frac{n!}{k!(n-k)!}$$

Notice that $\binom{n}{n} = 1$ and $\binom{n}{0} = 1$ since there is only one way to choose a set of (all) n elements or of no elements, and $\binom{n}{1} = n$ since there are n subsets of size 1.

Example 2.22 A bridge hand consists of any 13 cards selected from a 52-card deck without regard to order. There are $\binom{52}{13} = 52!/13!/39!$ different bridge hands, which works out to approximately 635 billion. Since there are 13 cards in each suit, the number of hands consisting entirely of clubs and/or spades (no red cards) is $\binom{26}{13} = 26!/13!13! = 10{,}400{,}600$. One of these $\binom{26}{13}$ hands consists entirely of spades, and one consists entirely of clubs, so there are $[\binom{26}{13} - 2]$ hands that consist entirely of clubs and spades with both suits represented in the hand. Suppose a bridge hand is dealt from a well-shuffled deck (i.e., 13 cards are randomly selected from among the 52 possibilities) and let

$$A = \{\text{the hand consists entirely of spades and clubs with both suits represented}\}$$

$$B = \{\text{the hand consists of exactly two suits}\}$$

The $N = \binom{52}{13}$ possible outcomes are equally likely, so

$$P(A) = \frac{N(A)}{N} = \frac{\binom{26}{13} - 2}{\binom{52}{13}} = .0000164$$

Since there are $\binom{4}{2} = 6$ combinations consisting of two suits, of which spades and clubs is one such combination,

$$P(B) = \frac{6\left[\binom{26}{13} - 2\right]}{\binom{52}{13}} = .0000983$$

That is, a hand consisting entirely of cards from exactly two of the four suits will occur roughly once in every 10,000 hands. If you play bridge only once a month, it is likely that you will never be dealt such a hand. ■

Example 2.23 A university warehouse has received a shipment of 25 printers, of which 10 are laser printers and 15 are inkjet models. If 6 of these 25 are selected at random to be checked by a particular technician, what is the probability that exactly 3 of those selected are laser printers (so that the other 3 are inkjets)?

Let $D_3 = \{\text{exactly 3 of the 6 selected are inkjet printers}\}$. Assuming that any particular set of 6 printers is as likely to be chosen as is any other set of 6, we have equally likely outcomes, so $P(D_3) = N(D_3)/N$, where N is the number of ways of choosing 6 printers from the 25 and $N(D_3)$ is the number of ways of choosing 3 laser printers and 3 inkjet models. Thus $N = \binom{25}{6}$. To obtain $N(D_3)$, think of first choosing 3 of the 15 inkjet models and then 3 of the laser printers. There are $\binom{15}{3}$ ways of choosing the 3 inkjet models, and there are $\binom{10}{3}$ ways of choosing the 3 laser printers; $N(D_3)$ is now the product of these two numbers (visualize a tree diagram—we are really using a product rule argument here), so

$$P(D_3) = \frac{N(D_3)}{N} = \frac{\binom{15}{3}\binom{10}{3}}{\binom{25}{6}} = \frac{\frac{15!}{3!12!} \cdot \frac{10!}{3!7!}}{\frac{25!}{6!19!}} = .3083$$

Let D_4 = {exactly 4 of the 6 printers selected are inkjet models} and define D_5 and D_6 in an analogous manner. Then the probability that at least 3 inkjet printers are selected is

$$P(D_3 \cup D_4 \cup D_5 \cup D_6) = P(D_3) + P(D_4) + P(D_5) + P(D_6)$$

$$= \frac{\binom{15}{3}\binom{10}{3}}{\binom{25}{6}} + \frac{\binom{15}{4}\binom{10}{2}}{\binom{25}{6}}$$

$$+ \frac{\binom{15}{5}\binom{10}{1}}{\binom{25}{6}} + \frac{\binom{15}{6}\binom{10}{0}}{\binom{25}{6}} = .8530$$

■

Exercises Section 2.3 (31–44)

31. The College of Science Council has one student representative from each of the five science departments (biology, chemistry, geology, mathematics, physics). In how many ways can

a. Both a council president and a vice president be selected?

b. A president, a vice president, and a secretary be selected?

c. Two members be selected for the Dean's Council?

32. A friend is giving a dinner party. Her current wine supply includes 8 bottles of zinfandel, 10 of merlot, and 12 of cabernet (she drinks only red wine), all from different wineries.

a. If she wants to serve 3 bottles of zinfandel and serving order is important, how many ways are there to do this?

b. If 6 bottles of wine are to be randomly selected from the 30 for serving, how many ways are there to do this?

c. If 6 bottles are randomly selected, how many ways are there to obtain two bottles of each variety?

d. If 6 bottles are randomly selected, what is the probability that this results in two bottles of each variety being chosen?

e. If 6 bottles are randomly selected, what is the probability that all of them are the same variety?

33. a. Beethoven wrote 9 symphonies and Mozart wrote 27 piano concertos. If a university radio station announcer wishes to play first a Beethoven symphony and then a Mozart concerto, in how many ways can this be done?

b. The station manager decides that on each successive night (7 days per week), a Beethoven symphony will be played, followed by a Mozart piano concerto, followed by a Schubert string quartet (of which there are 15). For roughly how many years could this policy be continued before exactly the same program would have to be repeated?

34. A chain of stereo stores is offering a special price on a complete set of components (receiver, compact disc player, speakers). A purchaser is offered a choice of manufacturer for each component:

Receiver:	Kenwood, Onkyo, Pioneer, Sony, Yamaha
Compact disc player:	Onkyo, Pioneer, Sony, Panasonic
Speakers:	Boston, Infinity, Polk

A switchboard display in the store allows a customer to hook together any selection of components (consisting of one of each type). Use the product rules to answer the following questions:

a. In how many ways can one component of each type be selected?

b. In how many ways can components be selected if both the receiver and the compact disc player are to be Sony?

c. In how many ways can components be selected if none is to be Sony?

d. In how many ways can a selection be made if at least one Sony component is to be included?

e. If someone flips switches on the selection in a completely random fashion, what is the probability that the system selected contains at least one Sony component? Exactly one Sony component?

35. Shortly after being put into service, some buses manufactured by a certain company have developed cracks on the underside of the main frame. Suppose a particular city has 25 of these buses, and cracks have actually appeared in 8 of them.

a. How many ways are there to select a sample of 5 buses from the 25 for a thorough inspection?

b. In how many ways can a sample of 5 buses contain exactly 4 with visible cracks?

c. If a sample of 5 buses is chosen at random, what is the probability that exactly 4 of the 5 will have visible cracks?

d. If buses are selected as in part (c), what is the probability that at least 4 of those selected will have visible cracks?

36. A production facility employs 20 workers on the day shift, 15 workers on the swing shift, and 10 workers on the graveyard shift. A quality control consultant is to select 6 of these workers for in-depth interviews. Suppose the selection is made in such a way that any particular group of 6 workers has the same chance of being selected as does any other group (drawing 6 slips without replacement from among 45).

a. How many selections result in all 6 workers coming from the day shift? What is the probability that all 6 selected workers will be from the day shift?

b. What is the probability that all 6 selected workers will be from the same shift?

c. What is the probability that at least two different shifts will be represented among the selected workers?

d. What is the probability that at least one of the shifts will be unrepresented in the sample of workers?

37. An academic department with five faculty members narrowed its choice for department head to either candidate A or candidate B. Each member then voted on a slip of paper for one of the candidates. Suppose there are actually three votes for A and two for B. If the slips are selected for tallying in random order, what is the probability that A remains ahead of B throughout the vote count (e.g., this event occurs if the selected ordering is $AABAB$, but not for $ABBAA$)?

38. An experimenter is studying the effects of temperature, pressure, and type of catalyst on yield from a certain chemical reaction. Three different temperatures, four different pressures, and five different catalysts are under consideration.

a. If any particular experimental run involves the use of a single temperature, pressure, and catalyst, how many experimental runs are possible?

b. How many experimental runs are there that involve use of the lowest temperature and two lowest pressures?

39. Refer to Exercise 38 and suppose that five different experimental runs are to be made on the first day of experimentation. If the five are randomly selected from among all the possibilities, so that any group of five has the same probability of selection, what is the probability that a different catalyst is used on each run?

40. A box in a certain supply room contains four 40-W lightbulbs, five 60-W bulbs, and six 75-W bulbs. Suppose that three bulbs are randomly selected.

a. What is the probability that exactly two of the selected bulbs are rated 75 W?

b. What is the probability that all three of the selected bulbs have the same rating?

c. What is the probability that one bulb of each type is selected?

d. Suppose now that bulbs are to be selected one by one until a 75-W bulb is found. What is the probability that it is necessary to examine at least six bulbs?

41. Fifteen telephones have just been received at an authorized service center. Five of these telephones are cellular, five are cordless, and the other five are corded phones. Suppose that these components are randomly allocated the numbers $1, 2, \ldots, 15$ to establish the order in which they will be serviced.

a. What is the probability that all the cordless phones are among the first ten to be serviced?

b. What is the probability that after servicing ten of these phones, phones of only two of the three types remain to be serviced?

c. What is the probability that two phones of each type are among the first six serviced?

42. Three molecules of type A, three of type B, three of type C, and three of type D are to be linked together to form a chain molecule. One such chain molecule is *ABCDABCDABCD*, and another is *BCDDAAABDBCC*.

a. How many such chain molecules are there? (*Hint*: If the three *A*'s were distinguishable from one another—A_1, A_2, A_3—and the *B*'s, *C*'s, and *D*'s were also, how many molecules would there be? How is this number reduced when the subscripts are removed from the *A*'s?)

b. Suppose a chain molecule of the type described is randomly selected. What is the probability that all three molecules of each type end up next to one another (such as in *BBBAAADDDCCC*)?

43. Three married couples have purchased theater tickets and are seated in a row consisting of just six seats. If they take their seats in a completely random fashion (random order), what is the probability that Jim and Paula (husband and wife) sit in the two seats on the far left? What is the probability that Jim and Paula end up sitting next to one another? What is the probability that at least one of the wives ends up sitting next to her husband?

44. Show that $\binom{n}{k} = \binom{n}{n-k}$. Give an interpretation involving subsets.

2.4 Conditional Probability

The probabilities assigned to various events depend on what is known about the experimental situation when the assignment is made. Subsequent to the initial assignment, partial information about or relevant to the outcome of the experiment may become available. Such information may cause us to revise some of our probability assignments. For a particular event *A*, we have used $P(A)$ to represent the probability assigned to *A*; we now think of $P(A)$ as the original or unconditional probability of the event *A*.

In this section, we examine how the information "an event *B* has occurred" affects the probability assigned to *A*. For example, *A* might refer to an individual having a particular disease in the presence of certain symptoms. If a blood test is performed on the individual and the result is negative (B = negative blood test), then the probability of having the disease will change (it should decrease, but not usually to zero, since blood tests are not infallible). We will use the notation $P(A \mid B)$ to represent the **conditional probability of *A* given that the event *B* has occurred**.

Example 2.24

Complex components are assembled in a plant that uses two different assembly lines, *A* and *A′*. Line *A* uses older equipment than *A′*, so it is somewhat slower and less reliable. Suppose on a given day line *A* has assembled 8 components, of which 2 have been identified as defective (*B*) and 6 as nondefective (*B′*), whereas *A′* has produced 1 defective and 9 nondefective components. This information is summarized in the accompanying table.

	Condition	
Line	***B***	***B′***
A	2	6
A′	1	9

Unaware of this information, the sales manager randomly selects 1 of these 18 components for a demonstration. Prior to the demonstration

$$P(\text{line } A \text{ component selected}) = P(A) = \frac{N(A)}{N} = \frac{8}{18} = .44$$

However, if the chosen component turns out to be defective, then the event B has occurred, so the component must have been 1 of the 3 in the B column of the table. Since these 3 components are equally likely among themselves after B has occurred,

$$P(A \mid B) = \frac{2}{3} = \frac{\frac{2}{18}}{\frac{3}{18}} = \frac{P(A \cap B)}{P(B)} \tag{2.2}$$

■

In Equation (2.2), the conditional probability is expressed as a ratio of unconditional probabilities: The numerator is the probability of the intersection of the two events, whereas the denominator is the probability of the conditioning event B. A Venn diagram illuminates this relationship (Figure 2.8).

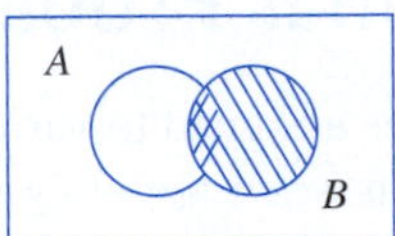

Figure 2.8 Motivating the definition of conditional probability

Given that B has occurred, the relevant sample space is no longer $\mathcal{S}$ but consists of outcomes in B; A has occurred if and only if one of the outcomes in the intersection occurred, so the conditional probability of A given B is proportional to $P(A \cap B)$. The proportionality constant $1/P(B)$ is used to ensure that the probability $P(B \mid B)$ of the new sample space B equals 1.

The Definition of Conditional Probability

Example 2.24 demonstrates that when outcomes are equally likely, computation of conditional probabilities can be based on intuition. When experiments are more complicated, though, intuition may fail us, so we want to have a general definition of conditional probability that will yield intuitive answers in simple problems. The Venn diagram and Equation (2.2) suggest the appropriate definition.

DEFINITION

For any two events A and B with $P(B) > 0$, the **conditional probability of A given that B has occurred** is defined by

$$P(A \mid B) = \frac{P(A \cap B)}{P(B)} \tag{2.3}$$

Example 2.25 Suppose that of all individuals buying a certain digital camera, 60% include an optional memory card in their purchase, 40% include an extra battery, and 30% include both a card and battery. Consider randomly selecting a buyer and let $A = \{\text{memory card purchased}\}$

and $B = \{\text{battery purchased}\}$. Then $P(A) = .60$, $P(B) = .40$, and $P(\text{both purchased}) = P(A \cap B) = .30$. Given that the selected individual purchased an extra battery, the probability that an optional card was also purchased is

$$P(A \mid B) = \frac{P(A \cap B)}{P(B)} = \frac{.30}{.40} = .75$$

That is, of all those purchasing an extra battery, 75% purchased an optional memory card. Similarly,

$$P(\text{battery} \mid \text{memory card}) = P(B \mid A) = \frac{P(A \cap B)}{P(A)} = \frac{.30}{.60} = .50$$

Notice that $P(A \mid B) \neq P(A)$ and $P(B \mid A) \neq P(B)$. ■

Example 2.26 A news magazine includes three columns entitled "Art" (A), "Books" (B), and "Cinema" (C). Reading habits of a randomly selected reader with respect to these columns are

Read regularly	A	B	C	$A \cap B$	$A \cap C$	$B \cap C$	$A \cap B \cap C$
Probability	.14	.23	.37	.08	.09	.13	.05

(See Figure 2.9.)

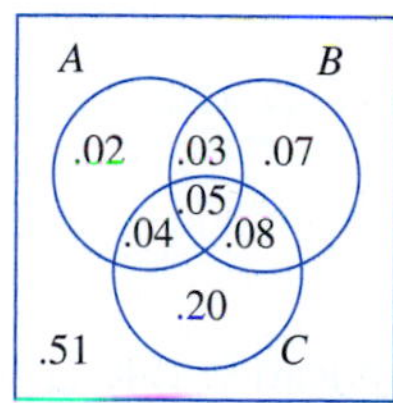

Figure 2.9 Venn diagram for Example 2.26

We thus have

$$P(A \mid B) = \frac{P(A \cap B)}{P(B)} = \frac{.08}{.23} = .348$$

$$P(A \mid B \cup C) = \frac{P(A \cap (B \cup C))}{P(B \cup C)} = \frac{.04 + .05 + .03}{.47} = \frac{.12}{.47} = .255$$

$$P(A \mid \text{reads at least one}) = P(A \mid A \cup B \cup C) = \frac{P(A \cap (A \cup B \cup C))}{P(A \cup B \cup C)}$$

$$= \frac{P(A)}{P(A \cup B \cup C)} = \frac{.14}{.49} = .286$$

and

$$P(A \cup B \mid C) = \frac{P((A \cup B) \cap C)}{P(C)} = \frac{.04 + .05 + .08}{.37} = .459$$ ■

The Multiplication Rule for $P(A \cap B)$

The definition of conditional probability yields the following result, obtained by multiplying both sides of Equation (2.3) by $P(B)$.

THE MULTIPLICATION RULE

$$P(A \cap B) = P(A \mid B) \cdot P(B)$$

This rule is important because it is often the case that $P(A \cap B)$ is desired, whereas both $P(B)$ and $P(A \mid B)$ can be specified from the problem description. Consideration of $P(B \mid A)$ gives $P(A \cap B) = P(B \mid A) \cdot P(A)$.

Example 2.27 Four individuals have responded to a request by a blood bank for blood donations. None of them has donated before, so their blood types are unknown. Suppose only type O+ is desired and only one of the four actually has this type. If the potential donors are selected in random order for typing, what is the probability that at least three individuals must be typed to obtain the desired type?

Making the identification $B = \{\text{first type not O+}\}$ and $A = \{\text{second type not O+}\}$, $P(B) = \frac{3}{4}$. Given that the first type is not O+, two of the three individuals left are not O+, so $P(A \mid B) = \frac{2}{3}$. The multiplication rule now gives

$$\begin{aligned} P(\text{at least three individuals are typed}) &= P(A \cap B) \\ &= P(A \mid B) \cdot P(B) \\ &= \frac{2}{3} \cdot \frac{3}{4} = \frac{6}{12} \\ &= .5 \end{aligned}$$

■

The multiplication rule is most useful when the experiment consists of several stages in succession. The conditioning event B then describes the outcome of the first stage and A the outcome of the second, so that $P(A \mid B)$—conditioning on what occurs first—will often be known. The rule is easily extended to experiments involving more than two stages. For example,

$$\begin{aligned} P(A_1 \cap A_2 \cap A_3) &= P(A_3 \mid A_1 \cap A_2) \cdot P(A_1 \cap A_2) \\ &= P(A_3 \mid A_1 \cap A_2) \cdot P(A_2 \mid A_1) \cdot P(A_1) \end{aligned} \tag{2.4}$$

where A_1 occurs first, followed by A_2, and finally A_3.

Example 2.28 For the blood typing experiment of Example 2.27,

$$\begin{aligned} P(\text{third type is O+}) &= P(\text{third is} \mid \text{first isn't} \cap \text{second isn't}) \\ &\quad \cdot P(\text{second isn't} \mid \text{first isn't}) \cdot P(\text{first isn't}) \\ &= \frac{1}{2} \cdot \frac{2}{3} \cdot \frac{3}{4} = \frac{1}{4} = .25 \end{aligned}$$

■

When the experiment of interest consists of a sequence of several stages, it is convenient to represent these with a tree diagram. Once we have an appropriate tree diagram, probabilities and conditional probabilities can be entered on the various branches; this will make repeated use of the multiplication rule quite straightforward.

Example 2.29 A chain of video stores sells three different brands of DVD players. Of its DVD player sales, 50% are brand 1 (the least expensive), 30% are brand 2, and 20% are brand 3. Each manufacturer offers a 1-year warranty on parts and labor. It is known that 25% of brand 1's DVD players require warranty repair work, whereas the corresponding percentages for brands 2 and 3 are 20% and 10%, respectively.

1. What is the probability that a randomly selected purchaser has bought a brand 1 DVD player that will need repair while under warranty?
2. What is the probability that a randomly selected purchaser has a DVD player that will need repair while under warranty?
3. If a customer returns to the store with a DVD player that needs warranty repair work, what is the probability that it is a brand 1 DVD player? A brand 2 DVD player? A brand 3 DVD player?

The first stage of the problem involves a customer selecting one of the three brands of DVD player. Let A_i = {brand i is purchased}, for $i = 1, 2,$ and 3. Then $P(A_1) = .50$, $P(A_2) = .30$, and $P(A_3) = .20$. Once a brand of DVD player is selected, the second stage involves observing whether the selected DVD player needs warranty repair. With B = {needs repair} and B' = {doesn't need repair}, the given information implies that $P(B|A_1) = .25$, $P(B|A_2) = .20$, and $P(B|A_3) = .10$.

The tree diagram representing this experimental situation is shown in Figure 2.10. The initial branches correspond to different brands of DVD players; there are two second-generation branches emanating from the tip of each initial branch, one for "needs repair"

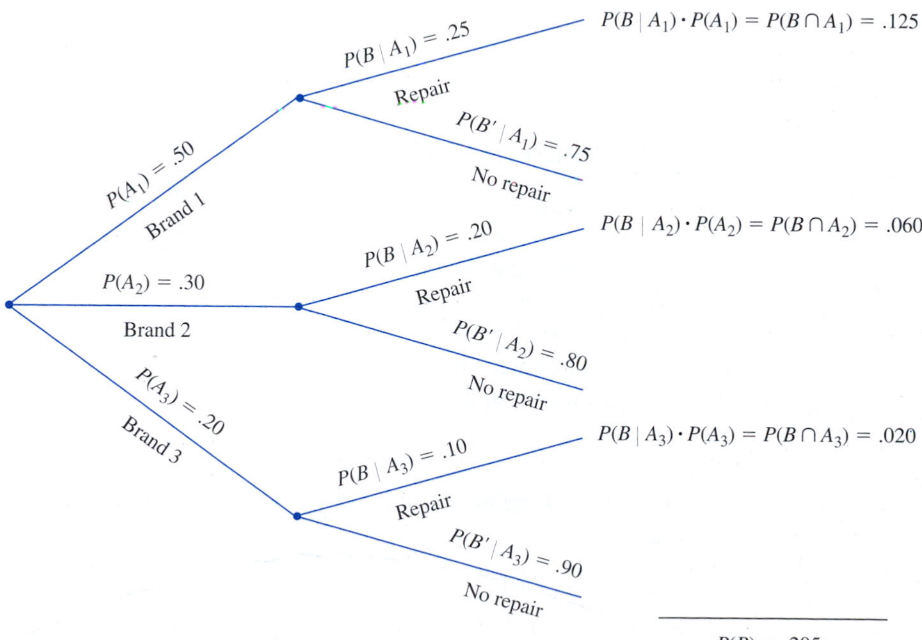

Figure 2.10 Tree diagram for Example 2.29

and the other for "doesn't need repair." The probability $P(A_i)$ appears on the ith initial branch, whereas the conditional probabilities $P(B|A_i)$ and $P(B'|A_i)$ appear on the second-generation branches. To the right of each second-generation branch corresponding to the occurrence of B, we display the product of probabilities on the branches leading out to that point. This is simply the multiplication rule in action. The answer to the question posed in 1 is thus $P(A_1 \cap B) = P(B|A_1) \cdot P(A_1) = .125$. The answer to question 2 is

$$\begin{aligned} P(B) &= P[(\text{brand 1 and repair}) \text{ or } (\text{brand 2 and repair}) \text{ or } (\text{brand 3 and repair})] \\ &= P(A_1 \cap B) + P(A_2 \cap B) + P(A_3 \cap B) \\ &= .125 + .060 + .020 = .205 \end{aligned}$$

Finally,

$$P(A_1|B) = \frac{P(A_1 \cap B)}{P(B)} = \frac{.125}{.205} = .61$$

$$P(A_2|B) = \frac{P(A_2 \cap B)}{P(B)} = \frac{.060}{.205} = .29$$

and

$$P(A_3|B) = 1 - P(A_1|B) - P(A_2|B) = .10$$

Notice that the initial or *prior probability* of brand 1 is .50, whereas once it is known that the selected DVD player needed repair, the *posterior probability* of brand 1 increases to .61. This is because brand 1 DVD players are more likely to need warranty repair than are the other brands. The posterior probability of brand 3 is $P(A_3|B) = .10$, which is much less than the prior probability $P(A_3) = .20$. ■

Bayes' Theorem

The computation of a posterior probability $P(A_j|B)$ from given prior probabilities $P(A_i)$ and conditional probabilities $P(B|A_i)$ occupies a central position in elementary probability. The general rule for such computations, which is really just a simple application of the multiplication rule, goes back to the Reverend Thomas Bayes, who lived in the eighteenth century. To state it we first need another result. Recall that events $A_1, \ldots, A_k$ are mutually exclusive if no two have any common outcomes. The events are *exhaustive* if one A_i must occur, so that $A_1 \cup \cdots \cup A_k = \mathcal{S}$.

THE LAW OF TOTAL PROBABILITY

Let $A_1, \ldots, A_k$ be mutually exclusive and exhaustive events. Then for any other event B,

$$\begin{aligned} P(B) &= P(B|A_1)P(A_1) + \cdots + P(B|A_k)P(A_k) \\ &= \sum_{i=1}^{k} P(B|A_i)P(A_i) \end{aligned} \qquad (2.5)$$

Proof Because the A_i's are mutually exclusive and exhaustive, if B occurs it must be in conjunction with exactly one of the A_i's. That is, $B = (A_1 \text{ and } B)$ or . . . or $(A_k \text{ and } B) = (A_1 \cap B) \cup \cdots \cup (A_k \cap B)$, where the events $(A_i \cap B)$ are mutually exclusive. This "partitioning of B" is illustrated in Figure 2.11. Thus

$$P(B) = \sum_{i=1}^{k} P(A_i \cap B) = \sum_{i=1}^{k} P(B \mid A_i)P(A_i)$$

as desired.

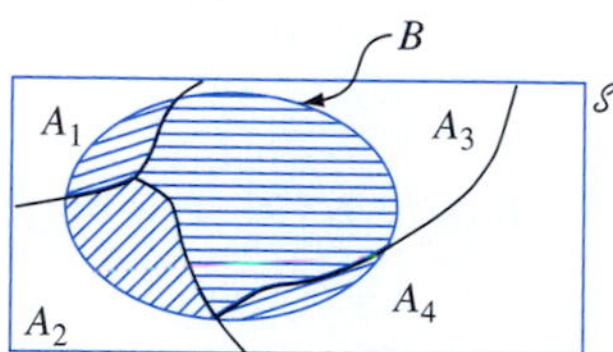

Figure 2.11 Partition of B by mutually exclusive and exhaustive A_i's ■

An example of the use of Equation (2.5) appeared in answering question 2 of Example 2.29, where $A_1 = \{\text{brand 1}\}, A_2 = \{\text{brand 2}\}, A_3 = \{\text{brand 3}\}$, and $B = \{\text{repair}\}$.

BAYES' THEOREM

Let $A_1, A_2, \ldots, A_k$ be a collection of k mutually exclusive and exhaustive events with $P(A_i) > 0$ for $i = 1, \ldots, k$. Then for any other event B for which $P(B) > 0$,

$$P(A_j \mid B) = \frac{P(A_j \cap B)}{P(B)} = \frac{P(B \mid A_j)P(A_j)}{\sum_{i=1}^{k} P(B \mid A_i) \cdot P(A_i)} \qquad j = 1, \ldots, k \tag{2.6}$$

The transition from the second to the third expression in (2.6) rests on using the multiplication rule in the numerator and the law of total probability in the denominator.

The proliferation of events and subscripts in (2.6) can be a bit intimidating to probability newcomers. As long as there are relatively few events in the partition, a tree diagram (as in Example 2.29) can be used as a basis for calculating posterior probabilities without ever referring explicitly to Bayes' theorem.

Example 2.30 INCIDENCE OF A RARE DISEASE Only 1 in 1000 adults is afflicted with a rare disease for which a diagnostic test has been developed. The test is such that when an individual actually has the disease, a positive result will occur 99% of the time, whereas an individual without the disease will show a positive test result only 2% of the time. If a randomly selected individual is tested and the result is positive, what is the probability that the individual has the disease?

To use Bayes' theorem, let $A_1 = \{\text{individual has the disease}\}, A_2 = \{\text{individual does not have the disease}\}$, and $B = \{\text{positive test result}\}$. Then $P(A_1) = .001$, $P(A_2) = .999$, $P(B \mid A_1) = .99$, and $P(B \mid A_2) = .02$. The tree diagram for this problem is in Figure 2.12.

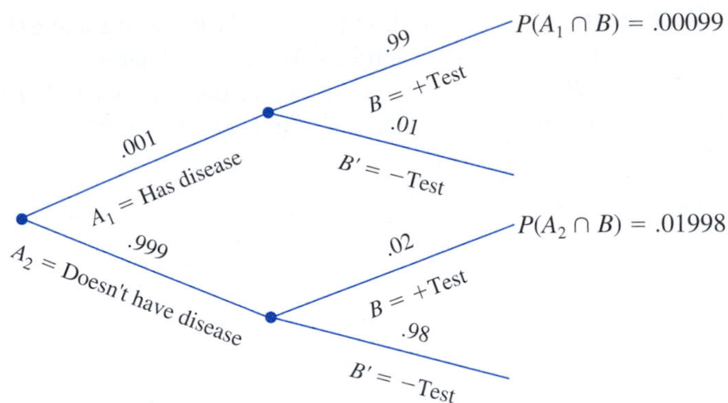

Figure 2.12 Tree diagram for the rare-disease problem

Next to each branch corresponding to a positive test result, the multiplication rule yields the recorded probabilities. Therefore, $P(B) = .00099 + .01998 = .02097$, from which we have

$$P(A_1 | B) = \frac{P(A_1 \cap B)}{P(B)} = \frac{.00099}{.02097} = .047$$

This result seems counterintuitive; the diagnostic test appears so accurate we expect someone with a positive test result to be highly likely to have the disease, whereas the computed conditional probability is only .047. However, because the disease is rare and the test only moderately reliable, most positive test results arise from errors rather than from diseased individuals. The probability of having the disease has increased by a multiplicative factor of 47 (from prior .001 to posterior .047); but to get a further increase in the posterior probability, a diagnostic test with much smaller error rates is needed. If the disease were not so rare (e.g., 25% incidence in the population), then the error rates for the present test would provide good diagnoses. ■

An important contemporary application of Bayes' theorem is in the identification of spam e-mail messages. A nice expository article on this appears in *Statistics: A Guide to the Unknown* (see the Chapter 1 bibliography).

Exercises Section 2.4 (45–65)

45. The population of a particular country consists of three ethnic groups. Each individual belongs to one of the four major blood groups. The accompanying *joint probability table* gives the proportions of individuals in the various ethnic group–blood group combinations.

	Blood Group			
Ethnic Group	**O**	**A**	**B**	**AB**
1	.082	.106	.008	.004
2	.135	.141	.018	.006
3	.215	.200	.065	.020

Suppose that an individual is randomly selected from the population, and define events by $A = \{\text{type A selected}\}$, $B = \{\text{type B selected}\}$, and $C = \{\text{ethnic group 3 selected}\}$.

a. Calculate $P(A)$, $P(C)$, and $P(A \cap C)$.

b. Calculate both $P(A|C)$ and $P(C|A)$, and explain in context what each of these probabilities represents.

c. If the selected individual does not have type B blood, what is the probability that he or she is from ethnic group 1?

46. Suppose an individual is randomly selected from the population of all adult males living in the United States. Let A be the event that the selected individual is over 6 ft in height, and let B be the event that the selected individual is a professional basketball player. Which do you think is larger, $P(A|B)$ or $P(B|A)$? Why?

47. Return to the credit card scenario of Exercise 14 (Section 2.2), where $A = \{\text{Visa}\}$, $B = \{\text{MasterCard}\}$, $P(A) = .5$, $P(B) = .4$, and $P(A \cap B) = .25$. Calculate and interpret each of the following probabilities (a Venn diagram might help).

a. $P(B|A)$ **b.** $P(B'|A)$

c. $P(A|B)$ **d.** $P(A'|B)$

e. Given that the selected individual has at least one card, what is the probability that he or she has a Visa card?

48. Reconsider the system defect situation described in Exercise 28 (Section 2.2).

a. Given that the system has a type 1 defect, what is the probability that it has a type 2 defect?

b. Given that the system has a type 1 defect, what is the probability that it has all three types of defects?

c. Given that the system has at least one type of defect, what is the probability that it has exactly one type of defect?

d. Given that the system has both of the first two types of defects, what is the probability that it does not have the third type of defect?

49. If two bulbs are randomly selected from the box of lightbulbs described in Exercise 40 (Section 2.3) and at least one of them is found to be rated 75 W, what is the probability that both of them are 75-W bulbs? Given that at least one of the two selected is not rated 75 W, what is the probability that both selected bulbs have the same rating?

50. A department store sells sport shirts in three sizes (small, medium, and large), three patterns (plaid, print, and stripe), and two sleeve lengths (long and short). The accompanying tables give the proportions of shirts sold in the various category combinations.

Short-sleeved

	Pattern		
Size	**Pl**	**Pr**	**St**
S	.04	.02	.05
M	.08	.07	.12
L	.03	.07	.08

Long-sleeved

	Pattern		
Size	**Pl**	**Pr**	**St**
S	.03	.02	.03
M	.10	.05	.07
L	.04	.02	.08

a. What is the probability that the next shirt sold is a medium, long-sleeved, print shirt?

b. What is the probability that the next shirt sold is a medium print shirt?

c. What is the probability that the next shirt sold is a short-sleeved shirt? A long-sleeved shirt?

d. What is the probability that the size of the next shirt sold is medium? That the pattern of the next shirt sold is a print?

e. Given that the shirt just sold was a short-sleeved plaid, what is the probability that its size was medium?

f. Given that the shirt just sold was a medium plaid, what is the probability that it was short-sleeved? Long-sleeved?

51. One box contains six red balls and four green balls, and a second box contains seven red balls and three green balls. A ball is randomly chosen from the first box and placed in the second box. Then a ball is randomly selected from the second box and placed in the first box.

a. What is the probability that a red ball is selected from the first box and a red ball is selected from the second box?

b. At the conclusion of the selection process, what is the probability that the numbers of red and green balls in the first box are identical to the numbers at the beginning?

52. A system consists of two identical pumps, #1 and #2. If one pump fails, the system will still operate. However, because of the added strain, the extra remaining pump is now more likely to fail than was originally the case. That is, $r = P(\#2 \text{ fails} \mid \#1 \text{ fails}) > P(\#2 \text{ fails}) = q$. If at least one pump fails by the end of the pump design life in 7% of all systems and both pumps fail during that period in only 1%, what is the probability that pump #1 will fail during the pump design life?

53. A certain shop repairs both audio and video components. Let A denote the event that the next component brought in for repair is an audio component, and let B be the event that the next component is a compact disc player (so the event B is contained in A). Suppose that $P(A) = .6$ and $P(B) = .05$. What is $P(B \mid A)$?

54. In Exercise 15, $A_i = \{\text{awarded project } i\}$, for $i = 1, 2, 3$. Use the probabilities given there to compute the following probabilities:

a. $P(A_2 \mid A_1)$

b. $P(A_2 \cap A_3 \mid A_1)$

c. $P(A_2 \cup A_3 \mid A_1)$

d. $P(A_1 \cap A_2 \cap A_3 \mid A_1 \cup A_2 \cup A_3)$
Express in words the probability you have calculated.

55. For any events A and B with $P(B) > 0$, show that $P(A \mid B) + P(A' \mid B) = 1$.

56. If $P(B \mid A) > P(B)$, show that $P(B' \mid A) < P(B')$. (*Hint*: Add $P(B' \mid A)$ to both sides of the given inequality and then use the result of Exercise 55.)

57. Show that for any three events A, B, and C with $P(C) > 0$, $P(A \cup B \mid C) = P(A \mid C) + P(B \mid C) - P(A \cap B \mid C)$.

58. At a certain gas station, 40% of the customers use regular unleaded gas (A_1), 35% use extra unleaded gas (A_2), and 25% use premium unleaded gas (A_3). Of those customers using regular gas, only 30% fill their tanks (event B). Of those customers using extra gas, 60% fill their tanks, whereas of those using premium, 50% fill their tanks.

a. What is the probability that the next customer will request extra unleaded gas and fill the tank ($A_2 \cap B$)?

b. What is the probability that the next customer fills the tank?

c. If the next customer fills the tank, what is the probability that regular gas is requested? Extra gas? Premium gas?

59. Seventy percent of the light aircraft that disappear while in flight in a certain country are subsequently discovered. Of the aircraft that are discovered, 60% have an emergency locator, whereas 90% of the aircraft not discovered do not have such a locator. Suppose a light aircraft has disappeared.

a. If it has an emergency locator, what is the probability that it will not be discovered?

b. If it does not have an emergency locator, what is the probability that it will be discovered?

60. Components of a certain type are shipped to a supplier in batches of ten. Suppose that 50% of all such batches contain no defective components, 30% contain one defective component, and 20% contain two defective components. Two components from a batch are randomly selected and tested. What are the probabilities associated with 0, 1, and 2 defective components being in the batch under each of the following conditions?

a. Neither tested component is defective.

b. One of the two tested components is defective. (*Hint*: Draw a tree diagram with three first-generation branches for the three different types of batches.)

61. A company that manufactures video cameras produces a basic model and a deluxe model. Over the past year, 40% of the cameras sold have been of the basic model. Of those buying the basic model, 30% purchase an extended warranty, whereas 50% of all deluxe purchasers do so. If you learn that a randomly selected purchaser has an extended warranty, how likely is it that he or she has a basic model?

62. For customers purchasing a full set of tires at a particular tire store, consider the events

$A = \{\text{tires purchased were made in the United States}\}$
$B = \{\text{purchaser has tires balanced immediately}\}$
$C = \{\text{purchaser requests front-end alignment}\}$

along with A', B', and C'. Assume the following unconditional and conditional probabilities:

$P(A) = .75 \quad P(B \mid A) = .9 \quad P(B \mid A') = .8$
$P(C \mid A \cap B) = .8 \quad P(C \mid A \cap B') = .6$
$P(C \mid A' \cap B) = .7 \quad P(C \mid A' \cap B') = .3$

a. Construct a tree diagram consisting of first-, second-, and third-generation branches and place an event label and appropriate probability next to each branch.

b. Compute $P(A \cap B \cap C)$.

c. Compute $P(B \cap C)$.

d. Compute $P(C)$.

e. Compute $P(A \mid B \cap C)$, the probability of a purchase of U.S. tires given that both balancing and an alignment were requested.

63. In Example 2.30, suppose that the incidence rate for the disease is 1 in 25 rather than 1 in 1000. What then is the probability of a positive test result? Given that the test result is positive, what is the probability that the individual has the disease? Given a negative test result, what is the probability that the individual does not have the disease?

64. At a large university, in the never-ending quest for a satisfactory textbook, the Statistics Department has tried a different text during each of the last three quarters. During the fall quarter, 500 students used the text by Professor Mean; during the winter quarter, 300 students used the text by Professor Median; and during the spring quarter, 200 students used the text by Professor Mode. A survey at the end of each quarter showed that 200 students were satisfied with Mean's book, 150 were satisfied with Median's book, and 160 were satisfied with Mode's book. If a student who took statistics during one of these quarters is selected at random and admits to having been satisfied with the text, is the student most likely to have used the book by Mean, Median, or Mode? Who is the least likely author? (*Hint*: Draw a tree-diagram or use Bayes' theorem.)

65. A friend who lives in Los Angeles makes frequent consulting trips to Washington, D.C.; 50% of the time she travels on airline #1, 30% of the time on airline #2, and the remaining 20% of the time on airline #3. For airline #1, flights are late into D.C. 30% of the time and late into L.A. 10% of the time. For airline #2, these percentages are 25% and 20%, whereas for airline #3 the percentages are 40% and 25%. If we learn that on a particular trip she arrived late at exactly one of the two destinations, what are the posterior probabilities of having flown on airlines #1, #2, and #3? Assume that the chance of a late arrival in L.A. is unaffected by what happens on the flight to D.C. (*Hint*: From the tip of each first-generation branch on a tree diagram, draw three second-generation branches labeled, respectively, 0 late, 1 late, and 2 late.)

2.5 Independence

The definition of conditional probability enables us to revise the probability $P(A)$ originally assigned to A when we are subsequently informed that another event B has occurred; the new probability of A is $P(A \mid B)$. In our examples, it was frequently the case that $P(A \mid B)$ was unequal to the unconditional probability $P(A)$, indicating that the information "B has occurred" resulted in a change in the chance of A occurring. There are other situations, though, in which the chance that A will occur or has occurred is not affected by knowledge that B has occurred, so that $P(A \mid B) = P(A)$. It is then natural to think of A and B as independent events, meaning that the occurrence or nonoccurrence of one event has no bearing on the chance that the other will occur.

DEFINITION Two events A and B are **independent** if $P(A \mid B) = P(A)$ and are **dependent** otherwise.

The definition of independence might seem "unsymmetric" because we do not demand that $P(B \mid A) = P(B)$ also. However, using the definition of conditional probability and the multiplication rule,

$$P(B \mid A) = \frac{P(A \cap B)}{P(A)} = \frac{P(A \mid B)P(B)}{P(A)} \tag{2.7}$$

The right-hand side of Equation (2.7) is $P(B)$ if and only if $P(A \mid B) = P(A)$ (independence), so the equality in the definition implies the other equality (and vice versa). It is also straightforward to show that if A and B are independent, then so are the following pairs of events: (1) A' and B, (2) A and B', and (3) A' and B'.

Example 2.31 Consider tossing a fair six-sided die once and define events $A = \{2, 4, 6\}$, $B = \{1, 2, 3\}$, and $C = \{1, 2, 3, 4\}$. We then have $P(A) = \frac{1}{2}$, $P(A \mid B) = \frac{1}{3}$, and $P(A \mid C) = \frac{1}{2}$. That is, events A and B are dependent, whereas events A and C are independent. Intuitively, if such a die is tossed and we are informed that the outcome was 1, 2, 3, or 4 (C has occurred), then the probability that A occurred is $\frac{1}{2}$, as it originally was, since two of the four relevant outcomes are even and the outcomes are still equally likely. ■

Example 2.32 Let A and B be any two mutually exclusive events with $P(A) > 0$. For example, for a randomly chosen automobile, let $A = \{\text{car is blue}\}$ and $B = \{\text{car is red}\}$. Since the events are mutually exclusive, if B occurs, then A cannot possibly have occurred, so $P(A \mid B) = 0 \neq P(A)$. The message here is that *if two events are mutually exclusive, they cannot be independent*. When A and B are mutually exclusive, the information that A occurred says something about B (it cannot have occurred), so independence is precluded. ■

$P(A \cap B)$ When Events Are Independent

Frequently the nature of an experiment suggests that two events A and B should be assumed independent. This is the case, for example, if a manufacturer receives a circuit board from each of two different suppliers, each board is tested on arrival, and $A = \{\text{first is defective}\}$ and $B = \{\text{second is defective}\}$. If $P(A) = .1$, it should also be the case that $P(A \mid B) = .1$; knowing the condition of the second board shouldn't provide information about the condition of the first. Our next result shows how to compute $P(A \cap B)$ when the events are independent.

PROPOSITION A and B are independent if and only if

$$P(A \cap B) = P(A) \cdot P(B) \tag{2.8}$$

To paraphrase the proposition, A and B are independent events iff* the probability that they both occur ($A \cap B$) is the product of the two individual probabilities. The verification is as follows:

$$P(A \cap B) = P(A \mid B) \cdot P(B) = P(A) \cdot P(B) \tag{2.9}$$

where the second equality in Equation (2.9) is valid iff A and B are independent. Because of the equivalence of independence with Equation (2.8), the latter can be used as a definition of independence.**

---

*iff is an abbreviation for "if and only if."

**However, the multiplication property is satisfied if $P(B) = 0$, yet $P(A \mid B)$ is not defined in this case. To make the multiplication property completely equivalent to the definition of independence, we should append to that definition that A and B are also independent if either $P(A) = 0$ or $P(B) = 0$.

Example 2.33 It is known that 30% of a certain company's washing machines require service while under warranty, whereas only 10% of its dryers need such service. If someone purchases both a washer and a dryer made by this company, what is the probability that both machines need warranty service?

Let A denote the event that the washer needs service while under warranty, and let B be defined analogously for the dryer. Then $P(A) = .30$ and $P(B) = .10$. Assuming that the two machines function independently of one another, the desired probability is

$$P(A \cap B) = P(A) \cdot P(B) = (.30)(.10) = .03$$

The probability that neither machine needs service is

$$P(A' \cap B') = P(A') \cdot P(B') = (.70)(.90) = .63$$

■

Example 2.34 Each day, Monday through Friday, a batch of components sent by a first supplier arrives at a certain inspection facility. Two days a week, a batch also arrives from a second supplier. Eighty percent of all supplier 1's batches pass inspection, and 90% of supplier 2's do likewise. What is the probability that, on a randomly selected day, two batches pass inspection? We will answer this assuming that on days when two batches are tested, whether the first batch passes is independent of whether the second batch does so. Figure 2.13 displays the relevant information.

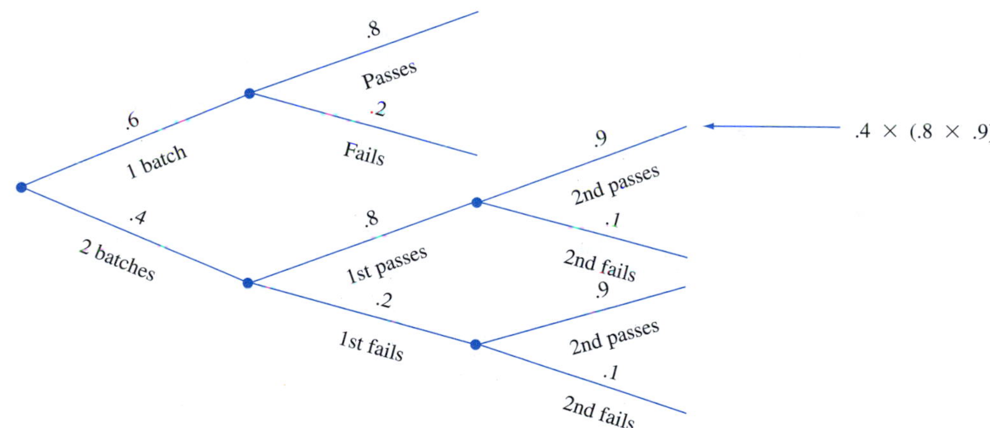

Figure 2.13 Tree diagram for Example 2.34

$$\begin{aligned} P(\text{two pass}) &= P(\text{two received} \cap \text{both pass}) \\ &= P(\text{both pass} \mid \text{two received}) \cdot P(\text{two received}) \\ &= [(.8)(.9)](.4) = .288 \end{aligned}$$

■

Independence of More Than Two Events

The notion of independence of two events can be extended to collections of more than two events. Although it is possible to extend the definition for two independent events by working in terms of conditional and unconditional probabilities, it is more direct and less cumbersome to proceed along the lines of the last proposition.

DEFINITION

Events $A_1, \ldots, A_n$ are **mutually independent** if for every k ($k = 2, 3, \ldots, n$) and every subset of indices $i_1, i_2, \ldots, i_k$,

$$P(A_{i_1} \cap A_{i_2} \cap \cdots \cap A_{i_k}) = P(A_{i_1}) \cdot P(A_{i_2}) \cdot \cdots \cdot P(A_{i_k})$$

To paraphrase the definition, the events are mutually independent if the probability of the intersection of any subset of the n events is equal to the product of the individual probabilities. As was the case with two events, we frequently specify at the outset of a problem the independence of certain events. The definition can then be used to calculate the probability of an intersection.

Example 2.35 The article "Reliability Evaluation of Solar Photovoltaic Arrays" (*Solar Energy*, 2002: 129–141) presents various configurations of solar photovoltaic arrays consisting of crystalline silicon solar cells. Consider first the system illustrated in Figure 2.14(a). There are two subsystems connected in parallel, each one containing three cells. In order for the system to function, at least one of the two parallel subsystems must work. Within each subsystem, the three cells are connected in series, so a subsystem will work only if all cells in the subsystem work. Consider a particular lifetime value t_0, and suppose we want to determine the probability that the system lifetime exceeds t_0. Let A_i denote the event that the lifetime of cell i exceeds t_0 ($i = 1, 2, \ldots, 6$). We assume that the A_i's are independent events (whether any particular cell lasts more than t_0 hours has no bearing on whether or not any other cell does) and that $P(A_i) = .9$ for every i since the cells are identical. Then

$$\begin{aligned} P(\text{system lifetime exceeds } t_0) &= P[(A_1 \cap A_2 \cap A_3) \cup (A_4 \cap A_5 \cap A_6)] \\ &= P(A_1 \cap A_2 \cap A_3) + P(A_4 \cap A_5 \cap A_6) \\ &\quad - P[(A_1 \cap A_2 \cap A_3) \cap (A_4 \cap A_5 \cap A_6)] \\ &= (.9)(.9)(.9) + (.9)(.9)(.9) \\ &\quad - (.9)(.9)(.9)(.9)(.9)(.9) = .927 \end{aligned}$$

Alternatively,

$$\begin{aligned} P(\text{system lifetime exceeds } t_0) &= 1 - P(\text{both subsystem lives are} \le t_0) \\ &= 1 - [P(\text{subsystem life is} \le t_0)]^2 \\ &= 1 - [1 - P(\text{subsystem life is} > t_0)]^2 \\ &= 1 - [1 - (.9)^3]^2 = .927 \end{aligned}$$

Next consider the total-cross-tied system shown in Figure 2.14(b), obtained from the series-parallel array by connecting ties across each column of junctions. Now the

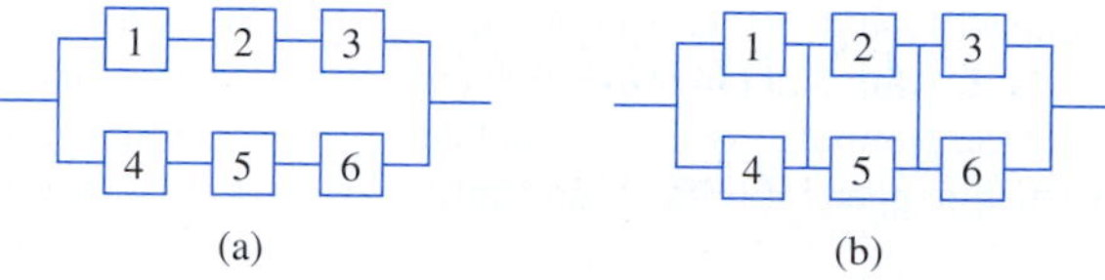

Figure 2.14 System configurations for Example 2.35: (a) series-parallel; (b) total-cross-tied

system fails as soon as an entire column fails, and system lifetime exceeds t_0 only if the life of every column does so. For this configuration,

$$\begin{aligned} P(\text{system lifetime is at least } t_0) &= [P(\text{column lifetime exceeds } t_0)]^3 \\ &= [1 - P(\text{column lifetime is} \le t_0)]^3 \\ &= [1 - P(\text{both cells in a column have lifetime} \le t_0)]^3 \\ &= [1 - (1 - .9)^2]^3 = .970 \end{aligned}$$

■

Exercises Section 2.5 (66–83)

66. Reconsider the credit card scenario of Exercise 47 (Section 2.4), and show that A and B are dependent first by using the definition of independence and then by verifying that the multiplication property does not hold.

67. An oil exploration company currently has two active projects, one in Asia and the other in Europe. Let A be the event that the Asian project is successful and B be the event that the European project is successful. Suppose that A and B are independent events with $P(A) = .4$ and $P(B) = .7$.

a. If the Asian project is not successful, what is the probability that the European project is also not successful? Explain your reasoning.

b. What is the probability that at least one of the two projects will be successful?

c. Given that at least one of the two projects is successful, what is the probability that only the Asian project is successful?

68. In Exercise 15, is any A_i independent of any other A_j? Answer using the multiplication property for independent events.

69. If A and B are independent events, show that A' and B are also independent. [*Hint*: First establish a relationship between $P(A' \cap B)$, $P(B)$, and $P(A \cap B)$.]

70. Suppose that the proportions of blood phenotypes in a particular population are as follows:

A	B	AB	O
.42	.10	.04	.44

Assuming that the phenotypes of two randomly selected individuals are independent of one another, what is the probability that both phenotypes are O? What is the probability that the phenotypes of two randomly selected individuals match?

71. The probability that a grader will make a marking error on any particular question of a multiple-choice exam is .1. If there are ten questions and questions are marked independently, what is the probability that no errors are made? That at least one error is made? If there are n questions and the probability of a marking error is p rather than .1, give expressions for these two probabilities.

72. An aircraft seam requires 25 rivets. The seam will have to be reworked if any of these rivets is defective. Suppose rivets are defective independently of one another, each with the same probability.

a. If 20% of all seams need reworking, what is the probability that a rivet is defective?

b. How small should the probability of a defective rivet be to ensure that only 10% of all seams need reworking?

73. A boiler has five identical relief valves. The probability that any particular valve will open on demand is .95. Assuming independent operation of the valves, calculate P(at least one valve opens) and P(at least one valve fails to open).

74. Two pumps connected in parallel fail independently of one another on any given day. The probability that only the older pump will fail is .10, and the probability that only the newer pump will fail is .05. What is the probability that the pumping system will fail on any given day (which happens if both pumps fail)?

75. Consider the system of components connected as in the accompanying picture. Components 1 and 2 are connected in parallel, so that subsystem works iff either 1 or 2 works; since 3 and 4 are connected in series, that subsystem works iff both 3 and 4 work. If components work independently of one another and P(component works) = .9, calculate P(system works).

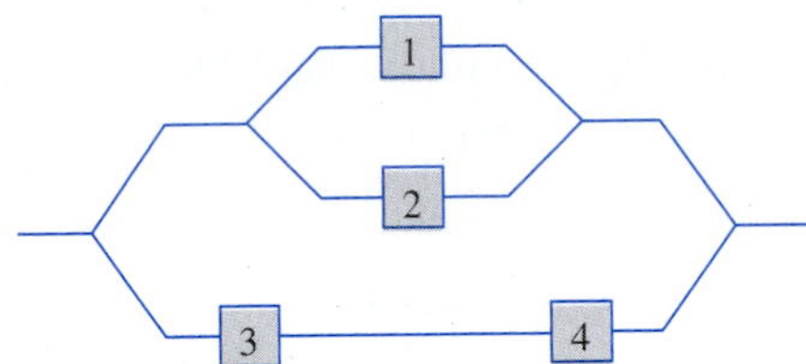

76. Refer back to the series-parallel system configuration introduced in Example 2.35, and suppose that there are only two cells rather than three in each parallel subsystem [in Figure 2.14(a), eliminate cells 3 and 6, and renumber cells 4 and 5 as 3 and 4]. Using $P(A_i) = .9$, the probability that system lifetime exceeds t_0 is easily seen to be .9639. To what value would .9 have to be changed in order to increase the system lifetime reliability from .9639 to .99? [*Hint*: Let $P(A_i) = p$, express system reliability in terms of p, and then let $x = p^2$.]

77. Consider independently rolling two fair dice, one red and the other green. Let A be the event that the red die shows 3 dots, B be the event that the green die shows 4 dots, and C be the event that the total number of dots showing on the two dice is 7. Are these events pairwise independent (i.e., are A and B independent events, are A and C independent, and are B and C independent)? Are the three events mutually independent?

78. Components arriving at a distributor are checked for defects by two different inspectors (each component is checked by both inspectors). The first inspector detects 90% of all defectives that are present, and the second inspector does likewise. At least one inspector does not detect a defect on 20% of all defective components. What is the probability that the following occur?

a. A defective component will be detected only by the first inspector? By exactly one of the two inspectors?

b. All three defective components in a batch escape detection by both inspectors (assuming inspections of different components are independent of one another)?

79. A quality control inspector is inspecting newly produced items for faults. The inspector searches an item for faults in a series of independent fixations, each of a fixed duration. Given that a flaw is actually present, let p denote the probability that the flaw is detected during any one fixation (this model is discussed in "Human Performance in Sampling Inspection," *Hum. Factors*, 1979: 99–105).

a. Assuming that an item has a flaw, what is the probability that it is detected by the end of the second fixation (once a flaw has been detected, the sequence of fixations terminates)?

b. Give an expression for the probability that a flaw will be detected by the end of the nth fixation.

c. If when a flaw has not been detected in three fixations, the item is passed, what is the probability that a flawed item will pass inspection?

d. Suppose 10% of all items contain a flaw [P(randomly chosen item is flawed) = .1]. With the assumption of part (c), what is the probability that a randomly chosen item will pass inspection (it will automatically pass if it is not flawed, but could also pass if it is flawed)?

e. Given that an item has passed inspection (no flaws in three fixations), what is the probability that it is actually flawed? Calculate for $p = .5$.

80. **a.** A lumber company has just taken delivery on a lot of 10,000 2 × 4 boards. Suppose that 20% of these boards (2000) are actually too green to be used in first-quality construction. Two boards are selected at random, one after the other. Let A = {the first board is green} and B = {the second board is green}. Compute $P(A)$, $P(B)$, and $P(A \cap B)$ (a tree diagram might help). Are A and B independent?

b. With A and B independent and $P(A) = P(B) = .2$, what is $P(A \cap B)$? How much difference is there between this answer and $P(A \cap B)$ in part (a)? For purposes of calculating $P(A \cap B)$, can we assume that A and B of part (a) are independent to obtain essentially the correct probability?

c. Suppose the lot consists of ten boards, of which two are green. Does the assumption of independence now yield approximately the correct answer for $P(A \cap B)$? What is the critical difference between the situation here and that of part (a)? When do you think that an independence assumption would be valid in obtaining an approximately correct answer to $P(A \cap B)$?

81. Refer to the assumptions stated in Exercise 75 and answer the question posed there for the system in the accompanying picture. How would the probability change if this were a subsystem connected in parallel to the subsystem pictured in Figure 2.14(a)?

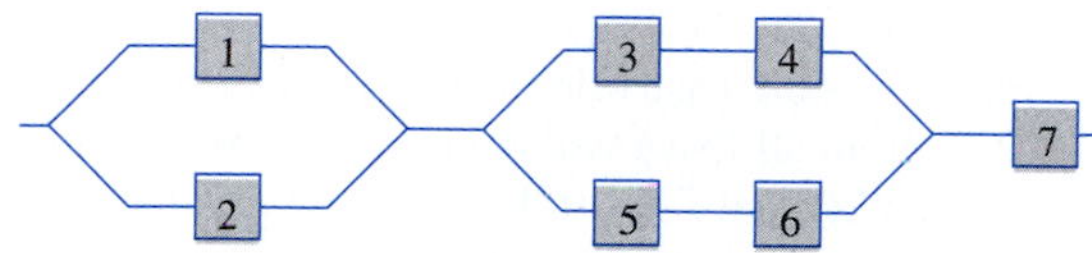

82. Professor Stan der Deviation can take one of two routes on his way home from work. On the first route, there are four railroad crossings. The probability that he will be stopped by a train at any particular one of the crossings is .1, and trains operate independently at the four crossings. The other route is longer but there are only two crossings, independent of one another, with the same stoppage probability for each as on the first route. On a particular day, Professor Deviation has a meeting scheduled at home for a certain time. Whichever route he takes, he calculates that he will be late if he is stopped by trains at at least half the crossings encountered.

a. Which route should he take to minimize the probability of being late to the meeting?

b. If he tosses a fair coin to decide on a route and he is late, what is the probability that he took the four-crossing route?

83. Suppose identical tags are placed on both the left ear and the right ear of a fox. The fox is then let loose for a period of time. Consider the two events C_1 = {left ear tag is lost} and C_2 = {right ear tag is lost}. Let $\pi = P(C_1) = P(C_2)$, and assume C_1 and C_2 are independent events. Derive an expression (involving π) for the probability that exactly one tag is lost given that at most one is lost ("Ear Tag Loss in Red Foxes," *J. Wildlife Manag.*, 1976: 164–167). (*Hint*: Draw a tree diagram in which the two initial branches refer to whether the left ear tag was lost.)

Supplementary Exercises (84–109)

84. A small manufacturing company will start operating a night shift. There are 20 machinists employed by the company.

a. If a night crew consists of 3 machinists, how many different crews are possible?

b. If the machinists are ranked 1, 2, . . . , 20 in order of competence, how many of these crews would not have the best machinist?

c. How many of the crews would have at least 1 of the 10 best machinists?

d. If one of these crews is selected at random to work on a particular night, what is the probability that the best machinist will not work that night?

85. A factory uses three production lines to manufacture cans of a certain type. The accompanying table gives percentages of nonconforming cans, categorized by type of nonconformance, for each of the three lines during a particular time period.

	Line 1	Line 2	Line 3
Blemish	15	12	20
Crack	50	44	40
Pull-Tab Problem	21	28	24
Surface Defect	10	8	15
Other	4	8	2

During this period, line 1 produced 500 nonconforming cans, line 2 produced 400 such cans, and line 3 was responsible for 600 nonconforming cans. Suppose that one of these 1500 cans is randomly selected.

a. What is the probability that the can was produced by line 1? That the reason for nonconformance is a crack?

b. If the selected can came from line 1, what is the probability that it had a blemish?

c. Given that the selected can had a surface defect, what is the probability that it came from line 1?

86. An employee of the records office at a certain university currently has ten forms on his desk awaiting processing. Six of these are withdrawal petitions and the other four are course substitution requests.

a. If he randomly selects six of these forms to give to a subordinate, what is the probability that only one of the two types of forms remains on his desk?

b. Suppose he has time to process only four of these forms before leaving for the day. If these four are randomly selected one by one, what is the probability that each succeeding form is of a different type from its predecessor?

87. One satellite is scheduled to be launched from Cape Canaveral in Florida, and another launching is scheduled for Vandenberg Air Force Base in California. Let A denote the event that the Vandenberg launch goes off on schedule, and let B represent the event that the Cape Canaveral launch goes off on schedule. If A and B are independent events with $P(A) > P(B)$ and $P(A \cup B) = .626$, $P(A \cap B) = .144$, determine the values of $P(A)$ and $P(B)$.

88. A transmitter is sending a message by using a binary code, namely, a sequence of 0's and 1's. Each transmitted bit (0 or 1) must pass through three relays to reach the receiver. At each relay, the probability is .20 that the bit sent will be different from the bit received (a reversal). Assume that the relays operate independently of one another.

Transmitter $\rightarrow$ Relay 1 $\rightarrow$ Relay 2 $\rightarrow$ Relay 3 $\rightarrow$ Receiver

a. If a 1 is sent from the transmitter, what is the probability that a 1 is sent by all three relays?

b. If a 1 is sent from the transmitter, what is the probability that a 1 is received by the receiver? (*Hint*: The eight experimental outcomes can be displayed on a tree diagram with three generations of branches, one generation for each relay.)

c. Suppose 70% of all bits sent from the transmitter are 1's. If a 1 is received by the receiver, what is the probability that a 1 was sent?

89. Individual A has a circle of five close friends (B, C, D, E, and F). A has heard a certain rumor from outside the circle and has invited the five friends to a party to circulate the rumor. To begin, A selects one of the five at random and tells the rumor to the chosen individual. That individual then selects at random one of the four remaining individuals and repeats the rumor. Continuing, a new individual is selected from those not already having heard the rumor by the individual who has just heard it, until everyone has been told.

a. What is the probability that the rumor is repeated in the order B, C, D, E, and F?

b. What is the probability that F is the third person at the party to be told the rumor?

c. What is the probability that F is the last person to hear the rumor?

90. Refer to Exercise 89. If at each stage the person who currently "has" the rumor does not know who has already heard it and selects the next recipient at random from all five possible individuals, what is the probability that F has still not heard the rumor after it has been told ten times at the party?

91. A chemist is interested in determining whether a certain trace impurity is present in a product. An experiment has a probability of .80 of detecting the impurity if it is present. The probability of not detecting the impurity if it is absent is .90. The prior probabilities of the impurity being present and being absent are .40 and .60, respectively. Three separate experiments result in only two detections. What is the posterior probability that the impurity is present?

92. Fasteners used in aircraft manufacturing are slightly crimped so that they lock enough to avoid loosening during vibration. Suppose that 95% of all fasteners pass an initial inspection. Of the 5% that fail, 20% are so seriously defective that they must be scrapped. The remaining fasteners are sent to a recrimping operation, where 40% cannot be salvaged and are discarded. The other 60% of these fasteners are corrected by the recrimping process and subsequently pass inspection.

a. What is the probability that a randomly selected incoming fastener will pass inspection either initially or after recrimping?

b. Given that a fastener passed inspection, what is the probability that it passed the initial inspection and did not need recrimping?

93. One percent of all individuals in a certain population are carriers of a particular disease. A diagnostic test for this disease has a 90% detection rate for carriers and a 5% detection rate for noncarriers. Suppose the test is applied independently to two different blood samples from the same randomly selected individual.

a. What is the probability that both tests yield the same result?

b. If both tests are positive, what is the probability that the selected individual is a carrier?

94. A system consists of two components. The probability that the second component functions in a satisfactory manner during its design life is .9, the probability that at least one of the two components does so is .96, and the probability that both components do so is .75. Given that the first component functions in a satisfactory manner throughout its design life, what is the probability that the second one does also?

95. A certain company sends 40% of its overnight mail parcels via express mail service E_1. Of these parcels, 2% arrive after the guaranteed delivery time (denote the event "late delivery" by L). If a record of an overnight mailing is randomly selected from the company's file, what is the probability that the parcel went via E_1 and was late?

96. Refer to Exercise 95. Suppose that 50% of the overnight parcels are sent via express mail service E_2 and the remaining 10% are sent via E_3. Of those sent via E_2, only 1% arrive late, whereas 5% of the parcels handled by E_3 arrive late.

a. What is the probability that a randomly selected parcel arrived late?

b. If a randomly selected parcel has arrived on time, what is the probability that it was not sent via E_1?

97. A company uses three different assembly lines—A_1, A_2, and A_3—to manufacture a particular component. Of those manufactured by line A_1, 5% need rework to remedy a defect, whereas 8% of A_2's components need rework and 10% of A_3's need rework. Suppose that 50% of all components are produced by line A_1, 30% are produced by line A_2, and 20% come from line A_3. If a randomly selected component needs rework, what is the probability that it came from line A_1? From line A_2? From line A_3?

98. Disregarding the possibility of a February 29 birthday, suppose a randomly selected individual is equally likely to have been born on any one of the other 365 days.

a. If ten people are randomly selected, what is the probability that all have different birthdays? That at least two have the same birthday?

b. With k replacing ten in part (a), what is the smallest k for which there is at least a 50–50 chance that two or more people will have the same birthday?

c. If ten people are randomly selected, what is the probability that either at least two have the same birthday or at least two have the same last three digits of their Social Security numbers? [*Note*: The article "Methods for Studying Coincidences" (F. Mosteller and P. Diaconis, *J. Amer. Statist. Assoc.*, 1989: 853–861) discusses problems of this type.]

99. One method used to distinguish between granitic (G) and basaltic (B) rocks is to examine a portion of the infrared spectrum of the sun's energy reflected from the rock surface. Let R_1, R_2, and R_3 denote measured spectrum intensities at three different wavelengths; typically, for granite $R_1 < R_2 < R_3$, whereas for basalt $R_3 < R_1 < R_2$. When measurements are made remotely (using aircraft), various orderings of the R_i's may arise whether the rock is basalt or granite. Flights over regions of known composition have yielded the following information:

	Granite	**Basalt**
$R_1 < R_2 < R_3$	60%	10%
$R_1 < R_3 < R_2$	25%	20%
$R_3 < R_1 < R_2$	15%	70%

Suppose that for a randomly selected rock in a certain region, $P(\text{granite}) = .25$ and $P(\text{basalt}) = .75$.

a. Show that $P(\text{granite} \mid R_1 < R_2 < R_3) > P(\text{basalt} \mid R_1 < R_2 < R_3)$. If measurements yielded $R_1 < R_2 < R_3$, would you classify the rock as granite or basalt?

b. If measurements yielded $R_1 < R_3 < R_2$, how would you classify the rock? Answer the same question for $R_3 < R_1 < R_2$.

c. Using the classification rules indicated in parts (a) and (b), when selecting a rock from this region, what is the probability of an erroneous classification? [*Hint*: Either G could be classified as B or B as G, and $P(B)$ and $P(G)$ are known.]

d. If $P(\text{granite}) = p$ rather than .25, are there values of p (other than 1) for which one would always classify a rock as granite?

100. In a Little League baseball game, team A's pitcher throws a strike 50% of the time and a ball 50% of the time, successive pitches are independent of one another, and the pitcher never hits a batter. Knowing this, team B's manager has instructed the first batter not to swing at anything. Calculate the probability that

a. The batter walks on the fourth pitch.

b. The batter walks on the sixth pitch (so two of the first five must be strikes), using a counting argument or constructing a tree diagram.

c. The batter walks.

d. The first batter up scores while no one is out (assuming that each batter pursues a no-swing strategy).

101. Four graduating seniors, A, B, C, and D, have been scheduled for job interviews at 10 a.m. on Friday, January 13, at Random Sampling, Inc. The personnel manager has scheduled the four for interview rooms 1, 2, 3, and 4, respectively. Unaware of this, the manager's secretary assigns them to the four rooms in a completely random fashion (what else!). What is the probability that

a. All four end up in the correct rooms?

b. None of the four ends up in the correct room?

102. A particular airline has 10 a.m. flights from Chicago to New York, Atlanta, and Los Angeles. Let A denote the event that the New York flight is full and define events B and C analogously for the other two flights. Suppose $P(A) = .6$, $P(B) = .5$, $P(C) = .4$ and the three events are independent. What is the probability that

a. All three flights are full? That at least one flight is not full?

b. Only the New York flight is full? That exactly one of the three flights is full?

103. A personnel manager is to interview four candidates for a job. These are ranked 1, 2, 3, and 4 in order of preference and will be interviewed in random order. However, at the conclusion of each interview, the manager will know only how the current candidate compares to those previously interviewed. For example, the interview order 3, 4, 1, 2 generates no information after the first interview, shows that the second candidate is worse than the first, and that the third is better than the first two. However, the order 3, 4, 2, 1 would generate the same information after each of the first three interviews. The manager wants to hire the best candidate but must make an irrevocable hire/no hire decision after each interview. Consider the following strategy: Automatically reject the first s candidates and then hire the first subsequent candidate who is best among those already interviewed (if no such candidate appears, the last one interviewed is hired).

For example, with $s = 2$, the order 3, 4, 1, 2 would result in the best being hired, whereas the order 3, 1, 2, 4 would not. Of the four possible s values (0, 1, 2, and 3), which one maximizes P(best is hired)? (*Hint*: Write out the 24 equally likely interview orderings: $s = 0$ means that the first candidate is automatically hired.)

104. Consider four independent events A_1, A_2, A_3, and A_4 and let $p_i = P(A_i)$ for $i = 1, 2, 3, 4$. Express the probability that at least one of these four events occurs in terms of the p_i's, and do the same for the probability that at least two of the events occur.

105. A box contains the following four slips of paper, each having exactly the same dimensions: (1) win prize 1; (2) win prize 2; (3) win prize 3; (4) win prizes 1, 2, and 3. One slip will be randomly selected. Let $A_1 = \{\text{win prize 1}\}$, $A_2 = \{\text{win prize 2}\}$, and $A_3 = \{\text{win prize 3}\}$. Show that A_1 and A_2 are independent, that A_1 and A_3 are independent, and that A_2 and A_3 are also independent (this is *pairwise* independence). However, show that $P(A_1 \cap A_2 \cap A_3) \neq P(A_1) \cdot P(A_2) \cdot P(A_3)$, so the three events are *not* mutually independent.

106. Consider a woman whose brother is afflicted with hemophilia, which implies that the woman's mother has the hemophilia gene on one of her two X chromosomes (almost surely not both, since that is generally fatal). Thus there is a 50–50 chance that the woman's mother has passed on the bad gene to her. The woman has two sons, each of whom will independently inherit the gene from one of her two chromosomes. If the woman herself has a bad gene, there is a 50–50 chance she will pass this on to a son. Suppose that neither of her two sons is afflicted with hemophilia. What then is the probability that the woman is indeed the carrier of the hemophilia gene? What is this probability if she has a third son who is also not afflicted?

107. Jurors may be a priori biased for or against the prosecution in a criminal trial. Each juror is questioned by both the prosecution and the defense (the "voir dire" process), but this may not reveal bias. Even if bias is revealed, the judge may not excuse the juror for cause because of the narrow legal definition of bias. For a randomly selected candidate for the jury, define events B_0, B_1, and B_2 as the juror being unbiased, biased against the prosecution, and biased against the defense, respectively. Also let C be the event that bias is revealed during the questioning and D be the event that the juror is eliminated for cause. Let $b_i = P(B_i)$ $(i = 0, 1, 2)$, $c = P(C|B_1) = P(C|B_2)$, and $d = P(D|B_1 \cap C) = P(D|B_2 \cap C)$ ["Fair Number of Peremptory Challenges in Jury Trials," *J. Amer. Statist. Assoc.*, 1979: 747–753].

a. If a juror survives the voir dire process, what is the probability that he/she is unbiased (in terms of the b_i's, c, and d)? What is the probability that he/she is biased against the prosecution? What is the probability that he/she is biased against the defense? *Hint*: Represent this situation using a tree diagram with three generations of branches.

b. What are the probabilities requested in (a) if $b_0 = .50$, $b_1 = .10$, $b_2 = .40$ (all based on data relating to the famous trial of the Florida murderer Ted Bundy), $c = .85$ (corresponding to the extensive questioning appropriate in a capital case), and $d = .7$ (a "moderate" judge)?

108. Allan and Beth currently have \$2 and \$3, respectively. A fair coin is tossed. If the result of the toss is H, Allan wins \$1 from Beth, whereas if the coin toss results in T, then Beth wins \$1 from Allan. This process is then repeated, with a coin toss followed by the exchange of \$1, until one of the two players goes broke (one of the two gamblers is ruined). We wish to determine

$$a_2 = P(\text{Allan is the winner} \mid \text{he starts with \$2})$$

To do so, let's also consider $a_i = P(\text{Allan wins} \mid \text{he starts with \$}i)$ for $i = 0, 1, 3, 4$, and 5.

a. What are the values of a_0 and a_5?

b. Use the law of total probability to obtain an equation relating a_2 to a_1 and a_3. *Hint*: Condition on the result of the first coin toss, realizing that if it is a H, then from that point Allan starts with \$3.

c. Using the logic described in (b), develop a system of equations relating a_i $(i = 1, 2, 3, 4)$ to a_{i-1} and a_{i+1}. Then solve these equations. *Hint*: Write each equation so that $a_i - a_{i-1}$ is on the left hand side. Then use the result of the first equation to express each other $a_i - a_{i-1}$ as a function of a_1, and add together all four of these expressions $(i = 2, 3, 4, 5)$.

d. Generalize the result to the situation in which Allan's initial fortune is \$$a$ and Beth's is \$$b$. *Note*: The solution is a bit more complicated if $p = P(\text{Allan wins \$1}) \neq .5$.

109. Prove that if $P(B|A) > P(B)$ [in which case we say that "A attracts B"], then $P(A|B) > P(A)$ ["B attracts A"].

110. Suppose a single gene determines whether the coloring of a certain animal is dark or light. The coloring will be dark if the genotype is either *AA* or *Aa* and will be light only if the genotype is *aa* (so *A* is dominant and *a* is recessive).

Consider two parents with genotypes *Aa* and *AA*. The first contributes *A* to an offspring with probability $\frac{1}{2}$ and *a* with probability $\frac{1}{2}$, whereas the second contributes *A* for sure. The resulting offspring will be either *AA* or *Aa*, and therefore will be dark colored. Assume that this child then mates with an *Aa* animal to produce a grandchild with dark coloring. In light of this information, what is the probability that the first-generation offspring has the *Aa* genotype (is heterozygous)? *Hint*: Construct an appropriate tree diagram.

Bibliography

Durrett, Richard, *The Essentials of Probability*, Duxbury Press, Belmont, CA, 1993. A concise presentation at a slightly higher level than this text.

Mosteller, Frederick, Robert Rourke, and George Thomas, *Probability with Statistical Applications* (2nd ed.), Addison-Wesley, Reading, MA, 1970. A very good precalculus introduction to probability, with many entertaining examples; especially good on counting rules and their application.

Olkin, Ingram, Cyrus Derman, and Leon Gleser, *Probability Models and Application* (2nd ed.), Macmillan, New York, 1994. A comprehensive introduction to probability, written at a slightly higher mathematical level than this text but containing many good examples.

Ross, Sheldon, *A First Course in Probability* (6th ed.), Macmillan, New York, 2002. Rather tightly written and more mathematically sophisticated than this text but contains a wealth of interesting examples and exercises.

Winkler, Robert, *Introduction to Bayesian Inference and Decision*, Holt, Rinehart & Winston, New York, 1972. A very good introduction to subjective probability.

CHAPTER THREE

Discrete Random Variables and Probability Distributions

Introduction

Whether an experiment yields qualitative or quantitative outcomes, methods of statistical analysis require that we focus on certain numerical aspects of the data (such as a sample proportion x/n, mean $\overline{x}$, or standard deviation s). The concept of a random variable allows us to pass from the experimental outcomes themselves to a numerical function of the outcomes. There are two fundamentally different types of random variables—discrete random variables and continuous random variables. In this chapter, we examine the basic properties and discuss the most important examples of discrete variables. Chapter 4 focuses on continuous random variables.

3.1 Random Variables

In any experiment, numerous characteristics can be observed or measured, but in most cases an experimenter will focus on some specific aspect or aspects of a sample. For example, in a study of commuting patterns in a metropolitan area, each individual in a sample might be asked about commuting distance and the number of people commuting in the same vehicle, but not about IQ, income, family size, and other such characteristics. Alternatively, a researcher may test a sample of components and record only the number that have failed within 1000 hours, rather than record the individual failure times.

In general, each outcome of an experiment can be associated with a number by specifying a rule of association (e.g., the number among the sample of ten components that fail to last 1000 hours or the total weight of baggage for a sample of 25 airline passengers). Such a rule of association is called a **random variable**—a variable because different numerical values are possible and random because the observed value depends on which of the possible experimental outcomes results (Figure 3.1).

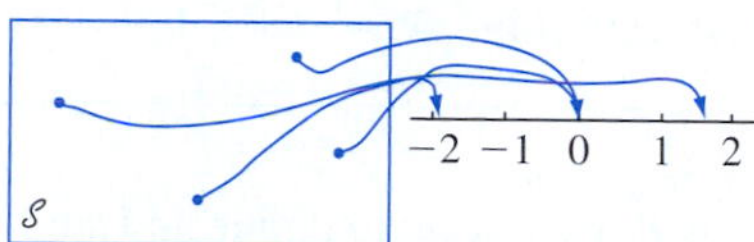

Figure 3.1 A random variable

DEFINITION For a given sample space $\mathcal{S}$ of some experiment, a **random variable (rv)** is any rule that associates a number with each outcome in $\mathcal{S}$. In mathematical language, a random variable is a function whose domain is the sample space and whose range is the set of real numbers.

Random variables are customarily denoted by uppercase letters, such as X and Y, near the end of our alphabet. In contrast to our previous use of a lowercase letter, such as x, to denote a variable, we will now use lowercase letters to represent some particular value of the corresponding random variable. The notation $X(s) = x$ means that x is the value associated with the outcome s by the rv X.

Example 3.1 When a student attempts to connect to a university computer system, either there is a failure (F), or there is a success (S). With $\mathcal{S} = \{S, F\}$, define an rv X by $X(S) = 1$, $X(F) = 0$. The rv X indicates whether (1) or not (0) the student can connect. ■

In Example 3.1, the rv X was specified by explicitly listing each element of $\mathcal{S}$ and the associated number. If $\mathcal{S}$ contains more than a few outcomes, such a listing is tedious, but it can frequently be avoided.

Example 3.2 Consider the experiment in which a telephone number in a certain area code is dialed using a random number dialer (such devices are used extensively by polling organizations), and define an rv Y by

$$Y = \begin{cases} 1 & \text{if the selected number is unlisted} \\ 0 & \text{if the selected number is listed in the directory} \end{cases}$$

For example, if 5282966 appears in the telephone directory, then $Y(5282966) = 0$, whereas $Y(7727350) = 1$ tells us that the number 7727350 is unlisted. A word description of this sort is more economical than a complete listing, so we will use such a description whenever possible. ■

In Examples 3.1 and 3.2, the only possible values of the random variable were 0 and 1. Such a random variable arises frequently enough to be given a special name, after the individual who first studied it.

DEFINITION Any random variable whose only possible values are 0 and 1 is called a **Bernoulli random variable**.

We will often want to define and study several different random variables from the same sample space.

Example 3.3 Example 2.3 described an experiment in which the number of pumps in use at each of two gas stations was determined. Define rv's X, Y, and U by

X = the total number of pumps in use at the two stations

Y = the difference between the number of pumps in use at station 1 and the number in use at station 2

U = the maximum of the numbers of pumps in use at the two stations

If this experiment is performed and $s = (2, 3)$ results, then $X((2, 3)) = 2 + 3 = 5$, so we say that the observed value of X is $x = 5$. Similarly, the observed value of Y would be $y = 2 - 3 = -1$, and the observed value of U would be $u = \max(2, 3) = 3$. ■

Each of the random variables of Examples 3.1–3.3 can assume only a finite number of possible values. This need not be the case.

Example 3.4 In Example 2.4, we considered the experiment in which batteries were examined until a good one (S) was obtained. The sample space was $\mathcal{S} = \{S, FS, FFS, \ldots\}$. Define an rv X by

X = the number of batteries examined before the experiment terminates

Then $X(S) = 1$, $X(FS) = 2$, $X(FFS) = 3, \ldots, X(FFFFFFS) = 7$, and so on. Any positive integer is a possible value of X, so the set of possible values is infinite. ■

Example 3.5 Suppose that in some random fashion, a location (latitude and longitude) in the continental United States is selected. Define an rv Y by

$$Y = \text{the height above sea level at the selected location}$$

For example, if the selected location were (39°50′N, 98°35′W), then we might have Y((39°50′N, 98°35′W)) = 1748.26 ft. The largest possible value of Y is 14,494 (Mt. Whitney), and the smallest possible value is −282 (Death Valley). The set of all possible values of Y is the set of all numbers in the interval between −282 and 14,494 — that is,

$$\{y: y \text{ is a number, } -282 \leq y \leq 14{,}494\}$$

and there are an infinite number of numbers in this interval. ■

Two Types of Random Variables

In Section 1.2 we distinguished between data resulting from observations on a counting variable and data obtained by observing values of a measurement variable. A slightly more formal distinction characterizes two different types of random variables.

DEFINITION

A discrete random variable is an rv whose possible values either constitute a finite set or else can be listed in an infinite sequence in which there is a first element, a second element, and so on.

A random variable is **continuous** if *both* of the following apply:

1. Its set of possible values consists either of all numbers in a single interval on the number line (possibly infinite in extent, e.g., from $-\infty$ to ∞) or all numbers in a disjoint union of such intervals (e.g., $[0, 10] \cup [20, 30]$).
2. No possible value of the variable has positive probability, that is, $P(X = c) = 0$ for any possible value c.

Although any interval on the number line contains an infinite number of numbers, it can be shown that there is no way to create an infinite listing of all these values — there are just too many of them. The second condition describing a continuous random variable is perhaps counterintuitive, since it would seem to imply a total probability of zero for all possible values. But we shall see in Chapter 4 that *intervals* of values have positive probability; the probability of an interval will decrease to zero as the width of the interval shrinks to zero.

Example 3.6 All random variables in Examples 3.1–3.4 are discrete. As another example, suppose we select married couples at random and do a blood test on each person until we find a husband and wife who both have the same Rh factor. With X = the number of blood tests to be performed, possible values of X are $D = \{2, 4, 6, 8, \ldots\}$. Since the possible values have been listed in sequence, X is a discrete rv. ■

To study basic properties of discrete rv's, only the tools of discrete mathematics — summation and differences — are required. The study of continuous variables requires the continuous mathematics of the calculus — integrals and derivatives.

Exercises Section 3.1 (1–10)

1. A concrete beam may fail either by shear (S) or flexure (F). Suppose that three failed beams are randomly selected and the type of failure is determined for each one. Let X = the number of beams among the three selected that failed by shear. List each outcome in the sample space along with the associated value of X.

2. Give three examples of Bernoulli rv's (other than those in the text).

3. Using the experiment in Example 3.3, define two more random variables and list the possible values of each.

4. Let X = the number of nonzero digits in a randomly selected zip code. What are the possible values of X? Give three possible outcomes and their associated X values.

5. If the sample space $\mathcal{S}$ is an infinite set, does this necessarily imply that any rv X defined from $\mathcal{S}$ will have an infinite set of possible values? If yes, say why. If no, give an example.

6. Starting at a fixed time, each car entering an intersection is observed to see whether it turns left (L), right (R), or goes straight ahead (A). The experiment terminates as soon as a car is observed to turn left. Let X = the number of cars observed. What are possible X values? List five outcomes and their associated X values.

7. For each random variable defined here, describe the set of possible values for the variable, and state whether the variable is discrete.

a. X = the number of unbroken eggs in a randomly chosen standard egg carton

b. Y = the number of students on a class list for a particular course who are absent on the first day of classes

c. U = the number of times a duffer has to swing at a golf ball before hitting it

d. X = the length of a randomly selected rattlesnake

e. Z = the amount of royalties earned from the sale of a first edition of 10,000 textbooks

f. Y = the pH of a randomly chosen soil sample

g. X = the tension (psi) at which a randomly selected tennis racket has been strung

h. X = the total number of coin tosses required for three individuals to obtain a match (*HHH* or *TTT*)

8. Each time a component is tested, the trial is a success (S) or failure (F). Suppose the component is tested repeatedly until a success occurs on three *consecutive* trials. Let Y denote the number of trials necessary to achieve this. List all outcomes corresponding to the five smallest possible values of Y, and state which Y value is associated with each one.

9. An individual named Claudius is located at the point 0 in the accompanying diagram.

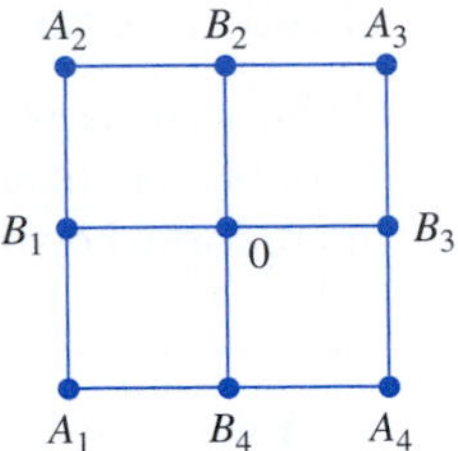

Using an appropriate randomization device (such as a tetrahedral die, one having four sides), Claudius first moves to one of the four locations B_1, B_2, B_3, B_4. Once at one of these locations, he uses another randomization device to decide whether he next returns to 0 or next visits one of the other two adjacent points. This process then continues; after each move, another move to one of the (new) adjacent points is determined by tossing an appropriate die or coin.

a. Let X = the number of moves that Claudius makes before first returning to 0. What are possible values of X? Is X discrete or continuous?

b. If moves are allowed also along the diagonal paths connecting 0 to A_1, A_2, A_3, and A_4, respectively, answer the questions in part (a).

10. The number of pumps in use at both a six-pump station and a four-pump station will be determined. Give the possible values for each of the following random variables:

a. T = the total number of pumps in use

b. X = the difference between the numbers in use at stations 1 and 2

c. U = the maximum number of pumps in use at either station

d. Z = the number of stations having exactly two pumps in use

3.2 Probability Distributions for Discrete Random Variables

When probabilities are assigned to various outcomes in $\mathcal{S}$, these in turn determine probabilities associated with the values of any particular rv X. The *probability distribution of* X says how the total probability of 1 is distributed among (allocated to) the various possible X values.

Example 3.7 Six lots of components are ready to be shipped by a certain supplier. The number of defective components in each lot is as follows:

Lot	1	2	3	4	5	6
Number of defectives	0	2	0	1	2	0

One of these lots is to be randomly selected for shipment to a particular customer. Let X be the number of defectives in the selected lot. The three possible X values are 0, 1, and 2. Of the six equally likely simple events, three result in $X = 0$, one in $X = 1$, and the other two in $X = 2$. Let $p(0)$ denote the probability that $X = 0$ and $p(1)$ and $p(2)$ represent the probabilities of the other two possible values of X. Then

$$p(0) = P(X = 0) = P(\text{lot 1 or 3 or 6 is sent}) = \frac{3}{6} = .500$$

$$p(1) = P(X = 1) = P(\text{lot 4 is sent}) = \frac{1}{6} = .167$$

$$p(2) = P(X = 2) = P(\text{lot 2 or 5 is sent}) = \frac{2}{6} = .333$$

That is, a probability of .500 is distributed to the X value 0, a probability of .167 is placed on the X value 1, and the remaining probability, .333, is associated with the X value 2. The values of X along with their probabilities collectively specify the probability distribution or *probability mass function of* X. If this experiment were repeated over and over again, in the long run $X = 0$ would occur one-half of the time, $X = 1$ one-sixth of the time, and $X = 2$ one-third of the time. ■

DEFINITION The **probability distribution** or **probability mass function** (pmf) of a discrete rv is defined for every number x by $p(x) = P(X = x) = P(\text{all } s \in \mathcal{S}\text{: } X(s) = x)$.*

In words, for every possible value x of the random variable, the pmf specifies the probability of observing that value when the experiment is performed. The conditions $p(x) \geq 0$ and $\Sigma_{\text{all possible } x} p(x) = 1$ are required of any pmf.

*$P(X = x)$ is read "the probability that the rv X assumes the value x." For example, $P(X = 2)$ denotes the probability that the resulting X value is 2.

Example 3.8 Suppose we go to a university bookstore during the first week of classes and observe whether the next person buying a computer buys a laptop or a desktop model. Let

$$X = \begin{cases} 1 & \text{if the customer purchases a laptop computer} \\ 0 & \text{if the customer purchases a desktop computer} \end{cases}$$

If 20% of all purchasers during that week select a laptop, the pmf for X is

$$\begin{aligned} p(0) &= P(X = 0) = P(\text{next customer purchases a desktop model}) = .8 \\ p(1) &= P(X = 1) = P(\text{next customer purchases a laptop model}) = .2 \\ p(x) &= P(X = x) = 0 \quad \text{for } x \neq 0 \text{ or } 1 \end{aligned}$$

An equivalent description is

$$p(x) = \begin{cases} .8 & \text{if } x = 0 \\ .2 & \text{if } x = 1 \\ 0 & \text{if } x \neq 0 \text{ or } 1 \end{cases}$$

Figure 3.2 is a picture of this pmf, called a *line graph*.

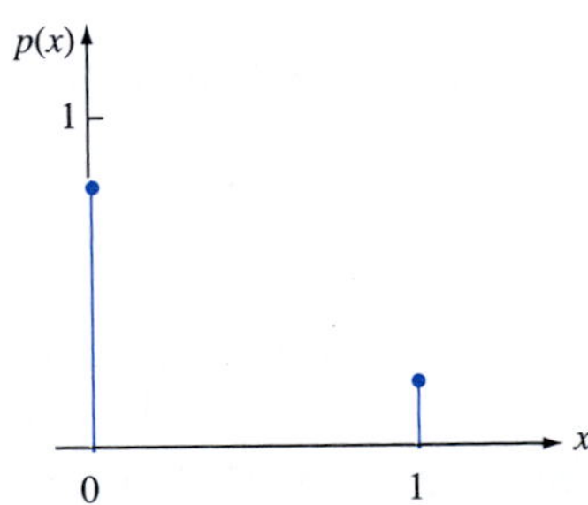

Figure 3.2 The line graph for the pmf in Example 3.8 ■

Example 3.9 Consider a group of five potential blood donors—A, B, C, D, and E—of whom only A and B have type O+ blood. Five blood samples, one from each individual, will be typed in random order until an O+ individual is identified. Let the rv Y = the number of typings necessary to identify an O+ individual. Then the pmf of Y is

$$\begin{aligned} p(1) &= P(Y = 1) = P(\text{A or B typed first}) = \frac{2}{5} = .4 \\ p(2) &= P(Y = 2) = P(\text{C, D, or E first, and then A or B}) \\ &= P(\text{C, D, or E first}) \cdot P(\text{A or B next} \mid \text{C, D, or E first}) = \frac{3}{5} \cdot \frac{2}{4} = .3 \\ p(3) &= P(Y = 3) = P(\text{C, D, or E first and second, and then A or B}) \\ &= \left(\frac{3}{5}\right)\left(\frac{2}{4}\right)\left(\frac{2}{3}\right) = .2 \end{aligned}$$

$$p(4) = P(Y = 4) = P(\text{C, D, and E all done first}) = \left(\frac{3}{5}\right)\left(\frac{2}{4}\right)\left(\frac{1}{3}\right) = .1$$

$$p(y) = 0 \quad \text{if } y \neq 1, 2, 3, 4$$

The pmf can be presented compactly in tabular form:

y	1	2	3	4
$p(y)$	.4	.3	.2	.1

where any y value not listed receives zero probability. This pmf can also be displayed in a line graph (Figure 3.3).

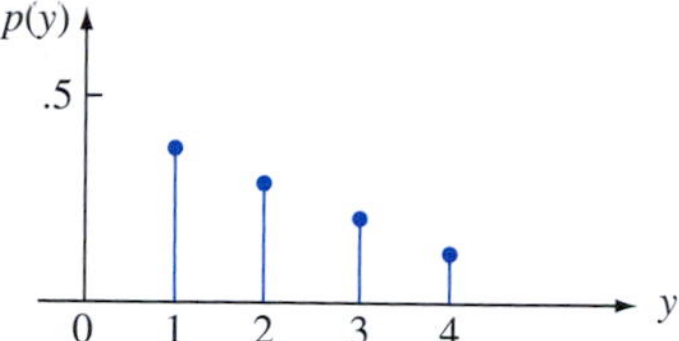

Figure 3.3 The line graph for the pmf in Example 3.9 ■

The name "probability mass function" is suggested by a model used in physics for a system of "point masses." In this model, masses are distributed at various locations x along a one-dimensional axis. Our pmf describes how the total probability mass of 1 is distributed at various points along the axis of possible values of the random variable (where and how much mass at each x).

Another useful pictorial representation of a pmf, called a **probability histogram**, is similar to histograms discussed in Chapter 1. Above each y with $p(y) > 0$, construct a rectangle centered at y. The height of each rectangle is proportional to $p(y)$, and the base is the same for all rectangles. When possible values are equally spaced, the base is frequently chosen as the distance between successive y values (though it could be smaller). Figure 3.4 shows two probability histograms.

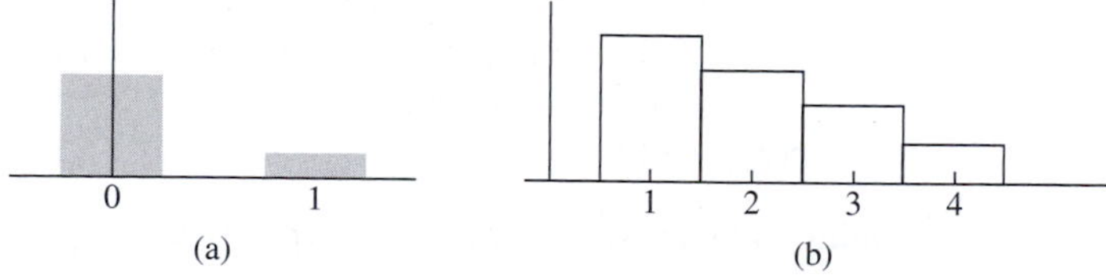

Figure 3.4 Probability histograms: (a) Example 3.8; (b) Example 3.9

A Parameter of a Probability Distribution

In Example 3.8, we had $p(0) = .8$ and $p(1) = .2$ because 20% of all purchasers selected a laptop computer. At another bookstore, it may be the case that $p(0) = .9$ and $p(1) = .1$. More generally, the pmf of any Bernoulli rv can be expressed in the form $p(1) = \alpha$ and

$p(0) = 1 - \alpha$, where $0 < \alpha < 1$. Because the pmf depends on the particular value of α, we often write $p(x; \alpha)$ rather than just $p(x)$:

$$p(x; \alpha) = \begin{cases} 1 - \alpha & \text{if } x = 0 \\ \alpha & \text{if } x = 1 \\ 0 & \text{otherwise} \end{cases} \tag{3.1}$$

Then each choice of α in Expression (3.1) yields a different pmf.

DEFINITION Suppose $p(x)$ depends on a quantity that can be assigned any one of a number of possible values, with each different value determining a different probability distribution. Such a quantity is called a **parameter** of the distribution. The collection of all probability distributions for different values of the parameter is called a **family** of probability distributions.

The quantity α in Expression (3.1) is a parameter. Each different number α between 0 and 1 determines a different member of a family of distributions; two such members are

$$p(x; .6) = \begin{cases} .4 & \text{if } x = 0 \\ .6 & \text{if } x = 1 \\ 0 & \text{otherwise} \end{cases} \quad \text{and} \quad p(x; .5) = \begin{cases} .5 & \text{if } x = 0 \\ .5 & \text{if } x = 1 \\ 0 & \text{otherwise} \end{cases}$$

Every probability distribution for a Bernoulli rv has the form of Expression (3.1), so it is called the *family of Bernoulli distributions*.

Example 3.10 Starting at a fixed time, we observe the gender of each newborn child at a certain hospital until a boy (B) is born. Let $p = P(B)$, assume that successive births are independent, and define the rv X by X = number of births observed. Then

$$p(1) = P(X = 1) = P(B) = p$$
$$p(2) = P(X = 2) = P(GB) = P(G) \cdot P(B) = (1 - p)p$$

and

$$p(3) = P(X = 3) = P(GGB) = P(G) \cdot P(G) \cdot P(B) = (1 - p)^2 p$$

Continuing in this way, a general formula emerges:

$$p(x) = \begin{cases} (1 - p)^{x-1} p & x = 1, 2, 3, \ldots \\ 0 & \text{otherwise} \end{cases} \tag{3.2}$$

The quantity p in Expression (3.2) represents a number between 0 and 1 and is a parameter of the probability distribution. In the gender example, $p = .51$ might be appropriate, but if we were looking for the first child with Rh-positive blood, then we might have $p = .85$. ■

The Cumulative Distribution Function

For some fixed value x, we often wish to compute the probability that the observed value of X will be at most x. For example, the pmf in Example 3.7 was

$$p(x) = \begin{cases} .500 & x = 0 \\ .167 & x = 1 \\ .333 & x = 2 \\ 0 & \text{otherwise} \end{cases}$$

The probability that X is at most 1 is then

$$P(X \leq 1) = p(0) + p(1) = .500 + .167 = .667$$

In this example, $X \leq 1.5$ iff $X \leq 1$, so $P(X \leq 1.5) = P(X \leq 1) = .667$. Similarly, $P(X \leq 0) = P(X = 0) = .5$, and $P(X \leq .75) = .5$ also. Since 0 is the smallest possible value of X, $P(X \leq -1.7) = 0$, $P(X \leq -.0001) = 0$, and so on. The largest possible X value is 2, so $P(X \leq 2) = 1$, and if x is any number larger than 2, $P(X \leq x) = 1$; that is, $P(X \leq 5) = 1$, $P(X \leq 10.23) = 1$, and so on. Notice that $P(X < 1) = .5 \neq P(X \leq 1)$, since the probability of the X value 1 is included in the latter probability but not in the former. When X is a discrete random variable and x is a possible value of X, $P(X < x) < P(X \leq x)$.

DEFINITION

The **cumulative distribution function** (cdf) $F(x)$ of a discrete rv variable X with pmf $p(x)$ is defined for every number x by

$$F(x) = P(X \leq x) = \sum_{y: y \leq x} p(y) \tag{3.3}$$

For any number x, $F(x)$ is the probability that the observed value of X will be at most x.

Example 3.11 The pmf of Y (the number of blood typings) in Example 3.9 was

y	1	2	3	4
$p(y)$	.4	.3	.2	.1

We first determine $F(y)$ for each value in the set $\{1, 2, 3, 4\}$ of possible values:

$$F(1) = P(Y \leq 1) = P(Y = 1) = p(1) = .4$$
$$F(2) = P(Y \leq 2) = P(Y = 1 \text{ or } 2) = p(1) + p(2) = .7$$
$$F(3) = P(Y \leq 3) = P(Y = 1 \text{ or } 2 \text{ or } 3) = p(1) + p(2) + p(3) = .9$$
$$F(4) = P(Y \leq 4) = P(Y = 1 \text{ or } 2 \text{ or } 3 \text{ or } 4) = 1$$

Now for any other number y, $F(y)$ will equal the value of F at the closest possible value of Y to the left of y. For example, $F(2.7) = P(Y \le 2.7) = P(Y \le 2) = .7$, and $F(3.999) = F(3) = .9$. The cdf is thus

$$F(y) = \begin{cases} 0 & \text{if } y < 1 \\ .4 & \text{if } 1 \le y < 2 \\ .7 & \text{if } 2 \le y < 3 \\ .9 & \text{if } 3 \le y < 4 \\ 1 & \text{if } 4 \le y \end{cases}$$

A graph of $F(y)$ is shown in Figure 3.5.

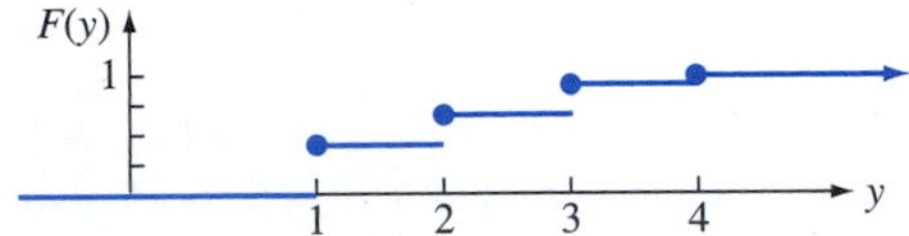

Figure 3.5 A graph of the cdf of Example 3.11 ■

For X a discrete rv, the graph of $F(x)$ will have a jump at every possible value of X and will be flat between possible values. Such a graph is called a **step function**.

Example 3.12 In Example 3.10, any positive integer was a possible X value, and the pmf was

$$p(x) = \begin{cases} (1-p)^{x-1}p & x = 1, 2, 3, \dots \\ 0 & \text{otherwise} \end{cases}$$

For any positive integer x,

$$F(x) = \sum_{y \le x} p(y) = \sum_{y=1}^{x} (1-p)^{y-1} p = p \sum_{y=0}^{x-1} (1-p)^y \tag{3.4}$$

To evaluate this sum, we use the fact that the partial sum of a geometric series is

$$\sum_{y=0}^{k} a^y = \frac{1 - a^{k+1}}{1 - a}$$

Using this in Equation (3.4), with $a = 1 - p$ and $k = x - 1$, gives

$$F(x) = p \cdot \frac{1 - (1-p)^x}{1 - (1-p)} = 1 - (1-p)^x \qquad x \text{ a positive integer}$$

Since F is constant in between positive integers,

$$F(x) = \begin{cases} 0 & x < 1 \\ 1 - (1-p)^{[x]} & x \ge 1 \end{cases} \tag{3.5}$$

where $[x]$ is the largest integer $\le x$ (e.g., $[2.7] = 2$). Thus if $p = .51$ as in the birth example, then the probability of having to examine at most five births to see the first boy is $F(5) = 1 - (.49)^5 = 1 - .0282 = .9718$, whereas $F(10) \approx 1.0000$. This cdf is graphed in Figure 3.6.

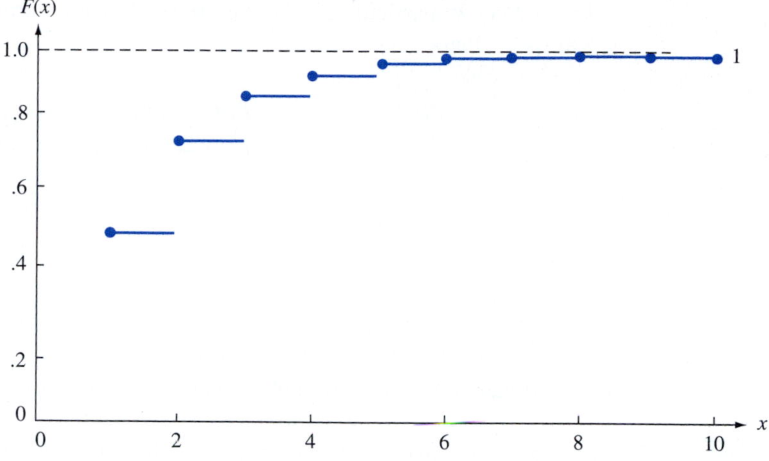

Figure 3.6 A graph of $F(x)$ for Example 3.12 ■

In our examples thus far, the cdf has been derived from the pmf. This process can be reversed to obtain the pmf from the cdf whenever the latter function is available. Suppose, for example, that X represents the number of defective components in a shipment consisting of six components, so that possible X values are 0, 1, . . . , 6. Then

$$\begin{aligned} p(3) &= P(X = 3) \\ &= [p(0) + p(1) + p(2) + p(3)] - [p(0) + p(1) + p(2)] \\ &= P(X \leq 3) - P(X \leq 2) \\ &= F(3) - F(2) \end{aligned}$$

More generally, the probability that X falls in a specified interval is easily obtained from the cdf. For example,

$$\begin{aligned} P(2 \leq X \leq 4) &= p(2) + p(3) + p(4) \\ &= [p(0) + \cdots + p(4)] - [p(0) + p(1)] \\ &= P(X \leq 4) - P(X \leq 1) \\ &= F(4) - F(1) \end{aligned}$$

Notice that $P(2 \leq X \leq 4) \neq F(4) - F(2)$. This is because the X value 2 is included in $2 \leq X \leq 4$, so we do not want to subtract out its probability. However, $P(2 < X \leq 4) = F(4) - F(2)$ because $X = 2$ is not included in the interval $2 < X \leq 4$.

PROPOSITION For any two numbers a and b with $a \leq b$,

$$P(a \leq X \leq b) = F(b) - F(a-)$$

where $F(a-)$ represents the maximum of $F(x)$ values to the left of a. Equivalently, if a is the limit of values of x approaching from the left, then $F(a-)$ is the limiting value of $F(x)$. In particular, if the only possible values are integers and if a and b are integers, then

$$P(a \le X \le b) = P(X = a \text{ or } a + 1 \text{ or} \dots \text{or } b)$$
$$= F(b) - F(a - 1)$$

Taking $a = b$ yields $P(X = a) = F(a) - F(a - 1)$ in this case.

The reason for subtracting $F(a-)$ rather than $F(a)$ is that we want to include $P(X = a)$; $F(b) - F(a)$ gives $P(a < X \le b)$. This proposition will be used extensively when computing binomial and Poisson probabilities in Sections 3.5 and 3.7.

Example 3.13 Let X = the number of days of sick leave taken by a randomly selected employee of a large company during a particular year. If the maximum number of allowable sick days per year is 14, possible values of X are 0, 1, . . . , 14. With $F(0) = .58$, $F(1) = .72$, $F(2) = .76$, $F(3) = .81$, $F(4) = .88$, and $F(5) = .94$,

$$P(2 \le X \le 5) = P(X = 2, 3, 4, \text{ or } 5) = F(5) - F(1) = .22$$

and

$$P(X = 3) = F(3) - F(2) = .05$$ ■

Another View of Probability Mass Functions

It is often helpful to think of a pmf as specifying a mathematical model for a discrete population.

Example 3.14 Consider selecting at random a student who is among the 15,000 registered for the current term at Mega University. Let X = the number of courses for which the selected student is registered, and suppose that X has the following pmf:

x	1	2	3	4	5	6	7
$p(x)$	.01	.03	.13	.25	.39	.17	.02

One way to view this situation is to think of the population as consisting of 15,000 individuals, each having his or her own X value; the proportion with each X value is given by $p(x)$. An alternative viewpoint is to forget about the students and think of the population itself as consisting of the X values: There are some 1's in the population, some 2's, . . . , and finally some 7's. The population then consists of the numbers 1, 2, . . . , 7 (so is discrete), and $p(x)$ gives a model for the distribution of population values. ■

Once we have such a population model, we will use it to compute values of population characteristics (e.g., the mean μ) and make inferences about such characteristics.

Exercises Section 3.2 (11–27)

11. An automobile service facility specializing in engine tune-ups knows that 45% of all tune-ups are done on four-cylinder automobiles, 40% on six-cylinder automobiles, and 15% on eight-cylinder automobiles. Let X = the number of cylinders on the next car to be tuned.

a. What is the pmf of X?

b. Draw both a line graph and a probability histogram for the pmf of part (a).

c. What is the probability that the next car tuned has at least six cylinders? More than six cylinders?

12. Airlines sometimes overbook flights. Suppose that for a plane with 50 seats, 55 passengers have tickets. Define the random variable Y as the number of ticketed passengers who actually show up for the flight. The probability mass function of Y appears in the accompanying table.

y	45	46	47	48	49	50	51	52	53	54	55
$p(y)$	.05	.10	.12	.14	.25	.17	.06	.05	.03	.02	.01

a. What is the probability that the flight will accommodate all ticketed passengers who show up?

b. What is the probability that not all ticketed passengers who show up can be accommodated?

c. If you are the first person on the standby list (which means you will be the first one to get on the plane if there are any seats available after all ticketed passengers have been accommodated), what is the probability that you will be able to take the flight? What is this probability if you are the third person on the standby list?

13. A mail-order computer business has six telephone lines. Let X denote the number of lines in use at a specified time. Suppose the pmf of X is as given in the accompanying table.

x	0	1	2	3	4	5	6
$p(x)$	.10	.15	.20	.25	.20	.06	.04

Calculate the probability of each of the following events.

a. {at most three lines are in use}

b. {fewer than three lines are in use}

c. {at least three lines are in use}

d. {between two and five lines, inclusive, are in use}

e. {between two and four lines, inclusive, are not in use}

f. {at least four lines are not in use}

14. A contractor is required by a county planning department to submit one, two, three, four, or five forms (depending on the nature of the project) in applying for a building permit. Let Y = the number of forms required of the next applicant. The probability that y forms are required is known to be proportional to y—that is, $p(y) = ky$ for $y = 1, \dots, 5$.

a. What is the value of k? [*Hint*: $\sum_{y=1}^{5} p(y) = 1$.]

b. What is the probability that at most three forms are required?

c. What is the probability that between two and four forms (inclusive) are required?

d. Could $p(y) = y^2/50$ for $y = 1, \dots, 5$ be the pmf of Y?

15. Many manufacturers have quality control programs that include inspection of incoming materials for defects. Suppose a computer manufacturer receives computer boards in lots of five. Two boards are selected from each lot for inspection. We can represent possible outcomes of the selection process by pairs. For example, the pair (1, 2) represents the selection of boards 1 and 2 for inspection.

a. List the ten different possible outcomes.

b. Suppose that boards 1 and 2 are the only defective boards in a lot of five. Two boards are to be chosen at random. Define X to be the number of defective boards observed among those inspected. Find the probability distribution of X.

c. Let $F(x)$ denote the cdf of X. First determine $F(0) = P(X \le 0)$, $F(1)$, and $F(2)$, and then obtain $F(x)$ for all other x.

16. Some parts of California are particularly earthquake-prone. Suppose that in one such area, 30% of all homeowners are insured against earthquake damage. Four homeowners are to be selected at random; let X denote the number among the four who have earthquake insurance.

a. Find the probability distribution of X. [*Hint*: Let S denote a homeowner who has insurance and F one who does not. Then one possible outcome is $SFSS$, with probability (.3)(.7)(.3)(.3) and associated X value 3. There are 15 other outcomes.]

b. Draw the corresponding probability histogram.

c. What is the most likely value for X?

d. What is the probability that at least two of the four selected have earthquake insurance?

17. A new battery's voltage may be acceptable (A) or unacceptable (U). A certain flashlight requires two batteries, so batteries will be independently selected and tested until two acceptable ones have been found. Suppose that 90% of all batteries have acceptable voltages. Let Y denote the number of batteries that must be tested.

a. What is $p(2)$, that is, $P(Y = 2)$?

b. What is $p(3)$? (*Hint*: There are two different outcomes that result in $Y = 3$.)

c. To have $Y = 5$, what must be true of the fifth battery selected? List the four outcomes for which $Y = 5$ and then determine $p(5)$.

d. Use the pattern in your answers for parts (a)–(c) to obtain a general formula for $p(y)$.

18. Two fair six-sided dice are tossed independently. Let M = the maximum of the two tosses [so $M(1, 5) = 5$, $M(3, 3) = 3$, etc.].

a. What is the pmf of M? [*Hint*: First determine $p(1)$, then $p(2)$, and so on.]

b. Determine the cdf of M and graph it.

19. In Example 3.9, suppose there are only four potential blood donors, of whom only one has type O+ blood. Compute the pmf of Y.

20. A library subscribes to two different weekly news magazines, each of which is supposed to arrive in Wednesday's mail. In actuality, each one may arrive on Wednesday, Thursday, Friday, or Saturday. Suppose the two arrive independently of one another, and for each one $P(\text{Wed.}) = .3$, $P(\text{Thurs.}) = .4$, $P(\text{Fri.}) = .2$, and $P(\text{Sat.}) = .1$. Let Y = the number of days beyond Wednesday that it takes for both magazines to arrive (so possible Y values are 0, 1, 2, or 3). Compute the pmf of Y. [*Hint*: There are 16 possible outcomes; $Y(W, W) = 0$, $Y(F, Th) = 2$, and so on.]

21. Refer to Exercise 13, and calculate and graph the cdf $F(x)$. Then use it to calculate the probabilities of the events given in parts (a)–(d) of that problem.

22. A consumer organization that evaluates new automobiles customarily reports the number of major defects in each car examined. Let X denote the number of major defects in a randomly selected car of a certain type. The cdf of X is as follows:

$$F(x) = \begin{cases} 0 & x < 0 \\ .06 & 0 \le x < 1 \\ .19 & 1 \le x < 2 \\ .39 & 2 \le x < 3 \\ .67 & 3 \le x < 4 \\ .92 & 4 \le x < 5 \\ .97 & 5 \le x < 6 \\ 1 & 6 \le x \end{cases}$$

Calculate the following probabilities directly from the cdf:

a. $p(2)$, that is, $P(X = 2)$

b. $P(X > 3)$

c. $P(2 \le X \le 5)$

d. $P(2 < X < 5)$

23. An insurance company offers its policyholders a number of different premium payment options. For a randomly selected policyholder, let X = the number of months between successive payments. The cdf of X is as follows:

$$F(x) = \begin{cases} 0 & x < 1 \\ .30 & 1 \le x < 3 \\ .40 & 3 \le x < 4 \\ .45 & 4 \le x < 6 \\ .60 & 6 \le x < 12 \\ 1 & 12 \le x \end{cases}$$

a. What is the pmf of X?

b. Using just the cdf, compute $P(3 \le X \le 6)$ and $P(4 \le X)$.

24. In Example 3.10, let Y = the number of girls born before the experiment terminates. With $p = P(B)$ and $1 - p = P(G)$, what is the pmf of Y? (*Hint*: First list the possible values of Y, starting with the smallest, and proceed until you see a general formula.)

25. Alvie Singer lives at 0 in the accompanying diagram and has four friends who live at A, B, C, and D. One day Alvie decides to go visiting, so he tosses a fair coin twice to decide which of the four to visit. Once at a friend's house, he will either return home or else proceed to one of the two adjacent houses (such as 0, A, or C when at B), with each of the three possibilities having probability $\frac{1}{3}$. In this way, Alvie continues to visit friends until he returns home.

a. Let X = the number of times that Alvie visits a friend. Derive the pmf of X.

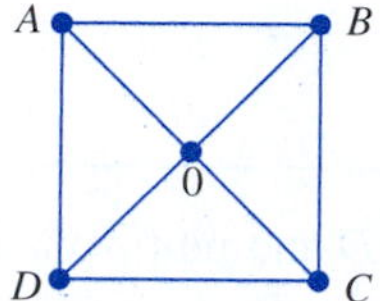

b. Let Y = the number of straight-line segments that Alvie traverses (including those leading to and from 0). What is the pmf of Y?

c. Suppose that female friends live at A and C and male friends at B and D. If Z = the number of visits to female friends, what is the pmf of Z?

26. After all students have left the classroom, a statistics professor notices that four copies of the text were left under desks. At the beginning of the next lecture, the professor distributes the four books in a completely random fashion to each of the four students (1, 2, 3, and 4) who claim to have left books. One possible outcome is that 1 receives 2's book, 2 receives 4's book, 3 receives his or her own book, and 4 receives 1's book. This outcome can be abbreviated as (2, 4, 3, 1).

a. List the other 23 possible outcomes.

b. Let X denote the number of students who receive their own book. Determine the pmf of X.

27. Show that the cdf $F(x)$ is a nondecreasing function; that is, $x_1 < x_2$ implies that $F(x_1) \le F(x_2)$. Under what condition will $F(x_1) = F(x_2)$?

3.3 Expected Values of Discrete Random Variables

In Example 3.14, we considered a university having 15,000 students and let X = the number of courses for which a randomly selected student is registered. The pmf of X follows. Since $p(1) = .01$, we know that $(.01) \cdot (15{,}000) = 150$ of the students are registered for one course, and similarly for the other x values.

x	1	2	3	4	5	6	7
$p(x)$	.01	.03	.13	.25	.39	.17	.02
Number registered	150	450	1950	3750	5850	2550	300

(3.6)

To compute the average number of courses per student, or the average value of X in the population, we should calculate the total number of courses and divide by the total number of students. Since each of 150 students is taking one course, these 150 contribute 150 courses to the total. Similarly, 450 students contribute 2(450) courses, and so on. The population average value of X is then

$$\frac{1(150) + 2(450) + 3(1950) + \cdots + 7(300)}{15{,}000} = 4.57 \tag{3.7}$$

Since $150/15{,}000 = .01 = p(1)$, $450/15{,}000 = .03 = p(2)$, and so on, an alternative expression for (3.7) is

$$1 \cdot p(1) + 2 \cdot p(2) + \cdots + 7 \cdot p(7) \tag{3.8}$$

Expression (3.8) shows that to compute the population average value of X, we need only the possible values of X along with their probabilities (proportions). In particular, the population size is irrelevant as long as the pmf is given by (3.6). The average or mean value of X is then a *weighted* average of the possible values $1, \dots, 7$, where the weights are the probabilities of those values.

The Expected Value of X

DEFINITION Let X be a discrete rv with set of possible values D and pmf $p(x)$. The **expected value** or **mean value** of X, denoted by $E(X)$ or μ_X, is

$$E(X) = \mu_X = \sum_{x \in D} x \cdot p(x)$$

This expected value will exist provided that $\sum_{x \in D} |x| \cdot p(x) < \infty$.

When it is clear to which X the expected value refers, μ rather than μ_X is often used.

Example 3.15 For the pmf in (3.6),

$$\begin{aligned}\mu &= 1 \cdot p(1) + 2 \cdot p(2) + \cdots + 7 \cdot p(7) \\ &= (1)(.01) + 2(.03) + \cdots + (7)(.02) \\ &= .01 + .06 + .39 + 1.00 + 1.95 + 1.02 + .14 = 4.57\end{aligned}$$

If we think of the population as consisting of the X values 1, 2, . . . , 7, then $\mu = 4.57$ is the population mean. In the sequel, we will often refer to μ as the *population mean* rather than the mean of X in the population. ■

In Example 3.15, the expected value μ was 4.57, which is not a possible value of X. The word *expected* should be interpreted with caution because one would not expect to see an X value of 4.57 when a single student is selected.

Example 3.16 Just after birth, each newborn child is rated on a scale called the Apgar scale. The possible ratings are 0, 1, . . . , 10, with the child's rating determined by color, muscle tone, respiratory effort, heartbeat, and reflex irritability (the best possible score is 10). Let X be the Apgar score of a randomly selected child born at a certain hospital during the next year, and suppose that the pmf of X is

x	0	1	2	3	4	5	6	7	8	9	10
$p(x)$	.002	.001	.002	.005	.02	.04	.18	.37	.25	.12	.01

Then the mean value of X is

$$\begin{aligned}E(X) = \mu &= 0(.002) + 1(.001) + 2(.002) + \cdots + 8(.25) + 9(.12) + 10(.01) \\ &= 7.15\end{aligned}$$

Again, μ is not a possible value of the variable X. Also, because the variable refers to a future child, there is no concrete existing population to which μ refers. Instead, we think of the pmf as a model for a conceptual population consisting of the values 0, 1, 2, . . . , 10. The mean value of this conceptual population is then $\mu = 7.15$. ■

Example 3.17 Let $X = 1$ if a randomly selected component needs warranty service and $= 0$ otherwise. Then X is a Bernoulli rv with pmf

$$p(x) = \begin{cases} 1 - p & x = 0 \\ p & x = 1 \\ 0 & x \neq 0, 1 \end{cases}$$

from which $E(X) = 0 \cdot p(0) + 1 \cdot p(1) = 0(1 - p) + 1(p) = p$. That is, the expected value of X is just the probability that X takes on the value 1. If we conceptualize a population consisting of 0's in proportion $1 - p$ and 1's in proportion p, then the population average is $\mu = p$. ■

Example 3.18 From Example 3.10 the general form for the pmf of X = the number of children born up to and including the first boy is

$$p(x) = \begin{cases} p(1 - p)^{x-1} & x = 1, 2, 3, \ldots \\ 0 & \text{otherwise} \end{cases}$$

From the definition,

$$E(X) = \sum_D x \cdot p(x) = \sum_{x=1}^{\infty} xp(1 - p)^{x-1} = p \sum_{x=1}^{\infty} \left[-\frac{d}{dp}(1 - p)^x \right] \tag{3.9}$$

If we interchange the order of taking the derivative and the summation, the sum is that of a geometric series. After the sum is computed, the derivative is taken, and the final result is $E(X) = 1/p$. If p is near 1, we expect to see a boy very soon, whereas if p is near 0, we expect many births before the first boy. For $p = .5$, $E(X) = 2$. ■

There is another frequently used interpretation of μ. Consider the pmf

$$p(x) = \begin{cases} (.5) \cdot (.5)^{x-1} & \text{if } x = 1, 2, 3, \ldots \\ 0 & \text{otherwise} \end{cases}$$

This is the pmf of X = the number of tosses of a fair coin necessary to obtain the first H (a special case of Example 3.18). Suppose we observe a value x from this pmf (toss a coin until an H appears), then observe independently another value (keep tossing), then another, and so on. If after observing a very large number of x values, we average them, the resulting sample average will be very near to $\mu = 2$. That is, μ can be interpreted as the long-run average observed value of X when the experiment is performed repeatedly.

Example 3.19 Let X, the number of interviews a student has prior to getting a job, have pmf

$$p(x) = \begin{cases} k/x^2 & x = 1, 2, 3, \ldots \\ 0 & \text{otherwise} \end{cases}$$

where k is chosen so that $\sum_{x=1}^{\infty}(k/x^2) = 1$. (Because $\sum_{x=1}^{\infty}(1/x^2) = \pi^2/6$, the value of k is $6/\pi^2$.) The expected value of X is

$$\mu = E(X) = \sum_{x=1}^{\infty} x \cdot \frac{k}{x^2} = k \sum_{x=1}^{\infty} \frac{1}{x} \tag{3.10}$$

The sum on the right of Equation (3.10) is the famous harmonic series of mathematics and can be shown to equal ∞. $E(X)$ is not finite here because $p(x)$ does not decrease sufficiently fast as x increases; statisticians say that the probability distribution of X has "a heavy tail." If a sequence of X values is chosen using this distribution, the sample average will not settle down to some finite number but will tend to grow without bound.

Statisticians use the phrase "heavy tails" in connection with any distribution having a large amount of probability far from μ (so heavy tails do not require $\mu = \infty$). Such heavy tails make it difficult to make inferences about μ. ■

The Expected Value of a Function

Often we will be interested in the expected value of some function $h(X)$ rather than X itself.

Example 3.20 Suppose a bookstore purchases ten copies of a book at \$6.00 each to sell at \$12.00 with the understanding that at the end of a 3-month period any unsold copies can be redeemed for \$2.00. If X represents the number of copies sold, then net revenue $= h(X) = 12X + 2(10 - X) - 60 = 10X - 40$. ■

An easy way of computing the expected value of $h(X)$ is suggested by the following example.

Example 3.21 Let $X =$ the number of cylinders in the engine of the next car to be tuned up at a certain facility. The cost of a tune-up is related to X by $h(X) = 20 + 3X + .5X^2$. Since X is a random variable, so is $h(X)$; denote this latter rv by Y. The pmf's of X and Y are as follows:

x	4	6	8
$p(x)$	.5	.3	.2

y	40	56	76
$p(y)$	.5	.3	.2

With D^* denoting possible values of Y,

$$E(Y) = E[h(X)] = \sum_{D^*} y \cdot p(y) \tag{3.11}$$
$$= (40)(.5) + (56)(.3) + (76)(.2)$$
$$= h(4) \cdot (.5) + h(6) \cdot (.3) + h(8) \cdot (.2)$$
$$= \sum_{D} h(x) \cdot p(x)$$

According to Equation (3.11), it was not necessary to determine the pmf of Y to obtain $E(Y)$; instead, the desired expected value is a weighted average of the possible $h(x)$ (rather than x) values. ■

PROPOSITION

If the rv X has a set of possible values D and pmf $p(x)$, then the expected value of any function $h(X)$, denoted by $E[h(X)]$ or $\mu_{h(X)}$, is computed by

$$E[h(X)] = \sum_D h(x) \cdot p(x)$$

assuming that $\sum_D |h(x)| \cdot p(x)$ is finite.

According to this proposition, $E[h(X)]$ is computed in the same way that $E(X)$ itself is, except that $h(x)$ is substituted in place of x.

Example 3.22

A computer store has purchased three computers of a certain type at \$500 apiece. It will sell them for \$1000 apiece. The manufacturer has agreed to repurchase any computers still unsold after a specified period at \$200 apiece. Let X denote the number of computers sold, and suppose that $p(0) = .1$, $p(1) = .2$, $p(2) = .3$, and $p(3) = .4$. With $h(X)$ denoting the profit associated with selling X units, the given information implies that $h(X) =$ revenue $-$ cost $= 1000X + 200(3 - X) - 1500 = 800X - 900$. The expected profit is then

$$\begin{aligned} E[h(X)] &= h(0) \cdot p(0) + h(1) \cdot p(1) + h(2) \cdot p(2) + h(3) \cdot p(3) \\ &= (-900)(.1) + (-100)(.2) + (700)(.3) + (1500)(.4) \\ &= \$700 \end{aligned}$$

■

The $h(X)$ function of interest is quite frequently a linear function $aX + b$. In this case, $E[h(X)]$ is easily computed from $E(X)$.

PROPOSITION

$$E(aX + b) = a \cdot E(X) + b \tag{3.12}$$

(Or, using alternative notation, $\mu_{aX+b} = a \cdot \mu_X + b$.)

To paraphrase, the expected value of a linear function equals the linear function evaluated at the expected value $E(X)$. Since $h(X)$ in Example 3.22 is linear and $E(X) = 2$, $E[h(X)] = 800(2) - 900 = \$700$, as before.

Proof

$$\begin{aligned} E(aX + b) &= \sum_D (ax + b) \cdot p(x) = a \sum_D x \cdot p(x) + b \sum_D p(x) \\ &= aE(X) + b \end{aligned}$$

■

Two special cases of the proposition yield two important rules of expected value.

1. For any constant a, $E(aX) = a \cdot E(X)$ [take $b = 0$ in (3.12)].
2. For any constant b, $E(X + b) = E(X) + b$ [take $a = 1$ in (3.12)].

Multiplication of X by a constant a changes the unit of measurement (from dollars to cents, where $a = 100$, inches to cm, where $a = 2.54$, etc.). Rule 1 says that the expected value in the new units equals the expected value in the old units multiplied by the conversion factor a. Similarly, if a constant b is added to each possible value of X, then the expected value will be shifted by that same constant amount.

The Variance of X

The expected value of X describes where the probability distribution is centered. Using the physical analogy of placing point mass $p(x)$ at the value x on a one-dimensional axis, if the axis were then supported by a fulcrum placed at μ, there would be no tendency for the axis to tilt. This is illustrated for two different distributions in Figure 3.7.

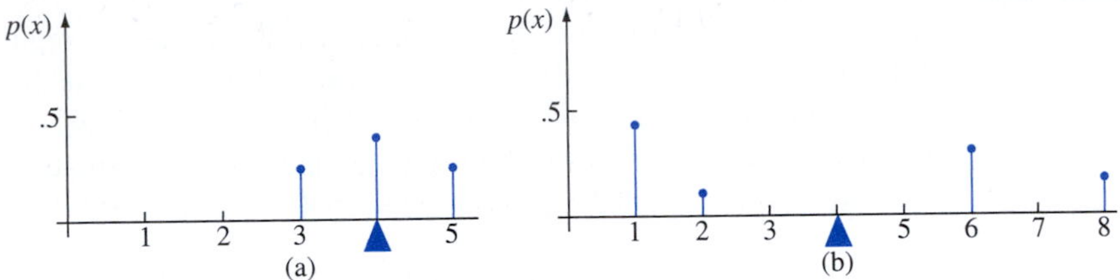

Figure 3.7 Two different probability distributions with $\mu = 4$

Although both distributions pictured in Figure 3.7 have the same center μ, the distribution of Figure 3.7(b) has greater spread or variability or dispersion than does that of Figure 3.7(a). We will use the variance of X to assess the amount of variability in (the distribution of) X, just as s^2 was used in Chapter 1 to measure variability in a sample.

DEFINITION Let X have pmf $p(x)$ and expected value μ. Then the **variance** of X, denoted by $V(X)$ or σ_X^2, or just σ^2, is

$$V(X) = \sum_D (x - \mu)^2 \cdot p(x) = E[(X - \mu)^2]$$

The **standard deviation** (SD) of X is

$$\sigma_X = \sqrt{\sigma_X^2}$$

The quantity $h(X) = (X - \mu)^2$ is the squared deviation of X from its mean, and σ^2 is the expected squared deviation. If most of the probability distribution is close to μ, then σ^2 will typically be relatively small. However, if there are x values far from μ that have large $p(x)$, then σ^2 will be quite large.

Example 3.23 If X is the number of cylinders on the next car to be tuned at a service facility, with pmf as given in Example 3.21 [$p(4) = .5$, $p(6) = .3$, $p(8) = .2$, from which $\mu = 5.4$], then

$$V(X) = \sigma^2 = \sum_{x=4}^{8} (x - 5.4)^2 \cdot p(x)$$
$$= (4 - 5.4)^2(.5) + (6 - 5.4)^2(.3) + (8 - 5.4)^2(.2) = 2.44$$

The standard deviation of X is $\sigma = \sqrt{2.44} = 1.562$. ■

When the pmf $p(x)$ specifies a mathematical model for the distribution of population values, both σ^2 and σ measure the spread of values in the population; σ^2 is the population variance, and σ is the population standard deviation.

A Shortcut Formula for σ^2

The number of arithmetic operations necessary to compute σ^2 can be reduced by using an alternative computing formula.

PROPOSITION

$$V(X) = \sigma^2 = \left[\sum_D x^2 \cdot p(x)\right] - \mu^2 = E(X^2) - [E(X)]^2$$

In using this formula, $E(X^2)$ is computed first without any subtraction; then $E(X)$ is computed, squared, and subtracted (once) from $E(X^2)$.

Example 3.24 The pmf of the number of cylinders X on the next car to be tuned at a certain facility was given in Example 3.23 as $p(4) = .5$, $p(6) = .3$, and $p(8) = .2$, from which $\mu = 5.4$ and

$$E(X^2) = (4^2)(.5) + (6^2)(.3) + (8^2)(.2) = 31.6$$

Thus $\sigma^2 = 31.6 - (5.4)^2 = 2.44$ as in Example 3.23. ■

Proof of the Shortcut Formula Expand $(x - \mu)^2$ in the definition of σ^2 to obtain $x^2 - 2\mu x + \mu^2$, and then carry Σ through to each of the three terms:

$$\begin{aligned}\sigma^2 &= \sum_D x^2 \cdot p(x) - 2\mu \cdot \sum_D x \cdot p(x) + \mu^2 \sum_D p(x) \\ &= E(X^2) - 2\mu \cdot \mu + \mu^2 = E(X^2) - \mu^2\end{aligned}$$

■

Rules of Variance

The variance of $h(X)$ is the expected value of the squared difference between $h(X)$ and its expected value:

$$V[h(X)] = \sigma^2_{h(X)} = \sum_D \{h(x) - E[h(X)]\}^2 \cdot p(x) \tag{3.13}$$

When $h(x)$ is a linear function, $V[h(X)]$ is easily related to $V(X)$ (Exercise 40).

PROPOSITION

$$V(aX + b) = \sigma^2_{aX+b} = a^2 \cdot \sigma^2_X \quad \text{and} \quad \sigma_{aX+b} = |a| \cdot \sigma_X$$

This result says that the addition of the constant b does not affect the variance, which is intuitive, because the addition of b changes the location (mean value) but not the spread of values. In particular,

$$\textbf{1. } \sigma_{aX}^2 = a^2 \cdot \sigma_X^2 \qquad \sigma_{aX} = |a| \cdot \sigma_X \qquad (3.14)$$

$$\textbf{2. } \sigma_{X+b}^2 = \sigma_X^2$$

The reason for the absolute value in σ_{aX} is that a may be negative, whereas a standard deviation cannot be negative; a^2 results when a is brought outside the term being squared in Equation (3.13).

Example 3.25 In the computer sales scenario of Example 3.22, $E(X) = 2$ and

$$E(X^2) = (0)^2(.1) + (1)^2(.2) + (2)^2(.3) + (3)^2(.4) = 5$$

so $V(X) = 5 - (2)^2 = 1$. The profit function $h(X) = 800X - 900$ then has variance $(800)^2 \cdot V(X) = (640{,}000)(1) = 640{,}000$ and standard deviation 800. ■

Exercises Section 3.3 (28–43)

28. The pmf for X = the number of major defects on a randomly selected appliance of a certain type is

x	0	1	2	3	4
$p(x)$	.08	.15	.45	.27	.05

Compute the following:

a. $E(X)$

b. $V(X)$ directly from the definition

c. The standard deviation of X

d. $V(X)$ using the shortcut formula

29. An individual who has automobile insurance from a certain company is randomly selected. Let Y be the number of moving violations for which the individual was cited during the last 3 years. The pmf of Y is

y	0	1	2	3
$p(y)$	.60	.25	.10	.05

a. Compute $E(Y)$.

b. Suppose an individual with Y violations incurs a surcharge of $\$100Y^2$. Calculate the expected amount of the surcharge.

30. Refer to Exercise 12 and calculate $V(Y)$ and σ_Y. Then determine the probability that Y is within 1 standard deviation of its mean value.

31. An appliance dealer sells three different models of upright freezers having 13.5, 15.9, and 19.1 cubic feet of storage space, respectively. Let X = the amount of storage space purchased by the next customer to buy a freezer. Suppose that X has pmf

x	13.5	15.9	19.1
$p(x)$	.2	.5	.3

a. Compute $E(X)$, $E(X^2)$, and $V(X)$.

b. If the price of a freezer having capacity X cubic feet is $25X - 8.5$, what is the expected price paid by the next customer to buy a freezer?

c. What is the variance of the price $25X - 8.5$ paid by the next customer?

d. Suppose that although the rated capacity of a freezer is X, the actual capacity is $h(X) = X - .01X^2$. What is the expected actual capacity of the freezer purchased by the next customer?

32. Let X be a Bernoulli rv with pmf as in Example 3.17.

a. Compute $E(X^2)$.

b. Show that $V(X) = p(1 - p)$.

c. Compute $E(X^{79})$.

33. Suppose that the number of plants of a particular type found in a rectangular region (called a quadrat by ecologists) in a certain geographic area is an rv X with pmf

$$p(x) = \begin{cases} c/x^3 & x = 1, 2, 3, \ldots \\ 0 & \text{otherwise} \end{cases}$$

Is $E(X)$ finite? Justify your answer (this is another distribution that statisticians would call heavy-tailed).

34. A small market orders copies of a certain magazine for its magazine rack each week. Let X = demand for the magazine, with pmf

x	1	2	3	4	5	6
$p(x)$	$\frac{1}{15}$	$\frac{2}{15}$	$\frac{3}{15}$	$\frac{4}{15}$	$\frac{3}{15}$	$\frac{2}{15}$

Suppose the store owner actually pays \$1.00 for each copy of the magazine and the price to customers is \$2.00. If magazines left at the end of the week have no salvage value, is it better to order three or four copies of the magazine? (*Hint*: For both three and four copies ordered, express net revenue as a function of demand X, and then compute the expected revenue.)

35. Let X be the damage incurred (in \$) in a certain type of accident during a given year. Possible X values are 0, 1000, 5000, and 10,000, with probabilities .8, .1, .08, and .02, respectively. A particular company offers a \$500 deductible policy. If the company wishes its expected profit to be \$100, what premium amount should it charge?

36. The n candidates for a job have been ranked 1, 2, 3, . . . , n. Let X = the rank of a randomly selected candidate, so that X has pmf

$$p(x) = \begin{cases} 1/n & x = 1, 2, 3, \ldots, n \\ 0 & \text{otherwise} \end{cases}$$

(this is called the *discrete uniform distribution*). Compute $E(X)$ and $V(X)$ using the shortcut formula. [*Hint*: The sum of the first n positive integers is $n(n + 1)/2$, whereas the sum of their squares is $n(n + 1)(2n + 1)/6$.]

37. Let X = the outcome when a fair die is rolled once. If before the die is rolled you are offered either (1/3.5) dollars or $h(X) = 1/X$ dollars, would you accept the guaranteed amount or would you gamble? [*Note*: It is not generally true that $1/E(X) = E(1/X)$.]

38. A chemical supply company currently has in stock 100 lb of a certain chemical, which it sells to customers in 5-lb containers. Let X = the number of containers ordered by a randomly chosen customer, and suppose that X has pmf

x	1	2	3	4
$p(x)$	.2	.4	.3	.1

Compute $E(X)$ and $V(X)$. Then compute the expected number of pounds left after the next customer's order is shipped and the variance of the number of pounds left. (*Hint*: The number of pounds left is a linear function of X.)

39. **a.** Draw a line graph of the pmf of X in Exercise 34. Then determine the pmf of $-X$ and draw its line graph. From these two pictures, what can you say about $V(X)$ and $V(-X)$?

b. Use the proposition involving $V(aX + b)$ to establish a general relationship between $V(X)$ and $V(-X)$.

40. Use the definition in Expression (3.13) to prove that $V(aX + b) = a^2 \cdot \sigma_X^2$. [*Hint*: With $h(X) = aX + b$, $E[h(X)] = a\mu + b$ where $\mu = E(X)$.]

41. Suppose $E(X) = 5$ and $E[X(X - 1)] = 27.5$. What is

a. $E(X^2)$? [*Hint*: $E[X(X - 1)] = E[X^2 - X] = E(X^2) - E(X)$.]

b. $V(X)$?

c. The general relationship among the quantities $E(X)$, $E[X(X - 1)]$, and $V(X)$?

42. Write a general rule for $E(X - c)$ where c is a constant. What happens when you let $c = \mu$, the expected value of X?

43. A result called **Chebyshev's inequality** states that for any probability distribution of an rv X and any number k that is at least 1, $P(|X - \mu| \geq k\sigma) \leq 1/k^2$. In words, the probability that the value of X lies at least k standard deviations from its mean is at most $1/k^2$.

a. What is the value of the upper bound for $k = 2$? $k = 3$? $k = 4$? $k = 5$? $k = 10$?

b. Compute μ and σ for the distribution of Exercise 13. Then evaluate $P(|X - \mu| \geq k\sigma)$ for the values of k given in part (a). What does this suggest about the upper bound relative to the corresponding probability?

c. Let X have three possible values, -1, 0, and 1, with probabilities $\frac{1}{18}$, $\frac{8}{9}$, and $\frac{1}{18}$, respectively. What is $P(|X - \mu| \geq 3\sigma)$, and how does it compare to the corresponding bound?

d. Give a distribution for which $P(|X - \mu| \geq 5\sigma) = .04$.

3.4 Moments and Moment Generating Functions

Sometimes the expected values of integer powers of X and $X - \mu$ are called **moments**, terminology borrowed from physics. Expected values of powers of X are called **moments about 0** and powers of $X - \mu$ are called **moments about the mean**. For example, $E(X^2)$ is the second moment about 0, and $E[(X - \mu)^3]$ is the third moment about the mean. Moments about 0 are sometimes simply called moments.

Example 3.26 Suppose the pmf of X, the number of points earned on a short quiz, is given by

x	0	1	2	3
$p(x)$	.1	.2	.3	.4

The first moment about 0 is the mean

$$\mu = E(X) = \sum_{x \in D} xp(x) = 0(.1) + 1(.2) + 2(.3) + 3(.4) = 2$$

The second moment about the mean is the variance

$$\begin{aligned} V(X) = \sigma^2 = E[(X - \mu)^2] &= \sum_{x \in D} (x - \mu)^2 p(x) \\ &= (0 - 2)^2(.1) + (1 - 2)^2(.2) + (2 - 2)^2(.3) + (3 - 2)^2(.4) = 1 \end{aligned}$$

The third moment about the mean is also important.

$$\begin{aligned} E[(X - \mu)^3] &= \sum_{x \in D} (x - \mu)^3 p(x) \\ &= (0 - 2)^3(.1) + (1 - 2)^3(.2) + (2 - 2)^3(.3) + (3 - 2)^3(.4) = -.6 \end{aligned}$$

We would like to use this as a measure of lack of symmetry, but $E[(X - \mu)^3]$ depends on the scale of measurement. That is, if X is measured in feet, the value is different from what would be obtained if X were measured in inches. Scale independence results from dividing the third moment about the mean by σ^3:

$$\frac{E[(X - \mu)^3]}{\sigma^3} = E\left[\left(\frac{X - \mu}{\sigma}\right)^3\right]$$

This is our measure of departure from symmetry, called the **skewness**. For a symmetric distribution the third moment about the mean would be 0, so the skewness in that case is 0. However, in the present example the skewness is $E[(X - \mu)^3]/\sigma^3 = -.6/1 = -.6$. When the skewness is negative, as it is here, we say that the distribution is negatively skewed or that it is skewed to the left. Generally speaking, it means that the distribution stretches farther to the left of the mean than to the right.

If the skewness were positive then we would say that the distribution is positively skewed or that it is skewed to the right. For example, suppose that $p(x)$ is reversed relative to the table above, so $p(x)$ is given by

x	0	1	2	3
$p(x)$	.4	.3	.2	.1

In Exercise 57 you are asked to show that this changes the sign of the skewness, so it becomes .6, and the distribution is skewed to the right. ■

Moments are not always easy to obtain, as shown by the calculation of $E(X)$ in Example 3.18. We now introduce the **moment generating function**, which will help in the calculation of moments and the understanding of statistical distributions. We have already discussed the expected value of a function, $E[h(X)]$. In particular, let e denote the base of the natural logarithms, with approximate value 2.71828. Then we may wish to calculate $E(e^{2X}) = \Sigma e^{2x}p(x)$, $E(e^{3.75X})$, or $E(e^{-2.56X})$. That is, for any particular number t, the expected value $E(e^{tX})$ is meaningful. When we consider this expected value as a function of t, the result is called the moment generating function.

DEFINITION

The moment generating function (mgf) of a discrete random variable X is defined to be

$$M_X(t) = E(e^{tX}) = \sum_{x \in D} e^{tx}p(x)$$

where D is the set of possible X values. We will say that the moment generating function exists if $M_X(t)$ is defined for an interval of numbers that includes zero as well as positive and negative values of t (an interval including 0 in its interior).

If the mgf exists, it will be defined on a symmetric interval of the form $(-t_0, t_0)$, where $t_0 > 0$, because t_0 can be chosen small enough so the symmetric interval is contained in the interval of the definition.

When $t = 0$, for any random variable X

$$M_X(0) = E(e^{0X}) = \sum_{x \in D} e^{0x}p(x) = \sum_{x \in D} 1p(x) = 1$$

See also Example 3.30. That is, $M_X(0)$ is the sum of all the probabilities, so it must always be 1. However, in order for the mgf to be useful in generating moments, it will need to be defined for an interval of values of t including 0 in its interior, and that is why we do not bother with the mgf otherwise. As you might guess, the moment generating function fails to exist in cases when moments themselves fail to exist, as in Example 3.19. See Example 3.30 below.

The simplest example of an mgf is for a Bernoulli distribution, where only the X values 0 and 1 receive positive probability.

Example 3.27 Let X be a Bernoulli random variable with $p(0) = \frac{1}{3}$ and $p(1) = \frac{2}{3}$. Then

$$M_X(t) = E(e^{tX}) = \sum_{x \in D} e^{tx}p(x) = e^{t \cdot 0}\frac{1}{3} + e^{t \cdot 1}\frac{2}{3} = \frac{1}{3} + e^{t}\frac{2}{3}$$

It should be clear that a Bernoulli random variable will always have an mgf of the form $p(0) + p(1)e^{t}$. This mgf exists because it is defined for all t. ■

The idea of the mgf is to have an alternate view of the distribution based on an infinite number of values of t. That is, the mgf for X is a function of t, and we get a different function for each different distribution. When the function is of the form of one constant plus another constant times e^t, we know that it corresponds to a Bernoulli random variable, and the constants tell us the probabilities. This is an example of the following "uniqueness property."

PROPOSITION If the mgf exists and is the same for two distributions, then the two distributions are the same. That is, the moment generating function uniquely specifies the probability distribution; there is a one-to-one correspondence between distributions and mgf's.

Example 3.28 Let X be the number of claims in a year by someone holding an automobile insurance policy with a company. The mgf for X is $M_X(t) = .7 + .2e^t + .1e^{2t}$. Then we can say that the pmf of X is given by

x	0	1	2
$p(x)$	.7	.2	.1

Why? If we compute $E(e^{tX})$ based on this table, we get the correct mgf. Because X and the random variable described by the table have the same mgf, the uniqueness property requires them to have the same distribution. Therefore, X has the given pmf. ■

Example 3.29 This is a continuation of Example 3.18, except that here we do not consider the number of births needed to produce a male child. Instead we are looking for a person whose blood type is Rh+. Set $p = .85$, which is the approximate probability that a random person has blood type Rh+. If X is the number of people we need to check until we find someone who is Rh+, then $p(x) = p(1 - p)^{x-1} = .85(.15)^{x-1}$ for $x = 1, 2, 3, \ldots$.

Determination of the moment generating function here requires using the formula for the sum of a geometric series:

$$a + ar + ar^2 + \cdots = \frac{a}{1 - r}$$

where a is the first term, r is the ratio of successive terms, and $|r| < 1$. The moment generating function is

$$M_X(t) = E(e^{tX}) = \sum_{x=1}^{\infty} e^{tx}.85(.15)^{x-1} = .85e^t \sum_{x=1}^{\infty} e^{t(x-1)}(.15)^{x-1}$$

$$= .85e^t \sum_{x=1}^{\infty} [e^t(.15)]^{x-1} = \frac{.85e^t}{1 - .15e^t}$$

The condition on r requires $|.15e^t| < 1$. Dividing by .15 and taking logs, this gives $t < -\ln(.15) \approx 1.90$. The result is an interval of values that includes 0 in its interior, so the mgf exists.

What about the value of the mgf at 0? Recall that $M_X(0) = 1$ always, because the value at 0 amounts to summing the probabilities. As a check, after computing an mgf we should make sure that this condition is satisfied. Here $M_X(0) = .85/(1 - .15) = 1$. ■

Example 3.30 Reconsider Example 3.19, where $p(x) = k/x^2$, $x = 1, 2, 3, \ldots$. Recall that $E(X)$ does not exist, so there might be problems with the mgf, too:

$$M_X(t) = E(e^{tX}) = \sum_{x=1}^{\infty} e^{tx}\frac{1}{x^2}$$

With the help of tests for convergence such as the ratio test, we find that the series converges if and only if $e^t \le 1$, which means that $t \le 0$. Because zero is on the boundary of this interval, not the interior of the interval (the interval must include both positive and negative values), this mgf does not exist. Of course, it could not be useful for finding moments, because X does not have even a first moment (mean). ■

How does the mgf produce moments? We will need various derivatives of $M_X(t)$. For any positive integer r, let $M_X^{(r)}(t)$ denote the rth derivative of $M_X(t)$. By computing this and then setting $t = 0$, we get the rth moment about 0.

THEOREM If the mgf exists,

$$E(X^r) = M_X^{(r)}(0)$$

Proof We show that the theorem is true for $r = 1$ and $r = 2$. A proof by mathematical induction can be used for general r. Differentiate

$$\frac{d}{dt}M_X(t) = \frac{d}{dt}\sum_{x\in D} e^{xt}p(x) = \sum_{x\in D}\frac{d}{dt}e^{xt}p(x) = \sum_{x\in D} xe^{xt}p(x)$$

where we have interchanged the order of summation and differentiation. This is justified inside the interval of convergence, which includes 0 in its interior. Next we set $t = 0$ and get the first moment

$$M_X'(0) = M_X^{(1)}(0) = \sum_{x\in D} xp(x) = E(X)$$

Differentiate again:

$$\frac{d^2}{dt^2}M_X(t) = \frac{d}{dt}\sum_{x\in D} xe^{xt}p(x) = \sum_{x\in D} x\frac{d}{dt}e^{xt}p(x) = \sum_{x\in D} x^2e^{xt}p(x)$$

Set $t = 0$ to get the second moment

$$M_X''(0) = M_X^{(2)}(0) = \sum_{x\in D} x^2p(x) = E(X^2)$$ ■

Example 3.31 This is a continuation of Example 3.28, where X represents the number of claims in a year with pmf and mgf

x	0	1	2
$p(x)$	.7	.2	.1

$$M_X(t) = .7 + .2e^t + .1e^{2t}$$

First, find the derivatives

$$M'_X(t) = .2e^t + .1(2)e^{2t}$$
$$M''_X(t) = .2e^t + .1(2)(2)e^{2t}$$

Setting t to 0 in the first derivative gives the first moment

$$E(X) = M'_X(0) = M_X^{(1)}(0) = .2e^0 + .1(2)e^{2(0)} = .2 + .1(2) = .4$$

Setting t to 0 in the second derivative gives the second moment

$$E(X^2) = M''_X(0) = M_X^{(2)}(0) = .2e^0 + .1(2)(2)e^{2(0)} = .2 + .1(2)(2) = .6$$

To get the variance recall the shortcut formula from the previous section:

$$V(X) = \sigma^2 = E(X^2) - [E(X)]^2 = .6 - .4^2 = .6 - .16 = .44$$

Taking the square root gives $\sigma = .66$ approximately. Do a mean of .4 and a standard deviation of .66 seem about right for a distribution concentrated mainly on 0 and 1? ■

Example 3.32 (Example 3.29 continued) Recall that $p = .85$ is the probability of a person having Rh+ blood and we keep checking people until we find one with this blood type. If X is the number of people we need to check, then $p(x) = .85(.15)^{x-1}$, $x = 1, 2, 3, \ldots$, and the mgf is

$$M_X(t) = E(e^{tX}) = \frac{.85e^t}{1 - .15e^t}$$

Differentiating with the help of the quotient rule,

$$M'_X(t) = \frac{.85e^t}{(1 - .15e^t)^2}$$

Setting $t = 0$,

$$\mu = E(X) = M'_X(0) = \frac{1}{.85}$$

Recalling that .85 corresponds to p, we see that this agrees with Example 3.18.

To get the second moment, differentiate again:

$$M''_X(t) = \frac{.85e^t(1 + .15e^t)}{(1 - .15e^t)^3}$$

Setting $t = 0$,

$$E(X^2) = M''_X(0) = \frac{1.15}{.85^2}$$

Now use the shortcut formula for the variance from the previous section:

$$V(X) = \sigma^2 = E(X^2) - [E(X)]^2 = \frac{1.15}{.85^2} - \frac{1}{.85^2} = \frac{.15}{.85^2} = .2076$$ ■

There is an alternate way of doing the differentiation that can sometimes make the effort easier. Define $R_X(t) = \ln[M_X(t)]$, where $\ln(u)$ is the natural log of u. In Exercise 54 you are requested to verify that if the moment generating function exists,

$$\mu = E(X) = R'_X(0)$$
$$\sigma^2 = V(X) = R''_X(0)$$

Example 3.33 Here we apply $R_X(t)$ to Example 3.32. Using $\ln(e^t) = t$,

$$R_X(t) = \ln[M_X(t)] = \ln\left(\frac{.85e^t}{1 - .15e^t}\right) = \ln .85 + t - \ln(1 - .15e^t)$$

The first derivative is

$$R'_X(t) = \frac{1}{1 - .15e^t}$$

and the second derivative is

$$R''_X(t) = \frac{.15e^t}{(1 - .15e^t)^2}$$

Setting t to 0 gives

$$\mu = E(X) = R'_X(0) = \frac{1}{.85}$$
$$\sigma^2 = V(X) = R''_X(0) = \frac{.15}{.85^2}$$

These are in agreement with the results of Example 3.32. ■

As mentioned at the end of the previous section, it is common to transform X using a linear function $Y = aX + b$. What happens to the mgf when we do this?

PROPOSITION Let X have the mgf $M_X(t)$ and let $Y = aX + b$. Then $M_Y(t) = e^{bt}M_X(at)$.

Example 3.34 Let X be a Bernoulli random variable with $p(0) = \frac{20}{38}$ and $p(1) = \frac{18}{38}$. Think of X as the number of wins, 0 or 1, in a single play of roulette. If you play roulette at an American casino and you bet red, then your chances of winning are $\frac{18}{38}$ because 18 of the 38 possible outcomes are red. Then from Example 3.27 $M_X(t) = (\frac{20}{38}) + e^t(\frac{18}{38})$. Let your bet be \$5 and let Y be your winnings. If $X = 0$ then $Y = -5$, and if $X = 1$ then $Y = +5$. The linear equation $Y = 10X - 5$ gives the appropriate relationship, as shown in the table.

$p(x)$	x	$y = 10x - 5$
$\frac{20}{38}$	0	−5
$\frac{18}{38}$	1	5

The equation is of the form $Y = aX + b$ with $a = 10$ and $b = -5$, so by the proposition

$$M_Y(t) = e^{bt}M_X(at) = e^{-5t}M_X(10t)$$
$$= e^{-5t}\left[\left(\frac{20}{38}\right) + e^{10t}\left(\frac{18}{38}\right)\right] = e^{-5t}\left(\frac{20}{38}\right) + e^{5t}\left(\frac{18}{38}\right)$$

From this we can read off the probabilities for Y: $p(-5) = \frac{20}{38}$ and $p(5) = \frac{18}{38}$. ■

Exercises Section 3.4 (44–57)

44. For a new car the number of defects X has the distribution given by the accompanying table. Find $M_X(t)$ and use it to find $E(X)$ and $V(X)$.

x	0	1	2	3	4	5	6
$p(x)$	.04	.20	.34	.20	.15	.04	.03

45. In flipping a fair coin let X be the number of tosses to get the first head. Then $p(x) = .5^x$ for $x = 1, 2, 3, \ldots$. Find $M_X(t)$ and use it to get $E(X)$ and $V(X)$.

46. Given $M_X(t) = .2 + .3e^t + .5e^{3t}$, find $p(x)$, $E(X)$, $V(X)$.

47. Using a calculation similar to the one in Example 3.29 show that, if X has the distribution of Example 3.18, then its mgf is

$$M_X(t) = \frac{pe^t}{1 - (1 - p)e^t}$$

Assuming that Y has mgf $M_Y(t) = .75e^t/(1 - .25e^t)$, determine the probability mass function $p_Y(y)$ with the help of the uniqueness property.

48. Let X have the moment generating function of Example 3.29 and let $Y = X - 1$. Recall that X is the number of people who need to be checked to get someone who is Rh+, so Y is the number of people checked before the first Rh+ person is found. Find $M_Y(t)$ using the second proposition.

49. If $M_X(t) = e^{5t+2t^2}$ then find $E(X)$ and $V(X)$ by differentiating

a. $M_X(t)$
b. $R_X(t)$

50. Prove the result in the second proposition, $M_{aX+b}(t) = e^{bt}M_X(at)$.

51. Let $M_X(t) = e^{5t+2t^2}$ and let $Y = (X - 5)/2$. Find $M_Y(t)$ and use it to find $E(Y)$ and $V(Y)$.

52. If you toss a fair die with outcome X, $p(x) = \frac{1}{6}$ for $x = 1, 2, 3, 4, 5, 6$. Find $M_X(t)$.

53. If $M_X(t) = 1/(1-t^2)$, find $E(X)$ and $V(X)$ by differentiating $M_X(t)$.

54. Prove that the mean and variance are obtainable from $R_X(t) = \ln(M_X(t))$:

$$\mu = E(X) = R'_X(0)$$
$$\sigma^2 = V(X) = R''_X(0)$$

55. Show that $g(t) = te^t$ cannot be a moment generating function.

56. If $M_X(t) = e^{5(e^t-1)}$ then find $E(X)$ and $V(X)$ by differentiating

a. $M_X(t)$
b. $R_X(t)$

57. Let X have the following distribution. Show that the skewness is .6.

x	0	1	2	3
$p(x)$	.4	.3	.2	.1

3.5 The Binomial Probability Distribution

Many experiments conform either exactly or approximately to the following list of requirements:

1. The experiment consists of a sequence of n smaller experiments called *trials*, where n is fixed in advance of the experiment.
2. Each trial can result in one of the same two possible outcomes (dichotomous trials), which we denote by success (S) or failure (F).
3. The trials are independent, so that the outcome on any particular trial does not influence the outcome on any other trial.
4. The probability of success is constant from trial to trial; we denote this probability by p.

DEFINITION An experiment for which Conditions 1–4 are satisfied is called **a binomial experiment**.

Example 3.35 The same coin is tossed successively and independently n times. We arbitrarily use S to denote the outcome H (heads) and F to denote the outcome T (tails). Then this experiment satisfies Conditions 1–4. Tossing a thumbtack n times, with $S =$ point up and $F =$ point down, also results in a binomial experiment. ■

Some experiments involve a sequence of independent trials for which there are more than two possible outcomes on any one trial. A binomial experiment can then be created by dividing the possible outcomes into two groups.

Example 3.36 The color of pea seeds is determined by a single genetic locus. If the two alleles at this locus are AA or Aa (the genotype), then the pea will be yellow (the phenotype), and if the allele is aa, the pea will be green. Suppose we pair off 20 Aa seeds and cross the two seeds in each of the ten pairs to obtain ten new genotypes. Call each new genotype a success S if it is aa and a failure otherwise. Then with this identification of S and F, the experiment is binomial with $n = 10$ and $p = P(\text{aa genotype})$. If each member of the pair is equally likely to contribute a or A, then $p = P(\text{a}) \cdot P(\text{a}) = (\frac{1}{2})(\frac{1}{2}) = \frac{1}{4}$. ■

Example 3.37 Suppose a certain city has 50 licensed restaurants, of which 15 currently have at least one serious health code violation and the other 35 have no serious violations. There are five inspectors, each of whom will inspect one restaurant during the coming week. The name of each restaurant is written on a different slip of paper, and after the slips are thoroughly mixed, each inspector in turn draws one of the slips *without replacement.* Label the ith trial as a success if the ith restaurant selected ($i = 1, \ldots, 5$) has no serious violations. Then

$$P(S \text{ on first trial}) = \frac{35}{50} = .70$$

and

$$\begin{aligned} P(S \text{ on second trial}) &= P(SS) + P(FS) \\ &= P(\text{second } S \mid \text{first } S)\, P(\text{first } S) + P(\text{second } S \mid \text{first } F)\, P(\text{first } F) \\ &= \frac{34}{49} \cdot \frac{35}{50} + \frac{35}{49} \cdot \frac{15}{50} = \frac{35}{50}\left(\frac{34}{49} + \frac{15}{49}\right) = \frac{35}{50} = .70 \end{aligned}$$

Similarly, it can be shown that $P(S \text{ on } i\text{th trial}) = .70$ for $i = 3, 4, 5$. However,

$$P(S \text{ on fifth trial} \mid SSSS) = \frac{31}{46} = .67$$

whereas

$$P(S \text{ on fifth trial} \mid FFFF) = \frac{35}{46} = .76$$

The experiment is not binomial because the trials are not independent. In general, if sampling is without replacement, the experiment will not yield independent trials. If each slip had been replaced after being drawn, then trials would have been independent, but this might have resulted in the same restaurant being inspected by more than one inspector. ■

Example 3.38 Suppose a certain state has 500,000 licensed drivers, of whom 400,000 are insured. A sample of 10 drivers is chosen without replacement. The ith trial is labeled S if the ith driver chosen is insured. Although this situation would seem identical to that of Example 3.37, the important difference is that the size of the population being sampled is very large relative to the sample size. In this case

$$P(S \text{ on } 2 \mid S \text{ on } 1) = \frac{399{,}999}{499{,}999} = .80000$$

and

$$P(S \text{ on } 10 \mid S \text{ on first } 9) = \frac{399{,}991}{499{,}991} = .799996 \approx .80000$$

These calculations suggest that although the trials are not exactly independent, the conditional probabilities differ so slightly from one another that for practical purposes the trials can be regarded as independent with constant $P(S) = .8$. Thus, to a very good approximation, the experiment is binomial with $n = 10$ and $p = .8$. ■

We will use the following rule of thumb in deciding whether a "without-replacement" experiment can be treated as a binomial experiment.

RULE Consider sampling without replacement from a dichotomous population of size N. If the sample size (number of trials) n is at most 5% of the population size, the experiment can be analyzed as though it were exactly a binomial experiment.

By "analyzed," we mean that probabilities based on the binomial experiment assumptions will be quite close to the actual "without-replacement" probabilities, which are typically more difficult to calculate. In Example 3.37, $n/N = 5/50 = .1 > .05$, so the binomial experiment is not a good approximation, but in Example 3.38, $n/N = 10/500{,}000 < .05$.

The Binomial Random Variable and Distribution

In most binomial experiments, it is the total number of S's, rather than knowledge of exactly which trials yielded S's, that is of interest.

DEFINITION Given a binomial experiment consisting of n trials, the **binomial random variable X** associated with this experiment is defined as

$$X = \text{the number of } S\text{'s among the } n \text{ trials}$$

Suppose, for example, that $n = 3$. Then there are eight possible outcomes for the experiment:

SSS SSF SFS SFF FSS FSF FFS FFF

From the definition of X, $X(SSF) = 2$, $X(SFF) = 1$, and so on. Possible values for X in an n-trial experiment are $x = 0, 1, 2, \ldots, n$. We will often write $X \sim \text{Bin}(n, p)$ to indicate that X is a binomial rv based on n trials with success probability p.

NOTATION Because the pmf of a binomial rv X depends on the two parameters n and p, we denote the pmf by $b(x; n, p)$.

Consider first the case $n = 4$ for which each outcome, its probability, and corresponding x value are listed in Table 3.1. For example,

$$\begin{aligned} P(SSFS) &= P(S) \cdot P(S) \cdot P(F) \cdot P(S) && \text{(independent trials)} \\ &= p \cdot p \cdot (1 - p) \cdot p && [\text{constant } P(S)] \\ &= p^3 \cdot (1 - p) \end{aligned}$$

Table 3.1 Outcomes and probabilities for a binomial experiment with four trials

Outcome	x	Probability	Outcome	x	Probability
SSSS	4	p^4	*FSSS*	3	$p^3(1-p)$
SSSF	3	$p^3(1-p)$	*FSSF*	2	$p^2(1-p)^2$
SSFS	3	$p^3(1-p)$	*FSFS*	2	$p^2(1-p)^2$
SSFF	2	$p^2(1-p)^2$	*FSFF*	1	$p(1-p)^3$
SFSS	3	$p^3(1-p)$	*FFSS*	2	$p^2(1-p)^2$
SFSF	2	$p^2(1-p)^2$	*FFSF*	1	$p(1-p)^3$
SFFS	2	$p^2(1-p)^2$	*FFFS*	1	$p(1-p)^3$
SFFF	1	$p(1-p)^3$	*FFFF*	0	$(1-p)^4$

In this special case, we wish $b(x; 4, p)$ for $x = 0, 1, 2, 3$, and 4. For $b(3; 4, p)$, we identify which of the 16 outcomes yield an x value of 3 and sum the probabilities associated with each such outcome:

$$b(3; 4, p) = P(FSSS) + P(SFSS) + P(SSFS) + P(SSSF) = 4p^3(1 - p)$$

There are four outcomes with $x = 3$ and each has probability $p^3(1 - \mathrm{p})$ (the probability depends only on the number of S's, *not* the order of S's and F's), so

$$b(3; 4, p) = \left\{\begin{matrix}\text{number of outcomes} \\ \text{with } X = 3\end{matrix}\right\} \cdot \left\{\begin{matrix}\text{probability of any particular} \\ \text{outcome with } X = 3\end{matrix}\right\}$$

Similarly, $b(2; 4, p) = 6p^2(1 - p)^2$, which is also the product of the number of outcomes with $X = 2$ and the probability of any such outcome.

In general,

$$b(x; n, p) = \left\{\begin{matrix}\text{number of sequences of} \\ \text{length } n \text{ consisting of } x\ S\text{'s}\end{matrix}\right\} \cdot \left\{\begin{matrix}\text{probability of any} \\ \text{particular such sequence}\end{matrix}\right\}$$

Since the ordering of S's and F's is not important, the second factor in the previous equation is $p^x(1 - p)^{n-x}$ (e.g., the first x trials resulting in S and the last $n - x$ resulting in F). The first factor is the number of ways of choosing x of the n trials to be S's—that is, the number of combinations of size x that can be constructed from n distinct objects (trials here).

THEOREM

$$b(x; n, p) = \begin{cases} \binom{n}{x} p^x(1 - p)^{n-x} & x = 0, 1, 2, \ldots, n \\ 0 & \text{otherwise} \end{cases}$$

Example 3.39 Each of six randomly selected cola drinkers is given a glass containing cola S and one containing cola F. The glasses are identical in appearance except for a code on the bottom to identify the cola. Suppose there is actually no tendency among cola drinkers to prefer one cola to the other. Then $p = P(\text{a selected individual prefers } S) = .5$, so with $X =$ the number among the six who prefer S, $X \sim \text{Bin}(6, .5)$.

Thus

$$P(X = 3) = b(3; 6, .5) = \binom{6}{3}(.5)^3(.5)^3 = 20(.5)^6 = .313$$

The probability that at least three prefer S is

$$P(3 \le X) = \sum_{x=3}^{6} b(x; 6, .5) = \sum_{x=3}^{6} \binom{6}{x}(.5)^x(.5)^{6-x} = .656$$

and the probability that at most one prefers S is

$$P(X \le 1) = \sum_{x=0}^{1} b(x; 6, .5) = .109$$

■

Using Binomial Tables

Even for a relatively small value of n, the computation of binomial probabilities can be tedious. Appendix Table A.1 tabulates the cdf $F(x) = P(X \le x)$ for $n = 5, 10, 15, 20, 25$ in combination with selected values of p. Various other probabilities can then be calculated using the proposition on cdf's from Section 3.2.

NOTATION For $X \sim \text{Bin}(n, p)$, the cdf will be denoted by

$$P(X \le x) = B(x; n, p) = \sum_{y=0}^{x} b(y; n, p) \qquad x = 0, 1, \ldots, n$$

Example 3.40 Suppose that 20% of all copies of a particular textbook fail a certain binding strength test. Let X denote the number among 15 randomly selected copies that fail the test. Then X has a binomial distribution with $n = 15$ and $p = .2$.

1. The probability that at most 8 fail the test is

$$P(X \le 8) = \sum_{y=0}^{8} b(y; 15, .2) = B(8; 15, .2)$$

which is the entry in the $x = 8$ row and the $p = .2$ column of the $n = 15$ binomial table. From Appendix Table A.1, the probability is $B(8; 15, .2) = .999$.

2. The probability that exactly 8 fail is

$$P(X = 8) = P(X \le 8) - P(X \le 7) = B(8; 15, .2) - B(7; 15, .2)$$

which is the difference between two consecutive entries in the $p = .2$ column. The result is $.999 - .996 = .003$.

3. The probability that at least 8 fail is

$$\begin{aligned} P(X \ge 8) &= 1 - P(X \le 7) = 1 - B(7; 15, .2) \\ &= 1 - \begin{pmatrix} \text{entry in } x = 7 \text{ row} \\ \text{of } p = .2 \text{ column} \end{pmatrix} \\ &= 1 - .996 = .004 \end{aligned}$$

4. Finally, the probability that between 4 and 7, inclusive, fail is

$$\begin{aligned} P(4 \le X \le 7) &= P(X = 4, 5, 6, \text{ or } 7) = P(X \le 7) - P(X \le 3) \\ &= B(7; 15, .2) - B(3; 15, .2) = .996 - .648 = .348 \end{aligned}$$

Notice that this latter probability is the difference between entries in the $x = 7$ and $x = 3$ rows, *not* the $x = 7$ and $x = 4$ rows. ■

Example 3.41 An electronics manufacturer claims that at most 10% of its power supply units need service during the warranty period. To investigate this claim, technicians at a testing laboratory purchase 20 units and subject each one to accelerated testing to simulate use during the warranty period. Let p denote the probability that a power supply unit needs repair during the period (the proportion of all such units that need repair). The laboratory technicians must decide whether the data resulting from the experiment supports the claim that $p \leq .10$. Let X denote the number among the 20 sampled that need repair, so $X \sim \text{Bin}(20, p)$. Consider the decision rule

> Reject the claim that $p \leq .10$ in favor of the conclusion that $p > .10$ if $x \geq 5$ (where x is the observed value of X), and consider the claim plausible if $x \leq 4$.

The probability that the claim is rejected when $p = .10$ (an incorrect conclusion) is

$$P(X \geq 5 \text{ when } p = .10) = 1 - B(4; 20, .1) = 1 - .957 = .043$$

The probability that the claim is not rejected when $p = .20$ (a different type of incorrect conclusion) is

$$P(X \leq 4 \text{ when } p = .2) = B(4; 20, .2) = .630$$

The first probability is rather small, but the second is intolerably large. When $p = .20$, so that the manufacturer has grossly understated the percentage of units that need service, and the stated decision rule is used, 63% of all samples will result in the manufacturer's claim being judged plausible!

One might think that the probability of this second type of erroneous conclusion could be made smaller by changing the cutoff value 5 in the decision rule to something else. However, although replacing 5 by a smaller number would yield a probability smaller than .630, the other probability would then increase. The only way to make both "error probabilities" small is to base the decision rule on an experiment involving many more units. ■

Note that a table entry of 0 signifies only that a probability is 0 to three significant digits, for all entries in the table are actually positive. Statistical computer packages such as MINITAB will generate either $b(x; n, p)$ or $B(x; n, p)$ once values of n and p are specified. In Chapter 4, we will present a method for obtaining quick and accurate approximations to binomial probabilities when n is large.

The Mean and Variance of X

For $n = 1$, the binomial distribution becomes the Bernoulli distribution. From Example 3.17, the mean value of a Bernoulli variable is $\mu = p$, so the expected number of S's on any single trial is p. Since a binomial experiment consists of n trials, intuition suggests that for $X \sim \text{Bin}(n, p)$, $E(X) = np$, the product of the number of trials and the probability of success on a single trial. The expression for $V(X)$ is not so intuitive.

PROPOSITION If $X \sim \text{Bin}(n, p)$, then $E(X) = np$, $V(X) = np(1 - p) = npq$, and $\sigma_X = \sqrt{npq}$ (where $q = 1 - p$).

Thus, calculating the mean and variance of a binomial rv does not necessitate evaluating summations. The proof of the result for $E(X)$ is sketched in Exercise 74, and both the mean and the variance are obtained below using the moment generating function.

Example 3.42 If 75% of all purchases at a certain store are made with a credit card and X is the number among ten randomly selected purchases made with a credit card, then $X \sim \text{Bin}(10, .75)$. Thus $E(X) = np = (10)(.75) = 7.5$, $V(X) = npq = 10(.75)(.25) = 1.875$, and $\sigma = \sqrt{1.875}$. Again, even though X can take on only integer values, $E(X)$ need not be an integer. If we perform a large number of independent binomial experiments, each with $n = 10$ trials and $p = .75$, then the average number of S's per experiment will be close to 7.5. ■

The Moment Generating Function of X

Let's find the moment generating function of a binomial random variable. Using the definition, $M_X(t) = E(e^{tX})$,

$$\begin{aligned} M_X(t) &= \sum_{x \in D} e^{tx} p(x) = \sum_{x=0}^{n} e^{tx} \binom{n}{x} p^x (1-p)^{n-x} \\ &= \sum_{x=0}^{n} \binom{n}{x} (pe^t)^x (1-p)^{n-x} = (pe^t + 1 - p)^n \end{aligned}$$

Here we have used the binomial theorem, $\sum_{x=0}^{n} a^x b^{n-x} = (a+b)^n$.

Notice that the mgf satisfies the property required of all moment generating functions, $M_X(0) = 1$, because the sum of the probabilities is 1. The mean and variance can be obtained by differentiating $M_X(t)$:

$$M'_X(t) = n(pe^t + 1 - p)^{n-1} pe^t \quad \text{and} \quad \mu = M'_X(0) = np$$

Then the second derivative is

$$M''_X(t) = n(n-1)(pe^t + 1 - p)^{n-2} pe^t pe^t + n(pe^t + 1 - p)^{n-1} pe^t$$

and

$$E(X^2) = M''_X(0) = n(n-1)p^2 + np$$

Therefore,

$$\begin{aligned} \sigma^2 = V(X) &= E(X^2) - [E(X)]^2 \\ &= n(n-1)p^2 + np - n^2p^2 = np - np^2 = np(1-p) \end{aligned}$$

in accord with the foregoing proposition.

Exercises Section 3.5 (58–79)

58. Compute the following binomial probabilities directly from the formula for $b(x; n, p)$:
a. $b(3; 8, .6)$
b. $b(5; 8, .6)$
c. $P(3 \le X \le 5)$ when $n = 8$ and $p = .6$
d. $P(1 \le X)$ when $n = 12$ and $p = .1$

59. Use Appendix Table A.1 to obtain the following probabilities:
a. $B(4; 10, .3)$
b. $b(4; 10, .3)$
c. $b(6; 10, .7)$
d. $P(2 \le X \le 4)$ when $X \sim \text{Bin}(10, .3)$
e. $P(2 \le X)$ when $X \sim \text{Bin}(10, .3)$
f. $P(X \le 1)$ when $X \sim \text{Bin}(10, .7)$
g. $P(2 < X < 6)$ when $X \sim \text{Bin}(10, .3)$

60. When circuit boards used in the manufacture of compact disc players are tested, the long-run percentage of defectives is 5%. Let X = the number of defective boards in a random sample of size $n = 25$, so $X \sim \text{Bin}(25, .05)$.
a. Determine $P(X \le 2)$.
b. Determine $P(X \ge 5)$.
c. Determine $P(1 \le X \le 4)$.
d. What is the probability that none of the 25 boards is defective?
e. Calculate the expected value and standard deviation of X.

61. A company that produces fine crystal knows from experience that 10% of its goblets have cosmetic flaws and must be classified as "seconds."
a. Among six randomly selected goblets, how likely is it that only one is a second?
b. Among six randomly selected goblets, what is the probability that at least two are seconds?
c. If goblets are examined one by one, what is the probability that at most five must be selected to find four that are not seconds?

62. Suppose that only 25% of all drivers come to a complete stop at an intersection having flashing red lights in all directions when no other cars are visible. What is the probability that, of 20 randomly chosen drivers coming to an intersection under these conditions,
a. At most 6 will come to a complete stop?
b. Exactly 6 will come to a complete stop?
c. At least 6 will come to a complete stop?
d. How many of the next 20 drivers do you expect to come to a complete stop?

63. Exercise 29 (Section 3.3) gave the pmf of Y, the number of traffic citations for a randomly selected individual insured by a particular company. What is the probability that among 15 randomly chosen such individuals
a. At least 10 have no citations?
b. Fewer than half have at least one citation?
c. The number that have at least one citation is between 5 and 10, inclusive?*

64. A particular type of tennis racket comes in a midsize version and an oversize version. Sixty percent of all customers at a certain store want the oversize version.
a. Among ten randomly selected customers who want this type of racket, what is the probability that at least six want the oversize version?
b. Among ten randomly selected customers, what is the probability that the number who want the oversize version is within 1 standard deviation of the mean value?
c. The store currently has seven rackets of each version. What is the probability that all of the next ten customers who want this racket can get the version they want from current stock?

65. Twenty percent of all telephones of a certain type are submitted for service while under warranty. Of these, 60% can be repaired, whereas the other 40% must be replaced with new units. If a company purchases ten of these telephones, what is the probability that exactly two will end up being replaced under warranty?

66. The College Board reports that 2% of the 2 million high school students who take the SAT each year receive special accommodations because of documented disabilities (*Los Angeles Times*, July 16, 2002). Consider a random sample of 25 students who have recently taken the test.
a. What is the probability that exactly 1 received a special accommodation?
b. What is the probability that at least 1 received a special accommodation?
c. What is the probability that at least 2 received a special accommodation?
d. What is the probability that the number among the 25 who received a special accommodation is

* "Between a and b, inclusive" is equivalent to $(a \le X \le b)$.

within 2 standard deviations of the number you would expect to be accommodated?

e. Suppose that a student who does not receive a special accommodation is allowed 3 hours for the exam, whereas an accommodated student is allowed 4.5 hours. What would you expect the average time allowed the 25 selected students to be?

67. Suppose that 90% of all batteries from a certain supplier have acceptable voltages. A certain type of flashlight requires two type-D batteries, and the flashlight will work only if both its batteries have acceptable voltages. Among ten randomly selected flashlights, what is the probability that at least nine will work? What assumptions did you make in the course of answering the question posed?

68. A very large batch of components has arrived at a distributor. The batch can be characterized as acceptable only if the proportion of defective components is at most .10. The distributor decides to randomly select 10 components and to accept the batch only if the number of defective components in the sample is at most 2.

a. What is the probability that the batch will be accepted when the actual proportion of defectives is .01? .05? .10? .20? .25?

b. Let p denote the actual proportion of defectives in the batch. A graph of P(batch is accepted) as a function of p, with p on the horizontal axis and P(batch is accepted) on the vertical axis, is called the *operating characteristic curve* for the acceptance sampling plan. Use the results of part (a) to sketch this curve for $0 \le p \le 1$.

c. Repeat parts (a) and (b) with "1" replacing "2" in the acceptance sampling plan.

d. Repeat parts (a) and (b) with "15" replacing "10" in the acceptance sampling plan.

e. Which of the three sampling plans, that of part (a), (c), or (d), appears most satisfactory, and why?

69. An ordinance requiring that a smoke detector be installed in all previously constructed houses has been in effect in a particular city for 1 year. The fire department is concerned that many houses remain without detectors. Let p = the true proportion of such houses having detectors, and suppose that a random sample of 25 homes is inspected. If the sample strongly indicates that fewer than 80% of all houses have a detector, the fire department will campaign for a mandatory inspection program. Because of the costliness of the program, the department prefers not to call for such inspections unless sample evidence strongly argues for their necessity. Let X denote the number of homes with detectors among the 25 sampled. Consider rejecting the claim that $p \ge .8$ if $x \le 15$.

a. What is the probability that the claim is rejected when the actual value of p is .8?

b. What is the probability of not rejecting the claim when $p = .7$? When $p = .6$?

c. How do the "error probabilities" of parts (a) and (b) change if the value 15 in the decision rule is replaced by 14?

70. A toll bridge charges $1.00 for passenger cars and $2.50 for other vehicles. Suppose that during daytime hours, 60% of all vehicles are passenger cars. If 25 vehicles cross the bridge during a particular daytime period, what is the resulting expected toll revenue? [*Hint*: Let X = the number of passenger cars; then the toll revenue $h(X)$ is a linear function of X.]

71. A student who is trying to write a paper for a course has a choice of two topics, A and B. If topic A is chosen, the student will order two books through interlibrary loan, whereas if topic B is chosen, the student will order four books. The student believes that a good paper necessitates receiving and using at least half the books ordered for either topic chosen. If the probability that a book ordered through interlibrary loan actually arrives in time is .9 and books arrive independently of one another, which topic should the student choose to maximize the probability of writing a good paper? What if the arrival probability is only .5 instead of .9?

72. **a.** For fixed n, are there values of p $(0 \le p \le 1)$ for which $V(X) = 0$? Explain why this is so.

b. For what value of p is $V(X)$ maximized? [*Hint*: Either graph $V(X)$ as a function of p or else take a derivative.]

73. **a.** Show that $b(x; n, 1 - p) = b(n - x; n, p)$.

b. Show that $B(x; n, 1 - p) = 1 - B(n - x - 1; n, p)$. [*Hint*: At most x S's is equivalent to at least $(n - x)$ F's.]

c. What do parts (a) and (b) imply about the necessity of including values of p greater than .5 in Appendix Table A.1?

74. Show that $E(X) = np$ when X is a binomial random variable. [*Hint*: First express $E(X)$ as a sum with lower limit $x = 1$. Then factor out np, let $y = x - 1$ so that the sum is from $y = 0$ to $y = n - 1$, and show that the sum equals 1.]

75. Customers at a gas station pay with a credit card (A), debit card (B), or cash (C). Assume that successive customers make independent choices, with $P(A) = .5$, $P(B) = .2$, and $P(C) = .3$.

a. Among the next 100 customers, what are the mean and variance of the number who pay with a debit card? Explain your reasoning.

b. Answer part (a) for the number among the 100 who don't pay with cash.

76. An airport limousine can accommodate up to four passengers on any one trip. The company will accept a maximum of six reservations for a trip, and a passenger must have a reservation. From previous records, 20% of all those making reservations do not appear for the trip. Answer the following questions, assuming independence wherever appropriate.

a. If six reservations are made, what is the probability that at least one individual with a reservation cannot be accommodated on the trip?

b. If six reservations are made, what is the expected number of available places when the limousine departs?

c. Suppose the probability distribution of the number of reservations made is given in the accompanying table.

Number of reservations	3	4	5	6
Probability	.1	.2	.3	.4

Let X denote the number of passengers on a randomly selected trip. Obtain the probability mass function of X.

77. Refer to Chebyshev's inequality given in Exercise 43 (Section 3.3). Calculate $P(|X - \mu| \geq k\sigma)$ for $k = 2$ and $k = 3$ when $X \sim \text{Bin}(20, .5)$, and compare to the corresponding upper bounds. Repeat for $X \sim \text{Bin}(20, .75)$.

78. At the end of this section we obtained the mean and variance of a binomial rv using the mgf. Obtain the mean and variance instead from $R_X(t) = \ln[M_X(t)]$.

79. Obtain the moment generating function of the number of failures $n - X$ in a binomial experiment, and use it to determine the expected number of failures and the variance of the number of failures. Are the expected value and variance intuitively consistent with the expressions for $E(X)$ and $V(X)$? Explain.

3.6 *Hypergeometric and Negative Binomial Distributions

The hypergeometric and negative binomial distributions are both closely related to the binomial distribution. Whereas the binomial distribution is the approximate probability model for sampling without replacement from a finite dichotomous (S–F) population, the hypergeometric distribution is the exact probability model for the number of S's in the sample. The binomial rv X is the number of S's when the number n of trials is fixed, whereas the negative binomial distribution arises from fixing the number of S's desired and letting the number of trials be random.

The Hypergeometric Distribution

The assumptions leading to the hypergeometric distribution are as follows:

1. The population or set to be sampled consists of N individuals, objects, or elements (a *finite* population).
2. Each individual can be characterized as a success (S) or a failure (F), and there are M successes in the population.

3. A sample of n individuals is selected without replacement in such a way that each subset of size n is equally likely to be chosen.

The random variable of interest is X = the number of S's in the sample. The probability distribution of X depends on the parameters n, M, and N, so we wish to obtain $P(X = x) = h(x; n, M, N)$.

Example 3.43 During a particular period a university's information technology office received 20 service orders for problems with printers, of which 8 were laser printers and 12 were inkjet models. A sample of 5 of these service orders is to be selected for inclusion in a customer satisfaction survey. Suppose that the 5 are selected in a completely random fashion, so that any particular subset of size 5 has the same chance of being selected as does any other subset (think of putting the numbers 1, 2, . . . , 20 on 20 identical slips of paper, mixing up the slips, and choosing 5 of them). What then is the probability that exactly x ($x = 0, 1, 2, 3, 4$, or 5) of the selected service orders were for inkjet printers?

In this example, the population size is $N = 20$, the sample size is $n = 5$, and the number of S's (inkjet = S) and F's in the population are $M = 12$ and $N - M = 8$, respectively. Consider the value $x = 2$. Because all outcomes (each consisting of 5 particular orders) are equally likely,

$$P(X = 2) = h(2; 5, 12, 20) = \frac{\text{number of outcomes having } X = 2}{\text{number of possible outcomes}}$$

The number of possible outcomes in the experiment is the number of ways of selecting 5 from the 20 objects without regard to order—that is, $\binom{20}{5}$. To count the number of outcomes having $X = 2$, note that there are $\binom{12}{2}$ ways of selecting 2 of the inkjet orders, and for each such way there are $\binom{8}{3}$ ways of selecting the 3 laser orders to fill out the sample. The product rule from Chapter 2 then gives $\binom{12}{2}\binom{8}{3}$ as the number of outcomes with $X = 2$, so

$$h(2; 5, 12, 20) = \frac{\binom{12}{2}\binom{8}{3}}{\binom{20}{5}} = \frac{77}{323} = .238$$ ■

In general, if the sample size n is smaller than the number of successes in the population (M), then the largest possible X value is n. However, if $M < n$ (e.g., a sample size of 25 and only 15 successes in the population), then X can be at most M. Similarly, whenever the number of population failures ($N - M$) exceeds the sample size, the smallest possible X value is 0 (since all sampled individuals might then be failures). However, if $N - M < n$, the smallest possible X value is $n - (N - M)$. Summarizing, the possible values of X satisfy the restriction $\max[0, n - (N - M)] \le x \le \min(n, M)$. An argument parallel to that of the previous example gives the pmf of X.

PROPOSITION If X is the number of S's in a completely random sample of size n drawn from a population consisting of M S's and $(N - M)$ F's, then the probability distribution of X, called the **hypergeometric distribution**, is given by

$$P(X = x) = h(x; n, M, N) = \frac{\binom{M}{x}\binom{N-M}{n-x}}{\binom{N}{n}} \tag{3.15}$$

for x an integer satisfying $\max(0, n - N + M) \le x \le \min(n, M)$.

In Example 3.43, $n = 5$, $M = 12$, and $N = 20$, so $h(x; 5, 12, 20)$ for $x = 0, 1, 2, 3, 4, 5$ can be obtained by substituting these numbers into Equation (3.15).

Example 3.44 Five individuals from an animal population thought to be near extinction in a certain region have been caught, tagged, and released to mix into the population. After they have had an opportunity to mix, a random sample of 10 of these animals is selected. Let $X =$ the number of tagged animals in the second sample. If there are actually 25 animals of this type in the region, what is the probability that (a) $X = 2$? (b) $X \le 2$?

Application of the hypergeometric distribution here requires assuming that every subset of 10 animals has the same chance of being captured. This in turn implies that released animals are no easier or harder to catch than are those not initially captured. Then the parameter values are $n = 10$, $M = 5$ (5 tagged animals in the population), and $N = 25$, so

$$h(x; 10, 5, 25) = \frac{\binom{5}{x}\binom{20}{10-x}}{\binom{25}{10}} \qquad x = 0, 1, 2, 3, 4, 5$$

For part (a),

$$P(X = 2) = h(2; 10, 5, 25) = \frac{\binom{5}{2}\binom{20}{8}}{\binom{25}{10}} = .385$$

For part (b),

$$\begin{aligned} P(X \le 2) &= P(X = 0, 1, \text{ or } 2) = \sum_{x=0}^{2} h(x; 10, 5, 25) \\ &= .057 + .257 + .385 = .699 \end{aligned}$$

■

Comprehensive tables of the hypergeometric distribution are available, but because the distribution has three parameters, these tables require much more space than tables for the binomial distribution. MINITAB and other statistical software packages will easily generate hypergeometric probabilities.

As in the binomial case, there are simple expressions for $E(X)$ and $V(X)$ for hypergeometric rv's.

PROPOSITION The mean and variance of the hypergeometric rv X having pmf $h(x; n, M, N)$ are

$$E(X) = n \cdot \frac{M}{N} \qquad V(X) = \left(\frac{N-n}{N-1}\right) \cdot n \cdot \frac{M}{N} \cdot \left(1 - \frac{M}{N}\right)$$

The proof will be given in Section 6.3. We do not give the moment generating function for the hypergeometric distribution, because the mgf is more trouble than it is worth here.

The ratio M/N is the proportion of S's in the population. Replacing M/N by p in $E(X)$ and $V(X)$ gives

$$E(X) = np \tag{3.16}$$

$$V(X) = \left(\frac{N-n}{N-1}\right) \cdot np(1-p)$$

Expression (3.16) shows that the means of the binomial and hypergeometric rv's are equal, whereas the variances of the two rv's differ by the factor $(N - n)/(N - 1)$, often called the **finite population correction factor**. This factor is less than 1, so the hypergeometric variable has smaller variance than does the binomial rv. The correction factor can be written $(1 - n/N)/(1 - 1/N)$, which is approximately 1 when n is small relative to N.

Example 3.45 (Example 3.44 continued)

In the animal-tagging example, $n = 10$, $M = 5$, and $N = 25$, so $p = \frac{5}{25} = .2$ and

$$E(X) = 10(.2) = 2$$

$$V(X) = \frac{15}{24}(10)(.2)(.8) = (.625)(1.6) = 1$$

If the sampling was carried out with replacement, $V(X) = 1.6$.

Suppose the population size N is not actually known, so the value x is observed and we wish to estimate N. It is reasonable to equate the observed sample proportion of S's, x/n, with the population proportion, M/N, giving the estimate

$$\hat{N} = \frac{M \cdot n}{x}$$

If $M = 100$, $n = 40$, and $x = 16$, then $\hat{N} = 250$. ■

Our general rule of thumb in Section 3.5 stated that if sampling was without replacement but n/N was at most .05, then the binomial distribution could be used to compute approximate probabilities involving the number of S's in the sample. A more precise statement is as follows: Let the population size, N, and number of population S's, M, get large with the ratio M/N approaching p. Then $h(x; n, M, N)$ approaches $b(x; n, p)$; so for n/N small, the two are approximately equal provided that p is not too near either 0 or 1. This is the rationale for our rule of thumb.

The Negative Binomial Distribution

The negative binomial rv and distribution are based on an experiment satisfying the following conditions:

1. The experiment consists of a sequence of independent trials.
2. Each trial can result in either a success (S) or a failure (F).
3. The probability of success is constant from trial to trial, so $P(S \text{ on trial } i) = p$ for $i = 1, 2, 3 \ldots$.
4. The experiment continues (trials are performed) until a total of r successes have been observed, where r is a specified positive integer.

The random variable of interest is $X =$ the number of failures that precede the rth success, and X is called a **negative binomial random variable**. In contrast to the binomial rv, the number of successes is fixed and the number of trials is random. Why the name "negative binomial?" Binomial probabilities are related to the terms in the binomial theorem, and negative binomial probabilities are related to the terms in the binomial theorem when the exponent is a negative integer. For details see the proof for the last proposition of this section.

Possible values of X are 0, 1, 2, ... Let $nb(x; r, p)$ denote the pmf of X. The event $\{X = x\}$ is equivalent to $\{r - 1$ S's in the first $(x + r - 1)$ trials and an S on the $(x + r)$th trial$\}$ (e.g., if $r = 5$ and $x = 10$, then there must be four S's in the first 14 trials and trial 15 must be an S). Since trials are independent,

$$\begin{aligned} nb(x; r, p) &= P(X = x) \\ &= P(r - 1 \; S\text{'s on the first } x + r - 1 \text{ trials}) \cdot P(S) \end{aligned} \tag{3.17}$$

The first probability on the far right of Expression (3.17) is the binomial probability

$$\binom{x + r - 1}{r - 1} p^{r-1}(1 - p)^x \quad \text{where } P(S) = p$$

PROPOSITION The pmf of the negative binomial rv X with parameters $r =$ number of S's and $p = P(S)$ is

$$nb(x; r, p) = \binom{x + r - 1}{r - 1} p^r(1 - p)^x \qquad x = 0, 1, 2, \ldots$$

Example 3.46 A pediatrician wishes to recruit 5 couples, each of whom is expecting their first child, to participate in a new natural childbirth regimen. Let $p = P(\text{a randomly selected couple agrees to participate})$. If $p = .2$, what is the probability that 15 couples must be asked before 5 are found who agree to participate? That is, with $S = \{\text{agrees to participate}\}$, what is the probability that 10 F's occur before the fifth S? Substituting $r = 5$, $p = .2$, and $x = 10$ into $nb(x; r, p)$ gives

$$nb(10; 5, .2) = \binom{14}{4}(.2)^5(.8)^{10} = .034$$

The probability that at most 10 F's are observed (at most 15 couples are asked) is

$$P(X \le 10) = \sum_{x=0}^{10} nb(x; 5, .2) = (.2)^5 \sum_{x=0}^{10} \binom{x+4}{4}(.8)^x = .164$$ ■

In some sources, the negative binomial rv is taken to be the number of trials $X + r$ rather than the number of failures.

In the special case $r = 1$, the pmf is

$$nb(x; 1, p) = (1 - p)^x p \qquad x = 0, 1, 2, \ldots \tag{3.18}$$

In Example 3.10, we derived the pmf for the number of trials necessary to obtain the first S, and the pmf there is similar to Expression (3.18). Both X = number of F's and Y = number of trials ($= 1 + X$) are referred to in the literature as **geometric random variables**, and the pmf in (3.18) is called the **geometric distribution**. The name is appropriate because the probabilities form a geometric series: $p, (1 - p)p, (1 - p)^2p, \ldots$. To see that the sum of the probabilities is 1, recall that the sum of a geometric series is $a + ar + ar^2 + \cdots = a/(1 - r)$ if $|r| < 1$, so for $p > 0$,

$$p + (1 - p)p + (1 - p)^2 p + \cdots = \frac{p}{1 - (1 - p)} = 1$$

In Example 3.18, the expected number of trials until the first S was shown to be $1/p$, so that the expected number of F's until the first S is $(1/p) - 1 = (1 - p)/p$. Intuitively, we would expect to see $r \cdot (1 - p)/p$ F's before the rth S, and this is indeed $E(X)$. There is also a simple formula for $V(X)$.

PROPOSITION If X is a negative binomial rv with pmf $nb(x; r, p)$, then

$$M_X(t) = \frac{p^r}{[1 - e^t(1 - p)]^r} \qquad E(X) = \frac{r(1 - p)}{p} \qquad V(X) = \frac{r(1 - p)}{p^2}$$

Proof In order to derive the moment generating function, we will use the binomial theorem as generalized by Isaac Newton to allow negative exponents, and this will help to explain the name of the distribution. If n is any real number, not necessarily a positive integer,

$$(a + b)^n = \sum_{x=0}^{\infty} \binom{n}{x} b^x a^{n-x}$$

where

$$\binom{n}{x} = \frac{n(n - 1) \cdot \cdots \cdot (n - x + 1)}{x!} \qquad \text{except that } \binom{n}{0} = 1$$

In the special case that $x > 0$ and n is a negative integer, $n = -r$,

$$\binom{-r}{x} = \frac{-r(-r-1)\cdot\cdots\cdot(-r-x+1)}{x!}$$
$$= \frac{(r+x-1)(r+x-2)\cdot\cdots\cdot r}{x!}(-1)^x = \binom{r+x-1}{r-1}(-1)^x$$

Using this in the generalized binomial theorem with $a = 1$ and $b = -u$,

$$(1-u)^{-r} = \sum_{x=0}^{\infty}\binom{r+x-1}{r-1}(-1)^x(-u)^x = \sum_{x=0}^{\infty}\binom{r+x-1}{r-1}u^x$$

Now we can find the moment generating function for the negative binomial distribution:

$$M_X(t) = \sum_{x=0}^{\infty} e^{tx}\binom{r+x-1}{r-1}p^r(1-p)^x$$
$$= p^r\sum_{x=0}^{\infty}\binom{r+x-1}{r-1}[e^t(1-p)]^x = \frac{p^r}{[1-e^t(1-p)]^r}$$

The mean and variance of X can now be obtained from the moment generating function (Exercise 91). ■

Finally, by expanding the binomial coefficient in front of $p^r(1-p)^x$ and doing some cancellation, it can be seen that $nb(x; r, p)$ is well defined even when r is not an integer. This *generalized negative binomial distribution* has been found to fit observed data quite well in a wide variety of applications.

Exercises Section 3.6 (80–92)

80. A certain type of digital camera comes in either a 3-megapixel version or a 4-megapixel version. A camera store has received a shipment of 15 of these cameras, of which 6 have 3-megapixel resolution. Suppose that 5 of these cameras are randomly selected to be stored behind the counter; the other 10 are placed in a storeroom. Let X = the number of 3-megapixel cameras among the 5 selected for behind-the-counter storage.

a. What kind of a distribution does X have (name and values of all parameters)?

b. Compute $P(X = 2)$, $P(X \le 2)$, and $P(X \ge 2)$.

c. Calculate the mean value and standard deviation of X.

81. Each of 12 refrigerators of a certain type has been returned to a distributor because of an audible, high-pitched, oscillating noise when the refrigerator is running. Suppose that 7 of these refrigerators have a defective compressor and the other 5 have less serious problems. If the refrigerators are examined in random order, let X be the number among the first 6 examined that have a defective compressor. Compute the following:

a. $P(X = 5)$

b. $P(X \le 4)$

c. The probability that X exceeds its mean value by more than 1 standard deviation.

d. Consider a large shipment of 400 refrigerators, of which 40 have defective compressors. If X is the number among 15 randomly selected refrigerators that have defective compressors, describe a less tedious way to calculate (at least approximately) $P(X \le 5)$ than to use the hypergeometric pmf.

82. An instructor who taught two sections of statistics last term, the first with 20 students and the second with 30, decided to assign a term project. After all projects had been turned in, the instructor randomly ordered them before grading. Consider the first 15 graded projects.

a. What is the probability that exactly 10 of these are from the second section?

b. What is the probability that at least 10 of these are from the second section?

c. What is the probability that at least 10 of these are from the same section?

d. What are the mean value and standard deviation of the number among these 15 that are from the second section?

e. What are the mean value and standard deviation of the number of projects not among these first 15 that are from the second section?

83. A geologist has collected 10 specimens of basaltic rock and 10 specimens of granite. The geologist instructs a laboratory assistant to randomly select 15 of the specimens for analysis.

a. What is the pmf of the number of granite specimens selected for analysis?

b. What is the probability that all specimens of one of the two types of rock are selected for analysis?

c. What is the probability that the number of granite specimens selected for analysis is within 1 standard deviation of its mean value?

84. Suppose that 20% of all individuals have an adverse reaction to a particular drug. A medical researcher will administer the drug to one individual after another until the first adverse reaction occurs. Define an appropriate random variable and use its distribution to answer the following questions.

a. What is the probability that when the experiment terminates, four individuals have not had adverse reactions?

b. What is the probability that the drug is administered to exactly five individuals?

c. What is the probability that at most four individuals do not have an adverse reaction?

d. How many individuals would you expect to not have an adverse reaction, and to how many individuals would you expect the drug to be given?

e. What is the probability that the number of individuals given the drug is within one standard deviation of what you expect?

85. Twenty pairs of individuals playing in a bridge tournament have been seeded 1, . . . , 20. In the first part of the tournament, the 20 are randomly divided into 10 east–west pairs and 10 north–south pairs.

a. What is the probability that x of the top 10 pairs end up playing east–west?

b. What is the probability that all of the top five pairs end up playing the same direction?

c. If there are $2n$ pairs, what is the pmf of $X =$ the number among the top n pairs who end up playing east–west? What are $E(X)$ and $V(X)$?

86. A second-stage smog alert has been called in a certain area of Los Angeles County in which there are 50 industrial firms. An inspector will visit 10 randomly selected firms to check for violations of regulations.

a. If 15 of the firms are actually violating at least one regulation, what is the pmf of the number of firms visited by the inspector that are in violation of at least one regulation?

b. If there are 500 firms in the area, of which 150 are in violation, approximate the pmf of part (a) by a simpler pmf.

c. For $X =$ the number among the 10 visited that are in violation, compute $E(X)$ and $V(X)$ both for the exact pmf and the approximating pmf in part (b).

87. Suppose that $p = P(\text{male birth}) = .5$. A couple wishes to have exactly two female children in their family. They will have children until this condition is fulfilled.

a. What is the probability that the family has x male children?

b. What is the probability that the family has four children?

c. What is the probability that the family has at most four children?

d. How many male children would you expect this family to have? How many children would you expect this family to have?

88. A family decides to have children until it has three children of the same gender. Assuming $P(B) = P(G) = .5$, what is the pmf of $X =$ the number of children in the family?

89. Three brothers and their wives decide to have children until each family has two female children. What is the pmf of $X =$ the total number of male children born to the brothers? What is $E(X)$, and how does it compare to the expected number of male children born to each brother?

90. Individual A has a red die and B has a green die (both fair). If they each roll until they obtain five "doubles" (1–1, . . . , 6–6), what is the pmf of $X =$

the total number of times a die is rolled? What are $E(X)$ and $V(X)$?

91. For the negative binomial distribution use the moment generating function to derive

a. The mean

b. The variance

92. If X is a negative binomial rv, then $Y = r + X$ is the total number of trials necessary to obtain r S's. Obtain the mgf of Y and then its mean value and variance. Are the mean and variance intuitively consistent with the expressions for $E(X)$ and $V(X)$? Explain.

3.7 *The Poisson Probability Distribution

The binomial, hypergeometric, and negative binomial distributions were all derived by starting with an experiment consisting of trials or draws and applying the laws of probability to various outcomes of the experiment. There is no simple experiment on which the Poisson distribution is based, though we will shortly describe how it can be obtained by certain limiting operations.

DEFINITION A random variable X is said to have a **Poisson distribution** with parameter λ ($\lambda > 0$) if the pmf of X is

$$p(x; \lambda) = \frac{e^{-\lambda}\lambda^x}{x!} \qquad x = 0, 1, 2, \ldots$$

The value of λ is frequently a rate per unit time or per unit area. Because λ must be positive, $p(x; \lambda) > 0$ for all possible x values. The fact that $\sum_{x=0}^{\infty} p(x; \lambda) = 1$ is a consequence of the Maclaurin infinite series expansion of e^{λ}, which appears in most calculus texts:

$$e^{\lambda} = 1 + \lambda + \frac{\lambda^2}{2!} + \frac{\lambda^3}{3!} + \cdots = \sum_{x=0}^{\infty} \frac{\lambda^x}{x!} \tag{3.19}$$

If the two extreme terms in Expression (3.19) are multiplied by $e^{-\lambda}$ and then $e^{-\lambda}$ is placed inside the summation, the result is

$$1 = \sum_{x=0}^{\infty} e^{-\lambda}\frac{\lambda^x}{x!}$$

which shows that $p(x; \lambda)$ fulfills the second condition necessary for specifying a pmf.

Example 3.47 Let X denote the number of creatures of a particular type captured in a trap during a given time period. Suppose that X has a Poisson distribution with $\lambda = 4.5$, so on average traps will contain 4.5 creatures. [The article "Dispersal Dynamics of the Bivalve *Gemma gemma* in a Patchy Environment (*Ecological Monographs*, 1995: 1–20) suggests this model; the bivalve *Gemma gemma* is a small clam.] The probability that a trap contains exactly five creatures is

$$P(X = 5) = \frac{e^{-4.5}(4.5)^5}{5!} = .1708$$

The probability that a trap has at most five creatures is

$$P(X \le 5) = \sum_{x=0}^{5} \frac{e^{-4.5}(4.5)^x}{x!} = e^{-4.5}\left[1 + 4.5 + \frac{(4.5)^2}{2!} + \cdots + \frac{(4.5)^5}{5!}\right] = .7029 \quad \blacksquare$$

The Poisson Distribution as a Limit

The rationale for using the Poisson distribution in many situations is provided by the following proposition.

PROPOSITION Suppose that in the binomial pmf $b(x; n, p)$ we let $n \to \infty$ and $p \to 0$ in such a way that np approaches a value $\lambda > 0$. Then $b(x; n, p) \to p(x; \lambda)$.

Proof Begin with the binomial pmf:

$$b(x; n, p) = \binom{n}{x} p^x(1-p)^{n-x} = \frac{n!}{x!(n-x)!}p^x(1-p)^{n-x}$$

$$= \frac{n(n-1)\cdot \cdots \cdot(n-x+1)}{x!}p^x(1-p)^{n-x}$$

Include n^x in both the numerator and denominator:

$$b(x; n, p) = \frac{n}{n}\frac{n-1}{n}\cdots\frac{n-x+1}{n}\cdot\frac{(np)^x}{x!}\cdot\frac{(1-p)^n}{(1-p)^x}$$

Taking the limit as $n \to \infty$ and $p \to 0$ with $np \to \lambda$,

$$\lim_{n\to\infty} b(x; n, p) = 1 \cdot 1 \cdot \cdots \cdot 1 \cdot \frac{\lambda^x}{x!}\left(\lim_{n\to\infty}\frac{(1-np/n)^n}{1}\right) \quad \blacksquare$$

The limit on the right can be obtained from the calculus theorem that says the limit of $(1 - a_n/n)^n$ is e^{-a} if $a_n \to a$. Because $np \to \lambda$,

$$\lim_{n\to\infty} b(x; n, p) = \frac{\lambda^x}{x!}\cdot\lim_{n\to\infty}\left(1 - \frac{np}{n}\right)^n = \frac{\lambda^x e^{-\lambda}}{x!} = p(x; \lambda)$$

It is interesting that Siméon Poisson discovered his distribution by this approach in the 1830s, as a limit of the binomial distribution. According to the proposition, *in any binomial experiment for which n is large and p is small,* $b(x; n, p) \approx p(x; \lambda)$ *where* $\lambda = np$. As a rule of thumb, this approximation can safely be applied if $n > 50$ and $np < 5$.

Example 3.48 If a publisher of nontechnical books takes great pains to ensure that its books are free of typographical errors, so that the probability of any given page containing at least one such error is .005 and errors are independent from page to page, what is the probability that one of its 400-page novels will contain exactly one page with errors? At most three pages with errors?

With S denoting a page containing at least one error and F an error-free page, the number X of pages containing at least one error is a binomial rv with $n = 400$ and $p = .005$, so $np = 2$. We wish

$$P(X = 1) = b(1; 400, .005) \approx p(1; 2) = \frac{e^{-2}(2)^1}{1!} = .270671$$

The binomial value is $b(1; 400, .005) = .270669$, so the approximation is good to five decimals here.

Similarly,

$$\begin{aligned} P(X \le 3) &\approx \sum_{x=0}^{3} p(x, 2) = \sum_{x=0}^{3} e^{-2}\frac{2^x}{x!} \\ &= .135335 + .270671 + .270671 + .180447 \\ &= .8571 \end{aligned}$$

and this again is quite close to the binomial value $P(X \le 3) = .8576$. ■

Table 3.2 shows the Poisson distribution for $\lambda = 3$ along with three binomial distributions with $np = 3$, and Figure 3.8 (from S-Plus) plots the Poisson along with the first two binomial distributions. The approximation is of limited use for $n = 30$, but of course the accuracy is better for $n = 100$ and much better for $n = 300$.

Table 3.2 Comparing the Poisson and three binomial distributions

x	$n = 30, p = .1$	$n = 100, p = .03$	$n = 300, p = .01$	Poisson, $\lambda = 3$
0	0.042391	0.047553	0.049041	0.049787
1	0.141304	0.147070	0.148609	0.149361
2	0.227656	0.225153	0.224414	0.224042
3	0.236088	0.227474	0.225170	0.224042
4	0.177066	0.170606	0.168877	0.168031
5	0.102305	0.101308	0.100985	0.100819
6	0.047363	0.049610	0.050153	0.050409
7	0.018043	0.020604	0.021277	0.021604
8	0.005764	0.007408	0.007871	0.008102
9	0.001565	0.002342	0.002580	0.002701
10	0.000365	0.000659	0.000758	0.000810

Appendix Table A.2 exhibits the cdf $F(x; \lambda)$ for $\lambda = .1, .2, \ldots, 1, 2, \ldots, 10, 15$, and 20. For example, if $\lambda = 2$, then $P(X \le 3) = F(3; 2) = .857$ as in Example 3.48, whereas $P(X = 3) = F(3; 2) - F(2; 2) = .180$. Alternatively, many statistical computer packages will generate $p(x; \lambda)$ and $F(x; \lambda)$ upon request.

The Mean, Variance and MGF of X

Since $b(x; n, p) \to p(x; \lambda)$ as $n \to \infty$, $p \to 0$, $np \to \lambda$, the mean and variance of a binomial variable should approach those of a Poisson variable. These limits are $np \to \lambda$ and $np(1 - p) \to \lambda$.

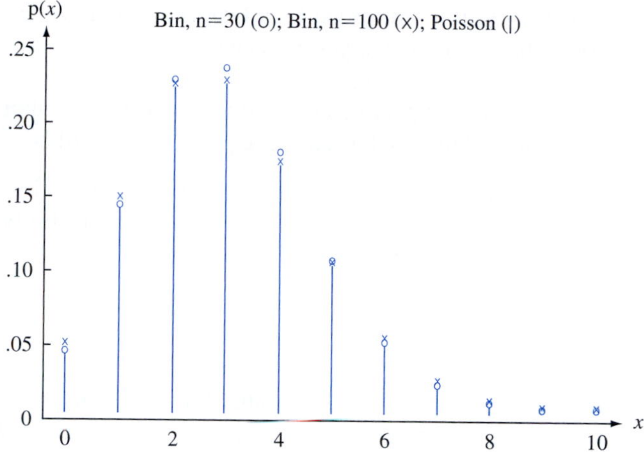

Figure 3.8 Comparing a Poisson and two binomial distributions

---

PROPOSITION

If X has a Poisson distribution with parameter λ, then $E(X) = V(X) = \lambda$.

These results can also be derived directly from the definitions of mean and variance (see Exercise 104 for the mean).

Example 3.49 (Example 3.47 continued)

Both the expected number of creatures trapped and the variance of the number trapped equal 4.5, and $\sigma_X = \sqrt{\lambda} = \sqrt{4.5} = 2.12$. ■

The moment generating function of the Poisson distribution is easy to derive, and it gives a direct route to the mean and variance (Exercise 108).

PROPOSITION

The Poisson moment generating function is

$$M_X(t) = e^{\lambda(e^t - 1)}$$

Proof The mgf is by definition

$$M_X(t) = E(e^{tX}) = \sum_{x=0}^{\infty} e^{tx} e^{-\lambda} \frac{\lambda^x}{x!} = e^{-\lambda} \sum_{x=0}^{\infty} \frac{(\lambda e^t)^x}{x!} = e^{-\lambda} e^{\lambda e^t} = e^{\lambda e^t - \lambda}$$

This uses the series expansion $\sum u^x/x! = e^u$. ■

The Poisson Process

A very important application of the Poisson distribution arises in connection with the occurrence of events of a particular type over time. As an example, suppose that starting from a time point that we label $t = 0$, we are interested in counting the number of radioactive pulses recorded by a Geiger counter. We make the following assumptions about the way in which pulses occur:

1. There exists a parameter $\alpha > 0$ such that for any short time interval of length Δt, the probability that exactly one pulse is received is $\alpha \cdot \Delta t + o(\Delta t)$.*
2. The probability of more than one pulse being received during Δt is $o(\Delta t)$ [which, along with Assumption 1, implies that the probability of no pulses during Δt is $1 - \alpha \cdot \Delta t - o(\Delta t)$].
3. The number of pulses received during the time interval Δt is independent of the number received prior to this time interval.

Informally, Assumption 1 says that for a short interval of time, the probability of receiving a single pulse is approximately proportional to the length of the time interval, where α is the constant of proportionality. Now let $P_k(t)$ denote the probability that k pulses will be received by the counter during any particular time interval of length t.

PROPOSITION

$P_k(t) = e^{-\alpha t}(\alpha t)^k/k!$, so that the number of pulses during a time interval of length t is a Poisson rv with parameter $\lambda = \alpha t$. The expected number of pulses during any such time interval is then αt, so the expected number during a unit interval of time is α.

See Exercise 107 for a derivation.

Example 3.50 Suppose pulses arrive at the counter at an average rate of six per minute, so that $\alpha = 6$. To find the probability that in a .5-min interval at least one pulse is received, note that the number of pulses in such an interval has a Poisson distribution with parameter $\alpha t = 6(.5) = 3$ (.5 min is used because α is expressed as a rate per minute). Then with $X =$ the number of pulses received in the 30-sec interval,

$$P(1 \le X) = 1 - P(X = 0) = 1 - \frac{e^{-3}(3)^0}{0!} = .950$$ ■

If in Assumptions 1–3 we replace "pulse" by "event," then the number of events occurring during a fixed time interval of length t has a Poisson distribution with parameter αt. Any process that has this distribution is called a **Poisson process**, and α is called the *rate of the process*. Other examples of situations giving rise to a Poisson process include monitoring the status of a computer system over time, with breakdowns constituting the events of interest; recording the number of accidents in an industrial facility over time; answering calls at a telephone switchboard; and observing the number of cosmic-ray showers from a particular observatory over time.

*A quantity is $o(\Delta t)$ (read "little o of delta t") if, as Δt approaches 0, so does $o(\Delta t)/\Delta t$. That is, $o(\Delta t)$ is even more negligible than Δt itself. The quantity $(\Delta t)^2$ has this property, but $\sin(\Delta t)$ does not.

Instead of observing events over time, consider observing events of some type that occur in a two- or three-dimensional region. For example, we might select on a map a certain region R of a forest, go to that region, and count the number of trees. Each tree would represent an event occurring at a particular point in space. Under assumptions similar to 1–3, it can be shown that the number of events occurring in a region R has a Poisson distribution with parameter $\alpha \cdot a(R)$, where $a(R)$ is the area of R. The quantity α is the expected number of events per unit area or volume.

Exercises Section 3.7 (93–109)

93. Let X, the number of flaws on the surface of a randomly selected carpet of a particular type, have a Poisson distribution with parameter $\lambda = 5$. Use Appendix Table A.2 to compute the following probabilities:
a. $P(X \leq 8)$ **b.** $P(X = 8)$ **c.** $P(9 \leq X)$
d. $P(5 \leq X \leq 8)$ **e.** $P(5 < X < 8)$

94. Suppose the number X of tornadoes observed in a particular region during a 1-year period has a Poisson distribution with $\lambda = 8$.
a. Compute $P(X \leq 5)$.
b. Compute $P(6 \leq X \leq 9)$.
c. Compute $P(10 \leq X)$.
d. What is the probability that the observed number of tornadoes exceeds the expected number by more than 1 standard deviation?

95. Suppose that the number of drivers who travel between a particular origin and destination during a designated time period has a Poisson distribution with parameter $\lambda = 20$ (suggested in the article "Dynamic Ride Sharing: Theory and Practice," *J. Transp. Engrg.*, 1997: 308–312). What is the probability that the number of drivers will
a. Be at most 10?
b. Exceed 20?
c. Be between 10 and 20, inclusive? Be strictly between 10 and 20?
d. Be within 2 standard deviations of the mean value?

96. Consider writing onto a computer disk and then sending it through a certifier that counts the number of missing pulses. Suppose this number X has a Poisson distribution with parameter $\lambda = .2$. (Suggested in "Average Sample Number for Semi-Curtailed Sampling Using the Poisson Distribution," *J. Qual. Tech.*, 1983: 126–129.)
a. What is the probability that a disk has exactly one missing pulse?
b. What is the probability that a disk has at least two missing pulses?
c. If two disks are independently selected, what is the probability that neither contains a missing pulse?

97. An article in the *Los Angeles Times* (Dec. 3, 1993) reports that 1 in 200 people carry the defective gene that causes inherited colon cancer. In a sample of 1000 individuals, what is the approximate distribution of the number who carry this gene? Use this distribution to calculate the approximate probability that
a. Between 5 and 8 (inclusive) carry the gene.
b. At least 8 carry the gene.

98. Suppose that only .10% of all computers of a certain type experience CPU failure during the warranty period. Consider a sample of 10,000 computers.
a. What are the expected value and standard deviation of the number of computers in the sample that have the defect?
b. What is the (approximate) probability that more than 10 sampled computers have the defect?
c. What is the (approximate) probability that no sampled computers have the defect?

99. Suppose small aircraft arrive at a certain airport according to a Poisson process with rate $\alpha = 8$ per hour, so that the number of arrivals during a time period of t hours is a Poisson rv with parameter $\lambda = 8t$.
a. What is the probability that exactly 6 small aircraft arrive during a 1-hour period? At least 6? At least 10?
b. What are the expected value and standard deviation of the number of small aircraft that arrive during a 90-min period?

c. What is the probability that at least 20 small aircraft arrive during a $2\frac{1}{2}$-hour period? That at most 10 arrive during this period?

100. The number of people arriving for treatment at an emergency room can be modeled by a Poisson process with a rate parameter of five per hour.

a. What is the probability that exactly four arrivals occur during a particular hour?

b. What is the probability that at least four people arrive during a particular hour?

c. How many people do you expect to arrive during a 45-min period?

101. The number of requests for assistance received by a towing service is a Poisson process with rate $\alpha = 4$ per hour.

a. Compute the probability that exactly ten requests are received during a particular 2-hour period.

b. If the operators of the towing service take a 30-min break for lunch, what is the probability that they do not miss any calls for assistance?

c. How many calls would you expect during their break?

102. In proof testing of circuit boards, the probability that any particular diode will fail is .01. Suppose a circuit board contains 200 diodes.

a. How many diodes would you expect to fail, and what is the standard deviation of the number that are expected to fail?

b. What is the (approximate) probability that at least four diodes will fail on a randomly selected board?

c. If five boards are shipped to a particular customer, how likely is it that at least four of them will work properly? (A board works properly only if all its diodes work.)

103. The article "Reliability-Based Service-Life Assessment of Aging Concrete Structures" (*J. Struct. Engrg.*, 1993: 1600–1621) suggests that a Poisson process can be used to represent the occurrence of structural loads over time. Suppose the mean time between occurrences of loads is .5 year.

a. How many loads can be expected to occur during a 2-year period?

b. What is the probability that more than five loads occur during a 2-year period?

c. How long must a time period be so that the probability of no loads occurring during that period is at most .1?

104. Let X have a Poisson distribution with parameter λ. Show that $E(X) = \lambda$ directly from the definition of expected value. (*Hint*: The first term in the sum equals 0, and then x can be canceled. Now factor out λ and show that what is left sums to 1.)

105. Suppose that trees are distributed in a forest according to a two-dimensional Poisson process with parameter α, the expected number of trees per acre, equal to 80.

a. What is the probability that in a certain quarter-acre plot, there will be at most 16 trees?

b. If the forest covers 85,000 acres, what is the expected number of trees in the forest?

c. Suppose you select a point in the forest and construct a circle of radius .1 mile. Let X = the number of trees within that circular region. What is the pmf of X? (*Hint*: 1 sq mile = 640 acres.)

106. Automobiles arrive at a vehicle equipment inspection station according to a Poisson process with rate $\alpha = 10$ per hour. Suppose that with probability .5 an arriving vehicle will have no equipment violations.

a. What is the probability that exactly ten arrive during the hour and all ten have no violations?

b. For any fixed $y \geq 10$, what is the probability that y arrive during the hour, of which ten have no violations?

c. What is the probability that ten "no-violation" cars arrive during the next hour? [*Hint*: Sum the probabilities in part (b) from $y = 10$ to ∞.]

107. a. In a Poisson process, what has to happen in both the time interval $(0, t)$ and the interval $(t, t + \Delta t)$ so that no events occur in the entire interval $(0, t + \Delta t)$? Use this and Assumptions 1–3 to write a relationship between $P_0(t + \Delta t)$ and $P_0(t)$.

b. Use the result of part (a) to write an expression for the difference $P_0(t + \Delta t) - P_0(t)$. Then divide by Δt and let $\Delta t \to 0$ to obtain an equation involving $(d/dt)P_0(t)$, the derivative of $P_0(t)$ with respect to t.

c. Verify that $P_0(t) = e^{-\alpha t}$ satisfies the equation of part (b).

d. It can be shown in a manner similar to parts (a) and (b) that the $P_k(t)$'s must satisfy the system of differential equations

$$\frac{d}{dt}P_k(t) = \alpha P_{k-1}(t) - \alpha P_k(t) \quad k = 1, 2, 3, \ldots$$

Verify that $P_k(t) = e^{-\alpha t}(\alpha t)^k/k!$ satisfies the system. (This is actually the only solution.)

108. **a.** Use derivatives of the moment generating function to obtain the mean and variance for the Poisson distribution.

b. As discussed in Section 3.4, obtain the Poisson mean and variance from $R_X(t) = \ln[M_X(t)]$. In terms of effort, how does this method compare with the one in part (a)?

109. Show that the binomial moment generating function converges to the Poisson moment generating function if we let $n \rightarrow \infty$ and $p \rightarrow 0$ in such a way that np approaches a value $\lambda > 0$. [*Hint*: Use the calculus theorem that was used in showing that the binomial probabilities converge to the Poisson probabilities.] There is in fact a theorem saying that convergence of the mgf implies convergence of the probability distribution. In particular, convergence of the binomial mgf to the Poisson mgf implies $b(x; n, p) \rightarrow p(x; \lambda)$.

Supplementary Exercises (110–139)

110. Consider a deck consisting of seven cards, marked 1, 2, . . . , 7. Three of these cards are selected at random. Define an rv W by W = the sum of the resulting numbers, and compute the pmf of W. Then compute μ and σ^2. [*Hint*: Consider outcomes as unordered, so that (1, 3, 7) and (3, 1, 7) are not different outcomes. Then there are 35 outcomes, and they can be listed. (This type of rv actually arises in connection with a hypothesis test called Wilcoxon's rank-sum test, in which there is an x sample and a y sample and W is the sum of the ranks of the x's in the combined sample.)]

111. After shuffling a deck of 52 cards, a dealer deals out 5. Let X = the number of suits represented in the five-card hand.

a. Show that the pmf of X is

x	1	2	3	4
$p(x)$	.002	.146	.588	.264

[*Hint*: $p(1) = 4P$(all are spades), $p(2) = 6P$(only spades and hearts with at least one of each), and $p(4) = 4P$(2 spades $\cap$ one of each other suit).]

b. Compute μ, σ^2, and σ.

112. The negative binomial rv X was defined as the number of F's preceding the rth S. Let Y = the number of trials necessary to obtain the rth S. In the same manner in which the pmf of X was derived, derive the pmf of Y.

113. Of all customers purchasing automatic garage-door openers, 75% purchase a chain-driven model. Let X = the number among the next 15 purchasers who select the chain-driven model.

a. What is the pmf of X?

b. Compute $P(X > 10)$.

c. Compute $P(6 \leq X \leq 10)$.

d. Compute μ and σ^2.

e. If the store currently has in stock 10 chain-driven models and 8 shaft-driven models, what is the probability that the requests of these 15 customers can all be met from existing stock?

114. A friend recently planned a camping trip. He had two flashlights, one that required a single 6-V battery and another that used two size-D batteries. He had previously packed two 6-V and four size-D batteries in his camper. Suppose the probability that any particular battery works is p and that batteries work or fail independently of one another. Our friend wants to take just one flashlight. For what values of p should he take the 6-V flashlight?

115. A *k-out-of-n system* is one that will function if and only if at least k of the n individual components in the system function. If individual components function independently of one another, each with probability .9, what is the probability that a 3-out-of-5 system functions?

116. A manufacturer of flashlight batteries wishes to control the quality of its product by rejecting any lot in which the proportion of batteries having unacceptable voltage appears to be too high. To this end, out of each large lot (10,000 batteries), 25 will be selected and tested. If at least 5 of these generate an unacceptable voltage, the entire lot will be rejected. What is the probability that a lot will be rejected if

a. 5% of the batteries in the lot have unacceptable voltages?

b. 10% of the batteries in the lot have unacceptable voltages?

c. 20% of the batteries in the lot have unacceptable voltages?

d. What would happen to the probabilities in parts (a)–(c) if the critical rejection number were increased from 5 to 6?

117. Of the people passing through an airport metal detector, .5% activate it; let $X =$ the number among a randomly selected group of 500 who activate the detector.

a. What is the (approximate) pmf of X?

b. Compute $P(X = 5)$.

c. Compute $P(5 \le X)$.

118. An educational consulting firm is trying to decide whether high school students who have never before used a hand-held calculator can solve a certain type of problem more easily with a calculator that uses reverse Polish logic or one that does not use this logic. A sample of 25 students is selected and allowed to practice on both calculators. Then each student is asked to work one problem on the reverse Polish calculator and a similar problem on the other. Let $p = P(S)$, where S indicates that a student worked the problem more quickly using reverse Polish logic than without, and let $X =$ number of S's.

a. If $p = .5$, what is $P(7 \le X \le 18)$?

b. If $p = .8$, what is $P(7 \le X \le 18)$?

c. If the claim that $p = .5$ is to be rejected when either $X \le 7$ or $X \ge 18$, what is the probability of rejecting the claim when it is actually correct?

d. If the decision to reject the claim $p = .5$ is made as in part (c), what is the probability that the claim is not rejected when $p = .6$? When $p = .8$?

e. What decision rule would you choose for rejecting the claim $p = .5$ if you wanted the probability in part (c) to be at most .01?

119. Consider a disease whose presence can be identified by carrying out a blood test. Let p denote the probability that a randomly selected individual has the disease. Suppose n individuals are independently selected for testing. One way to proceed is to carry out a separate test on each of the n blood samples. A potentially more economical approach, group testing, was introduced during World War II to identify syphilitic men among army inductees. First, take a part of each blood sample, combine these specimens, and carry out a single test. If no one has the disease, the result will be negative, and only the one test is required. If at least one individual is diseased, the test on the combined sample will yield a positive result, in which case the n individual tests are then carried out. If $p = .1$ and $n = 3$, what is the expected number of tests using this procedure? What is the expected number when $n = 5$? [The article "Random Multiple-Access Communication and Group Testing" (*IEEE Trans. Commun.*, 1984: 769–774) applied these ideas to a communication system in which the dichotomy was active/idle user rather than diseased/nondiseased.]

120. Let p_1 denote the probability that any particular code symbol is erroneously transmitted through a communication system. Assume that on different symbols, errors occur independently of one another. Suppose also that with probability p_2 an erroneous symbol is corrected upon receipt. Let X denote the number of correct symbols in a message block consisting of n symbols (after the correction process has ended). What is the probability distribution of X?

121. The purchaser of a power-generating unit requires c consecutive successful start-ups before the unit will be accepted. Assume that the outcomes of individual start-ups are independent of one another. Let p denote the probability that any particular start-up is successful. The random variable of interest is $X =$ the number of start-ups that must be made prior to acceptance. Give the pmf of X for the case $c = 2$. If $p = .9$, what is $P(X \le 8)$? [*Hint*: For $x \ge 5$, express $p(x)$ "recursively" in terms of the pmf evaluated at the smaller values $x - 3, x - 4, \ldots, 2$.] (This problem was suggested by the article "Evaluation of a Start-Up Demonstration Test," *J. Qual. Tech.*, 1983: 103–106.)

122. A plan for an executive travelers' club has been developed by an airline on the premise that 10% of its current customers would qualify for membership.

a. Assuming the validity of this premise, among 25 randomly selected current customers, what is the probability that between 2 and 6 (inclusive) qualify for membership?

b. Again assuming the validity of the premise, what are the expected number of customers who qualify and the standard deviation of the number who qualify in a random sample of 100 current customers?

c. Let X denote the number in a random sample of 25 current customers who qualify for membership. Consider rejecting the company's premise in favor of the claim that $p > .10$ if $x \geq 7$. What is the probability that the company's premise is rejected when it is actually valid?

d. Refer to the decision rule introduced in part (c). What is the probability that the company's premise is not rejected even though $p = .20$ (i.e., 20% qualify)?

123. Forty percent of seeds from maize (modern-day corn) ears carry single spikelets, and the other 60% carry paired spikelets. A seed with single spikelets will produce an ear with single spikelets 29% of the time, whereas a seed with paired spikelets will produce an ear with single spikelets 26% of the time. Consider randomly selecting ten seeds.

a. What is the probability that exactly five of these seeds carry a single spikelet and produce an ear with a single spikelet?

b. What is the probability that exactly five of the ears produced by these seeds have single spikelets? What is the probability that at most five ears have single spikelets?

124. A trial has just resulted in a hung jury because eight members of the jury were in favor of a guilty verdict and the other four were for acquittal. If the jurors leave the jury room in random order and each of the first four leaving the room is accosted by a reporter in quest of an interview, what is the pmf of X = the number of jurors favoring acquittal among those interviewed? How many of those favoring acquittal do you expect to be interviewed?

125. A reservation service employs five information operators who receive requests for information independently of one another, each according to a Poisson process with rate $\alpha = 2$ per minute.

a. What is the probability that during a given 1-min period, the first operator receives no requests?

b. What is the probability that during a given 1-min period, exactly four of the five operators receive no requests?

c. Write an expression for the probability that during a given 1-min period, all of the operators receive exactly the same number of requests.

126. Grasshoppers are distributed at random in a large field according to a Poisson distribution with parameter $\alpha = 2$ per square yard. How large should the radius R of a circular sampling region be taken so that the probability of finding at least one in the region equals .99?

127. A newsstand has ordered five copies of a certain issue of a photography magazine. Let X = the number of individuals who come in to purchase this magazine. If X has a Poisson distribution with parameter $\lambda = 4$, what is the expected number of copies that are sold?

128. Individuals A and B begin to play a sequence of chess games. Let $S = \{\text{A wins a game}\}$, and suppose that outcomes of successive games are independent with $P(S) = p$ and $P(F) = 1 - p$ (they never draw). They will play until one of them wins ten games. Let X = the number of games played (with possible values 10, 11, . . . , 19).

a. For $x = 10, 11, \ldots, 19$, obtain an expression for $p(x) = P(X = x)$.

b. If a draw is possible, with $p = P(S)$, $q = P(F)$, $1 - p - q = P(\text{draw})$, what are the possible values of X? What is $P(20 \leq X)$? [*Hint*: $P(20 \leq X) = 1 - P(X < 20)$.]

129. A test for the presence of a certain disease has probability .20 of giving a false-positive reading (indicating that an individual has the disease when this is not the case) and probability .10 of giving a false-negative result. Suppose that ten individuals are tested, five of whom have the disease and five of whom do not. Let X = the number of positive readings that result.

a. Does X have a binomial distribution? Explain your reasoning.

b. What is the probability that exactly three of the ten test results are positive?

130. The generalized negative binomial pmf is given by

$$nb(x; r, p) = k(r, x) \cdot p^r(1 - p)^x \qquad x = 0, 1, 2, \ldots$$

Let X, the number of plants of a certain species found in a particular region, have this distribution with $p = .3$ and $r = 2.5$. What is $P(X = 4)$? What is the probability that at least one plant is found?

131. Define a function $p(x; \lambda, \mu)$ by

$$p(x; \lambda, \mu) = \begin{cases} \frac{1}{2}e^{-\lambda}\frac{\lambda^x}{x!} + \frac{1}{2}e^{-\mu}\frac{\mu^x}{x!} & x = 0, 1, 2, \ldots \\ 0 & \text{otherwise} \end{cases}$$

a. Show that $p(x; \lambda, \mu)$ satisfies the two conditions necessary for specifying a pmf. [*Note*: If a firm employs two typists, one of whom makes typographical errors at the rate of λ per page and the other at rate μ per page and they each do half the firm's typing, then $p(x; \lambda, \mu)$ is the pmf of $X =$ the number of errors on a randomly chosen page.]

b. If the first typist (rate λ) types 60% of all pages, what is the pmf of X of part (a)?

c. What is $E(X)$ for $p(x; \lambda, \mu)$ given by the displayed expression?

d. What is σ^2 for $p(x; \lambda, \mu)$ given by that expression?

132. The *mode* of a discrete random variable X with pmf $p(x)$ is that value x^* for which $p(x)$ is largest (the most probable x value).

a. Let $X \sim \text{Bin}(n, p)$. By considering the ratio $b(x + 1; n, p)/b(x; n, p)$, show that $b(x; n, p)$ increases with x as long as $x < np - (1 - p)$. Conclude that the mode x^* is the integer satisfying $(n + 1)p - 1 \le x^* \le (n + 1)p$.

b. Show that if X has a Poisson distribution with parameter λ, the mode is the largest integer less than λ. If λ is an integer, show that both $\lambda - 1$ and λ are modes.

133. A computer disk storage device has ten concentric tracks, numbered 1, 2, . . . , 10 from outermost to innermost, and a single access arm. Let $p_i =$ the probability that any particular request for data will take the arm to track i ($i = 1, \ldots, 10$). Assume that the tracks accessed in successive seeks are independent. Let $X =$ the number of tracks over which the access arm passes during two successive requests (excluding the track that the arm has just left, so possible X values are $x = 0, 1, \ldots, 9$). Compute the pmf of X. [*Hint*: P(the arm is now on track i and $X = j$) $= P(X = j \mid \text{arm now on } i) \cdot p_i$. After the conditional probability is written in terms of $p_1, \ldots, p_{10}$, by the law of total probability, the desired probability is obtained by summing over i.]

134. If X is a hypergeometric rv, show directly from the definition that $E(X) = nM/N$ (consider only the case $n < M$). [*Hint*: Factor nM/N out of the sum for $E(X)$, and show that the terms inside the sum are of the form $h(y; n - 1, M - 1, N - 1)$, where $y = x - 1$.]

135. Use the fact that

$$\sum_{\text{all } x} (x - \mu)^2 p(x) \ge \sum_{x: |x - \mu| \ge k\sigma} (x - \mu)^2 p(x)$$

to prove Chebyshev's inequality, given in Exercise 43 (Section 3.3).

136. The simple Poisson process of Section 3.7 is characterized by a constant rate α at which events occur per unit time. A generalization of this is to suppose that the probability of exactly one event occurring in the interval $(t, t + \Delta t)$ is $\alpha(t) \cdot \Delta t + o(\Delta t)$. It can then be shown that the number of events occurring during an interval $[t_1, t_2]$ has a Poisson distribution with parameter

$$\lambda = \int_{t_1}^{t_2} \alpha(t)\, dt$$

The occurrence of events over time in this situation is called a *nonhomogeneous Poisson process*. The article "Inference Based on Retrospective Ascertainment," *J. Amer. Statist. Assoc.*, 1989: 360–372, considers the intensity function

$$\alpha(t) = e^{a + bt}$$

as appropriate for events involving transmission of HIV (the AIDS virus) via blood transfusions. Suppose that $a = 2$ and $b = .6$ (close to values suggested in the paper), with time in years.

a. What is the expected number of events in the interval [0, 4]? In [2, 6]?

b. What is the probability that at most 15 events occur in the interval [0, .9907]?

137. Suppose a store sells two different coffee makers of a particular brand, a basic model selling for $30 and a fancy one selling for $50. Let X be the number of people among the next 25 purchasing this brand who choose the fancy one. Then $h(X) =$ revenue $= 50X + 30(25 - X) = 20X + 750$, a linear function. If the choices are independent and have the same probability, then how is X distributed? Find the mean and standard deviation of $h(X)$. Explain why the choices might not be independent with the same probability.

138. Let X be a discrete rv with possible values 0, 1, 2, . . . or some subset of these. The function

$$h(s) = E(s^X) = \sum_{x=0}^{\infty} s^x \cdot p(x)$$

is called the probability generating function [e.g., $h(2) = \Sigma 2^x p(x)$, $h(3.7) = \Sigma(3.7)^x p(x)$, etc.].

a. Suppose X is the number of children born to a family, and $p(0) = .2$, $p(1) = .5$, and $p(2) = .3$. Determine the pgf of X.

b. Determine the pgf when X has a Poisson distribution with parameter λ.

c. Show that $h(1) = 1$.

d. Show that $h'(s)|_{s=0} = p(1)$ (assuming that the derivative can be brought inside the summation, which is justified). What results from taking the second derivative with respect to s and evaluating at $s = 0$? The third derivative? Explain how successive differentiation of $h(s)$ and evaluation at $s = 0$ "generates the probabilities in the distribution." Use this to recapture the probabilities of (a) from the pgf. *Note*: This shows that the pgf contains all the information about the distribution—knowing $h(s)$ is equivalent to knowing $p(x)$.

139. Three couples and two single individuals have been invited to a dinner party. Assume independence of arrivals to the party, and suppose that the probability of any particular individual or any particular couple arriving late is .4 (the two members of a couple arrive together). Let X = the number of people who show up late for the party. Determine the pmf of X.

Bibliography

Durrett, Richard, *Probability: Theory and Examples*, Duxbury Press, Belmont, CA, 1994.

Johnson, Norman, Samuel Kotz, and Adrienne Kemp, *Discrete Univariate Distributions*, Wiley, New York, 1992. An encyclopedia of information on discrete distributions.

Olkin, Ingram, Cyrus Derman, and Leon Gleser, *Probability Models and Applications* (2nd ed.), Macmillan, New York, 1994. Contains an in-depth discussion of both general properties of discrete and continuous distributions and results for specific distributions.

Pitman, Jim, *Probability*, Springer-Verlag, New York, 1993.

Ross, Sheldon, *Introduction to Probability Models* (7th ed.), Academic Press, New York, 2003. A good source of material on the Poisson process and generalizations and a nice introduction to other topics in applied probability.

CHAPTER FOUR

Continuous Random Variables and Probability Distributions

Introduction

As mentioned at the beginning of Chapter 3, the two important types of random variables are discrete and continuous. In this chapter, we study the second general type of random variable that arises in many applied problems. Sections 4.1 and 4.2 present the basic definitions and properties of continuous random variables, their probability distributions, and their moment generating functions. In Section 4.3, we study in detail the normal random variable and distribution, unquestionably the most important and useful in probability and statistics. Sections 4.4 and 4.5 discuss some other continuous distributions that are often used in applied work. In Section 4.6, we introduce a method for assessing whether given sample data is consistent with a specified distribution. Section 4.7 discusses methods for finding the pdf of a transformed random variable.

4.1 Probability Density Functions and Cumulative Distribution Functions

A discrete random variable (rv) is one whose possible values either constitute a finite set or else can be listed in an infinite sequence (a list in which there is a first element, a second element, etc.). A random variable whose set of possible values is an entire interval of numbers is not discrete.

Recall from Chapter 3 that a random variable X is continuous if (1) possible values comprise either a single interval on the number line (for some $A < B$, any number x between A and B is a possible value) or a union of disjoint intervals, and (2) $P(X = c) = 0$ for any number c that is a possible value of X.

Example 4.1 If in the study of the ecology of a lake, we make depth measurements at randomly chosen locations, then $X =$ the depth at such a location is a continuous rv. Here A is the minimum depth in the region being sampled, and B is the maximum depth. ■

Example 4.2 If a chemical compound is randomly selected and its pH X is determined, then X is a continuous rv because any pH value between 0 and 14 is possible. If more is known about the compound selected for analysis, then the set of possible values might be a subinterval of [0, 14], such as $5.5 \le x \le 6.5$, but X would still be continuous. ■

Example 4.3 Let X represent the amount of time a randomly selected customer spends waiting for a haircut before his/her haircut commences. Your first thought might be that X is a continuous random variable, since a measurement is required to determine its value. However, there are customers lucky enough to have no wait whatsoever before climbing into the barber's chair. So it must be the case that $P(X = 0) > 0$. Conditional on no chairs being empty, though, the waiting time will be continuous since X could then assume any value between some minimum possible time A and a maximum possible time B. This random variable is neither purely discrete nor purely continuous but instead is a mixture of the two types. ■

One might argue that although in principle variables such as height, weight, and temperature are continuous, in practice the limitations of our measuring instruments restrict us to a discrete (though sometimes very finely subdivided) world. However, continuous models often approximate real-world situations very well, and continuous mathematics (the calculus) is frequently easier to work with than the mathematics of discrete variables and distributions.

Probability Distributions for Continuous Variables

Suppose the variable X of interest is the depth of a lake at a randomly chosen point on the surface. Let $M =$ the maximum depth (in meters), so that any number in the interval $[0, M]$ is a possible value of X. If we "discretize" X by measuring depth to the nearest meter, then possible values are nonnegative integers less than or equal to M. The resulting discrete distribution of depth can be pictured using a probability histogram. If we draw the histogram so that the area of the rectangle above any possible integer k is the proportion of the lake whose depth is (to the nearest meter) k, then the total area of all rectangles is 1. A possible histogram appears in Figure 4.1(a).

If depth is measured much more accurately and the same measurement axis as in Figure 4.1(a) is used, each rectangle in the resulting probability histogram is much narrower, though the total area of all rectangles is still 1. A possible histogram is pictured in Figure 4.1(b); it has a much smoother appearance than the histogram in Figure 4.1(a). If we continue in this way to measure depth more and more finely, the resulting sequence of histograms approaches a smooth curve, such as is pictured in Figure 4.1(c). Because for each histogram the total area of all rectangles equals 1, the total area under the smooth curve is also 1. The probability that the depth at a randomly chosen point is between a and b is just the area under the smooth curve between a and b. It is exactly a smooth curve of the type pictured in Figure 4.1(c) that specifies a continuous probability distribution.

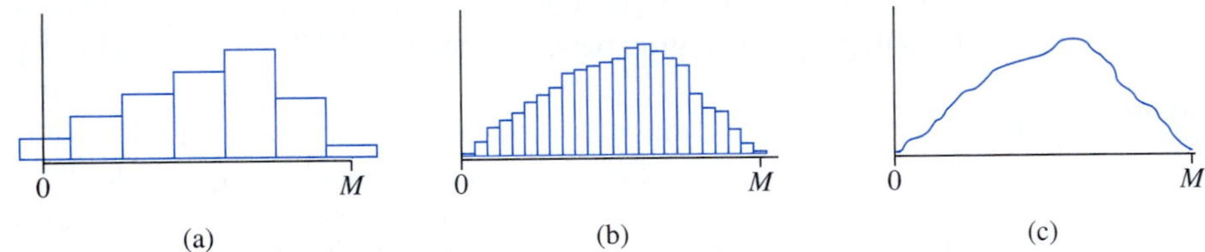

Figure 4.1 (a) Probability histogram of depth measured to the nearest meter; (b) probability histogram of depth measured to the nearest centimeter; (c) a limit of a sequence of discrete histograms

---

DEFINITION

Let X be a continuous rv. Then a **probability distribution** or **probability density function** (pdf) of X is a function $f(x)$ such that for any two numbers a and b with $a \le b$,

$$P(a \le X \le b) = \int_a^b f(x)\,dx$$

That is, the probability that X takes on a value in the interval $[a, b]$ is the area above this interval and under the graph of the density function, as illustrated in Figure 4.2. The graph of $f(x)$ is often referred to as the *density curve*.

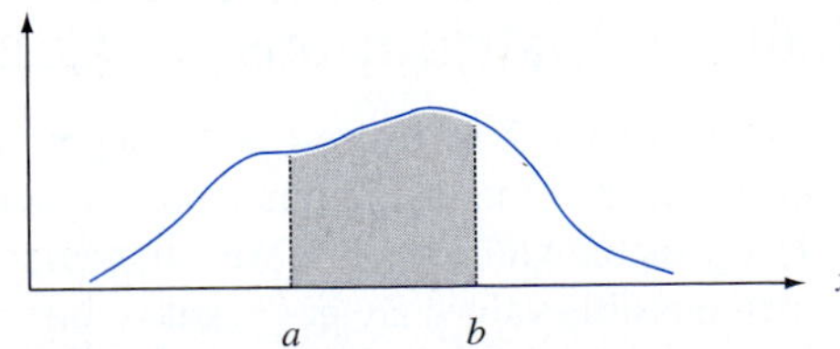

Figure 4.2 $P(a \le X \le b) =$ the area under the density curve between a and b

For $f(x)$ to be a legitimate pdf, it must satisfy the following two conditions:

1. $f(x) \geq 0$ for all x
2. $\int_{-\infty}^{\infty} f(x)\,dx = [\text{area under the entire graph of } f(x)] = 1$

Example 4.4 The direction of an imperfection with respect to a reference line on a circular object such as a tire, brake rotor, or flywheel is, in general, subject to uncertainty. Consider the reference line connecting the valve stem on a tire to the center point, and let X be the angle measured clockwise to the location of an imperfection. One possible pdf for X is

$$f(x) = \begin{cases} \dfrac{1}{360} & 0 \leq x < 360 \\ 0 & \text{otherwise} \end{cases}$$

The pdf is graphed in Figure 4.3. Clearly $f(x) \geq 0$. The area under the density curve is just the area of a rectangle: (height)(base) $= (\frac{1}{360})(360) = 1$. The probability that the angle is between 90° and 180° is

$$P(90 \leq X \leq 180) = \int_{90}^{180} \frac{1}{360}\,dx = \frac{x}{360}\Big|_{x=90}^{x=180} = \frac{1}{4} = .25$$

The probability that the angle of occurrence is within 90° of the reference line is

$$P(0 \leq X \leq 90) + P(270 \leq X < 360) = .25 + .25 = .50$$

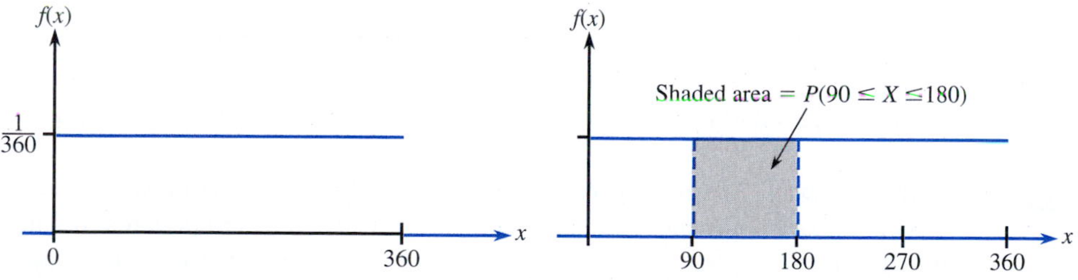

Figure 4.3 The pdf and probability for Example 4.4 ■

Because whenever $0 \leq a \leq b \leq 360$ in Example 4.4, $P(a \leq X \leq b)$ depends only on the width $b - a$ of the interval, X is said to have a uniform distribution.

DEFINITION A continuous rv X is said to have a **uniform distribution** on the interval $[A, B]$ if the pdf of X is

$$f(x; A, B) = \begin{cases} \dfrac{1}{B - A} & A \leq x \leq B \\ 0 & \text{otherwise} \end{cases}$$

The graph of any uniform pdf looks like the graph in Figure 4.3 except that the interval of positive density is $[A, B]$ rather than $[0, 360]$.

In the discrete case, a probability mass function (pmf) tells us how little "blobs" of probability mass of various magnitudes are distributed along the measurement axis. In the continuous case, probability density is "smeared" in a continuous fashion along the interval of possible values. When density is smeared uniformly over the interval, a uniform pdf, as in Figure 4.3, results.

When X is a discrete random variable, each possible value is assigned positive probability. This is not true of a continuous random variable (that is, the second condition of the definition is satisfied) because the area under a density curve that lies above any single value is zero:

$$P(X = c) = \int_c^c f(x)\,dx = \lim_{\varepsilon \to 0} \int_{c-\varepsilon}^{c+\varepsilon} f(x)\,dx = 0$$

The fact that $P(X = c) = 0$ when X is continuous has an important practical consequence: The probability that X lies in some interval between a and b does not depend on whether the lower limit a or the upper limit b is included in the probability calculation:

$$P(a \le X \le b) = P(a < X < b) = P(a < X \le b) = P(a \le X < b) \qquad (4.1)$$

If X is discrete and both a and b are possible values (e.g., X is binomial with $n = 20$ and $a = 5$, $b = 10$), then all four of these probabilities are different.

The zero probability condition has a physical analog. Consider a solid circular rod with cross-sectional area $= 1$ in^2. Place the rod alongside a measurement axis and suppose that the density of the rod at any point x is given by the value $f(x)$ of a density function. Then if the rod is sliced at points a and b and this segment is removed, the amount of mass removed is $\int_a^b f(x)dx$; if the rod is sliced just at the point c, no mass is removed. Mass is assigned to interval segments of the rod but not to individual points.

Example 4.5

"Time headway" in traffic flow is the elapsed time between the time that one car finishes passing a fixed point and the instant that the next car begins to pass that point. Let $X =$ the time headway for two randomly chosen consecutive cars on a freeway during a period of heavy flow. The following pdf of X is essentially the one suggested in "The Statistical Properties of Freeway Traffic" (*Transp. Res.*, vol. 11: 221–228):

$$f(x) = \begin{cases} .15e^{-.15(x-.5)} & x \ge .5 \\ 0 & \text{otherwise} \end{cases}$$

The graph of $f(x)$ is given in Figure 4.4; there is no density associated with headway times less than .5, and headway density decreases rapidly (exponentially fast) as x increases from .5. Clearly, $f(x) \ge 0$; to show that $\int_{-\infty}^{\infty} f(x)\,dx = 1$, we use the calculus result $\int_a^{\infty} e^{-kx}dx = (1/k)e^{-k \cdot a}$. Then

$$\int_{-\infty}^{\infty} f(x)\,dx = \int_{.5}^{\infty} .15e^{-.15(x-.5)}\,dx = .15e^{.075} \int_{.5}^{\infty} e^{-.15x}\,dx$$

$$= .15e^{.075} \cdot \frac{1}{.15} e^{-(.15)(.5)} = 1$$

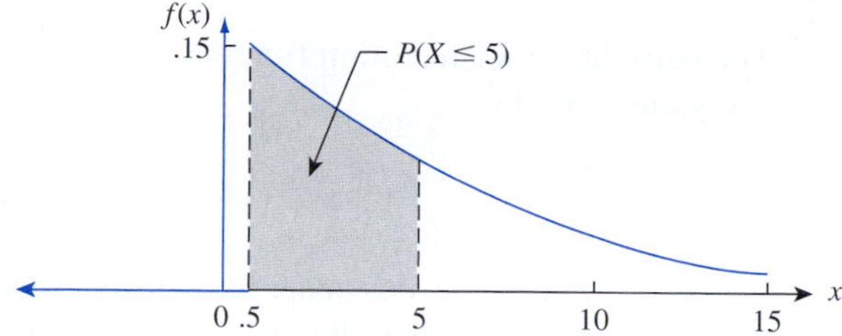

Figure 4.4 The density curve for headway time in Example 4.5

The probability that headway time is at most 5 sec is

$$P(X \le 5) = \int_{-\infty}^{5} f(x)\,dx = \int_{.5}^{5} .15e^{-.15(x-.5)}\,dx$$

$$= .15e^{.075}\int_{.5}^{5} e^{-.15x}\,dx = .15e^{.075} \cdot -\frac{1}{.15}e^{-.15x}\Big|_{x=.5}^{x=5}$$

$$= e^{.075}(-e^{-.75} + e^{-.075}) = 1.078(-.472 + .928) = .491$$

$$= P(\text{less than 5 sec}) = P(X < 5)$$ ■

Unlike discrete distributions such as the binomial, hypergeometric, and negative binomial, the distribution of any given continuous rv cannot usually be derived using simple probabilistic arguments. Instead, one must make a judicious choice of pdf based on prior knowledge and available data. Fortunately, some general families of pdf's have been found to fit well in a wide variety of experimental situations; several of these are discussed later in the chapter.

Just as in the discrete case, it is often helpful to think of the population of interest as consisting of X values rather than individuals or objects. The pdf is then a model for the distribution of values in this numerical population, and from this model various population characteristics (such as the mean) can be calculated.

Several of the most important concepts introduced in the study of discrete distributions also play an important role for continuous distributions. Definitions analogous to those in Chapter 3 involve replacing summation by integration.

The Cumulative Distribution Function

The cumulative distribution function (cdf) $F(x)$ for a discrete rv X gives, for any specified number x, the probability $P(X \le x)$. It is obtained by summing the pmf $p(y)$ over all possible values y satisfying $y \le x$. The cdf of a continuous rv gives the same probabilities $P(X \le x)$ and is obtained by integrating the pdf $f(y)$ between the limits $-\infty$ and x.

DEFINITION The **cumulative distribution function** $F(x)$ for a continuous rv X is defined for every number x by

$$F(x) = P(X \le x) = \int_{-\infty}^{x} f(y)\,dy$$

For each x, $F(x)$ is the area under the density curve to the left of x. This is illustrated in Figure 4.5, where $F(x)$ increases smoothly as x increases.

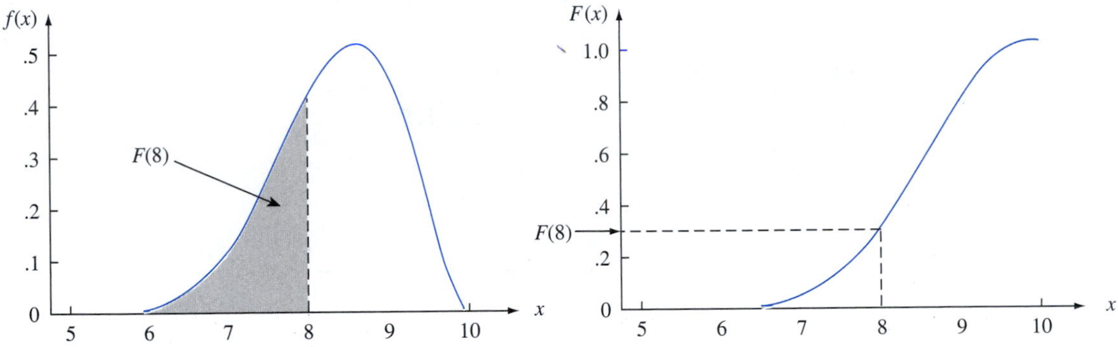

Figure 4.5 A pdf and associated cdf

Example 4.6 Let X, the thickness of a certain metal sheet, have a uniform distribution on $[A, B]$. The density function is shown in Figure 4.6. For $x < A$, $F(x) = 0$, since there is no area under the graph of the density function to the left of such an x. For $x \ge B$, $F(x) = 1$, since all the area is accumulated to the left of such an x. Finally, for $A \le x \le B$,

$$F(x) = \int_{-\infty}^{x} f(y)\,dy = \int_{A}^{x} \frac{1}{B-A}\,dy = \frac{1}{B-A} \cdot y \Big|_{y=A}^{y=x} = \frac{x-A}{B-A}$$

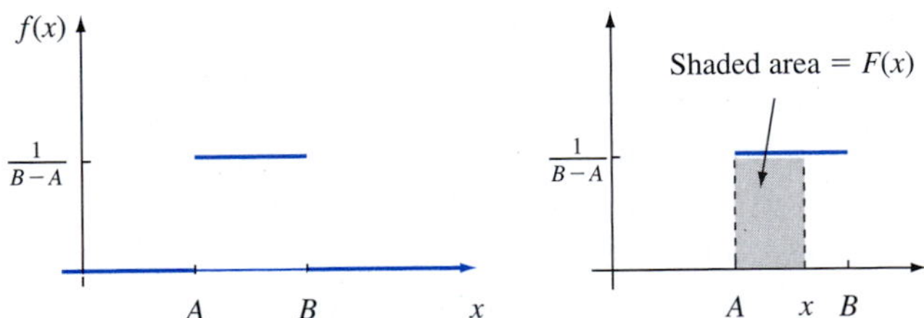

Figure 4.6 The pdf for a uniform distribution

The entire cdf is

$$F(x) = \begin{cases} 0 & x < A \\ \dfrac{x-A}{B-A} & A \le x < B \\ 1 & x \ge B \end{cases}$$

The graph of this cdf appears in Figure 4.7.

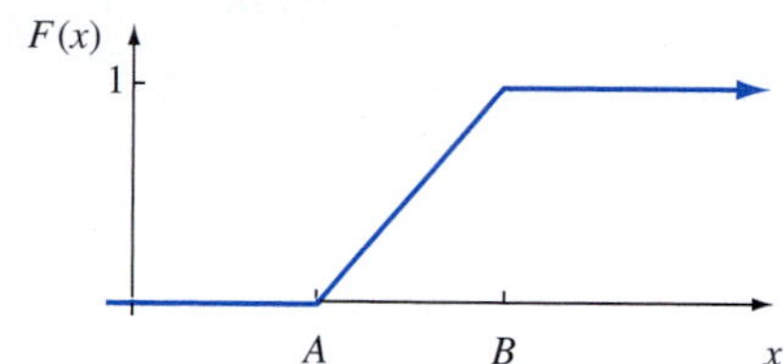

Figure 4.7 The cdf for a uniform distribution ■

Using F(x) to Compute Probabilities

The importance of the cdf here, just as for discrete rv's, is that probabilities of various intervals can be computed from a formula for or table of $F(x)$.

PROPOSITION Let X be a continuous rv with pdf $f(x)$ and cdf $F(x)$. Then for any number a,

$$P(X > a) = 1 - F(a)$$

and for any two numbers a and b with $a < b$,

$$P(a \le X \le b) = F(b) - F(a)$$

Figure 4.8 illustrates the second part of this proposition; the desired probability is the shaded area under the density curve between a and b, and it equals the difference between the two shaded cumulative areas.

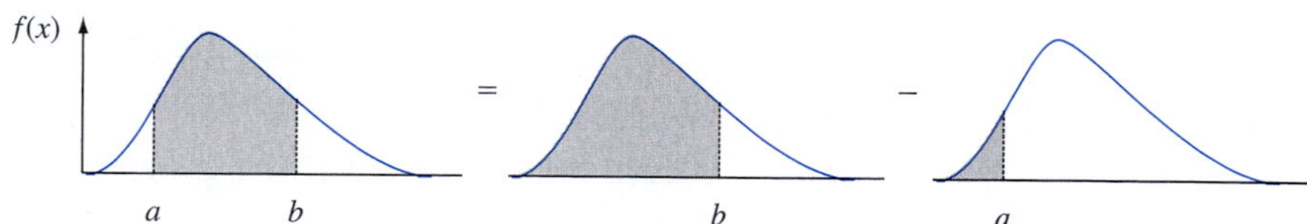

Figure 4.8 Computing $P(a \le X \le b)$ from cumulative probabilities

Example 4.7 Suppose the pdf of the magnitude X of a dynamic load on a bridge (in newtons) is given by

$$f(x) = \begin{cases} \frac{1}{8} + \frac{3}{8}x & 0 \le x \le 2 \\ 0 & \text{otherwise} \end{cases}$$

For any number x between 0 and 2,

$$F(x) = \int_{-\infty}^{x} f(y)\,dy = \int_{0}^{x} \left(\frac{1}{8} + \frac{3}{8}y\right) dy = \frac{x}{8} + \frac{3}{16}x^2$$

Thus

$$F(x) = \begin{cases} 0 & x < 0 \\ \dfrac{x}{8} + \dfrac{3}{16}x^2 & 0 \le x \le 2 \\ 1 & 2 < x \end{cases}$$

The graphs of $f(x)$ and $F(x)$ are shown in Figure 4.9. The probability that the load is between 1 and 1.5 is

$$\begin{aligned} P(1 \le X \le 1.5) &= F(1.5) - F(1) \\ &= \left[\frac{1}{8}(1.5) + \frac{3}{16}(1.5)^2\right] - \left[\frac{1}{8}(1) + \frac{3}{16}(1)^2\right] \\ &= \frac{19}{64} = .297 \end{aligned}$$

The probability that the load exceeds 1 is

$$P(X > 1) = 1 - P(X \le 1) = 1 - F(1) = 1 - \left[\frac{1}{8}(1) + \frac{3}{16}(1)^2\right]$$

$$= \frac{11}{16} = .688$$

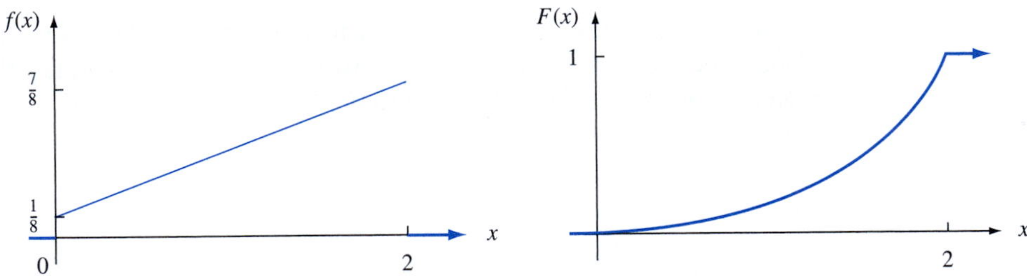

Figure 4.9 The pdf and cdf for Example 4.7 ■

Once the cdf has been obtained, any probability involving X can easily be calculated without any further integration.

Obtaining $f(x)$ from $F(x)$

For X discrete, the pmf is obtained from the cdf by taking the difference between two $F(x)$ values. The continuous analog of a difference is a derivative. The following result is a consequence of the Fundamental Theorem of Calculus.

PROPOSITION

If X is a continuous rv with pdf $f(x)$ and cdf $F(x)$, then at every x at which the derivative $F'(x)$ exists, $F'(x) = f(x)$.

Example 4.8 (Example 4.6 continued)

When X has a uniform distribution, $F(x)$ is differentiable except at $x = A$ and $x = B$, where the graph of $F(x)$ has sharp corners. Since $F(x) = 0$ for $x < A$ and $F(x) = 1$ for $x > B$, $F'(x) = 0 = f(x)$ for such x. For $A < x < B$,

$$F'(x) = \frac{d}{dx}\left(\frac{x - A}{B - A}\right) = \frac{1}{B - A} = f(x)$$

■

Percentiles of a Continuous Distribution

When we say that an individual's test score was at the 85th percentile of the population, we mean that 85% of all population scores were below that score and 15% were above. Similarly, the 40th percentile is the score that exceeds 40% of all scores and is exceeded by 60% of all scores.

DEFINITION

Let p be a number between 0 and 1. The **(100*p*)th percentile** of the distribution of a continuous rv X, denoted by $\eta(p)$, is defined by

$$p = F[\eta(p)] = \int_{-\infty}^{\eta(p)} f(y)\,dy \tag{4.2}$$

According to Expression (4.2), $\eta(p)$ is that value on the measurement axis such that $100p\%$ of the area under the graph of $f(x)$ lies to the left of $\eta(p)$ and $100(1 - p)\%$ lies to the right. Thus $\eta(.75)$, the 75th percentile, is such that the area under the graph of $f(x)$ to the left of $\eta(.75)$ is .75. Figure 4.10 illustrates the definition.

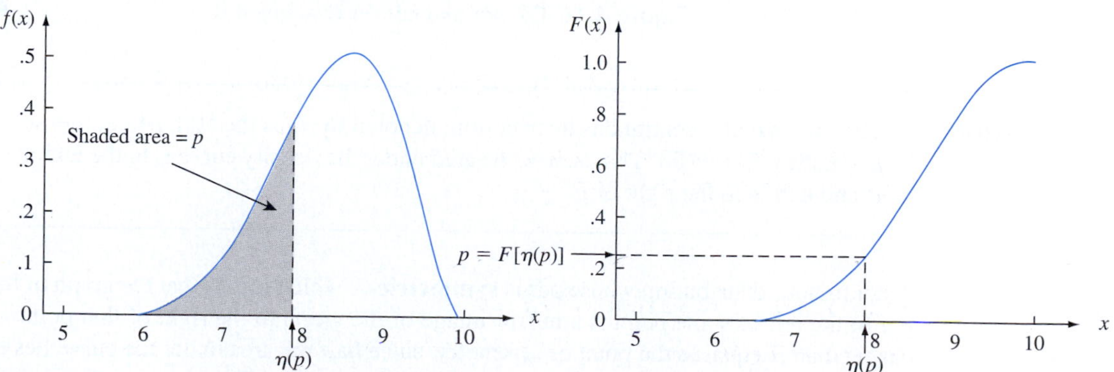

Figure 4.10 The (100*p*)th percentile of a continuous distribution

Example 4.9 The distribution of the amount of gravel (in tons) sold by a particular construction supply company in a given week is a continuous rv X with pdf

$$f(x) = \begin{cases} \frac{3}{2}(1 - x^2) & 0 \le x \le 1 \\ 0 & \text{otherwise} \end{cases}$$

The cdf of sales for any x between 0 and 1 is

$$F(x) = \int_0^x \frac{3}{2}(1 - y^2)\,dy = \frac{3}{2}\left(y - \frac{y^3}{3}\right)\Bigg|_{y=0}^{y=x} = \frac{3}{2}\left(x - \frac{x^3}{3}\right)$$

The graphs of both $f(x)$ and $F(x)$ appear in Figure 4.11. The $(100p)$th percentile of this distribution satisfies the equation

$$p = F[\eta(p)] = \frac{3}{2}\left[\eta(p) - \frac{[\eta(p)]^3}{3}\right]$$

that is,

$$[\eta(p)]^3 - 3\eta(p) + 2p = 0$$

For the 50th percentile, $p = .5$, and the equation to be solved is $\eta^3 - 3\eta + 1 = 0$; the solution is $\eta = \eta(.5) = .347$. If the distribution remains the same from week to week, then in the long run 50% of all weeks will result in sales of less than .347 ton and 50% in more than .347 ton.

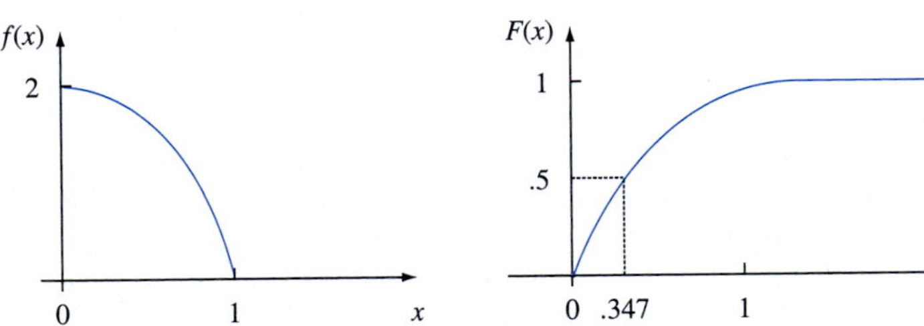

Figure 4.11 The pdf and cdf for Example 4.9 ■

DEFINITION The **median** of a continuous distribution, denoted by $\tilde{\mu}$, is the 50th percentile, so $\tilde{\mu}$ satisfies $.5 = F(\tilde{\mu})$. That is, half the area under the density curve is to the left of $\tilde{\mu}$ and half is to the right of $\tilde{\mu}$.

A continuous distribution whose pdf is **symmetric**—which means that the graph of the pdf to the left of some point is a mirror image of the graph to the right of that point—has median $\tilde{\mu}$ equal to the point of symmetry, since half the area under the curve lies to either side of this point. Figure 4.12 gives several examples. The amount of error in a measurement of a physical quantity is often assumed to have a symmetric distribution.

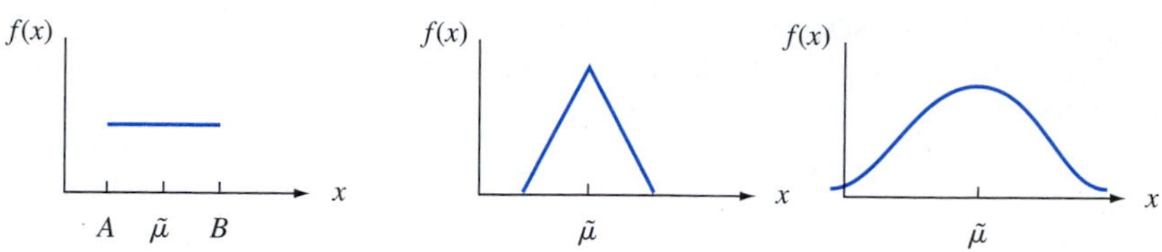

Figure 4.12 Medians of symmetric distributions

Exercises Section 4.1 (1–17)

1. Let X denote the amount of time for which a book on 2-hour reserve at a college library is checked out by a randomly selected student and suppose that X has density function

$$f(x) = \begin{cases} .5x & 0 \le x \le 2 \\ 0 & \text{otherwise} \end{cases}$$

Calculate the following probabilities:
a. $P(X \le 1)$
b. $P(.5 \le X \le 1.5)$
c. $P(1.5 < X)$

2. Suppose the reaction temperature X (in °C) in a certain chemical process has a uniform distribution with $A = -5$ and $B = 5$.
a. Compute $P(X < 0)$.
b. Compute $P(-2.5 < X < 2.5)$.
c. Compute $P(-2 \le X \le 3)$.
d. For k satisfying $-5 < k < k + 4 < 5$, compute $P(k < X < k + 4)$.

3. Suppose the error involved in making a certain measurement is a continuous rv X with pdf

$$f(x) = \begin{cases} .09375(4 - x^2) & -2 \le x \le 2 \\ 0 & \text{otherwise} \end{cases}$$

a. Sketch the graph of $f(x)$.
b. Compute $P(X > 0)$.
c. Compute $P(-1 < X < 1)$.
d. Compute $P(X < -.5 \text{ or } X > .5)$.

4. Let X denote the vibratory stress (psi) on a wind turbine blade at a particular wind speed in a wind tunnel. The article "Blade Fatigue Life Assessment with Application to VAWTS" (*J. Solar Energy Engrg.*, 1982: 107–111) proposes the Rayleigh distribution, with pdf

$$f(x, \theta) = \begin{cases} \dfrac{x}{\theta^2} \cdot e^{-x^2/(2\theta^2)} & x > 0 \\ 0 & \text{otherwise} \end{cases}$$

as a model for the X distribution.
a. Verify that $f(x; \theta)$ is a legitimate pdf.
b. Suppose $\theta = 100$ (a value suggested by a graph in the article). What is the probability that X is at most 200? Less than 200? At least 200?
c. What is the probability that X is between 100 and 200 (again assuming $\theta = 100$)?
d. Give an expression for $P(X \le x)$.

5. A college professor never finishes his lecture before the end of the hour and always finishes his lectures within 2 min after the hour. Let X = the time that elapses between the end of the hour and the end of the lecture and suppose the pdf of X is

$$f(x) = \begin{cases} kx^2 & 0 \le x \le 2 \\ 0 & \text{otherwise} \end{cases}$$

a. Find the value of k. [*Hint*: Total area under the graph of $f(x)$ is 1.]
b. What is the probability that the lecture ends within 1 min of the end of the hour?
c. What is the probability that the lecture continues beyond the hour for between 60 and 90 sec?
d. What is the probability that the lecture continues for at least 90 sec beyond the end of the hour?

6. The grade point averages (GPA's) for graduating seniors at a college are distributed as a continuous rv X with pdf

$$f(x) = \begin{cases} k[1 - (x - 3)^2] & 2 \le x \le 4 \\ 0 & \text{otherwise} \end{cases}$$

a. Sketch the graph of $f(x)$.
b. Find the value of k.
c. Find the probability that a GPA exceeds 3.

d. Find the probability that a GPA is within .25 of 3.
e. Find the probability that a GPA differs from 3 by more than .5.

7. The time X (min) for a lab assistant to prepare the equipment for a certain experiment is believed to have a uniform distribution with $A = 25$ and $B = 35$.
a. Write the pdf of X and sketch its graph.
b. What is the probability that preparation time exceeds 33 min?
c. What is the probability that preparation time is within 2 min of the mean time? [*Hint:* Identify μ from the graph of $f(x)$.]
d. For any a such that $25 < a < a + 2 < 35$, what is the probability that preparation time is between a and $a + 2$ min?

8. Commuting to work requires getting on a bus near home and then transferring to a second bus. If the waiting time (in minutes) at each stop has a uniform distribution with $A = 0$ and $B = 5$, then it can be shown that the total waiting time Y has the pdf

$$f(y) = \begin{cases} \frac{1}{25}y & 0 \le y < 5 \\ \frac{2}{5} - \frac{1}{25}y & 5 \le y \le 10 \\ 0 & y < 0 \text{ or } y > 10 \end{cases}$$

a. Sketch a graph of the pdf of Y.
b. Verify that $\int_{-\infty}^{\infty} f(y)\,dy = 1$.
c. What is the probability that total waiting time is at most 3 min?
d. What is the probability that total waiting time is at most 8 min?
e. What is the probability that total waiting time is between 3 and 8 min?
f. What is the probability that total waiting time is either less than 2 min or more than 6 min?

9. Consider again the pdf of X = time headway given in Example 4.5. What is the probability that time headway is
a. At most 6 sec?
b. More than 6 sec? At least 6 sec?
c. Between 5 and 6 sec?

10. A family of pdf's that has been used to approximate the distribution of income, city population size, and size of firms is the Pareto family. The family has two parameters, k and θ, both > 0, and the pdf is

$$f(x; k, \theta) = \begin{cases} \frac{k \cdot \theta^k}{x^{k+1}} & x \ge \theta \\ 0 & x < \theta \end{cases}$$

a. Sketch the graph of $f(x; k, \theta)$.
b. Verify that the total area under the graph equals 1.
c. If the rv X has pdf $f(x; k, \theta)$, for any fixed $b > \theta$, obtain an expression for $P(X \le b)$.
d. For $\theta < a < b$, obtain an expression for the probability $P(a \le X \le b)$.

11. The cdf of checkout duration X as described in Exercise 1 is

$$F(x) = \begin{cases} 0 & x < 0 \\ \frac{x^2}{4} & 0 \le x < 2 \\ 1 & 2 \le x \end{cases}$$

Use this to compute the following:
a. $P(X \le 1)$
b. $P(.5 \le X \le 1)$
c. $P(X > .5)$
d. The median checkout duration $\tilde{\mu}$ [solve $.5 = F(\tilde{\mu})$]
e. $F'(x)$ to obtain the density function $f(x)$

12. The cdf for X (= measurement error) of Exercise 3 is

$$F(x) = \begin{cases} 0 & x < -2 \\ \frac{1}{2} + \frac{3}{32}\left(4x - \frac{x^3}{3}\right) & -2 \le x < 2 \\ 1 & 2 \le x \end{cases}$$

a. Compute $P(X < 0)$.
b. Compute $P(-1 < X < 1)$.
c. Compute $P(.5 < X)$.
d. Verify that $f(x)$ is as given in Exercise 3 by obtaining $F'(x)$.
e. Verify that $\tilde{\mu} = 0$.

13. Example 4.5 introduced the concept of time headway in traffic flow and proposed a particular distribution for X = the headway between two randomly selected consecutive cars (sec). Suppose that in a different traffic environment, the distribution of time headway has the form

$$f(x) = \begin{cases} \frac{k}{x^4} & x > 1 \\ 0 & x \le 1 \end{cases}$$

a. Determine the value of k for which $f(x)$ is a legitimate pdf.
b. Obtain the cumulative distribution function.
c. Use the cdf from (b) to determine the probability that headway exceeds 2 sec and also the probability that headway is between 2 and 3 sec.

14. Let X denote the amount of space occupied by an article placed in a 1-ft^3 packing container. The pdf of X is

$$f(x) = \begin{cases} 90x^8(1 - x) & 0 < x < 1 \\ 0 & \text{otherwise} \end{cases}$$

a. Graph the pdf. Then obtain the cdf of X and graph it.
b. What is $P(X \le .5)$ [i.e., $F(.5)$]?
c. Using part (a), what is $P(.25 < X \le .5)$? What is $P(.25 \le X \le .5)$?
d. What is the 75th percentile of the distribution?

15. Answer parts (a)–(d) of Exercise 14 for the random variable X, lecture time past the hour, given in Exercise 5.

16. Let X be a continuous rv with cdf

$$F(x) = \begin{cases} 0 & x \le 0 \\ \dfrac{x}{4}\left[1 + \ln\left(\dfrac{4}{x}\right)\right] & 0 < x \le 4 \\ 1 & x > 4 \end{cases}$$

[This type of cdf is suggested in the article "Variability in Measured Bedload-Transport Rates" (*Water Resources Bull.*, 1985: 39–48) as a model for a certain hydrologic variable.] What is
a. $P(X \le 1)$?
b. $P(1 \le X \le 3)$?
c. The pdf of X?

17. Let X be the temperature in °C at which a certain chemical reaction takes place, and let Y be the temperature in °F (so $Y = 1.8X + 32$).
a. If the median of the X distribution is $\tilde{\mu}$, show that $1.8\tilde{\mu} + 32$ is the median of the Y distribution.
b. How is the 90th percentile of the Y distribution related to the 90th percentile of the X distribution? Verify your conjecture.
c. More generally, if $Y = aX + b$, how is any particular percentile of the Y distribution related to the corresponding percentile of the X distribution?

4.2 Expected Values and Moment Generating Functions

In Section 4.1 we saw that the transition from a discrete cdf to a continuous cdf entails replacing summation by integration. The same thing is true in moving from expected values and mgf's of discrete variables to those of continuous variables.

Expected Values

For a discrete random variable X, $E(X)$ was obtained by summing $x \cdot p(x)$ over possible X values. Here we replace summation by integration and the pmf by the pdf to get a continuous weighted average.

DEFINITION

The **expected** or **mean value** of a continuous rv X with pdf $f(x)$ is

$$\mu_X = E(X) = \int_{-\infty}^{\infty} x \cdot f(x)\,dx$$

This expected value will exist provided that $\int_{-\infty}^{\infty} |x| f(x)\,dx < \infty$.

Example 4.10 (Example 4.9 continued)

The pdf of weekly gravel sales X was

$$f(x) = \begin{cases} \frac{3}{2}(1 - x^2) & 0 \le x \le 1 \\ 0 & \text{otherwise} \end{cases}$$

so

$$E(X) = \int_{-\infty}^{\infty} x \cdot f(x)\,dx = \int_0^1 x \cdot \frac{3}{2}(1 - x^2)\,dx$$

$$= \frac{3}{2}\int_0^1 (x - x^3)\,dx = \frac{3}{2}\left(\frac{x^2}{2} - \frac{x^4}{4}\right)\Bigg|_{x=0}^{x=1} = \frac{3}{8}$$

If gravel sales are determined week after week according to the given pdf, then the long-run average value of sales per week will be .375 ton. ■

When the pdf $f(x)$ specifies a model for the distribution of values in a numerical population, then μ is the population mean, which is the most frequently used measure of population location or center.

Often we wish to compute the expected value of some function $h(X)$ of the rv X. If we think of $h(X)$ as a new rv Y, methods from Section 4.7 can be used to derive the pdf of Y, and $E(Y)$ can be computed from the definition. Fortunately, as in the discrete case, there is an easier way to compute $E[h(X)]$.

PROPOSITION

If X is a continuous rv with pdf $f(x)$ and $h(X)$ is any function of X, then

$$E[h(X)] = \mu_{h(X)} = \int_{-\infty}^{\infty} h(x) \cdot f(x)\,dx$$

Example 4.11

Two species are competing in a region for control of a limited amount of a certain resource. Let X = the proportion of the resource controlled by species 1 and suppose X has pdf

$$f(x) = \begin{cases} 1 & 0 \le x \le 1 \\ 0 & \text{otherwise} \end{cases}$$

which is a uniform distribution on [0, 1]. (In her book *Ecological Diversity*, E. C. Pielou calls this the "broken-stick" model for resource allocation, since it is analogous to breaking a stick at a randomly chosen point.) Then the species that controls the majority of this resource controls the amount

$$h(X) = \max(X, 1 - X) = \begin{cases} 1 - X & \text{if } 0 \le X < \frac{1}{2} \\ X & \text{if } \frac{1}{2} \le X \le 1 \end{cases}$$

The expected amount controlled by the species having majority control is then

$$E[h(X)] = \int_{-\infty}^{\infty} \max(x, 1 - x) \cdot f(x)\,dx = \int_{0}^{1} \max(x, 1 - x) \cdot 1\,dx$$

$$= \int_{0}^{1/2} (1 - x) \cdot 1\,dx + \int_{1/2}^{1} x \cdot 1\,dx = \frac{3}{4}$$

■

The Variance and Standard Deviation

DEFINITION

The **variance** of a continuous random variable X with pdf $f(x)$ and mean value μ is

$$\sigma_X^2 = V(X) = \int_{-\infty}^{\infty} (x - \mu)^2 \cdot f(x)\,dx = E[(X - \mu)^2]$$

The **standard deviation** (SD) of X is $\sigma_X = \sqrt{V(X)}$.

As in the discrete case, σ_X^2 is the expected or average squared deviation about the mean μ, and σ_X can be interpreted roughly as the size of a representative deviation from the mean value μ. The easiest way to compute σ^2 is again to use a shortcut formula.

PROPOSITION

$$V(X) = E(X^2) - [E(X)]^2$$

The derivation is similar to the derivation for the discrete case in Section 3.3.

Example 4.12 (Example 4.10 continued)

For X = weekly gravel sales, we computed $E(X) = \frac{3}{8}$. Since

$$E(X^2) = \int_{-\infty}^{\infty} x^2 \cdot f(x)\,dx = \int_{0}^{1} x^2 \cdot \frac{3}{2}(1 - x^2)\,dx$$

$$= \int_{0}^{1} \frac{3}{2}(x^2 - x^4)\,dx = \frac{1}{5}$$

$$V(X) = \frac{1}{5} - \left(\frac{3}{8}\right)^2 = \frac{19}{320} = .059 \quad \text{and} \quad \sigma_X = .244$$

■

Often in applications it is the case that $h(X) = aX + b$, a linear function of X. For example, $h(X) = 1.8X + 32$ gives the transformation of temperature from the Celsius scale to the Fahrenheit scale. When $h(X)$ is linear, its mean and variance are easily related to those of X itself, as discussed for the discrete case in Section 3.3. The derivations in the continuous case are the same. We have

$$E(aX + b) = aE(X) + b \qquad V(aX + b) = a^2\sigma_X^2 \qquad \sigma_{aX+b} = |a|\sigma_X$$

Example 4.13 When a dart is thrown at a circular target, consider the location of the landing point relative to the bull's eye. Let X be the angle in degrees measured from the horizontal, and assume that X is uniformly distributed on [0, 360]. By Exercise 23, $E(X) = 180$ and $\sigma_X = 360/\sqrt{12}$. Define Y to be the transformed variable $Y = h(X) = (2\pi/360)X - \pi$, so Y is the angle measured in radians and Y is between $-\pi$ and π. Then

$$E(Y) = \frac{2\pi}{360}E(X) - \pi = \frac{2\pi}{360}180 - \pi = 0$$

and

$$\sigma_Y = \frac{2\pi}{360}\sigma_X = \frac{2\pi}{360}\frac{360}{\sqrt{12}} = \frac{2\pi}{\sqrt{12}}$$

■

As a special case of the result $E(aX + b) = aE(X) + b$, set $a = 1$ and $b = -\mu$, giving $E(X - \mu) = E(X) - \mu = 0$. This can be interpreted as saying that the expected deviation from μ is 0; $\int_{-\infty}^{\infty}(x - \mu)f(x)\,dx = 0$. The integral suggests a physical interpretation: With $(x - \mu)$ as the lever arm and $f(x)$ as the weight function, the total torque is 0. Using a seesaw as a model with weight distributed in accord with $f(x)$, the seesaw will balance at μ. Alternatively, if the region bounded by the pdf curve and the x-axis is cut out of cardboard, then it will balance if supported at μ. If $f(x)$ is symmetric, then it will balance at its point of symmetry, which must be the mean μ, assuming that the mean exists. The point of symmetry for X in Example 4.13 is 180, so it follows that $\mu = 180$. Recall from Section 4.1 that the median is also the point of symmetry, so the median of X in Example 4.13 is also 180. In general, if the distribution is symmetric and the mean exists, then it is equal to the median.

Moment Generating Functions

Moments and moment generating functions for discrete random variables were introduced in Section 3.4. These concepts carry over to the continuous case.

DEFINITION The **moment generating function** (mgf) of a continuous random variable X is

$$M_X(t) = E(e^{tX}) = \int_{-\infty}^{\infty} e^{tx}f(x)\,dx$$

As in the discrete case, we will say that the moment generating function exists if $M_X(t)$ is defined for an interval of numbers that includes zero in its interior, which means that it includes both positive and negative values of t.

Just as before, when $t = 0$ the value of the mgf is always 1:

$$M_X(0) = E(e^{0X}) = \int_{-\infty}^{\infty} e^{0x}f(x)\,dx = \int_{-\infty}^{\infty} f(x)\,dx = 1$$

Example 4.14 At a store the checkout time X in minutes has the pdf $f(x) = 2e^{-2x}$, $x \geq 0$; $f(x) = 0$ otherwise. Then

$$M_X(t) = \int_{-\infty}^{\infty} e^{tx} f(x)\,dx = \int_0^{\infty} e^{tx}(2e^{-2x})\,dx = \int_0^{\infty} 2e^{-(2-t)x}\,dx$$

$$= \frac{-2}{2-t} e^{-(2-t)x}\Big|_0^{\infty} = \frac{2}{2-t} \quad \text{if } t < 2$$

This mgf exists because it is defined for an interval of values including 0 in its interior.

Notice that $M_X(0) = 2/(2 - 0) = 1$. Of course, from the calculation preceding this example we know that $M_X(0) = 1$ always, but it is useful as a check to set $t = 0$ and see if the result is 1. ■

Recall that in the discrete case we had a proposition stating the uniqueness principle: The mgf uniquely identifies the distribution. This proposition is equally valid in the continuous case. Two distributions have the same pdf if and only if they have the same moment generating function, assuming that the mgf exists.

Example 4.15 Let X be a random variable with mgf $M_X(t) = 2/(2 - t)$, $t < 2$. Can we find the pdf $f(x)$? Yes, because we know from Example 4.14 that if $f(x) = 2e^{-2x}$ when $x \geq 0$, and $f(x) = 0$ otherwise, then $M_X(t) = 2/(2 - t)$, $t < 2$. The uniqueness principle implies that this is the only pdf with the given mgf, and therefore $f(x) = 2e^{-2x}$, $x \geq 0$, $f(x) = 0$ otherwise. ■

In the discrete case we had a theorem on how to get moments from the mgf, and this theorem applies also in the continuous case: $E(X^r) = M_X^{(r)}(0)$, the rth derivative of the mgf with respect to t evaluated at $t = 0$, if the mgf exists.

Example 4.16 In Example 4.14, for the pdf $f(x) = 2e^{-2x}$ when $x \geq 0$, and $f(x) = 0$ otherwise, we found $M_X(t) = 2/(2 - t) = 2(2 - t)^{-1}$, $t < 2$. To find the mean and variance, first compute the derivatives.

$$M'_X(t) = -2(2 - t)^{-2}(-1) = \frac{2}{(2 - t)^2}$$

$$M''_X(t) = (-2)(-2)(2 - t)^{-3}(-1)(-1) = \frac{4}{(2 - t)^3}$$

Setting t to 0 in the first derivative gives the expected checkout time as

$$E(X) = M'_X(0) = M_X^{(1)}(0) = .5$$

Setting t to 0 in the second derivative gives the second moment:

$$E(X^2) = M'_X(0) = M_X^{(2)}(0) = .5$$

The variance of the checkout time is then

$$V(X) = \sigma^2 = E(X^2) - [E(X)]^2 = .5 - .5^2 = .25$$ ■

As mentioned in Section 3.4, there is another way of doing the differentiation that is sometimes more straightforward. Define $R_X(t) = \ln[M_X(t)]$, where $\ln(u)$ is the natural log of u. Then if the moment generating function exists,

$$\mu = E(X) = R_X'(0)$$

$$\sigma^2 = V(X) = R_X''(0)$$

The derivation for the discrete case in Exercise 54 of Section 3.4 also applies here in the continuous case.

We will sometimes need to transform X using a linear function $Y = aX + b$. As discussed in the discrete case, if X has the mgf $M_X(t)$ and $Y = aX + b$, then $M_Y(t) = e^{bt}M_X(at)$.

Example 4.17 Let X have a uniform distribution on the interval $[A, B]$, so its pdf is $f(x) = 1/(B - A)$, $A \le x \le B$; $f(x) = 0$ otherwise. As verified in Exercise 31, the moment generating function of X is

$$M_X(t) = \begin{cases} \dfrac{e^{Bt} - e^{At}}{(B - A)t} & t \ne 0 \\ 1 & t = 0 \end{cases}$$

In particular, consider the situation in Example 4.13. Let X, the angle measured in degrees, be uniform on $[0, 360]$, so $A = 0$ and $B = 360$. Then

$$M_X(t) = \frac{e^{360t} - 1}{360t} \qquad t \ne 0 \qquad M_X(0) = 1$$

Now let $Y = (2\pi/360)X - \pi$, so Y is the angle measured in radians and Y is between $-\pi$ and π. Using the foregoing property with $a = 2\pi/360$ and $b = -\pi$, we get

$$M_Y(t) = e^{bt}M_X(at) = e^{-\pi t}M_X\left(\frac{2\pi}{360}t\right)$$

$$= e^{-\pi t}\frac{e^{360(2\pi/360)t} - 1}{360\left(\dfrac{2\pi}{360}t\right)}$$

$$= \frac{e^{\pi t} - e^{-\pi t}}{2\pi t} \qquad t \ne 0 \qquad M_Y(0) = 1$$

This matches the general form of the moment generating function for a uniform random variable with $A = -\pi$ and $B = \pi$. Thus, by the uniqueness principle, Y is uniformly distributed on $[-\pi, \pi]$. ■

Exercises Section 4.2 (18–38)

18. Reconsider the distribution of checkout duration X described in Exercises 1 and 11. Compute the following:

a. $E(X)$

b. $V(X)$ and σ_X

c. If the borrower is charged an amount $h(X) = X^2$ when checkout duration is X, compute the expected charge $E[h(X)]$.

19. Recall the distribution of time headway used in Example 4.5.

a. Obtain the mean value of headway and the standard deviation of headway.

b. What is the probability that headway is within 1 standard deviation of the mean value?

20. The article "Modeling Sediment and Water Column Interactions for Hydrophobic Pollutants" (*Water Res.*, 1984: 1169–1174) suggests the uniform distribution on the interval (7.5, 20) as a model for depth (cm) of the bioturbation layer in sediment in a certain region.

a. What are the mean and variance of depth?

b. What is the cdf of depth?

c. What is the probability that observed depth is at most 10? Between 10 and 15?

d. What is the probability that the observed depth is within 1 standard deviation of the mean value? Within 2 standard deviations?

21. For the distribution of Exercise 14,

a. Compute $E(X)$ and σ_X.

b. What is the probability that X is more than 2 standard deviations from its mean value?

22. Consider the pdf of X = grade point average given in Exercise 6.

a. Obtain and graph the cdf of X.

b. From the graph of $f(x)$, what is $\tilde{\mu}$?

c. Compute $E(X)$ and $V(X)$.

23. Let X have a uniform distribution on the interval $[A, B]$.

a. Obtain an expression for the $(100p)$th percentile.

b. Compute $E(X)$, $V(X)$, and σ_X.

c. For n a positive integer, compute $E(X^n)$.

24. Consider the pdf for total waiting time Y for two buses

$$f(y) = \begin{cases} \frac{1}{25}y & 0 \le y < 5 \\ \frac{2}{5} - \frac{1}{25}y & 5 \le y \le 10 \\ 0 & \text{otherwise} \end{cases}$$

introduced in Exercise 8.

a. Compute and sketch the cdf of Y. [*Hint*: Consider separately $0 \le y < 5$ and $5 \le y \le 10$ in computing $F(y)$. A graph of the pdf should be helpful.]

b. Obtain an expression for the $(100p)$th percentile. (*Hint*: Consider separately $0 < p < .5$ and $.5 < p < 1$.)

c. Compute $E(Y)$ and $V(Y)$. How do these compare with the expected waiting time and variance for a single bus when the time is uniformly distributed on $[0, 5]$?

d. Explain how symmetry can be used to obtain $E(Y)$.

25. An ecologist wishes to mark off a circular sampling region having radius 10 m. However, the radius of the resulting region is actually a random variable R with pdf

$$f(r) = \begin{cases} \frac{3}{4}[1 - (10 - r)^2] & 9 \le r \le 11 \\ 0 & \text{otherwise} \end{cases}$$

What is the expected area of the resulting circular region?

26. The weekly demand for propane gas (in 1000's of gallons) from a particular facility is an rv X with pdf

$$f(x) = \begin{cases} 2\left(1 - \frac{1}{x^2}\right) & 1 \le x \le 2 \\ 0 & \text{otherwise} \end{cases}$$

a. Compute the cdf of X.

b. Obtain an expression for the $(100p)$th percentile. What is the value of $\tilde{\mu}$?

c. Compute $E(X)$ and $V(X)$.

d. If 1.5 thousand gallons are in stock at the beginning of the week and no new supply is due in during the week, how much of the 1.5 thousand gallons is expected to be left at the end of the week? *Hint*: Let $h(x)$ = amount left when demand = x.

27. If the temperature at which a certain compound melts is a random variable with mean value 120°C and standard deviation 2°C, what are the mean temperature and standard deviation measured in °F? (*Hint*: °F = 1.8°C + 32.)

28. Let X have the Pareto pdf

$$f(x; k, \theta) = \begin{cases} \dfrac{k \cdot \theta^k}{x^{k+1}} & x \geq \theta \\ 0 & x < \theta \end{cases}$$

introduced in Exercise 10.

a. If $k > 1$, compute $E(X)$.
b. What can you say about $E(X)$ if $k = 1$?
c. If $k > 2$, show that $V(X) = k\theta^2(k-1)^{-2}(k-2)^{-1}$.
d. If $k = 2$, what can you say about $V(X)$?
e. What conditions on k are necessary to ensure that $E(X^n)$ is finite?

29. At a Website, the waiting time X (in minutes) between hits has pdf $f(x) = 4e^{-4x}$, $x \geq 0$; $f(x) = 0$ otherwise. Find $M_X(t)$ and use it to obtain $E(X)$ and $V(X)$.

30. Suppose that the pdf of X is

$$f(x) = \begin{cases} .5 - \dfrac{x}{8} & 0 \leq x \leq 4 \\ 0 & \text{otherwise} \end{cases}$$

a. Show that $E(X) = \frac{4}{3}$, $V(X) = \frac{8}{9}$.
b. The coefficient of skewness is $E[(X - \mu)^3]/\sigma^3$. Show that its value for the given pdf is .566. What would the skewness be for a perfectly symmetric pdf?

31. Let X have a uniform distribution on the interval $[A, B]$, so its pdf is $f(x) = 1/(B - A)$, $A \leq x \leq B$, $f(x) = 0$ otherwise. Show that the moment generating function of X is

$$M_X(t) = \begin{cases} \dfrac{e^{Bt} - e^{At}}{(B - A)t} & t \neq 0 \\ 1 & t = 0 \end{cases}$$

32. Use Exercise 31 to find the pdf $f(x)$ of X if its moment generating function is

$$M_X(t) = \begin{cases} \dfrac{e^{5t} - e^{-5t}}{10t} & t \neq 0 \\ 1 & t = 0 \end{cases}$$

Explain why you know that your $f(x)$ is uniquely determined by $M_X(t)$.

33. If the pdf of a measurement error X is $f(x) = .5e^{-|x|}$, $-\infty < x < \infty$, show that

$$M_X(t) = \frac{1}{1 - t^2} \quad \text{for } |t| < 1$$

34. In Example 4.5 the pdf of X is given as

$$f(x) = \begin{cases} .15e^{-.15(x-.5)} & x \geq .5 \\ 0 & \text{otherwise} \end{cases}$$

Find the moment generating function and use it to find the mean and variance.

35. For the mgf of Exercise 34, obtain the mean and variance by differentiating $R_X(t)$. Compare the answers with the results of Exercise 34.

36. Let X be uniformly distributed on [0, 1]. Find a linear function $Y = g(X)$ such that the interval [0, 1] is transformed into [−5, 5]. Use the relationship for linear functions $M_{aX+b}(t) = e^{bt}M_X(at)$ to obtain the mgf of Y from the mgf of X. Compare your answer with the result of Exercise 31, and use this to obtain the pdf of Y.

37. Suppose the pdf of X is

$$f(x) = \begin{cases} .15e^{-.15x} & x \geq 0 \\ 0 & \text{otherwise} \end{cases}$$

Find the moment generating function and use it to find the mean and variance. Compare with Exercise 34, and explain the similarities and differences.

38. Let X be the random variable of Exercise 34. Let $Y = X - .5$ and use the relationship $M_{aX+b}(t) = e^{bt}M_X(at)$ to obtain the mgf of Y from the mgf of Exercise 34. Compare with the result of Exercise 37 and explain.

4.3 The Normal Distribution

The normal distribution is the most important one in all of probability and statistics. Many numerical populations have distributions that can be fit very closely by an appropriate normal curve. Examples include heights, weights, and other physical characteristics, measurement errors in scientific experiments, anthropometric measurements on fossils, reaction times in psychological experiments, measurements of intelligence and aptitude, scores on various tests, and numerous economic measures and indicators. Even when the underlying distribution is discrete, the normal curve often gives an excellent approximation. In addition, even when individual variables themselves are not normally distributed, sums and averages of the variables will under suitable conditions have approximately a normal distribution; this is the content of the Central Limit Theorem discussed in Chapter 6.

DEFINITION

A continuous rv X is said to have a **normal distribution** with parameters μ and σ (or μ and σ^2), where $-\infty < \mu < \infty$ and $0 < \sigma$, if the pdf of X is

$$f(x; \mu, \sigma) = \frac{1}{\sqrt{2\pi}\sigma} e^{-(x-\mu)^2/(2\sigma^2)} \qquad -\infty < x < \infty \tag{4.3}$$

Again e denotes the base of the natural logarithm system and equals approximately 2.71828, and π represents the familiar mathematical constant with approximate value 3.14159. The statement that X is normally distributed with parameters μ and σ^2 is often abbreviated $X \sim N(\mu, \sigma^2)$.

Here is a proof that the normal curve satisfies the requirement $\int_{-\infty}^{\infty} f(x)\,dx = 1$ (courtesy of Professor Robert Young of Oberlin College). Consider the special case where $\mu = 0$ and $\sigma = 1$, so $f(x) = (1/\sqrt{2\pi})e^{-x^2/2}$, and define $\int_{-\infty}^{\infty}(1/\sqrt{2\pi})e^{-x^2/2}\,dx = A$. Let $g(x, y)$ be the function of two variables

$$g(x, y) = f(x) \cdot f(y) = \frac{1}{\sqrt{2\pi}} e^{-x^2/2} \frac{1}{\sqrt{2\pi}} e^{-y^2/2} = \frac{1}{2\pi} e^{-(x^2+y^2)/2}$$

Using the rotational symmetry of $g(x, y)$, let's evaluate the volume under it by the shell method, which adds up the volumes of shells from rotation about the y-axis:

$$V = \int_0^{\infty} 2\pi x \frac{1}{2\pi} e^{-x^2/2}\,dx = \left[-e^{-x^2/2}\right]_0^{\infty} = 1$$

Now evaluate V by the usual double integral

$$V = \int_{-\infty}^{\infty}\int_{-\infty}^{\infty} f(x)f(y)\,dx\,dy = \int_{-\infty}^{\infty} f(x)\,dx \cdot \int_{-\infty}^{\infty} f(y)\,dy = \left[\int_{-\infty}^{\infty} f(x)\,dx\right]^2 = A^2$$

Because $1 = V = A^2$, we have $A = 1$ in this special case where $\mu = 0$ and $\sigma = 1$. How about the general case? Using a change of variables, $z = (x - \mu)/\sigma$,

$$\int_{-\infty}^{\infty} f(x)\,dx = \int_{-\infty}^{\infty} \frac{1}{\sqrt{2\pi}\sigma} e^{-(x-\mu)^2/(2\sigma^2)}\,dx = \int_{-\infty}^{\infty} \frac{1}{\sqrt{2\pi}} e^{-z^2/2}\,dz = 1$$ ■

It can be shown (Exercise 68) that $E(X) = \mu$ and $V(X) = \sigma^2$, so the parameters are the mean and the standard deviation of X. Figure 4.13 presents graphs of $f(x; \mu, \sigma)$ for several different (μ, σ^2) pairs. Each resulting density curve is symmetric about μ and bell-shaped, so the center of the bell (point of symmetry) is both the mean of the distribution and the median. The value of σ is the distance from μ to the inflection points of the curve (the points at which the curve changes from turning downward to turning upward). Large values of σ yield density curves that are quite spread out about μ, whereas small values of σ yield density curves with a high peak above μ and most of the area under the density curve quite close to μ. Thus a large σ implies that a value of X far from μ may well be observed, whereas such a value is quite unlikely when σ is small.

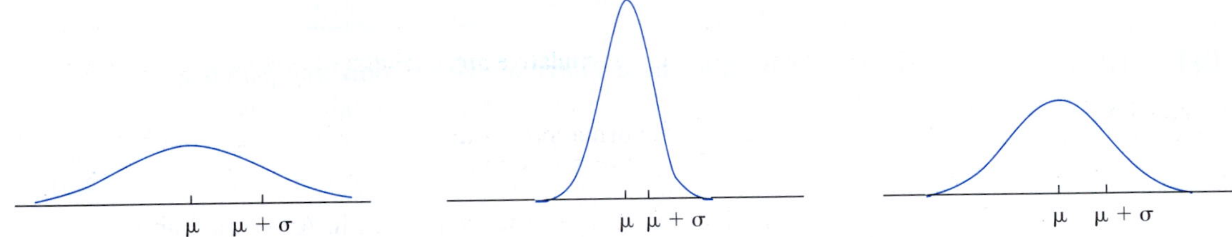

Figure 4.13 Normal density curves

The Standard Normal Distribution

To compute $P(a \le X \le b)$ when X is a normal rv with parameters μ and σ, we must evaluate

$$\int_a^b \frac{1}{\sqrt{2\pi}\sigma} e^{-(x-\mu)^2/(2\sigma^2)}\,dx \qquad (4.4)$$

None of the standard integration techniques can be used to evaluate Expression (4.4). Instead, for $\mu = 0$ and $\sigma = 1$, Expression (4.4) has been numerically evaluated and tabulated for certain values of a and b. This table can also be used to compute probabilities for any other values of μ and σ under consideration.

DEFINITION

The normal distribution with parameter values $\mu = 0$ and $\sigma = 1$ is called the **standard normal distribution**. A random variable that has a standard normal distribution is called a **standard normal random variable** and will be denoted by Z. The pdf of Z is

$$f(z; 0, 1) = \frac{1}{\sqrt{2\pi}} e^{-z^2/2} \qquad -\infty < z < \infty$$

The cdf of Z is $P(Z \le z) = \int_{-\infty}^{z} f(y; 0, 1)\,dy$, which we will denote by $\Phi(z)$.

The standard normal distribution does not usually model a naturally arising population. Instead, it is a reference distribution from which information about other normal distributions can be obtained. Appendix Table A.3 gives $\Phi(z) = P(Z \le z)$, the area under the graph of the standard normal pdf to the left of z, for $z = -3.49, -3.48, \ldots, 3.48, 3.49$. Figure 4.14 illustrates the type of cumulative area (probability) tabulated in Table A.3. From this table, various other probabilities involving Z can be calculated.

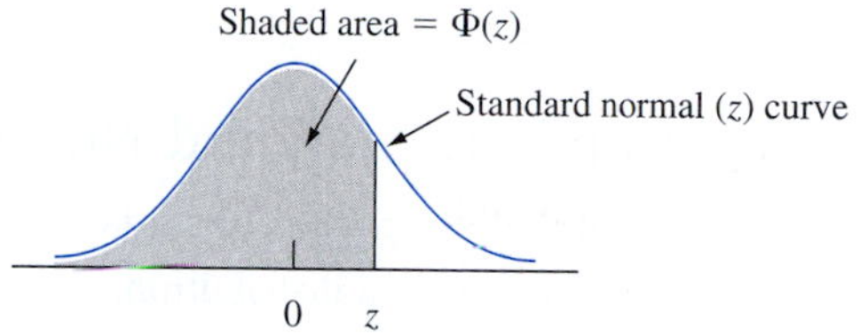

Figure 4.14 Standard normal cumulative areas tabulated in Appendix Table A.3

Example 4.18 Compute the following standard normal probabilities: (a) $P(Z \le 1.25)$, (b) $P(Z > 1.25)$, (c) $P(Z \le -1.25)$, and (d) $P(-.38 \le Z \le 1.25)$.

a. $P(Z \le 1.25) = \Phi(1.25)$, a probability that is tabulated in Appendix Table A.3 at the intersection of the row marked 1.2 and the column marked .05. The number there is .8944, so $P(Z \le 1.25) = .8944$. See Figure 4.15(a).

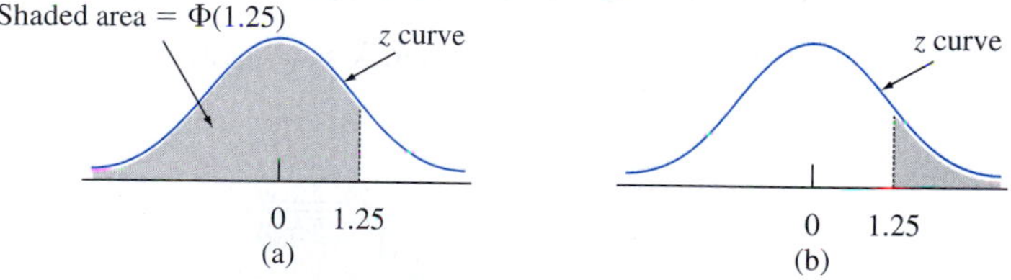

Figure 4.15 Normal curve areas (probabilities) for Example 4.18

b. $P(Z > 1.25) = 1 - P(Z \le 1.25) = 1 - \Phi(1.25)$, the area under the standard normal curve to the right of 1.25 (an upper-tail area). Since $\Phi(1.25) = .8944$, it follows that $P(Z > 1.25) = .1056$. Since Z is a continuous rv, $P(Z \ge 1.25)$ also equals .1056. See Figure 4.15(b).

c. $P(Z \le -1.25) = \Phi(-1.25)$, a lower-tail area. Directly from Appendix Table A.3, $\Phi(-1.25) = .1056$. By symmetry of the normal curve, this is the same answer as in part (b).

d. $P(-.38 \le Z \le 1.25)$ is the area under the standard normal curve above the interval whose left endpoint is $-.38$ and whose right endpoint is 1.25. From Section 4.1, if X is a continuous rv with cdf $F(x)$, then $P(a \le X \le b) = F(b) - F(a)$. This gives $P(-.38 \le Z \le 1.25) = \Phi(1.25) - \Phi(-.38) = .8944 - .3520 = .5424$. (See Figure 4.16.)

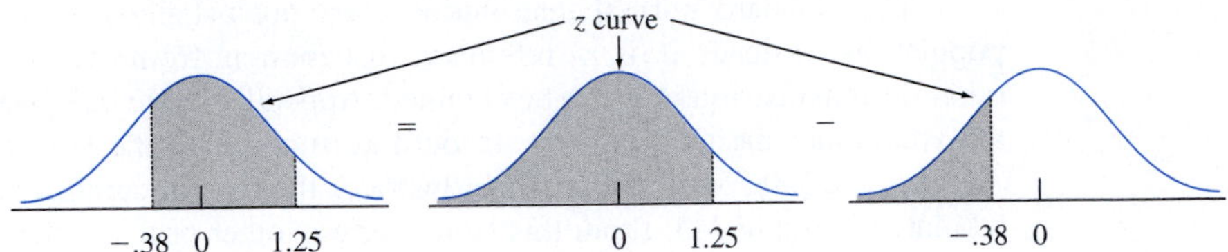

Figure 4.16 $P(-.38 \leq Z \leq 1.25)$ as the difference between two cumulative areas ■

Percentiles of the Standard Normal Distribution

For any p between 0 and 1, Appendix Table A.3 can be used to obtain the $(100p)$th percentile of the standard normal distribution.

Example 4.19 The 99th percentile of the standard normal distribution is that value on the horizontal axis such that the area under the curve to the left of the value is .9900. Now Appendix Table A.3 gives for fixed z the area under the standard normal curve to the left of z, whereas here we have the area and want the value of z. This is the "inverse" problem to $P(Z \leq z) = ?$ so the table is used in an inverse fashion: Find in the middle of the table .9900; the row and column in which it lies identify the 99th z percentile. Here .9901 lies in the row marked 2.3 and column marked .03, so the 99th percentile is (approximately) $z = 2.33$. (See Figure 4.17.) By symmetry, the first percentile is the negative of the 99th percentile, so it equals -2.33 (1% lies below the first and above the 99th). (See Figure 4.18.)

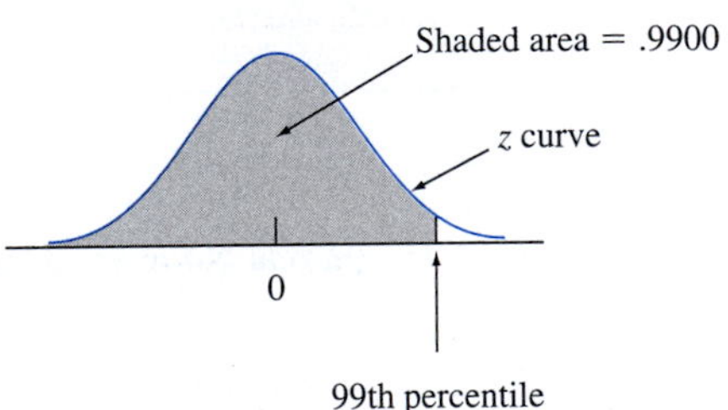

Figure 4.17 Finding the 99th percentile

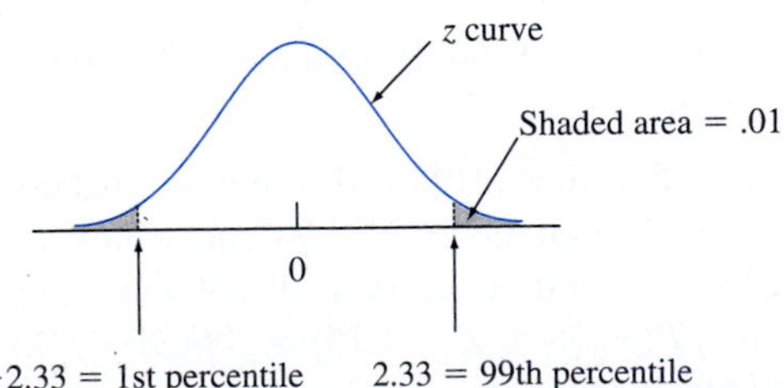

Figure 4.18 The relationship between the 1st and 99th percentiles ■

In general, the $(100p)$th percentile is identified by the row and column of Appendix Table A.3 in which the entry p is found (e.g., the 67th percentile is obtained by finding .6700 in the body of the table, which gives $z = .44$). If p does not appear, the number closest to it is often used, although linear interpolation gives a more accurate answer. For example, to find the 95th percentile, we look for .9500 inside the table. Although .9500 does not appear, both .9495 and .9505 do, corresponding to $z = 1.64$ and 1.65, respectively. Since .9500 is halfway between the two probabilities that do appear, we will use 1.645 as the 95th percentile and -1.645 as the 5th percentile.

z_α Notation

In statistical inference, we will need the values on the measurement axis that capture certain small tail areas under the standard normal curve.

---

z_α will denote the value on the measurement axis for which α of the area under the z curve lies to the right of z_α. (See Figure 4.19.)

For example, $z_{.10}$ captures upper-tail area .10 and $z_{.01}$ captures upper-tail area .01.

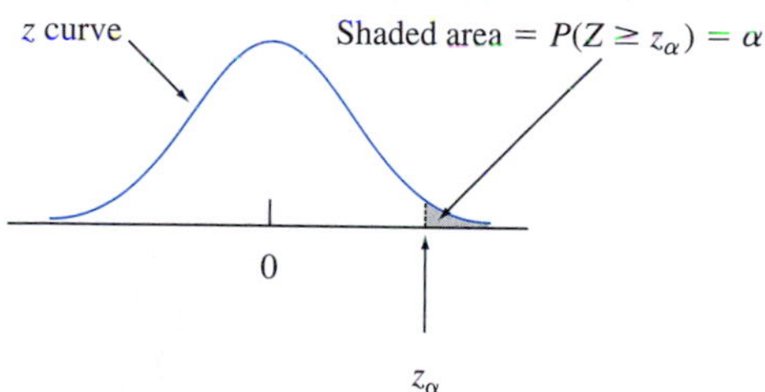

Figure 4.19 z_α notation illustrated

Since α of the area under the standard normal curve lies to the right of z_α, $1 - \alpha$ of the area lies to the left of z_α. Thus *z_α is the $100(1 - \alpha)$th percentile of the standard normal distribution.* By symmetry the area under the standard normal curve to the left of $-z_\alpha$ is also α. The z_α's are usually referred to as ***z* critical values**. Table 4.1 lists the most useful standard normal percentiles and z_α values.

Table 4.1 Standard normal percentiles and critical values

Percentile	90	95	97.5	99	99.5	99.9	99.95
α (tail area)	.1	.05	.025	.01	.005	.001	.0005
$z_\alpha = 100(1 - \alpha)$th percentile	1.28	1.645	1.96	2.33	2.58	3.08	3.27

Example 4.20 $z_{.05}$ is the $100(1 - .05)$th $=$ 95th percentile of the standard normal distribution, so $z_{.05} = 1.645$. The area under the standard normal curve to the left of $-z_{.05}$ is also .05. (See Figure 4.20.)

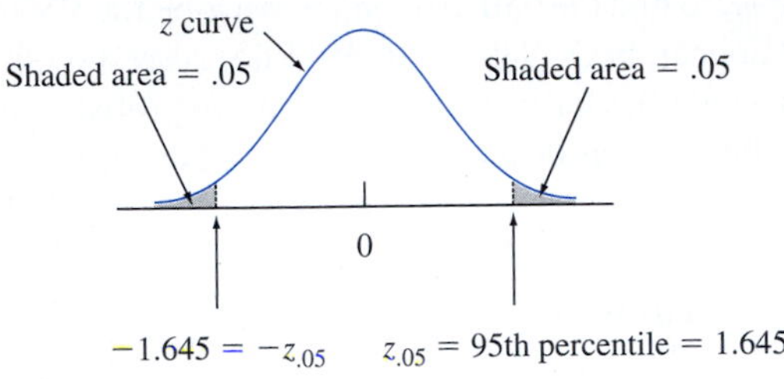

Figure 4.20 Finding $z_{.05}$ ■

Nonstandard Normal Distributions

When $X \sim N(\mu, \sigma^2)$, probabilities involving X are computed by "standardizing." The **standardized variable** is $(X - \mu)/\sigma$. Subtracting μ shifts the mean from μ to zero, and then dividing by σ scales the variable so that the standard deviation is 1 rather than σ.

PROPOSITION If X has a normal distribution with mean μ and standard deviation σ, then

$$Z = \frac{X - \mu}{\sigma}$$

has a standard normal distribution. Thus

$$P(a \le X \le b) = P\left(\frac{a - \mu}{\sigma} \le Z \le \frac{b - \mu}{\sigma}\right)$$
$$= \Phi\left(\frac{b - \mu}{\sigma}\right) - \Phi\left(\frac{a - \mu}{\sigma}\right)$$

$$P(X \le a) = \Phi\left(\frac{a - \mu}{\sigma}\right) \qquad P(X \ge b) = 1 - \Phi\left(\frac{b - \mu}{\sigma}\right)$$

The key idea of the proposition is that by standardizing, any probability involving X can be expressed as a probability involving a standard normal rv Z, so that Appendix Table A.3 can be used. This is illustrated in Figure 4.21. The proposition can be proved by writing the cdf of $Z = (X - \mu)/\sigma$ as

$$P(Z \le z) = P(X \le \sigma z + \mu) = \int_{-\infty}^{\sigma z + \mu} f(x; \mu, \sigma)\, dx$$

Using a result from calculus, this integral can be differentiated with respect to z to yield the desired pdf $f(z; 0, 1)$.

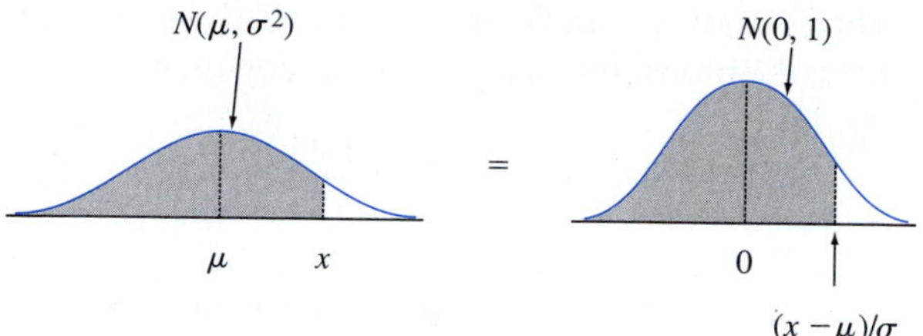

Figure 4.21 Equality of nonstandard and standard normal curve areas

Example 4.21 The time that it takes a driver to react to the brake lights on a decelerating vehicle is critical in helping to avoid rear-end collisions. The article "Fast-Rise Brake Lamp as a Collision-Prevention Device" (*Ergonomics*, 1993: 391–395) suggests that reaction time for an in-traffic response to a brake signal from standard brake lights can be modeled with a normal distribution having mean value 1.25 sec and standard deviation of .46 sec. What is the probability that reaction time is between 1.00 sec and 1.75 sec? If we let X denote reaction time, then standardizing gives

$$1.00 \le X \le 1.75$$

if and only if

$$\frac{1.00 - 1.25}{.46} \le \frac{X - 1.25}{.46} \le \frac{1.75 - 1.25}{.46}$$

Thus

$$\begin{aligned} P(1.00 \le X \le 1.75) &= P\left(\frac{1.00 - 1.25}{.46} \le Z \le \frac{1.75 - 1.25}{.46}\right) \\ &= P(-.54 \le Z \le 1.09) = \Phi(1.09) - \Phi(-.54) \\ &= .8621 - .2946 = .5675 \end{aligned}$$

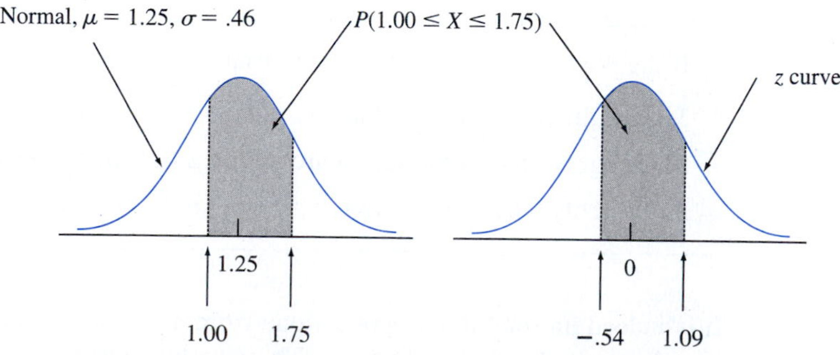

Figure 4.22 Normal curves for Example 4.21

This is illustrated in Figure 4.22. Similarly, if we view 2 sec as a critically long reaction time, the probability that actual reaction time will exceed this value is

$$P(X > 2) = P\left(Z > \frac{2 - 1.25}{.46}\right) = P(Z > 1.63) = 1 - \Phi(1.63) = .0516$$ ■

Standardizing amounts to nothing more than calculating a distance from the mean value and then reexpressing the distance as some number of standard deviations. For example, if $\mu = 100$ and $\sigma = 15$, then $x = 130$ corresponds to $z = (130 - 100)/15 = 30/15 = 2.00$. That is, 130 is 2 standard deviations above (to the right of) the mean value. Similarly, standardizing 85 gives $(85 - 100)/15 = -1$, so 85 is 1 standard deviation below the mean. The z table applies to *any* normal distribution, provided that we think in terms of number of standard deviations away from the mean value.

Example 4.22 The breakdown voltage of a randomly chosen diode of a particular type is known to be normally distributed. What is the probability that a diode's breakdown voltage is within 1 standard deviation of its mean value? This question can be answered without knowing either μ or σ, as long as the distribution is known to be normal; in other words, the answer is the same for *any* normal distribution:

$$\begin{aligned}
P\left(\begin{matrix}X \text{ is within 1 standard} \\ \text{deviation of its mean}\end{matrix}\right) &= P(\mu - \sigma \le X \le \mu + \sigma) \\
&= P\left(\frac{\mu - \sigma - \mu}{\sigma} \le Z \le \frac{\mu + \sigma - \mu}{\sigma}\right) \\
&= P(-1.00 \le Z \le 1.00) \\
&= \Phi(1.00) - \Phi(-1.00) = .6826
\end{aligned}$$

The probability that X is within 2 standard deviation is $P(-2.00 \le Z \le 2.00) = .9544$ and within 3 standard deviations is $P(-3.00 \le Z \le 3.00) = .9974$. ■

The results of Example 4.22 are often reported in percentage form and referred to as the *empirical rule* (because empirical evidence has shown that histograms of real data can very frequently be approximated by normal curves).

If the population distribution of a variable is (approximately) normal, then

1. Roughly 68% of the values are within 1 SD of the mean.
2. Roughly 95% of the values are within 2 SDs of the mean.
3. Roughly 99.7% of the values are within 3 SDs of the mean.

It is indeed unusual to observe a value from a normal population that is much farther than 2 standard deviations from μ. These results will be important in the development of hypothesis-testing procedures in later chapters.

Percentiles of an Arbitrary Normal Distribution

The $(100p)$th percentile of a normal distribution with mean μ and standard deviation σ is easily related to the $(100p)$th percentile of the standard normal distribution.

PROPOSITION

$$\begin{matrix}(100p)\text{th percentile} \\ \text{for normal } (\mu, \sigma)\end{matrix} = \mu + \left[\begin{matrix}(100p)\text{th percentile for} \\ \text{standard normal}\end{matrix}\right] \cdot \sigma$$

Another way of saying this is that if z is the desired percentile for the standard normal distribution, then the desired percentile for the normal (μ, σ) distribution is z standard deviations from μ. For justification, see Exercise 65.

Example 4.23 The amount of distilled water dispensed by a certain machine is normally distributed with mean value 64 oz and standard deviation .78 oz. What container size c will ensure that overflow occurs only .5% of the time? If X denotes the amount dispensed, the desired condition is that $P(X > c) = .005$, or, equivalently, that $P(X \leq c) = .995$. Thus c is the 99.5th percentile of the normal distribution with $\mu = 64$ and $\sigma = .78$. The 99.5th percentile of the standard normal distribution is 2.58, so

$$c = \eta(.995) = 64 + (2.58)(.78) = 64 + 2.0 = 66 \text{ oz}$$

This is illustrated in Figure 4.23.

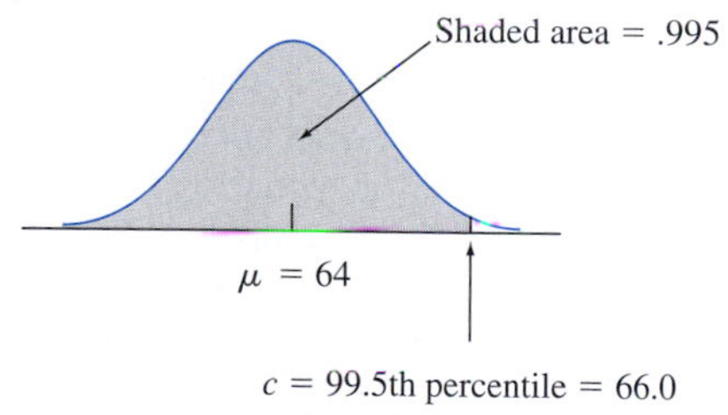

Figure 4.23 Distribution of amount dispensed for Example 4.23 ■

The Normal Distribution and Discrete Populations

The normal distribution is often used as an approximation to the distribution of values in a discrete population. In such situations, extra care must be taken to ensure that probabilities are computed in an accurate manner.

Example 4.24 IQ in a particular population (as measured by a standard test) is known to be approximately normally distributed with $\mu = 100$ and $\sigma = 15$. What is the probability that a randomly selected individual has an IQ of at least 125? Letting $X =$ the IQ of a randomly chosen person, we wish $P(X \geq 125)$. The temptation here is to standardize $X \geq 125$ immediately as in previous examples. However, the IQ population is actually discrete, since IQs are integer-valued, so the normal curve is an approximation to a discrete probability histogram, as pictured in Figure 4.24.

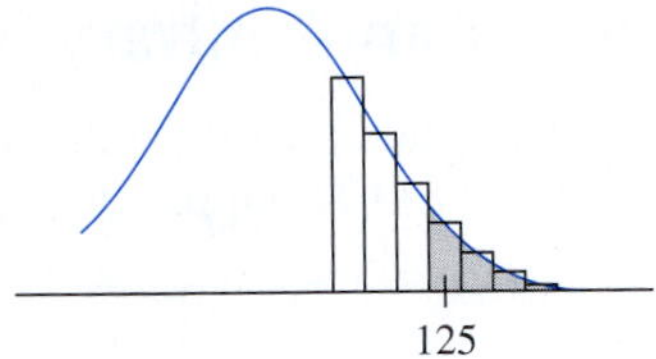

Figure 4.24 A normal approximation to a discrete distribution

The rectangles of the histogram are *centered* at integers, so IQs of at least 125 correspond to rectangles beginning at 124.5, as shaded in Figure 4.24. Thus we really want the area under the approximating normal curve to the right of 124.5. Standardizing this value gives $P(Z \geq 1.63) = .0516$. If we had standardized $X \geq 125$, we would have obtained $P(Z \geq 1.67) = .0475$. The difference is not great, but the answer .0516 is more accurate. Similarly, $P(X = 125)$ would be approximated by the area between 124.5 and 125.5, since the area under the normal curve above the single value 125 is zero. ■

The correction for discreteness of the underlying distribution in Example 4.24 is often called a **continuity correction**. It is useful in the following application of the normal distribution to the computation of binomial probabilities. The normal distribution was actually created as an approximation to the binomial distribution (by Abraham de Moivre in the 1730s).

Approximating the Binomial Distribution

Recall that the mean value and standard deviation of a binomial random variable X are $\mu_x = np$ and $\sigma_x = \sqrt{npq}$, respectively. Figure 4.25 displays a binomial probability histogram for the binomial distribution with $n = 20$, $p = .6$ [so $\mu = 20(.6) = 12$ and $\sigma = \sqrt{20(.6)(.4)} = 2.19$]. A normal curve with mean value and standard deviation equal to the corresponding values for the binomial distribution has been superimposed

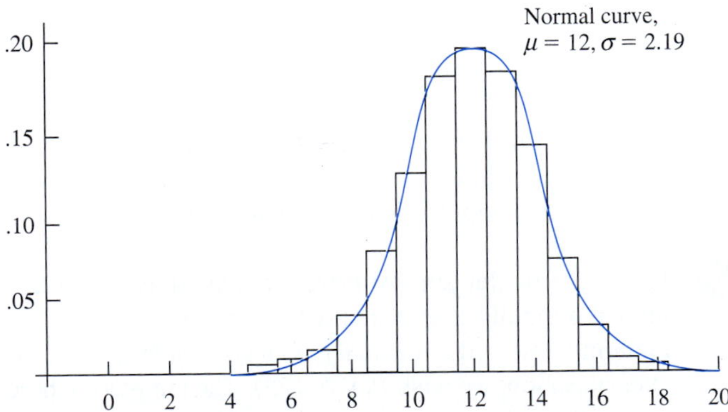

Figure 4.25 Binomial probability histogram for $n = 20$, $p = .6$ with normal approximation curve superimposed

on the probability histogram. Although the probability histogram is a bit skewed (because $p \neq .5$), the normal curve gives a very good approximation, especially in the middle part of the picture. The area of any rectangle (probability of any particular X value) except those in the extreme tails can be accurately approximated by the corresponding normal curve area. For example, $P(X = 10) = B(10; 20, .6) - B(9; 20, .6) = .117$, whereas the area under the normal curve between 9.5 and 10.5 is $P(-1.14 \leq Z \leq -.68) = .1212$.

More generally, as long as the binomial probability histogram is not too skewed, binomial probabilities can be well approximated by normal curve areas. It is then customary to say that X has approximately a normal distribution.

PROPOSITION

Let X be a binomial rv based on n trials with success probability p. Then if the binomial probability histogram is not too skewed, X has approximately a normal distribution with $\mu = np$ and $\sigma = \sqrt{npq}$. In particular, for $x =$ a possible value of X,

$$P(X \leq x) = B(x; n, p) \approx \text{(area under the normal curve to the left of } x + .5)$$

$$= \Phi\left(\frac{x + .5 - np}{\sqrt{npq}}\right)$$

In practice, the approximation is adequate provided that both $np \geq 10$ and $nq \geq 10$.

If either $np < 10$ or $nq < 10$, the binomial distribution is too skewed for the (symmetric) normal curve to give accurate approximations.

Example 4.25

Suppose that 25% of all licensed drivers in a particular state do not have insurance. Let X be the number of uninsured drivers in a random sample of size 50 (somewhat perversely, a success is an uninsured driver), so that $p = .25$. Then $\mu = 12.5$ and $\sigma = 3.06$. Since $np = 50(.25) = 12.5 \geq 10$ and $nq = 37.5 \geq 10$, the approximation can safely be applied:

$$P(X \leq 10) = B(10; 50, .25) \approx \Phi\left(\frac{10 + .5 - 12.5}{3.06}\right)$$

$$= \Phi(-.65) = .2578$$

Similarly, the probability that between 5 and 15 (inclusive) of the selected drivers are uninsured is

$$P(5 \leq X \leq 15) = B(15; 50, .25) - B(4; 50, .25)$$

$$\approx \Phi\left(\frac{15.5 - 12.5}{3.06}\right) - \Phi\left(\frac{4.5 - 12.5}{3.06}\right) = .8320$$

The exact probabilities are .2622 and .8348, respectively, so the approximations are quite good. In the last calculation, the probability $P(5 \le X \le 15)$ is being approximated by the area under the normal curve between 4.5 and 15.5—the continuity correction is used for both the upper and lower limits. ■

When the objective of our investigation is to make an inference about a population proportion p, interest will focus on the sample proportion of successes X/n rather than on X itself. Because this proportion is just X multiplied by the constant $1/n$, it will also have approximately a normal distribution (with mean $\mu = p$ and standard deviation $\sigma = \sqrt{pq/n}$) provided that both $np \ge 10$ and $nq \ge 10$. This normal approximation is the basis for several inferential procedures to be discussed in later chapters.

It is quite difficult to give a direct proof of the validity of this normal approximation (the first one goes back about 270 years to de Moivre). In Chapter 6, we'll see that it is a consequence of an important general result called the Central Limit Theorem.

The Normal Moment Generating Function

The moment generating function provides a straightforward way to verify that the parameters μ and σ^2 are indeed the mean and variance of X (Exercise 68).

PROPOSITION The moment generating function of a normally distributed random variable X is $M_X(t) = e^{\mu t + \sigma^2 t^2/2}$.

Proof Consider first the special case of a standard normal rv Z. Then

$$M_Z(t) = E(e^{tZ}) = \int_{-\infty}^{\infty} e^{tz} \frac{1}{\sqrt{2\pi}} e^{-z^2/2} dz = \int_{-\infty}^{\infty} \frac{1}{\sqrt{2\pi}} e^{-(z^2 - 2tz)/2} dz$$

Completing the square in the exponent, we have

$$M_Z(t) = e^{t^2/2} \int_{-\infty}^{\infty} \frac{1}{\sqrt{2\pi}} e^{-(z^2 - 2tz + t^2)/2} dz = e^{t^2/2} \int_{-\infty}^{\infty} \frac{1}{\sqrt{2\pi}} e^{-(z-t)^2/2} dz$$

The last integral is the area under a normal density with mean t and standard deviation 1, so the value of the integral is 1. Therefore, $M_Z(t) = e^{t^2/2}$.

Now let X be any normal rv with mean μ and standard deviation σ. Then, by the first proposition in this section, $(X - \mu)/\sigma = Z$, where Z is standard normal. That is, $X = \mu + \sigma Z$. Now use the property $M_{aY+b}(t) = e^{bt} M_Y(at)$:

$$M_X(t) = M_{\mu + \sigma Z}(t) = e^{\mu t} M_Z(\sigma t) = e^{\mu t} e^{\sigma^2 t^2/2} = e^{\mu t + \sigma^2 t^2/2}$$

■

Exercises Section 4.3 (39–68)

39. Let Z be a standard normal random variable and calculate the following probabilities, drawing pictures wherever appropriate.

a. $P(0 \le Z \le 2.17)$
b. $P(0 \le Z \le 1)$
c. $P(-2.50 \le Z \le 0)$
d. $P(-2.50 \le Z \le 2.50)$
e. $P(Z \le 1.37)$
f. $P(-1.75 \le Z)$
g. $P(-1.50 \le Z \le 2.00)$
h. $P(1.37 \le Z \le 2.50)$
i. $P(1.50 \le Z)$
j. $P(|Z| \le 2.50)$

40. In each case, determine the value of the constant c that makes the probability statement correct.

a. $\Phi(c) = .9838$
b. $P(0 \le Z \le c) = .291$
c. $P(c \le Z) = .121$
d. $P(-c \le Z \le c) = .668$
e. $P(c \le |Z|) = .016$

41. Find the following percentiles for the standard normal distribution. Interpolate where appropriate.

a. 91st
b. 9th
c. 75th
d. 25th
e. 6th

42. Determine z_α for the following:

a. $\alpha = .0055$
b. $\alpha = .09$
c. $\alpha = .663$

43. If X is a normal rv with mean 80 and standard deviation 10, compute the following probabilities by standardizing:

a. $P(X \le 100)$
b. $P(X \le 80)$
c. $P(65 \le X \le 100)$
d. $P(70 \le X)$
e. $P(85 \le X \le 95)$
f. $P(|X - 80| \le 10)$

44. The plasma cholesterol level (mg/dL) for patients with no prior evidence of heart disease who experience chest pain is normally distributed with mean 200 and standard deviation 35. Consider randomly selecting an individual of this type. What is the probability that the plasma cholesterol level

a. Is at most 250?
b. Is between 300 and 400?
c. Differs from the mean by at least 1.5 standard deviations?

45. The article "Reliability of Domestic-Waste Biofilm Reactors" (*J. Envir. Engrg.*, 1995: 785–790) suggests that substrate concentration (mg/cm^3) of influent to a reactor is normally distributed with $\mu = .30$ and $\sigma = .06$.

a. What is the probability that the concentration exceeds .25?
b. What is the probability that the concentration is at most .10?
c. How would you characterize the largest 5% of all concentration values?

46. Suppose the diameter at breast height (in.) of trees of a certain type is normally distributed with $\mu = 8.8$ and $\sigma = 2.8$, as suggested in the article "Simulating a Harvester-Forwarder Softwood Thinning" (*Forest Products J.*, May 1997: 36–41).

a. What is the probability that the diameter of a randomly selected tree will be at least 10 in.? Will exceed 10 in.?
b. What is the probability that the diameter of a randomly selected tree will exceed 20 in.?
c. What is the probability that the diameter of a randomly selected tree will be between 5 and 10 in.?
d. What value c is such that the interval $(8.8 - c, 8.8 + c)$ includes 98% of all diameter values?
e. If four trees are independently selected, what is the probability that at least one has a diameter exceeding 10 in.?

47. There are two machines available for cutting corks intended for use in wine bottles. The first produces corks with diameters that are normally distributed with mean 3 cm and standard deviation .1 cm. The second machine produces corks with diameters that have a normal distribution with mean 3.04 cm and standard deviation .02 cm. Acceptable corks have diameters between 2.9 cm and 3.1 cm. Which machine is more likely to produce an acceptable cork?

48. Human body temperatures for healthy individuals have approximately a normal distribution with mean

98.25°F and standard deviation .75°F. (The past accepted value of 98.6° Fahrenheit was obtained by converting the Celsius value of 37°, which is correct to the nearest integer.)

a. Find the 90th percentile of the distribution.

b. Find the 5th percentile of the distribution.

c. What temperature separates the coolest 25% from the others?

49. The article "Monte Carlo Simulation—Tool for Better Understanding of LRFD" (*J. Struct. Engrg.*, 1993: 1586–1599) suggests that yield strength (ksi) for A36 grade steel is normally distributed with $\mu = 43$ and $\sigma = 4.5$.

a. What is the probability that yield strength is at most 40? Greater than 60?

b. What yield strength value separates the strongest 75% from the others?

50. The automatic opening device of a military cargo parachute has been designed to open when the parachute is 200 m above the ground. Suppose opening altitude actually has a normal distribution with mean value 200 m and standard deviation 30 m. Equipment damage will occur if the parachute opens at an altitude of less than 100 m. What is the probability that there is equipment damage to the payload of at least one of five independently dropped parachutes?

51. The temperature reading from a thermocouple placed in a constant-temperature medium is normally distributed with mean μ, the actual temperature of the medium, and standard deviation σ. What would the value of σ have to be to ensure that 95% of all readings are within .1° of μ?

52. The distribution of resistance for resistors of a certain type is known to be normal, with 10% of all resistors having a resistance exceeding 10.256 ohms and 5% having a resistance smaller than 9.671 ohms. What are the mean value and standard deviation of the resistance distribution?

53. If adult female heights are normally distributed, what is the probability that the height of a randomly selected woman is

a. Within 1.5 SDs of its mean value?

b. Farther than 2.5 SDs from its mean value?

c. Between 1 and 2 SDs from its mean value?

54. A machine that produces ball bearings has initially been set so that the true average diameter of the bearings it produces is .500 in. A bearing is acceptable if its diameter is within .004 in. of this target value. Suppose, however, that the setting has changed during the course of production, so that the bearings have normally distributed diameters with mean value .499 in. and standard deviation .002 in. What percentage of the bearings produced will not be acceptable?

55. The Rockwell hardness of a metal is determined by impressing a hardened point into the surface of the metal and then measuring the depth of penetration of the point. Suppose the Rockwell hardness of a particular alloy is normally distributed with mean 70 and standard deviation 3. (Rockwell hardness is measured on a continuous scale.)

a. If a specimen is acceptable only if its hardness is between 67 and 75, what is the probability that a randomly chosen specimen has an acceptable hardness?

b. If the acceptable range of hardness is $(70 - c, 70 + c)$, for what value of c would 95% of all specimens have acceptable hardness?

c. If the acceptable range is as in part (a) and the hardness of each of ten randomly selected specimens is independently determined, what is the expected number of acceptable specimens among the ten?

d. What is the probability that at most eight of ten independently selected specimens have a hardness of less than 73.84? (*Hint*: Y = the number among the ten specimens with hardness less than 73.84 is a binomial variable; what is p?)

56. The weight distribution of parcels sent in a certain manner is normal with mean value 12 lb and standard deviation 3.5 lb. The parcel service wishes to establish a weight value c beyond which there will be a surcharge. What value of c is such that 99% of all parcels are at least 1 lb under the surcharge weight?

57. Suppose Appendix Table A.3 contained $\Phi(z)$ only for $z \geq 0$. Explain how you could still compute

a. $P(-1.72 \leq Z \leq -.55)$

b. $P(-1.72 \leq Z \leq .55)$

Is it necessary to table $\Phi(z)$ for z negative? What property of the standard normal curve justifies your answer?

58. Consider babies born in the "normal" range of 37–43 weeks of gestational age. Extensive data supports the assumption that for such babies born in the

United States, birth weight is normally distributed with mean 3432 g and standard deviation 482 g. [The article "Are Babies Normal?" (*Amer. Statist.*, 1999: 298–302) analyzed data from a particular year. A histogram with a sensible choice of class intervals did not look at all normal, but further investigation revealed this was because some hospitals measured weight in grams and others measured to the nearest ounce and then converted to grams. Modifying the class intervals to allow for this gave a histogram that was well described by a normal distribution.]

a. What is the probability that the birth weight of a randomly selected baby of this type exceeds 4000 grams? Is between 3000 and 4000 grams?

b. What is the probability that the birth weight of a randomly selected baby of this type is either less than 2000 grams or greater than 5000 grams?

c. What is the probability that the birth weight of a randomly selected baby of this type exceeds 7 lb?

d. How would you characterize the most extreme .1% of all birth weights?

e. If X is a random variable with a normal distribution and a is a numerical constant ($a \neq 0$), then $Y = aX$ also has a normal distribution. Use this to determine the distribution of birth weight expressed in pounds (shape, mean, and standard deviation), and then recalculate the probability from part (c). How does this compare to your previous answer?

59. In response to concerns about nutritional contents of fast foods, McDonald's has announced that it will use a new cooking oil for its french fries that will decrease substantially trans fatty acid levels and increase the amount of more beneficial polyunsaturated fat. The company claims that 97 out of 100 people cannot detect a difference in taste between the new and old oils. Assuming that this figure is correct (as a long-run proportion), what is the approximate probability that in a random sample of 1000 individuals who have purchased fries at McDonald's,

a. At least 40 can taste the difference between the two oils?

b. At most 5% can taste the difference between the two oils?

60. Chebyshev's inequality, introduced in Exercise 43 (Chapter 3), is valid for continuous as well as discrete distributions. It states that for any number k satisfying $k \geq 1$, $P(|X - \mu| \geq k\sigma) \leq 1/k^2$ (see Exercise 43 in Section 3.3 for an interpretation and Exercise 135 in Chapter 3 Supplementary Exercises for a proof). Obtain this probability in the case of a normal distribution for $k = 1, 2$, and 3, and compare to the upper bound.

61. Let X denote the number of flaws along a 100-m reel of magnetic tape (an integer-valued variable). Suppose X has approximately a normal distribution with $\mu = 25$ and $\sigma = 5$. Use the continuity correction to calculate the probability that the number of flaws is

a. Between 20 and 30, inclusive.

b. At most 30. Less than 30.

62. Let X have a binomial distribution with parameters $n = 25$ and p. Calculate each of the following probabilities using the normal approximation (with the continuity correction) for the cases $p = .5, .6$, and $.8$ and compare to the exact probabilities calculated from Appendix Table A.1.

a. $P(15 \leq X \leq 20)$

b. $P(X \leq 15)$

c. $P(20 \leq X)$

63. Suppose that 10% of all steel shafts produced by a certain process are nonconforming but can be reworked (rather than having to be scrapped). Consider a random sample of 200 shafts, and let X denote the number among these that are nonconforming and can be reworked. What is the (approximate) probability that X is

a. At most 30?

b. Less than 30?

c. Between 15 and 25 (inclusive)?

64. Suppose only 70% of all drivers in a certain state regularly wear a seat belt. A random sample of 500 drivers is selected. What is the probability that

a. Between 320 and 370 (inclusive) of the drivers in the sample regularly wear a seat belt?

b. Fewer than 325 of those in the sample regularly wear a seat belt? Fewer than 315?

65. Show that the relationship between a general normal percentile and the corresponding z percentile is as stated in this section.

66. a. Show that if X has a normal distribution with parameters μ and σ, then $Y = aX + b$ (a linear function of X) also has a normal distribution. What are the parameters of the distribution of Y [i.e., $E(Y)$ and $V(Y)$]? [*Hint*: Write the cdf of Y, $P(Y \leq y)$, as an integral involving the pdf of X, and then differentiate with respect to y to get the pdf of Y.]

b. If when measured in °C, temperature is normally distributed with mean 115 and standard deviation 2, what can be said about the distribution of temperature measured in °F?

67. There is no nice formula for the standard normal cdf $\Phi(z)$, but several good approximations have been published in articles. The following is from "Approximations for Hand Calculators Using Small Integer Coefficients" (*Math. Comput.*, 1977: 214–222). For $0 < z \le 5.5$,

$$P(Z \ge z) = 1 - \Phi(z) \approx .5 \exp\left\{-\left[\frac{(83z + 351)z + 562}{703/z + 165}\right]\right\}$$

The relative error of this approximation is less than .042%. Use this to calculate approximations to the following probabilities, and compare whenever possible to the probabilities obtained from Appendix Table A.3.

a. $P(Z \ge 1)$
b. $P(Z < -3)$
c. $P(-4 < Z < 4)$
d. $P(Z > 5)$

68. The moment generating function can be used to find the mean and variance of the normal distribution.

a. Use derivatives of $M_X(t)$ to verify that $E(X) = \mu$ and $V(X) = \sigma^2$.
b. Repeat (a) using $R_X(t) = \ln[M_X(t)]$, and compare with part (a) in terms of effort.

4.4 *The Gamma Distribution and Its Relatives

The graph of any normal pdf is bell-shaped and thus symmetric. In many practical situations, the variable of interest to the experimenter might have a skewed distribution. A family of pdf's that yields a wide variety of skewed distributional shapes is the gamma family. To define the family of gamma distributions, we first need to introduce a function that plays an important role in many branches of mathematics.

DEFINITION For $\alpha > 0$, the **gamma function** $\Gamma(\alpha)$ is defined by

$$\Gamma(\alpha) = \int_0^\infty x^{\alpha-1}e^{-x}\,dx \tag{4.5}$$

The most important properties of the gamma function are the following:

1. For any $\alpha > 1$, $\Gamma(\alpha) = (\alpha - 1) \cdot \Gamma(\alpha - 1)$ [via integration by parts]
2. For any positive integer, n, $\Gamma(n) = (n - 1)!$
3. $\Gamma(\frac{1}{2}) = \sqrt{\pi}$

By Expression (4.5), if we let

$$f(x;\alpha) = \begin{cases} \dfrac{x^{\alpha-1}e^{-x}}{\Gamma(\alpha)} & x \ge 0 \\ 0 & \text{otherwise} \end{cases} \tag{4.6}$$

then $f(x;\alpha) \ge 0$ and $\int_0^\infty f(x;\alpha)\,dx = \Gamma(\alpha)/\Gamma(\alpha) = 1$, so $f(x;\alpha)$ satisfies the two basic properties of a pdf.

The Family of Gamma Distributions

DEFINITION A continuous random variable X is said to have a **gamma distribution** if the pdf of X is

$$f(x; \alpha, \beta) = \begin{cases} \dfrac{1}{\beta^{\alpha}\Gamma(\alpha)} x^{\alpha-1} e^{-x/\beta} & x \ge 0 \\ 0 & \text{otherwise} \end{cases} \tag{4.7}$$

where the parameters α and β satisfy $\alpha > 0$, $\beta > 0$. The **standard gamma distribution** has $\beta = 1$, so the pdf of a standard gamma rv is given by (4.6).

Figure 4.26(a) illustrates the graphs of the gamma pdf for several (α, β) pairs, whereas Figure 4.26(b) presents graphs of the standard gamma pdf. For the standard pdf, when $\alpha \le 1$, $f(x; \alpha)$ is strictly decreasing as x increases from 0; when $\alpha > 1$, $f(x; \alpha)$ rises from 0 at $x = 0$ to a maximum and then decreases. The parameter β in (4.7) is called the *scale parameter* because values other than 1 either stretch or compress the pdf in the x direction.

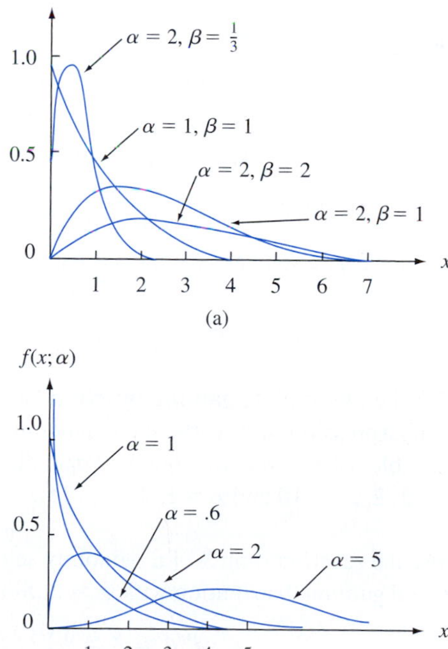

Figure 4.26 (a) Gamma density curves; (b) standard gamma density curves

PROPOSITION

The moment generating function of a gamma random variable is

$$M_X(t) = \frac{1}{(1 - \beta t)^\alpha}$$

Proof By definition, the mgf is

$$M_X(t) = E(e^{tX}) = \int_0^\infty e^{tx} \frac{x^{\alpha-1}}{\Gamma(\alpha)\beta^\alpha} e^{-x/\beta}\, dx = \int_0^\infty \frac{x^{\alpha-1}}{\Gamma(\alpha)\beta^\alpha} e^{-x(-t+1/\beta)}\, dx$$

One way to evaluate the integral is to express the integrand in terms of a gamma density. This means writing the exponent in the form $-x/b$ and having b take the place of β. We have $-x(-t + 1/\beta) = -x[(-\beta t + 1)/\beta] = -x/[\beta/(1 - \beta t)]$. Now multiplying and at the same time dividing the integrand by $1/(1 - \beta t)^\alpha$ gives

$$M_X(t) = \frac{1}{(1 - \beta t)^\alpha} \int_0^\infty \frac{x^{\alpha-1}}{\Gamma(\alpha)[\beta/(1 - \beta t)]^\alpha} e^{-x/[\beta/(1-\beta t)]}\, dx$$

But now the integrand is a gamma pdf, so it integrates to 1. This establishes the result. ■

The mean and variance can be obtained from the moment generating function (Exercise 80), but they can also be obtained directly through integration (Exercise 81).

PROPOSITION

The mean and variance of a random variable X having the gamma distribution $f(x; \alpha, \beta)$ are

$$E(X) = \mu = \alpha\beta \qquad V(X) = \sigma^2 = \alpha\beta^2$$

When X is a standard gamma rv, the cdf of X, which is

$$F(x; \alpha) = \int_0^x \frac{y^{\alpha-1}e^{-y}}{\Gamma(\alpha)}\, dy \qquad x > 0 \tag{4.8}$$

is called the **incomplete gamma function** [sometimes the incomplete gamma function refers to Expression (4.8) without the denominator $\Gamma(\alpha)$ in the integrand]. There are extensive tables of $F(x; \alpha)$ available; in Appendix Table A.4, we present a small tabulation for $\alpha = 1, 2, \ldots, 10$ and $x = 1, 2, \ldots, 15$.

Example 4.26 Suppose the reaction time X of a randomly selected individual to a certain stimulus has a standard gamma distribution with $\alpha = 2$. Since

$$P(a \le X \le b) = F(b) - F(a)$$

when X is continuous,

$$P(3 \le X \le 5) = F(5; 2) - F(3; 2) = .960 - .801 = .159$$

The probability that the reaction time is more than 4 sec is

$$P(X > 4) = 1 - P(X \le 4) = 1 - F(4; 2) = 1 - .908 = .092$$ ■

The incomplete gamma function can also be used to compute probabilities involving nonstandard gamma distributions.

PROPOSITION

Let X have a gamma distribution with parameters α and β. Then for any $x > 0$, the cdf of X is given by

$$P(X \le x) = F(x; \alpha, \beta) = F\left(\frac{x}{\beta}; \alpha\right)$$

the incomplete gamma function evaluated at x/β.*

Proof Calculate, with the help of the substitution $y = u/\beta$,

$$P(X \le x) = \int_0^x \frac{u^{\alpha-1}}{\Gamma(\alpha)\beta^\alpha} e^{-u/\beta}\, du = \int_0^{x/\beta} \frac{y^{\alpha-1}}{\Gamma(\alpha)} e^{-y}\, dy = F\left(\frac{x}{\beta}; \alpha\right)$$ ■

Example 4.27 Suppose the survival time X in weeks of a randomly selected male mouse exposed to 240 rads of gamma radiation has a gamma distribution with $\alpha = 8$ and $\beta = 15$. (Data in *Survival Distributions: Reliability Applications in the Biomedical Services*, by A. J. Gross and V. Clark, suggests $\alpha \approx 8.5$ and $\beta \approx 13.3$.) The expected survival time is $E(X) = (8)(15) = 120$ weeks, whereas $V(X) = (8)(15)^2 = 1800$ and $\sigma_X = \sqrt{1800} = 42.43$ weeks. The probability that a mouse survives between 60 and 120 weeks is

$$\begin{aligned} P(60 \le X \le 120) &= P(X \le 120) - P(X \le 60) \\ &= F(120/15; 8) - F(60/15; 8) \\ &= F(8; 8) - F(4; 8) = .547 - .051 = .496 \end{aligned}$$

The probability that a mouse survives at least 30 weeks is

$$\begin{aligned} P(X \ge 30) &= 1 - P(X < 30) = 1 - P(X \le 30) \\ &= 1 - F(30/15; 8) = .999 \end{aligned}$$ ■

The Exponential Distribution

The family of exponential distributions provides probability models that are widely used in engineering and science disciplines.

---

*MINITAB and other statistical packages calculate $F(x; \alpha, \beta)$ once values of x, α, and β are specified.

DEFINITION X is said to have an **exponential distribution** with parameter λ ($\lambda > 0$) if the pdf of X is

$$f(x; \lambda) = \begin{cases} \lambda e^{-\lambda x} & x \ge 0 \\ 0 & \text{otherwise} \end{cases} \tag{4.9}$$

The exponential pdf is a special case of the general gamma pdf (4.7) in which $\alpha = 1$ and β has been replaced by $1/\lambda$ [some authors use the form $(1/\beta)e^{-x/\beta}$]. The mean and variance of X are then

$$\mu = \alpha\beta = \frac{1}{\lambda} \qquad \sigma^2 = \alpha\beta^2 = \frac{1}{\lambda^2}$$

Both the mean and standard deviation of the exponential distribution equal $1/\lambda$. Graphs of several exponential pdf's appear in Figure 4.27.

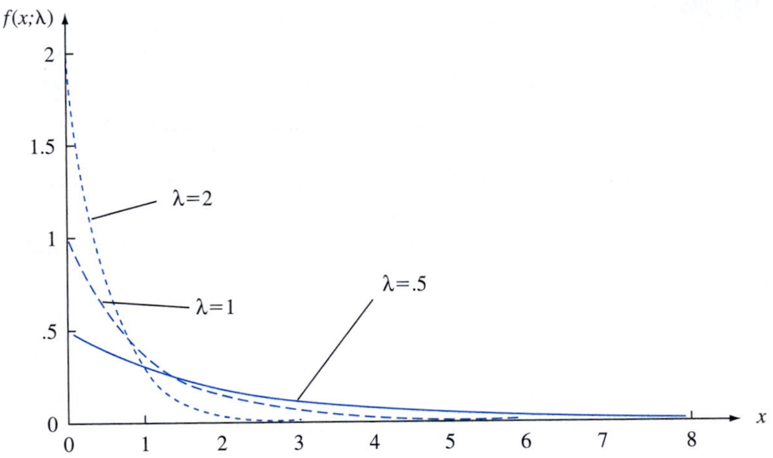

Figure 4.27 Exponential density curves

Unlike the general gamma pdf, the exponential pdf can be easily integrated. In particular, the cdf of X is

$$F(x; \lambda) = \begin{cases} 0 & x < 0 \\ 1 - e^{-\lambda x} & x \ge 0 \end{cases}$$

Example 4.28 The response time X at a certain on-line computer terminal (the elapsed time between the end of a user's inquiry and the beginning of the system's response to that inquiry) has an exponential distribution with expected response time equal to 5 sec. Then $E(X) = 1/\lambda = 5$, so $\lambda = .2$. The probability that the response time is at most 10 sec is

$$P(X \le 10) = F(10; .2) = 1 - e^{-(.2)(10)} = 1 - e^{-2} = 1 - .135 = .865$$

The probability that response time is between 5 and 10 sec is

$$\begin{aligned} P(5 \le X \le 10) &= F(10; .2) - F(5; .2) \\ &= (1 - e^{-2}) - (1 - e^{-1}) = .233 \end{aligned}$$ ■

The exponential distribution is frequently used as a model for the distribution of times between the occurrence of successive events, such as customers arriving at a service facility or calls coming in to a switchboard. The reason for this is that the exponential distribution is closely related to the Poisson process discussed in Chapter 3.

PROPOSITION Suppose that the number of events occurring in any time interval of length t has a Poisson distribution with parameter αt (where α, the rate of the event process, is the expected number of events occurring in 1 unit of time) and that numbers of occurrences in nonoverlapping intervals are independent of one another. Then the distribution of elapsed time between the occurrence of two successive events is exponential with parameter $\lambda = \alpha$.

Although a complete proof is beyond the scope of the text, the result is easily verified for the time X_1 until the first event occurs:

$$\begin{aligned} P(X_1 \le t) &= 1 - P(X_1 > t) = 1 - P[\text{no events in } (0, t)] \\ &= 1 - \frac{e^{-\alpha t} \cdot (\alpha t)^0}{0!} = 1 - e^{-\alpha t} \end{aligned}$$

which is exactly the cdf of the exponential distribution.

Example 4.29 Calls are received at a 24-hour "suicide hotline" according to a Poisson process with rate $\alpha = .5$ call per day. Then the number of days X between successive calls has an exponential distribution with parameter value .5, so the probability that more than 2 days elapse between calls is

$$P(X > 2) = 1 - P(X \le 2) = 1 - F(2; .5) = e^{-(.5)(2)} = .368$$

The expected time between successive calls is $1/.5 = 2$ days. ■

Another important application of the exponential distribution is to model the distribution of component lifetime. A partial reason for the popularity of such applications is the **"memoryless" property** of the exponential distribution. Suppose component lifetime is exponentially distributed with parameter λ. After putting the component into service, we leave for a period of t_0 hours and then return to find the component still working; what now is the probability that it lasts at least an additional t hours? In symbols, we wish $P(X \geq t + t_0 \mid X \geq t_0)$. By the definition of conditional probability,

$$P(X \geq t + t_0 | X \geq t_0) = \frac{P[(X \geq t + t_0) \cap (X \geq t_0)]}{P(X \geq t_0)}$$

But the event $X \geq t_0$ in the numerator is redundant, since both events can occur if and only if $X \geq t + t_0$. Therefore,

$$P(X \geq t + t_0 | X \geq t_0) = \frac{P(X \geq t + t_0)}{P(X \geq t_0)} = \frac{1 - F(t + t_0; \lambda)}{1 - F(t_0; \lambda)} = e^{-\lambda t}$$

This conditional probability is identical to the original probability $P(X \geq t)$ that the component lasted t hours. Thus *the distribution of additional lifetime is exactly the same as the original distribution of lifetime*, so at each point in time the component shows no effect of wear. In other words, the distribution of remaining lifetime is independent of current age.

Although the memoryless property can be justified at least approximately in many applied problems, in other situations components deteriorate with age or occasionally improve with age (at least up to a certain point). More general lifetime models are then furnished by the gamma, Weibull, and lognormal distributions (the latter two are discussed in the next section).

The Chi-Squared Distribution

DEFINITION

Let ν be a positive integer. Then a random variable X is said to have a **chi-squared distribution** with parameter ν if the pdf of X is the gamma density with $\alpha = \nu/2$ and $\beta = 2$. The pdf of a chi-squared rv is thus

$$f(x; \nu) = \begin{cases} \dfrac{1}{2^{\nu/2}\Gamma(\nu/2)} x^{(\nu/2)-1} e^{-x/2} & x \geq 0 \\ 0 & x < 0 \end{cases} \tag{4.10}$$

The parameter ν is called the **number of degrees of freedom** (df) of X. The symbol χ^2 is often used in place of "chi-squared."

The chi-squared distribution is important because it is the basis for a number of procedures in statistical inference. The reason for this is that chi-squared distributions are intimately related to normal distributions (see Exercise 79). We will discuss the chi-squared distribution in more detail in Section 6.4 and the chapters on inference.

Exercises Section 4.4 (69–81)

69. Evaluate the following:

a. $\Gamma(6)$

b. $\Gamma(5/2)$

c. $F(4; 5)$ (the incomplete gamma function)

d. $F(5; 4)$

e. $F(0; 4)$

70. Let X have a standard gamma distribution with $\alpha = 7$. Evaluate the following:

a. $P(X \le 5)$

b. $P(X < 5)$

c. $P(X > 8)$

d. $P(3 \le X \le 8)$

e. $P(3 < X < 8)$

f. $P(X < 4 \text{ or } X > 6)$

71. Suppose the time spent by a randomly selected student at a campus computer lab has a gamma distribution with mean 20 minutes and variance 80 minutes^2.

a. What are the values of α and β?

b. What is the probability that a student uses the lab for at most 24 minutes?

c. What is the probability that a student spends between 20 and 40 minutes at the lab?

72. Suppose that when a transistor of a certain type is subjected to an accelerated life test, the lifetime X (in weeks) has a gamma distribution with mean 24 weeks and standard deviation 12 weeks.

a. What is the probability that a transistor will last between 12 and 24 weeks?

b. What is the probability that a transistor will last at most 24 weeks? Is the median of the lifetime distribution less than 24? Why or why not?

c. What is the 99th percentile of the lifetime distribution?

d. Suppose the test will actually be terminated after t weeks. What value of t is such that only .5% of all transistors would still be operating at termination?

73. Let X = the time between two successive arrivals at the drive-up window of a local bank. If X has an exponential distribution with $\lambda = 1$ (which is identical to a standard gamma distribution with $\alpha = 1$), compute the following:

a. The expected time between two successive arrivals

b. The standard deviation of the time between successive arrivals

c. $P(X \le 4)$

d. $P(2 \le X \le 5)$

74. Let X denote the distance (m) that an animal moves from its birth site to the first territorial vacancy it encounters. Suppose that for banner-tailed kangaroo rats, X has an exponential distribution with parameter $\lambda = .01386$ (as suggested in the article "Competition and Dispersal from Multiple Nests," *Ecology*, 1997: 873–883).

a. What is the probability that the distance is at most 100 m? At most 200 m? Between 100 and 200 m?

b. What is the probability that distance exceeds the mean distance by more than 2 standard deviations?

c. What is the value of the median distance?

75. Extensive experience with fans of a certain type used in diesel engines has suggested that the exponential distribution provides a good model for time until failure. Suppose the mean time until failure is 25,000 hours. What is the probability that

a. A randomly selected fan will last at least 20,000 hours? At most 30,000 hours? Between 20,000 and 30,000 hours?

b. The lifetime of a fan exceeds the mean value by more than 2 standard deviations? More than 3 standard deviations?

76. The special case of the gamma distribution in which α is a positive integer n is called an Erlang distribution. If we replace β by $1/\lambda$ in Expression (4.7), the Erlang pdf is

$$f(x; \lambda, n) = \begin{cases} \dfrac{\lambda(\lambda x)^{n-1} e^{-\lambda x}}{(n-1)!} & x \ge 0 \\ 0 & x < 0 \end{cases}$$

It can be shown that if the times between successive events are independent, each with an exponential distribution with parameter λ, then the total time X that elapses before all of the next n events occur has pdf $f(x; \lambda, n)$.

a. What is the expected value of X? If the time (in minutes) between arrivals of successive customers is exponentially distributed with $\lambda = .5$, how much time can be expected to elapse before the tenth customer arrives?

b. If customer interarrival time is exponentially distributed with $\lambda = .5$, what is the probability that the tenth customer (after the one who has just arrived) will arrive within the next 30 min?

c. The event $\{X \le t\}$ occurs iff at least n events occur in the next t units of time. Use the fact that the number of events occurring in an interval of length t has a Poisson distribution with parameter λt to write an expression (involving Poisson probabilities) for the Erlang cumulative distribution function $F(t; \lambda, n) = P(X \le t)$.

77. A system consists of five identical components connected in series as shown:

As soon as one component fails, the entire system will fail. Suppose each component has a lifetime that is exponentially distributed with $\lambda = .01$ and that components fail independently of one another. Define events $A_i = \{i\text{th component lasts at least } t \text{ hours}\}$, $i = 1, \dots, 5$, so that the A_i's are independent events. Let X = the time at which the system fails—that is, the shortest (minimum) lifetime among the five components.

a. The event $\{X \ge t\}$ is equivalent to what event involving $A_1, \dots, A_5$?

b. Using the independence of the five A_i's, compute $P(X \ge t)$. Then obtain $F(t) = P(X \le t)$ and the pdf of X. What type of distribution does X have?

c. Suppose there are n components, each having exponential lifetime with parameter λ. What type of distribution does X have?

78. If X has an exponential distribution with parameter λ, derive a general expression for the $(100p)$th percentile of the distribution. Then specialize to obtain the median.

79. a. The event $\{X^2 \le y\}$ is equivalent to what event involving X itself?

b. If X has a standard normal distribution, use part (a) to write the integral that equals $P(X^2 \le y)$. Then differentiate this with respect to y to obtain the pdf of X^2 [the square of a $N(0, 1)$ variable]. Finally, show that X^2 has a chi-squared distribution with $\nu = 1$ df [see Expression (4.10)]. (*Hint*: Use the following identity.)

$$\frac{d}{dy}\left\{\int_{a(y)}^{b(y)} f(x)\,dx\right\} = f[b(y)]\cdot b'(y) - f[a(y)]\cdot a'(y)$$

80. a. Find the mean and variance of the gamma distribution by differentiating the moment generating function $M_X(t)$.

b. Find the mean and variance of the gamma distribution by differentiating $R_X(t) = \ln[M_X(t)]$.

81. Find the mean and variance of the gamma distribution using integration to obtain $E(X)$ and $E(X^2)$. [*Hint*: Express the integrand in terms of a gamma density.]

4.5 *Other Continuous Distributions

The normal, gamma (including exponential), and uniform families of distributions provide a wide variety of probability models for continuous variables, but there are many practical situations in which no member of these families fits a set of observed data very well. Statisticians and other investigators have developed other families of distributions that are often appropriate in practice.

The Weibull Distribution

The family of Weibull distributions was introduced by the Swedish physicist Waloddi Weibull in 1939; his 1951 article "A Statistical Distribution Function of Wide Applicability" (*J. Appl. Mech.*, vol. 18: 293–297) discusses a number of applications.

DEFINITION

A random variable X is said to have a **Weibull distribution** with parameters α and β ($\alpha > 0, \beta > 0$) if the pdf of X is

$$f(x; \alpha, \beta) = \begin{cases} \dfrac{\alpha}{\beta^\alpha} x^{\alpha-1} e^{-(x/\beta)^\alpha} & x \ge 0 \\ 0 & x < 0 \end{cases} \tag{4.11}$$

In some situations there are theoretical justifications for the appropriateness of the Weibull distribution, but in many applications $f(x; \alpha, \beta)$ simply provides a good fit to observed data for particular values of α and β. When $\alpha = 1$, the pdf reduces to the exponential distribution (with $\lambda = 1/\beta$), so the exponential distribution is a special case of both the gamma and Weibull distributions. However, there are gamma distributions that are not Weibull distributions and vice versa, so one family is not a subset of the other. Both α and β can be varied to obtain a number of different distributional shapes, as illustrated in Figure 4.28. Note that β is a scale parameter, so different values stretch or compress the graph in the x direction.

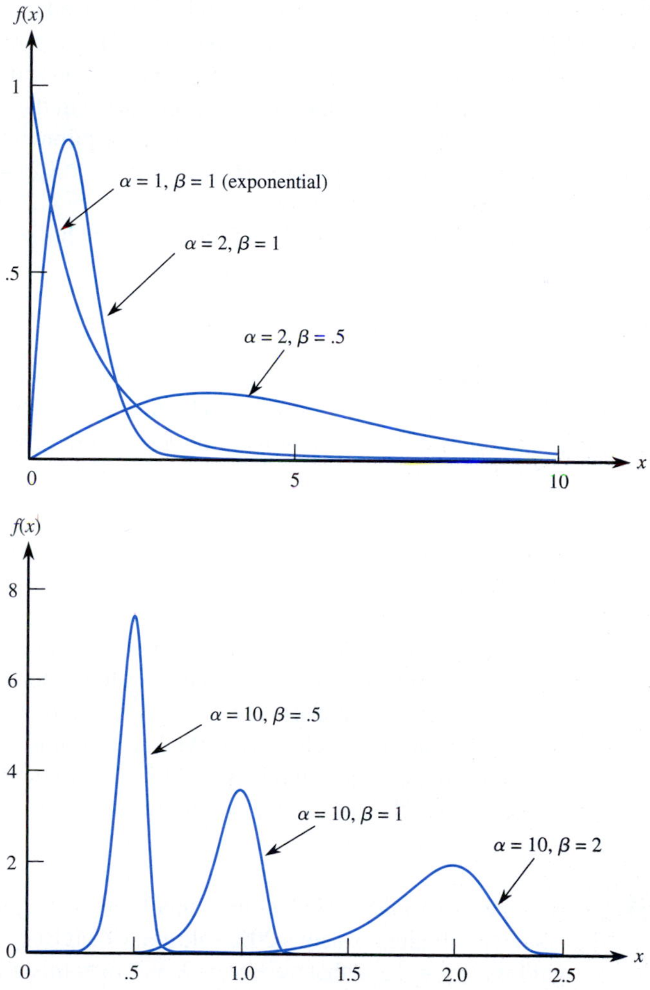

Figure 4.28 Weibull density curves

Integrating to obtain $E(X)$ and $E(X^2)$ yields

$$\mu = \beta\Gamma\left(1 + \frac{1}{\alpha}\right) \qquad \sigma^2 = \beta^2\left\{\Gamma\left(1 + \frac{2}{\alpha}\right) - \left[\Gamma\left(1 + \frac{1}{\alpha}\right)\right]^2\right\}$$

The computation of μ and σ^2 thus necessitates using the gamma function.

The integration $\int_0^x f(y; \alpha, \beta)\, dy$ is easily carried out to obtain the cdf of X. The cdf of a Weibull rv having parameters α and β is

$$F(x; \alpha, \beta) = \begin{cases} 0 & x < 0 \\ 1 - e^{-(x/\beta)^\alpha} & x \geq 0 \end{cases} \tag{4.12}$$

Example 4.30 In recent years the Weibull distribution has been used to model engine emissions of various pollutants. Let X denote the amount of NO_x emission (g/gal) from a randomly selected four-stroke engine of a certain type, and suppose that X has a Weibull distribution with $\alpha = 2$ and $\beta = 10$ (suggested by information in the article "Quantification of Variability and Uncertainty in Lawn and Garden Equipment NO_x and Total Hydrocarbon Emission Factors," *J. Air Waste Manag. Assoc.*, 2002: 435–448). The corresponding density curve looks exactly like the one in Figure 4.28 for $\alpha = 2$, $\beta = 1$ except that now the values 50 and 100 replace 5 and 10 on the horizontal axis (because β is a "scale parameter"). Then

$$P(X \leq 10) = F(10; 2, 10) = 1 - e^{-(10/10)^2} = 1 - e^{-1} = .632$$

Similarly, $P(X \leq 25) = .998$, so the distribution is almost entirely concentrated on values between 0 and 25. The value c, which separates the 5% of all engines having the largest amounts of NO_x emissions from the remaining 95%, satisfies

$$.95 = 1 - e^{-(c/10)^2}$$

Isolating the exponential term on one side, taking logarithms, and solving the resulting equation gives $c \approx 17.3$ as the 95th percentile of the emission distribution. ■

Frequently, in practical situations, a Weibull model may be reasonable except that the smallest possible X value may be some value γ not assumed to be zero (this would also apply to a gamma model). The quantity γ can then be regarded as a third parameter of the distribution, which is what Weibull did in his original work. For, say, $\gamma = 3$, all curves in Figure 4.28 would be shifted 3 units to the right. This is equivalent to saying that $X - \gamma$ has the pdf (4.11), so that the cdf of X is obtained by replacing x in (4.12) by $x - \gamma$.

Example 4.31 Let X = the corrosion weight loss for a small square magnesium alloy plate immersed for 7 days in an inhibited aqueous 20% solution of $MgBr_2$. Suppose the minimum possible weight loss is $\gamma = 3$ and that the excess $X - 3$ over this minimum has a Weibull distribution with $\alpha = 2$ and $\beta = 4$. (This example was considered in "Practical Applications of the Weibull Distribution," *Indust. Qual. Control*, Aug. 1964: 71–78; values for α and β were taken to be 1.8 and 3.67, respectively, though a slightly different choice of parameters was used in the article.) The cdf of X is then

$$F(x; \alpha, \beta, \gamma) = F(x; 2, 4, 3) = \begin{cases} 0 & x < 3 \\ 1 - e^{-[(x-3)/4]^2} & x \geq 3 \end{cases}$$

Therefore,

$$P(X > 3.5) = 1 - F(3.5; 2, 4, 3) = e^{-.0156} = .985$$

and

$$P(7 \le X \le 9) = 1 - e^{-2.25} - (1 - e^{-1}) = .895 - .632 = .263$$ ■

The Lognormal Distribution

Lognormal distributions have been used extensively in engineering, medicine, and more recently, finance.

DEFINITION A nonnegative rv X is said to have a **lognormal distribution** if the rv $Y = \ln(X)$ has a normal distribution. The resulting pdf of a lognormal rv when $\ln(X)$ is normally distributed with parameters μ and σ is

$$f(x; \mu, \sigma) = \begin{cases} \dfrac{1}{\sqrt{2\pi}\sigma x} e^{-[\ln(x) - \mu]^2/(2\sigma^2)} & x \ge 0 \\ 0 & x < 0 \end{cases}$$

Be careful here; the parameters μ and σ are not the mean and standard deviation of X but of $\ln(X)$. The mean and variance of X can be shown to be

$$E(X) = e^{\mu + \sigma^2/2} \qquad V(X) = e^{2\mu + \sigma^2} \cdot (e^{\sigma^2} - 1)$$

In Chapter 6, we will present a theoretical justification for this distribution in connection with the Central Limit Theorem, but as with other distributions, the lognormal can be used as a model even in the absence of such justification. Figure 4.29 illustrates

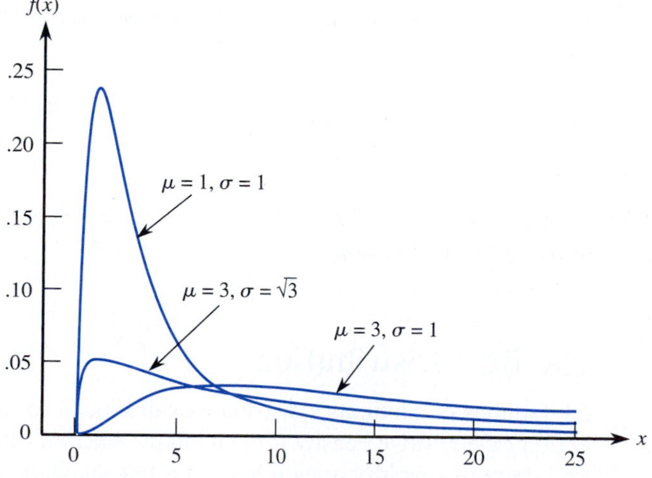

Figure 4.29 Lognormal density curves

graphs of the lognormal pdf; although a normal curve is symmetric, a lognormal curve has a positive skew.

Because ln(X) has a normal distribution, the cdf of X can be expressed in terms of the cdf $\Phi(z)$ of a standard normal rv Z. For $x \geq 0$,

$$F(x; \mu, \sigma) = P(X \leq x) = P[\ln(X) \leq \ln(x)] \tag{4.13}$$
$$= P\left[Z \leq \frac{\ln(x) - \mu}{\sigma}\right] = \Phi\left[\frac{\ln(x) - \mu}{\sigma}\right]$$

Example 4.32 The lognormal distribution is frequently used as a model for various material properties. The article "Reliability of Wood Joist Floor Systems with Creep" (*J. Struct. Engrg.* 1995: 946–954) suggests that the lognormal distribution with $\mu = .375$ and $\sigma = .25$ is a plausible model for X = the modulus of elasticity (MOE, in 10^6 psi) of wood joist floor systems constructed from #2 grade hem-fir. The mean value and variance of MOE are

$$E(X) = e^{.375 + (.25)^2/2} = e^{.40625} = 1.50$$
$$V(X) = e^{.8125}(e^{.0625} - 1) = .1453$$

The probability that MOE is between 1 and 2 is

$$P(1 \leq X \leq 2) = P[\ln(1) \leq \ln(X) \leq \ln(2)]$$
$$= P[0 \leq \ln(X) \leq .693]$$
$$= P\left(\frac{0 - .375}{.25} \leq Z \leq \frac{.693 - .375}{.25}\right)$$
$$= \Phi(1.27) - \Phi(-1.50) = .8312$$

What value c is such that only 1% of all systems have an MOE exceeding c? We wish the c for which

$$.99 = P(X \leq c) = P\left[Z \leq \frac{\ln(c) - .375}{.25}\right]$$

from which $[\ln(c) - .375]/.25 = 2.33$ and $c = 2.605$. Thus 2.605 is the 99th percentile of the MOE distribution. ■

The Beta Distribution

All families of continuous distributions discussed so far except for the uniform distribution have positive density over an infinite interval (though typically the density function decreases rapidly to zero beyond a few standard deviations from the mean). The beta distribution provides positive density only for X in an interval of finite length.

DEFINITION A random variable X is said to have a **beta distribution** with parameters α, β (both positive), A, and B if the pdf of X is

$$f(x; \alpha, \beta, A, B) = \begin{cases} \dfrac{1}{B - A} \cdot \dfrac{\Gamma(\alpha + \beta)}{\Gamma(\alpha) \cdot \Gamma(\beta)} \left(\dfrac{x - A}{B - A}\right)^{\alpha - 1} \left(\dfrac{B - x}{B - A}\right)^{\beta - 1} & A \le x \le B \\ 0 & \text{otherwise} \end{cases}$$

The case $A = 0$, $B = 1$ gives the **standard beta distribution.**

Figure 4.30 illustrates several standard beta pdf's. Graphs of the general pdf are similar, except they are shifted and then stretched or compressed to fit over $[A, B]$. Unless α and β are integers, integration of the pdf to calculate probabilities is difficult, so either a table of the incomplete beta function or software is generally used. The mean and variance of X are

$$\mu = A + (B - A) \cdot \frac{\alpha}{\alpha + \beta} \qquad \sigma^2 = \frac{(B - A)^2 \alpha\beta}{(\alpha + \beta)^2(\alpha + \beta + 1)}$$

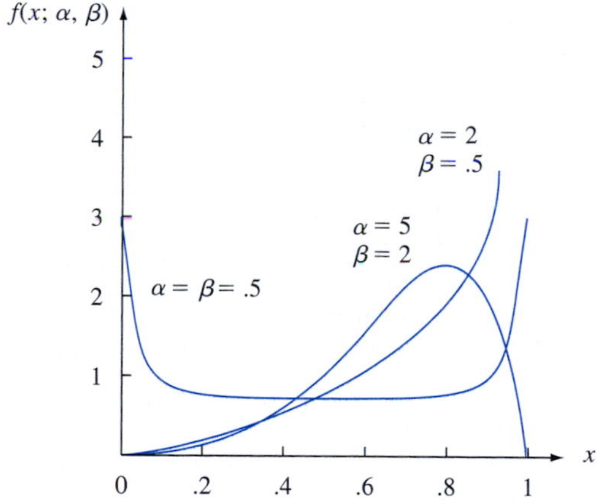

Figure 4.30 Standard beta density curves

Example 4.33 Project managers often use a method labeled PERT—for program evaluation and review technique—to coordinate the various activities making up a large project. (One successful application was in the construction of the *Apollo* spacecraft.) A standard assumption in PERT analysis is that the time necessary to complete any particular activity once it has been started has a beta distribution with A = the optimistic time (if everything goes well) and B = the pessimistic time (if everything goes badly). Suppose that in constructing a single-family house, the time X (in days) necessary for laying the foundation has a beta distribution with $A = 2$, $B = 5$, $\alpha = 2$, and $\beta = 3$. Then $\alpha/(\alpha + \beta) = .4$, so $E(X) = 2 + (3)(.4) = 3.2$. For these values of α and β, the pdf of

X is a simple polynomial function. The probability that it takes at most 3 days to lay the foundation is

$$P(X \le 3) = \int_2^3 \frac{1}{3} \cdot \frac{4!}{1!2!}\left(\frac{x-2}{3}\right)\left(\frac{5-x}{3}\right)^2 dx$$

$$= \frac{4}{27}\int_2^3 (x-2)(5-x)^2\,dx = \frac{4}{27}\cdot\frac{11}{4} = \frac{11}{27} = .407 \qquad \blacksquare$$

The standard beta distribution is commonly used to model variation in the proportion or percentage of a quantity occurring in different samples, such as the proportion of a 24-hour day that an individual is asleep or the proportion of a certain element in a chemical compound.

Exercises Section 4.5 (82–96)

82. The lifetime X (in hundreds of hours) of a certain type of vacuum tube has a Weibull distribution with parameters $\alpha = 2$ and $\beta = 3$. Compute the following:

a. $E(X)$ and $V(X)$

b. $P(X \le 6)$

c. $P(1.5 \le X \le 6)$

(This Weibull distribution is suggested as a model for time in service in "On the Assessment of Equipment Reliability: Trading Data Collection Costs for Precision," *J. Engrg. Manuf.*, 1991: 105–109).

83. The authors of the article "A Probabilistic Insulation Life Model for Combined Thermal-Electrical Stresses" (*IEEE Trans. Electr. Insul.*, 1985: 519–522) state that "the Weibull distribution is widely used in statistical problems relating to aging of solid insulating materials subjected to aging and stress." They propose the use of the distribution as a model for time (in hours) to failure of solid insulating specimens subjected to ac voltage. The values of the parameters depend on the voltage and temperature; suppose $\alpha = 2.5$ and $\beta = 200$ (values suggested by data in the article).

a. What is the probability that a specimen's lifetime is at most 250? Less than 250? More than 300?

b. What is the probability that a specimen's lifetime is between 100 and 250?

c. What value is such that exactly 50% of all specimens have lifetimes exceeding that value?

84. Let X = the time (in 10^{-1} weeks) from shipment of a defective product until the customer returns the product. Suppose that the minimum return time is $\gamma = 3.5$ and that the excess $X - 3.5$ over the minimum has a Weibull distribution with parameters $\alpha = 2$ and $\beta = 1.5$ (see the *Indust. Qual. Control* article referenced in Example 4.31).

a. What is the cdf of X?

b. What are the expected return time and variance of return time? [*Hint*: First obtain $E(X - 3.5)$ and $V(X - 3.5)$.]

c. Compute $P(X > 5)$.

d. Compute $P(5 \le X \le 8)$.

85. Let X have a Weibull distribution with the pdf from Expression (4.11). Verify that $\mu = \beta\Gamma(1 + 1/\alpha)$. (*Hint*: In the integral for $E(X)$, make the change of variable $y = (x/\beta)^\alpha$, so that $x = \beta y^{1/\alpha}$.)

86. **a.** In Exercise 82, what is the median lifetime of such tubes? [*Hint*: Use Expression (4.12).]

b. In Exercise 84, what is the median return time?

c. If X has a Weibull distribution with the cdf from Expression (4.12), obtain a general expression for the $(100p)$th percentile of the distribution.

d. In Exercise 84, the company wants to refuse to accept returns after t weeks. For what value of t will only 10% of all returns be refused?

87. Let X denote the ultimate tensile strength (ksi) at $-200°$ of a randomly selected steel specimen of a certain type that exhibits "cold brittleness" at low temperatures. Suppose that X has a Weibull distribution with $\alpha = 20$ and $\beta = 100$.

a. What is the probability that X is at most 105 ksi?

b. If specimen after specimen is selected, what is the long-run proportion having strength values between 100 and 105 ksi?

c. What is the median of the strength distribution?

88. The authors of a paper from which the data in Exercise 25 of Chapter 1 was extracted suggested that a reasonable probability model for drill lifetime was a lognormal distribution with $\mu = 4.5$ and $\sigma = .8$.

a. What are the mean value and standard deviation of lifetime?

b. What is the probability that lifetime is at most 100?

c. What is the probability that lifetime is at least 200? Greater than 200?

89. Let X = the hourly median power (in decibels) of received radio signals transmitted between two cities. The authors of the article "Families of Distributions for Hourly Median Power and Instantaneous Power of Received Radio Signals" (*J. Res. Nat. Bureau Standards*, vol. 67D, 1963: 753–762) argue that the lognormal distribution provides a reasonable probability model for X. If the parameter values are $\mu = 3.5$ and $\sigma = 1.2$, calculate the following:

a. The mean value and standard deviation of received power.

b. The probability that received power is between 50 and 250 dB.

c. The probability that X is less than its mean value. Why is this probability not .5?

90. a. Use Equation (4.13) to write a formula for the median $\tilde{\mu}$ of the lognormal distribution. What is the median for the power distribution of Exercise 89?

b. Recalling that z_α is our notation for the $100(1 - \alpha)$ percentile of the standard normal distribution, write an expression for the $100(1 - \alpha)$ percentile of the lognormal distribution. In Exercise 89, what value will received power exceed only 5% of the time?

91. A theoretical justification based on a certain material failure mechanism underlies the assumption that ductile strength X of a material has a lognormal distribution. Suppose the parameters are $\mu = 5$ and $\sigma = .1$.

a. Compute $E(X)$ and $V(X)$.

b. Compute $P(X > 125)$.

c. Compute $P(110 \le X \le 125)$.

d. What is the value of median ductile strength?

e. If ten different samples of an alloy steel of this type were subjected to a strength test, how many would you expect to have strength of at least 125?

f. If the smallest 5% of strength values were unacceptable, what would the minimum acceptable strength be?

92. The article "The Statistics of Phytotoxic Air Pollutants" (*J. Roy. Statist Soc.*, 1989: 183–198) suggests the lognormal distribution as a model for SO_2 concentration above a certain forest. Suppose the parameter values are $\mu = 1.9$ and $\sigma = .9$.

a. What are the mean value and standard deviation of concentration?

b. What is the probability that concentration is at most 10? Between 5 and 10?

93. What condition on α and β is necessary for the standard beta pdf to be symmetric?

94. Suppose the proportion X of surface area in a randomly selected quadrate that is covered by a certain plant has a standard beta distribution with $\alpha = 5$ and $\beta = 2$.

a. Compute $E(X)$ and $V(X)$.

b. Compute $P(X \le .2)$.

c. Compute $P(.2 \le X \le .4)$.

d. What is the expected proportion of the sampling region not covered by the plant?

95. Let X have a standard beta density with parameters α and β.

a. Verify the formula for $E(X)$ given in the section.

b. Compute $E[(1 - X)^m]$. If X represents the proportion of a substance consisting of a particular ingredient, what is the expected proportion that does not consist of this ingredient?

96. Stress is applied to a 20-in. steel bar that is clamped in a fixed position at each end. Let Y = the distance from the left end at which the bar snaps. Suppose $Y/20$ has a standard beta distribution with $E(Y) = 10$ and $V(Y) = \frac{100}{7}$.

a. What are the parameters of the relevant standard beta distribution?

b. Compute $P(8 \le Y \le 12)$.

c. Compute the probability that the bar snaps more than 2 in. from where you expect it to.

4.6 *Probability Plots

An investigator will often have obtained a numerical sample $x_1, x_2, \ldots, x_n$ and wish to know whether it is plausible that it came from a population distribution of some particular type (e.g., from a normal distribution). For one thing, many formal procedures from statistical inference are based on the assumption that the population distribution is of a specified type. The use of such a procedure is inappropriate if the actual underlying probability distribution differs greatly from the assumed type. Additionally, understanding the underlying distribution can sometimes give insight into the physical mechanisms involved in generating the data. An effective way to check a distributional assumption is to construct what is called a **probability plot**. The essence of such a plot is that if the distribution on which the plot is based is correct, the points in the plot will fall close to a straight line. If the actual distribution is quite different from the one used to construct the plot, the points should depart substantially from a linear pattern.

Sample Percentiles

The details involved in constructing probability plots differ a bit from source to source. The basis for our construction is a comparison between percentiles of the sample data and the corresponding percentiles of the distribution under consideration. Recall that the $(100p)$th percentile of a continuous distribution with cdf $F(x)$ is the number $\eta(p)$ that satisfies $F[\eta(p)] = p$. That is, $\eta(p)$ is the number on the measurement scale such that the area under the density curve to the left of $\eta(p)$ is p. Thus the 50th percentile $\eta(.5)$ satisfies $F[\eta(.5)] = .5$, and the 90th percentile satisfies $F[\eta(.9)] = .9$. Consider as an example the standard normal distribution, for which we have denoted the cdf by $\Phi(z)$. From Appendix Table A.3, we find the 20th percentile by locating the row and column in which .2000 (or a number as close to it as possible) appears inside the table. Since .2005 appears at the intersection of the $-.8$ row and the .04 column, the 20th percentile is approximately $-.84$. Similarly, the 25th percentile of the standard normal distribution is (using linear interpolation) approximately $-.675$.

Roughly speaking, sample percentiles are defined in the same way that percentiles of a population distribution are defined. The 50th-sample percentile should separate the smallest 50% of the sample from the largest 50%, the 90th percentile should be such that 90% of the sample lies below that value and 10% lies above, and so on. Unfortunately, we run into problems when we actually try to compute the sample percentiles for a particular sample of n observations. If, for example, $n = 10$, we can split off 20% of these values or 30% of the data, but there is no value that will split off exactly 23% of these ten observations. To proceed further, we need an operational definition of sample percentiles (this is one place where different people do slightly different things). Recall that when n is odd, the sample median or 50th-sample percentile is the middle value in the ordered list, for example, the sixth largest value when $n = 11$. This amounts to regarding the middle observation as being half in the lower half of the data and half in the upper half. Similarly, suppose $n = 10$. Then if we call the third smallest value the 25th percentile, we are regarding that value as being half in the lower group (consisting of the two smallest observations) and half in the upper group (the seven largest observations). This leads to the following general definition of sample percentiles.

DEFINITION Order the n sample observations from smallest to largest. Then the ith smallest observation in the list is taken to be the **$[100(i - .5)/n]$th sample percentile**.

Once the percentage values $100(i - .5)/n$ ($i = 1, 2, \ldots, n$) have been calculated, sample percentiles corresponding to intermediate percentages can be obtained by linear interpolation. For example, if $n = 10$, the percentages corresponding to the ordered sample observations are $100(1 - .5)/10 = 5\%$, $100(2 - .5)/10 = 15\%$, $25\%, \ldots$, and $100(10 - .5)/10 = 95\%$. The 10th percentile is then halfway between the 5th percentile (smallest sample observation) and the 15th percentile (second smallest observation). For our purposes, such interpolation is not necessary because a probability plot will be based only on the percentages $100(i - .5)/n$ corresponding to the n sample observations.

A Probability Plot

Suppose now that for percentages $100(i - .5)/n$ ($i = 1, \ldots, n$) the percentiles are determined for a specified population distribution whose plausibility is being investigated. If the sample was actually selected from the specified distribution, the sample percentiles (ordered sample observations) should be reasonably close to the corresponding population distribution percentiles. That is, for $i = 1, 2, \ldots, n$ there should be reasonable agreement between the ith smallest sample observation and the $[100(i - .5)/n]$th percentile for the specified distribution. Consider the (population percentile, sample percentile) pairs—that is, the pairs

$$\left(\begin{matrix} [100(i - .5)/n]\text{th percentile} \\ \text{of the distribution} \end{matrix}, \begin{matrix} i\text{th smallest sample} \\ \text{observation} \end{matrix} \right)$$

for $i = 1, \ldots, n$. Each such pair can be plotted as a point on a two-dimensional coordinate system. If the sample percentiles are close to the corresponding population distribution percentiles, the first number in each pair will be roughly equal to the second number. The plotted points will then fall close to a 45° line. Substantial deviations of the plotted points from a 45° line cast doubt on the assumption that the distribution under consideration is the correct one.

Example 4.34 The value of a certain physical constant is known to an experimenter. The experimenter makes $n = 10$ independent measurements of this value using a particular measurement device and records the resulting measurement errors (error = observed value − true value). These observations appear in the accompanying table.

Percentage	5	15	25	35	45
z percentile	−1.645	−1.037	−.675	−.385	−.126
Sample observation	−1.91	−1.25	−.75	−.53	.20

Percentage	55	65	75	85	95
z percentile	.126	.385	.675	1.037	1.645
Sample observation	.35	.72	.87	1.40	1.56

Is it plausible that the random variable *measurement error* has a standard normal distribution? The needed standard normal (z) percentiles are also displayed in the table. Thus the points in the probability plot are $(-1.645, -1.91)$, $(-1.037, -1.25)$, . . . , and $(1.645, 1.56)$. Figure 4.31 shows the resulting plot. Although the points deviate a bit from the 45° line, the predominant impression is that this line fits the points very well. The plot suggests that the standard normal distribution is a reasonable probability model for measurement error.

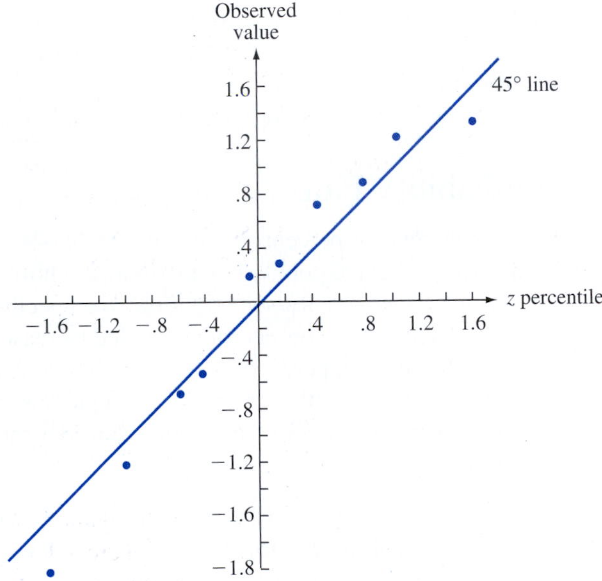

Figure 4.31 Plots of pairs (z percentile, observed value) for the data of Example 4.34: first sample

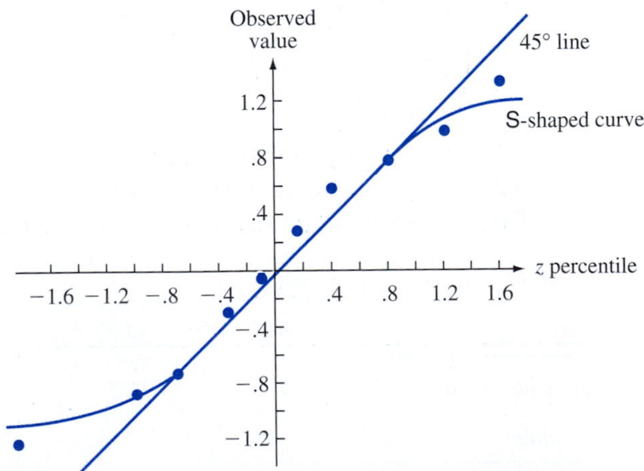

Figure 4.32 Plots of pairs (z percentile, observed value) for the data of Example 4.34: second sample

Figure 4.32 shows a plot of pairs (z percentile, observation) for a second sample of ten observations. The 45° line gives a good fit to the middle part of the sample but not to the extremes. The plot has a well-defined S-shaped appearance. The two smallest sample observations are considerably larger than the corresponding z percentiles (the points on the far left of the plot are well above the 45° line). Similarly, the two largest sample observations are much smaller than the associated z percentiles. This plot indicates that the standard normal distribution would not be a plausible choice for the probability model that gave rise to these observed measurement errors. ■

An investigator is typically not interested in knowing whether a specified probability distribution, such as the standard normal distribution (normal with $\mu = 0$ and $\sigma = 1$) or the exponential distribution with $\lambda = .1$, is a plausible model for the population distribution from which the sample was selected. Instead, the investigator will want to know whether *some* member of a family of probability distributions specifies a plausible model—the family of normal distributions, the family of exponential distributions, the family of Weibull distributions, and so on. The values of the parameters of a distribution are usually not specified at the outset. If the family of Weibull distributions is under consideration as a model for lifetime data, the issue is whether there are *any* values of the parameters α and β for which the corresponding Weibull distribution gives a good fit to the data. Fortunately, it is almost always the case that just one probability plot will suffice for assessing the plausibility of an entire family. If the plot deviates substantially from a straight line, no member of the family is plausible. When the plot is quite straight, further work is necessary to estimate values of the parameters (e.g., find values for μ and σ) that yield the most reasonable distribution of the specified type.

Let's focus on a plot for checking normality. Such a plot can be very useful in applied work because many formal statistical procedures are appropriate (give accurate inferences) only when the population distribution is at least approximately normal. These procedures should generally not be used if the normal probability plot shows a very pronounced departure from linearity. The key to constructing an omnibus normal probability plot is the relationship between standard normal (z) percentiles and those for any other normal distribution:

$$\begin{array}{l}\text{percentile for a normal}\\ (\mu, \sigma)\text{ distribution}\end{array} = \mu + \sigma \cdot (\text{corresponding } z \text{ percentile})$$

Consider first the case $\mu = 0$. Then if each observation is exactly equal to the corresponding normal percentile for a particular value of σ, the pairs ($\sigma \cdot$ [z percentile], observation) fall on a 45° line, which has slope 1. This implies that the pairs (z percentile, observation) fall on a line passing through (0, 0) (i.e., one with y-intercept 0) but having slope σ rather than 1. The effect of a nonzero value of μ is simply to change the y-intercept from 0 to μ.

A plot of the n pairs

$$([100(i - .5)/n]\text{th } z \text{ percentile, } i\text{th smallest observation})$$

on a two-dimensional coordinate system is called a **normal probability plot**. If the sample observations are in fact drawn from a normal distribution with mean value

μ and standard deviation σ, the points should fall close to a straight line with slope σ and intercept μ. Thus a plot for which the points fall close to some straight line suggests that the assumption of a normal population distribution is plausible.

Example 4.35 The accompanying sample consisting of $n = 20$ observations on dielectric breakdown voltage of a piece of epoxy resin appeared in the article "Maximum Likelihood Estimation in the 3-Parameter Weibull Distribution (*IEEE Trans. Dielectrics Electr. Insul.*, 1996: 43–55). Values of $(i - .5)/n$ for which z percentiles are needed are $(1 - .5)/20 = .025$, $(2 - .5)/20 = .075, \ldots,$ and .975.

Observation	24.46	25.61	26.25	26.42	26.66	27.15	27.31	27.54	27.74	27.94
z percentile	−1.96	−1.44	−1.15	−.93	−.76	−.60	−.45	−.32	−.19	−.06

Observation	27.98	28.04	28.28	28.49	28.50	28.87	29.11	29.13	29.50	30.88
z percentile	.06	.19	.32	.45	.60	.76	.93	1.15	1.44	1.96

Figure 4.33 shows the resulting normal probability plot. The pattern in the plot is quite straight, indicating it is plausible that the population distribution of dielectric breakdown voltage is normal.

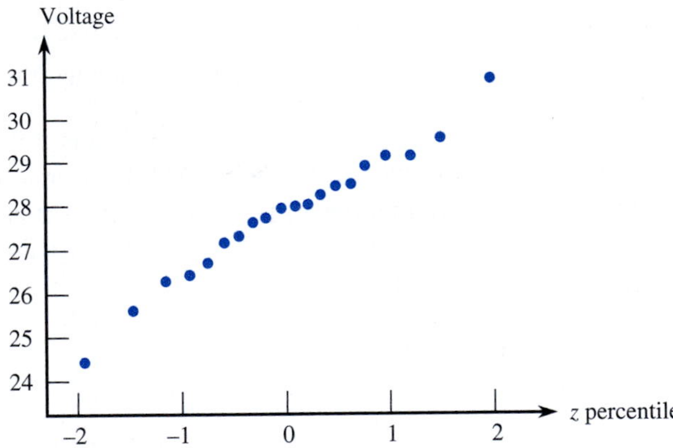

Figure 4.33 Normal probability plot for the dielectric breakdown voltage sample ■

There is an alternative version of a normal probability plot in which the z percentile axis is replaced by a nonlinear probability axis. The scaling on this axis is constructed so that plotted points should again fall close to a line when the sampled distribution is normal. Figure 4.34 shows such a plot from MINITAB for the breakdown voltage data of Example 4.35. Here the z values are replaced by the corresponding normal percentiles. The plot remains the same, and it is just the labeling of the axis that changes. Note that MINITAB and various other software packages use the refinement $(i - .375)/(n + .25)$ of the formula $(i - .5)/n$ in order to get a better approximation to

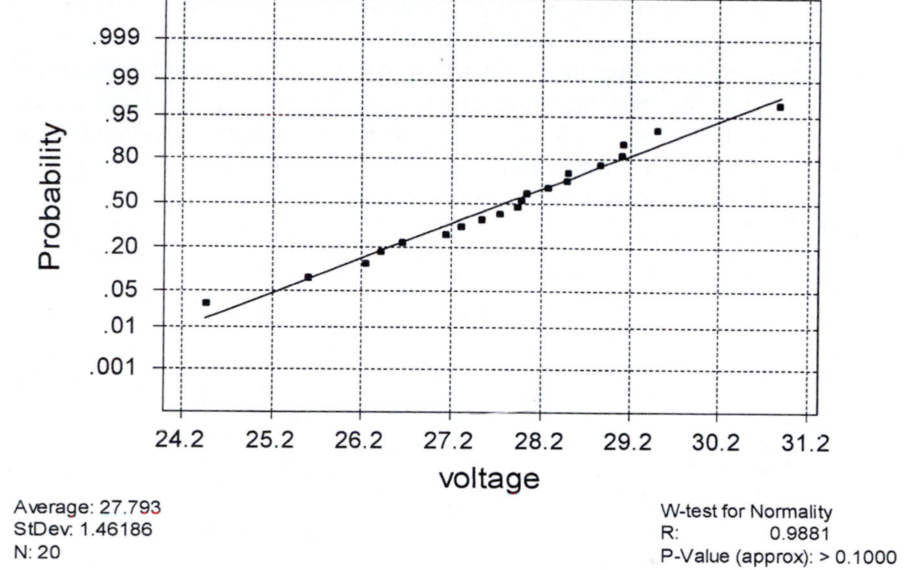

Figure 4.34 Normal probability plot of the breakdown voltage data from MINITAB

what is expected for the ordered values of the standard normal distribution. Also notice that the axes in Figure 4.34 are reversed relative to those in Figure 4.33. A nonnormal population distribution can often be placed in one of the following three categories:

1. It is symmetric and has "lighter tails" than does a normal distribution; that is, the density curve declines more rapidly out in the tails than does a normal curve.
2. It is symmetric and heavy-tailed compared to a normal distribution.
3. It is skewed.

A uniform distribution is light-tailed, since its density function drops to zero outside a finite interval. The density function $f(x) = 1/[\pi(1+x^2)]$ for $-\infty < x < \infty$ is one example of a heavy-tailed distribution, since $1/(1 + x^2)$ declines much less rapidly than does $e^{-x^2/2}$. Lognormal and Weibull distributions are among those that are skewed. When the points in a normal probability plot do not adhere to a straight line, the pattern will frequently suggest that the population distribution is in a particular one of these three categories.

When the distribution from which the sample is selected is light-tailed, the largest and smallest observations are usually not as extreme as would be expected from a normal random sample. Visualize a straight line drawn through the middle part of the plot; points on the far right tend to be below the line (observed value $< z$ percentile), whereas points on the left end of the plot tend to fall above the straight line (observed value $> z$ percentile). The result is an S-shaped pattern of the type pictured in Figure 4.32.

A sample from a heavy-tailed distribution also tends to produce an S-shaped plot. However, in contrast to the light-tailed case, the left end of the plot curves downward (observed < z percentile), as shown in Figure 4.35(a). If the underlying distribution is positively skewed (a short left tail and a long right tail), the smallest sample observations will be larger than expected from a normal sample and so will the largest observations. In this case, points on both ends of the plot will fall above a straight line through the middle part, yielding a curved pattern, as illustrated in Figure 4.35(b). A sample from a lognormal distribution will usually produce such a pattern. A plot of (z percentile, $\ln(x)$) pairs should then resemble a straight line.

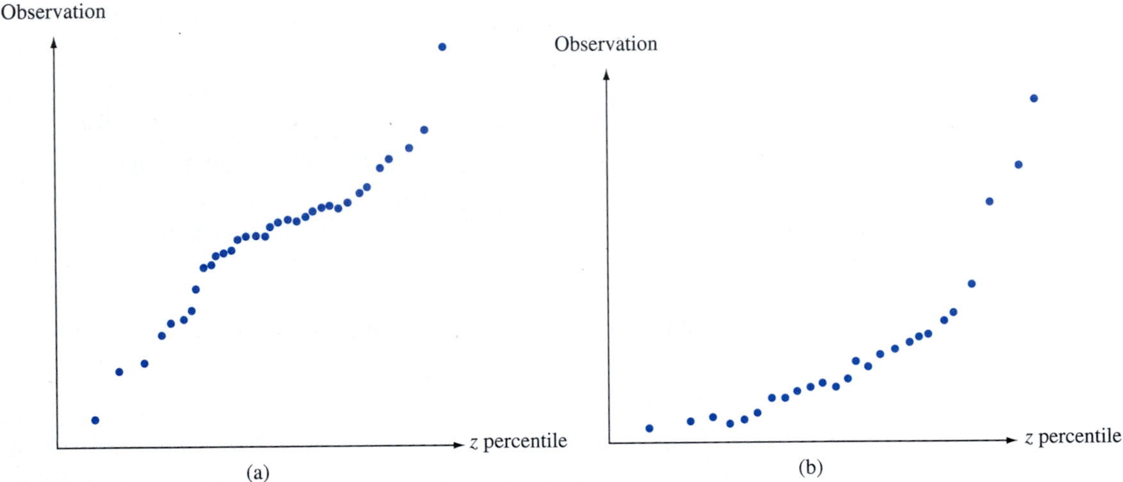

Figure 4.35 Probability plots that suggest a nonnormal distribution: (a) a plot consistent with a heavy-tailed distribution; (b) a plot consistent with a positively skewed distribution

Even when the population distribution is normal, the sample percentiles will not coincide exactly with the theoretical percentiles because of sampling variability. How much can the points in the probability plot deviate from a straight-line pattern before the assumption of population normality is no longer plausible? This is not an easy question to answer. Generally speaking, a small sample from a normal distribution is more likely to yield a plot with a nonlinear pattern than is a large sample. The book *Fitting Equations to Data* (see the Chapter 12 bibliography) presents the results of a simulation study in which numerous samples of different sizes were selected from normal distributions. The authors concluded that there is typically greater variation in the appearance of the probability plot for sample sizes smaller than 30, and only for much larger sample sizes does a linear pattern generally predominate. When a plot is based on a small sample size, only a very substantial departure from linearity should be taken as conclusive evidence of nonnormality. A similar comment applies to probability plots for checking the plausibility of other types of distributions.

Given the limitations of probability plots, there is need for an alternative. In Section 13.2 we introduce a formal procedure for judging whether the pattern of points in a normal probability plot is far enough from linear to cast doubt on population normality.

Beyond Normality

Consider a family of probability distributions involving two parameters, θ_1 and θ_2, and let $F(x; \theta_1, \theta_2)$ denote the corresponding cdf's. The family of normal distributions is one such family, with $\theta_1 = \mu$, $\theta_2 = \sigma$, and $F(x; \mu, \sigma) = \Phi[(x - \mu)/\sigma]$. Another example is the Weibull family, with $\theta_1 = \alpha$, $\theta_2 = \beta$, and

$$F(x; \alpha, \beta) = 1 - e^{-(x/\beta)^{\alpha}}$$

Still another family of this type is the gamma family, for which the cdf is an integral involving the incomplete gamma function that cannot be expressed in any simpler form.

The parameters θ_1 and θ_2 are said to be **location** and **scale parameters**, respectively, if $F(x; \theta_1, \theta_2)$ is a function of $(x - \theta_1)/\theta_2$. The parameters μ and σ of the normal family are location and scale parameters, respectively. Changing μ shifts the location of the bell-shaped density curve to the right or left, and changing σ amounts to stretching or compressing the measurement scale (the scale on the horizontal axis when the density function is graphed). Another example is given by the cdf

$$F(x; \theta_1, \theta_2) = 1 - e^{-e^{(x-\theta_1)/\theta_2}} \qquad -\infty < x < \infty$$

A random variable with this cdf is said to have an *extreme value distribution*. It is used in applications involving component lifetime and material strength.

Although the form of the extreme value cdf might at first glance suggest that θ_1 is the point of symmetry for the density function, and therefore the mean and median, this is not the case. Instead, $P(X \le \theta_1) = F(\theta_1; \theta_1, \theta_2) = 1 - e^{-1} = .632$, and the density function $f(x; \theta_1, \theta_2) = F'(x; \theta_1, \theta_2)$ is negatively skewed (a long lower tail). Similarly, the scale parameter θ_2 is not the standard deviation ($\mu = \theta_1 - .5772\theta_2$ and $\sigma = 1.283\theta_2$). However, changing the value of θ_1 does change the location of the density curve, whereas a change in θ_2 rescales the measurement axis.

The parameter β of the Weibull distribution is a scale parameter, but α is not a location parameter. The parameter α is usually referred to as a **shape parameter**. A similar comment applies to the parameters α and β of the gamma distribution. In the usual form, the density function for any member of either the gamma or Weibull distribution is positive for $x > 0$ and zero otherwise. A location parameter can be introduced as a third parameter γ (we did this for the Weibull distribution) to shift the density function so that it is positive if $x > \gamma$ and zero otherwise.

When the family under consideration has only location and scale parameters, the issue of whether any member of the family is a plausible population distribution can be addressed via a single, easily constructed probability plot. One first obtains the percentiles of the *standard distribution*, the one with $\theta_1 = 0$ and $\theta_2 = 1$, for percentages $100(i - .5)/n$ $(i = 1, \ldots, n)$. The n (standardized percentile, observation) pairs give the points in the plot. This is, of course, exactly what we did to obtain an omnibus normal probability plot. Somewhat surprisingly, this methodology can be applied to yield an omnibus Weibull probability plot. The key result is that if X has a Weibull distribution with shape parameter α and scale parameter β, then the transformed variable $\ln(X)$ has an extreme value distribution with location parameter $\theta_1 = \ln(\beta)$ and scale parameter $1/\alpha$. Thus a plot of the (extreme value standardized percentile, $\ln(x)$) pairs that shows a strong linear pattern provides support for choosing the Weibull distribution as a population model.

Example 4.36 The accompanying observations are on lifetime (in hours) of power apparatus insulation when thermal and electrical stress acceleration were fixed at particular values ("On the Estimation of Life of Power Apparatus Insulation Under Combined Electrical and Thermal Stress," *IEEE Trans. Electr. Insul.*, 1985: 70–78). A Weibull probability plot necessitates first computing the 5th, 15th, . . . , and 95th percentiles of the standard extreme value distribution. The $(100p)$th percentile $\eta(p)$ satisfies

$$p = F[\eta(p)] = 1 - e^{-e^{\eta(p)}}$$

from which $\eta(p) = \ln[-\ln(1 - p)]$.

Percentile	−2.97	−1.82	−1.25	−.84	−.51
x	282	501	741	851	1072
ln(x)	5.64	6.22	6.61	6.75	6.98

Percentile	−.23	.05	.33	.64	1.10
x	1122	1202	1585	1905	2138
ln(x)	7.02	7.09	7.37	7.55	7.67

The pairs (−2.97, 5.64), (−1.82, 6.22), . . . , (1.10, 7.67) are plotted as points in Figure 4.36. The straightness of the plot argues strongly for using the Weibull distribution as a model for insulation life, a conclusion also reached by the author of the cited article.

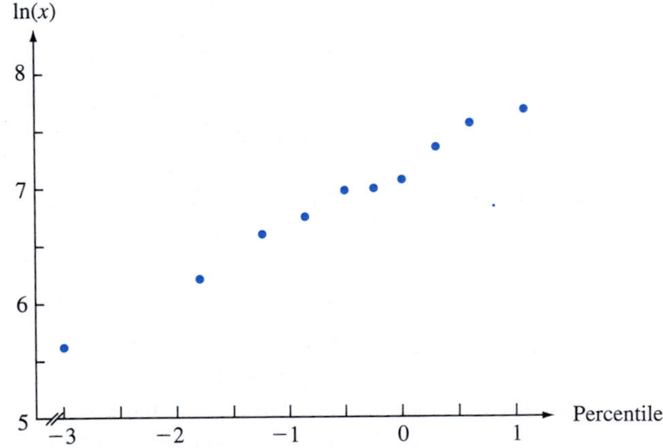

Figure 4.36 A Weibull probability plot of the insulation lifetime data ■

The gamma distribution is an example of a family involving a shape parameter for which there is no transformation $h(x)$ such that $h(X)$ has a distribution that depends only on location and scale parameters. Construction of a probability plot necessitates first estimating the shape parameter from sample data (some methods for doing this are described in Chapter 7). Sometimes an investigator wishes to know whether the transformed variable X^θ has a normal distribution for some value of θ (by convention, $\theta = 0$ is identified with the logarithmic transformation, in which case X has a lognormal distribution). The book *Graphical Methods for Data Analysis,* listed in the Chapter 1 bibliography, discusses this type of problem as well as other refinements of probability plotting.

Exercises Section 4.6 (97–107)

97. The accompanying normal probability plot was constructed from a sample of 30 readings on tension for mesh screens behind the surface of video display tubes used in computer monitors. Does it appear plausible that the tension distribution is normal?

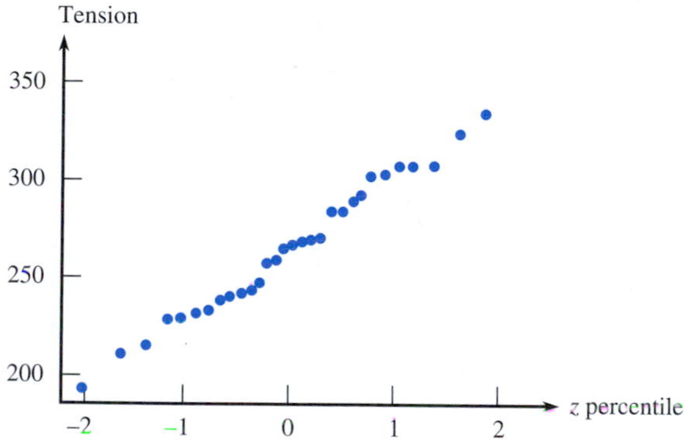

98. Consider the following ten observations on bearing lifetime (in hours):

152.7	172.0	172.5	173.3	193.0
204.7	216.5	234.9	262.6	422.6

Construct a normal probability plot and comment on the plausibility of the normal distribution as a model for bearing lifetime (data from "Modified Moment Estimation for the Three-Parameter Lognormal Distribution," *J. Qual. Tech.*, 1985: 92–99).

99. Construct a normal probability plot for the following sample of observations on coating thickness for low-viscosity paint ("Achieving a Target Value for a Manufacturing Process: A Case Study," *J. Qual. Tech.*, 1992: 22–26). Would you feel comfortable estimating population mean thickness using a method that assumed a normal population distribution?

.83	.88	.88	1.04	1.09	1.12	1.29	1.31
1.48	1.49	1.59	1.62	1.65	1.71	1.76	1.83

100. The article "A Probabilistic Model of Fracture in Concrete and Size Effects on Fracture Toughness" (*Mag. Concrete Res.*, 1996: 311–320) gives arguments for why the distribution of fracture toughness in concrete specimens should have a Weibull distribution and presents several histograms of data that appear well fit by superimposed Weibull curves. Consider the following sample of size $n = 18$ observations on toughness for high-strength concrete (consistent with one of the histograms); values of $p_i = (i - .5)/18$ are also given.

Observation	.47	.58	.65	.69	.72	.74
p_i	.0278	.0833	.1389	.1944	.2500	.3056
Observation	.77	.79	.80	.81	.82	.84
p_i	.3611	.4167	.4722	.5278	.5833	.6389
Observation	.86	.89	.91	.95	1.01	1.04
p_i	.6944	.7500	.8056	.8611	.9167	.9722

Construct a Weibull probability plot and comment.

101. Construct a normal probability plot for the fatigue crack propagation data given in Exercise 36 of Chapter 1. Does it appear plausible that propagation life has a normal distribution? Explain.

102. The article "The Load-Life Relationship for M50 Bearings with Silicon Nitride Ceramic Balls" (*Lubricat. Engrg.*, 1984: 153–159) reports the accompanying data on bearing load life (million revs.) for bearings tested at a 6.45-kN load.

47.1	68.1	68.1	90.8	103.6	106.0	115.0
126.0	146.6	229.0	240.0	240.0	278.0	278.0
289.0	289.0	367.0	385.9	392.0	505.0	

a. Construct a normal probability plot. Is normality plausible?
b. Construct a Weibull probability plot. Is the Weibull distribution family plausible?

103. Construct a probability plot that will allow you to assess the plausibility of the lognormal distribution as a model for the rainfall data of Exercise 80 in Chapter 1.

104. The accompanying observations are precipitation values during March over a 30-year period in Minneapolis–St. Paul.

.77	1.20	3.00	1.62	2.81	2.48
1.74	.47	3.09	1.31	1.87	.96
.81	1.43	1.51	.32	1.18	1.89
1.20	3.37	2.10	.59	1.35	.90
1.95	2.20	.52	.81	4.75	2.05

a. Construct and interpret a normal probability plot for this data set.
b. Calculate the square root of each value and then construct a normal probability plot based on this transformed data. Does it seem plausible that the square root of precipitation is normally distributed?
c. Repeat part (b) after transforming by cube roots.

105. Use a statistical software package to construct a normal probability plot of the shower-flow rate data given in Exercise 13 of Chapter 1, and comment.

106. Let the *ordered* sample observations be denoted by $y_1, y_2, \ldots, y_n$ (y_1 being the smallest and y_n the largest). Our suggested check for normality is to plot the $(\Phi^{-1}[(i - .5)/n], y_i)$ pairs. Suppose we believe that the observations come from a distribution with mean 0, and let $w_1, \ldots, w_n$ be the *ordered absolute* values of the x_i's. A **half-normal** plot is a probability plot of the w_i's. More specifically, since $P(|Z| \le w) = P(-w \le Z \le w) = 2\Phi(w) - 1$, a half-normal plot is a plot of the $(\Phi^{-1}[(p_i + 1)/2], w_i)$ pairs, where $p_i = (i - .5)/n$. The virtue of this plot is that small or large outliers in the original sample will now appear only at the upper end of the plot rather than at both ends. Construct a half-normal plot for the following sample of measurement errors, and comment: −3.78, −1.27, 1.44, −.39, 12.38, −43.40, 1.15, −3.96, −2.34, 30.84.

107. The following failure time observations (1000's of hours) resulted from accelerated life testing of 16 integrated circuit chips of a certain type:

82.8	11.6	359.5	502.5	307.8	179.7
242.0	26.5	244.8	304.3	379.1	212.6
229.9	558.9	366.7	204.6		

Use the corresponding percentiles of the exponential distribution with $\lambda = 1$ to construct a probability plot. Then explain why the plot assesses the plausibility of the sample having been generated from *any* exponential distribution.

4.7 *Transformations of a Random Variable

Often we need to deal with a transformation $Y = g(X)$ of the random variable X. Here $g(X)$ could be a simple change of time scale. If X is in hours and Y is in minutes, then $Y = 60X$. What happens to the pdf when we do this? Can we get the pdf of Y from the pdf of X? Consider first a simple example.

Example 4.37 The interval X in minutes between calls to a 911 center is exponentially distributed with mean 2 min, so its pdf $f_X(x) = \frac{1}{2}e^{-x/2}$ for $x > 0$. Can we find the pdf of $Y = 60X$, so Y is the number of seconds? In order to get the pdf, we first find the cdf. The cdf of Y is

$$F_Y(y) = P(Y \le y) = P(60X \le y) = P(X \le y/60) = F_X(y/60)$$

$$= \int_0^{y/60} \frac{1}{2}e^{-u/2}\,du = 1 - e^{-y/120}$$

Differentiating this with respect to y gives $f_Y(y) = \frac{1}{120}e^{-y/120}$ for $y > 0$. The distribution of Y is exponential with mean 120 sec (2 min).

Sometimes it isn't possible to evaluate the cdf in closed form. Could we still find the pdf of Y without evaluating the integral? Yes, and it involves differentiating the integral with respect to the upper limit of integration. The rule, which is sometimes presented as part of the Fundamental Theorem of Calculus, is

$$\frac{d}{dx}\int_a^x h(u)\,du = h(x)$$

Now, setting $x = y/60$ and using the chain rule, we get the pdf using the rule for differentiating integrals:

$$\begin{aligned} f_Y(y) &= \frac{d}{dy}F_Y(y) = \frac{d}{dy}F_X(x)\Big|_{x=y/60} = \frac{dx}{dy}\frac{d}{dx}F_X(x)\Big|_{x=y/60} \qquad y > 0 \\ &= \frac{1}{60}\frac{d}{dx}\int_0^x \frac{1}{2}e^{-u/2}\,du = \frac{1}{60}\frac{1}{2}e^{-x/2} = \frac{1}{120}e^{-y/120} \end{aligned}$$

Although it is useful to have the integral expression of the cdf here for clarity, it is not necessary. A more abstract approach is just to use differentiation of the cdf to get the pdf. That is, with $x = y/60$ and again using the chain rule,

$$\begin{aligned} f_Y(y) &= \frac{d}{dy}F_Y(y) = \frac{d}{dy}F_X(x)\Big|_{x=y/60} = \frac{dx}{dy}\frac{d}{dx}F_X(x) \\ &= \frac{1}{60}f_X(x) = \frac{1}{60}\frac{1}{2}e^{-x/2} = \frac{1}{120}e^{-y/120} \qquad y > 0 \end{aligned}$$ ■

Is it plausible that, if $X \sim$ exponential with mean 2, then $60X \sim$ exponential with mean 120? In terms of time between calls, if it is exponential with mean 2 minutes, then this should be the same as exponential with mean 120 seconds. Generalizing, there is nothing special here about 2 and 60, so it should be clear that if we multiply an exponential random variable with mean μ by a positive constant c we get another exponential random variable with mean $c\mu$. This is also easily verified using a moment generating function argument.

The method illustrated here can be applied to other transformations.

THEOREM Let X have pdf $f_X(x)$ and let $Y = g(X)$, where g is monotonic (either strictly increasing or strictly decreasing) so it has an inverse function $X = h(Y)$. Then $f_Y(y) = f_X[h(y)]|h'(y)|$.

Proof Here is the proof assuming that g is monotonically increasing. The proof for g monotonically decreasing is similar. We follow the last method in Example 4.37. First find the cdf.

$$F_Y(y) = P(Y \le y) = P[g(X) \le y] = P[X \le h(y)] = F_X[h(y)]$$

Now differentiate the cdf, letting $x = h(y)$.

$$f_Y(y) = \frac{d}{dy}F_Y(y) = \frac{d}{dy}F_X[h(y)] = \frac{dx}{dy}\frac{d}{dx}F_X(x) = h'(y)f_X(x) = h'(y)f_X[h(y)]$$

The absolute value is needed on the derivative only in the other case where g is decreasing. The set of possible values for Y is obtained by applying g to the set of possible values for X. ■

A heuristic view of the theorem (and a good way to remember it) is to say that

$$f_X(x)\,dx = f_Y(y)\,dy, \text{ so}$$

$$f_Y(y) = f_X(x)\frac{dx}{dy} = f_X[h(y)]h'(y)$$

Of course, because the pdf's must be nonnegative, the absolute value will be needed on the derivative if it is negative.

Sometimes it is easier to find the derivative of g than to find the derivative of h. In this case, remember that

$$\frac{dx}{dy} = \frac{1}{\dfrac{dy}{dx}}$$

Example 4.38 Let's apply the theorem to the situation introduced in Example 4.37. There $Y = g(X) = 60X$ and $X = h(Y) = Y/60$.

$$f_Y(y) = f_X[h(y)]|h'(y)| = \frac{1}{2}e^{-x/2}\frac{1}{60} = \frac{1}{120}e^{-y/120} \qquad y > 0$$ ■

Example 4.39 Here is an even simpler example. Suppose the arrival time of a delivery truck will be somewhere between noon and 2:00. We model this with a random variable X that is uniform on $[0, 2]$, so $f_X(x) = \frac{1}{2}$ on that interval. Let Y be the time in minutes, starting at noon, $Y = g(X) = 60X$ so $X = h(Y) = Y/60$. Then

$$f_Y(y) = f_X[h(y)]|h'(y)| = \frac{1}{2}\frac{1}{60} = \frac{1}{120} \qquad 0 < y < 120$$

Is this intuitively reasonable? Beginning with a uniform distribution on [0, 2], we multiply it by 60, and this spreads it out over the interval [0, 120]. Notice that the pdf is divided by 60, not multiplied by 60. Because the distribution is spread over a wider interval, the density curve must be lower if the total area under the curve is to be 1. ■

Example 4.40 This being a special day (an A in statistics!), you plan to buy a steak (substitute five portobello mushrooms if you are a vegetarian) for dinner. The weight X of the steak is normally distributed with mean μ and variance σ^2. The steak costs a dollars per pound, and your other purchases total b dollars. Let Y be the total bill at the cash register, so $Y = aX + b$. What is the distribution of the new variable Y? Let $X \sim N(\mu, \sigma^2)$ and $Y = aX + b$, where $a \neq 0$. In our example a is positive, but we will do a more general calculation that allows negative a. Then the inverse function is $x = h(y) = (y - b)/a$.

$$f_Y(y) = f_X[h(y)]|h'(y)| = \frac{1}{\sqrt{2\pi}\sigma} e^{-(\{[(y-b)/a]-\mu\}/\sigma)^2} \frac{1}{|a|} = \frac{1}{\sqrt{2\pi}\sigma|a|} e^{-[(y-b-\mu a)/\sigma|a|]^2}$$

Thus Y is normally distributed with mean $\mu a + b$ and standard deviation $\sigma|a|$. The mean and standard deviation did not require the new theory of this section, because we could have just calculated $E(Y) = E(aX + b) = a\mu + b$, $V(Y) = V(aX + b) = a^2\sigma^2$, and therefore $\sigma_Y = |a|\sigma$.

As a special case, take $Y = (X - \mu)/\sigma$, so $b = -\mu/\sigma$ and $a = 1/\sigma$. Then Y is normal with mean $a\mu + b = \mu/\sigma - \mu/\sigma = 0$ and standard deviation $|a|\sigma = |1/\sigma|\sigma = 1$. Thus the transformation $Y = (X - \mu)/\sigma$ creates a new normal random variable with mean 0 and standard deviation 1. That is, Y is standard normal. This is the first proposition in Section 4.3.

On the other hand, suppose that X is already standard normal, $X \sim N(0, 1)$. If we let $Y = \mu + \sigma X$, then $a = \sigma$ and $b = \mu$, so Y will have mean $0 \cdot \sigma + \mu = \mu$, and standard deviation $|a| \cdot 1 = \sigma$. If we start with a standard normal, we can obtain any other normal distribution by means of a linear transformation. ■

Example 4.41 Here we want to see what can be done with the simple uniform distribution. Let X have uniform distribution on [0, 1], so $f_X(x) = 1$ for $0 < x < 1$. We want to transform X so that $g(X) = Y$ has a specified distribution. Let's specify that $f_Y(y) = y/2$ for $0 < y < 2$. Integrating this, we get the cdf $F_Y(y) = y^2/4$, $0 < y < 2$. The trick is to set this equal to the inverse function $h(y)$. That is, $x = h(y) = y^2/4$. Inverting this (solving for y, and using the positive root), we get $y = g(x) = F_Y^{-1}(x) = \sqrt{4x} = 2\sqrt{x}$. Let's apply the foregoing theorem to see if $Y = g(X) = 2\sqrt{X}$ has the desired pdf:

$$f_Y(y) = f_X[h(y)]|h'(y)| = 1\frac{2y}{4} = \frac{y}{2} \qquad 0 < y < 2$$

A graphical approach may help in understanding why the transform $Y = 2\sqrt{X}$ yields $f_Y(y) = y/2$ if X is uniform on [0, 1]. Figure 4.37(a) shows the uniform distribution with [0, 1] partitioned into ten subintervals. In Figure 4.37(b) the endpoints of these intervals are shown after transforming according to $y = 2\sqrt{x}$. The heights of the rectangles are arranged so each rectangle still has area .1, and therefore the probability in each interval is preserved. Notice the close fit of the dashed line, which has the equation $f_Y(y) = y/2$.

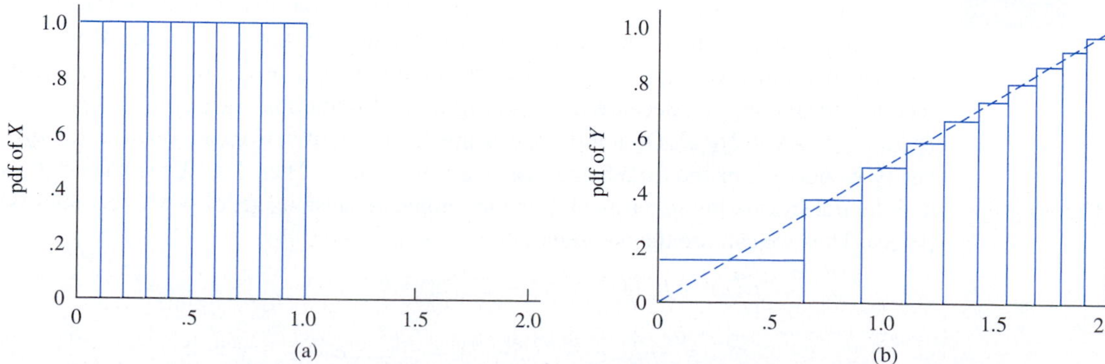

Figure 4.37 The effect on the pdf if X is uniform on [0, 1] and $Y = 2\sqrt{X}$

Can the method be generalized to produce a random variable with any desired pdf? Let the pdf $f_Y(y)$ be specified along with the corresponding cdf $F_Y(y)$. Define g to be the inverse function of F_Y, so $h(y) = F_Y(y)$. If X is uniformly distributed on [0, 1], then using the theorem, the pdf of $Y = g(X) = F_Y^{-1}(X)$ is

$$f_X[h(y)]|h'(y)| = (1)f_Y(y) = f_Y(y)$$

This says that you can build any random variable you want from humble uniform variates. Values of uniformly distributed random variables are available from almost any calculator or computer language, so our method enables you to produce values of any continuous random variable, as long as you know its cdf.

To get a sequence of random values with the pdf $f_Y(y) = y/2$, $0 < y < 2$, start with a sequence of random values from the uniform distribution on [0, 1]: .529, .043, .294, ... Then take $Y = g(X) = F_Y^{-1}(X) = 2\sqrt{X}$ to get 1.455, .415, 1.084, ... ■

Can the process be reversed, so we start with any continuous random variable and transform to a uniform variable? Let X have pdf $f_X(x)$ and cdf $F_X(x)$. Transform X to $Y = g(X) = F_X(X)$, so g is F_X. The inverse function of $g = F_X$ is h. Again apply the theorem to show that Y is uniform:

$$f_Y(y) = f_X[h(y)]|h'(y)| = f_X(x)/f_X(x) = 1 \qquad 0 \le x \le 1$$

This works because h and F are inverse functions, so their derivatives are reciprocals.

Example 4.42 To illustrate the transformation to uniformity, assume that X has pdf $f_X(x) = x/2$, $0 < x < 2$. Integrating this, we get the cdf $F_X(x) = x^2/4$, $0 < x < 2$. Let $Y = g(X) = F_X(X) = X^2/4$. Then the inverse function is $h(y) = \sqrt{4y} = 2\sqrt{y}$ and

$$f_Y(y) = f_X(x)\left|\frac{dx}{dy}\right| = \frac{f_X(x)}{\left|\frac{dy}{dx}\right|} = \frac{x/2}{x/2} = 1 \qquad 0 < y < 1$$

■

The foregoing theorem requires a monotonic transformation, but there are important applications in which the transformation is not monotonic. Nevertheless, it may be possible to use the theorem anyway with a little trickery.

Example 4.43 In this example, we start with a standard normal random variable X, and we transform to $Y = X^2$. Of course, this is not monotonic over the interval for X, $(-\infty, \infty)$. However, consider the transformation $U = |X|$. Can we obtain the pdf of this intuitively, without recourse to any theory? Because X has a symmetric distribution, the pdf of U is $f_U(u) = f_X(u) + f_X(-u) = 2f_X(u)$. Do not despair if this is not intuitively clear, because we will verify it shortly. For the time being, assume it to be true. Then $Y = X^2 = |X|^2 = U^2$, and the transformation in terms of U is monotonic because its set of possible values is $[0, \infty)$. Thus we can use the theorem with $h(y) = y^{.5}$:

$$\begin{aligned} f_Y(y) &= f_U[h(y)]h'(y) = 2f_X[h(y)]h'(y) \\ &= \frac{2}{\sqrt{2\pi}}e^{-.5(y^{.5})^2}(.5y^{-.5}) = \frac{1}{\sqrt{2\pi y}}e^{-y/2} \qquad y > 0 \end{aligned}$$

This is the chi-squared distribution (with 1 degree of freedom) introduced in Section 4.4. The squares of normal random variables are important because the sample variance is built from squares, and we will need the distribution of the variance. The variance for normal data is proportional to a chi-squared rv.

You were asked to believe that $f_U(u) = 2f_X(u)$ on an intuitive basis. Here is a derivation that works as long as f_X is an even function, that is, $f_X(-x) = f_X(x)$. If $u > 0$,

$$\begin{aligned} F_U(u) &= P(U < u) = P(|X| < u) = P(-u < X < u) \\ &= 2P(0 < X < u) = 2[F_X(u) - F_X(0)] \end{aligned}$$

Differentiating this with respect to u gives $f_U(u) = 2f_X(u)$. ■

Example 4.44 Sometimes the theorem cannot be used at all, and you need to use the cdf. Let $f_X(x) = (x + 1)/8$, $-1 < x < 3$, and $Y = X^2$. The transformation is not monotonic and $f_X(x)$ is not an even function. Possible values of Y are $\{y\colon 0 \le y \le 9\}$. Consider first $0 \le y \le 1$:

$$F_Y(y) = P(Y \le y) = P(X^2 \le y) = P(-\sqrt{y} \le X \le \sqrt{y}) = \int_{-\sqrt{y}}^{\sqrt{y}} \frac{u + 1}{8}\, du = \frac{\sqrt{y}}{4}$$

Then, on the other subinterval, $1 < y \le 9$,

$$\begin{aligned} F_Y(y) &= P(Y \le y) = P(X^2 \le y) = P(-\sqrt{y} \le X \le \sqrt{y}) \\ &= P(-1 \le X \le \sqrt{y}) = \int_{-1}^{\sqrt{y}} \frac{u + 1}{8}\, du = (1 + y + 2\sqrt{y})/16 \end{aligned}$$

Differentiating, we get

$$f_Y(y) = \begin{cases} \dfrac{1}{8\sqrt{y}} & 0 < y < 1 \\ \dfrac{y + \sqrt{y}}{16y} & 1 < y < 9 \\ 0 & \text{otherwise} \end{cases}$$

■

If X is discrete, what happens to the pmf when we do a monotonic transformation?

Example 4.45 Let X have the geometric distribution, with pmf $p_X(x) = (1 - p)^x p$, $x = 0, 1, 2, \ldots$, and define $Y = X/3$. Then the pmf of Y is

$$p_Y(y) = P(Y = y) = P\left(\frac{X}{3} = y\right) = P(X = 3y) = p_X(3y) = (1 - p)^{3y}p$$

$$y = 0, \frac{1}{3}, \frac{2}{3}, \ldots$$

Notice that there is no need for a derivative in finding the pmf for transformations of discrete random variables.

To put this on a more general basis in the discrete case, if $Y = g(X)$ with inverse $X = h(Y)$, then

$$p_Y(y) = P(Y = y) = P[g(X) = y] = P[X = h(y)] = p_X[h(y)]$$

and the set of possible values of Y is obtained by applying g to the set of possible values of X. ■

Exercises Section 4.7 (108–126)

108. Relative to the winning time, the time X of another runner in a 10-km race has pdf $f_X(x) = 2/x^3$, $x > 1$. The reciprocal $Y = 1/X$ represents the ratio of the time for the winner divided by the time of the other runner. Find the pdf of Y. Explain why Y also represents the speed of the other runner relative to the winner.

109. If X has the pdf $f_X(x) = 2x$, $0 < x < 1$, find the pdf of $Y = 1/X$. The distribution of Y is a special case of the Pareto distribution (see Exercise 10).

110. Let X have the pdf $f_X(x) = 2/x^3$, $x > 1$. Find the pdf of $Y = \sqrt{X}$.

111. Let X have the chi-squared distribution with 2 degrees of freedom, so $f_X(x) = \frac{1}{2}e^{-(x/2)}$, $x > 0$. Find the pdf of $Y = \sqrt{X}$. Suppose you choose a point in two dimensions randomly, with the horizontal and vertical coordinates chosen independently from the standard normal distribution. Then X has the distribution of the squared distance from the origin and Y has the distribution of the distance from the origin. Because Y is the length of a vector with normal components, there are lots of applications in physics, and its distribution has the name *Rayleigh*.

112. If X is distributed as $N(\mu, \sigma^2)$, find the pdf of $Y = e^X$. The distribution of Y is lognormal, as discussed in Section 4.5.

113. If the side of a square X is random with the pdf $f_X(x) = x/8$, $0 < x < 4$, and Y is the area of the square, find the pdf of Y.

114. Let X have the uniform distribution on [0, 1]. Find the pdf of $Y = -\ln(X)$.

115. Let X be uniformly distributed on [0, 1]. Find the pdf of $Y = \tan[\pi(X - .5)]$. This is called the *Cauchy distribution* after the famous mathematician.

116. If X is uniformly distributed on [0, 1], find a linear transformation $Y = cX + d$ such that Y is uniformly distributed on $[a, b]$, where a and b are any two numbers such that $a < b$. Is there another solution? Explain.

117. If X has the pdf $f_X(x) = x/8$, $0 < x < 4$, find a transformation $Y = g(X)$ such that Y is uniformly distributed on [0, 1].

118. If X is uniformly distributed on [−1, 1], find the pdf of $Y = |X|$.

119. If X is uniformly distributed on [−1, 1], find the pdf of $Y = X^2$.

120. Ann is expected at 7:00 pm after an all-day drive. She may be as much as one hour early or as much as three hours late. Assuming that her arrival time X is uniformly distributed over that interval, find the pdf of $|X - 7|$, the unsigned difference between her actual and predicted arrival times.

121. If X is uniformly distributed on [−1, 3], find the pdf of $Y = X^2$.

122. If X is distributed as $N(0, 1)$, find the pdf of $|X|$.

123. A circular target has radius 1 foot. Assume that you hit the target (we shall ignore misses) and that the probability of hitting any region of the target is proportional to the region's area. If you hit the target at a distance Y from the center, then let $X = \pi Y^2$ be the corresponding circular area. Show that

a. X is uniformly distributed on $[0, \pi]$. *Hint*: Show that $F_X(x) = P(X \le x) = x/\pi$.

b. Y has pdf $f_Y(y) = 2y$, $0 < y < 1$.

124. In Exercise 123, suppose instead that Y is uniformly distributed on [0, 1]. Find the pdf of $X = \pi Y^2$. Geometrically speaking, why should X have a pdf that is unbounded near 0?

125. Let X have the geometric distribution with pmf $p_X(x) = (1 - p)^x p$, $x = 0, 1, 2, \ldots$. Find the pmf of $Y = X + 1$. The resulting distribution is also referred to as *geometric* (see Example 3.10).

126. Let X have a binomial distribution with $n = 1$ (Bernoulli distribution). That is, X has pmf $b(x; 1, p)$. If $Y = 2X - 1$, find the pmf of Y.

Supplementary Exercises (127–155)

127. Let X = the time it takes a read/write head to locate a desired record on a computer disk memory device once the head has been positioned over the correct track. If the disks rotate once every 25 msec, a reasonable assumption is that X is uniformly distributed on the interval [0, 25].

a. Compute $P(10 \le X \le 20)$.

b. Compute $P(X \ge 10)$.

c. Obtain the cdf $F(X)$.

d. Compute $E(X)$ and σ_X.

128. A 12-in. bar that is clamped at both ends is to be subjected to an increasing amount of stress until it snaps. Let Y = the distance from the left end at which the break occurs. Suppose Y has pdf

$$f(y) = \begin{cases} \left(\frac{1}{24}\right) y \left(1 - \frac{y}{12}\right) & 0 \le y \le 12 \\ 0 & \text{otherwise} \end{cases}$$

Compute the following:

a. The cdf of Y, and graph it.

b. $P(Y \le 4)$, $P(Y > 6)$, and $P(4 \le Y \le 6)$.

c. $E(Y)$, $E(Y^2)$, and $V(Y)$.

d. The probability that the break point occurs more than 2 in. from the expected break point.

e. The expected length of the shorter segment when the break occurs.

129. Let X denote the time to failure (in years) of a certain hydraulic component. Suppose the pdf of X is $f(x) = 32/(x + 4)^3$ for $x > 0$.

a. Verify that $f(x)$ is a legitimate pdf.

b. Determine the cdf.

c. Use the result of part (b) to calculate the probability that time to failure is between 2 and 5 years.

d. What is the expected time to failure?

e. If the component has a salvage value equal to $100/(4 + x)$ when its time to failure is x, what is the expected salvage value?

130. The completion time X for a certain task has cdf $F(x)$ given by

$$F(x) = \begin{cases} 0 & x < 0 \\ \frac{x^3}{3} & 0 \le x < 1 \\ 1 - \frac{1}{2}\left(\frac{7}{3} - x\right)\left(\frac{7}{4} - \frac{3}{4}x\right) & 1 \le x \le \frac{7}{3} \\ 1 & x \ge \frac{7}{3} \end{cases}$$

a. Obtain the pdf $f(x)$ and sketch its graph.

b. Compute $P(.5 \le X \le 2)$.

c. Compute $E(X)$.

131. The breakdown voltage of a randomly chosen diode of a certain type is known to be normally distributed with mean value 40 V and standard deviation 1.5 V.

a. What is the probability that the voltage of a single diode is between 39 and 42?

b. What value is such that only 15% of all diodes have voltages exceeding that value?

c. If four diodes are independently selected, what is the probability that at least one has a voltage exceeding 42?

132. The article "Computer Assisted Net Weight Control" (*Qual. Prog.*, 1983: 22–25) suggests a normal distribution with mean 137.2 oz and standard deviation 1.6 oz, for the actual contents of jars of a certain type. The stated contents was 135 oz.

a. What is the probability that a single jar contains more than the stated contents?

b. Among ten randomly selected jars, what is the probability that at least eight contain more than the stated contents?

c. Assuming that the mean remains at 137.2, to what value would the standard deviation have to be changed so that 95% of all jars contain more than the stated contents?

133. When circuit boards used in the manufacture of compact disc players are tested, the long-run percentage of defectives is 5%. Suppose that a batch of

250 boards has been received and that the condition of any particular board is independent of that of any other board.

a. What is the approximate probability that at least 10% of the boards in the batch are defective?

b. What is the approximate probability that there are exactly 10 defectives in the batch?

134. The article "Characterization of Room Temperature Damping in Aluminum-Indium Alloys" (*Metallurgical Trans.*, 1993: 1611–1619) suggests that Al matrix grain size (μm) for an alloy consisting of 2% indium could be modeled with a normal distribution with a mean value 96 and standard deviation 14.

a. What is the probability that grain size exceeds 100?

b. What is the probability that grain size is between 50 and 80?

c. What interval (a, b) includes the central 90% of all grain sizes (so that 5% are below a and 5% are above b)?

135. The reaction time (in seconds) to a certain stimulus is a continuous random variable with pdf

$$f(x) = \begin{cases} \dfrac{3}{2} \cdot \dfrac{1}{x^2} & 1 \le x \le 3 \\ 0 & \text{otherwise} \end{cases}$$

a. Obtain the cdf.

b. What is the probability that reaction time is at most 2.5 sec? Between 1.5 and 2.5 sec?

c. Compute the expected reaction time.

d. Compute the standard deviation of reaction time.

e. If an individual takes more than 1.5 sec to react, a light comes on and stays on either until one further second has elapsed or until the person reacts (whichever happens first). Determine the expected amount of time that the light remains lit. [*Hint*: Let $h(X)$ = the time that the light is on as a function of reaction time X.]

136. Let X denote the temperature at which a certain chemical reaction takes place. Suppose that X has pdf

$$f(x) = \begin{cases} \dfrac{1}{9}(4 - x^2) & -1 \le x \le 2 \\ 0 & \text{otherwise} \end{cases}$$

a. Sketch the graph of $f(x)$.

b. Determine the cdf and sketch it.

c. Is 0 the median temperature at which the reaction takes place? If not, is the median temperature smaller or larger than 0?

d. Suppose this reaction is independently carried out once in each of ten different labs and that the pdf of reaction time in each lab is as given. Let Y = the number among the ten labs at which the temperature exceeds 1. What kind of distribution does Y have? (Give the name and values of any parameters.)

137. The article "Determination of the MTF of Positive Photoresists Using the Monte Carlo Method" (*Photographic Sci. Engrg.*, 1983: 254–260) proposes the exponential distribution with parameter $\lambda = .93$ as a model for the distribution of a photon's free path length (μm) under certain circumstances. Suppose this is the correct model.

a. What is the expected path length, and what is the standard deviation of path length?

b. What is the probability that path length exceeds 3.0? What is the probability that path length is between 1.0 and 3.0?

c. What value is exceeded by only 10% of all path lengths?

138. The article "The Prediction of Corrosion by Statistical Analysis of Corrosion Profiles" (*Corrosion Sci.*, 1985: 305–315) suggests the following cdf for the depth X of the deepest pit in an experiment involving the exposure of carbon manganese steel to acidified seawater.

$$F(x; \alpha, \beta) = e^{-e^{-(x-\alpha)/\beta}} \qquad -\infty < x < \infty$$

The authors propose the values $\alpha = 150$ and $\beta = 90$. Assume this to be the correct model.

a. What is the probability that the depth of the deepest pit is at most 150? At most 300? Between 150 and 300?

b. Below what value will the depth of the maximum pit be observed in 90% of all such experiments?

c. What is the density function of X?

d. The density function can be shown to be unimodal (a single peak). Above what value on the measurement axis does this peak occur? (This value is the mode.)

e. It can be shown that $E(X) \approx .5772\beta + \alpha$. What is the mean for the given values of α and β, and how does it compare to the median and mode? Sketch the graph of the density function.

(*Note*: This is called the *largest extreme value distribution*.)

139. A component has lifetime X that is exponentially distributed with parameter λ.
   a. If the cost of operation per unit time is c, what is the expected cost of operating this component over its lifetime?
   b. Instead of a constant cost rate c as in part (a), suppose the cost rate is $c(1 - .5e^{ax})$ with $a < 0$, so that the cost per unit time is less than c when the component is new and gets more expensive as the component ages. Now compute the expected cost of operation over the lifetime of the component.

140. The *mode* of a continuous distribution is the value x^* that maximizes $f(x)$.
   a. What is the mode of a normal distribution with parameters μ and σ?
   b. Does the uniform distribution with parameters A and B have a single mode? Why or why not?
   c. What is the mode of an exponential distribution with parameter λ? (Draw a picture.)
   d. If X has a gamma distribution with parameters α and β, and $\alpha > 1$, find the mode. [*Hint*: $\ln[f(x)]$ will be maximized iff $f(x)$ is, and it may be simpler to take the derivative of $\ln[f(x)]$.]
   e. What is the mode of a chi-squared distribution having ν degrees of freedom?

141. The article "Error Distribution in Navigation" (*J. Institut. Navigation*, 1971: 429–442) suggests that the frequency distribution of positive errors (magnitudes of errors) is well approximated by an exponential distribution. Let X = the lateral position error (nautical miles), which can be either negative or positive. Suppose the pdf of X is

$$f(x) = (.1)e^{-.2|x|} \qquad -\infty < x < \infty$$

   a. Sketch a graph of $f(x)$ and verify that $f(x)$ is a legitimate pdf (show that it integrates to 1).
   b. Obtain the cdf of X and sketch it.
   c. Compute $P(X \le 0)$, $P(X \le 2)$, $P(-1 \le X \le 2)$, and the probability that an error of more than 2 miles is made.

142. In some systems, a customer is allocated to one of two service facilities. If the service time for a customer served by facility i has an exponential distribution with parameter λ_i ($i = 1, 2$) and p is the proportion of all customers served by facility 1, then the pdf of X = the service time of a randomly selected customer is

$$f(x; \lambda_1, \lambda_2, p) = \begin{cases} p\lambda_1 e^{-\lambda_1 x} + (1-p)\lambda_2 e^{-\lambda_2 x} & x \ge 0 \\ 0 & \text{otherwise} \end{cases}$$

This is often called the hyperexponential or mixed exponential distribution. This distribution is also proposed as a model for rainfall amount in "Modeling Monsoon Affected Rainfall of Pakistan by Point Processes" (*J. Water Resources Planning Manag.*, 1992: 671–688).
   a. Verify that $f(x; \lambda_1, \lambda_2, p)$ is indeed a pdf.
   b. What is the cdf $F(x; \lambda_1, \lambda_2, p)$?
   c. If X has $f(x; \lambda_1, \lambda_2, p)$ as its pdf, what is $E(X)$?
   d. Using the fact that $E(X^2) = 2/\lambda^2$ when X has an exponential distribution with parameter λ, compute $E(X^2)$ when X has pdf $f(x; \lambda_1, \lambda_2, p)$. Then compute $V(X)$.
   e. The coefficient of variation of a random variable (or distribution) is $CV = \sigma/\mu$. What is CV for an exponential rv? What can you say about the value of CV when X has a hyperexponential distribution?
   f. What is CV for an Erlang distribution with parameters λ and n as defined in Exercise 76? (*Note*: In applied work, the sample CV is used to decide which of the three distributions might be appropriate.)

143. Suppose a particular state allows individuals filing tax returns to itemize deductions only if the total of all itemized deductions is at least \$5000. Let X (in 1000's of dollars) be the total of itemized deductions on a randomly chosen form. Assume that X has the pdf

$$f(x; \alpha) = \begin{cases} k/x^\alpha & x \ge 5 \\ 0 & \text{otherwise} \end{cases}$$

   a. Find the value of k. What restriction on α is necessary?
   b. What is the cdf of X?
   c. What is the expected total deduction on a randomly chosen form? What restriction on α is necessary for $E(X)$ to be finite?
   d. Show that $\ln(X/5)$ has an exponential distribution with parameter $\alpha - 1$.

144. Let I_i be the input current to a transistor and I_o be the output current. Then the current gain is proportional to $\ln(I_o/I_i)$. Suppose the constant of proportionality is 1 (which amounts to choosing a particular unit of measurement), so that current gain $= X = \ln(I_o/I_i)$. Assume X is normally distributed with $\mu = 1$ and $\sigma = .05$.

a. What type of distribution does the ratio I_o/I_i have?

b. What is the probability that the output current is more than twice the input current?

c. What are the expected value and variance of the ratio of output to input current?

145. The article "Response of SiC_f/Si_3N_4 Composites Under Static and Cyclic Loading—An Experimental and Statistical Analysis" (*J. Engrg. Materials Tech.*, 1997: 186–193) suggests that tensile strength (MPa) of composites under specified conditions can be modeled by a Weibull distribution with $\alpha = 9$ and $\beta = 180$.

a. Sketch a graph of the density function.

b. What is the probability that the strength of a randomly selected specimen will exceed 175? Will be between 150 and 175?

c. If two randomly selected specimens are chosen and their strengths are independent of one another, what is the probability that at least one has a strength between 150 and 175?

d. What strength value separates the weakest 10% of all specimens from the remaining 90%?

146. a. Suppose the lifetime X of a component, when measured in hours, has a gamma distribution with parameters α and β. Let $Y =$ lifetime measured in minutes. Derive the pdf of Y.

b. If X has a gamma distribution with parameters α and β, what is the probability distribution of $Y = cX$?

147. Based on data from a dart-throwing experiment, the article "Shooting Darts" (*Chance*, Summer 1997: 16–19) proposed that the horizontal and vertical errors from aiming at a point target should be independent of one another, each with a normal distribution having mean 0 and variance σ^2. It can then be shown that the pdf of the distance V from the target to the landing point is

$$f(v) = \frac{v}{\sigma^2} \cdot e^{-v^2/2\sigma^2} \qquad v > 0$$

a. This pdf is a member of what family introduced in this chapter?

b. If $\sigma = 20$ mm (close to the value suggested in the paper), what is the probability that a dart will land within 25 mm (roughly 1 in.) of the target?

148. The article "Three Sisters Give Birth on the Same Day"(*Chance*, Spring 2001: 23–25) used the fact that three Utah sisters had all given birth on March 11, 1998, as a basis for posing some interesting questions regarding birth coincidences.

a. Disregarding leap year and assuming that the other 365 days are equally likely, what is the probability that three randomly selected births all occur on March 11? Be sure to indicate what, if any, extra assumptions you are making.

b. With the assumptions used in part (a), what is the probability that three randomly selected births all occur on the same day?

c. The author suggested that, based on extensive data, the length of gestation (time between conception and birth) could be modeled as having a normal distribution with mean value 280 days and standard deviation 19.88 days. The due dates for the three Utah sisters were March 15, April 1, and April 4, respectively. Assuming that all three due dates are at the mean of the distribution, what is the probability that all births occurred on March 11? (*Hint*: The deviation of birth date from due date is normally distributed with mean 0.)

d. Explain how you would use the information in part (c) to calculate the probability of a common birth date.

149. Let X denote the lifetime of a component, with $f(x)$ and $F(x)$ the pdf and cdf of X. The probability that the component fails in the interval $(x, x + \Delta x)$ is approximately $f(x) \cdot \Delta x$. The conditional probability that it fails in $(x, x + \Delta x)$ given that it has lasted at least x is $f(x) \cdot \Delta x/[1 - F(x)]$. Dividing this by Δx produces the failure rate function:

$$r(x) = \frac{f(x)}{1 - F(x)}$$

An increasing failure rate function indicates that older components are increasingly likely to wear out, whereas a decreasing failure rate is evidence of increasing reliability with age. In practice, a "bathtub-shaped" failure is often assumed.

a. If X is exponentially distributed, what is $r(x)$?

b. If X has a Weibull distribution with parameters α and β, what is $r(x)$? For what parameter values will $r(x)$ be increasing? For what parameter values will $r(x)$ decrease with x?

c. Since $r(x) = -(d/dx) \ln[1 - F(x)]$, $\ln[1 - F(x)] = -\int r(x)\, dx$. Suppose

$$r(x) = \begin{cases} \alpha\left(1 - \dfrac{x}{\beta}\right) & 0 \le x \le \beta \\ 0 & \text{otherwise} \end{cases}$$

so that if a component lasts β hours, it will last forever (while seemingly unreasonable, this model can be used to study just "initial wearout"). What are the cdf and pdf of X?

150. Let U have a uniform distribution on the interval [0, 1]. Then observed values having this distribution can be obtained from a computer's random number generator. Let $X = -(1/\lambda)\ln(1 - U)$.

a. Show that X has an exponential distribution with parameter λ.

b. How would you use part (a) and a random number generator to obtain observed values from an exponential distribution with parameter $\lambda = 10$?

151. Consider an rv X with mean μ and standard deviation σ, and let $g(X)$ be a specified function of X. The first-order Taylor series approximation to $g(X)$ in the neighborhood of μ is

$$g(X) \approx g(\mu) + g'(\mu)\cdot(X - \mu)$$

The right-hand side of this equation is a linear function of X. If the distribution of X is concentrated in an interval over which $g(X)$ is approximately linear [e.g., $\sqrt{x}$ is approximately linear in (1, 2)], then the equation yields approximations to $E[g(X)]$ and $V[g(X)]$.

a. Give expressions for these approximations. (*Hint*: Use rules of expected value and variance for a linear function $aX + b$.)

b. If the voltage v across a medium is fixed but current I is random, then resistance will also be a random variable related to I by $R = v/I$. If $\mu_I = 20$ and $\sigma_I = .5$, calculate approximations to μ_R and σ_R.

152. A function $g(x)$ is *convex* if the chord connecting any two points on the function's graph lies above the graph. When $g(x)$ is differentiable, an equivalent condition is that for every x, the tangent line at x lies entirely on or below the graph. (See the figures below.) How does $g(\mu) = g[E(X)]$ compare to $E[g(X)]$? [*Hint*: The equation of the tangent line at $x = \mu$ is $y = g(\mu) + g'(\mu)\cdot(x - \mu)$. Use the condition of convexity, substitute X for x, and take expected values. *Note*: Unless $g(x)$ is linear, the resulting inequality (usually called Jensen's inequality) is strict (< rather than $\le$); it is valid for both continuous and discrete rv's.]

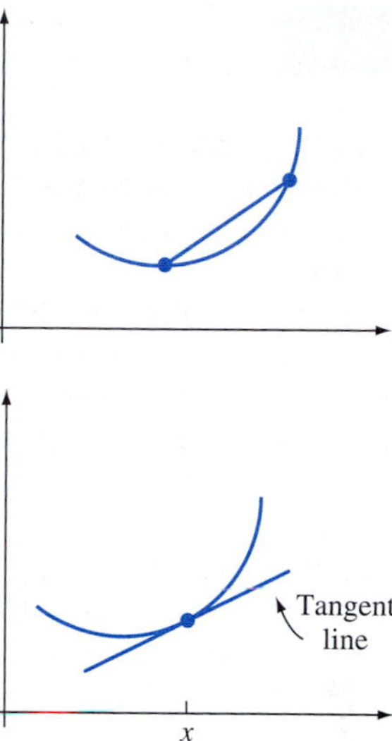

153. Let X have a Weibull distribution with parameters $\alpha = 2$ and β. Show that $Y = 2X^2/\beta^2$ has a chi-squared distribution with $\nu = 2$.

154. Let X have the pdf $f(x) = 1/[\pi(1 + x^2)]$ for $-\infty < x < \infty$ (a central Cauchy distribution), and show that $Y = 1/X$ has the same distribution. *Hint*: Consider $P(|Y| \le y)$, the cdf of $|Y|$, then obtain its pdf and show it is identical to the pdf of $|X|$.

155. A store will order q gallons of a certain liquid product to meet demand during a particular time period. This product can be dispensed to customers in any amount desired, so demand during the period is a continuous random variable X with cdf $F(x)$. There is a fixed cost c_0 for ordering the product plus a cost of c_1 per gallon purchased. The per-gallon sale price of the product is d. Liquid left unsold at the end of the time period has a salvage value of e per gallon. Finally, if demand exceeds q, there will be a shortage cost for loss of goodwill and future business; this cost is f per gallon of unfulfilled demand. Show that the value of q that maximizes expected profit, denoted by q^*, satisfies

$$P(\text{satisfying demand}) = F(q^*) = \frac{d - c_1 + f}{d - e + f}$$

Then determine the value of $F(q^*)$ if $d = \$35$, $c_0 = \$25$, $c_1 = \$15$, $e = \$5$, and $f = \$25$. *Hint*: Let x denote a particular value of X. Develop an expression for profit when $x \le q$ and another expression for profit when $x > q$. Now write an integral expression for expected profit (as a function of q) and differentiate.

Bibliography

Bury, Karl, *Statistical Distributions in Engineering*, Cambridge Univ. Press, Cambridge, England, 1999. A readable and informative survey of distributions and their properties.

Johnson, Norman, Samuel Kotz, and N. Balakrishnan, *Continuous Univariate Distributions*, vols. 1–2, Wiley, New York, 1994. These two volumes together present an exhaustive survey of various continuous distributions.

Nelson, Wayne, *Applied Life Data Analysis*, Wiley, New York, 1982. Gives a comprehensive discussion of distributions and methods that are used in the analysis of lifetime data.

Olkin, Ingram, Cyrus Derman, and Leon Gleser, *Probability Models and Applications* (2nd ed.), Macmillan, New York, 1994. Good coverage of general properties and specific distributions.

CHAPTER FIVE

Joint Probability Distributions

Introduction

In Chapters 3 and 4, we studied probability models for a single random variable. Many problems in probability and statistics lead to models involving several random variables simultaneously. In this chapter, we first discuss probability models for the joint behavior of several random variables, putting special emphasis on the case in which the variables are independent of one another. We then study expected values of functions of several random variables, including covariance and correlation as measures of the degree of association between two variables.

The third section considers conditional distributions, the distributions of random variables given the values of other random variables. The next section is about transformations of two or more random variables, generalizing the results of Section 4.7. In the last section of this chapter we discuss the distribution of order statistics: the minimum, maximum, median, and other statistics that can be found by arranging the observations in order.

5.1 Jointly Distributed Random Variables

There are many experimental situations in which more than one random variable (rv) will be of interest to an investigator. We shall first consider joint probability distributions for two discrete rv's, then for two continuous variables, and finally for more than two variables.

The Joint Probability Mass Function for Two Discrete Random Variables

The probability mass function (pmf) of a single discrete rv X specifies how much probability mass is placed on each possible X value. The joint pmf of two discrete rv's X and Y describes how much probability mass is placed on each possible pair of values (x, y).

DEFINITION Let X and Y be two discrete rv's defined on the sample space $\mathcal{S}$ of an experiment. The **joint probability mass function** $p(x, y)$ is defined for each pair of numbers (x, y) by

$$p(x, y) = P(X = x \text{ and } Y = y)$$

Let A be any set consisting of pairs of (x, y) values. Then the probability that the random pair (X, Y) lies in A is obtained by summing the joint pmf over pairs in A:

$$P[(X, Y) \in A] = \sum_{(x,y) \in A}\sum p(x, y)$$

Example 5.1 A large insurance agency services a number of customers who have purchased both a homeowner's policy and an automobile policy from the agency. For each type of policy, a deductible amount must be specified. For an automobile policy, the choices are \$100 and \$250, whereas for a homeowner's policy, the choices are 0, \$100, and \$200. Suppose an individual with both types of policy is selected at random from the agency's files. Let X = the deductible amount on the auto policy and Y = the deductible amount on the homeowner's policy. Possible (X, Y) pairs are then (100, 0), (100, 100), (100, 200), (250, 0), (250, 100), and (250, 200); the joint pmf specifies the probability associated with each one of these pairs, with any other pair having probability zero. Suppose the joint pmf is given in the accompanying **joint probability table**:

$p(x, y)$		y = 0	100	200
x	100	.20	.10	.20
	250	.05	.15	.30

Then $p(100, 100) = P(X = 100 \text{ and } Y = 100) = P(\$100 \text{ deductible on both policies}) = .10$. The probability $P(Y \geq 100)$ is computed by summing probabilities of all (x, y) pairs for which $y \geq 100$:

$$P(Y \geq 100) = p(100, 100) + p(250, 100) + p(100, 200) + p(250, 200)$$
$$= .75$$

■

A function $p(x, y)$ can be used as a joint pmf provided that $p(x, y) \geq 0$ for all x and y and $\sum_x \sum_y p(x, y) = 1$.

The pmf of one of the variables alone is obtained by summing $p(x, y)$ over values of the other variable. The result is called a *marginal pmf* because when the $p(x, y)$ values appear in a rectangular table, the sums are just marginal (row or column) totals.

DEFINITION

The **marginal probability mass functions** of X and of Y, denoted by $p_X(x)$ and $p_Y(y)$, respectively, are given by

$$p_X(x) = \sum_y p(x, y) \qquad p_Y(y) = \sum_x p(x, y)$$

Thus to obtain the marginal pmf of X evaluated at, say, $x = 100$, the probabilities $p(100, y)$ are added over all possible y values. Doing this for each possible X value gives the marginal pmf of X alone (without reference to Y). From the marginal pmf's, probabilities of events involving only X or only Y can be computed.

Example 5.2

(Example 5.1 continued)

The possible X values are $x = 100$ and $x = 250$, so computing row totals in the joint probability table yields

$$p_X(100) = p(100, 0) + p(100, 100) + p(100, 200) = .50$$

and

$$p_X(250) = p(250, 0) + p(250, 100) + p(250, 200) = .50$$

The marginal pmf of X is then

$$p_X(x) = \begin{cases} .5 & x = 100, 250 \\ 0 & \text{otherwise} \end{cases}$$

Similarly, the marginal pmf of Y is obtained from column totals as

$$p_Y(y) = \begin{cases} .25 & y = 0, 100 \\ .50 & y = 20 \\ 0 & \text{otherwise} \end{cases}$$

so $P(Y \geq 100) = p_Y(100) + p_Y(200) = .75$ as before. ■

The Joint Probability Density Function for Two Continuous Random Variables

The probability that the observed value of a continuous rv X lies in a one-dimensional set A (such as an interval) is obtained by integrating the pdf $f(x)$ over the set A. Similarly, the probability that the pair (X, Y) of continuous rv's falls in a two-dimensional set A (such as a rectangle) is obtained by integrating a function called the *joint density function*.

DEFINITION

Let X and Y be continuous rv's. Then $f(x, y)$ is the **joint probability density function** for X and Y if for any two-dimensional set A

$$P[(X, Y) \in A] = \int_A\int f(x, y)\,dx\,dy$$

In particular, if A is the two-dimensional rectangle $\{(x, y): a \le x \le b, c \le y \le d\}$, then

$$P[(X, Y) \in A] = P(a \le X \le b, c \le Y \le d) = \int_a^b \int_c^d f(x, y)\,dy\,dx$$

For $f(x, y)$ to be a candidate for a joint pdf, it must satisfy $f(x, y) \ge 0$ and $\int_{-\infty}^{\infty}\int_{-\infty}^{\infty} f(x, y)\,dx\,dy = 1$. We can think of $f(x, y)$ as specifying a surface at height $f(x, y)$ above the point (x, y) in a three-dimensional coordinate system. Then $P[(X, Y) \in A]$ is the volume underneath this surface and above the region A, analogous to the area under a curve in the one-dimensional case. This is illustrated in Figure 5.1.

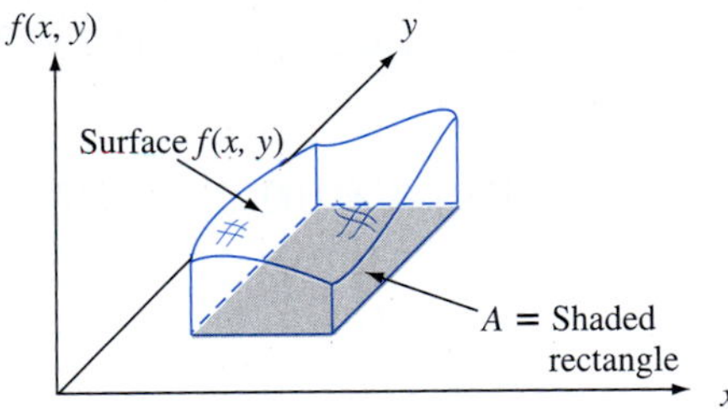

Figure 5.1 $P[(X, Y) \in A]$ = volume under density surface above A

Example 5.3 A bank operates both a drive-up facility and a walk-up window. On a randomly selected day, let X = the proportion of time that the drive-up facility is in use (at least one customer is being served or waiting to be served) and Y = the proportion of time that the walk-up window is in use. Then the set of possible values for (X, Y) is the rectangle $D = \{(x, y): 0 \le x \le 1, 0 \le y \le 1\}$. Suppose the joint pdf of (X, Y) is given by

$$f(x, y) = \begin{cases} \frac{6}{5}(x + y^2) & 0 \le x \le 1, 0 \le y \le 1 \\ 0 & \text{otherwise} \end{cases}$$

To verify that this is a legitimate pdf, note that $f(x, y) \geq 0$ and

$$\int_{-\infty}^{\infty}\int_{-\infty}^{\infty} f(x, y)\,dx\,dy = \int_0^1\int_0^1 \frac{6}{5}(x + y^2)\,dx\,dy$$

$$= \int_0^1\int_0^1 \frac{6}{5}x\,dx\,dy + \int_0^1\int_0^1 \frac{6}{5}y^2\,dx\,dy$$

$$= \int_0^1 \frac{6}{5}x\,dx + \int_0^1 \frac{6}{5}y^2\,dy = \frac{6}{10} + \frac{6}{15} = 1$$

The probability that neither facility is busy more than one-quarter of the time is

$$P\left(0 \leq X \leq \frac{1}{4}, 0 \leq Y \leq \frac{1}{4}\right) = \int_0^{1/4}\int_0^{1/4} \frac{6}{5}(x + y^2)\,dx\,dy$$

$$= \frac{6}{5}\int_0^{1/4}\int_0^{1/4} x\,dx\,dy + \frac{6}{5}\int_0^{1/4}\int_0^{1/4} y^2\,dx\,dy$$

$$= \frac{6}{20}\cdot\frac{x^2}{2}\bigg|_{x=0}^{x=1/4} + \frac{6}{20}\cdot\frac{y^3}{3}\bigg|_{y=0}^{y=1/4} = \frac{7}{640}$$

$$= .0109$$

■

As with joint pmf's, from the joint pdf of X and Y, each of the two marginal density functions can be computed.

DEFINITION

The **marginal probability density functions** of X and Y, denoted by $f_X(x)$ and $f_Y(y)$, respectively, are given by

$$f_X(x) = \int_{-\infty}^{\infty} f(x, y)\,dy \quad \text{for } -\infty < x < \infty$$

$$f_Y(y) = \int_{-\infty}^{\infty} f(x, y)\,dx \quad \text{for } -\infty < y < \infty$$

Example 5.4
(Example 5.3 continued)

The marginal pdf of X, which gives the probability distribution of busy time for the drive-up facility without reference to the walk-up window, is

$$f_X(x) = \int_{-\infty}^{\infty} f(x, y)\,dy = \int_0^1 \frac{6}{5}(x + y^2)\,dy = \frac{6}{5}x + \frac{2}{5}$$

for $0 \leq x \leq 1$ and 0 otherwise. The marginal pdf of Y is

$$f_Y(y) = \begin{cases} \frac{6}{5}y^2 + \frac{3}{5} & 0 \leq y \leq 1 \\ 0 & \text{otherwise} \end{cases}$$

Then

$$P\left(\frac{1}{4} \le Y \le \frac{3}{4}\right) = \int_{1/4}^{3/4} f_Y(y)\,dy = \frac{37}{80} = .4625$$

■

In Example 5.3, the region of positive joint density was a rectangle, which made computation of the marginal pdf's relatively easy. Consider now an example in which the region of positive density is a more complicated figure.

Example 5.5 A nut company markets cans of deluxe mixed nuts containing almonds, cashews, and peanuts. Suppose the net weight of each can is exactly 1 lb, but the weight contribution of each type of nut is random. Because the three weights sum to 1, a joint probability model for any two gives all necessary information about the weight of the third type. Let X = the weight of almonds in a selected can and Y = the weight of cashews. Then the region of positive density is $D = \{(x, y): 0 \le x \le 1, 0 \le y \le 1, x + y \le 1\}$, the shaded region pictured in Figure 5.2.

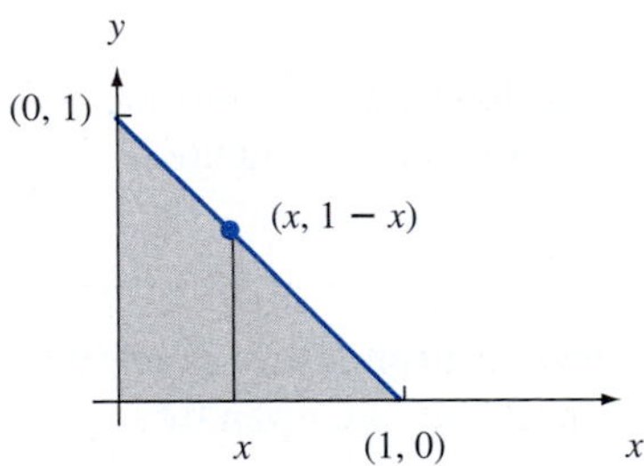

Figure 5.2 Region of positive density for Example 5.5

Now let the joint pdf for (X, Y) be

$$f(x, y) = \begin{cases} 24xy & 0 \le x \le 1, 0 \le y \le 1, x + y \le 1 \\ 0 & \text{otherwise} \end{cases}$$

For any fixed x, $f(x, y)$ increases with y; for fixed y, $f(x, y)$ increases with x. This is appropriate because the word *deluxe* implies that most of the can should consist of almonds and cashews rather than peanuts, so that the density function should be large near the upper boundary and small near the origin. The surface determined by $f(x, y)$ slopes upward from zero as (x, y) moves away from either axis.

Clearly, $f(x, y) \ge 0$. To verify the second condition on a joint pdf, recall that a double integral is computed as an iterated integral by holding one variable fixed (such as x as in Figure 5.2), integrating over values of the other variable lying along the straight line passing through the value of the fixed variable, and finally integrating over all possible values of the fixed variable. Thus

$$\int_{-\infty}^{\infty}\int_{-\infty}^{\infty} f(x, y)\, dy\, dx = \iint_D f(x, y)\, dy\, dx = \int_0^1 \left\{ \int_0^{1-x} 24xy\, dy \right\} dx$$

$$= \int_0^1 24x \left\{ \frac{y^2}{2} \Bigg|_{y=0}^{y=1-x} \right\} dx = \int_0^1 12x(1-x)^2\, dx = 1$$

To compute the probability that the two types of nuts together make up at most 50% of the can, let $A = \{(x, y): 0 \le x \le 1, 0 \le y \le 1, \text{ and } x + y \le .5\}$, as shown in Figure 5.3. Then

$$P[(X, Y) \in A] = \iint_A f(x, y)\, dx\, dy = \int_0^{.5}\int_0^{.5-x} 24xy\, dy\, dx = .0625$$

The marginal pdf for almonds is obtained by holding X fixed at x and integrating $f(x, y)$ along the vertical line through x:

$$f_X(x) = \int_{-\infty}^{\infty} f(x, y)\, dy = \begin{cases} \int_0^{1-x} 24xy\, dy = 12x(1-x)^2 & 0 \le x \le 1 \\ 0 & \text{otherwise} \end{cases}$$

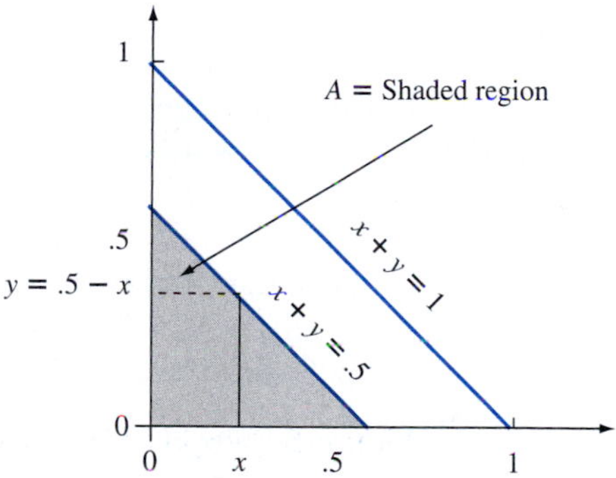

Figure 5.3 Computing $P[(X, Y) \in A]$ for Example 5.5

By symmetry of $f(x, y)$ and the region D, the marginal pdf of Y is obtained by replacing x and X in $f_X(x)$ by y and Y, respectively. ■

Independent Random Variables

In many situations, information about the observed value of one of the two variables X and Y gives information about the value of the other variable. In Example 5.1, the marginal probability of X at $x = 250$ was .5, as was the probability that $X = 100$. If, however, we are told that the selected individual had $Y = 0$, then $X = 100$ is four times as likely as $X = 250$. Thus there is a dependence between the two variables.

In Chapter 2 we pointed out that one way of defining independence of two events is to say that A and B are independent if $P(A \cap B) = P(A) \cdot P(B)$. Here is an analogous definition for the independence of two rv's.

DEFINITION Two random variables X and Y are said to be **independent** if for every pair of x and y values,

$$p(x, y) = p_X(x) \cdot p_Y(y) \quad \text{when } X \text{ and } Y \text{ are discrete} \tag{5.1}$$

or

$$f(x, y) = f_X(x) \cdot f_Y(y) \quad \text{when } X \text{ and } Y \text{ are continuous}$$

If (5.1) is not satisfied for all (x, y), then X and Y are said to be **dependent**.

The definition says that two variables are independent if their joint pmf or pdf is the product of the two marginal pmf's or pdf's.

Example 5.6 In the insurance situation of Examples 5.1 and 5.2,

$$p(100, 100) = .10 \neq (.5)(.25) = p_X(100) \cdot p_Y(100)$$

so X and Y are not independent. Independence of X and Y requires that *every* entry in the joint probability table be the product of the corresponding row and column marginal probabilities. ■

Example 5.7 (Example 5.5 continued) Because $f(x, y)$ in the nut scenario has the form of a product, X and Y would appear to be independent. However, although $f_X(\frac{3}{4}) = f_Y(\frac{3}{4}) = \frac{9}{16}, f(\frac{3}{4}, \frac{3}{4}) = 0 \neq \frac{9}{16} \cdot \frac{9}{16}$, so the variables are not in fact independent. To be independent, $f(x, y)$ must have the form $g(x) \cdot h(y)$ *and* the region of positive density must be a rectangle whose sides are parallel to the coordinate axes. ■

Independence of two random variables is most useful when the description of the experiment under study tells us that X and Y have no effect on one another. Then once the marginal pmf's or pdf's have been specified, the joint pmf or pdf is simply the product of the two marginal functions. It follows that

$$P(a \leq X \leq b, c \leq Y \leq d) = P(a \leq X \leq b) \cdot P(c \leq Y \leq d)$$

Example 5.8 Suppose that the lifetimes of two components are independent of one another and that the first lifetime, X_1, has an exponential distribution with parameter λ_1 whereas the second, X_2, has an exponential distribution with parameter λ_2. Then the joint pdf is

$$\begin{aligned} f(x_1, x_2) &= f_{X_1}(x_1) \cdot f_{X_2}(x_2) \\ &= \begin{cases} \lambda_1 e^{-\lambda_1 x_1} \cdot \lambda_2 e^{-\lambda_2 x_2} = \lambda_1 \lambda_2 e^{-\lambda_1 x_1 - \lambda_2 x_2} & x_1 > 0, x_2 > 0 \\ 0 & \text{otherwise} \end{cases} \end{aligned}$$

Let $\lambda_1 = 1/1000$ and $\lambda_2 = 1/1200$, so that the expected lifetimes are 1000 hours and 1200 hours, respectively. The probability that both component lifetimes are at least 1500 hours is

$$\begin{aligned} P(1500 \leq X_1, 1500 \leq X_2) &= P(1500 \leq X_1) \cdot P(1500 \leq X_2) \\ &= e^{-\lambda_1(1500)} \cdot e^{-\lambda_2(1500)} \\ &= (.2231)(.2865) = .0639 \end{aligned}$$

■

More Than Two Random Variables

To model the joint behavior of more than two random variables, we extend the concept of a joint distribution of two variables.

DEFINITION If $X_1, X_2, \ldots, X_n$ are all discrete random variables, the **joint pmf** of the variables is the function

$$p(x_1, x_2, \ldots, x_n) = P(X_1 = x_1, X_2 = x_2, \ldots, X_n = x_n)$$

If the variables are continuous, the **joint pdf** of $X_1, X_2, \ldots, X_n$ is the function $f(x_1, x_2, \ldots, x_n)$ such that for any n intervals $[a_1, b_1], \ldots, [a_n, b_n]$,

$$P(a_1 \le X_1 \le b_1, \ldots, a_n \le X_n \le b_n) = \int_{a_1}^{b_1} \cdots \int_{a_n}^{b_n} f(x_1, \ldots, x_n)\, dx_n \ldots dx_1$$

In a binomial experiment, each trial could result in one of only two possible outcomes. Consider now an experiment consisting of n independent and identical trials, in which each trial can result in any one of r possible outcomes. Let $p_i = P(\text{outcome } i$ on any particular trial), and define random variables by $X_i =$ the number of trials resulting in outcome i ($i = 1, \ldots, r$). Such an experiment is called a **multinomial experiment**, and the joint pmf of $X_1, \ldots, X_r$ is called the **multinomial distribution**. By using a counting argument analogous to the one used in deriving the binomial distribution, the joint pmf of $X_1, \ldots, X_r$ can be shown to be

$$p(x_1, \ldots, x_r) = \begin{cases} \dfrac{n!}{(x_1!)(x_2!)\cdot \cdots \cdot (x_r!)} p_1^{x_1} \cdot \cdots \cdot p_r^{x_r}, & x_i = 0, 1, 2, \ldots, \text{ with } x_1 + \cdots + x_r = n \\ 0 & \text{otherwise} \end{cases}$$

The case $r = 2$ gives the binomial distribution, with $X_1 =$ number of successes and $X_2 = n - X_1 =$ number of failures.

Example 5.9 If the allele of each of ten independently obtained pea sections is determined and $p_1 = P(\text{AA})$, $p_2 = P(\text{Aa})$, $p_3 = P(\text{aa})$, $X_1 =$ number of AA's, $X_2 =$ number of Aa's, and $X_3 =$ number of aa's, then

$$p(x_1, x_2, x_3) = \frac{10!}{(x_1!)(x_2!)(x_3!)} p_1^{x_1} p_2^{x_2} p_3^{x_3} \qquad x_i = 0, 1, \ldots \text{ and } x_1 + x_2 + x_3 = 10$$

If $p_1 = p_3 = .25$, $p_2 = .5$, then

$$\begin{aligned} P(X_1 = 2, X_2 = 5, X_3 = 3) &= p(2, 5, 3) \\ &= \frac{10!}{2!\,5!\,3!}(.25)^2(.5)^5(.25)^3 = .0769 \end{aligned}$$

■

Example 5.10 When a certain method is used to collect a fixed volume of rock samples in a region, there are four resulting rock types. Let X_1, X_2, and X_3 denote the proportion by volume of rock types 1, 2, and 3 in a randomly selected sample (the proportion of rock type 4 is $1 - X_1 - X_2 - X_3$, so a variable X_4 would be redundant). If the joint pdf of X_1, X_2, X_3 is

$$f(x_1, x_2, x_3) = \begin{cases} kx_1x_2(1 - x_3) & 0 \le x_1 \le 1, 0 \le x_2 \le 1, 0 \le x_3 \le 1, \\ & x_1 + x_2 + x_3 \le 1 \\ 0 & \text{otherwise} \end{cases}$$

then k is determined by

$$\begin{aligned} 1 &= \int_{-\infty}^{\infty}\int_{-\infty}^{\infty}\int_{-\infty}^{\infty} f(x_1, x_2, x_3)\, dx_3\, dx_2\, dx_1 \\ &= \int_0^1 \left\{ \int_0^{1-x_1} \left[\int_0^{1-x_1-x_2} kx_1x_2(1 - x_3)\, dx_3 \right] dx_2 \right\} dx_1 \end{aligned}$$

This iterated integral has value $k/144$, so $k = 144$. The probability that rocks of types 1 and 2 together account for at most 50% of the sample is

$$\begin{aligned} P(X_1 + X_2 \le .5) &= \underset{\substack{\{0 \le x_i \le 1 \text{ for } i=1,2,3 \\ x_1+x_2+x_3 \le 1,\, x_1+x_2 \le .5\}}}{\int\int\int} f(x_1, x_2, x_3)\, dx_3\, dx_2\, dx_1 \\ &= \int_0^{.5} \left\{ \int_0^{.5-x_1} \left[\int_0^{1-x_1-x_2} 144x_1x_2(1 - x_3)\, dx_3 \right] dx_2 \right\} dx_1 \\ &= .6066 \end{aligned}$$

■

The notion of independence of more than two random variables is similar to the notion of independence of more than two events.

DEFINITION The random variables $X_1, X_2, \ldots, X_n$ are said to be **independent** if for *every* subset $X_{i_1}, X_{i_2}, \ldots, X_{i_k}$ of the variables (each pair, each triple, and so on), the joint pmf or pdf of the subset is equal to the product of the marginal pmf's or pdf's.

Thus if the variables are independent with $n = 4$, then the joint pmf or pdf of any two variables is the product of the two marginals, and similarly for any three variables and all four variables together. Most important, once we are told that n variables are independent, then the joint pmf or pdf is the product of the n marginals.

Example 5.11 If $X_1, \ldots, X_n$ represent the lifetimes of n components, the components operate independently of one another, and each lifetime is exponentially distributed with parameter λ, then

$$\begin{aligned} f(x_1, x_2, \ldots, x_n) &= (\lambda e^{-\lambda x_1}) \cdot (\lambda e^{-\lambda x_2}) \cdot \cdots \cdot (\lambda e^{-\lambda x_n}) \\ &= \begin{cases} \lambda^n e^{-\lambda \Sigma x_i} & x_1 \ge 0, x_2 \ge 0, \ldots, x_n \ge 0 \\ 0 & \text{otherwise} \end{cases} \end{aligned}$$

If these n components are connected in series, so that the system will fail as soon as a single component fails, then the probability that the system lasts past time t is

$$P(X_1 > t, \ldots, X_n > t) = \int_t^{\infty} \cdots \int_t^{\infty} f(x_1, \ldots, x_n)\, dx_1 \ldots dx_n$$
$$= \left(\int_t^{\infty} \lambda e^{-\lambda x_1}\, dx_1\right) \cdots \left(\int_t^{\infty} \lambda e^{-\lambda x_n}\, dx_n\right)$$
$$= (e^{-\lambda t})^n = e^{-n\lambda t}$$

Therefore,

$$P(\text{system lifetime} \le t) = 1 - e^{-n\lambda t} \quad \text{for } t \ge 0$$

which shows that *system* lifetime has an exponential distribution with parameter $n\lambda$; the expected value of system lifetime is $1/n\lambda$. ■

In many experimental situations to be considered in this book, independence is a reasonable assumption, so that specifying the joint distribution reduces to deciding on appropriate marginal distributions.

Exercises Section 5.1 (1–17)

1. A service station has both self-service and full-service islands. On each island, there is a single regular unleaded pump with two hoses. Let X denote the number of hoses being used on the self-service island at a particular time, and let Y denote the number of hoses on the full-service island in use at that time. The joint pmf of X and Y appears in the accompanying tabulation.

$p(x, y)$		y = 0	1	2
x	0	.10	.04	.02
	1	.08	.20	.06
	2	.06	.14	.30

a. What is $P(X = 1 \text{ and } Y = 1)$?
b. Compute $P(X \le 1 \text{ and } Y \le 1)$.
c. Give a word description of the event $\{X \ne 0 \text{ and } Y \ne 0\}$, and compute the probability of this event.
d. Compute the marginal pmf of X and of Y. Using $p_X(x)$, what is $P(X \le 1)$?
e. Are X and Y independent rv's? Explain.

2. When an automobile is stopped by a roving safety patrol, each tire is checked for tire wear, and each headlight is checked to see whether it is properly aimed. Let X denote the number of headlights that need adjustment, and let Y denote the number of defective tires.

a. If X and Y are independent with $p_X(0) = .5$, $p_X(1) = .3$, $p_X(2) = .2$, and $p_Y(0) = .6$, $p_Y(1) = .1$, $p_Y(2) = p_Y(3) = .05$, $p_Y(4) = .2$, display the joint pmf of (X, Y) in a joint probability table.
b. Compute $P(X \le 1 \text{ and } Y \le 1)$ from the joint probability table, and verify that it equals the product $P(X \le 1) \cdot P(Y \le 1)$.
c. What is $P(X + Y = 0)$ (the probability of no violations)?
d. Compute $P(X + Y \le 1)$.

3. A certain market has both an express checkout line and a superexpress checkout line. Let X_1 denote the number of customers in line at the express checkout at a particular time of day, and let X_2 denote the number of customers in line at the superexpress checkout at the same time. Suppose the joint pmf of X_1 and X_2 is as given in the accompanying table.

		x_2 = 0	1	2	3
x_1	0	.08	.07	.04	.00
	1	.06	.15	.05	.04
	2	.05	.04	.10	.06
	3	.00	.03	.04	.07
	4	.00	.01	.05	.06

a. What is $P(X_1 = 1, X_2 = 1)$, that is, the probability that there is exactly one customer in each line?

b. What is $P(X_1 = X_2)$, that is, the probability that the numbers of customers in the two lines are identical?

c. Let A denote the event that there are at least two more customers in one line than in the other line. Express A in terms of X_1 and X_2, and calculate the probability of this event.

d. What is the probability that the total number of customers in the two lines is exactly four? At least four?

4. Return to the situation described in Exercise 3.

a. Determine the marginal pmf of X_1, and then calculate the expected number of customers in line at the express checkout.

b. Determine the marginal pmf of X_2.

c. By inspection of the probabilities $P(X_1 = 4)$, $P(X_2 = 0)$, and $P(X_1 = 4, X_2 = 0)$, are X_1 and X_2 independent random variables? Explain.

5. The number of customers waiting for gift-wrap service at a department store is an rv X with possible values 0, 1, 2, 3, 4 and corresponding probabilities .1, .2, .3, .25, .15. A randomly selected customer will have 1, 2, or 3 packages for wrapping with probabilities .6, .3, and .1, respectively. Let Y = the total number of packages to be wrapped for the customers waiting in line (assume that the number of packages submitted by one customer is independent of the number submitted by any other customer).

a. Determine $P(X = 3, Y = 3)$, that is, $p(3, 3)$.

b. Determine $p(4, 11)$.

6. Let X denote the number of Canon digital cameras sold during a particular week by a certain store. The pmf of X is

x	0	1	2	3	4
$p_X(x)$	.1	.2	.3	.25	.15

Sixty percent of all customers who purchase these cameras also buy an extended warranty. Let Y denote the number of purchasers during this week who buy an extended warranty.

a. What is $P(X = 4, Y = 2)$? [*Hint*: This probability equals $P(Y = 2 \mid X = 4) \cdot P(X = 4)$; now think of the four purchases as four trials of a binomial experiment, with success on a trial corresponding to buying an extended warranty.]

b. Calculate $P(X = Y)$.

c. Determine the joint pmf of X and Y and then the marginal pmf of Y.

7. The joint probability distribution of the number X of cars and the number Y of buses per signal cycle at a proposed left-turn lane is displayed in the accompanying joint probability table.

$p(x, y)$		y = 0	1	2
x	0	.025	.015	.010
	1	.050	.030	.020
	2	.125	.075	.050
	3	.150	.090	.060
	4	.100	.060	.040
	5	.050	.030	.020

a. What is the probability that there is exactly one car and exactly one bus during a cycle?

b. What is the probability that there is at most one car and at most one bus during a cycle?

c. What is the probability that there is exactly one car during a cycle? Exactly one bus?

d. Suppose the left-turn lane is to have a capacity of five cars, and one bus is equivalent to three cars. What is the probability of an overflow during a cycle?

e. Are X and Y independent rv's? Explain.

8. A stockroom currently has 30 components of a certain type, of which 8 were provided by supplier 1, 10 by supplier 2, and 12 by supplier 3. Six of these are to be randomly selected for a particular assembly. Let X = the number of supplier 1's components selected, Y = the number of supplier 2's components selected, and $p(x, y)$ denote the joint pmf of X and Y.

a. What is $p(3, 2)$? [*Hint*: Each sample of size 6 is equally likely to be selected. Therefore, $p(3, 2)$ = (number of outcomes with $X = 3$ and $Y = 2$)/(total number of outcomes). Now use the product rule for counting to obtain the numerator and denominator.]

b. Using the logic of part (a), obtain $p(x, y)$. (This can be thought of as a multivariate hypergeometric distribution—sampling without replacement from a finite population consisting of more than two categories.)

9. Each front tire on a particular type of vehicle is supposed to be filled to a pressure of 26 psi. Suppose the actual air pressure in each tire is a random variable—X for the right tire and Y for the left tire, with joint pdf

$$f(x, y) = \begin{cases} K(x^2 + y^2) & 20 \le x \le 30,\ 20 \le y \le 30 \\ 0 & \text{otherwise} \end{cases}$$

a. What is the value of K?

b. What is the probability that both tires are underfilled?

c. What is the probability that the difference in air pressure between the two tires is at most 2 psi?

d. Determine the (marginal) distribution of air pressure in the right tire alone.

e. Are X and Y independent rv's?

10. Annie and Alvie have agreed to meet between 5:00 p.m. and 6:00 p.m. for dinner at a local health-food restaurant. Let X = Annie's arrival time and Y = Alvie's arrival time. Suppose X and Y are independent with each uniformly distributed on the interval [5, 6].

a. What is the joint pdf of X and Y?

b. What is the probability that they both arrive between 5:15 and 5:45?

c. If the first one to arrive will wait only 10 min before leaving to eat elsewhere, what is the probability that they have dinner at the health-food restaurant? [*Hint*: The event of interest is $A = \{(x, y): |x - y| \le \frac{1}{6}\}$.]

11. Two different professors have just submitted final exams for duplication. Let X denote the number of typographical errors on the first professor's exam and Y denote the number of such errors on the second exam. Suppose X has a Poisson distribution with parameter λ, Y has a Poisson distribution with parameter θ, and X and Y are independent.

a. What is the joint pmf of X and Y?

b. What is the probability that at most one error is made on both exams combined?

c. Obtain a general expression for the probability that the total number of errors in the two exams is m (where m is a nonnegative integer). [*Hint*: $A = \{(x, y): x + y = m\} = \{(m, 0), (m - 1, 1), \ldots, (1, m - 1), (0, m)\}$. Now sum the joint pmf over $(x, y) \in A$ and use the binomial theorem, which says that

$$\sum_{k=0}^{m} \binom{m}{k} a^k b^{m-k} = (a + b)^m$$

for any a, b.]

12. Two components of a minicomputer have the following joint pdf for their useful lifetimes X and Y:

$$f(x, y) = \begin{cases} xe^{-x(1+y)} & x \ge 0 \text{ and } y \ge 0 \\ 0 & \text{otherwise} \end{cases}$$

a. What is the probability that the lifetime X of the first component exceeds 3?

b. What are the marginal pdf's of X and Y? Are the two lifetimes independent? Explain.

c. What is the probability that the lifetime of at least one component exceeds 3?

13. You have two lightbulbs for a particular lamp. Let X = the lifetime of the first bulb and Y = the lifetime of the second bulb (both in 1000's of hours). Suppose that X and Y are independent and that each has an exponential distribution with parameter $\lambda = 1$.

a. What is the joint pdf of X and Y?

b. What is the probability that each bulb lasts at most 1000 hours (i.e., $X \le 1$ and $Y \le 1$)?

c. What is the probability that the total lifetime of the two bulbs is at most 2? [*Hint*: Draw a picture of the region $A = \{(x, y): x \ge 0, y \ge 0, x + y \le 2\}$ before integrating.]

d. What is the probability that the total lifetime is between 1 and 2?

14. Suppose that you have ten lightbulbs, that the lifetime of each is independent of all the other lifetimes, and that each lifetime has an exponential distribution with parameter λ.

a. What is the probability that all ten bulbs fail before time t?

b. What is the probability that exactly k of the ten bulbs fail before time t?

c. Suppose that nine of the bulbs have lifetimes that are exponentially distributed with parameter λ and that the remaining bulb has a lifetime that is exponentially distributed with parameter θ (it is made by another manufacturer). What is the probability that exactly five of the ten bulbs fail before time t?

15. Consider a system consisting of three components as pictured. The system will continue to function as long as the first component functions and either component 2 or component 3 functions. Let X_1, X_2, and X_3 denote the lifetimes of components 1, 2, and 3, respectively. Suppose the X_i's are independent of one another and each X_i has an exponential distribution with parameter λ.

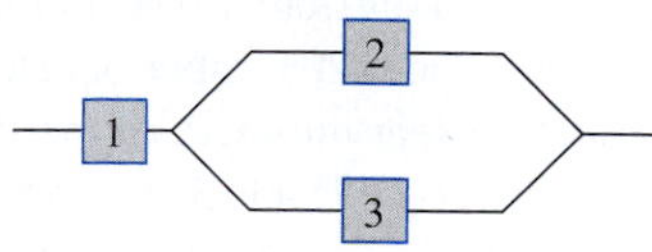

a. Let Y denote the system lifetime. Obtain the cumulative distribution function of Y and differentiate to obtain the pdf. [*Hint*: $F(y) = P(Y \le y)$;

express the event $\{Y \le y\}$ in terms of unions and/or intersections of the three events $\{X_1 \le y\}$, $\{X_2 \le y\}$, and $\{X_3 \le y\}$.]

b. Compute the expected system lifetime.

16. **a.** For $f(x_1, x_2, x_3)$ as given in Example 5.10, compute the **joint marginal density function** of X_1 and X_3 alone (by integrating over x_2).

b. What is the probability that rocks of types 1 and 3 together make up at most 50% of the sample? [*Hint*: Use the result of part (a).]

c. Compute the marginal pdf of X_1 alone. [*Hint*: Use the result of part (a).]

17. An ecologist wishes to select a point inside a circular sampling region according to a uniform distribution (in practice this could be done by first selecting a direction and then a distance from the center in that direction). Let X = the x coordinate of the point selected and Y = the y coordinate of the point selected. If the circle is centered at (0, 0) and has radius R, then the joint pdf of X and Y is

$$f(x, y) = \begin{cases} \dfrac{1}{\pi R^2} & x^2 + y^2 \le R^2 \\ 0 & \text{otherwise} \end{cases}$$

a. What is the probability that the selected point is within $R/2$ of the center of the circular region? [*Hint*: Draw a picture of the region of positive density D. Because $f(x, y)$ is constant on D, computing a probability reduces to computing an area.]

b. What is the probability that both X and Y differ from 0 by at most $R/2$?

c. Answer part (b) for $R/\sqrt{2}$ replacing $R/2$.

d. What is the marginal pdf of X? Of Y? Are X and Y independent?

5.2 Expected Values, Covariance, and Correlation

We previously saw that any function $h(X)$ of a single rv X is itself a random variable. However, to compute $E[h(X)]$, it was not necessary to obtain the probability distribution of $h(X)$; instead, $E[h(X)]$ was computed as a weighted average of $h(x)$ values, where the weight function was the pmf $p(x)$ or pdf $f(x)$ of X. A similar result holds for a function $h(X, Y)$ of two jointly distributed random variables.

PROPOSITION

Let X and Y be jointly distributed rv's with pmf $p(x, y)$ or pdf $f(x, y)$ according to whether the variables are discrete or continuous. Then the expected value of a function $h(X, Y)$, denoted by $E[h(X, Y)]$ or $\mu_{h(X,Y)}$, is given by

$$E[h(X, Y)] = \begin{cases} \displaystyle\sum_x \sum_y h(x, y) \cdot p(x, y) & \text{if } X \text{ and } Y \text{ are discrete} \\ \displaystyle\int_{-\infty}^{\infty} \int_{-\infty}^{\infty} h(x, y) \cdot f(x, y)\, dx\, dy & \text{if } X \text{ and } Y \text{ are continuous} \end{cases}$$

Example 5.12 Five friends have purchased tickets to a certain concert. If the tickets are for seats 1–5 in a particular row and the tickets are randomly distributed among the five, what is the expected number of seats separating any particular two of the five? Let X and Y denote the seat numbers of the first and second individuals, respectively. Possible (X, Y) pairs are $\{(1, 2), (1, 3), \ldots, (5, 4)\}$, and the joint pmf of (X, Y) is

$$p(x, y) = \begin{cases} \dfrac{1}{20} & x = 1, \ldots, 5;\ y = 1, \ldots, 5;\ x \ne y \\ 0 & \text{otherwise} \end{cases}$$

The number of seats separating the two individuals is $h(X, Y) = |X - Y| - 1$. The accompanying table gives $h(x, y)$ for each possible (x, y) pair.

$h(x, y)$		x: 1	2	3	4	5
	1	—	0	1	2	3
	2	0	—	0	1	2
y	3	1	0	—	0	1
	4	2	1	0	—	0
	5	3	2	1	0	—

Thus

$$E[h(X, Y)] = \sum_{(x,y)}\sum h(x, y) \cdot p(x, y) = \sum_{x=1}^{5} \sum_{\substack{y=1 \\ x \neq y}}^{5} (|x - y| - 1) \cdot \frac{1}{20} = 1$$

■

Example 5.13 In Example 5.5, the joint pdf of the amount X of almonds and amount Y of cashews in a 1-lb can of nuts was

$$f(x, y) = \begin{cases} 24xy & 0 \leq x \leq 1, 0 \leq y \leq 1, x + y \leq 1 \\ 0 & \text{otherwise} \end{cases}$$

If 1 lb of almonds costs the company \$2.00, 1 lb of cashews costs \$3.00, and 1 lb of peanuts costs \$1.00, then the total cost of the contents of a can is

$$h(X, Y) = 2X + 3Y + 1(1 - X - Y) = 1 + X + 2Y$$

(since $1 - X - Y$ of the weight consists of peanuts). The expected total cost is

$$E[h(X, Y)] = \int_{-\infty}^{\infty}\int_{-\infty}^{\infty} h(x, y) \cdot f(x, y)\, dx\, dy$$

$$= \int_{0}^{1}\int_{0}^{1-x} (1 + x + 2y) \cdot 24xy\, dy\, dx = \$2.20$$

■

The method of computing the expected value of a function $h(X_1, \ldots, X_n)$ of n random variables is similar to that for two random variables. If the X_i's are discrete, $E[h(X_1, \ldots, X_n)]$ is an n-dimensional sum; if the X_i's are continuous, it is an n-dimensional integral.

When $h(X, Y)$ is a product of a function of X and a function of Y, the expected value simplifies in the case of independence. In particular, let X and Y be continuous independent random variables and suppose $h(X, Y) = XY$. Then

$$E(XY) = \int_{-\infty}^{\infty}\int_{-\infty}^{\infty} xyf(x, y)\, dx\, dy = \int_{-\infty}^{\infty}\int_{-\infty}^{\infty} xyf_X(x)f_Y(y)\, dx\, dy$$

$$= \int_{-\infty}^{\infty} yf_Y(y)\left[\int_{-\infty}^{\infty} xf_X(x)\, dx\right] dy = E(X)E(Y)$$

The discrete case is similar. More generally, essentially the same derivation works for several functions of random variables, as stated in this proposition:

PROPOSITION Let $X_1, X_2, \ldots, X_n$ be independent random variables and assume that the expected values of $h_1(X_1), h_2(X_2), \ldots, h_n(X_n)$ all exist. Then

$$E[h_1(X_1) \cdot h_2(X_2) \cdot \cdots \cdot h_n(X_n)] = E[h_1(X_1)] \cdot E[h_2(X_2)] \cdot \cdots \cdot E[h_n(X_n)]$$

Covariance

When two random variables X and Y are not independent, it is frequently of interest to assess how strongly they are related to one another.

DEFINITION The **covariance** between two rv's X and Y is

$$\text{Cov}(X, Y) = E[(X - \mu_X)(Y - \mu_Y)]$$

$$= \begin{cases} \sum_x \sum_y (x - \mu_X)(y - \mu_Y)p(x, y) & X, Y \text{ discrete} \\ \int_{-\infty}^{\infty}\int_{-\infty}^{\infty} (x - \mu_X)(y - \mu_Y)f(x, y)\,dx\,dy & X, Y \text{ continuous} \end{cases}$$

The rationale for the definition is as follows. Suppose X and Y have a strong positive relationship to one another, by which we mean that large values of X tend to occur with large values of Y and small values of X with small values of Y. Then most of the probability mass or density will be associated with $(x - \mu_X)$ and $(y - \mu_Y)$ either both positive (both X and Y above their respective means) or both negative, so the product $(x - \mu_X) \cdot (y - \mu_Y)$ will tend to be positive. Thus for a strong positive relationship, Cov(X, Y) should be quite positive. For a strong negative relationship, the signs of $(x - \mu_X)$ and $(y - \mu_Y)$ will tend to be opposite, yielding a negative product. Thus for a strong negative relationship, Cov(X, Y) should be quite negative. If X and Y are not strongly related, positive and negative products will tend to cancel one another, yielding a covariance near 0. Figure 5.4

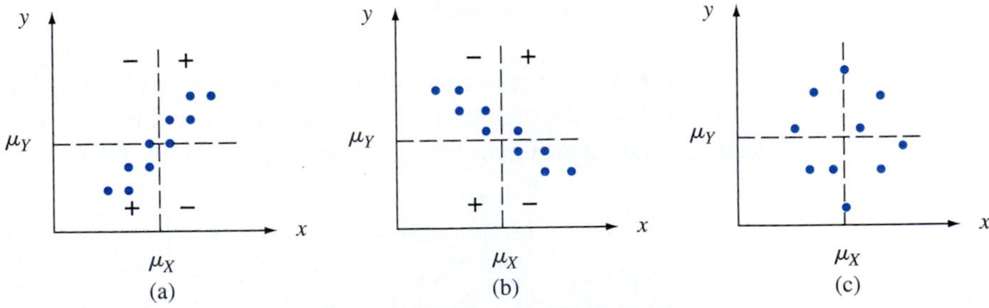

Figure 5.4 $p(x, y) = \frac{1}{10}$ for each of ten pairs corresponding to indicated points; (a) positive covariance; (b) negative covariance; (c) covariance near zero

illustrates the different possibilities. The covariance depends on *both* the set of possible pairs and the probabilities. In Figure 5.4, the probabilities could be changed without altering the set of possible pairs, and this could drastically change the value of Cov(X, Y).

Example 5.14 The joint and marginal pmf's for X = automobile policy deductible amount and Y = homeowner policy deductible amount in Example 5.1 were

$p(x, y)$		y: 0	100	200
x	100	.20	.10	.20
	250	.05	.15	.30

x	100	250
$p_X(x)$	.5	.5

y	0	100	200
$p_Y(y)$	.25	.25	.5

from which $\mu_X = \Sigma x p_X(x) = 175$ and $\mu_Y = 125$. Therefore,

$$\begin{aligned}\text{Cov}(X, Y) &= \sum_{(x, y)}\sum (x - 175)(y - 125)p(x, y) \\ &= (100 - 175)(0 - 125)(.20) + \cdots \\ &\quad + (250 - 175)(200 - 125)(.30) \\ &= 1875\end{aligned}$$

■

The following shortcut formula for Cov(X, Y) simplifies the computations.

PROPOSITION

$$\text{Cov}(X, Y) = E(XY) - \mu_X \cdot \mu_Y$$

According to this formula, no intermediate subtractions are necessary; only at the end of the computation is $\mu_X \cdot \mu_Y$ subtracted from $E(XY)$. The proof involves expanding $(X - \mu_X)(Y - \mu_Y)$ and then taking the expected value of each term separately. Note that Cov$(X, X) = E(X^2) - \mu_X^2 = V(X)$.

Example 5.15 (Example 5.5 continued) The joint and marginal pdf's of X = amount of almonds and Y = amount of cashews were

$$f(x, y) = \begin{cases} 24xy & 0 \le x \le 1, 0 \le y \le 1, x + y \le 1 \\ 0 & \text{otherwise} \end{cases}$$

$$f_X(x) = \begin{cases} 12x(1 - x)^2 & 0 \le x \le 1 \\ 0 & \text{otherwise} \end{cases}$$

with $f_Y(y)$ obtained by replacing x by y in $f_X(x)$. It is easily verified that $\mu_X = \mu_Y = \frac{2}{5}$, and

$$\begin{aligned}E(XY) &= \int_{-\infty}^{\infty}\int_{-\infty}^{\infty} xy\, f(x, y)\, dx\, dy = \int_0^1\int_0^{1-x} xy \cdot 24xy\, dy\, dx \\ &= 8\int_0^1 x^2(1 - x)^3\, dx = \frac{2}{15}\end{aligned}$$

Thus Cov$(X, Y) = \frac{2}{15} - \left(\frac{2}{5}\right)\left(\frac{2}{5}\right) = \frac{2}{15} - \frac{4}{25} = -\frac{2}{75}$. A negative covariance is reasonable here because more almonds in the can implies fewer cashews. ■

The covariance satisfies a useful linearity property (Exercise 33).

PROPOSITION If X, Y, and Z are rv's and a and b are constants then

$$\text{Cov}(aX + bY, Z) = a\,\text{Cov}(X, Z) + b\,\text{Cov}(Y, Z)$$

It would appear that the relationship in the insurance example is quite strong since $\text{Cov}(X, Y) = 1875$, whereas in the nut example $\text{Cov}(X, Y) = -\frac{2}{75}$ would seem to imply quite a weak relationship. Unfortunately, the covariance has a serious defect that makes it impossible to interpret a computed value of the covariance. In the insurance example, suppose we had expressed the deductible amount in cents rather than in dollars. Then $100X$ would replace X, $100Y$ would replace Y, and the resulting covariance would be $\text{Cov}(100X, 100Y) = (100)(100)\,\text{Cov}(X, Y) = 18{,}750{,}000$. If, on the other hand, the deductible amount had been expressed in hundreds of dollars, the computed covariance would have been $(.01)(.01)(1875) = .1875$. *The defect of covariance is that its computed value depends critically on the units of measurement*. Ideally, the choice of units should have no effect on a measure of strength of relationship. This is achieved by scaling the covariance.

Correlation

DEFINITION The **correlation coefficient** of X and Y, denoted by $\text{Corr}(X, Y)$, or $\rho_{X,Y}$, or just ρ, is defined by

$$\rho_{X,Y} = \frac{\text{Cov}(X, Y)}{\sigma_X \cdot \sigma_Y}$$

Example 5.16 It is easily verified that in the insurance problem of Example 5.14, $E(X^2) = 36{,}250$, $\sigma_X^2 = 36{,}250 - (175)^2 = 5625$, $\sigma_X = 75$, $E(Y^2) = 22{,}500$, $\sigma_Y^2 = 6875$, and $\sigma_Y = 82.92$. This gives

$$\rho = \frac{1875}{(75)(82.92)} = .301$$

■

The following proposition shows that ρ remedies the defect of $\text{Cov}(X, Y)$ and also suggests how to recognize the existence of a strong (linear) relationship.

PROPOSITION

1. If a and c are either both positive or both negative,

$$\text{Corr}(aX + b, cY + d) = \text{Corr}(X, Y)$$

2. For any two rv's X and Y, $-1 \le \text{Corr}(X, Y) \le 1$.

Statement 1 says precisely that the correlation coefficient is not affected by a linear change in the units of measurement (if, say, X = temperature in °C, then $9X/5 + 32$ = temperature in °F). According to Statement 2, the strongest possible positive relationship is evidenced by $\rho = +1$, whereas the strongest possible negative relationship corresponds to $\rho = -1$. The proof of the first statement is sketched in Exercise 31, and that of the second appears in Exercise 35 and also Supplementary Exercise 76 in Chapter 6. For descriptive purposes, the relationship will be described as strong if $|\rho| \geq .8$, moderate if $.5 < |\rho| < .8$, and weak if $|\rho| \leq .5$.

If we think of $p(x, y)$ or $f(x, y)$ as prescribing a mathematical model for how the two numerical variables X and Y are distributed in some population (height and weight, verbal SAT score and quantitative SAT score, etc.), then ρ is a population characteristic or parameter that measures how strongly X and Y are related in the population. In Chapter 12, we will consider taking a sample of pairs $(x_1, y_1), \ldots, (x_n, y_n)$ from the population. The sample correlation coefficient r will then be defined and used to make inferences about ρ.

The correlation coefficient ρ is actually not a completely general measure of the strength of a relationship.

PROPOSITION

1. If X and Y are independent, then $\rho = 0$, but $\rho = 0$ does not imply independence.
2. $\rho = 1$ or -1 iff $Y = aX + b$ for some numbers a and b with $a \neq 0$.

Exercise 29 and Example 5.17 relate to Property 1, and Property 2 is investigated in Exercises 32 and 35.

This proposition says that ρ is a measure of the degree of **linear** relationship between X and Y, and only when the two variables are perfectly related in a linear manner will ρ be as positive or negative as it can be. A ρ less than 1 in absolute value indicates only that the relationship is not completely linear, but there may still be a very strong nonlinear relation. Also, $\rho = 0$ does not imply that X and Y are independent, but only that there is complete absence of a linear relationship. When $\rho = 0$, X and Y are said to be **uncorrelated**. Two variables could be uncorrelated yet highly dependent because there is a strong nonlinear relationship, so be careful not to conclude too much from knowing that $\rho = 0$.

Example 5.17 Let X and Y be discrete rv's with joint pmf

$$p(x, y) = \begin{cases} \dfrac{1}{4} & (x, y) = (-4, 1), (4, -1), (2, 2), (-2, -2) \\ 0 & \text{otherwise} \end{cases}$$

The points that receive positive probability mass are identified on the (x, y) coordinate system in Figure 5.5. It is evident from the figure that the value of X is completely determined by the value of Y and vice versa, so the two variables are completely dependent. However, by symmetry $\mu_X = \mu_Y = 0$ and $E(XY) = (-4)\frac{1}{4} + (-4)\frac{1}{4} + (4)\frac{1}{4} + (4)\frac{1}{4} = 0$, so $\text{Cov}(X, Y) = E(XY) - \mu_X \cdot \mu_Y = 0$ and thus $\rho_{X,Y} = 0$. Although there is perfect dependence, there is also complete absence of any linear relationship!

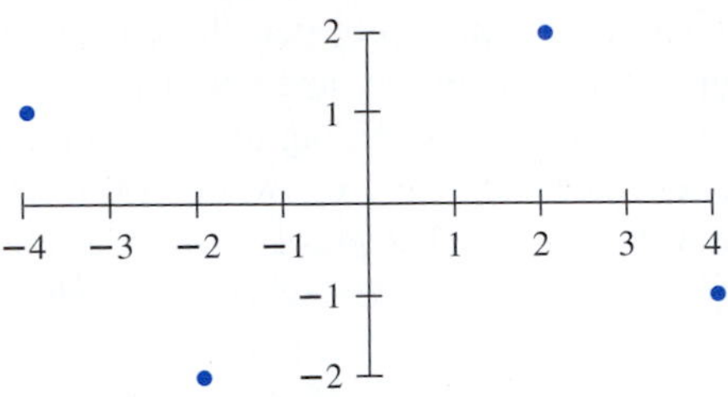

Figure 5.5 The population of pairs for Example 5.17 ■

A value of ρ near 1 does not necessarily imply that increasing the value of X *causes* Y to increase. It implies only that large X values are *associated* with large Y values. For example, in the population of children, vocabulary size and number of cavities are quite positively correlated, but it is certainly not true that cavities cause vocabulary to grow. Instead, the values of both these variables tend to increase as the value of age, a third variable, increases. For children of a fixed age, there is probably a very low correlation between number of cavities and vocabulary size. In summary, association (a high correlation) is not the same as causation.

Exercises Section 5.2 (18–35)

18. An instructor has given a short quiz consisting of two parts. For a randomly selected student, let X = the number of points earned on the first part and Y = the number of points earned on the second part. Suppose that the joint pmf of X and Y is given in the accompanying table.

$p(x, y)$		y: 0	5	10	15
x	0	.02	.06	.02	.10
	5	.04	.15	.20	.10
	10	.01	.15	.14	.01

a. If the score recorded in the grade book is the total number of points earned on the two parts, what is the expected recorded score $E(X + Y)$?

b. If the maximum of the two scores is recorded, what is the expected recorded score?

19. The difference between the number of customers in line at the express checkout and the number in line at the superexpress checkout in Exercise 3 is $X_1 - X_2$. Calculate the expected difference.

20. Six individuals, including A and B, take seats around a circular table in a completely random fashion. Suppose the seats are numbered 1, . . . , 6. Let X = A's seat number and Y = B's seat number. If A sends a written message around the table to B in the direction in which they are closest, how many individuals (including A and B) would you expect to handle the message?

21. A surveyor wishes to lay out a square region with each side having length L. However, because of measurement error, he instead lays out a rectangle in which the north–south sides both have length X and the east–west sides both have length Y. Suppose that X and Y are independent and that each is uniformly distributed on the interval $[L - A, L + A]$ (where $0 < A < L$). What is the expected area of the resulting rectangle?

22. Consider a small ferry that can accommodate cars and buses. The toll for cars is $3, and the toll for buses is $10. Let X and Y denote the number of cars and buses, respectively, carried on a single trip. Suppose the joint distribution of X and Y is as given in the table of Exercise 7. Compute the expected revenue from a single trip.

23. Annie and Alvie have agreed to meet for lunch between noon (0:00 p.m.) and 1:00 p.m. Denote

Annie's arrival time by X, Alvie's by Y, and suppose X and Y are independent with pdf's

$$f_X(x) = \begin{cases} 3x^2 & 0 \le x \le 1 \\ 0 & \text{otherwise} \end{cases}$$

$$f_Y(y) = \begin{cases} 2y & 0 \le y \le 1 \\ 0 & \text{otherwise} \end{cases}$$

What is the expected amount of time that the one who arrives first must wait for the other person? [*Hint*: $h(X, Y) = |X - Y|$.]

24. Suppose that X and Y are independent rv's with moment generating functions $M_X(t)$ and $M_Y(t)$, respectively. If $Z = X + Y$, show that $M_Z(t) = M_X(t)\, M_Y(t)$. (*Hint*: Use the proposition on the expected value of a product.)

25. Compute the correlation coefficient ρ for X and Y of Example 5.15 (the covariance has already been computed).

26. a. Compute the covariance for X and Y in Exercise 18.

b. Compute ρ for X and Y in the same exercise.

27. a. Compute the covariance between X and Y in Exercise 9.

b. Compute the correlation coefficient ρ for this X and Y.

28. Reconsider the minicomputer component lifetimes X and Y as described in Exercise 12. Determine $E(XY)$. What can be said about $\text{Cov}(X, Y)$ and ρ?

29. Show that when X and Y are independent variables, $\text{Cov}(X, Y) = \text{Corr}(X, Y) = 0$.

30. a. Recalling the definition of σ^2 for a single rv X, write a formula that would be appropriate for computing the variance of a function $h(X, Y)$ of two random variables. (*Hint*: Remember that variance is just a special expected value.)

b. Use this formula to compute the variance of the recorded score $h(X, Y)$ [$= \max(X, Y)$] in part (b) of Exercise 18.

31. a. Use the rules of expected value to show that $\text{Cov}(aX + b, cY + d) = ac\,\text{Cov}(X, Y)$.

b. Use part (a) along with the rules of variance and standard deviation to show that $\text{Corr}(aX + b, cY + d) = \text{Corr}(X, Y)$ when a and c have the same sign.

c. What happens if a and c have opposite signs?

32. Show that if $Y = aX + b$ ($a \ne 0$), then $\text{Corr}(X, Y) = +1$ or -1. Under what conditions will $\rho = +1$?

33. Show that if X, Y, and Z are rv's and a and b are constants, then $\text{Cov}(aX + bY, Z) = a\,\text{Cov}(X, Z) + b\,\text{Cov}(Y, Z)$

34. Let Z_X be the standardized X, $Z_X = (X - \mu_X)/\sigma_X$, and let Z_Y be the standardized Y, $Z_Y = (Y - \mu_Y)/\sigma_Y$. Use the results of Exercise 31 to show that $\text{Corr}(X, Y) = \text{Cov}(Z_X, Z_Y) = E(Z_X Z_Y)$.

35. Let Z_X be the standardized X, $Z_X = (X - \mu_X)/\sigma_X$, and let Z_Y be the standardized Y, $Z_Y = (Y - \mu_Y)/\sigma_Y$.

a. Show with the help of the previous exercise that $E[(Z_Y - \rho Z_X)]^2 = 1 - \rho^2$.

b. Use part (a) to show that $-1 \le \rho \le 1$.

c. Use part (a) to show that $\rho = 1$ implies that $Y = aX + b$ where $a > 0$, and $\rho = -1$ implies that $Y = aX + b$ where $a < 0$.

5.3 *Conditional Distributions

The distribution of Y can depend strongly on the value of another variable X. For example, if X is height and Y is weight, the distribution of weight for men who are 6 feet tall is very different from the distribution of weight for short men. The conditional distribution of Y given $X = x$ describes for each possible x how probability is distributed over the set of possible y values. We define the conditional distribution of Y given X, but the conditional distribution of X given Y can be obtained by just reversing the roles of X and Y. Both definitions are analogous to that of the conditional probability $P(A \mid B)$ as the ratio $P(A \cap B)/P(B)$.

DEFINITION

Let X and Y be two discrete random variables with joint pmf $p(x, y)$ and marginal X pmf $p_X(x)$. Then for any x value such that $p_X(x) > 0$, the **conditional probability mass function of Y given $X = x$** is

$$p_{Y|X}(y \mid x) = \frac{p(x, y)}{p_X(x)}$$

An analogous formula holds in the continuous case. Let X and Y be two continuous random variables with joint pdf $f(x, y)$ and marginal X pdf $f_X(x)$. Then for any x value such that $f_X(x) > 0$, the **conditional probability density function of Y given $X = x$** is

$$f_{Y|X}(y \mid x) = \frac{f(x, y)}{f_X(x)}$$

Example 5.18

For a discrete example, reconsider Example 5.1, where X represents the deductible amount on an automobile policy and Y represents the deductible amount on a homeowner's policy. Here is the joint distribution again.

$p(x, y)$		y: 0	100	200
	100	.20	.10	.20
x	250	.05	.15	.30

The distribution of Y depends on X. In particular, let's find the conditional probability that Y is 200, given that X is 250, using the definition of conditional probability from Section 2.4.

$$P(Y = 200 \mid X = 250) = \frac{P(Y = 200 \text{ and } X = 250)}{P(X = 250)} = \frac{.3}{.05 + .15 + .3} = .6$$

With our new definition we obtain the same result:

$$p_{Y|X}(200 \mid 250) = \frac{p(250, 200)}{p_X(250)} = \frac{.3}{.05 + .15 + .3} = .6$$

Continuing with this example, we get

$$p_{Y|X}(0 \mid 250) = \frac{p(250, 0)}{p_X(250)} = \frac{.05}{.05 + .15 + .3} = .1$$

$$p_{Y|X}(100 \mid 250) = \frac{p(250, 100)}{p_X(250)} = \frac{.15}{.05 + .15 + .3} = .3$$

Thus, $p_{Y|X}(0 \mid 250) + p_{Y|X}(100 \mid 250) + p_{Y|X}(200 \mid 250) = .1 + .3 + .6 = 1$. This is no coincidence; conditional probabilities satisfy the properties of ordinary probabilities.

They are nonnegative and they sum to 1. Essentially, the denominator in the definition of conditional probability is designed to make the total be 1.

Reversing the roles of X and Y, we find the conditional probabilities for X, given that $Y = 0$:

$$p_{X|Y}(100\,|\,0) = \frac{p(100, 0)}{p_Y(0)} = \frac{.20}{.20 + .05} = .8$$

$$p_{X|Y}(250\,|\,0) = \frac{p(250, 0)}{p_Y(0)} = \frac{.05}{.20 + .05} = .2$$

Again, the conditional probabilities add to 1. ■

Example 5.19 For a continuous example, recall Example 5.5, where X is the weight of almonds and Y is the weight of cashews in a can of mixed nuts. The sum $X + Y$ is at most 1 lb, the total weight of the can of nuts. The joint pdf of X and Y is

$$f(x, y) = \begin{cases} 24xy & 0 \le x \le 1, 0 \le y \le 1, x + y \le 1 \\ 0 & \text{otherwise} \end{cases}$$

In Example 5.5 it was shown that

$$f_X(x) = \begin{cases} 12x(1 - x)^2 & 0 \le x \le 1 \\ 0 & \text{otherwise} \end{cases}$$

Compute

$$f_{Y|X}(y\,|\,x) = \frac{f(x, y)}{f_X(x)} = \frac{24xy}{12x(1 - x)^2} = \frac{2y}{(1 - x)^2} \qquad 0 \le y \le 1 - x$$

This can be used to get conditional probabilities for Y. For example,

$$P(Y \le .25\,|\,X = .5) = \int_{-\infty}^{.25} f_{Y|X}(y\,|\,.5)\, dy = \int_0^{.25} \frac{2y}{(1 - .5)^2}\, dy = [4y^2]_0^{.25} = .25$$

Recall that X is the weight of almonds and Y is the weight of cashews, so this says that, given that the weight of almonds is .5 lb, the probability is $\frac{1}{4}$ for the weight of cashews to be less than .25 lb.

Just as in the discrete case, the total conditional probability here should be 1. That is, integrating the conditional density over its set of possible values should yield 1.

$$\int_{-\infty}^{\infty} f_{Y|X}(y\,|\,x)\, dy = \int_0^{1-x} \frac{2y}{(1 - x)^2}\, dy = \left[\frac{y^2}{(1 - x)^2}\right]_0^{1-x} = 1$$

Whenever you calculate a conditional density it is a good idea to do this integration as a validity check. ■

Because the conditional distribution is a valid probability distribution, it makes sense to define the conditional mean and variance.

DEFINITION

Let X and Y be two discrete random variables with conditional probability mass function $p_{Y|X}(y|x)$. Then the **conditional mean** or **expected value of Y given that $X = x$** is

$$\mu_{Y|X=x} = E(Y|X = x) = \sum_{y \in D_Y} y p_{Y|X}(y|x)$$

An analogous formula holds in the continuous case. Let X and Y be two continuous random variables with conditional probability density function $f_{Y|X}(y|x)$. Then

$$\mu_{Y|X=x} = E(Y|X = x) = \int_{-\infty}^{\infty} y f_{Y|X}(y|x)\,dy$$

The conditional mean of any function $g(Y)$ can be obtained similarly. In the discrete case,

$$E[g(Y)|X = x] = \sum_{y \in D_Y} g(y) p_{Y|X}(y|x)$$

In the continuous case,

$$E[g(Y)|X = x] = \int_{-\infty}^{\infty} g(y) f_{Y|X}(y|x)\,dy$$

The **conditional variance of Y given $X = x$** is

$$\sigma^2_{Y|X=x} = V(Y|X = x) = E\{[Y - E(Y|X = x)]^2|X = x\}$$

There is a shortcut formula for the conditional variance analogous to that for $V(Y)$:

$$\sigma^2_{Y|X=x} = V(Y|X = x) = E(Y^2|X = x) - \mu^2_{Y|X=x}$$

Example 5.20

Having found the conditional distribution of Y given $X = 250$ in Example 5.18, we compute the conditional mean and variance.

$$\mu_{Y|X=250} = E(Y|X = 250) = 0p_{Y|X}(0|250) + 100p_{Y|X}(100|250) + 200p_{Y|X}(200|250) = 0(.1) + 100(.3) + 200(.6) = 150$$

Given that the possibilities for Y are 0, 100, and 200 and most of the probability is on 100 and 200, it is reasonable that the conditional mean should be between 100 and 200.

Let's use the alternative formula for the conditional variance:

$$E(Y^2|X = 250) = 0^2p_{Y|X}(0|250) + 100^2p_{Y|X}(100|250) + 200^2P_{Y|X}(200|250) = 0^2(.1) + 100^2(.3) + 200^2(.6) = 27{,}000$$

Thus,

$$\sigma^2_{Y|X=250} = V(Y|X = 250) = E(Y^2|X = 250) - \mu^2_{Y|X=250} = 27{,}000 - 150^2 = 4500$$

Taking the square root, we get $\sigma_{Y|X=250} = 67.08$, which is in the right ballpark when we recall that the possible values of Y are 0, 100, and 200.

It is important to realize that $E(Y|X = x)$ is one particular possible value of a random variable $E(Y|X)$, which is a function of X. Similarly, the conditional variance $V(Y|X = x)$ is a value of the rv $V(Y|X)$. The value of X might be 100 or 250. So far, we have just $E(Y|X = 250) = 150$ and $V(Y|X = 250) = 4500$. If the calculations are repeated for $X = 100$, the results are $E(Y|X = 100) = 100$ and $V(Y|X = 100) = 8000$. Here is a summary in the form of a table:

x	$P(X = x)$	$E(Y\|X = x)$	$V(Y\|X = x)$
100	.5	100	8000
250	.5	150	4500

Similarly, the conditional mean and variance of X can be computed for specific Y. Taking the conditional probabilities from Example 5.18,

$$\mu_{X|Y=0} = E(X|Y = 0) = 100p_{X|Y}(100|0) + 250p_{X|Y}(250|0) = 100(.8) + 250(.2) = 130$$

$$\begin{aligned}\sigma^2_{X|Y=0} &= V(X|Y = 0) = E([X - E(X|Y = 0)]^2|Y = 0) \\ &= (100 - 130)^2p_{X|Y}(100|0) + (250 - 130)^2p_{X|Y}(250|0) \\ &= 30^2(.8) + 120^2(.2) = 3600\end{aligned}$$

Similar calculations give the other entries in this table:

y	$P(Y = y)$	$E(X\|Y = y)$	$V(X\|Y = y)$
0	.25	130	3600
100	.25	190	5400
200	.50	190	5400

Again, the conditional mean and variance are random because they depend on the random value of Y. ■

Example 5.21 (Example 5.19 continued)

For any given weight of almonds, let's find the expected weight of cashews. Using the definition of conditional mean,

$$\begin{aligned}\mu_{Y|X=x} &= E(Y|X = x) = \int_{-\infty}^{\infty} yf_{Y|X}(y|x)\,dy \\ &= \int_0^{1-x} y\frac{2y}{(1 - x)^2}\,dy = \frac{2}{3}(1 - x) \qquad 0 \le x \le 1\end{aligned}$$

This is a linear decreasing function of x. When there are more almonds, we expect less cashews. This is in accord with Figure 5.2, which shows that for large X the domain of Y is restricted to small values. To get the corresponding variance, compute first

$$\begin{aligned}E(Y^2|X = x) &= \int_{-\infty}^{\infty} y^2f_{Y|X}(y|x)\,dy \\ &= \int_0^{1-x} y^2\frac{2y}{(1 - x)^2}\,dy = .5(1 - x)^2 \qquad 0 \le x \le 1\end{aligned}$$

Then the conditional variance is

$$\begin{aligned}\sigma^2_{Y|X=x} &= V(Y|X = x) = E(Y^2|X = x) - \mu^2_{Y|X=x} \\ &= .5(1 - x)^2 - \frac{4}{9}(1 - x)^2 = \frac{1}{18}(1 - x)^2 \\ \sigma_{Y|X=x} &= \left(\frac{1}{18}\right)^{.5}(1 - x)\end{aligned}$$

This says that the variance gets smaller as the weight of almonds approaches 1. Does this make sense? When the weight of almonds is 1, the weight of cashews is guaranteed to be 0, implying that the variance is 0. This is clarified by Figure 5.2, which shows that the set of y values narrows to 0 as x approaches 1. ■

Independence

Recall that in Section 5.1 two random variables were defined to be independent if their joint pmf or pdf factors into the product of the marginal pmf's or pdf's. We can understand this definition better with the help of conditional distributions. For example, suppose there is independence in the discrete case. Then

$$p_{Y|X}(y|x) = \frac{p(x, y)}{p_X(x)} = \frac{p_X(x)p_Y(y)}{p_X(x)} = p_Y(y)$$

That is, independence implies that the conditional distribution of Y is the same as the unconditional distribution. The implication works in the other direction, too. If

$$p_{Y|X}(y|x) = p_Y(y)$$

then

$$\frac{p(x, y)}{p_X(x)} = p_Y(y)$$

so

$$p(x, y) = p_X(x)p_Y(y)$$

and therefore X and Y are independent. Is this intuitively reasonable? Yes, because independence means that knowing X does not change our probabilities for Y.

In Example 5.7 we said that independence requires a rectangular region of support for the joint distribution. In terms of conditional distribution this region tells us the domain of Y for each X. For independence we need to have the domain of Y not be dependent on X. That is, the conditional distributions must all be the same, so the domains must all be the same, which requires a rectangle.

The Bivariate Normal Distribution

Perhaps the most useful example of a joint distribution is the bivariate normal. Although the formula may seem rather messy, it is based on a simple quadratic expression in the standardized variables (subtract the mean and then divide by the standard deviation). The bivariate normal density is

$$f(x, y) = \frac{1}{2\pi\sigma_1\sigma_2\sqrt{1-\rho^2}} e^{-\{[(x-\mu_1)/\sigma_1]^2 - 2\rho(x-\mu_1)(y-\mu_2)/\sigma_1\sigma_2 + [(y-\mu_2)/\sigma_2]^2\}/[2(1-\rho^2)]}$$

There are five parameters, including the mean μ_1 and the standard deviation σ_1 of X, and the mean μ_2 and the standard deviation σ_2 of Y. The fifth parameter ρ is the correlation between X and Y. The integration required to do bivariate normal probability calculations is quite difficult. Computer code is available for calculating $P(X < x, Y < y)$ approximately using numerical integration, and some statistical software packages (e.g., SAS, S-Plus, Stata) include this feature.

What does the density look like when plotted as a function of x and y? If we set $f(x, y)$ to a constant to investigate the contours, this is setting the exponent to a constant, and it will give ellipses centered at $(x, y) = (\mu_1, \mu_2)$. That is, all of the contours are concentric ellipses. The plot in three dimensions looks like a mountain with elliptical cross-sections. The vertical cross-sections are all proportional to normal densities. See Figure 5.6.

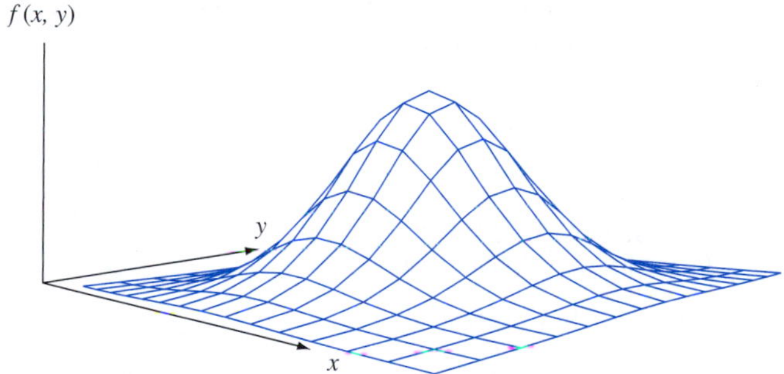

Figure 5.6 A graph of the bivariate normal pdf

If $\rho = 0$, then $f(x, y) = f_X(x) f_Y(y)$, where X is normal with mean μ_1 and standard deviation σ_1, and Y is normal with mean μ_2 and standard deviation σ_2. That is, X and Y have independent normal distributions. In this case the plot in three dimensions has elliptical contours that reduce to circles. Recall that in Section 5.2 we emphasized that independence of X and Y implies $\rho = 0$ but, in general, $\rho = 0$ does not imply independence. However, we have just seen that when X and Y are bivariate normal, $\rho = 0$ does imply independence. Therefore, in the bivariate normal case $\rho = 0$ if and only if the two rv's are independent.

What do we get for the marginal distributions? As you might guess, the marginal distribution $f_X(x)$ is just a normal distribution with mean μ_1 and standard deviation σ_1:

$$f_X(x) = \frac{1}{\sigma_1\sqrt{2\pi}} e^{-.5\{[(x-\mu_1)/\sigma_1]^2\}}$$

The integration to show this [integrating $f(x, y)$ on y from $-\infty$ to ∞] is rather messy. More generally, any linear combination of the form $aX + bY$, where a and b are constants, is normally distributed.

We get the conditional density by dividing the marginal density of X into $f(x, y)$. Unfortunately, the algebra is again a mess, but the result is fairly simple. The conditional density $f_{Y|X}(y|x)$ is a normal density with mean and variance given by

$$\mu_{Y|X=x} = E(Y|X = x) = \mu_2 + \rho\sigma_2\frac{x - \mu_1}{\sigma_1}$$

$$\sigma^2_{Y|X=x} = V(Y|X = x) = \sigma_2^2(1 - \rho^2)$$

Notice that the conditional mean is a linear function of x and the conditional variance doesn't depend on x at all. When $\rho = 0$, the conditional mean is the mean of Y and the conditional variance is just the variance of Y. In other words, if $\rho = 0$, then the conditional distribution of Y is the same as the unconditional distribution of Y. This says that if $\rho = 0$ then X and Y are independent, but we already saw that previously in terms of the factorization of $f(x, y)$ into the product of the marginal densities.

When ρ is close to 1 or -1 the conditional variance will be much smaller than $V(Y)$, which says that knowledge of X will be very helpful in predicting Y. If ρ is near 0 then X and Y are nearly independent and knowledge of X is not very useful in predicting Y.

Example 5.22 Let X be mother's height and Y be daughter's height. A similar situation was one of the first applications of the bivariate normal distribution (Galton, 1886), and the data was found to fit the distribution very well. Suppose a bivariate normal distribution with mean $\mu_1 = 64$ inches and standard deviation $\sigma_1 = 3$ inches for X and mean $\mu_2 = 65$ inches and standard deviation $\sigma_2 = 3$ inches for Y. Here $\mu_2 > \mu_1$, which is in accord with the increase in height from one generation to the next. Assume $\rho = .4$. Then

$$\mu_{Y|X=x} = \mu_2 + \rho\sigma_2\frac{x - \mu_1}{\sigma_1} = 65 + .4(3)\frac{x - 64}{3} = 65 + .4\,(x - 64) = .4x + 39.4$$

$$\sigma^2_{Y|X=x} = V(Y|X = x) = \sigma_2^2(1 - \rho^2) = 9(1 - .4^2) = 7.56 \quad \text{and} \quad \sigma_{Y|X=x} = 2.75$$

Notice that the conditional variance is 16% less than the variance of Y. Squaring the correlation gives the percentage by which the conditional variance is reduced relative to the variance of Y. ■

Regression to the Mean

The formula for the conditional mean can be reexpressed as

$$\frac{\mu_{Y|X=x} - \mu_2}{\sigma_2} = \rho \cdot \frac{x - \mu_1}{\sigma_1}$$

In words, when the formula is expressed in terms of standardized variables, the standardized conditional mean is just ρ times the standardized x. In particular, for the example of heights,

$$\frac{\mu_{Y|X=x} - 65}{3} = .4 \cdot \frac{x - 64}{3}$$

If the mother is 5 inches above the mean of 64 inches for mothers, then the daughter's conditional expected height is just 2 inches above the mean for daughters. The daughter's conditional expected height is always closer to its mean than the mother's height is to its mean. One can think of the conditional expectation as falling back toward the mean, and that is why Galton called this *regression to the mean*.

Regression to the mean occurs in many contexts. For example, let X be a baseball player's average for the first half of the season and let Y be the average for the second half. Most of the players with a high X (above .300) will not have such a high Y. The same kind of reasoning applies to the "sophomore jinx," which says that if a player has a very good first season, then the player is unlikely to do as well in the second season.

The Mean and Variance via the Conditional Mean and Variance

From the conditional mean we can obtain the mean of Y. From the conditional mean and the conditional variance, the variance of Y can be obtained. The following theorem uses the idea that the conditional mean and variance are themselves random variables, as shown in the tables of Example 5.20.

THEOREM

a. $E(Y) = E[E(Y|X)]$

b. $V(Y) = V[E(Y|X)] + E[V(Y|X)]$

The result in (a) says that $E(Y)$ is a weighted average of the conditional means $E(Y|X = x)$, where the weights are given by the pmf or pdf of X. We give the proof of just part (a) in the discrete case:

$$E[E(Y|X)] = \sum_{x \in D_X} E(Y|X = x)p_X(x) = \sum_{x \in D_X} \sum_{y \in D_Y} y p_{Y|X}(y|x)p_X(x)$$

$$= \sum_{x \in D_X} \sum_{y \in D_Y} y \frac{p(x, y)}{p_X(x)} p_X(x) = \sum_{y \in D_Y} y \sum_{x \in D_X} p(x, y) = \sum_{y \in D_Y} y p_Y(y) = E(Y)$$

Example 5.23 To try to get a feel for the theorem, let's apply it to Example 5.20. Here again is the table for the conditional mean and variance of Y given X.

x	$P(X = x)$	$E(Y\|X = x)$	$V(Y\|X = x)$
100	.5	100	8000
250	.5	150	4500

Compute

$$E[E(Y|X)] = E(Y|X = 100)P(X = 100) + E(Y|X = 250)P(X = 250)$$
$$= 100(.5) + 150(.5) = 125$$

Compare this with $E(Y)$ computed directly:

$$E(Y) = 0P(Y = 0) + 100P(Y = 100) + 200P(Y = 200)$$
$$= 0(.25) + 100(.25) + 200(.5) = 125$$

For the variance, first compute the mean of the conditional variance:

$$E[V(Y|X)] = V(Y|X = 100)P(X = 100) + V(Y|X = 250)P(X = 250)$$
$$= 4500(.5) + 8000(.5) = 6250$$

Then comes the variance of the conditional mean. We have already computed the mean of this random variable to be 125. The variance is

$$V[E(Y|X)] = .5(100 - 125)^2 + .5(150 - 125)^2 = 625$$

Finally, do the sum in part (b) of the theorem:

$$V(Y) = V[E(Y|X)] + E[V(Y|X)] = 625 + 6250 = 6875$$

To compare this with $V(Y)$ calculated from the pmf of Y, compute first

$$E(Y^2) = 0^2P(Y = 0) + 100^2P(Y = 100) + 200^2P(Y = 200)$$
$$= 0(.25) + 10{,}000(.25) + 40{,}000(.5) = 22{,}500$$

Thus, $V(Y) = E(Y^2) - [E(Y)]^2 = 22{,}500 - 125^2 = 6875$, in agreement with the calculation based on the theorem. ■

Here is an example where the theorem is helpful because we are finding the mean and variance of a random variable that is neither discrete nor continuous.

Example 5.24 The probability of a claim being filed on an insurance policy is .1, and only one claim can be filed. If a claim is filed, the amount is exponentially distributed with mean \$1000. Recall from Section 4.4 that the mean and standard deviation of the exponential distribution are the same, so the variance is the square of this value. We want to find the mean and variance of the amount paid. Let X be the number of claims (0 or 1) and let Y be the payment. We know that $E(Y|X = 0) = 0$ and $E(Y|X = 1) = 1000$. Also, $V(Y|X = 0) = 0$ and $V(Y|X = 1) = 1000^2 = 1{,}000{,}000$. Here is a table for the distribution of $E(Y|X = x)$ and $V(Y|X = x)$:s

x	$P(X = x)$	$E(Y\|X = x)$	$V(Y\|X = x)$
0	.9	0	0
1	.1	1000	1,000,000

Therefore,

$$E(Y) = E[E(Y|X)] = E(Y|X = 0)P(X = 0) + E(Y|X = 1)P(X = 1)$$
$$= 0(.9) + 1000(.1) = 100$$

The variance of the conditional mean is

$$V[E(Y|X)] = .9(0 - 100)^2 + .1(1000 - 100)^2 = 90{,}000$$

The expected value of the conditional variance is

$$E[V(Y|X)] = .9(0) + .1(1{,}000{,}000) = 100{,}000$$

Finally, use part (b) of the theorem to get $V(Y)$:

$$V(Y) = V[E(Y|X)] + E[V(Y|X)] = 90{,}000 + 100{,}000 = 190{,}000$$

Taking the square root gives the standard deviation, $\sigma_Y = \$435.89$.

Suppose that we want to compute the mean and variance of Y directly. Notice that X is discrete, but the conditional distribution of Y given $X = 1$ is continuous. The random variable Y itself is neither discrete nor continuous, because it has probability .9 of being 0, but the other .1 of its probability is spread out from 0 to ∞. Such "mixed" distributions may require a little extra effort to evaluate means and variances, although it is not especially hard in this case. Compute

$$\mu_Y = E(Y) = (.1)\int_0^\infty y\frac{1}{1000}e^{-y/1000}\,dy = (.1)(1000) = 100$$

$$E(Y^2) = (.1)\int_0^\infty y^2\frac{1}{1000}e^{-y/1000}\,dy = (.1)2(1000^2) = 200{,}000$$

$$V(Y) = E(Y^2) - [E(Y)]^2 = 200{,}000 - 10{,}000 = 190{,}000$$

These agree with what we found using the theorem. ■

Exercises Section 5.3 (36–57)

36. According to an article in the August 30, 2002, issue of the *Chron. Higher Ed.*, 30% of first-year college students are liberals, 20% are conservatives, and 50% characterize themselves as middle-of-the-road. Choose two students at random, let X be the number of liberals, and let Y be the number of conservatives.

a. Using the multinomial distribution from Section 5.1, give the joint probability mass function $p(x, y)$ of X and Y. Give the joint probability table showing all nine values, of which three should be 0.

b. Find the marginal probability mass functions by summing $p(x, y)$ numerically. How could these be obtained directly? *Hint*: What are the univariate distributions of X and Y?

c. Find the conditional probability mass function of Y given $X = x$ for $x = 0, 1, 2$. Compare with the Bin$[2 - x, .2/(.2 + .5)]$ distribution. Why should this work?

d. Are X and Y independent? Explain.

e. Find $E(Y|X = x)$ for $x = 0, 1, 2$. Do this numerically and then compare with the use of the formula for the binomial mean, using the binomial distribution given in part (c). Is $E(Y|X = x)$ a linear function of x?

f. Find $V(Y|X = x)$ for $x = 0, 1, 2$. Do this numerically and then compare with the use of the formula for the binomial variance, using the binomial distribution given in part (c).

37. Teresa and Allison each have arrival times uniformly distributed between 12:00 and 1:00. Their times do not influence each other. If Y is the first of the two times and X is the second, on a scale of 0 to 1, then the joint pdf of X and Y is $f(x, y) = 2$ for $0 < y < x < 1$.

a. Find the marginal density of X.

b. Find the conditional density of Y given $X = x$.

c. Find the conditional probability that Y is between 0 and .3, given that X is .5.

d. Are X and Y independent? Explain.

e. Find the conditional mean of Y given $X = x$. Is $E(Y|X = x)$ a linear function of x?

f. Find the conditional variance of Y given $X = x$.

38. In Exercise 37,

a. Find the marginal density of Y.

b. Find the conditional density of X given $Y = y$.

c. Find the conditional mean of X given $Y = y$. Is $E(X|Y = y)$ a linear function of y?

d. Find the conditional variance of X given $Y = y$.

39. A photographic supply business accepts orders on each of two different phone lines. On each line the waiting time until the first call is exponentially distributed with mean 1 minute, and the two times are independent of one another. Let X be the shorter of the two waiting times and let Y be the longer. It can be shown that the joint pdf of X and Y is $f(x, y) = 2e^{-(x+y)}, 0 < x < y < \infty$.

a. Find the marginal density of X.

b. Find the conditional density of Y given $X = x$.

c. Find the probability that Y is greater than 2, given that $X = 1$.

d. Are X and Y independent? Explain.

e. Find the conditional mean of Y given $X = x$. Is $E(Y|X = x)$ a linear function of x?

f. Find the conditional variance of Y given $X = x$.

40. A class has 10 mathematics majors, 6 computer science majors, and 4 statistics majors. A committee of two is selected at random to work on a problem. Let X be the number of mathematics majors and let Y be the number of computer science majors chosen.

a. Find the joint probability mass function $p(x, y)$. This generalizes the hypergeometric distribution studied in Section 3.6. Give the joint probability table showing all nine values, of which three should be 0.

b. Find the marginal probability mass functions by summing numerically. How could these be obtained directly? *Hint*: What are the univariate distributions of X and Y?

c. Find the conditional probability mass function of Y given $X = x$ for $x = 0, 1, 2$. Compare with the $h(y; 2 - x, 6, 10)$ distribution. Intuitively, why should this work?

d. Are X and Y independent? Explain.

e. Find $E(Y|X = x)$, $x = 0, 1, 2$. Do this numerically and then compare with the use of the formula for the hypergeometric mean, using the hypergeometric distribution given in part (c). Is $E(Y|X = x)$ a linear function of x?

f. Find $V(Y|X = x)$, $x = 0, 1, 2$. Do this numerically and then compare with the use of the formula for the hypergeometric variance, using the hypergeometric distribution given in part (c).

41. A stick is 1 foot long. You break it at a point X (measured from the left end) chosen randomly uniformly along its length. Then you break the left part at a point Y chosen randomly uniformly along its length. In other words, X is uniformly distributed between 0 and 1 and, given $X = x$, Y is uniformly distributed between 0 and x.

a. Determine $E(Y|X = x)$ and then $V(Y|X = x)$. Is $E(Y|X = x)$ a linear function of x?

b. Find $f(x, y)$ using $f_X(x)$ and $f_{Y|X}(y|x)$.

c. Find $f_Y(y)$.

42. This is a continuation of the previous exercise.

a. Use $f_Y(y)$ from Exercise 41(c) to get $E(Y)$ and $V(Y)$.

b. Use Exercise 41(a) and the theorem of this section to get $E(Y)$ and $V(Y)$.

43. Refer to Exercise 1 and answer the following questions:

a. Given that $X = 1$, determine the conditional pmf of Y—that is, $p_{Y|X}(0|1)$, $p_{Y|X}(1|1)$, and $p_{Y|X}(2|1)$.

b. Given that two hoses are in use at the self-service island, what is the conditional pmf of the number of hoses in use on the full-service island?

c. Use the result of part (b) to calculate the conditional probability $P(Y \le 1|X = 2)$.

d. Given that two hoses are in use at the full-service island, what is the conditional pmf of the number in use at the self-service island?

44. The joint pdf of pressures for right and left front tires is given in Exercise 9.

a. Determine the conditional pdf of Y given that $X = x$ and the conditional pdf of X given that $Y = y$.

b. If the pressure in the right tire is found to be 22 psi, what is the probability that the left tire has a pressure of at least 25 psi? Compare this to $P(Y \ge 25)$.

c. If the pressure in the right tire is found to be 22 psi, what is the expected pressure in the left tire, and what is the standard deviation of pressure in this tire?

45. Suppose that X is uniformly distributed between 0 and 1. Given $X = x$, Y is uniformly distributed between 0 and x^2.

a. Determine $E(Y|X = x)$ and then $V(Y|X = x)$. Is $E(Y|X = x)$ a linear function of x?

b. Find $f(x, y)$ using $f_X(x)$ and $f_{Y|X}(y|x)$.

c. Find $f_Y(y)$.

46. This is a continuation of the previous exercise.

a. Use $f_Y(y)$ from Exercise 45(c) to get $E(Y)$ and $V(Y)$.

b. Use Exercise 45(a) and the theorem of this section to get $E(Y)$ and $V(Y)$.

47. David and Peter independently choose at random a number from 1, 2, 3, with each possibility equally likely. Let X be the larger of the two numbers, and let Y be the smaller.

a. Find $p(x, y)$.
b. Find $p_X(x)$, $x = 1, 2, 3$.
c. Find $p_{Y|X}(y|x)$.
d. Find $E(Y|X = x)$. Is this a linear function of x?
e. Find $V(Y|X = x)$.

48. In Exercise 47 find

a. $E(X)$.
b. $p_Y(y)$.
c. $E(Y)$ using $p_Y(y)$.
d. $E(Y)$ using $E(Y|X)$.
e. $E(X) + E(Y)$. Why should this be 4, intuitively?

49. In Exercise 47 find

a. $p_{X|Y}(x|y)$.
b. $E(X|Y = y)$. Is this a linear function of y?
c. $V(X|Y = y)$.

50. For a Calculus I class, the final exam score Y and the average of the four earlier tests X are bivariate normal with mean $\mu_1 = 73$, standard deviation $\sigma_1 = 12$ and mean $\mu_2 = 70$, standard deviation $\sigma_2 = 15$. The correlation is $\rho = .71$. Find

a. $\mu_{Y|X=x}$
b. $\sigma^2_{Y|X=x}$
c. $\sigma_{Y|X=x}$
d. $P(Y > 90|X = 80)$, that is, the probability that the final exam score exceeds 90 given that the average of the four earlier tests is 80

51. Let X and Y, reaction times (sec) to two different stimuli, have a bivariate normal distribution with mean $\mu_1 = 20$ and standard deviation $\sigma_1 = 2$ for X and mean $\mu_2 = 30$ and standard deviation $\sigma_2 = 5$ for Y. Assume $\rho = .8$. Find

a. $\mu_{Y|X=x}$
b. $\sigma^2_{Y|X=x}$
c. $\sigma_{Y|X=x}$
d. $P(Y > 46|X = 25)$

52. Consider three ping pong balls numbered 1, 2, and 3. Two balls are randomly selected with replacement. If the sum of the two resulting numbers exceeds 4, two balls are again selected. This process continues until the sum is at most 4. Let X and Y denote the last two numbers selected. Possible (X, Y) pairs are $\{(1, 1), (1, 2), (1, 3), (2, 1), (2, 2), (3, 1)\}$.

a. Find $p_{X,Y}(x, y)$.
b. Find $p_{Y|X}(y|x)$.
c. Find $E(Y|X = x)$. Is this a linear function of x?
d. Find $E(X|Y = y)$. What special property of $p(x, y)$ allows us to get this from (c)?
e. Find $V(Y|X = x)$.

53. Let X be a random digit (0, 1, 2, . . . , 9 are equally likely) and let Y be a random digit not equal to X. That is, the nine digits other than X are equally likely for Y.

a. Find $p_X(x)$, $p_{Y|X}(y|x)$, $p_{X,Y}(x, y)$.
b. Find a formula for $E(Y|X = x)$. Is this a linear function of x?

54. In our discussion of the bivariate normal distribution, there is an expression for $E(Y|X = x)$.

a. By reversing the roles of X and Y give a similar formula for $E(X|Y = y)$.
b. Both $E(Y|X = x)$ and $E(X|Y = y)$ are linear functions. Show that the product of the two slopes is ρ^2.

55. This week the number X of claims coming into an insurance office has a Poisson distribution with mean 100. The probability that any particular claim relates to automobile insurance is .6, independent of any other claim. If Y is the number of automobile claims, then Y is binomial with X trials, each with "success" probability .6.

a. Find $E(Y|X = x)$ and $V(Y|X = x)$.
b. Use part (a) to find $E(Y)$.
c. Use part (a) to find $V(Y)$.

56. In Exercise 55 show that the distribution of Y is Poisson with mean 60. You will need to recognize the Maclaurin series expansion for the exponential function. Use the knowledge that Y is Poisson with mean 60 to find $E(Y)$ and $V(Y)$.

57. Let X and Y be the times for a randomly selected individual to complete two different tasks, and assume that (X, Y) has a bivariate normal distribution with $\mu_X = 100$, $\sigma_X = 50$, $\mu_Y = 25$, $\sigma_Y = 5$, $\rho = .5$. From statistical software we obtain $P(X < 100, Y < 25) = .3333$, $P(X < 50, Y < 20) = .0625$, $P(X < 50, Y < 25) = .1274$, and $P(X < 100, Y < 20) = .1274$.

a. Find $P(50 < X < 100, 20 < Y < 25)$.
b. Leave the other parameters the same but change the correlation to $\rho = 0$ (independence). Now recompute the answer to part (a). Intuitively, why should the answer to part (a) be larger?

5.4 *Transformations of Random Variables

In the previous chapter we discussed the problem of starting with a single random variable X, forming some function of X, such as X^2 or e^X, to obtain a new random variable $Y = h(X)$, and investigating the distribution of this new random variable. We now generalize this scenario by starting with more than a single random variable. Consider as an example a system having a component that can be replaced just once before the system itself expires. Let X_1 denote the lifetime of the original component and X_2 the lifetime of the replacement component. Then any of the following functions of X_1 and X_2 may be of interest to an investigator:

1. The total lifetime $X_1 + X_2$
2. The ratio of lifetimes X_1/X_2; for example, if the value of this ratio is 2, the original component lasted twice as long as its replacement
3. The ratio $X_1/(X_1 + X_2)$, which represents the proportion of system lifetime during which the original component operated

The Joint Distribution of Two New Random Variables

Given two random variables X_1 and X_2, consider forming two new random variables $Y_1 = u_1(X_1, X_2)$ and $Y_2 = u_2(X_1, X_2)$. We now focus on finding the joint distribution of these two new variables. Since most applications assume that the X_i's are continuous we restrict ourselves to that case. Some notation is needed before a general result can be given. Let

$$f(x_1, x_2) = \text{the joint pdf of the two original variables}$$

$$g(y_1, y_2) = \text{the joint pdf of the two new variables}$$

The $u_1(\cdot)$ and $u_2(\cdot)$ functions express the new variables in terms of the original ones. The general result presumes that these functions can be inverted to solve for the original variables in terms of the new ones:

$$X_1 = v_1(Y_1, Y_2) \qquad X_2 = v_2(Y_1, Y_2)$$

For example, if

$$y_1 = x_1 + x_2 \quad \text{and} \quad y_2 = \frac{x_1}{x_1 + x_2}$$

then multiplying y_2 by y_1 gives an expression for x_1, and then we can substitute this into the expression for y_1 and solve for x_2:

$$x_1 = y_1 y_2 = v_1(y_1, y_2) \qquad x_2 = y_1(1 - y_2) = v_2(y_1, y_2)$$

In a final burst of notation, let

$$S = \{(x_1, x_2): f(x_1, x_2) > 0\} \qquad T = \{(y_1, y_2): g(y_1, y_2) > 0\}$$

That is, S is the region of positive density for the original variables and T is the region of positive density for the new variables; T is the "image" of S under the transformation.

THEOREM

Suppose that the partial derivative of each $v_i(y_1, y_2)$ with respect to both y_1 and y_2 exists for every (y_1, y_2) pair in T and is continuous. Form the 2×2 matrix

$$M = \begin{pmatrix} \dfrac{\partial v_1(y_1, y_2)}{\partial y_1} & \dfrac{\partial v_1(y_1, y_2)}{\partial y_2} \\ \dfrac{\partial v_2(y_1, y_2)}{\partial y_1} & \dfrac{\partial v_2(y_1, y_2)}{\partial y_2} \end{pmatrix}$$

The determinant of this matrix, called the *Jacobian*, is

$$\det(M) = \frac{\partial v_1}{\partial y_1} \cdot \frac{\partial v_2}{\partial y_2} - \frac{\partial v_1}{\partial y_2} \cdot \frac{\partial v_2}{\partial y_1}$$

The joint pdf for the new variables then results from taking the joint pdf $f(x_1, x_2)$ for the original variables, replacing x_1 and x_2 by their expressions in terms of y_1 and y_2, and finally multiplying this by the absolute value of the Jacobian:

$$g(y_1, y_2) = f[v_1(y_1, y_2), v_2(y_1, y_2)] \cdot |\det(M)| \qquad (y_1, y_2) \in T$$

The theorem can be rewritten slightly by using the notation

$$|\det(M)| = \left| \frac{\partial(x_1, x_2)}{\partial(y_1, y_2)} \right|$$

Then we have

$$g(y_1, y_2) = f(x_1, x_2) \cdot \left| \frac{\partial(x_1, x_2)}{\partial(y_1, y_2)} \right|$$

which is the natural extension of the univariate result (transforming a single rv X to obtain a single new rv Y) $g(y) = f(x) \cdot |dx/dy|$ discussed in Chapter 4.

Example 5.25 Continuing with the component lifetime situation, suppose that X_1 and X_2 are independent, each having an exponential distribution with parameter λ. Let's determine the joint pdf of

$$Y_1 = u_1(X_1, X_2) = X_1 + X_2 \quad \text{and} \quad Y_2 = u_2(X_1, X_2) = \frac{X_1}{X_1 + X_2}$$

We have already inverted this transformation:

$$x_1 = v_1(y_1, y_2) = y_1 y_2 \qquad x_2 = v_2(y_1, y_2) = y_1(1 - y_2)$$

The image of the transformation, that is, the set of (y_1, y_2) pairs with positive density, is $0 < y_1$ and $0 < y_2 < 1$. The four relevant partial derivatives are

$$\frac{\partial v_1}{\partial y_1} = y_2 \qquad \frac{\partial v_1}{\partial y_2} = y_1 \qquad \frac{\partial v_2}{\partial y_1} = 1 - y_2 \qquad \frac{\partial v_2}{\partial y_2} = -y_1$$

from which the Jacobian is $-y_1 y_2 - y_1(1 - y_2) = -y_1$.

Since the joint pdf of X_1 and X_2 is

$$f(x_1, x_2) = \lambda e^{-\lambda x_1} \cdot \lambda e^{-\lambda x_2} = \lambda^2 e^{-\lambda(x_1+x_2)} \qquad x_1 > 0,\ x_2 > 0$$

we have

$$g(y_1, y_2) = \lambda^2 e^{-\lambda y_1} \cdot y_1 = \lambda^2 y_1 e^{-\lambda y_1} \cdot 1 \qquad 0 < y_1, 0 < y_2 < 1$$

The joint pdf thus factors into two parts. The first part is a gamma pdf with parameters $\alpha = 2$ and $\beta = 1/\lambda$, and the second part is a uniform pdf on (0, 1). Since the pdf factors and the region of positive density is rectangular, we have demonstrated that

1. The distribution of system lifetime $X_1 + X_2$ is gamma with $\alpha = 2$, $\beta = 1/\lambda$.
2. The distribution of the proportion of system lifetime during which the original component functions is uniform on (0, 1).
3. $Y_1 = X_1 + X_2$ and $Y_2 = X_1/(X_1 + X_2)$ are independent of one another. ■

In the foregoing example, because the joint pdf factored into one pdf involving y_1 alone and another pdf involving y_2 alone, the individual (i.e., marginal) pdf's of the two new variables were obtained from the joint pdf without any further effort. Often this will not be the case — that is, Y_1 and Y_2 will not be independent. Then to obtain the marginal pdf of Y_1, the joint pdf must be integrated over all values of the second variable. In fact, in many applications an investigator wishes to obtain the distribution of a single function $u_1(X_1, X_2)$ of the original variables. To accomplish this, a second function $u_2(X_1, X_2)$ is selected, the joint pdf is obtained, and then y_2 is integrated out. There are of course many ways to select the second function. The choice should be made so that the transformation can be easily inverted *and* the integration in the last step is straightforward.

Example 5.26 Consider a rectangular coordinate system with a horizontal x_1 axis and a vertical x_2 axis as shown in Figure 5.7(a). First a point (X_1, X_2) is randomly selected, where the joint pdf of X_1, X_2 is

$$f(x_1, x_2) = \begin{cases} x_1 + x_2 & 0 < x_1 < 1, 0 < x_2 < 1 \\ 0 & \text{otherwise} \end{cases}$$

Then a rectangle with vertices (0, 0), $(X_1, 0)$, $(0, X_2)$, and (X_1, X_2) is formed. What is the distribution of X_1X_2, the area of this rectangle? To answer this question, let

$$Y_1 = X_1X_2 \qquad Y_2 = X_2$$

so

$$y_1 = u_1(x_1, x_2) = x_1x_2 \qquad y_2 = u_2(x_1, x_2) = x_2$$

Then

$$x_1 = v_1(y_1, y_2) = \frac{y_1}{y_2} \qquad x_2 = v_2(y_1, y_2) = y_2$$

Notice that because $x_2\ (= y_2)$ is between 0 and 1 and y_1 is the product of the two x_i's, it must be the case that $0 < y_1 < y_2$. The region of positive density for the new variables is then

$$T = \{(y_1, y_2)\colon 0 < y_1 < y_2, 0 < y_2 < 1\}$$

which is the triangular region shown in Figure 5.7(b).

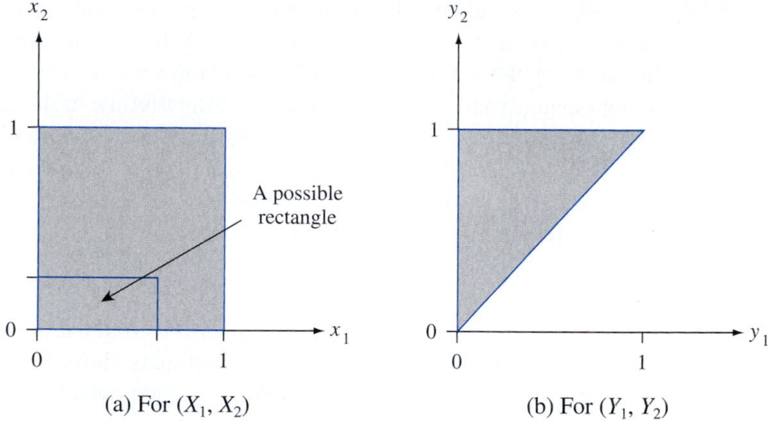

(a) For (X_1, X_2) (b) For (Y_1, Y_2)

Figure 5.7 Regions of positive density for Example 5.26

Since $\partial v_2/\partial y_1 = 0$, the product of the two off-diagonal elements in the matrix M will be 0, so only the two diagonal elements contribute to the Jacobian:

$$M = \begin{pmatrix} \frac{1}{y_2} & ? \\ 0 & 1 \end{pmatrix} \qquad |\det(M)| = \frac{1}{y_2}$$

The joint pdf of the two new variables is now

$$g(y_1, y_2) = f\left(\frac{y_1}{y_2}, y_2\right) \cdot |\det(M)| = \begin{cases} \frac{y_1}{y_2} + y_2 \cdot \left(\frac{1}{y_2}\right) & 0 < y_1 < y_2, 0 < y_2 < 1 \\ 0 & \text{otherwise} \end{cases}$$

To obtain the marginal pdf of Y_1 alone, we now fix y_1 at some arbitrary value between 0 and 1, and integrate out y_2. Figure 5.7(b) shows that we must integrate along the vertical line segment passing through y_1 whose lower limit is y_1 and whose upper limit is 1:

$$g_1(y_1) = \int_{y_1}^{1} \left(\frac{y_1}{y_2} + y_2\right) \cdot \frac{1}{y_2}\, dy_2 = 2(1 - y_1) \qquad 0 < y_1 < 1$$

This marginal pdf can now be integrated to obtain any desired probability involving the area. For example, integrating from 0 to .5 gives $P(\text{area} < .5) = .75$. ■

The Joint Distribution of More Than Two New Variables

Consider now starting with three random variables X_1, X_2, and X_3, and forming three new variables Y_1, Y_2, and Y_3. Suppose again that the transformation can be inverted to express the original variables in terms of the new ones:

$$x_1 = v_1(y_1, y_2, y_3) \qquad x_2 = v_2(y_1, y_2, y_3) \qquad x_3 = v_3(y_1, y_2, y_3)$$

Then the foregoing proposition can be extended to this new situation. The Jacobian matrix has dimension 3×3, with the entry in the ith row and jth column being $\partial v_i/\partial y_j$. The joint pdf of the new variables results from replacing each x_i in the original pdf $f(\cdot)$ by its expression in terms of the y_j's and multiplying by the absolute value of the Jacobian.

Example 5.27 Consider $n = 3$ identical components with independent lifetimes X_1, X_2, X_3, each having an exponential distribution with parameter λ. If the first component is used until it fails, replaced by the second one which remains in service until it fails, and finally the third component is used until failure, then the total lifetime of these components is $Y_3 = X_1 + X_2 + X_3$. To find the distribution of total lifetime, let's first define two other new variables: $Y_1 = X_1$ and $Y_2 = X_1 + X_2$ (so that $Y_1 < Y_2 < Y_3$). After finding the joint pdf of all three variables, we integrate out the first two variables to obtain the desired information. Solving for the old variables in terms of the new gives

$$x_1 = y_1 \qquad x_2 = y_2 - y_1 \qquad x_3 = y_3 - y_2$$

It is obvious by inspection of these expressions that the three diagonal elements of the Jacobian matrix are all 1's and that the elements above the diagonal are all 0's, so the determinant is 1, the product of the diagonal elements. Since

$$f(x_1, x_2, x_3) = \lambda^3 e^{-\lambda(x_1+x_2+x_3)} \qquad x_1 > 0, x_2 > 0, x_3 > 0$$

by substitution,

$$g(y_1, y_2, y_3) = \lambda^3 e^{-\lambda y_3} \qquad 0 < y_1 < y_2 < y_3$$

Integrating this joint pdf first with respect to y_1 between 0 and y_2 and then with respect to y_2 between 0 and y_3 (try it!) gives

$$g_3(y_3) = \frac{\lambda^3}{2} y_3^2 e^{-\lambda y_3} \qquad y_3 > 0$$

This is a gamma pdf. The result is easily extended to n components. It can also be obtained (more easily) by using a moment generating function argument. ■

Exercises Section 5.4 (58–64)

58. Consider two components whose lifetimes X_1 and X_2 are independent and exponentially distributed with parameters λ_1 and λ_2, respectively. Obtain the joint pdf of total lifetime $X_1 + X_2$ and the proportion of total lifetime $X_1/(X_1 + X_2)$ during which the first component operates.

59. Let X_1 denote the time (hr) it takes to perform a first task and X_2 denote the time it takes to perform a second one. The second task always takes at least as long to perform as the first task. The joint pdf of these variables is

$$f(x_1, x_2) = \begin{cases} 2(x_1 + x_2) & 0 \le x_1 \le x_2 \le 1 \\ 0 & \text{otherwise} \end{cases}$$

a. Obtain the pdf of the total completion time for the two tasks.

b. Obtain the pdf of the difference $X_2 - X_1$ between the longer completion time and the shorter time.

60. An exam consists of a problem section and a short-answer section. Let X_1 denote the amount of time (hr) that a student spends on the problem section and X_2 represent the amount of time the same student spends on the short-answer section. Suppose the joint pdf of these two times is

$$f(x_1, x_2) = \begin{cases} cx_1x_2 & \frac{x_1}{3} < x_2 < \frac{x_1}{2}, 0 < x_1 < 1 \\ 0 & \text{otherwise} \end{cases}$$

a. What is the value of c?

b. If the student spends exactly .25 hr on the short-answer section, what is the probability that at most .60 hr was spent on the problem section? *Hint*: First obtain the relevant conditional distribution.

c. What is the probability that the amount of time spent on the problem part of the exam exceeds the amount of time spent on the short-answer part by at least .5 hr?

d. Obtain the joint distribution of $Y_1 = X_2/X_1$, the ratio of the two times, and $Y_2 = X_2$. Then obtain the marginal distribution of the ratio.

61. Consider randomly selecting a point (X_1, X_2, X_3) in the unit cube $\{(x_1, x_2, x_3): 0 < x_1 < 1, 0 < x_2 < 1, 0 < x_3 < 1\}$ according to the joint pdf

$$f(x_1, x_2, x_3) = \begin{cases} 8x_1x_2x_3 & 0 < x_1 < 1, 0 < x_2 < 1, 0 < x_3 < 1 \\ 0 & \text{otherwise} \end{cases}$$

(so the three variables are independent). Then form a rectangular solid whose vertices are (0, 0, 0), $(X_1, 0, 0)$, $(0, X_2, 0)$, $(X_1, X_2, 0)$, $(0, 0, X_3)$, $(X_1, 0, X_3)$, $(0, X_2, X_3)$, and (X_1, X_2, X_3). The volume of this cube is $Y_3 = X_1X_2X_3$. Obtain the pdf of this volume. *Hint*: Let $Y_1 = X_1$ and $Y_2 = X_1X_2$.

62. Let X_1 and X_2 be independent, each having a standard normal distribution. The pair (X_1, X_2) corresponds to a point in a two-dimensional coordinate system. Consider now changing to polar coordinates via the transformation

$$Y_1 = X_1^2 + X_2^2$$

$$Y_2 = \begin{cases} \arctan\left(\dfrac{X_2}{X_1}\right) & X_1 > 0, X_2 \ge 0 \\ \arctan\left(\dfrac{X_2}{X_1}\right) + 2\pi & X_1 > 0, X_2 < 0 \\ \arctan\left(\dfrac{X_2}{X_1}\right) + \pi & X_1 < 0 \\ 0 & X_1 = 0 \end{cases}$$

from which $x_1 = \sqrt{y_1}\cos(y_2)$, $x_2 = \sqrt{y_1}\sin(y_2)$. Obtain the joint pdf of the new variables and then the marginal distribution of each one. *Note*: It would be nice if we could simply let $Y_2 = \arctan(X_2/X_1)$, but in order to ensure invertibility of the arctan function, it is defined to take on values only between $-\pi/2$ and $\pi/2$. Our specification of Y_2 allows it to assume any value between 0 and 2π.

63. The result of the previous exercise suggests how observed values of two independent standard normal variables can be generated by first generating their polar coordinates with an exponential rv with $\lambda = \frac{1}{2}$ and an independent uniform $(0, 2\pi)$ rv: Let U_1 and U_2 be independent uniform (0, 1) rv's, and then let

$$Y_1 = -2\ln(U_1) \qquad Y_2 = 2\pi U_2$$
$$Z_1 = \sqrt{Y_1}\cos(Y_2) \qquad Z_2 = \sqrt{Y_1}\sin(Y_2)$$

Show that the Z_i's are independent standard normal. *Note*: This is called the *Box-Muller transformation* after the two individuals who discovered it. Now that statistical software packages will generate almost instantaneously observations from a normal distribution with any mean and variance, it is thankfully no longer necessary for us to carry out the transformations just described—let the software do it!

64. Let X_1 and X_2 be independent random variables, each having a standard normal distribution. Show that the pdf of the ratio $Y = X_1/X_2$ is given by $f(y) = 1/[\pi(1 + y^2)]$ for $-\infty < y < \infty$ (this is called the *standard Cauchy distribution*).

5.5 *Order Statistics

Many statistical procedures involve ordering the sample observations from smallest to largest and then manipulating these ordered values in various ways. For example, the sample median is either the middle value in the ordered list or the average of the two middle values depending on whether the sample size n is odd or even. The sample range is the difference between the largest and smallest values. And a trimmed mean results from deleting the same number of observations from each end of the ordered list and averaging the remaining values.

Suppose that $X_1, X_2, \ldots, X_n$ is a random sample from a continuous distribution with cumulative distribution function $F(x)$ and density function $f(x)$. Because of continuity, for any i, j with $i \neq j$, $P(X_i = X_j) = 0$. This implies that with probability 1, the

n sample observations will all be different (of course, in practice all measuring instruments have accuracy limitations, so tied values may in fact result).

DEFINITION The **order statistics** from a random sample are the random variables $Y_1, \ldots, Y_n$ given by

$$\begin{aligned} Y_1 &= \text{the smallest among } X_1, X_2, \ldots, X_n \\ Y_2 &= \text{the second smallest among } X_1, X_2, \ldots, X_n \\ &\vdots \\ Y_n &= \text{the largest among } X_1, X_2, \ldots, X_n \end{aligned}$$

so that with probability 1, $Y_1 < Y_2 < \cdots < Y_{n-1} < Y_n$.

The sample median is then $Y_{(n+1)/2}$ when n is odd, the sample range is $Y_n - Y_1$, and for $n = 10$ the 20% trimmed mean is $\sum_{i=3}^{8} Y_i/6$. The order statistics are defined as random variables (hence the use of uppercase letters); observed values are denoted by $y_1, \ldots, y_n$.

The Distributions of Y_n and Y_1

The key idea in obtaining the distribution of the largest order statistic is that Y_n is at most y if and only if every one of the X_i's is at most y. Similarly, the distribution of Y_1 is based on the fact that it will be at least y if and only if all X_i's are at least y.

Example 5.28 Consider five identical components connected in parallel, as illustrated in Figure 5.8(a). Let X_i denote the lifetime (hr) of the ith component ($i = 1, 2, 3, 4, 5$). Suppose that the X_i's are independent and that each has an exponential distribution with $\lambda = .01$, so the expected lifetime of any particular component is $1/\lambda = 100$ hr. Because of the parallel configuration, the system will continue to function as long as at least one component is still working, and will fail as soon as the last component functioning ceases to do so. That is, the system lifetime is just Y_5, the largest order statistic in a sample of size 5 from the specified exponential distribution. Now Y_5 will be at most y if and only if every one of the five X_i's is at most y. With $G_5(y)$ denoting the cumulative distribution function of Y_5,

$$\begin{aligned} G_5(y) &= P(Y_5 \le y) = P(X_1 \le y, X_2 \le y, \ldots, X_5 \le y) \\ &= P(X_1 \le y)\, P(X_2 \le y) \cdot \cdots \cdot P(X_5 \le y) = [F(y)]^5 = [1 - e^{-.01y}]^5 \end{aligned}$$

The pdf of Y_5 can now be obtained by differentiating the cdf with respect to y.

Suppose instead that the five components are connected in series rather than in parallel [Figure 5.8(b)]. In this case the system lifetime will be Y_1, the *smallest* of the five order statistics, since the system will crash as soon as a single one of the individual components fails. Note that system lifetime will exceed y hr if and only if the lifetime of every component exceeds y hr. Thus

$$\begin{aligned} G_1(y) &= P(Y_1 \le y) = 1 - P(Y_1 > y) = 1 - P(X_1 > y, X_2 > y, \ldots, X_5 > y) \\ &= 1 - P(X_1 > y) \cdot P(X_2 > y) \cdot \cdots \cdot P(X_5 > y) = 1 - [e^{-.01y}]^5 = 1 - e^{-.05y} \end{aligned}$$

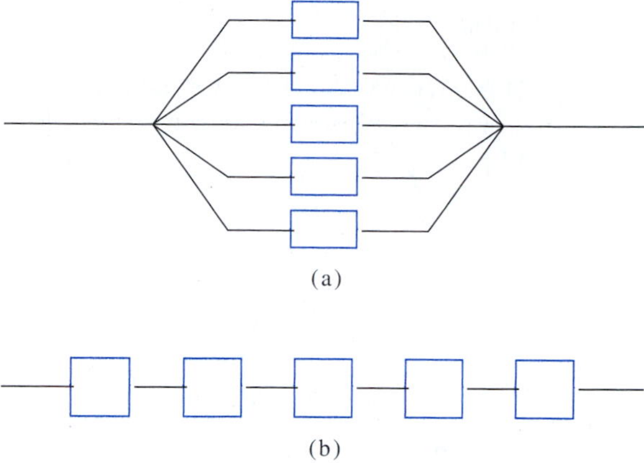

Figure 5.8 Systems of components for Example 5.28: (a) parallel connection; (b) series connection

This is the form of an exponential cdf with parameter .05. More generally, if the n components in a series connection have lifetimes which are independent, each exponentially distributed with the same parameter λ, then system lifetime will be exponentially distributed with parameter $n\lambda$. The expected system lifetime will then be $1/n\lambda$, much smaller than the expected lifetime of an individual component. ■

An argument parallel to that of the previous example for a general sample size n and an arbitrary pdf $f(x)$ gives the following general results.

PROPOSITION Let Y_1 and Y_n denote the smallest and largest order statistics, respectively, based on a random sample from a continuous distribution with cdf $F(x)$ and pdf $f(x)$. Then the cdf and pdf of Y_n are

$$G_n(y) = [F(y)]^n \qquad g_n(y) = n\,[F(y)]^{n-1}\cdot f(y)$$

The cdf and pdf of Y_1 are

$$G_1(y) = 1 - [1 - F(y)]^n \qquad g_1(y) = n\,[1 - F(y)]^{n-1}\cdot f(y)$$

Example 5.29 Let X denote the contents of a 1-gallon container of a particular type, and suppose that its pdf $f(x) = 2x$ for $0 \le x \le 1$ (and 0 otherwise) with corresponding cdf $F(x) = x^2$ in the interval of positive density. Consider a random sample of four such containers. Let's determine the expected value of $Y_4 - Y_1$, the difference between the contents of the most-filled container and the least-filled container; $Y_4 - Y_1$ is just the sample range. The pdf's of Y_4 and Y_1 are

$$g_4(y) = 4(y^2)^3\cdot 2y \qquad 0 \le y \le 1$$
$$g_1(y) = 4(1 - y^2)^3\cdot 2y \qquad 0 \le y \le 1$$

The corresponding density curves appear in Figure 5.9.

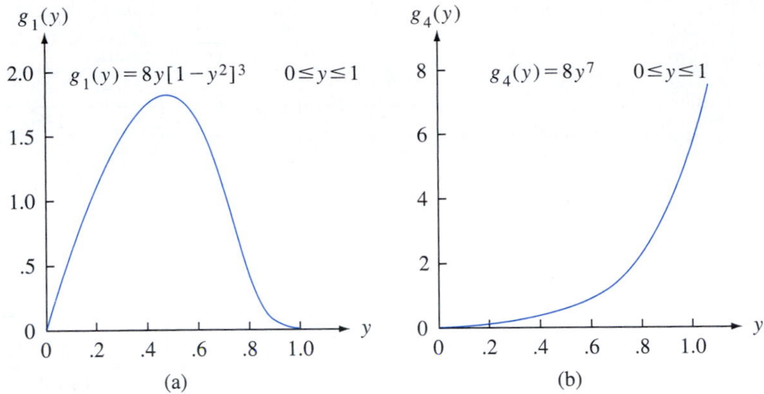

Figure 5.9 Density curves for the order statistics (a) Y_1 and (b) Y_4 in Example 5.29

$$E(Y_4 - Y_1) = E(Y_4) - E(Y_1) = \int_0^1 y \cdot 8y^7\,dy - \int_0^1 y \cdot 8y(1 - y^2)^3\,dy$$

$$= \frac{8}{9} - \frac{384}{945} = .889 - .406 = .483$$

If random samples of four containers were repeatedly selected and the sample range of contents determined for each one, the long-run average value of the range would be .483. ■

The Joint Distribution of the n Order Statistics

We now develop the joint pdf of $Y_1, Y_2, \ldots, Y_n$. Consider first a random sample X_1, X_2, X_3 of fuel efficiency measurements (mpg). The joint pdf of this random sample is

$$f(x_1, x_2, x_3) = f(x_1) \cdot f(x_2) \cdot f(x_3)$$

The joint pdf of Y_1, Y_2, Y_3 will be positive only for values of y_1, y_2, y_3 satisfying $y_1 < y_2 < y_3$. What is this joint pdf at the values $y_1 = 28.4$, $y_2 = 29.0$, $y_3 = 30.5$? There are six different ways to obtain these ordered values:

$$\begin{array}{lll}
X_1 = 28.4, & X_2 = 29.0, & X_3 = 30.5 \\
X_1 = 28.4, & X_2 = 30.5, & X_3 = 29.0 \\
X_1 = 29.0, & X_2 = 28.4, & X_3 = 30.5 \\
X_1 = 29.0, & X_2 = 30.5, & X_3 = 28.4 \\
X_1 = 30.5, & X_2 = 28.4, & X_3 = 29.0 \\
X_1 = 30.5, & X_2 = 29.0, & X_3 = 28.4
\end{array}$$

These six possibilities come from the 3! ways to order the three numerical observations once their values are fixed. Thus

$$g(28.4, 29.0, 30.5) = f(28.4) \cdot f(29.0) \cdot f(30.5) + \cdots + f(30.5) \cdot f(29.0) \cdot f(28.4)$$
$$= 3!f(28.4) \cdot f(29.0) \cdot f(30.5)$$

Analogous reasoning with a sample of size n yields the following result:

PROPOSITION Let $g(y_1, y_2, \ldots, y_n)$ denote the joint pdf of the order statistics $Y_1, Y_2, \ldots, Y_n$ resulting from a random sample of X_i's from a pdf $f(x)$. Then

$$g(y_1, y_2, \ldots, y_n) = \begin{cases} n!f(y_1)\cdot f(y_2)\cdot \cdots \cdot f(y_n) & y_1 < y_2 < \cdots < y_n \\ 0 & \text{otherwise} \end{cases}$$

For example, if we have a random sample of component lifetimes and the lifetime distribution is exponential with parameter λ, then the joint pdf of the order statistics is

$$g(y_1, \ldots, y_n) = n!\lambda^n e^{-\lambda(y_1 + \cdots + y_n)} \qquad 0 < y_1 < \cdots < y_n$$

The Distribution of a Single Order Statistic

We have already obtained the (marginal) distribution of the largest order statistic Y_n and also that of the smallest order statistic Y_1. Let's now focus on an intermediate order statistic Y_i where $1 < i < n$. For concreteness, consider a random sample $X_1, X_2, \ldots, X_6$ of $n = 6$ component lifetimes, and suppose we wish the distribution of the third smallest lifetime Y_3. Now the joint pdf of all six order statistics is

$$g(y_1, y_2, \ldots, y_6) = 6!f(y_1)\cdot \cdots \cdot f(y_6) \qquad y_1 < y_2 < y_3 < y_4 < y_5 < y_6$$

To obtain the pdf of Y_3 alone, we must hold y_3 fixed in the joint pdf and integrate out all the other y_i's. One way to do this is to

1. Integrate y_1 from $-\infty$ to y_2, and then integrate y_2 from $-\infty$ to y_3.
2. Integrate y_6 from y_5 to ∞, then integrate y_5 from y_4 to ∞, and finally integrate y_4 from y_3 to ∞.

That is,

$$\begin{aligned} g(y_3) &= \int_{y_3}^{\infty}\int_{y_4}^{\infty}\int_{y_5}^{\infty}\int_{-\infty}^{y_3}\int_{-\infty}^{y_2} 6!f(y_1)\cdot f(y_2)\cdot \cdots \cdot f(y_6)\,dy_1\,dy_2\,dy_6\,dy_5\,dy_4 \\ &= 6!\left[\int_{-\infty}^{y_3}\int_{-\infty}^{y_2} f(y_1)f(y_2)\,dy_1\,dy_2\right]\cdot\left[\int_{y_3}^{\infty}\int_{y_4}^{\infty}\int_{y_5}^{\infty} f(y_4)f(y_5)f(y_6)\,dy_6\,dy_5\,dy_4\right]\cdot f(y_3) \end{aligned}$$

In these integrations we use the following general results:

$$\int [F(x)]^k f(x)\,dx = \frac{1}{k+1}[F(x)]^{k+1} + c \qquad [\text{let } u = F(x)]$$

$$\int [1 - F(x)]^k f(x)\,dx = -\frac{1}{k+1}[1 - F(x)]^{k+1} + c \qquad [\text{let } u = 1 - F(x)]$$

Therefore

$$\int_{-\infty}^{y_3}\int_{-\infty}^{y_2} f(y_1)f(y_2)\,dy_1\,dy_2 = \int_{-\infty}^{y_3} F(y_2)f(y_2)\,dy_2 = \frac{1}{2}[F(y_3)]^2$$

and

$$\int_{y_3}^{\infty}\int_{y_4}^{\infty}\int_{y_5}^{\infty} f(y_6)f(y_5)f(y_4)\,dy_6\,dy_5\,dy_4 = \int_{y_3}^{\infty}\int_{y_4}^{\infty} [1 - F(y_5)]f(y_5)f(y_4)\,dy_5\,dy_4$$

$$= -\int_{y_3}^{\infty}\frac{1}{2}[1 - F(y_4)]^2 f(y_4)\,dy_4$$

$$= \frac{1}{3 \cdot 2}[1 - F(y_3)]^3$$

Thus

$$g(y_3) = \frac{6!}{2!3!}[F(y_3)]^2[1 - F(y_3)]^3 f(y_3) \qquad -\infty < y_3 < \infty$$

A generalization of the foregoing argument gives the following expression for the pdf of any single order statistic.

PROPOSITION The pdf of the ith smallest order statistic Y_i is

$$g(y_i) = \frac{n!}{(i - 1)! \cdot (n - i)!}[F(y_i)]^{i-1}[1 - F(y_i)]^{n-i}f(y_i) \qquad -\infty < y_i < \infty$$

Example 5.30 Suppose that component lifetime is exponentially distributed with parameter λ. For a random sample of $n = 5$ components, the expected value of the sample median lifetime is

$$E(Y_3) = \int_0^{\infty} y \cdot \frac{5!}{2! \cdot 2!}(1 - e^{-\lambda y})^2(e^{-\lambda y})^2 \cdot \lambda e^{-\lambda y}\,dy$$

Expanding out the integrand and integrating term by term, the expected value is $.783/\lambda$. The median of the exponential distribution is, from solving $F(\tilde{\mu}) = .5$, $\tilde{\mu} = .693/\lambda$. Thus if sample after sample of five components is selected, the long-run average value of the sample median will be somewhat larger than the value of the lifetime population distribution median. This is because the exponential distribution has a positive skew. ■

The Joint Distribution of Two Order Statistics

We now focus on the joint distribution of two order statistics Y_i and Y_j with $i < j$. Consider first $n = 6$ and the two order statistics Y_3 and Y_5. We must then take the joint pdf of all six order statistics, hold y_3 and y_5 fixed, and integrate out y_1, y_2, y_4, and y_6. That is,

$$g_{3,5}(y_3, y_5) = \int_{y_5}^{\infty}\int_{y_3}^{y_5}\int_{-\infty}^{y_3}\int_{y_1}^{y_3} 6!f(y_1) \cdot \cdots \cdot f(y_6)\,dy_2\,dy_1\,dy_4\,dy_6$$

The result of this integration is

$$g_{3,5}(y_3, y_5) = \frac{6!}{2!1!1!}[F(y_3)]^2[F(y_5) - F(y_3)]^1[1 - F(y_5)]^1 f(y_3)f(y_5)$$

$$-\infty < y_3 < y_5 < \infty$$

In the general case, the numerator in the leading expression involving factorials becomes $n!$ and the denominator becomes $(i-1)!(j-i-1)!(n-j)!$. The three exponents on bracketed terms change in a corresponding way.

An Intuitive Derivation of Order Statistic PDF's

Let Δ be a number quite close to 0, and consider the three class intervals $[-\infty, y]$, $[y, y+\Delta]$, and $[y+\Delta, \infty]$. For a single X, the probabilities of these three classes are

$$p_1 = F(y) \qquad p_2 = \int_y^{y+\Delta} f(x)\,dx \approx f(y)\cdot\Delta \qquad p_3 = 1 - F(y+\Delta)$$

For a random sample of size n, it is very unlikely that two or more X's will fall in the second interval. The probability that the ith order statistic falls in the second interval is then approximately the probability that $i-1$ of the X's are in the first interval, one is in the second, and the remaining $n-i$ X's are in the third class. This is just a multinomial probability:

$$P(y < Y_i \le y + \Delta) \approx \frac{n!}{(i-1)!1!(n-i)!}\cdot[F(y)]^{i-1}\cdot f(y)\cdot\Delta[1 - F(y+\Delta)]^{n-i}$$

Dividing both sides by Δ and taking the limit as $\Delta \to 0$ gives exactly the pdf of Y_i obtained earlier via integration.

Similar reasoning works with the joint pdf of Y_i and Y_j ($i < j$). In this case there are five relevant class intervals: $(-\infty, y_i]$, $(y_i, y_i + \Delta_1]$, $(y_i + \Delta_1, y_j]$, $(y_j, y_j + \Delta_2]$, and $(y_j + \Delta_2, \infty)$.

Exercises Section 5.5 (65–77)

65. A friend of ours takes the bus five days per week to her job. The five waiting times until she can board the bus are a random sample from a uniform distribution on the interval from 0 to 10 min.

a. Determine the pdf and then the expected value of the largest of the five waiting times.

b. Determine the expected value of the difference between the largest and smallest times.

c. What is the expected value of the sample median waiting time?

d. What is the standard deviation of the largest time?

66. Refer back to Example 5.29. Because $n = 4$, the sample median is $(Y_2 + Y_3)/2$. What is the expected value of the sample median, and how does it compare to the median of the population distribution?

67. Referring back to Exercise 65, suppose you learn that the smallest of the five waiting times is 4 min. What is the conditional density function of the largest waiting time, and what is the expected value of the largest waiting time in light of this information?

68. Let X represent a measurement error of some sort. It is natural to assume that the pdf $f(x)$ is symmetric about 0, so that the density at a value $-c$ is the same as the density at c (an error of a given magnitude is equally likely to be positive or negative). Consider a random sample of n measurements, where $n = 2k + 1$, so that Y_{k+1} is the sample median. What can be said about $E(Y_{k+1})$? If the X distribution is symmetric about some other value, so that value is the median of the distribution, what does this imply about $E(Y_{k+1})$? *Hints*: For the first question, symmetry implies that $1 - F(x) = P(X > x) = P(X < -x) = F(-x)$. For the second question, consider $W = X - \tilde{\mu}$; what is the median of the distribution of W?

69. A store is expecting n deliveries between the hours of noon and 1 p.m. Suppose the arrival time of each delivery truck is uniformly distributed on this 1-hour interval and that the times are independent of one another. What are the expected values of the ordered arrival times?

70. Suppose the cdf $F(x)$ is strictly increasing and let $F^{-1}(y)$ denote the inverse function for $0 < y < 1$. Show that the distribution of $F[Y_i]$ is the same as the distribution of the ith smallest order statistic from a uniform distribution on (0, 1). *Hint*: Start with $P[F(Y_i) \le y]$ and apply the inverse function to both sides of the inequality. *Note*: This result should not be surprising to you, since we have already noted that $F(X)$ has a uniform distribution on (0, 1). The result also holds when the cdf is not strictly increasing, but then extra care is necessary in defining the inverse function.

71. Let X be the amount of time an ATM is in use during a particular 1-hour period, and suppose that X has the cdf $F(x) = x^\theta$ for $0 < x < 1$ (where $\theta > 1$). Give expressions involving the gamma function for both the mean and variance of the ith smallest amount of time Y_i from a random sample of n such time periods.

72. The logistic pdf $f(x) = e^{-x}/(1 + e^{-x})^2$ for $-\infty < x < \infty$ is sometimes used to describe the distribution of measurement errors.

a. Graph the pdf. Does the appearance of the graph surprise you?

b. For a random sample of size n, obtain an expression involving the gamma function of the moment generating function of the ith smallest order statistic Y_i. This expression can then be differentiated to obtain moments of the order statistics. *Hint*: Set up the appropriate integral, and then let $u = 1/(1 + e^{-x})$.

73. Consider a random sample of 10 waiting times from a uniform distribution on the interval from 0 to 5 min and determine the joint pdf of the third smallest and third largest times.

74. Conjecture the form of the joint pdf of three order statistics Y_i, Y_j, Y_k in a random sample of size n.

75. Use the intuitive argument sketched in this section to obtain a general formula for the joint pdf of two order statistics.

76. Consider a sample of size $n = 3$ from the standard normal distribution, and obtain the expected value of the largest order statistic. What does this say about the expected value of the largest order statistic in a sample of this size from *any* normal distribution? *Hint*: With $\phi(x)$ denoting the standard normal pdf, use the fact that $(d/dx)\phi(x) = -x\phi(x)$ along with integration by parts.

77. Let Y_1 and Y_n be the smallest and largest order statistics, respectively, from a random sample of size n, and let $W_2 = Y_n - Y_1$ (this is the sample range).

a. Let $W_1 = Y_1$, obtain the joint pdf of the W_i's (use the method of Section 5.4), and then derive an expression involving an integral for the pdf of the sample range.

b. For the case in which the random sample is from a uniform (0, 1) distribution, carry out the integration of (a) to obtain an explicit formula for the pdf of the sample range.

Supplementary Exercises (78–91)

78. A restaurant serves three fixed-price dinners costing \$12, \$15, and \$20. For a randomly selected couple dining at this restaurant, let X = the cost of the man's dinner and Y = the cost of the woman's dinner. The joint pmf of X and Y is given in the following table:

$p(x, y)$		y = 12	15	20
x	12	.05	.05	.10
	15	.05	.10	.35
	20	0	.20	.10

a. Compute the marginal pmf's of X and Y.
b. What is the probability that the man's and the woman's dinner cost at most \$15 each?
c. Are X and Y independent? Justify your answer.
d. What is the expected total cost of the dinner for the two people?
e. Suppose that when a couple opens fortune cookies at the conclusion of the meal, they find the message "You will receive as a refund the difference between the cost of the more expensive and the less expensive meal that you have chosen." How much does the restaurant expect to refund?

79. A health-food store stocks two different brands of a certain type of grain. Let X = the amount (lb) of brand A on hand and Y = the amount of brand B on hand. Suppose the joint pdf of X and Y is

$$f(x, y) = \begin{cases} kxy & x \geq 0, y \geq 0, 20 \leq x + y \leq 30 \\ 0 & \text{otherwise} \end{cases}$$

a. Draw the region of positive density and determine the value of k.
b. Are X and Y independent? Answer by first deriving the marginal pdf of each variable.
c. Compute $P(X + Y \leq 25)$.
d. What is the expected total amount of this grain on hand?
e. Compute $\text{Cov}(X, Y)$ and $\text{Corr}(X, Y)$.
f. What is the variance of the total amount of grain on hand?

80. Let $X_1, X_2, \ldots, X_n$ be random variables denoting n independent bids for an item that is for sale. Suppose each X_i is uniformly distributed on the interval [100, 200]. If the seller sells to the highest bidder, how much can he expect to earn on the sale? [*Hint*: Let $Y = \max(X_1, X_2, \ldots, X_n)$. Find $F_Y(y)$ by using the results of Section 5.5 or else by noting that $Y \leq y$ iff each X_i is $\leq y$. Then obtain the pdf and $E(Y)$.]

81. Suppose a randomly chosen individual's verbal score X and quantitative score Y on a nationally administered aptitude examination have joint pdf

$$f(x, y) = \begin{cases} \frac{2}{5}(2x + 3y) & 0 \leq x \leq 1, 0 \leq y \leq 1 \\ 0 & \text{otherwise} \end{cases}$$

You are asked to provide a prediction t of the individual's total score $X + Y$. The error of prediction is the mean squared error $E[(X + Y - t)^2]$. What value of t minimizes the error of prediction?

82. Let X_1 and X_2 be quantitative and verbal scores on one aptitude exam, and let Y_1 and Y_2 be corresponding scores on another exam. If $\text{Cov}(X_1, Y_1) = 5$, $\text{Cov}(X_1, Y_2) = 1$, $\text{Cov}(X_2, Y_1) = 2$, and $\text{Cov}(X_2, Y_2) = 8$, what is the covariance between the two total scores $X_1 + X_2$ and $Y_1 + Y_2$?

83. Simulation studies are important in investigating various characteristics of a system or process. They are generally employed when the mathematical analysis necessary to answer important questions is too complicated to yield closed form solutions. For example, in a system where the time between successive customer arrivals has a particular pdf and the service time of any particular customer has another particular pdf, simulation can provide information about the probability that the system is empty when a customer arrives, the expected number of customers in the system, and the expected waiting time in queue. Such studies depend on being able to generate observations from a specified probability distribution.

The *rejection method* gives a way of generating an observation from a pdf $f(\cdot)$ when we have a way of generating an observation from $g(\cdot)$ and the ratio $f(x)/g(x)$ is bounded, that is, $\leq c$ for some finite c. The steps are as follows:
1. Use a software package's random number generator to obtain a value u from a uniform distribution on the interval from 0 to 1.
2. Generate a value y from the distribution with pdf $g(y)$.
3. If $u \leq f(y)/cg(y)$, set $x = y$ ("accept" x); otherwise return to step 1. That is, the procedure is repeated until at some stage $u \leq f(y)/cg(y)$.

a. Argue that $c \geq 1$. *Hint*: If $c < 1$, then $f(y) < g(y)$ for all y; why is this bad?
b. Show that this procedure does result in an observation from the pdf $f(\bullet)$; that is, $P(\text{accepted value} \leq x) = F(x)$. *Hint:* This probability is $P(\{U \leq f(Y)/cg(Y)\} \cap \{Y \leq x\})$; to calculate, first integrate with respect to u for fixed y and then integrate with respect to y.
c. Show that the probability of "accepting" at any particular stage is $1/c$. What does this imply about the expected number of stages necessary to obtain an acceptable value? What kind of value of c is desirable?
d. Let $f(x) = 20x(1 - x)^3$ for $0 < x < 1$, a particular beta distribution. Show that taking $g(y)$ to be a uniform pdf on (0, 1) works. What is the best value of c in this situation?

84. You are driving on a highway at speed X_1. Cars entering this highway after you travel at speeds X_2, X_3, ... Suppose these X_i's are independent and identically distributed with pdf $f(x)$ and cdf $F(x)$. Unfortunately there is no way for a faster car to pass a slower one—it will catch up to the slower one and then travel at the same speed. For example, if $X_1 =$ 52.3, $X_2 = 37.5$, and $X_3 = 42.8$, then no car will catch up to yours, but the third car will catch up to the second. Let N = the number of cars that ultimately travel at your speed (in your "cohort"), including your own car. Possible values of N are 1, 2, 3, ... Show that the pmf of N is $p(n) = 1/[n(n + 1)]$, and then determine the expected number of cars in your cohort. *Hint*: $N = 3$ requires that $X_1 < X_2$, $X_1 < X_3$, $X_4 < X_1$.

85. Suppose the number of children born to an individual has pmf $p(x)$. A *Galton–Watson branching process* unfolds as follows: At time $t = 0$, the population consists of a single individual. Just prior to time $t = 1$, this individual gives birth to X_1 individuals according to the pmf $p(x)$, so there are X_1 individuals in the first generation. Just prior to time $t = 2$, each of these X_1 individuals gives birth independently of the others according to the pmf $p(x)$, resulting in X_2 individuals in the second generation (e.g., if $X_1 = 3$, then $X_2 = Y_1 + Y_2 + Y_3$, where Y_i is the number of progeny of the ith individual in the first generation). This process then continues to yield a third generation of size X_3, and so on.

a. If $X_1 = 3$, $Y_1 = 4$, $Y_2 = 0$, $Y_3 = 1$, draw a tree diagram with two generations of branches to represent this situation.

b. Let A be the event that the process ultimately becomes extinct (one way for A to occur would be to have $X_1 = 3$ with none of these three second-generation individuals having any progeny) and let $p^* = P(A)$. Argue that p^* satisfies the equation

$$p^* = \sum (p^*)^x \cdot p(x)$$

That is, $p^* = h(p^*)$ where $h(s)$ is the probability generating function introduced in Exercise 138 from Chapter 3. *Hint*: $A = \cup (A \cap \{X_1 = x\})$, so the law of total probability can be applied. Now given that $X_1 = 3$, A will occur if and only if each of the three separate branching processes starting from the first generation ultimately becomes extinct; what is the probability of this happening?

c. Verify that one solution to the equation in (b) is $p^* = 1$. It can be shown that this equation has just one other solution, and that the probability of ultimate extinction is in fact the *smaller* of the two roots. If $p(0) = .3$, $p(1) = .5$, and $p(2) = .2$, what is p^*? Is this consistent with the value of μ, the expected number of progeny from a single individual? What happens if $p(0) = .2$, $p(1) = .5$, and $p(2) = .3$?

86. Let $f(x)$ and $g(y)$ be pdf's with corresponding cdf's $F(x)$ and $G(y)$, respectively. With c denoting a numerical constant satisfying $|c| \le 1$, consider

$$f(x, y) = f(x)g(y)\{1 + c[2F(x) - 1][2G(y) - 1]\}$$

Show that $f(x, y)$ satisfies the conditions necessary to specify a joint pdf for two continuous rv's. What is the marginal pdf of the first variable X? Of the second variable Y? For what values of c are X and Y independent? If $f(x)$ and $g(y)$ are normal pdf's, is the joint distribution of X and Y bivariate normal?

87. The **joint cumulative distribution function** of two random variables X and Y, denoted by $F(x, y)$, is defined by

$$F(x, y) = P(X \le x \cap Y \le y) \qquad -\infty < x < \infty, -\infty < y < \infty$$

a. Suppose that X and Y are both continuous variables. Once the joint cdf is available, explain how it can be used to determine the probability $P[(X, Y) \in A]$, where A is the rectangular region $\{(x, y): a \le x \le b, c \le y \le d\}$.

b. Suppose the only possible values of X and Y are 0, 1, 2, . . . and consider the values $a = 5$, $b = 10$, $c = 2$, and $d = 6$ for the rectangle specified in (a). Describe how you would use the joint cdf to calculate the probability that the pair (X, Y) falls in the rectangle. More generally, how can the rectangular probability be calculated from the joint cdf if a, b, c, and d are all integers?

c. Determine the joint cdf for the scenario of Example 5.1. *Hint*: First determine $F(x, y)$ for $x =$ 100, 250 and $y = 0$, 100, and 200. Then describe the joint cdf for various other (x, y) pairs.

d. Determine the joint cdf for the scenario of Example 5.3 and use it to calculate the probability that X and Y are both between .25 and .75. *Hint*: For $0 \le x \le 1$ and $0 \le y \le 1$,

$$F(x, y) = \int_0^x \int_0^y f(u, v)\, dv\, du$$

e. Determine the joint cdf for the scenario of Example 5.5. *Hint*: Proceed as in (d), but be careful about the order of integration and consider separately (x, y) points that lie inside the triangular region of positive density and then points that lie outside this region.

88. A circular sampling region with radius X is chosen by a biologist, where X has an exponential distribution with mean value 10 ft. Plants of a certain type occur in this region according to a (spatial) Poisson process with "rate" .5 plant per square foot. Let Y denote the number of plants in the region.
a. Find $E(Y|X = x)$ and $V(Y|X = x)$.
b. Use part (a) to find $E(Y)$.
c. Use part (a) to find $V(Y)$.

89. The number of individuals arriving at a post office to mail packages during a certain period is a Poisson random variable X with mean value 20. Independently of one another, any particular customer will mail either 1, 2, 3, or 4 packages with probabilities .4, .3, .2, and .1, respectively. Let Y denote the total number of packages mailed during this time period.
a. Find $E(Y|X = x)$ and $V(Y|X = x)$.
b. Use part (a) to find $E(Y)$.
c. Use part (a) to find $V(Y)$.

90. Consider a sealed-bid auction in which each of the n bidders has his/her valuation (assessment of inherent worth) of the item being auctioned. The valuation of any particular bidder is not known to the other bidders. Suppose these valuations constitute a random sample $X_1, \ldots, X_n$ from a distribution with cdf $F(x)$, with corresponding order statistics $Y_1 \le Y_2 \le \cdots \le Y_n$. The *rent* of the winning bidder is the difference between the winner's valuation and the price. The article "Mean Sample Spacings, Sample Size and Variability in an Auction-Theoretic Framework" (*Oper. Res. Lett.*, 2004: 103–108) argues that the rent is just $Y_n - Y_{n-1}$ (why?).
a. Suppose that the valuation distribution is uniform on [0, 100]. What is the expected rent when there are $n = 10$ bidders?
b. Referring back to (a), what happens when there are 11 bidders? More generally, what is the relationship between the expected rent for n bidders and for $n + 1$ bidders? Is this intuitive? *Note*: The cited article presents a counterexample.

91. Suppose two identical components are connected in parallel, so the system continues to function as long as at least one of the components does so. The two lifetimes are independent of one another, each having an exponential distribution with mean 1000 hr. Let W denote system lifetime. Obtain the moment generating function of W, and use it to calculate the expected lifetime.

Bibliography

Larsen, Richard, and Morris Marx, *An Introduction to Mathematical Statistics and Its Applications* (3rd ed.), Prentice Hall, Englewood Cliffs, NJ, 2000. More limited coverage than in the book by Olkin et al., but well written and readable.

Olkin, Ingram, Cyrus Derman, and Leon Gleser, *Probability Models and Applications* (2nd ed.), Macmillan, New York, 1994. Contains a careful and comprehensive exposition of joint distributions and rules of expectation.

CHAPTER SIX

Statistics and Sampling Distributions

Introduction

This chapter helps make the transition between probability and inferential statistics. Given a sample of n observations from a population, we will be calculating *estimates* of the population mean, median, standard deviation, and various other population characteristics (parameters). Prior to obtaining data, there is uncertainty as to which of all possible samples will occur. Because of this, estimates such as $\bar{x}$, $\tilde{x}$, and s will vary from one sample to another. The behavior of such estimates in repeated sampling is described by what are called *sampling distributions.* Any particular sampling distribution will give an indication of how close the estimate is likely to be to the value of the parameter being estimated.

The first three sections use probability results to study sampling distributions. A particularly important result is the Central Limit Theorem, which shows how the behavior of the sample mean can be described by a particular normal distribution when the sample size is large. The last section introduces several distributions related to normal samples that will be used as a basis for numerous inferential procedures.

6.1 Statistics and Their Distributions

The observations in a single sample were denoted in Chapter 1 by $x_1, x_2, \ldots, x_n$. Consider selecting two different samples of size n from the same population distribution. The x_i's in the second sample will virtually always differ at least a bit from those in the first sample. For example, a first sample of $n = 3$ cars of a particular type might result in fuel efficiencies $x_1 = 30.7$, $x_2 = 29.4$, $x_3 = 31.1$, whereas a second sample may give $x_1 = 28.8$, $x_2 = 30.0$, and $x_3 = 31.1$. Before we obtain data, there is uncertainty about the value of each x_i. Because of this uncertainty, *before* the data becomes available we view each observation as a random variable and denote the sample by $X_1, X_2, \ldots, X_n$ (uppercase letters for random variables).

This variation in observed values in turn implies that the value of any function of the sample observations—such as the sample mean, sample standard deviation, or sample fourth spread—also varies from sample to sample. That is, prior to obtaining $x_1, \ldots, x_n$, there is uncertainty as to the value of $\bar{x}$, the value of s, and so on.

Example 6.1 Suppose that material strength for a randomly selected specimen of a particular type has a Weibull distribution with parameter values $\alpha = 2$ (shape) and $\beta = 5$ (scale). The corresponding density curve is shown in Figure 6.1. Formulas from Section 4.5 give

$$\mu = E(X) = 4.4311 \qquad \tilde{\mu} = 4.1628 \qquad \sigma^2 = V(X) = 5.365 \qquad \sigma = 2.316$$

The mean exceeds the median because of the distribution's positive skew.

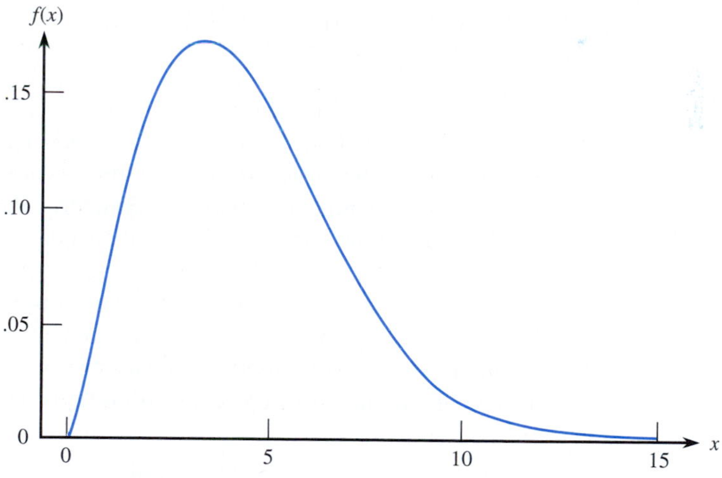

Figure 6.1 The Weibull density curve for Example 6.1

We used MINITAB to generate six different samples, each with $n = 10$, from this distribution (material strengths for six different groups of ten specimens each). The results appear in Table 6.1, followed by the values of the sample mean, sample median, and sample standard deviation for each sample. Notice first that the ten observations in any particular sample are all different from those in any other sample. Second, the six values of the sample mean are all different from one another, as are the six values of the

Table 6.1 Samples from the Weibull distribution of Example 6.1

	Sample					
	1	2	3	4	5	6
Observation						
1	6.1171	5.07611	3.46710	1.55601	3.12372	8.93795
2	4.1600	6.79279	2.71938	4.56941	6.09685	3.92487
3	3.1950	4.43259	5.88129	4.79870	3.41181	8.76202
4	0.6694	8.55752	5.14915	2.49759	1.65409	7.05569
5	1.8552	6.82487	4.99635	2.33267	2.29512	2.30932
6	5.2316	7.39958	5.86887	4.01295	2.12583	5.94195
7	2.7609	2.14755	6.05918	9.08845	3.20938	6.74166
8	10.2185	8.50628	1.80119	3.25728	3.23209	1.75468
9	5.2438	5.49510	4.21994	3.70132	6.84426	4.91827
10	4.5590	4.04525	2.12934	5.50134	4.20694	7.26081
Statistic						
Mean	4.401	5.928	4.229	4.132	3.620	5.761
Median	4.360	6.144	4.608	3.857	3.221	6.342
SD	2.642	2.062	1.611	2.124	1.678	2.496

sample median and the six values of the sample standard deviation. The same is true of the sample 10% trimmed means, sample fourth spreads, and so on.

Furthermore, the value of the sample mean from any particular sample can be regarded as a *point estimate* ("point" because it is a single number, corresponding to a single point on the number line) of the population mean μ, whose value is known to be 4.4311. None of the estimates from these six samples is identical to what is being estimated. The estimates from the second and sixth samples are much too large, whereas the fifth sample gives a substantial underestimate. Similarly, the sample standard deviation gives a point estimate of the population standard deviation. All six of the resulting estimates are in error by at least a small amount.

In summary, the values of the individual sample observations vary from sample to sample, so in general the value of any quantity computed from sample data, and the value of a sample characteristic used as an estimate of the corresponding population characteristic, will virtually never coincide with what is being estimated. ■

DEFINITION

A **statistic** is any quantity whose value can be calculated from sample data. Prior to obtaining data, there is uncertainty as to what value of any particular statistic will result. Therefore, a statistic is a random variable and will be denoted by an uppercase letter; a lowercase letter is used to represent the calculated or observed value of the statistic.

Thus the sample mean, regarded as a statistic (before a sample has been selected or an experiment has been carried out), is denoted by $\overline{X}$; the calculated value of this statistic is $\overline{x}$. Similarly, S represents the sample standard deviation thought of as a statistic, and its computed value is s. Suppose a drug is given to a sample of patients, another drug is given to a second sample, and the cholesterol levels are denoted by $X_1, \ldots, X_m$ and $Y_1, \ldots, Y_n$, respectively. Then the statistic $\overline{X} - \overline{Y}$, the difference between the two sample mean cholesterol levels, may be important.

Any statistic, being a random variable, has a probability distribution. In particular, the sample mean $\overline{X}$ has a probability distribution. Suppose, for example, that $n = 2$ components are randomly selected and the number of breakdowns while under warranty is determined for each one. Possible values for the sample mean number of breakdowns $\overline{X}$ are 0 (if $X_1 = X_2 = 0$), .5 (if either $X_1 = 0$ and $X_2 = 1$ or $X_1 = 1$ and $X_2 = 0$), 1, 1.5, ... The probability distribution of $\overline{X}$ specifies $P(\overline{X} = 0)$, $P(\overline{X} = .5)$, and so on, from which other probabilities such as $P(1 \le \overline{X} \le 3)$ and $P(\overline{X} \ge 2.5)$ can be calculated. Similarly, if for a sample of size $n = 2$, the only possible values of the sample variance are 0, 12.5, and 50 (which is the case if X_1 and X_2 can each take on only the values 40, 45, and 50), then the probability distribution of S^2 gives $P(S^2 = 0)$, $P(S^2 = 12.5)$, and $P(S^2 = 50)$. The probability distribution of a statistic is sometimes referred to as its **sampling distribution** to emphasize that it describes how the statistic varies in value across all samples that might be selected.

Random Samples

The probability distribution of any particular statistic depends not only on the population distribution (normal, uniform, etc.) and the sample size n but also on the method of sampling. Consider selecting a sample of size $n = 2$ from a population consisting of just the three values 1, 5, and 10, and suppose that the statistic of interest is the sample variance. If sampling is done "with replacement," then $S^2 = 0$ will result if $X_1 = X_2$. However, S^2 cannot equal 0 if sampling is "without replacement." So $P(S^2 = 0) = 0$ for one sampling method, and this probability is positive for the other method. Our next definition describes a sampling method often encountered (at least approximately) in practice.

DEFINITION

The rv's $X_1, X_2, \ldots, X_n$ are said to form a (simple) **random sample** of size n if

1. The X_i's are independent rv's.
2. Every X_i has the same probability distribution.

Conditions 1 and 2 can be paraphrased by saying that the X_i's are *independent and identically distributed* (iid). If sampling is either with replacement or from an infinite (conceptual) population, Conditions 1 and 2 are satisfied exactly. These conditions will be approximately satisfied if sampling is without replacement, yet the sample size n is much smaller than the population size N. In practice, if $n/N \le .05$ (at most 5% of the population is sampled), we can proceed as if the X_i's form a random sample. The virtue of this sampling method is that the probability distribution of any statistic can be more easily obtained than for any other sampling method.

There are two general methods for obtaining information about a statistic's sampling distribution. One method involves calculations based on probability rules, and the other involves carrying out a simulation experiment.

Deriving the Sampling Distribution of a Statistic

Probability rules can be used to obtain the distribution of a statistic provided that it is a "fairly simple" function of the X_i's and either there are relatively few different X values in the population or else the population distribution has a "nice" form. Our next two examples illustrate such situations.

Example 6.2 A large automobile service center charges \$40, \$45, and \$50 for a tune-up of four-, six-, and eight-cylinder cars, respectively. If 20% of its tune-ups are done on four-cylinder cars, 30% on six-cylinder cars, and 50% on eight-cylinder cars, then the probability distribution of revenue from a single randomly selected tune-up is given by

x	40	45	50
$p(x)$	.2	.3	.5

with $\mu = 46.5$, $\sigma^2 = 15.25$ (6.1)

Suppose on a particular day only two servicing jobs involve tune-ups. Let X_1 = the revenue from the first tune-up and X_2 = the revenue from the second. Suppose that X_1 and X_2 are independent, each with the probability distribution shown in (6.1) [so that X_1 and X_2 constitute a random sample from the distribution (6.1)]. Table 6.2 lists possible (x_1, x_2) pairs, the probability of each [computed using (6.1) and the assumption of independence], and the resulting $\bar{x}$ and s^2 values. Now to obtain the probability distribution of $\bar{X}$, the sample average revenue per tune-up, we must consider each possible value $\bar{x}$ and

Table 6.2 Outcomes, probabilities, and values of $\bar{x}$ and s^2 for Example 6.2

x_1	x_2	$p(x_1, x_2)$	$\bar{x}$	s^2
40	40	.04	40	0
40	45	.06	42.5	12.5
40	50	.10	45	50
45	40	.06	42.5	12.5
45	45	.09	45	0
45	50	.15	47.5	12.5
50	40	.10	45	50
50	45	.15	47.5	12.5
50	50	.25	50	0

compute its probability. For example, $\bar{x} = 45$ occurs three times in the table with probabilities .10, .09, and .10, so

$$p_{\bar{X}}(45) = P(\bar{X} = 45) = .10 + .09 + .10 = .29$$

Similarly,

$$\begin{aligned} p_{s^2}(50) &= P(S^2 = 50) = P(X_1 = 40, X_2 = 50 \quad \text{or} \quad X_1 = 50, X_2 = 40) \\ &= .10 + .10 = .20 \end{aligned}$$

The complete sampling distributions of $\bar{X}$ and S^2 appear in (6.2) and (6.3).

$\bar{x}$	40	42.5	45	47.5	50
$p_{\bar{X}}(\bar{x})$	.04	.12	.29	.30	.25

(6.2)

s^2	0	12.5	50
$p_{S^2}(s^2)$	.38	.42	.20

(6.3)

Figure 6.2 pictures a probability histogram for both the original distribution (6.1) and the $\bar{X}$ distribution (6.2). The figure suggests first that the mean (expected value) of the $\bar{X}$ distribution is equal to the mean 46.5 of the original distribution, since both histograms appear to be centered at the same place.

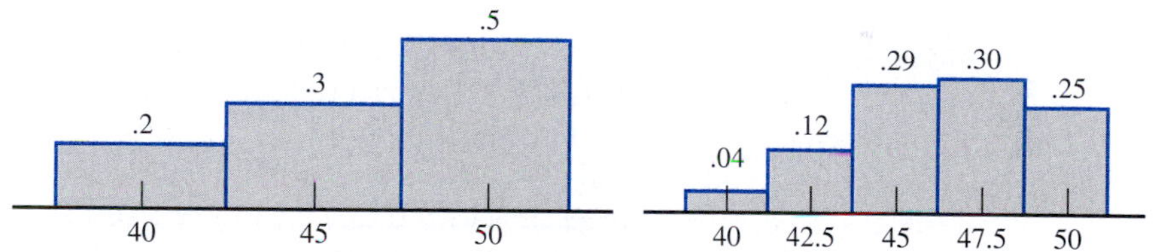

Figure 6.2 Probability histograms for the underlying distribution and $\bar{X}$ distribution in Example 6.2

From (6.2),

$$\mu_{\bar{X}} = E(\bar{X}) = \sum \bar{x} p_{\bar{X}}(\bar{x}) = (40)(.04) + \cdots + (50)(.25) = 46.5 = \mu$$

Second, it appears that the $\bar{X}$ distribution has smaller spread (variability) than the original distribution, since probability mass has moved in toward the mean. Again from (6.2),

$$\begin{aligned} \sigma_{\bar{X}}^2 = V(\bar{X}) &= \sum \bar{x}^2 \cdot p_{\bar{X}}(\bar{x}) - \mu_{\bar{X}}^2 \\ &= (40)^2(.04) + \cdots + (50)^2(.25) - (46.5)^2 \\ &= 7.625 = \frac{15.25}{2} = \frac{\sigma^2}{2} \end{aligned}$$

The variance of $\overline{X}$ is precisely half that of the original variance (because $n = 2$).

The mean value of S^2 is

$$\mu_{S^2} = E(S^2) = \sum s^2 \cdot p_{S^2}(s^2)$$

$$= (0)(.38) + (12.5)(.42) + (50)(.20) = 15.25 = \sigma^2$$

That is, the $\overline{X}$ sampling distribution is centered at the population mean μ, and the S^2 sampling distribution is centered at the population variance σ^2.

If four tune-ups had been done on the day of interest, the sample average revenue $\overline{X}$ would be based on a random sample of four X_i's, each having the distribution (6.1). More calculation eventually yields the pmf of $\overline{X}$ for $n = 4$ as

$\bar{x}$	40	41.25	42.5	43.75	45	46.25	47.5	48.75	50
$p_{\overline{X}}(\bar{x})$	.0016	.0096	.0376	.0936	.1761	.2340	.2350	.1500	.0625

From this, for $n = 4$, $\mu_{\overline{X}} = 46.50 = \mu$ and $\sigma^2_{\overline{X}} = 3.8125 = \sigma^2/4$. Figure 6.3 is a probability histogram of this pmf.

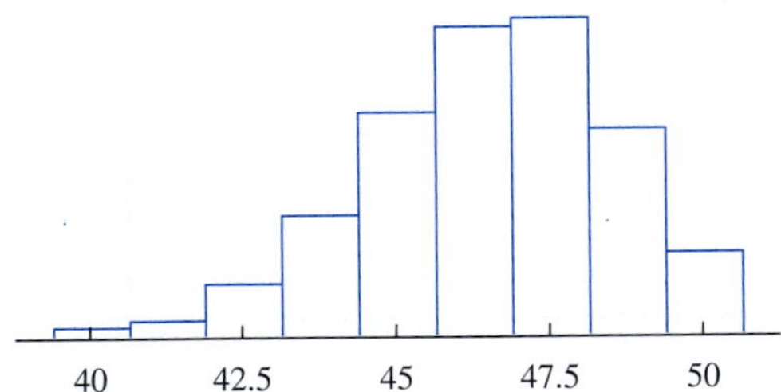

Figure 6.3 Probability histogram for $\overline{X}$ based on $n = 4$ in Example 6.2 ■

Example 6.2 should suggest first of all that the computation of $p_{\overline{X}}(\bar{x})$ and $p_{S^2}(s^2)$ can be tedious. If the original distribution (6.1) had allowed for more than three possible values 40, 45, and 50, then even for $n = 2$ the computations would have been more involved. The example should also suggest, however, that there are some general relationships between $E(\overline{X})$, $V(\overline{X})$, $E(S^2)$, and the mean μ and variance σ^2 of the original distribution. These are stated in the next section. Now consider an example in which the random sample is drawn from a continuous distribution.

Example 6.3 The time that it takes to serve a customer at the cash register in a minimarket is a random variable having an exponential distribution with parameter λ. Suppose X_1 and X_2 are service times for two different customers, assumed independent of each other.

Consider the total service time $T_o = X_1 + X_2$ for the two customers, also a statistic. The cdf of T_o is, for $t \geq 0$,

$$F_{T_o}(t) = P(X_1 + X_2 \leq t) = \iint\limits_{\{(x_1, x_2): x_1 + x_2 \leq t\}} f(x_1, x_2)\, dx_1\, dx_2$$

$$= \int_0^t \int_0^{t - x_1} \lambda e^{-\lambda x_1} \cdot \lambda e^{-\lambda x_2}\, dx_2\, dx_1 = \int_0^t (\lambda e^{-\lambda x_1} - \lambda e^{-\lambda t})\, dx_1$$

$$= 1 - e^{-\lambda t} - \lambda t e^{-\lambda t}$$

The region of integration is pictured in Figure 6.4.

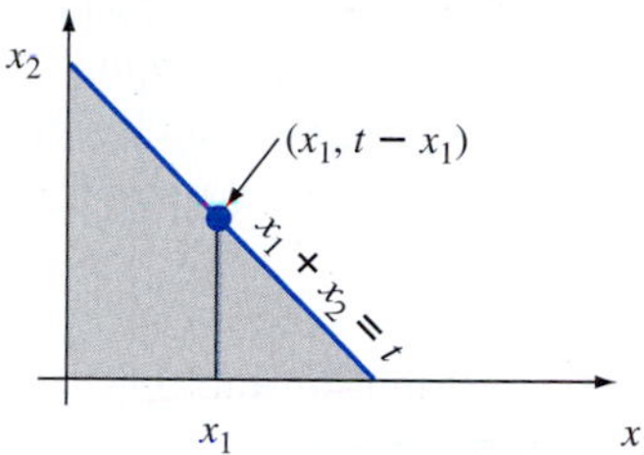

Figure 6.4 Region of integration to obtain cdf of T_o in Example 6.3

The pdf of T_o is obtained by differentiating $F_{T_o}(t)$:

$$f_{T_o}(t) = \begin{cases} \lambda^2 t e^{-\lambda t} & t \geq 0 \\ 0 & t < 0 \end{cases} \tag{6.4}$$

This is a gamma pdf ($\alpha = 2$ and $\beta = 1/\lambda$). This distribution for T_o can also be derived by a moment generating function argument.

The pdf of $\overline{X} = T_o/2$ can be obtained by the method of Section 4.7 as

$$f_{\overline{X}}(\overline{x}) = \begin{cases} 4\lambda^2 \overline{x} e^{-2\lambda \overline{x}} & \overline{x} \geq 0 \\ 0 & \overline{x} < 0 \end{cases} \tag{6.5}$$

The mean and variance of the underlying exponential distribution are $\mu = 1/\lambda$ and $\sigma^2 = 1/\lambda^2$. Using Expressions (6.4) and (6.5), it can be verified that $E(\overline{X}) = 1/\lambda$, $V(\overline{X}) = 1/(2\lambda^2)$, $E(T_o) = 2/\lambda$, and $V(T_o) = 2/\lambda^2$. These results again suggest some general relationships between means and variances of $\overline{X}$, T_o, and the underlying distribution. ■

Simulation Experiments

The second method of obtaining information about a statistic's sampling distribution is to perform a simulation experiment. This method is usually used when a derivation via probability rules is too difficult or complicated to be carried out. Such an experiment is virtually always done with the aid of a computer. The following characteristics of an experiment must be specified:

1. The statistic of interest ($\overline{X}$, S, a particular trimmed mean, etc.)
2. The population distribution (normal with $\mu = 100$ and $\sigma = 15$, uniform with lower limit $A = 5$ and upper limit $B = 10$, etc.)
3. The sample size n (e.g., $n = 10$ or $n = 50$)
4. The number of replications k (e.g., $k = 500$)

Then use a computer to obtain k different random samples, each of size n, from the designated population distribution. For each such sample, calculate the value of the statistic and construct a histogram of the k calculated values. This histogram gives the *approximate* sampling distribution of the statistic. The larger the value of k, the better the approximation will tend to be (the actual sampling distribution emerges as $k \to \infty$). In practice, $k = 500$ or 1000 is usually enough if the statistic is "fairly simple."

Example 6.4 The population distribution for our first simulation study is normal with $\mu = 8.25$ and $\sigma = .75$, as pictured in Figure 6.5. [The article "Platelet Size in Myocardial Infarction" (*British Med. J.*, 1983: 449–451) suggests this distribution for platelet volume in individuals with no history of serious heart problems.]

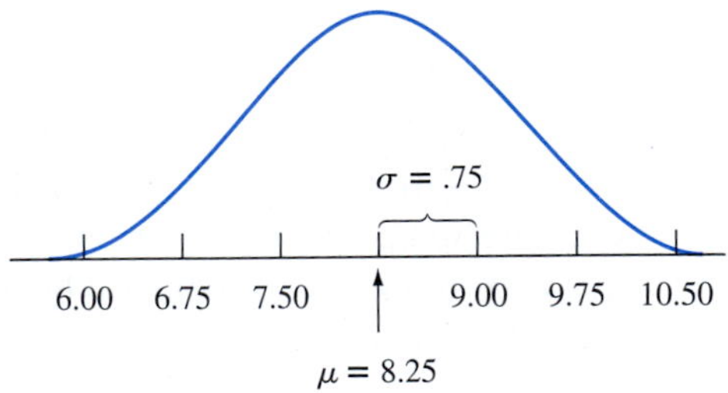

Figure 6.5 Normal distribution, with $\mu = 8.25$ and $\sigma = .75$

We actually performed four different experiments, with 500 replications for each one. In the first experiment, 500 samples of $n = 5$ observations each were generated using MINITAB, and the sample sizes for the other three were $n = 10$, $n = 20$, and $n = 30$, respectively. The sample mean was calculated for each sample, and the resulting histograms of $\bar{x}$ values appear in Figure 6.6.

The first thing to notice about the histograms is their shape. To a reasonable approximation, each of the four looks like a normal curve. The resemblance would be even more striking if each histogram had been based on many more than 500 $\bar{x}$ values. Second, each histogram is centered approximately at 8.25, the mean of the population being sampled. Had the histograms been based on an unending sequence of $\bar{x}$ values, their centers would have been exactly the population mean, 8.25.

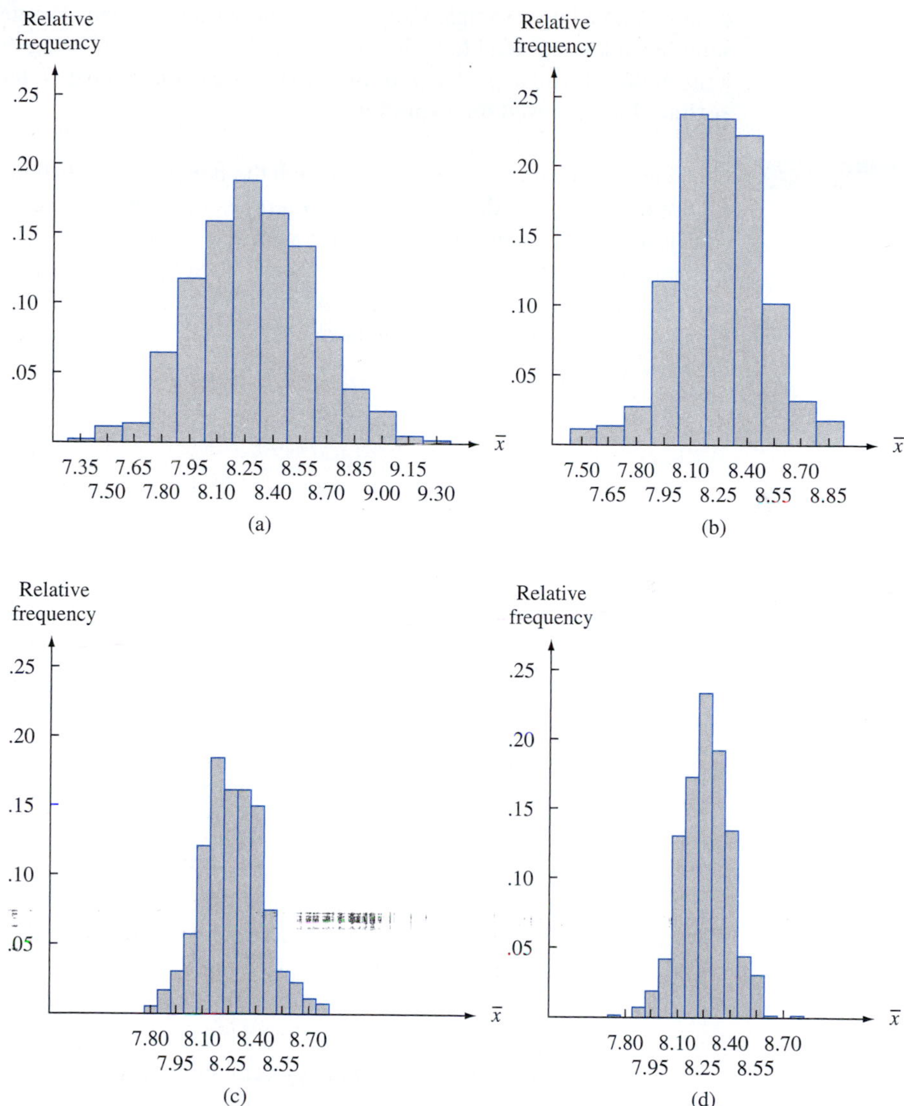

Figure 6.6 Sample histograms for $\bar{X}$ based on 500 samples, each consisting of n observations: (a) $n = 5$; (b) $n = 10$; (c) $n = 20$; (d) $n = 30$

The final aspect of the histograms to note is their spread relative to one another. The smaller the value of n, the greater the extent to which the sampling distribution spreads out about the mean value. This is why the histograms for $n = 20$ and $n = 30$ are based on narrower class intervals than those for the two smaller sample sizes. For the larger sample sizes, most of the $\bar{x}$ values are quite close to 8.25. This is the effect of averaging. When n is small, a single unusual x value can result in an $\bar{x}$ value far from the center. With a larger sample size, any unusual x values, when averaged in with the other sample values, still tend to yield an $\bar{x}$ value close to μ. Combining these insights yields a result that should appeal to your intuition: **$\bar{X}$ based on a large n tends to be closer to μ than does $\bar{X}$ based on a small n.** ■

Example 6.5 Consider a simulation experiment in which the population distribution is quite skewed. Figure 6.7 shows the density curve for lifetimes of a certain type of electronic control (this is actually a lognormal distribution with $E[\ln(X)] = 3$ and $V[\ln(X)] = .16$). Again the statistic of interest is the sample mean $\bar{X}$. The experiment utilized 500 replications and considered the same four sample sizes as in Example 6.4. The resulting histograms along with a normal probability plot from MINITAB for the 500 $\bar{x}$ values based on $n = 30$ are shown in Figure 6.8.

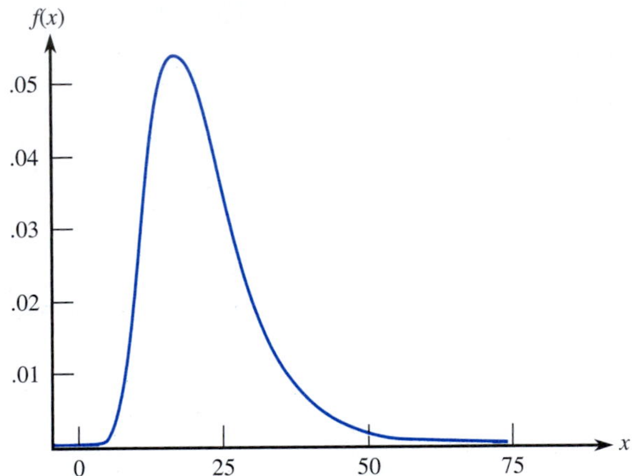

Figure 6.7 Density curve for the simulation experiment of Example 6.5 $[E(X) = 21.7584,\ V(X) = 82.1449]$

Unlike the normal case, these histograms all differ in shape. In particular, they become progressively less skewed as the sample size n increases. The average of the 500 $\bar{x}$ values for the four different sample sizes are all quite close to the mean value of the population distribution. If each histogram had been based on an unending sequence of $\bar{x}$ values rather than just 500, all four would have been centered at exactly 21.7584. Thus different values of n change the shape but not the center of the sampling distribution of $\bar{X}$. Comparison of the four histograms in Figure 6.8 also shows that as n increases, the spread of the histograms decreases. Increasing n results in a greater degree of concentration about the population mean value and makes the histogram look more like a normal curve. The histogram of Figure 6.8(d) and the normal probability plot in Figure 6.8(e) provide convincing evidence that a sample size of $n = 30$ is sufficient to overcome the skewness of the population distribution and give an approximately normal $\bar{X}$ sampling distribution.

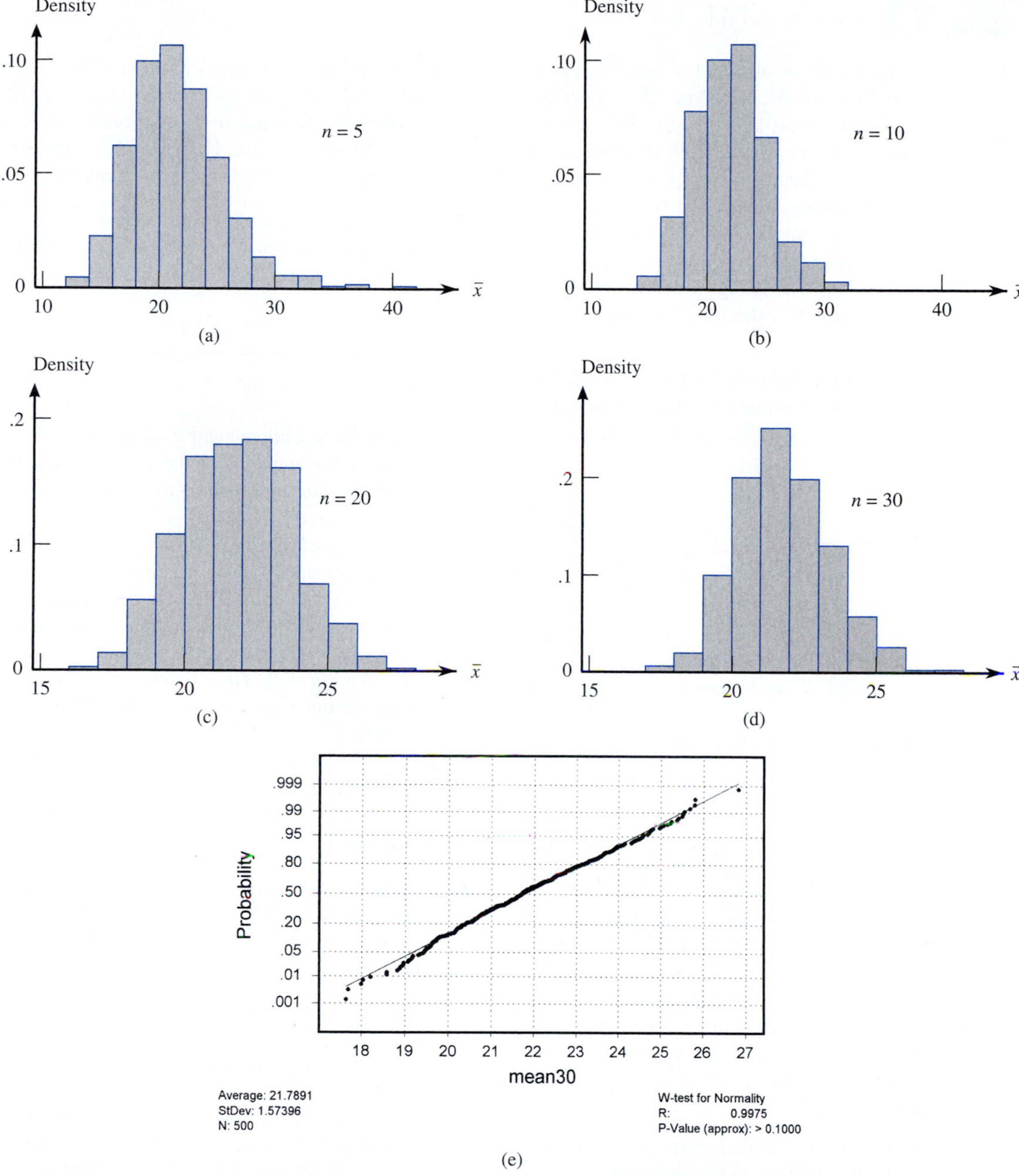

Figure 6.8 Results of the simulation experiment of Example 6.5: (a) $\overline{X}$ histogram for $n = 5$; (b) $\overline{X}$ histogram for $n = 10$; (c) $\overline{X}$ histogram for $n = 20$; (d) $\overline{X}$ histogram for $n = 30$; (e) normal probability plot for $n = 30$ (from MINITAB) ■

Exercises Section 6.1 (1–10)

1. A particular brand of dishwasher soap is sold in three sizes: 25 oz, 40 oz, and 65 oz. Twenty percent of all purchasers select a 25-oz box, 50% select a 40-oz box, and the remaining 30% choose a 65-oz box. Let X_1 and X_2 denote the package sizes selected by two independently selected purchasers.

a. Determine the sampling distribution of $\overline{X}$, calculate $E(\overline{X})$, and compare to μ.

b. Determine the sampling distribution of the sample variance S^2, calculate $E(S^2)$, and compare to σ^2.

2. There are two traffic lights on the way to work. Let X_1 be the number of lights that are red, requiring a stop, and suppose that the distribution of X_1 is as follows:

x_1	0	1	2
$p(x_1)$	.2	.5	.3

$\mu = 1.1, \sigma^2 = .49$

Let X_2 be the number of lights that are red on the way home; X_2 is independent of X_1. Assume that X_2 has the same distribution as X_1, so that X_1, X_2 is a random sample of size $n = 2$.

a. Let $T_o = X_1 + X_2$, and determine the probability distribution of T_o.

b. Calculate μ_{T_o}. How does it relate to μ, the population mean?

c. Calculate $\sigma^2_{T_o}$. How does it relate to σ^2, the population variance?

3. It is known that 80% of all brand A DVD players work in a satisfactory manner throughout the warranty period (are "successes"). Suppose that $n = 10$ players are randomly selected. Let $X =$ the number of successes in the sample. The statistic X/n is the sample proportion (fraction) of successes. Obtain the sampling distribution of this statistic. [*Hint*: One possible value of X/n is .3, corresponding to $X = 3$. What is the probability of this value (what kind of random variable is X)?]

4. A box contains ten sealed envelopes numbered 1, . . . , 10. The first five contain no money, the next three each contain \$5, and there is a \$10 bill in each of the last two. A sample of size 3 is selected *with* replacement (so we have a random sample), and you get the largest amount in any of the envelopes selected. If X_1, X_2, and X_3 denote the amounts in the selected envelopes, the statistic of interest is $M =$ the maximum of X_1, X_2, and X_3.

a. Obtain the probability distribution of this statistic.

b. Describe how you would carry out a simulation experiment to compare the distributions of M for various sample sizes. How would you guess the distribution would change as n increases?

5. Let X be the number of packages being mailed by a randomly selected customer at a certain shipping facility. Suppose the distribution of X is as follows:

x	1	2	3	4
$p(x)$	.4	.3	.2	.1

a. Consider a random sample of size $n = 2$ (two customers), and let $\overline{X}$ be the sample mean number of packages shipped. Obtain the probability distribution of $\overline{X}$.

b. Refer to part (a) and calculate $P(\overline{X} \le 2.5)$.

c. Again consider a random sample of size $n = 2$, but now focus on the statistic $R =$ the sample range (difference between the largest and smallest values in the sample). Obtain the distribution of R. [*Hint*: Calculate the value of R for each outcome and use the probabilities from part (a).]

d. If a random sample of size $n = 4$ is selected, what is $P(\overline{X} \le 1.5)$? (*Hint*: You should not have to list all possible outcomes, only those for which $\overline{x} \le 1.5$.)

6. A company maintains three offices in a certain region, each staffed by two employees. Information concerning yearly salaries (1000's of dollars) is as follows:

Office	1	1	2	2	3	3
Employee	1	2	3	4	5	6
Salary	29.7	33.6	30.2	33.6	25.8	29.7

a. Suppose two of these employees are randomly selected from among the six (without replacement). Determine the sampling distribution of the sample mean salary $\overline{X}$.

b. Suppose one of the three offices is randomly selected. Let X_1 and X_2 denote the salaries of the two employees. Determine the sampling distribution of $\overline{X}$.

c. How does $E(\overline{X})$ from parts (a) and (b) compare to the population mean salary μ?

7. The number of dirt specks on a randomly selected square yard of polyethylene film of a certain type has a Poisson distribution with a mean value of 2 specks per square yard. Consider a random sample of $n = 5$ film specimens, each having area 1 square yard, and let $\overline{X}$ be the resulting sample mean number of dirt specks. Obtain the first 21 probabilities in the $\overline{X}$ sampling distribution. *Hint*: What does a moment generating function argument say about the distribution of $X_1 + \cdots + X_5$?

8. Suppose the amount of liquid dispensed by a certain machine is uniformly distributed with lower limit $A = 8$ oz and upper limit $B = 10$ oz. Describe how you would carry out simulation experiments to compare the sampling distribution of the (sample) fourth spread for sample sizes $n = 5$, 10, 20, and 30.

9. Carry out a simulation experiment using a statistical computer package or other software to study the sampling distribution of $\overline{X}$ when the population distribution is Weibull with $\alpha = 2$ and $\beta = 5$, as in Example 6.1. Consider the four sample sizes $n = 5$, 10, 20, and 30, and in each case use 500 replications. For which of these sample sizes does the $\overline{X}$ sampling distribution appear to be approximately normal?

10. Carry out a simulation experiment using a statistical computer package or other software to study the sampling distribution of $\overline{X}$ when the population distribution is lognormal with $E[\ln(X)] = 3$ and $V[\ln(X)] = 1$. Consider the four sample sizes $n = 10$, 20, 30, and 50, and in each case use 500 replications. For which of these sample sizes does the $\overline{X}$ sampling distribution appear to be approximately normal?

6.2 The Distribution of the Sample Mean

The importance of the sample mean $\overline{X}$ springs from its use in drawing conclusions about the population mean μ. Some of the most frequently used inferential procedures are based on properties of the sampling distribution of $\overline{X}$. A preview of these properties appeared in the calculations and simulation experiments of the previous section, where we noted relationships between $E(\overline{X})$ and μ and also among $V(\overline{X})$, σ^2, and n.

PROPOSITION Let $X_1, X_2, \ldots, X_n$ be a random sample from a distribution with mean value μ and standard deviation σ. Then

1. $E(\overline{X}) = \mu_{\overline{X}} = \mu$

2. $V(\overline{X}) = \sigma^2_{\overline{X}} = \sigma^2/n$ and $\sigma_{\overline{X}} = \sigma/\sqrt{n}$

In addition, with $T_o = X_1 + \cdots + X_n$ (the sample total), $E(T_o) = n\mu$, $V(T_o) = n\sigma^2$, and $\sigma_{T_o} = \sqrt{n}\sigma$.

Proofs of these results are deferred to the next section. According to Result 1, the sampling (i.e., probability) distribution of $\overline{X}$ is centered precisely at the mean of the population from which the sample has been selected. Result 2 shows that the $\overline{X}$ distribution becomes more concentrated about μ as the sample size n increases. In marked contrast, the distribution of T_o becomes more spread out as n increases. Averaging moves probability in toward the middle, whereas totaling spreads probability out over a wider and wider range of values.

Example 6.6 At an automobile body shop the expected number of days in the shop for an American car is 4.5 and the standard deviation is 2 days. Let $X_1, X_2, \ldots, X_{25}$ be a random sample of size 25, where each X_i is the number of days for an American car to be fixed at the shop. Then the expected value of the sample mean number of days in the shop is $E(\overline{X}) = \mu = 4.5$, and the expected total number of days in the shop for the 25 cars is $E(T_o) = n\mu = 25(4.5) = 112.5$. The standard deviations of $\overline{X}$ and T_o are

$$\sigma_{\overline{X}} = \frac{\sigma}{\sqrt{n}} = \frac{2}{\sqrt{25}} = .4 \qquad \sigma_{T_o} = \sqrt{n}\sigma = \sqrt{25}(2) = 10$$

If the sample size increases to $n = 100$, $E(\overline{X})$ is unchanged, but $\sigma_{\overline{X}} = .2$, half of its previous value (the sample size must be quadrupled to halve the standard deviation of $\overline{X}$). ■

The Case of a Normal Population Distribution

Looking back to the simulation experiment of Example 6.4, we see that when the population distribution is normal, each histogram of $\bar{x}$ values is well approximated by a normal curve. The precise result follows (see the next section for a derivation).

PROPOSITION Let $X_1, X_2, \ldots, X_n$ be a random sample from a *normal* distribution with mean μ and standard deviation σ. Then for *any* n, $\overline{X}$ is normally distributed (with mean μ and standard deviation $\sigma/\sqrt{n}$), as is T_o (with mean $n\mu$ and standard deviation $\sqrt{n}\sigma$).

We know everything there is to know about the $\overline{X}$ and T_o distributions when the population distribution is normal. In particular, probabilities such as $P(a \le \overline{X} \le b)$ and $P(c \le T_o \le d)$ can be obtained simply by standardizing. Figure 6.9 illustrates the proposition.

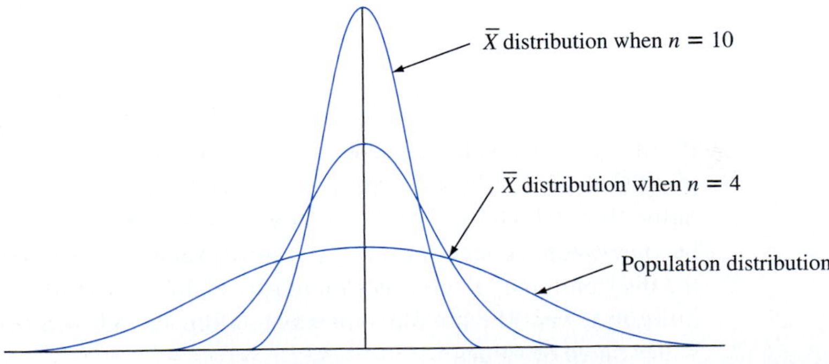

Figure 6.9 A normal population distribution and $\overline{X}$ sampling distributions

Example 6.7 The time that it takes a randomly selected rat of a certain subspecies to find its way through a maze is a normally distributed rv with $\mu = 1.5$ min and $\sigma = .35$ min. Suppose five rats are selected. Let $X_1, \ldots, X_5$ denote their times in the maze. Assuming the X_i's to be a random sample from this normal distribution, what is the probability that the total time $T_o = X_1 + \cdots + X_5$ for the five is between 6 and 8 min? By the proposition, T_o has a normal distribution with $\mu_{T_o} = n\mu = 5(1.5) = 7.5$ and variance $\sigma^2_{T_o} = n\sigma^2 = 5(.1225) = .6125$, so $\sigma_{T_o} = .783$. To standardize T_o, subtract μ_{T_o} and divide by σ_{T_o}:

$$P(6 \le T_o \le 8) = P\left(\frac{6 - 7.5}{.783} \le Z \le \frac{8 - 7.5}{.783}\right)$$

$$= P(-1.92 \le Z \le .64) = \Phi(.64) - \Phi(-1.92) = .7115$$

Determination of the probability that the sample average time $\overline{X}$ (a normally distributed variable) is at most 2.0 min requires $\mu_{\overline{X}} = \mu = 1.5$ and $\sigma_{\overline{X}} = \sigma/\sqrt{n} = .35/\sqrt{5} = .1565$. Then

$$P(\overline{X} \le 2.0) = P\left(Z \le \frac{2.0 - 1.5}{.1565}\right) = P(Z \le 3.19) = \Phi(3.19) = .9993$$ ■

The Central Limit Theorem

When the X_i's are normally distributed, so is $\overline{X}$ for every sample size n. The simulation experiment of Example 6.5 suggests that even when the population distribution is highly nonnormal, averaging produces a distribution more bell-shaped than the one being sampled. A reasonable conjecture is that if n is large, a suitable normal curve will approximate the actual distribution of $\overline{X}$. The formal statement of this result is the most important theorem of probability.

THEOREM **The Central Limit Theorem (CLT)**

Let $X_1, X_2, \ldots, X_n$ be a random sample from a distribution with mean μ and variance σ^2. Then, in the limit as $n \to \infty$, the standardized versions of $\overline{X}$ and T_o have the standard normal distribution. That is,

$$\lim_{n\to\infty} P\left(\frac{\overline{X} - \mu}{\sigma/\sqrt{n}} \le z\right) = P(Z \le z) = \Phi(z)$$

and

$$\lim_{n\to\infty} P\left(\frac{T_o - n\mu}{\sqrt{n}\sigma} \le z\right) = P(Z \le z) = \Phi(z)$$

where Z is a standard normal rv. As an alternative to saying that the standardized versions of $\overline{X}$ and T_o have limiting standard normal distributions, it is customary to say that $\overline{X}$ and T_o are **asymptotically normal**. Thus when n is sufficiently large, $\overline{X}$ has approximately a normal distribution with mean $\mu_{\overline{X}} = \mu$ and variance $\sigma^2_{\overline{X}} = \sigma^2/n$. Equivalently, for large n the sum T_o has approximately a normal distribution with mean $\mu_{T_o} = n\mu$ and variance $\sigma^2_{T_o} = n\sigma^2$.

A partial proof of the CLT appears in the appendix to this chapter. It is shown that, if the moment generating function exists, then the mgf of the standardized $\overline{X}$ (and T_o) approaches the standard normal mgf. With the aid of an advanced probability theorem, this implies the CLT statement about convergence of probabilities.

Figure 6.10 illustrates the Central Limit Theorem. According to the CLT, when n is large and we wish to calculate a probability such as $P(a \le \overline{X} \le b)$, we need only "pretend" that $\overline{X}$ is normal, standardize it, and use the normal table. The resulting answer will be approximately correct. The exact answer could be obtained only by first finding the distribution of $\overline{X}$, so the CLT provides a truly impressive shortcut.

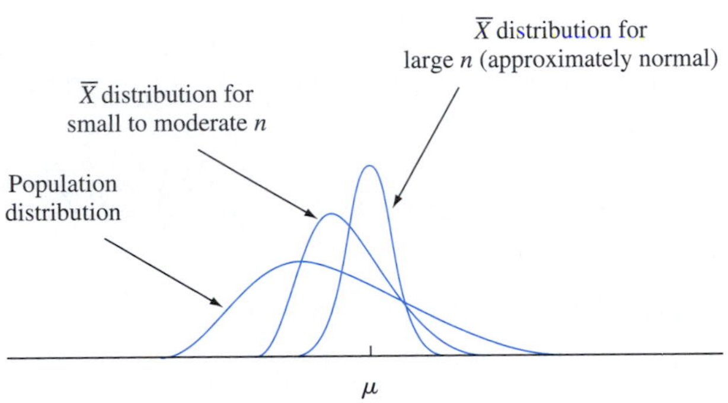

Figure 6.10 The Central Limit Theorem illustrated

Example 6.8 When a batch of a certain chemical product is prepared, the amount of a particular impurity in the batch is a random variable with mean value 4.0 g and standard deviation 1.5 g. If 50 batches are independently prepared, what is the (approximate) probability that the sample average amount of impurity $\overline{X}$ is between 3.5 and 3.8 g? According to the rule of thumb to be stated shortly, $n = 50$ is large enough for the CLT to be applicable. $\overline{X}$ then has approximately a normal distribution with mean value $\mu_{\overline{X}} = 4.0$ and $\sigma_{\overline{X}} = 1.5/\sqrt{50} = .2121$, so

$$P(3.5 \le \overline{X} \le 3.8) \approx P\left(\frac{3.5 - 4.0}{.2121} \le Z \le \frac{3.8 - 4.0}{.2121}\right)$$

$$= \Phi(-.94) - \Phi(-2.36) = .1645$$

■

Example 6.9 A certain consumer organization customarily reports the number of major defects for each new automobile that it tests. Suppose the number of such defects for a certain model is a random variable with mean value 3.2 and standard deviation 2.4. Among 100 randomly selected cars of this model, how likely is it that the sample average number of major defects exceeds 4? Let X_i denote the number of major defects for the ith car in the random sample. Notice that X_i is a discrete rv, but the CLT is not limited to continuous random variables. Also, although the fact that the standard deviation of this nonnegative variable is quite large relative to the mean value suggests that its distribution is positively skewed, the large sample size implies that $\overline{X}$ does have approximately a normal distribution. Using $\mu_{\overline{X}} = 3.2$ and $\sigma_{\overline{X}} = .24$,

$$P(\overline{X} > 4) \approx P\left(Z > \frac{4 - 3.2}{.24}\right) = 1 - \Phi(3.33) = .0004$$

The CLT provides insight into why many random variables have probability distributions that are approximately normal. For example, the measurement error in a scientific experiment can be thought of as the sum of a number of underlying perturbations and errors of small magnitude.

Although the usefulness of the CLT for inference will soon be apparent, the intuitive content of the result gives many beginning students difficulty. Again looking back to Figure 6.2, the probability histogram on the left is a picture of the distribution being sampled. It is discrete and quite skewed, so it does not look at all like a normal distribution. The distribution of $\overline{X}$ for $n = 2$ starts to exhibit some symmetry, and this is even more pronounced for $n = 4$ in Figure 6.3. Figure 6.11 contains the probability distribution of $\overline{X}$ for $n = 8$, as well as a probability histogram for this distribution. With

$\bar{x}$	40	40.625	41.25	41.875	42.5	43.125
$p(\bar{x})$	.0000	.0000	.0003	.0012	.0038	.0112

$\bar{x}$	43.75	44.375	45	45.625	46.25	46.875
$p(\bar{x})$	.0274	.0556	.0954	.1378	.1704	.1746

$\bar{x}$	47.5	48.125	48.75	49.375	50
$p(\bar{x})$	.1474	.0998	.0519	.0188	.0039

(a)

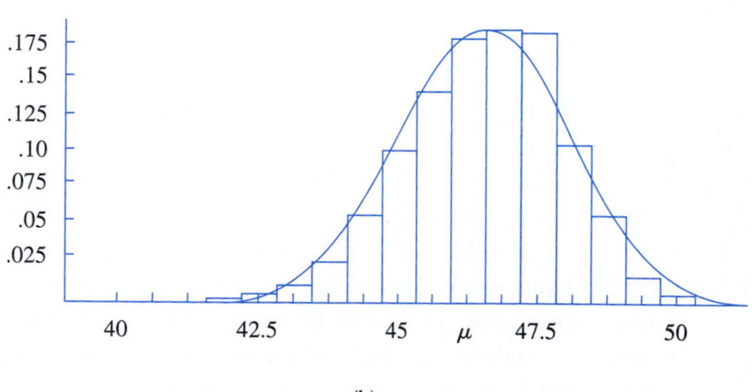

(b)

Figure 6.11 (a) Probability distribution of $\overline{X}$ for $n = 8$; (b) probability histogram and normal approximation to the distribution of $\overline{X}$ when the original distribution is as in Example 6.2

$\mu_{\overline{X}} = \mu = 46.5$ and $\sigma_{\overline{X}} = \sigma/\sqrt{n} = 3.905/\sqrt{8} = 1.38$, if we fit a normal curve with this mean and standard deviation through the histogram of $\overline{X}$, the areas of rectangles in the probability histogram are reasonably well approximated by the normal curve areas, at least in the central part of the distribution. The picture for T_o is similar except that the horizontal scale is much more spread out, with T_o ranging from 320 ($\overline{x} = 40$) to 400 ($\overline{x} = 50$).

A practical difficulty in applying the CLT is in knowing when n is sufficiently large. The problem is that the accuracy of the approximation for a particular n depends on the shape of the original underlying distribution being sampled. If the underlying distribution is symmetric and there is not much probability in the tails, then the approximation will be good even for a small n, whereas if it is highly skewed or there is a lot of probability in the tails, then a large n will be required. For example, if the distribution is uniform on an interval, then it is symmetric with no probability in the tails, and the normal approximation is very good for n as small as 10. However, at the other extreme, a distribution can have such fat tails that the mean fails to exist and the Central Limit Theorem does not apply, so no n is big enough. We will use the following rule of thumb, which is frequently somewhat conservative.

RULE OF THUMB

If $n > 30$, the Central Limit Theorem can be used.

Of course, there are exceptions, but this rule applies to most distributions of real data.

Other Applications of the Central Limit Theorem

The CLT can be used to justify the normal approximation to the binomial distribution discussed in Chapter 4. Recall that a binomial variable X is the number of successes in a binomial experiment consisting of n independent success/failure trials with $p = P(S)$ for any particular trial. Define new rv's $X_1, X_2, \ldots, X_n$ by

$$X_i = \begin{cases} 1 & \text{if the } i\text{th trial results in a success} \\ 0 & \text{if the } i\text{th trial results in a failure} \end{cases} \qquad (i = 1, \ldots, n)$$

Because the trials are independent and $P(S)$ is constant from trial to trial, the X_i's are iid (a random sample from a Bernoulli distribution). The CLT then implies that if n is sufficiently large, both the sum and the average of the X_i's have approximately normal distributions. When the X_i's are summed, a 1 is added for every S that occurs and a 0 for every F, so $X_1 + \cdots + X_n = X = T_o$. The sample mean of the X_i's is $\overline{X} = X/n$, the sample proportion of successes. That is, both X and X/n are approximately normal when n is large. The necessary sample size for this approximation depends on the value of p: When p is close to .5, the distribution of each X_i is reasonably symmetric (see Figure 6.12), whereas the distribution is quite skewed when p is near 0 or 1. Using the approximation only if both $np \geq 10$ and $n(1 - p) \geq 10$ ensures that n is large enough to overcome any skewness in the underlying Bernoulli distribution.

Recall from Section 4.5 that X has a lognormal distribution if $\ln(X)$ has a normal distribution.

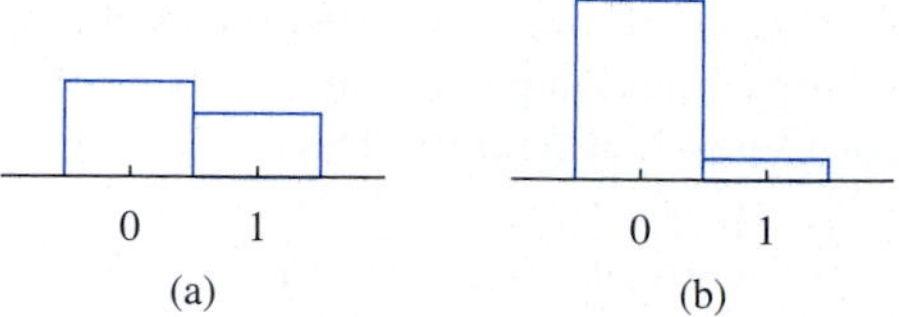

Figure 6.12 Two Bernoulli distributions: (a) p = .4 (reasonably symmetric); (b) p = .1 (very skewed)

PROPOSITION

Let $X_1, X_2, \ldots, X_n$ be a random sample from a distribution for which only positive values are possible $[P(X_i > 0) = 1]$. Then if n is sufficiently large, the product $Y = X_1 X_2 \cdot \cdots \cdot X_n$ has approximately a lognormal distribution.

To verify this, note that

$$\ln(Y) = \ln(X_1) + \ln(X_2) + \cdots + \ln(X_n)$$

Since $\ln(Y)$ is a sum of independent and identically distributed rv's [the $\ln(X_i)$'s], it is approximately normal when n is large, so Y itself has approximately a lognormal distribution. As an example of the applicability of this result, it has been argued that the damage process in plastic flow and crack propagation is a multiplicative process, so that variables such as percentage elongation and rupture strength have approximately lognormal distributions.

The Law of Large Numbers

Recall the first proposition in this section: If $X_1, X_2, \ldots, X_n$ is a random sample from a distribution with mean μ and variance σ^2, then $E(\overline{X}) = \mu$ and $V(\overline{X}) = \sigma^2/n$. What happens to $\overline{X}$ as the number of observations becomes large? The expected value of $\overline{X}$ remains at μ but the variance approaches zero. That is, $V(\overline{X}) = E[(\overline{X} - \mu)]^2 \rightarrow 0$. We say that $\overline{X}$ converges *in mean square* to μ because the mean of the squared difference between $\overline{X}$ and μ goes to zero. This is one form of the *Law of Large Numbers*, which says that $\overline{X} \rightarrow \mu$ as $n \rightarrow \infty$.

The Law of Large Numbers should be intuitively reasonable. For example, consider a fair die with equal probabilities for the values 1, 2, . . . , 6 so $\mu = 3.5$. After many repeated throws of the die $x_1, x_2, \ldots, x_n$, we should be surprised if $\bar{x}$ is not close to 3.5.

Another form of convergence can be shown with the help of Chebyshev's inequality (Exercises 43 and 135 in Chapter 3), which states that for any random variable Y, $P(|Y - \mu| \geq k\sigma) \leq 1/k^2$ whenever $k \geq 1$. In words, the probability that Y is at least k standard deviations away from its mean value is at most $1/k^2$; as k increases, the probability gets closer to 0. Apply this to the mean $Y = \overline{X}$ of a random sample $X_1, X_2, \ldots, X_n$ from a distribution with mean μ and variance σ^2. Then $E(Y) = E(\overline{X}) = \mu$ and $V(Y) = V(\overline{X}) = \sigma^2/n$, so the σ in Chebyshev's inequality needs to be replaced by $\sigma/\sqrt{n}$. Now let ε be a positive number close to 0, such as .01 or .001, and consider

$P(|\overline{X} - \mu| \geq \varepsilon)$, the probability that $\overline{X}$ differs from μ by at least ε (at least .01, at least .001, etc.). What happens to this probability as $n \to \infty$? Setting $\varepsilon = k\sigma/\sqrt{n}$ and solving for k gives $k = \varepsilon\sqrt{n}/\sigma$. Thus

$$P(|\overline{X} - \mu| \geq \varepsilon) = P\left[|\overline{X} - \mu| \geq \left(\varepsilon\frac{\sqrt{n}}{\sigma}\right)\frac{\sigma}{\sqrt{n}}\right] \leq \frac{1}{\left(\varepsilon\frac{\sqrt{n}}{\sigma}\right)^2} = \frac{\sigma^2}{n\varepsilon^2}$$

so as n gets arbitrarily large, the probability will approach 0 regardless of how small ε is. That is, for *any* ε, the chance that $\overline{X}$ will differ from μ by at least ε decreases to 0 as the sample size increases. Because of this, statisticians say that $\overline{X}$ *converges to* μ *in probability*.

We can summarize the two forms of the Law of Large Numbers in the following theorem.

THEOREM If $X_1, X_2, \ldots, X_n$ is a random sample from a distribution with mean μ and variance σ^2, then $\overline{X}$ converges to μ

a. In mean square $\quad E[(\overline{X} - \mu)^2] \to 0$ as $n \to \infty$

b. In probability $\quad P(|\overline{X} - \mu| \geq \varepsilon) \to 0$ as $n \to \infty$

Example 6.10 Let's apply the Law of Large Numbers to the repeated flipping of a fair coin. Intuitively, the fraction of heads should approach $\frac{1}{2}$ as we get more and more coin flips. For $i = 1, \ldots, n$, let $X_i = 1$ if the *ith* toss is a head and $= 0$ if it is a tail. Then the X_i's are independent and each X_i is a Bernoulli rv with $\mu = .5$ and standard deviation $\sigma = .5$. Furthermore, the sum $X_1 + X_2 + \cdots + X_n$ is the total number of heads, so $\overline{X}$ is the fraction of heads. Thus, the fraction of heads approaches the mean, $\mu = .5$, by the Law of Large Numbers. ■

Exercises Section 6.2 (11–26)

11. The inside diameter of a randomly selected piston ring is a random variable with mean value 12 cm and standard deviation .04 cm.

a. If $\overline{X}$ is the sample mean diameter for a random sample of $n = 16$ rings, where is the sampling distribution of $\overline{X}$ centered, and what is the standard deviation of the $\overline{X}$ distribution?

b. Answer the questions posed in part (a) for a sample size of $n = 64$ rings.

c. For which of the two random samples, the one of part (a) or the one of part (b), is $\overline{X}$ more likely to be within .01 cm of 12 cm? Explain your reasoning.

12. Refer to Exercise 11. Suppose the distribution of diameter is normal.

a. Calculate $P(11.99 \leq \overline{X} \leq 12.01)$ when $n = 16$.

b. How likely is it that the sample mean diameter exceeds 12.01 when $n = 25$?

13. Let $X_1, X_2, \ldots, X_{100}$ denote the actual net weights of 100 randomly selected 50-lb bags of fertilizer.

a. If the expected weight of each bag is 50 and the variance is 1, calculate $P(49.75 \leq \overline{X} \leq 50.25)$ (approximately) using the CLT.

b. If the expected weight is 49.8 lb rather than 50 lb so that on average bags are underfilled, calculate $P(49.75 \leq \overline{X} \leq 50.25)$.

14. There are 40 students in an elementary statistics class. On the basis of years of experience, the instructor knows that the time needed to grade a randomly chosen first examination paper is a random variable with an expected value of 6 min and a standard deviation of 6 min.

a. If grading times are independent and the instructor begins grading at 6:50 p.m. and grades continuously, what is the (approximate) probability that he is through grading before the 11:00 p.m. TV news begins?

b. If the sports report begins at 11:10, what is the probability that he misses part of the report if he waits until grading is done before turning on the TV?

15. The tip percentage at a restaurant has a mean value of 18% and a standard deviation of 6%.

a. What is the approximate probability that the sample mean tip percentage for a random sample of 40 bills is between 16% and 19%?

b. If the sample size had been 15 rather than 40, could the probability requested in part (a) be calculated from the given information?

16. The time taken by a randomly selected applicant for a mortgage to fill out a certain form has a normal distribution with mean value 10 min and standard deviation 2 min. If five individuals fill out a form on one day and six on another, what is the probability that the sample average amount of time taken on each day is at most 11 min?

17. The lifetime of a certain type of battery is normally distributed with mean value 10 hours and standard deviation 1 hour. There are four batteries in a package. What lifetime value is such that the total lifetime of all batteries in a package exceeds that value for only 5% of all packages?

18. Rockwell hardness of pins of a certain type is known to have a mean value of 50 and a standard deviation of 1.2.

a. If the distribution is normal, what is the probability that the sample mean hardness for a random sample of 9 pins is at least 51?

b. What is the (approximate) probability that the sample mean hardness for a random sample of 40 pins is at least 51?

19. Suppose the sediment density (g/cm) of a randomly selected specimen from a certain region is normally distributed with mean 2.65 and standard deviation .85 (suggested in "Modeling Sediment and Water Column Interactions for Hydrophobic Pollutants," *Water Res.*, 1984: 1169–1174).

a. If a random sample of 25 specimens is selected, what is the probability that the sample average sediment density is at most 3.00? Between 2.65 and 3.00?

b. How large a sample size would be required to ensure that the first probability in part (a) is at least .99?

20. The first assignment in a statistical computing class involves running a short program. If past experience indicates that 40% of all students will make no programming errors, compute the (approximate) probability that in a class of 50 students

a. At least 25 will make no errors (*Hint*: Normal approximation to the binomial)

b. Between 15 and 25 (inclusive) will make no errors

21. The number of parking tickets issued in a certain city on any given weekday has a Poisson distribution with parameter $\lambda = 50$. What is the approximate probability that

a. Between 35 and 70 tickets are given out on a particular day? (*Hint*: When λ is large, a Poisson rv has approximately a normal distribution.)

b. The total number of tickets given out during a 5-day week is between 225 and 275?

22. Suppose the distribution of the time X (in hours) spent by students at a certain university on a particular project is gamma with parameters $\alpha = 50$ and $\beta = 2$. Because α is large, it can be shown that X has approximately a normal distribution. Use this fact to compute the probability that a randomly selected student spends at most 125 hours on the project.

23. The Central Limit Theorem says that $\overline{X}$ is approximately normal if the sample size is large. More specifically, the theorem states that the standardized $\overline{X}$ has a limiting standard normal distribution. That is, $(\overline{X} - \mu)/(\sigma/\sqrt{n})$ has a distribution approaching the standard normal. Can you reconcile this with the Law of Large Numbers? If the standardized $\overline{X}$ is approximately standard normal, then what about $\overline{X}$ itself?

24. Assume a sequence of independent trials, each with probability p of success. Use the Law of Large Numbers to show that the proportion of successes approaches p as the number of trials becomes large.

25. Let Y_n be the largest order statistic in a sample of size n from the uniform distribution on $[0, \theta]$. Show that Y_n converges in probability to θ, that is, that $P(|Y_n - \theta| \geq \varepsilon) \to 0$ as n approaches ∞. *Hint*: The pdf of the largest order statistic appears in Section 5.5, so the relevant probability can be obtained by integration (Chebyshev's inequality is not needed).

26. A friend commutes by bus to and from work six days per week. Suppose that waiting time is uniformly distributed between 0 and 10 min, and that waiting times going and returning on various days are independent of one another. What is the approximate probability that total waiting time for an entire week is at most 75 min? *Hint*: Carry out a simulation experiment using statistical software to investigate the sampling distribution of T_o under these circumstances. The idea of this problem is that even for an n as small as 12, T_o and $\overline{X}$ should be approximately normal when the parent distribution is uniform. What do you think?

6.3 The Distribution of a Linear Combination

The sample mean $\overline{X}$ and sample total T_o are special cases of a type of random variable that arises very frequently in statistical applications.

DEFINITION

Given a collection of n random variables $X_1, \ldots, X_n$ and n numerical constants $a_1, \ldots, a_n$, the rv

$$Y = a_1X_1 + \cdots + a_nX_n = \sum_{i=1}^{n} a_iX_i \tag{6.6}$$

is called a **linear combination** of the X_i's.

Taking $a_1 = a_2 = \cdots = a_n = 1$ gives $Y = X_1 + \cdots + X_n = T_o$, and $a_1 = a_2 = \cdots = a_n = \frac{1}{n}$ yields $Y = \frac{1}{n}X_1 + \cdots + \frac{1}{n}X_n = \frac{1}{n}(X_1 + \cdots + X_n) = \frac{1}{n}T_o = \overline{X}$. Notice that we are not requiring the X_i's to be independent or identically distributed. All the X_i's could have different distributions and therefore different mean values and variances. We first consider the expected value and variance of a linear combination.

PROPOSITION

Let $X_1, X_2, \ldots, X_n$ have mean values $\mu_1, \ldots, \mu_n$, respectively, and variances $\sigma_1^2, \ldots, \sigma_n^2$, respectively.

1. Whether or not the X_i's are independent,

$$\begin{aligned} E(a_1X_1 + a_2X_2 + \cdots + a_nX_n) &= a_1E(X_1) + a_2E(X_2) + \cdots + a_nE(X_n) \\ &= a_1\mu_1 + \cdots + a_n\mu_n \end{aligned} \tag{6.7}$$

2. If $X_1, \ldots, X_n$ are independent,

$$\begin{aligned} V(a_1X_1 + a_2X_2 + \cdots + a_nX_n) &= a_1^2V(X_1) + a_2^2V(X_2) + \cdots + a_n^2V(X_n) \\ &= a_1^2\sigma_1^2 + \cdots + a_n^2\sigma_n^2 \end{aligned} \tag{6.8}$$

and

$$\sigma_{a_1X_1+\cdots+a_nX_n} = \sqrt{a_1^2\sigma_1^2 + \cdots + a_n^2\sigma_n^2} \tag{6.9}$$

3. For any $X_1, \ldots, X_n$,

$$V(a_1X_1 + \cdots + a_nX_n) = \sum_{i=1}^{n}\sum_{j=1}^{n} a_ia_j\text{Cov}(X_i, X_j) \tag{6.10}$$

Proofs are sketched out later in the section. A paraphrase of (6.7) is that the expected value of a linear combination is the same linear combination of the expected values—for example, $E(2X_1 + 5X_2) = 2\mu_1 + 5\mu_2$. The result (6.8) in Statement 2 is a special case of (6.10) in Statement 3; when the X_i's are independent, $\text{Cov}(X_i, X_j) = 0$ for $i \neq j$ and $= V(X_i)$ for $i = j$ (this simplification actually occurs when the X_i's are uncorrelated, a weaker condition than independence). Specializing to the case of a random sample (X_i's iid) with $a_i = 1/n$ for every i gives $E(\overline{X}) = \mu$ and $V(\overline{X}) = \sigma^2/n$, as discussed in Section 6.2. A similar comment applies to the rules for T_o.

Example 6.11 A gas station sells three grades of gasoline: regular unleaded, extra unleaded, and super unleaded. These are priced at \$2.20, \$2.35, and \$2.50 per gallon, respectively. Let X_1, X_2, and X_3 denote the amounts of these grades purchased (gallons) on a particular day. Suppose the X_i's are independent with $\mu_1 = 1000$, $\mu_2 = 500$, $\mu_3 = 300$, $\sigma_1 = 100$, $\sigma_2 = 80$, and $\sigma_3 = 50$. The revenue from sales is $Y = 2.2X_1 + 2.35X_2 + 2.5X_3$, and

$$E(Y) = 2.2\mu_1 + 2.35\mu_2 + 2.5\mu_3 = \$4125$$
$$V(Y) = (2.2)^2\sigma_1^2 + (2.35)^2\sigma_2^2 + (2.5)^2\sigma_3^2 = 99{,}369$$
$$\sigma_Y = \sqrt{99{,}369} = \$315.23$$

■

Example 6.12 The results of the previous proposition allow for a straightforward derivation of the mean and variance of a hypergeometric rv, which were given without proof in Section 3.6. Recall that the distribution is defined in terms of a population with N items, of which M are successes and $N - M$ are failures. A sample of size n is drawn, of which X are successes. It is equivalent to view this as random arrangement of all N items, followed by selection of the first n. Let X_i be 1 if the ith item is a success and 0 if it is a failure, $i = 1, 2, \ldots, N$. Then

$$X = X_1 + X_2 + \cdots + X_n$$

According to the proposition, we can find the mean and variance of X if we can find the means, variances, and covariances of the terms in the sum. By symmetry, all N of the X_i's have the same mean and variance, and all of their covariances are the same. Because each X_i is a Bernoulli random variable with success probability $p = M/N$,

$$E(X_i) = p = \frac{M}{N} \qquad V(X_i) = p(1 - p) = \left(\frac{M}{N}\right)\left(1 - \frac{M}{N}\right)$$

Therefore,

$$E(X) = E\left(\sum_{i=1}^{n} X_i\right) = np$$

Here is a trick for finding the covariances $\text{Cov}(X_i, X_j)$, all of which equal $\text{Cov}(X_1, X_2)$. The sum of all N of the X_i's is M, which is a constant, so its variance is 0. We can use Statement 3 of the proposition to express the variance in terms of N identical variances and $N(N - 1)$ identical covariances:

$$\begin{aligned} 0 = V(M) = V\left(\sum_{i=1}^{N} X_i\right) &= NV(X_1) + N(N-1)\text{Cov}(X_1, X_2) \\ &= Np(1-p) + N(N-1)\text{Cov}(X_1, X_2) \end{aligned}$$

Solving this equation for the covariance,

$$\text{Cov}(X_1, X_2) = \frac{-p(1-p)}{N-1}$$

Thus, using Statement 3 of the proposition with n identical variances and $n(n - 1)$ identical covariances,

$$\begin{aligned} V(X) = V\left(\sum_{i=1}^{n} X_i\right) &= nV(X_1) + n(n-1)\text{Cov}(X_1, X_2) \\ &= np(1-p) + n(n-1)\frac{-p(1-p)}{N-1} \\ &= np(1-p)\left(1 - \frac{n-1}{N-1}\right) \\ &= np(1-p)\left(\frac{N-n}{N-1}\right) \end{aligned}$$

■

The Difference Between Two Random Variables

An important special case of a linear combination results from taking $n = 2$, $a_1 = 1$, and $a_2 = -1$:

$$Y = a_1X_1 + a_2X_2 = X_1 - X_2$$

We then have the following corollary to the proposition.

COROLLARY $E(X_1 - X_2) = E(X_1) - E(X_2)$ and, if X_1 and X_2 are independent, $V(X_1 - X_2) = V(X_1) + V(X_2)$.

The expected value of a difference is the difference of the two expected values, but the variance of a difference between two independent variables is the *sum, not* the difference, of the two variances. There is just as much variability in $X_1 - X_2$ as in $X_1 + X_2$ [writing $X_1 - X_2 = X_1 + (-1)X_2$, $(-1)X_2$ has the same amount of variability as X_2 itself].

Example 6.13 A certain automobile manufacturer equips a particular model with either a six-cylinder engine or a four-cylinder engine. Let X_1 and X_2 be fuel efficiencies for independently and randomly selected six-cylinder and four-cylinder cars, respectively. With $\mu_1 = 22$, $\mu_2 = 26$, $\sigma_1 = 1.2$, and $\sigma_2 = 1.5$,

$$E(X_1 - X_2) = \mu_1 - \mu_2 = 22 - 26 = -4$$

$$V(X_1 - X_2) = \sigma_1^2 + \sigma_2^2 = (1.2)^2 + (1.5)^2 = 3.69$$

$$\sigma_{X_1 - X_2} = \sqrt{3.69} = 1.92$$

If we relabel so that X_1 refers to the four-cylinder car, then $E(X_1 - X_2) = 4$, but the variance of the difference is still 3.69. ■

The Case of Normal Random Variables

When the X_i's form a random sample from a normal distribution, $\overline{X}$ and T_o are both normally distributed. Here is a more general result concerning linear combinations. The proof will be given toward the end of the section.

PROPOSITION If $X_1, X_2, \ldots, X_n$ are independent, normally distributed rv's (with possibly different means and/or variances), then any linear combination of the X_i's also has a normal distribution. In particular, the difference $X_1 - X_2$ between two independent, normally distributed variables is itself normally distributed.

Example 6.14 (Example 6.11 continued) The total revenue from the sale of the three grades of gasoline on a particular day was $Y = 2.2X_1 + 2.35X_2 + 2.5X_3$, and we calculated $\mu_Y = 4125$ and (assuming independence) $\sigma_Y = 315.23$. If the X_i's are normally distributed, the probability that revenue exceeds 4500 is

$$P(Y > 4500) = P\left(Z > \frac{4500 - 4125}{315.23}\right)$$

$$= P(Z > 1.19) = 1 - \Phi(1.19) = .1170$$ ■

The CLT can also be generalized so it applies to certain linear combinations. Roughly speaking, if n is large and no individual term is likely to contribute too much to the overall value, then Y has approximately a normal distribution.

Proofs for the Case $n = 2$ For the result concerning expected values, suppose that X_1 and X_2 are continuous with joint pdf $f(x_1, x_2)$. Then

$$\begin{aligned} E(a_1X_1 + a_2X_2) &= \int_{-\infty}^{\infty}\int_{-\infty}^{\infty} (a_1x_1 + a_2x_2)f(x_1, x_2)\,dx_1\,dx_2 \\ &= a_1\int_{-\infty}^{\infty}\int_{-\infty}^{\infty} x_1f(x_1, x_2)\,dx_2\,dx_1 + a_2\int_{-\infty}^{\infty}\int_{-\infty}^{\infty} x_2f(x_1, x_2)\,dx_1\,dx_2 \\ &= a_1\int_{-\infty}^{\infty} x_1f_{X_1}(x_1)\,dx_1 + a_2\int_{-\infty}^{\infty} x_2f_{X_2}(x_2)\,dx_2 \\ &= a_1E(X_1) + a_2E(X_2) \end{aligned}$$

Summation replaces integration in the discrete case. The argument for the variance result does not require specifying whether either variable is discrete or continuous. Recalling that $V(Y) = E[(Y - \mu_Y)^2]$,

$$\begin{aligned} V(a_1X_1 + a_2X_2) &= E\{[a_1X_1 + a_2X_2 - (a_1\mu_1 + a_2\mu_2)]^2\} \\ &= E\{a_1^2(X_1 - \mu_1)^2 + a_2^2(X_2 - \mu_2)^2 + 2a_1a_2(X_1 - \mu_1)(X_2 - \mu_2)\} \end{aligned}$$

The expression inside the braces is a linear combination of the variables $Y_1 = (X_1 - \mu_1)^2$, $Y_2 = (X_2 - \mu_2)^2$, and $Y_3 = (X_1 - \mu_1)(X_2 - \mu_2)$, so carrying the E operation through to the three terms gives $a_1^2V(X_1) + a_2^2V(X_2) + 2a_1a_2\text{Cov}(X_1, X_2)$ as required. ■

The previous proposition has a generalization to two linear combinations:

PROPOSITION Let U and V be linear combinations of the independent normal rv's $X_1, X_2, \ldots, X_n$. Then the joint distribution of U and V is bivariate normal. The converse is also true: If U and V have a bivariate normal distribution, then they can be expressed as linear combinations of independent normal rv's.

The proof uses the methods of Section 5.4 together with a little matrix theory.

Example 6.15 How can we create two bivariate normal rv's X and Y with a specified correlation ρ? Let Z_1 and Z_2 be independent standard normal rv's and let

$$X = Z_1 \qquad Y = \rho \cdot Z_1 + \sqrt{1 - \rho^2}Z_2$$

Then X and Y are linear combinations of independent normal random variables, so their joint distribution is bivariate normal. Furthermore, they each have standard deviation 1 (verify this for Y) and their covariance is ρ, so their correlation is ρ. ■

Moment Generating Functions for Linear Combinations

We shall use moment generating functions to prove the proposition on linear combinations of normal random variables, but we first need a general proposition on the distribution of linear combinations. This will be useful for normal random variables and others.

Recall that the second proposition in Section 5.2 shows how to simplify the expected value of a product of functions of independent random variables. We now use this to simplify the moment generating function of a linear combination of independent random variables.

PROPOSITION

Let $X_1, X_2, \ldots, X_n$ be independent random variables with moment generating functions $M_{X_1}(t), M_{X_2}(t), \ldots, M_{X_n}(t)$, respectively. Define $Y = a_1X_1 + a_2X_2 + \cdots + a_nX_n$, where $a_1, a_2, \ldots, a_n$ are constants. Then

$$M_Y(t) = M_{X_1}(a_1t) \cdot M_{X_2}(a_2t) \cdot \cdots \cdot M_{X_n}(a_nt)$$

In the special case that $a_1 = a_2 = \cdots = a_n = 1$,

$$M_Y(t) = M_{X_1}(t) \cdot M_{X_2}(t) \cdot \cdots \cdot M_{X_n}(t)$$

That is, the mgf of a sum of independent rv's is the product of the individual mgf's.

Proof First, we write the moment generating function of Y as the expected value of a product:

$$\begin{aligned} M_Y(t) &= E(e^{tY}) = E(e^{t(a_1X_1+a_2X_2+\cdots+a_nX_n)}) \\ &= E(e^{ta_1X_1+ta_2X_2+\cdots+ta_nX_n}) = E(e^{ta_1X_1} \cdot e^{ta_2X_2} \cdot \cdots \cdot e^{ta_nX_n}) \end{aligned}$$

Next, we use the second proposition in Section 5.2, which says that the expected value of a product of functions of independent random variables is the product of the expected values:

$$\begin{aligned} E(e^{ta_1X_1} \cdot e^{ta_2X_2} \cdot \cdots \cdot e^{ta_nX_n}) &= E(e^{ta_1X_1}) \cdot E(e^{ta_2X_2}) \cdot \cdots \cdot E(e^{ta_nX_n}) \\ &= M_{X_1}(a_1t) \cdot M_{X_2}(a_2t) \cdot \cdots \cdot M_{X_n}(a_nt) \end{aligned}$$ ■

Now let's apply this to prove the previous proposition about normality for a linear combination of independent normal random variables. If $Y = a_1X_1 + a_2X_2 + \cdots + a_nX_n$, where X_i is normally distributed with mean μ_i and standard deviation σ_i, and a_i is a constant, $i = 1, 2, \ldots, n$, then $M_{X_i}(t) = e^{\mu_i t+\sigma_i^2 t^2/2}$. Therefore,

$$\begin{aligned} M_Y(t) &= M_{X_1}(a_1t) \cdot M_{X_2}(a_2t) \cdot \cdots \cdot M_{X_n}(a_nt) \\ &= e^{\mu_1 a_1 t+\sigma_1^2 a_1^2 t^2/2} e^{\mu_2 a_2 t+\sigma_2^2 a_2^2 t^2/2} \cdot \cdots \cdot e^{\mu_n a_n t+\sigma_n^2 a_n^2 t^2/2} \\ &= e^{(\mu_1 a_1+\mu_2 a_2+\cdots+\mu_n a_n)t+(\sigma_1^2 a_1^2+\sigma_2^2 a_2^2+\cdots+\sigma_n^2 a_n^2)t^2/2} \end{aligned}$$

Because the moment generating function of Y is the moment generating function of a normal random variable, it follows that Y is normally distributed by the uniqueness principle for moment generating functions. In agreement with the first proposition in this section, the mean is the coefficient of t,

$$E(Y) = a_1\mu_1 + a_2\mu_2 + \cdots + a_n\mu_n$$

and the variance is the coefficient of $t^2/2$,

$$V(Y) = a_1^2\sigma_1^2 + a_2^2\sigma_2^2 + \cdots + a_n^2\sigma_n^2$$

Example 6.16 Suppose X and Y are independent Poisson random variables, where X has mean λ and Y has mean ν. We can show that $X + Y$ also has the Poisson distribution, with the help of the proposition on the moment generating function of a linear combination.

According to the proposition,

$$M_{X+Y}(t) = M_X(t) \cdot M_Y(t) = e^{\lambda(e^t-1)}e^{\nu(e^t-1)} = e^{(\lambda+\nu)(e^t-1)}$$

Here we have used for both X and Y the moment generating function of the Poisson distribution from Section 3.7. The resulting moment generating function for $X + Y$ is the moment generating function of a Poisson random variable with mean $\lambda + \nu$. By the uniqueness property of moment generating functions, $X + Y$ is Poisson distributed with mean $\lambda + \nu$.

In words, if X and Y are independent Poisson random variables, then their sum is also Poisson, and the mean of $X + Y$ is the sum of the two means. ■

Exercises Section 6.3 (27–45)

27. A shipping company handles containers in three different sizes: (1) 27 ft^3 ($3 \times 3 \times 3$), (2) 125 ft^3, and (3) 512 ft^3. Let X_i ($i = 1, 2, 3$) denote the number of type i containers shipped during a given week. With $\mu_i = E(X_i)$ and $\sigma_i^2 = V(X_i)$, suppose that the mean values and standard deviations are as follows:

$$\mu_1 = 200 \qquad \mu_2 = 250 \qquad \mu_3 = 100$$
$$\sigma_1 = 10 \qquad \sigma_2 = 12 \qquad \sigma_3 = 8$$

a. Assuming that X_1, X_2, X_3 are independent, calculate the expected value and variance of the total volume shipped. [*Hint*: Volume $= 27X_1 + 125X_2 + 512X_3$.]

b. Would your calculations necessarily be correct if the X_i's were not independent? Explain.

28. Let X_1, X_2, and X_3 represent the times necessary to perform three successive repair tasks at a certain service facility. Suppose they are independent, normal rv's with expected values μ_1, μ_2, and μ_3 and variances σ_1^2, σ_2^2, and σ_3^2, respectively.

a. If $\mu = \mu_2 = \mu_3 = 60$ and $\sigma_1^2 = \sigma_2^2 = \sigma_3^2 = 15$, calculate $P(X_1 + X_2 + X_3 \le 200)$. What is $P(150 \le X_1 + X_2 + X_3 \le 200)$?

b. Using the μ_i's and σ_i's given in part (a), calculate $P(55 \le \overline{X})$ and $P(58 \le \overline{X} \le 62)$.

c. Using the μ_i's and σ_i's given in part (a), calculate $P(-10 \le X_1 - .5X_2 - .5X_3 \le 5)$.

d. If $\mu_1 = 40$, $\mu_2 = 50$, $\mu_3 = 60$, $\sigma_1^2 = 10$, $\sigma_2^2 = 12$, and $\sigma_3^2 = 14$, calculate $P(X_1 + X_2 + X_3 \le 160)$ and $P(X_1 + X_2 \ge 2X_3)$.

29. Five automobiles of the same type are to be driven on a 300-mile trip. The first two will use an economy brand of gasoline, and the other three will use a name brand. Let X_1, X_2, X_3, X_4, and X_5 be the observed fuel efficiencies (mpg) for the five cars. Suppose these variables are independent and normally distributed with $\mu_1 = \mu_2 = 20$, $\mu_3 = \mu_4 = \mu_5 = 21$, and $\sigma^2 = 4$ for the economy brand and 3.5 for the name brand. Define an rv Y by

$$Y = \frac{X_1 + X_2}{2} - \frac{X_3 + X_4 + X_5}{3}$$

so that Y is a measure of the difference in efficiency between economy gas and name-brand gas. Compute $P(0 \le Y)$ and $P(-1 \le Y \le 1)$. (*Hint*: $Y = a_1X_1 + \cdots + a_5X_5$, with $a_1 = \frac{1}{2}, \ldots, a_5 = -\frac{1}{3}$.)

30. Exercise 22 in Chapter 5 introduced random variables X and Y, the number of cars and buses, respectively, carried by a ferry on a single trip. The joint pmf of X and Y is given in the table in Exercise 7 of Chapter 5. It is readily verified that X and Y are independent.

a. Compute the expected value, variance, and standard deviation of the total number of vehicles on a single trip.

b. If each car is charged \$3 and each bus \$10, compute the expected value, variance, and standard deviation of the revenue resulting from a single trip.

31. A concert has three pieces of music to be played before intermission. The time taken to play each piece has a normal distribution. Assume that the three times are independent of one another. The mean times are 15, 30, and 20 minutes, respectively, and the standard deviations are 1, 2, and 1.5 minutes, respectively. What is the probability that this part of the concert takes at most one hour? Are there reasons to question the independence assumption? Explain.

32. Refer to Exercise 3 in Chapter 5.

a. Calculate the covariance between X_1 = the number of customers in the express checkout and X_2 = the number of customers in the superexpress checkout.

b. Calculate $V(X_1 + X_2)$. How does this compare to $V(X_1) + V(X_2)$?

33. Suppose your waiting time for a bus in the morning is uniformly distributed on [0, 8], whereas waiting time in the evening is uniformly distributed on [0, 10] independent of morning waiting time.

a. If you take the bus each morning and evening for a week, what is your total expected waiting time? (*Hint*: Define rv's $X_1, \ldots, X_{10}$ and use a rule of expected value.)

b. What is the variance of your total waiting time?

c. What are the expected value and variance of the difference between morning and evening waiting times on a given day?

d. What are the expected value and variance of the difference between total morning waiting time and total evening waiting time for a particular week?

34. Suppose that when the pH of a certain chemical compound is 5.00, the pH measured by a randomly selected beginning chemistry student is a random variable with mean 5.00 and standard deviation .2. A large batch of the compound is subdivided and a sample given to each student in a morning lab and each student in an afternoon lab. Let $\overline{X}$ = the average pH as determined by the morning students and $\overline{Y}$ = the average pH as determined by the afternoon students.

a. If pH is a normally distributed random variable and there are 25 students in each lab, compute $P(-.1 \le \overline{X} - \overline{Y} \le .1)$. (*Hint*: $\overline{X} - \overline{Y}$ is a linear combination of normal variables, so it is normally distributed. Compute $\mu_{\overline{X}-\overline{Y}}$ and $\sigma_{\overline{X}-\overline{Y}}$.)

b. If there are 36 students in each lab, but pH determinations are not assumed normal, calculate (approximately) $P(-.1 \le \overline{X} - \overline{Y} \le .1)$.

35. If two loads are applied to a cantilever beam as shown in the accompanying drawing, the bending moment at 0 due to the loads is $a_1X_1 + a_2X_2$.

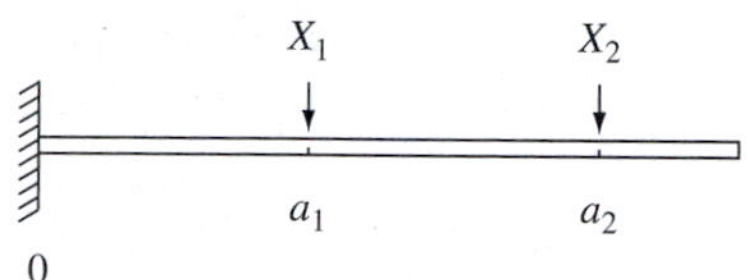

a. Suppose that X_1 and X_2 are independent rv's with means 2 and 4 kips, respectively, and standard deviations .5 and 1.0 kip, respectively. If a_1 = 5 ft and a_2 = 10 ft, what is the expected bending moment and what is the standard deviation of the bending moment?

b. If X_1 and X_2 are normally distributed, what is the probability that the bending moment will exceed 75 kip-ft?

c. Suppose the positions of the two loads are random variables. Denoting them by A_1 and A_2, assume that these variables have means of 5 and 10 ft, respectively, that each has a standard deviation of .5, and that all A_i's and X_i's are independent of one another. What is the expected moment now?

d. For the situation of part (c), what is the variance of the bending moment?

e. If the situation is as described in part (a) except that $\text{Corr}(X_1, X_2) = .5$ (so that the two loads are not independent), what is the variance of the bending moment?

36. One piece of PVC pipe is to be inserted inside another piece. The length of the first piece is normally distributed with mean value 20 in. and standard deviation .5 in. The length of the second piece is a normal rv with mean and standard deviation 15 in. and .4 in., respectively. The amount of overlap is normally distributed with mean value 1 in. and standard deviation .1 in. Assuming that the lengths and amount of overlap are independent of one another, what is the probability that the total length after insertion is between 34.5 in. and 35 in.?

37. Two airplanes are flying in the same direction in adjacent parallel corridors. At time $t = 0$, the first airplane is 10 km ahead of the second one. Suppose the speed of the first plane (km/hr) is normally distributed with mean 520 and standard deviation 10 and the second plane's speed, independent of the first, is also normally distributed with mean and standard deviation 500 and 10, respectively.

a. What is the probability that after 2 hr of flying, the second plane has not caught up to the first plane?

b. Determine the probability that the planes are separated by at most 10 km after 2 hr.

38. Three different roads feed into a particular freeway entrance. Suppose that during a fixed time period, the number of cars coming from each road onto the freeway is a random variable, with expected value and standard deviation as given in the table.

	Road 1	Road 2	Road 3
Expected value	800	1000	600
Standard deviation	16	25	18

a. What is the expected total number of cars entering the freeway at this point during the period? (*Hint*: Let X_i = the number from road i.)

b. What is the variance of the total number of entering cars? Have you made any assumptions about the relationship between the numbers of cars on the different roads?

c. With X_i denoting the number of cars entering from road i during the period, suppose that $\text{Cov}(X_1, X_2) = 80$, $\text{Cov}(X_1, X_3) = 90$, and $\text{Cov}(X_2, X_3) = 100$ (so that the three streams of traffic are not independent). Compute the expected total number of entering cars and the standard deviation of the total.

39. Suppose we take a random sample of size n from a continuous distribution having median 0 so that the probability of any one observation being positive is .5. We now disregard the signs of the observations, rank them from smallest to largest in absolute value, and then let W = the sum of the ranks of the observations having positive signs. For example, if the observations are $-.3$, $+.7$, $+2.1$, and -2.5, then the ranks of positive observations are 2 and 3, so $W = 5$. In Chapter 14, W will be called *Wilcoxon's signed-rank statistic*. W can be represented as follows:

$$W = 1 \cdot Y_1 + 2 \cdot Y_2 + 3 \cdot Y_3 + \cdots + n \cdot Y_n$$
$$= \sum_{i=1}^{n} i \cdot Y_i$$

where the Y_i's are independent Bernoulli rv's, each with $p = .5$ ($Y_i = 1$ corresponds to the observation with rank i being positive). Compute the following:

a. $E(Y_i)$ and then $E(W)$ using the equation for W [*Hint*: The first n positive integers sum to $n(n+1)/2$.]

b. $V(Y_i)$ and then $V(W)$ [*Hint*: The sum of the squares of the first n positive integers is $n(n+1) \cdot (2n+1)/6$.]

40. In Exercise 35, the weight of the beam itself contributes to the bending moment. Assume that the beam is of uniform thickness and density so that the resulting load is uniformly distributed on the beam. If the weight of the beam is random, the resulting load from the weight is also random; denote this load by W (kip-ft).

a. If the beam is 12 ft long, W has mean 1.5 and standard deviation .25, and the fixed loads are as described in part (a) of Exercise 35, what are the expected value and variance of the bending moment? (*Hint:* If the load due to the beam were w kip-ft, the contribution to the bending moment would be $w \int_0^{12} x\,dx$.)

b. If all three variables (X_1, X_2, and W) are normally distributed, what is the probability that the bending moment will be at most 200 kip-ft?

41. A professor has three errands to take care of in the Administration Building. Let X_i = the time that it takes for the ith errand ($i = 1, 2, 3$), and let X_4 = the total time in minutes that she spends walking to and from the building and between each errand. Suppose the X_i's are independent, normally distributed, with the following means and standard deviations: $\mu_1 = 15$, $\sigma_1 = 4$, $\mu_2 = 5$, $\sigma_2 = 1$, $\mu_3 = 8$, $\sigma_3 = 2$, $\mu_4 = 12$, $\sigma_4 = 3$. She plans to leave her office at precisely 10:00 a.m. and wishes to post a note on her door that reads, "I will return by t a.m." What time t should she write down if she wants the probability of her arriving after t to be .01?

42. For males the expected pulse rate is 70 per second and the standard deviation is 10 per second. For women the expected pulse rate is 77 per second and the standard deviation is 12 per second. Let $\overline{X}$ = the sample average pulse rate for a random sample of 40 men and let $\overline{Y}$ = the sample average pulse rate for a random sample of 36 women.

a. What is the approximate distribution of $\overline{X}$? Of $\overline{Y}$?

b. What is the approximate distribution of $\overline{X} - \overline{Y}$? Justify your answer.

c. Calculate (approximately) the probability $P(-2 \le \overline{X} - \overline{Y} \le 1)$.

d. Calculate (approximately) $P(\overline{X} - \overline{Y} \le -15)$. If you actually observed $\overline{X} - \overline{Y} \le -15$, would you doubt that $\mu_1 - \mu_2 = -7$? Explain.

43. In an area having sandy soil, 50 small trees of a certain type were planted, and another 50 trees were planted in an area having clay soil. Let X = the number of trees planted in sandy soil that survive 1 year and Y = the number of trees planted in clay soil that survive 1 year. If the probability that a tree planted in sandy soil will survive 1 year is .7 and the probability of 1-year survival in clay soil is .6, compute an approximation to $P(-5 \le X - Y \le 5)$ (do not bother with the continuity correction).

44. Let X and Y be independent gamma random variables, both with the same scale parameter β. The value of the other parameter is α_1 for X and α_2 for Y. Use moment generating functions to show that $X + Y$ is also gamma distributed with scale parameter β, and with the other parameter equal to $\alpha_1 + \alpha_2$. Is $X + Y$ gamma distributed if the scale parameters are different? Explain.

45. The proof of the Central Limit Theorem requires calculating the moment generating function for the standardized mean from a random sample of any distribution, and showing that it approaches the moment generating function of the standard normal distribution. Here we look at a particular case of the Laplace distribution, for which the calculation is simpler.

a. Letting X have pdf $f(x) = \frac{1}{2}e^{-|x|}$, $-\infty < x < \infty$, show that $M_X(t) = 1/(1 - t^2)$, $-1 < t < 1$.

b. Find the moment generating function $M_Y(t)$ for the standardized mean Y of a random sample from this distribution.

c. Show that the limit of $M_Y(t)$ is $e^{t^2/2}$, the moment generating function of a standard normal random variable. [*Hint*: Notice that the denominator of $M_Y(t)$ is of the form $(1 + a/n)^n$ and recall that the limit of this is e^a.]

6.4 Distributions Based on a Normal Random Sample

This section is about three distributions that are related to the sample variance S^2. The chi-squared, t, and F distributions all play important roles in statistics. For normal data we need to be able to work with the distribution of the sample variance, which is built from squares, and this will require finding the distribution for sums of squares of normal variables. The chi-squared distribution, defined in Section 4.4 as a special case of the gamma distribution, turns out to be just what is needed. Also, in order to use the sample standard deviation in a measure of precision for the mean $\overline{X}$, we will need a distribution that combines the square root of a chi-squared variable with a normal variable, and this is the t distribution. Finally, we will need a distribution to compare two independent sample variances, and for this we will define the F distribution in terms of the ratio of two independent chi-squared variables.

The Chi-Squared Distribution

Recall from Section 4.4 that the chi-squared distribution is a special case of the gamma distribution. It has one parameter ν called the *number of degrees of freedom* of the distribution. Possible values of ν are 1, 2, 3, ... The chi-squared pdf is

$$f(x) = \frac{1}{2^{\nu/2}\Gamma(\nu/2)} x^{(\nu/2)-1} e^{-x/2} \quad \text{if } x > 0,\ f(x) = 0 \quad \text{if } x \le 0$$

We use the notation χ^2_ν to indicate a chi-squared variable with ν df (degrees of freedom).

The mean, variance, and moment generating function of a chi-squared rv follow from the fact that the chi-squared distribution is a special case of the gamma distribution with $\alpha = \nu/2$ and $\beta = 2$:

$$\mu = \alpha\beta = \nu \qquad \sigma^2 = \alpha\beta^2 = 2\nu \qquad M_X(t) = (1 - 2t)^{-\nu/2}$$

Here is a result that is not at all obvious, a proposition showing that the square of a standard normal variable has the chi-squared distribution.

PROPOSITION If Z has a standard normal distribution and $X = Z^2$, then the pdf of X is

$$f(x) = \frac{1}{2^{1/2}\Gamma(1/2)} x^{(1/2)-1} e^{-x/2}$$

where $x > 0$ and $f(x) = 0$ if $x \le 0$. That is, X is chi-squared with 1 df, $X \sim \chi_1$.

Proof The proof involves determining the cdf of X and differentiating to get the pdf. If $x > 0$,

$$\begin{aligned} P(X \le x) &= P(Z^2 \le x) = P(-\sqrt{x} \le Z \le \sqrt{x}) \\ &= 2P(0 \le Z \le \sqrt{x}) = 2\Phi(\sqrt{x}) - 2\Phi(0) \end{aligned}$$

where Φ is the cdf of the standard normal distribution. Differentiating, and using ϕ for the pdf of the standard normal distribution, we obtain the pdf

$$f(x) = 2\phi(\sqrt{x})(.5x^{-.5}) = 2\frac{1}{\sqrt{2\pi}} e^{-.5x}(.5x^{-.5}) = \frac{1}{2^{1/2}\Gamma(1/2)} x^{(1/2)-1} e^{-x/2}$$

The last equality makes use of the relationship $\Gamma(1/2) = \sqrt{\pi}$. ■

For another proof, see Example 4.43.

The next proposition allows us to combine two independent chi-squared variables to get another.

PROPOSITION If $X_1 \sim \chi^2_{\nu_1}$, $X_2 \sim \chi^2_{\nu_2}$, and they are independent, then $X_1 + X_2 \sim \chi^2_{\nu_1+\nu_2}$.

Proof The proof uses moment generating functions. Recall from Section 6.3 that, if random variables are independent, then the moment generating function of their sum is the product of their moment generating functions. Therefore,

$$M_{X_1+X_2}(t) = M_{X_1}(t)M_{X_2}(t) = (1 - 2t)^{-\nu_1/2}(1 - 2t)^{-\nu_2/2} = (1 - 2t)^{-(\nu_1+\nu_2)/2}$$

Because the sum has the moment generating function of a chi-squared variable with $\nu_1 + \nu_2$ degrees of freedom, the uniqueness principle implies that the sum has the chi-squared distribution with $\nu_1 + \nu_2$ degrees of freedom. ■

By combining these two propositions we can see that the sum of two independent standard normal squares is chi-squared with 2 degrees of freedom, the sum of three independent standard normal squares is chi-squared with 3 degrees of freedom, and so on.

PROPOSITION If $Z_1, Z_2, \ldots, Z_n$ are independent and each has the standard normal distribution, then $Z_1^2 + Z_2^2 + \cdots + Z_n^2 \sim \chi_n^2$.

Now the meaning of the degrees of freedom parameter is clear. It is the number of independent standard normal squares that are added to build a chi-squared variable.

Figure 6.13 shows plots of the chi-squared pdf for 1, 2, 3, and 5 degrees of freedom. Notice that the pdf is unbounded for 1 df and the pdf is exponentially decreasing for 2 df. Indeed, the chi-squared distribution for 2 df is exponential with mean 2, $f(x) = \frac{1}{2}e^{-x/2}$ for $x > 0$. If $\nu > 2$ the pdf is unimodal with a peak at $x = \nu - 2$, as shown in Exercise 49. The distribution is skewed, but it becomes more symmetric as the degrees of freedom increase, and for large df values the distribution is approximately normal (see Exercise 47).

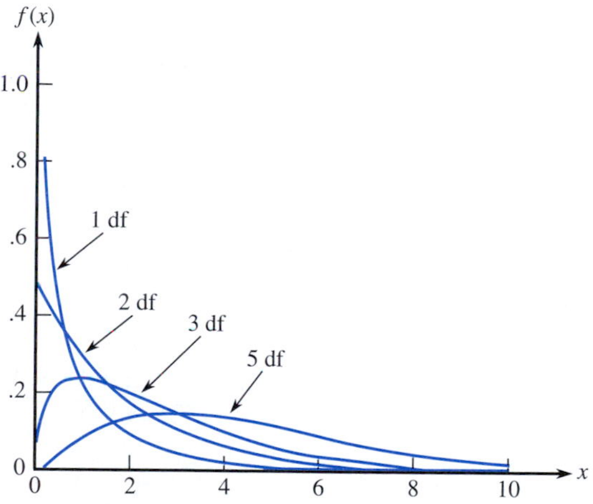

Figure 6.13 Chi-squared density curves

Except for a few special cases, it is difficult to integrate the pdf, so Table A.7 in the appendix has critical values for chi-squared distributions. For example, the second row of the table is for 2 df, and under the heading .01 the value 9.210 indicates that $P(\chi_2^2 > 9.210) = .01$. We use the notation $\chi^2_{.01,2} = 9.210$, where in general $\chi^2_{\alpha,\nu} = c$ means that $P(\chi_\nu^2 > c) = \alpha$.

In Section 1.4 we defined the sample variance in terms of $\bar{x}$,

$$s^2 = \frac{1}{n-1}\sum_{i=1}^{n}(x_i - \bar{x})^2$$

which gives an estimate of σ^2 when the population mean μ is unknown. If we happen to know the value of μ, then the appropriate estimate is

$$\hat{\sigma}^2 = \frac{1}{n}\sum_{i=1}^{n}(x_i - \mu)^2$$

Replacing x_i's by X_i's results in S^2 and $\hat{\sigma}^2$ becoming statistics (and therefore random variables). A simple function of $\hat{\sigma}^2$ is a chi-squared rv. First recall that if X is normally distributed, then $(X - \mu)/\sigma$ is a standard normal rv. Thus

$$\frac{n\hat{\sigma}^2}{\sigma^2} = \sum_{i=1}^{n}\left(\frac{X_i - \mu}{\sigma}\right)^2$$

is the sum of n independent standard normal squares, so it is χ^2_n.

A similar relationship connects the sample variance S^2 to the chi-squared distribution. First, compute

$$\begin{aligned}\sum (X_i - \mu)^2 &= \sum [(X_i - \overline{X}) + (\overline{X} - \mu)]^2 \\ &= \sum (X_i - \overline{X})^2 + 2(\overline{X} - \mu)\sum (X_i - \overline{X}) + \sum (\overline{X} - \mu)^2\end{aligned}$$

The middle term on the second line vanishes (why?). Dividing through by σ^2,

$$\begin{aligned}\sum\left(\frac{X_i - \mu}{\sigma}\right)^2 &= \sum\left(\frac{X_i - \overline{X}}{\sigma}\right)^2 + \sum\left(\frac{\overline{X} - \mu}{\sigma}\right)^2 \\ &= \sum\left(\frac{X_i - \overline{X}}{\sigma}\right)^2 + n\left(\frac{\overline{X} - \mu}{\sigma}\right)^2\end{aligned}$$

The last term can be written as the square of a standard normal rv, and therefore as a χ^2_1 rv.

$$\begin{aligned}\sum\left(\frac{X_i - \mu}{\sigma}\right)^2 &= \sum\left(\frac{X_i - \overline{X}}{\sigma}\right)^2 + n\left(\frac{\overline{X} - \mu}{\sigma}\right)^2 \\ &= \sum\left(\frac{X_i - \overline{X}}{\sigma}\right)^2 + \left(\frac{\overline{X} - \mu}{\sigma/\sqrt{n}}\right)^2\end{aligned} \qquad (6.11)$$

It is crucial here that the two terms on the right be independent. This is equivalent to saying that S^2 and $\overline{X}$ are independent. Although it is a bit much to show rigorously, one approach is based on the covariances between the sample mean and the deviations from the sample mean. Using the linearity of the covariance operator,

$$\begin{aligned}\text{Cov}(X_i - \overline{X}, \overline{X}) &= \text{Cov}(X_i, \overline{X}) - \text{Cov}(\overline{X}, \overline{X}) = \text{Cov}\left(X_i, \frac{1}{n}\sum X_i\right) - V(\overline{X}) \\ &= \frac{\sigma^2}{n} - \frac{\sigma^2}{n} = 0\end{aligned}$$

This shows that $\overline{X}$ is uncorrelated with all the deviations of the observations from their mean. In general, this does not imply independence, but in the special case of the bivariate normal distribution, being uncorrelated is equivalent to independence. Both $\overline{X}$ and $X_i - \overline{X}$ are linear combinations of the independent normal observations, so they are bivariate normal, as discussed in Section 5.3. In the special case of the bivariate normal, being uncorrelated implies independence. Because the sample variance S^2 is composed of the deviations $X_i - \overline{X}$, we have this result.

PROPOSITION If $X_1, X_2, \ldots, X_n$ are a random sample from a normal distribution, then $\overline{X}$ and S^2 are independent.

To understand this proposition better we can look at the relationship between the sample standard deviation and mean for a large number of samples. In particular, suppose we select sample after sample of size n from a particular population distribution, calculate $\bar{x}$ and s for each one, and then plot the resulting $(\bar{x}, s)$ pairs. Figure 6.14a shows the result for 1000 samples of size $n = 5$ from a standard normal population distribution. The elliptical pattern, with axes parallel to the coordinate axes, suggests no relationship between $\bar{x}$ and s—that is, independence of the statistics $\overline{X}$ and S (equivalently, $\overline{X}$ and S^2). However, this independence fails for data from a nonnormal distribution, and Figure 6.14b illustrates what happens for samples of size 5 from an exponential distribution with mean 1. This graph shows a strong relationship between the two statistics, which is what might be expected for data from a highly skewed distribution.

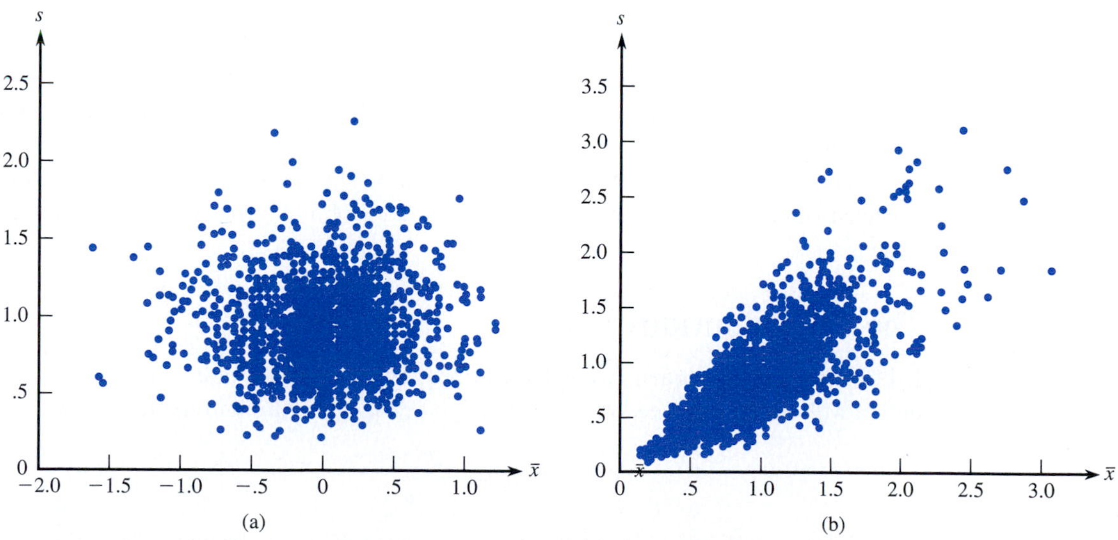

Figure 6.14 Scatter plot of $(\bar{x}, s)$ pairs

We will use the independence of $\overline{X}$ and S^2 together with the following proposition to show that S^2 is proportional to a chi-squared random variable.

PROPOSITION If $X_3 = X_1 + X_2$, with $X_1 \sim \chi^2_{\nu_1}$, $X_3 \sim \chi^2_{\nu_3}$, $\nu_3 > \nu_1$, and X_1 and X_2 are independent, then $X_2 \sim \chi^2_{\nu_3 - \nu_1}$.

The proof is similar to that of the proposition involving the sum of independent chi-squared variables, and it is left as an exercise (Exercise 51).

From Equation (6.11),

$$\sum\left(\frac{X_i - \mu}{\sigma}\right)^2 = \sum\left(\frac{X_i - \overline{X}}{\sigma}\right)^2 + \left(\frac{\overline{X} - \mu}{\sigma/\sqrt{n}}\right)^2 = \frac{(n-1)S^2}{\sigma^2} + \left(\frac{\overline{X} - \mu}{\sigma/\sqrt{n}}\right)^2$$

Assuming a random sample from the normal distribution, the term on the left is χ^2_n, and the last term is the square of a standard normal variable, so it is χ^2_1.

Putting the last two propositions together gives the following.

PROPOSITION If $X_1, X_2, \ldots, X_n$ are a random sample from a normal distribution, then $(n-1)S^2/\sigma^2 \sim \chi^2_{n-1}$.

Intuitively, the degrees of freedom make sense because s^2 is built from the deviations $(x_1 - \bar{x}), (x_2 - \bar{x}), \ldots, (x_n - \bar{x})$, which sum to zero:

$$\sum(x_i - \bar{x}) = \sum x_i - \sum \bar{x} = n\bar{x} - n\bar{x} = 0$$

The last deviation is determined by the first $(n-1)$ deviations, so it is reasonable that s^2 has only $(n-1)$ degrees of freedom.

The degrees of freedom helps to explain why the definition of s^2 has $(n-1)$ and not n in the denominator.

Knowing that $(n-1)S^2/\sigma^2 \sim \chi^2_{n-1}$, it can be shown (see Exercise 50) that the expected value of S^2 is σ^2, and also that the variance of S^2 approaches 0 as n becomes large.

The t Distribution

Let Z be a standard normal rv and let X be a χ^2_ν rv independent of Z. Then the t distribution with degrees of freedom ν is defined to be the distribution of the ratio

$$T = \frac{Z}{\sqrt{X/\nu}}$$

Sometimes we will include a subscript to indicate the df: $t = t_\nu$. From the definition it is not obvious how the t distribution can be applied to data, but the next result puts the distribution in more directly usable form.

THEOREM If $X_1, X_2, \ldots, X_n$ is a random sample from a normal distribution $N(\mu, \sigma^2)$, then the distribution of

$$T = \frac{\overline{X} - \mu}{S/\sqrt{n}}$$

is the t distribution with $(n - 1)$ degrees of freedom, t_{n-1}.

Proof First we express T in a slightly different way:

$$T = \frac{\overline{X} - \mu}{S/\sqrt{n}} = \frac{(\overline{X} - \mu)/(\sigma/\sqrt{n})}{\sqrt{\left[\dfrac{(n-1)S^2}{\sigma^2}/(n-1)\right]}}$$

The numerator on the right is standard normal because the mean of a random sample from $N(\mu, \sigma^2)$ is normal with population mean μ and variance σ^2/n. The denominator is the square root of a chi-squared variable with $(n - 1)$ degrees of freedom, divided by its degrees of freedom. This chi-squared variable is independent of the numerator, so the ratio has the t distribution with $(n - 1)$ degrees of freedom. ■

It is not hard to obtain the pdf for t.

PROPOSITION The pdf of the t distribution with ν degrees of freedom is

$$f(t) = \frac{1}{\sqrt{\pi\nu}} \frac{\Gamma[(\nu+1)/2]}{\Gamma(\nu/2)} \frac{1}{(1 + t^2/\nu)^{(\nu+1)/2}} \qquad -\infty < t < \infty$$

Proof We first find the cdf of T and then differentiate to obtain the pdf. A t variable is defined in terms of a standard normal Z and a chi-squared variable X with ν degrees of freedom. They are independent, so their joint pdf $f(x, z)$ is the product of their individual pdf's. Thus

$$P(T \le t) = P\left(\frac{Z}{\sqrt{X/\nu}} \le t\right) = P\left(Z \le t\sqrt{\frac{X}{\nu}}\right) = \int_0^{\infty}\int_{-\infty}^{t\sqrt{x/\nu}} f(x, z)\, dz\, dx$$

Differentiating with respect to t using the Fundamental Theorem of Calculus,

$$f(t) = \frac{d}{dt}P(T \le t) = \int_0^{\infty} \frac{d}{dt}\int_{-\infty}^{t\sqrt{x/\nu}} f(x, z)\, dz\, dx = \int_0^{\infty} \sqrt{\frac{x}{\nu}}\, f\left(x, t\sqrt{\frac{x}{\nu}}\right) dx$$

Now substitute the joint pdf and integrate:

$$f(t) = \int_0^\infty \sqrt{\frac{x}{\nu}} \frac{x^{\nu/2-1}}{2^{\nu/2}\Gamma(\nu/2)} e^{-x/2} \frac{1}{\sqrt{2\pi}} e^{-t^2x/(2\nu)}\, dx$$

The integral can be evaluated by writing the integrand in terms of a gamma pdf.

$$f(t) = \frac{\Gamma[(\nu+1)/2]}{\sqrt{2\pi\nu}\Gamma(\nu/2)[1/2 + t^2/(2\nu)]^{[(\nu+1)/2]}2^{\nu/2}} \cdot$$

$$\int_0^\infty \left(\frac{1}{2} + \frac{t^2}{2\nu}\right)^{(\nu+1)/2} \cdot \frac{x^{(\nu+1)/2-1}}{\Gamma[(\nu+1)/2]} e^{-[1/2+t^2/(2\nu)]x}\, dx$$

The integral of the gamma pdf is 1, so

$$f(t) = \frac{\Gamma[(\nu+1)/2]}{\sqrt{2\pi\nu}\Gamma(\nu/2)[1/2 + t^2/(2\nu)]^{[(\nu+1)/2]}2^{\nu/2}} \qquad -\infty < t < \infty$$

$$= \frac{\Gamma[(\nu+1)/2]}{\sqrt{\pi\nu}\Gamma(\nu/2)} \frac{1}{(1 + t^2/\nu)^{[(\nu+1)/2]}} \qquad \blacksquare$$

The pdf has a maximum at 0 and decreases symmetrically as $|t|$ increases. As ν becomes large, the t pdf approaches the standard normal pdf, as shown in Exercise 54. It makes sense that the t distribution would be close to the standard normal for large ν, because $T = Z/\sqrt{\chi_\nu^2/\nu}$, and χ_ν^2/ν converges to 1 by the Law of Large Numbers, as shown in Exercise 48.

Figure 6.15 shows t density curves for $\nu = 1$, 5, and 20 along with the standard normal curve. Notice how fat the tails are for 1 df, as compared to the standard normal. As the degrees of freedom increase, the t pdf becomes more like the standard normal. For 20 df there is not much difference.

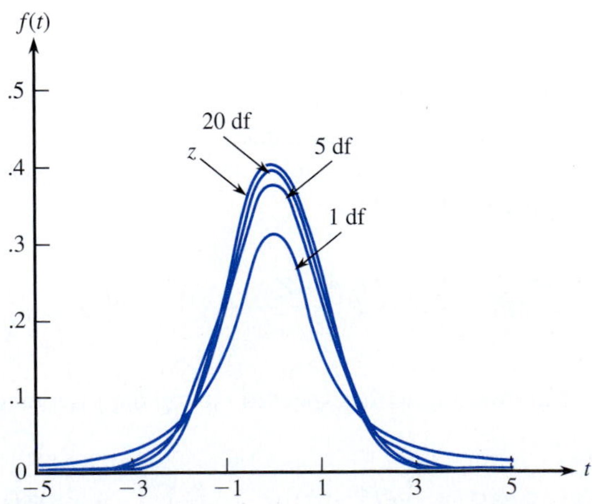

Figure 6.15 Comparison of t curves to the z curve

Integration of the t pdf is difficult except for low degrees of freedom, so values of upper-tail areas are given in Table A.8. For example, the value in the column labeled 2 and the row labeled 3.0 is .048, meaning that for 2 degrees of freedom $P(T > 3.0) = .048$. We write this as $t_{.048,2} = 3.0$, and in general we write $t_{\alpha,\nu} = c$ if $P(T_\nu > c) = \alpha$. A tabulation of these t critical values (i.e., $t_{\alpha,\nu}$) for frequently used tail areas α appears in Table A.5.

Using $\nu = 1$ and $\Gamma(1/2) = \sqrt{\pi}$ in the chi-squared pdf, we obtain the pdf for the t distribution with 1 degree of freedom as $1/[\pi(1 + t^2)]$. It has another name, the Cauchy distribution. This distribution has such fat tails that the mean does not exist (Exercise 55).

The mean and variance of a t variable can be obtained directly from the pdf, but there is another route, through the definition in terms of independent standard normal and chi-squared variables, $T = Z/\sqrt{X/\nu}$. Recall from Section 5.2 that $E(UV) = E(U)E(V)$ if U and V are independent. Thus, $E(T) = E(Z)E(1/\sqrt{X/\nu})$. Of course, $E(Z) = 0$, so $E(T) = 0$ if the second expected value on the right exists. Let's compute it from a more general expectation, $E(X^k)$ for any k if X is chi-squared:

$$\begin{aligned} E(X^k) &= \int_0^\infty x^k \frac{x^{(\nu/2)-1}}{2^{\nu/2}\Gamma(\nu/2)} e^{-x/2}\,dx \\ &= \frac{2^{k+\nu/2}\Gamma(k+\nu/2)}{2^{\nu/2}\Gamma(\nu/2)} \int_0^\infty \frac{x^{(k+\nu/2)-1}}{2^{k+\nu/2}\Gamma(k+\nu/2)} e^{-x/2}\,dx \end{aligned}$$

The second integrand is a gamma pdf, so its integral is 1 if $k + \nu/2 > 0$, and otherwise the integral does not exist. Therefore

$$E(X^k) = \frac{2^k\Gamma(k+\nu/2)}{\Gamma(\nu/2)} \tag{6.12}$$

if $k + \nu/2 > 0$, and otherwise the expectation does not exist. The requirement $k + \nu/2 > 0$ translates when $k = -\frac{1}{2}$ into $\nu > 1$. The mean of T fails to exist if $\nu = 1$ and the mean is indeed 0 otherwise.

For the variance of T we need $E(T^2) = E(Z^2)E[1/(X/\nu)] = 1 \cdot \nu/E(1/X)$. Using $k = -1$ in Equation (6.12), we obtain, with the help of $\Gamma(a+1) = a\Gamma(a)$,

$$E(X^{-1}) = \frac{2^{-1}\Gamma(-1+\nu/2)}{\Gamma(\nu/2)} = \frac{2^{-1}}{\nu/2 - 1} = \frac{1}{\nu - 2} \quad \text{if } \nu > 2$$

and therefore $V(T) = \nu/(\nu - 2)$. For 1 or 2 df the variance does not exist. The variance always exceeds 1, and for large df the variance is close to 1. This is appropriate because any t curve spreads out wider than the z curve, but for large df the t curve approaches the z curve.

The *F* Distribution

Let X_1 and X_2 be independent chi-squared random variables with ν_1 and ν_2 degrees of freedom, respectively. The F distribution with ν_1 numerator degrees of freedom and ν_2 denominator degrees of freedom is defined to be the distribution of the ratio

$$F = \frac{X_1/\nu_1}{X_2/\nu_2} \tag{6.13}$$

Sometimes the degrees of freedom will be indicated with subscripts, F_{ν_1,ν_2}.

Suppose that we have a random sample of m observations from the normal population $N(\mu_1, \sigma_1^2)$ and an independent random sample of n observations from a second normal population $N(\mu_2, \sigma_2^2)$. Then for the sample variance from the first group we know $(m-1)S_1^2/\sigma_1^2$ is χ^2_{m-1}, and similarly for the second group $(n-1)S_2^2/\sigma_2^2$ is χ^2_{n-1}. Thus, according to Equation (6.13),

$$F_{m-1,n-1} = \frac{\dfrac{(m-1)S_1^2/\sigma_1^2}{m-1}}{\dfrac{(n-1)S_2^2/\sigma_2^2}{n-1}} = \frac{S_1^2/\sigma_1^2}{S_2^2/\sigma_2^2} \tag{6.14}$$

The F distribution, via Equation (6.14), will be used in Chapter 10 to compare the variances from two independent groups. Also, for several independent groups, in Chapter 11 we will use the F distribution to see if the differences among sample means are bigger than would be expected by chance.

What happens to F if the degrees of freedom are large? Suppose that ν_2 is large. Then, using the Law of Large Numbers we can see (Exercise 48) that the denominator of Equation (6.13) will be close to 1, and F will be just the numerator chi-squared over its degrees of freedom. Similarly, if both ν_1 and ν_2 are large, then both the numerator and denominator will be close to 1, and the F ratio therefore will be close to 1.

Here is the pdf of the F distribution:

$$g(x) = \begin{cases} \dfrac{\Gamma[(\nu_1+\nu_2)/2]}{\Gamma(\nu_1/2)\Gamma(\nu_2/2)}\left(\dfrac{\nu_1}{\nu_2}\right)^{\nu_1/2} \cdot \dfrac{x^{\nu_1/2-1}}{(1+\nu_1 x/\nu_2)^{(\nu_1+\nu_2)/2}} & \text{for } x > 0 \\ 0 & \text{if } x \le 0 \end{cases}$$

Its derivation (Exercise 60) is similar to the derivation of the t pdf. Figure 6.16 shows the F density curves for several choices of ν_1 and ν_2. It should be clear by comparison with Figure 6.13 that the numerator degrees of freedom determines a lot about the shapes in Figure 6.16. For example, with $\nu_1 = 1$, the pdf is unbounded at $x = 0$, just as in Figure 6.13 with $\nu = 1$. For $\nu_1 = 2$, the pdf is positive at $x = 0$, just as in Figure 6.13 with $\nu = 2$. For $\nu_1 > 2$, the pdf is 0 at $x = 0$, just as in Figure 6.13 with $\nu > 2$. However, the

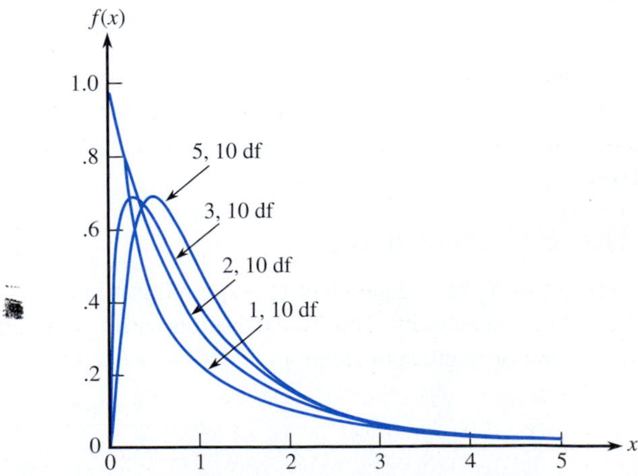

Figure 6.16 *F* **density curves**

F pdf has a fatter tail, especially for low values of ν_2. This should be evident because the F pdf does not decrease to 0 exponentially as the chi-squared pdf does.

Except for a few special choices of degrees of freedom, integration of the F pdf is difficult, so F critical values (values that capture specified F distribution tail areas) are given in Table A.9. For example, the value in the column labeled 1 and the row labeled 2 and .100 is 8.53, meaning that for 1 numerator df and 2 denominator df $P(F > 8.53) = .100$. We can express this as $F_{.1,1,2} = 8.53$, where $F_{\alpha,\nu_1,\nu_2} = c$ means that $P(F_{\nu_1,\nu_2} > c) = \alpha$.

What about lower-tail areas? Since $1/F = (X_2/\nu_2)/(X_1/\nu_1)$, the reciprocal of an F variable also has an F distribution, but with the degrees of freedom reversed, and this can be used to obtain lower-tail critical values. For example, $.100 = P(F_{1,2} > 8.53) = P(1/F_{1,2} < 1/8.53) = P(F_{2,1} < .117)$. This can be written as $F_{.9,2,1} = .117$ because $.9 = P(F_{2,1} > .117)$. In general we have

$$F_{p,\nu_1,\nu_2} = \frac{1}{F_{1-p,\nu_2,\nu_1}} \tag{6.15}$$

Recalling that $T = Z/\sqrt{X/\nu}$, it follows that the square of this t random variable is an F random variable with 1 numerator degree of freedom and ν denominator degrees of freedom, $t_\nu^2 = F_{1,\nu}$. We can use this to obtain tail areas. For example,

$$.100 = P(F_{1,2} > 8.53) = P(T_2^2 > 8.53) = P(|T_2| > \sqrt{8.53}) = 2P(T_2 > 2.92)$$

and therefore $.05 = P(T_2 > 2.92)$. We previously determined $.048 = P(T_2 > 3.0)$, which is very nearly the same statement. In terms of our notation, $t_{.05,2} = \sqrt{F_{.10,1,2}}$, and we can similarly show that in general $t_{\alpha,\nu} = \sqrt{F_{2\alpha,1,\nu}}$.

The mean of the F distribution can be obtained with the help of Equation (6.12): $E(F) = \nu_2/(\nu_2 - 2)$ if $\nu_2 > 2$, and the mean does not exist if $\nu_2 \le 2$ (Exercise 57).

Summary of Relationships

Is it clear how the standard normal, chi-squared, t, and F distributions are related? Starting with a sequence of n independent standard normal random variables (let's use five, $Z_1, Z_2, \ldots, Z_5$, to be specific), can we construct random variables having the other distributions? For example, the chi-squared distribution with n degrees of freedom is the sum of n independent standard normal squares, so $Z_1^2 + Z_2^2 + Z_3^2$ has the chi-squared distribution with 3 degrees of freedom.

Recall that the ratio of a standard normal rv to the square root of an independent chi-squared rv, divided by its df ν, has the t distribution with ν df. This implies that $Z_4/\sqrt{(Z_1^2 + Z_2^2 + Z_3^2)/3}$ has the t distribution with 3 degrees of freedom. Why would it be wrong to use Z_1 in place of Z_4?

Building a random variable having the F distribution requires two independent chi-squared rv's. We already have $Z_1^2 + Z_2^2 + Z_3^2$ chi-squared with 3 df, and similarly we obtain $Z_4^2 + Z_5^2$ chi-squared with 2 df. Dividing each chi-square rv by its df and taking the ratio gives an $F_{2,3}$ random variable, $[(Z_4^2 + Z_5^2)/2]/[(Z_1^2 + Z_2^2 + Z_3^2)/3]$.

Exercises Section 6.4 (46–66)

46. **a.** Use Table A.7 to find $\chi^2_{.05,2}$.
b. Verify the answer to (a) by integrating the pdf.
c. Verify the answer to (a) by using software (e.g., TI-89 calculator or MINITAB).

47. Why should χ^2_ν be approximately normal for large ν? What theorem applies here, and why?

48. Apply the Law of Large Numbers to show that χ^2_ν/ν approaches 1 as ν becomes large.

49. Show that the χ^2_ν pdf has a maximum at $\nu - 2$ if $\nu > 2$.

50. Knowing that $(n-1)S^2/\sigma^2 \sim \chi^2_{n-1}$ for a normal random sample,
a. Show that $E(S^2) = \sigma^2$.
b. Show that $V(S^2) = 2\sigma^4/(n-1)$. What happens to this variance as n gets large?
c. Apply Equation (6.12) to show that

$$E(S) = \sigma \frac{\sqrt{2}\Gamma(n/2)}{\sqrt{n-1}\Gamma[(n-1)/2]}$$

Then show that $E(S) = \sigma\sqrt{2/\pi}$ if $n = 2$. Is it true that $E(S) = \sigma$ for normal data?

51. Use moment generating functions to show that if $X_3 = X_1 + X_2$, with $X_1 \sim \chi^2_{\nu_1}$, $X_3 \sim \chi^2_{\nu_3}$, $\nu_3 > \nu_1$, and X_1 and X_2 are independent, then $X_2 \sim \chi^2_{\nu_3-\nu_1}$.

52. **a.** Use Table A.8 to find $t_{.102,1}$.
b. Verify the answer to part (a) by integrating the pdf.
c. Verify the answer to part (a) using software (e.g., TI-89 calculator or MINITAB).

53. **a.** Use Table A.8 to find $t_{.005,10}$.
b. Use Table A.9 to find $F_{.01,1,10}$ and relate this to the value you obtained in part (a).
c. Verify the answer to part (b) using software (e.g., TI-89 calculator or MINITAB).

54. Show that the t pdf approaches the standard normal pdf for large df values. *Hint*: Use $(1 + a/x)^x \to e^a$ and $\Gamma(x + 1/2)/[\sqrt{x}\Gamma(x)] \to 1$ as $x \to \infty$.

55. Show directly from the pdf that the mean of a t_1 (Cauchy) random variable does not exist.

56. Show that the ratio of two independent standard normal random variables has the t_1 distribution. Apply the method used to derive the t pdf in this section. *Hint*: Split the domain of the denominator into positive and negative parts.

57. Let X have an F distribution with ν_1 numerator df and ν_2 denominator df.
a. Determine the mean value of X.
b. Determine the variance of X.

58. Is it true that $E(F_{\nu_1,\nu_2}) = E(\chi^2_{\nu_1}/\nu_1)/E(\chi^2_{\nu_2}/\nu_2)$? Explain.

59. Show that $F_{p,\nu_1,\nu_2} = 1/F_{1-p,\nu_2,\nu_1}$.

60. Derive the F pdf by applying the method used to derive the t pdf.

61. **a.** Use Table A.9 to find $F_{.1,2,4}$.
b. Verify the answer to part (a) using the pdf.
c. Verify the answer to part (a) using software (e.g., TI-89 calculator or MINITAB).

62. **a.** Use Table A.8 to find $t_{.25,10}$.
b. Use (a) to find the median of $F_{1,10}$.
c. Verify the answer to part (b) using software (e.g., TI-89 calculator or MINITAB).

63. Show that if X is gamma distributed and $c(> 0)$ is a constant, then cX is gamma distributed. In particular, if X is chi-squared distributed, then cX is gamma distributed.

64. Let $Z_1, Z_2, \ldots, Z_{10}$ be independent standard normal. Use these to construct
a. A χ^2_4 random variable
b. A t_4 random variable
c. An $F_{4,6}$ random variable
d. A Cauchy random variable
e. An exponential random variable with mean 2
f. An exponential random variable with mean 1
g. A gamma random variable with mean 1 and variance $\frac{1}{2}$ (*Hint*: Use part (a) and Exercise 63.)

65. **a.** Use Exercise 47 to approximate $P(\chi^2_{50} > 70)$, and compare the result with the answer given by software, .03237.
b. Use the formula given at the bottom of Table A.7, $\chi^2_\nu \approx \nu(1 - 2/9\nu + Z\sqrt{2/9\nu})^3$, to approximate $P(\chi^2_{50} > 70)$, and compare with part (a).

66. The difference of two independent normal variables itself has a normal distribution. Is it true that the difference between two independent chi-squared variables has a chi-squared distribution? Explain.

Supplementary Exercises (67–81)

67. In cost estimation, the total cost of a project is the sum of component task costs. Each of these costs is a random variable with a probability distribution. It is customary to obtain information about the total cost distribution by adding together characteristics of the individual component cost distributions—this is called the "roll-up" procedure. For example, $E(X_1 + \cdots + X_n) = E(X_1) + \cdots + E(X_n)$, so the roll-up procedure is valid for mean cost. Suppose that there are two component tasks and that X_1 and X_2 are independent, normally distributed random variables. Is the roll-up procedure valid for the 75th percentile? That is, is the 75th percentile of the distribution of $X_1 + X_2$ the same as the sum of the 75th percentiles of the two individual distributions? If not, what is the relationship between the percentile of the sum and the sum of percentiles? For what percentiles is the roll-up procedure valid in this case?

68. Suppose that for a certain individual, calorie intake at breakfast is a random variable with expected value 500 and standard deviation 50, calorie intake at lunch is random with expected value 900 and standard deviation 100, and calorie intake at dinner is a random variable with expected value 2000 and standard deviation 180. Assuming that intakes at different meals are independent of one another, what is the probability that average calorie intake per day over the next (365-day) year is at most 3500? [*Hint*: Let X_i, Y_i, and Z_i denote the three calorie intakes on day i. Then total intake is given by $\Sigma(X_i + Y_i + Z_i)$.]

69. The mean weight of luggage checked by a randomly selected tourist-class passenger flying between two cities on a certain airline is 40 lb, and the standard deviation is 10 lb. The mean and standard deviation for a business-class passenger are 30 lb and 6 lb, respectively.

a. If there are 12 business-class passengers and 50 tourist-class passengers on a particular flight, what are the expected value of total luggage weight and the standard deviation of total luggage weight?

b. If individual luggage weights are independent, normally distributed rv's, what is the probability that total luggage weight is at most 2500 lb?

70. We have seen that if $E(X_1) = E(X_2) = \cdots = E(X_n) = \mu$, then $E(X_1 + \cdots + X_n) = n\mu$. In some applications, the number of X_i's under consideration is not a fixed number n but instead is an rv N. For example, let N = the number of components that are brought into a repair shop on a particular day, and let X_i denote the repair shop time for the ith component. Then the total repair time is $X_1 + X_2 + \cdots + X_N$, the sum of a *random* number of random variables. When N is independent of the X_i's, it can be shown that

$$E(X_1 + \cdots + X_N) = E(N) \cdot \mu$$

a. If the expected number of components brought in on a particular day is 10 and expected repair time for a randomly submitted component is 40 min, what is the expected total repair time for components submitted on any particular day?

b. Suppose components of a certain type come in for repair according to a Poisson process with a rate of 5 per hour. The expected number of defects per component is 3.5. What is the expected value of the total number of defects on components submitted for repair during a 4-hour period? Be sure to indicate how your answer follows from the general result just given.

71. Suppose the proportion of rural voters in a certain state who favor a particular gubernatorial candidate is .45 and the proportion of suburban and urban voters favoring the candidate is .60. If a sample of 200 rural voters and 300 urban and suburban voters is obtained, what is the approximate probability that at least 250 of these voters favor this candidate?

72. Let μ denote the true pH of a chemical compound. A sequence of n independent sample pH determinations will be made. Suppose each sample pH is a random variable with expected value μ and standard deviation .1. How many determinations are required if we wish the probability that the sample average is within .02 of the true pH to be at least .95? What theorem justifies your probability calculation?

73. The amount of soft drink that Ann consumes on any given day is independent of consumption on any other day and is normally distributed with $\mu = 13$ oz and $\sigma = 2$. If she currently has two six-packs of 16-oz bottles, what is the probability that she still has some soft drink left at the end of 2 weeks (14 days)? Why should we worry about the validity of the independence assumption here?

74. Refer to Exercise 27, and suppose that the X_i's are independent with each one having a normal distribution. What is the probability that the total volume shipped is at most 100,000 ft^3?

75. A student has a class that is supposed to end at 9:00 a.m. and another that is supposed to begin at 9:10 a.m. Suppose the actual ending time of the 9 a.m. class is a normally distributed rv X_1 with mean 9:02 and standard deviation 1.5 min and that the starting time of the next class is also a normally distributed rv X_2 with mean 9:10 and standard deviation 1 min. Suppose also that the time necessary to get from one classroom to the other is a normally distributed rv X_3 with mean 6 min and standard deviation 1 min. What is the probability that the student makes it to the second class before the lecture starts? (Assume independence of X_1, X_2, and X_3, which is reasonable if the student pays no attention to the finishing time of the first class.)

76. **a.** Use the general formula for the variance of a linear combination to write an expression for $V(aX + Y)$. Then let $a = \sigma_Y/\sigma_X$, and show that $\rho \geq -1$. [*Hint*: Variance is always ≥ 0, and $\text{Cov}(X, Y) = \sigma_X \cdot \sigma_Y \cdot \rho$.]

b. By considering $V(aX - Y)$, conclude that $\rho \leq 1$.

c. Use the fact that $V(W) = 0$ only if W is a constant to show that $\rho = 1$ only if $Y = aX + b$.

77. A rock specimen from a particular area is randomly selected and weighed two different times. Let W denote the actual weight and X_1 and X_2 the two measured weights. Then $X_1 = W + E_1$ and $X_2 = W + E_2$, where E_1 and E_2 are the two measurement errors. Suppose that the E_i's are independent of one another and of W and that $V(E_1) = V(E_2) = \sigma_E^2$.

a. Express ρ, the correlation coefficient between the two measured weights X_1 and X_2, in terms of σ_W^2, the variance of actual weight, and σ_X^2, the variance of measured weight.

b. Compute ρ when $\sigma_W = 1$ kg and $\sigma_E = .01$ kg.

78. Let A denote the percentage of one constituent in a randomly selected rock specimen, and let B denote the percentage of a second constituent in that same specimen. Suppose D and E are measurement errors in determining the values of A and B so that measured values are $X = A + D$ and $Y = B + E$, respectively. Assume that measurement errors are independent of one another and of actual values.

a. Show that

$$\text{Corr}(X, Y) = \text{Corr}(A, B) \cdot \sqrt{\text{Corr}(X_1, X_2)} \cdot \sqrt{\text{Corr}(Y_1, Y_2)}$$

where X_1 and X_2 are replicate measurements on the value of A, and Y_1 and Y_2 are defined analogously with respect to B. What effect does the presence of measurement error have on the correlation?

b. What is the maximum value of $\text{Corr}(X, Y)$ when $\text{Corr}(X_1, X_2) = .8100$ and $\text{Corr}(Y_1, Y_2) = .9025$? Is this disturbing?

79. Let $X_1, \ldots, X_n$ be independent rv's with mean values $\mu_1, \ldots, \mu_n$ and variances $\sigma_1^2, \ldots, \sigma_n^2$. Consider a function $h(x_1, \ldots, x_n)$, and use it to define a new rv $Y = h(X_1, \ldots, X_n)$. Under rather general conditions on the h function, if the σ_i's are all small relative to the corresponding μ_i's, it can be shown that $E(Y) \approx h(\mu_1, \ldots, \mu_n)$ and

$$V(Y) \approx \left(\frac{\partial h}{\partial x_1}\right)^2 \cdot \sigma_1^2 + \cdots + \left(\frac{\partial h}{\partial x_n}\right)^2 \cdot \sigma_n^2$$

where each partial derivative is evaluated at $(x_1, \ldots, x_n) = (\mu_1, \ldots, \mu_n)$. Suppose three resistors with resistances X_1, X_2, X_3 are connected in parallel across a battery with voltage X_4. Then by Ohm's law, the current is

$$Y = X_4\left(\frac{1}{X_1} + \frac{1}{X_2} + \frac{1}{X_3}\right)$$

Let $\mu_1 = 10$ ohms, $\sigma_1 = 1.0$ ohm, $\mu_2 = 15$ ohms, $\sigma_2 = 1.0$ ohm, $\mu_3 = 20$ ohms, $\sigma_3 = 1.5$ ohms, $\mu_4 = 120$ V, $\sigma_4 = 4.0$ V. Calculate the approximate expected value and standard deviation of the current (suggested by "Random Samplings," *CHEMTECH*, 1984: 696–697).

80. A more accurate approximation to $E[h(X_1, \ldots, X_n)]$ in Exercise 79 is

$$h(\mu_1, \ldots, \mu_n) + \frac{1}{2}\sigma_1^2\left(\frac{\partial^2 h}{\partial x_1^2}\right) + \cdots + \frac{1}{2}\sigma_n^2\left(\frac{\partial^2 h}{\partial x_n^2}\right)$$

Compute this for $Y = h(X_1, X_2, X_3, X_4)$ given in Exercise 79, and compare it to the leading term $h(\mu_1, \ldots, \mu_n)$.

81. Explain how you would use a statistical software package capable of generating independent standard normal observations to obtain observed values of (X, Y), where X and Y are bivariate normal with means 100 and 50, standard deviations 5 and 2, and correlation .5. *Hint*: Example 6.15.

Bibliography

Larsen, Richard, and Morris Marx, *An Introduction to Mathematical Statistics and Its Applications* (3rd ed.), Prentice Hall, Englewood Cliffs, NJ, 2000. More limited coverage than in the book by Olkin et al., but well written and readable.

Olkin, Ingram, Cyrus Derman, and Leon Gleser, *Probability Models and Applications* (2nd ed.), Macmillan, New York, 1994. Contains a careful and comprehensive exposition of limit theorems.

Appendix: Proof of the Central Limit Theorem

First, here is a restatement of the theorem. Let $X_1, X_2, \ldots, X_n$ be a random sample from a distribution with mean μ and variance σ^2. Then, if Z is a standard normal random variable,

$$\lim_{n\to\infty} P\left(\frac{\overline{X} - \mu}{\sigma/\sqrt{n}} < z\right) = P(Z < z)$$

The theorem says that the distribution of the standardized $\overline{X}$ approaches the standard normal distribution. Our proof is only for the special case in which the moment generating function exists, which implies also that all its derivatives exist and that they are continuous. We will show that the moment generating function of the standardized $\overline{X}$ approaches the moment generating function of the standard normal distribution. However, convergence of the moment generating function does not by itself imply the desired convergence of the distribution. This requires a theorem, which we will not prove, showing that convergence of the moment generating function implies the convergence of the distribution.

The standardized $\overline{X}$ can be written as

$$Y = \frac{\overline{X} - \mu}{\sigma/\sqrt{n}} = \frac{(1/n)[(X_1 - \mu)/\sigma + (X_2 - \mu)/\sigma + \cdots + (X_n - \mu)/\sigma] - 0}{1/\sqrt{n}}$$

The mean and standard deviation for the first ratio come from the first proposition of Section 6.2, and the second ratio is algebraically equivalent to the first. Thus, if we define W to be the standardized X, so $W_i = (X_i - \mu)/\sigma$, $i = 1, 2, \ldots, n$, then the standardized $\overline{X}$ can be written as the standardized $\overline{W}$,

$$Y = \frac{\overline{X} - \mu}{\sigma/\sqrt{n}} = \frac{\overline{W} - 0}{1/\sqrt{n}}$$

This allows a simplification of the proof because we can work with the simpler variable W, which has mean 0 and variance 1. We need to obtain the moment generating function of

$$Y = \frac{\overline{W} - 0}{1/\sqrt{n}} = \sqrt{n}\,\overline{W} = (W_1 + W_2 + \cdots + W_n)/\sqrt{n}$$

from the moment generating function $M(t)$ of W. With the help of the Section 6.3 proposition on moment generating functions for linear combinations of independent random variables, we get $M_Y(t) = M(t/\sqrt{n})^n$. We want to show that this converges to the moment generating function of a standard normal random variable, $M_Z(t) = e^{t^2/2}$. It is easier to take the logarithm of both sides and show instead that $\ln[M_Y(t)] = n\ln[M(t/\sqrt{n})] \to t^2/2$. This is equivalent because the logarithm and its inverse are continuous functions.

The limit can be obtained from two applications of L'Hôpital's rule if we set $x = 1/\sqrt{n}$, $\ln[M_Y(t)] = n\ln[M(t/\sqrt{n})] = \ln[M(tx)]/x^2$. Both the numerator and the denominator approach 0 as n gets large and x gets small [recall that $M(0) = 1$ and $M(t)$ is continuous], so L'Hôpital's rule is applicable. Thus, differentiating the numerator and denominator with respect to x,

$$\lim_{x\to 0}\frac{\ln[M(tx)]}{x^2} = \lim_{x\to 0}\frac{M'(tx)t/M(tx)}{2x} = \lim_{x\to 0}\frac{M'(tx)t}{2xM(tx)}$$

Recall that $M(0) = 1$, $M'(0) = E(W) = 0$, and $M(t)$ and its derivative $M'(t)$ are continuous, so both the numerator and denominator of the limit on the right approach 0. Thus we can use L'Hôpital's rule again.

$$\lim_{x\to 0}\frac{M'(tx)t}{2xM(tx)} = \lim_{x\to 0}\frac{M''(tx)t^2}{2M(tx) + 2xM'(tx)t} = \frac{1(t^2)}{2(1) + 2(0)(0)t} = \frac{t^2}{2}$$

In evaluating the limit we have used the continuity of $M(t)$ and its derivatives and $M(0) = 1$, $M'(0) = E(W) = 0$, $M''(0) = E(W^2) = 1$. We conclude that the mgf converges to the mgf of a standard normal random variable.

CHAPTER SEVEN

Point Estimation

Introduction

Given a parameter of interest, such as a population mean μ or population proportion p, the objective of point estimation is to use a sample to compute a number that represents in some sense a good guess for the true value of the parameter. The resulting number is called *a point estimate*. In Section 7.1, we present some general concepts of point estimation. In Section 7.2, we describe and illustrate two important methods for obtaining point estimates: the method of moments and the method of maximum likelihood.

Obtaining a point estimate entails calculating the value of a statistic such as the sample mean $\overline{X}$ or sample standard deviation S. We should therefore be concerned that the chosen statistic contains all the relevant information about the parameter of interest. The idea of no information loss is made precise by the concept of sufficiency, which is developed in Section 7.3. Finally, Section 7.4 further explores the meaning of efficient estimation and properties of maximum likelihood.

7.1 General Concepts and Criteria

Statistical inference is frequently directed toward drawing some type of conclusions about one or more parameters (population characteristics). To do so requires that an investigator obtain sample data from each of the populations under study. Conclusions can then be based on the computed values of various sample quantities. For example, let μ (a parameter) denote the true average breaking strength of wire connections used in bonding semiconductor wafers. A random sample of $n = 10$ connections might be made, and the breaking strength of each one determined, resulting in observed strengths x_1, $x_2, \ldots, x_{10}$. The sample mean breaking strength $\bar{x}$ could then be used to draw a conclusion about the value of μ. Similarly, if σ^2 is the variance of the breaking strength distribution (population variance, another parameter), the value of the sample variance s^2 can be used to infer something about σ^2.

When discussing general concepts and methods of inference, it is convenient to have a generic symbol for the parameter of interest. We will use the Greek letter θ for this purpose. The objective of point estimation is to select a single number, based on sample data, that represents a sensible value for θ. Suppose, for example, that the parameter of interest is μ, the true average lifetime of batteries of a certain type. A random sample of $n = 3$ batteries might yield observed lifetimes (hours) $x_1 = 5.0$, $x_2 = 6.4$, $x_3 = 5.9$. The computed value of the sample mean lifetime is $\bar{x} = 5.77$, and it is reasonable to regard 5.77 as a very plausible value of μ—our "best guess" for the value of μ based on the available sample information.

Suppose we want to estimate a parameter of a single population (e.g., μ or σ) based on a random sample of size n. Recall from the previous chapter that before data is available, the sample observations must be considered random variables (rv's) $X_1, X_2, \ldots, X_n$. It follows that any function of the X_i's—that is, any statistic—such as the sample mean $\overline{X}$ or sample standard deviation S is also a random variable. The same is true if available data consists of more than one sample. For example, we can represent tensile strengths of m type 1 specimens and n type 2 specimens by $X_1, \ldots, X_m$ and $Y_1, \ldots, Y_n$, respectively. The difference between the two sample mean strengths is $\overline{X} - \overline{Y}$, the natural statistic for making inferences about $\mu_1 - \mu_2$, the difference between the population mean strengths.

DEFINITION

A point estimate of a parameter θ is a single number that can be regarded as a sensible value for θ. A point estimate is obtained by selecting a suitable statistic and computing its value from the given sample data. The selected statistic is called the **point estimator** of θ.

In the battery example just given, the estimator used to obtain the point estimate of μ was $\overline{X}$, and the point estimate of μ was 5.77. If the three observed lifetimes had instead been $x_1 = 5.6$, $x_2 = 4.5$, and $x_3 = 6.1$, use of the estimator $\overline{X}$ would have resulted in the estimate $\bar{x} = (5.6 + 4.5 + 6.1)/3 = 5.40$. The symbol $\hat{\theta}$ ("theta hat") is customarily used to denote both the estimator of θ and the point estimate resulting from a given sample.*

---

*Following earlier notation, we could use $\hat{\Theta}$ (an uppercase theta) for the estimator, but this is cumbersome to write.

Thus $\hat{\mu} = \overline{X}$ is read as "the point estimator of μ is the sample mean $\overline{X}$." The statement "the point estimate of μ is 5.77" can be written concisely as $\hat{\mu} = 5.77$. Notice that in writing $\hat{\theta} = 72.5$, there is no indication of how this point estimate was obtained (what statistic was used). It is recommended that both the estimator and the resulting estimate be reported.

Example 7.1 An automobile manufacturer has developed a new type of bumper, which is supposed to absorb impacts with less damage than previous bumpers. The manufacturer has used this bumper in a sequence of 25 controlled crashes against a wall, each at 10 mph, using one of its compact car models. Let X = the number of crashes that result in no visible damage to the automobile. The parameter to be estimated is p = the proportion of all such crashes that result in no damage [alternatively, $p = P$(no damage in a single crash)]. If X is observed to be $x = 15$, the most reasonable estimator and estimate are

$$\text{estimator } \hat{p} = \frac{X}{n} \qquad \text{estimate} = \frac{x}{n} = \frac{15}{25} = .60$$

■

If for each parameter of interest there were only one reasonable point estimator, there would not be much to point estimation. In most problems, though, there will be more than one reasonable estimator.

Example 7.2 Reconsider the accompanying 20 observations on dielectric breakdown voltage for pieces of epoxy resin first introduced in Example 4.35 (Section 4.6).

24.46	25.61	26.25	26.42	26.66	27.15	27.31	27.54	27.74	27.94
27.98	28.04	28.28	28.49	28.50	28.87	29.11	29.13	29.50	30.88

The pattern in the normal probability plot given there is quite straight, so we now assume that the distribution of breakdown voltage is normal with mean value μ. Because normal distributions are symmetric, μ is also the median lifetime of the distribution. The given observations are then assumed to be the result of a random sample $X_1, X_2, \ldots, X_{20}$ from this normal distribution. Consider the following estimators and resulting estimates for μ:

a. Estimator $= \overline{X}$, estimate $= \bar{x} = \Sigma x_i/n = 555.86/20 = 27.793$

b. Estimator $= \widetilde{X}$, estimate $= \tilde{x} = (27.94 + 27.98)/2 = 27.960$

c. Estimator $= [\min(X_i) + \max(X_i)]/2$ = the average of the two extreme lifetimes, estimate $= [\min(x_i) + \max(x_i)]/2 = (24.46 + 30.88)/2 = 27.670$

d. Estimator $= \overline{X}_{\text{tr}(10)}$, the 10% trimmed mean (discard the smallest and largest 10% of the sample and then average),

$$\begin{aligned} \text{estimate} &= \bar{x}_{\text{tr}(10)} \\ &= \frac{555.86 - 24.46 - 25.61 - 29.50 - 30.88}{16} \\ &= 27.838 \end{aligned}$$

Each one of the estimators (a)–(d) uses a different measure of the center of the sample to estimate μ. Which of the estimates is closest to the true value? We cannot answer this without knowing the true value. A question that can be answered is, "Which estimator, when used on other samples of X_i's, will tend to produce estimates closest to the true value?" We will shortly consider this type of question. ■

Example 7.3 Studies have shown that a calorie-restricted diet can prolong life. Of course, controlled studies are much easier to do with lab animals. Here is a random sample of eight lifetimes of rats that were fed a restricted diet (from "Tests and Confidence Sets for Comparing Two Mean Residual Life Functions," *Biometrics*, 1988: 103–115):

716 1144 1017 1138 389 1221 530 958

Label the observations $X_1, \ldots, X_8$. We want to estimate the population variance σ^2. A natural estimator is the sample variance:

$$\hat{\sigma}^2 = S^2 = \frac{\sum (X_i - \overline{X})^2}{n-1} = \frac{\sum X_i^2 - \left(\sum X_i\right)^2/n}{n-1}$$

The corresponding estimate is

$$\hat{\sigma}^2 = s^2 = \frac{\sum x_i^2 - \left(\sum x_i\right)^2/8}{7} = \frac{6{,}991{,}551 - (7113)^2/8}{7} = \frac{667{,}205}{7} = 95{,}315$$

The estimate of σ would then be $\hat{\sigma} = s = \sqrt{95{,}315} = 309$.

An alternative estimator would result from using divisor n instead of $n - 1$ (i.e., the average squared deviation):

$$\hat{\sigma}^2 = \frac{\sum (X_i - \overline{X})^2}{n} \qquad \text{estimate} = \frac{667{,}205}{8} = 83{,}401$$

We will shortly indicate why many statisticians prefer S^2 to the estimator with divisor n. ■

In the best of all possible worlds, we could find an estimator $\hat{\theta}$ for which $\hat{\theta} = \theta$ always. However, $\hat{\theta}$ is a function of the sample X_i's, so it is a random variable. For some samples, $\hat{\theta}$ will yield a value larger than θ, whereas for other samples $\hat{\theta}$ will underestimate θ. If we write

$$\hat{\theta} = \theta + \text{error of estimation}$$

then an accurate estimator would be one resulting in small estimation errors, so that estimated values will be near the true value.

Mean Square Error

A popular way to quantify the idea of $\hat{\theta}$ being close to θ is to consider the squared error $(\hat{\theta} - \theta)^2$. Another possibility is the absolute error $|\hat{\theta} - \theta|$, but this is more difficult to work with mathematically. For some samples, $\hat{\theta}$ will be quite close to θ and the resulting squared error will be very small, whereas the squared error will be quite large whenever a sample produces an estimate $\hat{\theta}$ that is far from the target. An omnibus measure of accuracy is the mean square error (expected square error), which entails averaging the squared error over all possible samples and resulting estimates.

DEFINITION The **mean square error** of an estimator $\hat{\theta}$ is $E[(\hat{\theta} - \theta)^2]$.

A useful result when evaluating mean square error is a consequence of the following rearrangement of the shortcut for evaluating a variance $V(Y)$:

$$V(Y) = E(Y^2) - [E(Y)]^2 \Rightarrow E(Y^2) = V(Y) + [E(Y)]^2$$

That is, the expected value of the square of Y is the variance plus the square of the mean value. Letting $Y = \hat{\theta} - \theta$, the estimation error, the left-hand side is just the mean square error. The first term on the right-hand side is $V(\hat{\theta} - \theta) = V(\hat{\theta})$ since θ is just a constant. The second term involves $E(\hat{\theta} - \theta) = E(\hat{\theta}) - \theta$, the difference between the expected value of the estimator and the value of the parameter. This difference is called the **bias** of the estimator. Thus

$$\text{MSE} = V(\hat{\theta}) + [E(\hat{\theta}) - \theta]^2 = \text{variance of estimator} + (\text{bias})^2$$

Example 7.4 (Example 7.1 continued) Consider once again estimating a population proportion of "successes" p. The natural estimator of p is the sample proportion of successes $\hat{p} = X/n$. The number of successes X in the sample has a binomial distribution with parameters n and p, so $E(X) = np$ and $V(X) = np(1 - p)$. The expected value of the estimator is

$$E(\hat{p}) = E\left(\frac{X}{n}\right) = \frac{1}{n}E(X) = \frac{1}{n}np = p$$

Thus the bias of $\hat{p}$ is $p - p = 0$, giving the mean square error as

$$E[(\hat{p} - p)^2] = V(\hat{p}) + 0^2 = V\left(\frac{X}{n}\right) = \frac{1}{n^2}V(X) = \frac{p(1 - p)}{n}$$

Now consider the alternative estimator $\hat{p} = (X + 2)/(n + 4)$. That is, add two successes and two failures to the sample and then calculate the sample proportion of successes. One intuitive justification for this estimator is that

$$\left|\frac{X}{n} - .5\right| = \left|\frac{X - .5n}{n}\right| \qquad \left|\frac{X + 2}{n + 4} - .5\right| = \left|\frac{X - .5n}{n + 4}\right|$$

from which we see that the alternative estimator is always somewhat closer to .5 than is the usual estimator. It seems particularly reasonable to move the estimate toward .5 when the number of successes in the sample is close to 0 or n. For example, if there are no successes at all in the sample, is it sensible to estimate the population proportion of successes as 0, especially if n is small?

The bias of the alternative estimator is

$$E\left(\frac{X+2}{n+4}\right) - p = \frac{1}{n+4}E(X+2) - p = \frac{np+2}{n+4} - p = \frac{2/n - 4p/n}{1 + 4/n}$$

This bias is not zero unless $p = .5$. However, as n increases, the numerator approaches 0 and the denominator approaches 1, so the bias approaches 0. The variance of the estimator is

$$V\left(\frac{X+2}{n+4}\right) = \frac{1}{(n+4)^2}V(X+2) = \frac{V(X)}{(n+4)^2} = \frac{np(1-p)}{(n+4)^2} = \frac{p(1-p)}{n+8+16/n}$$

This variance approaches 0 as the sample size increases. The mean square error of the alternative estimator is

$$\text{MSE} = \frac{p(1-p)}{n+8+16/n} + \left(\frac{2/n - 4p/n}{1+4/n}\right)^2$$

So how does the mean square error of the usual estimator, the sample proportion, compare to that of the alternative estimator? If one MSE were smaller than the other for all values of p, then we could say that one estimator is always preferred to the other (using MSE as our criterion). But as Figure 7.1 shows, this is not the case at least for the sample sizes $n = 10$ and $n = 100$, and in fact is not true for any other sample size.

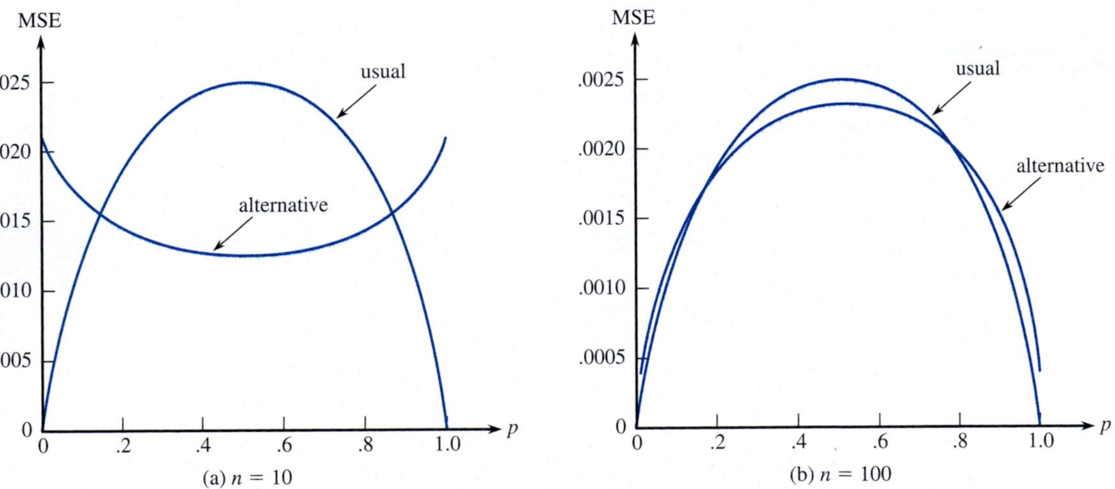

Figure 7.1 Graphs of MSE for the usual and alternative estimators of p

According to Figure 7.1, the two MSE's are quite different when n is small. In this case the alternative estimator is better for values of p near .5 (since it moves the sample proportion toward .5) but not for extreme values of p. For a large n the two MSE's are quite similar, but again neither dominates the other. ■

Seeking an estimator whose mean square error is smaller than that of every other estimator for all values of the parameter is generally too ambitious a goal. One common approach is to restrict the class of estimators under consideration in some way, and then seek the estimator that is best in that restricted class. A very popular restriction is to impose the condition of unbiasedness.

Unbiased Estimators

Suppose we have two measuring instruments; one instrument has been accurately calibrated, but the other systematically gives readings smaller than the true value being measured. When each instrument is used repeatedly on the same object, because of measurement error, the observed measurements will not be identical. However, the measurements produced by the first instrument will be distributed about the true value in such a way that on average this instrument measures what it purports to measure, so it is called an unbiased instrument. The second instrument yields observations that have a systematic error component or bias.

DEFINITION A point estimator $\hat{\theta}$ is said to be an **unbiased estimator** of θ if $E(\hat{\theta}) = \theta$ for every possible value of θ. If $\hat{\theta}$ is not unbiased, the difference $E(\hat{\theta}) - \theta$ is called the **bias** of $\hat{\theta}$.

That is, $\hat{\theta}$ is unbiased if its probability (i.e., sampling) distribution is always "centered" at the true value of the parameter. Suppose $\hat{\theta}$ is an unbiased estimator; then if $\theta = 100$, the $\hat{\theta}$ sampling distribution is centered at 100; if $\theta = 27.5$, then the $\hat{\theta}$ sampling distribution is centered at 27.5, and so on. Figure 7.2 pictures the distributions of several biased and unbiased estimators. Note that "centered" here means that the expected value, not the median, of the distribution of $\hat{\theta}$ is equal to θ.

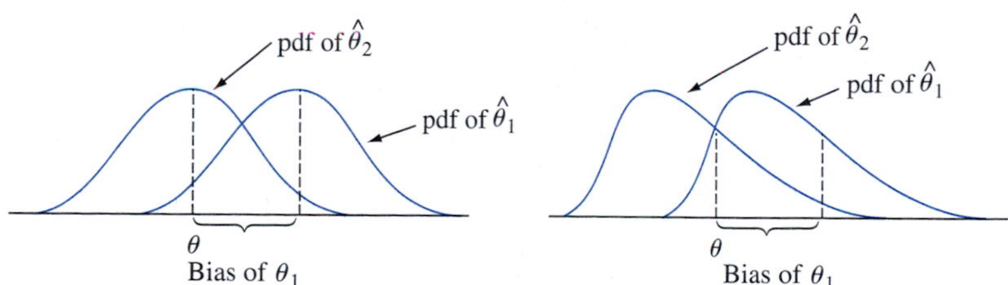

Figure 7.2 The pdf's of a biased estimator $\hat{\theta}_1$ and an unbiased estimator $\hat{\theta}_2$ for a parameter θ

It may seem as though it is necessary to know the value of θ (in which case estimation is unnecessary) to see whether $\hat{\theta}$ is unbiased. This is usually not the case, though, because unbiasedness is a general property of the estimator's sampling distribution—where it is centered—which is typically not dependent on any particular parameter value. For example, in Example 7.4 we showed that $E(\hat{p}) = p$ when $\hat{p}$ is the sample proportion of successes. Thus if $p = .25$, the sampling distribution of $\hat{p}$ is centered at .25 (centered in the sense of mean value), when $p = .9$ the sampling distribution is centered at .9, and so on. It is not necessary to know the value of p to know that $\hat{p}$ is unbiased.

PROPOSITION When X is a binomial rv with parameters n and p, the sample proportion $\hat{p} = X/n$ is an unbiased estimator of p.

Example 7.5 Suppose that X, the reaction time to a certain stimulus, has a uniform distribution on the interval from 0 to an unknown upper limit θ (so the density function of X is rectangular in shape with height $1/\theta$ for $0 \le x \le \theta$). An investigator wants to estimate θ on the basis of a random sample $X_1, X_2, \ldots, X_n$ of reaction times. Since θ is the largest possible time in the entire population of reaction times, consider as a first estimator the largest sample reaction time: $\hat{\theta}_b = \max(X_1, \ldots, X_n)$. If $n = 5$ and $x_1 = 4.2$, $x_2 = 1.7$, $x_3 = 2.4$, $x_4 = 3.9$, $x_5 = 1.3$, the point estimate of θ is $\hat{\theta}_b = \max(4.2, 1.7, 2.4, 3.9, 1.3) = 4.2$.

Unbiasedness implies that some samples will yield estimates that exceed θ and other samples will yield estimates smaller than θ—otherwise θ could not possibly be the center (balance point) of $\hat{\theta}_b$'s distribution. However, our proposed estimator will never overestimate θ (the largest sample value cannot exceed the largest population value) and will underestimate θ unless the largest sample value equals θ. This intuitive argument shows that $\hat{\theta}_b$ is a biased estimator. More precisely, using our earlier results on order statistics, it can be shown (see Exercise 50) that

$$E(\hat{\theta}_b) = \frac{n}{n+1} \cdot \theta < \theta \quad \left(\text{since } \frac{n}{n+1} < 1\right)$$

The bias of $\hat{\theta}_b$ is given by $n\theta/(n+1) - \theta = -\theta/(n+1)$, which approaches 0 as n gets large.

It is easy to modify $\hat{\theta}_b$ to obtain an unbiased estimator of θ. Consider the estimator

$$\hat{\theta}_u = \frac{n+1}{n} \cdot \hat{\theta}_b = \frac{n+1}{n} \cdot \max(X_1, \ldots, X_n)$$

Using this estimator on the given data gives the estimate $(6/5)(4.2) = 5.04$. The fact that $(n+1)/n > 1$ implies that $\hat{\theta}_u$ will overestimate θ for some samples and underestimate it for others. The mean value of this estimator is

$$E(\hat{\theta}_u) = E\left[\frac{n+1}{n} \cdot \max(X_1, \ldots, X_n)\right] = \frac{n+1}{n} \cdot E[\max(X_1, \ldots, X_n)]$$

$$= \frac{n+1}{n} \cdot \frac{n}{n+1}\theta = \theta$$

If $\hat{\theta}_u$ is used repeatedly on different samples to estimate θ, some estimates will be too large and others will be too small, but in the long run there will be no systematic tendency to underestimate or overestimate θ. ■

Statistical practitioners who buy into the **Principle of Unbiased Estimation** would employ an unbiased estimator in preference to a biased estimator. On this basis, the sample proportion of successes should be preferred to the alternative estimator of p, and the unbiased estimator $\hat{\theta}_u$ should be preferred to the biased estimator $\hat{\theta}_b$.

Example 7.6 Let's turn now to the problem of estimating σ^2 based on a random sample $X_1, \ldots, X_n$. First consider the estimator $S^2 = \sum(X_i - \overline{X})^2/(n-1)$, the sample variance as we have defined it. Applying the result $E(Y^2) = V(Y) + [E(Y)]^2$ to

$$S^2 = \frac{1}{n-1}\left[\sum X_i^2 - \frac{\left(\sum X_i\right)^2}{n}\right]$$

gives

$$\begin{aligned} E(S^2) &= \frac{1}{n-1}\left\{\sum E(X_i^2) - \frac{1}{n}E\left[\left(\sum X_i\right)^2\right]\right\} \\ &= \frac{1}{n-1}\left\{\sum(\sigma^2 + \mu^2) - \frac{1}{n}\left\{V\left(\sum X_i\right) + \left[E\left(\sum X_i\right)\right]^2\right\}\right\} \\ &= \frac{1}{n-1}\left\{n\sigma^2 + n\mu^2 - \frac{1}{n}n\sigma^2 - \frac{1}{n}(n\mu)^2\right\} \\ &= \frac{1}{n-1}\{n\sigma^2 - \sigma^2\} = \sigma^2 \end{aligned}$$

Thus we have shown that **the sample variance S^2 is an unbiased estimator of σ^2.**

The estimator that uses divisor n can be expressed as $(n-1)S^2/n$, so

$$E\left[\frac{(n-1)S^2}{n}\right] = \frac{n-1}{n}E(S^2) = \frac{n-1}{n}\sigma^2$$

This estimator is therefore biased. The bias is $(n-1)\sigma^2/n - \sigma^2 = -\sigma^2/n$. Because the bias is negative, the estimator with divisor n tends to underestimate σ^2, and this is why the divisor $n - 1$ is preferred by many statisticians (though when n is large, the bias is small and there is little difference between the two).

This is not quite the whole story, though. Suppose the random sample has come from a normal distribution. Then from Section 6.4, we know that $(n-1)S^2/\sigma^2$ has a chi-squared distribution with $n - 1$ degrees of freedom. The mean and variance of a chi-squared variable are df and 2 df, respectively. Let's now consider estimators of the form

$$\hat{\sigma}^2 = c\sum(X_i - \overline{X})^2$$

The expected value of the estimator is

$$E\left[c\sum(X_i - \overline{X})^2\right] = c(n-1)E(S^2) = c(n-1)\sigma^2$$

so the bias is $c(n-1)\sigma^2 - \sigma^2$. The only unbiased estimator of this type is the sample variance, with $c = 1/(n-1)$.

Similarly, the variance of the estimator is

$$V\left[c\sum(X_i - \overline{X})^2\right] = V\left[c\sigma^2\frac{(n-1)S^2}{\sigma^2}\right] = c^2\sigma^4[2(n-1)]$$

Substituting these expressions into the relationship MSE = variance + (bias)^2, the value of c for which MSE is minimized can be found by taking the derivative with respect to c, equating the resulting expression to zero, and solving for c. The result is $c = 1/(n + 1)$. So in this situation, the principle of unbiasedness and the principle of minimum MSE are at loggerheads.

As a final blow, even though S^2 is unbiased for estimating σ^2, *it is not true* that the sample standard deviation S is unbiased for estimating σ. This is because the square root function is not linear, so the expected value of the square root is not the square root of the expected value. Well, if S is biased, why not find an unbiased estimator for σ and use it rather than S? Unfortunately there is no estimator of σ that is unbiased irrespective of the nature of the population distribution (though in special cases, e.g. a normal distribution, an unbiased estimator does exist). Fortunately the bias of S is not serious unless n is quite small. So we shall generally employ it as an estimator. ■

In Example 7.2, we proposed several different estimators for the mean μ of a normal distribution. If there were a unique unbiased estimator for μ, the estimation dilemma could be resolved by using that estimator. Unfortunately, this is not the case.

PROPOSITION If $X_1, X_2, \ldots, X_n$ is a random sample from a distribution with mean μ, then $\overline{X}$ is an unbiased estimator of μ. If in addition the distribution is continuous and symmetric, then $\widetilde{X}$ and any trimmed mean are also unbiased estimators of μ.

The fact that $\overline{X}$ is unbiased is just a restatement of one of our rules of expected value: $E(\overline{X}) = \mu$ for every possible value of μ (for discrete as well as continuous distributions). The unbiasedness of the other estimators is more difficult to verify; the argument requires invoking results on distributions of order statistics from Section 5.5.

According to this proposition, the principle of unbiasedness by itself does not always allow us to select a single estimator. When the underlying population is normal, even the third estimator in Example 7.2 is unbiased, and there are many other unbiased estimators. What we now need is a way of selecting among unbiased estimators.

Estimators with Minimum Variance

Suppose $\hat{\theta}_1$ and $\hat{\theta}_2$ are two estimators of θ that are both unbiased. Then, although the distribution of each estimator is centered at the true value of θ, the spreads of the distributions about the true value may be different.

PRINCIPLE OF MINIMUM VARIANCE UNBIASED ESTIMATION Among all estimators of θ that are unbiased, choose the one that has minimum variance. The resulting $\hat{\theta}$ is called the **minimum variance unbiased estimator (MVUE)** of θ. Since MSE = variance + $(\text{bias})^2$, seeking an unbiased estimator with minimum variance is the same as seeking an unbiased estimator that has minimum mean square error.

Figure 7.3 pictures the pdf's of two unbiased estimators, with the first $\hat{\theta}$ having smaller variance than the second estimator. Then the first $\hat{\theta}$ is more likely than the second one to produce an estimate close to the true θ. The MVUE is, in a certain sense, the most likely among all unbiased estimators to produce an estimate close to the true θ.

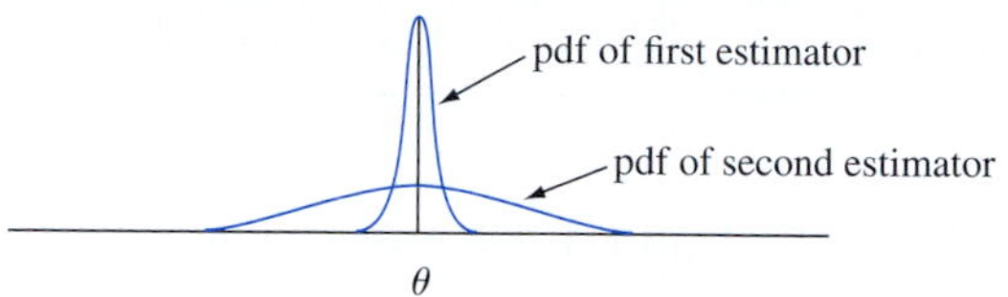

Figure 7.3 Graphs of the pdf's of two different unbiased estimators

Example 7.7 We argued in Example 7.5 that when $X_1, \ldots, X_n$ is a random sample from a uniform distribution on $[0, \theta]$, the estimator

$$\hat{\theta}_1 = \frac{n+1}{n} \cdot \max(X_1, \ldots, X_n)$$

is unbiased for θ (we previously denoted this estimator by $\hat{\theta}_u$). This is not the only unbiased estimator of θ. The expected value of a uniformly distributed rv is just the midpoint of the interval of positive density, so $E(X_i) = \theta/2$. This implies that $E(\overline{X}) = \theta/2$, from which $E(2\overline{X}) = \theta$. That is, the estimator $\hat{\theta}_2 = 2\overline{X}$ is unbiased for θ.

If X is uniformly distributed on the interval $[A, B]$, then $V(X) = \sigma^2 = (B - A)^2/12$. Thus, in our situation, $V(X_i) = \theta^2/12$, $V(\overline{X}) = \sigma^2/n = \theta^2/(12n)$, and $V(\hat{\theta}_2) = V(2\overline{X}) = 4V(\overline{X}) = \theta^2/(3n)$. The results of Exercise 50 can be used to show that $V(\hat{\theta}_1) = \theta^2/[n(n+2)]$. The estimator $\hat{\theta}_1$ has smaller variance than does $\hat{\theta}_2$ if $3n < n(n+2)$—that is, if $0 < n^2 - n = n(n-1)$. As long as $n > 1$, $V(\hat{\theta}_1) < V(\hat{\theta}_2)$, so $\hat{\theta}_1$ is a better estimator than $\hat{\theta}_2$. More advanced methods can be used to show that $\hat{\theta}_1$ is the MVUE of θ—every other unbiased estimator of θ has variance that exceeds $\theta^2/[n(n+2)]$. ■

One of the triumphs of mathematical statistics has been the development of methodology for identifying the MVUE in a wide variety of situations. The most important result of this type for our purposes concerns estimating the mean μ of a normal distribution. For a proof in the special case that σ is known, see Exercise 45.

THEOREM Let $X_1, \ldots, X_n$ be a random sample from a normal distribution with parameters μ and σ. Then the estimator $\hat{\mu} = \overline{X}$ is the MVUE for μ.

Whenever we are convinced that the population being sampled is normal, the result says that $\overline{X}$ should be used to estimate μ. In Example 7.2, then, our estimate would be $\bar{x} = 27.793$.

Once again, in some situations such as the one in Example 7.6, it is possible to obtain an estimator with small bias that would be preferred to the best unbiased estimator. This is illustrated in Figure 7.4. However, MVUEs are often easier to obtain than the type of biased estimator whose distribution is pictured.

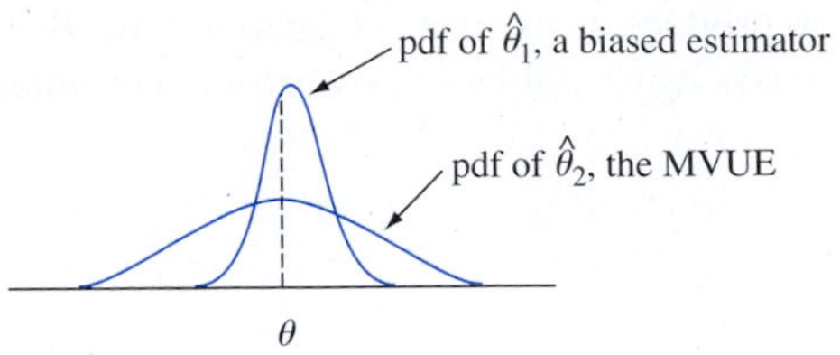

Figure 7.4 A biased estimator that is preferable to the MVUE

More Complications

The last theorem does not say that in estimating a population mean μ, the estimator $\overline{X}$ should be used irrespective of the distribution being sampled.

Example 7.8 Suppose we wish to estimate the number of calories θ in a certain food. Using standard measurement techniques, we will obtain a random sample $X_1, \ldots, X_n$ of n calorie measurements. Let's assume that the population distribution is a member of one of the following three families:

$$f(x) = \frac{1}{\sqrt{2\pi\sigma^2}} e^{-(x-\theta)^2/(2\sigma^2)} \qquad -\infty < x < \infty \tag{7.1}$$

$$f(x) = \frac{1}{\pi[1 + (x - \theta)^2]} \qquad -\infty < x < \infty \tag{7.2}$$

$$f(x) = \begin{cases} \dfrac{1}{2c} & -c \le x - \theta \le c \\ 0 & \text{otherwise} \end{cases} \tag{7.3}$$

The pdf (7.1) is the normal distribution, (7.2) is called the Cauchy distribution, and (7.3) is a uniform distribution. All three distributions are symmetric about θ, which is therefore the median of each distribution. The value θ is also the mean for the normal and uniform distributions, but the mean of the Cauchy distribution fails to exist. This happens because, even though the Cauchy distribution is bell-shaped like the normal distribution, it has much heavier tails (more probability far out) than the normal curve. The uniform distribution has no tails. The four estimators for μ considered earlier are $\overline{X}$, $\widetilde{X}$, $\overline{X}_e$ (the average of the two extreme observations), and $\overline{X}_{\text{tr}(10)}$, a trimmed mean.

The very important moral here is that the best estimator for μ depends crucially on which distribution is being sampled. In particular,

1. If the random sample comes from a normal distribution, then $\overline{X}$ is the best of the four estimators, since it has minimum variance among all unbiased estimators.
2. If the random sample comes from a Cauchy distribution, then $\overline{X}$ and $\overline{X}_e$ are terrible estimators for μ, whereas $\widetilde{X}$ is quite good (the MVUE is not known); $\overline{X}$ is bad because it is very sensitive to outlying observations, and the heavy tails of the Cauchy distribution make a few such observations likely to appear in any sample.

3. If the underlying distribution is the particular uniform distribution in (7.3), then the best estimator is $\overline{X}_e$; this estimator is greatly influenced by outlying observations, but the lack of tails makes such observations impossible.
4. *The trimmed mean is best in none of these three situations but works reasonably well in all three*. That is, $\overline{X}_{tr(10)}$ does not suffer too much in comparison with the best procedure in any of the three situations. ■

More generally, recent research in statistics has established that when estimating a point of symmetry μ of a continuous probability distribution, a trimmed mean with trimming proportion 10% or 20% (from each end of the sample) produces reasonably behaved estimates over a very wide range of possible models. For this reason, a trimmed mean with small trimming percentage is said to be a **robust estimator**.

Until now, we have focused on comparing several estimators based on the same data, such as $\overline{X}$ and $\widetilde{X}$ for estimating μ when a sample of size n is selected from a normal population distribution. Sometimes an investigator is faced with a choice between alternative ways of gathering data; the form of an appropriate estimator then may well depend on how the experiment was carried out.

Example 7.9 Suppose a certain type of component has a lifetime distribution that is exponential with parameter λ so that expected lifetime is $\mu = 1/\lambda$. A sample of n such components is selected, and each is put into operation. If the experiment is continued until all n lifetimes, $X_1, \ldots, X_n$, have been observed, then $\overline{X}$ is an unbiased estimator of μ.

In some experiments, though, the components are left in operation only until the time of the rth failure, where $r < n$. This procedure is referred to as **censoring**. Let Y_1 denote the time of the first failure (the minimum lifetime among the n components), Y_2 denote the time at which the second failure occurs (the second smallest lifetime), and so on. Since the experiment terminates at time Y_r, the total accumulated lifetime at termination is

$$T_r = \sum_{i=1}^{r} Y_i + (n - r)Y_r$$

We now demonstrate that $\hat{\mu} = T_r/r$ is an unbiased estimator for μ. To do so, we need two properties of exponential variables:

1. The memoryless property (see Section 4.4) says that at any time point, remaining lifetime has the same exponential distribution as original lifetime.
2. If $X_1, \ldots, X_k$ are independent, each exponentially distributed with parameter λ, then min $(X_1, \ldots, X_k)$ is exponential with parameter $k\lambda$ and has expected value $1/(k\lambda)$. See Example 5.28.

Since all n components last until Y_1, $n - 1$ last an additional $Y_2 - Y_1$, $n - 2$ an additional $Y_3 - Y_2$ amount of time, and so on, another expression for T_r is

$$T_r = nY_1 + (n - 1)(Y_2 - Y_1) + (n - 2)(Y_3 - Y_2) + \cdots + (n - r + 1)(Y_r - Y_{r-1})$$

But Y_1 is the minimum of n exponential variables, so $E(Y_1) = 1/(n\lambda)$. Similarly, $Y_2 - Y_1$ is the smallest of the $n - 1$ remaining lifetimes, each exponential with parameter λ (by the memoryless property), so $E(Y_2 - Y_1) = 1/[(n-1)\lambda]$. Continuing, $E(Y_{i+1} - Y_i) = 1/[(n-i)\lambda]$, so

$$E(T_r) = nE(Y_1) + (n-1)E(Y_2 - Y_1) + \cdots + (n-r+1)E(Y_r - Y_{r-1})$$
$$= n \cdot \frac{1}{n\lambda} + (n-1) \cdot \frac{1}{(n-1)\lambda} + \cdots + (n-r+1) \cdot \frac{1}{(n-r+1)\lambda}$$
$$= \frac{r}{\lambda}$$

Therefore, $E(T_r/r) = (1/r)E(T_r) = (1/r) \cdot (r/\lambda) = 1/\lambda = \mu$ as claimed.

As an example, suppose 20 components are put on test and $r = 10$. Then if the first ten failure times are 11, 15, 29, 33, 35, 40, 47, 55, 58, and 72, the estimate of μ is

$$\hat{\mu} = \frac{11 + 15 + \cdots + 72 + (10)(72)}{10} = 111.5$$

The advantage of the experiment with censoring is that it terminates more quickly than the uncensored experiment. However, it can be shown that $V(T_r/r) = 1/(\lambda^2 r)$, which is larger than $1/(\lambda^2 n)$, the variance of $\overline{X}$ in the uncensored experiment. ■

Reporting a Point Estimate: The Standard Error

Besides reporting the value of a point estimate, some indication of its precision should be given. The usual measure of precision is the standard error of the estimator used.

DEFINITION

The **standard error** of an estimator $\hat{\theta}$ is its standard deviation $\sigma_{\hat{\theta}} = \sqrt{V(\hat{\theta})}$. If the standard error itself involves unknown parameters whose values can be estimated, substitution of these estimates into $\sigma_{\hat{\theta}}$ yields the **estimated standard error** (estimated standard deviation) of the estimator. The estimated standard error can be denoted either by $\hat{\sigma}_{\hat{\theta}}$ (the ^ over σ emphasizes that $\sigma_{\hat{\theta}}$ is being estimated) or by $s_{\hat{\theta}}$.

Example 7.10 (Example 7.2 continued)

Assuming that breakdown voltage is normally distributed, $\hat{\mu} = \overline{X}$ is the best estimator of μ. If the value of σ is known to be 1.5, the standard error of $\overline{X}$ is $\sigma_{\overline{X}} = \sigma/\sqrt{n} = 1.5/\sqrt{20} = .335$. If, as is usually the case, the value of σ is unknown, the estimate $\hat{\sigma} = s = 1.462$ is substituted into $\sigma_{\overline{X}}$ to obtain the estimated standard error $\hat{\sigma}_{\overline{X}} = s_{\overline{X}} = s/\sqrt{n} = 1.462/\sqrt{20} = .327$. ■

Example 7.11 (Example 7.1 continued)

The standard error of $\hat{p} = X/n$ is

$$\sigma_{\hat{p}} = \sqrt{V(X/n)} = \sqrt{\frac{V(X)}{n^2}} = \sqrt{\frac{npq}{n^2}} = \sqrt{\frac{pq}{n}}$$

Since p and $q = 1 - p$ are unknown (else why estimate?), we substitute $\hat{p} = x/n$ and $\hat{q} = 1 - x/n$ into $\sigma_{\hat{p}}$, yielding the estimated standard error $\hat{\sigma}_{\hat{p}} = \sqrt{\hat{p}\hat{q}/n} = \sqrt{(.6)(.4)/25} = .098$. Alternatively, since the largest value of pq is attained when $p = q = .5$, an upper bound on the standard error is $\sqrt{1/(4n)} = .10$. ■

When the point estimator $\hat{\theta}$ has approximately a normal distribution, which will often be the case when n is large, then we can be reasonably confident that the true value of θ lies within approximately 2 standard errors (standard deviations) of $\hat{\theta}$. Thus if a sample of $n = 36$ component lifetimes gives $\hat{\mu} = \bar{x} = 28.50$ and $s = 3.60$, then $s/\sqrt{n} = .60$, so "within 2 estimated standard errors of $\hat{\mu}$" translates to the interval $28.50 \pm (2)(.60) = (27.30, 29.70)$.

If $\hat{\theta}$ is not necessarily approximately normal but is unbiased, then it can be shown (using Chebyshev's inequality, introduced in Exercises 43, 77, and 135 of Chapter 3) that the estimate will deviate from θ by as much as 4 standard errors at most 6% of the time. We would then expect the true value to lie within 4 standard errors of $\hat{\theta}$ (and this is a very conservative statement, since it applies to *any* unbiased $\hat{\theta}$). Summarizing, the standard error tells us roughly within what distance of $\hat{\theta}$ we can expect the true value of θ to lie.

The Bootstrap

The form of the estimator $\hat{\theta}$ may be sufficiently complicated so that standard statistical theory cannot be applied to obtain an expression for $\sigma_{\hat{\theta}}$. This is true, for example, in the case $\theta = \sigma$, $\hat{\theta} = S$; the standard deviation of the statistic S, σ_S, cannot in general be determined. In recent years, a new computer-intensive method called the **bootstrap** has been introduced to address this problem. Suppose that the population pdf is $f(x; \theta)$, a member of a particular parametric family, and that data $x_1, x_2, \ldots, x_n$ gives $\hat{\theta} = 21.7$. We now use the computer to obtain "bootstrap samples" from the pdf $f(x; 21.7)$, and for each sample we calculate a "bootstrap estimate" $\hat{\theta}^*$:

First bootstrap sample: $x_1^*, x_2^*, \ldots, x_n^*$; estimate $= \hat{\theta}_1^*$

Second bootstrap sample: $x_1^*, x_2^*, \ldots, x_n^*$; estimate $= \hat{\theta}_2^*$

⋮

Bth bootstrap sample: $x_1^*, x_2^*, \ldots, x_n^*$; estimate $= \hat{\theta}_B^*$

$B = 100$ or 200 is often used. Now let $\bar{\theta}^* = \Sigma\hat{\theta}_i^*/B$, the sample mean of the bootstrap estimates. The **bootstrap estimate** of $\hat{\theta}$'s standard error is now just the sample standard deviation of the $\hat{\theta}_i^*$'s:

$$S_{\hat{\theta}} = \sqrt{\frac{1}{B-1}\sum(\hat{\theta}_i^* - \bar{\theta}^*)^2}$$

(In the bootstrap literature, B is often used in place of $B - 1$; for typical values of B, there is usually little difference between the resulting estimates.)

Example 7.12 A theoretical model suggests that X, the time to breakdown of an insulating fluid between electrodes at a particular voltage, has $f(x; \lambda) = \lambda e^{-\lambda x}$, an exponential distribution. A random sample of $n = 10$ breakdown times (min) gives the following data:

41.53 18.73 2.99 30.34 12.33 117.52 73.02 223.63 4.00 26.78

Since $E(X) = 1/\lambda$, $E(\overline{X}) = 1/\lambda$, so a reasonable estimate of λ is $\hat{\lambda} = 1/\overline{x} = 1/55.087 = .018153$. We then used a statistical computer package to obtain $B = 100$ bootstrap samples, each of size 10, from $f(x; 018153)$. The first such sample was 41.00, 109.70, 16.78, 6.31, 6.76, 5.62, 60.96, 78.81, 192.25, 27.61, from which $\Sigma x_i^* = 545.8$ and $\hat{\lambda}_1^* = 1/54.58 = .01832$. The average of the 100 bootstrap estimates is $\lambda^* = .02153$, and the sample standard deviation of these 100 estimates is $s_{\hat{\lambda}} = .0091$, the bootstrap estimate of $\hat{\lambda}$'s standard error. A histogram of the 100 $\hat{\lambda}_i^*$'s was somewhat positively skewed, suggesting that the sampling distribution of $\hat{\lambda}$ also has this property. ■

Sometimes an investigator wishes to estimate a population characteristic without assuming that the population distribution belongs to a particular parametric family. An instance of this occurred in Example 7.8, where a 10% trimmed mean was proposed for estimating a symmetric population distribution's center θ. The data of Example 7.2 gave $\hat{\theta} = \overline{X}_{tr(10)} = 27.838$, but now there is no assumed $f(x; \theta)$, so how can we obtain a bootstrap sample? The answer is to regard the sample itself as constituting the population (the $n = 20$ observations in Example 7.2) and take B different samples, each of size n, *with* replacement from this population. See Section 8.5.

Exercises Section 7.1 (1–20)

1. The accompanying data on IQ for first graders in a particular school was introduced in Example 1.2.

82 96 99 102 103 103 106 107 108 108 108
108 109 110 110 111 113 113 113 113 115 115
118 118 119 121 122 122 127 132 136 140 146

a. Calculate a point estimate of the mean value of IQ for the conceptual population of all first graders in this school, and state which estimator you used. (*Hint*: $\Sigma x_i = 3753$.)

b. Calculate a point estimate of the IQ value that separates the lowest 50% of all such students from the highest 50%, and state which estimator you used.

c. Calculate and interpret a point estimate of the population standard deviation σ. Which estimator did you use? (*Hint*: $\Sigma x_i^2 = 432{,}015$.)

d. Calculate a point estimate of the proportion of all such students whose IQ exceeds 100. (*Hint*: Think of an observation as a "success" if it exceeds 100.)

e. Calculate a point estimate of the population coefficient of variation σ/μ, and state which estimator you used.

2. A sample of 20 students who had recently taken elementary statistics yielded the following information on brand of calculator owned (T = Texas Instruments, H = Hewlett-Packard, C = Casio, S = Sharp):

T T H T C T T S C H
S S T H C T T T H T

a. Estimate the true proportion of all such students who own a Texas Instruments calculator.

b. Of the 10 students who owned a TI calculator, 4 had graphing calculators. Estimate the proportion of students who do not own a TI graphing calculator.

3. Consider the following sample of observations on coating thickness for low-viscosity paint ("Achieving a Target Value for a Manufacturing Process: A Case Study," *J. Qual. Tech.*, 1992: 22–26):

.83	.88	.88	1.04	1.09	1.12	1.29	1.31
1.48	1.49	1.59	1.62	1.65	1.71	1.76	1.83

Assume that the distribution of coating thickness is normal (a normal probability plot strongly supports this assumption).

a. Calculate a point estimate of the mean value of coating thickness, and state which estimator you used.

b. Calculate a point estimate of the median of the coating thickness distribution, and state which estimator you used.

c. Calculate a point estimate of the value that separates the largest 10% of all values in the thickness distribution from the remaining 90%, and state which estimator you used. (*Hint*: Express what you are trying to estimate in terms of μ and σ.)

d. Estimate $P(X < 1.5)$, that is, the proportion of all thickness values less than 1.5. (*Hint*: If you knew the values of μ and σ, you could calculate this probability. These values are not available, but they can be estimated.)

e. What is the estimated standard error of the estimator that you used in part (b)?

4. The data set of Exercise 1 also includes these third-grade verbal IQ observations for males:

117	103	121	112	120	132	113	117	132
149	125	131	136	107	108	113	136	114

and females:

114	102	113	131	124	117	120	90
114	109	102	114	127	127	103	

Prior to obtaining data, denote the male values by $X_1, \ldots, X_m$ and the female values by $Y_1, \ldots, Y_n$. Suppose that the X_i's constitute a random sample from a distribution with mean μ_1 and standard deviation σ_1 and that the Y_i's form a random sample (independent of the X_i's) from another distribution with mean μ_2 and standard deviation σ_2.

a. Use rules of expected value to show that $\overline{X} - \overline{Y}$ is an unbiased estimator of $\mu_1 - \mu_2$. Calculate the estimate for the given data.

b. Use rules of variance from Chapter 6 to obtain an expression for the variance and standard deviation (standard error) of the estimator in part (a), and then compute the estimated standard error.

c. Calculate a point estimate of the ratio σ_1/σ_2 of the two standard deviations.

d. Suppose one male third-grader and one female third-grader are randomly selected. Calculate a point estimate of the variance of the difference $X - Y$ between male and female IQ.

5. As an example of a situation in which several different statistics could reasonably be used to calculate a point estimate, consider a population of N invoices. Associated with each invoice is its "book value," the recorded amount of that invoice. Let T denote the total book value, a known amount. Some of these book values are erroneous. An audit will be carried out by randomly selecting n invoices and determining the audited (correct) value for each one. Suppose that the sample gives the following results (in dollars).

	Invoice				
	1	**2**	**3**	**4**	**5**
Book value	300	720	526	200	127
Audited value	300	520	526	200	157
Error	0	200	0	0	−30

Let

$\overline{Y}$ = sample mean book value
$\overline{X}$ = sample mean audited value
$\overline{D}$ = sample mean error

Several different statistics for estimating the total audited (correct) value have been proposed (see "Statistical Models and Analysis in Auditing," *Statistical Sci.*, 1989: 2–33). These include

Mean per unit statistic $= N\overline{X}$
Difference statistic $= T - N\overline{D}$
Ratio statistic $= T \cdot (\overline{X}/\overline{Y})$

If $N = 5000$ and $T = 1{,}761{,}300$, calculate the three corresponding point estimates. (The cited article discusses properties of these estimators.)

6. Consider the accompanying observations on stream flow (1000's of acre-feet) recorded at a station in Colorado for the period April 1–August 31 over a 31-year span (from an article in the 1974 volume of *Water Resources Res.*).

127.96	210.07	203.24	108.91	178.21
285.37	100.85	89.59	185.36	126.94
200.19	66.24	247.11	299.87	109.64

125.86	114.79	109.11	330.33	85.54
117.64	302.74	280.55	145.11	95.36
204.91	311.13	150.58	262.09	477.08
94.33				

An appropriate probability plot supports the use of the lognormal distribution (see Section 4.5) as a reasonable model for stream flow.

a. Estimate the parameters of the distribution. [*Hint*: Remember that X has a lognormal distribution with parameters μ and σ^2 if $\ln(X)$ is normally distributed with mean μ and variance σ^2.]

b. Use the estimates of part (a) to calculate an estimate of the expected value of stream flow. [*Hint*: What is $E(X)$?]

7. a. A random sample of 10 houses in a particular area, each of which is heated with natural gas, is selected and the amount of gas (therms) used during the month of January is determined for each house. The resulting observations are 103, 156, 118, 89, 125, 147, 122, 109, 138, 99. Let μ denote the average gas usage during January by all houses in this area. Compute a point estimate of μ.

b. Suppose there are 10,000 houses in this area that use natural gas for heating. Let τ denote the total amount of gas used by all of these houses during January. Estimate τ using the data of part (a). What estimator did you use in computing your estimate?

c. Use the data in part (a) to estimate p, the proportion of all houses that used at least 100 therms.

d. Give a point estimate of the population median usage (the middle value in the population of all houses) based on the sample of part (a). What estimator did you use?

8. In a random sample of 80 components of a certain type, 12 are found to be defective.

a. Give a point estimate of the proportion of all such components that are *not* defective.

b. A system is to be constructed by randomly selecting two of these components and connecting them in series, as shown here.

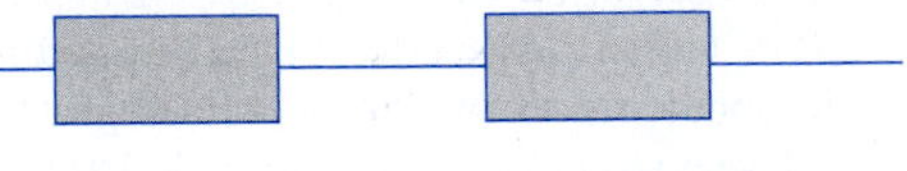

The series connection implies that the system will function if and only if neither component is defective (i.e., both components work properly). Estimate the proportion of all such systems that work properly. [*Hint*: If p denotes the probability that a component works properly, how can $P(\text{system works})$ be expressed in terms of p?]

c. Let $\hat{p}$ be the sample proportion of successes. Is $\hat{p}^2$ an unbiased estimator for p^2? *Hint*: For any rv Y, $E(Y^2) = V(Y) + [E(Y)]^2$.

9. Each of 150 newly manufactured items is examined and the number of scratches per item is recorded (the items are supposed to be free of scratches), yielding the following data:

Number of scratches per item	0	1	2	3	4	5	6	7
Observed frequency	18	37	42	30	13	7	2	1

Let X = the number of scratches on a randomly chosen item, and assume that X has a Poisson distribution with parameter λ.

a. Find an unbiased estimator of λ and compute the estimate for the data. [*Hint*: $E(X) = \lambda$ for X Poisson, so $E(\overline{X}) = ?$]

b. What is the standard deviation (standard error) of your estimator? Compute the estimated standard error. (*Hint*: $\sigma_X^2 = \lambda$ for X Poisson.)

10. Using a long rod that has length μ, you are going to lay out a square plot in which the length of each side is μ. Thus the area of the plot will be μ^2. However, you do not know the value of μ, so you decide to make n independent measurements $X_1, X_2, \ldots X_n$ of the length. Assume that each X_i has mean μ (unbiased measurements) and variance σ^2.

a. Show that $(\overline{X})^2$ is not an unbiased estimator for μ^2. (*Hint*: For any rv Y, $E(Y^2) = V(Y) + [E(Y)]^2$. Apply this with $Y = \overline{X}$.)

b. For what value of k is the estimator $(\overline{X})^2 - kS^2$ unbiased for μ^2? [*Hint*: Compute $E((\overline{X})^2 - kS^2)$.]

11. Of n_1 randomly selected male smokers, X_1 smoked filter cigarettes, whereas of n_2 randomly selected female smokers, X_2 smoked filter cigarettes. Let p_1 and p_2 denote the probabilities that a randomly selected male and female, respectively, smoke filter cigarettes.

a. Show that $(X_1/n_1) - (X_2/n_2)$ is an unbiased estimator for $p_1 - p_2$. [*Hint*: $E(X_i) = n_ip_i$ for $i = 1, 2$.]

b. What is the standard error of the estimator in part (a)?

c. How would you use the observed values x_1 and x_2 to estimate the standard error of your estimator?

d. If $n_1 = n_2 = 200$, $x_1 = 127$, and $x_2 = 176$, use the estimator of part (a) to obtain an estimate of $p_1 - p_2$.

e. Use the result of part (c) and the data of part (d) to estimate the standard error of the estimator.

12. Suppose a certain type of fertilizer has an expected yield per acre of μ_1 with variance σ^2, whereas the expected yield for a second type of fertilizer is μ_2 with the same variance σ^2. Let S_1^2 and S_2^2 denote the sample variances of yields based on sample sizes n_1 and n_2, respectively, of the two fertilizers. Show that the pooled (combined) estimator

$$\hat{\sigma}^2 = \frac{(n_1 - 1)S_1^2 + (n_2 - 1)S_2^2}{n_1 + n_2 - 2}$$

is an unbiased estimator of σ^2.

13. Consider a random sample $X_1, \ldots, X_n$ from the pdf

$$f(x; \theta) = .5(1 + \theta x) \qquad -1 \le x \le 1$$

where $-1 \le \theta \le 1$ (this distribution arises in particle physics). Show that $\hat{\theta} = 3\overline{X}$ is an unbiased estimator of θ. [*Hint*: First determine $\mu = E(X) = E(\overline{X})$.]

14. A sample of n captured Pandemonium jet fighters results in serial numbers $x_1, x_2, x_3, \ldots, x_n$. The CIA knows that the aircraft were numbered consecutively at the factory starting with α and ending with β, so that the total number of planes manufactured is $\beta - \alpha + 1$ (e.g., if $\alpha = 17$ and $\beta = 29$, then $29 - 17 + 1 = 13$ planes having serial numbers 17, 18, 19, . . . , 28, 29 were manufactured). However, the CIA does not know the values of α or β. A CIA statistician suggests using the estimator $\max(X_i) - \min(X_i) + 1$ to estimate the total number of planes manufactured.

a. If $n = 5$, $x_1 = 237$, $x_2 = 375$, $x_3 = 202$, $x_4 = 525$, and $x_5 = 418$, what is the corresponding estimate?

b. Under what conditions on the sample will the value of the estimate be exactly equal to the true total number of planes? Will the estimate ever be larger than the true total? Do you think the estimator is unbiased for estimating $\beta - \alpha + 1$? Explain in one or two sentences.

(A similar method was used to estimate German tank production in World War II.)

15. Let $X_1, X_2, \ldots, X_n$ represent a random sample from a Rayleigh distribution with pdf

$$f(x; \theta) = \frac{x}{\theta} e^{-x^2/(2\theta)} \qquad x > 0$$

a. It can be shown that $E(X^2) = 2\theta$. Use this fact to construct an unbiased estimator of θ based on $\sum X_i^2$ (and use rules of expected value to show that it is unbiased).

b. Estimate θ from the following measurements of blood plasma beta concentration (in pmol/L) for $n = 10$ men.

16.88	10.23	4.59	6.66	13.68
14.23	19.87	9.40	6.51	10.95

16. Suppose the true average growth μ of one type of plant during a 1-year period is identical to that of a second type, but the variance of growth for the first type is σ^2, whereas for the second type, the variance is $4\sigma^2$. Let $X_1, \ldots, X_m$ be m independent growth observations on the first type [so $E(X_i) = \mu$, $V(X_i) = \sigma^2$], and let $Y_1, \ldots, Y_n$ be n independent growth observations on the second type [$E(Y_i) = \mu$, $V(Y_i) = 4\sigma^2$].

a. Show that for any constant δ between 0 and 1, the estimator $\hat{\mu} = \delta\overline{X} + (1 - \delta)\overline{Y}$ is unbiased for μ.

b. For fixed m and n, compute $V(\hat{\mu})$, and then find the value of δ that minimizes $V(\hat{\mu})$. [*Hint*: Differentiate $V(\hat{\mu})$ with respect to δ.]

17. In Chapter 3, we defined a negative binomial rv as the number of failures that occur before the rth success in a sequence of independent and identical success/failure trials. The probability mass function (pmf) of X is

$$nb(x; r, p) = \begin{cases} \dbinom{x+r-1}{x} p^r (1-p)^x & x = 0, 1, 2, \ldots \\ 0 & \text{otherwise} \end{cases}$$

a. Suppose that $r \ge 2$. Show that

$$\hat{p} = (r - 1)/(X + r - 1)$$

is an unbiased estimator for p. [*Hint*: Write out $E(\hat{p})$ and cancel $x + r - 1$ inside the sum.]

b. A reporter wishing to interview five individuals who support a certain candidate begins asking people whether (S) or not (F) they support the candidate. If the sequence of responses is *SFFSFFFSSS*, estimate p = the true proportion who support the candidate.

18. Let $X_1, X_2, \ldots, X_n$ be a random sample from a pdf $f(x)$ that is symmetric about μ, so that $\widetilde{X}$ is an unbiased estimator of μ. If n is large, it can be shown that $V(\widetilde{X}) \approx 1/\{4n[f(\mu)]^2\}$. When the underlying pdf is Cauchy (see Example 7.8), $V(\overline{X}) = \infty$, so $\overline{X}$ is a terrible estimator. What is $V(\widetilde{X})$ in this case when n is large?

19. An investigator wishes to estimate the proportion of students at a certain university who have violated the honor code. Having obtained a random sample of n students, she realizes that asking each, "Have you violated the honor code?" will probably result in some untruthful responses. Consider the following scheme, called a **randomized response** technique. The investigator makes up a deck of 100 cards, of which 50 are of type I and 50 are of type II.

Type I: Have you violated the honor code (yes or no)?

Type II: Is the last digit of your telephone number a 0, 1, or 2 (yes or no)?

Each student in the random sample is asked to mix the deck, draw a card, and answer the resulting question truthfully. Because of the irrelevant question on type II cards, a yes response no longer stigmatizes the respondent, so we assume that responses are truthful. Let p denote the proportion of honor-code violators (i.e., the probability of a randomly selected student being a violator), and let $\lambda = P(\text{yes response})$. Then λ and p are related by $\lambda = .5p + (.5)(.3)$.

a. Let Y denote the number of yes responses, so $Y \sim \text{Bin}(n, \lambda)$. Thus Y/n is an unbiased estimator of λ. Derive an estimator for p based on Y. If $n = 80$ and $y = 20$, what is your estimate? (*Hint*: Solve $\lambda = .5p + .15$ for p and then substitute Y/n for λ.)

b. Use the fact that $E(Y/n) = \lambda$ to show that your estimator $\hat{p}$ is unbiased.

c. If there were 70 type I and 30 type II cards, what would be your estimator for p?

20. Return to the problem of estimating the population proportion p and consider another adjusted estimator, namely

$$\hat{p} = \frac{X + \sqrt{n/4}}{n + \sqrt{n}}$$

The justification for this estimator comes from the Bayesian approach to point estimation to be introduced in Section 14.4.

a. Determine the mean square error of this estimator. What do you find interesting about this MSE?

b. Compare the MSE of this estimator to the MSE of the usual estimator (the sample proportion).

7.2 *Methods of Point Estimation

The definition of unbiasedness does not in general indicate how unbiased estimators can be derived. We now discuss two "constructive" methods for obtaining point estimators: the method of moments and the method of maximum likelihood. By constructive we mean that the general definition of each type of estimator suggests explicitly how to obtain the estimator in any specific problem. Although maximum likelihood estimators are generally preferable to moment estimators because of certain efficiency properties, they often require significantly more computation than do moment estimators. It is sometimes the case that these methods yield unbiased estimators.

The Method of Moments

The basic idea of this method is to equate certain sample characteristics, such as the mean, to the corresponding population expected values. Then solving these equations for unknown parameter values yields the estimators.

DEFINITION

Let $X_1, \ldots, X_n$ be a random sample from a pmf or pdf $f(x)$. For $k = 1, 2, 3, \ldots,$ the ***k*th population moment**, or ***k*th moment of the distribution $f(x)$**, is $E(X^k)$. The ***k*th sample moment** is $(1/n)\sum_{i=1}^{n} X_i^k$.

Thus the first population moment is $E(X) = \mu$ and the first sample moment is $\sum X_i/n = \overline{X}$. The second population and sample moments are $E(X^2)$ and $\sum X_i^2/n$, respectively. The population moments will be functions of any unknown parameters $\theta_1, \theta_2, \ldots$.

DEFINITION Let $X_1, X_2, \ldots, X_n$ be a random sample from a distribution with pmf or pdf $f(x; \theta_1, \ldots, \theta_m)$, where $\theta_1, \ldots, \theta_m$ are parameters whose values are unknown. Then the **moment estimators** $\hat{\theta}_1, \ldots, \hat{\theta}_m$ are obtained by equating the first m sample moments to the corresponding first m population moments and solving for $\theta_1, \ldots, \theta_m$.

If, for example, $m = 2$, $E(X)$ and $E(X^2)$ will be functions of θ_1 and θ_2. Setting $E(X) = (1/n)\sum X_i \; (= \overline{X})$ and $E(X^2) = (1/n)\sum X_i^2$ gives two equations in θ_1 and θ_2. The solution then defines the estimators. For estimating a population mean μ, the method gives $\mu = \overline{X}$, so the estimator is the sample mean.

Example 7.13 Let $X_1, X_2, \ldots, X_n$ represent a random sample of service times of n customers at a certain facility, where the underlying distribution is assumed exponential with parameter λ. Since there is only one parameter to be estimated, the estimator is obtained by equating $E(X)$ to $\overline{X}$. Since $E(X) = 1/\lambda$ for an exponential distribution, this gives $1/\lambda = \overline{X}$ or $\lambda = 1/\overline{X}$. The moment estimator of λ is then $\hat{\lambda} = 1/\overline{X}$. ■

Example 7.14 Let $X_1, \ldots, X_n$ be a random sample from a gamma distribution with parameters α and β. From Section 4.4, $E(X) = \alpha\beta$ and $E(X^2) = \beta^2\Gamma(\alpha + 2)/\Gamma(\alpha) = \beta^2(\alpha + 1)\alpha$. The moment estimators of α and β are obtained by solving

$$\overline{X} = \alpha\beta \qquad \frac{1}{n}\sum X_i^2 = \alpha(\alpha + 1)\beta^2$$

Since $\alpha(\alpha + 1)\beta^2 = \alpha^2\beta^2 + \alpha\beta^2$ and the first equation implies $\alpha^2\beta^2 = (\overline{X})^2$, the second equation becomes

$$\frac{1}{n}\sum X_i^2 = (\overline{X})^2 + \alpha\beta^2$$

Now dividing each side of this second equation by the corresponding side of the first equation and substituting back gives the estimators

$$\hat{\alpha} = \frac{(\overline{X})^2}{(1/n)\sum X_i^2 - (\overline{X})^2} \qquad \hat{\beta} = \frac{(1/n)\sum X_i^2 - (\overline{X})^2}{\overline{X}}$$

To illustrate, the survival time data mentioned in Example 4.27 is

152	115	109	94	88	137	152	77	160	165
125	40	128	123	136	101	62	153	83	69

with $\bar{x} = 113.5$ and $(1/20)\sum x_i^2 = 14{,}087.8$. The estimates are

$$\hat{\alpha} = \frac{(113.5)^2}{14{,}087.8 - (113.5)^2} = 10.7 \qquad \hat{\beta} = \frac{14{,}087.8 - (113.5)^2}{113.5} = 10.6$$

These estimates of α and β differ from the values suggested by Gross and Clark because they used a different estimation technique. ■

Example 7.15 Let $X_1, \ldots, X_n$ be a random sample from a generalized negative binomial distribution with parameters r and p (Section 3.6). Since $E(X) = r(1 - p)/p$ and $V(X) = r(1 - p)/p^2$, $E(X^2) = V(X) + [E(X)]^2 = r(1 - p)(r - rp + 1)/p^2$. Equating $E(X)$ to $\overline{X}$ and $E(X^2)$ to $(1/n)\sum X_i^2$ eventually gives

$$\hat{p} = \frac{\overline{X}}{(1/n)\sum X_i^2 - (\overline{X})^2} \qquad \hat{r} = \frac{(\overline{X})^2}{(1/n)\sum X_i^2 - (\overline{X})^2 - \overline{X}}$$

As an illustration, Reep, Pollard, and Benjamin ("Skill and Chance in Ball Games," *J. Royal Statist. Soc.*, 1971: 623–629) consider the negative binomial distribution as a model for the number of goals per game scored by National Hockey League teams. The data for 1966–1967 follows (420 games):

Goals	0	1	2	3	4	5	6	7	8	9	10
Frequency	29	71	82	89	65	45	24	7	4	1	3

Then,

$$\bar{x} = \sum x_i/420 = [(0)(29) + (1)(71) + \cdots + (10)(3)]/420 = 2.98$$

and

$$\sum x_i^2/420 = [(0)^2(29) + (1)^2(71) + \cdots + (10)^2(3)]/420 = 12.40$$

Thus,

$$\hat{p} = \frac{2.98}{12.40 - (2.98)^2} = .85 \qquad \hat{r} = \frac{(2.98)^2}{12.40 - (2.98)^2 - 2.98} = 16.5$$

Although r by definition must be positive, the denominator of $\hat{r}$ could be negative, indicating that the negative binomial distribution is not appropriate (or that the moment estimator is flawed). ■

Maximum Likelihood Estimation

The method of maximum likelihood was first introduced by R. A. Fisher, a geneticist and statistician, in the 1920s. Most statisticians recommend this method, at least when the sample size is large, since the resulting estimators have certain desirable efficiency properties (see the proposition on large sample behavior toward the end of this section).

Example 7.16 A sample of ten new bike helmets manufactured by a certain company is obtained. Upon testing, it is found that the first, third, and tenth helmets are flawed, whereas the others are not. Let $p = P$ (flawed helmet) and define $X_1, \ldots, X_{10}$ by $X_i = 1$ if the ith helmet is flawed and zero otherwise. Then the observed x_i's are 1, 0, 1, 0, 0, 0, 0, 0, 0, 1, so the joint pmf of the sample is

$$f(x_1, x_2, \ldots, x_{10}; p) = p(1 - p)p \cdot \cdots \cdot p = p^3(1 - p)^7 \tag{7.4}$$

We now ask, "For what value of p is the observed sample most likely to have occurred?" That is, we wish to find the value of p that maximizes the pmf (7.4) or, equivalently, maximizes the natural log of (7.4).* Since

$$\ln[f(x_1, \ldots, x_{10}; p)] = 3\ln(p) + 7\ln(1-p) \tag{7.5}$$

and this is a differentiable function of p, equating the derivative of (7.5) to zero gives the maximizing value[†]:

$$\frac{d}{dp}\ln[f(x_1, \ldots, x_{10}; p)] = \frac{3}{p} - \frac{7}{1-p} = 0 \Rightarrow p = \frac{3}{10} = \frac{x}{n}$$

where x is the observed number of successes (flawed helmets). The estimate of p is now $\hat{p} = \frac{3}{10}$. It is called the maximum likelihood estimate because for fixed $x_1, \ldots, x_{10}$, it is the parameter value that maximizes the likelihood (joint pmf) of the observed sample. The likelihood and log likelihood are graphed in Figure 7.5. Of course, the maximum on both graphs occurs at the same value, $p = .3$.

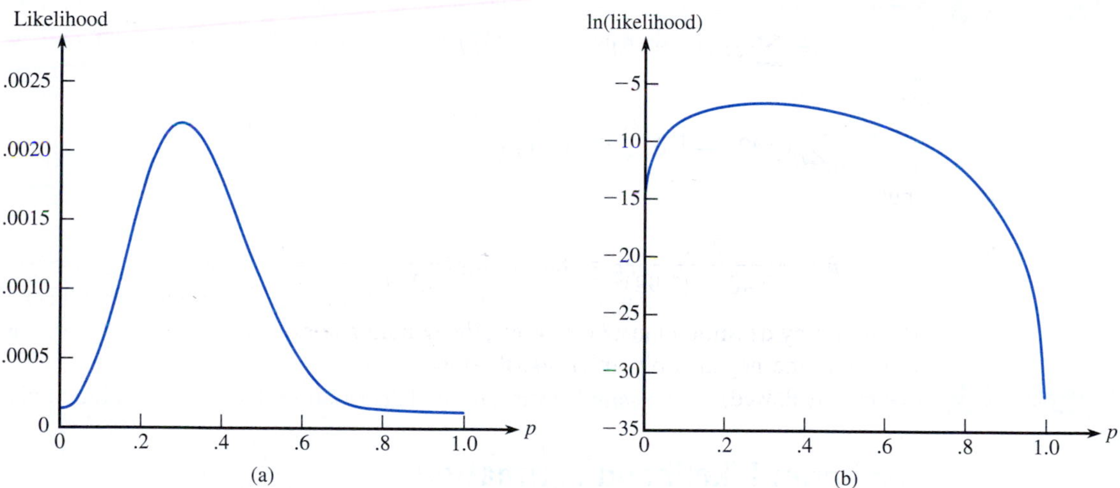

Figure 7.5 Likelihood and log likelihood plotted against p

Note that if we had been told only that among the ten helmets there were three that were flawed, Equation (7.4) would be replaced by the binomial pmf $\binom{10}{3}p^3(1-p)^7$, which is also maximized for $\hat{p} = \frac{3}{10}$. ■

---

* Since $\ln[g(x)]$ is a monotonic function of $g(x)$, finding x to maximize $\ln[g(x)]$ is equivalent to maximizing $g(x)$ itself. In statistics, taking the logarithm frequently changes a product to a sum, which is easier to work with.

[†] This conclusion requires checking the second derivative, but the details are omitted.

DEFINITION

Let $X_1, X_2, \ldots, X_n$ have joint pmf or pdf

$$f(x_1, x_2, \ldots, x_n; \theta_1, \ldots, \theta_m) \tag{7.6}$$

where the parameters $\theta_1, \ldots, \theta_m$ have unknown values. When $x_1, \ldots, x_n$ are the observed sample values and (7.6) is regarded as a function of $\theta_1, \ldots, \theta_m$, it is called the **likelihood function**. The maximum likelihood estimates $\hat{\theta}_1, \ldots, \hat{\theta}_m$ are those values of the θ_i's that maximize the likelihood function, so that

$$f(x_1, \ldots, x_n; \hat{\theta}_1, \ldots, \hat{\theta}_m) \ge f(x_1, \ldots, x_n; \theta_1, \ldots, \theta_m) \quad \text{for all } \theta_1, \ldots, \theta_m$$

When the X_i's are substituted in place of the x_i's, the **maximum likelihood estimators** (mle's) result.

The likelihood function tells us how likely the observed sample is as a function of the possible parameter values. Maximizing the likelihood gives the parameter values for which the observed sample is most likely to have been generated—that is, the parameter values that "agree most closely" with the observed data.

Example 7.17

Suppose $X_1, X_2, \ldots, X_n$ is a random sample from an exponential distribution with parameter λ. Because of independence, the likelihood function is a product of the individual pdf's:

$$f(x_1, \ldots, x_n; \lambda) = (\lambda e^{-\lambda x_1}) \cdot \cdots \cdot (\lambda e^{-\lambda x_n}) = \lambda^n e^{-\lambda \Sigma x_i}$$

The ln(likelihood) is

$$\ln[f(x_1, \ldots, x_n; \lambda)] = n \ln(\lambda) - \lambda \sum x_i$$

Equating $(d/d\lambda)$[ln(likelihood)] to zero results in $n/\lambda - \Sigma x_i = 0$, or $\lambda = n/\Sigma x_i = 1/\overline{x}$. Thus the mle is $\hat{\lambda} = 1/\overline{X}$; it is identical to the method of moments estimator [but it is not an unbiased estimator, since $E(1/\overline{X}) \neq 1/E(\overline{X})$]. ■

Example 7.18

Let $X_1, \ldots, X_n$ be a random sample from a normal distribution. The likelihood function is

$$\begin{aligned} f(x_1, \ldots, x_n; \mu, \sigma^2) &= \frac{1}{\sqrt{2\pi\sigma^2}} e^{-(x_1-\mu)^2/(2\sigma^2)} \cdot \cdots \cdot \frac{1}{\sqrt{2\pi\sigma^2}} e^{-(x_n-\mu)^2/(2\sigma^2)} \\ &= \left(\frac{1}{2\pi\sigma^2}\right)^{n/2} e^{-\Sigma(x_i-\mu)^2/(2\sigma^2)} \end{aligned}$$

so

$$\ln[f(x_1, \ldots, x_n; \mu, \sigma^2)] = -\frac{n}{2}\ln(2\pi\sigma^2) - \frac{1}{2\sigma^2}\sum (x_i - \mu)^2$$

To find the maximizing values of μ and σ^2, we must take the partial derivatives of $\ln(f)$ with respect to μ and σ^2, equate them to zero, and solve the resulting two equations. Omitting the details, the resulting mle's are

$$\hat{\mu} = \overline{X} \qquad \hat{\sigma}^2 = \frac{\sum (X_i - \overline{X})^2}{n}$$

The mle of σ^2 is not the unbiased estimator, so two different principles of estimation (unbiasedness and maximum likelihood) yield two different estimators. ■

Example 7.19 In Chapter 3, we discussed the use of the Poisson distribution for modeling the number of "events" that occur in a two-dimensional region. Assume that when the region R being sampled has area $a(R)$, the number X of events occurring in R has a Poisson distribution with parameter $\lambda a(R)$ (where λ is the expected number of events per unit area) and that nonoverlapping regions yield independent X's.

Suppose an ecologist selects n nonoverlapping regions $R_1, \ldots, R_n$ and counts the number of plants of a certain species found in each region. The joint pmf (likelihood) is then

$$p(x_1, \ldots, x_n; \lambda) = \frac{[\lambda \cdot a(R_1)]^{x_1} e^{-\lambda \cdot a(R_1)}}{x_1!} \cdot \ldots \cdot \frac{[\lambda \cdot a(R_n)]^{x_n} e^{-\lambda \cdot a(R_n)}}{x_n!}$$
$$= \frac{[a(R_1)]^{x_1} \cdot \ldots \cdot [a(R_n)]^{x_n} \cdot \lambda^{\Sigma x_i} \cdot e^{-\lambda \Sigma a(R_i)}}{x_1! \cdot \ldots \cdot x_n!}$$

The ln(likelihood) is

$$\ln[p(x_1, \ldots, x_n; \lambda)] = \sum x_i \cdot \ln[a(R_i)] + \ln(\lambda) \cdot \sum x_i - \lambda \sum a(R_i) - \sum \ln(x_i!)$$

Taking $d/d\lambda \ln(p)$ and equating it to zero yields

$$\frac{\sum x_i}{\lambda} - \sum a(R_i) = 0$$

so

$$\lambda = \frac{\sum x_i}{\sum a(R_i)}$$

The mle is then $\hat{\lambda} = \Sigma X_i / \Sigma a(R_i)$. This is intuitively reasonable because λ is the true density (plants per unit area), whereas $\hat{\lambda}$ is the sample density since $\Sigma a(R_i)$ is just the total area sampled. Because $E(X_i) = \lambda \cdot a(R_i)$, the estimator is unbiased.

Sometimes an alternative sampling procedure is used. Instead of fixing regions to be sampled, the ecologist will select n points in the entire region of interest and let $y_i =$ the distance from the ith point to the nearest plant. The cumulative distribution function (cdf) of $Y =$ distance to the nearest plant is

$$F_Y(y) = P(Y \le y) = 1 - P(Y > y) = 1 - P\begin{pmatrix}\text{no plants in a} \\ \text{circle of radius } y\end{pmatrix}$$
$$= 1 - \frac{e^{-\lambda \pi y^2}(\lambda \pi y^2)^0}{0!} = 1 - e^{-\lambda \pi y^2}$$

Taking the derivative of $F_Y(y)$ with respect to y yields

$$f_Y(y; \lambda) = \begin{cases} 2\pi \lambda y e^{-\lambda \pi y^2} & y \ge 0 \\ 0 & \text{otherwise} \end{cases}$$

If we now form the likelihood $f_Y(y_1; \lambda) \cdot \cdots \cdot f_Y(y_n; \lambda)$, differentiate ln(likelihood), and so on, the resulting mle is

$$\hat{\lambda} = \frac{n}{\pi \sum Y_i^2} = \frac{\text{number of plants observed}}{\text{total area sampled}}$$

which is also a sample density. It can be shown that in a sparse environment (small λ), the distance method is in a certain sense better, whereas in a dense environment, the first sampling method is better. ■

Example 7.20 Let $X_1, \ldots, X_n$ be a random sample from a Weibull pdf

$$f(x; \alpha, \beta) = \begin{cases} \dfrac{\alpha}{\beta^\alpha} \cdot x^{\alpha-1} \cdot e^{-(x/\beta)^\alpha} & x \geq 0 \\ 0 & \text{otherwise} \end{cases}$$

Writing the likelihood and ln(likelihood), then setting both $(\partial/\partial\alpha)[\ln(f)] = 0$ and $(\partial/\partial\beta)[\ln(f)] = 0$ yields the equations

$$\alpha = \left[\frac{\sum x_i^\alpha \cdot \ln(x_i)}{\sum x_i^\alpha} - \frac{\sum \ln(x_i)}{n} \right]^{-1} \qquad \beta = \left(\frac{\sum x_i^\alpha}{n} \right)^{1/\alpha}$$

These two equations cannot be solved explicitly to give general formulas for the mle's $\hat{\alpha}$ and $\hat{\beta}$. Instead, for each sample $x_1, \ldots, x_n$, the equations must be solved using an iterative numerical procedure. Even moment estimators of α and β are somewhat complicated (see Exercise 22).

The iterative mle computations can be done on a computer, and they are available in some statistical packages. MINITAB gives maximum likelihood estimates for both the Weibull and the gamma distributions (under "Quality Tools"). Stata has a general procedure that can be used for these and other distributions. For the data of Example 7.14 the maximum likelihood estimates for the Weibull distribution are $\hat{\alpha} = 3.799$ and $\hat{\beta} = 125.88$. (The mle's for the gamma distribution are $\hat{\alpha} = 8.799$ and $\hat{\beta} = 12.893$, a little different from the moment estimates in Example 7.14.) Figure 7.6 shows the Weibull log likelihood as a function of α and β. The surface near the top has a rounded shape, allowing the maximum to be found easily, but for some distributions the surface can be much more irregular, and the maximum may be hard to find.

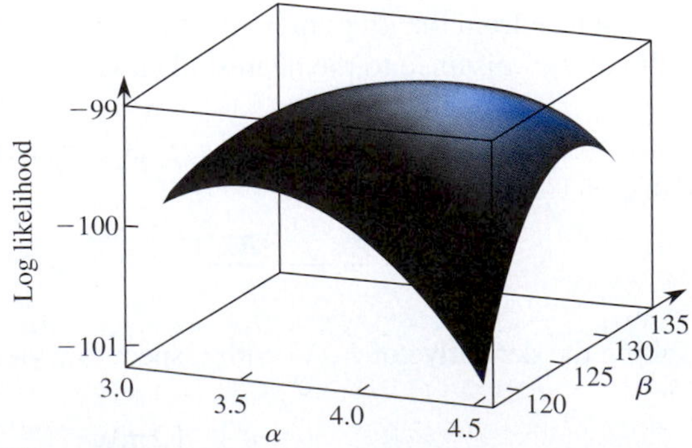

Figure 7.6 Weibull log likelihood for Example 7.20 ■

Some Properties of MLEs

In Example 7.18, we obtained the mle of σ^2 when the underlying distribution is normal. The mle of $\sigma = \sqrt{\sigma^2}$, as well as many other mle's, can be easily derived using the following proposition.

PROPOSITION **The Invariance Principle**

Let $\hat{\theta}_1, \hat{\theta}_2, \ldots, \hat{\theta}_m$ be the mle's of the parameters $\theta_1, \theta_2, \ldots, \theta_m$. Then the mle of any function $h(\theta_1, \theta_2, \ldots, \theta_m)$ of these parameters is the function $h(\hat{\theta}_1, \hat{\theta}_2, \ldots, \hat{\theta}_m)$ of the mle's.

Proof For an intuitive idea of the proof, consider the special case $m = 1$, with $\theta_1 = \theta$, and assume that $h(\cdot)$ is a one-to-one function. On the graph of the likelihood as a function of the parameter θ, the highest point occurs where $\theta = \hat{\theta}$. Now consider the graph of the likelihood as a function of $h(\theta)$. In the new graph the same heights occur, but the height that was previously plotted at $\theta = a$ is now plotted at $h(\theta) = h(a)$, and the highest point is now plotted at $h(\theta) = h(\hat{\theta})$. Thus, the maximum remains the same, but it now occurs at $h(\hat{\theta})$. ■

Example 7.21 (Example 7.18 continued)

In the normal case, the mle's of μ and σ^2 are $\hat{\mu} = \overline{X}$ and $\hat{\sigma}^2 = \Sigma(X_i - \overline{X})^2/n$. To obtain the mle of the function $h(\mu, \sigma^2) = \sqrt{\sigma^2} = \sigma$, substitute the mle's into the function:

$$\hat{\sigma} = \sqrt{\hat{\sigma}^2} = \left[\frac{1}{n}\sum (X_i - \overline{X})^2\right]^{1/2}$$

The mle of σ is not the sample standard deviation S, though they are close unless n is quite small. ■

Example 7.22 (Example 7.20 continued)

The mean value of an rv X that has a Weibull distribution is

$$\mu = \beta \cdot \Gamma(1 + 1/\alpha)$$

The mle of μ is therefore $\hat{\mu} = \hat{\beta}\Gamma(1 + 1/\hat{\alpha})$, where $\hat{\alpha}$ and $\hat{\beta}$ are the mle's of α and β. In particular, $\overline{X}$ is not the mle of μ, though it is an unbiased estimator. At least for large n, $\hat{\mu}$ is a better estimator than $\overline{X}$. ■

Large-Sample Behavior of the MLE

Although the principle of maximum likelihood estimation has considerable intuitive appeal, the following proposition provides additional rationale for the use of mle's. (See Section 7.4 for more details.)

PROPOSITION Under very general conditions on the joint distribution of the sample, when the sample size is large, the maximum likelihood estimator of any parameter θ is close to θ (consistency), is approximately unbiased $[E(\hat{\theta}) \approx \theta]$, and has variance that is nearly as small as can be achieved by any unbiased estimator. Stated another way, the mle $\hat{\theta}$ is approximately the MVUE of θ.

Because of this result and the fact that calculus-based techniques can usually be used to derive the mle's (though often numerical methods, such as Newton's method, are necessary), maximum likelihood estimation is the most widely used estimation technique among statisticians. Many of the estimators used in the remainder of the book are mle's. Obtaining an mle, however, does require that the underlying distribution be specified.

Note that there is no similar result for method of moments estimators. In general, if there is a choice between maximum likelihood and moment estimators, the mle is preferable. For example, the maximum likelihood method applied to estimating gamma distribution parameters tends to give better estimates (closer to the parameter values) than does the method of moments, so the extra computation is worth the price.

Some Complications

Sometimes calculus cannot be used to obtain mle's.

Example 7.23 Suppose the waiting time for a bus is uniformly distributed on $[0, \theta]$ and the results $x_1, \ldots, x_n$ of a random sample from this distribution have been observed. Since $f(x; \theta) = 1/\theta$ for $0 \le x \le \theta$ and 0 otherwise,

$$f(x_1, \ldots, x_n; \theta) = \begin{cases} \dfrac{1}{\theta^n} & 0 \le x_1 \le \theta, \ldots, 0 \le x_n \le \theta \\ 0 & \text{otherwise} \end{cases}$$

As long as $\max(x_i) \le \theta$, the likelihood is $1/\theta^n$, which is positive, but as soon as $\theta < \max(x_i)$, the likelihood drops to 0. This is illustrated in Figure 7.7. Calculus will not work because the maximum of the likelihood occurs at a point of discontinuity, but the figure shows that $\hat{\theta} = \max(X_i)$. Thus if my waiting times are 2.3, 3.7, 1.5, .4, and 3.2, then the mle is $\hat{\theta} = 3.7$. Note that the mle is not unbiased (see Example 7.5).

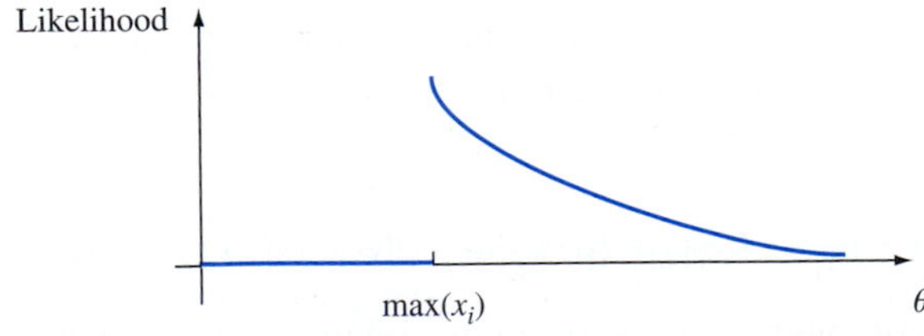

Figure 7.7 The likelihood function for Example 7.23 ■

Example 7.24 A method that is often used to estimate the size of a wildlife population involves performing a capture/recapture experiment. In this experiment, an initial sample of M animals is captured, each of these animals is tagged, and the animals are then returned to the population. After allowing enough time for the tagged individuals to mix into the population, another sample of size n is captured. With $X =$ the number of tagged animals in the second sample, the objective is to use the observed x to estimate the population size N.

The parameter of interest is $\theta = N$, which can assume only integer values, so even after determining the likelihood function (pmf of X here), using calculus to obtain N would present difficulties. If we think of a success as a previously tagged animal being recaptured, then sampling is without replacement from a population containing M successes and $N - M$ failures, so that X is a hypergeometric rv and the likelihood function is

$$p(x; N) = h(x; n, M, N) = \frac{\binom{M}{x} \cdot \binom{N-M}{n-x}}{\binom{N}{n}}$$

The integer-valued nature of N notwithstanding, it would be difficult to take the derivative of $p(x; N)$. However, if we consider the ratio of $p(x; N)$ to $p(x; N - 1)$, we have

$$\frac{p(x; N)}{p(x; N-1)} = \frac{(N-M) \cdot (N-n)}{N(N-M-n+x)}$$

This ratio is larger than 1 if and only if (iff) $N < Mn/x$. The value of N for which $p(x; N)$ is maximized is therefore the largest integer less than Mn/x. If we use standard mathematical notation $[r]$ for the largest integer less than or equal to r, the mle of N is $\hat{N} = [Mn/x]$. As an illustration, if $M = 200$ fish are taken from a lake and tagged, subsequently $n = 100$ fish are recaptured, and among the 100 there are $x = 11$ tagged fish, then $\hat{N} = [(200)(100)/11] = [1818.18] = 1818$. The estimate is actually rather intuitive; x/n is the proportion of the recaptured sample that is tagged, whereas M/N is the proportion of the entire population that is tagged. The estimate is obtained by equating these two proportions (estimating a population proportion by a sample proportion). ■

Suppose $X_1, X_2, \ldots, X_n$ is a random sample from a pdf $f(x; \theta)$ that is symmetric about θ, but the investigator is unsure of the form of the f function. It is then desirable to use an estimator $\hat{\theta}$ that is *robust*—that is, one that performs well for a wide variety of underlying pdf's. One such estimator is a trimmed mean. In recent years, statisticians have proposed another type of estimator, called an *M-estimator*, based on a generalization of maximum likelihood estimation. Instead of maximizing the log likelihood $\sum \ln[f(x; \theta)]$ for a specified f, one maximizes $\sum \rho(x_i; \theta)$. The "objective function" ρ is selected to yield an estimator with good robustness properties. The book by David Hoaglin et al. (see the bibliography) contains a good exposition on this subject.

Exercises Section 7.2 (21–31)

21. A random sample of n bike helmets manufactured by a certain company is selected. Let X = the number among the n that are flawed, and let p = P(flawed). Assume that only X is observed, rather than the sequence of S's and F's.

a. Derive the maximum likelihood estimator of p. If $n = 20$ and $x = 3$, what is the estimate?

b. Is the estimator of part (a) unbiased?

c. If $n = 20$ and $x = 3$, what is the mle of the probability $(1 - p)^5$ that none of the next five helmets examined is flawed?

22. Let X have a Weibull distribution with parameters α and β, so

$$E(X) = \beta \cdot \Gamma(1 + 1/\alpha)$$
$$V(X) = \beta^2\{\Gamma(1 + 2/\alpha) - [\Gamma(1 + 1/\alpha)]^2\}$$

a. Based on a random sample $X_1, \ldots, X_n$, write equations for the method of moments estimators of β and α. Show that, once the estimate of α has been obtained, the estimate of β can be found from a table of the gamma function and that the estimate of α is the solution to a complicated equation involving the gamma function.

b. If $n = 20$, $\bar{x} = 28.0$, and $\Sigma x_i^2 = 16{,}500$, compute the estimates. (*Hint*: $[\Gamma(1.2)]^2/\Gamma(1.4) = .95$.)

23. Let X denote the proportion of allotted time that a randomly selected student spends working on a certain aptitude test. Suppose the pdf of X is

$$f(x; \theta) = \begin{cases} (\theta + 1)x^\theta & 0 \le x \le 1 \\ 0 & \text{otherwise} \end{cases}$$

where $-1 < \theta$. A random sample of ten students yields data $x_1 = .92, x_2 = .79, x_3 = .90, x_4 = .65, x_5 = .86, x_6 = .47, x_7 = .73, x_8 = .97, x_9 = .94, x_{10} = .77$.

a. Use the method of moments to obtain an estimator of θ, and then compute the estimate for this data.

b. Obtain the maximum likelihood estimator of θ, and then compute the estimate for the given data.

24. Two different computer systems are monitored for a total of n weeks. Let X_i denote the number of breakdowns of the first system during the ith week, and suppose the X_i's are independent and drawn from a Poisson distribution with parameter λ_1. Similarly, let Y_i denote the number of breakdowns of the second system during the ith week, and assume independence with each Y_i Poisson with parameter λ_2. Derive the mle's of λ_1, λ_2, and $\lambda_1 - \lambda_2$. [*Hint*: Using independence, write the joint pmf (likelihood) of the X_i's and Y_i's together.]

25. Refer to Exercise 21. Instead of selecting $n = 20$ helmets to examine, suppose we examine helmets in succession until we have found $r = 3$ flawed ones. If the 20th helmet is the third flawed one (so that the number of helmets examined that were not flawed is $x = 17$), what is the mle of p? Is this the same as the estimate in Exercise 21? Why or why not? Is it the same as the estimate computed from the unbiased estimator of Exercise 17?

26. Six Pepperidge Farm bagels were weighed, yielding the following data (grams):

117.6 109.5 111.6 109.2 119.1 110.8

(*Note*: 4 ounces = 113.4 grams)

a. Assuming that the six bagels are a random sample and the weight is normally distributed, estimate the true average weight and standard deviation of the weight using maximum likelihood.

b. Again assuming a normal distribution, estimate the weight below which 95% of all bagels will have their weights. (*Hint*: What is the 95th percentile in terms of μ and σ? Now use the invariance principle.)

27. Refer to Exercise 26. Suppose we choose another bagel and weigh it. Let X = weight of the bagel. Use the given data to obtain the mle of $P(X \le 113.4)$. (*Hint*: $P(X \le 113.4) = \Phi[(113.4 - \mu)/\sigma)]$.)

28. Let $X_1, \ldots, X_n$ be a random sample from a gamma distribution with parameters α and β.

a. Derive the equations whose solution yields the maximum likelihood estimators of α and β. Do you think they can be solved explicitly?

b. Show that the mle of $\mu = \alpha\beta$ is $\hat{\mu} = \bar{X}$.

29. Let $X_1, X_2, \ldots, X_n$ represent a random sample from the Rayleigh distribution with density function given in Exercise 15. Determine

a. The maximum likelihood estimator of θ and then calculate the estimate for the vibratory stress data given in that exercise. Is this estimator the same as the unbiased estimator suggested in Exercise 15?

b. The mle of the median of the vibratory stress distribution. (*Hint*: First express the median in terms of θ.)

30. Consider a random sample $X_1, X_2, \dots, X_n$ from the shifted exponential pdf

$$f(x; \lambda, \theta) = \begin{cases} \lambda e^{-\lambda(x-\theta)} & x \ge \theta \\ 0 & \text{otherwise} \end{cases}$$

Taking $\theta = 0$ gives the pdf of the exponential distribution considered previously (with positive density to the right of zero). An example of the shifted exponential distribution appeared in Example 4.4, in which the variable of interest was time headway in traffic flow and $\theta = .5$ was the minimum possible time headway.

a. Obtain the maximum likelihood estimators of θ and λ.

b. If $n = 10$ time headway observations are made, resulting in the values 3.11, .64, 2.55, 2.20, 5.44, 3.42, 10.39, 8.93, 17.82, and 1.30, calculate the estimates of θ and λ.

31. At time $t = 0$, 20 identical components are put on test. The lifetime distribution of each is exponential with parameter λ. The experimenter then leaves the test facility unmonitored. On his return 24 hours later, the experimenter immediately terminates the test after noticing that $y = 15$ of the 20 components are still in operation (so 5 have failed). Derive the mle of λ. [*Hint*: Let Y = the number that survive 24 hours. Then $Y \sim \text{Bin}(n, p)$. What is the mle of p? Now notice that $p = P(X_i \ge 24)$, where X_i is exponentially distributed. This relates λ to p, so the former can be estimated once the latter has been.]

7.3 *Sufficiency

An investigator who wishes to make an inference about some parameter θ will base conclusions on the value of one or more statistics—the sample mean $\overline{X}$, the sample variance S^2, the sample range $Y_n - Y_1$, and so on. Intuitively, some statistics will contain more information about θ than will others. Sufficiency, the topic of this section, will help us decide which functions of the data are most informative for making inferences.

As a first point, we note that a statistic $T = t(X_1, \dots, X_n)$ will not be useful for drawing conclusions about θ unless the distribution of T depends on θ. Consider, for example, a random sample of size $n = 2$ from a normal distribution with mean μ and variance σ^2, and let $T = X_1 - X_2$. Then T has a normal distribution with mean 0 and variance $2\sigma^2$, which does not depend on μ. Thus this statistic cannot be used as a basis for drawing any conclusions about μ, though it certainly does carry information about the variance σ^2.

The relevance of this observation to sufficiency is as follows. Suppose an investigator is given the value of some statistic T, and then examines the *conditional distribution of the sample* $X_1, X_2, \dots, X_n$ given the value of the statistic—for example, the conditional distribution given that $\overline{X} = 28.7$. If this conditional distribution does not depend upon θ, then it can be concluded that there is no additional information about θ in the data over and above what is provided by T. In this sense, for purposes of making inferences about θ, it is *sufficient* to know the value of T, which contains all the information in the data relevant to θ.

Example 7.25 An investigation of major defects on new vehicles of a certain type involved selecting a random sample of $n = 3$ vehicles and determining for each one the value of X = the number of major defects. This resulted in observations $x_1 = 1$, $x_2 = 0$, and $x_3 = 3$. You, as a consulting statistician, have been provided with a description of the experiment, from which it is reasonable to assume that X has a Poisson distribution, and told only that the total number of defects for the three sampled vehicles was four.

Knowing that $T = \Sigma X_i = 4$, would there be any additional advantage in having the observed values of the individual X_i's when making an inference about the Poisson parameter λ? Or rather is it the case that the statistic T contains all relevant information about λ in the data? To address this issue, consider the conditional distribution of X_1, X_2, X_3 given that $\Sigma X_i = 4$. First of all, there are only a few possible (x_1, x_2, x_3) triples for which $x_1 + x_2 + x_3 = 4$. For example, (0, 4, 0) is a possibility, as are (2, 2, 0) and (1, 0, 3), but not (1, 2, 3) or (5, 0, 2). That is,

$$P\left(X_1 = x_1, X_2 = x_2, X_3 = x_3 \,\Big|\, \sum_{i=1}^{3} X_i = 4\right) = 0 \quad \text{unless } x_1 + x_2 + x_3 = 4$$

Now consider the triple (2, 1, 1), which is consistent with $\Sigma X_i = 4$. If we let A denote the event that $X_1 = 2$, $X_2 = 1$, and $X_3 = 1$ and B denote the event that $\Sigma X_i = 4$, then the event A implies the event B (i.e., A is contained in B), so the intersection of the two events is just the smaller event A. Thus

$$P\left(X_1 = 2, X_2 = 1, X_3 = 1 \,\Big|\, \sum_{i=1}^{3} X_i = 4\right) = P(A \mid B) = \frac{P(A \cap B)}{P(B)}$$

$$= \frac{P(X_1 = 2, X_2 = 1, X_3 = 1)}{P(\Sigma X_i = 4)}$$

A moment generating function argument shows that ΣX_i has a Poisson distribution with parameter 3λ. Thus the desired conditional probability is

$$\frac{\dfrac{e^{-\lambda}\cdot\lambda^2}{2!}\cdot\dfrac{e^{-\lambda}\cdot\lambda^1}{1!}\cdot\dfrac{e^{-\lambda}\cdot\lambda^1}{1!}}{\dfrac{e^{-3\lambda}\cdot(3\lambda)^4}{4!}} = \frac{4!}{3^4\cdot 2!} = \frac{4}{27}$$

Similarly,

$$P\left(X_1 = 1, X_2 = 0, X_3 = 3 \,\Big|\, \sum X_i = 4\right) = \frac{4!}{3^4\cdot 3!} = \frac{4}{81}$$

The complete conditional distribution is as follows:

$$P\left(X_1 = x_1, X_2 = x_2, X_3 = x_3 \,\Big|\, \sum_{i=1}^{3} X_i = 4\right)$$

$$= \begin{cases} \frac{6}{81} & (x_1, x_2, x_3) = (2, 2, 0), (2, 0, 2), (0, 2, 2) \\ \frac{12}{81} & (x_1, x_2, x_3) = (2, 1, 1), (1, 2, 1), (1, 1, 2) \\ \frac{1}{81} & (x_1, x_2, x_3) = (4, 0, 0), (0, 4, 0), (0, 0, 4) \\ \frac{4}{81} & (x_1, x_2, x_3) = (3, 1, 0), (1, 3, 0), (3, 0, 1), (1, 0, 3), (0, 1, 3), (0, 3, 1) \end{cases}$$

This conditional distribution does not involve λ. Thus once the value of the statistic ΣX_i has been provided, there is no additional information about λ in the individual observations.

To put this another way, think of obtaining the data from the experiment in two stages:

1. Observe the value of $T = X_1 + X_2 + X_3$ from a Poisson distribution with parameter 3λ.
2. Having observed $T = 4$, now obtain the individual x_i's from the conditional distribution

$$P\left(X_1 = x_1, X_2 = x_2, X_3 = x_3 \,\middle|\, \sum_{i=1}^{3} X_i = 4\right)$$

Since the conditional distribution in step 2 does not involve λ, there is no additional information about λ resulting from the second stage of the data generation process. This argument holds more generally for any sample size n and any value t other than 4 (e.g., the total number of defects among 10 randomly selected vehicles might be $\sum X_i = 16$). Once the value of $\sum X_i$ is known, there is no further information in the data about the Poisson parameter. ■

DEFINITION

A statistic $T = t(X_1, \ldots, X_n)$ is said to be **sufficient** for making inferences about a parameter θ if the joint distribution of $X_1, X_2, \ldots, X_n$ given that $T = t$ does not depend upon θ for every possible value t of the statistic T.

The notion of sufficiency formalizes the idea that a statistic T contains all relevant information about θ. Once the value of T for the given data is available, it is of no benefit to know anything else about the sample.

The Factorization Theorem

How can a sufficient statistic be identified? It may seem as though one would have to select a statistic, determine the conditional distribution of the X_i's given any particular value of the statistic, and keep doing this until hitting paydirt by finding one that satisfies the defining condition. This would be terribly time-consuming, and when the X_i's are continuous there are additional technical difficulties in obtaining the relevant conditional distribution. Fortunately, the next result provides a relatively straightforward way of proceeding.

THE NEYMAN FACTORIZATION THEOREM

Let $f(x_1, x_2, \ldots, x_n; \theta)$ denote the joint pmf or pdf of $X_1, X_2, \ldots, X_n$. Then $T = t(X_1, \ldots, X_n)$ is a sufficient statistic for θ if and only if the joint pmf or pdf can be represented as a product of two factors in which the first factor involves θ and the data only through $t(x_1, \ldots, x_n)$, whereas the second factor involves $x_1, \ldots, x_n$ but does not depend on θ:

$$f(x_1, x_2, \ldots, x_n; \theta) = g[t(x_1, \ldots, x_n); \theta] \cdot h(x_1, \ldots, x_n)$$

Before sketching a proof of this theorem, we consider several examples.

Example 7.26 Let's generalize the previous example by considering a random sample $X_1, X_2, \ldots, X_n$ from a Poisson distribution with parameter λ— for example, the numbers of blemishes on n independently selected reels of high-quality recording tape or the numbers of errors in n batches of invoices where each batch consists of 200 invoices. The joint pmf of these variables is

$$f(x_1, \ldots, x_n; \lambda) = \frac{e^{-\lambda}\lambda^{x_1}}{x_1!} \cdot \frac{e^{-\lambda}\lambda^{x_2}}{x_2!} \cdot \cdots \cdot \frac{e^{-\lambda}\lambda^{x_n}}{x_n!} = \frac{e^{-n\lambda} \cdot \lambda^{x_1+x_2+\cdots+x_n}}{x_1! \cdot x_2! \cdot \cdots \cdot x_n!}$$

$$= (e^{-n\lambda} \cdot \lambda^{\Sigma x_i}) \cdot \left(\frac{1}{x_1! \cdot x_2! \cdot \cdots \cdot x_n!}\right)$$

The factor inside the first set of parentheses involves the parameter λ and the data only through Σx_i, whereas the factor inside the second set of parentheses involves the data but not λ. So we have the desired factorization, and the sufficient statistic is $T = \Sigma X_i$, as we previously ascertained directly from the definition of sufficiency. ■

A sufficient statistic is not unique; any one-to-one function of a sufficient statistic is itself sufficient. In the Poisson example, the sample mean $\overline{X} = (1/n)\,\Sigma X_i$ is a one-to-one function of ΣX_i (knowing the value of the sum of the n observations is equivalent to knowing their mean), so the sample mean is also a sufficient statistic.

Example 7.27 Suppose that the waiting time for a bus on a weekday morning is uniformly distributed on the interval from 0 to θ, and consider a random sample $X_1, \ldots, X_n$ of waiting times (i.e., times on n independently selected mornings). The joint pdf of these times is

$$f(x_1, \ldots, x_n; \theta) = \begin{cases} \dfrac{1}{\theta} \cdot \dfrac{1}{\theta} \cdot \cdots \cdot \dfrac{1}{\theta} = \dfrac{1}{\theta^n} & 0 \le x_1 \le \theta, \ldots, 0 \le x_n \le \theta \\ 0 & \text{otherwise} \end{cases}$$

To obtain the desired factorization, we introduce notation for an indicator function of an event A: $I(A) = 1$ if $(x_1, x_2, \ldots, x_n)$ lies in A and $I(A) = 0$ otherwise. Now let

$$A = \{(x_1, x_2, \ldots, x_n): 0 \le x_1 \le \theta, 0 \le x_2 \le \theta, \ldots, 0 \le x_n \le \theta\}$$

That is, A is the indicator for the event that all x_i's are between 0 and θ. All n of the x_i's will be between 0 and θ if and only if the smallest of the x_i's is at least 0 and the largest is at most θ. Thus

$$I(A) = I(0 \le \min\{x_1, \ldots, x_n\}) \cdot I(\max\{x_1, \ldots, x_n\} \le \theta)$$

We can now use this indicator function notation to write a one-line expression for the joint pdf:

$$f(x_1, x_2, \ldots, x_n; \theta) = \left[\frac{1}{\theta^n} \cdot I(\max\{x_1, \ldots, x_n\} \le \theta)\right] \cdot [I(0 \le \min\{x_1, \ldots, x_n\})]$$

The factor inside the first set of square brackets involves θ and the x_i's only through $t(x_1, \ldots, x_n) = \max\{x_1, \ldots, x_n\}$. Voila, we have our desired factorization, and the sufficient statistic for the uniform parameter θ is $T = \max\{X_1, \ldots, X_n\}$, the largest order statistic.

All the information about θ in this uniform random sample is contained in the largest of the n observations. This result is much more difficult to obtain directly from the definition of sufficiency. ■

Proof of the Factorization Theorem A general proof when the X_i's constitute a random sample from a continuous distribution is fraught with technical details that are beyond the level of our text. So we content ourselves with a proof in the discrete case. For the sake of concise notation, denote $X_1, X_2, \ldots, X_n$ by $\mathbf{X}$ and $x_1, x_2, \ldots, x_n$ by $\mathbf{x}$.

Suppose first that $T = t(\mathbf{x})$ is sufficient, so that $P(\mathbf{X} = \mathbf{x} \mid T = t)$ does not depend upon θ. Focus on a value t for which $t(\mathbf{x}) = t$ (e.g., $\mathbf{x} = 3, 0, 1$, $t(\mathbf{x}) = \sum x_i$, so $t = 4$). The event that $\mathbf{X} = \mathbf{x}$ is then identical to the event that both $\mathbf{X} = \mathbf{x}$ and $T = t$ because the former equality implies the latter one. Thus

$$\begin{aligned} f(\mathbf{x}; \theta) &= P(\mathbf{X} = \mathbf{x}; \theta) = P(\mathbf{X} = \mathbf{x}, T = t; \theta) \\ &= P(\mathbf{X} = \mathbf{x} \mid T = t; \theta) \cdot P(T = t; \theta) = P(\mathbf{X} = \mathbf{x} \mid T = t) \cdot P(T = t; \theta) \end{aligned}$$

Since the first factor in this latter product does not involve θ and the second one involves the data only through t, we have our desired factorization.

Now let's go the other way: Assume a factorization, and show that T is sufficient, that is, that the conditional probability that $\mathbf{X} = \mathbf{x}$ given that $T = t$ does not involve θ.

$$\begin{aligned} P(\mathbf{X} = \mathbf{x} \mid T = t; \theta) &= \frac{P(\mathbf{X} = \mathbf{x}, T = t; \theta)}{P(T = t; \theta)} = \frac{P(\mathbf{X} = \mathbf{x}; \theta)}{P(T = t; \theta)} \\ &= \frac{g(t; \theta) \cdot h(\mathbf{x})}{\sum_{\mathbf{u}: t(\mathbf{u}) = t} P(\mathbf{X} = u; \theta)} = \frac{g(t; \theta) \cdot h(\mathbf{x})}{\sum_{\mathbf{u}: t(\mathbf{u}) = t} g[t(\mathbf{u}); \theta] \cdot h(\mathbf{u})} = \frac{h(\mathbf{x})}{\sum_{\mathbf{u}: t(\mathbf{u}) = t} h(\mathbf{u})} \end{aligned}$$

Sure enough, this latter ratio does not involve θ. ■

Jointly Sufficient Statistics

When the joint pmf or pdf of the data involves a single unknown parameter θ, there is frequently a single statistic (single function of the data) that is sufficient. However, when there are several unknown parameters—for example the mean μ and standard deviation σ of a normal distribution, or the shape parameter α and scale parameter β of a gamma distribution—we must expand our notion of sufficiency.

DEFINITION

Suppose the joint pmf or pdf of the data involves k unknown parameters $\theta_1, \theta_2, \ldots, \theta_k$. The m statistics $T_1 = t_1(X_1, \ldots, X_n)$, $T_2 = t_2(X_1, \ldots, X_n), \ldots, T_m = t_m(X_1, \ldots, X_n)$ are said to be **jointly sufficient** for the parameters if the conditional distribution of the X_i's given that $T_1 = t_1, T_2 = t_2, \ldots, T_m = t_m$ does not depend on any of the unknown parameters, and this is true for all possible values $t_1, t_2, \ldots, t_m$ of the statistics.

Example 7.28 Consider a random sample of size $n = 3$ from a continuous distribution, and let T_1, T_2, and T_3 be the three order statistics, that is, T_1 = the smallest of the three X_i's, T_2 = the second smallest X_i, and T_3 = the largest X_i (these order statistics were previously denoted by Y_1, Y_2, and Y_3). Then for any values t_1, t_2, and t_3 satisfying $t_1 < t_2 < t_3$,

$$P(X_1 = x_1, X_2 = x_2, X_3 = x_3 \mid T_1 = t_1, T_2 = t_2, T_3 = t_3)$$

$$= \begin{cases} \dfrac{1}{3!} & x_1, x_2, x_3 = t_1, t_2, t_3;\ t_1, t_3, t_2;\ t_2, t_1, t_3;\ t_2, t_3, t_1;\ t_3, t_1, t_2;\ t_3, t_2, t_1 \\ 0 & \text{otherwise} \end{cases}$$

For example, if the three ordered values are 21.4, 23.8, and 26.0, then the conditional probability distribution of the three X_i's places probability $\frac{1}{6}$ on each of the 6 permutations of these three numbers (23.8, 21.4, 26.0, and so on). This conditional distribution clearly does not involve any unknown parameters.

Generalizing this argument to a sample of size n, we see that for a random sample from a continuous distribution, the order statistics are jointly sufficient for $\theta_1, \theta_2, \ldots, \theta_k$ regardless of whether $k = 1$ (e.g., the exponential distribution has a single parameter) or 2 (the normal distribution) or even $k > 2$. ■

The factorization theorem extends to the case of jointly sufficient statistics: T_1, $T_2, \ldots, T_m$ are jointly sufficient for $\theta_1, \theta_2, \ldots, \theta_k$ if and only if the joint pmf or pdf of the X_i's can be represented as a product of two factors, where the first involves the θ_i's and the data only through $t_1, t_2, \ldots, t_m$ and the second does not involve the θ_i's.

Example 7.29 Let $X_1, \ldots, X_n$ be a random sample from a normal distribution with mean μ and variance σ^2. The joint pdf is

$$f(x_1, \ldots, x_n; \mu, \sigma^2) = \prod_{i=1}^{n} \frac{1}{\sqrt{2\pi\sigma^2}} e^{-(x_i-\mu)^2/2\sigma^2}$$

$$= \left[\frac{1}{\sigma^n} \cdot e^{-(\Sigma x_i^2 - 2\mu\Sigma x_i + n\mu^2)/2\sigma^2}\right] \cdot \left(\frac{1}{2\pi}\right)^{n/2}$$

This factorization shows that the two statistics ΣX_i and ΣX_i^2 are jointly sufficient for the two parameters μ and σ^2. Since $\Sigma(X_i - \overline{X})^2 = \Sigma X_i^2 - n(\overline{X})^2$, there is a one-to-one correspondence between the two sufficient statistics and the statistics $\overline{X}$ and $\Sigma(X_i - \overline{X})^2$—that is, values of the two original sufficient statistics uniquely determine values of the latter two statistics, and vice versa. This implies that the latter two statistics are also jointly sufficient, which in turn implies that the sample mean and sample variance (or sample standard deviation) are jointly sufficient statistics. The sample mean and sample variance encapsulate all the information about μ and σ^2 that is contained in the sample data. ■

Minimal Sufficiency

When $X_1, \ldots, X_n$ constitute a random sample from a normal distribution, the n order statistics $Y_1, \ldots, Y_n$ are jointly sufficient for μ and σ^2, and the sample mean and sample variance are also jointly sufficient. Both the order statistics and the pair $(\overline{X}, S^2)$ reduce the data without any information loss, but the sample mean and variance represent a greater reduction. In general, we would like the greatest possible reduction without information loss. A **minimal** (possibly jointly) **sufficient statistic** is a function of every other sufficient statistic. That is, given the value(s) of any other sufficient statistic(s), the value(s) of the minimal sufficient statistic(s) can be calculated. The minimal sufficient statistic is the sufficient statistic having the smallest dimensionality, and thus represents the greatest possible reduction of the data without any information loss.

A general discussion of minimal sufficiency is beyond the scope of our text. In the case of a normal distribution with values of both μ and σ^2 unknown, it can be shown that the sample mean and sample variance are jointly minimal sufficient (so the same is true of $\sum X_i$ and $\sum X_i^2$). It is intuitively reasonable that because there are two unknown parameters, there should be a pair of sufficient statistics. It is indeed often the case that the number of the (jointly) sufficient statistic(s) matches the number of unknown parameters. But this is not always true. Consider a random sample $X_1, \ldots, X_n$ from a Cauchy distribution, one with pdf $f(x; \theta) = 1/\{\pi[1 + (x - \theta)]^2\}$ for $-\infty < x < \infty$. The graph of this pdf is bell-shaped and centered at θ, but its tails decrease much more slowly than those of a normal density curve. Because the Cauchy distribution is continuous, the order statistics are jointly sufficient for θ. It would seem, though, that a single sufficient statistic (one-dimensional) could be found for the single parameter. Unfortunately this is not the case; it can be shown that the order statistics are *minimal* sufficient! So going beyond the order statistics to any single function of the X_i's as a point estimator of θ entails a loss of information from the original data.

Improving an Estimator

Because a sufficient statistic contains all the information the data has to offer about the value of θ, it is reasonable that an estimator of θ or any function of θ should depend on the data only through the sufficient statistic. A general result due to Rao and Blackwell shows how to start with an unbiased statistic that is not a function of sufficient statistics and create an improved estimator that is sufficient.

THEOREM Suppose that the joint distribution of $X_1, \ldots, X_n$ depends on some unknown parameter θ and that T is sufficient for θ. Consider estimating $h(\theta)$, a specified function of θ. If U is an unbiased statistic for estimating $h(\theta)$, then the estimator $U^* = E(U|T)$ is also unbiased for $h(\theta)$ and has variance no greater than the original unbiased estimator U.

Proof First of all, we must show that U^* is indeed an estimator—that it is a function of the X_i's which does not depend on θ. This follows because since T is sufficient, the distribution of U conditional on T does not involve θ, so the expected value calculated from the conditional distribution will of course not involve θ. The fact that U^* has variance no greater than U is a consequence of a conditional expectation–conditional variance formula for $V(U)$ introduced in Section 5.3:

$$V(U) = V[E(U|T)] + E[V(U|T)] = V(U^*) + E[V(U|T)]$$

Because $V(U|T)$, being a variance, is nonnegative, it follows that $V(U) \geq V(U^*)$ as desired. ■

Example 7.30 Suppose that the number of major defects on a randomly selected new vehicle of a certain type has a Poisson distribution with parameter λ. Consider estimating $e^{-\lambda}$, the probability that a vehicle has no such defects, based on a random sample of n vehicles. Let's start with the estimator $U = I(X_1 = 0)$, the indicator function of the event that the first vehicle in the sample has no defects. That is,

$$U = \begin{cases} 1 & \text{if } X_1 = 0 \\ 0 & \text{if } X_1 > 0 \end{cases}$$

Then

$$E(U) = 1 \cdot P(X_1 = 0) + 0 \cdot P(X_1 = 0) > P(X_1 = 0) = e^{-\lambda} \cdot \lambda^0/0! = e^{-\lambda}$$

Our estimator is therefore unbiased for estimating the probability of no defects. The sufficient statistic here is $T = \Sigma X_i$, so of course the estimator U is not a function of T. The improved estimator is $U^* = E(U|\,\Sigma X_i) = P(X_1 = 0|\,\Sigma X_i)$. Let's consider $P(X_1 = 0|\,\Sigma X_i = t)$ where t is some nonnegative integer. The event that $X_1 = 0$ and $\Sigma X_i = t$ is identical to the event that the first vehicle has no defects and the total number of defects on the last $n - 1$ vehicles is t. Thus

$$P\left(X_1 = 0 \,\Big|\, \sum_{i=1}^{n} X_i = t\right) = \frac{P\left(\{X_1 = 0\} \cap \left\{\sum_{i=1}^{n} X_i = t\right\}\right)}{P\left(\sum_{i=1}^{n} X_i = t\right)} = \frac{P\left(\{X_1 = 0\} \cap \left\{\sum_{i=2}^{n} X_i = t\right\}\right)}{P\left(\sum_{i=1}^{n} X_i = t\right)}$$

A moment generating function argument shows that the sum of all n X_i's has a Poisson distribution with parameter $n\lambda$ and the sum of the last $n - 1$ X_i's has a Poisson distribution with parameter $(n - 1)\lambda$. Furthermore, X_1 is independent of the other $n - 1$ X_i's so it is independent of their sum, from which

$$P\left(X_1 = 0 \,\Big|\, \sum_{i=1}^{n} X_i = t\right) = \frac{\dfrac{e^{-\lambda}\lambda^0}{0!} \cdot \dfrac{e^{-(n-1)\lambda}[(n-1)\lambda]^t}{t!}}{\dfrac{e^{-n\lambda}(n\lambda)^t}{t!}} = \left(\frac{n-1}{n}\right)^t$$

The improved unbiased estimator is then $U^* = (1 - 1/n)^T$. If, for example, there are a total of 15 defects among 10 randomly selected vehicles, then the estimate is $(1 - \frac{1}{10})^{15} = .206$. For this sample, $\hat{\lambda} = \bar{x} = 1.5$, so the maximum likelihood estimate of $e^{-\lambda}$ is $e^{-1.5} = .223$. Here as in some other situations the principles of unbiasedness and maximum likelihood are in conflict. However, if n is large, the improved estimate is $(1 - 1/n)^t = [(1 - 1/n)^n]^{\bar{x}} \approx e^{-\bar{x}}$, which is the mle. That is, the unbiased and maximum likelihood estimators are "asymptotically equivalent." ■

We have emphasized that in general there will not be a unique sufficient statistic. Suppose there are two different sufficient statistics T_1 and T_2 such that the first is not a one-to-one function of the second (e.g., we are not considering $T_1 = \Sigma X_i$ and $T_2 = \bar{X}$). Then it would be distressing if we started with an unbiased estimator U and found that

$E(U|T_1) \neq E(U|T_2)$, so our improved estimator depended on which sufficient statistic we used. Fortunately there are general conditions under which, starting with a minimal sufficient statistic T, the improved estimator is the MVUE (minimum variance unbiased estimator). That is, the new estimator is unbiased and has smaller variance than any other unbiased estimator. Please consult one of the references in the chapter bibliography for more detail.

Further Comments

Maximum likelihood is by far the most popular method for obtaining point estimates, so it would be disappointing if maximum likelihood estimators did not make full use of sample information. Fortunately the mle's do not suffer from this defect. If $T_1, \ldots, T_m$ are jointly sufficient statistics for parameters $\theta_1, \ldots, \theta_k$, then the joint pmf or pdf factors as follows:

$$f(x_1, \ldots, x_n; \theta_1, \ldots, \theta_k) = g(t_1, \ldots, t_m; \theta_1, \ldots, \theta_k) \cdot h(x_1, \ldots, x_n)$$

The maximum likelihood estimates result from maximizing $f(\cdot)$ with respect to the θ_i's. Since the $h(\cdot)$ factor does not involve the parameters, this is equivalent to maximizing the $g(\cdot)$ factor with respect to the θ_i's. The resulting $\hat{\theta}_i$'s will involve the data only through the t_i's. Thus it is always possible to find a maximum likelihood estimator that is a function of just the sufficient statistic(s). There are contrived examples of situations where the mle is not unique, in which case an mle that is not a function of the sufficient statistics can be constructed—but there is also an mle that *is* a function of the sufficient statistics.

The concept of sufficiency is compelling when an investigator is sure that the underlying distribution that generated the data is a member of some particular family (normal, exponential, etc.). However, two different families of distributions might each furnish plausible models for the data in a particular application, and yet the sufficient statistics for these two families might be different (an analogous comment applies to maximum likelihood estimation). For example, there are data sets for which a gamma probability plot suggests that a member of the gamma family would give a reasonable model, and a lognormal probability plot (normal probability plot of the logs of the observations) also indicates that lognormality is plausible. Yet the jointly sufficient statistics for the parameters of the gamma family are not the same as those for the parameters of the lognormal family. When estimating some parameter θ in such situations (e.g., the mean μ or median $\tilde{\mu}$), one would look for a *robust estimator* that performs well for a wide variety of underlying distributions, as discussed in Section 7.1. Please consult a more advanced source for additional information.

Exercises Section 7.3 (32–41)

32. The long-run proportion of vehicles that pass a certain emissions test is p. Suppose that three vehicles are independently selected for testing. Let $X_i = 1$ if the ith vehicle passes the test and $X_i = 0$ otherwise ($i = 1, 2, 3$), and let $X = X_1 + X_2 + X_3$. Use the definition of sufficiency to show that X is sufficient for p by obtaining the conditional distribution of the X_i's given that $X = x$ for each possible value x. Then generalize by giving an analogous argument for the case of n vehicles.

33. Components of a certain type are shipped in batches of size k. Suppose that whether or not any particular component is satisfactory is independent of the condition of any other component, and that the long-run proportion of satisfactory components is p. Consider n batches, and let X_i denote the number of satisfactory components in the ith batch ($i = 1, 2, \ldots, n$). Statistician A is provided with the values of all the X_i's, whereas statistician B is given only the value of $X = \sum X_i$. Use a conditional probability argument to decide whether statistician A has more information about p than does statistician B.

34. Let $X_1, \ldots, X_n$ be a random sample of component lifetimes from an exponential distribution with parameter λ. Use the factorization theorem to show that $\sum X_i$ is a sufficient statistic for λ.

35. Identify a pair of jointly sufficient statistics for the two parameters of a gamma distribution based on a random sample of size n from that distribution.

36. Suppose waiting time for delivery of an item is uniform on the interval from θ_1 to θ_2 [so $f(x; \theta_1, \theta_2) = 1/(\theta_2 - \theta_1)$ for $\theta_1 < x < \theta_2$ and is 0 otherwise]. Consider a random sample of n waiting times, and use the factorization theorem to show that $\min(X_i)$, $\max(X_i)$ is a pair of jointly sufficient statistics for θ_1 and θ_2. (*Hint*: Introduce an appropriate indicator function as we did in Example 7.27.)

37. For $\theta > 0$ consider a random sample from a uniform distribution on the interval from θ to 2θ (pdf $1/\theta$ for $\theta < x < 2\theta$), and use the factorization theorem to determine a sufficient statistic for θ.

38. Suppose that material strength X has a lognormal distribution with parameters μ and σ [which are the mean and standard deviation of $\ln(X)$, not of X itself]. Are $\sum X_i$ and $\sum X_i^2$ jointly sufficient for the two parameters? If not, what is a pair of jointly sufficient statistics?

39. The probability that any particular component of a certain type works in a satisfactory manner is p. If n of these components are independently selected, then the statistic X, the number among the selected components that perform in a satisfactory manner, is sufficient for p. You must purchase two of these components for a particular system. Obtain an unbiased statistic for the probability that exactly one of your purchased components will perform in a satisfactory manner. (*Hint*: Start with the statistic U, the indicator function of the event that exactly one of the first two components in the sample of size n performs as desired, and improve on it by conditioning on the sufficient statistic.)

40. In Example 7.30, we started with $U = I(X_1 = 0)$ and used a conditional expectation argument to obtain an unbiased estimator of the zero-defect probability based on the sufficient statistic. Consider starting with a different statistic: $U = [\sum I(X_i = 0)]/n$. Show that the improved estimator based on the sufficient statistic is identical to the one obtained in the cited example. [*Hint*: Use the general property $E(Y + Z|T) = E(Y|T) + E(Z|T)$.]

41. A particular quality characteristic of items produced using a certain process is known to be normally distributed with mean μ and standard deviation 1. Let X denote the value of the characteristic for a randomly selected item. An unbiased estimator for the parameter $\theta = P(X \le c)$, where c is a critical threshold, is desired. The estimator will be based on a random sample $X_1, \ldots, X_n$.

a. Obtain a sufficient statistic for μ.

b. Consider the estimator $\hat{\theta} = I(X_1 \le c)$. Obtain an improved unbiased estimator based on the sufficient statistic (it is actually the minimum variance unbiased estimator). [*Hint*: You may use the following facts: (1) The joint distribution of X_1 and $\overline{X}$ is bivariate normal with means μ and μ, respectively, variances 1 and $1/n$, respectively, and correlation ρ (which you should verify). (2) If Y_1 and Y_2 have a bivariate normal distribution, then the conditional distribution of Y_1 given that $Y_2 = y_2$ is normal with mean $\mu_1 + (\rho\sigma_1/\sigma_2)(y_2 - \mu_2)$ and variance $\sigma_1^2(1 - \rho^2)$.]

7.4 *Information and Efficiency

In this section we introduce the idea of *Fisher information* and two of its applications. The first application is to find the minimum possible variance for an unbiased estimator. The second application is to show that the maximum likelihood estimator is asymptotically unbiased and normal (that is, for large n it has expected value approximately θ and it has approximately a normal distribution) with the minimum possible variance.

Here the notation $f(x; \theta)$ will be used for a probability mass function or a probability density function with unknown parameter θ. The Fisher information is intended to measure the precision in a single observation. Consider the random variable U obtained by taking the partial derivative of $\ln[f(x; \theta)]$ with respect to θ and then replacing x by X: $U = \partial[\ln f(X; \theta)]/\partial\theta$. For example, if the pdf is $\theta x^{\theta-1}$ for $0 < x < 1$ $(\theta > 0)$, then $\partial[\ln(\theta x^{\theta-1})]/\partial\theta = \partial[\ln(\theta) + (\theta - 1)\ln(x)]/\partial\theta = 1/\theta + \ln(x)$, so $U = \ln(X) + 1/\theta$.

DEFINITION

The **Fisher information $I(\theta)$ in a single observation from a pmf or pdf $f(x; \theta)$** is the variance of the random variable $U = \partial\{\ln[f(X; \theta)]\}/\partial\theta$:

$$I(\theta) = V\left[\frac{\partial}{\partial\theta}\ln f(X; \theta)\right] \tag{7.7}$$

It may seem strange to differentiate the logarithm of the pmf or pdf, but this is exactly what is often done in maximum likelihood estimation. In what follows we will assume that $f(x; \theta)$ is a pmf, but everything that we do will apply also in the continuous case if appropriate assumptions are made. In particular, it is important to assume that the set of possible x's does not depend on the parameter.

When $f(x; \theta)$ is a pmf, we know that $1 = \sum_x f(x; \theta)$. Therefore, differentiating both sides with respect to θ and using the fact that $[\ln(f)]' = f'/f$, we find that the mean of U is 0:

$$\begin{aligned} 0 &= \frac{\partial}{\partial\theta}\sum_x f(x; \theta) = \sum_x \frac{\partial}{\partial\theta} f(x; \theta) \\ &= \sum_x \frac{\partial}{\partial\theta}[\ln f(x; \theta)]f(x; \theta) = E\left[\frac{\partial}{\partial\theta}\ln f(X; \theta)\right] = E(U) \end{aligned} \tag{7.8}$$

This involves interchanging the order of differentiation and summation, which requires certain technical assumptions if the set of possible x values is infinite. We will omit those assumptions here and elsewhere in this section, but we emphasize that switching differentiation and summation (or integration) is not allowed if the set of possible values depends on the parameter. For example, if the summation were from $-\theta$ to θ there would be additional variability, and therefore terms for the limits of summation would be needed.

There is an alternative expression for $I(\theta)$ that is sometimes easier to compute than the variance in the definition:

$$I(\theta) = -E\left[\frac{\partial^2}{\partial\theta^2}\ln f(X; \theta)\right] \tag{7.9}$$

This is a consequence of taking another derivative in (7.8):

$$\begin{aligned} 0 &= \sum_x \frac{\partial^2}{\partial\theta^2}[\ln f(x; \theta)]f(x; \theta) + \sum_x \frac{\partial}{\partial\theta}[\ln f(x; \theta)]\frac{\partial}{\partial\theta}[\ln f(x; \theta)]f(x; \theta) \\ &= E\left\{\frac{\partial^2}{\partial\theta^2}[\ln f(X; \theta)]\right\} + E\left\{\left[\frac{\partial}{\partial\theta}\ln f(X; \theta)\right]^2\right\} \end{aligned} \tag{7.10}$$

To complete the derivation of (7.9), recall that U has mean 0, so its variance is

$$I(\theta) = V\left\{\frac{\partial}{\partial\theta}[\ln f(X;\theta)]\right\} = E\left\{\left[\frac{\partial}{\partial\theta}\ln f(X;\theta)\right]^2\right\} = -E\left\{\frac{\partial^2}{\partial\theta^2}[\ln f(X;\theta)]\right\}$$

where Equation (7.10) is used in the last step.

Example 7.31 Let X be a Bernoulli rv, so $f(x;p) = p^x(1-p)^{1-x}$, $x = 0, 1$. Then

$$\frac{\partial}{\partial p}\ln f(X;p) = \frac{\partial}{\partial p}[X\ln p + (1-X)\ln(1-p)] = \frac{X}{p} - \frac{1-X}{1-p} = \frac{X-p}{p(1-p)} \quad (7.11)$$

This has mean 0, in accord with Equation (7.8), because $E(X) = p$. Computing the variance of the partial derivative, we get the Fisher information:

$$I(p) = V\left[\frac{\partial}{\partial p}\ln f(X;p)\right] = \frac{V(X-p)}{[p(1-p)]^2} = \frac{V(X)}{[p(1-p)]^2} = \frac{p(1-p)}{[p(1-p)]^2} = \frac{1}{p(1-p)} \quad (7.12)$$

The alternative method uses Equation (7.9). Differentiating Equation (7.11) with respect to p gives

$$\frac{\partial^2}{\partial p^2}\ln f(X;p) = \frac{-X}{p^2} - \frac{1-X}{(1-p)^2} \quad (7.13)$$

Taking the negative of the expected value in Equation (7.13) gives the information in an observation:

$$I(p) = -E\left[\frac{\partial^2}{\partial p^2}\ln f(X;p)\right] = \frac{p}{p^2} + \frac{1-p}{(1-p)^2} = \frac{1}{p} + \frac{1}{(1-p)} = \frac{1}{p(1-p)} \quad (7.14)$$

Both methods yield the answer $I(p) = 1/[p(1-p)]$, which says that the information is the reciprocal of $V(X)$. It is reasonable that the information is greatest when the variance is smallest. ■

Information in a Random Sample

Now assume a random sample $X_1, X_2, \ldots, X_n$ from a distribution with pmf or pdf $f(x;\theta)$. Let $f(X_1, X_2, \ldots, X_n;\theta) = f(X_1;\theta)\cdot f(X_2;\theta)\cdot\cdots\cdot f(X_n;\theta)$ be the likelihood function. The Fisher information $I_n(\theta)$ for the random sample is the variance of the **score function**

$$\frac{\partial}{\partial\theta}\ln f(X_1, X_2, \ldots, X_n;\theta) = \frac{\partial}{\partial\theta}\ln[f(X_1;\theta)\cdot f(X_2;\theta)\cdot\cdots\cdot f(X_n;\theta)]$$

The log of a product is the sum of the logs, so the score function is a sum:

$$\frac{\partial}{\partial\theta}\ln f(X_1, X_2, \ldots, X_n;\theta) = \frac{\partial}{\partial\theta}\ln f(X_1;\theta) + \frac{\partial}{\partial\theta}\ln f(X_2;\theta) + \cdots + \frac{\partial}{\partial\theta}\ln f(X_n;\theta) \quad (7.15)$$

This is a sum of terms for which the mean is zero, by Equation (7.8), and therefore

$$E\left[\frac{\partial}{\partial\theta}\ln f(X_1, X_2, \ldots, X_n; \theta)\right] = 0 \tag{7.16}$$

The right-hand side of Equation (7.15) is a sum of independent identically distributed random variables, and each has variance $I(\theta)$. Taking the variance of both sides of Equation (7.15) gives the information $I_n(\theta)$ in the random sample

$$I_n(\theta) = V\left[\frac{\partial}{\partial\theta}\ln f(X_1, X_2, \ldots, X_n; \theta)\right] = nV\left[\frac{\partial}{\partial\theta}\ln f(X_1; \theta)\right] = nI(\theta) \tag{7.17}$$

Therefore, the Fisher information in a random sample is just n times the information in a single observation. This should make sense intuitively, because it says that twice as many observations yield twice as much information.

Example 7.32 Continuing with Example 7.31, let $X_1, X_2, \ldots, X_n$ be a random sample from the Bernoulli distribution with $f(x; p) = p^x(1-p)^{1-x}$, $x = 0, 1$. Suppose the purpose is to estimate the proportion p of drivers who are wearing seat belts. We saw that the information in a single observation is $I(p) = 1/[p(1-p)]$, and therefore the Fisher information in the random sample is $I_n(p) = nI(p) = n/[p(1-p)]$. ■

The Cramér–Rao Inequality

We will use the concept of Fisher information to show that, if $t(X_1, X_2, \ldots, X_n)$ is an unbiased estimator of θ, then its minimum possible variance is the reciprocal of $I_n(\theta)$. Harald Cramér in Sweden and C. R. Rao in India independently derived this inequality during World War II, but R. A. Fisher had some notion of it 20 years previously.

THEOREM (CRAMÉR–RAO INEQUALITY) Assume a random sample $X_1, X_2, \ldots, X_n$ from a distribution with pmf or pdf $f(x; \theta)$ such that the set of possible values does not depend on θ. If $T = t(X_1, X_2, \ldots, X_n)$ is an unbiased estimator for the parameter θ, then

$$V(T) \geq \frac{1}{V\left\{\frac{\partial}{\partial\theta}[\ln f(X_1, \ldots, X_n; \theta)]\right\}} = \frac{1}{nI(\theta)} = \frac{1}{I_n(\theta)}$$

Proof The basic idea here is to consider the correlation ρ between T and the score function, and the desired inequality will result from $-1 \leq \rho \leq 1$. If $T = t(X_1, X_2, \ldots, X_n)$ is an unbiased estimator of θ, then

$$\theta = E(T) = \sum_{x_1, \ldots, x_n} t(x_1, \ldots, x_n) f(x_1, \ldots, x_n; \theta)$$

Differentiating this with respect to θ,

$$1 = \frac{\partial}{\partial\theta}\sum_{x_1, \ldots, x_n} t(x_1, \ldots, x_n)\, f(x_1, \ldots, x_n; \theta) = \sum_{x_1, \ldots, x_n} t(x_1, \ldots, x_n)\frac{\partial}{\partial\theta} f(x_1, \ldots, x_n; \theta)$$

Multiplying and dividing the last term by the likelihood $f(x_1, \ldots, x_n; \theta)$ gives

$$1 = \sum_{x_1, \ldots, x_n} t(x_1, \ldots, x_n) \frac{\frac{\partial}{\partial\theta} f(x_1, \ldots, x_n; \theta)}{f(x_1, \ldots, x_n; \theta)} f(x_1, \ldots, x_n; \theta)$$

which is equivalent to

$$1 = \sum_{x_1, \ldots, x_n} t(x_1, \ldots, x_n) \frac{\partial}{\partial\theta}[\ln f(x_1, \ldots, x_n; \theta)] f(x_1, \ldots, x_n; \theta)$$

$$= E\left\{ t(X_1, \ldots, X_n) \frac{\partial}{\partial\theta}[\ln f(X_1, \ldots, X_n; \theta)] \right\}$$

Therefore, because of Equation (7.16), the covariance of T with the score function is 1:

$$1 = \text{Cov}\left\{ T, \frac{\partial}{\partial\theta}[\ln f(X_1, \ldots, X_n; \theta)] \right\} \tag{7.18}$$

Recall from Section 5.2 that the correlation between two rv's is $\rho_{X,Y} = \text{Cov}(X, Y)/\sigma_X\sigma_Y$, and that $-1 \le \rho_{X,Y} \le 1$. Therefore,

$$\text{Cov}(X, Y)^2 = \rho_{X,Y}^2 \sigma_X^2 \sigma_Y^2 \le \sigma_X^2 \sigma_Y^2$$

Apply this to Equation (7.18):

$$1 = \left(\text{Cov}\left\{ T, \frac{\partial}{\partial\theta}[\ln f(X_1, \ldots, X_n; \theta)] \right\} \right)^2 \tag{7.19}$$

$$\le V(T) \cdot V\left\{ \frac{\partial}{\partial\theta}[\ln f(X_1, \ldots, X_n; \theta)] \right\}$$

Dividing both sides by the variance of the score function and using the fact that this variance equals $nI(\theta)$, we obtain the desired result. ■

Because the variance of T must be at least $1/nI(\theta)$, it is natural to call T an efficient estimator of θ if $V(T) = 1/nI(\theta)$.

DEFINITION Let T be an unbiased estimator of θ. The ratio of the lower bound to the variance of T is its **efficiency**. Then T is said to be an **efficient** estimator if T achieves the Cramér–Rao lower bound (the efficiency is 1). An efficient estimator is a minimum variance unbiased estimator (MVUE), as discussed in Section 7.1.

Example 7.33 Continuing with Example 7.32, let $X_1, X_2, \ldots, X_n$ be a random sample from the Bernoulli distribution, where the purpose is to estimate the proportion p of drivers who are wearing seat belts. We saw that the information in the sample is $I_n(p) = n/[p(1 - p)]$, and therefore the Cramér–Rao lower bound is $1/I_n(p) = p(1 - p)/n$. Let $T(X_1, X_2, \ldots, X_n) = \hat{p} = \bar{x} = \sum x_i/n$. Then $E(T) = E(\sum X_i)/n = np/n = p$, so T is unbiased, and $V(T) = V(\sum X_i)/n^2 = np(1 - p)/n^2 = p(1 - p)/n$. Because T is unbiased and $V(T)$ is equal to the lower bound, T has efficiency 1 and therefore is an efficient estimator. ■

Large-Sample Properties of the MLE

As discussed in Section 7.2, the maximum likelihood estimator $\hat{\theta}$ has some nice properties. First of all it is *consistent*, which means that it converges in probability to the parameter θ as the sample size increases. A verification of this is beyond the level of this book, but we can use it as a basis for showing that the mle is asymptotically normal with mean θ (asymptotic unbiasedness) and variance equal to the Cramér–Rao lower bound.

THEOREM Given a random sample $X_1, X_2, \ldots, X_n$ from a distribution with pmf or pdf $f(x; \theta)$, assume that the set of possible values does not depend on θ. Then for large n the maximum likelihood estimator $\hat{\theta}$ has approximately a normal distribution with mean θ and variance $1/[nI(\theta)]$. More precisely, the limiting distribution of $\sqrt{n}(\hat{\theta} - \theta)$ is normal with mean 0 and variance $1/I(\theta)$.

Proof Consider the score function

$$S(\theta) = \frac{\partial}{\partial\theta}\ln[f(X_1, X_2, \ldots, X_n; \theta)]$$

Its derivative $S'(\theta)$ at the true θ is approximately equal to the difference quotient

$$\frac{S(\hat{\theta}) - S(\theta)}{\hat{\theta} - \theta} \tag{7.20}$$

and the error approaches zero asymptotically because $\hat{\theta}$ approaches θ (consistency). Equation 7.20 connects the mle $\hat{\theta}$ to the score function, so the asymptotic behavior of the score function can be applied to $\hat{\theta}$. Because $\hat{\theta}$ is the maximum likelihood estimate, $S(\hat{\theta}) = 0$, so in the limit,

$$\hat{\theta} - \theta = \frac{S(\theta)}{-S'(\theta)}$$

Multiplying both sides by $\sqrt{n}$, then dividing numerator and denominator by $n\sqrt{I(\theta)}$,

$$\sqrt{n}(\hat{\theta} - \theta) = \frac{\{\sqrt{n}/[n\sqrt{I(\theta)}]\}S(\theta)}{-\{1/[n\sqrt{I(\theta)}]\}S'(\theta)} = \frac{S(\theta)/\sqrt{nI(\theta)}}{-(1/n)S'(\theta)/\sqrt{I(\theta)}}$$

Now rewrite $S(\theta)$ and $S'(\theta)$ as sums using Equation (7.15):

$$\sqrt{n}(\hat{\theta} - \theta) = \frac{\dfrac{\dfrac{1}{n}\left\{\dfrac{\partial}{\partial\theta}\ln[f(X_1; \theta)] + \cdots + \dfrac{\partial}{\partial\theta}\ln[f(X_n; \theta)]\right\}}{\sqrt{I(\theta)/n}}}{\dfrac{\dfrac{1}{n}\left\{-\dfrac{\partial^2}{\partial\theta^2}\ln[f(X_1; \theta)] - \cdots - \dfrac{\partial^2}{\partial\theta^2}\ln[f(X_n; \theta)]\right\}}{\sqrt{I(\theta)}}} \tag{7.21}$$

The denominator braces contain a sum of independent identically distributed rv's each with mean

$$I(\theta) = -E\left\{\frac{\partial^2}{\partial\theta^2}[\ln f(X;\theta)]\right\}$$

by Equation (7.9). Therefore, by the Law of Large Numbers, the average in the top part of the main denominator converges to $I(\theta)$. Thus the denominator converges to $\sqrt{I(\theta)}$. The top part of the main numerator contains an average of independent identically distributed rv's with mean 0 [by Equation (7.8)] and variance $I(\theta)$, so the numerator ratio is an average minus its expected value, divided by its standard deviation. Therefore, by the Central Limit Theorem it is approximately normal with mean 0 and variance 1. Thus, the ratio in Equation (7.21) has a numerator that is approximately $N(0, 1)$ and a denominator that is approximately $\sqrt{I(\theta)}$, so the ratio is approximately $N(0, 1/\sqrt{I(\theta)}^2) = N(0, 1/I(\theta))$. That is, $\sqrt{n}(\hat{\theta} - \theta)$ is approximately $N(0, 1/I(\theta))$, and it follows that $\hat{\theta}$ is approximately normal with mean θ and variance $1/[nI(\theta)]$, the Cramér–Rao lower bound. ■

Example 7.34 Continue the previous example, with $X_1, X_2, \ldots, X_n$ a random sample from the Bernoulli distribution, and the object is to estimate the proportion p of drivers who are wearing seat belts. The pmf is $f(x; p) = p^x(1-p)^{1-x}$, $x = 0, 1$ so the likelihood is

$$f(x_1, x_2, \ldots, x_n; p) = p^{x_1+x_2+\cdots+x_n}(1-p)^{n-(x_1+x_2+\cdots+x_n)}$$

Then the log likelihood is

$$\ln[f(x_1, x_2, \ldots, x_n; p)] = \sum x_i \ln(p) + (n - \sum x_i)\ln(1-p)$$

and therefore its derivative, the score function, is

$$\frac{\partial}{\partial p}\ln[f(x_1, x_2, \ldots, x_n; p)] = \frac{\sum x_i}{p} - \frac{n - \sum x_i}{1-p} = \frac{\sum x_i - np}{p(1-p)}$$

Conclude that the maximum likelihood estimate is $\hat{p} = \bar{x} = \sum x_i/n$. Recall from Example 7.33 that this is unbiased and efficient with the minimum variance of the Cramér–Rao inequality. It is also asymptotically normal by the Central Limit Theorem. These properties are in accord with the asymptotic distribution given by the theorem, $\hat{p} \sim N(p, 1/[nI(p)])$. ■

Example 7.35 Let $X_1, X_2, \ldots, X_n$ be a random sample from the distribution with pdf $f(x;\theta) = \theta x^{\theta-1}$ for $0 < x < 1$, assuming $\theta > 0$. Here X_i, $i = 1, 2, \ldots, n$, represents the fraction of a perfect score assigned to the ith applicant by a recruiting team. The Fisher information is the variance of

$$U = \frac{\partial}{\partial\theta}\ln[f(X;\theta)] = \frac{\partial}{\partial\theta}[\ln\theta + (\theta-1)\ln(X)] = \frac{1}{\theta} + \ln(X)$$

However, it is easier to use the alternative method of Equation (7.9):

$$I(\theta) = -E\left\{\frac{\partial^2}{\partial\theta^2}\ln[f(X;\theta)]\right\} = -E\left\{\frac{\partial}{\partial\theta}\left[\frac{1}{\theta} + \ln(X)\right]\right\} = -E\left\{\frac{-1}{\theta^2}\right\} = \frac{1}{\theta^2}$$

To obtain the maximum likelihood estimator, we first find the log likelihood:

$$\ln[f(x_1, x_2, \ldots, x_n; \theta)] = \ln(\theta^n \Pi x_i^{\theta-1}) = n\ln(\theta) + (\theta - 1)\sum \ln(x_i)$$

Its derivative, the score function, is

$$\frac{\partial}{\partial\theta}\ln[f(x_1, x_2, \ldots, x_n; \theta)] = \frac{n}{\theta} + \sum \ln(x_i)$$

Setting this to 0, we find that the maximum likelihood estimate is

$$\hat{\theta} = \frac{-1}{\sum \ln(x_i)/n} \tag{7.22}$$

The expected value of $\ln(X)$ is $-1/\theta$, because $E(U) = 0$, so the denominator of (7.22) converges in probability to $-1/\theta$ by the Law of Large Numbers. Therefore $\hat{\theta}$ converges in probability to θ, which means that $\hat{\theta}$ is consistent. By the theorem, the asymptotic distribution of $\hat{\theta}$ is normal with mean θ and variance $1/[nI(\theta)] = \theta^2/n$. ■

Exercises Section 7.4 (42–48)

42. Assume that the number of defects in a car has a Poisson distribution with parameter λ. To estimate λ we obtain the random sample $X_1, X_2, \ldots, X_n$.

a. Find the Fisher information in a single observation using two methods.

b. Find the Cramér–Rao lower bound for the variance of an unbiased estimator of λ.

c. Use the score function to find the mle of λ and show that the mle is an efficient estimator.

d. Is the asymptotic distribution of the mle in accord with the second theorem? Explain.

43. In Example 7.23 $f(x; \theta) = 1/\theta$ for $0 \le x \le \theta$ and 0 otherwise. Given a random sample, the maximum likelihood estimate $\hat{\theta}$ is the largest observation.

a. Letting $\tilde{\theta} = [(n + 1)/n]\hat{\theta}$, show that $\tilde{\theta}$ is unbiased and find its variance.

b. Find the Cramér–Rao lower bound for the variance of an unbiased estimator of θ.

c. Compare the answers in parts (a) and (b) and explain why it is apparent that they disagree. What assumption is violated, causing the theorem not to apply here?

44. Component lifetimes have the exponential distribution with pdf $f(x; \lambda) = \lambda e^{-\lambda x}$, $x \ge 0$, and $f(x; \lambda) = 0$ otherwise, where $\lambda > 0$. However, we wish to estimate the mean $\mu = 1/\lambda$ based on the random sample $X_1, X_2, \ldots, X_n$ so let's reexpress the pdf in the form $(1/\mu)e^{-x/\mu}$.

a. Find the information in a single observation and the Cramér–Rao lower bound.

b. Use the score function to find the mle of μ.

c. Find the mean and variance of the mle.

d. Is the mle an efficient estimator? Explain.

45. Let $X_1, X_2, \ldots, X_n$ be a random sample from the normal distribution with known standard deviation σ.

a. Find the mle of μ.

b. Find the distribution of the mle.

c. Is the mle an efficient estimator? Explain.

d. How does the answer to part (b) compare with the asymptotic distribution given by the second theorem?

46. Let $X_1, X_2, \ldots, X_n$ be a random sample from the normal distribution with known mean μ but with the variance σ^2 as the unknown parameter.

a. Find the information in a single observation and the Cramér–Rao lower bound.

b. Find the mle of σ^2.

c. Find the distribution of the mle.

d. Is the mle an efficient estimator? Explain.

e. Is the answer to part (c) in conflict with the asymptotic distribution of the mle given by the second theorem? Explain.

47. Let $X_1, X_2, \ldots, X_n$ be a random sample from the normal distribution with known mean μ but with the standard deviation σ as the unknown parameter.

a. Find the information in a single observation.

b. Compare the answer in part (a) to the answer in part (a) of Exercise 46. Does the information depend on the parameterization?

48. Let $X_1, X_2, \ldots, X_n$ be a random sample from a continuous distribution with pdf $f(x; \theta)$. For large n, the variance of the sample median is approximately $1/\{4n[f(\tilde{\mu}; \theta)]^2\}$. If $X_1, X_2, \ldots, X_n$ is a random sample from the normal distribution with known standard deviation σ and unknown μ, determine the efficiency of the sample median.

Supplementary Exercises (49–63)

49. At time $t = 0$, there is one individual alive in a certain population. A **pure birth process** then unfolds as follows. The time until the first birth is exponentially distributed with parameter λ. After the first birth, there are two individuals alive. The time until the first gives birth again is exponential with parameter λ, and similarly for the second individual. Therefore, the time until the next birth is the minimum of two exponential (λ) variables, which is exponential with parameter 2λ. Similarly, once the second birth has occurred, there are three individuals alive, so the time until the next birth is an exponential rv with parameter 3λ, and so on (the memoryless property of the exponential distribution is being used here). Suppose the process is observed until the sixth birth has occurred and the successive birth times are 25.2, 41.7, 51.2, 55.5, 59.5, 61.8 (from which you should calculate the times between successive births). Derive the mle of λ. (*Hint*: The likelihood is a product of exponential terms.)

50. a. Let $X_1, \ldots, X_n$ be a random sample from a uniform distribution on $[0, \theta]$. Then the mle of θ is $\hat{\theta} = Y = \max(X_i)$. Use the fact that $Y \le y$ iff each $X_i \le y$ to derive the cdf of Y. Then show that the pdf of $Y = \max(X_i)$ is

$$f_Y(y) = \begin{cases} \dfrac{ny^{n-1}}{\theta^n} & 0 \le y \le \theta \\ 0 & \text{otherwise} \end{cases}$$

b. Use the result of part (a) to show that the mle is biased but that $(n + 1)\max(X_i)/n$ is unbiased.

51. The **mean square error** of an estimator $\hat{\theta}$ is $\text{MSE}(\hat{\theta}) = E(\hat{\theta} - \theta)^2$. If $\hat{\theta}$ is unbiased, then $\text{MSE}(\hat{\theta}) = V(\hat{\theta})$, but in general $\text{MSE}(\hat{\theta}) = V(\hat{\theta}) + (\text{bias})^2$. Consider the estimator $\hat{\sigma}^2 = KS^2$, where $S^2 =$ sample variance. What value of K minimizes the mean square error of this estimator when the population distribution is normal? [*Hint*: Assuming normality, it can be shown that

$$E[(S^2)^2] = (n + 1)\sigma^4/(n - 1)$$

In general, it is difficult to find $\hat{\theta}$ to minimize $\text{MSE}(\hat{\theta})$, which is why we look only at unbiased estimators and minimize $V(\hat{\theta})$.]

52. Let $X_1, \ldots, X_n$ be a random sample from a pdf that is symmetric about μ. An estimator for μ that has been found to perform well for a variety of underlying distributions is the *Hodges–Lehmann estimator*. To define it, first compute for each $i \le j$ and each $j = 1, 2, \ldots, n$ the pairwise average $\overline{X}_{i,j} = (X_i + X_j)/2$. Then the estimator is $\hat{\mu} =$ the median of the $\overline{X}_{i,j}$'s. Compute the value of this estimate using the data of Exercise 41 of Chapter 1. (*Hint*: Construct a square table with the x_i's listed on the left margin and on top. Then compute averages on and above the diagonal.)

53. When the population distribution is normal, the statistic median $\{|X_1 - \widetilde{X}|, \ldots, |X_n - \widetilde{X}|\}/.6745$ can be used to estimate σ. This estimator is more resistant to the effects of outliers (observations far from the bulk of the data) than is the sample standard deviation. Compute both the corresponding point estimate and s for the data of Example 7.2.

54. When the sample standard deviation S is based on a random sample from a normal population distribution, it can be shown that

$$E(S) = \sqrt{2/(n - 1)}\Gamma(n/2)\sigma/\Gamma[(n - 1)/2]$$

Use this to obtain an unbiased estimator for σ of the form cS. What is c when $n = 20$?

55. Each of n specimens is to be weighed twice on the same scale. Let X_i and Y_i denote the two observed weights for the ith specimen. Suppose X_i and Y_i are independent of one another, each normally distributed with mean value μ_i (the true weight of specimen i) and variance σ^2.

a. Show that the maximum likelihood estimator of σ^2 is $\hat{\sigma}^2 = \Sigma(X_i - Y_i)^2/(4n)$. [*Hint*: If $\bar{z} = (z_1 + z_2)/2$, then $\Sigma(z_i - \bar{z})^2 = (z_1 - z_2)^2/2$.]

b. Is the mle $\hat{\sigma}^2$ an unbiased estimator of σ^2? Find an unbiased estimator of σ^2. [*Hint*: For any rv Z, $E(Z^2) = V(Z) + [E(Z)]^2$. Apply this to $Z = X_i - Y_i$.]

56. For $0 < \theta < 1$ consider a random sample from a uniform distribution on the interval from θ to $1/\theta$. Identify a sufficient statistic for θ.

57. Let p denote the proportion of all individuals who are allergic to a particular medication. An investigator tests individual after individual to obtain a group of r individuals who have the allergy. Let $X_i = 1$ if the ith individual tested has the allergy and $X_i = 0$ otherwise ($i = 1, 2, 3, \ldots$). Recall that in this situation, X = the number of nonallergic individuals tested prior to obtaining the desired group has a negative binomial distribution. Use the definition of sufficiency to show that X is a sufficient statistic for p.

58. The fraction of a bottle that is filled with a particular liquid is a continuous random variable X with pdf $f(x; \theta) = \theta x^{\theta-1}$ for $0 < x < 1$ (where $\theta > 0$).

a. Obtain the method of moments estimator for θ.

b. Is the estimator of (a) a sufficient statistic? If not, what is a sufficient statistic, and what is an estimator of θ (not necessarily unbiased) based on a sufficient statistic?

59. Let $X_1, \ldots, X_n$ be a random sample from a normal distribution with both μ and σ unknown. An unbiased estimator of $\theta = P(X \le c)$ based on the jointly sufficient statistics is desired. Let $k = \sqrt{n/(n-1)}$ and $w = (c - \bar{x})/s$. Then it can be shown that the minimum variance unbiased estimator for θ is

$$\hat{\theta} = \begin{cases} 0 & kw \le -1 \\ P\left(T < \dfrac{kw\sqrt{n-2}}{\sqrt{1-k^2w^2}}\right) & -1 < kw < 1 \\ 1 & kw \ge 1 \end{cases}$$

where T has a t distribution with $n - 2$ df. The article "Big and Bad: How the S.U.V. Ran over Automobile Safety" (*The New Yorker*, Jan. 24, 2004) reported that when an engineer with Consumer Union (the product testing and rating organization that publishes *Consumers Reports*) performed three different trials in which a Chevrolet Blazer was accelerated to 60 mph and then suddenly braked, the stopping distances (ft) were 146.2, 151.6, and 153.4, respectively. Assuming that braking distance is normally distributed, obtain the minimum variance unbiased estimate for the probability that distance is at most 150 ft, and compare to the maximum likelihood estimate of this probability.

60. Here is a result that allows for easy identification of a *minimal* sufficient statistic: Suppose there is a function $t(x_1, \ldots, x_n)$ such that for any two sets of observations $x_1, \ldots, x_n$ and $y_1, \ldots, y_n$, the likelihood ratio $f(x_1, \ldots, x_n; \theta)/f(y_1, \ldots, y_n; \theta)$ does not depend on θ if and only if $t(x_1, \ldots, x_n) = t(y_1, \ldots, y_n)$. Then $T = t(X_1, \ldots, X_n)$ is a minimal sufficient statistic. The result is also valid if θ is replaced by $\theta_1, \ldots, \theta_k$, in which case there will typically be several jointly minimal sufficient statistics. For example, if the underlying pdf is exponential with parameter λ, then the likelihood ratio is $\lambda^{\Sigma x_i - \Sigma y_i}$, which will not depend on λ if and only if $\Sigma x_i = \Sigma y_i$, so $T = \Sigma x_i$, is a minimal sufficient statistic for λ (and so is the sample mean).

a. Identify a minimal sufficient statistic when the X_i's are a random sample from a Poisson distribution.

b. Identify a minimal sufficient statistic or jointly minimal sufficient statistics when the X_i's are a random sample from a normal distribution with mean θ and variance θ.

c. Identify a minimal sufficient statistic or jointly minimal sufficient statistics when the X_i's are a random sample from a normal distribution with mean θ and standard deviation θ.

61. The principle of unbiasedness (prefer an unbiased estimator to any other) has been criticized on the grounds that in some situations the only unbiased estimator is patently ridiculous. Here is one such example. Suppose that the number of major defects X on a randomly selected vehicle has a Poisson distribution with parameter λ. You are going to purchase two such vehicles and wish to estimate $\theta = P(X_1 = 0, X_2 = 0) = e^{-2\lambda}$, the probability that neither of these vehicles has any major defects. Your estimate is based on observing the value of X for a single vehicle. Denote this estimator by $\hat{\theta} = \delta(X)$. Write the equation implied by the condition of unbiasedness, $E[\delta(X)] = e^{-2\lambda}$, cancel $e^{-\lambda}$ from both sides, then expand what remains on the right-hand side in an infinite series, and compare the two sides to determine $\delta(X)$. If $X = 200$, what is the estimate? Does this seem reasonable? What is the estimate if $X = 199$? Is this reasonable?

is quite narrow, our knowledge of the value of the parameter is reasonably precise. A very wide confidence interval, however, gives the message that there is a great deal of uncertainty concerning the value of what we are estimating. Figure 8.1 shows 95% confidence intervals for true average breaking strengths of two different brands of paper towels. One of these intervals suggests precise knowledge about μ, whereas the other suggests a very wide range of plausible values.

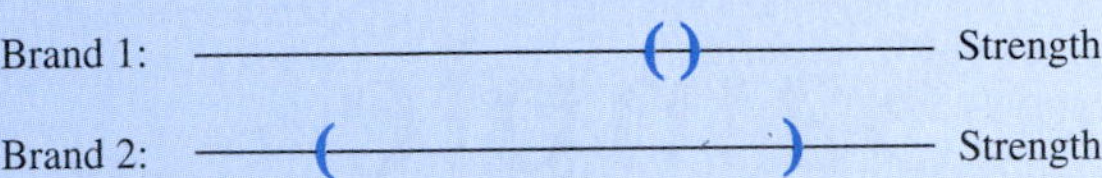

Figure 8.1 Confidence intervals indicating precise (brand 1) and imprecise (brand 2) information about μ

8.1 Basic Properties of Confidence Intervals

The basic concepts and properties of confidence intervals (CIs) are most easily introduced by first focusing on a simple, albeit somewhat unrealistic, problem situation. Suppose that the parameter of interest is a population mean μ and that

1. The population distribution is normal.
2. The value of the population standard deviation σ is known.

Normality of the population distribution is often a reasonable assumption. However, if the value of μ is unknown, it is implausible that the value of σ would be available (knowledge of a population's center typically precedes information concerning spread). In later sections, we will develop methods based on less restrictive assumptions.

Example 8.1 Industrial engineers who specialize in ergonomics are concerned with designing workspace and devices operated by workers so as to achieve high productivity and comfort. The article "Studies on Ergonomically Designed Alphanumeric Keyboards" (*Hum. Factors*, 1985: 175–187) reports on a study of preferred height for an experimental keyboard with large forearm–wrist support. A sample of $n = 31$ trained typists was selected, and the preferred keyboard height was determined for each typist. The resulting sample average preferred height was $\bar{x} = 80.0$ cm. Assuming that the preferred height is normally distributed with $\sigma = 2.0$ cm (a value suggested by data in the article), obtain a CI for μ, the true average preferred height for the population of all experienced typists. ■

The actual sample observations $x_1, x_2, \ldots, x_n$ are assumed to be the result of a random sample $X_1, \ldots, X_n$ from a normal distribution with mean value μ and standard deviation σ. The results of Chapter 6 then imply that irrespective of the sample size n, the sample mean $\overline{X}$ is normally distributed with expected value μ and standard deviation

$\sigma/\sqrt{n}$. Standardizing $\overline{X}$ by first subtracting its expected value and then dividing by its standard deviation yields the variable

$$Z = \frac{\overline{X} - \mu}{\sigma/\sqrt{n}} \tag{8.1}$$

which has a standard normal distribution. Because the area under the standard normal curve between -1.96 and 1.96 is .95,

$$P\left(-1.96 < \frac{\overline{X} - \mu}{\sigma/\sqrt{n}} < 1.96\right) = .95 \tag{8.2}$$

The next step in the development is to manipulate the inequalities inside the parentheses in (8.2) so that they appear in the equivalent form $l < \mu < u$, where the endpoints l and u involve $\overline{X}$ and $\sigma/\sqrt{n}$. This is achieved through the following sequence of operations, each one yielding inequalities equivalent to those we started with:

1. Multiply through by $\sigma/\sqrt{n}$ to obtain

$$-1.96 \cdot \frac{\sigma}{\sqrt{n}} < \overline{X} - \mu < 1.96 \cdot \frac{\sigma}{\sqrt{n}}$$

2. Subtract $\overline{X}$ from each term to obtain

$$-\overline{X} - 1.96 \cdot \frac{\sigma}{\sqrt{n}} < -\mu < -\overline{X} + 1.96 \cdot \frac{\sigma}{\sqrt{n}}$$

3. Multiply through by -1 to eliminate the minus sign in front of μ (which reverses the direction of each inequality) to obtain

$$\overline{X} + 1.96 \cdot \frac{\sigma}{\sqrt{n}} > \mu > \overline{X} - 1.96 \cdot \frac{\sigma}{\sqrt{n}}$$

that is,

$$\overline{X} - 1.96 \cdot \frac{\sigma}{\sqrt{n}} < \mu < \overline{X} + 1.96 \cdot \frac{\sigma}{\sqrt{n}}$$

Because each set of inequalities in the sequence is equivalent to the original one, the probability associated with each is .95. In particular,

$$P\left(\overline{X} - 1.96\frac{\sigma}{\sqrt{n}} < \mu < \overline{X} + 1.96\frac{\sigma}{\sqrt{n}}\right) = .95 \tag{8.3}$$

The event inside the parentheses in (8.3) has a somewhat unfamiliar appearance; always before, the random quantity has appeared in the middle with constants on both ends, as in $a \le Y \le b$. In (8.3) the random quantity appears on the two ends, whereas the unknown constant μ appears in the middle. To interpret (8.3), think of a **random interval** having left endpoint $\overline{X} - 1.96 \cdot \sigma/\sqrt{n}$ and right endpoint $\overline{X} + 1.96 \cdot \sigma/\sqrt{n}$, which in interval notation is

$$\left(\overline{X} - 1.96 \cdot \frac{\sigma}{\sqrt{n}}, \overline{X} + 1.96 \cdot \frac{\sigma}{\sqrt{n}}\right) \tag{8.4}$$

The interval (8.4) is random because the two endpoints of the interval involve a random variable (rv). Note that the interval is centered at the sample mean $\overline{X}$ and extends $1.96\sigma/\sqrt{n}$ to each side of $\overline{X}$. Thus the interval's width is $2 \cdot (1.96) \cdot \sigma/\sqrt{n}$, which is not random; only the location of the interval (its midpoint $\overline{X}$) is random (Figure 8.2). Now (8.3) can be paraphrased as "*the probability is* .95 *that the random interval* (8.4) *includes or covers the true value of* μ." Before any experiment is performed and any data is gathered, it is quite likely (probability .95) that μ will lie inside the interval in Expression (8.4).

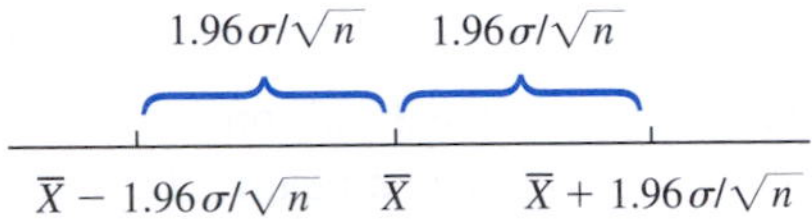

Figure 8.2 The random interval (8.4) centered at $\overline{X}$

DEFINITION

If after observing $X_1 = x_1, X_2 = x_2, \ldots, X_n = x_n$, we compute the observed sample mean $\bar{x}$ and then substitute $\bar{x}$ into (8.4) in place of $\overline{X}$, the resulting fixed interval is called a **95% confidence interval for μ**. This CI can be expressed either as

$$\left(\bar{x} - 1.96 \cdot \frac{\sigma}{\sqrt{n}}, \bar{x} + 1.96 \cdot \frac{\sigma}{\sqrt{n}}\right) \quad \text{is a 95\% CI for } \mu$$

or as

$$\bar{x} - 1.96 \cdot \frac{\sigma}{\sqrt{n}} < \mu < \bar{x} + 1.96 \cdot \frac{\sigma}{\sqrt{n}} \quad \text{with 95\% confidence}$$

A concise expression for the interval is $\bar{x} \pm 1.96 \cdot \sigma/\sqrt{n}$, where $-$ gives the left endpoint (lower limit) and $+$ gives the right endpoint (upper limit).

Example 8.2 (Example 8.1 continued)

The quantities needed for computation of the 95% CI for true average preferred height are $\sigma = 2.0$, $n = 31$, and $\bar{x} = 80.0$. The resulting interval is

$$\bar{x} \pm 1.96 \cdot \frac{\sigma}{\sqrt{n}} = 80.0 \pm (1.96)\frac{2.0}{\sqrt{31}} = 80.0 \pm .7 = (79.3, 80.7)$$

That is, we can be highly confident, at the 95% confidence level, that $79.3 < \mu < 80.7$. This interval is relatively narrow, indicating that μ has been rather precisely estimated. ■

Interpreting a Confidence Level

The confidence level 95% for the interval just defined was inherited from the probability .95 for the random interval (8.4). Intervals having other levels of confidence will be introduced shortly. For now, though, consider how 95% confidence can be interpreted.

Because we started with an event whose probability was .95 — that the random interval (8.4) would capture the true value of μ — and then used the data in Example 8.1 to compute the fixed interval (79.3, 80.7), it is tempting to conclude that μ is within this fixed interval with probability .95. But by substituting $\bar{x} = 80.0$ for $\bar{X}$, all randomness disappears; the interval (79.3, 80.7) is not a random interval, and μ is a constant (unfortunately unknown to us), so it is *incorrect* to write the statement $P[\mu \text{ lies in } (79.3, 80.7)] = .95$.

A correct interpretation of "95% confidence" relies on the long-run relative frequency interpretation of probability: To say that an event A has probability .95 is to say that if the experiment on which A is defined is performed over and over again, in the long run A will occur 95% of the time. Suppose we obtain another sample of typists' preferred heights and compute another 95% interval. Then we consider repeating this for a third sample, a fourth sample, and so on. Let A be the event that $\bar{X} - 1.96 \cdot \sigma/\sqrt{n} < \mu < \bar{X} + 1.96 \cdot \sigma/\sqrt{n}$. Since $P(A) = .95$, in the long run 95% of our computed CIs will contain μ. This is illustrated in Figure 8.3, where the vertical line cuts the measurement axis at the true (but unknown) value of μ. Notice that of the 11 intervals pictured, only intervals 3 and 11 fail to contain μ. In the long run, only 5% of the intervals so constructed would fail to contain μ.

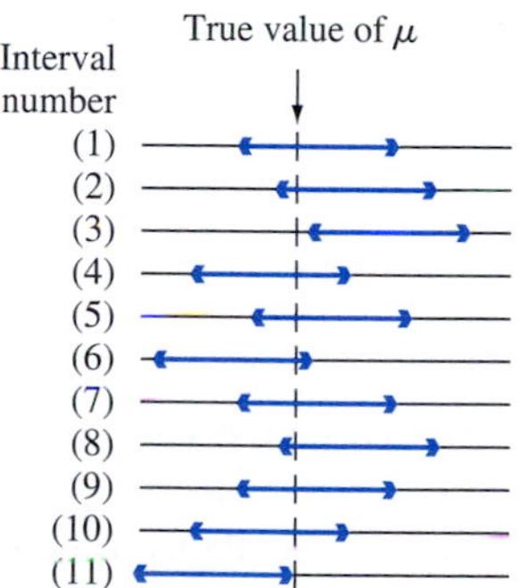

Figure 8.3 Repeated construction of 95% CIs

According to this interpretation, the confidence level 95% is not so much a statement about any particular interval such as (79.3, 80.7), but pertains to what would happen if a very large number of like intervals were to be constructed. Although this may seem unsatisfactory, the root of the difficulty lies with our interpretation of probability — it applies to a long sequence of replications of an experiment rather than just a single replication. There is another approach to the construction and interpretation of CIs that uses the notion of subjective probability and Bayes' theorem, as discussed in Section 14.4. The interval presented here (as well as each interval presented subsequently) is called a "classical" CI because its interpretation rests on the classical notion of probability (though the main ideas were developed as recently as the 1930s).

Other Levels of Confidence

The confidence level of 95% was inherited from the probability .95 for the initial inequalities in (8.2). If a confidence level of 99% is desired, the initial probability of .95 must be replaced by .99, which necessitates changing the z critical value from 1.96

to 2.58. A 99% CI then results from using 2.58 in place of 1.96 in the formula for the 95% CI.

This suggests that any desired level of confidence can be achieved by replacing 1.96 or 2.58 with the appropriate standard normal critical value. As Figure 8.4 shows, a probability of $1 - \alpha$ is achieved by using $z_{\alpha/2}$ in place of 1.96.

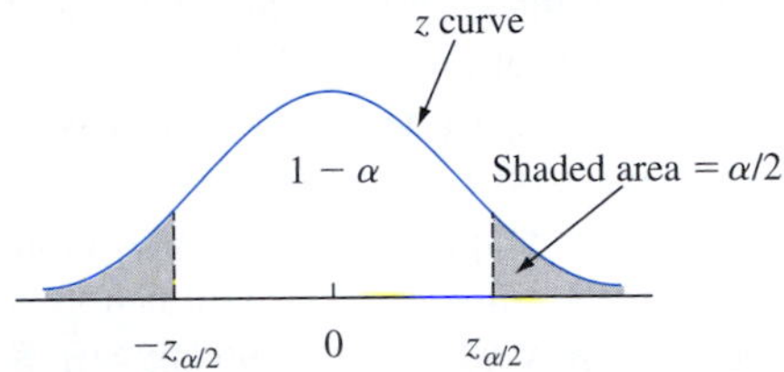

Figure 8.4 $P(-z_{\alpha/2} \le Z \le z_{\alpha/2}) = 1 - \alpha$

DEFINITION A **100(1 − α)% confidence interval** for the mean μ of a normal population when the value of σ is known is given by

$$\left(\bar{x} - z_{\alpha/2} \cdot \frac{\sigma}{\sqrt{n}},\ \bar{x} + z_{\alpha/2} \cdot \frac{\sigma}{\sqrt{n}}\right) \tag{8.5}$$

or, equivalently, by $\bar{x} \pm z_{\alpha/2} \cdot \sigma/\sqrt{n}$.

Example 8.3 A finite mathematics course has recently been changed, and the homework is now done online via computer instead of from the textbook exercises. Past experience suggests that the distribution of final exam scores is normally distributed with standard deviation 13. It is believed that the distribution is still normal with standard deviation 13, but the mean has likely changed. A sample of 40 students has a mean final exam score of 70.7. Let's calculate a confidence interval for the population mean using a confidence level of 90%. This requires that $100(1 - \alpha) = 90$, from which $\alpha = .10$ and $z_{\alpha/2} = z_{.05} = 1.645$ (corresponding to a cumulative z-curve area of .9500). The desired interval is then

$$70.7 \pm (1.645)\frac{13}{\sqrt{40}} = 70.7 \pm 3.4 = (67.3, 74.1)$$

With a reasonably high degree of confidence, we can say that $67.3 < \mu < 74.1$. ■

Confidence Level, Precision, and Choice of Sample Size

Why settle for a confidence level of 95% when a level of 99% is achievable? Because the price paid for the higher confidence level is a wider interval. The 95% interval extends $1.96 \cdot \sigma/\sqrt{n}$ to each side of $\bar{x}$, so the width of the interval is $2(1.96) \cdot \sigma/\sqrt{n} = 3.92 \cdot \sigma/\sqrt{n}$. Similarly, the width of the 99% interval is $2(2.58) \cdot \sigma/\sqrt{n} = 5.16 \cdot \sigma/\sqrt{n}$. That is, we have more confidence in the 99% interval precisely because it is wider. The higher the desired degree of confidence, the wider the resulting interval. In fact, the only

100% CI for μ is $(-\infty, \infty)$, which is not terribly informative because, even before sampling, we knew that this interval covers μ.

If we think of the width of the interval as specifying its precision or accuracy, then the confidence level (or reliability) of the interval is inversely related to its precision. A highly reliable interval estimate may be imprecise in that the endpoints of the interval may be far apart, whereas a precise interval may entail relatively low reliability. Thus it cannot be said unequivocally that a 99% interval is to be preferred to a 95% interval; the gain in reliability entails a loss in precision.

An appealing strategy is to specify both the desired confidence level and interval width and then determine the necessary sample size.

Example 8.4 Extensive monitoring of a computer time-sharing system has suggested that response time to a particular editing command is normally distributed with standard deviation 25 millisec. A new operating system has been installed, and we wish to estimate the true average response time μ for the new environment. Assuming that response times are still normally distributed with $\sigma = 25$, what sample size is necessary to ensure that the resulting 95% CI has a width of (at most) 10? The sample size n must satisfy

$$10 = 2 \cdot (1.96)(25/\sqrt{n})$$

Rearranging this equation gives

$$\sqrt{n} = 2 \cdot (1.96)(25)/10 = 9.80$$

so

$$n = (9.80)^2 = 96.04$$

Since n must be an integer, a sample size of 97 is required. ■

The general formula for the sample size n necessary to ensure an interval width w is obtained from $w = 2 \cdot z_{\alpha/2} \cdot \sigma/\sqrt{n}$ as

$$n = \left(2z_{\alpha/2} \cdot \frac{\sigma}{w}\right)^2 \qquad (8.6)$$

The smaller the desired width w, the larger n must be. In addition, n is an increasing function of σ (more population variability necessitates a larger sample size) and of the confidence level $100(1 - \alpha)$ (as α decreases, $z_{\alpha/2}$ increases).

The half-width $1.96\sigma/\sqrt{n}$ of the 95% CI is sometimes called the **bound on the error of estimation** associated with a 95% confidence level; that is, with 95% confidence, the point estimate $\bar{x}$ will be no farther than this from μ. Before obtaining data, an investigator may wish to determine a sample size for which a particular value of the bound is achieved. For example, with μ representing the average fuel efficiency (mpg) for all cars of a certain type, the objective of an investigation may be to estimate μ to within 1 mpg with 95% confidence. More generally, if we wish to estimate μ to within an amount B

(the specified bound on the error of estimation) with $100(1 - \alpha)\%$ confidence, the necessary sample size results from replacing $2/w$ by $1/B$ in (8.6).

Deriving a Confidence Interval

Let $X_1, X_2, \ldots, X_n$ denote the sample on which the CI for a parameter θ is to be based. Suppose a random variable satisfying the following two properties can be found:

1. The variable depends functionally on both $X_1, \ldots, X_n$ and θ.
2. The probability distribution of the variable does not depend on θ or on any other unknown parameters.

Let $h(X_1, X_2, \ldots, X_n; \theta)$ denote this random variable. For example, if the population distribution is normal with known σ and $\theta = \mu$, the variable $h(X_1, \ldots, X_n; \mu) = (\overline{X} - \mu)/(\sigma/\sqrt{n})$ satisfies both properties; it clearly depends functionally on μ, yet has the standard normal probability distribution, which does not depend on μ. In general, the form of the h function is usually suggested by examining the distribution of an appropriate estimator $\hat{\theta}$.

For any α between 0 and 1, constants a and b can be found to satisfy

$$P[a < h(X_1, \ldots, X_n; \theta) < b] = 1 - \alpha \qquad (8.7)$$

Because of the second property, a and b do not depend on θ. In the normal example, $a = -z_{\alpha/2}$ and $b = z_{\alpha/2}$. Now suppose that the inequalities in (8.7) can be manipulated to isolate θ, giving the equivalent probability statement

$$P[l(X_1, X_2, \ldots, X_n) < \theta < u(X_1, X_2, \ldots, X_n)] = 1 - \alpha$$

Then $l(x_1, x_2, \ldots, x_n)$ and $u(x_1, \ldots, x_n)$ are the lower and upper confidence limits, respectively, for a $100(1 - \alpha)\%$ CI. In the normal example, we saw that $l(X_1, \ldots, X_n) = \overline{X} - z_{\alpha/2} \cdot \sigma/\sqrt{n}$ and $u(X_1, \ldots, X_n) = \overline{X} + z_{\alpha/2} \cdot \sigma/\sqrt{n}$.

Example 8.5 A theoretical model suggests that the time to breakdown of an insulating fluid between electrodes at a particular voltage has an exponential distribution with parameter λ (see Section 4.4). A random sample of $n = 10$ breakdown times yields the following sample data (in min): $x_1 = 41.53$, $x_2 = 18.73$, $x_3 = 2.99$, $x_4 = 30.34$, $x_5 = 12.33$, $x_6 = 117.52$, $x_7 = 73.02$, $x_8 = 223.63$, $x_9 = 4.00$, $x_{10} = 26.78$. A 95% CI for λ and for the true average breakdown time are desired.

Let $h(X_1, X_2, \ldots, X_n; \lambda) = 2\lambda\Sigma X_i$. Using a moment generating function argument, it can be shown that this random variable has a chi-squared distribution with $2n$ degrees of freedom (df) ($\nu = 2n$, as discussed in Section 6.4). Appendix Table A.7 pictures a typical chi-squared density curve and tabulates critical values that capture specified tail areas. The relevant number of degrees of freedom here is $2(10) = 20$. The $\nu = 20$ row of the table shows that 34.170 captures upper-tail area .025 and 9.591 captures lower-tail area .025 (upper-tail area .975). Thus for $n = 10$,

$$P(9.591 < 2\lambda\, \Sigma X_i < 34.170) = .95$$

Division by $2\Sigma X_i$ isolates λ, yielding

$$P[9.591/(2\,\Sigma X_i) < \lambda < 34.170/(2\,\Sigma X_i)] = .95$$

The lower limit of the 95% CI for λ is $9.591/(2\Sigma x_i)$, and the upper limit is $34.170/(2\Sigma x_i)$. For the given data, $\Sigma x_i = 550.87$, giving the interval (.00871, .03101).

The expected value of an exponential rv is $\mu = 1/\lambda$. Since

$$P(2\,\Sigma X_i/34.170 < 1/\lambda < 2\,\Sigma X_i/9.591) = .95$$

the 95% CI for true average breakdown time is $(2\Sigma x_i/34.170,\ 2\Sigma x_i/9.591) = (32.24, 114.87)$. This interval is obviously quite wide, reflecting substantial variability in breakdown times and a small sample size. ■

In general, the upper and lower confidence limits result from replacing each $<$ in (8.7) by $=$ and solving for θ. In the insulating fluid example just considered, $2\lambda\Sigma x_i = 34.170$ gives $\lambda = 34.170/(2\Sigma x_i)$ as the upper confidence limit, and the lower limit is obtained from the other equation. Notice that the two interval limits are not equidistant from the point estimate, since the interval is not of the form $\hat{\theta} \pm c$.

Exercises Section 8.1 (1–11)

1. Consider a normal population distribution with the value of σ known.

a. What is the confidence level for the interval $\bar{x} \pm 2.81\sigma/\sqrt{n}$?

b. What is the confidence level for the interval $\bar{x} \pm 1.44\sigma/\sqrt{n}$?

c. What value of $z_{\alpha/2}$ in the CI formula (8.5) results in a confidence level of 99.7%?

d. Answer the question posed in part (c) for a confidence level of 75%.

2. Each of the following is a confidence interval for μ = true average (i.e., population mean) resonance frequency (Hz) for all tennis rackets of a certain type:

(114.4, 115.6) (114.1, 115.9)

a. What is the value of the sample mean resonance frequency?

b. Both intervals were calculated from the same sample data. The confidence level for one of these intervals is 90% and for the other is 99%. Which of the intervals has the 90% confidence level, and why?

3. Suppose that a random sample of 50 bottles of a particular brand of cough syrup is selected and the alcohol content of each bottle is determined. Let μ denote the average alcohol content for the population of all bottles of the brand under study. Suppose that the resulting 95% confidence interval is (7.8, 9.4).

a. Would a 90% confidence interval calculated from this same sample have been narrower or wider than the given interval? Explain your reasoning.

b. Consider the following statement: There is a 95% chance that μ is between 7.8 and 9.4. Is this statement correct? Why or why not?

c. Consider the following statement: We can be highly confident that 95% of all bottles of this type of cough syrup have an alcohol content that is between 7.8 and 9.4. Is this statement correct? Why or why not?

d. Consider the following statement: If the process of selecting a sample of size 50 and then computing the corresponding 95% interval is repeated 100 times, 95 of the resulting intervals will include μ. Is this statement correct? Why or why not?

4. A CI is desired for the true average stray-load loss μ (watts) for a certain type of induction motor when the line current is held at 10 amps for a speed of 1500 rpm. Assume that stray-load loss is normally distributed with $\sigma = 3.0$.

a. Compute a 95% CI for μ when $n = 25$ and $\bar{x} = 58.3$.

b. Compute a 95% CI for μ when $n = 100$ and $\bar{x} = 58.3$.

c. Compute a 99% CI for μ when $n = 100$ and $\bar{x} = 58.3$.

d. Compute an 82% CI for μ when $n = 100$ and $\bar{x} = 58.3$.

e. How large must n be if the width of the 99% interval for μ is to be 1.0?

5. Assume that the helium porosity (in percentage) of coal samples taken from any particular seam is normally distributed with true standard deviation .75.

a. Compute a 95% CI for the true average porosity of a certain seam if the average porosity for 20 specimens from the seam was 4.85.

b. Compute a 98% CI for true average porosity of another seam based on 16 specimens with a sample average porosity of 4.56.

c. How large a sample size is necessary if the width of the 95% interval is to be .40?

d. What sample size is necessary to estimate true average porosity to within .2 with 99% confidence?

6. On the basis of extensive tests, the yield point of a particular type of mild steel reinforcing bar is known to be normally distributed with $\sigma = 100$. The composition of the bar has been slightly modified, but the modification is not believed to have affected either the normality or the value of σ.

a. Assuming this to be the case, if a sample of 25 modified bars resulted in a sample average yield point of 8439 lb, compute a 90% CI for the true average yield point of the modified bar.

b. How would you modify the interval in part (a) to obtain a confidence level of 92%?

7. By how much must the sample size n be increased if the width of the CI (8.5) is to be halved? If the sample size is increased by a factor of 25, what effect will this have on the width of the interval? Justify your assertions.

8. Let $\alpha_1 > 0$, $\alpha_2 > 0$, with $\alpha_1 + \alpha_2 = \alpha$. Then

$$P\left(-z_{\alpha_1} < \frac{\bar{X} - \mu}{\sigma/\sqrt{n}} < z_{\alpha_2}\right) = 1 - \alpha$$

a. Use this equation to derive a more general expression for a $100(1 - \alpha)\%$ CI for μ of which the interval (8.5) is a special case.

b. Let $\alpha = .05$ and $\alpha_1 = \alpha/4$, $\alpha_2 = 3\alpha/4$. Does this result in a narrower or wider interval than the interval (8.5)?

9. **a.** Under the same conditions as those leading to the interval (8.5), $P[(\bar{X} - \mu)/(\sigma/\sqrt{n}) < 1.645] = .95$. Use this to derive a one-sided interval for μ that has infinite width and provides a lower confidence bound on μ. What is this interval for the data in Exercise 5(a)?

b. Generalize the result of part (a) to obtain a lower bound with confidence level $100(1 - \alpha)\%$.

c. What is an analogous interval to that of part (b) that provides an upper bound on μ? Compute this 99% interval for the data of Exercise 4(a).

10. A random sample of $n = 15$ heat pumps of a certain type yielded the following observations on lifetime (in years):

2.0	1.3	6.0	1.9	5.1	.4	1.0	5.3
15.7	.7	4.8	.9	12.2	5.3	.6	

a. Assume that the lifetime distribution is exponential and use an argument parallel to that of Example 8.5 to obtain a 95% CI for expected (true average) lifetime.

b. How should the interval of part (a) be altered to achieve a confidence level of 99%?

c. What is a 95% CI for the standard deviation of the lifetime distribution? (*Hint*: What is the standard deviation of an exponential random variable?)

11. Consider the next 1000 95% CIs for μ that a statistical consultant will obtain for various clients. Suppose the data sets on which the intervals are based are selected independently of one another. How many of these 1000 intervals do you expect to capture the corresponding value of μ? What is the probability that between 940 and 960 of these intervals contain the corresponding value of μ? (*Hint*: Let Y = the number among the 1000 intervals that contain μ. What kind of random variable is Y?)

8.2 Large-Sample Confidence Intervals for a Population Mean and Proportion

The CI for μ given in the previous section assumed that the population distribution is normal and that the value of σ is known. We now present a large-sample CI whose validity does not require these assumptions. After showing how the argument leading to this interval generalizes to yield other large-sample intervals, we focus on an interval for a population proportion p.

A Large-Sample Interval for μ

Let $X_1, X_2, \ldots, X_n$ be a random sample from a population having a mean μ and standard deviation σ. Provided that n is large, the Central Limit Theorem (CLT) implies that $\overline{X}$ has approximately a normal distribution whatever the nature of the population distribution. It then follows that $Z = (\overline{X} - \mu)/(\sigma/\sqrt{n})$ has approximately a standard normal distribution, so that

$$P\left(-z_{\alpha/2} < \frac{\overline{X} - \mu}{\sigma/\sqrt{n}} < z_{\alpha/2}\right) \approx 1 - \alpha$$

An argument parallel with that given in Section 8.1 yields $\bar{x} \pm z_{\alpha/2} \cdot \sigma/\sqrt{n}$ as a large-sample CI for μ with a confidence level of *approximately* $100(1 - \alpha)\%$. That is, when n is large, the CI for μ given previously remains valid whatever the population distribution, provided that the qualifier "approximately" is inserted in front of the confidence level.

One practical difficulty with this development is that computation of the interval requires the value of σ, which will almost never be known. Consider the standardized variable

$$Z = \frac{\overline{X} - \mu}{S/\sqrt{n}}$$

in which the sample standard deviation S replaces σ. Previously there was randomness only in the numerator of Z (by virtue of $\overline{X}$). Now there is randomness in both the numerator and the denominator — the values of both $\overline{X}$ and S vary from sample to sample. However, when n is large, the use of S rather than σ adds very little extra variability to Z. More specifically, in this case the new Z also has approximately a standard normal distribution. Manipulation of the inequalities in a probability statement involving this new Z yields a general large-sample interval for μ.

PROPOSITION If n is sufficiently large, the standardized variable

$$Z = \frac{\overline{X} - \mu}{S/\sqrt{n}}$$

has approximately a standard normal distribution. This implies that

$$\bar{x} \pm z_{\alpha/2} \cdot \frac{s}{\sqrt{n}} \tag{8.8}$$

is a **large-sample confidence interval for μ** with confidence level approximately $100(1 - \alpha)\%$. This formula is valid regardless of the shape of the population distribution.

Generally speaking, $n > 40$ will be sufficient to justify the use of this interval. This is somewhat more conservative than the rule of thumb for the CLT because of the additional variability introduced by using S in place of σ.

Example 8.6 An algebra placement test was used to determine placement in mathematics courses, as described in the article "Factors Affecting Achievement in the First Course in Calculus" (*J. Experiment. Ed.*, 1981: 136–140). A sample of calculus students gave the following 50 scores:

29	21	23	24	22	24	22	23	15	21	22	17	15	23	17	18	23	18
19	17	14	19	16	22	23	14	19	19	22	16	21	12	28	20	17	24
12	18	18	10	21	22	26	24	14	27	15	24	28	13				

A boxplot of the data (Figure 8.5) shows that most of the scores are between 15 and 30, there are no outliers, and the distribution looks fairly symmetric.

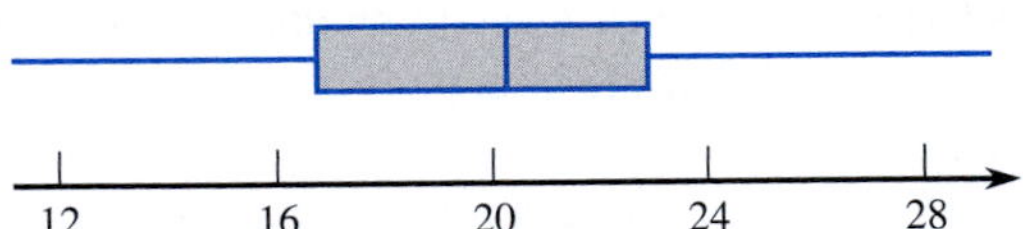

Figure 8.5 A boxplot for the algebra data of Example 8.6

Summary quantities include $n = 50$, $\Sigma x_i = 991$, and $\Sigma x_i^2 = 20{,}635$, from which $\bar{x} = 19.82$ and $s = 4.50$. The 95% confidence interval is then

$$19.82 \pm 1.96\frac{4.50}{\sqrt{50}} = 19.82 \pm 1.25 = (18.6, 21.1)$$

That is,

$$18.6 < \mu < 21.1$$

with a confidence level of approximately 95%. The interval has a reasonably narrow width of 2.5, indicating fairly precise estimation of μ. ■

Unfortunately, the choice of sample size to yield a desired interval width is not as straightforward here as it was for the case of known σ. This is because the width of (8.8) is $2z_{\alpha/2}s/\sqrt{n}$. Since the value of s is not available before data collection, the width of the interval cannot be determined solely by the choice of n. The only option for an investigator who wishes to specify a desired width is to make an educated guess as to what the value of s might be. By being conservative and guessing a larger value of s, an n larger than necessary will be chosen. The investigator may be able to specify a reasonably accurate value of the population range (the difference between the largest and smallest values). Then if the population distribution is not too skewed, dividing the range by 4 gives a ballpark value of what s might be. The idea is that roughly 95% of the data lie within $\pm 2\sigma$ of the mean, so the range is roughly 4σ.

Example 8.7 Refer to Example 8.6 on algebra scores. Suppose the investigator believes that virtually all values in the population are between 10 and 30. Then $(30 - 10)/4 = 5$ gives a reasonable value for s. The appropriate sample size for estimating the true average algebra score to within 1 with confidence level 95%—that is, for the 95% CI to have a width of 2—is

$$n = [(1.96)(5)/1]^2 \approx 96$$

■

A General Large-Sample Confidence Interval

The large-sample intervals $\bar{x} \pm z_{\alpha/2} \cdot \sigma/\sqrt{n}$ and $\bar{x} \pm z_{\alpha/2} \cdot s/\sqrt{n}$ are special cases of a general large-sample CI for a parameter θ. Suppose that $\hat{\theta}$ is an estimator satisfying the following properties: (1) It has approximately a normal distribution; (2) it is (at least approximately) unbiased; and (3) an expression for $\sigma_{\hat{\theta}}$, the standard deviation of $\hat{\theta}$, is available. For example, in the case $\theta = \mu$, $\hat{\mu} = \bar{X}$ is an unbiased estimator whose distribution is approximately normal when n is large and $\sigma_{\hat{\mu}} = \sigma_{\bar{X}} = \sigma/\sqrt{n}$. Standardizing $\hat{\theta}$ yields the rv $Z = (\hat{\theta} - \theta)/\sigma_{\hat{\theta}}$, which has approximately a standard normal distribution. This justifies the probability statement

$$P\left(-z_{\alpha/2} < \frac{\hat{\theta} - \theta}{\sigma_{\hat{\theta}}} < z_{\alpha/2}\right) \approx 1 - \alpha \qquad (8.9)$$

Suppose, first, that $\sigma_{\hat{\theta}}$ does not involve any unknown parameters (e.g., known σ in the case $\theta = \mu$). Then replacing each $<$ in (8.9) by $=$ results in $\theta = \hat{\theta} \pm z_{\alpha/2}\sigma_{\hat{\theta}}$, so the lower and upper confidence limits are $\hat{\theta} - z_{\alpha/2} \cdot \sigma_{\hat{\theta}}$ and $\hat{\theta} + z_{\alpha/2} \cdot \sigma_{\hat{\theta}}$, respectively. Now suppose that $\sigma_{\hat{\theta}}$ does not involve θ but does involve at least one other unknown parameter. Let $s_{\hat{\theta}}$ be the estimate of $\sigma_{\hat{\theta}}$ obtained by using estimates in place of the unknown parameters (e.g., $s/\sqrt{n}$ estimates $\sigma/\sqrt{n}$). Under general conditions (essentially that $s_{\hat{\theta}}$ be close to $\sigma_{\hat{\theta}}$ for most samples), a valid CI is $\hat{\theta} \pm z_{\alpha/2} \cdot s_{\hat{\theta}}$. The interval $\bar{x} \pm z_{\alpha/2} \cdot s/\sqrt{n}$ is an example.

Finally, suppose that $\sigma_{\hat{\theta}}$ does involve the unknown θ. This is the case, for example, when $\theta = p$, a population proportion. Then $(\hat{\theta} - \theta)/\sigma_{\hat{\theta}} = z_{\alpha/2}$ can be difficult to solve. An approximate solution can often be obtained by replacing θ in $\sigma_{\hat{\theta}}$ by its estimate $\hat{\theta}$. This results in an estimated standard deviation $s_{\hat{\theta}}$, and the corresponding interval is again $\hat{\theta} \pm z_{\alpha/2} \cdot s_{\hat{\theta}}$.

A Confidence Interval for a Population Proportion

Let p denote the proportion of "successes" in a population, where *success* identifies an individual or object that has a specified property. A random sample of n individuals is to be selected, and X is the number of successes in the sample. Provided that n is small compared to the population size, X can be regarded as a binomial rv with $E(X) = np$ and $\sigma_X = \sqrt{np(1-p)}$. Furthermore, if n is large ($np \geq 10$ and $nq \geq 10$), X has approximately a normal distribution.

The natural estimator of p is $\hat{p} = X/n$, the sample fraction of successes. Since $\hat{p}$ is just X multiplied by a constant $1/n$, $\hat{p}$ also has approximately a normal distribution. As shown in Section 7.1, $E(\hat{p}) = p$ (unbiasedness) and $\sigma_{\hat{p}} = \sqrt{p(1-p)/n}$. The standard deviation $\sigma_{\hat{p}}$ involves the unknown parameter p. Standardizing $\hat{p}$ by subtracting p and dividing by $\sigma_{\hat{p}}$ then implies that

$$P\left(-z_{\alpha/2} < \frac{\hat{p} - p}{\sqrt{p(1-p)/n}} < z_{\alpha/2}\right) \approx 1 - \alpha$$

Proceeding as suggested in the subsection "Deriving a Confidence Interval" (Section 8.1), the confidence limits result from replacing each $<$ by $=$ and solving the resulting quadratic equation for p. With $\hat{q} = 1 - \hat{p}$, this gives the two roots

$$p = \frac{\hat{p} + \dfrac{z_{\alpha/2}^2}{2n} \pm z_{\alpha/2}\sqrt{\dfrac{\hat{p}\hat{q}}{n} + \dfrac{z_{\alpha/2}^2}{4n^2}}}{1 + (z_{\alpha/2}^2)/n}$$

PROPOSITION

A confidence interval for a population proportion p with confidence level approximately $100(1-\alpha)\%$ has

$$\text{lower confidence limit} = \frac{\hat{p} + \dfrac{z_{\alpha/2}^2}{2n} - z_{\alpha/2}\sqrt{\dfrac{\hat{p}\hat{q}}{n} + \dfrac{z_{\alpha/2}^2}{4n^2}}}{1 + (z_{\alpha/2}^2)/n}$$

and (8.10)

$$\text{upper confidence limit} = \frac{\hat{p} + \dfrac{z_{\alpha/2}^2}{2n} + z_{\alpha/2}\sqrt{\dfrac{\hat{p}\hat{q}}{n} + \dfrac{z_{\alpha/2}^2}{4n^2}}}{1 + (z_{\alpha/2}^2)/n}$$

If the sample size is quite large, $z^2/(2n)$ is negligible compared to $\hat{p}$, $z^2/(4n^2)$ under the square root is negligible compared to $\hat{p}\hat{q}/n$, and z^2/n is negligible compared to 1. Discarding these negligible terms gives approximate confidence limits

$$\hat{p} \pm z_{\alpha/2}\sqrt{\hat{p}\hat{q}/n} \tag{8.11}$$

This is of the general form $\hat{\theta} \pm z_{\alpha/2}\hat{\sigma}_{\hat{\theta}}$ of a large-sample interval suggested in the last subsection. For decades this latter interval has been recommended as long as the normal approximation for $\hat{p}$ is justified. However, recent research has shown that the somewhat more complicated interval given in the proposition has an actual confidence level that tends to be closer to the nominal level than does the traditional interval (Agresti and Coull, "Approximate Is Better Than 'Exact' for Interval Estimation of a Binomial Proportion," *Amer. Statist.*, 1998: 119–126). That is, if $z_{\alpha/2} = 1.96$ is used, the confidence level for the "new" interval tends to be closer to 95% for almost all values of p than is the case for the traditional interval; this is also true for other confidence levels. In addition, Agresti and Coull state that the interval "can be recommended for use with nearly all sample sizes and parameter values," so the conditions $n\hat{p} \geq 10$ and $n\hat{q} \geq 10$ need not be checked.

Example 8.8 The article "Repeatability and Reproducibility for Pass/Fail Data" (*J. Testing Eval.*, 1997: 151–153) reported that in $n = 48$ trials in a particular laboratory, 16 resulted in ignition of a particular type of substrate by a lighted cigarette. Let p denote the long-run proportion of all such trials that would result in ignition. A point estimate for p is $\hat{p} = 16/48 = .333$. A confidence interval for p with a confidence level of approximately 95% is

$$\frac{.333 + (1.96)^2/96 \pm 1.96\sqrt{(.333)(.667)/48 + (1.96)^2/9216}}{1 + (1.96)^2/48}$$

$$= \frac{.373 \pm .139}{1.08} = (.217, .474)$$

The traditional interval is

$$.333 \pm 1.96\sqrt{(.333)(.667)/48} = .333 \pm .133 = (.200, .466)$$

These two intervals would be in much closer agreement were the sample size substantially larger. ■

Equating the width of the CI for p to a prespecified width w gives a quadratic equation for the sample size n necessary to give an interval with a desired degree of precision. Suppressing the subscript in $z_{\alpha/2}$, the solution is

$$n = \frac{2z^2\hat{p}\hat{q} - z^2w^2 \pm \sqrt{4z^4\hat{p}\hat{q}(\hat{p}\hat{q} - w^2) + w^2z^4}}{w^2} \tag{8.12}$$

Neglecting the terms in the numerator involving w^2 gives

$$n \approx \frac{4z^2\hat{p}\hat{q}}{w^2}$$

This latter expression is what results from equating the width of the traditional interval to w.

These formulas unfortunately involve the unknown p. The most conservative approach is to take advantage of the fact that $\hat{p}\hat{q}[= \hat{p}(1 - \hat{p})]$ is a maximum when $\hat{p} = .5$. Thus if $\hat{p} = \hat{q} = .5$ is used in (8.12), the width will be at most w regardless of what value of $\hat{p}$ results from the sample. Alternatively, if the investigator believes strongly, based on prior information, that $p \leq p_0 \leq .5$, then p_0 can be used in place of $\hat{p}$. A similar comment applies when $p \geq p_0 \geq .5$.

Example 8.9 The width of the 95% CI in Example 8.8 is .257. The value of n necessary to ensure a width of .10 irrespective of the value of p is

$$n = \frac{2(1.96)^2(.25) - (1.96)^2(.01) \pm \sqrt{4(1.96)^4(.25)(.25 - .01) + (.01)(1.96)^4}}{.01}$$

$$= 380.3$$

Thus a sample size of 381 should be used. The expression for n based on the traditional CI gives a slightly larger value of 385. ■

One-Sided Confidence Intervals (Confidence Bounds)

The confidence intervals discussed thus far give both a lower confidence bound *and* an upper confidence bound for the parameter being estimated. In some circumstances, an investigator will want only one of these two types of bounds. For example, a psychologist may wish to calculate a 95% upper confidence bound for true average reaction time to a particular stimulus, or a surgeon may want only a lower confidence bound for true average remission time after colon cancer surgery. Because the cumulative area under the standard normal curve to the left of 1.645 is .95,

$$P\left(\frac{\overline{X} - \mu}{S/\sqrt{n}} < 1.645\right) \approx .95$$

Manipulating the inequality inside the parentheses to isolate μ on one side and replacing rv's by calculated values gives the inequality $\mu > \overline{x} - 1.645s/\sqrt{n}$; the expression on the right is the desired lower confidence bound. Starting with $P(-1.645 < Z) \approx .95$ and manipulating the inequality results in the upper confidence bound. A similar argument gives a one-sided bound associated with any other confidence level.

PROPOSITION

A **large-sample upper confidence bound for μ** is

$$\mu < \overline{x} + z_\alpha \cdot \frac{s}{\sqrt{n}}$$

and a **large-sample lower confidence bound for μ** is

$$\mu > \overline{x} - z_\alpha \cdot \frac{s}{\sqrt{n}}$$

A **one-sided confidence bound for p** results from replacing $z_{\alpha/2}$ by z_α and $\pm$ by either $+$ or $-$ in the CI formula (8.10) for p. In all cases the confidence level is approximately $100(1 - \alpha)\%$.

Example 8.10 Recall the algebra test data of Example 8.6. A sample of 50 scores gave a sample mean $\overline{x} = 19.82$ and a standard deviation $s = 4.50$. A lower confidence bound for true average score μ with confidence level 95% is

$$19.82 - (1.645) \cdot \frac{4.50}{\sqrt{50}} = 19.82 - 1.05 = 18.77$$

That is, with a confidence level of 95%, the value of μ lies in the interval $(18.77, \infty)$. Of course, the lower confidence bound here is above the lower bound in the two-sided interval of Example 8.6. ■

Exercises Section 8.2 (12–28)

12. A random sample of 110 lightning flashes in a certain region resulted in a sample average radar echo duration of .81 sec and a sample standard deviation of .34 sec ("Lightning Strikes to an Airplane in a Thunderstorm," *J. Aircraft*, 1984: 607–611). Calculate a 99% (two-sided) confidence interval for the true average echo duration μ, and interpret the resulting interval.

13. The article "Extravisual Damage Detection? Defining the Standard Normal Tree" (*Photogrammetric Engrg. Remote Sensing*, 1981: 515–522) discusses the use of color infrared photography in identification of normal trees in Douglas fir stands. Among data reported were summary statistics for green-filter analytic optical densitometric measurements on samples of both healthy and diseased trees. For a sample of 69 healthy trees, the sample mean dye-layer density was 1.028, and the sample standard deviation was .163.

a. Calculate a 95% (two-sided) CI for the true average dye-layer density for all such trees.

b. Suppose the investigators had made a rough guess of .16 for the value of s before collecting data. What sample size would be necessary to obtain an interval width of .05 for a confidence level of 95%?

14. The article "Evaluating Tunnel Kiln Performance" (*Amer. Ceramic Soc. Bull.*, Aug. 1997: 59–63) gave the following summary information for fracture strengths (MPa) of $n = 169$ ceramic bars fired in a particular kiln: $\bar{x} = 89.10$, $s = 3.73$.

a. Calculate a (two-sided) confidence interval for true average fracture strength using a confidence level of 95%. Does it appear that true average fracture strength has been precisely estimated?

b. Suppose the investigators had believed a priori that the population standard deviation was about 4 MPa. Based on this supposition, how large a sample would have been required to estimate μ to within .5 MPa with 95% confidence?

15. Determine the confidence level for each of the following large-sample one-sided confidence bounds:

a. Upper bound: $\bar{x} + .84s/\sqrt{n}$

b. Lower bound: $\bar{x} - 2.05s/\sqrt{n}$

c. Upper bound: $\bar{x} + .67s/\sqrt{n}$

16. A sample of 66 obese adults was put on a low-carbohydrate diet for a year. The average weight loss was 11 pounds and the standard deviation was 19 pounds. Calculate a 99% lower confidence bound for the true average weight loss. What does the bound say about confidence that the mean weight loss is positive?

17. A study was done on 41 first-year medical students to see if their anxiety levels changed during the first semester. One measure used was the level of serum cortisol, which is associated with stress. For each of the 41 students the level was compared during finals at the end of the semester against the level in the first week of classes. The average difference was 2.08 with a standard deviation of 7.88. Find a 95% lower confidence bound for the population mean difference μ. Does the bound suggest that the mean population stress change is necessarily positive?

18. The article "Ultimate Load Capacities of Expansion Anchor Bolts" (*J. Energy Engrg.*, 1993: 139–158) gave the following summary data on shear strength (kip) for a sample of 3/8-in. anchor bolts: $n = 78$, $\bar{x} = 4.25$, $s = 1.30$. Calculate a lower confidence bound using a confidence level of 90% for true average shear strength.

19. The article "Limited Yield Estimation for Visual Defect Sources" (*IEEE Trans. Semiconductor Manuf.*, 1997: 17–23) reported that, in a study of a particular wafer inspection process, 356 dies were examined by an inspection probe and 201 of these passed the probe. Assuming a stable process, calculate a 95% (two-sided) confidence interval for the proportion of all dies that pass the probe.

20. The Associated Press (October 9, 2002) reported that in a survey of 4722 American youngsters aged 6 to 19, 15% were seriously overweight (a body mass index of at least 30; this index is a measure of weight relative to height). Calculate and interpret a confidence interval using a 99% confidence level for the proportion of all American youngsters who are seriously overweight.

21. A random sample of 539 households from a certain midwestern city was selected, and it was determined that 133 of these households owned at least one firearm ("The Social Determinants of Gun Ownership: Self-Protection in an Urban Environment," *Criminology*, 1997: 629–640). Using a 95% confidence level, calculate a lower confidence bound for the proportion of all households in this city that own at least one firearm.

22. A random sample of 487 nonsmoking women of normal weight (body mass index between 19.8 and 26.0) who had given birth at a large metropolitan medical center was selected ("The Effects of Cigarette Smoking and Gestational Weight Change on Birth Outcomes in Obese and Normal-Weight Women," *Amer. J. Public Health*, 1997: 591–596). It was determined that 7.2% of these births resulted in children of low birth weight (less than 2500 g). Calculate an upper confidence bound using a confidence level of 99% for the proportion of all such births that result in children of low birth weight.

23. The article "An Evaluation of Football Helmets Under Impact Conditions" (*Amer. J. Sports Medi.*, 1984: 233–237) reports that when each football helmet in a random sample of 37 suspension-type helmets was subjected to a certain impact test, 24 showed damage. Let p denote the proportion of all helmets of this type that would show damage when tested in the prescribed manner.

a. Calculate a 99% CI for p.

b. What sample size would be required for the width of a 99% CI to be at most .10, irrespective of $\hat{p}$?

24. A sample of 56 research cotton samples resulted in a sample average percentage elongation of 8.17 and a sample standard deviation of 1.42 ("An Apparent Relation Between the Spiral Angle ϕ, the Percent Elongation E_1, and the Dimensions of the Cotton Fiber," *Textile Res. J.*, 1978: 407–410). Calculate a 95% large-sample CI for the true average percentage elongation μ. What assumptions are you making about the distribution of percentage elongation?

25. A state legislator wishes to survey residents of her district to see what proportion of the electorate is aware of her position on using state funds to pay for abortions.

a. What sample size is necessary if the 95% CI for p is to have width of at most .10 irrespective of p?

b. If the legislator has strong reason to believe that at least $\frac{2}{3}$ of the electorate know of her position, how large a sample size would you recommend?

26. The superintendent of a large school district, having once had a course in probability and statistics, believes that the number of teachers absent on any given day has a Poisson distribution with parameter λ. Use the accompanying data on absences for 50 days to derive a large-sample CI for λ. (*Hint*: The mean and variance of a Poisson variable both equal λ, so

$$Z = \frac{\overline{X} - \lambda}{\sqrt{\lambda/n}}$$

has approximately a standard normal distribution. Now proceed as in the derivation of the interval for p by making a probability statement [with probability $1 - \alpha$] and solving the resulting inequalities for λ [see the argument just after (8.10)].

Number of absences	0	1	2	3	4	5	6	7	8	9	10
Frequency	1	4	8	10	8	7	5	3	2	1	1

27. Reconsider the CI (8.10) for p, and focus on a confidence level of 95%. Show that the confidence limits agree quite well with those of the traditional interval (8.11) once two successes and two failures have been appended to the sample [i.e., (8.11) based on $(x + 2)$ S's in $(n + 4)$ trials]. (*Hint*: $1.96 \approx 2$. *Note*: Agresti and Coull showed that this adjustment of the traditional interval also has actual confidence level close to the nominal level.)

28. Young people may feel they are carrying the weight of the world on their shoulders, when what they are actually carrying too often is an excessively heavy backpack. The article "Effectiveness of a School-Based Backpack Health Promotion Program" (*Work*, 2003: 113–123) reported the following data for a sample of 131 sixth graders: for backpack weight (lb), $\bar{x} = 13.83$, $s = 5.05$; for backpack weight as a percentage of body weight, a 95% CI for the population mean was (13.62, 15.89).

a. Calculate and interpret a 99% CI for population mean backpack weight.

b. Obtain a 99% CI for population mean weight as a percentage of body weight.

c. The American Academy of Orthopedic Surgeons recommends that backpack weight be at most 10% of body weight. What does your calculation of (b) suggest, and why?

8.3 Intervals Based on a Normal Population Distribution

The CI for μ presented in Section 8.2 is valid provided that n is large. The resulting interval can be used whatever the nature of the population distribution. The CLT cannot be invoked, however, when n is small. In this case, one way to proceed is to make a specific assumption about the form of the population distribution and then derive a CI tailored to that assumption. For example, we could develop a CI for μ when the population is described by a gamma distribution, another interval for the case of a Weibull population, and so on. Statisticians have indeed carried out this program for a number of different distributional families. Because the normal distribution is more frequently appropriate as a population model than is any other type of distribution, we will focus here on a CI for this situation.

ASSUMPTION The population of interest is normal, so that $X_1, \ldots, X_n$ constitutes a random sample from a normal distribution with both μ and σ unknown.

The key result underlying the interval in Section 8.2 is that for large n, the rv $Z = (\overline{X} - \mu)/(S/\sqrt{n})$ has approximately a standard normal distribution. When n is small, S is no longer likely to be close to σ, so the variability in the distribution of Z arises from randomness in both the numerator and the denominator. This implies that the probability distribution of $(\overline{X} - \mu)/(S/\sqrt{n})$ will be more spread out than the standard normal distribution. Inferences are based on the following result from Section 6.4 using the family of *t distributions*:

THEOREM When $\overline{X}$ is the mean of a random sample of size n from a normal distribution with mean μ, the rv

$$T = \frac{\overline{X} - \mu}{S/\sqrt{n}} \tag{8.13}$$

has the t distribution with $n - 1$ degrees of freedom (df).

Properties of t Distributions

Before applying this theorem, a review of properties of t distributions is in order. Although the variable of interest is still $(\overline{X} - \mu)/(S/\sqrt{n})$, we now denote it by T to emphasize that it does not have a standard normal distribution when n is small. Recall that a normal distribution is governed by two parameters, the mean μ and the standard deviation σ. A t distribution is governed by only one parameter, the **number of degrees of freedom** of the distribution, abbreviated df and denoted by ν. Possible values of ν are the positive integers 1, 2, 3, ... Each different value of ν corresponds to a different t distribution.

For any fixed value of the parameter ν, the density function that specifies the associated t curve has an even more complicated appearance than the normal density function. Fortunately, we need concern ourselves only with several of the more important features of these curves.

PROPERTIES OF t DISTRIBUTIONS

Let t_ν denote the density function curve for ν df.

1. Each t_ν curve is bell-shaped and centered at 0.
2. Each t_ν curve is more spread out than the standard normal (z) curve.
3. As ν increases, the spread of the corresponding t_ν curve decreases.
4. As $\nu \to \infty$, the sequence of t_ν curves approaches the standard normal curve (so the z curve is often called the t curve with df $= \infty$).

Recall the notation for values that capture particular upper-tail areas under the t pdf.

NOTATION

Let $t_{\alpha,\nu}$ = the number on the measurement axis for which the area under the t curve with ν df to the right of $t_{\alpha,\nu}$, is α; $t_{\alpha,\nu}$ is called a ***t* critical value**.

This notation is illustrated in Figure 8.6. Appendix Table A.5 gives $t_{\alpha,\nu}$ for selected values of α and ν. This table also appears inside the back cover. The columns of the table correspond to different values of α. To obtain $t_{.05,15}$, go to the $\alpha = .05$ column, look down to the $\nu = 15$ row, and read $t_{.05,15} = 1.753$. Similarly, $t_{.05,22} = 1.717$ (.05 column, $\nu = 22$ row), and $t_{.01,22} = 2.508$.

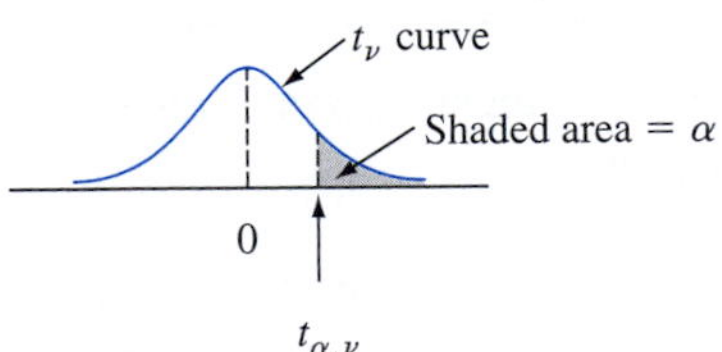

Figure 8.6 A pictorial definition of $t_{\alpha,\nu}$

The values of $t_{\alpha,\nu}$ exhibit regular behavior as we move across a row or down a column. For fixed ν, $t_{\alpha,\nu}$ increases as α decreases, since we must move farther to the right of zero to capture area α in the tail. For fixed α, as ν is increased (i.e., as we look down any particular column of the t table) the value of $t_{\alpha,\nu}$ decreases. This is because a larger value of ν implies a t distribution with smaller spread, so it is not necessary to go so far from zero to capture tail area α. Furthermore, $t_{\alpha,\nu}$ decreases more slowly as ν increases. Consequently, the table values are shown in increments of 2 between 30 df and 40 df and then jump to $\nu = 50$, 60, 120, and finally ∞. Because t_∞ is the standard normal curve, the familiar z_α values appear in the last row of the table. The rule of thumb suggested earlier for use of the large-sample CI (if $n > 40$) comes from the approximate equality of the standard normal and t distributions for $\nu \geq 40$.

The One-Sample t Confidence Interval

The standardized variable T has a t distribution with $n - 1$ df, and the area under the corresponding t density curve between $-t_{\alpha/2,n-1}$ and $t_{\alpha/2,n-1}$ is $1 - \alpha$ (area $\alpha/2$ lies in each tail), so

$$P(-t_{\alpha/2,n-1} < T < t_{\alpha 2,n-1}) = 1-\alpha \qquad (8.14)$$

Expression (8.14) differs from expressions in previous sections in that T and $t_{\alpha/2,n-1}$ are used in place of Z and $z_{\alpha/2}$, but it can be manipulated in the same manner to obtain a confidence interval for μ.

PROPOSITION

Let $\bar{x}$ and s be the sample mean and sample standard deviation computed from the results of a random sample from a normal population with mean μ. Then a **100(1 − α)% confidence interval for μ** is

$$\left(\bar{x} - t_{\alpha/2,n-1} \cdot \frac{s}{\sqrt{n}}, \bar{x} + t_{\alpha/2,n-1} \cdot \frac{s}{\sqrt{n}}\right) \qquad (8.15)$$

or, more compactly, $\bar{x} \pm t_{\alpha/2,n-1} \cdot s/\sqrt{n}$.

An **upper confidence bound for μ** is

$$\bar{x} + t_{\alpha,n-1} \cdot \frac{s}{\sqrt{n}}$$

and replacing + by − in this latter expression gives a **lower confidence bound for μ**, both with confidence level 100(1 − α)%.

Example 8.11 Here are the alcohol percentages for a sample of 16 beers (light beers excluded):

4.68	4.13	4.80	4.63	5.08	5.79	6.29	6.79
4.93	4.25	5.70	4.74	5.88	6.77	6.04	4.95

Figure 8.7 shows a normal probability plot obtained from SAS. The plot is sufficiently straight for the percentage to be assumed approximately normal.

The mean is $\bar{x} = 5.34$ and the standard deviation is $s = .8483$. The sample size is 16, so a confidence interval for the population mean percentage is based on 15 df. A confidence level of 95% for a two-sided interval requires the t critical value of 2.131. The resulting interval is

$$\begin{aligned}\bar{x} \pm t_{.025,15} \cdot \frac{s}{\sqrt{n}} &= 5.34 \pm 2.131 \frac{.8483}{\sqrt{16}} \\ &= 5.34 \pm .45 = (4.89, 5.79)\end{aligned}$$

A 95% lower bound would use −1.753 in place of ±2.131. It is interesting that the 95% confidence interval is consistent with the usual statement about the equivalence of wine and beer in terms of alcohol content. That is, assuming an alcohol percentage of 13% for wine, a 5-ounce serving yields .65 ounce of alcohol, and assuming 5.34% alcohol for beer, a 12-ounce serving has .64 ounce of alcohol.

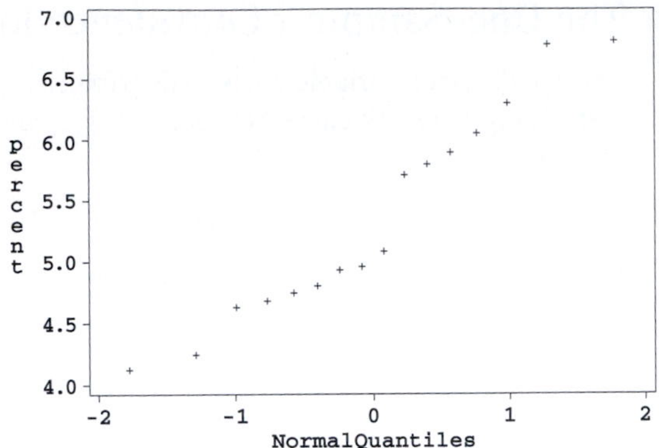

Figure 8.7 A normal probability plot of the alcohol percentage data ■

Unfortunately, it is not easy to select n to control the width of the t interval. This is because the width involves the unknown (before data collection) s and because n enters not only through $1/\sqrt{n}$ but also through $t_{\alpha/2,n-1}$. As a result, an appropriate n can be obtained only by trial and error.

In Chapter 14, we will discuss a small-sample CI for μ that is valid provided only that the population distribution is symmetric, a weaker assumption than normality. However, when the population distribution is normal, the t interval tends to be shorter than would be *any* other interval with the same confidence level.

A Prediction Interval for a Single Future Value

In many applications, an investigator wishes to *predict a* single value of a variable to be observed at some future time, rather than to *estimate* the mean value of that variable.

Example 8.12 Consider the following sample of fat content (in percentage) of $n = 10$ randomly selected hot dogs ("Sensory and Mechanical Assessment of the Quality of Frankfurters," *J. Texture Stud.*, 1990: 395–409):

25.2 21.3 22.8 17.0 29.8 21.0 25.5 16.0 20.9 19.5

Assuming that these were selected from a normal population distribution, a 95% CI for (interval estimate of) the population mean fat content is

$$\bar{x} \pm t_{.025,9} \cdot \frac{s}{\sqrt{n}} = 21.90 \pm 2.262 \cdot \frac{4.134}{\sqrt{10}} = 21.90 \pm 2.96$$
$$= (18.94, 24.86)$$

Suppose, however, you are going to eat a single hot dog of this type and want *a prediction* for the resulting fat content. *A point* prediction, analogous to *a point* estimate, is just $\bar{x} = 21.90$. This prediction unfortunately gives no information about reliability or precision. ■

The general setup is as follows: We will have available a random sample X_1, $X_2, \ldots, X_n$ from a normal population distribution, and we wish to predict the value of X_{n+1}, a single future observation. A point predictor is $\overline{X}$, and the resulting prediction error is $\overline{X} - X_{n+1}$. The expected value of the prediction error is

$$E(\overline{X} - X_{n+1}) = E(\overline{X}) - E(X_{n+1}) = \mu - \mu = 0$$

Since X_{n+1} is independent of $X_1, \ldots, X_n$, it is independent of $\overline{X}$, so the variance of the prediction error is

$$V(\overline{X} - X_{n+1}) = V(\overline{X}) + V(X_{n+1}) = \frac{\sigma^2}{n} + \sigma^2 = \sigma^2\left(1 + \frac{1}{n}\right)$$

The prediction error is a linear combination of independent normally distributed rv's, so itself is normally distributed. Thus

$$Z = \frac{(\overline{X} - X_{n+1}) - 0}{\sqrt{\sigma^2\left(1 + \frac{1}{n}\right)}} = \frac{\overline{X} - X_{n+1}}{\sqrt{\sigma^2\left(1 + \frac{1}{n}\right)}}$$

has a standard normal distribution. As in the derivation of the distribution of $(\overline{X} - \mu)/(S/\sqrt{n})$ in Section 6.4, it can be shown (Exercise 43) that replacing σ by the sample standard deviation S (of $X_1, \ldots, X_n$) results in

$$T = \frac{\overline{X} - X_{n+1}}{S\sqrt{1 + \frac{1}{n}}} \sim t \text{ distribution with } n - 1 \text{ df}$$

Manipulating this T variable as $T = (\overline{X} - \mu)/(S/\sqrt{n})$ was manipulated in the development of a CI gives the following result.

PROPOSITION

A **prediction interval (PI)** for a single observation to be selected from a normal population distribution is

$$\bar{x} \pm t_{\alpha/2,n-1} \cdot s\sqrt{1 + \frac{1}{n}} \tag{8.16}$$

The *prediction level* is $100(1 - \alpha)\%$.

The interpretation of a 95% prediction level is similar to that of a 95% confidence level; if the interval (8.16) is calculated for sample after sample, in the long run 95% of these intervals will include the corresponding future values of X.

Example 8.13 (Example 8.12 continued)

With $n = 10$, $\bar{x} = 21.90$, $s = 4.134$, and $t_{.025,9} = 2.262$, a 95% PI for the fat content of a single hot dog is

$$21.90 \pm (2.262)(4.134)\sqrt{1 + \frac{1}{10}} = 21.90 \pm 9.81$$
$$= (12.09, 31.71)$$

This interval is quite wide, indicating substantial uncertainty about fat content. Notice that the width of the PI is more than three times that of the CI. ■

The error of prediction is $\overline{X} - X_{n+1}$, a difference between two random variables, whereas the estimation error is $\overline{X} - \mu$, the difference between a random variable and a fixed (but unknown) value. The PI is wider than the CI because there is more variability in the prediction error (due to X_{n+1}) than in the estimation error. In fact, as n gets arbitrarily large, the CI shrinks to the single value μ, and the PI approaches $\mu \pm z_{\alpha/2} \cdot \sigma$. There is uncertainty about a single X value even when there is no need to estimate.

Tolerance Intervals

Consider a population of automobiles of a certain type, and suppose that under specified conditions, fuel efficiency (mpg) has a normal distribution with $\mu = 30$ and $\sigma = 2$. Then since the interval from -1.645 to 1.645 captures 90% of the area under the z curve, 90% of all these automobiles will have fuel efficiency values between $\mu - 1.645\sigma = 26.71$ and $\mu + 1.645\sigma = 33.29$. But what if the values of μ and σ are not known? We can take a sample of size n, determine the fuel efficiencies, $\bar{x}$, and s, and form the interval whose lower limit is $\bar{x} - 1.645s$ and whose upper limit is $\bar{x} + 1.645s$. However, because of sampling variability in the estimates of μ and σ, there is a good chance that the resulting interval will include less than 90% of the population values. Intuitively, to have an a priori 95% chance of the resulting interval including at least 90% of the population values, when $\bar{x}$ and s are used in place of μ and σ, we should also replace 1.645 by some larger number. For example, when $n = 20$, the value 2.310 is such that we can be 95% confident that the interval $\bar{x} \pm 2.310s$ will include at least 90% of the fuel efficiency values in the population.

Let k be a number between 0 and 100. A **tolerance interval** for capturing at least k% of the values in a normal population distribution with a confidence level 95% has the form

$$\bar{x} \pm (\text{tolerance critical value}) \cdot s$$

Tolerance critical values for $k = 90$, 95, and 99 in combination with various sample sizes are given in Appendix Table A.6. This table also includes critical values for a confidence level of 99% (these values are larger than the corresponding 95% values). Replacing $\pm$ by $+$ gives an upper tolerance bound, and using $-$ in place of $\pm$ results in a lower tolerance bound. Critical values for obtaining these one-sided bounds also appear in Appendix Table A.6.

Example 8.14 Data on the modulus of elasticity (MPa) for $n = 16$ Scotch pine lumber specimens resulted in $\bar{x} = 14{,}532.5$, $s = 2055.67$, and a normal probability plot of the data indicated that population normality was quite plausible. For a confidence level of 95%, a two-sided tolerance interval for capturing at least 95% of the modulus of elasticity values for specimens of lumber in the population sampled uses the tolerance critical value of 2.903: The resulting interval is

$$14{,}532.5 \pm (2.903)(2055.67) = 14{,}532.5 \pm 5967.6 = (8{,}564.9, 20{,}500.1)$$

We can be highly confident that at least 95% of all lumber specimens have modulus of elasticity values between 8,564.9 and 20,500.1.

The 95% CI for μ is (13,437.3, 15,627.7), and the 95% prediction interval for the modulus of elasticity of a single lumber specimen is (10,017.0, 19,048.0). Both the prediction interval and the tolerance interval are substantially wider than the confidence interval. ■

Intervals Based on Nonnormal Population Distributions

The one-sample t CI for μ is robust to small or even moderate departures from normality unless n is quite small. By this we mean that if a critical value for 95% confidence, for example, is used in calculating the interval, the actual confidence level will be reasonably close to the nominal 95% level. If, however, n is small and the population distribution is highly nonnormal, then the actual confidence level may be considerably different from the one you think you are using when you obtain a particular critical value from the t table. It would certainly be distressing to believe that your confidence level is about 95% when in fact it was really more like 88%! The bootstrap technique, discussed in the last section of this chapter, has been found to be quite successful at estimating parameters in a wide variety of nonnormal situations.

In contrast to the confidence interval, the validity of the prediction intervals described in this section is closely tied to the normality assumption. These latter intervals should not be used in the absence of compelling evidence for normality. The excellent reference *Statistical Intervals*, cited in the bibliography at the end of this chapter, discusses alternative procedures of this sort for various other situations.

Exercises Section 8.3 (29–43)

29. Determine the values of the following quantities:
a. $t_{.1,15}$ **b.** $t_{.05,15}$ **c.** $t_{.05,25}$
d. $t_{.05,40}$ **e.** $t_{.005,40}$

30. Determine the t critical value that will capture the desired t curve area in each of the following cases:
a. Central area = .95, df = 10
b. Central area = .95, df = 20
c. Central area = .99, df = 20
d. Central area = .99, df = 50
e. Upper-tail area = .01, df = 25
f. Lower-tail area = .025, df = 5

31. Determine the t critical value for a two-sided confidence interval in each of the following situations:
a. Confidence level = 95%, df = 10
b. Confidence level = 95%, df = 15
c. Confidence level = 99%, df = 15
d. Confidence level = 99%, n = 5
e. Confidence level = 98%, df = 24
f. Confidence level = 99%, n = 38

32. Determine the t critical value for a lower or an upper confidence bound for each of the situations described in Exercise 31.

33. A sample of ten guinea pigs yielded the following measurements of body temperature in degrees Celsius (*Statistical Exercises in Medical Research*, New York: Wiley, 1979, p. 26):

38.1 38.4 38.3 38.2 38.2 37.9 38.7 38.6 38.0 38.2

a. Verify graphically that it is reasonable to assume the normal distribution.
b. Compute a 95% confidence interval for the population mean temperature.
c. What is the CI if temperature is reexpressed in degrees Fahrenheit? Are guinea pigs warmer on average than humans?

34. Here is a sample of ACT scores (average of the Math, English, Social Science, and Natural Science scores) for students taking college freshman calculus:

24.00	28.00	27.75	27.00	24.25	23.50	26.25
24.00	25.00	30.00	23.25	26.25	21.50	26.00
28.00	24.50	22.50	28.25	21.25	19.75	

a. Using an appropriate graph, see if it is plausible that the observations were selected from a normal distribution.

b. Calculate a two-sided 95% confidence interval for the population mean.

c. The university ACT average for entering freshmen that year was about 21. Are the calculus students better than average, as measured by the ACT?

35. A sample of 14 joint specimens of a particular type gave a sample mean proportional limit stress of 8.48 MPa and a sample standard deviation of .79 MPa ("Characterization of Bearing Strength Factors in Pegged Timber Connections," *J. Struct. Engrg.*, 1997: 326–332).

a. Calculate and interpret a 95% lower confidence bound for the true average proportional limit stress of all such joints. What, if any, assumptions did you make about the distribution of proportional limit stress?

b. Calculate and interpret a 95% lower prediction bound for the proportional limit stress of a single joint of this type.

36. Exercise 43 in Chapter 1 introduced the following sample observations on stabilized viscosity of asphalt specimens: 2781, 2900, 3013, 2856, 2888. A normal probability plot supports the assumption that viscosity is at least approximately normally distributed.

a. Estimate true average viscosity in a way that conveys information about precision and reliability.

b. Predict the viscosity for a single asphalt specimen in a way that conveys information about precision and reliability. How does the prediction compare to the estimate calculated in part (a)?

37. The $n = 26$ observations on escape time given in Exercise 33 of Chapter 1 give a sample mean and sample standard deviation of 370.69 and 24.36, respectively.

a. Calculate an upper confidence bound for population mean escape time using a confidence level of 95%.

b. Calculate an upper prediction bound for the escape time of a single additional worker using a prediction level of 95%. How does this bound compare with the confidence bound of part (a)?

c. Suppose that two additional workers will be chosen to participate in the simulated escape exercise. Denote their escape times by X_{27} and X_{28}, and let $\overline{X}_{new}$ denote the average of these two values. Modify the formula for a PI for a single x value to obtain a PI for $\overline{X}_{new}$, and calculate a 95% two-sided interval based on the given escape data.

38. A study of the ability of individuals to walk in a straight line ("Can We Really Walk Straight?" *Amer. J. Phys. Anthropol.*, 1992: 19–27) reported the accompanying data on cadence (strides per second) for a sample of $n = 20$ randomly selected healthy men.

.95	.85	.92	.95	.93	.86	1.00	.92	.85	.81
.78	.93	.93	1.05	.93	1.06	1.06	.96	.81	.96

A normal probability plot gives substantial support to the assumption that the population distribution of cadence is approximately normal. A descriptive summary of the data from MINITAB follows:

```
Variable  N   Mean    Median  TrMean  StDev   SEMean
cadence   20  0.9255  0.9300  0.9261  0.0809  0.0181

Variable      Min      Max       Q1       Q3
cadence    0.7800   1.0600   0.8525   0.9600
```

a. Calculate and interpret a 95% confidence interval for population mean cadence.

b. Calculate and interpret a 95% prediction interval for the cadence of a single individual randomly selected from this population.

c. Calculate an interval that includes at least 99% of the cadences in the population distribution using a confidence level of 95%.

39. A sample of 25 pieces of laminate used in the manufacture of circuit boards was selected and the amount of warpage (in.) under particular conditions was determined for each piece, resulting in a sample mean warpage of .0635 and a sample standard deviation of .0065.

a. Calculate a prediction for the amount of warpage of a single piece of laminate in a way that provides information about precision and reliability.

b. Calculate an interval for which you can have a high degree of confidence that at least 95% of all pieces of laminate result in amounts of warpage that are between the two limits of the interval.

40. Exercise 69 of Chapter 1 gave the following observations on a receptor binding measure (adjusted distribution volume) for a sample of 13 healthy individuals: 23, 39, 40, 41, 43, 47, 51, 58, 63, 66, 67, 69, 72.

a. Is it plausible that the population distribution from which this sample was selected is normal?

b. Calculate an interval for which you can be 95% confident that at least 95% of all healthy individuals in the population have adjusted distribution volumes lying between the limits of the interval.

c. Predict the adjusted distribution volume of a single healthy individual by calculating a 95% prediction interval. How does this interval's width compare to the width of the interval calculated in part (b)?

41. Here are the lengths (in minutes) of the 63 nine-inning games from the first week of the 2001 major league baseball season:

194 160 176 203 187 163 162 183 152 177
177 151 173 188 179 194 149 165 186 187
187 177 187 186 187 173 136 150 173 173
136 153 152 149 152 180 186 166 174 176
198 193 218 173 144 148 174 163 184 155
151 172 216 149 207 212 216 166 190 165
176 158 198

Assume that this is a random sample of nine-inning games (the mean differs by 12 seconds from the mean for the whole season).

a. Give a 95% confidence interval for the population mean.

b. Give a 95% prediction interval for the length of the next nine-inning game. On the first day of the next week, Boston beat Tampa Bay 3-0 in a nine-inning game of 152 minutes. Is this within the prediction interval?

c. Compare the two intervals and explain why one is much wider than the other.

d. Explore the issue of normality for the data and explain how this is relevant to parts (a) and (b).

42. A more extensive tabulation of t critical values than what appears in this book shows that for the t distribution with 20 df, the areas to the right of the values .687, .860, and 1.064 are .25, .20, and .15, respectively. What is the confidence level for each of the following three confidence intervals for the mean μ of a normal population distribution? Which of the three intervals would you recommend be used, and why?

a. $(\bar{x} - .687s/\sqrt{21}, \bar{x} + 1.725s/\sqrt{21})$

b. $(\bar{x} - .860s/\sqrt{21}, \bar{x} + 1.325s/\sqrt{21})$

c. $(\bar{x} - 1.064s/\sqrt{21}, \bar{x} + 1.064s/\sqrt{21})$

43. Use the results of Section 6.4 to show that the variable T on which the PI is based does in fact have a t distribution with $n - 1$ df.

8.4 *Confidence Intervals for the Variance and Standard Deviation of a Normal Population

Although inferences concerning a population variance σ^2 or standard deviation σ are usually of less interest than those about a mean or proportion, there are occasions when such procedures are needed. In the case of a normal population distribution, inferences are based on the following result from Section 6.4 concerning the sample variance S^2.

THEOREM Let $X_1, X_2, \ldots, X_n$ be a random sample from a normal distribution with parameters μ and σ^2. Then the rv

$$\frac{(n-1)S^2}{\sigma^2} = \frac{\sum (X_i - \bar{X})^2}{\sigma^2}$$

has a chi-squared (χ^2) probability distribution with $n - 1$ df.

As discussed in Sections 4.4 and 6.4, the chi-squared distribution is a continuous probability distribution with a single parameter ν, the number of degrees of freedom, with possible values 1, 2, 3, ... To specify inferential procedures that use the chi-squared distribution, recall the notation for critical values from Section 6.4.

NOTATION Let $\chi^2_{\alpha,\nu}$, called a **chi-squared critical value**, denote the number on the measurement axis such that α of the area under the chi-squared curve with ν df lies to the right of $\chi^2_{\alpha,\nu}$.

Because the t distribution is symmetric, it was necessary to tabulate only upper-tail critical values ($t_{\alpha,\nu}$ for small values of α). The chi-squared distribution is not symmetric, so Appendix Table A.7 contains values of $\chi^2_{\alpha,\nu}$ for α both near 0 and near 1, as illustrated in Figure 8.8(b). For example, $\chi^2_{.025,14} = 26.119$ and $\chi^2_{.95,20}$ (the 5th percentile) $= 10.851$.

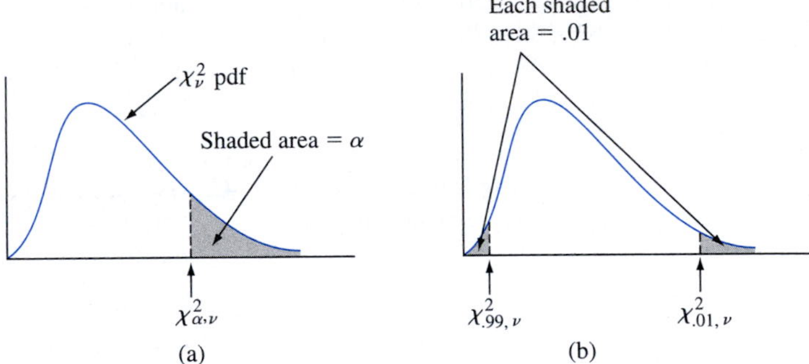

Figure 8.8 $\chi^2_{\alpha,\nu}$ notation illustrated

The rv $(n-1)S^2/\sigma^2$ satisfies the two properties on which the general method for obtaining a CI is based: It is a function of the parameter of interest σ^2, yet its probability distribution (chi-squared) does not depend on this parameter. The area under a chi-squared curve with ν df to the right of $\chi^2_{\alpha/2,\nu}$ is $\alpha/2$, as is the area to the left of $\chi^2_{1-\alpha/2,\nu}$. Thus the area captured between these two critical values is $1-\alpha$. As a consequence of this and the theorem just stated,

$$P\left(\chi^2_{1-\alpha/2,n-1} < \frac{(n-1)S^2}{\sigma^2} < \chi^2_{\alpha/2,n-1}\right) = 1-\alpha \tag{8.17}$$

The inequalities in (8.17) are equivalent to

$$\frac{(n-1)S^2}{\chi^2_{\alpha/2,n-1}} < \sigma^2 < \frac{(n-1)S^2}{\chi^2_{1-\alpha/2,n-1}}$$

Substituting the computed value s^2 into the limits gives a CI for σ^2, and taking square roots gives an interval for σ.

A **100(1 − α)% confidence interval for the variance σ^2 of a normal population** has lower limit

$$(n - 1)s^2/\chi^2_{\alpha/2,n-1}$$

and upper limit

$$(n - 1)s^2/\chi^2_{1-\alpha/2,n-1}$$

A **confidence interval for σ** has lower and upper limits that are the square roots of the corresponding limits in the interval for σ^2.

Example 8.15 Recall the beer alcohol percentage data from Example 8.11, where the normal plot was acceptably straight and the standard deviation was found to be $s = .8483$. Then the estimated variance is $s^2 = .8483^2 = .7196$, and we wish to estimate the population variance σ^2. With df $= n - 1 = 15$, a 95% confidence interval requires $\chi^2_{.975,15} = 6.262$ and $\chi^2_{.025,15} = 27.488$. The interval for σ^2 is

$$\left(\frac{15(.7196)}{27.488}, \frac{15(.7196)}{6.262}\right) = (.393, 1.724)$$

Taking the square root of each endpoint yields (.627, 1.313) as the 95% confidence interval for σ. With lower and upper limits differing by more than a factor of two, this interval is quite wide. Precise estimates of variability require large samples. ■

Unfortunately, our confidence interval requires that the data be normal or nearly normal. In the case of nonnormal data the interval could be very far from valid; for example, the true confidence level could be 70% where 95% is intended. See Exercise 57 in the next section for a method that does not require the normal distribution.

Exercises Section 8.4 (44–48)

44. Determine the values of the following quantities:
a. $\chi^2_{.1,15}$ **b.** $\chi^2_{.1,25}$
c. $\chi^2_{.01,25}$ **d.** $\chi^2_{.005,25}$
e. $\chi^2_{.99,25}$ **f.** $\chi^2_{.995,25}$

45. Determine the following:
a. The 95th percentile of the chi-squared distribution with $\nu = 10$
b. The 5th percentile of the chi-squared distribution with $\nu = 10$
c. $P(10.98 \le \chi^2 \le 36.78)$, where χ^2 is a chi-squared rv with $\nu = 22$
d. $P(\chi^2 < 14.611$ or $\chi^2 > 37.652)$, where χ^2 is a chi-squared rv with $\nu = 25$

46. Exercise 34 gave a random sample of 20 ACT scores from students taking college freshman calculus. Calculate a 99% CI for the standard deviation of the population distribution. Is this interval valid whatever the nature of the distribution? Explain.

47. Here are the names of 12 orchestra conductors and their performance times in minutes for Beethoven's Ninth Symphony:

Bernstein	71.03	Furtwängler	74.38
Leinsdorf	65.78	Ormandy	64.72
Solti	74.70	Szell	66.22
Bohm	72.68	Karajan	66.90
Mazur	69.45	Rattle	69.93
Steinberg	68.62	Tennstedt	68.40

a. Check to see that normality is a reasonable assumption for the performance time distribution.

b. Compute a 95% CI for the population standard deviation, and interpret the interval.

c. Supposedly, classical music is 100% determined by the composer's notation, including all timings. Based on your results, is this true or false?

48. Refer to the baseball game times in Exercise 41. Calculate an upper confidence bound with confidence level 95% for the population standard deviation of game time. Interpret your interval. Explore the issue of normality for the data and explain how this is relevant to your interval.

8.5 *Bootstrap Confidence Intervals

How can we find a confidence interval for the mean if the population distribution is not normal and the sample size n is not large? Can we find confidence intervals for other parameters such as the population median or the 90th percentile of the population distribution? The bootstrap, developed by Bradley Efron in the late 1970s, allows us to calculate estimates in situations where there is no adequate statistical theory. The method substitutes heavy computation for theory, and it has been feasible only fairly recently with the availability of fast computers. The bootstrap was introduced in Section 7.1 for applications with known distribution (the *parametric bootstrap*), but here we are concerned with the case of unknown distribution (the *nonparametric bootstrap*).

Bootstrapping the Mean

Perhaps the best way to appreciate the ideas of the bootstrap is through an example. In a student project, Erich Brandt studied tips at a restaurant. Here is a random sample of 30 observed tip percentages:

22.7	16.3	13.6	16.8	29.9	15.9	14.0	15.0	14.1	18.1	22.8	27.6	16.4	16.1	19.0
13.5	18.9	20.2	19.7	18.2	15.4	15.7	19.0	11.5	18.4	16.0	16.9	12.0	40.1	19.2

We would like to get a confidence interval for the population mean tip percentage at this restaurant. However, this is not a large sample and there is a problem with positive skewness, as shown in the normal probability plot of Figure 8.9. Notice that, of the plots

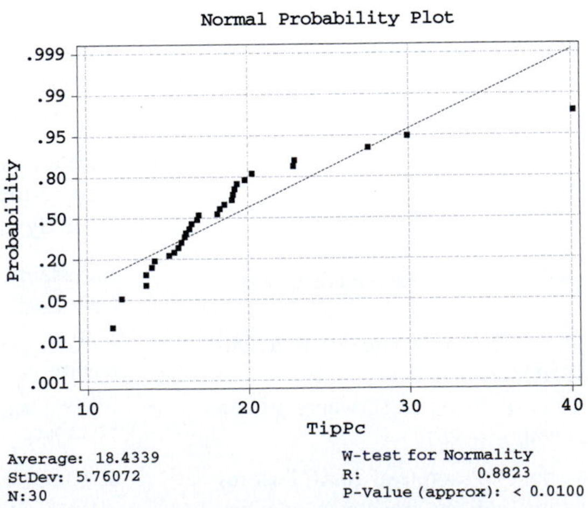

Figure 8.9 Normal probability plot from MINITAB of the tip percentages

discussed in Section 4.6, this is like Figure 4.34 and not Figure 4.31 in the sense that the data values here are plotted on the horizontal axis instead of the vertical axis. Therefore, positive skewness is shown by downward curvature instead of upward curvature.

Most of the tips are between 10% and 20%, but a few big tips cause enough skewness to invalidate the normality assumption. The sample mean is 18.43% and the sample standard deviation is 5.76%.

If population normality were plausible, then we could form a confidence interval using the mean and standard deviation calculated from the sample. From Section 8.3, the resulting 95% confidence interval for the population mean would be

$$\bar{x} \pm t_{.025,n-1}\frac{s}{\sqrt{n}} = 18.43 \pm 2.045\frac{5.76}{\sqrt{30}} = 18.43 \pm 2.15 = (16.3, 20.6)$$

How does the bootstrap approach differ from this? For the moment, we regard the 30 observations as constituting a population, and take a large number of random samples (1000 is a common choice), each of size 30, from this population. These are samples with replacement, so repetitions are allowed. For each of these samples we compute the mean (or the median or whatever statistic estimates the population parameter). Then we use the distribution of these 1000 means to get a confidence interval for the population mean. To help get a feeling for how this works, here is the first of the 1000 samples:

22.8	16.8	16.0	19.0	19.2	20.2	13.6	15.9	22.8	11.5	15.9	14.0	29.9	19.2	16.0
27.6	14.1	13.5	16.8	15.4	20.2	16.4	20.2	16.9	16.8	22.8	19.7	18.2	22.7	18.2

This sample has mean $\bar{x}_1^* = 18.41$, where the asterisk emphasizes that this is the mean of a bootstrap sample.

Of course, when we take a random sample with replacement, repetitions usually occur as they do here, and this implies that not all of the 30 observations will appear in each sample. After doing this 999 more times and computing the means $\bar{x}_2^*, \ldots, \bar{x}_{1000}^*$ for these 999 samples, we construct Figure 8.10, the histogram of the 1000 $\bar{x}^*$ values.

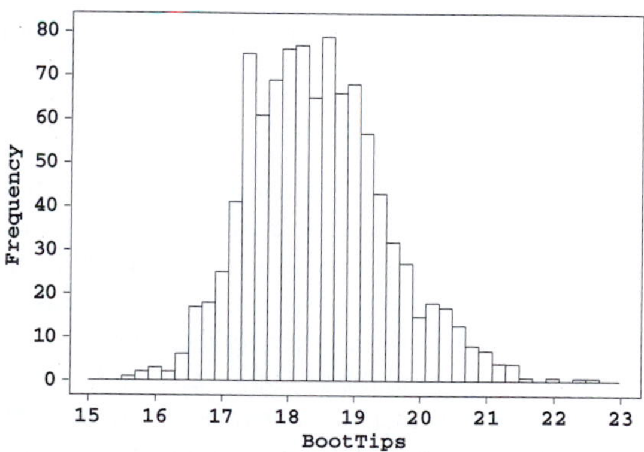

Figure 8.10 Histogram of the tip bootstrap distribution, from MINITAB

This describes approximately the sampling distribution of $\bar{X}$ for samples of 30 from the true tip population. That is, if we could draw the pdf for the true population distribution of $\bar{x}$ values, then it should look something like the histogram in Figure 8.10. Does the distribution appear to be normal? The histogram is not exactly symmetric, and the distribution looks skewed to the right. Figure 8.11 has the normal probability plot from MINITAB.

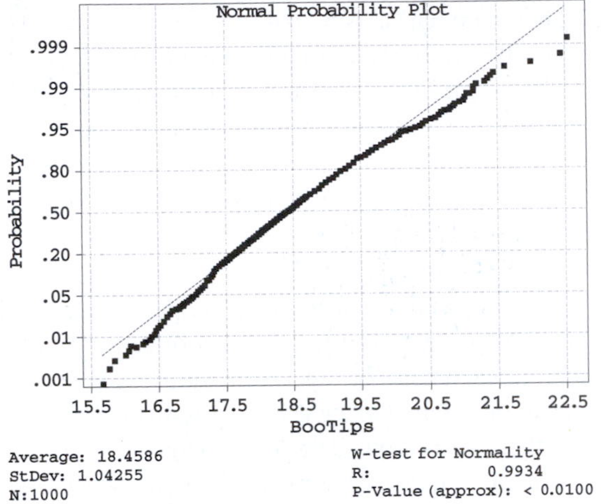

Figure 8.11 Normal probability plot of the tip bootstrap distribution

The pattern in this plot gives evidence of slight positive skewness (see Section 4.6). If this plot were straighter, then we could form a 95% confidence interval for the population mean in the following way. Let s_{boot} denote the sample standard deviation of the 1000 bootstrap means. That is, defining $\bar{x}^*$ to be the mean of the 1000 bootstrap means,

$$s_{\text{boot}}^2 = \frac{\sum (\bar{x}_i^* - \bar{x}^*)^2}{1000 - 1}$$

The value of s_{boot} turns out to be 1.043. The sample mean of the original 30 tip percentages is $\bar{x} = 18.43$, and $t_{.025,n-1} = 2.045$ is the t critical value with $n - 1 = 30 - 1 = 29$ degrees of freedom, such that 2.5% of the probability is beyond this point. Then the 95% confidence interval is

$$\bar{x} \pm t_{0.25,n-1}s_{\text{boot}} = 18.43 \pm 2.045(1.043) = 18.43 \pm 2.13 = (16.3, 20.6)$$

Notice that this is very similar to the previous interval based on the method of Section 8.3. The reason for using the t critical value instead of the z critical value is that the bootstrap interval should agree with the t interval from Section 8.3 if the bootstrap standard deviation s_{boot} agrees with the estimated standard error $s/\sqrt{n}$. There should be good agreement if the original data set looks normal. Even if the normality assumption is not satisfied, there should be good agreement if the sample size n is big enough.

In the case that the bootstrap distribution (as represented here by the histogram of Figure 8.10) is normal, the foregoing interval uses the middle 95% of the bootstrap distribution. Because the 1000 bootstrap means do not fit a normal curve, we need an alternative approach to finding a confidence interval. To allow for a nonnormal bootstrap distribution, we need to use something other than the standard deviation and the t table to determine the confidence limits. The *percentile interval* uses the 2.5th percentile and the 97.5th percentile of the bootstrap distribution for confidence limits of a 95% confidence interval. Computationally, one way to find the two percentiles is to sort the 1000 means and then use the 25th value from each end. These turn out to be 16.7 and 20.8, so the interval is (16.7, 20.8). Because the bootstrap distribution is positively skewed, the percentile interval is shifted slightly to the right compared to the interval based on a normal bootstrap distribution.

DEFINITION

The **bootstrap percentile interval** with a confidence level of $100(1 - \alpha)\%$ for a specified parameter is obtained by first generating B bootstrap samples, for each one calculating the value of some particular statistic that estimates the parameter, and sorting these values from smallest to largest. Then we compute $k = \alpha B/2$ and choose the kth value from each end of the sorted list. These two values form the confidence limits for the confidence interval. If k is not an integer, then interpolation can be used, but this is not crucial. As an example, if $\alpha = .05$ and $B = 1000$ then $k = \alpha B/2 = (.05)(1000)/2 = 25$.

When the percentile method is used to obtain a confidence interval, under some circumstances the actual confidence level may differ substantially from the nominal level (the level you think you are getting); in our example, the nominal level was 95%, and the actual level could be quite different from this. There are refined bootstrap intervals that often yield an improvement in this respect. In particular, the BCa (bias-corrected and accelerated) interval, implemented in the S-Plus and Stata software packages, is a method that corrects for bias. Here bias refers to the difference between the median (note that this differs from the usual definition of bias, which would use the mean) of the bootstrap distribution and the value of the estimate based on the original sample. For example, in estimating the mean for the tip data, the mean of the 30 tips in the original sample is 18.43 but the median of the 1000 bootstrap sample means is 18.33, so there is just a slight bias of $18.33 - 18.43 = -.10$.

The acceleration aspect of the BCa interval is an adjustment for dependence of the standard error of the estimator on the parameter that is being estimated. For example, suppose we are trying to estimate the mean in the case of exponential data. In this case the standard deviation is equal to the mean, and the standard error of $\overline{X}$ is $\sigma/\sqrt{n} = \mu/\sqrt{n}$, so the standard error of the estimator $\overline{X}$ depends strongly on the parameter μ that is being estimated. If the histogram in Figure 8.10 resembled the exponential pdf, we would expect the BCa method to make a substantial correction to the percentile interval. Recall that the percentile interval for the mean of the tip data is (16.7, 20.8). Compared to this, the BCa interval (16.9, 21.8) is shifted a little to the right.

Is the bootstrap guaranteed to work, or is it possible that the method can give grossly incorrect estimates? The key here is how closely the original sample represents the whole distribution of the random variable X. When the sample is small, then there is a possibility that important features of the distribution are not included in the data set. In terms of our 30 observations, the value 40.1% is highly influential. If we drew another sample of 30 observations independent of this sample, the luck of the draw might give no values above 25, and the sample would yield very different conclusions. The bootstrap is a useful method for making inferences from data, but it is dependent on a good sample. If this is all the data that we can get, we will never know how well our sample represents the distribution, and therefore how good our answer is. Of course, no statistical method will give good answers if the sample is not representative of the population.

Bootstrapping the Median

We do have a statistic that is less sensitive than the mean to the influence of individual observations. For the 30 tip percentages, the median is 16.85, substantially less than the mean of 18.43. The mean is pulled upward by the few large values, but these extremes have little effect on the median. However, it is more difficult to get a confidence interval for the median. There is a nice statistic to estimate the standard deviation of the mean $(S/\sqrt{n})$, but unfortunately there is nothing like this for the median.

Let's use the bootstrap method to get a confidence interval for the median. We can use the same 1000 samples of 30 as we did previously, but now we instead look at the 1000 medians. The first sample has mean $\bar{x}_1^* = 18.41$, whereas its median is $\tilde{x}_1^* = 17.55$. The histogram of this and the other 999 bootstrap medians $\tilde{x}_2^*, \ldots, \tilde{x}_{1000}^*$ is shown in Figure 8.12.

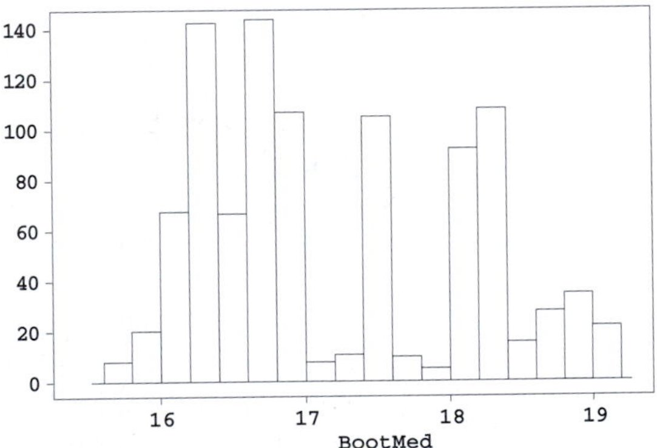

Figure 8.12 Histogram of the bootstrap medians from S-Plus

It should be apparent that the distribution of the 1000 bootstrap medians is not normal. As is often the case with the median, the bootstrap distribution takes on just a few values and there are many repeats. Instead of 1000 different values, as would be expected if we took 1000 samples from a true continuous distribution, here there are only 72 values, and some appear more than 50 times. These are apparent in the normal probability plot, shown in Figure 8.13. In contrast to what MINITAB does, the values here are plotted vertically, so the horizontal segments indicate repeats.

The mean of the 1000 bootstrap medians is 17.20 with standard deviation .917. Even though the procedure is inappropriate because of nonnormality, we can for comparative purposes use the median $\tilde{x} = 16.85$ of the original 30 observations together with the bootstrap standard deviation $\tilde{s}_{\text{boot}} = .917$ to get a confidence interval based on the normal distribution:

$$\tilde{x} \pm t_{.025,n-1}\tilde{s}_{\text{boot}} = 16.85 \pm 2.045(.917) = 16.85 \pm 1.88 = (15.0, 18.7)$$

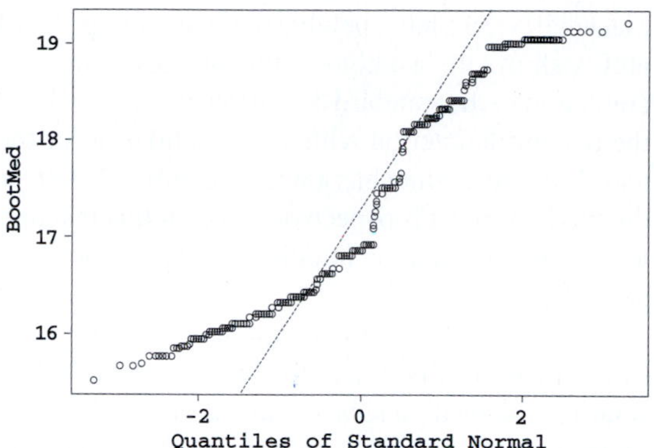

Figure 8.13 Normal probability plot of the bootstrap medians from S-Plus

Because the bootstrap distribution is so nonnormal, it is more appropriate to use the percentile interval in which the confidence limits for a 95% confidence interval are taken from the 2.5th and 97.5th percentiles of the bootstrap distribution. When the 1000 bootstrap medians are sorted, the 25th value is 15.9 and the 25th value from the top is 19.0, so the 95% confidence interval for the population median is (15.9, 19.0). In accord with the nonnormal bootstrap distribution, this interval differs from the interval that assumes normality.

We should be a bit uncomfortable with the results of bootstrapping the median. Given that the bootstrap distribution takes on just a few values but the true sampling distribution is continuous, we should worry a little about how well the bootstrap distribution approximates the true sampling distribution. On the other hand, the situation here is nowhere near as bad as it could be. Sometimes, especially when the sample size is smaller, the bootstrap distribution has far fewer values.

What can be done to see if the bootstrap results are valid for the median? We performed a simulation experiment with data from the exponential distribution, a distribution that is more strongly skewed than the tip percentages. We generated 100 samples, each of size 30, and then took 1000 bootstrap samples from each of them. In this way we obtained percentile intervals with a 95% confidence level for the mean and the median from each of the 100 samples. We used the exponential distribution with mean $\mu = 1/\lambda = 1$, for which the median $\tilde{\mu} = \ln(2) = .693$. In checking each of the 100 confidence intervals for the mean, we found that 93 of them contained the true mean. Similarly, we found that 93 of the confidence intervals for the median contained the true median. It is gratifying to see that, in spite of the strange distribution of the bootstrapped medians, the performance of the percentile confidence intervals is reasonably on target.

The bias-corrected and accelerated BCa refinement makes only a slight change to the percentile interval for the median. Here there is no bias because the median of the original sample of 30 is 16.85 and this is also the median of the 1000 bootstrap medians. The percentile interval is refined from (15.94, 18.99) to (15.87, 18.94).

The Mean Versus the Median

For the tip percentages, is it better to use the mean or the median? The median is much less affected by the extreme observations in this skewed data set. This suggests that the mean will vary a lot depending on whether a particular sample has outliers. Here, the variability shows up in a higher standard deviation of 1.043 for the 1000 bootstrap means as compared to the standard deviation of .917 for the 1000 bootstrap medians. Furthermore, the percentile interval with 95% confidence for the mean has width 4.2 whereas the interval for the median has a width of only 3.1. In terms of precision, we are better off with the median. For a prospective server at this restaurant, it might also be more meaningful to give the median, the middle tip value in the sense that roughly half are above and half are below.

Of course, it is not always necessary to choose one statistic over the other. Sometimes a case can be made for presenting both the mean and the median. In the case of salaries, the median salary may be more relevant to an employee, but the mean may be more useful to the employer because the mean is proportional to the total payroll.

Exercises Section 8.5 (49–57)

49. In a survey, students gave their study time per week (hr), and here are the 22 values:

15.0	10.0	10.0	15.0	25.0	7.0	3.0	8.0	10.0
10.0	11.0	7.0	5.0	15.0	7.5	7.5	12.0	7.0
10.5	6.0	10.0	7.5					

We would like to get a 95% confidence interval for the population mean.

a. Compute the t-based confidence interval of Section 8.3.

b. Display a normal plot. Is it apparent that the data set is not normal, so the t-based interval is of questionable validity?

c. Generate a bootstrap sample of 1000 means.

d. Use the standard deviation for part (c) to get a 95% confidence interval for the population mean.

e. Investigate the distribution of the bootstrap means to see if part (d) is valid.

f. Use part (c) to form the 95% confidence interval using the percentile method.

g. Say which interval should be used and explain why.

50. We would like to obtain a 95% confidence interval for the population median of the study hours data in Exercise 49.

a. Obtain a bootstrap sample of 1000 medians.

b. Use the standard deviation for part (a) to get a 95% confidence interval for the population median.

c. Investigate the distribution of the bootstrap medians and discuss the validity of part (b). Does the distribution take on just a few values?

d. Use part (a) to form a 95% confidence interval for the median using the percentile method.

e. For the study hours data, state your preference between the median and the mean and explain your reasoning.

51. Here are 68 weight gains in pounds for pregnant women from conception to delivery ("Classifying Data Displays with an Assessment of Displays Found in Popular Software," *Teaching Statist.*, Autumn 2002: 96–101).

25	14	20	38	21	22	36	38	35	37
35	24	31	28	25	32	23	30	39	26
38	20	21	11	35	42	31	25	59	23
43	38	21	76	22	26	10	19	25	25
15	31	34	36	35	33	24	44	35	43
7	32	25	27	31	14	25	16	25	47
35	−14	65	40	35	45	27	24		

We would like to get a 95% confidence interval for the population mean.

a. Compute the t-based confidence interval of Section 8.3.

b. Check for normality to see if part (a) is valid. Is the sample large enough that the interval might be valid anyway?

c. Generate a bootstrap sample of 1000 means.

d. Use the standard deviation for part (c) to get a 95% confidence interval for the population mean.

e. Investigate the distribution of the bootstrap means to see if part (d) is valid.

f. Use part (c) to form the 95% confidence interval using the percentile method.

g. Compare the intervals. If they are all close, then the bootstrap supports the use of part (a).

52. We would like to obtain a 95% confidence interval for the population median weight gain using the data in Exercise 51.

a. Obtain a bootstrap sample of 1000 medians.

b. Use the standard deviation for part (a) to get a 95% confidence interval for the population median.

c. Investigate the distribution of the bootstrap medians and discuss the validity of part (b). Does the distribution take on just a few values?

d. Use part (a) to form a 95% confidence interval for the median using the percentile method.

e. For the weight gain data, state your preference between the median and the mean and explain your reasoning.

53. Nine Australian soldiers were subjected to extreme conditions, which involved a 100-minute walk with a 25-lb pack when the temperature was 40°C (104°F). One of them overheated (above 39°C) and was removed from the study. Here are the rectal Celsius temperatures of the other eight at the end of the walk ("Neural Network Training on Human Body Core Temperature Data," Combatant Protection and Nutrition Branch, Aeronautical and Maritime Research Laboratory of Australia, DSTO TN-0241, 1999):

38.4 38.7 39.0 38.5 38.5 39.0 38.5 38.6

We would like to get a 95% confidence interval for the population mean.

a. Compute the t-based confidence interval of Section 8.3.

b. Check for the validity of part (a).

c. Generate a bootstrap sample of 1000 means.

d. Use the standard deviation for part (c) to get a 95% confidence interval for the population mean.

e. Investigate the distribution of the bootstrap means to see if part (d) is valid.

f. Use part (c) to form the 95% confidence interval using the percentile method.

g. Compare the intervals and explain your preference.

h. Based on your knowledge of normal body temperature, would you say that body temperature can be influenced by environment?

54. We would like to obtain a 95% confidence interval for the population median temperature using the data in Exercise 53.

a. Obtain a bootstrap sample of 1000 medians.

b. Use the standard deviation for part (a) to get a 95% confidence interval for the population median.

c. Investigate the distribution of the bootstrap medians and discuss the validity of part (b). Does the distribution take on just a few values?

d. Use part (a) to form a 95% confidence interval for the median using the percentile method.

e. Compare all the intervals for the mean and median. Are they fairly similar? How do you explain that?

55. If you go to a major league baseball game, how long do you expect the game to be? From the 2429 games played in 2001, here is a random sample of 25 times in minutes:

352 150 164 167 225 159 142 182 229 163
188 197 189 235 161 195 177 166 195 160
154 130 189 188 225

This is one of those rare instances in which we can do a confidence interval and compare with the true population mean. The mean of all 2429 lengths is 178.29 (almost three hours).

a. Compute the t-based confidence interval of Section 8.3.

b. Use a normal plot to see if part (a) is valid.

c. Generate a bootstrap sample of 1000 means.

d. Use the standard deviation for part (c) to get a 95% confidence interval for the population mean.

e. Investigate the distribution of the bootstrap means to see if part (d) is valid.

f. Use part (c) to form the 95% confidence interval using the percentile method.

g. Say which interval should be used and explain why. Does your interval include the true value, 178.29?

56. The median might be a more meaningful statistic for the length-of-game data in Exercise 55. The median of all 2429 lengths is 175 minutes.

a. Obtain a bootstrap sample of 1000 medians.

b. Use the standard deviation for part (a) to get a 95% confidence interval for the population median.

c. Investigate the distribution of the bootstrap medians and discuss the validity of part (b). Does the distribution take on just a few values?

d. Use part (a) to form a 95% confidence interval for the median using the percentile method. Compare your answer with the population median, 175.

e. Comparing the percentile intervals for the mean and the median, is there much difference in their widths? If not, and you are forced to choose between them for the length-of-game data, which do you choose and why?

57. We would like to obtain a 95% confidence interval for the population standard deviation study time using the data in Exercise 49.

a. Obtain a bootstrap sample of 1000 standard deviations and use it to form a 95% confidence interval for the population standard deviation using the percentile method.

b. Recalling that it requires normal data, use the method of Section 8.4 to obtain a 95% confidence interval for the population standard deviation. Discuss normality for the study hours data. How does this interval compare with the percentile interval?

Supplementary Exercises (58–79)

58. A triathlon consisting of swimming, cycling, and running is one of the more strenuous amateur sporting events. The article "Cardiovascular and Thermal Response of Triathlon Performance" (*Med. Sci. Sports Exercise*, 1988: 385–389) reports on a research study involving nine male triathletes. Maximum heart rate (beats/min) was recorded during performance of each of the three events. For swimming, the sample mean and sample standard deviation were 188.0 and 7.2, respectively. Assuming that the heart rate distribution is (approximately) normal, construct a 98% CI for true mean heart rate of triathletes while swimming.

59. The reaction time (RT) to a stimulus is the interval of time commencing with stimulus presentation and ending with the first discernible movement of a certain type. The article "Relationship of Reaction Time and Movement Time in a Gross Motor Skill" (*Percept. Motor Skills*, 1973: 453–454) reports that the sample average RT for 16 experienced swimmers to a pistol start was .214 sec and the sample standard deviation was .036 sec.

a. Making any necessary assumptions, derive a 90% CI for true average RT for all experienced swimmers.

b. Calculate a 90% upper confidence bound for the standard deviation of the reaction time distribution.

c. Predict RT for another such individual in a way that conveys information about precision and reliability.

60. For each of 18 preserved cores from oil-wet carbonate reservoirs, the amount of residual gas saturation after a solvent injection was measured at water flood-out. Observations, in percentage of pore volume, were

23.5	31.5	34.0	46.7	45.6	32.5
41.4	37.2	42.5	46.9	51.5	36.4
44.5	35.7	33.5	39.3	22.0	51.2

(See "Relative Permeability Studies of Gas-Water Flow Following Solvent Injection in Carbonate Rocks," *Soc. Petroleum Engineers J.*, 1976: 23–30.)

a. Construct a boxplot of this data, and comment on any interesting features.

b. Is it plausible that the sample was selected from a normal population distribution?

c. Calculate a 98% CI for the true average amount of residual gas saturation.

61. A manufacturer of college textbooks is interested in estimating the strength of the bindings produced by a particular binding machine. Strength can be measured by recording the force required to pull the pages from the binding. If this force is measured in pounds, how many books should be tested to estimate the average force required to break the binding to within .1 lb with 95% confidence? Assume that σ is known to be .8.

62. The financial manager of a large department store chain selected a random sample of 200 of its credit card customers and found that 136 had incurred an interest charge during the previous year because of an unpaid balance.

a. Compute a 90% CI for the true proportion of credit card customers who incurred an interest charge during the previous year.

b. If the desired width of the 90% interval is .05, what sample size is necessary to ensure this?

c. Does the upper limit of the interval in part (a) specify a 90% upper confidence bound for the proportion being estimated? Explain.

63. There were 12 first-round heats in the men's 100-meter race at the 1996 Atlanta Summer Olympics. Here are the reaction times in seconds (time to first movement) of the top four finishers of each heat. The first 12 are the 12 winners, then the second-place finishers, and so on.

1st	.187	.152	.137	.175	.172	.165
	.184	.185	.147	.189	.172	.156

2nd	.168	.140	.214	.163	.202	.173
	.175	.154	.160	.169	.148	.144
3rd	.159	.145	.187	.222	.190	.158
	.202	.162	.156	.141	.167	.155
4th	.156	.164	.160	.145	.163	.170
	.182	.187	.148	.183	.162	.186

Because reaction time has little if any relationship to the order of finish, it is reasonable to view the times as coming from a single population.

a. Estimate the population mean in a way that conveys information about precision and reliability. (*Note*: $\Sigma x_i = 8.08100$, $\Sigma x_i^2 = 1.37813$.) Do the runners seem to react faster than the swimmers in Exercise 59?

b. Calculate a 95% confidence interval for the population proportion of reaction times that are below .15. Reaction times below .10 are regarded as false starts, meaning that the runner anticipates the starter's gun, because such times are considered physically impossible. Linford Christie, who had a reaction time of .160 in placing second in his first-round heat, had two such false starts in the finals and was disqualified.

64. Aphid infestation of fruit trees can be controlled either by spraying with pesticide or by inundation with ladybugs. In a particular area, four different groves of fruit trees are selected for experimentation. The first three groves are sprayed with pesticides 1, 2, and 3, respectively, and the fourth is treated with ladybugs, with the following results on yield:

Treatment	n_i (number of trees)	$\bar{x}_i$ (bushels/tree)	s_i
1	100	10.5	1.5
2	90	10.0	1.3
3	100	10.1	1.8
4	120	10.7	1.6

Let μ_i = the true average yield (bushels/tree) after receiving the ith treatment. Then

$$\theta = \frac{1}{3}(\mu_1 + \mu_2 + \mu_3) - \mu_4$$

measures the difference in true average yields between treatment with pesticides and treatment with ladybugs. When n_1, n_2, n_3, and n_4 are all large, the estimator $\hat{\theta}$ obtained by replacing each μ_i by $\overline{X}_i$ is approximately normal. Use this to derive a large-sample $100(1 - \alpha)\%$ CI for θ, and compute the 95% interval for the given data.

65. It is important that face masks used by firefighters be able to withstand high temperatures because firefighters commonly work in temperatures of 200–500°F. In a test of one type of mask, 11 of 55 masks had lenses pop out at 250°. Construct a 90% CI for the true proportion of masks of this type whose lenses would pop out at 250°.

66. A journal article reports that a sample of size 5 was used as a basis for calculating a 95% CI for the true average natural frequency (Hz) of delaminated beams of a certain type. The resulting interval was (229.764, 233.504). You decide that a confidence level of 99% is more appropriate than the 95% level used. What are the limits of the 99% interval? (*Hint*: Use the center of the interval and its width to determine $\bar{x}$ and s.)

67. Chronic exposure to asbestos fiber is a well-known health hazard. The article "The Acute Effects of Chrysotile Asbestos Exposure on Lung Function" (*Envir. Res.*, 1978: 360–372) reports results of a study based on a sample of construction workers who had been exposed to asbestos over a prolonged period. Among the data given in the article were the following (ordered) values of pulmonary compliance ($cm^3/cm\ H_2O$) for each of 16 subjects eight months after the exposure period (pulmonary compliance is a measure of lung elasticity, or how effectively the lungs are able to inhale and exhale):

167.9	180.8	184.8	189.8	194.8	200.2
201.9	206.9	207.2	208.4	226.3	227.7
228.5	232.4	239.8	258.6		

a. Is it plausible that the population distribution is normal?

b. Compute a 95% CI for the true average pulmonary compliance after such exposure.

c. Calculate an interval that, with a confidence level of 95%, includes at least 95% of the pulmonary compliance values in the population distribution.

68. In Example 7.9, we introduced the concept of a censored experiment in which n components are put on test and the experiment terminates as soon as r of the components have failed. Suppose component lifetimes are independent, each having an exponential distribution with parameter λ. Let Y_1 denote the time at which the first failure occurs, Y_2 the time at which the second failure occurs, and so on, so that

$T_r = Y_1 + \cdots + Y_r + (n - r)Y_r$ is the total accumulated lifetime at termination. Then it can be shown that $2\lambda T_r$ has a chi-squared distribution with $2r$ df. Use this fact to develop a $100(1 - \alpha)\%$ CI formula for true average lifetime $1/\lambda$. Compute a 95% CI from the data in Example 7.9.

69. Exercise 63 from Chapter 7 introduced "regression through the origin" to relate a dependent variable y to an independent variable x. The assumption there was that for any fixed x value, the dependent variable is a random variable Y with mean value βx and variance σ^2 (so that Y has mean value zero when $x = 0$). The data consists of n independent (x_i, Y_i) pairs, where each Y_i is normally distributed with mean βx_i and variance σ^2. The likelihood is then a product of normal pdf's with different mean values but the same variance.

a. Show that the mle of β is $\hat{\beta} = \Sigma x_i Y_i / \Sigma x_i^2$.

b. Verify that the mle of (a) is unbiased.

c. Obtain an expression for $V(\hat{\beta})$ and then for $\sigma_{\hat{\beta}}$.

d. For purposes of obtaining a precise estimate of β, is it better to have the x_i's all close to 0 (the origin) or spread out quite far above 0? Explain your reasoning.

e. The natural prediction of Y_i is $\hat{\beta}x_i$. Let $S^2 = \Sigma(Y_i - \hat{\beta}x_i)^2/(n - 1)$ which is analogous to our earlier sample variance $S^2 = \Sigma(X_i - \overline{X})^2/(n - 1)$ for a univariate sample $X_1, \ldots, X_n$ (in which case $\overline{X}$ is a natural prediction for each X_i). Then it can be shown that $T = (\hat{\beta} - \beta)/(S/\sqrt{\Sigma x_i^2})$ has a t distribution based on $n - 1$ df. Use this to obtain a CI formula for estimating β, and calculate a 95% CI using the data from the cited exercise.

70. Let $X_1, X_2, \ldots, X_n$ be a random sample from a uniform distribution on the interval $[0, \theta]$, so that

$$f(x) = \begin{cases} \dfrac{1}{\theta} & 0 \le x \le \theta \\ 0 & \text{otherwise} \end{cases}$$

Then if $Y = \max(X_i)$, by the first proposition in Section 5.5, $U = Y/\theta$ has density function

$$f_U(u) = \begin{cases} nu^{n-1} & 0 \le u \le 1 \\ 0 & \text{otherwise} \end{cases}$$

a. Use $f_U(u)$ to verify that

$$P\left[(\alpha/2)^{1/n} < \frac{Y}{\theta} \le (1 - \alpha/2)^{1/n}\right] = 1 - \alpha$$

and use this to derive a $100(1 - \alpha)\%$ CI for θ.

b. Verify that $P(\alpha^{1/n} \le Y/\theta \le 1) = 1 - \alpha$, and derive a $100(1 - \alpha)\%$ CI for θ based on this probability statement.

c. Which of the two intervals derived previously is shorter? If my waiting time for a morning bus is uniformly distributed and observed waiting times are $x_1 = 4.2$, $x_2 = 3.5$, $x_3 = 1.7$, $x_4 = 1.2$, and $x_5 = 2.4$, derive a 95% CI for θ by using the shorter of the two intervals.

71. Let $0 \le \gamma \le \alpha$. Then a $100(1 - \alpha)\%$ CI for μ when n is large is

$$\left(\bar{x} - z_\gamma \cdot \frac{s}{\sqrt{n}},\ \bar{x} + z_{\alpha-\gamma} \cdot \frac{s}{\sqrt{n}}\right)$$

The choice $\gamma = \alpha/2$ yields the usual interval derived in Section 8.2; if $\gamma \ne \alpha/2$, this confidence interval is not symmetric about $\bar{x}$. The width of this interval is $w = s(z_\gamma + z_{\alpha-\gamma})/\sqrt{n}$. Show that w is minimized for the choice $\gamma = \alpha/2$, so that the symmetric interval is the shortest. [*Hints*: (a) By definition of z_α, $\Phi(z_\alpha) = 1 - \alpha$, so that $z_\alpha = \Phi^{-1}(1 - \alpha)$; (b) the relationship between the derivative of a function $y = f(x)$ and the inverse function $x = f^{-1}(y)$ is $(d/dy)f^{-1}(y) = 1/f'(x)$.]

72. Suppose $x_1, x_2, \ldots, x_n$ are observed values resulting from a random sample from a symmetric but possibly heavy-tailed distribution. Let $\tilde{x}$ and f_s denote the sample median and fourth spread, respectively. Chapter 11 of *Understanding Robust and Exploratory Data Analysis* (see the bibliography in Chapter 7) suggests the following robust 95% CI for the population mean (point of symmetry):

$$\tilde{x} \pm \left(\frac{\text{conservative } t \text{ critical value}}{1.075}\right) \cdot \frac{f_s}{\sqrt{n}}$$

The value of the quantity in parentheses is 2.10 for $n = 10$, 1.94 for $n = 20$, and 1.91 for $n = 30$. Compute this CI for the restaurant tip data of Section 8.5, and compare to the t CI appropriate for a normal population distribution.

73. **a.** Use the results of Example 8.5 to obtain a 95% lower confidence bound for the parameter λ of an exponential distribution, and calculate the bound based on the data given in the example.

b. If lifetime X has an exponential distribution, the probability that lifetime exceeds t is $P(X > t) = e^{-\lambda t}$. Use the result of part (a) to obtain a 95% lower confidence bound for the probability that breakdown time exceeds 100 min.

74. Let θ_1 and θ_2 denote the mean weights for animals of two different species. An investigator wishes to estimate the ratio θ_1/θ_2. Unfortunately the species are extremely rare, so the estimate will be based on finding a single animal of each species. Let X_i denote the weight of the species i animal ($i = 1, 2$), assumed to be normally distributed with mean θ_i and standard deviation 1.

a. What is the distribution of the variable $h(X_1, X_2; \theta_1, \theta_2) = (\theta_2 X_1 - \theta_1 X_2)/\sqrt{\theta_1^2 + \theta_2^2}$? Show that this variable depends on θ_1 and θ_2 only through θ_1/θ_2 (divide numerator and denominator by θ_2).

b. Consider Expression (8.6) from the first section of this chapter with $a = -1.96$ and $b = 1.96$. Now replace $<$ by $=$ and solve for θ_1/θ_2. Then show that a confidence interval results if $x_1^2 + x_2^2 \geq (1.96)^2$, whereas if this inequality is not satisfied, the resulting *confidence set* is the complement of an interval.

75. The one-sample CI for a normal mean and PI for a single observation from a normal distribution were both based on the *central* t distribution. A CI for a particular percentile (e.g., the 1st percentile or the 95th percentile) of a normal population distribution is based on the *noncentral* t distribution. A particular distribution of this type is specified by both df and the value of the noncentrality parameter δ ($\delta = 0$ gives the central t distribution). The key result is that the variable

$$T = \frac{\dfrac{\overline{X} - \mu}{\sigma/\sqrt{n}} - (z\text{ percentile})\sqrt{n}}{S/\sigma}$$

has a noncentral t distribution with df $= n - 1$ and $\delta = -(z\text{ percentile})\sqrt{n}$.

Let $t_{.025,\nu,\delta}$ and $t_{.975,\nu,\delta}$ denote the critical values that capture upper-tail area .025 and lower-tail area .025, respectively, under the noncentral t curve with ν df and noncentrality parameter δ (when $\delta = 0$, $t_{.975} = -t_{.025}$, since central t distributions are symmetric about 0).

a. Use the given information to obtain a formula for a 95% confidence interval for the $(100p)$th percentile of a normal population distribution.

b. For $\delta = 6.58$ and df $= 15$, $t_{.975}$ and $t_{.025}$ are (from MINITAB) 4.1690 and 10.9684, respectively. Use this information to obtain a 95% CI for the 5th percentile of the beer alcohol distribution considered in Example 8.11.

76. The one-sample t CI for μ is also a confidence interval for the population median $\tilde{\mu}$ when the population distribution is normal. We now develop a CI for $\tilde{\mu}$ that is valid whatever the shape of the population distribution as long as it is continuous. Let $X_1, \ldots, X_n$ be a random sample from the distribution and $Y_1, \ldots, Y_n$ denote the corresponding order statistics (smallest observation, second smallest, and so on).

a. What is $P(X_1 < \tilde{\mu})$? What is $P(\{X_1 < \tilde{\mu}\} \cap \{X_2 < \tilde{\mu}\})$?

b. What is $P(Y_n < \tilde{\mu})$? What is $P(Y_1 > \tilde{\mu})$? (*Hint*: What condition involving all of the X_i's is equivalent to the largest being smaller than the population median?)

c. What is $P(Y_1 < \tilde{\mu} < Y_n)$? What does this imply about the confidence level associated with the CI (y_1, y_n) for $\tilde{\mu}$?

d. An experiment carried out to study the time (min) necessary for an anesthetic to produce the desired result yielded the following data: 31.2, 36.0, 31.5, 28.7, 37.2, 35.4, 33.3, 39.3, 42.0, 29.9. Determine the confidence interval of (c) and the associated confidence level. Also calculate the one-sample t CI using the same level and compare the two intervals.

77. Consider the situation described in the previous exercise.

a. What is $P(\{X_1 < \tilde{\mu}\} \cap \{X_2 > \tilde{\mu}\} \cap \cdots \cap \{X_n > \tilde{\mu}\})$, that is, the probability that only the first observation is smaller than the median?

b. What is the probability that exactly one of the n observations is smaller than the median?

c. What is $P(\tilde{\mu} < Y_2)$? (*Hint*: The event in parentheses occurs if all n of the observations exceed the median. How else can it occur? What does this imply about the confidence level associated with the CI (y_2, y_{n-1}) for $\tilde{\mu}$? Determine the confidence level and CI for the data given in the previous exercise.)

78. The previous two exercises considered a CI for a population median $\tilde{\mu}$ based on the n order statistics from a random sample. Let's now consider a prediction interval for the next observation X_{n+1}.

a. What is $P(X_{n+1} < X_1)$? What is $P(\{X_{n+1} < X_1\} \cap \{X_{n+1} < X_2\})$?

b. What is $P(X_{n+1} < Y_1)$? What is $P(X_{n+1} > Y_n)$?

c. What is $P(Y_1 < X_{n+1} < Y_n)$? What does this say about the prediction level for the PI (y_1, y_n)? Determine the prediction level and interval for the data given in the previous exercise.

79. Consider 95% CI's for two different parameters θ_1 and θ_2, and let A_i $(i = 1, 2)$ denote the event that the value of θ_i is included in the random interval that results in the CI. Thus $P(A_i) = .95$.

a. Suppose that the data on which the CI for θ_1 is based is independent of the data used to obtain the CI for θ_2 (e.g., we might have $\theta_1 = \mu$, the population mean height for American females, and $\theta_2 = p$, the proportion of all Kodak digital cameras that don't need warranty service). What can be said about the *simultaneous* (i.e., *joint*) confidence level for the two intervals? That is, how confident can we be that the first interval contains the value of θ_1 and that the second contains the value of θ_2? [*Hint*: Consider $P(A_1 \cap A_2)$.]

b. Now suppose the data for the first CI is not independent of that for the second one. What now can be said about the simultaneous confidence level for both intervals? (*Hint*: Consider $P(A_1' \cup A_2')$, the probability that at least one interval fails to include the value of what it is estimating. Now use the fact that $P(A_1' \cup A_2') \le P(A_1') + P(A_2')$ [why?] to show that the probability that *both* random intervals include what they are estimating is at least .90. The generalization of the bound on $P(A_1' \cup A_2')$ to the probability of a k-fold union is one version of the *Bonferroni* inequality.)

c. What can be said about the simultaneous confidence level if the confidence level for each interval separately is $100(1 - \alpha)\%$? What can be said about the simultaneous confidence level if a $100(1 - \alpha)\%$ CI is computed separately for each of k parameters $\theta_1, \ldots, \theta_k$?

Bibliography

DeGroot, Morris, and Mark Schervish, *Probability and Statistics* (3rd ed.), Addison-Wesley, Reading, MA, 2002. A very good exposition of the general principles of statistical inference.

Efron, Bradley, and Robert Tibshirani, *An Introduction to the Bootstrap*, Chapman and Hall, New York, 1993. The bible of the bootstrap.

Hahn, Gerald, and William Meeker, *Statistical Intervals*, Wiley, New York, 1991. Everything you ever wanted to know about statistical intervals (confidence, prediction, tolerance, and others).

Larsen, Richard, and Morris Marx, *Introduction to Mathematical Statistics* (2nd ed.), Prentice Hall, Englewood Cliffs, NJ, 1986. Similar to DeGroot's presentation, but slightly less mathematical.

CHAPTER NINE

Tests of Hypotheses Based on a Single Sample

Introduction

A parameter can be estimated from sample data either by a single number (a point estimate) or an entire interval of plausible values (a confidence interval). Frequently, however, the objective of an investigation is not to estimate a parameter but to decide which of two contradictory claims about the parameter is correct. Methods for accomplishing this comprise the part of statistical inference called *hypothesis testing*. In this chapter, we first discuss some of the basic concepts and terminology in hypothesis testing and then develop decision procedures for the most frequently encountered testing problems based on a sample from a single population.

9.1 Hypotheses and Test Procedures

A **statistical hypothesis**, or just *hypothesis*, is a claim or assertion either about the value of a single parameter (population characteristic or characteristic of a probability distribution), about the values of several parameters, or about the form of an entire probability distribution. One example of a hypothesis is the claim $\mu = \$311$, where μ is the true average one-term textbook expenditure for students at a university. Another example is the statement $p < .50$, where p is the proportion of adults who approve of the job that the President is doing. If μ_1 and μ_2 denote the true average decreases in systolic blood pressure for two different drugs, one hypothesis is the assertion that $\mu_1 - \mu_2 = 0$, and another is the statement $\mu_1 - \mu_2 > 5$. Yet another example of a hypothesis is the assertion that the stopping distance for a car under particular conditions has a normal distribution. Hypotheses of this latter sort will be considered in Chapter 13. In this and the next several chapters, we concentrate on hypotheses about parameters.

In any hypothesis-testing problem, there are two contradictory hypotheses under consideration. One hypothesis might be the claim $\mu = \$311$ and the other $\mu \neq \$311$, or the two contradictory statements might be $p \geq .50$ and $p < .50$. The objective is to decide, based on sample information, which of the two hypotheses is correct. There is a familiar analogy to this in a criminal trial. One claim is the assertion that the accused individual is innocent. In the U.S. judicial system, this is the claim that is initially believed to be true. Only in the face of strong evidence to the contrary should the jury reject this claim in favor of the alternative assertion that the accused is guilty. In this sense, the claim of innocence is the favored or protected hypothesis, and the burden of proof is placed on those who believe in the alternative claim.

Similarly, in testing statistical hypotheses, the problem will be formulated so that one of the claims is initially favored. This initially favored claim will not be rejected in favor of the alternative claim unless sample evidence contradicts it and provides strong support for the alternative assertion.

DEFINITION

The **null hypothesis**, denoted by H_0, is the claim that is initially assumed to be true (the "prior belief" claim). The **alternative hypothesis**, denoted by H_a, is the assertion that is contradictory to H_0.

The null hypothesis will be rejected in favor of the alternative hypothesis only if sample evidence suggests that H_0 is false. If the sample does not strongly contradict H_0, we will continue to believe in the probability of the null hypothesis. The two possible conclusions from a hypothesis-testing analysis are then *reject* H_0 or *fail to reject* H_0.

A **test of hypotheses** is a method for using sample data to decide whether the null hypothesis should be rejected. Thus we might test H_0: $\mu = .75$ against the alternative H_a: $\mu \neq .75$. Only if sample data strongly suggests that μ is something other than .75 should the null hypothesis be rejected. In the absence of such evidence, H_0 should not be rejected, since it is still quite plausible.

Sometimes an investigator does not want to accept a particular assertion unless and until data can provide strong support for the assertion. As an example, suppose a company is considering putting a new additive in the dried fruit that it produces. The true average shelf life with the current additive is known to be 200 days. With μ denoting the true average life for the new additive, the company would not want to make a change unless evidence strongly suggested that μ exceeds 200. An appropriate problem formulation would involve testing H_0: $\mu = 200$ against H_a: $\mu > 200$. The conclusion that a change is justified is identified with H_a, and it would take conclusive evidence to justify rejecting H_0 and switching to the new additive.

Scientific research often involves trying to decide whether a current theory should be replaced by a more plausible and satisfactory explanation of the phenomenon under investigation. A conservative approach is to identify the current theory with H_0 and the researcher's alternative explanation with H_a. Rejection of the current theory will then occur only when evidence is much more consistent with the new theory. In many situations, H_a is referred to as the "research hypothesis," since it is the claim that the researcher would really like to validate. The word *null* means "of no value, effect, or consequence," which suggests that H_0 should be identified with the hypothesis of no change (from current opinion), no difference, no improvement, and so on. Suppose, for example, that 10% of all computer circuit boards produced by a certain manufacturer during a recent period were defective. An engineer has suggested a change in the production process in the belief that it will result in a reduced defective rate. Let p denote the true proportion of defective boards resulting from the changed process. Then the research hypothesis, on which the burden of proof is placed, is the assertion that $p < .10$. Thus the alternative hypothesis is H_a: $p < .10$.

In our treatment of hypothesis testing, H_0 will always be stated as an equality claim. If θ denotes the parameter of interest, the null hypothesis will have the form H_0: $\theta = \theta_0$, where θ_0 is a specified number called the *null value* of the parameter (value claimed for θ by the null hypothesis). As an example, consider the circuit board situation just discussed. The suggested alternative hypothesis was H_a: $p < .10$, the claim that the defective rate is reduced by the process modification. A natural choice of H_0 in this situation is the claim that $p \geq .10$, according to which the new process is either no better *or* worse than the one currently used. We will instead consider H_0: $p = .10$ versus H_a: $p < .10$. The rationale for using this simplified null hypothesis is that any reasonable decision procedure for deciding between H_0: $p = .10$ and H_a: $p < .10$ will also be reasonable for deciding between the claim that $p \geq .10$ and H_a. The use of a simplified H_0 is preferred because it has certain technical benefits, which will be apparent shortly.

The alternative to the null hypothesis H_0: $\theta = \theta_0$ will look like one of the following three assertions: (1) H_a: $\theta > \theta_0$ (in which case the implicit null hypothesis is $\theta \leq \theta_0$), (2) H_a: $\theta < \theta_0$ (so the implicit null hypothesis states that $\theta \geq \theta_0$), or (3) H_a: $\theta \neq \theta_0$. For example, let σ denote the standard deviation of the distribution of outside diameters (inches) for an engine piston. If the decision was made to use the piston unless sample evidence conclusively demonstrated that $\sigma > .0001$ inch, the appropriate hypotheses would be H_0: $\sigma = .0001$ versus H_a: $\sigma > .0001$. The number θ_0 that appears in both H_0 and H_a (separates the alternative from the null) is called the **null value**.

Test Procedures

A test procedure is a rule, based on sample data, for deciding whether to reject H_0. A test of H_0: $p = .10$ versus H_a: $p < .10$ in the circuit board problem might be based on examining a random sample of $n = 200$ boards. Let X denote the number of defective boards in the sample, a binomial random variable; x represents the observed value of X. If H_0 is true, $E(X) = np = 200(.10) = 20$, whereas we can expect fewer than 20 defective boards if H_a is true. A value x just a bit below 20 does not strongly contradict H_0, so it is reasonable to reject H_0 only if x is substantially less than 20. One such test procedure is to reject H_0 if $x \leq 15$ and not reject H_0 otherwise. This procedure has two constituents: (1) a *test statistic* or function of the sample data used to make a decision and (2) a *rejection region* consisting of those x values for which H_0 will be rejected in favor of H_a. For the rule just suggested, the rejection region consists of $x = 0, 1, 2, \ldots, 15$. H_0 will not be rejected if $x = 16, 17, \ldots, 199$, or 200.

A test procedure is specified by the following:

1. A **test statistic**, a function of the sample data on which the decision (reject H_0 or do not reject H_0) is to be based
2. A **rejection region**, the set of all test statistic values for which H_0 will be rejected

The null hypothesis will then be rejected if and only if the observed or computed test statistic value falls in the rejection region.

As another example, suppose a cigarette manufacturer claims that the average nicotine content μ of brand B cigarettes is (at most) 1.5 mg. It would be unwise to reject the manufacturer's claim without strong contradictory evidence, so an appropriate problem formulation is to test H_0: $\mu = 1.5$ versus H_a: $\mu > 1.5$. Consider a decision rule based on analyzing a random sample of 32 cigarettes. Let $\overline{X}$ denote the sample average nicotine content. If H_0 is true, $E(\overline{X}) = \mu = 1.5$, whereas if H_0 is false, we expect $\overline{X}$ to exceed 1.5. Strong evidence against H_0 is provided by a value $\overline{x}$ that considerably exceeds 1.5. Thus we might use $\overline{X}$ as a test statistic along with the rejection region $\overline{x} \geq 1.6$.

In both the circuit board and nicotine examples, the choice of test statistic and form of the rejection region make sense intuitively. However, the choice of cutoff value used to specify the rejection region is somewhat arbitrary. Instead of rejecting H_0: $p = .10$ in favor of H_a: $p < .10$ when $x \leq 15$, we could use the rejection region $x \leq 14$. For this region, H_0 would not be rejected if 15 defective boards are observed, whereas this occurrence would lead to rejection of H_0 if the initially suggested region is employed. Similarly, the rejection region $\overline{x} \geq 1.55$ might be used in the nicotine problem in place of the region $\overline{x} \geq 1.60$.

Errors in Hypothesis Testing

The basis for choosing a particular rejection region lies in an understanding of the errors that one might be faced with in drawing a conclusion. Consider the rejection region $x \leq 15$ in the circuit board problem. Even when H_0: $p = .10$ is true, it might happen that an unusual sample results in $x = 13$, so that H_0 is erroneously rejected. On the other hand, even when H_a: $p < .10$ is true, an unusual sample might yield $x = 20$, in which case H_0

would not be rejected, again an incorrect conclusion. Thus it is possible that H_0 may be rejected when it is true or that H_0 may not be rejected when it is false. These possible errors are not consequences of a foolishly chosen rejection region. Either one of these two errors might result when the region $x \leq 14$ is employed, or indeed when any other region is used.

DEFINITION

A **type I error** consists of rejecting the null hypothesis H_0 when it is true.
A **type II error** involves not rejecting H_0 when H_0 is false.

In the nicotine problem, a type I error consists of rejecting the manufacturer's claim that $\mu = 1.5$ when it is actually true. If the rejection region $\bar{x} \geq 1.6$ is employed, it might happen that $\bar{x} = 1.63$ even when $\mu = 1.5$, resulting in a type I error. Alternatively, it may be that H_0 is false and yet $\bar{x} = 1.52$ is observed, leading to H_0 not being rejected (a type II error).

In the best of all possible worlds, test procedures for which neither type of error is possible could be developed. However, this ideal can be achieved only by basing a decision on an examination of the entire population, which is almost always impractical. The difficulty with using a procedure based on sample data is that because of sampling variability, an unrepresentative sample may result. Even though $E(\bar{X}) = \mu$, the observed value $\bar{x}$ may differ substantially from μ (at least if n is small). Thus when $\mu = 1.5$ in the nicotine situation, $\bar{x}$ may be much larger than 1.5, resulting in erroneous rejection of H_0. Alternatively, it may be that $\mu = 1.6$ yet an $\bar{x}$ much smaller than this is observed, leading to a type II error.

Instead of demanding error-free procedures, we must look for procedures for which either type of error is unlikely to occur. That is, a good procedure is one for which the probability of making either type of error is small. The choice of a particular rejection region cutoff value fixes the probabilities of type I and type II errors. These error probabilities are traditionally denoted by α and β, respectively. Because H_0 specifies a unique value of the parameter, there is a single value of α. However, there is a different value of β for each value of the parameter consistent with H_a.

Example 9.1 A certain type of automobile is known to sustain no visible damage 25% of the time in 10-mph crash tests. A modified bumper design has been proposed in an effort to increase this percentage. Let p denote the proportion of all 10-mph crashes with this new bumper that result in no visible damage. The hypotheses to be tested are H_0: $p = .25$ (no improvement) versus H_a: $p > .25$. The test will be based on an experiment involving $n = 20$ independent crashes with prototypes of the new design. Intuitively, H_0 should be rejected if a substantial number of the crashes show no damage. Consider the following test procedure:

Test statistic: $X =$ the number of crashes with no visible damage

Rejection region: $R_8 = \{8, 9, 10, \ldots, 19, 20\}$; that is, reject H_0 if $x \geq 8$, where x is the observed value of the test statistic

This rejection region is called *upper-tailed* because it consists only of large values of the test statistic.

When H_0 is true, X has a binomial probability distribution with $n = 20$ and $p = .25$. Then

$$\begin{aligned}\alpha &= P(\text{type I error}) = P(H_0 \text{ is rejected when it is true}) \\ &= P[X \geq 8 \text{ when } X \sim \text{Bin}(20, .25)] = 1 - B(7; 20, .25) \\ &= 1 - .898 = .102\end{aligned}$$

That is, when H_0 is actually true, roughly 10% of all experiments consisting of 20 crashes would result in H_0 being incorrectly rejected (a type I error).

In contrast to α, there is not a single β. Instead, there is a different β for each different p that exceeds .25. Thus there is a value of β for $p = .3$ [in which case $X \sim \text{Bin}(20, .3)$], another value of β for $p = .5$, and so on. For example,

$$\begin{aligned}\beta(.3) &= P(\text{type II error when } p = .3) \\ &= P(H_0 \text{ is not rejected when it is false because } p = .3) \\ &= P[X \leq 7 \text{ when } X \sim \text{Bin}(20, .3)] = B(7; 20, .3) = .772\end{aligned}$$

When p is actually .3 rather than .25 (a "small" departure from H_0), roughly 77% of all experiments of this type would result in H_0 being incorrectly not rejected!

The accompanying table displays β for selected values of p (each calculated for the rejection region R_8). Clearly, β decreases as the value of p moves farther to the right of the null value .25. Intuitively, the greater the departure from H_0, the less likely it is that such a departure will not be detected.

p	.3	.4	.5	.6	.7	.8
$\beta(p)$	.772	.416	.132	.021	.001	.000

The proposed test procedure is still reasonable for testing the more realistic null hypothesis that $p \leq .25$. In this case, there is no longer a single α, but instead there is an α for each p that is at most .25: $\alpha(.25)$, $\alpha(.23)$, $\alpha(.20)$, $\alpha(.15)$, and so on. It is easily verified, though, that $\alpha(p) < \alpha(.25) = .102$ if $p < .25$. That is, the largest value of α occurs for the boundary value .25 between H_0 and H_a. Thus if α is small for the simplified null hypothesis, it will also be as small as or smaller for the more realistic H_0. ■

Example 9.2 The drying time of a certain type of paint under specified test conditions is known to be normally distributed with mean value 75 min and standard deviation 9 min. Chemists have proposed a new additive designed to decrease average drying time. It is believed that drying times with this additive will remain normally distributed with $\sigma = 9$. Because of the expense associated with the additive, evidence should strongly suggest an improvement in average drying time before such a conclusion is adopted. Let μ denote the true average drying time when the additive is used. The appropriate hypotheses are H_0: $\mu = 75$ versus H_a: $\mu < 75$. Only if H_0 can be rejected will the additive be declared successful and used.

Experimental data is to consist of drying times from $n = 25$ test specimens. Let $X_1, \ldots, X_{25}$ denote the 25 drying times—a random sample of size 25 from a normal distribution with mean value μ and standard deviation $\sigma = 9$. The sample mean drying time $\overline{X}$ then has a normal distribution with expected value $\mu_{\overline{X}} = \mu$ and standard deviation $\sigma_{\overline{X}} = \sigma/\sqrt{n} = 9/\sqrt{25} = 1.80$. When H_0 is true, $\mu_{\overline{X}} = 75$, so an $\overline{x}$ value somewhat less than 75 would not strongly contradict H_0. A reasonable rejection region has the form $\bar{x} \le c$, where the cutoff value c is suitably chosen. Consider the choice $c = 70.8$, so that the test procedure consists of test statistic $\bar{X}$ and rejection region $\bar{x} \le 70.8$. Because the rejection region consists only of small values of the test statistic, the test is said to be *lower-tailed*. Calculation of α and β now involves a routine standardization of $\bar{X}$ followed by reference to the standard normal probabilities of Appendix Table A.3:

$$\begin{aligned}
\alpha &= P(\text{type I error}) = P(H_0 \text{ is rejected when it is true}) \\
&= P(\bar{X} \le 70.8 \text{ when } \bar{X} \sim \text{normal with } \mu_{\bar{X}} = 75, \sigma_{\bar{X}} = 1.8) \\
&= \Phi\left(\frac{70.8 - 75}{1.8}\right) = \Phi(-2.33) = .01
\end{aligned}$$

$$\begin{aligned}
\beta(72) &= P(\text{type II error when } \mu = 72) \\
&= P(H_0 \text{ is not rejected when it is false because } \mu = 72) \\
&= P(\bar{X} > 70.8 \text{ when } \bar{X} \sim \text{normal with } \mu_{\bar{X}} = 72 \text{ and } \sigma_{\bar{X}} = 1.8) \\
&= 1 - \Phi\left(\frac{70.8 - 72}{1.8}\right) = 1 - \Phi(-.67) = 1 - .2514 = .7486 \\
\beta(70) &= 1 - \Phi\left(\frac{70.8 - 70}{1.8}\right) = .3300 \qquad \beta(67) = .0174
\end{aligned}$$

For the specified test procedure, only 1% of all experiments carried out as described will result in H_0 being rejected when it is actually true. However, the chance of a type II error is very large when $\mu = 72$ (only a small departure from H_0), somewhat less when $\mu = 70$, and quite small when $\mu = 67$ (a very substantial departure from H_0). These error probabilities are illustrated in Figure 9.1. Notice that α is computed using the probability distribution of the test statistic when H_0 is true, whereas determination of β requires knowing the test statistic's distribution when H_0 is false.

As in Example 9.1, if the more realistic null hypothesis $\mu \ge 75$ is considered, there is an α for each parameter value for which H_0 is true: $\alpha(75)$, $\alpha(75.8)$, $\alpha(76.5)$, and so on. It is easily verified, though, that $\alpha(75)$ is the largest of all these type I error probabilities. Focusing on the boundary value amounts to working explicitly with the "worst case." ■

The specification of a cutoff value for the rejection region in the examples just considered was somewhat arbitrary. Use of the rejection region $R_8 = \{8, 9, \ldots, 20\}$ in Example 9.1 resulted in $\alpha = .102$, $\beta(.3) = .772$, and $\beta(.5) = .132$. Many would think these error probabilities intolerably large. Perhaps they can be decreased by changing the cutoff value.

Example 9.3 (Example 9.1 continued)

Let us use the same experiment and test statistic X as previously described in the automobile bumper problem but now consider the rejection region $R_9 = \{9, 10, \ldots, 20\}$. Since X still has a binomial distribution with parameters $n = 20$ and p,

$$\begin{aligned}
\alpha &= P(H_0 \text{ is rejected when } p = .25) \\
&= P[X \ge 9 \text{ when } X \sim \text{Bin}(20, .25)] = 1 - B(8; 20, .25) = .041
\end{aligned}$$

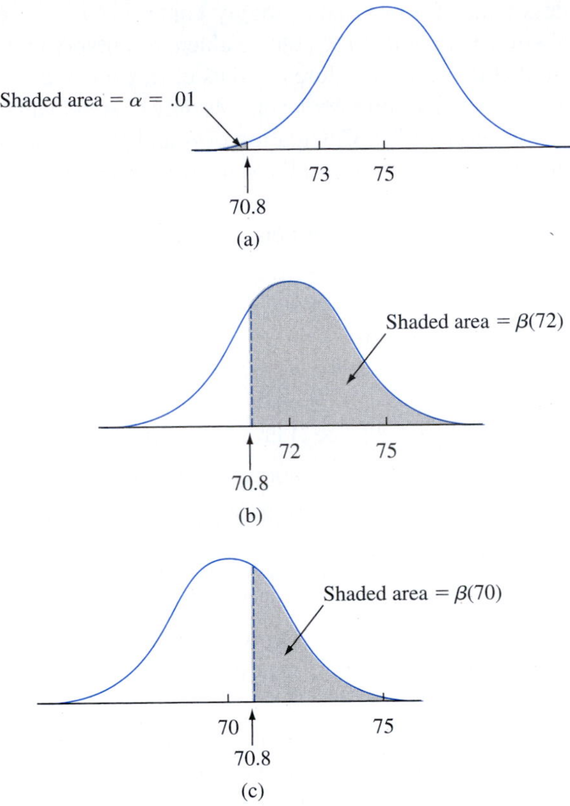

Figure 9.1 α and β illustrated for Example 9.2: (a) the distribution of $\overline{X}$ when $\mu = 75$ (H_0 true); (b) the distribution of $\overline{X}$ when $\mu = 72$ (H_0 false); (c) the distribution of $\overline{X}$ when $\mu = 70$ (H_0 false)

The type I error probability has been decreased by using the new rejection region. However, a price has been paid for this decrease:

$$\begin{aligned}\beta(.3) &= P(H_0 \text{ is not rejected when } p = .3)\\ &= P[X \le 8 \text{ when } X \sim \text{Bin}(20, .3)] = B(8; 20, .3) = .887\\ \beta(.5) &= B(8; 20, .5) = .252\end{aligned}$$

Both these β's are larger than the corresponding error probabilities .772 and .132 for the region R_8. In retrospect, this is not surprising; α is computed by summing over probabilities of test statistic values *in the rejection region*, whereas β is the probability that X falls *in the complement* of the rejection region. Making the rejection region smaller must therefore decrease α while increasing β for any fixed alternative value of the parameter. ■

Example 9.4
(Example 9.2 continued)

The use of cutoff value $c = 70.8$ in the paint-drying example resulted in a very small value of α (.01) but rather large β's. Consider the same experiment and test statistic $\overline{X}$ with the new rejection region $\bar{x} \le 72$. Because $\overline{X}$ is still normally distributed with mean value $\mu_{\overline{X}} = \mu$ and $\sigma_{\overline{X}} = 1.8$,

$$\begin{aligned}\alpha &= P(H_0 \text{ is rejected when it is true}) \\ &= P[\overline{X} \le 72 \text{ when } \overline{X} \sim N(75, 1.8^2)] \\ &= \Phi\left(\frac{72 - 75}{1.8}\right) = \Phi(-1.67) = .0475 \approx .05\end{aligned}$$

$$\begin{aligned}\beta(72) &= P(H_0 \text{ is not rejected when } \mu = 72) \\ &= P(\overline{X} > 72 \text{ when } \overline{X} \text{ is a normal rv with mean 72 and standard deviation 1.8}) \\ &= 1 - \Phi\left(\frac{72 - 72}{1.8}\right) = 1 - \Phi(0) = .5\end{aligned}$$

$$\beta(70) = 1 - \Phi\left(\frac{72 - 70}{1.8}\right) = .1335 \qquad \beta(67) = .0027$$

The change in cutoff value has made the rejection region larger (it includes more $\bar{x}$ values), resulting in a decrease in β for each fixed μ less than 75. However, α for this new region has increased from the previous value .01 to approximately .05. If a type I error probability this large can be tolerated, though, the second region ($c = 72$) is preferable to the first ($c = 70.8$) because of the smaller β's. ■

The results of these examples can be generalized in the following manner.

PROPOSITION Suppose an experiment and a sample size are fixed and a test statistic is chosen. Then decreasing the size of the rejection region to obtain a smaller value of α results in a larger value of β for any particular parameter value consistent with H_a.

This proposition says that once the test statistic and n are fixed, there is no rejection region that will simultaneously make both α and all β's small. A region must be chosen to effect a compromise between α and β.

Because of the suggested guidelines for specifying H_0 and H_a, a type I error is usually more serious than a type II error (this can always be achieved by proper choice of the hypotheses). The approach adhered to by most statistical practitioners is then to specify the largest value of α that can be tolerated and find a rejection region having that value of α rather than anything smaller. This makes β as small as possible subject to the bound on α. The resulting value of α is often referred to as the **significance level** of the test. Traditional levels of significance are .10, .05, and .01, though the level in any particular problem will depend on the seriousness of a type I error—the more serious this error, the smaller should be the significance level. The corresponding test procedure is called a **level $\boldsymbol{\alpha}$ test** (e.g., a level .05 test or a level .01 test). A test with significance level α is one for which the type I error probability is controlled at the specified level.

Example 9.5 Consider the situation mentioned previously in which μ was the true average nicotine content of brand B cigarettes. The objective is to test H_0: $\mu = 1.5$ versus H_a: $\mu > 1.5$ based on a random sample $X_1, X_2, \ldots, X_{32}$ of nicotine contents. Suppose the distribution of nicotine content is known to be normal with $\sigma = .20$. Then $\overline{X}$ is normally distributed with mean value $\mu_{\overline{X}} = \mu$ and standard deviation $\sigma_{\overline{X}} = .20/\sqrt{32} = .0354$.

Rather than use $\overline{X}$ itself as the test statistic, let's standardize $\overline{X}$ assuming that H_0 is true.

$$\text{Test statistic:} \quad Z = \frac{\overline{X} - 1.5}{\sigma/\sqrt{n}} = \frac{\overline{X} - 1.5}{.0354}$$

Z expresses the distance between $\overline{X}$ and its expected value when H_0 is true as some number of standard deviations. For example, $z = 3$ results from an $\bar{x}$ that is 3 standard deviations larger than we would have expected it to be were H_0 true.

Rejecting H_0 when $\bar{x}$ "considerably" exceeds 1.5 is equivalent to rejecting H_0 when z "considerably" exceeds 0. That is, the form of the rejection region is $z \geq c$. Let's now determine c so that $\alpha = .05$. When H_0 is true, Z has a standard normal distribution. Thus

$$\begin{aligned}\alpha &= P(\text{type I error}) = P(\text{rejecting } H_0 \text{ when } H_0 \text{ is true})\\ &= P[Z \geq c \text{ when } Z \sim N(0, 1)]\end{aligned}$$

The value c must capture upper-tail area .05 under the z curve. Either from Section 4.3 or directly from Appendix Table A.3, $c = z_{.05} = 1.645$.

Notice that $z \geq 1.645$ is equivalent to $\bar{x} - 1.5 \geq (.0354)(1.645)$, that is, $\bar{x} \geq 1.56$. Then β is the probability that $\overline{X} < 1.56$ and can be calculated for any μ greater than 1.5. ■

Exercises Section 9.1 (1–14)

1. For each of the following assertions, state whether it is a legitimate statistical hypothesis and why:

a. $H: \sigma > 100$ **b.** $H: \tilde{x} = 45$

c. $H: s \leq .20$ **d.** $H: \sigma_1/\sigma_2 < 1$

e. $H: \overline{X} - \overline{Y} = 5$

f. $H: \lambda \leq .01$, where λ is the parameter of an exponential distribution used to model component lifetime

2. For the following pairs of assertions, indicate which do not comply with our rules for setting up hypotheses and why (the subscripts 1 and 2 differentiate between quantities for two different populations or samples):

a. $H_0: \mu = 100, H_a: \mu > 100$

b. $H_0: \sigma = 20, H_a: \sigma \leq 20$

c. $H_0: p \neq .25, H_a: p = .25$

d. $H_0: \mu_1 - \mu_2 = 25, H_a: \mu_1 - \mu_2 > 100$

e. $H_0: S_1^2 = S_2^2, H_a: S_1^2 \neq S_2^2$

f. $H_0: \mu = 120, H_a: \mu = 150$

g. $H_0: \sigma_1/\sigma_2 = 1, H_a: \sigma_1/\sigma_2 \neq 1$

h. $H_0: p_1 - p_2 = -.1, H_a: p_1 - p_2 < -.1$

3. To determine whether the girder welds in a new performing arts center meet specifications, a random sample of welds is selected, and tests are conducted on each weld in the sample. Weld strength is measured as the force required to break the weld. Suppose the specifications state that mean strength of welds should exceed 100 lb/in^2; the inspection team decides to test $H_0: \mu = 100$ versus $H_a: \mu > 100$. Explain why it might be preferable to use this H_a rather than $\mu < 100$.

4. Let μ denote the true average radioactivity level (picocuries per liter). The value 5 pCi/L is considered the dividing line between safe and unsafe water. Would you recommend testing $H_0: \mu = 5$ versus $H_a: \mu > 5$ or $H_0: \mu = 5$ versus $H_a: \mu < 5$? Explain your reasoning. (*Hint*: Think about the consequences of a type I and type II error for each possibility.)

5. Before agreeing to purchase a large order of polyethylene sheaths for a particular type of high-pressure oil-filled submarine power cable, a company wants to see conclusive evidence that the true standard deviation of sheath thickness is less than .05 mm. What hypotheses should be tested, and why? In this context, what are the type I and type II errors?

6. Many older homes have electrical systems that use fuses rather than circuit breakers. A manufacturer of 40-amp fuses wants to make sure that the mean amperage at which its fuses burn out is in fact 40. If the mean amperage is lower than 40, customers will complain because the fuses require replacement too often. If the mean amperage is higher than 40, the manufacturer might be liable for damage to an electrical system due to fuse malfunction. To verify the amperage of the fuses, a sample of fuses is to be selected and inspected. If a hypothesis test were to be performed on the resulting data, what null and alternative hypotheses would be of interest to the manufacturer? Describe type I and type II errors in the context of this problem situation.

7. Water samples are taken from water used for cooling as it is being discharged from a power plant into a river. It has been determined that as long as the mean temperature of the discharged water is at most 150°F, there will be no negative effects on the river's ecosystem. To investigate whether the plant is in compliance with regulations that prohibit a mean discharge-water temperature above 150°, 50 water samples will be taken at randomly selected times, and the temperature of each sample recorded. The resulting data will be used to test the hypotheses H_0: $\mu = 150°$ versus H_a: $\mu > 150°$. In the context of this situation, describe type I and type II errors. Which type of error would you consider more serious? Explain.

8. A regular type of laminate is currently being used by a manufacturer of circuit boards. A special laminate has been developed to reduce warpage. The regular laminate will be used on one sample of specimens and the special laminate on another sample, and the amount of warpage will then be determined for each specimen. The manufacturer will then switch to the special laminate only if it can be demonstrated that the true average amount of warpage for that laminate is less than for the regular laminate. State the relevant hypotheses, and describe the type I and type II errors in the context of this situation.

9. Two different companies have applied to provide cable television service in a certain region. Let p denote the proportion of all potential subscribers who favor the first company over the second. Consider testing H_0: $p = .5$ versus H_a: $p \neq .5$ based on a random sample of 25 individuals. Let X denote the number in the sample who favor the first company and x represent the observed value of X.

   a. Which of the following rejection regions is most appropriate and why?

   $R_1 = \{x: x \leq 7 \text{ or } x \geq 18\}$, $R_2 = \{x: x \leq 8\}$,
   $R_3 = \{x: x \geq 17\}$

   b. In the context of this problem situation, describe what type I and type II errors are.
   c. What is the probability distribution of the test statistic X when H_0 is true? Use it to compute the probability of a type I error.
   d. Compute the probability of a type II error for the selected region when $p = .3$, again when $p = .4$, and also for both $p = .6$ and $p = .7$.
   e. Using the selected region, what would you conclude if 6 of the 25 queried favored company 1?

10. For healthy individuals the level of prothrombin in the blood is approximately normally distributed with mean 20 mg/100 mL and standard deviation 4 mg/100 mL. Low levels indicate low clotting ability. In studying the effect of gallstones on prothrombin, the level of each patient in a sample is measured to see if there is a deficiency. Let μ be the true average level of prothrombin for gallstone patients.
   a. What are the appropriate null and alternative hypotheses?
   b. Let $\bar{X}$ denote the sample average level of prothrombin in a sample of $n = 20$ randomly selected gallstone patients. Consider the test procedure with test statistic $\bar{X}$ and rejection region $\bar{x} \leq 17.92$. What is the probability distribution of the test statistic when H_0 is true? What is the probability of a type I error for the test procedure?
   c. What is the probability distribution of the test statistic when $\mu = 16.7$? Using the test procedure of part (b), what is the probability that gallstone patients will be judged not deficient in prothrombin, when in fact $\mu = 16.7$ (a type II error)?
   d. How would you change the test procedure of part (b) to obtain a test with significance level .05? What impact would this change have on the error probability of part (c)?
   e. Consider the standardized test statistic $Z = (\bar{X} - 20)/(\sigma/\sqrt{n}) = (\bar{X} - 20)/.8944$. What are the values of Z corresponding to the rejection region of part (b)?

11. The calibration of a scale is to be checked by weighing a 10-kg test specimen 25 times. Suppose that the results of different weighings are independent of one another and that the weight on each trial is

normally distributed with $\sigma = .200$ kg. Let μ denote the true average weight reading on the scale.

a. What hypotheses should be tested?

b. Suppose the scale is to be recalibrated if either $\bar{x} \geq 10.1032$ or $\bar{x} \leq 9.8968$. What is the probability that recalibration is carried out when it is actually unnecessary?

c. What is the probability that recalibration is judged unnecessary when in fact $\mu = 10.1$? When $\mu = 9.8$?

d. Let $z = (\bar{x} - 10)/(\sigma/\sqrt{n})$. For what value c is the rejection region of part (b) equivalent to the "two-tailed" region *either* $z \geq c$ *or* $z \leq -c$?

e. If the sample size were only 10 rather than 25, how should the procedure of part (d) be altered so that $\alpha = .05$?

f. Using the test of part (e), what would you conclude from the following sample data?

9.981	10.006	9.857	10.107	9.888
9.728	10.439	10.214	10.190	9.793

g. Reexpress the test procedure of part (b) in terms of the standardized test statistic $Z = (\bar{X} - 10)/(\sigma/\sqrt{n})$.

12. A new design for the braking system on a certain type of car has been proposed. For the current system, the true average braking distance at 40 mph under specified conditions is known to be 120 ft. It is proposed that the new design be implemented only if sample data strongly indicates a reduction in true average braking distance for the new design.

a. Define the parameter of interest and state the relevant hypotheses.

b. Suppose braking distance for the new system is normally distributed with $\sigma = 10$. Let $\bar{X}$ denote the sample average braking distance for a random sample of 36 observations. Which of the following rejection regions is appropriate: $R_1 = \{\bar{x}: \bar{x} \geq 124.80\}$, $R_2 = \{\bar{x}: \bar{x} \leq 115.20\}$, $R_3 = \{\bar{x}: \text{either } \bar{x} \geq 125.13 \text{ or } \bar{x} \leq 114.87\}$?

c. What is the significance level for the appropriate region of part (b)? How would you change the region to obtain a test with $\alpha = .001$?

d. What is the probability that the new design is not implemented when its true average braking distance is actually 115 ft and the appropriate region from part (b) is used?

e. Let $Z = (\bar{X} - 120)/(\sigma/\sqrt{n})$. What is the significance level for the rejection region $\{z: z \leq -2.33\}$? For the region $\{z: z \leq -2.88\}$?

13. Let $X_1, \ldots, X_n$ denote a random sample from a normal population distribution with a known value of σ.

a. For testing the hypotheses H_0: $\mu = \mu_0$ versus H_a: $\mu > \mu_0$ (where μ_0 is a fixed number), show that the test with test statistic $\bar{X}$ and rejection region $\bar{x} \geq \mu_0 + 2.33\sigma/\sqrt{n}$ has significance level .01.

b. Suppose the procedure of part (a) is used to test H_0: $\mu \leq \mu_0$ versus H_a: $\mu > \mu_0$. If $\mu_0 = 100$, $n = 25$, and $\sigma = 5$, what is the probability of committing a type I error when $\mu = 99$? When $\mu = 98$? In general, what can be said about the probability of a type I error when the actual value of μ is less than μ_0? Verify your assertion.

14. Reconsider the situation of Exercise 11 and suppose the rejection region is $\{\bar{x}: \bar{x} \geq 10.1004 \text{ or } \bar{x} \leq 9.8940\} = \{z: z \geq 2.51 \text{ or } z \leq -2.65\}$.

a. What is α for this procedure?

b. What is β when $\mu = 10.1$? When $\mu = 9.9$? Is this desirable?

9.2 Tests About a Population Mean

The general discussion in Chapter 8 of confidence intervals for a population mean μ focused on three different cases. We now develop test procedures for these same three cases.

Case I: A Normal Population with Known σ

Although the assumption that the value of σ is known is rarely met in practice, this case provides a good starting point because of the ease with which general procedures and their properties can be developed. The null hypothesis in all three cases will state that μ has a particular numerical value, the *null value*, which we will denote by μ_0. Let $X_1, \ldots,$

X_n represent a random sample of size n from the normal population. Then the sample mean $\overline{X}$ has a normal distribution with expected value $\mu_{\overline{X}} = \mu$ and standard deviation $\sigma_{\overline{X}} = \sigma/\sqrt{n}$. When H_0 is true, $\mu_{\overline{X}} = \mu_0$. Consider now the statistic Z obtained by standardizing $\overline{X}$ under the assumption that H_0 is true:

$$Z = \frac{\overline{X} - \mu_0}{\sigma/\sqrt{n}}$$

Substitution of the computed sample mean $\bar{x}$ gives z, the distance between $\bar{x}$ and μ_0 expressed in "standard deviation units." For example, if the null hypothesis is H_0: $\mu = 100$, $\sigma_{\overline{X}} = \sigma/\sqrt{n} = 10/\sqrt{25} = 2.0$ and $\bar{x} = 103$, then the test statistic value is given by $z = (103 - 100)/2.0 = 1.5$. That is, the observed value of $\bar{x}$ is 1.5 standard deviations (of $\overline{X}$) above what we expect it to be when H_0 is true. The statistic Z is a natural measure of the distance between $\overline{X}$, the estimator of μ, and its expected value when H_0 is true. If this distance is too great in a direction consistent with H_a, the null hypothesis should be rejected.

Suppose first that the alternative hypothesis has the form H_a: $\mu > \mu_0$. Then an $\bar{x}$ value less than μ_0 certainly does not provide support for H_a. Such an $\bar{x}$ corresponds to a negative value of z (since $\bar{x} - \mu_0$ is negative and the divisor $\sigma/\sqrt{n}$ is positive). Similarly, an $\bar{x}$ value that exceeds μ_0 by only a small amount (corresponding to z which is positive but small) does not suggest that H_0 should be rejected in favor of H_a. The rejection of H_0 is appropriate only when $\bar{x}$ considerably exceeds μ_0—that is, when the z value is positive and large. In summary, the appropriate rejection region, based on the test statistic Z rather than $\overline{X}$, has the form $z \geq c$.

As discussed in Section 9.1, the cutoff value c should be chosen to control the probability of a type I error at the desired level α. This is easily accomplished because the distribution of the test statistic Z when H_0 is true is the standard normal distribution (that's why μ_0 was subtracted in standardizing). The required cutoff c is the z critical value that captures upper-tail area α under the standard normal curve. As an example, let $c = 1.645$, the value that captures tail area .05 ($z_{.05} = 1.645$). Then,

$$\begin{aligned}\alpha &= P(\text{type I error}) = P(H_0 \text{ is rejected when } H_0 \text{ is true}) \\ &= P[Z \geq 1.645 \text{ when } Z \sim N(0, 1)] = 1 - \Phi(1.645) = .05\end{aligned}$$

More generally, the rejection region $z \geq z_\alpha$ has type I error probability α. The test procedure is *upper-tailed* because the rejection region consists only of large values of the test statistic.

Analogous reasoning for the alternative hypothesis H_a: $\mu < \mu_0$ suggests a rejection region of the form $z \leq c$, where c is a suitably chosen negative number ($\bar{x}$ is far below μ_0 if and only if z is quite negative). Because Z has a standard normal distribution when H_0 is true, taking $c = -z_\alpha$ yields $P(\text{type I error}) = \alpha$. This is a *lower-tailed* test. For example, $z_{.10} = 1.28$ implies that the rejection region $z \leq -1.28$ specifies a test with significance level .10.

Finally, when the alternative hypothesis is H_a: $\mu \neq \mu_0$, H_0 should be rejected if $\bar{x}$ is too far to either side of μ_0. This is equivalent to rejecting H_0 either if $z \geq c$ or if $z \leq -c$. Suppose we desire $\alpha = .05$. Then,

$$\begin{aligned}.05 &= P(Z \geq c \text{ or } Z \leq -c \text{ when } Z \text{ has a standard normal distribution}) \\ &= \Phi(-c) + 1 - \Phi(c) = 2[1 - \Phi(c)]\end{aligned}$$

Thus c is such that $1 - \Phi(c)$, the area under the standard normal curve to the right of c, is .025 (and not .05!). From Section 4.3 or Appendix Table A.3, $c = 1.96$, and the rejection region is $z \geq 1.96$ *or* $z \leq -1.96$. For any α, the *two-tailed* rejection region $z \geq z_{\alpha/2}$ *or* $z \leq -z_{\alpha/2}$ has type I error probability α (since area $\alpha/2$ is captured under each of the two tails of the z curve). Again, the key reason for using the standardized test statistic Z is that because Z has a known distribution when H_0 is true (standard normal), a rejection region with desired type I error probability is easily obtained by using an appropriate critical value.

The test procedure for case I is summarized in the accompanying box, and the corresponding rejection regions are illustrated in Figure 9.2.

Null hypothesis: $H_0\colon \mu = \mu_0$

Test statistic value: $z = \dfrac{\bar{x} - \mu_0}{\sigma/\sqrt{n}}$

Alternative Hypothesis	**Rejection Region for Level α Test**
$H_a\colon \mu > \mu_0$	$z \geq z_\alpha$ (upper-tailed test)
$H_a\colon \mu < \mu_0$	$z \leq -z_\alpha$ (lower-tailed test)
$H_a\colon \mu \neq \mu_0$	either $z \geq z_{\alpha/2}$ or $z \leq -z_{\alpha/2}$ (two-tailed test)

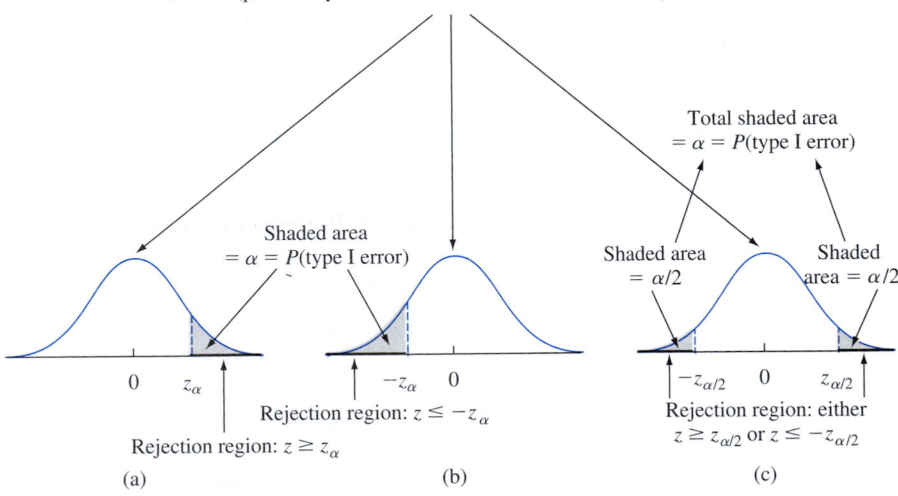

Figure 9.2 Rejection regions for z tests: (a) upper-tailed test; (b) lower-tailed test; (c) two-tailed test

Use of the following sequence of steps is recommended when testing hypotheses about a parameter.

1. Identify the parameter of interest and describe it in the context of the problem situation.

2. Determine the null value and state the null hypothesis.
3. State the appropriate alternative hypothesis.
4. Give the formula for the computed value of the test statistic (substituting the null value and the known values of any other parameters, but *not* those of any sample-based quantities).
5. State the rejection region for the selected significance level α.
6. Compute any necessary sample quantities, substitute into the formula for the test statistic value, and compute that value.
7. Decide whether H_0 should be rejected and state this conclusion in the problem context.

The formulation of hypotheses (steps 2 and 3) should be done before examining the data.

Example 9.6 A manufacturer of sprinkler systems used for fire protection in office buildings claims that the true average system-activation temperature is 130°. A sample of $n = 9$ systems, when tested, yields a sample average activation temperature of 131.08°F. If the distribution of activation times is normal with standard deviation 1.5°F, does the data contradict the manufacturer's claim at significance level $\alpha = .01$?

1. Parameter of interest: μ = true average activation temperature.
2. Null hypothesis: H_0: $\mu = 130$ (null value $= \mu_0 = 130$).
3. Alternative hypothesis: H_a: $\mu \neq 130$ (a departure from the claimed value in *either* direction is of concern).
4. Test statistic value:

$$z = \frac{\bar{x} - \mu_0}{\sigma/\sqrt{n}} = \frac{\bar{x} - 130}{1.5/\sqrt{n}}$$

5. Rejection region: The form of H_a implies use of a two-tailed test with rejection region *either* $z \geq z_{.005}$ *or* $z \leq -z_{.005}$. From Section 4.3 or Appendix Table A.3, $z_{.005} = 2.58$, so we reject H_0 if either $z \geq 2.58$ or $z \leq -2.58$.
6. Substituting $n = 9$ and $\bar{x} = 131.08$,

$$z = \frac{131.08 - 130}{1.5/\sqrt{9}} = \frac{1.08}{.5} = 2.16$$

That is, the observed sample mean is a bit more than 2 standard deviations above what would have been expected were H_0 true.

7. The computed value $z = 2.16$ does not fall in the rejection region ($-2.58 < 2.16 < 2.58$), so H_0 cannot be rejected at significance level .01. The data does not give strong support to the claim that the true average differs from the design value of 130. ■

Another view of the analysis in the previous example involves calculating a 99% CI for μ based on Equation (8.5):

$$\bar{x} \pm 2.58\sigma/\sqrt{n} = 131.08 \pm 2.58(1.5/\sqrt{9}) = 131.08 \pm 1.29 = (129.79, 132.37)$$

Notice that the interval includes $\mu_0 = 130$, and it is not hard to see that the 99% CI excludes μ_0 if and only if the two-tailed hypothesis test rejects H_0 at level .01. In general, the $100(1 - \alpha)\%$ CI excludes μ_0 if and only if the two-tailed hypothesis test rejects H_0 at level α. Although we will not always call attention to it, this kind of relationship between hypothesis tests and confidence intervals will occur over and over in the remainder of the book. It should be intuitively reasonable that the CI will exclude a value when the corresponding test rejects the value. There is a similar relationship between lower-tailed tests and upper confidence bounds, and also between upper-tailed tests and lower confidence bounds.

β and Sample Size Determination The z tests for case I are among the few in statistics for which there are simple formulas available for β, the probability of a type II error. Consider first the upper-tailed test with rejection region $z \geq z_\alpha$. This is equivalent to $\bar{x} \geq \mu_0 + z_\alpha \cdot \sigma/\sqrt{n}$, so H_0 will not be rejected if $\bar{x} < \mu_0 + z_\alpha \cdot \sigma/\sqrt{n}$. Now let μ' denote a particular value of μ that exceeds the null value μ_0. Then,

$$\begin{aligned}
\beta(\mu') &= P(H_0 \text{ is not rejected when } \mu = \mu') \\
&= P(\overline{X} < \mu_0 + z_\alpha \cdot \sigma/\sqrt{n} \text{ when } \mu = \mu') \\
&= P\left(\frac{\overline{X} - \mu'}{\sigma/\sqrt{n}} < z_\alpha + \frac{\mu_0 - \mu'}{\sigma/\sqrt{n}} \text{ when } \mu = \mu'\right) \\
&= \Phi\left(z_\alpha + \frac{\mu_0 - \mu'}{\sigma/\sqrt{n}}\right)
\end{aligned}$$

As μ' increases, $\mu_0 - \mu'$ becomes more negative, so $\beta(\mu')$ will be small when μ' greatly exceeds μ_0 (because the value at which Φ is evaluated will then be quite negative). Error probabilities for the lower-tailed and two-tailed tests are derived in an analogous manner.

If σ is large, the probability of a type II error can be large at an alternative value μ' that is of particular concern to an investigator. Suppose we fix α and also specify β for such an alternative value. In the sprinkler example, company officials might view $\mu' = 132$ as a very substantial departure from H_0: $\mu = 130$ and therefore wish $\beta(132) = .10$ in addition to $\alpha = .01$. More generally, consider the two restrictions $P(\text{type I error}) = \alpha$ and $\beta(\mu') = \beta$ for specified α, μ', and β. Then for an upper-tailed test, the sample size n should be chosen to satisfy

$$\Phi\left(z_\alpha + \frac{\mu_0 - \mu'}{\sigma/\sqrt{n}}\right) = \beta$$

This implies that

$$-z_\beta = \begin{array}{c} z \text{ critical value that} \\ \text{captures lower } = \text{ tail area } \beta \end{array} = z_\alpha + \frac{\mu_0 - \mu'}{\sigma/\sqrt{n}}$$

It is easy to solve this equation for the desired n. A parallel argument yields the necessary sample size for lower- and two-tailed tests as summarized in the next box.

Alternative Hypothesis	Type II Error Probability $\beta(\mu')$ for a Level α Test
H_a: $\mu > \mu_0$	$\Phi\left(z_\alpha + \frac{\mu_0 - \mu'}{\sigma/\sqrt{n}}\right)$
H_a: $\mu < \mu_0$	$1 - \Phi\left(-z_\alpha + \frac{\mu_0 - \mu'}{\sigma/\sqrt{n}}\right)$
H_a: $\mu \neq \mu_0$	$\Phi\left(z_{\alpha/2} + \frac{\mu_0 - \mu'}{\sigma/\sqrt{n}}\right) - \Phi\left(-z_{\alpha/2} + \frac{\mu_0 - \mu'}{\sigma/\sqrt{n}}\right)$

where $\Phi(z)$ = the standard normal cdf.

The sample size n for which a level α test also has $\beta(\mu') = \beta$ at the alternative value μ' is

$$n = \begin{cases} \left[\dfrac{\sigma(z_\alpha + z_\beta)}{\mu_0 - \mu'}\right]^2 & \text{for a one-tailed (upper or lower) test} \\ \left[\dfrac{\sigma(z_{\alpha/2} + z_\beta)}{\mu_0 - \mu'}\right]^2 & \text{for a two-tailed test (an approximate solution)} \end{cases}$$

Example 9.7 Let μ denote the true average tread life of a certain type of tire. Consider testing H_0: $\mu = 30{,}000$ versus H_a: $\mu > 30{,}000$ based on a sample of size $n = 16$ from a normal population distribution with $\sigma = 1500$. A test with $\alpha = .01$ requires $z_\alpha = z_{.01} = 2.33$. The probability of making a type II error when $\mu = 31{,}000$ is

$$\beta(31{,}000) = \Phi\left(2.33 + \frac{30{,}000 - 31{,}000}{1500/\sqrt{16}}\right) = \Phi(-.34) = .3669$$

Since $z_{.1} = 1.28$, the requirement that the level .01 test also have $\beta(31{,}000) = .1$ necessitates

$$n = \left[\frac{1500(2.33 + 1.28)}{30{,}000 - 31{,}000}\right]^2 = (-5.42)^2 = 29.32$$

The sample size must be an integer, so $n = 30$ tires should be used. ■

Case II: Large-Sample Tests

When the sample size is large, the z tests for case I are easily modified to yield valid test procedures without requiring either a normal population distribution or known σ. The key result was used in Chapter 8 to justify large-sample confidence intervals: A large n implies that the sample standard deviation s will be close to σ for most samples, so that the standardized variable

$$Z = \frac{\overline{X} - \mu}{S/\sqrt{n}}$$

has *approximately* a standard normal distribution. Substitution of the null value μ_0 in place of μ yields the test statistic

$$Z = \frac{\overline{X} - \mu_0}{S/\sqrt{n}}$$

which has approximately a standard normal distribution when H_0 is true. The use of rejection regions given previously for case I (e.g., $z \geq z_\alpha$ when the alternative hypothesis is H_a: $\mu > \mu_0$) then results in test procedures for which the significance level is approximately (rather than exactly) α. The rule of thumb $n > 40$ will again be used to characterize a large sample size.

Example 9.8 A sample of bills for meals was obtained at a restaurant (by Erich Brandt). For each of 70 bills the tip was found as a percentage of the raw bill (before taxes). Does it appear that the population mean tip percentage for this restaurant exceeds the standard 15%? Here are the 70 tip percentages:

14.21	20.24	20.10	14.94	15.69	15.04	12.04	20.16	17.85	16.35
19.12	20.37	15.29	18.39	27.55	16.01	10.94	13.52	17.42	14.48
29.87	17.92	19.74	22.73	14.56	15.16	16.09	16.42	19.07	13.74
13.46	16.79	19.03	19.19	19.23	12.39	16.89	18.93	13.56	17.70
11.48	13.96	21.58	11.94	19.02	17.73	20.07	40.09	19.88	22.79
15.23	16.09	19.19	11.91	18.21	15.37	16.31	16.03	48.77	12.31
21.53	12.76	18.07	14.11	15.86	20.67	15.66	18.54	27.88	13.81

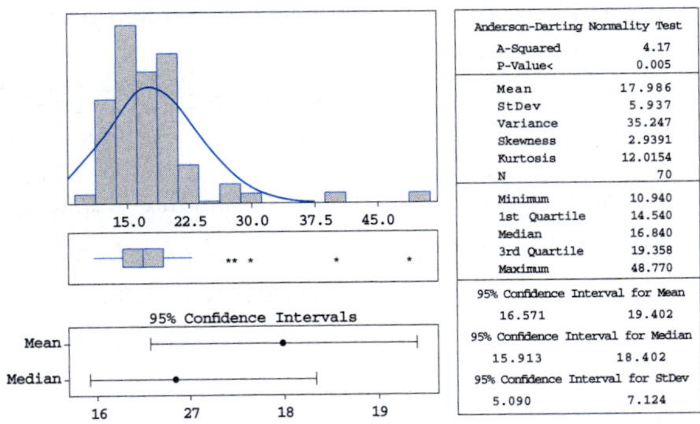

Figure 9.3 MINITAB descriptive summary for the tip data of Example 9.8

Figure 9.3 shows a descriptive summary obtained from MINITAB. The sample mean tip percentage is greater than 15. Notice that the distribution is positively skewed because there are some very large tips (and a normal probability plot therefore does not exhibit a linear pattern), but the large-sample z tests do not require a normal population distribution.

1. μ = true average tip percentage
2. H_0: $\mu = 15$
3. H_a: $\mu > 15$

4. $z = \dfrac{\bar{x} - 15}{s/\sqrt{n}}$

5. Using a test with a significance level .05, H_0 will be rejected if $z \geq 1.645$ (an upper-tailed test).

6. With $n = 70$, $\bar{x} = 17.99$, and $s = 5.937$,

$$z = \frac{17.99 - 15}{5.937/\sqrt{70}} = \frac{2.99}{.7096} = 4.21$$

7. Since $4.21 \geq 1.645$, H_0 is rejected. There is evidence that the true average tip percentage exceeds 15%. ■

Determination of β and the necessary sample size for these large-sample tests can be based either on specifying a plausible value of σ and using the case I formulas (even though s is used in the test) or on using the curves to be introduced shortly in connection with case III.

Case III: A Normal Population Distribution with Unknown σ

When n is small, the Central Limit Theorem (CLT) can no longer be invoked to justify the use of a large-sample test. We faced this same difficulty in obtaining a small-sample confidence interval (CI) for μ in Chapter 8. Our approach here will be the same one used there: We will assume that the population distribution is at least approximately normal and describe test procedures whose validity rests on this assumption. If an investigator has good reason to believe that the population distribution is quite nonnormal, a distribution-free test from Chapter 14 can be used. Alternatively, a statistician can be consulted regarding procedures valid for specific families of population distributions other than the normal family. Or a bootstrap procedure can be developed.

The key result on which tests for a normal population mean are based was used in Chapter 8 to derive the one-sample t CI: If $X_1, X_2, \ldots, X_n$ is a random sample from a normal distribution, the standardized variable

$$T = \frac{\bar{X} - \mu}{S/\sqrt{n}}$$

has a t distribution with $n - 1$ degrees of freedom (df). Consider testing H_0: $\mu = \mu_0$ against H_a: $\mu > \mu_0$ by using the test statistic $T = (\bar{X} - \mu_0)/(S/\sqrt{n})$. That is, the test statistic results from standardizing $\bar{X}$ under the assumption that H_0 is true (using $S/\sqrt{n}$, the estimated standard deviation of $\bar{X}$, rather than $\sigma/\sqrt{n}$). When H_0 is true, the test statistic has a t distribution with $n - 1$ df. Knowledge of the test statistic's distribution when H_0 is true (the "null distribution") allows us to construct a rejection region for which the type I error probability is controlled at the desired level. In particular, use of the upper-tail t critical value $t_{\alpha,n-1}$ to specify the rejection region $t \geq t_{\alpha,n-1}$ implies that

$$\begin{aligned} P(\text{type I error}) &= P(H_0 \text{ is rejected when it is true}) \\ &= P(T \geq t_{\alpha,n-1} \text{ when } T \text{ has a } t \text{ distribution with } n - 1 \text{ df}) \\ &= \alpha \end{aligned}$$

The test statistic is really the same here as in the large-sample case but is labeled T to emphasize that its null distribution is a t distribution with $n - 1$ df rather than the standard normal (z) distribution. The rejection region for the t test differs from that for the z test only in that a t critical value $t_{\alpha,n-1}$ replaces the z critical value z_α. Similar comments apply to alternatives for which a lower-tailed or two-tailed test is appropriate.

THE ONE-SAMPLE t TEST

Null hypothesis: H_0: $\mu = \mu_0$

Test statistic value: $t = \dfrac{\bar{x} - \mu_0}{s/\sqrt{n}}$

Alternative Hypothesis	**Rejection Region for a Level α Test**
H_a: $\mu > \mu_0$	$t \geq t_{\alpha,n-1}$ (upper-tailed)
H_a: $\mu < \mu_0$	$t \leq -t_{\alpha,n-1}$ (lower-tailed)
H_a: $\mu \neq \mu_0$	either $t \geq t_{\alpha/2,n-1}$ or $t \leq -t_{\alpha/2,n-1}$ (two-tailed)

Example 9.9 A well-designed and safe workplace can contribute greatly to increased productivity. It is especially important that workers not be asked to perform tasks, such as lifting, that exceed their capabilities. The accompanying data on maximum weight of lift (MAWL, in kg) for a frequency of four lifts/min was reported in the article "The Effects of Speed, Frequency, and Load on Measured Hand Forces for a Floor-to-Knuckle Lifting Task" (*Ergonomics*, 1992: 833–843); subjects were randomly selected from the population of healthy males age 18–30. Assuming that MAWL is normally distributed, does the following data suggest that the population mean MAWL exceeds 25?

25.8 36.6 26.3 21.8 27.2

Let's carry out a test using a significance level of .05.

1. μ = population mean MAWL

2. H_0: $\mu = 25$

3. H_a: $\mu > 25$

4. $t = \dfrac{\bar{x} - 25}{s/\sqrt{n}}$

5. Reject H_0 if $t \geq t_{\alpha,\,n-1} = t_{.05,4} = 2.132$.

6. $\Sigma x_i = 137.7$ and $\Sigma x_i^2 = 3911.97$, from which $\bar{x} = 27.54$, $s = 5.47$, and

$$t = \frac{27.54 - 25}{5.47/\sqrt{5}} = \frac{2.54}{2.45} = 1.04$$

The accompanying MINITAB output from a request for a one-sample t test has the same calculated values (the P-value is discussed in Section 9.4).

```
Test of mu = 25.00 vs mu > 25.00
Variable    N    Mean    StDev    SE Mean      T      P-Value
mawl        5   27.54     5.47      2.45     1.04      0.18
```

7. Since 1.04 does not fall in the rejection region ($1.04 < 2.132$), H_0 cannot be rejected at significance level .05. It is still plausible that μ is (at most) 25. ■

β and Sample Size Determination The calculation of β at the alternative value μ' in case I was carried out by expressing the rejection region in terms of $\bar{x}$ (e.g., $\bar{x} \geq \mu_0 + z_\alpha \cdot \sigma/\sqrt{n}$) and then subtracting μ' to standardize correctly. An equivalent approach involves noting that when $\mu = \mu'$, the test statistic $Z = (\bar{X} - \mu_0)/(\sigma/\sqrt{n})$ still has a normal distribution with variance 1, but now the mean value of Z is given by $(\mu' - \mu_0)/(\sigma/\sqrt{n})$. That is, when $\mu = \mu'$, the test statistic still has a normal distribution though not the standard normal distribution. Because of this, $\beta(\mu')$ is an area under the normal curve corresponding to mean value $(\mu' - \mu_0)/(\sigma/\sqrt{n})$ and variance 1. Both α and β involve working with normally distributed variables.

The calculation of $\beta(\mu')$ for the t test is much less straightforward. This is because the distribution of the test statistic $T = (\bar{X} - \mu_0)/(S/\sqrt{n})$ is quite complicated when H_0 is false and H_a is true. Thus, for an upper-tailed test, determining

$$\beta(\mu') = P(T < t_{\alpha,n-1} \text{ when } \mu = \mu' \text{ rather than } \mu_0)$$

involves integrating a very unpleasant density function. This must be done numerically, but fortunately it has been done by research statisticians for both one- and two-tailed t tests. The results are summarized in graphs of β that appear in Appendix Table A.17. There are four sets of graphs, corresponding to one-tailed tests at level .05 and level .01 and two-tailed tests at the same levels.

To understand how these graphs are used, note first that both β and the necessary sample size n in case I are functions not just of the absolute difference $|\mu_0 - \mu'|$ but of $d = |\mu_0 - \mu'|/\sigma$. Suppose, for example, that $|\mu_0 - \mu'| = 10$. This departure from H_0 will be much easier to detect (smaller β) when $\sigma = 2$, in which case μ_0 and μ' are 5 population standard deviations apart, than when $\sigma = 10$. The fact that β for the t test depends on d rather than just $|\mu_0 - \mu'|$ is unfortunate, since to use the graphs one must have some idea of the true value of σ. A conservative (large) guess for σ will yield a conservative (large) value of $\beta(\mu')$ and a conservative estimate of the sample size necessary for prescribed α and $\beta(\mu')$.

Once the alternative μ' and value of σ are selected, d is calculated and its value located on the horizontal axis of the relevant set of curves. The value of β is the height of the $n - 1$ df curve above the value of d (visual interpolation is necessary if $n - 1$ is not a value for which the corresponding curve appears), as illustrated in Figure 9.4.

Rather than fixing n (i.e., $n - 1$, and thus the particular curve from which β is read), one might prescribe both α (.05 or .01 here) and a value of β for the chosen μ' and σ. After computing d, the point (d, β) is located on the relevant set of graphs. The curve below and closest to this point gives $n - 1$ and thus n (again, interpolation is often necessary).

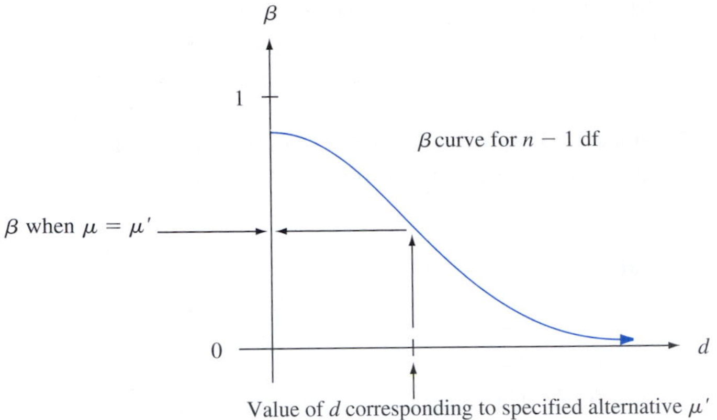

Figure 9.4 A typical β curve for the t test

Example 9.10 The true average voltage drop from collector to emitter of insulated gate bipolar transistors of a certain type is supposed to be at most 2.5 volts. An investigator selects a sample of $n = 10$ such transistors and uses the resulting voltages as a basis for testing H_0: $\mu = 2.5$ versus H_a: $\mu > 2.5$ using a t test with significance level $\alpha = .05$. If the standard deviation of the voltage distribution is $\sigma = .100$, how likely is it that H_0 will not be rejected when in fact $\mu = 2.6$? With $d = |2.5 - 2.6|/.100 = 1.0$, the point on the β curve at 9 df for a one-tailed test with $\alpha = .05$ above 1.0 has height approximately .1, so $\beta \approx .1$. The investigator might think that this is too large a value of β for such a substantial departure from H_0 and may wish to have $\beta = .05$ for this alternative value of μ. Since $d = 1.0$, the point $(d, \beta) = (1.0, .05)$ must be located. This point is very close to the 14 df curve, so using $n = 15$ will give both $\alpha = .05$ and $\beta = .05$ when the value of μ is 2.6 and $\sigma = .10$. A larger value of σ would give a larger β for this alternative, and an alternative value of μ closer to 2.5 would also result in an increased value of β. ■

Most of the widely used statistical computer packages will also calculate type II error probabilities and determine necessary sample sizes. As an example, we asked MINITAB to do the calculations from Example 9.10. Its computations are based on **power**, which is simply $1 - \beta$. We want β to be small, which is equivalent to asking that the power of the test be large. For example, $\beta = .05$ corresponds to a value of .95 for power. Here is the resulting MINITAB output.

```
Power and Sample Size

Testing mean = null (versus > null)
Calculating power for mean = null + 0.1
Alpha = 0.05  Sigma = 0.1

Sample
Size    Power
10      0.8975
```

```
Power and Sample Size

1-Sample t Test

Testing mean = null (versus > null)
Calculating power for mean = null + 0.1
Alpha = 0.05  Sigma = 0.1

Sample  Target  Actual
Size    Power   Power
13      0.9500  0.9597
```

Notice from the second part of the output that the sample size necessary to obtain a power of .95 ($\beta = .05$) for an upper-tailed test with $\alpha = .05$ when $\sigma = .1$ and μ' is .1 larger than μ_0 is only $n = 13$, whereas eyeballing our β curves gave 15. When available, this type of software is more trustworthy than the curves.

Exercises Section 9.2 (15–35)

15. Let the test statistic Z have a standard normal distribution when H_0 is true. Give the significance level for each of the following situations:

a. H_a: $\mu > \mu_0$, rejection region $z \geq 1.88$

b. H_a: $\mu < \mu_0$, rejection region $z \leq -2.75$

c. H_a: $\mu \neq \mu_0$, rejection region $z \geq 2.88$ or $z \leq -2.88$

16. Let the test statistic T have a t distribution when H_0 is true. Give the significance level for each of the following situations:

a. H_a: $\mu > \mu_0$, df $= 15$, rejection region $t \geq 3.733$

b. H_a: $\mu < \mu_0$, $n = 24$, rejection region $t \leq -2.500$

c. H_a: $\mu \neq \mu_0$, $n = 31$, rejection region $t \geq 1.697$ or $t \leq -1.697$

17. Answer the following questions for the tire problem in Example 9.7.

a. If $\bar{x} = 30{,}960$ and a level $\alpha = .01$ test is used, what is the decision?

b. If a level .01 test is used, what is $\beta(30{,}500)$?

c. If a level .01 test is used and it is also required that $\beta(30{,}500) = .05$, what sample size n is necessary?

d. If $\bar{x} = 30{,}960$, what is the smallest α at which H_0 can be rejected (based on $n = 16$)?

18. Reconsider the paint-drying situation of Example 9.2, in which drying time for a test specimen is normally distributed with $\sigma = 9$. The hypotheses H_0: $\mu = 75$ versus H_a: $\mu < 75$ are to be tested using a random sample of $n = 25$ observations.

a. How many standard deviations (of $\bar{X}$) below the null value is $\bar{x} = 72.3$?

b. If $\bar{x} = 72.3$, what is the conclusion using $\alpha = .01$?

c. What is α for the test procedure that rejects H_0 when $z \leq -2.88$?

d. For the test procedure of part (c), what is $\beta(70)$?

e. If the test procedure of part (c) is used, what n is necessary to ensure that $\beta(70) = .01$?

f. If a level .01 test is used with $n = 100$, what is the probability of a type I error when $\mu = 76$?

19. The melting point of each of 16 samples of a certain brand of hydrogenated vegetable oil was determined, resulting in $\bar{x} = 94.32$. Assume that the distribution of melting point is normal with $\sigma = 1.20$.

a. Test H_0: $\mu = 95$ versus H_a: $\mu \neq 95$ using a two-tailed level .01 test.

b. If a level .01 test is used, what is $\beta(94)$, the probability of a type II error when $\mu = 94$?

c. What value of n is necessary to ensure that $\beta(94) = .1$ when $\alpha = .01$?

20. Lightbulbs of a certain type are advertised as having an average lifetime of 750 hours. The price of these bulbs is very favorable, so a potential customer has decided to go ahead with a purchase arrangement unless it can be conclusively demonstrated that the true average lifetime is smaller than what is advertised. A random sample of 50 bulbs was selected, the lifetime of each bulb determined, and the appropriate hypotheses were tested using MINITAB, resulting in the accompanying output.

```
Variable  N   Mean    StDev  SEMean   Z     P-Value
lifetime  50  738.44  38.20  5.40    -2.14  0.016
```

What conclusion would be appropriate for a significance level of .05? A significance level of .01? What significance level and conclusion would you recommend?

21. The true average diameter of ball bearings of a certain type is supposed to be .5 in. A one-sample t test will be carried out to see whether this is the case. What conclusion is appropriate in each of the following situations?

a. $n = 13, t = 1.6, \alpha = .05$
b. $n = 13, t = -1.6, \alpha = .05$
c. $n = 25, t = -2.6, \alpha = .01$
d. $n = 25, t = -3.9$

22. The article "The Foreman's View of Quality Control" (*Quality Engrg.*, 1990: 257–280) described an investigation into the coating weights for large pipes resulting from a galvanized coating process. Production standards call for a true average weight of 200 lb per pipe. The accompanying descriptive summary and boxplot are from MINITAB.

```
Variable   N   Mean    Median  TrMean  StDev  SEMean
ctg wt    30  206.73  206.00  206.81   6.35    1.16

Variable       Min        Max        Q1         Q3
ctg wt       193.00     218.00     202.75     212.00
```

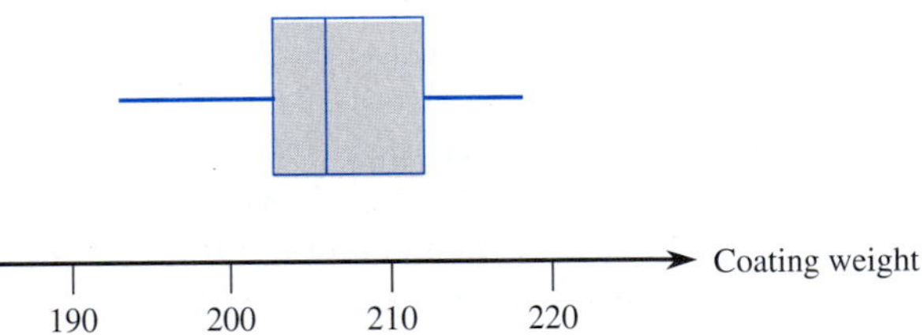

a. What does the boxplot suggest about the status of the specification for true average coating weight?
b. A normal probability plot of the data was quite straight. Use the descriptive output to test the appropriate hypotheses.

23. Exercise 33 in Chapter 1 gave $n = 26$ observations on escape time (sec) for oil workers in a simulated exercise, from which the sample mean and sample standard deviation are 370.69 and 24.36, respectively. Suppose the investigators had believed a priori that true average escape time would be at most 6 min. Does the data contradict this prior belief? Assuming normality, test the appropriate hypotheses using a significance level of .05.

24. Reconsider the sample observations on stabilized viscosity of asphalt specimens introduced in Exercise 43 in Chapter 1 (2781, 2900, 3013, 2856, and 2888). Suppose that for a particular application, it is required that true average viscosity be 3000. Does this requirement appear to have been satisfied? State and test the appropriate hypotheses.

25. Recall the first-grade IQ scores of Example 1.2. Here is a random sample of 10 of those scores:

107 113 108 127 146 103 108 118 111 119

The IQ test score has approximately a normal distribution with mean 100 and standard deviation 15 for the entire U.S. population of first graders. Here we are interested in seeing whether the population of first graders at this school is different from the national population. Assume that the normal distribution with standard deviation 15 is valid for the school, and test at the .05 level to see whether the school mean differs from the national mean. Summarize your conclusion in a sentence about these first graders.

26. In recent years major league baseball games have averaged three hours in duration. However, because games in Denver tend to be high-scoring, it might be expected that the games would be longer there. In 2001, the 81 games in Denver averaged 185.54 minutes with standard deviation 24.6 minutes. What would you conclude?

27. On the label, Pepperidge Farm bagels are said to weigh four ounces each (113 grams). A random sample of six bagels resulted in the following weights (in grams):

117.6 109.5 111.6 109.2 119.1 110.8

a. Based on this sample, is there any reason to doubt that the population mean is at least 113 grams?
b. Assume that the population mean is actually 110 grams and that the distribution is normal with standard deviation 4 grams. In a z test of H_0: $\mu = 113$ against H_a: $\mu < 113$ with $\alpha = .05$, find the probability of rejecting H_0 with six observations.
c. Under the conditions of part (b) with $\alpha = .05$, how many more observations would be needed in order for the power to be at least .95?

28. Minor surgery on horses under field conditions requires a reliable short-term anesthetic producing good muscle relaxation, minimal cardiovascular and respiratory changes, and a quick, smooth recovery with minimal aftereffects so that horses can be left unattended. The article "A Field Trial of Ketamine Anesthesia in the Horse" (*Equine Vet. J.*, 1984: 176–179) reports that for a sample of $n = 73$ horses to which ketamine was administered under certain conditions, the sample average lateral recumbency (lying-down) time was 18.86 min and the standard deviation was 8.6 min. Does this data suggest that true average lateral recumbency time under these conditions is less than 20 min? Test the appropriate hypotheses at level of significance .10.

29. The amount of shaft wear (.0001 in.) after a fixed mileage was determined for each of $n = 8$ internal combustion engines having copper lead as a bearing material, resulting in $\bar{x} = 3.72$ and $s = 1.25$.

a. Assuming that the distribution of shaft wear is normal with mean μ, use the t test at level .05 to test H_0: $\mu = 3.50$ versus H_a: $\mu > 3.50$.

b. Using $\sigma = 1.25$, what is the type II error probability $\beta(\mu')$ of the test for the alternative $\mu' = 4.00$?

30. The recommended daily dietary allowance for zinc among males older than age 50 years is 15 mg/day. The article "Nutrient Intakes and Dietary Patterns of Older Americans: A National Study" (*J. Gerontology*, 1992: M145–150) reports the following summary data on intake for a sample of males age 65–74 years: $n = 115$, $\bar{x} = 11.3$, and $s = 6.43$. Does this data indicate that average daily zinc intake in the population of all males age 65–74 falls below the recommended allowance?

31. In an experiment designed to measure the time necessary for an inspector's eyes to become used to the reduced amount of light necessary for penetrant inspection, the sample average time for $n = 9$ inspectors was 6.32 sec and the sample standard deviation was 1.65 sec. It has previously been assumed that the average adaptation time was at least 7 sec. Assuming adaptation time to be normally distributed, does the data contradict prior belief? Use the t test with $\alpha = .1$.

32. A sample of 12 radon detectors of a certain type was selected, and each was exposed to 100 pCi/L of radon. The resulting readings were as follows:

105.6	90.9	91.2	96.9	96.5	91.3
100.1	105.0	99.6	107.7	103.3	92.4

a. Does this data suggest that the population mean reading under these conditions differs from 100? State and test the appropriate hypotheses using $\alpha = .05$.

b. Suppose that prior to the experiment, a value of $\sigma = 7.5$ had been assumed. How many determinations would then have been appropriate to obtain $\beta = .10$ for the alternative $\mu = 95$?

33. Show that for any $\Delta > 0$, when the population distribution is normal and σ is known, the two-tailed test satisfies $\beta(\mu_0 - \Delta) = \beta(\mu_0 + \Delta)$, so that $\beta(\mu')$ is symmetric about μ_0.

34. For a fixed alternative value μ', show that $\beta(\mu') \rightarrow 0$ as $n \rightarrow \infty$ for either a one-tailed or a two-tailed z test in the case of a normal population distribution with known σ.

35. The industry standard for the amount of alcohol poured into many types of drinks (e.g., gin for a gin and tonic, whiskey on the rocks) is 1.5 oz. Each individual in a sample of 8 bartenders with at least five years of experience was asked to pour rum for a rum and coke into a short, wide (tumbler) glass, resulting in the following data:

2.00 1.78 2.16 1.91 1.70 1.67 1.83 1.48

(summary quantities agree with those given in the article "Bottoms Up! The Influence of Elongation on Pouring and Consumption Volume," *J. Consumer Res.*, 2003: 455–463).

a. What does a boxplot suggest about the distribution of the amount poured?

b. Carry out a test of hypotheses to decide whether there is strong evidence for concluding that the true average amount poured differs from the industry standard.

c. Does the validity of the test you carried out in (b) depend on any assumptions about the population distribution? If so, check the plausibility of such assumptions.

d. Suppose you walk into a bar and ask for a rum and coke. If the bartender has the requisite experience, predict the amount of rum you will get in a way that conveys information about precision and reliability.

e. Suppose the actual standard deviation of the amount poured is .20 oz. Determine the probability of a type II error for the test of (b) when the true average amount poured is actually (1) 1.6, (2) 1.7, (3) 1.8.

9.3 Tests Concerning a Population Proportion

Let p denote the proportion of individuals or objects in a population who possess a specified property (e.g., cars with manual transmissions or smokers who smoke a filter cigarette). If an individual or object with the property is labeled a success (S), then p is the population proportion of successes. Tests concerning p will be based on a random sample of size n from the population. Provided that n is small relative to the population size, X (the number of S's in the sample) has (approximately) a binomial distribution. Furthermore, if n itself is large, both X and the estimator $\hat{p} = X/n$ are approximately normally distributed. We first consider large-sample tests based on this latter fact and then turn to the small-sample case that directly uses the binomial distribution.

Large-Sample Tests

Large-sample tests concerning p are a special case of the more general large-sample procedures for a parameter θ. Let $\hat{\theta}$ be an estimator of θ that is (at least approximately) unbiased and has approximately a normal distribution. The null hypothesis has the form H_0: $\theta = \theta_0$, where θ_0 denotes a number (the null value) appropriate to the problem context. Suppose that when H_0 is true, the standard deviation of $\hat{\theta}$, $\sigma_{\hat{\theta}}$, involves no unknown parameters. For example, if $\theta = \mu$ and $\hat{\theta} = \overline{X}$, $\sigma_{\hat{\theta}} = \sigma_{\overline{X}} = \sigma/\sqrt{n}$, which involves no unknown parameters only if the value of σ is known. A large-sample test statistic results from standardizing $\hat{\theta}$ under the assumption that H_0 is true [so that $E(\hat{\theta}) = \theta_0$]:

$$\text{Test statistic:} \quad Z = \frac{\hat{\theta} - \theta_0}{\sigma_{\hat{\theta}}}$$

If the alternative hypothesis is H_a: $\theta > \theta_0$, an upper-tailed test whose significance level is approximately α is specified by the rejection region $z \geq z_\alpha$. The other two alternatives, H_a: $\theta < \theta_0$ and H_a: $\theta \neq \theta_0$, are tested using a lower-tailed z test and a two-tailed z test, respectively.

In the case $\theta = p$, $\sigma_{\hat{\theta}}$ will not involve any unknown parameters when H_0 is true, but this is atypical. When $\sigma_{\hat{\theta}}$ does involve unknown parameters, it is often possible to use an estimated standard deviation $S_{\hat{\theta}}$ in place of $\sigma_{\hat{\theta}}$ and still have Z approximately normally distributed when H_0 is true (because when n is large, $s_{\hat{\theta}} \approx \sigma_{\hat{\theta}}$ for most samples). The large-sample test of the previous section furnishes an example of this: Because σ is usually unknown, we use $s_{\hat{\theta}} = s_{\overline{X}} = s/\sqrt{n}$ in place of $\sigma/\sqrt{n}$ in the denominator of z.

The estimator $\hat{p} = X/n$ is unbiased $[E(\hat{p}) = p]$, has approximately a normal distribution, and its standard deviation is $\sigma_{\hat{p}} = \sqrt{p(1-p)/n}$. These facts were used in Section 8.2 to obtain a confidence interval for p. When H_0 is true, $E(\hat{p}) = p_0$ and $\sigma_{\hat{p}} = \sqrt{p_0(1-p_0)/n}$, so $\sigma_{\hat{p}}$ does not involve any unknown parameters. It then follows that when n is large and H_0 is true, the test statistic

$$Z = \frac{\hat{p} - p_0}{\sqrt{p_0(1-p_0)/n}}$$

has approximately a standard normal distribution. If the alternative hypothesis is H_a: $p > p_0$ and the upper-tailed rejection region $z \geq z_\alpha$ is used, then

$$\begin{aligned} P(\text{type I error}) &= P(H_0 \text{ is rejected when it is true}) \\ &= P(Z \geq z_\alpha \text{when } Z \text{ has approximately a standard normal} \\ &\qquad \text{distribution}) \approx \alpha \end{aligned}$$

Thus the desired level of significance α is attained by using the critical value that captures area α in the upper tail of the z curve. Rejection regions for the other two alternative hypotheses, lower-tailed for H_a: $p < p_0$ and two-tailed for H_a: $p \neq p_0$, are justified in an analogous manner.

Null hypothesis: H_0: $p = p_0$

Test statistic value: $z = \dfrac{\hat{p} - p_0}{\sqrt{p_0(1 - p_0)/n}}$

Alternative Hypothesis	**Rejection Region**
H_a: $p > p_0$	$z \geq z_\alpha$ (upper-tailed)
H_a: $p < p_0$	$z \leq -z_\alpha$ (lower-tailed)
H_a: $p \neq p_0$	either $z \geq z_{\alpha/2}$ or $z \leq -z_{\alpha/2}$ (two-tailed)

These test procedures are valid provided that $np_0 \geq 10$ and $n(1 - p_0) \geq 10$.

Example 9.11 Recent information suggests that obesity is an increasing problem in America among all age groups. The Associated Press (Oct. 9, 2002) reported that 1276 individuals in a sample of 4115 adults were found to be obese (a body mass index exceeding 30; this index is a measure of weight relative to height). A 1998 survey based on people's own assessment revealed that 20% of adult Americans considered themselves obese. Does the recent data suggest that the true proportion of adults who are obese is more than 1.5 times the percentage from the self-assessment survey? Let's carry out a test of hypotheses using a significance level of .10.

1. p = the proportion of all American adults who are obese.
2. Saying that the current percentage is 1.5 times the self-assessment percentage is equivalent to the assertion that the current percentage is 30%, from which we have the null hypothesis as H_0: $p = .30$.
3. The phrase "more than" in the problem description implies that the alternative hypothesis is H_a: $p > .30$.
4. Since $np_0 = 4115(.3) \geq 10$ and $nq_0 = 4115(.7) \geq 10$, the large-sample z test can certainly be used. The test statistic value is

$$z = (\hat{p} - .3)/\sqrt{(.3)(.7)/n}$$

5. The form of H_a implies that an upper-tailed test is appropriate: Reject H_0 if $z \geq z_{.10} = 1.28$.

6. $\hat{p} = 1276/4115 = .310$, from which $z = (.310 - .3)/\sqrt{(.3)(.7)/4115} = .010/.0071 = 1.40$.

7. Since 1.40 exceeds the critical value 1.28, z lies in the rejection region. This justifies rejecting the null hypothesis. Using a significance level of .10, it does appear that more than 30% of American adults are obese. ■

β and Sample Size Determination When H_0 is true, the test statistic Z has approximately a standard normal distribution. Now suppose that H_0 is *not* true and that $p = p'$. Then Z still has approximately a normal distribution (because it is a linear function of $\hat{p}$), but its mean value and variance are no longer 0 and 1, respectively. Instead,

$$E(Z) = \frac{p' - p_0}{\sqrt{p_0(1 - p_0)/n}} \qquad V(Z) = \frac{p'(1 - p')/n}{p_0(1 - p_0)/n}$$

The probability of a type II error for an upper-tailed test is $\beta(p') = P(Z < z_\alpha$ when $p = p')$. This can be computed by using the given mean and variance to standardize and then referring to the standard normal cdf. In addition, if it is desired that the level α test also have $\beta(p') = \beta$ for a specified value of β, this equation can be solved for the necessary n as in Section 9.2. General expressions for $\beta(p')$ and n are given in the accompanying box.

Alternative Hypothesis	**$\beta(p')$**
H_a: $p > p_0$	$\Phi\left[\dfrac{p_0 - p' + z_\alpha\sqrt{p_0(1 - p_0)/n}}{\sqrt{p'(1 - p')/n}}\right]$
H_a: $p < p_0$	$1 - \Phi\left[\dfrac{p_0 - p' - z_\alpha\sqrt{p_0(1 - p_0)/n}}{\sqrt{p'(1 - p')/n}}\right]$
H_a: $p \neq p_0$	$\Phi\left[\dfrac{p_0 - p' + z_{\alpha/2}\sqrt{p_0(1 - p_0)/n}}{\sqrt{p'(1 - p')/n}}\right] - \Phi\left[\dfrac{p_0 - p' - z_{\alpha/2}\sqrt{p_0(1 - p_0)/n}}{\sqrt{p'(1 - p')/n}}\right]$

The sample size n for which the level α test also satisfies $\beta(p') = \beta$ is

$$n = \begin{cases} \left[\dfrac{z_\alpha\sqrt{p_0(1 - p_0)} + z_\beta\sqrt{p'(1 - p')}}{p' - p_0}\right]^2 & \text{one-tailed test} \\ \left[\dfrac{z_{\alpha/2}\sqrt{p_0(1 - p_0)} + z_\beta\sqrt{p'(1 - p')}}{p' - p_0}\right]^2 & \text{two-tailed test (an approximate solution)} \end{cases}$$

Example 9.12 A package-delivery service advertises that at least 90% of all packages brought to its office by 9 a.m. for delivery in the same city are delivered by noon that day. Let p denote the true proportion of such packages that are delivered as advertised and consider the hypotheses H_0: $p = .9$ versus H_a: $p < .9$. If only 80% of the packages are delivered as advertised, how likely is it that a level .01 test based on $n = 225$ packages will detect such a departure from H_0? What should the sample size be to ensure that $\beta(.8) = .01$? With $\alpha = .01$, $p_0 = .9$, $p' = .8$, and $n = 225$,

$$\begin{aligned}\beta(.8) &= 1 - \Phi\left[\frac{.9 - .8 - 2.33\sqrt{(.9)(.1)/225}}{\sqrt{(.8)(.2)/225}}\right] \\ &= 1 - \Phi(2.00) = .0228\end{aligned}$$

Thus the probability that H_0 will be rejected using the test when $p = .8$ is .9772—roughly 98% of all samples will result in correct rejection of H_0.

Using $z_\alpha = z_\beta = 2.33$ in the sample size formula yields

$$n = \left[\frac{2.33\sqrt{(.9)(.1)} + 2.33\sqrt{(.8)(.2)}}{.8 - .9}\right]^2 \approx 266$$

■

Small-Sample Tests

Test procedures when the sample size n is small are based directly on the binomial distribution rather than the normal approximation. Consider the alternative hypothesis H_a: $p > p_0$ and again let X be the number of successes in the sample. Then X is the test statistic, and the upper-tailed rejection region has the form $x \geq c$. When H_0 is true, X has a binomial distribution with parameters n and p_0, so

$$\begin{aligned}P(\text{type I error}) &= P(H_0 \text{ is rejected when it is true}) \\ &= P[X \geq c \text{ when } X \sim \text{Bin}(n, p_0)] \\ &= 1 - P[X \leq c - 1 \text{ when } X \sim \text{Bin}(n, p_0)] \\ &= 1 - B(c - 1; n, p_0)\end{aligned}$$

As the critical value c decreases, more x values are included in the rejection region and P(type I error) increases. Because X has a discrete probability distribution, it is usually not possible to find a value of c for which P(type I error) is exactly the desired significance level α (e.g., .05 or .01). Instead, the largest rejection region of the form $\{c, c + 1, \ldots, n\}$ satisfying $1 - B(c - 1; n, p_0) \leq \alpha$ is used.

Let p' denote an alternative value of p ($p' > p_0$). When $p = p'$, $X \sim \text{Bin}(n, p')$, so

$$\begin{aligned}\beta(p') &= P(\text{type II error when } p = p') \\ &= P[X < c \text{ when } X \sim \text{Bin}(n, p')] = B(c - 1; n, p')\end{aligned}$$

That is, $\beta(p')$ is the result of a straightforward binomial probability calculation. The sample size n necessary to ensure that a level α test also has specified β at a particular alternative value p' must be determined by trial and error using the binomial cdf.

Test procedures for H_a: $p < p_0$ and for H_a: $p \neq p_0$ are constructed in a similar manner. In the former case, the appropriate rejection region has the form $x \leq c$ (a lower-tailed test). The critical value c is the largest number satisfying $B(c; n, p_0) \leq \alpha$.

The rejection region when the alternative hypothesis is H_a: $p \neq p_0$ consists of both large and small x values.

Example 9.13 A plastics manufacturer has developed a new type of plastic trash can and proposes to sell them with an unconditional 6-year warranty. To see whether this is economically feasible, 20 prototype cans are subjected to an accelerated life test to simulate 6 years of use. The proposed warranty will be modified only if the sample data strongly suggests that fewer than 90% of such cans would survive the 6-year period. Let p denote the proportion of all cans that survive the accelerated test. The relevant hypotheses are then H_0: $p = .9$ versus H_a: $p < .9$. A decision will be based on the test statistic X, the number among the 20 that survive. If the desired significance level is $\alpha = .05$, c must satisfy $B(c; 20, .9) \leq .05$. From Appendix Table A.1, $B(15; 20, .9) = .043$, and $B(16; 20, .9) = .133$. The appropriate rejection region is therefore $x \leq 15$. If the accelerated test results in $x = 14$, H_0 would be rejected in favor of H_a, necessitating a modification of the proposed warranty. The probability of a type II error for the alternative value $p' = .8$ is

$$\begin{aligned}\beta(.8) &= P[H_0 \text{ is not rejected when } X \sim \text{Bin}(20, .8)] \\ &= P[X \geq 16 \text{ when } X \sim \text{Bin}(20, .8)] \\ &= 1 - B(15; 20, .8) = 1 - .370 = .630\end{aligned}$$

That is, when $p = .8$, 63% of all samples consisting of $n = 20$ cans would result in H_0 being incorrectly not rejected. This error probability is high because 20 is a small sample size and $p' = .8$ is close to the null value $p_0 = .9$. ■

Exercises Section 9.3 (36–44)

36. State DMV records indicate that of all vehicles undergoing emissions testing during the previous year, 70% passed on the first try. A random sample of 200 cars tested in a particular county during the current year yields 124 that passed on the initial test. Does this suggest that the true proportion for this county during the current year differs from the previous statewide proportion? Test the relevant hypotheses using $\alpha = .05$.

37. A manufacturer of nickel–hydrogen batteries randomly selects 100 nickel plates for test cells, cycles them a specified number of times, and determines that 14 of the plates have blistered.

a. Does this provide compelling evidence for concluding that more than 10% of all plates blister under such circumstances? State and test the appropriate hypotheses using a significance level of .05. In reaching your conclusion, what type of error might you have committed?

b. If it is really the case that 15% of all plates blister under these circumstances and a sample size of 100 is used, how likely is it that the null hypothesis of part (a) will not be rejected by the level .05 test? Answer this question for a sample size of 200.

c. How many plates would have to be tested to have $\beta(.15) = .10$ for the test of part (a)?

38. A random sample of 150 recent donations at a certain blood bank reveals that 82 were type A blood. Does this suggest that the actual percentage of type A donations differs from 40%, the percentage of the population having type A blood? Carry out a test of the appropriate hypotheses using a significance level of .01. Would your conclusion have been different if a significance level of .05 had been used?

39. A university library ordinarily has a complete shelf inventory done once every year. Because of new shelving rules instituted the previous year, the head librarian believes it may be possible to save money by postponing the inventory. The librarian decides to select at random 1000 books from the library's collection and have them searched in a preliminary manner. If evidence indicates strongly that the true proportion of misshelved or unlocatable books is less than .02, then the inventory will be postponed.

a. Among the 1000 books searched, 15 were misshelved or unlocatable. Test the relevant hypotheses and advise the librarian what to do (use $\alpha = .05$).

b. If the true proportion of misshelved and lost books is actually .01, what is the probability that the inventory will be (unnecessarily) taken?

c. If the true proportion is .05, what is the probability that the inventory will be postponed?

40. The article "Statistical Evidence of Discrimination" (*J. Amer. Statist. Assoc.*, 1982: 773–783) discusses the court case *Swain v. Alabama* (1965), in which it was alleged that there was discrimination against blacks in grand jury selection. Census data suggested that 25% of those eligible for grand jury service were black, yet a random sample of 1050 people called to appear for possible duty yielded only 177 blacks. Using a level .01 test, does this data argue strongly for a conclusion of discrimination?

41. A plan for an executive traveler's club has been developed by an airline on the premise that 5% of its current customers would qualify for membership. A random sample of 500 customers yielded 40 who would qualify.

a. Using this data, test at level .01 the null hypothesis that the company's premise is correct against the alternative that it is not correct.

b. What is the probability that when the test of part (a) is used, the company's premise will be judged correct when in fact 10% of all current customers qualify?

42. Each of a group of 20 intermediate tennis players is given two rackets, one having nylon strings and the other synthetic gut strings. After several weeks of playing with the two rackets, each player will be asked to state a preference for one of the two types of strings. Let p denote the proportion of all such players who would prefer gut to nylon, and let X be the number of players in the sample who prefer gut. Because gut strings are more expensive, consider the null hypothesis that at most 50% of all such players prefer gut. We simplify this to H_0: $p = .5$, planning to reject H_0 only if sample evidence strongly favors gut strings.

a. Which of the rejection regions {15, 16, 17, 18, 19, 20}, {0, 1, 2, 3, 4, 5}, or {0, 1, 2, 3, 17, 18, 19, 20} is most appropriate, and why are the other two not appropriate?

b. What is the probability of a type I error for the chosen region of part (a)? Does the region specify a level .05 test? Is it the best level .05 test?

c. If 60% of all enthusiasts prefer gut, calculate the probability of a type II error using the appropriate region from part (a). Repeat if 80% of all enthusiasts prefer gut.

d. If 13 out of the 20 players prefer gut, should H_0 be rejected using a significance level of .10?

43. A manufacturer of plumbing fixtures has developed a new type of washerless faucet. Let $p = P$(a randomly selected faucet of this type will develop a leak within 2 years under normal use). The manufacturer has decided to proceed with production unless it can be determined that p is too large; the borderline acceptable value of p is specified as .10. The manufacturer decides to subject n of these faucets to accelerated testing (approximating 2 years of normal use). With X = the number among the n faucets that leak before the test concludes, production will commence unless the observed X is too large. It is decided that if $p = .10$, the probability of not proceeding should be at most .10, whereas if $p = .30$ the probability of proceeding should be at most .10. Can $n = 10$ be used? $n = 20$? $n = 25$? What is the appropriate rejection region for the chosen n, and what are the actual error probabilities when this region is used?

44. Scientists have recently become concerned about the safety of Teflon cookware and various food containers because perfluorooctanoic acid (PFOA) is used in the manufacturing process. An article in the July 27, 2005, *New York Times* reported that of 600 children tested, 96% had PFOA in their blood. According to the FDA, 90% of all Americans have PFOA in their blood.

a. Does the data on PFOA incidence among children suggest that the percentage of all children who have PFOA in their blood exceeds the FDA percentage for all Americans? Carry out an appropriate test of hypotheses.

b. If 95% of all children have PFOA in their blood, how likely is it that the null hypothesis tested in (a) will be rejected when a significance level of .01 is employed?

c. Referring back to (b), what sample size would be necessary for the relevant probability to be .10?

9.4 *P*-Values

One way to report the result of a hypothesis-testing analysis is to simply say whether the null hypothesis was rejected at a specified level of significance. Thus an investigator might state that H_0 was rejected at level of significance .05 or that use of a level .01 test resulted in not rejecting H_0. This type of statement is somewhat inadequate because it says nothing about whether the computed value of the test statistic just barely fell into the rejection region or whether it exceeded the critical value by a large amount. A related difficulty is that such a report imposes the specified significance level on other decision makers. In many decision situations, individuals may have different views concerning the consequences of a type I or type II error. Each individual would then want to select his or her own significance level — some selecting $\alpha = .05$, others .01, and so on — and reach a conclusion accordingly. This could result in some individuals rejecting H_0 while others conclude that the data does not show a strong enough contradiction of H_0 to justify its rejection.

Example 9.14 The true average time to initial relief of pain for a best-selling pain reliever is known to be 10 min. Let μ denote the true average time to relief for a company's newly developed reliever. The company wishes to produce and market this reliever only if it provides quicker relief than the best-seller, so it wishes to test H_0: $\mu = 10$ versus H_a: $\mu < 10$. Only if experimental evidence leads to rejection of H_0 will the new reliever be introduced. After weighing the relative seriousness of each type of error, a single level of significance must be agreed on and a decision — to reject H_0 and introduce the reliever or not to do so — made at that level.

Suppose the new reliever has been introduced. The company supports its claim of quicker relief by stating that, based on an analysis of experimental data, H_0: $\mu = 10$ was rejected in favor of H_a: $\mu < 10$ using level of significance $\alpha = .10$. Any individuals contemplating a switch to this new reliever would naturally want to reach their own conclusions concerning the validity of the claim. Individuals who are satisfied with the best-seller would view a type I error (concluding that the new product provides quicker relief when it actually does not) as serious, so they might wish to use $\alpha = .05$, .01, or even smaller levels. Unfortunately, the nature of the company's statement prevents an individual decision maker from reaching a conclusion at such a level. The company has imposed its own choice of significance level on others. The report could have been done in a manner that allowed each individual flexibility in drawing a conclusion at a personally selected α. ■

A *P-value* conveys much information about the strength of evidence against H_0 and allows an individual decision maker to draw a conclusion at any specified level α. Before we give a general definition, consider how the conclusion in a hypothesis-testing problem depends on the selected level α.

Example 9.15 The nicotine content problem discussed in Example 9.5 involved testing H_0: $\mu = 1.5$ versus H_a: $\mu > 1.5$. Because of the inequality in H_a, the rejection region is upper-tailed, with H_0 rejected if $z \geq z_\alpha$. Suppose $z = 2.10$. The accompanying table displays the rejection region for each of four different α's along with the resulting conclusion.

Level of Significance α	Rejection Region	Conclusion
.05	$z \geq 1.645$	Reject H_0
.025	$z \geq 1.96$	Reject H_0
.01	$z \geq 2.33$	Do not reject H_0
.005	$z \geq 2.58$	Do not reject H_0

For α relatively large, the z critical value z_α is not very far out in the upper tail; 2.10 exceeds the critical value, and so H_0 is rejected. However, as α decreases, the critical value increases. For small α, the z critical value is large, 2.10 is less than z_α, and H_0 is not rejected.

Recall that for an upper-tailed z test, α is just the area under the z curve to the right of the critical value z_α. That is, once α is specified, the critical value is chosen to capture upper-tail area α. Appendix Table A.3 shows that the area to the right of 2.10 is .0179. Using an α larger than .0179 corresponds to $z_\alpha < 2.10$. An α less than .0179 necessitates using a z critical value that exceeds 2.10. The decision at a particular level α thus depends on how the selected α compares to the tail area captured by the computed z. This is illustrated in Figure 9.5. Notice in particular that .0179, the captured tail area, is the smallest level α at which H_0 would be rejected, because using any smaller α results in a z critical value that exceeds 2.10, so that 2.10 is not in the rejection region.

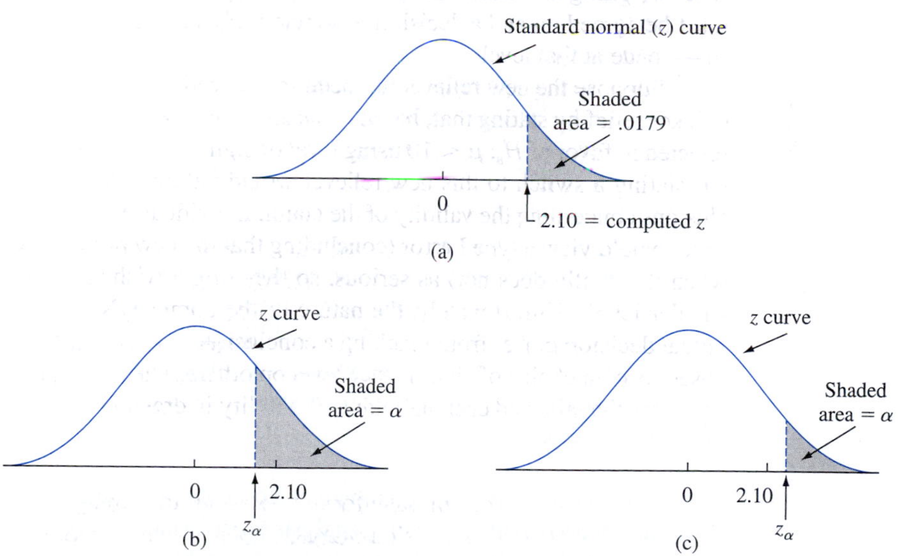

Figure 9.5 Relationship between α and tail area captured by computed z: (a) tail area captured by computed z; (b) when $\alpha > .0179$, $z_\alpha < 2.10$ and H_0 is rejected; (c) when $\alpha < .0179$, $z_\alpha > 2.10$ and H_0 is not rejected ■

In general, suppose the probability distribution of a test statistic when H_0 is true has been determined. Then, for specified α, the rejection region is determined by finding a critical value or values that capture tail area α (upper-, lower-, or two-tailed, whichever is appropriate) under the probability distribution curve. The smallest α for which H_0 would be rejected is the tail area captured by the computed value of the test statistic. This smallest α is the P-value.

DEFINITION

The ***P*-value** (or *observed significance level*) is the smallest level of significance at which H_0 would be rejected when a specified test procedure is used on a given data set. Once the P-value has been determined, the conclusion at any particular level α results from comparing the P-value to α:

1. P-value $\leq \alpha \Rightarrow$ reject H_0 at level α.
2. P-value $> \alpha \Rightarrow$ do not reject H_0 at level α.

It is customary to call the data *significant* when H_0 is rejected and *not significant* otherwise. The P-value is then the smallest level at which the data is significant. An easy way to visualize the comparison of the P-value with the chosen α is to draw a picture like that of Figure 9.6. The calculation of the P-value depends on whether the test is upper-, lower-, or two-tailed. However, once it has been calculated, the comparison with α does not depend on which type of test was used.

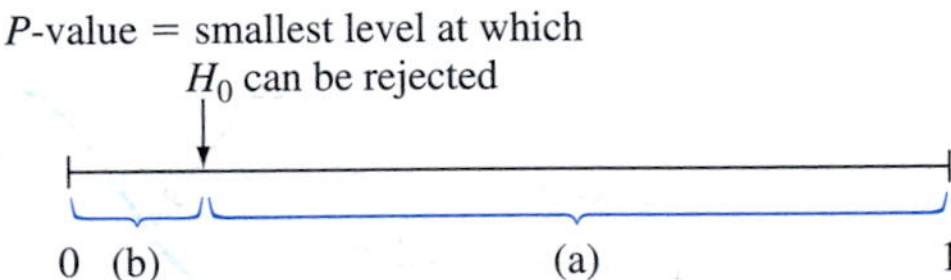

Figure 9.6 Comparing α and the P-value: (a) reject H_0 when α lies here; (b) do not reject H_0 when α lies here

Example 9.16 (Example 9.14 continued)

Suppose that when data from an experiment involving the new pain reliever was analyzed, the P-value for testing H_0: $\mu = 10$ versus H_a: $\mu < 10$ was calculated as .0384. Since $\alpha = .05$ is larger than the P-value [.05 lies in the interval (a) of Figure 9.6], H_0 would be rejected by anyone carrying out the test at level .05. However, at level .01, H_0 would not be rejected because .01 is smaller than the smallest level (.0384) at which H_0 can be rejected. ■

The most widely used statistical computer packages automatically include a P-value when a hypothesis-testing analysis is performed. A conclusion can then be drawn directly from the output, without reference to a table of critical values.

A useful alternative definition equivalent to the one just given is as follows:

DEFINITION

The ***P*-value** is the probability, calculated assuming H_0 is true, of obtaining a test statistic value at least as contradictory to H_0 as the value that actually resulted. The smaller the P-value, the more contradictory is the data to H_0.

Thus if $z = 2.10$ for an upper-tailed z test, P-value $= P(Z \geq 2.10$, when H_0 is true$) = 1 - \Phi(2.10) = .0179$, as before. Beware: The P-value is not the probability that H_0 is true, nor is it an error probability!

P-Values for z Tests

The P-value for a z test (one based on a test statistic whose distribution when H_0 is true is at least approximately standard normal) is easily determined from the information in Appendix Table A.3. Consider an upper-tailed test and let z denote the computed value of the test statistic Z. The null hypothesis is rejected if $z \geq z_\alpha$, and the P-value is the smallest α for which this is the case. Since z_α increases as α decreases, the P-value is the value of α for which $z = z_\alpha$. That is, the P-value is just the area captured by the computed value z in the upper tail of the standard normal curve. The corresponding cumulative area is $\Phi(z)$, so in this case P-value $= 1 - \Phi(z)$.

An analogous argument for a lower-tailed test shows that the P-value is the area captured by the computed value z in the lower tail of the standard normal curve. More care must be exercised in the case of a two-tailed test. Suppose first that z is positive. Then the P-value is the value of α satisfying $z = z_{\alpha/2}$ (i.e., computed z = upper-tail critical value). This says that the area captured in the upper tail is half the P-value, so that P-value $= 2[1 - \Phi(z)]$. If z is negative, the P-value is the α for which $z = -z_{\alpha/2}$, or, equivalently,

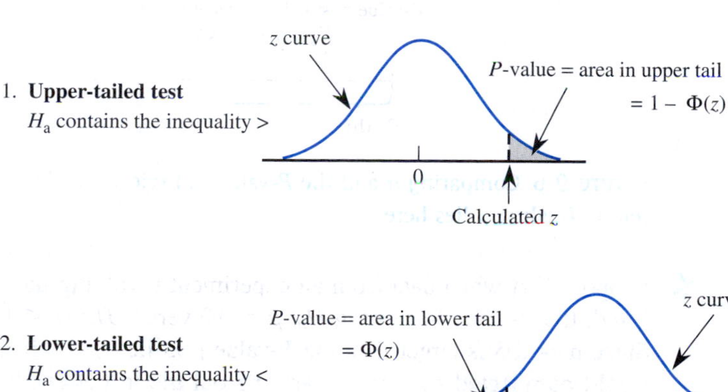

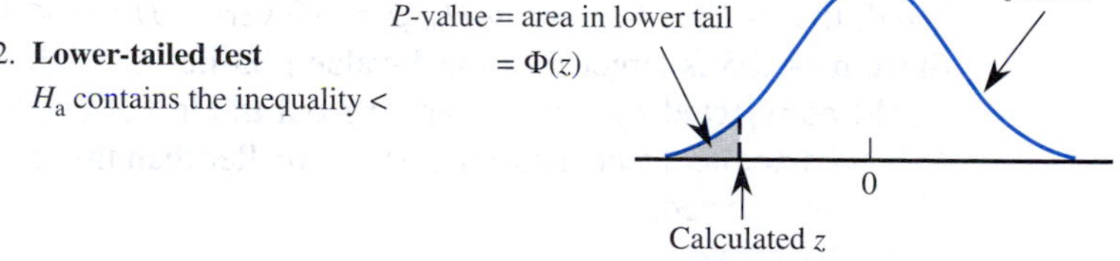

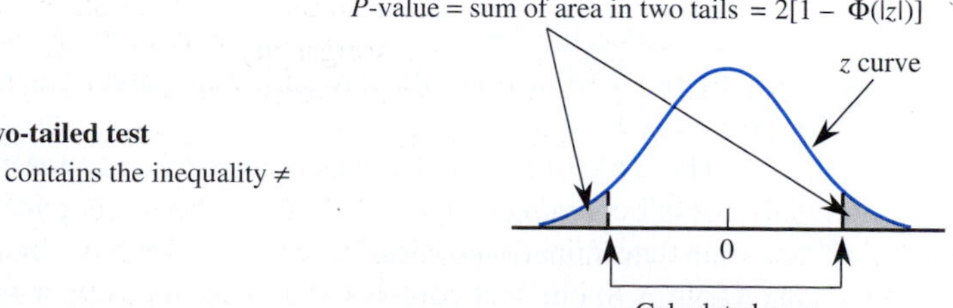

Figure 9.7 Determination of the P-value for a z test

$-z = z_{\alpha/2}$, so P-value $= 2[1 - \Phi(-z)]$. Since $-z = |z|$ when z is negative, P-value $=$ $2[1 - \Phi(|z|)]$ for either positive or negative z.

$$P\text{-value:} \quad P = \begin{cases} 1 - \Phi(z) & \text{for an upper-tailed test} \\ \Phi(z) & \text{for a lower-tailed test} \\ 2[1 - \Phi(|z|)] & \text{for a two-tailed test} \end{cases}$$

Each of these is the probability of getting a value at least as extreme as what was obtained (assuming H_0 true). The three cases are illustrated in Figure 9.7.

The next example illustrates the use of the P-value approach to hypothesis testing by means of a sequence of steps modified from our previously recommended sequence.

Example 9.17 The target thickness for silicon wafers used in a certain type of integrated circuit is 245 μm. A sample of 50 wafers is obtained and the thickness of each one is determined, resulting in a sample mean thickness of 246.18 μm and a sample standard deviation of 3.60 μm. Does this data suggest that true average wafer thickness is something other than the target value?

1. Parameter of interest: μ = true average wafer thickness
2. Null hypothesis: H_0: $\mu = 245$
3. Alternative hypothesis: H_a: $\mu \neq 245$
4. Formula for test statistic value: $z = \dfrac{\bar{x} - 245}{s/\sqrt{n}}$
5. Calculation of test statistic value: $z = \dfrac{246.18 - 245}{3.60/\sqrt{50}} = 2.32$
6. Determination of P-value: Because the test is two-tailed,
$$P\text{-value} = 2[1 - \Phi(2.32)] = .0204$$
7. Conclusion: Using a significance level of .01, H_0 would not be rejected since $.0204 > .01$. At this significance level, there is insufficient evidence to conclude that true average thickness differs from the target value. ■

P-Values for *t* Tests

Just as the P-value for a z test is a z curve area, the P-value for a t test will be a t curve area. Figure 9.8 illustrates the three different cases. The number of df for the one-sample t test is $n - 1$.

The table of t critical values used previously for confidence and prediction intervals doesn't contain enough information about any particular t distribution to allow for accurate determination of desired areas. So we have included another t table in Appendix Table A.8, one that contains a tabulation of upper-tail t curve areas. Each different column of the table is for a different number of df, and the rows are for calculated values of the test statistic t ranging from 0.0 to 4.0 in increments of .1. For example, the number .074 appears at the intersection of the 1.6 row and the 8 df column, so the area under

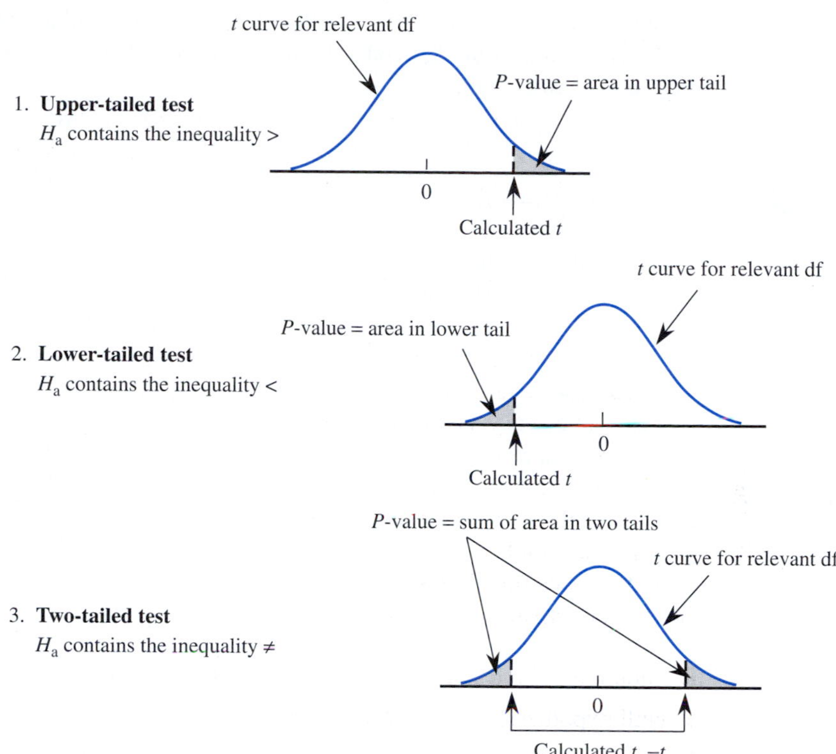

Figure 9.8 *P*-values for *t* tests

the 8 df curve to the right of 1.6 (an upper-tail area) is .074. Because t curves are symmetric, .074 is also the area under the 8 df curve to the left of -1.6 (a lower-tail area).

Suppose, for example, that a test of H_0: $\mu = 100$ versus H_a: $\mu > 100$ is based on the 8 df t distribution. If the calculated value of the test statistic is $t = 1.6$, then the P-value for this upper-tailed test is .074. Because .074 exceeds .05, we would not be able to reject H_0 at a significance level of .05. If the alternative hypothesis is H_a: $\mu < 100$ and a test based on 20 df yields $t = -3.2$, then Appendix Table A.8 shows that the P-value is the captured lower-tail area .002. The null hypothesis can be rejected at either level .05 or .01. Consider testing H_0: $\mu_1 - \mu_2 = 0$ versus H_a: $\mu_1 - \mu_2 \neq 0$; the null hypothesis states that the means of the two populations are identical, whereas the alternative hypothesis states that they are different without specifying a direction of departure from H_0. If a t test is based on 20 df and $t = 3.2$, then the P-value for this two-tailed test is $2(.002) = .004$. This would also be the P-value for $t = -3.2$. The tail area is doubled because values both larger than 3.2 and smaller than -3.2 are more contradictory to H_0 than what was calculated (values farther out in *either* tail of the t curve).

Example 9.18 In Example 9.9, we carried out a test of H_0: $\mu = 25$ versus H_a: $\mu > 25$ based on 4 df. The calculated value of t was 1.04. Looking to the 4 df column of Appendix Table A.8 and down to the 1.0 row, we see that the entry is .187, so the P-value $\approx$ .187. This P-value is clearly larger than any reasonable significance level α (.01, .05, and even .10), so there is no reason to reject the null hypothesis. The MINITAB output included in

Example 9.9 has P-value $= .18$. P-values from software packages will be more accurate than what results from Appendix Table A.8 since values of t in our table are accurate only to the tenths digit. ■

Exercises Section 9.4 (45–59)

45. For which of the given P-values would the null hypothesis be rejected when performing a level .05 test?
a. .001 **b.** .021 **c.** .078
d. .047 **e.** .148

46. Pairs of P-values and significance levels, α, are given. For each pair, state whether the observed P-value would lead to rejection of H_0 at the given significance level.
a. P-value $= .084$, $\alpha = .05$
b. P-value $= .003$, $\alpha = .001$
c. P-value $= .498$, $\alpha = .05$
d. P-value $= .084$, $\alpha = .10$
e. P-value $= .039$, $\alpha = .01$
f. P-value $= .218$, $\alpha = .10$

47. Let μ denote the mean reaction time to a certain stimulus. For a large-sample z test of H_0: $\mu = 5$ versus H_a: $\mu > 5$, find the P-value associated with each of the given values of the z test statistic.
a. 1.42 **b.** .90 **c.** 1.96
d. 2.48 **e.** $-.11$

48. Newly purchased tires of a certain type are supposed to be filled to a pressure of 30 lb/in^2. Let μ denote the true average pressure. Find the P-value associated with each given z statistic value for testing H_0: $\mu = 30$ versus H_a: $\mu \neq 30$.
a. 2.10 **b.** -1.75 **c.** $-.55$
d. 1.41 **e.** -5.3

49. Give as much information as you can about the P-value of a t test in each of the following situations:
a. Upper-tailed test, df $= 8$, $t = 2.0$
b. Lower-tailed test, df $= 11$, $t = -2.4$
c. Two-tailed test, df $= 15$, $t = -1.6$
d. Upper-tailed test, df $= 19$, $t = -.4$
e. Upper-tailed test, df $= 5$, $t = 5.0$
f. Two-tailed test, df $= 40$, $t = -4.8$

50. The paint used to make lines on roads must reflect enough light to be clearly visible at night. Let μ denote the true average reflectometer reading for a new type of paint under consideration. A test of H_0: $\mu = 20$ versus H_a: $\mu > 20$ will be based on a random sample of size n from a normal population distribution. What conclusion is appropriate in each of the following situations?
a. $n = 15$, $t = 3.2$, $\alpha = .05$
b. $n = 9$, $t = 1.8$, $\alpha = .01$
c. $n = 24$, $t = -.2$

51. Let μ denote true average serum receptor concentration for all pregnant women. The average for all women is known to be 5.63. The article "Serum Transferrin Receptor for the Detection of Iron Deficiency in Pregnancy" (*Amer. J. Clin. Nutrit.*, 1991: 1077–1081) reports that P-value $> .10$ for a test of H_0: $\mu = 5.63$ versus H_a: $\mu \neq 5.63$ based on $n = 176$ pregnant women. Using a significance level of .01, what would you conclude?

52. An aspirin manufacturer fills bottles by weight rather than by count. Since each bottle should contain 100 tablets, the average weight per tablet should be 5 grains. Each of 100 tablets taken from a very large lot is weighed, resulting in a sample average weight per tablet of 4.87 grains and a sample standard deviation of .35 grain. Does this information provide strong evidence for concluding that the company is not filling its bottles as advertised? Test the appropriate hypotheses using $\alpha = .01$ by first computing the P-value and then comparing it to the specified significance level.

53. Because of variability in the manufacturing process, the actual yielding point of a sample of mild steel subjected to increasing stress will usually differ from the theoretical yielding point. Let p denote the true proportion of samples that yield before their theoretical yielding point. If on the basis of a sample it can be concluded that more than 20% of all specimens yield before the theoretical point, the production process will have to be modified.
a. If 15 of 60 specimens yield before the theoretical point, what is the P-value when the appropriate test is used, and what would you advise the company to do?
b. If the true percentage of "early yields" is actually 50% (so that the theoretical point is the median

of the yield distribution) and a level .01 test is used, what is the probability that the company concludes a modification of the process is necessary?

54. Many consumers are turning to generics as a way of reducing the cost of prescription medications. The article "Commercial Information on Drugs: Confusing to the Physician?" (*J. Drug Issues*, 1988: 245–257) gives the results of a survey of 102 doctors. Only 47 of those surveyed knew the generic name for the drug methadone. Does this provide strong evidence for concluding that fewer than half of all physicians know the generic name for methadone? Carry out a test of hypotheses using a significance level of .01 using the *P*-value method.

55. A random sample of soil specimens was obtained, and the amount of organic matter (%) in the soil was determined for each specimen, resulting in the accompanying data (from "Engineering Properties of Soil," *Soil Sci.*, 1998: 93–102).

1.10	5.09	0.97	1.59	4.60	0.32	0.55	1.45
0.14	4.47	1.20	3.50	5.02	4.67	5.22	2.69
3.98	3.17	3.03	2.21	0.69	4.47	3.31	1.17
0.76	1.17	1.57	2.62	1.66	2.05		

The values of the sample mean, sample standard deviation, and (estimated) standard error of the mean are 2.481, 1.616, and .295, respectively. Does this data suggest that the true average percentage of organic matter in such soil is something other than 3%? Carry out a test of the appropriate hypotheses at significance level .10 by first determining the *P*-value. Would your conclusion be different if $\alpha = .05$ had been used? (*Note*: A normal probability plot of the data shows an acceptable pattern in light of the reasonably large sample size.)

56. The times of first sprinkler activation for a series of tests with fire prevention sprinkler systems using an aqueous film-forming foam were (in sec)

27 41 22 27 23 35 30 33 24 27 28 22 24

(see "Use of AFFF in Sprinkler Systems," *Fire Tech.*, 1976: 5). The system has been designed so that true average activation time is at most 25 sec under such conditions. Does the data strongly contradict the validity of this design specification? Test the relevant hypotheses at significance level .05 using the *P*-value approach.

57. A certain pen has been designed so that true average writing lifetime under controlled conditions (involving the use of a writing machine) is at least 10 hours. A random sample of 18 pens is selected, the writing lifetime of each is determined, and a normal probability plot of the resulting data supports the use of a one-sample *t* test.

a. What hypotheses should be tested if the investigators believe a priori that the design specification has been satisfied?

b. What conclusion is appropriate if the hypotheses of part (a) are tested, $t = -2.3$, and $\alpha = .05$?

c. What conclusion is appropriate if the hypotheses of part (a) are tested, $t = -1.8$, and $\alpha = .01$?

d. What should be concluded if the hypotheses of part (a) are tested and $t = -3.6$?

58. A spectrophotometer used for measuring CO concentration [ppm (parts per million) by volume] is checked for accuracy by taking readings on a manufactured gas (called span gas) in which the CO concentration is very precisely controlled at 70 ppm. If the readings suggest that the spectrophotometer is not working properly, it will have to be recalibrated. Assume that if it is properly calibrated, measured concentration for span gas samples is normally distributed. On the basis of the six readings—85, 77, 82, 68, 72, and 69—is recalibration necessary? Carry out a test of the relevant hypotheses using the *P*-value approach with $\alpha = .05$.

59. The relative conductivity of a semiconductor device is determined by the amount of impurity "doped" into the device during its manufacture. A silicon diode to be used for a specific purpose requires an average cut-on voltage of .60 V, and if this is not achieved, the amount of impurity must be adjusted. A sample of diodes was selected and the cut-on voltage was determined. The accompanying SAS output resulted from a request to test the appropriate hypotheses.

```
N        Mean    Std Dev              T  Prob>|T|
15 0.0453333 0.0899100 1.9527887      0.0711
```

[*Note*: SAS explicitly tests H_0: $\mu = 0$, so to test H_0: $\mu = .60$, the null value .60 must be subtracted from each x_i; the reported mean is then the average of the $(x_i - .60)$ values. Also, SAS's *P*-value is always for a two-tailed test.] What would be concluded for a significance level of .01? .05? .10?

9.5 *Some Comments on Selecting a Test Procedure

Once the experimenter has decided on the question of interest and the method for gathering data (the design of the experiment), construction of an appropriate test procedure consists of three distinct steps:

1. Specify a test statistic (the decision is based on this function of the data).
2. Decide on the general form of the rejection region (typically, reject H_0 for suitably large values of the test statistic, reject for suitably small values, or reject for either small or large values).
3. Select the specific numerical critical value or values that will separate the rejection region from the acceptance region (by obtaining the distribution of the test statistic when H_0 is true, and then selecting a level of significance).

In the examples thus far, both steps 1 and 2 were carried out in an ad hoc manner through intuition. For example, when the underlying population was assumed normal with mean μ and known σ, we were led from $\overline{X}$ to the standardized test statistic

$$Z = \frac{\overline{X} - \mu_0}{\sigma/\sqrt{n}}$$

For testing H_0: $\mu = \mu_0$ versus H_a: $\mu > \mu_0$, intuition then suggested rejecting H_0 when z was large. Finally, the critical value was determined by specifying the level of significance α and using the fact that Z has a standard normal distribution when H_0 is true. The reliability of the test in reaching a correct decision can be assessed by studying type II error probabilities.

Issues to be considered in carrying out steps 1–3 encompass the following questions:

1. What are the practical implications and consequences of choosing a particular level of significance once the other aspects of a test procedure have been determined?
2. Does there exist a general principle, not dependent just on intuition, that can be used to obtain best or good test procedures?
3. When two or more tests are appropriate in a given situation, how can the tests be compared to decide which should be used?
4. If a test is derived under specific assumptions about the distribution or population being sampled, how well will the test procedure work when the assumptions are violated?

Statistical Versus Practical Significance

Although the process of reaching a decision by using the methodology of classical hypothesis testing involves selecting a level of significance and then rejecting or not rejecting H_0 at that level, simply reporting the α used and the decision reached conveys little of the information contained in the sample data. Especially when the results of an experiment are to be communicated to a large audience, rejection of H_0 at level .05 will be much more convincing if the observed value of the test statistic greatly exceeds the 5% critical value than if it barely exceeds that value. This is precisely what led to the notion of P-value as a way of reporting significance without imposing a particular α on others who might wish to draw their own conclusions.

Even if a P-value is included in a summary of results, however, there may be difficulty in interpreting this value and in making a decision. This is because a small P-value, which would ordinarily indicate **statistical significance** in that it would strongly suggest rejection of H_0 in favor of H_a, may be the result of a large sample size in combination with a departure from H_0 that has little **practical significance**. In many experimental situations, only departures from H_0 of large magnitude would be worthy of detection, whereas a small departure from H_0 would have little practical significance.

Consider as an example testing H_0: $\mu = 100$ versus H_a: $\mu > 100$ where μ is the mean of a normal population with $\sigma = 10$. Suppose a true value of $\mu = 101$ would not represent a serious departure from H_0 in the sense that not rejecting H_0 when $\mu = 101$ would be a relatively inexpensive error. For a reasonably large sample size n, this μ would lead to an $\bar{x}$ value near 101, so we would not want this sample evidence to argue strongly for rejection of H_0 when $\bar{x} = 101$ is observed. For various sample sizes, Table 9.1 records both the P-value when $\bar{x} = 101$ and also the probability of not rejecting H_0 at level .01 when $\mu = 101$.

The second column in Table 9.1 shows that even for moderately large sample sizes, the P-value of $\bar{x} = 101$ argues very strongly for rejection of H_0, whereas the observed $\bar{x}$ itself suggests that in practical terms the true value of μ differs little from the null value $\mu_0 = 100$. The third column points out that even when there is little practical difference between the true μ and the null value, for a fixed level of significance a large sample size will almost always lead to rejection of the null hypothesis at that level. To summarize, *one must be especially careful in interpreting evidence when the sample size is large, since any small departure from H_0 will almost surely be detected by a test, yet such a departure may have little practical significance.*

Table 9.1 An illustration of the effect of sample size on P-values and β

n	P-Value When $\bar{x} = 101$	$\beta(101)$ for Level .01 Test
25	.3085	.9664
100	.1587	.9082
400	.0228	.6293
900	.0013	.2514
1600	.0000335	.0475
2500	.000000297	.0038
10,000	7.69×10^{-24}	.0000

Best Tests for Simple Hypotheses

The test procedures presented thus far are (hopefully) intuitively reasonable, but they have not been shown to be best in any sense. How can an optimal test be obtained, one for which the type II error probability is as small as possible, subject to controlling the type I error probability at the desired level? Our starting point here will be a rather unrealistic situation from a practical viewpoint: testing a *simple* null hypothesis against a *simple* alternative hypothesis. A **simple hypothesis** is one which, when true, completely specifies the distribution of the sample X_i's. Suppose, for example, that the X_i's form a random sample from an exponential distribution with parameter λ. Then the hypothesis H: $\lambda = 1$ is simple, since when H is true each X_i has an exponential distribution with parameter $\lambda = 1$. We might then consider H_0: $\lambda = 1$ versus H_a: $\lambda = 2$, both of which are simple hypotheses. The hypothesis H: $\lambda \leq 1$ is not simple, because when H is true, the distribution of each X_i might be exponential with $\lambda = 1$ or with $\lambda = .8$ or . . . . Similarly, if the X_i's constitute a random sample from a normal distribution with *known* σ, then H: $\mu = 100$ is a simple hypothesis. But if the value of σ is unknown, this hypothesis is not simple because the distribution of each X_i is then not completely specified—it could be normal with $\mu = 100$ and $\sigma = 15$ or normal with $\mu = 100$ and $\sigma = 12$ or normal with $\mu = 100$ and any other positive value of σ. For a hypothesis to be simple, the value of *every* parameter in the pmf or pdf of the X_i's must be specified.

The next result was a milestone in the theory of hypothesis testing—a method for constructing a best test for a simple null hypothesis versus a simple alternative hypothesis. Let $f(x_1, \ldots, x_n; \theta)$ be the joint pmf or pdf of the X_i's. Then our null hypothesis will assert that $\theta = \theta_0$ and the relevant alternative hypothesis will claim that $\theta = \theta_a$. The result will carry over to the case of more than one parameter as long as the value of each parameter is completely specified in both H_0 and H_a.

THE NEYMAN–PEARSON THEOREM

For testing a simple null hypothesis H_0: $\theta = \theta_0$ versus a simple alternative hypothesis H_a: $\theta = \theta_a$, let k be a positive fixed number and form the rejection region

$$R^* = \left\{(x_1, \ldots, x_n): \frac{f(x_1, \ldots, x_n; \theta_a)}{f(x_1, \ldots, x_n; \theta_0)} \geq k\right\}$$

Thus R^* is the set of all observations for which the likelihood ratio—ratio of the alternative likelihood to the null likelihood—is at least k. The probability of a type I error for the test with this rejection region is $\alpha^* = P[(X_1, \ldots, X_n) \in R^*$ when $\theta = \theta_0]$, whereas the type II error probability β^* is the probability that the X_i's lie in the complement of R^* (in the "acceptance" region) when $\theta = \theta_a$.

Then for any other test procedure with type I error probability α satisfying $\alpha \leq \alpha^*$, the probability of a type II error must satisfy $\beta \geq \beta^*$. Thus the test with rejection region R^* has the smallest type II error probability among all tests for which the type I error probability is at most α^*.

The choice of the constant k in the rejection region will determine the type I error probability α^*. In the continuous case, k can be selected to give one of the traditional significance levels .05, .01, and so on, whereas in the discrete case $\alpha^* = .057$ or $.039$ may be as close as one can get to .05.

Example 9.19 Consider randomly selecting $n = 5$ new vehicles of a certain type and determining the number of major defects on each one. Letting X_i denote the number of such defects for the ith selected vehicle ($i = 1, \ldots, 5$), suppose that the X_i's form a random sample from a Poisson distribution with parameter λ. Let's find the best test for testing H_0: $\lambda = 1$ versus H_a: $\lambda = 2$. The Poisson likelihood is $f(x_1, \ldots, x_5; \lambda) = e^{-5\lambda}\lambda^{\Sigma x_i}/\Pi x_i!$. Substituting first $\lambda = 2$, then $\lambda = 1$, and then taking the ratio of these two likelihoods gives the rejection region

$$R^* = \{(x_1, \ldots, x_5): e^{-5}2^{\Sigma x_i} \geq k\}$$

Multiplying both sides of the inequality by e^5 and letting $k' = ke^5$ gives the rejection region $2^{\Sigma x_i} \geq k'$. Now, take the natural logarithm of both sides and let $c = \ln(k')/\ln(2)$ to obtain the rejection region $\sum x_i \geq c$.

This latter rejection region is completely equivalent to R^*: For any particular value k there will be a corresponding value c, and vice versa. It is much easier to express the rejection region in this latter form and then select c to obtain a desired significance level than it is to determine an appropriate value of k for the likelihood ratio. In particular, $T = \sum X_i$ has a Poisson distribution with parameter 5λ (via a moment generating function argument), so when H_0 is true, T has a Poisson distribution with parameter 5. From the 5.0 column of our Poisson table (Table A.2), the cumulative probabilities for the values 8 and 9 are .932 and .968, respectively. Thus if we use $c = 9$ in the rejection region,

$$\alpha^* = P(\text{Poisson rv with parameter 5 is} \geq 9) = 1 - .932 = .068$$

Choosing instead $c = 10$ gives $\alpha^* = .032$. If we insist that the significance level be at most .05, then the optimal rejection region is $\sum x_i \geq 10$.

When H_a is true, the test statistic has a Poisson distribution with parameter 10. Thus

$$\begin{aligned}\beta^* &= P(H_0 \text{ is } \textit{not} \text{ rejected when } H_a \text{ is true}) \\ &= P(\text{Poisson rv with parameter 10 is} \leq 9) = .458\end{aligned}$$

Obviously this type II error probability is quite large. This is because the sample size $n = 5$ is too small to allow for effective discrimination between $\lambda = 1$ and $\lambda = 2$. For a sample size of 10, the Poisson table reveals that the best test having significance level at most .05 uses $c = 16$, for which $\alpha^* = .049$ (Poisson parameter $= 10$) and $\beta^* = .157$ (Poisson parameter $= 20$).

Finally, returning to a sample size of 5, $c = 10$ implies that $10 = \ln(ke^5)/\ln(2)$, from which $k = 2^{10}/e^5 \approx 6.9$. For the best test to have a significance level of at most .05, the null hypothesis should be rejected only when the likelihood for the alternative value of λ is more than about 7 times what it is for the null value. ■

Example 9.20 Let $X_1, \ldots, X_n$ be a random sample from a normal distribution with mean μ and variance 1 (the argument to be given will work for any other known value of σ^2). Consider testing H_0: $\mu = \mu_0$ versus H_a: $\mu = \mu_a$ where $\mu_a > \mu_0$. The likelihood ratio is

$$\frac{\left(\dfrac{1}{2\pi}\right)^{n/2} e^{-(1/2)\Sigma(x_i-\mu_a)^2}}{\left(\dfrac{1}{2\pi}\right)^{n/2} e^{-(1/2)\Sigma(x_i-\mu_0)^2}} = e^{\mu_a\Sigma x_i - \mu_0\Sigma x_i - (n/2)(\mu_a^2 - \mu_0^2)}$$

$$= \left[e^{-n(\mu_a^2-\mu_0^2)/2}\right] \cdot \left[e^{(\mu_a-\mu_0)\Sigma x_i}\right]$$

The term in the first set of brackets is a numerical constant. The fact that $\mu_a - \mu_0 > 0$ then implies that the likelihood ratio will be at least k if and only if $\sum x_i \geq k'$, that is, if and only if $\bar{x} \geq k''$, which means if and only if

$$z = \frac{\bar{x} - \mu_0}{1/\sqrt{n}} \geq c$$

If we now let $c = z_{.01} = 2.33$, this z test (one for which the test statistic has a standard normal distribution when H_0 is true) will have minimum β among all tests for which $\alpha \leq .01$. ■

The key idea in these last two examples cannot be overemphasized: Write an expression for the likelihood ratio, and then manipulate the inequality *likelihood ratio* $\geq$ k so it is equivalent to an inequality involving a test statistic whose distribution when H_0 is true is known or can be derived. Then this known or derived distribution can be used to obtain a test with the desired α. In the first example the distribution was Poisson with parameter 5, and in the second it was the standard normal distribution.

Proof of the Neyman–Pearson Theorem We shall consider the case in which the X_i's have a discrete distribution, so that type I and type II error probabilities are obtained by summation. In the continuous case, integration replaces summation. Then

$$R^* = \{(x_1, \ldots, x_n): f(x_1, \ldots, x_n; \theta_a) \geq k \cdot f(x_1, \ldots, x_n; \theta_0)\}$$

$$\alpha^* = P[(X_1, \ldots, X_n) \in R^* \text{ when } \theta = \theta_0] = \sum_{R^*} f(x_1, \ldots, x_n; \theta_0)$$

$$\beta^* = P[(X_1, \ldots, X_n) \in R^{*\prime} \text{ when } \theta = \theta_a] = \sum_{R^{*\prime}} f(x_1, \ldots, x_n; \theta_a)$$

(β^* is the sum over values in the *complement* of the rejection region.) Suppose that R is a rejection region different from R^* whose type I error probability is at most α^*, that is,

$$\alpha = P[(X_1, \ldots, X_n) \in R \text{ when } \theta = \theta_0] = \sum_{R} f(x_1, \ldots, x_n; \theta_0) \leq \alpha^*$$

We then wish to show that β for this rejection region must be at least as large as β^*. Consider the difference

$$\begin{aligned}\Delta &= \sum_{R^*} [f(x_1, \ldots, x_n; \theta_a) - k \cdot f(x_1, \ldots, x_n; \theta_0)] \\ &\quad - \sum_{R} [f(x_1, \ldots, x_n; \theta_a) - k \cdot f(x_1, \ldots, x_n; \theta_0)] \\ &= \sum_{R^* \cap R} [\ldots] + \sum_{R^* \cap R'} [\ldots] - \left\{ \sum_{R \cap R^*} [\ldots] + \sum_{R \cap R^{*\prime}} [\ldots] \right\} \\ &= \sum_{R^* \cap R'} [\ldots] - \sum_{R \cap R^{*\prime}} [\ldots]\end{aligned}$$

This last difference is nonnegative (i.e., ≥ 0) because the term in the square brackets is ≥ 0 for any set of x_i's in R^* and is negative for any set of x_i's not in R^*. It then follows that

$$
\begin{aligned}
0 \leq & \sum_{R^*} f(x_1, \ldots, x_n; \theta_a) - k \sum_{R^*} f(x_1, \ldots, x_n; \theta_0) \\
& - \sum_{R} f(x_1, \ldots, x_n; \theta_a) + k \sum_{R} f(x_1, \ldots, x_n; \theta_0) \\
& = (1 - \beta^*) - k\alpha^* - (1 - \beta) + k\alpha \\
& = \beta - \beta^* - k(\alpha^* - \alpha) \leq \beta - \beta^* \quad \text{(since } \alpha \leq \alpha^* \text{ implies that the term being subtracted is nonnegative)}
\end{aligned}
$$

Thus we have shown that $\beta^* \leq \beta$ as desired. ■

Power and Uniformly Most Powerful Tests

The Neyman–Pearson theorem can be restated in a slightly different way by considering the *power of a test*, first introduced in Section 9.2.

DEFINITION

Let Ω_0 and Ω_a be two disjoint sets of possible values of θ, and consider testing H_0: $\theta \in \Omega_0$ versus H_a: $\theta \in \Omega_a$ using a test with rejection region R. Then the **power function** of the test, denoted by $\pi(\cdot)$, is the probability of rejecting H_0 considered as a function of θ:

$$\pi(\theta') = P[(X_1, \ldots, X_n) \in R \text{ when } \theta = \theta']$$

Since we don't want to reject the null hypothesis when $\theta \in \Omega_0$ and do want to reject it when $\theta \in \Omega_a$, we wish a test for which the power function is close to 0 whenever θ' is in Ω_0 and close to 1 whenever θ' is in Ω_a. The power is easily related to the type I and type II error probabilities:

$$\pi(\theta') = \begin{cases} P(\text{type I error when } \theta = \theta') = \alpha(\theta') & \text{when } \theta' \in \Omega_0 \\ 1 - P(\text{type II error when } \theta = \theta') = 1 - \beta(\theta') & \text{when } \theta' \in \Omega_a \end{cases}$$

Thus large power when $\theta' \in \Omega_a$ is equivalent to small β for such parameter values.

Example 9.21 The drying time (min) of a particular brand and type of paint on a test board under controlled conditions is known to be normally distributed with $\mu = 75$ and $\sigma = 9.4$. A new additive has been developed for the purpose of improving drying time. Assume that drying time with the additive is still normally distributed with the same standard deviation, and consider testing H_0: $\mu \geq 75$ versus H_a: $\mu < 75$ based on a sample of size $n = 100$. A test with significance level .01 rejects the null hypothesis if $z \leq -2.33$, where $z = (\bar{x} - 75)/(9.4/\sqrt{100}) = (\bar{x} - 75)/.94$. Manipulating the inequality in the rejection region to isolate $\bar{x}$ gives the equivalent rejection region $\bar{x} \leq 72.81$. Thus the power of the test when $\mu = 70$ (a substantial departure from the null hypothesis) is

$$
\begin{aligned}
\pi(70) &= P(\bar{X} \leq 72.81 \text{ when } \mu = 70) = \Phi\left(\frac{72.81 - 70}{9.4/\sqrt{100}}\right) \\
&= \Phi(2.99) = .9986
\end{aligned}
$$

so $\beta = .0014$. It is easily verified that $\pi(75) = .01$, the significance level. The power when $\mu = 76$ (a parameter value for which H_0 is true) is

$$\pi(76) = P(\overline{X} \leq 72.81 \text{ when } \mu = 76) = \Phi\left(\frac{72.81 - 76}{9.4/\sqrt{100}}\right)$$
$$= \Phi(-3.39) = .0003$$

which is quite small as it should be. By repeating this calculation for various other values of μ we obtain the entire power function. A graph of the ideal power function appears in Figure 9.9(a) and the actual power function is graphed in Figure 9.9(b). The maximum power for $\mu \geq 75$ (i.e., in Ω_0) occurs at $\mu = 75$, on the boundary between Ω_0 and Ω_a. Because the power function is continuous, there are values of μ smaller than 75 for which the power is quite small. Even with a large sample size, it is difficult to detect a very small departure from the null hypothesis.

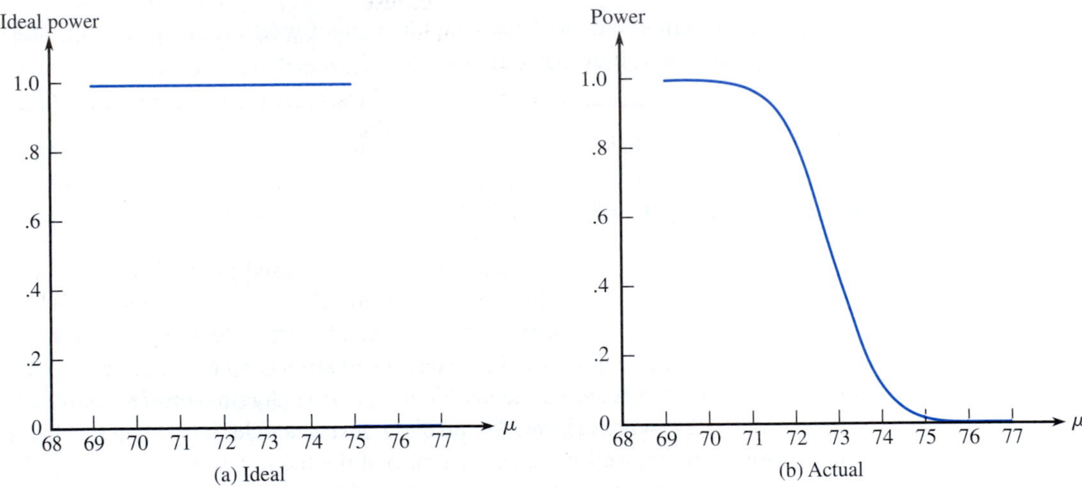

Figure 9.9 Graphs of power functions for Example 9.21 ■

The Neyman–Pearson theorem says that when Ω_0 consists of a single value θ_0 and Ω_a also consists of a single value θ_a, the rejection region R^* specifies a test for which the power $\pi(\theta_a)$ at the alternative value θ_a (which is just $1 - \beta$) is maximized subject to $\pi(\theta_0) \leq \alpha$ for some specified value of α. That is, R^* specifies a *most powerful test* subject to the restriction on the power when the null hypothesis is true.

What about best tests when at least one of the two hypotheses is **composite**, that is, Ω_0 or Ω_a or both consist of more than a single value?

Example 9.22 (Example 9.19 continued)

Consider again a random sample of size $n = 5$ from a Poisson distribution, and suppose we now wish to test H_0: $\lambda \leq 1$ versus H_a: $\lambda > 1$. Both of these hypotheses are composite. Arguing as in Example 9.19, for any value λ_a exceeding 1, a most powerful test of H_0: $\lambda = 1$ versus H_a: $\lambda = \lambda_a$ with significance level (power when $\lambda = 1$) .032 rejects the null hypothesis when $\sum x_i \geq 10$. Furthermore, it is easily verified that the power of this test at λ' is smaller than .032 if $\lambda' < 1$. Thus the test that rejects H_0: $\lambda \leq 1$ in favor of H_0: $\lambda > 1$ when $\sum x_i \geq 10$ has maximum power for *any* $\lambda' > 1$ subject to the condition that $\pi(\lambda') \leq .032$. This test is *uniformly most powerful*. ■

More generally, a **uniformly most powerful (UMP) level α test** is one for which $\pi(\theta')$ is maximized for any $\theta \in \Omega_a$ subject to $\pi(\theta') \le \alpha$ for any $\theta' \in \Omega_0$. Unfortunately UMP tests are fairly rare, especially in commonly encountered situations when H_0 and H_a are assertions about a single parameter θ_1 whereas the distribution of the X_i's involves not only θ_1 but also at least one other "nuisance parameter." For example, when the population distribution is normal with values of both μ and σ unknown, σ is a nuisance parameter when testing H_0: $\mu = \mu_0$ versus H_a: $\mu \neq \mu_0$. Be careful here—the null hypothesis is not simple because Ω_0 consists of all pairs (μ, σ) for which $\mu = \mu_0$ and $\sigma > 0$, and there is certainly more than one such pair. In this situation, the one-sample t test is not UMP.

However, suppose we restrict attention to **unbiased** tests, those for which the smallest value of $\pi(\theta')$ for $\theta' \in \Omega_a$ is at least as large as the largest value of $\pi(\theta')$ for $\theta' \in \Omega_0$. Unbiasedness simply says that we are at least as likely to reject the null hypothesis when H_0 is false as we are to reject it when H_0 is true. The test proposed in Example 9.21 involving paint drying times is unbiased because, as Figure 9.9(b) shows, the power function at or to the right of 75 is smaller than it is to the left of 75. It can be shown that the one-sample t test is *UMP unbiased*—that is, it is uniformly most powerful among all tests that are unbiased. Several other commonly used tests also have this property. Please consult one of the chapter references for more details.

Likelihood Ratio Tests

The likelihood ratio (LR) principle is the most frequently used method for finding an appropriate test statistic in a new situation. As before, denote the joint pmf or pdf of $X_1, \ldots, X_n$ by $f(x_1, \ldots, x_n; \theta)$. In the usual case of a random sample, it will be a product $f(x_1; \theta) \cdot \cdots \cdot f(x_n; \theta)$. When the x_i's are the actual observations and $f(x_1, \ldots, x_n; \theta)$ is regarded as a function of θ, it is called the *likelihood function*. Again consider testing H_0: $\theta \in \Omega_0$ versus H_a: $\theta \in \Omega_a$, where Ω_0 and Ω_a are disjoint sets, and let $\Omega = \Omega_0 \cup \Omega_a$. In the Neyman–Pearson theorem, we focused on the ratio of the likelihood when $\theta \in \Omega_a$ to the likelihood when $\theta \in \Omega_0$, rejecting H_0 when the value of the ratio was "sufficiently large." Now we consider the ratio of the likelihood when $\theta \in \Omega_0$ to the likelihood when $\theta \in \Omega$. A very *small* value of this ratio argues against the null hypothesis, since a small value arises when the data is much more consistent with the alternative hypothesis than with the null hypothesis. More formally,

1. Find the largest value of the likelihood for any $\theta \in \Omega_0$ by finding the maximum likelihood estimate of θ within Ω_0 and substituting this mle into the likelihood function to obtain $L(\hat{\Omega}_0)$.
2. Find the largest value of the likelihood for any $\theta \in \Omega$ by finding the maximum likelihood estimate of θ within Ω and substituting this mle into the likelihood function to obtain $L(\hat{\Omega})$. Because Ω_0 is a subset of Ω, this likelihood $L(\hat{\Omega})$ can't be any smaller than the likelihood $L(\hat{\Omega}_0)$ obtained in the first step, and will be much larger when the data is much more consistent with H_a than with H_0.
3. Form the *likelihood ratio* $L(\hat{\Omega}_0)/L(\hat{\Omega})$ and reject the null hypothesis in favor of the alternative when this ratio is $\le k$. The critical value k is chosen to give a test with the desired significance level. In practice, the inequality $L(\hat{\Omega}_0)/L(\hat{\Omega}) \le k$ is often reexpressed in terms of a more convenient statistic (such as the sum of the observations) whose distribution is known or can be derived.

The above prescription remains valid if the single parameter θ is replaced by several parameters $\theta_1, \ldots, \theta_k$. The mle's of all parameters must be obtained in both steps 1 and 2 and substituted back into the likelihood function.

Example 9.23 Consider a random sample from a normal distribution with the values of both parameters unknown. We wish to test H_0: $\mu = \mu_0$ versus H_a: $\mu \neq \mu_0$. Here Ω consists of all values of μ and σ^2 for which $-\infty < \mu < \infty$ and $\sigma^2 > 0$, and the likelihood function is

$$\left(\frac{1}{2\pi\sigma^2}\right)^{n/2} e^{(1/2\sigma^2)\Sigma(x_i-\mu)^2}$$

In Section 7.2 we obtained the mle's as $\hat{\mu} = \bar{x}$, $\hat{\sigma}^2 = \Sigma(x_i - \bar{x})^2/n$. Substituting these estimates back into the likelihood function gives

$$L(\hat{\Omega}) = \left[\frac{1}{2\pi \sum (x_i - \bar{x})^2/n}\right]^{n/2} e^{-n/2}$$

Within Ω_0, μ in the foregoing likelihood is replaced by μ_0, so that only σ^2 must be estimated. It is easily verified that the mle is $\hat{\sigma}^2 = \Sigma(x_i - \mu_0)^2/n$. Substitution of this estimate in the likelihood function yields

$$L(\hat{\Omega}_0) = \left[\frac{1}{2\pi \sum (x_i - \mu_0)^2/n}\right]^{n/2} e^{-n/2}$$

Thus we reject H_0 in favor of H_a when

$$\frac{L(\hat{\Omega}_0)}{L(\hat{\Omega})} = \left[\frac{\sum (x_i - \bar{x})^2}{\sum (x_i - \mu_0)^2}\right]^{n/2} \leq k$$

Raising both sides of this inequality to the power $2/n$, we reject H_0 whenever

$$\frac{\sum (x_i - \bar{x})^2}{\sum (x_i - \mu_0)^2} \leq k^{2/n} = k'$$

This is intuitively quite reasonable: the value μ_0 is implausible for μ if the sum of squared deviations about the sample mean is much smaller than the sum of squared deviations about μ_0. The denominator of this latter ratio can be expressed as

$$\sum [(x_i - \bar{x}) + (\bar{x} - \mu_0)]^2 = \sum (x_i - \bar{x})^2 + 2\sum (\bar{x} - \mu_0)(x_i - \bar{x}) + n(\bar{x} - \mu_0)^2$$

The middle (i.e., cross-product) term in this expression is 0, because the constant $\bar{x} - \mu_0$ can be moved outside the summation, and then the sum of deviations from the sample mean is 0. Thus we should reject H_0 when

$$\frac{\sum (x_i - \bar{x})^2}{\sum (x_i - \bar{x})^2 + n(\bar{x} - \mu_0)^2} = \frac{1}{1 + n(\bar{x} - \mu_0)^2 / \sum (x_i - \bar{x})^2} \leq k'$$

This latter ratio will be small when the second term in the denominator is large, so the condition for rejection becomes

$$\frac{n(\bar{x} - \mu_0)^2}{\sum (x_i - \bar{x})^2} \geq k''$$

Dividing both sides by $n - 1$ and taking square roots gives the rejection region

$$\text{either} \quad \frac{\bar{x} - \mu_0}{s/\sqrt{n}} \geq c \quad \text{or} \quad \frac{\bar{x} - \mu_0}{s/\sqrt{n}} \leq -c$$

If we now let $c = t_{\alpha/2,n-1}$, we have exactly the two-tailed one-sample t test. The bottom line is that when testing H_0: $\mu = \mu_0$ against the two-sided ($\neq$) alternative, the one-sample t test is the likelihood ratio test. This is also true of the upper-tailed version of the t test when the alternative is H_a: $\mu > \mu_0$ and of the lower-tailed test when the alternative is H_a: $\mu < \mu_0$. We could trace back through the argument to recover the critical constant k from c, but there is no point in doing this; the rejection region in terms of t is much more convenient than the rejection region in terms of the likelihood ratio. ■

A number of tests discussed subsequently, including the "pooled" t test from the next chapter and various tests from ANOVA (the analysis of variance) and regression analysis, can be derived by the likelihood ratio principle. Rather frequently the inequality for the rejection region of a likelihood ratio test cannot be manipulated to express the test procedure in terms of a simple statistic whose distribution can be ascertained. The following large-sample result, valid under fairly general conditions, can then be used: If the sample size n is sufficiently large, then the statistic -2[ln(likelihood ratio)] has approximately a chi-squared distribution with ν degrees of freedom, where ν is the difference between the number of "freely varying" parameters in Ω and the number of such parameters in Ω_0. For example, if the distribution sampled is bivariate normal with the 5 parameters μ_1, μ_2, σ_1, σ_2, and ρ and the null hypothesis asserts that $\mu_1 = \mu_2$ and $\sigma_1 = \sigma_2$, then $\nu = 5 - 3 = 2$. By definition $L(\hat{\Omega}_0)/L(\hat{\Omega}) \leq 1$, and the likelihood ratio test rejects H_0 when this likelihood ratio is much less than 1. This is equivalent to rejecting when the logarithm of the likelihood ratio is quite negative, that is, when $-\ln(\text{LR})$ is quite positive. The large-sample version of the test is thus upper-tailed: H_0 should be rejected if $-2\ln$(likelihood ratio)] $\geq \chi^2_{\alpha,\nu}$ (an upper-tail critical value extracted from Table A.7).

Example 9.24 Suppose a scientist makes n measurements of some physical characteristic, such as the specific gravity of a certain liquid. Let $X_1, \ldots, X_n$ denote the resulting measurement errors. Assume that these X_i's are independent and identically distributed according to the double exponential (Laplace) distribution with pdf $f(x) = .5e^{-|x-\theta|}$ for $-\infty < x < \infty$. This pdf is symmetric about θ with somewhat heavier tails than the normal pdf. If $\theta = 0$ then the measurements are unbiased, so it is natural to test H_0: $\theta = 0$ versus H_a: $\theta \neq 0$. Here $\nu = 1 - 0 = 1$. The likelihood is

$$L(\theta) = (.5)^n e^{-\Sigma|x_i - \theta|}$$

Because of the minus sign preceding the summation, the likelihood is maximized when $\Sigma|x_i - \theta|$ is minimized. The absolute value function is not differentiable, and therefore differential calculus cannot be used. Instead, consider for a moment the case $n = 5$ and let $y_1, \ldots, y_5$ denote the values of the x_i's ordered from smallest to largest, so the y_i's are the observed values of the order statistics. For example, a random sample of size five from the Laplace distribution with $\theta = 0$ is $-.24998, .75446, -.19053, 1.16237, .83229$, so $(y_1, \ldots, y_5) = (-.24998, -.19053, .75446, .83229, 1.16237)$. Then

$$\sum \left| x_i - \theta \right| = \sum \left| y_i - \theta \right| = \begin{cases} y_1 + y_2 + y_3 + y_4 + y_5 - 5\theta & \theta < y_1 \\ -y_1 + y_2 + y_3 + y_4 + y_5 - 3\theta & y_1 \le \theta < y_2 \\ -y_1 - y_2 + y_3 + y_4 + y_5 - \theta & y_2 \le \theta < y_3 \\ -y_1 - y_2 - y_3 + y_4 + y_5 + \theta & y_3 \le \theta < y_4 \\ -y_1 - y_2 - y_3 - y_4 + y_5 + 3\theta & y_4 \le \theta < y_5 \\ -y_1 - y_2 - y_3 - y_4 - y_5 + 5\theta & \theta \ge y_5 \end{cases}$$

The graph of this expression as a function of θ appears in Figure 9.10, from which it is apparent that the minimum occurs at $y_3 = \tilde{x} = .75446$, the sample median. The situation is similar whenever n is odd. When n is even, the function achieves its minimum for any θ between $y_{n/2}$ and $y_{(n/2)+1}$; one such θ *is* $(y_{n/2} + y_{(n/2)+1})/2 = \tilde{x}$. In summary, the mle of θ is the sample median.

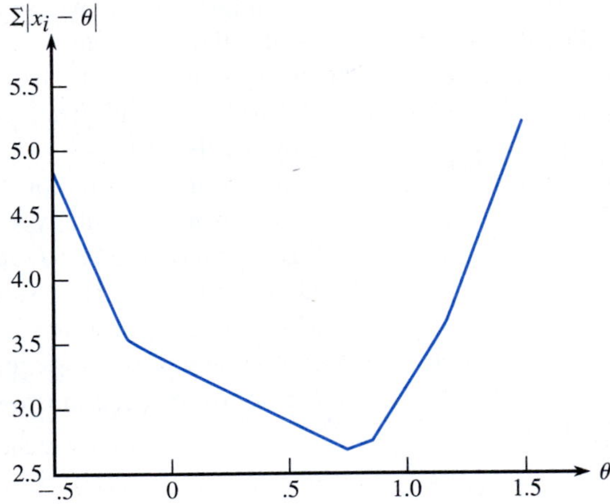

Figure 9.10 Determining the mle of the double exponential parameter by minimizing $\Sigma|x_i - \theta|$

The LR statistic for testing the relevant hypotheses is $(.5)^n e^{-\Sigma|x_i|}/(.5)^n e^{-\Sigma|x_i - \tilde{x}|}$. Taking the natural log of the likelihood ratio and multiplying by -2 gives the rejection region $2\Sigma|x_i| - 2\Sigma|x_i - \tilde{x}| \ge \chi^2_{\alpha,1}$ for the large-sample version of the LR test.

Suppose that a sample of $n = 30$ errors results in $\Sigma|x_i| = 38.6$ and $\Sigma|x_i - \tilde{x}| = 37.3$. Then

$$-2\ln(LR) = 2\left(\sum |x_i| - \sum |x_i - \tilde{x}|\right) = 2.6$$

Comparing this to $\chi^2_{.05,1} = 3.84$, we would not reject the null hypothesis at the 5% significance level. It is plausible that the measurement process is indeed unbiased. ■

Exercises Section 9.5 (60–71)

60. Reconsider the paint-drying problem discussed in Example 9.2. The hypotheses were H_0: $\mu = 75$ versus H_a: $\mu < 75$, with σ assumed to have value 9.0. Consider the alternative value $\mu = 74$, which in the context of the problem would presumably not be a practically significant departure from H_0.

a. For a level .01 test, compute β at this alternative for sample sizes $n = 100$, 900, and 2500.

b. If the observed value of $\overline{X}$ is $\overline{x} = 74$, what can you say about the resulting P-value when $n = 2500$? Is the data statistically significant at any of the standard values of α?

c. Would you really want to use a sample size of 2500 along with a level .01 test (disregarding the cost of such an experiment)? Explain.

61. Consider the large-sample level .01 test in Section 9.3 for testing H_0: $p = .2$ against H_a: $p > .2$.

a. For the alternative value $p = .21$, compute $\beta(.21)$ for sample sizes $n = 100$, 2500, 10,000, 40,000, and 90,000.

b. For $\hat{p} = x/n = .21$, compute the P-value when $n = 100$, 2500, 10,000, and 40,000.

c. In most situations, would it be reasonable to use a level .01 test in conjunction with a sample size of 40,000? Why or why not?

62. For a random sample of n individuals taking a licensing exam, let $X_i = 1$ if the ith individual in the sample passes the exam and $X_i = 0$ otherwise ($i = 1, \ldots, n$).

a. With p denoting the proportion of all exam takers who pass, show that the most powerful test of H_0: $p = .5$ versus H_a: $p = .75$ rejects H_0 when $\sum x_i \geq c$.

b. If $n = 20$ and you want $\alpha \leq .05$ for the test described in (a), would you reject H_0 if 15 of the 20 individuals in the sample pass the exam?

c. What is the power of the test you used in (b) when $p = .75$ [i.e., what is $\pi(.75)$]?

d. Is the test derived in (a) UMP for testing H_0: $p = .5$ versus H_a: $p > .5$? Explain your reasoning.

e. Graph the power function $\pi(p)$ of the test for the hypotheses of (d) when $n = 20$ and $\alpha \leq .05$.

63. The error X in a measurement has a normal distribution with mean value 0 and variance σ^2. Consider testing H_0: $\sigma^2 = 2$ versus H_a: $\sigma^2 = 3$ based on a random sample $X_1, \ldots, X_n$ of errors.

a. Show that a most powerful test rejects H_0 when $\sum x_i^2 \geq c$.

b. For $n = 10$, find the value of c for the test in (a) that results in $\alpha = .05$.

c. Is the test described in (a) UMP for H_0: $\sigma^2 = 2$ versus H_a: $\sigma^2 > 2$? Justify your assertion.

64. Suppose that X, the fraction of a container that is filled, has pdf $f(x; \theta) = \theta x^{\theta-1}$ for $0 < x < 1$ (where $\theta > 0$), and let $X_1, \ldots, X_n$ be a random sample from this distribution.

a. Show that the most powerful test for H_0: $\theta = 1$ versus H_a: $\theta = 2$ rejects the null hypothesis if $\sum \ln(x_i) \geq c$.

b. Is the test described in (a) UMP for testing H_0: $\theta = 1$ versus H_a: $\theta > 1$? Explain your reasoning.

c. If $n = 50$, what is the (approximate) value of c for which the test has significance level .05?

65. Consider a random sample of n component lifetimes, where the distribution of lifetime is exponential with parameter λ.

a. Obtain a most powerful test for H_0: $\lambda = 1$ versus H_a: $\lambda = .5$, and express the rejection region in terms of a "simple" statistic.

b. Is the test found in (a) uniformly most powerful for H_0: $\lambda = 1$ versus H_a: $\lambda < 1$? Justify your answer.

66. Consider a random sample of size n from the "shifted exponential" distribution with pdf $f(x; \theta) = e^{-(x-\theta)}$ for $x > \theta$ and 0 otherwise (the graph is that of the ordinary exponential pdf with $\lambda = 1$ shifted so that it begins its descent at θ rather than at 0). Let Y_1 denote the smallest order statistic, and show that the likelihood ratio test of H_0: $\theta \leq 1$ versus H_a: $\theta > 1$ rejects the null hypothesis if y_1, the observed value of Y_1, is $\geq c$.

67. Suppose that each of n randomly selected individuals is classified according to his/her genotype with respect to a particular characteristic and that the three possible genotypes are AA, Aa, and aa with long-run proportions (probabilities) θ^2, $2\theta(1-\theta)$, and $(1-\theta)^2$, respectively $(0 < \theta < 1)$. It is then straightforward to show that the likelihood is

$$\theta^{2x_1} \cdot [2\theta(1-\theta)]^{x_2} \cdot (1-\theta)^{2x_3}$$

where x_1, x_2, and x_3 are the number of individuals in the sample who have the AA, Aa, and aa genotypes, respectively. Show that the most powerful test for testing H_0: $\theta = .5$ versus H_a: $\theta = .8$ rejects the null hypothesis when $2x_1 + x_2 \geq c$. Is this test UMP for the alternative H_a: $\theta > .5$? Explain. *Note*: The fact that the joint distribution of X_1, X_2, and X_3 is multinomial can be used to obtain the value of c that yields a test with any desired significance level when n is large.

68. The error in a measurement is normally distributed with mean μ and standard deviation 1. Consider a random sample of n errors, and show that the likelihood ratio test for H_0: $\mu = 0$ versus H_a: $\mu \neq 0$ rejects the null hypothesis when either $\bar{x} \geq c$ or $\bar{x} \leq -c$. What is c for a test with $\alpha = .05$? How does the test change if the standard deviation of an error is σ_0 (known) and the relevant hypotheses are H_0: $\mu = \mu_0$ versus H_a: $\mu \neq \mu_0$?

69. Measurement error in a particular situation is normally distributed with mean value μ and standard deviation 4. Consider testing H_0: $\mu = 0$ versus H_a: $\mu \neq 0$ based on a sample of $n = 16$ measurements.

a. Verify that the usual test with significance level .05 rejects H_0 if either $\bar{x} \geq 1.96$ or $\bar{x} \leq -1.96$. *Note*: That this test is unbiased follows from the fact that the way to capture the largest area under the z curve above an interval having width 3.92 is to center that interval at 0 (so it extends from -1.96 to 1.96).

b. Consider the test which rejects H_0 if either $\bar{x} \geq 2.17$ or $\bar{x} \leq -1.81$. What is α, that is, $\pi(0)$?

c. What is the power of the test proposed in (b) when $\mu = .1$ and when $\mu = -.1$? (Note that .1 and $-.1$ are very close to the null value, so one would not expect large power for such values.) Is the test unbiased?

d. Calculate the power of the usual test when $\mu = .1$ and when $\mu = -.1$. Is the usual test a most powerful test? *Hint*: Refer to your calculations in (c). *Note*: It can be shown that the usual test is most powerful among all unbiased tests.

70. A test of whether or not a coin is fair will be based on $n = 50$ tosses. Let X be the resulting number of heads. Consider two rejection regions: $R_1 = \{x$: either $x \leq 17$ or $x \geq 33\}$ and $R_2 = \{x$: either $x \leq 18$ or $x \geq 37\}$.

a. Determine the significance level (type I error probability) for each rejection region.

b. Determine the power of each test when $p = .49$. Is the test with rejection region R_1 a uniformly most powerful level .033 test? Explain.

c. Determine the power of the test with rejection region R_2. Is this test unbiased? Explain.

d. Sketch the power function for the test with rejection region R_1, and then do so for the test with the rejection region R_2. What does your intuition suggest about the desirability of using the rejection region R_2?

71. Consider Example 9.23.

a. With $t = (\bar{x} - \mu_0)/(s/\sqrt{n})$, show that the likelihood ratio is equal to $\lambda = [1 + t^2/(n-1)]^{-n/2}$, where $t = (\bar{x} - \mu_0)/(s/\sqrt{n})$, and therefore the approximate chi-squared statistic is $-2[\ln(\lambda)] = n \cdot \ln[1 + t^2/(n-1)]$.

b. Apply part (a) to test the hypotheses of Exercise 55, using the data given there. Compare your results with the answers found in Exercise 55.

Supplementary Exercises (72–94)

72. A sample of 50 lenses used in eyeglasses yields a sample mean thickness of 3.05 mm and a sample standard deviation of .34 mm. The desired true average thickness of such lenses is 3.20 mm. Does the data strongly suggest that the true average thickness of such lenses is something other than what is desired? Test using $\alpha = .05$.

73. In Exercise 72, suppose the experimenter had believed before collecting the data that the value of σ was approximately .30. If the experimenter wished the probability of a type II error to be .05 when $\mu = 3.00$, was a sample size of 50 unnecessarily large?

74. It is specified that a certain type of iron should contain .85 g of silicon per 100 g of iron (.85%). The silicon content of each of 25 randomly selected iron specimens was determined, and the accompanying MINITAB output resulted from a test of the appropriate hypotheses.

```
Variable   N    Mean    StDev   SE Mean     T      P
sil cont  25  0.8880  0.1807   0.0361   1.05   0.30
```

a. What hypotheses were tested?

b. What conclusion would be reached for a significance level of .05, and why? Answer the same question for a significance level of .10.

75. One method for straightening wire before coiling it to make a spring is called "roller straightening." The article "The Effect of Roller and Spinner Wire Straightening on Coiling Performance and Wire Properties" (*Springs*, 1987: 27–28) reports on the tensile properties of wire. Suppose a sample of 16 wires is selected and each is tested to determine tensile strength (N/mm^2). The resulting sample mean and standard deviation are 2160 and 30, respectively.

a. The mean tensile strength for springs made using spinner straightening is 2150 N/mm^2. What hypotheses should be tested to determine whether the mean tensile strength for the roller method exceeds 2150?

b. Assuming that the tensile strength distribution is approximately normal, what test statistic would you use to test the hypotheses in part (a)?

c. What is the value of the test statistic for this data?

d. What is the P-value for the value of the test statistic computed in part (c)?

e. For a level .05 test, what conclusion would you reach?

76. A new method for measuring phosphorus levels in soil is described in the article "A Rapid Method to Determine Total Phosphorus in Soils" (*Soil Sci. Amer. J.*, 1988: 1301–1304). Suppose a sample of 11 soil specimens, each with a true phosphorus content of 548 mg/kg, is analyzed using the new method. The resulting sample mean and standard deviation for phosphorus level are 587 and 10, respectively.

a. Is there evidence that the mean phosphorus level reported by the new method differs significantly from the true value of 548 mg/kg? Use $\alpha = .05$.

b. What assumptions must you make for the test in part (a) to be appropriate?

77. The article "Orchard Floor Management Utilizing Soil-Applied Coal Dust for Frost Protection" (*Agric. Forest Meteorology*, 1988: 71–82) reports the following values for soil heat flux of eight plots covered with coal dust.

34.7 35.4 34.7 37.7 32.5 28.0 18.4 24.9

The mean soil heat flux for plots covered only with grass is 29.0. Assuming that the heat-flux distribution is approximately normal, does the data suggest that the coal dust is effective in increasing the mean heat flux over that for grass? Test the appropriate hypotheses using $\alpha = .05$.

78. The article "Caffeine Knowledge, Attitudes, and Consumption in Adult Women" (*J. Nutrit. Ed.*, 1992: 179–184) reports the following summary data on daily caffeine consumption for a sample of adult women: $n = 47$, $\bar{x} = 215$ mg, $s = 235$ mg, and range = 5–1176.

a. Does it appear plausible that the population distribution of daily caffeine consumption is normal? Is it necessary to assume a normal population distribution to test hypotheses about the value of the population mean consumption? Explain your reasoning.

b. Suppose it had previously been believed that mean consumption was at most 200 mg. Does the given data contradict this prior belief? Test the appropriate hypotheses at significance level .10 and include a P-value in your analysis.

79. The accompanying output resulted when MINITAB was used to test the appropriate hypotheses about true average activation time based on the data in Exercise 56. Use this information to reach a conclusion at significance level .05 and also at level .01.

```
TEST OF MU = 25.000 VS MU G.T. 25.000
          N    MEAN   STDEV   SE MEAN      T   P VALUE
time     13  27.923   5.619     1.559   1.88     0.043
```

80. The true average breaking strength of ceramic insulators of a certain type is supposed to be at least 10 psi. They will be used for a particular application unless sample data indicates conclusively that this specification has not been met. A test of hypotheses using $\alpha = .01$ is to be based on a random sample of ten insulators. Assume that the breaking-strength distribution is normal with unknown standard deviation.

a. If the true standard deviation is .80, how likely is it that insulators will be judged satisfactory when true average breaking strength is actually only 9.5? Only 9.0?

b. What sample size would be necessary to have a 75% chance of detecting that true average breaking strength is 9.5 when the true standard deviation is .80?

81. The accompanying observations on residual flame time (sec) for strips of treated children's nightwear were given in the article "An Introduction to Some Precision and Accuracy of Measurement Problems" (*J. Testing Eval.*, 1982: 132–140). Suppose a true average flame time of at most 9.75 had been mandated. Does the data suggest that this condition has not been met? Carry out an appropriate test after first investigating the plausibility of assumptions that underlie your method of inference.

9.85	9.93	9.75	9.77	9.67	9.87	9.67
9.94	9.85	9.75	9.83	9.92	9.74	9.99
9.88	9.95	9.95	9.93	9.92	9.89	

82. The incidence of a certain type of chromosome defect in the U.S. adult male population is believed to be 1 in 75. A random sample of 800 individuals in U.S. penal institutions reveals 16 who have such defects. Can it be concluded that the incidence rate of this defect among prisoners differs from the presumed rate for the entire adult male population?

a. State and test the relevant hypotheses using $\alpha = .05$. What type of error might you have made in reaching a conclusion?

b. What P-value is associated with this test? Based on this P-value, could H_0 be rejected at significance level .20?

83. In an investigation of the toxin produced by a certain poisonous snake, a researcher prepared 26 different vials, each containing 1 g of the toxin, and then determined the amount of antitoxin needed to neutralize the toxin. The sample average amount of antitoxin necessary was found to be 1.89 mg, and the sample standard deviation was .42. Previous research had indicated that the true average neutralizing amount was 1.75 mg/g of toxin. Does the new data contradict the value suggested by prior research? Test the relevant hypotheses using the P-value approach. Does the validity of your analysis depend on any assumptions about the population distribution of neutralizing amount? Explain.

84. The sample average unrestrained compressive strength for 45 specimens of a particular type of brick was computed to be 3107 psi, and the sample standard deviation was 188. The distribution of unrestrained compressive strength may be somewhat skewed. Does the data strongly indicate that the true average unrestrained compressive strength is less than the design value of 3200? Test using $\alpha = .001$.

85. To test the ability of auto mechanics to identify simple engine problems, an automobile with a single such problem was taken in turn to 72 different car repair facilities. Only 42 of the 72 mechanics who worked on the car correctly identified the problem. Does this strongly indicate that the true proportion of mechanics who could identify this problem is less than .75? Compute the P-value and reach a conclusion accordingly.

86. When $X_1, X_2, \ldots, X_n$ are independent Poisson variables, each with parameter λ, and n is large, the sample mean $\overline{X}$ has approximately a normal distribution with $\mu = E(\overline{X}) = \lambda$ and $\sigma^2 = V(\overline{X}) = \lambda/n$. This implies that

$$Z = \frac{\overline{X} - \lambda}{\sqrt{\lambda/n}}$$

has approximately a standard normal distribution. For testing H_0: $\lambda = \lambda_0$, we can replace λ by λ_0 in the equation for Z to obtain a test statistic. This statistic is actually preferred to the large-sample statistic with denominator $S/\sqrt{n}$ (when the X_i's are Poisson) because it is tailored explicitly to the Poisson assumption. If the number of requests for consulting received by a certain statistician during a 5-day work week has a Poisson distribution and the total number of consulting requests during a 36-week period is 160, does this suggest that the true average number of weekly requests exceeds 4.0? Test using $\alpha = .02$.

87. A hot-tub manufacturer advertises that with its heating equipment, a temperature of 100°F can be achieved in at most 15 min. A random sample of 32 tubs is selected, and the time necessary to achieve a 100°F temperature is determined for each tub. The sample average time and sample standard deviation are 17.5 min and 2.2 min, respectively. Does this data cast doubt on the company's claim? Compute the P-value and use it to reach a conclusion at level .05 (assume that the heating-time distribution is approximately normal).

88. Chapter 8 presented a CI for the variance σ^2 of a normal population distribution. The key result there was that the rv $\chi^2 = (n-1)S^2/\sigma^2$ has a chi-squared distribution with $n - 1$ df. Consider the null hypothesis H_0: $\sigma^2 = \sigma_0^2$ (equivalently, $\sigma = \sigma_0$). Then when H_0 is true, the test statistic $\chi^2 = (n-1)S^2/\sigma_0^2$ has a chi-squared distribution with $n - 1$ df. If the relevant alternative is H_a: $\sigma^2 > \sigma_0^2$, rejecting H_0 if $(n-1)s^2/\sigma_0^2 \geq \chi^2_{\alpha,n-1}$ gives a test with significance level α. To ensure reasonably uniform characteristics for a particular application, it is desired that the true standard deviation of the softening point of a certain type of petroleum pitch be at most .50°C. The softening points of ten different specimens were determined, yielding a sample standard deviation of .58°C. Does this strongly contradict the uniformity specification? Test the appropriate hypotheses using $\alpha = .01$.

89. Referring to Exercise 88, suppose an investigator wishes to test H_0: $\sigma^2 = .04$ versus H_a: $\sigma^2 < .04$

based on a sample of 21 observations. The computed value of $20s^2/.04$ is 8.58. Place bounds on the P-value and then reach a conclusion at level .01.

90. When the population distribution is normal and n is large, the sample standard deviation S has approximately a normal distribution with $E(S) \approx \sigma$ and $V(S) \approx \sigma^2/(2n)$. We already know that in this case, for any n, $\overline{X}$ is normal with $E(\overline{X}) = \mu$ and $V(\overline{X}) = \sigma^2/n$.

a. Assuming that the underlying distribution is normal, what is an approximately unbiased estimator of the 99th percentile $\theta = \mu + 2.33\sigma$?

b. As discussed in Section 6.4, when the X_i's are normal $\overline{X}$ and S are independent rv's (one measures location whereas the other measures spread). Use this to compute $V(\hat{\theta})$ and $\sigma_{\hat{\theta}}$ for the estimator $\hat{\theta}$ of part (a). What is the estimated standard error $\hat{\sigma}_{\hat{\theta}}$?

c. Write a test statistic for testing H_0: $\theta = \theta_0$ that has approximately a standard normal distribution when H_0 is true. If soil pH is normally distributed in a certain region and 64 soil samples yield $\bar{x} = 6.33$, $s = .16$, does this provide strong evidence for concluding that at most 99% of all possible samples would have a pH of less than 6.75? Test using $\alpha = .01$.

91. Let $X_1, X_2, \ldots, X_n$ be a random sample from an exponential distribution with parameter λ. Then it can be shown that $2\lambda\sum X_i$ has a chi-squared distribution with $\nu = 2n$ (by first showing that $2\lambda X_i$ has a chi-squared distribution with $\nu = 2$).

a. Use this fact to obtain a test statistic and rejection region that together specify a level α test for H_0: $\mu = \mu_0$ versus each of the three commonly encountered alternatives. [*Hint*: $E(X_i) = \mu = 1/\lambda$, so $\mu = \mu_0$ is equivalent to $\lambda = 1/\mu_0$.]

b. Suppose that ten identical components, each having exponentially distributed time until failure, are tested. The resulting failure times are

95 16 11 3 42 71 225 64 87 123

Use the test procedure of part (a) to decide whether the data strongly suggests that the true average lifetime is less than the previously claimed value of 75.

92. Suppose the population distribution is normal with known σ. Let γ be such that $0 < \gamma < \alpha$. For testing H_0: $\mu = \mu_0$ versus H_a: $\mu \neq \mu_0$, consider the test that rejects H_0 if either $z \geq z_\gamma$ or $z \leq -z_{\alpha-\gamma}$, where the test statistic is $Z = (\overline{X} - \mu_0)/(\sigma/\sqrt{n})$.

a. Show that P(type I error) $= \alpha$.

b. Derive an expression for $\beta(\mu')$. (*Hint*: Express the test in the form "reject H_0 if either $\bar{x} \geq c_1$ or $\leq c_2$.")

c. Let $\Delta > 0$. For what values of γ (relative to α) will $\beta(\mu_0 + \Delta) < \beta(\mu_0 - \Delta)$?

93. After a period of apprenticeship, an organization gives an exam that must be passed to be eligible for membership. Let $p = P$(randomly chosen apprentice passes). The organization wishes an exam that most but not all should be able to pass, so it decides that $p = .90$ is desirable. For a particular exam, the relevant hypotheses are H_0: $p = .90$ versus the alternative H_a: $p \neq .90$. Suppose ten people take the exam, and let X = the number who pass.

a. Does the lower-tailed region $\{0, 1, \ldots, 5\}$ specify a level .01 test?

b. Show that even though H_a is two-sided, no two-tailed test is a level .01 test.

c. Sketch a graph of $\beta(p')$ as a function of p' for this test. Is this desirable?

94. A service station has six gas pumps. When no vehicles are at the station, let p_i denote the probability that the next vehicle will select pump i ($i = 1, 2, \ldots, 6$). Based on a sample of size n, we wish to test H_0: $p_1 = \cdots = p_6$ versus H_a: $p_1 = p_3 = p_5$, $p_2 = p_4 = p_6$ (note that H_a is not a simple hypothesis). Let X be the number of customers in the sample that select an even-numbered pump.

a. Show that the likelihood ratio test rejects H_0 if either $X \geq c$ or $X \leq n - c$. (*Hint*: When H_a is true, let θ denote the common value of p_2, p_4, and p_6.)

b. Let $n = 10$ and $c = 9$. Determine the power of the test both when H_0 is true and also when $p_2 = p_4 = p_6 = \frac{1}{10}$, $p_1 = p_3 = p_5 = \frac{7}{30}$.

Bibliography

See the bibliographies for Chapter 7 and Chapter 8.

CHAPTER TEN

Inferences Based on Two Samples

Introduction

Chapters 8 and 9 presented confidence intervals (CIs) and hypothesis testing procedures for a single mean μ, single proportion p, and a single variance σ^2. Here we extend these methods to situations involving the means, proportions, and variances of two different population distributions. For example, let μ_1 and μ_2 denote true average decrease in cholesterol for two drugs. Then an investigator might wish to use results from patients assigned at random to two groups as a basis for testing the hypothesis H_0: $\mu_1 = \mu_2$ versus the alternative hypothesis H_a: $\mu_1 \neq \mu_2$. As another example, let p_1 denote the true proportion of all Catholics who plan to vote for the Republican candidate in the next presidential election, and let p_2 represent the true proportion of all Protestants who plan to vote Republican. Based on a survey of 500 Catholics and 500 Protestants we might like an interval estimate for the difference $p_1 - p_2$.

10.1 z Tests and Confidence Intervals for a Difference Between Two Population Means

The inferences discussed in this section concern a difference $\mu_1 - \mu_2$ between the means of two different population distributions. An investigator might, for example, wish to test hypotheses about the difference between the true average weight losses of two diets. One such hypothesis would state that $\mu_1 - \mu_2 = 0$, that is, that $\mu_1 = \mu_2$. Alternatively, it may be appropriate to estimate $\mu_1 - \mu_2$ by computing a 95% CI. Such inferences are based on a sample of weight losses for each diet.

BASIC ASSUMPTIONS

1. $X_1, X_2, \ldots, X_m$ is a random sample from a population with mean μ_1 and variance σ_1^2.
2. $Y_1, Y_2, \ldots, Y_n$ is a random sample from a population with mean μ_2 and variance σ_2^2.
3. The X and Y samples are independent of one another.

The natural estimator of $\mu_1 - \mu_2$ is $\overline{X} - \overline{Y}$, the difference between the corresponding sample means. The test statistic results from standardizing this estimator, so we need expressions for the expected value and standard deviation of $\overline{X} - \overline{Y}$.

PROPOSITION

The expected value of $\overline{X} - \overline{Y}$ is $\mu_1 - \mu_2$, so $\overline{X} - \overline{Y}$ is an unbiased estimator of $\mu_1 - \mu_2$. The standard deviation of $\overline{X} - \overline{Y}$ is

$$\sigma_{\overline{X}-\overline{Y}} = \sqrt{\frac{\sigma_1^2}{m} + \frac{\sigma_2^2}{n}}$$

Proof Both these results depend on the rules of expected value and variance presented in Chapter 6. Since the expected value of a difference is the difference of expected values,

$$E(\overline{X} - \overline{Y}) = E(\overline{X}) - E(\overline{Y}) = \mu_1 - \mu_2$$

Because the X and Y samples are independent, $\overline{X}$ and $\overline{Y}$ are independent quantities, so the variance of the difference is the *sum* of $V(\overline{X})$ and $V(\overline{Y})$:

$$V(\overline{X} - \overline{Y}) = V(\overline{X}) + V(\overline{Y}) = \frac{\sigma_1^2}{m} + \frac{\sigma_2^2}{n}$$

The standard deviation of $\overline{X} - \overline{Y}$ is the square root of this expression. ■

If we think of $\mu_1 - \mu_2$ as a parameter θ, then its estimator is $\hat{\theta} = \overline{X} - \overline{Y}$ with standard deviation $\sigma_{\hat{\theta}}$ given by the proposition. When σ_1^2 and σ_2^2 both have known values, the test statistic will have the form $(\hat{\theta} - \text{null value})/\sigma_{\hat{\theta}}$; this form of a test statistic was used in several one-sample problems in the previous chapter. When σ_1^2 and σ_2^2 are unknown, the sample variances must be used to estimate $\sigma_{\hat{\theta}}$.

Test Procedures for Normal Populations with Known Variances

In Chapters 8 and 9, the first CI and test procedure for a population mean μ were based on the assumption that the population distribution was normal with the value of the population variance σ^2 known to the investigator. Similarly, we first assume here that *both* population distributions are normal and that the values of *both* σ_1^2 and σ_1^2 are known. Situations in which one or both of these assumptions can be dispensed with will be presented shortly.

Because the population distributions are normal, both $\overline{X}$ and $\overline{Y}$ have normal distributions. This implies that $\overline{X} - \overline{Y}$ is normally distributed, with expected value $\mu_1 - \mu_2$ and standard deviation $\sigma_{\overline{X}-\overline{Y}}$ given in the foregoing proposition. Standardizing $\overline{X} - \overline{Y}$ gives the standard normal variable

$$Z = \frac{\overline{X} - \overline{Y} - (\mu_1 - \mu_2)}{\sqrt{\dfrac{\sigma_1^2}{m} + \dfrac{\sigma_2^2}{n}}} \tag{10.1}$$

In a hypothesis-testing problem, the null hypothesis will state that $\mu_1 - \mu_2$ has a specified value. Denoting this null value by Δ_0, the null hypothesis becomes H_0: $\mu_1 - \mu_2 = \Delta_0$. Often $\Delta_0 = 0$, in which case H_0 says that $\mu_1 = \mu_2$. A test statistic results from replacing $\mu_1 - \mu_2$ in Expression (10.1) by the null value Δ_0. Because the test statistic Z is obtained by standardizing $\overline{X} - \overline{Y}$ under the assumption that H_0 is true, it has a standard normal distribution in this case. Consider the alternative hypothesis H_a: $\mu_1 - \mu_2 > \Delta_0$. A value $\bar{x} - \bar{y}$ that considerably exceeds Δ_0 (the expected value of $\overline{X} - \overline{Y}$ when H_0 is true) provides evidence against H_0 and for H_a. Such a value of $\bar{x} - \bar{y}$ corresponds to a positive and large value of z. Thus H_0 should be rejected in favor of H_a if z is greater than or equal to an appropriately chosen critical value. Because the test statistic Z has a standard normal distribution when H_0 is true, the upper-tailed rejection region $z \geq z_\alpha$ gives a test with significance level (type I error probability) α. Rejection regions for the alternatives H_a: $\mu_1 - \mu_2 < \Delta_0$ and H_a: $\mu_1 - \mu_2 \neq \Delta_0$ that yield tests with desired significance level α are lower-tailed and two-tailed, respectively.

Null hypothesis: H_0: $\mu_1 - \mu_2 = \Delta_0$

Test statistic value: $$z = \frac{\bar{x} - \bar{y} - \Delta_0}{\sqrt{\dfrac{\sigma_1^2}{m} + \dfrac{\sigma_2^2}{n}}}$$

Alternative Hypothesis	**Rejection Region for Level α Test**
H_a: $\mu_1 - \mu_2 > \Delta_0$	$z \geq z_\alpha$ (upper-tailed)
H_a: $\mu_1 - \mu_2 < \Delta_0$	$z \leq -z_\alpha$ (lower-tailed)
H_a: $\mu_1 - \mu_2 \neq \Delta_0$	either $z \geq z_{\alpha/2}$ or $z \leq -z_{\alpha/2}$ (two-tailed)

Because these are z tests, a P-value is computed as it was for the z tests in Chapter 9 [e.g., P-value $= 1 - \Phi(z)$ for an upper-tailed test].

Example 10.1 Each student in a class of 21 responded to a questionnaire that requested their grade point average (GPA) and the number of hours each week that they studied. For those who studied less than 10 hours per week the GPAs were

2.80 3.40 4.00 3.60 2.00 3.00 3.47 2.80 2.60 2.00

and for those who studied at least 10 hours per week the GPAs were

3.00 3.00 2.20 2.40 4.00 2.96 3.41 3.27 3.80 3.10 2.50

Normal plots for both sets are reasonably linear, so the normality assumption is tenable. Because the standard deviation of GPAs for the whole campus is .6, it is reasonable to apply that value here. The sample means are 2.97 for the $<$10 study hours group and 3.06 for the $\geq$10 study hours group. Treating the two samples as random, is there evidence that true average GPA differs for the two study times? Let's carry out a test of significance at level .05.

1. The parameter of interest is $\mu_1 - \mu_2$, the difference between true mean GPA for the $<$10 (conceptual) population and true mean GPA for the $\geq$10 population.
2. The null hypothesis is H_0: $\mu_1 - \mu_2 = 0$.
3. The alternative hypothesis is H_a: $\mu_1 - \mu_2 \neq 0$; if H_a is true then μ_1 and μ_2 are different. Although it would seem unlikely that $\mu_1 - \mu_2 > 0$ (those with low study hours have higher mean GPA) we will allow it as a possibility and do a two-tailed test.
4. With $\Delta_0 = 0$, the test statistic value is

$$z = \frac{\bar{x} - \bar{y}}{\sqrt{\dfrac{\sigma_1^2}{m} + \dfrac{\sigma_2^2}{n}}}$$

5. The inequality in H_a implies that the test is two-tailed. For $\alpha = .05$, $\alpha/2 = .025$ and $z_{\alpha/2} = z_{.025} = 1.96$. H_0 will be rejected if $z \geq 1.96$ or $z \leq -1.96$.
6. Substituting $m = 10$, $\bar{x} = 2.97$, $\sigma_1^2 = .36$, $n = 11$, $\bar{y} = 3.06$, and $\sigma_2^2 = .36$ into the formula for z yields

$$z = \frac{2.97 - 3.06}{\sqrt{\dfrac{.36}{10} + \dfrac{.36}{11}}} = \frac{-.09}{.262} = -.34$$

   That is, the value of $\bar{x} - \bar{y}$ is only one-third of a standard deviation below what would be expected when H_0 is true.
7. Because the value of z is not even close to the rejection region, there is no reason to reject the null hypothesis. This test shows no evidence of any relationship between study hours and GPA. ■

Using a Comparison to Identify Causality

Investigators are often interested in comparing either the effects of two different treatments on a response or the response after treatment with the response after no treatment (treatment vs. control). If the individuals or objects to be used in the comparison are not assigned by the investigators to the two different conditions, the study is said to be **observational**. The difficulty with drawing conclusions based on an observational study is that although statistical analysis may indicate a significant difference in response between the two groups, the difference may be due to some underlying factors that had not been controlled rather than to any difference in treatments.

Example 10.2 A letter in the *Journal of the American Medical Association* (May 19, 1978) reports that of 215 male physicians who were Harvard graduates and died between November 1974 and October 1977, the 125 in full-time practice lived an average of 48.9 years beyond graduation, whereas the 90 with academic affiliations lived an average of 43.2 years beyond graduation. Does the data suggest that the mean lifetime after graduation for doctors in full-time practice exceeds the mean lifetime for those who have an academic affiliation (if so, those medical students who say that they are "dying to obtain an academic affiliation" may be closer to the truth than they realize; in other words, is "publish or perish" really "publish and perish")?

Let μ_1 denote the true average number of years lived beyond graduation for physicians in full-time practice, and let μ_2 denote the same quantity for physicians with academic affiliations. Assume the 125 and 90 physicians to be random samples from populations 1 and 2, respectively (which may not be reasonable if there is reason to believe that Harvard graduates have special characteristics that differentiate them from all other physicians—in this case inferences would be restricted just to the "Harvard populations"). The letter from which the data was taken gave no information about variances, so for illustration assume that $\sigma_1 = 14.6$ and $\sigma_2 = 14.4$. The relevant hypotheses are H_0: $\mu_1 - \mu_2 = 0$ versus H_a: $\mu_1 - \mu_2 > 0$, so Δ_0 is zero. The computed value of z is

$$z = \frac{48.9 - 43.2}{\sqrt{\frac{(14.6)^2}{125} + \frac{(14.4)^2}{90}}} = \frac{5.70}{\sqrt{1.70 + 2.30}} = 2.85$$

The P-value for an upper-tailed test is $1 - \Phi(2.85) = .0022$. At significance level .01, H_0 is rejected (because $\alpha > P$-value) in favor of the conclusion that $\mu_1 - \mu_2 > 0$ $(\mu_1 > \mu_2)$. This is consistent with the information reported in the letter.

This data resulted from a **retrospective** observational study; the investigator did not start out by selecting a sample of doctors and assigning some to the "academic affiliation" treatment and the others to the "full-time practice" treatment, but instead identified members of the two groups by looking backward in time (through obituaries!) to past records. Can the statistically significant result here really be attributed to a difference in the type of medical practice after graduation, or is there some other underlying factor (e.g., age at graduation, exercise regimens, etc.) that might also furnish a plausible explanation for the difference?

Once upon a time, it could be argued that the studies linking smoking and lung cancer were all observational, and therefore that nothing had been proved. This was the view of the great (perhaps the greatest) statistician R. A. Fisher, who maintained till his death in 1962 that the observational studies did not show causation. He said that people who choose to smoke might be more susceptible to lung cancer. This explanation for the relationship had plenty of opposition then, and few would support it now. At that time few women got lung cancer because few women had smoked, but when smoking increased among women, so did lung cancer. Furthermore, the incidence of lung cancer was higher for those who smoked more, and quitters had reduced incidence. Eventually, the physiological effects on the body were better understood, and nonobservational animal studies made it clear that smoking does cause lung cancer. ■

A **randomized controlled experiment** results when investigators assign subjects to the two treatments in a random fashion. When statistical significance is observed in such an experiment, the investigator and other interested parties will have more confidence in the conclusion that the difference in response has been caused by a difference in treatments. A famous example of this type of experiment and conclusion is the Salk polio vaccine experiment described in Section 10.4. These issues are discussed at greater length in the (nonmathematical) books by Moore and by Freedman et al., listed in the Chapter 1 bibliography.

β and the Choice of Sample Size

The probability of a type II error is easily calculated when both population distributions are normal with known values of σ_1 and σ_2. Consider the case in which the alternative hypothesis is H_a: $\mu_1 - \mu_2 > \Delta_0$. Let Δ' denote a value of $\mu_1 - \mu_2$ that exceeds Δ_0 (a value for which H_0 is false). The upper-tailed rejection region $z \geq z_\alpha$ can be reexpressed in the form $\bar{x} - \bar{y} \geq \Delta_0 + z_\alpha \sigma_{\bar{X}-\bar{Y}}$. Thus the probability of a type II error when $\mu_1 - \mu_2 = \Delta'$ is

$$\begin{aligned}\beta(\Delta') &= P(\text{not rejecting } H_0 \text{ when } \mu_1 - \mu_2 = \Delta') \\ &= P(\bar{X} - \bar{Y} < \Delta_0 + z_\alpha \sigma_{\bar{X}-\bar{Y}} \text{ when } \mu_1 - \mu_2 = \Delta')\end{aligned}$$

When $\mu_1 - \mu_2 = \Delta'$, $\bar{X} - \bar{Y}$ is normally distributed with mean value Δ' and standard deviation $\sigma_{\bar{X}-\bar{Y}}$ (the same standard deviation as when H_0 is true); using these values to standardize the inequality in parentheses gives β.

Alternative Hypothesis	$\boldsymbol{\beta(\Delta') = P}$**(type II error when** $\boldsymbol{\mu_1 - \mu_2 = \Delta'}$**)**
H_a: $\mu_1 - \mu_2 > \Delta_0$	$\Phi\left(z_\alpha - \dfrac{\Delta' - \Delta_0}{\sigma}\right)$
H_a: $\mu_1 - \mu_2 < \Delta_0$	$1 - \Phi\left(-z_\alpha - \dfrac{\Delta' - \Delta_0}{\sigma}\right)$
H_a: $\mu_1 - \mu_2 \neq \Delta_0$	$\Phi\left(z_{\alpha/2} - \dfrac{\Delta' - \Delta_0}{\sigma}\right) - \Phi\left(-z_{\alpha/2} - \dfrac{\Delta' - \Delta_0}{\sigma}\right)$

where $\sigma = \sigma_{\bar{X}-\bar{Y}} = \sqrt{(\sigma_1^2/m) + (\sigma_2^2/n)}$

Example 10.3
(Example 10.1 continued)

If μ_1 and μ_2 (the true average GPAs for the two levels of effort) differ by as much as .5, what is the probability of detecting such a departure from H_0 based on a level .05 test with sample sizes $m = 10$ and $n = 11$? The value of σ for these sample sizes (the denominator of z) was previously calculated as .262. The probability of a type II error for the two-tailed level .05 test when

$$\mu_1 - \mu_2 = \Delta' = .5$$

is

$$\begin{aligned}\beta(.5) &= \Phi\left(1.96 - \frac{.5 - 0}{.262}\right) - \Phi\left(-1.96 - \frac{.5 - 0}{.262}\right)\\ &= \Phi(.0516) - \Phi(-3.868) = .521\end{aligned}$$

By symmetry we also have $\beta(-.5) = .521$. Thus the probability of detecting such a departure is $1 - \beta(.5) = .479$. Clearly, we do not have a very good chance of detecting a difference of .5 with these sample sizes. We should not conclude from Example 10.1 that there is no relationship between study time and GPA, because the sample sizes were insufficient. ■

As in Chapter 9, sample sizes m and n can be determined that will satisfy both $P(\text{type I error}) =$ a specified α and $P(\text{type II error when } \mu_1 - \mu_2 = \Delta') =$ a specified β. For an upper-tailed test, equating the previous expression for $\beta(\Delta')$ to the specified value of β gives

$$\frac{\sigma_1^2}{m} + \frac{\sigma_2^2}{n} = \frac{(\Delta' - \Delta_0)^2}{(z_\alpha + z_\beta)^2}$$

When the two sample sizes are equal, this equation yields

$$m = n = \frac{(\sigma_1^2 + \sigma_2^2)(z_\alpha + z_\beta)^2}{(\Delta' - \Delta_0)^2}$$

These expressions are also correct for a lower-tailed test, whereas α is replaced by $\alpha/2$ for a two-tailed test.

Large-Sample Tests

The assumptions of normal population distributions and known values of σ_1 and σ_2 are unnecessary when both sample sizes are large. In this case, the Central Limit Theorem guarantees that $\overline{X} - \overline{Y}$ has approximately a normal distribution regardless of the underlying population distributions. Furthermore, using S_1^2 and S_2^2 in place of σ_1^2 and σ_2^2 in Expression (10.1) gives a variable whose distribution is approximately standard normal:

$$Z = \frac{\overline{X} - \overline{Y} - (\mu_1 - \mu_2)}{\sqrt{\dfrac{S_1^2}{m} + \dfrac{S_2^2}{n}}}$$

A large-sample test statistic results from replacing $\mu_1 - \mu_2$ by Δ_0, the expected value of $\overline{X} - \overline{Y}$ when H_0 is true. This statistic Z then has approximately a standard normal distribution when H_0 is true, so level α tests are obtained by using z critical values exactly as before.

Use of the test statistic value

$$z = \frac{\bar{x} - \bar{y} - \Delta_0}{\sqrt{\frac{s_1^2}{m} + \frac{s_2^2}{n}}}$$

along with the previously stated upper-, lower-, and two-tailed rejection regions based on z critical values gives large-sample tests whose significance levels are approximately α. These tests are usually appropriate if both $m > 40$ and $n > 40$. A P-value is computed exactly as it was for our earlier z tests.

Example 10.4 A study was carried out in an attempt to improve student performance in a low-level university mathematics course. Experience had shown that many students had fallen by the wayside, meaning that they had dropped out or completed the course with minimal effort and low grades. The study involved assigning the students to sections based on odd or even Social Security number. It is important that the assignment to sections not be on the basis of student choice, because then the differences in performance might be attributable to differences in student attitude or ability. Half of the sections were taught traditionally, whereas the other half were taught in a way that hopefully would keep the students involved. They were given frequent assignments that were collected and graded, they had frequent quizzes, and they were allowed retakes on exams. Lotus Hershberger conducted the experiment and he supplied the data. Here are the final exam scores for the 79 students taught traditionally (the control group) and for the 85 students taught with more involvement (the experimental group):

Control

37	22	29	29	33	22	32	36	29	06	04	37	00	36	00	32
27	07	19	35	26	22	28	28	32	35	28	33	35	24	21	00
32	28	27	08	30	37	09	33	30	36	28	03	08	31	29	09
00	00	35	25	29	03	33	33	28	32	39	20	32	22	24	20
32	07	08	33	29	09	00	30	26	25	32	38	22	29	29	

Experimental

34	27	26	33	23	37	24	34	22	23	32	05	30	35	28	25
37	28	26	29	22	33	31	23	37	29	00	30	34	26	28	27
32	29	31	33	28	21	34	29	33	06	08	29	36	07	21	30
28	34	28	35	30	34	09	38	09	27	25	33	09	23	32	25
37	28	23	26	34	32	34	00	24	30	36	28	38	35	16	37
25	34	38	34	31											

Table 10.1 summarizes the data. Does this information suggest that true mean for the experimental condition exceeds that for the control condition? Let's use a test with $\alpha = .05$.

Table 10.1 Summary results for Example 10.4

Group	Sample Size	Sample Mean	Sample SD
Control	79	23.87	11.60
Experimental	85	27.34	8.85

Let μ_1 and μ_2 denote the true mean scores for the control group and the experimental group, respectively. The two hypotheses are H_0: $\mu_1 - \mu_2 = 0$ versus H_a: $\mu_1 - \mu_2 < 0$. H_0 will be rejected if $z \le -z_{.05} = -1.645$. We calculate

$$z = \frac{23.87 - 27.34}{\sqrt{\dfrac{11.60^2}{79} + \dfrac{8.85^2}{85}}} = \frac{-3.47}{1.620} = -2.14$$

Since $-2.14 \le -1.645$, H_0 is rejected at significance level .05. Alternatively, the P-value for a lower-tailed z test is

$$P\text{-value} = \Phi(z) = \Phi(-2.14) = .016$$

which implies rejection at significance level .05. Also, if the test had been two-tailed, then the P-value would be $2(.016) = .032$, so the two-tailed test would reject H_0 at the .05 level.

We have shown fairly conclusively that the experimental method of instruction is an improvement. Nevertheless, there is more to be said. It is important to view the data graphically to see if anything is strange. Figure 10.1 shows a plot from Systat combining a boxplot and dotplot.

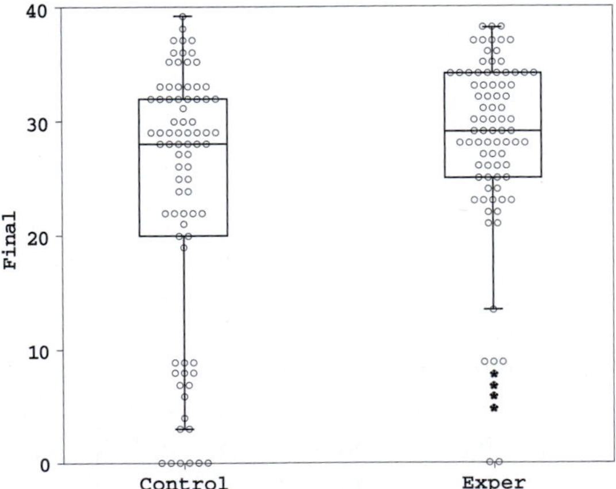

Figure 10.1 Boxplot/dotplot for the teaching experiment of Example 10.4

The plot shows that both groups have outlying observations at the low end; some students who showed up for the final performed very poorly. What happens if we compare the groups while ignoring the low performers whose scores are below 10? The resulting summary information is in Table 10.2.

Table 10.2 Summary results without poor performers

Group	Sample Size	Sample Mean	Sample SD
Control	61	29.59	5.005
Experimental	76	29.88	4.950

Notice that the means and standard deviations for the two groups are now very similar. Indeed, based on Table 10.2 the z statistic is $-.34$, giving no reason to reject the null hypothesis. For the majority of the students, there appears to be not much effect from the experimental treatment. It is the low performers who make a big difference in the results. There were 18 low performers in the control group but only 9 in the experimental group. The effect of the experimental instruction is to decrease the number of students who perform at the bottom of the scale. This is in accord with the goals of the experimental treatment, which was designed to keep students on track. ■

Confidence Intervals for $\mu_1 - \mu_2$

When both population distributions are normal, standardizing $\overline{X} - \overline{Y}$ gives a random variable Z with a standard normal distribution. Since the area under the z curve between $-z_{\alpha/2}$ and $z_{\alpha/2}$ is $1 - \alpha$, it follows that

$$P\left(-z_{\alpha/2} < \frac{\overline{X} - \overline{Y} - (\mu_1 - \mu_2)}{\sqrt{\dfrac{\sigma_1^2}{m} + \dfrac{\sigma_2^2}{n}}} < z_{\alpha/2}\right) = 1 - \alpha$$

Manipulation of the inequalities inside the parentheses to isolate $\mu_1 - \mu_2$ yields the equivalent probability statement

$$P\left(\overline{X} - \overline{Y} - z_{\alpha/2}\sqrt{\frac{\sigma_1^2}{m} + \frac{\sigma_2^2}{n}} < \mu_1 - \mu_2 < \overline{X} - \overline{Y} + z_{\alpha/2}\sqrt{\frac{\sigma_1^2}{m} + \frac{\sigma_2^2}{n}}\right) = 1 - \alpha$$

This implies that a $100(1 - \alpha)\%$ CI for $\mu_1 - \mu_2$ has lower limit $\overline{x} - \overline{y} - z_{\alpha/2} \cdot \sigma_{\overline{X}-\overline{Y}}$ and upper limit $\overline{x} - \overline{y} + z_{\alpha/2} \cdot \sigma_{\overline{X}-\overline{Y}}$, where $\sigma_{\overline{X}-\overline{Y}}$ is the square-root expression. This interval is a special case of the general formula $\hat{\theta} \pm z_{\alpha/2} \cdot \sigma_{\hat{\theta}}$.

If both m and n are large, the CLT implies that this interval is valid even without the assumption of normal populations; in this case, the confidence level is *approximately* $100(1 - \alpha)\%$. Furthermore, use of the sample variances S_1^2 and S_2^2 in the standardized variable Z yields a valid interval in which s_1^2 and s_2^2 replace σ_1^2 and σ_2^2.

Provided that m and n are both large, a CI for $\mu_1 - \mu_2$ with a confidence level of approximately $100(1 - \alpha)\%$ is

$$\overline{x} - \overline{y} \pm z_{\alpha/2}\sqrt{\frac{s_1^2}{m} + \frac{s_2^2}{n}}$$

where $-$ gives the lower limit and $+$ the upper limit of the interval. An upper or lower confidence bound can also be calculated by retaining the appropriate sign ($+$ or $-$) and replacing $z_{\alpha/2}$ by z_α.

Our standard rule of thumb for characterizing sample sizes as large is $m > 40$ and $n > 40$.

Example 10.5 For many calculus instructors it seems that students taking Calculus I in the fall semester are better prepared than are the students taking it in the spring. If so, it would be nice to have some measure of the difference. We use data from a study of the influence of various predictors on calculus performance, "Factors Affecting Achievement in the First Course in Calculus" (*J. Experiment. Ed.*, 1984: 136–140). Here are the ACT mathematics scores for both the fall students and the spring students:

Fall

27	29	30	34	29	30	29	28	28	31	25	34	27	28	31	26
24	30	25	25	27	27	28	27	27	27	26	33	27	26	35	27
32	30	27	30	30	28	28	30	26	31	28	26	23	28	31	28
33	24	32	20	28	34	33	30	29	16	30	30	26	29	26	27
26	25	31	18	29	29	30	29	29	30	33	29	29	27	28	28

Spring

29	26	25	24	14	31	25	33	27	30	27	29	26	27	29	31
25	28	26	23	28	27	27	19	28	25	23	20	34	25	33	30
26	19	18	25	17	26	24	29	20	27	26	26	27	20	28	26
27	24	28	28	30	27	27	27	14	25	27	32	35	13	28	25
29	25	19	27	30	15	28	27	28	32						

Figure 10.2 shows a graph from Systat combining a boxplot and dotplot.

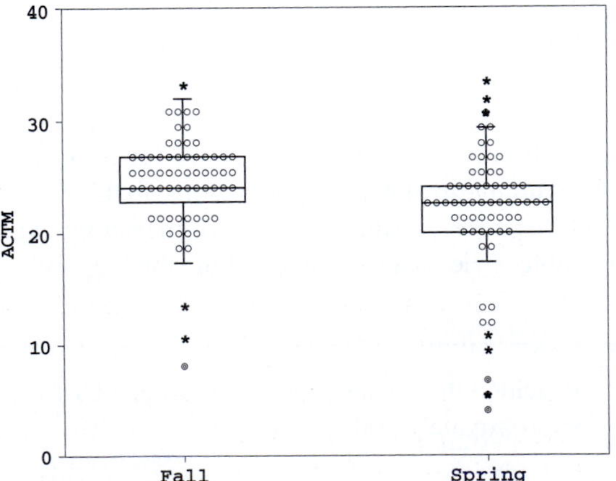

Figure 10.2 Boxplot/dotplot for fall and spring ACT mathematics scores

It is evident that there are more high scorers in the fall and more low scorers in the spring. Table 10.3 summarizes the data.

Table 10.3 Summary results for Example 10.5

Group	Sample Size	Sample Mean	Sample SD
Fall	80	28.25	3.25
Spring	74	25.88	4.59

Let's now calculate a confidence interval for the difference between true average fall ACT score and true average spring ACT score, using a confidence level of 95%:

$$28.25 - 25.88 \pm (1.96)\sqrt{\frac{3.25^2}{80} + \frac{4.59^2}{74}} = 2.37 \pm (1.96)(.6456)$$

$$= 2.37 \pm 1.265 = (1.10, 3.64)$$

That is, with 95% confidence, $1.10 < \mu_1 - \mu_2 < 3.64$. We can therefore be highly confident that the true fall average exceeds the true spring average by between 1.10 and 3.64. It makes sense that the fall average should be higher, because students who were less prepared in the fall (as judged by an algebra placement test) were required to take a fall-semester college algebra course before taking Calculus I in the spring. ■

If the variances σ_1^2 and σ_2^2 are at least approximately known and the investigator uses equal sample sizes, then the sample size n for each sample that yields a $100(1 - \alpha)\%$ interval of width w is

$$n = \frac{4z_{\alpha/2}^2(\sigma_1^2 + \sigma_2^2)}{w^2}$$

which will generally have to be rounded up to an integer.

Exercises Section 10.1 (1–19)

1. An article in the November 1983 *Consumer Reports* compared various types of batteries. The average lifetimes of Duracell Alkaline AA batteries and Eveready Energizer Alkaline AA batteries were given as 4.1 hours and 4.5 hours, respectively. Suppose these are the population average lifetimes.

a. Let $\overline{X}$ be the sample average lifetime of 100 Duracell batteries and $\overline{Y}$ be the sample average lifetime of 100 Eveready batteries. What is the mean value of $\overline{X} - \overline{Y}$ (i.e., where is the distribution of $\overline{X} - \overline{Y}$ centered)? How does your answer depend on the specified sample sizes?

b. Suppose the population standard deviations of lifetime are 1.8 hours for Duracell batteries and 2.0 hours for Eveready batteries. With the sample sizes given in part (a), what is the variance of the statistic $\overline{X} - \overline{Y}$, and what is its standard deviation?

c. For the sample sizes given in part (a), draw a picture of the approximate distribution curve of

$\overline{X} - \overline{Y}$ (include a measurement scale on the horizontal axis). Would the shape of the curve necessarily be the same for sample sizes of 10 batteries of each type? Explain.

2. Let μ_1 and μ_2 denote true average tread lives for two competing brands of size P205/65R15 radial tires. Test H_0: $\mu_1 - \mu_2 = 0$ versus H_a: $\mu_1 - \mu_2 \neq 0$ at level .05 using the following data: $m = 45$, $\overline{x} = 42{,}500$, $s_1 = 2200$, $n = 45$, $\overline{y} = 40{,}400$, and $s_2 = 1900$.

3. Let μ_1 denote true average tread life for a premium brand of P205/65R15 radial tire and let μ_2 denote the true average tread life for an economy brand of the same size. Test H_0: $\mu_1 - \mu_2 = 5000$ versus H_a: $\mu_1 - \mu_2 > 5000$ at level .01 using the following data: $m = 45$, $\overline{x} = 42{,}500$, $s_1 = 2200$, $n = 45$, $\overline{y} = 36{,}800$, and $s_2 = 1500$.

4. **a.** Use the data of Exercise 2 to compute a 95% CI for $\mu_1 - \mu_2$. Does the resulting interval suggest that $\mu_1 - \mu_2$ has been precisely estimated?

b. Use the data of Exercise 3 to compute a 95% upper confidence bound for $\mu_1 - \mu_2$.

5. Persons having Raynaud's syndrome are apt to suffer a sudden impairment of blood circulation in fingers and toes. In an experiment to study the extent of this impairment, each subject immersed a forefinger in water and the resulting heat output (cal/cm^2/min) was measured. For $m = 10$ subjects with the syndrome, the average heat output was $\overline{x} = .64$, and for $n = 10$ nonsufferers, the average output was 2.05. Let μ_1 and μ_2 denote the true average heat outputs for the two types of subjects. Assume that the two distributions of heat output are normal with $\sigma_1 = .2$ and $\sigma_2 = .4$.

a. Consider testing H_0: $\mu_1 - \mu_2 = -1.0$ versus H_a: $\mu_1 - \mu_2 < -1.0$ at level .01. Describe in words what H_a says, and then carry out the test.

b. Compute the P-value for the value of Z obtained in part (a).

c. What is the probability of a type II error when the actual difference between μ_1 and μ_2 is $\mu_1 - \mu_2 = -1.2$?

d. Assuming that $m = n$, what sample sizes are required to ensure that $\beta = .1$ when $\mu_1 - \mu_2 = -1.2$?

6. An experiment to compare the tension bond strength of polymer latex modified mortar (Portland cement mortar to which polymer latex emulsions have been added during mixing) to that of unmodified mortar resulted in $\overline{x} = 18.12$ kgf/cm^2 for the modified mortar ($m = 40$) and $\overline{y} = 16.87$ kgf/cm^2 for the unmodified mortar ($n = 32$). Let μ_1 and μ_2 be the true average tension bond strengths for the modified and unmodified mortars, respectively. Assume that the bond strength distributions are both normal.

a. Assuming that $\sigma_1 = 1.6$ and $\sigma_2 = 1.4$, test H_0: $\mu_1 - \mu_2 = 0$ versus H_a: $\mu_1 - \mu_2 > 0$ at level .01.

b. Compute the probability of a type II error for the test of part (a) when $\mu_1 - \mu_2 = 1$.

c. Suppose the investigator decided to use a level .05 test and wished $\beta = .10$ when $\mu_1 - \mu_2 = 1$. If $m = 40$, what value of n is necessary?

d. How would the analysis and conclusion of part (a) change if σ_1 and σ_2 were unknown but $s_1 = 1.6$ and $s_2 = 1.4$?

7. Are male college students more easily bored than their female counterparts? This question was examined in the article "Boredom in Young Adults—Gender and Cultural Comparisons" (*J. Cross-Cult. Psych.*, 1991: 209–223). The authors administered a scale called the Boredom Proneness Scale to 97 male and 148 female U.S. college students. Does the accompanying data support the research hypothesis that the mean Boredom Proneness Rating is higher for men than for women? Test the appropriate hypotheses using a .05 significance level.

Gender	Sample Size	Sample Mean	Sample SD
Male	97	10.40	4.83
Female	148	9.26	4.68

8. Is touching by a coworker sexual harassment? This question was included on a survey given to federal employees, who responded on a scale of 1 to 5, with 1 meaning a strong negative and 5 indicating a strong yes. The table summarizes the results.

Gender	Sample Size	Sample Mean	Sample SD
Female	4343	4.6056	.8659
Male	3903	4.1709	1.2157

Of course, with 1 to 5 being the only possible values, the normal distribution does not apply here, but the sample sizes are sufficient that it does not matter. Obtain a two-sided confidence interval for the difference in population means. Does your interval suggest that females are more likely than males to regard touching as harassment? Explain your reasoning.

9. The article "Evaluation of a Ventilation Strategy to Prevent Barotrauma in Patients at High Risk for Acute Respiratory Distress Syndrome" (*New Engl. J. Med.*, 1998: 355–358) reported on an experiment in which 120 patients with similar clinical features were randomly divided into a control group and a treatment group, each consisting of 60 patients. The sample mean ICU stay (days) and sample standard deviation for the treatment group were 19.9 and 39.1, respectively, whereas these values for the control group were 13.7 and 15.8.

a. Calculate a point estimate for the difference between true average ICU stay for the treatment and control groups. Does this estimate suggest that there is a significant difference between true average stays under the two conditions?

b. Answer the question posed in part (a) by carrying out a formal test of hypotheses. Is the result different from what you conjectured in part (a)?

c. Does it appear that ICU stay for patients given the ventilation treatment is normally distributed? Explain your reasoning.

d. Estimate true average length of stay for patients given the ventilation treatment in a way that conveys information about precision and reliability.

10. An experiment was performed to compare the fracture toughness of high-purity 18 Ni maraging steel with commercial-purity steel of the same type (*Corrosion Sci.*, 1971: 723–736). The sample average toughness was $\bar{x} = 65.6$ for $m = 32$ specimens of the high-purity steel, whereas for $n = 38$ specimens of commercial steel $\bar{y} = 59.8$. Because the high-purity steel is more expensive, its use for a certain application can be justified only if its fracture toughness exceeds that of commercial-purity steel by more than 5. Suppose that both toughness distributions are normal.

a. Assuming that $\sigma_1 = 1.2$ and $\sigma_2 = 1.1$, test the relevant hypotheses using $\alpha = .001$.

b. Compute β for the test conducted in part (a) when $\mu_1 - \mu_2 = 6$.

11. The level of lead in the blood was determined for a sample of 152 male hazardous-waste workers age 20–30 and also for a sample of 86 female workers, resulting in a mean ± standard error of 5.5 ± 0.3 for the men and 3.8 ± 0.2 for the women ("Temporal Changes in Blood Lead Levels of Hazardous Waste Workers in New Jersey, 1984–1987," *Envir. Monitoring Assessment*, 1993: 99–107). Calculate an estimate of the difference between true average blood lead levels for male and female workers in a way that provides information about reliability and precision.

12. A 3-year study was carried out to see if fluoride toothpaste helps to prevent cavities ("Clinical Testing of Fluoride and non-Fluoride Containing Dentifrices in Hounslow School Children," *British Dental J.*, Feb., 1971: 154–158). The dependent variable was the DMFS increment, the number of new Decayed, Missing, and Filled Surfaces. The table gives summary data.

Group	Sample Size	Sample Mean	Sample SD
Control	289	12.83	8.31
Fluoride	260	9.78	7.51

Calculate and interpret a 99% confidence interval for the difference between true means. Is fluoride toothpaste beneficial?

13. A study seeks to compare hospitals based on the performance of their intensive care units. The dependent variable is the mortality ratio, the ratio of the number of deaths over the predicted number of deaths based on the condition of the patients. The comparison will be between hospitals with nurse staffing problems and hospitals without such problems. Assume, based on past experience, that the standard deviation of the mortality ratio will be around .2 in both types of hospital. How many of each type of hospital should be included in the study in order to have both the type I and type II error probabilities be .05, if the true difference of mean mortality ratio for the two types of hospital is .2? If we conclude that hospitals with nurse staffing problems have a higher mortality ratio, does this imply a causal relationship? Explain.

14. The level of monoamine oxidase (MAO) activity in blood platelets (nm/mg protein/hr) was determined for each individual in a sample of 43 chronic schizophrenics, resulting in $\bar{x} = 2.69$ and $s_1 = 2.30$, as well as for 45 normal subjects, resulting in $\bar{y} = 6.35$ and $s_2 = 4.03$. Does this data strongly suggest that true average MAO activity for normal subjects is more than twice the activity level for schizophrenics? Derive a test procedure and carry out the test

Example 10.7 The deterioration of many municipal pipeline networks across the country is a growing concern. One technology proposed for pipeline rehabilitation uses a flexible liner threaded through existing pipe. The article "Effect of Welding on a High-Density Polyethylene Liner" (*J. Materials Civil Engrg.*, 1996: 94–100) reported the following data on tensile strength (psi) of liner specimens both when a certain fusion process was used and when this process was not used.

No fusion	2748	2700	2655	2822	2511			
	3149	3257	3213	3220	2753			
	$m = 10$	$\bar{x} = 2902.8$		$s_1 = 277.3$				
Fused	3027	3356	3359	3297	3125	2910	2889	2902
	$n = 8$	$\bar{y} = 3108.1$		$s_2 = 205.9$				

Figure 10.3 shows normal probability plots from MINITAB. The linear pattern in each plot supports the assumption that the tensile strength distributions under the two conditions are both normal.

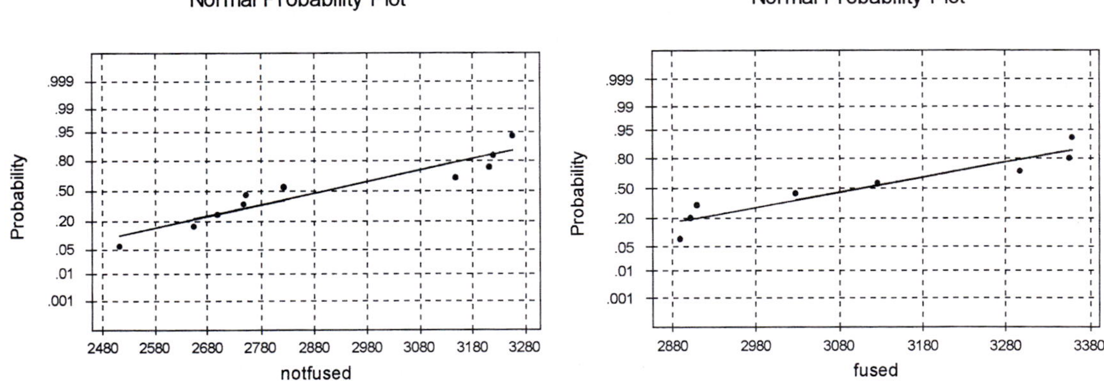

Figure 10.3 Normal probability plots from MINITAB for the tensile strength data

The authors of the article stated that the fusion process increased the average tensile strength. The message from the comparative boxplot of Figure 10.4 is not all that clear. Let's carry out a test of hypotheses to see whether the data supports this conclusion.

1. Let μ_1 be the true average tensile strength of specimens when the no-fusion treatment is used and μ_2 denote the true average tensile strength when the fusion treatment is used.
2. H_0: $\mu_1 - \mu_2 = 0$ (no difference in the true average tensile strengths for the two treatments)
3. H_a: $\mu_1 - \mu_2 < 0$ (true average tensile strength for the no-fusion treatment is less than that for the fusion treatment, so that the investigators' conclusion is correct)

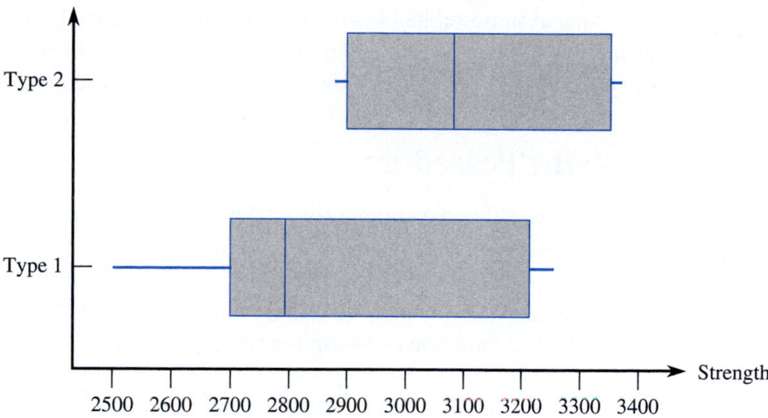

Figure 10.4 A comparative boxplot of the tensile strength data

4. The null value is $\Delta_0 = 0$, so the test statistic is

$$t = \frac{\bar{x} - \bar{y}}{\sqrt{\dfrac{s_1^2}{m} + \dfrac{s_2^2}{n}}}$$

5. We now compute both the test statistic value and the df for the test:

$$t = \frac{2902.8 - 3108.1}{\sqrt{\dfrac{(277.3)^2}{10} + \dfrac{(205.9)^2}{8}}} = \frac{-205.3}{113.97} = -1.8$$

Using $s_1^2/m = 7689.529$ and $s_2^2/n = 5299.351$,

$$\nu = \frac{(7689.529 + 5299.351)^2}{(7689.529)^2/9 + (5299.351)^2/7} = \frac{168{,}711{,}003.7}{10{,}581{,}747.35} = 15.94$$

so the test will be based on 15 df.

6. Appendix Table A.8 shows that the area under the 15 df t curve to the right of 1.8 is .046, so the P-value for a lower-tailed test is also .046. The following MINITAB output summarizes all the computations:

```
Twosample T for nofusion vs fused

              N       Mean      StDev   SE Mean
no fusion    10       2903        277        88
fused         8       3108        206        73
95% C.I. for mu nofusion-mu fused: (-488, 38)
T-Test mu nofusion = mu fused (vs <): T = -1.80 P = 0.046 DF = 15
```

7. Using a significance level of .05, we can barely reject the null hypothesis in favor of the alternative hypothesis, confirming the conclusion stated in the article. However, someone demanding more compelling evidence might select $\alpha = .01$, a level for which H_0 cannot be rejected.

If the question posed had been whether fusing increased true average strength by more than 100 psi, then the relevant hypotheses would have been H_0: $\mu_1 - \mu_2 = -100$ versus H_a: $\mu_1 - \mu_2 < -100$; that is, the null value would have been $\Delta_0 = -100$. ■

Pooled *t* Procedures

Alternatives to the two-sample *t* procedures just described result from assuming not only that the two population distributions are normal but also that they have equal variances ($\sigma_1^2 = \sigma_2^2$). That is, the two population distribution curves are assumed normal with equal spreads, the only possible difference between them being where they are centered.

Let σ^2 denote the common population variance. Then standardizing $\overline{X} - \overline{Y}$ gives

$$Z = \frac{\overline{X} - \overline{Y} - (\mu_1 - \mu_2)}{\sqrt{\dfrac{\sigma^2}{m} + \dfrac{\sigma^2}{n}}} = \frac{\overline{X} - \overline{Y} - (\mu_1 - \mu_2)}{\sqrt{\sigma^2\left(\dfrac{1}{m} + \dfrac{1}{n}\right)}}$$

which has a standard normal distribution. Before this variable can be used as a basis for making inferences about $\mu_1 - \mu_2$, the common variance must be estimated from sample data. One estimator of σ^2 is S_1^2, the variance of the m observations in the first sample, and another is S_2^2, the variance of the second sample. Intuitively, a better estimator than either individual sample variance results from combining the two sample variances. A first thought might be to use $(S_1^2 + S_2^2)/2$, the ordinary average of the two sample variances. However, if $m > n$, then the first sample contains more information about σ^2 than does the second sample, and an analogous comment applies if $m < n$. The following *weighted* average of the two sample variances, called the **pooled** (i.e., combined) **estimator of** σ^2, adjusts for any difference between the two sample sizes:

$$S_p^2 = \frac{m-1}{m+n-2}S_1^2 + \frac{n-1}{m+n-2}S_2^2$$

We can show that S_p^2 is proportional to a chi-squared rv with $m + n - 2$ df.

Recall that $(m-1)S_1^2/\sigma_1^2 \sim \chi^2_{m-1}$, $(n-1)S_2^2/\sigma_2^2 \sim \chi^2_{n-1}$. Furthermore, S_1^2 and S_2^2 are independent, so with $\sigma_1^2 = \sigma_2^2 = \sigma^2$,

$$\frac{(m+n-2)S_p^2}{\sigma^2} = \frac{m-1}{\sigma^2}S_1^2 + \frac{n-1}{\sigma^2}S_2^2$$

is the sum of two independent chi-squared rv's with $m - 1$ and $n - 1$ degrees of freedom, respectively, so the sum is a chi-squared rv with $(m - 1) + (n - 1) = m + n - 2$ degrees of freedom. Furthermore, it is also independent of $\overline{X}$ and $\overline{Y}$ because the sample means are independent of the sample variances. Now consider the ratio

$$\frac{\dfrac{\overline{X} - \overline{Y} - (\mu_1 - \mu_2)}{\sqrt{\sigma^2(1/m + 1/n)}}}{\sqrt{\dfrac{(m+n-2)S_p^2}{\sigma^2} \cdot \dfrac{1}{m+n-2}}} = \frac{\overline{X} - \overline{Y} - (\mu_1 - \mu_2)}{\sqrt{S_p^2\left(\dfrac{1}{m} + \dfrac{1}{n}\right)}}$$

On the left is the ratio of a standard normal rv to the square root of an independent chi-squared rv over its degrees of freedom, $m + n - 2$, so the ratio has the *t* distribution with $m + n - 2$ degrees of freedom. We see therefore that if S_p^2 replaces σ^2 in the expression for Z, the resulting standardized variable has a t distribution. In the same way that earlier standardized variables were used as a basis for deriving confidence intervals and test procedures, this t variable immediately leads to the pooled t confidence interval for estimating $\mu_1 - \mu_2$ and the pooled t test for testing hypotheses about a difference between means.

In the past, many statisticians recommended these pooled t procedures over the two-sample t procedures. The pooled t test, for example, can be derived from the likelihood ratio principle, whereas the two-sample t test is not a likelihood ratio test. Furthermore, the significance level for the pooled t test is exact, whereas it is only approximate for the two-sample t test. However, recent research has shown that although the pooled t test does outperform the two-sample t test by a bit (smaller β's for the same α) when $\sigma_1^2 = \sigma_2^2$, the former test can easily lead to erroneous conclusions if applied when the variances are different. Analogous comments apply to the behavior of the two confidence intervals. That is, the pooled t procedures are not robust to violations of the equal variance assumption.

It has been suggested that one could carry out a preliminary test of H_0: $\sigma_1^2 = \sigma_2^2$ and use a pooled t procedure if this null hypothesis is not rejected. Unfortunately, the usual "F test" of equal variances (Section 10.5) is quite sensitive to the assumption of normal population distributions, much more so than t procedures. We therefore recommend the conservative approach of using two-sample t procedures unless there is very compelling evidence for doing otherwise, particularly when the two sample sizes are different.

Type II Error Probabilities

Determining type II error probabilities (or equivalently, power $= 1 - \beta$) for the two-sample t test is complicated. There does not appear to be any simple way to use the β curves of Appendix Table A.17. The most recent version of MINITAB (Version 14) will calculate power for the pooled t test but not for the two-sample t test. However, the UCLA Statistics Department home page (http://www.stat.ucla.edu) permits access to a power calculator that will do this. For example, we specified $m = 10$, $n = 8$, $\sigma_1 = 300$, $\sigma_2 = 225$ (these are the sample sizes for Example 10.7, whose sample standard deviations are somewhat smaller than these values of σ_1 and σ_2) and asked for the power of a two-tailed level .05 test of H_0: $\mu_1 - \mu_2 = 0$ when $\mu_1 - \mu_2 = 100$, 250, and 500. The resulting values of the power were .1089, .4609, and .9635 (corresponding to $\beta = .89$, .54, and .04), respectively. In general, β will decrease as the sample sizes increase, as α increases, and as $\mu_1 - \mu_2$ moves farther from 0. The software will also calculate sample sizes necessary to obtain a specified value of power for a particular value of $\mu_1 - \mu_2$.

Exercises Section 10.2 (20–38)

20. Determine the number of degrees of freedom for the two-sample t test or CI in each of the following situations:

a. $m = 10, n = 10, s_1 = 5.0, s_2 = 6.0$
b. $m = 10, n = 15, s_1 = 5.0, s_2 = 6.0$
c. $m = 10, n = 15, s_1 = 2.0, s_2 = 6.0$
d. $m = 12, n = 24, s_1 = 5.0, s_2 = 6.0$

21. Expert and amateur pianists were compared in a study "Maintaining Excellence: Deliberate Practice and Elite Performance in Young and Older Pianists" (*J. Experiment Psych.: General*, 1996: 331–340). The researchers used a keyboard that allowed measurement of the force applied by a pianist in striking a key. All 48 pianists played Prelude Number 1 from

Bach's Well-Tempered Clavier. For 24 amateur pianists the mean force applied was 74.5 with standard deviation 6.29, and for 24 expert pianists the mean force was 81.8 with standard deviation 8.64. Do expert pianists hit the keys harder? Assuming normally distributed data, state and test the relevant hypotheses, and interpret the results.

22. Quantitative noninvasive techniques are needed for routinely assessing symptoms of peripheral neuropathies, such as carpal tunnel syndrome (CTS). The article "A Gap Detection Tactility Test for Sensory Deficits Associated with Carpal Tunnel Syndrome" (*Ergonomics*, 1995: 2588–2601) reported on a test that involved sensing a tiny gap in an otherwise smooth surface by probing with a finger; this functionally resembles many work-related tactile activities, such as detecting scratches or surface defects. When finger probing was not allowed, the sample average gap detection threshold for $m = 8$ normal subjects was 1.71 mm, and the sample standard deviation was .53; for $n = 10$ CTS subjects, the sample mean and sample standard deviation were 2.53 and .87, respectively. Does this data suggest that the true average gap detection threshold for CTS subjects exceeds that for normal subjects? State and test the relevant hypotheses using a significance level of .01.

23. Fusible interlinings are being used with increasing frequency to support outer fabrics and improve the shape and drape of various pieces of clothing. The article "Compatibility of Outer and Fusible Interlining Fabrics in Tailored Garments" (*Textile Res. J.*, 1997: 137–142) gave the accompanying data on extensibility (%) at 100 g/cm for both high-quality fabric (H) and poor-quality fabric (P) specimens.

H	1.2	.9	.7	1.0	1.7	1.7	1.1	.9	1.7
	1.9	1.3	2.1	1.6	1.8	1.4	1.3	1.9	1.6
	.8	2.0	1.7	1.6	2.3	2.0			
P	1.6	1.5	1.1	2.1	1.5	1.3	1.0	2.6	

a. Construct normal probability plots to verify the plausibility of both samples having been selected from normal population distributions.

b. Construct a comparative boxplot. Does it suggest that there is a difference between true average extensibility for high-quality fabric specimens and that for poor-quality specimens?

c. The sample mean and standard deviation for the high-quality sample are 1.508 and .444, respectively, and those for the poor-quality sample are 1.588 and .530. Use the two-sample t test to decide whether true average extensibility differs for the two types of fabric.

24. Low-back pain (LBP) is a serious health problem in many industrial settings. The article "Isodynamic Evaluation of Trunk Muscles and Low-Back Pain Among Workers in a Steel Factory" (*Ergonomics*, 1995: 2107–2117) reported the accompanying summary data on lateral range of motion (degrees) for a sample of workers without a history of LBP and another sample with a history of this malady.

Condition	Sample Size	Sample Mean	Sample SD
No LBP	28	91.5	5.5
LBP	31	88.3	7.8

Calculate a 90% confidence interval for the difference between population mean extent of lateral motion for the two conditions. Does the interval suggest that population mean lateral motion differs for the two conditions? Is the message different if we use a confidence level of 95%?

25. The article "The Influence of Corrosion Inhibitor and Surface Abrasion on the Failure of Aluminum-Wired Twist-on Connections" (*IEEE Trans. Components, Hybrids, Manuf. Tech.*, 1984: 20–25) reported data on potential drop measurements for one sample of connectors wired with alloy aluminum and another sample wired with EC aluminum. Does the accompanying SAS output suggest that the true average potential drop for alloy connections (type 1) is higher than that for EC connections (as stated in the article)? Carry out the appropriate test using a significance level of .01. In reaching your conclusion, what type of error might you have committed? (*Note*: SAS reports the P-value for a two-tailed test.)

```
Type  N         Mean      Std Dev   Std Error
1     20  17.49900000  0.55012821  0.12301241
2     20  16.90000000  0.48998389  0.10956373

  Variances         T          DF   Prob > |T|
    Unequal    3.6362        37.5       0.0008
      Equal    3.6362        38.0       0.0008
```

26. Tennis elbow is thought to be aggravated by the impact experienced when hitting the ball. The article "Forces on the Hand in the Tennis One-Handed Backhand" (*Internat. J. Sport Biomechanics*, 1991: 282–292) reported the force (N) on the hand just after impact on a one-handed backhand drive for six advanced players and for eight intermediate players.

Type of Player	Sample Size	Sample Mean	Sample SD
1. Advanced	6	40.3	11.3
2. Intermediate	8	21.4	8.3

In their analysis of the data, the authors assumed that both force distributions were normal. Calculate a 95% CI for the difference between true average force for advanced players (μ_1) and true average force for intermediate players (μ_2). Does your interval provide compelling evidence for concluding that the two μ's are different? Would you have reached the same conclusion by calculating a CI for $\mu_2 - \mu_1$ (i.e., by reversing the 1 and 2 labels on the two types of players)? Explain.

27. As the population ages, there is increasing concern about accident-related injuries to the elderly. The article "Age and Gender Differences in Single-Step Recovery from a Forward Fall" (*J. Gerontol.*, 1999: M44–M50) reported on an experiment in which the maximum lean angle—the furthest a subject is able to lean and still recover in one step—was determined for both a sample of younger females (21–29 years) and a sample of older females (67–81 years). The following observations are consistent with summary data given in the article:

YF: 29, 34, 33, 27, 28, 32, 31, 34, 32, 27
OF: 18, 15, 23, 13, 12

Does the data suggest that true average maximum lean angle for older females is more than 10 degrees smaller than it is for younger females? State and test the relevant hypotheses at significance level .10 by obtaining a P-value.

28. The article "Effect of Internal Gas Pressure on the Compression Strength of Beverage Cans and Plastic Bottles" (*J. Testing Eval.*, 1993: 129–131) includes the accompanying data on compression strength (lb) for a sample of 12-oz aluminum cans filled with strawberry drink and another sample filled with cola. Does the data suggest that the extra carbonation of cola results in a higher average compression strength? Base your answer on a P-value. What assumptions are necessary for your analysis?

Beverage	Sample Size	Sample Mean	Sample SD
Strawberry drink	15	540	21
Cola	15	554	15

29. Which foams more when you pour it, Coke or Pepsi? Here are measurements by Diane Warfield on the foam volume (mL) after pouring a 12-oz can of Coke, based on a sample of 12 cans:

312.2	292.6	331.7	355.1	362.9	331.7
292.6	245.8	280.9	320.0	273.1	288.7

and here are measurements for Pepsi, based on a sample of 12 cans:

148.3	210.7	152.2	117.1	89.7	140.5
128.8	167.8	156.1	136.6	124.9	136.6

a. Verify graphically that normality is an appropriate assumption.

b. Calculate a 99% confidence interval for the population difference in mean volumes.

c. Does the upper limit of your interval in (b) give a 99% lower confidence bound for the difference between the two μ's? If not, calculate such a bound and interpret it in terms of the relationship between the foam volumes of Coke and Pepsi.

d. Summarize in a sentence what you have learned about the foam volumes of Coke and Pepsi.

30. The article "Measuring and Understanding the Aging of Kraft Insulating Paper in Power Transformers" (*IEEE Electr. Insul. Mag.*, 1996: 28–34) contained observations on degree of polymerization for paper specimens. There were observations for two different ranges of viscosity times concentration:

Middle range

418 421 421 422 425 427 431 434 437
439 446 447 448 453 454 463 465

Higher range

429 430 430 431 436 437
440 441 445 446 447

a. Construct a comparative boxplot for the two samples, and comment on any interesting features.

b. Calculate a 95% confidence interval for the difference between true average degree of polymerization for the middle range and that for the high range. Does the interval suggest that μ_1 and μ_2 may in fact be different? Explain your reasoning.

31. The article "Characterization of Bearing Strength Factors in Pegged Timber Connections" (*J. Struct. Engrg.*, 1997: 326–332) gave the following summary data on proportional stress limits for specimens constructed using two different types of wood:

Type of Wood	Sample Size	Sample Mean	Sample SD
Red oak	14	8.48	.79
Douglas fir	10	6.65	1.28

Assuming that both samples were selected from normal distributions, carry out a test of hypotheses to decide whether the true average proportional stress limit for red oak joints exceeds that for Douglas fir joints by more than 1 MPa.

32. The accompanying table summarizes data on body weight gain (g) both for a sample of animals given a 1-mg/pellet dose of a certain soft steroid and for a sample of control animals ("The Soft Drug Approach," *CHEMTECH*, 1984: 28–38).

Treatment	Sample Size	Sample Mean	Sample SD
Steroid	8	32.8	2.6
Control	10	40.5	2.5

Does the data suggest that the true average weight gain in the control situation exceeds that for the steroid treatment by more than 5 g? State and test the appropriate hypotheses at a significance level of .01 using the *P*-value method.

33. Consider the pooled t variable

$$T = \frac{(\bar{X} - \bar{Y}) - (\mu_1 - \mu_2)}{S_p\sqrt{\dfrac{1}{m} + \dfrac{1}{n}}}$$

which has a t distribution with $m + n - 2$ df when both population distributions are normal with $\sigma_1 = \sigma_2$ (see the Pooled t Procedures subsection for a description of S_p).

a. Use this t variable to obtain a pooled t confidence interval formula for $\mu_1 - \mu_2$.

b. A sample of ultrasonic humidifiers of one particular brand was selected for which the observations on maximum output of moisture (oz) in a controlled chamber were 14.0, 14.3, 12.2, and 15.1. A sample of the second brand gave output values 12.1, 13.6, 11.9, and 11.2 ("Multiple Comparisons of Means Using Simultaneous Confidence Intervals," *J. Qual. Techn.*, 1989: 232–241). Use the pooled t formula from part (a) to estimate the difference between true average outputs for the two brands with a 95% confidence interval.

c. Estimate the difference between the two μ's using the two-sample t interval discussed in this section, and compare it to the interval of part (b).

34. Refer to Exercise 33. Describe the pooled t test for testing H_0: $\mu_1 - \mu_2 = 0$ when both population distributions are normal with $\sigma_1 = \sigma_2$. Then use this test procedure to test the hypotheses suggested in Exercise 32.

35. Exercise 35 from Chapter 9 gave the following data on amount (oz) of alcohol poured into a short, wide tumbler glass by a sample of experienced bartenders: 2.00, 1.78, 2.16, 1.91,1.70, 1.67, 1.83, 1.48. The cited article also gave summary data on the amount poured by a different sample of experienced bartenders into a tall, slender (highball) glass; the following observations are consistent with the reported summary data: 1.67, 1.57, 1.64, 1.69, 1.74, 1.75, 1.70, 1.60.

a. What does a comparative boxplot suggest about similarities and differences in the data?

b. Carry out a test of hypotheses to decide whether the true average amount poured is different for the two types of glasses; be sure to check the validity of any assumptions necessary to your analysis, and report a *P*-value for the test.

36. Is the incidence of head or neck pain among video display terminal users related to the monitor angle (degrees from horizontal)? The paper "An Analysis of VDT Monitor Placement and Daily Hours of Use for Female Bifocal Users" (*Work*, 2003: 77–80) reported the accompanying data. Carry out an appropriate test of hypotheses (be sure to include a *P*-value in your analysis).

Head/ Neck Pain	Sample Size	Sample Mean	Sample SD
Yes	32	2.20	3.42
No	40	3.20	2.52

37. The article "Gender Differences in Individuals with Comorbid Alcohol Dependence and Post-Traumatic Stress Disorder" (*Amer. J. Addiction*, 2003: 412–423) reported the accompanying data on total score on the Obsessive-Compulsive Drinking Scale (OCSD).

Gender	Sample Size	Sample Mean	Sample SD
Male	44	19.93	7.74
Female	40	16.26	7.58

Formulate hypotheses and carry out an appropriate analysis. Does your conclusion depend on whether a significance level of .05 or .01 was employed? (The cited paper reported P-value $< .05$; presumably .05 would have been replaced by .01 if the P-value were really that small.)

38. Which factors are relevant to the time a consumer spends looking at a product on the shelf prior to selection? The article "Effects of Base Price upon Search Behavior of Consumers in a Supermarket" (*J. Economic Psych.*, 2003: 637–652) reported the following data on elapsed time (sec) for fabric softener purchasers and washing-up liquid purchasers; the former product is significantly more expensive than the latter. These products were chosen because they are similar with respect to allocated shelf space and number of alternative brands.

Product	Sample Size	Sample Mean	Sample SD
Fabric softener	15	30.47	19.15
Washing-up liquid	19	26.53	15.37

a. What if any assumptions are needed before an inferential procedure can be used to compare true average elapsed times?

b. If just the two sample means had been reported, would they provide persuasive evidence for a significant difference between true average elapsed times for the two products?

c. Carry out an appropriate test of significance and state your conclusion.

10.3 Analysis of Paired Data

In Sections 10.1 and 10.2, we considered estimating or testing for a difference between two means μ_1 and μ_2. This was done by utilizing the results of a random sample X_1, $X_2, \ldots, X_m$ from the distribution with mean μ_1 and a completely independent (of the X's) sample $Y_1, \ldots, Y_n$ from the distribution with mean μ_2. That is, either m individuals were selected from population 1 and n different individuals from population 2, or m individuals (or experimental objects) were given one treatment and another set of n individuals were given the other treatment. In contrast, many experimental situations have only one set of n individuals or experimental objects, and two observations are made on each individual or object, resulting in a natural pairing of values.

Example 10.8 Trace metals in drinking water affect the flavor, and unusually high concentrations can pose a health hazard. The article "Trace Metals of South Indian River" (*Envir. Stud.*, 1982: 62–66) reports on a study in which six river locations were selected (six experimental objects) and the zinc concentration (mg/L) determined for both surface water and bottom water at each location. The six pairs of observations are displayed in the accompanying table. Does the data suggest that true average concentration in bottom water exceeds that of surface water?

	Location					
	1	2	3	4	5	6
Zinc concentration in bottom water (x)	.430	.266	.567	.531	.707	.716
Zinc concentration in surface water (y)	.415	.238	.390	.410	.605	.609
Difference	.015	.028	.177	.121	.102	.107

Figure 10.5(a) displays a plot of this data. At first glance, there appears to be little difference between the x and y samples. From location to location, there is a great deal of variability in each sample, and it looks as though any differences between the samples can be attributed to this variability. However, when the observations are identified by location, as in Figure 10.5(b), a different view emerges. At each location, bottom concentration exceeds surface concentration. This is confirmed by the fact that all $x - y$ differences (bottom water concentration − surface water concentration) displayed in the bottom row of the data table are positive. As we will see, a correct analysis of this data focuses on these differences.

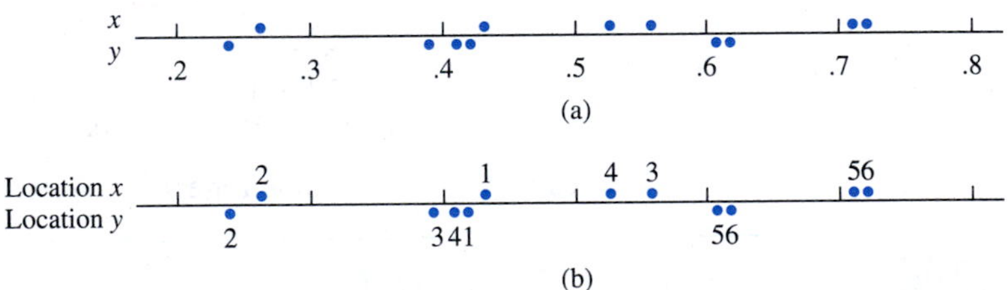

Figure 10.5 Plot of paired data from Example 10.8: (a) observations not identified by location; (b) observations identified by location ■

ASSUMPTIONS The data consists of n independently selected pairs $(X_1, Y_1), (X_2, Y_2), \ldots, (X_n, Y_n)$, with $E(X_i) = \mu_1$ and $E(Y_i) = \mu_2$. Let $D_1 = X_1 - Y_1$, $D_2 = X_2 - Y_2, \ldots, D_n = X_n - Y_n$, so the D_i's are the differences within pairs. Then the D_i's are assumed to be normally distributed with mean value μ_D and variance σ_D^2.

We are again interested in hypothesis testing or estimation for the difference $\mu_1 - \mu_2$. The denominator of the two-sample t statistic was obtained by first applying the rule $V(\overline{X} - \overline{Y}) = V(\overline{X}) + V(\overline{Y})$. However, with paired data, the X and Y observations within each pair are often not independent, so $\overline{X}$ and $\overline{Y}$ are not independent of one another, and the rule is not valid. We must therefore abandon the two-sample t procedures and look for an alternative method of analysis.

The Paired t Test

Because different pairs are independent, the D_i's are independent of one another. If we let $D = X - Y$, where X and Y are the first and second observations, respectively, within an arbitrary pair, then the expected difference is

$$\mu_D = E(X - Y) = E(X) - E(Y) = \mu_1 - \mu_2$$

(the rule of expected values used here is valid even when X and Y are dependent). Thus any hypothesis about $\mu_1 - \mu_2$ can be phrased as a hypothesis about the mean difference μ_D. But since the D_i's constitute a normal random sample (of differences) with mean μ_D, hypotheses about μ_D can be tested using a one-sample t test. That is, *to test hypotheses about* $\mu_1 - \mu_2$ *when data is paired, form the differences* $D_1, D_2, \ldots, D_n$ *and carry out a one-sample t test (based on* $n - 1$ *df) on the differences.*

THE PAIRED t TEST

Null hypothesis: H_0: $\mu_D = \Delta_0$ (where $D = X - Y$ is the difference between the first and second observations within a pair, and $\mu_D = \mu_1 - \mu_2$)

Test statistic value: $t = \dfrac{\bar{d} - \Delta_0}{s_D/\sqrt{n}}$ (where $\bar{d}$ and s_D are the sample mean and standard deviation, respectively, of the d_i's)

Alternative Hypothesis	Rejection Region for Level α Test
H_a: $\mu_D > \Delta_0$	$t \geq t_{\alpha,n-1}$
H_a: $\mu_D < \Delta_0$	$t \leq -t_{\alpha,n-1}$
H_a: $\mu_D \neq \Delta_0$	either $t \geq t_{\alpha/2,n-1}$ or $t \leq -t_{\alpha/2,n-1}$

A P-value can be calculated as was done for earlier t tests.

Example 10.9 Musculoskeletal neck-and-shoulder disorders are all too common among office staff who perform repetitive tasks using visual display units. The article "Upper-Arm Elevation During Office Work" (*Ergonomics*, 1996: 1221–1230) reported on a study to determine whether more varied work conditions would have any impact on arm movement. The accompanying data was obtained from a sample of $n = 16$ subjects. Each observation is the amount of time, expressed as a proportion of total time observed, during which arm elevation was below 30°. The two measurements from each subject were obtained 18 months apart. During this period, work conditions were changed, and subjects were allowed to engage in a wider variety of work tasks. Does the data suggest that true average time during which elevation is below 30° differs after the change from what it was before the change? This particular angle is important because in Sweden, where the research was conducted, workers' compensation regulations assert that arm elevation less than 30° is not harmful.

Subject	1	2	3	4	5	6	7	8
Before	81	87	86	82	90	86	96	73
After	78	91	78	78	84	67	92	70
Difference	3	−4	8	4	6	19	4	3

Subject	9	10	11	12	13	14	15	16
Before	74	75	72	80	66	72	56	82
After	58	62	70	58	66	60	65	73
Difference	16	13	2	22	0	12	−9	9

Figure 10.6 shows a normal probability plot of the 16 differences; the pattern in the plot is quite straight, supporting the normality assumption. A boxplot of these differences appears in Figure 10.7; the box is located considerably to the right of zero, suggesting that perhaps $\mu_D > 0$ (note also that 13 of the 16 differences are positive and only two are negative).

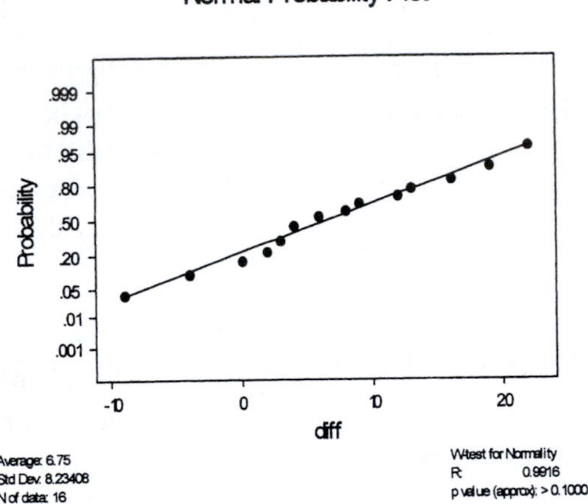

Figure 10.6 A normal probability plot from MINITAB of the differences in Example 10.9

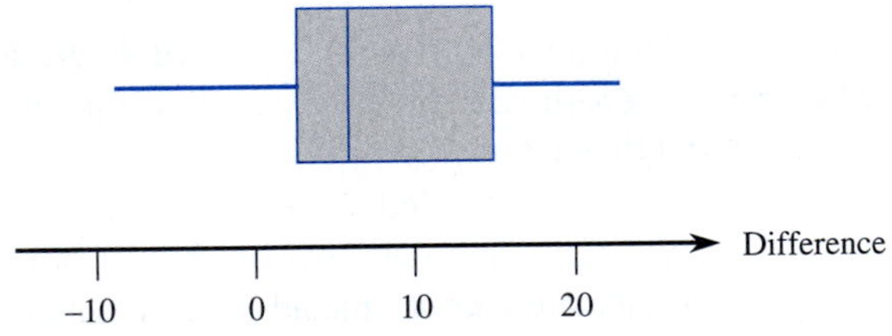

Figure 10.7 A boxplot of the differences in Example 10.9

Let's now use the recommended sequence of steps to test the appropriate hypotheses.

1. Let μ_D denote the true average difference between elevation time before the change in work conditions and time after the change.
2. H_0: $\mu_D = 0$ (there is no difference between true average time before the change and true average time after the change)
3. H_a: $\mu_D \neq 0$

4. $t = \dfrac{\bar{d} - 0}{s_D/\sqrt{n}} = \dfrac{\bar{d}}{s_D/\sqrt{n}}$

5. $n = 16$, $\Sigma d_i = 108$, $\Sigma d_i^2 = 1746$, from which $\bar{d} = 6.75$, $s_D = 8.234$, and

$$t = \frac{6.75}{8.234/\sqrt{16}} = 3.28 \approx 3.3$$

6. Appendix Table A.8 shows that the area to the right of 3.3 under the t curve with 15 df is .002. The inequality in H_a implies that a two-tailed test is appropriate, so the P-value is approximately $2(.002) = .004$ (MINITAB gives .0051).

7. Since $.004 < .01$, the null hypothesis can be rejected at either significance level .05 or .01. It does appear that the true average difference between times is something other than zero; that is, true average time after the change is different from that before the change. Recalling that arm elevation should be kept under 30°, we can conclude that the situation became worse because the amount of time below 30° decreased. ■

When the number of pairs is large, the assumption of a normal difference distribution is not necessary. The CLT validates the resulting z test.

A Confidence Interval for μ_D

In the same way that the t CI for a single population mean μ is based on the t variable $T = (\bar{X} - \mu)/(S/\sqrt{n})$, a t confidence interval for $\mu_D (= \mu_1 - \mu_2)$ is based on the fact that

$$T = \frac{\bar{D} - \mu_D}{S_D/\sqrt{n}}$$

has a t distribution with $n - 1$ df. Manipulation of this t variable, as in previous derivations of CIs, yields the following $100(1 - \alpha)\%$ CI:

The **paired t CI for μ_D** is

$$\bar{d} \pm t_{\alpha/2,n-1} \cdot s_D/\sqrt{n}$$

A one-sided confidence bound results from retaining the relevant sign and replacing $t_{\alpha/2}$ by t_α.

When n is small, the validity of this interval requires that the distribution of differences be at least approximately normal. For large n, the CLT ensures that the resulting z interval is valid without any restrictions on the distribution of differences.

Example 10.10 Adding computerized medical images to a database promises to provide great resources for physicians. However, there are other methods of obtaining such information, so the issue of efficiency of access needs to be investigated. The article "The Comparative Effectiveness of Conventional and Digital Image Libraries" (*J. Audiovisual Media*

Med., 2001: 8–15) reported on an experiment in which 13 computer-proficient medical professionals were timed both while retrieving an image from a library of slides and while retrieving the same image from a computer database with a Web front end.

Subject	1	2	3	4	5	6	7	8	9	10	11	12	13
Slide	30	35	40	25	20	30	35	62	40	51	25	42	33
Digital	25	16	15	15	10	20	7	16	15	13	11	19	19
Difference	5	19	25	10	10	10	28	46	25	38	14	23	14

Let μ_D denote the true mean difference between slide retrieval time (sec) and digital retrieval time. Using the paired t confidence interval to estimate μ_D requires that the difference distribution be at least approximately normal. The linear pattern of points in the normal probability plot from MINITAB (Figure 10.8) validates the normality assumption. (Only 9 points appear because of ties in the differences.)

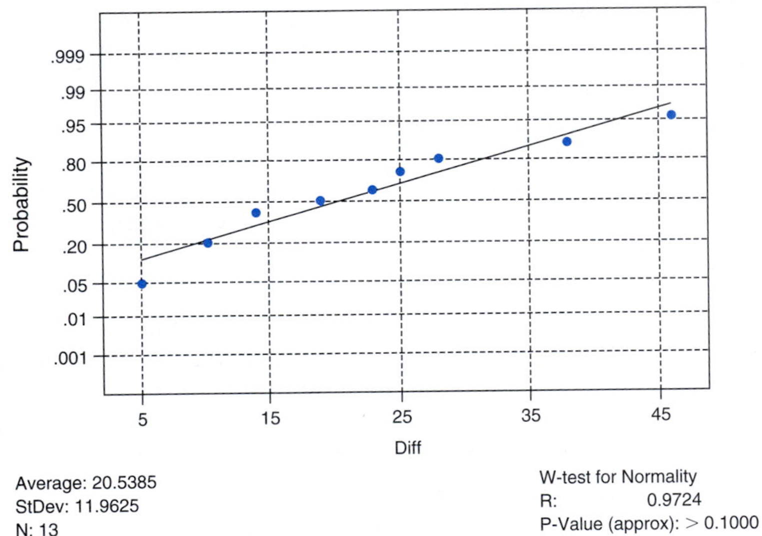

Figure 10.8 Normal probability plot of the differences in Example 10.10

Relevant summary quantities are $\sum d_i = 267$, $\sum d_i^2 = 7201$, from which $\bar{d} = 20.5$, $s_D = 11.96$. The t critical value required for a 95% confidence level is $t_{.025,12} = 2.179$, and the 95% CI is

$$\bar{d} \pm t_{\alpha/2,n-1} \cdot \frac{s_D}{\sqrt{n}} = 20.5 \pm (2.179) \cdot \frac{11.96}{\sqrt{13}} = 20.5 \pm 7.2 = (13.3, 27.7)$$

Thus we can be highly confident (at the 95% confidence level) that $13.3 < \mu_D < 27.7$. This interval of plausible values is rather wide, a consequence of the sample standard deviation being large relative to the sample mean. A sample size much larger than 13 would be required to estimate with substantially more precision. Notice, however, that 0 lies well outside the interval, suggesting that $\mu_D > 0$; this is confirmed by a formal hypothesis test. It is not hard to show that 0 is outside the 95% CI if and only if the two-tailed test rejects H_0: $\mu_D = 0$ at the .05 level. We can conclude from the experiment that computer retrieval appears to be faster on average. ■

Paired Data and Two-Sample *t* Procedures

Consider using the two-sample t test on paired data. The numerators of the paired t and two-sample t test statistics are identical, since $\bar{d} = \sum d_i/n = [\sum(x_i - y_i)]/n = (\sum x_i)/n - (\sum y_i)/n = \bar{x} - \bar{y}$. The difference between the two statistics is due entirely to the denominators. Each test statistic is obtained by standardizing $\bar{X} - \bar{Y}(=\bar{D})$, but in the presence of dependence the two-sample t standardization is incorrect. To see this, recall from Section 6.3 that

$$V(X \pm Y) = V(X) + V(Y) \pm 2\,\text{Cov}(X, Y)$$

Since the correlation between X and Y is

$$\rho = \text{Corr}(X, Y) = \text{Cov}(X, Y)/[\sqrt{V(X)} \cdot \sqrt{V(Y)}]$$

it follows that

$$V(X - Y) = \sigma_1^2 + \sigma_2^2 - 2\rho\sigma_1\sigma_2$$

Applying this to $\bar{X} - \bar{Y}$ yields

$$V(\bar{X} - \bar{Y}) = V(\bar{D}) = V\left(\frac{1}{n}\sum D_i\right) = \frac{V(D_i)}{n} = \frac{\sigma_1^2 + \sigma_2^2 - 2\rho\sigma_1\sigma_2}{n}$$

The two-sample t test is based on the assumption of independence, in which case $\rho = 0$. But in many paired experiments, there will be a strong *positive* dependence between X and Y (large X associated with large Y), so that ρ will be positive and the variance of $\bar{X} - \bar{Y}$ will be smaller than $\sigma_1^2/n + \sigma_2^2/n$. Thus *whenever there is positive dependence within pairs, the denominator for the paired t statistic should be smaller than for t of the independent-samples test.* Often two-sample t will be much closer to zero than paired t, considerably understating the significance of the data.

Similarly, when data is paired, the paired t CI will usually be narrower than the (incorrect) two-sample t CI. This is because there is typically much less variability in the differences than in the x and y values.

Paired Versus Unpaired Experiments

In our examples, paired data resulted from two observations on the same subject (Example 10.9) or experimental object (location in Example 10.8). Even when this cannot be done, paired data with dependence within pairs can be obtained by matching individuals or objects on one or more characteristics thought to influence responses. For example, in a medical experiment to compare the efficacy of two drugs for lowering blood pressure, the experimenter's budget might allow for the treatment of 20 patients. If 10 patients are randomly selected for treatment with the first drug and another 10 independently selected for treatment with the second drug, an independent-samples experiment results.

However, the experimenter, knowing that blood pressure is influenced by age and weight, might decide to pair off patients so that within each of the resulting 10 pairs, age and weight were approximately equal (though there might be sizable differences between pairs). Then each drug would be given to a different patient within each pair for a total of 10 observations on each drug.

Without this matching (or "blocking"), one drug might appear to outperform the other just because patients in one sample were lighter and younger and thus more susceptible to a decrease in blood pressure than the heavier and older patients in the second sample. However, there is a price to be paid for pairing—a smaller number of degrees of freedom for the paired analysis—so we must ask when one type of experiment should be preferred to the other.

There is no straightforward and precise answer to this question, but there are some useful guidelines. If we have a choice between two t tests that are both valid (and carried out at the same level of significance α), we should prefer the test that has the larger number of degrees of freedom. The reason for this is that a larger number of degrees of freedom means a smaller β for any fixed alternative value of the parameter or parameters. That is, for a fixed type I error probability, the probability of a type II error is decreased by increasing degrees of freedom.

However, if the experimental units are quite heterogeneous in their responses, it will be difficult to detect small but significant differences between two treatments. This is essentially what happened in the data set in Example 10.8; for both "treatments" (bottom water and surface water), there is great between-location variability, which tends to mask differences in treatments within locations. If there is a high positive correlation within experimental units or subjects, the variance of $\overline{D} = \overline{X} - \overline{Y}$ will be much smaller than the unpaired variance. Because of this reduced variance, it will be easier to detect a difference with paired samples than with independent samples. The pros and cons of pairing can now be summarized as follows.

---

1. If there is great heterogeneity between experimental units and a large correlation within experimental units (large positive ρ), then the loss in degrees of freedom will be compensated for by the increased precision associated with pairing, so a paired experiment is preferable to an independent-samples experiment.
2. If the experimental units are relatively homogeneous and the correlation within pairs is not large, the gain in precision due to pairing will be outweighed by the decrease in degrees of freedom, so an independent-samples experiment should be used.

Of course, values of σ_1^2, σ_2^2 and ρ will not usually be known very precisely, so an investigator will be required to make a seat-of-the-pants judgment as to whether Situation 1 or 2 obtains. In general, if the number of observations that can be obtained is large, then a loss in degrees of freedom (e.g., from 40 to 20) will not be serious; but if the number is small, then the loss (say, from 16 to 8) because of pairing may be serious if not compensated for by increased precision. Similar considerations apply when choosing between the two types of experiments to estimate $\mu_1 - \mu_2$ with a confidence interval.

Exercises Section 10.3 (39–47)

39. Consider the accompanying data on breaking load (kg/25-mm width) for various fabrics in both an unabraded condition and an abraded condition ("The Effect of Wet Abrasive Wear on the Tensile Properties of Cotton and Polyester-Cotton Fabrics," *J. Testing Eval.*, 1993: 84–93).

a. Use the paired *t* test, as did the authors of the cited article, to test H_0: $\mu_D = 0$ versus H_a: $\mu_D > 0$ at significance level .01.

	Fabric							
	1	**2**	**3**	**4**	**5**	**6**	**7**	**8**
U	36.4	55.0	51.5	38.7	43.2	48.8	25.6	49.8
A	28.5	20.0	46.0	34.5	36.5	52.5	26.5	46.5

b. Do the assumptions necessary for using the paired *t* test appear to be plausible? Explain.

c. Delete the observation on fabric 2 and use the paired *t* test on the remaining observations. How does your conclusion compare with that of part (a)?

40. Hexavalent chromium has been identified as an inhalation carcinogen and an air toxin of concern in a number of different locales. The article "Airborne Hexavalent Chromium in Southwestern Ontario" (*J. Air Waste Manag. Assoc.*, 1997: 905–910) gave the accompanying data on both indoor and outdoor concentration (nanograms/m^3) for a sample of houses selected from a certain region.

House	1	2	3	4	5	6	7	8	9
Indoor	.07	.08	.09	.12	.12	.12	.13	.14	.15
Outdoor	.29	.68	.47	.54	.97	.35	.49	.84	.86

House	10	11	12	13	14	15	16	17
Indoor	.15	.17	.17	.18	.18	.18	.18	.19
Outdoor	.28	.32	.32	1.55	.66	.29	.21	1.02

House	18	19	20	21	22	23	24	25
Indoor	.20	.22	.22	.23	.23	.25	.26	.28
Outdoor	1.59	.90	.52	.12	.54	.88	.49	1.24

House	26	27	28	29	30	31	32	33
Indoor	.28	.29	.34	.39	.40	.45	.54	.62
Outdoor	.48	.27	.37	1.26	.70	.76	.99	.36

a. Calculate a confidence interval for the population mean difference between indoor and outdoor concentrations using a confidence level of 95%, and interpret the resulting interval.

b. If a 34th house were to be randomly selected from the population, between what values would you predict the difference in concentrations to lie?

41. Are there drugs that will help a chronic insomniac get a good night's sleep? This problem was studied a hundred years ago ("The Action of Optical Isomers, II: Hyoscines," *J. Physiol.*, 1905: 501–510). Ten patients with chronic sleep loss were each tested on four treatments, of which one was no drug. Here are the average hours of sleep for each patient with no drug and the average amount of sleep with the three drugs:

Patient	1	2	3	4	5
No drug	0.6	1.1	2.5	2.8	2.9
Drug	1.97	4.20	7.47	4.1	5.87

Patient	6	7	8	9	10
No drug	3.0	3.2	4.7	5.5	6.2
Drug	3.2	7.5	5	4.9	6.3

a. Verify graphically that the relevant normal distribution assumption appears appropriate.

b. Obtain a confidence interval for the population mean difference. Interpret your result in terms of the advantage or disadvantage of the drugs.

42. Scientists and engineers frequently wish to compare two different techniques for measuring or determining the value of a variable. In such situations, interest centers on testing whether the mean difference in measurements is zero. The article "Evaluation of the Deuterium Dilution Technique Against the Test Weighing Procedure for the Determination of Breast Milk Intake" (*Amer. J. Clin. Nutrit.*, 1983: 996–1003) reports the accompanying data on measuring amount of milk ingested by each of 14 randomly selected infants.

a. Is it plausible that the population distribution of differences is normal?

b. Does it appear that the true average difference between intake values measured by the two methods is something other than zero? Determine the *P*-value of the test, and use it to reach a conclusion at significance level .05.

c. What happens if the two-sample *t* test is (incorrectly) used? *Hint*: $s_1 = 352.970$, $s_2 = 234.042$.

	Infant													
Method	**1**	**2**	**3**	**4**	**5**	**6**	**7**	**8**	**9**	**10**	**11**	**12**	**13**	**14**
Isotopic	1509	1418	1561	1556	2169	1760	1098	1198	1479	1281	1414	1954	2174	2058
Test	1498	1254	1336	1565	2000	1318	1410	1129	1342	1124	1468	1604	1722	1518
Difference	11	164	225	−9	169	442	−312	69	137	157	−54	350	452	540

43. In an experiment designed to study the effects of illumination level on task performance ("Performance of Complex Tasks Under Different Levels of Illumination," *J. Illuminating Engrg.*, 1976: 235–242), subjects were required to insert a fine-tipped probe into the eyeholes of ten needles in rapid succession both for a low light level with a black background and a higher level with a white background. Each data value is the time (sec) required to complete the task.

Subject	1	2	3	4	5
Black	25.85	28.84	32.05	25.74	20.89
White	18.23	20.84	22.96	19.68	19.50

Subject	6	7	8	9
Black	41.05	25.01	24.96	27.47
White	24.98	16.61	16.07	24.59

Does the data indicate that the higher level of illumination yields a decrease of more than 5 sec in true average task completion time? Test the appropriate hypotheses using the *P*-value approach.

44. It has been estimated that between 1945 and 1971, as many as 2 million children were born to mothers treated with diethylstilbestrol (DES), a nonsteroidal estrogen recommended for pregnancy maintenance. The FDA banned this drug in 1971 because research indicated a link with the incidence of cervical cancer. The article "Effects of Prenatal Exposure to Diethylstilbestrol (DES) on Hemispheric Laterality and Spatial Ability in Human Males" (*Hormones Behav.*, 1992: 62–75) discussed a study in which 10 males exposed to DES and their unexposed brothers underwent various tests. This is the summary data on the results of a spatial ability test: $\bar{x} = 12.6$ (exposed), $\bar{y} = 13.7$, and standard error of mean difference $= .5$. Test at level .05 to see whether exposure is associated with reduced spatial ability by obtaining the *P*-value.

45. Cushing's disease is characterized by muscular weakness due to adrenal or pituitary dysfunction. To provide effective treatment, it is important to detect childhood Cushing's disease as early as possible. Age at onset of symptoms and age at diagnosis for 15 children suffering from the disease were given in the article "Treatment of Cushing's Disease in Childhood and Adolescence by Transphenoidal Microadenomectomy" (*New Engl. J. Med.*, 1984: 889). Here are the values of the differences between age at onset of symptoms and age at diagnosis:

−24	−12	−55	−15	−30	−60	−14	−21
−48	−12	−25	−53	−61	−69	−80	

a. Does the accompanying normal probability plot cast strong doubt on the approximate normality of the population distribution of differences?

b. Calculate a lower 95% confidence bound for the population mean difference, and interpret the resulting bound.

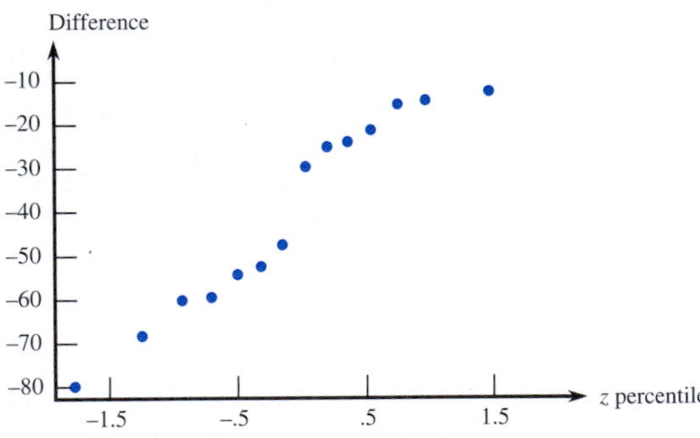

c. Suppose the (age at diagnosis) − (age at onset) differences had been calculated. What would be a 95% upper confidence bound for the corresponding population mean difference?

46. Example 1.2 describes a study of children's private speech (talking to themselves). The 33 children were each observed in about 100 ten-second intervals in the first grade, and again in the second and third grades. Because private speech occurs more in challenging circumstances, the children were observed while doing their mathematics. The speech was classified as on task (about the math lesson), off task, or mumbling (the observer could not tell what was said). Here are the 33 first-grade mumble scores:

20.8	24.4	19.4	33.3	26.0	56.6	39.5	24.7
21.6	32.1	48.1	19.5	19.2	43.0	26.3	22.7
49.4	35.4	56.8	45.4	28.7	42.2	20.3	20.0
34.0	26.9	48.4	27.6	52.6	5.9	38.5	22.1
22.2							

and here are the third-grade mumble scores:

28.8	57.0	23.9	46.9	50.0	64.6	54.2	55.3
21.4	38.3	78.5	38.1	44.3	11.7	58.6	76.1
76.4	48.6	37.2	69.8	29.1	60.4	57.8	38.7
46.5	50.0	69.6	69.8	59.4	22.7	84.9	42.0
67.2							

The numbers are in the same order for each grade; for example, the third student mumbled in 19.4% of the intervals in the first grade and 23.9% of the intervals in the third grade.

a. Verify graphically that normality is plausible for the population distribution of differences.

b. Find a 95% confidence interval for the difference of population means, and interpret the result.

47. Construct a paired data set for which $t = \infty$, so that the data is highly significant when the correct analysis is used, yet t for the two-sample t test is quite near zero, so the incorrect analysis yields an insignificant result.

10.4 Inferences About Two Population Proportions

Having presented methods for comparing the means of two different populations, we now turn to the comparison of two population proportions. The notation for this problem is an extension of the notation used in the corresponding one-population problem. We let p_1 and p_2 denote the proportions of individuals in populations 1 and 2, respectively, who possess a particular characteristic. Alternatively, if we use the label S for an individual who possesses the characteristic of interest (does favor a particular proposition, has read at least one book within the last month, etc.), then p_1 and p_2 represent the probabilities of seeing the label S on a randomly chosen individual from populations 1 and 2, respectively.

We will assume the availability of a sample of m individuals from the first population and n from the second. The variables X and Y will represent the number of individuals in each sample possessing the characteristic that defines p_1 and p_2. Provided the population sizes are much larger than the sample sizes, the distribution of X can be taken to be binomial with parameters m and p_1, and similarly, Y is taken to be a binomial variable with parameters n and p_2. Furthermore, the samples are assumed to be independent of one another, so that X and Y are independent rv's.

The obvious estimator for $p_1 - p_2$, the difference in population proportions, is the corresponding difference in sample proportions $X/m - Y/n$. With $\hat{p}_1 = X/m$ and $\hat{p}_2 = Y/n$, the estimator of $p_1 - p_2$ can be expressed as $\hat{p}_1 - \hat{p}_2$.

PROPOSITION Let $X \sim \text{Bin}(m, p_1)$ and $Y \sim \text{Bin}(n, p_2)$ with X and Y independent variables. Then

$$E(\hat{p}_1 - \hat{p}_2) = p_1 - p_2$$

so $\hat{p}_1 - \hat{p}_2$ is an unbiased estimator of $p_1 - p_2$, and

$$V(\hat{p}_1 - \hat{p}_2) = \frac{p_1 q_1}{m} + \frac{p_2 q_2}{n} \qquad \text{(where } q_i = 1 - p_i\text{)} \tag{10.3}$$

Proof Since $E(X) = mp_1$ and $E(Y) = np_2$,

$$E\left(\frac{X}{m} - \frac{Y}{n}\right) = \frac{1}{m}E(X) - \frac{1}{n}E(Y) = \frac{1}{m}mp_1 - \frac{1}{n}np_2 = p_1 - p_2$$

Since $V(X) = mp_1q_1$, $V(Y) = np_2q_2$, and X and Y are independent,

$$V\left(\frac{X}{m} - \frac{Y}{n}\right) = V\left(\frac{X}{m}\right) + V\left(\frac{Y}{n}\right) = \frac{1}{m^2}V(X) + \frac{1}{n^2}V(Y) = \frac{p_1q_1}{m} + \frac{p_2q_2}{n} \quad \blacksquare$$

We will focus first on situations in which both m and n are large. Then because $\hat{p}_1$ and $\hat{p}_2$ individually have approximately normal distributions, the estimator $\hat{p}_1 - \hat{p}_2$ also has approximately a normal distribution. Standardizing $\hat{p}_1 - \hat{p}_2$ yields a variable Z whose distribution is approximately standard normal:

$$Z = \frac{\hat{p}_1 - \hat{p}_2 - (p_1 - p_2)}{\sqrt{\dfrac{p_1q_1}{m} + \dfrac{p_2q_2}{n}}}$$

A Large-Sample Test Procedure

Analogously to the hypotheses for $\mu_1 - \mu_2$, the most general null hypothesis an investigator might consider would be of the form H_0: $p_1 - p_2 = \Delta_0$, where Δ_0 is again a specified number. Although for population means the case $\Delta_0 \neq 0$ presented no difficulties, for population proportions the cases $\Delta_0 = 0$ and $\Delta_0 \neq 0$ must be considered separately. Since the vast majority of actual problems of this sort involve $\Delta_0 = 0$ (i.e., the null hypothesis $p_1 = p_2$), we will concentrate on this case. When H_0: $p_1 - p_2 = 0$ is true, let p denote the common value of p_1 and p_2 (and similarly for q). Then the standardized variable

$$Z = \frac{\hat{p}_1 - \hat{p}_2 - 0}{\sqrt{pq\left(\dfrac{1}{m} + \dfrac{1}{n}\right)}} \tag{10.4}$$

has approximately a standard normal distribution when H_0 is true. However, this Z cannot serve as a test statistic because the value of p is unknown—H_0 asserts only that there is a common value of p, but does not say what that value is. To obtain a test statistic having approximately a standard normal distribution when H_0 is true (so that use of an appropriate z critical value specifies a level α test), p must be estimated from the sample data.

Assuming then that $p_1 = p_2 = p$, instead of separate samples of size m and n from two different populations (two different binomial distributions), we really have a single sample of size $m + n$ from one population with proportion p. Since the total number of individuals in this combined sample having the characteristic of interest is $X + Y$, the estimator of p is

$$\hat{p} = \frac{X + Y}{m + n} = \frac{m}{m + n}\hat{p}_1 + \frac{n}{m + n}\hat{p}_2 \tag{10.5}$$

The second expression for $\hat{p}$ shows that it is actually a weighted average of estimators $\hat{p}_1$ and $\hat{p}_2$ obtained from the two samples. If we take (10.5) (with $\hat{q} = 1 - \hat{p}$) and substitute back into (10.4), the resulting statistic has approximately a standard normal distribution when H_0 is true.

Null hypothesis: H_0: $p_1 - p_2 = 0$

Test statistic value (large samples): $z = \dfrac{\hat{p}_1 - \hat{p}_2}{\sqrt{\hat{p}\hat{q}\left(\dfrac{1}{m} + \dfrac{1}{n}\right)}}$

Alternative Hypothesis	**Rejection Region for Approximate Level α Test**
H_a: $p_1 - p_2 > 0$	$z \geq z_\alpha$
H_a: $p_1 - p_2 < 0$	$z \leq -z_\alpha$
H_a: $p_1 - p_2 \neq 0$	either $z \geq z_{\alpha/2}$ or $z \leq -z_{\alpha/2}$

A P-value is calculated in the same way as for previous z tests.

Example 10.11 Some defendants in criminal proceedings plead guilty and are sentenced without a trial, whereas others who plead innocent are subsequently found guilty and then are sentenced. In recent years, legal scholars have speculated as to whether sentences of those who plead guilty differ in severity from sentences for those who plead innocent and are subsequently judged guilty. Consider the accompanying data on defendants from San Francisco County accused of robbery, all of whom had previous prison records ("Does It Pay to Plead Guilty? Differential Sentencing and the Functioning of Criminal Courts," *Law and Society Rev.*, 1981–1982: 45–69). Does this data suggest that the proportion of all defendants in these circumstances who plead guilty and are sent to prison differs from the proportion who are sent to prison after pleading innocent and being found guilty?

	Plea	
	Guilty	**Not Guilty**
Number judged guilty	$m = 191$	$n = 64$
Number sentenced to prison	$x = 101$	$y = 56$
Sample proportion	$\hat{p}_1 = .529$	$\hat{p}_2 = .875$

Let p_1 and p_2 denote the two population proportions. The hypotheses of interest are H_0: $p_1 - p_2 = 0$ versus H_a: $p_1 - p_2 \neq 0$. At level .01, H_0 should be rejected if either $z \geq z_{.005} = 2.58$ or if $z \leq -2.58$. The combined estimate of the common success proportion is $\hat{p} = (101 + 56)/(191 + 64) = .616$. The value of the test statistic is then

$$z = \frac{.529 - .875}{\sqrt{(.616)(.384)\left(\dfrac{1}{191} + \dfrac{1}{64}\right)}} = \frac{-.346}{.070} = -4.94$$

Since $-4.94 \le -2.58$, H_0 must be rejected.

The P-value for a two-tailed z test is

$$P\text{-value} = 2[1 - \Phi(|z|)] = 2[1 - \Phi(4.94)] < 2[1 - \Phi(3.49)] = .0004$$

A more extensive standard normal table yields P-value $\approx .0000006$. This P-value is so minuscule that at any reasonable level α, H_0 should be rejected. The data very strongly suggests that $p_1 \ne p_2$ and, in particular, that initially pleading guilty may be a good strategy as far as avoiding prison is concerned.

The cited article also reported data on defendants in several other counties. The authors broke down the data by type of crime (burglary or robbery) and by nature of prior record (none, some but no prison, and prison). In every case, the conclusion was the same: Among defendants judged guilty, those who pleaded that way were less likely to receive prison sentences. ■

Type II Error Probabilities and Sample Sizes

Here the determination of β is a bit more cumbersome than it was for other large-sample tests. The reason is that the denominator of Z is an estimate of the standard deviation of $\hat{p}_1 - \hat{p}_2$, assuming that $p_1 = p_2 = p$. When H_0 is false, $\hat{p}_1 - \hat{p}_2$ must be restandardized using

$$\sigma_{\hat{p}_1 - \hat{p}_2} = \sqrt{\frac{p_1 q_1}{m} + \frac{p_2 q_2}{n}} \tag{10.6}$$

The form of σ implies that β is not a function of just $p_1 - p_2$, so we denote it by $\beta(p_1, p_2)$.

Alternative Hypothesis	$\beta(p_1, p_2)$
H_a: $p_1 - p_2 > 0$	$\Phi\left[\dfrac{z_\alpha\sqrt{\bar{p}\bar{q}\left(\frac{1}{m} + \frac{1}{n}\right)} - (p_1 - p_2)}{\sigma}\right]$
H_a: $p_1 - p_2 < 0$	$1 - \Phi\left[\dfrac{-z_\alpha\sqrt{\bar{p}\bar{q}\left(\frac{1}{m} + \frac{1}{n}\right)} - (p_1 - p_2)}{\sigma}\right]$
H_a: $p_1 - p_2 \ne 0$	$\Phi\left[\dfrac{z_{\alpha/2}\sqrt{\bar{p}\bar{q}\left(\frac{1}{m} + \frac{1}{n}\right)} - (p_1 - p_2)}{\sigma}\right]$
	$-\Phi\left[\dfrac{-z_{\alpha/2}\sqrt{\bar{p}\bar{q}\left(\frac{1}{m} + \frac{1}{n}\right)} - (p_1 - p_2)}{\sigma}\right]$

where $\bar{p} = (mp_1 + np_2)/(m + n)$, $\bar{q} = (mq_1 + nq_2)/(m + n)$, and σ is given by (10.6).

Proof For the upper-tailed test (H_a: $p_1 - p_2 > 0$),

$$\beta(p_1, p_2) = P\left[\hat{p}_1 - \hat{p}_2 < z_\alpha\sqrt{\hat{p}\hat{q}\left(\frac{1}{m} + \frac{1}{n}\right)}\right]$$

$$= P\left[\frac{\hat{p}_1 - \hat{p}_2 - (p_1 - p_2)}{\sigma} < \frac{z_\alpha\sqrt{\hat{p}\hat{q}\left(\frac{1}{m} + \frac{1}{n}\right)} - (p_1 - p_2)}{\sigma}\right]$$

When m and n are both large,

$$\hat{p} = \frac{m\hat{p}_1 + n\hat{p}_2}{m + n} \approx \frac{mp_1 + np_2}{m + n} = \bar{p}$$

and $\hat{q} \approx \bar{q}$, which yields the previous (approximate) expression for $\beta(p_1, p_2)$. ■

Alternatively, for specified p_1, p_2 with $p_1 - p_2 = d$, the sample sizes necessary to achieve $\beta(p_1, p_2) = \beta$ can be determined. For example, for the upper-tailed test, we equate $-z_\beta$ to the argument of $\Phi(\cdot)$ (i.e., what's inside the parentheses) in the foregoing box. If $m = n$, there is a simple expression for the common value.

For the case $m = n$, the level α test has type II error probability β at the alternative values p_1, p_2 with $p_1 - p_2 = d$ when

$$n = \frac{\left[z_\alpha\sqrt{(p_1 + p_2)(q_1 + q_2)/2} + z_\beta\sqrt{p_1q_1 + p_2q_2}\right]^2}{d^2} \tag{10.7}$$

for an upper- or lower-tailed test, with $\alpha/2$ replacing α for a two-tailed test.

Example 10.12 One of the truly impressive applications of statistics occurred in connection with the design of the 1954 Salk polio vaccine experiment and analysis of the resulting data. Part of the experiment focused on the efficacy of the vaccine in combating paralytic polio. Because it was thought that without a control group of children, there would be no sound basis for assessment of the vaccine, it was decided to administer the vaccine to one group and a placebo injection (visually indistinguishable from the vaccine but known to have no effect) to a control group. For ethical reasons and also because it was thought that the knowledge of vaccine administration might have an effect on treatment and diagnosis, the experiment was conducted in a **double-blind** manner. That is, neither the individuals receiving injections nor those administering them actually knew who was receiving vaccine and who was receiving the placebo (samples were numerically coded)—remember, at that point it was not at all clear whether the vaccine was beneficial.

Let p_1 and p_2 be the probabilities of a child getting paralytic polio for the control and treatment conditions, respectively. The objective was to test H_0: $p_1 - p_2 = 0$ versus H_a: $p_1 - p_2 > 0$ (the alternative hypothesis states that a vaccinated child is less likely to contract polio than an unvaccinated child). Supposing the true value of p_1 is .0003 (an incidence rate of 30 per 100,000), the vaccine would be a significant improvement if the incidence rate was halved—that is, $p_2 = .00015$. Using a level $\alpha = .05$ test, it would then be reasonable to ask for sample sizes for which $\beta = .1$ when $p_1 = .0003$ and $p_2 = .00015$. Assuming equal sample sizes, the required n is obtained from (10.7) as

$$n = \frac{\left[1.645\sqrt{(.5)(.00045)(1.99955)} + 1.28\sqrt{(.00015)(.99985) + (.0003)(.9997)}\right]^2}{(.0003 - .00015)^2}$$

$$= [(.0349 + .0271)/.00015]^2 \approx 171{,}000$$

The actual data for this experiment follows. Sample sizes of approximately 200,000 were used. The reader can easily verify that $z = 6.43$, a highly significant value. The vaccine was judged a resounding success!

Placebo: $m = 201{,}229$ $\quad x =$ number of cases of paralytic polio $= 110$

Vaccine: $n = 200{,}745$ $\quad y = 33$ ■

A Large-Sample Confidence Interval for $p_1 - p_2$

As with means, many two-sample problems involve the objective of comparison through hypothesis testing, but sometimes an interval estimate for $p_1 - p_2$ is appropriate. Both $\hat{p}_1 = X/m$ and $\hat{p}_2 = Y/n$ have approximate normal distributions when m and n are both large. If we identify θ with $p_1 - p_2$, then $\hat{\theta} = \hat{p}_1 - \hat{p}_2$ satisfies the conditions necessary for obtaining a large-sample CI. In particular, the estimated standard deviation of $\hat{\theta}$ is $\sqrt{(\hat{p}_1\hat{q}_1/m) + (\hat{p}_2\hat{q}_2/n)}$. The $100(1 - \alpha)\%$ interval $\hat{\theta} \pm z_{\alpha/2} \cdot \hat{\sigma}_{\hat{\theta}}$ then becomes

$$\hat{p}_1 - \hat{p}_2 \pm z_{\alpha/2}\sqrt{\frac{\hat{p}_1\hat{q}_1}{m} + \frac{\hat{p}_2\hat{q}_2}{n}}$$

Notice that the estimated standard deviation of $\hat{p}_1 - \hat{p}_2$ (the square-root expression) is different here from what it was for hypothesis testing when $\Delta_0 = 0$.

Recent research has shown that the actual confidence level for the traditional CI just given can sometimes deviate substantially from the nominal level (the level you think you are getting when you use a particular z critical value—e.g., 95% when $z_{\alpha/2} = 1.96$). The suggested improvement is to add one success and one failure to each of the two samples and then replace the $\hat{p}$'s and $\hat{q}$'s in the foregoing formula by $\tilde{p}$'s and $\tilde{q}$'s where $\tilde{p}_1 = (x + 1)/(m + 2)$, etc. This interval can also be used when sample sizes are quite small.

Example 10.13 The authors of the article "Adjuvant Radiotherapy and Chemotherapy in Node-Positive Premenopausal Women with Breast Cancer" (*New Engl. J. Med.*, 1997: 956–962) reported on the results of an experiment designed to compare treating cancer patients with only chemotherapy to treatment with a combination of chemotherapy and radiation. Of the 154 individuals who received the chemotherapy-only treatment, 76 survived at least 15 years, whereas 98 of the 164 patients who received the hybrid treatment survived at least that long. With p_1 denoting the proportion of all such women who, when treated with just chemotherapy, survive at least 15 years and p_2 denoting the analogous proportion for the hybrid treatment, $\hat{p}_1 = 76/154 = .494$ and $\hat{p}_2 = 98/164 = .598$. A confidence interval for the difference between proportions based on the traditional formula with a confidence level of approximately 99% is

$$.494 - .598 \pm (2.58)\sqrt{\frac{(.494)(.506)}{154} + \frac{(.598)(.402)}{164}} = -.104 \pm .143$$

$$= (-.247, .039)$$

At the 99% confidence level, it is plausible that $-.247 < p_1 - p_2 < .039$. This interval is reasonably wide, a reflection of the fact that the sample sizes are not terribly large for this type of interval. Notice that 0 is one of the plausible values of $p_1 - p_2$ suggesting that neither treatment can be judged superior to the other. Using $\tilde{p}_1 = 77/156 = .494$, $\tilde{q}_1 = 79/156 = .506$, $\tilde{p}_2 = .596$, $\tilde{q}_2 = .404$ based on sample sizes of 156 and 166, respectively, the "improved" interval here is identical to the earlier interval. ■

Small-Sample Inferences

On occasion an inference concerning $p_1 - p_2$ may have to be based on samples for which at least one sample size is small. Appropriate methods for such situations are not as straightforward as those for large samples, and there is more controversy among statisticians as to recommended procedures. One frequently used test, called the Fisher–Irwin test, is based on the hypergeometric distribution.

Exercises Section 10.4 (48–59)

48. Is someone who switches brands because of a financial inducement less likely to remain loyal than someone who switches without inducement? Let p_1 and p_2 denote the true proportions of switchers to a certain brand with and without inducement, respectively, who subsequently make a repeat purchase. Test H_0: $p_1 - p_2 = 0$ versus H_a: $p_1 - p_2 < 0$ using $\alpha = .01$ and the following data:

$m = 200$ number of successes = 30
$n = 600$ number of successes = 180

(Similar data is given in "Impact of Deals and Deal Retraction on Brand Switching," *J. Marketing*, 1980: 62–70.)

49. A sample of 300 urban adult residents of a particular state revealed 63 who favored increasing the highway speed limit from 55 to 65 mph, whereas a sample of 180 rural residents yielded 75 who favored the increase. Does this data indicate that the sentiment for increasing the speed limit is different for the two groups of residents?

a. Test H_0: $p_1 = p_2$ versus H_a: $p_1 \neq p_2$ using $\alpha = .05$, where p_1 refers to the urban population.

b. If the true proportions favoring the increase are actually $p_1 = .20$ (urban) and $p_2 = .40$ (rural), what is the probability that H_0 will be rejected using a level .05 test with $m = 300$, $n = 180$?

50. It is thought that the front cover and the nature of the first question on mail surveys influence the response rate. The article "The Impact of Cover Design and First Questions on Response Rates for a Mail Survey of Skydivers" (*Leisure Sci.*, 1991: 67–76) tested this theory by experimenting with different cover designs. One cover was plain; the other used a picture of a skydiver. The researchers speculated that the return rate would be lower for the plain cover.

Cover	Number Sent	Number Returned
Plain	207	104
Skydiver	213	109

Does this data support the researchers' hypothesis? Test the relevant hypotheses using $\alpha = .10$ by first calculating a P-value.

51. Do teachers find their work rewarding and satisfying? The article "Work-Related Attitudes" (*Psych. Rep.*, 1991: 443–450) reports the results of a survey of 395 elementary school teachers and 266 high school teachers. Of the elementary school teachers, 224 said they were very satisfied with their jobs, whereas 126 of the high school teachers were very satisfied with their work. Estimate the difference between the proportion of all elementary school teachers who are satisfied and all high school teachers who are satisfied by calculating a CI.

52. A random sample of 5726 telephone numbers from a certain region taken in March 2002 yielded 1105 that were unlisted, and 1 year later a sample of 5384 yielded 980 unlisted numbers.

a. Test at level .10 to see whether there is a difference in true proportions of unlisted numbers between the two years.

b. If $p_1 = .20$ and $p_2 = .18$, what sample sizes ($m = n$) would be necessary to detect such a difference with probability .90?

53. Ionizing radiation is being given increasing attention as a method for preserving horticultural products. The article "The Influence of Gamma-Irradiation on the Storage Life of Red Variety Garlic" (*J. Food Process. Preserv.*, 1983: 179–183) reports that 153 of 180 irradiated garlic bulbs were marketable (no external sprouting, rotting, or softening) 240 days after treatment, whereas only 119 of 180 untreated bulbs were marketable after this length of time. Does this data suggest that ionizing radiation is beneficial as far as marketability is concerned?

54. In medical investigations, the ratio $\theta = p_1/p_2$ is often of more interest than the difference $p_1 - p_2$ (e.g., individuals given treatment 1 are how many times as likely to recover as those given treatment 2?). Let $\hat{\theta} = \hat{p}_1/\hat{p}_2$. When m and n are both large, the statistic $\ln(\hat{\theta})$ has approximately a normal distribution with approximate mean value $\ln(\theta)$ and approximate standard deviation $[(m - x)/(mx) + (n - y)/(ny)]^{1/2}$.

a. Use these facts to obtain a large-sample 95% CI formula for estimating $\ln(\theta)$, and then a CI for θ itself.

b. Return to the heart attack data of Example 1.3, and calculate an interval of plausible values for θ at the 95% confidence level. What does this interval suggest about the efficacy of the aspirin treatment?

55. Sometimes experiments involving success or failure responses are run in a paired or before/after manner. Suppose that before a major policy speech by a political candidate, n individuals are selected and asked whether (S) or not (F) they favor the candidate. Then after the speech the same n people are asked the same question. The responses can be entered in a table as follows:

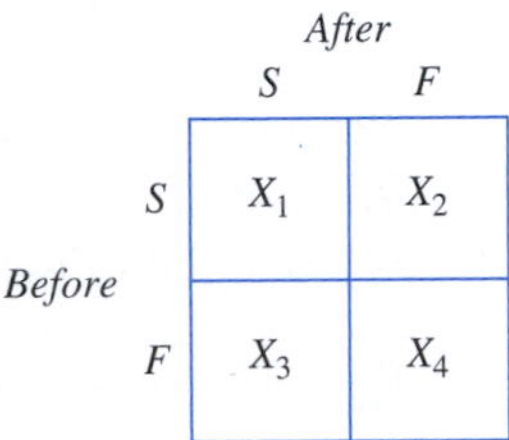

		After	
		S	F
Before	S	X_1	X_2
	F	X_3	X_4

where $X_1 + X_2 + X_3 + X_4 = n$. Let p_1, p_2, p_3, and p_4 denote the four cell probabilities, so that $p_1 = P(S$ before and S after), and so on. We wish to test the hypothesis that the true proportion of supporters (S) after the speech has not increased against the alternative that it has increased.

a. State the two hypotheses of interest in terms of p_1, p_2, p_3, and p_4.

b. Construct an estimator for the after/before difference in success probabilities.

c. When n is large, it can be shown that the rv $(X_i - X_j)/n$ has approximately a normal distribution with variance given by $[p_i + p_j - (p_i - p_j)^2]/n$. Use this to construct a test statistic with approximately a standard normal distribution when H_0 is true (the result is called McNemar's test).

d. If $x_1 = 350$, $x_2 = 150$, $x_3 = 200$, and $x_4 = 300$, what do you conclude?

56. The Chicago Cubs won 73 games and lost 71 in 1995. This was described as a much more successful season for them than 1994, when they won only 49 and lost 64.

a. Based on a binomial model with p_1 for 1994 and p_2 for 1995, carry out a two-tailed test for the difference. Based on your result, could the difference in sample proportions be attributed to luck (bad in 1994, good in 1995)?

b. Criticize the binomial model. Do baseball games satisfy the assumptions?

57. Using the traditional formula, a 95% CI for $p_1 - p_2$ is to be constructed based on equal sample sizes from the two populations. For what value of $n\ (= m)$ will the resulting interval have width at most .1 irrespective of the results of the sampling?

58. Statin drugs are used to decrease cholesterol levels, and therefore hopefully to decrease the chances of a heart attack. In a British study ("MRC/BHF Heart Protection Study of Cholesterol Lowering with Simvastin in 20,536 High-Risk Individuals: A Randomized Placebo-Controlled Trial," *Lancet,* 2002: 7–22) 20,536 at-risk adults were assigned randomly to take either a 40-mg statin pill or placebo. The subjects had coronary disease, artery blockage, or diabetes. After five years there were 1328 deaths (587 from heart attack) among the 10,269 in the statin group and 1507 deaths (707 from heart attack) among the 10,267 in the placebo group.

a. Give a 95% confidence interval for the difference in population death proportions.
b. Give a 95% confidence interval for the difference in population heart attack death proportions.
c. Is it reasonable to say that most of the difference in death proportions is due to heart attacks, as would be expected?

59. A study of male navy enlisted personnel was reported in the Bloomington, Illinois, *Daily Pantagraph*, Aug. 23, 1993. It was found that 90 of 231 left-handers had been hospitalized for injuries, whereas 623 of 2148 right-handers had been hospitalized for injuries. Test for equal population proportions at the .01 level, find the *P*-value for the test, and interpret your results. Can it be concluded that there is a causal relationship between handedness and proneness to injury? Explain.

10.5 *Inferences About Two Population Variances

Methods for comparing two population variances (or standard deviations) are occasionally needed, though such problems arise much less frequently than those involving means or proportions. For the case in which the populations under investigation are normal, the procedures are based on the F distribution, as discussed in Section 6.4.

Testing Hypotheses

A test procedure for hypotheses concerning the ratio σ_1^2/σ_2^2, as well as a CI for this ratio, are based on the following result from Section 6.4.

THEOREM Let $X_1, \ldots, X_m$ be a random sample from a normal distribution with variance σ_1^2, let $Y_1, \ldots, Y_n$ be another random sample (independent of the X_i's) from a normal distribution with variance σ_2^2, and let S_1^2 and S_2^2 denote the two sample variances. Then the rv

$$F = \frac{S_1^2/\sigma_1^2}{S_2^2/\sigma_2^2} \qquad (10.8)$$

has an F distribution with $\nu_1 = m - 1$ and $\nu_2 = n - 1$.

Under the null hypothesis of equal population variances, (10.8) reduces to the ratio of sample variances. For a test statistic we use this ratio of sample variances; and the claim that $\sigma_1^2 = \sigma_2^2$ is rejected if the ratio differs by too much from 1.

THE F TEST FOR EQUALITY OF VARIANCES

Null hypothesis: H_0: $\sigma_1^2 = \sigma_2^2$

Test statistic value: $f = s_1^2/s_2^2$

Alternative Hypothesis	Rejection Region for a Level α Test
H_a: $\sigma_1^2 > \sigma_2^2$	$f \geq F_{\alpha,m-1,n-1}$
H_a: $\sigma_1^2 < \sigma_2^2$	$f \leq F_{1-\alpha,m-1,n-1}$
H_a: $\sigma_1^2 \neq \sigma_2^2$	either $f \geq F_{\alpha/2,m-1,n-1}$ or $f \leq F_{1-\alpha/2,m-1,n-1}$

Since critical values are tabled only for α = .10, .05, .01, and .001, the two-tailed test can be performed only at levels .20, .10, .02, and .002. More extensive tabulations of F critical values are available elsewhere.

Example 10.14 Is there less variation in weights of some baked goods than others? Here are the weights (in grams) for a sample of Bruegger's bagels (their Iowa City shop) and another sample of Wolferman's muffins (made in Kansas City):

B:	99.8	105.4	94.7	107.8	114.3	106.3		
W:	99.0	98.2	98.1	102.1	102.9	104.1	98.8	99.5

The normality assumption is very important for the use of Expression (10.8) so we check the normal plot from MINITAB, shown in Figure 10.9. There is no apparent reason to doubt normality here.

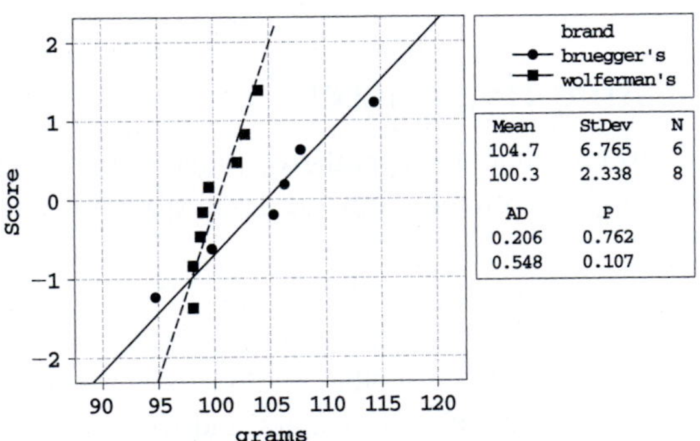

Figure 10.9 Normal plot for weights of baked goods

Notice the difference in slopes for the two sources in Figure 10.9. This suggests different variabilities because the vertical axis is the z-score and is related to the horizontal axis (grams) by z = (grams − mean)/(std dev). Thus, when score is plotted against grams the slope is the reciprocal of the standard deviation. Now let's test H_0: $\sigma_1^2 = \sigma_2^2$ against a two-tailed alternative with $\alpha = .02$. We need the critical values $F_{.01,5,7} = 7.46$ and $F_{.99,5,7} = 1/F_{.01,7,5} = 1/10.46 = .0956$. We have

$$f = \frac{s_1^2}{s_2^2} = \frac{6.765^2}{2.338^2} = 8.37$$

which exceeds 7.46, so the hypothesis of equal variances is rejected. We conclude that there is a difference in weight variation, and the English muffins are less variable.

Notice that it is not really necessary to use the lower-tail critical value here if the groups are chosen so the first group has the larger variance, and therefore the value of $f = s_1^2/s_2^2$ exceeds 1. Because $f > 1$, the only comparison is between the computed f and the upper critical value 7.46. It does not change the result of the test to fix things so $f > 1$, so it is not cheating to simplify the test in this way. ■

P-Values for *F* Tests

Recall that the *P*-value for an upper-tailed *t* test is the area under the relevant *t* curve (the one with appropriate df) to the right of the calculated *t*. In the same way, the *P*-value for an upper-tailed *F* test is the area under the *F* curve with appropriate numerator and denominator df to the right of the calculated *f*. Figure 10.10 illustrates this for a test based on $\nu_1 = 4$ and $\nu_2 = 6$.

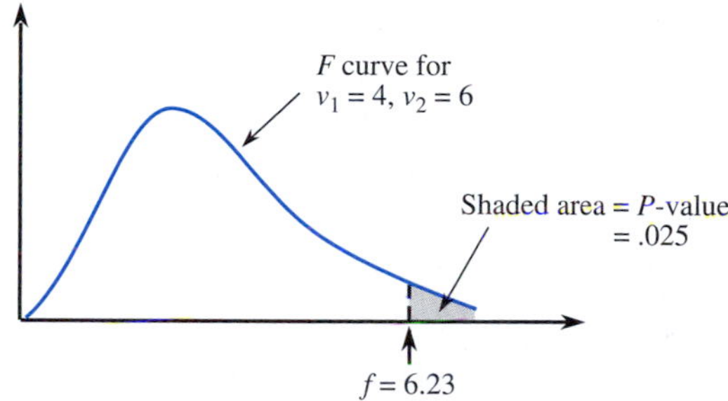

Figure 10.10 A *P*-value for an upper-tailed *F* test

Unfortunately, tabulation of *F* curve upper-tail areas is much more cumbersome than for *t* curves because two df's are involved. For each combination of ν_1 and ν_2, our *F* table gives only the four critical values that capture areas .10, .05, .01, and .001. Figure 10.11 shows what can be said about the *P*-value depending on where *f* falls relative to the four critical values. For example, for a test with $\nu_1 = 4$ and $\nu_2 = 6$,

$$f = 5.70 \quad \Rightarrow .01 < P\text{-value} < .05$$
$$f = 2.16 \quad \Rightarrow P\text{-value} > .10$$
$$f = 25.03 \quad \Rightarrow P\text{-value} < .001$$

Only if *f* equals a tabulated value do we obtain an exact *P*-value (e.g., if $f = 4.53$, then *P*-value = .05). Once we know that .01 < *P*-value < .05, H_0 would be rejected at a significance level of .05 but not at a level of .01. When *P*-value < .001, H_0 should be rejected at any reasonable significance level.

The *F* tests discussed in succeeding chapters will all be upper-tailed. If, however, a lower-tailed *F* test is appropriate, then (6.15) should be used to obtain lower-tailed critical values so that a bound or bounds on the *P*-value can be established. In the case of a two-tailed test, the bound or bounds from a one-tailed test should be multiplied by 2. For example, if $f = 5.82$ when $\nu_1 = 4$ and $\nu_2 = 6$, then since 5.82 falls between the .05 and .01 critical values, 2(.01) < *P*-value < 2(.05), giving .02 < *P*-value < .10. H_0 would

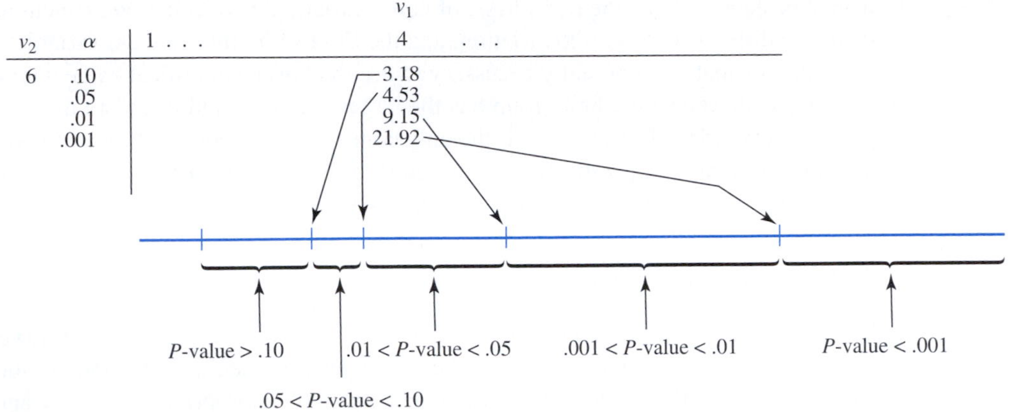

Figure 10.11 Obtaining *P*-value information from the *F* table for an upper-tailed *F* test

then be rejected if $\alpha = .10$ but not if $\alpha = .01$. In this case, we cannot say from our table what conclusion is appropriate when $\alpha = .05$ (since we don't know whether the *P*-value is smaller or larger than this). However, statistical software shows that the area to the right of 5.82 under this *F* curve is .029, so the *P*-value is .058 and the null hypothesis should therefore not be rejected at level .05 (.058 is the smallest α for which H_0 can be rejected and our chosen α is smaller than this).

A Confidence Interval for σ_1/σ_2

The CI for σ_1^2/σ_2^2 is based on replacing *F* in the probability statement

$$P(F_{1-\alpha/2,\nu_1,\nu_2} < F < F_{\alpha/2,\nu_1,\nu_2}) = 1 - \alpha$$

by the *F* variable (10.8) and manipulating the inequalities to isolate σ_1^2/σ_2^2:

$$\frac{s_1^2}{s_2^2} \cdot \frac{1}{F_{\alpha/2,\nu_1,\nu_2}} < \frac{\sigma_1^2}{\sigma_2^2} < \frac{s_1^2}{s_2^2} \cdot \frac{1}{F_{1-\alpha/2,\nu_1,\nu_2}} = \frac{s_1^2}{s_2^2} \cdot F_{\alpha/2,\nu_2,\nu_1}$$

Equation (6.15) has been used here to simplify the upper bound and enable use of Table A.9. Thus the confidence interval for σ_1^2/σ_2^2 is

$$\left(\frac{s_1^2}{s_2^2} \cdot \frac{1}{F_{\alpha/2,m-1,n-1}}, \frac{s_1^2}{s_2^2} \cdot F_{\alpha/2,n-1,m-1} \right)$$

An interval for σ_1/σ_2 results from taking the square root of each limit:

$$\left(\frac{s_1}{s_2} \cdot \frac{1}{\sqrt{F_{\alpha/2,m-1,n-1}}}, \frac{s_1}{s_2} \cdot \sqrt{F_{\alpha/2,n-1,m-1}} \right)$$

In the interval for the ratio of population variances, notice that the limits of the interval are proportional to the ratio of sample variances. Of course, the lower limit is less than the ratio of sample variances, and the upper limit is greater.

Example 10.15 Let's find a confidence interval using the data of Example 10.14. The sample standard deviations are $s_1 = 6.765$ for 6 Bruegger's bagels, and $s_2 = 2.338$ for 8 Wolferman English muffins. Then a 98% confidence interval for the ratio σ_1/σ_2 is

$$\left(\frac{6.765}{2.338}\cdot\frac{1}{\sqrt{F_{.01,5,7}}},\ \frac{6.765}{2.338}\cdot\sqrt{F_{.01,7,5}}\right) = \left(2.89\cdot\frac{1}{\sqrt{7.46}},\ 2.89\cdot\sqrt{10.46}\right)$$
$$= (1.06, 9.35)$$

Because 1 is not included in the interval, it suggests that the two standard deviations differ. By comparing the CI calculation with the hypothesis test calculation, it should be clear that a two-tailed test would reject equality at the 2% level, and this is consistent with the results of Example 10.14. ■

It is important to emphasize that the methods of this section are strongly dependent on the normality assumption. Expression (10.8) is valid only in the case of normal data or nearly normal data. Otherwise, the F distribution in (10.8) does not apply. The t procedures of this chapter are robust to the normality assumption, meaning that the procedures still work in the case of moderate departures from normality, but this is not true for comparison of variances based on (10.8).

Exercises Section 10.5 (60–68)

60. Obtain or compute the following quantities:
a. $F_{.05,5,8}$ **b.** $F_{.05,8,5}$ **c.** $F_{.95,5,8}$ **d.** $F_{.95,8,5}$
e. The 99th percentile of the F distribution with $\nu_1 = 10$, $\nu_2 = 12$
f. The 1st percentile of the F distribution with $\nu_1 = 10$, $\nu_2 = 12$
g. $P(F \le 6.16)$ for $\nu_1 = 6$, $\nu_2 = 4$
h. $P(.177 \le F \le 4.74)$ for $\nu_1 = 10$, $\nu_2 = 5$

61. Give as much information as you can about the P-value of the F test in each of the following situations:
a. $\nu_1 = 5$, $\nu_2 = 10$, upper-tailed test, $f = 4.75$
b. $\nu_1 = 5$, $\nu_2 = 10$, upper-tailed test, $f = 2.00$
c. $\nu_1 = 5$, $\nu_2 = 10$, two-tailed test, $f = 5.64$
d. $\nu_1 = 5$, $\nu_2 = 10$, lower-tailed test, $f = .200$
e. $\nu_1 = 35$, $\nu_2 = 20$, upper-tailed test, $f = 3.24$

62. Return to the data on maximum lean angle given in Exercise 27 of this chapter. Carry out a test at significance level .10 to see whether the population standard deviations for the two age groups are different (normal probability plots support the necessary normality assumption).

63. Refer to Example 10.7. Does the data suggest that the standard deviation of the strength distribution for fused specimens is smaller than that for not-fused specimens? Carry out a test at significance level .01 by obtaining as much information as you can about the P-value.

64. Toxaphene is an insecticide that has been identified as a pollutant in the Great Lakes ecosystem. To investigate the effect of toxaphene exposure on animals, groups of rats were given toxaphene in their diet. The article "Reproduction Study of Toxaphene in the Rat" (*J. Envir. Sci. Health*, 1988: 101–126) reports weight gains (in grams) for rats given a low dose (4 ppm) and for control rats whose diet did not include the insecticide. The sample standard deviation for 23 female control rats was 32 g and for 20 female low-dose rats was 54 g. Does this data suggest that there is more variability in low-dose weight gains than in control weight gains? Assuming normality, carry out a test of hypotheses at significance level .05.

65. In a study of copper deficiency in cattle, the copper values (μg/100 mL blood) were determined both for cattle grazing in an area known to have well-defined molybdenum anomalies (metal values in excess of the normal range of regional variation) and for cattle grazing in a nonanomalous area ("An Investigation into Copper Deficiency in Cattle in the Southern

Pennines," *J. Agric. Soc. Cambridge*, 1972: 157–163), resulting in $s_1 = 21.5$ ($m = 48$) for the anomalous condition and $s_2 = 19.45$ ($n = 45$) for the nonanomalous condition. Test for the equality versus inequality of population variances at significance level .10 by using the *P*-value approach.

66. The article "Enhancement of Compressive Properties of Failed Concrete Cylinders with Polymer Impregnation" (*J. Testing Eval.*, 1977: 333–337) reports the following data on impregnated compressive modulus (psi $\times 10^6$) when two different polymers were used to repair cracks in failed concrete.

Epoxy	1.75	2.12	2.05	1.97
MMA prepolymer	1.77	1.59	1.70	1.69

Obtain a 90% confidence interval for the ratio of variances.

67. Reconsider the data of Example 10.6, and calculate a 95% upper confidence bound for the ratio of the standard deviation of the triacetate porosity distribution to that of the cotton porosity distribution.

68. For the data of Exercise 27 find a 90% confidence interval for the ratio of population standard deviations, and relate your CI to the test of Exercise 62.

10.6 *Comparisons Using the Bootstrap and Permutation Methods

In this chapter we have discussed how to make comparisons based on normal data. We have also considered comparisons of means when the sample sizes are large enough for the sampling distributions of the sample means to be approximately normal. What about all other cases, especially small skewed data sets?

We now consider the bootstrap technique for forming confidence intervals and permutation tests for testing hypotheses. As described in Section 8.5, bootstrapping involves a lot of computation. The same will be true here for bootstrap confidence intervals and for permutation tests.

The Bootstrap for Two Samples

The bootstrap for two samples is similar to the one-sample bootstrap of Section 8.5, except that samples with replacement are taken from the two groups separately. That is, a sample is taken from the first group, a separate sample is taken from the second group, and then the difference of means or some other comparison statistic is computed. This process is repeated until there are 1000 (or another large number) values of the comparison statistic, and this constitutes the *bootstrap sample*. The distribution of the bootstrap sample is called the *bootstrap distribution*.

If the bootstrap distribution appears normal, then a confidence interval can be computed by using the standard deviation of the bootstrap distribution in place of the square root expression in the theorem of Section 10.2. That is, instead of estimating the standard error for the difference of means from the two sample standard deviations, we use the standard deviation of the bootstrap distribution. The idea is that the bootstrap distribution should represent the actual sampling distribution for the difference of means.

However, if the bootstrap distribution does not look normal, then the *percentile interval* should be calculated, just as was done in Section 8.5. Assuming a bootstrap sample of size 1000, this involves sorting the 1000 bootstrap values, finding the 25th from the bottom and the 25th from the top, and using these values as confidence limits for a 95% CI. The bias-corrected and accelerated interval is a further refinement available in some software, including Stata and S-Plus.

Example 10.16 As an example of the bootstrap for two samples, consider data from a study of children talking to themselves (private speech), introduced in Example 1.2. The children were each observed in many 10-second intervals (about 100) and the researchers computed the percentage of intervals in which private speech occurred. Because private speech tends to occur when there is a challenging task, the students were observed when they were doing arithmetic. The private speech is classified as on task if it is about arithmetic, off task if it is about something else, and mumbling if the subject is not clear.

Each child was observed in the first, second, and third grades, but we will consider here just the first-grade off-task private speech. For the 18 boys and 15 girls, here are the percentages:

B: 4.9 5.5 6.5 0.0 0.0 3.0 2.8 6.4 1.0 0.9 0.0 28.1 8.7 1.6 5.1 17.0 4.7 28.1

G: 0.0 1.3 2.2 0.0 1.3 0.0 0.0 0.0 0.0 3.9 0.0 10.1 5.2 3.2 0.0

With the large number of zeroes, a majority for the girls, the normality assumption of Section 10.2 is not plausible here. Also, the sample sizes for the two groups are not very large, so the two-sample z methods of Section 10.1 might not work for this data set. Nevertheless, it is useful to give the t CI for comparison purposes. The 95% interval is

$$\bar{x} - \bar{y} \pm t_{.025,\nu}\sqrt{\frac{s_1^2}{18} + \frac{s_2^2}{15}} = 6.906 - 1.813 \pm 2.080\sqrt{\frac{8.719^2}{18} + \frac{2.846^2}{15}}$$
$$= 5.093 \pm 2.080(2.1825)$$
$$= 5.093 \pm 4.540 = (.55, 9.63)$$

The degrees of freedom $\nu = 21$ come from the messy formula in the theorem of Section 10.2. The confidence interval does not include 0, which implies that we would reject the hypothesis $\mu_1 = \mu_2$ against a two-tailed alternative at the .05 level. This is in agreement with what we get in testing this hypothesis directly: $t = 2.33$, P-value .030.

The t method is of questionable validity, because the sample sizes might not be enough to compensate for the nonnormality. The bootstrap method involves drawing a random sample of size 18 with replacement from the 18 boys, drawing a random sample of size 15 with replacement from the 15 girls, and calculating the difference of means. Then this process is repeated to give a total of 1000 differences of means. The distribution of these 1000 differences of means is the bootstrap distribution.

To help clarify the procedure, here are the first random samples from the boys and girls:

B: 0.0 3.0 2.8 0.9 3.0 0.0 0.0 6.5 6.4 8.7 6.4 1.0 0.9 5.5 17.0 17.0 0.0 3.0

G: 1.3 0.0 0.0 0.0 0.0 1.3 1.3 0.0 3.2 0.0 1.3 5.2 0.0 0.0 0.0

Of course, in sampling with replacement some values will occur more than once and some will not occur at all. For these two samples, the difference of means is $4.56 - .91 = 3.65$. Doing this 1000 times gives the bootstrap sample summarized in Figure 10.12.

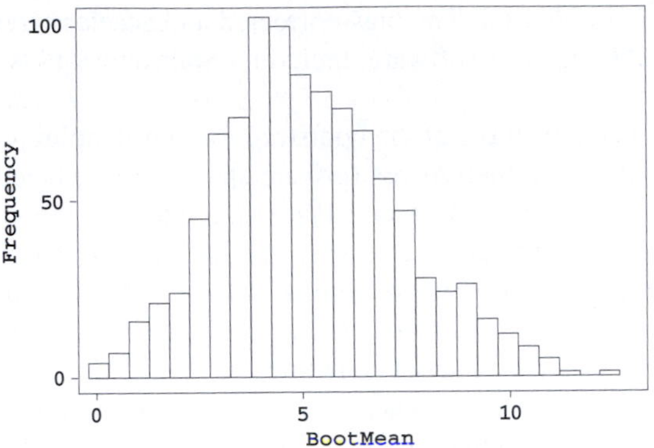

Figure 10.12 Histogram of the bootstrapped difference in means from MINITAB

The distribution looks almost normal, but with some positive skewness. The idea of the bootstrap, with its samples taken from the original samples of boys and girls, is for this histogram to resemble the true distribution of the difference of means. If the original samples of boys and girls are representative of their populations, then our histogram should be a reasonable imitation of the population distribution for the difference of means.

In spite of the nonnormality of the bootstrap distribution, we will use its standard deviation to compute a confidence interval to see how much it differs from the percentile interval. The standard deviation of the bootstrap distribution (i.e., of the 1000 $\bar{x} - \bar{y}$ values) is 2.1612, which is a little less than the 2.1825 that was computed for the square root in the t interval above. Using 2.1612 instead of 2.1825 gives the 95% confidence interval

$$\begin{aligned}\bar{x} - \bar{y} \pm t_{.025,\nu} s_{\text{boot}} &= 6.906 - 1.813 \pm 2.078(2.1612) \\ &= 5.093 \pm 4.491 = (.60, 9.58)\end{aligned}$$

which is very similar to the t interval, (.56, 9.63).

In the presence of a nonnormal bootstrap distribution, we now use the percentile interval, which for a 95% confidence interval finds the middle 95% of the bootstrap distribution. The confidence limits for a 95% confidence interval are the 2.5th percentile and the 97.5th percentile. When the 1000 bootstrap differences of means are sorted, the 25th value from the bottom is 1.199 and the 25th value from the top is 9.896. This gives a 95% CI (1.199, 9.896). The skewness of the bootstrap distribution pushes the endpoints a little to the right of the endpoints computed from s_{boot}. In addition, one can compute the bias-corrected and accelerated refinement, as discussed in Section 8.5. The improved interval (1.755, 10.61), obtained from S-Plus, is moved even farther to the right compared to the previous intervals. ■

In Example 10.16 we have used for the bootstrap t interval the t critical value from the theorem of Section 10.2, in order to get a similar answer if the two estimates of standard error (one based on the sample standard deviations and the other from the bootstrap) are close. Other critical values $t_{.025,\nu}$ can be used together with the bootstrap standard deviation s_{boot}. For example, with ν denoting df for the two-sample t procedure, it can be shown that $\nu \geq \min(m - 1, n - 1)$. For this reason, some sources, including

S-Plus, use df = $\min(m - 1, n - 1)$. Another natural choice would be df = $m + n - 2$, the value for the pooled t procedure. Although we do want to have the bootstrap interval be close to the t interval when we are working with the difference of means, note that we might also be bootstrapping a difference of medians or a ratio of standard deviations, and so on, where there is no concern about agreement with a t interval. A reasonable alternative (used by the Stata package) is to get the critical value instead from the standard normal distribution (df = ∞), which has the advantage of better agreement with the percentile interval if the bootstrap distribution is approximately normal.

Permutation Tests

How should we test hypotheses when the validity of the t test is in doubt? Permutation tests do not require any specific distribution for the data. The idea is that under the null hypothesis, every observation has the same distribution and thus the same expected value, so we can rearrange the group labels without changing the group population means. We look at all possible arrangements, compute the difference of means for each of these, and compute a P-value by seeing how extreme is our original difference of means. That is, the P-value is the fraction of arrangements that are at least as extreme as the value computed for the original data.

Example 10.17 Consider a small-scale version of the off-task private speech data. The first three values for the boys are 4.9, 5.5, 6.5 and the first two values for the girls are 0.0, 1.3. To demonstrate the permutation test, we will act as if this is the whole data set. First, we compute the difference of means of the boys versus the girls, $5.63 - .65 = 4.98$. Under the null hypothesis of equal population means, it should not matter if we reassign boys and girls. Therefore, we consider all ways of selecting three from among the five observations to be in the boys sample, leaving the other two for the girls sample. Under the null hypothesis, the following ten choices are equally likely.

Boys			$\bar{x}$	Girls		$\bar{y}$	$\bar{x} - \bar{y}$
4.9	5.5	6.5	5.63	0.0	1.3	.65	4.98
4.9	5.5	0.0	3.47	6.5	1.3	3.90	−.43
4.9	5.5	1.3	3.90	0.0	6.5	3.25	.65
4.9	6.5	0.0	3.80	5.5	1.3	3.40	.40
4.9	6.5	1.3	4.23	5.5	0.0	2.75	1.48
4.9	0.0	1.3	2.07	5.5	6.5	6.00	−3.93
5.5	6.5	0.0	4.00	4.9	1.3	3.10	.90
5.5	6.5	1.3	4.43	4.9	0.0	2.45	1.98
5.5	0.0	1.3	2.27	6.5	4.9	5.70	−3.43
6.5	0.0	1.3	2.60	5.5	4.9	5.20	−2.60

How extreme is our original difference of means (4.98) in this set of ten differences? Because 4.98 is the largest of ten, our P-value for an upper-tailed alternative hypothesis is $\frac{1}{10} = .10$. That is, for an upper-tailed test the P-value is the fraction of arrangements that give a difference at least as large as our original difference. For a two-tailed test we simply double the one-tailed P-value, giving $P = .20$ for this example. ■

When $m = 3$ and $n = 2$, it is simple enough to deal with all $\binom{5}{3} = 10$ arrangements. What happens when we try to use the whole set of 18 boys and 15 girls in the private speech data set?

Example 10.18 Consider a permutation test for the full private speech data. Here we are dealing with $\binom{33}{18} = 1{,}037{,}158{,}320$ arrangements of the 18 boys and 15 girls, more than a billion arrangements. Even on a reasonably fast computer it might take a while to generate this many differences and see how many are at least as big as the value $\bar{x} - \bar{y} = 6.906 - 1.813 = 5.093$ computed for the original data. It took around an hour on an 800-MHz Dell using the free statistical software BLOSSOM, which can be downloaded from the Internet. The two-tailed *P*-value is .0203, a little less than the *P*-value .030 from the *t* test. There is fairly strong evidence, at least at the 5% level, that the boys engage in more off task private speech than the girls.

We might have expected that the hypothesis test would reject the null hypothesis (of zero difference in means) at the 5% level with a two-tailed test. Recall that all three of our 95% confidence intervals in Example 10.16 consisted of only positive values, so none of the intervals included zero.

The number of arrangements goes up very quickly as the group sizes increase. If there are 20 boys and 20 girls, then the number of arrangements is more than 100 times as big as when there are 18 boys and 15 girls. Doing the test exactly, using all of the arrangements, becomes entirely impractical, but there is an approximate alternative. We can take a random sample of a few thousand arrangements and get quite close to the exact answer. For example, with our 18 boys and 15 girls, BLOSSOM gives (almost instantaneously) a *P*-value of .0204, which is certainly close enough to the exact answer of .0203. An approximate computation is also built into S-Plus and Stata and can easily be programmed in other software such as MINITAB. ■

PERMUTATION TESTS

Let θ_1 and θ_2 be the same parameters (means, medians, standard deviations, etc.) for two different populations, and consider testing H_0: $\theta_1 = \theta_2$ based on independent samples of size m and n, respectively. Suppose that when H_0 is true, the two population distributions are identical in all respects, so all $m + n$ observations have actually been selected from the same population distribution. In this case, the labels 1 and 2 are arbitrary, as any m of the $m + n$ observations have the same chance of ending up in the first sample (leaving the remaining n for the second sample). An exact permutation test computes a chosen comparison statistic for all possible rearrangements, and sets the *P*-value equal to the fraction of these that are at least as extreme as the statistic computed on the original samples. This is the *P*-value for a one-tailed test, and it needs to be doubled for a two-tailed test. For an approximate permutation test, instead of all possible arrangements, we take a random sample with replacement from the set of all possible arrangements.

Permutation tests are nonparametric, meaning that they do not assume a specific underlying distribution such as the normal distribution. However, this does not mean that there are no assumptions whatsoever. The null hypothesis in a permutation test is that the two distributions are the same, and any deviation can increase the probability of rejecting the null hypothesis. Thus, strictly speaking, we are doing a test for equal means only if the distributions are alike in all other respects, and this means that the two distributions have the same shape. In particular, it requires the distributions to have the same spread. See Exercise 84 for an example in which the permutation test underestimates the true *P*-value.

Inferences About Variability

Section 10.5 discussed the use of the *F* distribution for comparing two variances, but this inferential method is strongly dependent on normality. For highly skewed data the *F* test for equal variances will tend to reject the null hypothesis too often.

Example 10.19 Consider the off-task private speech data from Example 10.16. The sample standard deviations for boys and girls are 8.72 and 2.85, respectively. Then the method of Section 10.5 gives for the ratio of male to female variances the 95% confidence interval

$$\left(\frac{s_1^2}{s_2^2}\frac{1}{F_{.025,17,14}}, \frac{s_1^2}{s_2^2}\frac{1}{F_{.025,14,17}}\right) = \left(\frac{8.72^2}{2.85^2}\frac{1}{2.900}, \frac{8.72^2}{2.85^2}2.753\right) = (3.23, 25.77)$$

Taking the square root gives (1.80, 5.08) as the 95% confidence interval for the ratio of standard deviations. However, the legitimacy of this interval is seriously in question because of the skewed distributions.

What about a hypothesis test of equal population variances? The ratio of male variance to female variance is $s_1^2/s_2^2 = 9.385$. Comparing this to the *F* distribution with 17 numerator degrees of freedom and 14 denominator degrees of freedom, we find that the one-tailed *P*-value is .000061, and therefore the two-tailed *P*-value is .00012. This is consistent with the 95% confidence interval not including 1. It would be strong evidence for the male variance being greater than the female variance, except that the validity of the test is in doubt because of nonnormality.

Let's apply the bootstrap to this problem. We can use the same set of 1000 samples from the boys and 1000 samples from the girls that were used to compare means. The first sample from the boys has standard deviation 5.264 and the first sample from the girls has standard deviation 1.505, so the first ratio is 5.264/1.505 = 3.498. This value is included along with the 999 others in the histogram of Figure 10.13.

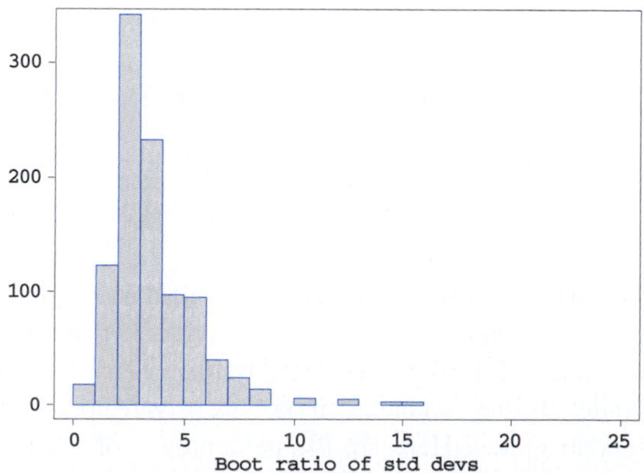

Figure 10.13 Histogram of bootstrap standard deviation ratios from S-Plus

The bootstrap distribution is strongly skewed to the right. For a 95% confidence interval, the percentile method uses the middle 95% of the bootstrap distribution. The 2.5th percentile is 1.115 and the 97.5th percentile is 8.17, so the 95% confidence interval for the population ratio of standard deviations is (1.115, 8.17). The bias-corrected and accelerated (BCa) refinement gives the interval (.755, 7.026). These two intervals differ in an important respect, that the percentile interval excludes 1 but the BCa refinement includes 1. In other words, the BCa interval allows the possibility that the two population standard deviations are the same, but the percentile interval does not. We expect the BCa method to be an improvement, and this is verified in the next example, where we see that the BCa result is consistent with the results of a permutation test. ■

Consider using a permutation test for H_0: $\sigma_1 = \sigma_2$.

Example 10.20 From Example 10.19 we know that the ratio of standard deviations for off-task private speech, males versus females, is 8.72/2.85 = 3.064. The idea of the permutation test is to find out how unusual this value is if we blur the distinction between males and females. That is, we remove the labels from the 18 males and 15 females and then consider all possible choices of 18 from the 33 children. For each of these possible choices we find the ratio of the standard deviation of the first 18 to the standard deviation of the last 15. The one-tailed P-value is the fraction that are at least as big as the original value, 3.064. Because there are more than a billion possible choices of 18 from 33, we instead selected 4999 random choices. This gives a total of 5000 when the original selection of males and females is included. Of these, 432 are at least as big as 3.064, so the one-tailed P-value is 432/5000 = .0864. For a two-tailed P-value we double this and get .1728. The permutation test does not reject at the 5% level (or the 10% level) the null hypothesis that the two population standard deviations are the same.

How does the permutation test result compare with the other results? Recall that the F interval and the percentile interval ruled out the possibility that the two standard deviations are the same, but the BCa refinement disagreed, because 1 is included in the BCa interval. Taking for granted that the permutation test is a valid approach and the permutation test does not reject the equality of standard deviations, the BCa interval is the only one of the three CIs consistent with this result. ■

The Analysis of Paired Data

The bootstrap can be used for paired data if we work with the paired differences, as in the paired t methods of Section 10.3.

Example 10.21 The private speech study was introduced in Examples 1.2 and 10.16. The study included the percentage of intervals with on-task private speech for 33 children in the first, second, and third grades. Here we will consider just the 15 girls in the first and second grades. Is there a change in on-task private speech when the girls go from the first to the second grade? Here are the percentages of intervals in which on-task private speech occurred, and also the differences.

Grade 1	Grade 2	Difference
25.7	18.6	7.1
36.0	17.4	18.6
27.6	2.6	25.0
29.7	0.9	28.8
36.0	1.5	34.5
35.1	14.1	21.0
42.0	3.3	38.7
7.6	1.6	6.0
14.1	0.0	14.1
25.0	1.5	23.5
20.2	0.0	20.2
24.4	2.1	22.3
10.4	18.4	−8.0
21.1	2.6	18.5
5.6	26.0	−20.4

Our null hypothesis is that the population mean difference between first- and second-grade percentages is zero. Figure 10.14 shows a histogram for the differences, and it shows a negatively skewed distribution.

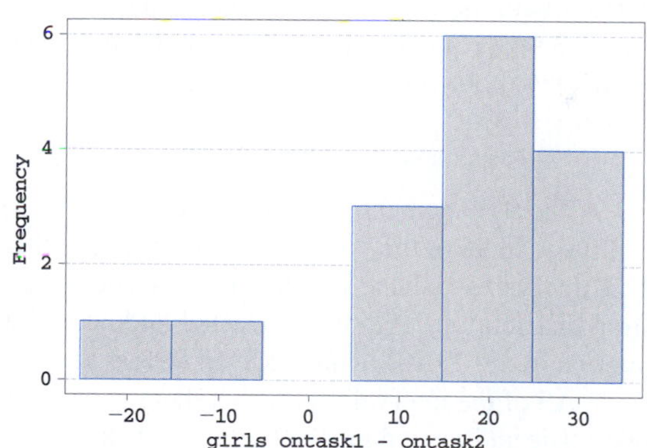

Figure 10.14 Histogram of differences for girls from Stata

The paired t method of Section 10.3 requires normality, so the skewness might invalidate this, but we will show the results here anyway for comparison purposes. The mean of the differences is $\bar{d} = 16.66$ with standard deviation $s_D = 15.43$, so the 95% confidence interval for the population mean difference is

$$\bar{d} \pm t_{.025,15-1}\frac{s_D}{\sqrt{15}} = 16.66 \pm 2.145\frac{15.43}{\sqrt{15}} = 16.66 \pm 8.54 = (8.12, 25.20)$$

What about the bootstrap for paired data? The bootstrap focuses on the 15 differences and uses the method of Section 8.5. We draw 1000 samples of size 15 with

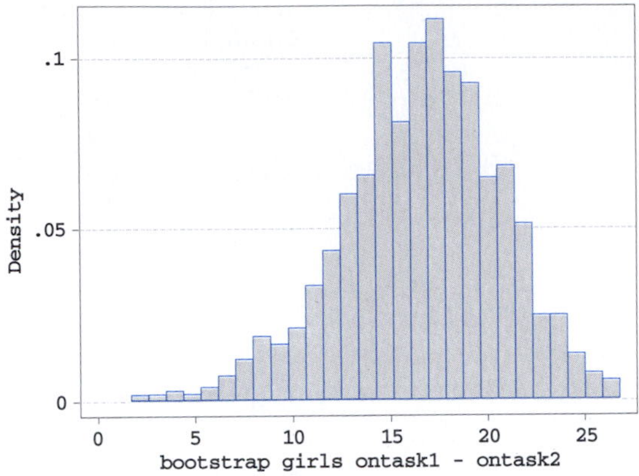

Figure 10.15 Histogram of bootstrap differences for girls from Stata

replacement from the 15 differences, and these 1000 samples constitute the bootstrap distribution. Figure 10.15 shows the histogram.

The histogram shows negative skewness, which is expected because of the negative skewness shown in Figure 10.14 for the original sample. The skewness implies that a symmetric confidence interval will not be entirely appropriate, but we show it for comparison with the other intervals. The standard deviation of the bootstrap distribution is $s_{\text{boot}} = 3.993$, compared to the estimated standard error $s_D/\sqrt{15} = 15.43/\sqrt{15} = 3.984$, so the 95% bootstrap t confidence interval is almost identical to the paired t interval:

$$\bar{d} \pm t_{.025,15-1}s_{\text{boot}} = 16.66 \pm 2.145(3.993) = 16.66 \pm 8.56 = (8.10, 25.22)$$

The 95% percentile interval uses the 2.5th percentile = 7.91 and the 97.5th percentile = 23.96 of the bootstrap distribution, so this confidence interval is (7.91, 23.96). This interval is to the left of the t intervals because of the negative skewness of the bootstrap distribution. The bias-corrected and accelerated refinement from Stata yields the interval (6.43, 23.12), which is even farther to the left.

All of the intervals agree that there is a substantial population difference between first grade and second grade. There is a strong reduction in the on-task private speech of girls between first and second grades. ■

A permutation test for paired data involves permutations within the pairs. Under the null hypothesis, the two observations in a pair have the same population mean, so the population mean difference is zero, even if the order is reversed. Therefore, we consider all possible orderings of the n pairs. Because there are two possible orderings within each pair, there are 2^n arrangements of n pairs. The one-tailed P-value is the fraction of the 2^n differences that are at least as extreme as the observed value, and the two-tailed P-value is double this.

Example 10.22 To see how the permutation test works for paired data, consider a scaled-down version of the data from Example 10.21 with only the first three pairs. These are (25.7, 18.6), (36.0, 17.4), (27.6, 2.6). They give a mean difference of (7.1 + 18.6 + 25.0)/3 = 16.9. Here are all 8 = 2^3 permutations with the corresponding means.

Arrangements			Mean Difference
(25.7, 18.6)	(36.0, 17.4)	(27.6, 2.6)	16.90
(25.7, 18.6)	(36.0, 17.4)	(2.6, 27.6)	.23
(25.7, 18.6)	(17.4, 36.0)	(27.6, 2.6)	4.50
(25.7, 18.6)	(17.4, 36.0)	(2.6, 27.6)	−12.17
(18.6, 25.7)	(36.0, 17.4)	(27.6, 2.6)	12.17
(18.6, 25.7)	(36.0, 17.4)	(2.6, 27.6)	−4.50
(18.6, 25.7)	(17.4, 36.0)	(27.6, 2.6)	−.23
(18.6, 25.7)	(17.4, 36.0)	(2.6, 27.6)	−16.90

Because the mean difference for the original sample is the highest value of eight, the one-tailed P-value is $\frac{1}{8} = .125$, and the two-tailed P-value is $2(\frac{1}{8}) = .25$. ■

Example 10.23 Let's now apply the permutation test to the paired data for the 15 girls of Example 10.21. In principle it is no harder to deal with the $2^n = 2^{15} = 32{,}768$ arrangements when all 15 pairs are included, but this exact approach is generally approximated using a random sample. We used Stata to draw an additional 4999 samples. Of the 4999, none yielded a mean difference as large as the value of 16.66 obtained for the original sample of 15 differences. Therefore, the one-tailed P-value is $\frac{1}{5000} = .0002$, and the two-tailed P-value is $2(.0002) = .0004$. Rejection of the null hypothesis at the 5% level was to be expected, given that none of the confidence intervals in Example 10.21 included 0.

It is interesting to compare the permutation test result with the t test of Section 10.3. For testing the null hypothesis of 0 population mean difference, the value of t is

$$\frac{\bar{d} - 0}{s_D/\sqrt{15}} = \frac{16.66}{15.425/\sqrt{15}} = 4.183$$

The two-tailed P-value for this is .0009, not very different from the result of the permutation test. ■

Exercises Section 10.6 (69–84)

69. A student project by Heather Kral studied students on "lifestyle floors" of a dormitory in comparison to students on other floors. On a lifestyle floor the students share a common major, and there are a faculty coordinator and resident assistant from that department. Here are the grade point averages of 30 students on lifestyle floors (L) and 30 students on other floors (N):

L: 2.00 2.25 2.60 2.90 3.00 3.00 3.00 3.00
3.00 3.20 3.20 3.25 3.30 3.30 3.32 3.50
3.50 3.60 3.60 3.70 3.75 3.75 3.79 3.80
3.80 3.90 4.00 4.00 4.00 4.00

N: 1.20 2.00 2.29 2.45 2.50 2.50 2.50 2.50
2.65 2.70 2.75 2.75 2.79 2.80 2.80 2.80
2.86 2.90 3.00 3.07 3.10 3.25 3.50 3.54
3.56 3.60 3.70 3.75 3.80 4.00

Notice that the GPA's from lifestyle floors have a large number of repeats and the distribution is skewed, so there is some question about normality.

a. Obtain a 95% confidence interval for the difference of population means using the method based on the theorem of Section 10.2.

b. Obtain a bootstrap sample of 1000 differences of means. Check the bootstrap distribution for normality using a normal probability plot.

c. Use the standard deviation of the bootstrap distribution along with the mean and t critical value from (a) to get a 95% confidence interval for the difference of means.

d. Use the bootstrap sample and the percentile method to obtain a 95% confidence interval for the difference of means.

e. Compare your three confidence intervals. If they are very similar, why do you think this is the case?

f. Interpret your results. Is there a substantial difference between lifestyle and other floors? Why do you think the difference is as big as it is?

70. In this application from Major League Baseball, the populations represent an abstraction of what the players can do, so the populations will vary from year to year. The Colorado Rockies and the Arizona Diamondbacks played nine games in Phoenix and ten games in Denver in 2001. The thinner air in Denver causes curve balls to curve less and it allows fly balls to travel farther. Does this mean that more runs are scored in Denver? The numbers of runs scored by the two teams in the nine Phoenix games (P) and ten Denver games (D) are

P: 5.09 15.88 3 8.47 11.65 6.48 11.65 7.41 9.53
D: 10 18 15.56 19 8.1 14 13.76 10 20.12 10.59

The fractions occur because the numbers have been adjusted for nine innings (54 outs). For example, in the third Denver game the Rockies won 10 to 7 on a home run with two out in the bottom of the tenth inning, so there were 59 outs instead of 54, and the number of runs is adjusted to $(54/59)(17) = 15.56$. We want to compare the average runs in Denver with the average runs in Phoenix.

a. Find a 95% confidence interval for the difference of population means using the method given in the theorem of Section 10.2.

b. Obtain a bootstrap sample of 1000 differences of means. Check the bootstrap distribution for normality using a normal probability plot.

c. Use the standard deviation of the bootstrap distribution along with the mean and t critical value from (a) to get a 95% confidence interval for the difference of means.

d. Use the bootstrap sample and the percentile method to obtain a 95% confidence interval for the difference of means.

e. Compare your three confidence intervals. If you used a standard normal critical value in place of the t critical value in (c), why would that make this interval more like the one in (d)? Why should the three intervals be fairly similar for this data set?

f. Interpret your results. Is there a substantial difference between the two locations? Compare the difference with what you thought it would be. If you were a major league pitcher, would you want to be traded to the Rockies?

71. For the data of Exercise 70 we want to compare population medians for the runs in Denver versus the runs in Phoenix.

a. Obtain a bootstrap sample of 1000 differences of medians. Check the bootstrap distribution for normality using a normal probability plot.

b. Use the standard deviation of the bootstrap distribution along with the difference of the medians in the original sample and the t critical value from Exercise 70(a) to get a 95% confidence interval for the difference of population medians.

c. Use the bootstrap sample and the percentile method to obtain a 95% confidence interval for the difference of population medians.

d. Compare the two confidence intervals.

e. How do the results for the median compare with the results for the mean? In terms of precision (measured by the width of the confidence interval), which gives the best results?

72. For the data of Exercise 69 now consider testing the hypothesis of equal population variances.

a. Carry out a two-tailed test using the method of Section 10.5. Recall that this method requires the data to be normal, and the method is sensitive to departures from normality. Check the data for normality to see if the F test is justified.

b. Carry out a two-tailed permutation test for the hypothesis of equal population variances (or standard deviations). Why does it not matter whether you use variances or standard deviations?

c. Compare the two results and summarize your conclusions.

73. For the data of Exercise 69 we want a 95% confidence interval for the ratio of population standard deviations.

a. Use the method of Section 10.5. Recall that this method requires the data to be normal, and the method is sensitive to departures from normality. Check the data for normality to see if the F distribution can be used for the ratio of sample variances.

b. With a bootstrap sample of size 1000 use the percentile method to obtain a 95% confidence interval for the ratio of standard deviations.

c. Compare the two results and discuss the relationship of the results to those of Exercise 72.

74. Can the right diet help us cope with aging-related diseases such as Alzheimer's disease? A study ("Reversals of Age-Related Declines in Neuronal Signal Transduction, Cognitive, and Motor Behavioral

Deficits with Blueberry, Spinach, or Strawberry Dietary Supplement," *J. Neurosci.*, 1999; 8114–8121) investigated the effects of fruit and vegetable supplements in the diet of rats. The rats were 19 months old, which is aged by rat standards. The 40 rats were randomly assigned to four diets, of which we will consider just the blueberry diet and the control diet here. After eight weeks on their diets, the rats were given a number of tests. We give the data for just one of the tests, which measured how many seconds they could walk on a rod. Here are the times for the ten control rats (C) and ten blueberry rats (B):

C: 15.00 7.00 2.44 5.60 3.63 6.24 4.12 8.21 3.90 0.95

B: 5.12 9.38 18.77 15.03 6.67 7.91 7.38 15.09 11.57 8.98

The objective is to obtain a 95% confidence interval for the difference of population means.

a. Determine a 95% confidence interval for the difference of population means using the method based on the theorem of Section 10.2.

b. Obtain a bootstrap sample of 1000 differences of means. Check the bootstrap distribution for normality using a normal probability plot.

c. Use the standard deviation of the bootstrap distribution along with the mean and t critical value from (a) to get a 95% confidence interval for the difference of means.

d. Use the bootstrap sample and the percentile method to obtain a 95% confidence interval for the difference of means.

e. Compare your three confidence intervals. If they are very similar, why do you think this is the case? If you had used a critical value from the normal table rather than the t table, would the result of (c) agree better with the result of (d)? Why?

f. Interpret your results. Do the blueberries make a substantial difference?

75. For the data of Exercise 74, we now want to test the hypothesis of equal population means.

a. Carry out a two-tailed test using the method based on the theorem of Section 10.2. Although this test requires normal data, it will work fairly well for moderately nonnormal data. Nevertheless, you should check the data for normality to see if the test is justified.

b. Carry out a two-tailed permutation test for the hypothesis of equal population means.

c. Compare the results of (a) and (b). Would you expect them to be similar for the data of this problem? Discuss their relationship to the results of Exercise 74. Summarize your conclusions about the effectiveness of blueberries.

76. Researchers at the University of Alaska have been trying to find inexpensive feed sources for Alaska reindeer growers ("Effects of Two Barley-Based Diets on Body Mass and Intake Rates of Captive Reindeer During Winter," Poster Presentation: School of Agriculture and Land Resources Management, University of Alaska Fairbanks, 2002). They are focusing on Alaska-grown barley because commercially available feed supplies are too expensive for farmers. Typically, reindeer lose weight in the fall and winter, and the researchers are trying to find a feed to minimize this loss. Thirteen pregnant reindeer were randomly divided into two groups to be fed on two different varieties of barley, thual and finaska. Here are the weight gains between October 1 and December 15 for the seven that were fed thual barley (T) and the six that were fed finaska barley (F).

T: -5.83 -11.5 -5.5 -1.33 -3.83 -3.33 -7.17

F: -0.17 -0.67 -4 -3 -1.33 -0.5

The weight gains are all negative, indicating that all of the animals lost weight. The thual barley is less fibrous and more digestible, and the intake rates for the two varieties of barley were very nearly the same, so the experimenters expected less weight loss for the thual variety.

a. Determine a 95% confidence interval for the difference of population means using the method given in the theorem of Section 10.2.

b. Obtain a bootstrap sample of 1000 differences of means. Check the bootstrap distribution for normality using a normal probability plot.

c. Use the standard deviation of the bootstrap distribution along with the mean and t critical value from (a) to get a 95% confidence interval for the difference of means.

d. Use the bootstrap sample and the percentile method to obtain a 95% confidence interval for the difference of means.

e. Compare your three confidence intervals. If they are very similar, why do you think this is the case?

f. Interpret your results. Is there a substantial difference? Is it in the direction anticipated by the experimenters?

77. For the data of Exercise 76, consider testing the hypothesis of equal population variances.

a. Carry out a two-tailed test using the method of Section 10.5. Recall that this method requires the data to be normal, and the method is sensitive to departures from normality. Check the data for normality to see if the F test is justified.

b. Carry out a two-tailed permutation test for the hypothesis of equal population variances (or standard deviations).

c. Compare the two results and summarize your conclusions.

78. Recall the data from Example 10.4 about the experiment in the low-level college mathematics course. Here again are the 85 final exam scores for those in the experimental group (E) and the 79 final exam scores for those in the control group (C):

E: 34 27 26 33 23 37 24 34 22 23 32 5
30 35 28 25 37 28 26 29 22 33 31 23
37 29 0 30 34 26 28 27 32 29 31 33
28 21 34 29 33 6 8 29 36 7 21 30
28 34 28 35 30 34 9 38 9 27 25 33
9 23 32 25 37 28 23 26 34 32 34 0
24 30 36 28 38 35 16 37 25 34 38 34
31

C: 37 22 29 29 33 22 32 36 29 6 4 37
0 36 0 32 27 7 19 35 26 22 28 28
32 35 28 33 35 24 21 0 32 28 27 8
30 37 9 33 30 36 28 3 8 31 29 9
0 0 35 25 29 3 33 33 28 32 39 20
32 22 24 20 32 7 8 33 29 9 0 30
26 25 32 38 22 29 29

a. Determine a 95% confidence interval for the difference of population means using the z method given in Section 10.1.

b. Obtain a bootstrap sample of 1000 differences of means. Check the bootstrap distribution for normality using a normal probability plot.

c. Use the standard deviation of the bootstrap distribution along with the mean and t critical value from (a) to get a 95% confidence interval for the difference of means.

d. Use the bootstrap sample and the percentile method to obtain a 95% confidence interval for the difference of means.

e. Compare your three confidence intervals. If they are very similar, why do you think this is the case? In the light of your results for (c) and (d), does the z method of (a) seem to work, regardless of normality? Explain.

f. Are your results consistent with the results of Example 10.4? Explain.

79. For the data of Example 10.4 we want to try a permutation test.

a. Carry out a two-tailed permutation test for the hypothesis of equal population means.

b. Compare the results for (a) and Example 10.4. Why should you have expected (a) and Example 10.4 to give similar results?

80. For the data of Example 10.4 it might be more appropriate to compare medians.

a. Find the medians for the two groups. With the help of a stem-and-leaf display for each group, explain why the medians are much closer than the means.

b. Do a two-tailed permutation test to compare the medians. Given what you found in (a), explain why the result of the permutation test was to be expected.

81. Two students, Miguel Melo and Cody Watson, did a study of textbook pricing. They compared prices at the campus bookstore and Amazon.com. To be fair, they included the sales tax for the local store and added shipping for Amazon. Here are the prices for a sample of 27 books.

Campus	Amazon	Campus	Amazon
100.41	106.94	59.50	69.24
99.34	113.94	87.66	73.84
51.53	61.44	26.56	33.98
20.45	31.59	44.63	40.39
28.69	29.89	96.69	117.99
70.66	83.94	18.06	27.94
98.81	107.74	103.06	115.74
111.56	115.99	14.61	24.69
97.22	108.29	77.03	88.04
61.89	78.44	99.34	113.94
70.39	82.94	81.81	90.74
58.17	65.74	48.88	58.94
108.38	122.09	76.50	91.94
61.63	63.49		

a. Determine a 95% confidence interval for the difference of population means using the t method of Section 10.3. Check the data for normality. Even if the normality assumption is not valid here, explain why the t method (or the z method of Section 10.1) might still be appropriate.

b. Based on the 27 differences, obtain a bootstrap sample of 1000 differences of means. Check the bootstrap distribution for normality.

c. Use the standard deviation of the bootstrap distribution along with the mean and t critical value from (a) to get a 95% confidence interval for the difference of means.

d. Use the bootstrap sample and the percentile method to obtain a 95% confidence interval for the difference of means.

e. Compare your three confidence intervals. In the light of your results for (d), does nonnormality invalidate the results of (a) and (c)? Explain.

f. Interpret your results. Is there a substantial difference between the two ways to buy books? Assuming that the populations remain unchanged and you have just these two sources, where would you buy?

82. Consider testing the hypothesis of equal population means based on the data in Exercise 81.

a. Carry out a two-tailed test using the method of Section 10.3. Is the normality assumption satisfied here? If not, why might the test be valid anyway?

b. Carry out a two-tailed permutation test for the hypothesis of equal population means.

c. Compare the results for (a) and (b). If the two results are similar, does that tend to validate (a), regardless of normality?

83. Compare bootstrapping with approximate permutation tests in which random permutations are used. Discuss the similarities and differences.

84. Assume that X is uniformly distributed on $(-1, 1)$ and Y is split evenly between a uniform distribution on $(-101, -100)$ and a uniform distribution on $(100, 101)$. Thus the means are both 0, but the variances differ strongly. We take random samples of three from each distribution and apply a permutation test for the null hypothesis $H_0\colon \mu_1 = \mu_2$ against the alternative $H_a\colon \mu_1 < \mu_2$.

a. Show that the probability is $\frac{1}{8}$ that all three of the Y values come from (100, 101).

b. Show that, if all three Y values come from (100, 101), then the P-value for the permutation test is .05.

c. Explain why (a) and (b) are in conflict. What is the true probability that the permutation test rejects the null hypothesis at the .05 level?

Supplementary Exercises (85–113)

85. A group of 115 University of Iowa students was randomly divided into a build-up condition group ($m = 56$) and a scale-down condition group ($n = 59$). The task for each subject was to build his/her own pizza from a menu of 12 ingredients. The build-up group was told that a basic cheese pizza costs \$5 and that each extra ingredient would cost 50 cents. The scale-down group was told that a pizza with all 12 ingredients (ugh!!!) would cost \$11 and that deleting an ingredient would save 50 cents. The article "A Tale of Two Pizzas: Building Up from a Basic Product Versus Scaling Down from a Fully Loaded Product" (*Marketing Lett.*, 2002: 335–344) reported that the mean number of ingredients selected by the scale-down group was significantly greater than the mean number for the build-up group: 5.29 versus 2.71. The calculated value of the appropriate t statistic was 6.07. Would you reject the null hypothesis of equality in favor of inequality at a significance level of .05? .01? .001? Can you think of other products aside from pizza where one could build up or scale down? *Note*: A separate experiment involved students from the University of Rome, but details were a bit different because there are typically not so many ingredient choices in Italy.

86. Is the number of export markets in which a firm sells its products related to the firm's return on sales? The article "Technology Industry Success: Strategic Options for Small and Medium Firms" (Gongming Qian, Lee Li, *Bus. Horizons*, Sept.–Oct. 2003: 41–46) gave the accompanying information on the number of export markets for one group of firms whose return on sales was less than 10% and another group whose return was at least 10%.

	Sample Size	Sample Mean	Sample SD
Less than 10%	36	5.12	.57
At least 10%	47	8.26	1.20

The investigators reported that an appropriate test of hypotheses resulted in a P-value between .01 and .05. What hypotheses do you think were tested, and do you agree with the stated P-value information?

What assumptions if any are needed in order to carry out the test? Can the plausibility of these assumptions be investigated based just on the foregoing summary data? Explain.

87. Suppose when using a two-sample t CI or test that $m < n$, and show that df $> m - 1$. This is why some authors suggest using min($m - 1$, $n - 1$) as df in place of the formula given in the text. What impact does this have on the CI and test procedure?

88. The accompanying summary data on compression strength (lb) for $12 \times 10 \times 8$ in. boxes appeared in the article "Compression of Single-Wall Corrugated Shipping Containers Using Fixed and Floating Test Platens" (*J. Testing Eval.*, 1992: 318–320). The authors stated that "the difference between the compression strength using fixed and floating platen method was found to be small compared to normal variation in compression strength between identical boxes." Do you agree?

Method	Sample Size	Sample Mean	Sample SD
Fixed	10	807	27
Floating	10	757	41

89. The authors of the article "Dynamics of Canopy Structure and Light Interception in *Pinus elliotti*, North Florida" (*Ecol. Monogr.*, 1991: 33–51) planned an experiment to determine the effect of fertilizer on a measure of leaf area. A number of plots were available for the study, and half were selected at random to be fertilized. To ensure that the plots to receive the fertilizer and the control plots were similar, before beginning the experiment tree density (the number of trees per hectare) was recorded for eight plots to be fertilized and eight control plots, resulting in the given data. MINITAB output follows.

Fertilizer plots	1024	1216	1312	1280
	1216	1312	992	1120
Control plots	1104	1072	1088	1328
	1376	1280	1120	1200

```
Two sample T for fertilizer vs control

             N     Mean  StDev  SE Mean
fertilize    8     1184    126       44
control      8     1196    118       42

95% CI for mu fertilize-mu control:
(-144, 120)
```

a. Construct a comparative boxplot and comment on any interesting features.

b. Would you conclude that there is a significant difference in the mean tree density for fertilizer and control plots? Use $\alpha = .05$.

c. Interpret the given confidence interval.

90. Is the response rate for questionnaires affected by including some sort of incentive to respond along with the questionnaire? In one experiment, 110 questionnaires with no incentive resulted in 75 being returned, whereas 98 questionnaires that included a chance to win a lottery yielded 66 responses ("Charities, No; Lotteries, No; Cash, Yes," *Public Opinion Q.*, 1996: 542–562). Does this data suggest that including an incentive increases the likelihood of a response? State and test the relevant hypotheses at significance level .10 by using the P-value method.

91. The article "Quantitative MRI and Electrophysiology of Preoperative Carpal Tunnel Syndrome in a Female Population" (*Ergonomics*, 1997: 642–649) reported that (−473.3, 1691.9) was a large-sample 95% confidence interval for the difference between true average thenar muscle volume (mm^3) for sufferers of carpal tunnel syndrome and true average volume for nonsufferers. Calculate and interpret a 90% confidence interval for this difference.

92. The following summary data on bending strength (lb-in/in) of joints is taken from the article "Bending Strength of Corner Joints Constructed with Injection Molded Splines" (*Forest Products J.*, April 1997: 89–92). Assume normal distributions.

Type	Sample Size	Sample Mean	Sample SD
Without side coating	10	80.95	9.59
With side coating	10	63.23	5.96

a. Calculate a 95% lower confidence bound for true average strength of joints with a side coating.

b. Calculate a 95% lower prediction bound for the strength of a single joint with a side coating.

c. Calculate an interval that, with 95% confidence, includes the strength values for at least 95% of the population of all joints with side coatings.

d. Calculate a 95% confidence interval for the difference between true average strengths for the two types of joints.

93. An experiment was carried out to compare various properties of cotton/polyester spun yarn finished with softener only and yarn finished with softener plus 5% DP-resin ("Properties of a Fabric Made with Tandem Spun Yarns," *Textile Res. J.*, 1996:

607–611). One particularly important characteristic of fabric is its durability, that is, its ability to resist wear. For a sample of 40 softener-only specimens, the sample mean stoll-flex abrasion resistance (cycles) in the filling direction of the yarn was 3975.0, with a sample standard deviation of 245.1. Another sample of 40 softener-plus specimens gave a sample mean and sample standard deviation of 2795.0 and 293.7, respectively. Calculate a confidence interval with confidence level 99% for the difference between true average abrasion resistances for the two types of fabrics. Does your interval provide convincing evidence that true average resistances differ for the two types of fabrics? Why or why not?

94. The derailment of a freight train due to the catastrophic failure of a traction motor armature bearing provided the impetus for a study reported in the article "Locomotive Traction Motor Armature Bearing Life Study" (*Lubricat. Engrg.*, Aug. 1997: 12–19). A sample of 17 high-mileage traction motors was selected, and the amount of cone penetration (mm/10) was determined both for the pinion bearing and for the commutator armature bearing, resulting in the following data:

Motor	1	2	3	4	5	6
Commutator	211	273	305	258	270	209
Pinion	226	278	259	244	273	236

Motor	7	8	9	10	11	12
Commutator	223	288	296	233	262	291
Pinion	290	287	315	242	288	242

Motor	13	14	15	16	17
Commutator	278	275	210	272	264
Pinion	278	208	281	274	268

Calculate an estimate of the population mean difference between penetration for the commutator armature bearing and penetration for the pinion bearing, and do so in a way that conveys information about the reliability and precision of the estimate. (*Note*: A normal probability plot validates the necessary normality assumption.) Would you say that the population mean difference has been precisely estimated? Does it look as though population mean penetration differs for the two types of bearings? Explain.

95. The article "Two Parameters Limiting the Sensitivity of Laboratory Tests of Condoms as Viral Barriers" (*J. Testing Eval.*, 1996: 279–286) reported that, in brand A condoms, among 16 tears produced by a puncturing needle, the sample mean tear length was 74.0 μm, whereas for the 14 brand B tears, the sample mean length was 61.0 μm (determined using light microscopy and scanning electron micrographs). Suppose the sample standard deviations are 14.8 and 12.5, respectively (consistent with the sample ranges given in the article). The authors commented that the thicker brand B condom displayed a smaller mean tear length than the thinner brand A condom. Is this difference in fact statistically significant? State the appropriate hypotheses and test at $\alpha = .05$.

96. Information about hand posture and forces generated by the fingers during manipulation of various daily objects is needed for designing high-tech hand prosthetic devices. The article "Grip Posture and Forces During Holding Cylindrical Objects with Circular Grips" (*Ergonomics*, 1996: 1163–1176) reported that for a sample of 11 females, the sample mean four-finger pinch strength (N) was 98.1 and the sample standard deviation was 14.2. For a sample of 15 males, the sample mean and sample standard deviation were 129.2 and 39.1, respectively.

a. A test carried out to see whether true average strengths for the two genders were different resulted in $t = 2.51$ and P-value $= .019$. Does the appropriate test procedure described in this chapter yield this value of t and the stated P-value?

b. Is there substantial evidence for concluding that true average strength for males exceeds that for females by more than 25 N? State and test the relevant hypotheses.

97. The article "Pine Needles as Sensors of Atmospheric Pollution" (*Environ. Monitoring*, 1982: 273–286) reported on the use of neutron-activity analysis to determine pollutant concentration in pine needles. According to the article's authors, "These observations strongly indicated that for those elements which are determined well by the analytical procedures, the distribution of concentration is lognormal. Accordingly, in tests of significance the logarithms of concentrations will be used." The given data refers to bromine concentration in needles taken from a site near an oil-fired steam plant and from a relatively clean site. The summary values are means and standard deviations of the log-transformed observations.

Site	**Sample Size**	**Mean Log Concentration**	**SD of Log Concentration**
Steam plant	8	18.0	4.9
Clean	9	11.0	4.6

Let μ_1^* be the true average *log* concentration at the first site, and define μ_2^* analogously for the second site.

a. Use the pooled t test (based on assuming normality *and* equal standard deviations) to decide at significance level .05 whether the two concentration distribution means are equal.

b. If σ_1^* and σ_2^*, the standard deviations of the two log concentration distributions, are not equal, would μ_1 and μ_2, the means of the concentration distributions, be the same if $\mu_1^* = \mu_2^*$? Explain your reasoning.

98. Long-term exposure of textile workers to cotton dust released during processing can result in substantial health problems, so textile researchers have been investigating methods that will result in reduced risks while preserving important fabric properties. The accompanying data on roving cohesion strength (kN · m/kg) for specimens produced at five different twist multiples is from the article "Heat Treatment of Cotton: Effect on Endotoxin Content, Fiber and Yarn Properties, and Processability" (*Textile Res. J.*, 1996: 727–738).

Twist multiple	1.054	1.141	1.245	1.370	1.481
Control strength	.45	.60	.61	.73	.69
Heated strength	.51	.59	.63	.73	.74

The authors of the cited article stated that strength for heated specimens appeared to be slightly higher on average than for the control specimens. Is the difference statistically significant? State and test the relevant hypotheses using $\alpha = .05$ by calculating the P-value.

99. The accompanying summary data on the ratio of strength to cross-sectional area for knee extensors is taken from the article "Knee Extensor and Knee Flexor Strength: Cross-Sectional Area Ratios in Young and Elderly Men" (*J. Gerontol.*, 1992: M204–M210).

Group	Sample Size	Sample Mean	Standard Error
Young	13	7.47	.22
Elderly men	12	6.71	.28

Does this data suggest that the true average ratio for young men exceeds that for elderly men? Carry out a test of appropriate hypotheses using $\alpha = .05$. Be sure to state any assumptions necessary for your analysis.

100. The accompanying data on response time appeared in the article "The Extinguishment of Fires Using Low-Flow Water Hose Streams—Part II" (*Fire Techn.*, 1991: 291–320). The samples are independent, not paired.

Good visibility .43 1.17 .37 .47 .68 .58 .50 2.75

Poor visibility 1.47 .80 1.58 1.53 4.33 4.23 3.25 3.22

The authors analyzed the data with the pooled t test. Does the use of this test appear justified? (*Hint*: Check for normality. The normal scores for $n = 8$ are $-1.53, -.89, -.49, -.15, .15, .49, .89$, and 1.53.)

101. The accompanying data on the alcohol content of wine is representative of that reported in a study in which wines from the years 1999 and 2000 were randomly selected and the actual content was determined by laboratory analysis (*London Times*, Aug. 5, 2001).

Wine	1	2	3	4	5	6
Actual	14.2	14.5	14.0	14.9	13.6	12.6
Label	14.0	14.0	13.5	15.0	13.0	12.5

The two-sample t test gives a test statistic value of .62 and a two-tailed P-value of .55. Does this convince you that there is no significant difference between true average actual alcohol content and true average content stated on the label? Explain.

102. In an experiment to compare bearing strengths of pegs inserted in two different types of mounts, a sample of 14 observations on stress limit for red oak mounts resulted in a sample mean and sample standard deviation of 8.48 MPa and .79 MPa, respectively, whereas a sample of 12 observations when Douglas fir mounts were used gave a mean of 9.36 and a standard deviation of 1.52 ("Bearing Strength of White Oak Pegs in Red Oak and Douglas Fir Timbers," *J. Testing Eval.*, 1998, 109–114). Consider testing whether or not true average stress limits are identical for the two types of mounts. Compare df's and P-values for the unpooled and pooled t tests.

103. How does energy intake compare to energy expenditure? One aspect of this issue was considered in the article "Measurement of Total Energy Expenditure by the Doubly Labelled Water Method in Professional Soccer Players" (*J. Sports Sci.*,

2002: 391–397), which contained the accompanying data (MJ/day).

Player	1	2	3	4	5	6	7
Expenditure	14.4	12.1	14.3	14.2	15.2	15.5	17.8
Intake	14.6	9.2	11.8	11.6	12.7	15.0	16.3

Test to see whether there is a significant difference between intake and expenditure. Does the conclusion depend on whether a significance level of .05, .01, or .001 is used?

104. An experimenter wishes to obtain a CI for the difference between true average breaking strength for cables manufactured by company I and by company II. Suppose breaking strength is normally distributed for both types of cable with $\sigma_1 = 30$ psi and $\sigma_2 = 20$ psi.

a. If costs dictate that the sample size for the type I cable should be three times the sample size for the type II cable, how many observations are required if the 99% CI is to be no wider than 20 psi?

b. Suppose a total of 400 observations is to be made. How many of the observations should be made on type I cable samples if the width of the resulting interval is to be a minimum?

105. An experiment to determine the effects of temperature on the survival of insect eggs was described in the article "Development Rates and a Temperature-Dependent Model of Pales Weevil" (*Environ. Entomology*, 1987: 956–962). At 11°C, 73 of 91 eggs survived to the next stage of development. At 30°C, 102 of 110 eggs survived. Do the results of this experiment suggest that the survival rate (proportion surviving) differs for the two temperatures? Calculate the *P*-value and use it to test the appropriate hypotheses.

106. The insulin-binding capacity (pmol/mg protein) was measured for four different groups of rats: (1) nondiabetic, (2) untreated diabetic, (3) diabetic treated with a low dose of insulin, (4) diabetic treated with a high dose of insulin. The accompanying table gives sample sizes and sample standard deviations. Denote the sample size for the *i*th treatment by n_i and the sample variance by S_i^2 ($i = 1, 2, 3, 4$). Assuming that the true variance for each treatment is σ^2, construct a pooled estimator of σ^2 that is unbiased, and verify using rules of expected value that it is indeed unbiased. What is your estimate for the following actual data? (*Hint*: Modify the pooled estimator S_p^2 from Section 10.2.)

	Treatment			
	1	**2**	**3**	**4**
Sample size	16	18	8	12
Sample SD	.64	.81	.51	.35

107. Suppose a level .05 test of H_0: $\mu_1 - \mu_2 = 0$ versus H_a: $\mu_1 - \mu_2 > 0$ is to be performed, assuming $\sigma_1 = \sigma_2 = 10$ and normality of both distributions, using equal sample sizes ($m = n$). Evaluate the probability of a type II error when $\mu_1 - \mu_2 = 1$ and $n = 25$, 100, 2500, and 10,000. Can you think of real problems in which the difference $\mu_1 - \mu_2 = 1$ has little practical significance? Would sample sizes of $n = 10{,}000$ be desirable in such problems?

108. The following data refers to airborne bacteria count (number of colonies/ft^3) both for $m = 8$ carpeted hospital rooms and for $n = 8$ uncarpeted rooms ("Microbial Air Sampling in a Carpeted Hospital," *J. Environ. Health*, 1968: 405). Does there appear to be a difference in true average bacteria count between carpeted and uncarpeted rooms?

Carpeted	11.8	8.2	7.1	13.0	10.8	10.1	14.6	14.0
Uncarpeted	12.1	8.3	3.8	7.2	12.0	11.1	10.1	13.7

Suppose you later learned that all carpeted rooms were in a veterans' hospital, whereas all uncarpeted rooms were in a children's hospital. Would you be able to assess the effect of carpeting? Comment.

109. Researchers sent 5000 resumes in response to job ads that appeared in the *Boston Globe* and *Chicago Tribune*. The resumes were identical except that 2500 of them had "white sounding" first names, such as Brett and Emily, whereas the other 2500 had "black sounding" names such as Tamika and Rasheed. The resumes of the first type elicited 250 responses and the resumes of the second type only 167 responses (these numbers are very consistent with information that appeared in a Jan. 15, 2003, report by the Associated Press). Does this data strongly suggest that a resume with a "black" name is less likely to result in a response than is a resume with a "white" name?

110. McNemar's test, developed in Exercise 55, can also be used when individuals are paired (matched) to yield *n* pairs and then one member of each pair is given treatment 1 and the other is given treatment 2. Then X_1 is the number of pairs in which both treatments were successful, and similarly for X_2, X_3, and X_4. The test statistic for testing equal efficacy of the two treatments is given by $(X_2 - X_3)/\sqrt{(X_2 + X_3)}$, which has approximately a standard normal distribution when H_0 is true. Use this to test whether the drug ergotamine is effective in the treatment of migraine headaches.

		Ergotamine	
		S	*F*
Placebo	*S*	44	34
	F	46	30

The data is fictitious, but the conclusion agrees with that in the article "Controlled Clinical Trial of Ergotamine Tartrate" (*British Med. J.*, 1970: 325–327).

111. Let $X_1, \ldots, X_m$ be a random sample from a Poisson distribution with parameter λ_1, and let $Y_1, \ldots, Y_n$ be a random sample from another Poisson distribution with parameter λ_2. We wish to test H_0: $\lambda_1 - \lambda_2 = 0$ against one of the three standard alternatives. Since $\mu = \lambda$ for a Poisson distribution, when m and n are large the large-sample z test of Section 10.1 can be used. However, the fact that $V(\overline{X}) = \lambda/n$ suggests that a different denominator should be used in standardizing $\overline{X} - \overline{Y}$. Develop a large-sample test procedure appropriate to this problem, and then apply it to the following data to test whether the plant densities for a particular species are equal in two different regions (where each observation is the number of plants found in a randomly located square sampling quadrate having area 1 m^2, so for region 1, there were 40 quadrates in which one plant was observed, etc.):

	Frequency								
	0	**1**	**2**	**3**	**4**	**5**	**6**	**7**	
Region 1	28	40	28	17	8	2	1	1	$m = 125$
Region 2	14	25	30	18	49	2	1	1	$n = 140$

112. Referring to Exercise 111, develop a large-sample confidence interval formula for $\lambda_1 - \lambda_2$. Calculate the interval for the data given there using a confidence level of 95%.

113. Let R_1 be a rejection region with significance level α for testing H_{01}: $\theta \in \Omega_1$ versus H_{a1}: $\theta \notin \Omega_1$, and let R_2 be a level α rejection region for testing H_{02}: $\theta \in \Omega_2$ versus H_{a2}: $\theta \notin \Omega_2$, where Ω_1 and Ω_2 are two disjoint sets of possible values of θ. Now consider testing H_0: $\theta \in \Omega_1 \cup \Omega_2$ versus the alternative H_a: $\theta \notin \Omega_1 \cup \Omega_2$. The proposed rejection region for this latter test is $R_1 \cap R_2$. That is, H_0 is rejected only if both H_{01} and H_{02} can be rejected. This procedure is called a *union–intersection test* (UIT).

a. Show that the UIT is a level α test.

b. As an example, let μ_T denote the mean value of a particular variable for a generic (test) drug, and μ_R denote the mean value of this variable for a brand-name (reference) drug. In *bioequivalence* testing, the relevant hypotheses are H_0: $\mu_T/\mu_R \le \delta_L$ or $\mu_T/\mu_R \ge \delta_U$ (not bioequivalent) versus H_a: $\delta_L < \mu_T/\mu_R < \delta_U$ (bioequivalent). δ_L and δ_U are standards set by regulatory agencies; for certain purposes the FDA uses .80 and $1.25 = 1/.8$, respectively. By taking logarithms and letting $\eta = \ln(\mu)$, $\tau = \ln(\delta)$, the hypotheses become H_0: either $\eta_T - \eta_R \le \tau_L$ or $\ge \tau_U$ versus H_a: $\tau_L < \eta_T - \eta_R < \tau_U$. With this setup, a type I error involves saying the drugs are bioequivalent when they are not. The FDA mandates $\alpha = .05$.

Let D be an estimator of $\eta_T - \eta_R$ with standard error S_D such that $T = [D - (\eta_T - \eta_R)]/S_D$ has a t distribution with ν df. The standard test procedure is referred to as *TOST* for "two one-sided tests," and is based on the two test statistics $T_U = (D - \tau_U)/S_D$ and $T_L = (D - \tau_L)/S_D$. If $\nu = 20$, state the appropriate conclusion in each of the following cases: (1) $t_L = 2.0$, $t_U = -1.5$; (2) $t_L = 1.5$, $t_U = -2.0$; (3) $t_L = 2.0$, $t_U = -2.0$.

Bibliography

See the bibliography at the end of Chapter 8.

CHAPTER ELEVEN

The Analysis of Variance

Introduction

In studying methods for the analysis of quantitative data, we first focused on problems involving a single sample of numbers and then turned to a comparative analysis of two different samples. Now we are ready for the analysis of several samples.

The **analysis of variance**, or more briefly ANOVA, refers broadly to a collection of statistical procedures for the analysis of quantitative responses. The simplest ANOVA problem is referred to variously as a **single-factor**, **single-classification**, or **one-way** ANOVA and involves the analysis of data sampled from two or more numerical populations (distributions). The characteristic that labels the populations is called the **factor** under study, and the populations are referred to as the **levels** of the factor. Examples of such situations include the following:

1. An experiment to study the effects of five different brands of gasoline on automobile engine operating efficiency (mpg)
2. An experiment to study the effects of four different sugar solutions (glucose, sucrose, fructose, and a mixture of the three) on bacterial growth
3. An experiment to investigate whether hardwood concentration in pulp (%) has an effect on tensile strength of bags made from the pulp
4. An experiment to decide whether the color density of fabric specimens depends on the amount of dye used

In (1) the factor of interest is gasoline brand, and there are five different levels of the factor. In (2) the factor is sugar, with four levels (or five, if a control solution containing no sugar is used). In both (1) and (2), the factor is qualitative in nature, and the levels correspond to possible categories of the factor. In (3) and (4), the factors are concentration of hardwood and amount of dye, respectively; both these factors are quantitative in nature, so the levels identify different settings of the factor. When the factor of interest is quantitative, statistical techniques from regression analysis (discussed in Chapter 12) can also be used to analyze the data.

In this chapter we first introduce single-factor ANOVA. Section 11.1 presents the F test for testing the null hypothesis that the population means are identical. Section 11.2 considers further analysis of the data when H_0 has been rejected. Section 11.3 covers some other aspects of single-factor ANOVA. Many experimental situations involve studying the simultaneous impact of more than one factor. Various aspects of two-factor ANOVA are considered in the last two sections of the chapter.

11.1 Single-Factor ANOVA

Single-factor ANOVA focuses on a comparison of two or more populations. Let

I = the number of treatments being compared

μ_1 = the mean of population 1 (the true average response when treatment 1 is applied)

$\vdots$

μ_I = the mean of population I (the true average response when treatment I is applied)

Then the hypotheses of interest are

$$H_0: \quad \mu_1 = \mu_2 = \cdots = \mu_I$$

versus

$$H_a: \quad \text{at least two of the } \mu_i\text{'s are different}$$

If $I = 4$, H_0 is true only if all four μ_i's are identical. H_a would be true, for example, if $\mu_1 = \mu_2 \neq \mu_3 = \mu_4$, if $\mu_1 = \mu_3 = \mu_4 \neq \mu_2$, or if all four μ_i's differ from one another.

A test of these hypotheses requires that we have available a random sample from each population.

Example 11.1 The article "Compression of Single-Wall Corrugated Shipping Containers Using Fixed and Floating Test Platens" (*J. Testing Eval.*, 1992: 318–320) describes an experiment in which several different types of boxes were compared with respect to compression strength (lb). Table 11.1 presents the results of a single-factor ANOVA experiment involving $I = 4$ types of boxes (the sample means and standard deviations are in good agreement with values given in the article).

Table 11.1 The data and summary quantities for Example 11.1

Type of Box	Compression Strength (lb)	Sample Mean	Sample SD
1	655.5 788.3 734.3 721.4 679.1 699.4	713.00	46.55
2	789.2 772.5 786.9 686.1 732.1 774.8	756.93	40.34
3	737.1 639.0 696.3 671.7 717.2 727.1	698.07	37.20
4	535.1 628.7 542.4 559.0 586.9 520.0	562.02	39.87
		Grand mean = 682.50	

With μ_i denoting the true average compression strength for boxes of type i ($i = 1, 2, 3, 4$), the null hypothesis is H_0: $\mu_1 = \mu_2 = \mu_3 = \mu_4$. Figure 11.1(a) shows a comparative boxplot for the four samples. There is a substantial amount of overlap among observations on the first three types of boxes, but compression strengths for the fourth type appear considerably smaller than for the other types. This suggests that H_0 is not true. The

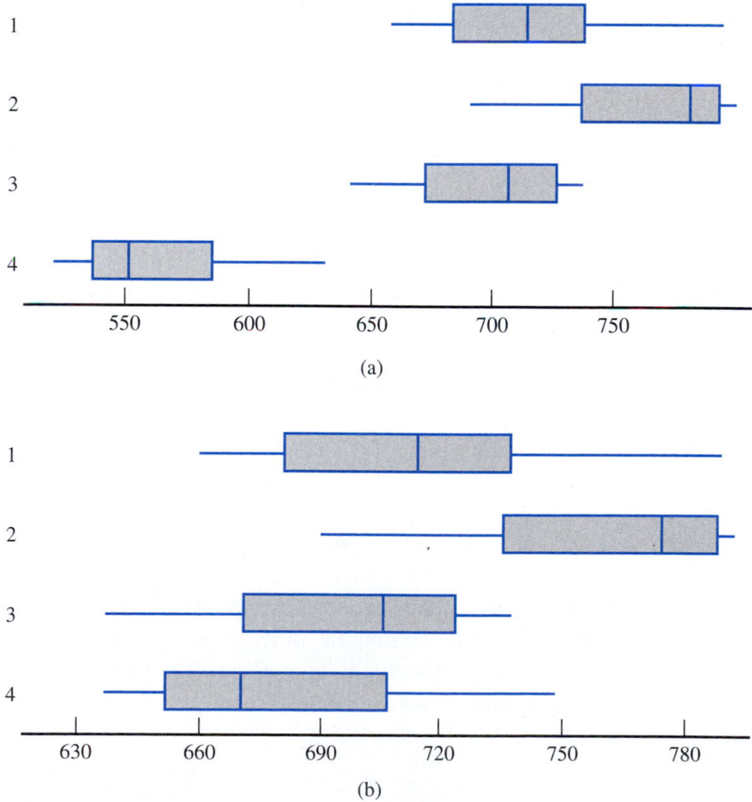

Figure 11.1 Boxplots for Example 11.1: (a) original data; (b) altered data

comparative boxplot in Figure 11.1(b) is based on adding 120 to each observation in the fourth sample (giving mean 682.02 and the same standard deviation) and leaving the other observations unaltered. It is no longer obvious whether H_0 is true or false. In situations such as this, we need a formal test procedure. ■

Notation and Assumptions

In two-sample problems, we used the letters X and Y to designate the observations in the two samples. Because this is cumbersome for three or more samples, it is customary to use a single letter with two subscripts. The first subscript identifies the sample number, corresponding to the population or treatment being sampled, and the second subscript denotes the position of the observation within that sample. Let

X_{ij} = the random variable (rv) denoting the jth measurement from the ith population

x_{ij} = the observed value of X_{ij} when the experiment is performed

The observed data is usually displayed in a rectangular table, such as Table 11.1. There samples from the different populations appear in different rows of the table, and $x_{i,j}$ is the jth number in the ith row. For example, $x_{2,3} = 786.9$ (the third observation from the second population), and $x_{4,1} = 535.1$. When there is no ambiguity, we will write x_{ij} rather than $x_{i,j}$ (e.g., if there were 15 observations on each of 12 treatments, x_{112} could mean $x_{1,12}$ or $x_{11,2}$). It is assumed that the X_{ij}'s within any particular sample are independent—a random sample from the ith population or treatment distribution—and that different samples are independent of one another.

In some experiments, different samples contain different numbers of observations. However, the concepts and methods of single-factor ANOVA are most easily developed for the case of equal sample sizes. Unequal sample sizes will be considered in Section 11.3. Restricting ourselves for the moment to equal sample sizes, let J denote the number of observations in each sample ($J = 6$ in Example 11.1). The data set consists of IJ observations. The individual sample means will be denoted by $\overline{X}_{1\cdot}, \overline{X}_{2\cdot}, \ldots, \overline{X}_{I\cdot}$. That is,

$$\overline{X}_{i\cdot} = \frac{\sum_{j=1}^{J} X_{ij}}{J} \qquad i = 1, 2, \ldots, I$$

The dot in place of the second subscript signifies that we have added over all values of that subscript while holding the other subscript value fixed, and the horizontal bar indicates division by J to obtain an average. Similarly, the average of all IJ observations, called the **grand mean**, is

$$\overline{X}_{\cdot\cdot} = \frac{\sum_{i=1}^{I} \sum_{j=1}^{J} X_{ij}}{IJ}$$

For the strength data in Table 11.1, $\bar{x}_{1\cdot} = 713.00$, $\bar{x}_{2\cdot} = 756.93$, $\bar{x}_{3\cdot} = 698.07$, $\bar{x}_{4\cdot} = 562.02$, and $\bar{x}_{\cdot\cdot} = 682.50$. Additionally, let $S_1^2, S_2^2, \ldots, S_I^2$ represent the sample variances:

$$S_i^2 = \frac{\sum_{j=1}^{J} (X_{ij} - \bar{X}_{i\cdot})^2}{J - 1} \qquad i = 1, 2, \ldots, I$$

From Example 11.1, $s_1 = 46.55$, $s_1^2 = 2166.90$, and so on.

ASSUMPTIONS The I population or treatment distributions are all normal with the same variance σ^2. That is, each X_{ij} is normally distributed with

$$E(X_{ij}) = \mu_i \qquad V(X_{ij}) = \sigma^2$$

In previous chapters, a normal probability plot was suggested for checking normality. The individual sample sizes in ANOVA are typically too small for I separate plots to be informative. A single plot can be constructed by subtracting $\bar{x}_{1\cdot}$ from each observation in the first sample, $\bar{x}_{2\cdot}$ from each observation in the second, and so on, and then plotting these IJ deviations against the z percentiles. The deviations are called **residuals**, so this plot is the normal plot of the residuals. Figure 11.2 gives the plot for the residuals of Example 11.1. The straightness of the pattern gives strong support to the normality assumption.

At the end of the section we discuss Levene's test for the equal variance assumption. For the moment, a rough rule of thumb is that if the largest s is not much more than twice the smallest s, it is reasonable to assume equal variances. This is especially true if the sample sizes are equal or close to equal. In Example 11.1, the largest s is only about 1.25 times the smallest.

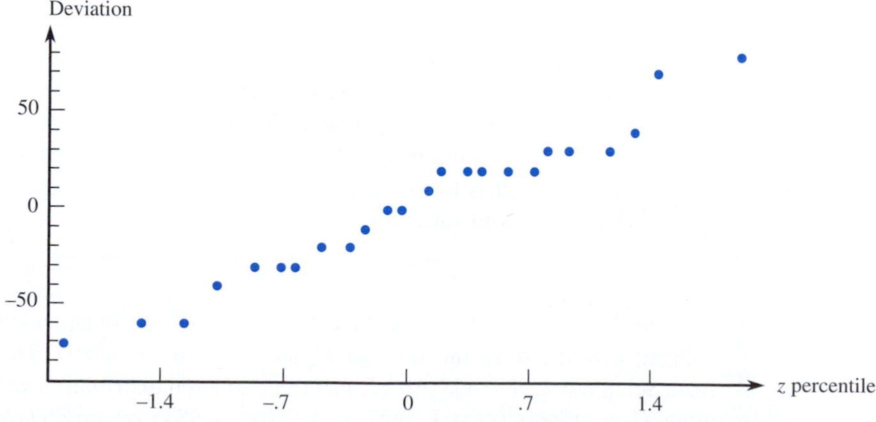

Figure 11.2 A normal probability plot based on the data of Example 11.1

Sums of Squares and Mean Squares

If H_0 is true, the J observations in each sample come from a normal population distribution with the *same* mean value μ, in which case the sample means $\bar{x}_{1\cdot}, \bar{x}_{2\cdot}, \ldots, \bar{x}_{I\cdot}$ should be reasonably close. The test procedure is based on comparing a measure of differences among these sample means ("between-samples" variation) to a measure of variation calculated from *within* each sample. These measures involve quantities called *sums of squares*.

DEFINITION

The **treatment sum of squares SSTr** is given by

$$\text{SSTr} = J\sum_i (\bar{X}_{i\cdot} - \bar{X}_{\cdot\cdot})^2 = J[(\bar{X}_{1\cdot} - \bar{X}_{\cdot\cdot})^2 + \cdots + (\bar{X}_{I\cdot} - \bar{X}_{\cdot\cdot})^2]$$

and the **error sum of squares SSE** is

$$\begin{aligned}\text{SSE} &= \sum_i \sum_j (X_{ij} - \bar{X}_{i\cdot})^2 \\ &= \sum_j (X_{1j} - \bar{X}_{1\cdot})^2 + \cdots + \sum_j (X_{Ij} - \bar{X}_{I\cdot})^2 \\ &= (J-1)S_1^2 + (J-1)S_2^2 + \cdots + (J-1)S_I^2 \\ &= (J-1)[S_1^2 + S_2^2 + \cdots + S_I^2]\end{aligned}$$

Now recall a key result from Section 6.4: If $X_1, \ldots, X_n$ is a random sample from a normal distribution with mean μ and variance σ^2, then the sample mean $\bar{X}$ and the sample variance S^2 are independent. Also, $\bar{X}$ is normally distributed, and $(n - 1)S^2/\sigma^2$ [i.e., $\Sigma(X_i - \bar{X})^2/\sigma^2$] has a chi-squared distribution with $n - 1$ df. That is, dividing the sum of squares $\Sigma(X_i - \bar{X})^2$ by σ^2 gives a chi-squared random variable. Similar results hold in our ANOVA situation.

THEOREM

When the basic assumptions of this section are satisfied, SSE/σ^2 has a chi-squared distribution with $I(J - 1)$ df (each sample contributes $J - 1$ df, and df's add because the samples are independent). Furthermore, when H_0 is true, SSTr/σ^2 has a chi-squared distribution with $I - 1$ df [there are I deviations $\bar{X}_{1\cdot} - \bar{X}_{\cdot\cdot}, \ldots, \bar{X}_{I\cdot} - \bar{X}_{\cdot\cdot}$, but 1 df is lost because $\Sigma_i(\bar{X}_{i\cdot} - \bar{X}_{\cdot\cdot}) = 0$]. Lastly, SSE and SSTr are independent random variables.

If we let $Y_i = \bar{X}_{i\cdot}$, $i = 1, \ldots, I$, then $Y_1, Y_2, \ldots, Y_I$ are independent and normally distributed with the same mean under H_0 and with variance σ^2/J. Thus, by the key result from Section 6.4, $(I - 1)S_Y^2/(\sigma^2/J)$ has a chi-squared distribution with $I - 1$ df. Furthermore, $(I - 1)S_Y^2/(\sigma^2/J) = J\,\Sigma(\bar{X}_{i\cdot} - \bar{X}_{\cdot\cdot})^2/\sigma^2 = \text{SSTr}/\sigma^2$, so $\text{SSTr}/\sigma^2 \sim \chi^2_{I-1}$. Independence of SSTr and SSE follows from the fact that SSTr is based on the individual sample means whereas SSE is based on the sample variances, and $\overline{X}_{i\cdot}$ is independent of S_i^2 for each i.

The expected value of a chi-squared variable with ν df is just ν. Thus

$$E\left(\frac{\text{SSE}}{\sigma^2}\right) = I(J-1) \Rightarrow E\left(\frac{\text{SSE}}{I(J-1)}\right) = \sigma^2$$

$$H_0 \text{ true} \Rightarrow E\left(\frac{\text{SSTr}}{\sigma^2}\right) = I - 1 \Rightarrow E\left(\frac{\text{SSTr}}{I-1}\right) = \sigma^2$$

Whenever the ratio of a sum of squares over σ^2 has a chi-squared distribution, we divide the sum of squares by its degrees of freedom to obtain a *mean square* ("mean" is used in the sense of "average").

DEFINITION

The **mean square for treatments** is MSTr = SSTr/(I − 1) and the **mean square for error** is MSE = SSE/[I(J − 1)].

Notice that uppercase X's and S's are used in defining the sums of squares and thus the mean squares, so the SS's and MS's are statistics (random variables). We will follow tradition and also use MSTr and MSE (rather than mstr and mse) to denote the calculated values of these statistics.

The foregoing results concerning expected values can now be restated:

$$E(\text{MSE}) = \sigma^2, \text{ that is, MSE is an unbiased estimator of } \sigma^2$$

$$H_0 \text{ true} \Rightarrow E(\text{MSTr}) = \sigma^2, \text{ so MSTr is an unbiased estimator of } \sigma^2$$

MSTr is unbiased for σ^2 when H_0 is true, but what about when H_0 is false? It can be shown (Exercise 10) that in this case, $E(\text{MSTr}) > \sigma^2$. This is because the $\overline{X}_i$'s tend to differ more from one another, and therefore from the grand mean, when the μ_i's are not identical than when they are the same.

The *F* Test

The test statistic is the ratio F = MSTr/MSE. F is a ratio of two estimators of σ^2. The numerator (the between-samples estimator), MSTr, is unbiased when H_0 is true but tends to overestimate σ^2 when H_0 is false, whereas the denominator (the within-samples estimator), MSE, is unbiased regardless of the status of H_0. Thus if H_0 is true the F ratio should be reasonably close to 1, but if the μ_i's differ considerably from one another, F should greatly exceed 1. Thus a value of F considerably exceeding 1 argues for rejection of H_0.

In Section 6.4 we introduced a family of probability distributions called F distributions. If Y_1 and Y_2 are two independent chi-squared random variables with ν_1 and ν_2 df, respectively, then the ratio $F = (Y_1/\nu_1)/(Y_2/\nu_2)$ has an F distribution with ν_1 numerator df and ν_2 denominator df. Figure 11.3 shows an F density curve and corresponding upper-tail critical value F_{α,ν_1,ν_2}. Appendix Table A.9 gives these critical values for α = .10, .05, .01, and .001. Values of ν_1 are identified with different columns of the table and the rows are labeled with various values of ν_2. For example, the F critical value that captures upper-tail area .05 under the F curve with $\nu_1 = 4$ and $\nu_2 = 6$ is $F_{.05,4,6} = 4.53$, whereas $F_{.05,6,4} = 6.16$ (so don't accidentally switch numerator and denominator df!). The key theoretical result that justifies the test procedure is that the test statistic F has an F distribution when H_0 is true.

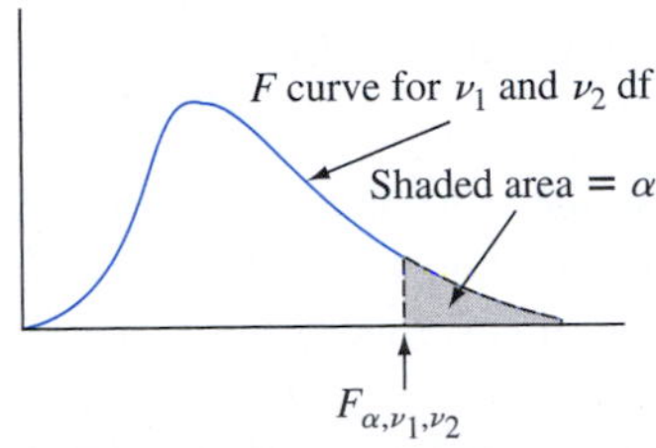

Figure 11.3 An F curve and critical value F_{α,ν_1,ν_2}

THEOREM

The test statistic in single-factor ANOVA is $F = \text{MSTr}/\text{MSE}$. We can write this as

$$F = \frac{\left[\dfrac{\text{SSTr}}{\sigma^2}\right] \Big/ (I-1)}{\left[\dfrac{\text{SSE}}{\sigma^2}\right] \Big/ [I(J-1)]}$$

When H_0 is true, the previous theorem implies that the numerator and denominator of F are independent chi-squared variables divided by their df's, in which case F has an F distribution with $I - 1$ numerator df and $I(J - 1)$ denominator df. The rejection region $f \geq F_{\alpha,I-1,I(J-1)}$ then specifies an upper-tailed test that has the desired significance level α. The P-value for an upper-tailed F test is the area under the relevant F curve (the one with correct numerator and denominator df's) to the right of the calculated f.

Refer to Section 10.5 to see how P-value information for F tests can be obtained from the table of F critical values. Alternatively, statistical software packages will automatically include the P-value with ANOVA output.

Computational Formulas

The calculations leading to f can be done efficiently by using formulas similar to the computing formula for the numerator of the sample variance s^2 from Section 1.4. The first two computational formulas here are essentially repetitions of that formula with new notation. Let $x_{i.}$ represent the *sum* (not the average, since there is no overbar) of the x_{ij}'s for fixed i (the total of the J observations in the ith sample). Similarly, let $x_{..}$ denote the sum of all IJ observations (the **grand total**). We also need a third sum of squares in addition to SSTr and SSE.

Sum of Squares	df	Definition	Computing Formula
total = SST	$IJ - 1$	$\sum_i \sum_j (x_{ij} - \bar{x}_{..})^2$	$\sum_i \sum_j x_{ij}^2 - x_{..}^2/IJ$
treatment = SSTr	$I - 1$	$\sum_i \sum_j (\bar{x}_{i.} - \bar{x}_{..})^2 = J \sum_i (\bar{x}_{i.} - \bar{x}_{..})^2$	$\dfrac{\sum_i x_{i.}^2}{J} - \dfrac{x_{..}^2}{IJ}$
error = SSE	$I(J - 1)$	$\sum_i \sum_j (x_{ij} - \bar{x}_{i.})^2$	SST − SSTr

Both SST and SSTr involve $x_{..}^2/IJ$, which is called either the *correction factor* or the *correction for the mean*. SST results from squaring each observation, adding these squares, and then subtracting the correction factor. Calculation of SSTr entails squaring each sample total (each row total from the data table), summing these squares, dividing the sum by J, and again subtracting the correction factor. SSTr is subtracted from SST to give SSE (it must be the case that SST $\geq$ SSTr), after which MSTr, MSE, and finally f are calculated.

The computational formula for SSE is a consequence of the **fundamental ANOVA identity**

$$\text{SST} = \text{SSTr} + \text{SSE} \tag{11.1}$$

The identity implies that once any two of the SS's have been calculated, the remaining one is easily obtained by addition or subtraction. The two that are most easily calculated are SST and SSTr. The proof of the identity follows from squaring both sides of the relationship

$$x_{ij} - \bar{x}_{..} = (x_{ij} - \bar{x}_{i.}) + (\bar{x}_{i.} - \bar{x}_{..}) \tag{11.2}$$

and summing over all i and j. This gives SST on the left and SSTr and SSE as the two extreme terms on the right; the cross-product term is easily seen to be zero (Exercise 9).

The interpretation of the fundamental identity is an important aid to understanding ANOVA. SST is a measure of total variation in the data—the sum of all squared deviations about the grand mean. The identity says that this total variation can be partitioned into two pieces; it is this decomposition of SST that gives rise to the name "analysis of variance" (more appropriately, "analysis of variation"). SSE measures variation that would be present (within samples) even if H_0 were true and is thus the part of total variation that is *unexplained* by the status of H_0 (true or false). SSTr is the part of total variation (between samples) that *can be* explained by possible differences in the μ_i's. If explained variation is large relative to unexplained variation, then H_0 is rejected in favor of H_a.

Once SSTr and SSE are computed, each is divided by its associated df to obtain a mean square (mean in the sense of average). Then F is the ratio of the two mean squares.

$$\text{MSTr} = \frac{\text{SSTr}}{I - 1} \qquad \text{MSE} = \frac{\text{SSE}}{I(J - 1)} \qquad F = \frac{\text{MSTr}}{\text{MSE}} \tag{11.3}$$

The computations are often summarized in a tabular format, called an **ANOVA table**, as displayed in Table 11.2. Tables produced by statistical software customarily include a *P*-value column to the right of *f*.

Table 11.2 An ANOVA table

Source of Variation	df	Sum of Squares	Mean Square	f
Treatments	$I - 1$	SSTr	$\text{MSTr} = \text{SSTr}/(I - 1)$	MSTr/MSE
Error	$I(J - 1)$	SSE	$\text{MSE} = \text{SSE}/[I(J - 1)]$	
Total	$IJ - 1$	SST		

Example 11.2

The accompanying data resulted from an experiment comparing the degree of soiling for fabric copolymerized with three different mixtures of methacrylic acid (similar data appeared in the article "Chemical Factors Affecting Soiling and Soil Release from Cotton DP Fabric," *American Dyestuff Reporter*, 1983: 25–30).

Mixture	Degree of Soiling					$x_{i\cdot}$	$\bar{x}_{i\cdot}$
1	.56	1.12	.90	1.07	.94	4.59	.918
2	.72	.69	.87	.78	.91	3.97	.794
3	.62	1.08	1.07	.99	.93	4.69	.938
						$x_{\cdot\cdot} = 13.25$	

Let μ_i denote the true average degree of soiling when mixture i is used ($i = 1, 2, 3$). The null hypothesis H_0: $\mu_1 = \mu_2 = \mu_3$ states that the true average degree of soiling is identical for the three mixtures. We will carry out a test at significance level .01 to see whether H_0 should be rejected in favor of the assertion that true average degree of soiling is not the same for all mixtures. Since $I - 1 = 2$ and $I(J - 1) = 12$, the F critical value for the rejection region is $F_{.01,2,12} = 6.93$. Squaring each of the 15 observations and summing gives $\sum\sum x_{ij}^2 = (.56)^2 + (1.12)^2 + \cdots + (.93)^2 = 12.1351$. The values of the three sums of squares are

$$\text{SST} = 12.1351 - (13.25)^2/15 = 12.1351 - 11.7042 = .4309$$

$$\text{SSTr} = \frac{1}{5}[(4.59)^2 + (3.97)^2 + (4.69)^2] - 11.7042$$

$$= 11.7650 - 11.7042 = .0608$$

$$\text{SSE} = .4309 - .0608 = .3701$$

The remaining computations are summarized in the accompanying ANOVA table. Because $f = .99$ is not at least $F_{.01,2,12} = 6.93$, H_0 is not rejected at significance level .01. The mixtures appear to be indistinguishable with respect to degree of soiling ($F_{.10,2,12} = 2.81 \Rightarrow P\text{-value} > .10$).

Source of Variation	df	Sum of Squares	Mean Square	f
Treatments	2	.0608	.0304	.99
Error	12	.3701	.0308	
Total	14	.4309		

■

When the F test causes H_0 to be rejected, the experimenter will often be interested in further analysis to decide which μ_i's differ from which others. Procedures for doing this are called multiple comparison procedures, and several are described in the next two sections.

Testing for the Assumption of Equal Variances

One of the two assumptions for ANOVA is that the populations have equal variances. If the likelihood ratio principle is applied to the problem of testing for equal variances for normal data, then the result is Bartlett's test. This is a generalization of the F test for equal variances given in Section 10.5, and it is very sensitive to the normality assumption.

The Levene test is much less sensitive to the assumption of normality. Essentially, this test involves performing an ANOVA on the absolute values of the residuals, which are the deviations $x_{ij} - \bar{x}_{i\cdot}, j = 1, 2, \ldots, J$ for each $i = 1, 2, \ldots, I$. That is, a residual is the difference between an observation and its row mean (mean for its sample). The Levene test performs an ANOVA F test using the absolute residuals $|x_{ij} - \bar{x}_{i\cdot}|$ in place of x_{ij}. The idea is to use absolute residuals to compare the variability of the samples.

Example 11.3 (Example 11.2 continued)

Consider the data of Example 11.2. Here are the observations again along with the means and the absolute values of the residuals.

						$\bar{x}_{i\cdot}$	$\Sigma\lvert x_{ij} - \bar{x}_{i\cdot}\rvert$
Mixture 1	.56	1.12	.90	1.07	.94	.918	
|residual 1|	.358	.202	.018	.152	.022		.752
Mixture 2	.72	.69	.87	.78	.91	.794	
|residual 2|	.074	.104	.076	.014	.116		.384
Mixture 3	.62	1.08	1.07	.99	.93	.938	
|residual 3|	.318	.142	.132	.052	.008		.652
							1.788

Now apply ANOVA to the absolute residuals. The sum of all 15 squared absolute residuals is .3701, so

$$\text{SST} = .3701 - 1.788^2/15 = .3701 - .2131 = .1570$$

$$\text{SSTr} = \frac{1}{5}[.752^2 + .384^2 + .652^2] - .2131 = .2276 - .2131 = .0145$$

$$\text{SSE} = .1570 - .0145 = .1425$$

$$f = \frac{.0145/2}{.1425/12} = .61$$

Compare .61 to the critical value $F_{.10,2,12} = 2.81$. Because .61 is much smaller than 2.81, there is no reason to doubt that the variances are equal. ■

Given that the absolute residuals are not normally distributed, it might seem unreasonable to do an ANOVA on them. However, the ANOVA F test is robust to the assumption of normality, meaning that the assumption can be relaxed somewhat. Thus, the Levene test works in spite of the normality assumption. Note also that the residuals are dependent because they sum to zero within each sample (row), but this again is not a problem if the samples are of sufficient size (if $J = 2$, why does each sample have both absolute residuals the same?). A sample size of 10 is sufficient for excellent accuracy in the Levene test, but smaller samples can still give useful results when only approximate critical values are needed. This occurs when the test value is either far beyond the nominal critical value or well below it, as in Example 11.3.

Some software packages perform the Levene test, but they will not necessarily get the same answer because they do not necessarily use absolute deviations from the mean. For example, MINITAB uses absolute residuals with respect to the median, an especially good idea in case of skewed data. By default, SAS uses the squared deviations from the mean, although the absolute deviations from the mean can be requested. SAS also allows absolute deviations from the median (known as the BF test, because Brown and Forsythe studied this procedure).

The ANOVA F test is pretty robust to both the normality and constant variance assumptions. The test will still work under moderate departures from these two assumptions. When the sample sizes are all the same, as we are assuming so far, the test is especially insensitive to unequal variances. Also, there is a generalization of the two sample t test of Section 10.2 for more than two samples, and it does not demand equal variances. This test is available in JMP, R, and SAS.

If there is a major violation of assumptions, the situation can sometimes be corrected by a data transformation, as discussed in Section 11.3. Alternatively, the bootstrap can be used, by generalizing the method of Section 10.6 from two groups to several. There is also a nonparametric test (no normality required), as discussed in Exercise 41 of Chapter 14.

Exercises Section 11.1 (1–10)

1. An experiment to compare $I = 5$ brands of golf balls involved using a robotic driver to hit $J = 7$ balls of each brand. The resulting between-sample and within-sample estimates of σ^2 were MSTr $= 123.50$ and MSE $= 22.16$, respectively.

a. State and test the relevant hypotheses using a significance level of .05.

b. What can be said about the P-value of the test?

2. The lumen output was determined for each of $I = 3$ different brands of 60-watt soft-white lightbulbs, with $J = 8$ bulbs of each brand tested. The sums of squares were computed as SSE $= 4773.3$ and SSTr $= 591.2$. State the hypotheses of interest (including word definitions of parameters), and use the F test of ANOVA ($\alpha = .05$) to decide whether there are any differences in true average lumen outputs among the three brands for this type of bulb by obtaining as much information as possible about the P-value.

3. In a study to assess the effects of malaria infection on mosquito hosts ("*Plasmodium cynomolgi*: Effects of Malaria Infection on Laboratory Flight Performance of *Anopheles stephensi* Mosquitos," *Experiment. Parasitol.*, 1977: 397–404), mosquitos were fed on either infective or noninfective rhesus monkeys. Subsequently the distance they flew during a 24-hour period was measured using a flight mill. The mosquitos were divided into four groups of eight mosquitos each: infective rhesus and sporozites present (IRS), infective rhesus and oocysts present (IRD), infective rhesus and no infection developed (IRN), and noninfective (C). The summary data values are $\bar{x}_{1\cdot} = 4.39$ (IRS), $\bar{x}_{2\cdot} = 4.52$ (IRD), $\bar{x}_{3\cdot} = 5.49$ (IRN), $\bar{x}_{4\cdot} = 6.36$ (C), $\bar{x}_{\cdot\cdot} = 5.19$, and $\Sigma\Sigma x_{ij}^2 = 911.91$. Use the ANOVA F test at level .05 to decide whether there are any differences between true average flight times for the four treatments.

4. Consider the following summary data on the modulus of elasticity ($\times 10^6$ psi) for lumber of three different grades (in close agreement with values in the article "Bending Strength and Stiffness of Second-Growth Douglas-Fir Dimension Lumber" (*Forest Products J.*, 1991: 35–43), except that the sample sizes there were larger):

Grade	J	$\bar{x}_{i\cdot}$	s_i
1	10	1.63	.27
2	10	1.56	.24
3	10	1.42	.26

Use this data and a significance level of .01 to test the null hypothesis of no difference in mean modulus of elasticity for the three grades.

5. The article "Origin of Precambrian Iron Formations" (*Econ. Geol.*, 1964: 1025–1057) reports the following data on total Fe for four types of iron formation (1 = carbonate, 2 = silicate, 3 = magnetite, 4 = hematite).

1:	20.5	28.1	27.8	27.0	28.0
	25.2	25.3	27.1	20.5	31.3
2:	26.3	24.0	26.2	20.2	23.7
	34.0	17.1	26.8	23.7	24.9
3:	29.5	34.0	27.5	29.4	27.9
	26.2	29.9	29.5	30.0	35.6
4:	36.5	44.2	34.1	30.3	31.4
	33.1	34.1	32.9	36.3	25.5

Carry out an analysis of variance F test at significance level .01, and summarize the results in an ANOVA table.

6. In an experiment to investigate the performance of four different brands of spark plugs intended for use on a 125-cc two-stroke motorcycle, five plugs of each brand were tested for the number of miles (at a constant speed) until failure. The partial ANOVA table for the data is given here. Fill in the missing entries, state the relevant hypotheses, and carry out a test by obtaining as much information as you can about the P-value.

Source	df	Sum of Squares	Mean Square	f
Brand				
Error			14,713.69	
Total		310,500.76		

7. A study of the properties of metal plate–connected trusses used for roof support ("Modeling Joints Made with Light-Gauge Metal Connector Plates," *Forest Products J.*, 1979: 39–44) yielded the following observations on axial stiffness index (kips/in.) for plate lengths 4, 6, 8, 10, and 12 in.:

4:	309.2	409.5	311.0	326.5	316.8	349.8	309.7
6:	402.1	347.2	361.0	404.5	331.0	348.9	381.7
8:	392.4	366.2	351.0	357.1	409.9	367.3	382.0
10:	346.7	452.9	461.4	433.1	410.6	384.2	362.6
12:	407.4	441.8	419.9	410.7	473.4	441.2	465.8

a. Check the ANOVA assumptions with a normal plot and a test for equal variances.

b. Does variation in plate length have any effect on true average axial stiffness? State and test the relevant hypotheses using analysis of variance with $\alpha = .01$. Display your results in an ANOVA table. (*Hint*: $\Sigma\Sigma x_{ij}^2 = 5{,}241{,}420.79$.)

8. Six samples of each of four types of cereal grain grown in a certain region were analyzed to determine thiamin content, resulting in the following data (μg/g):

Wheat	5.2	4.5	6.0	6.1	6.7	5.8
Barley	6.5	8.0	6.1	7.5	5.9	5.6
Maize	5.8	4.7	6.4	4.9	6.0	5.2
Oats	8.3	6.1	7.8	7.0	5.5	7.2

a. Check the ANOVA assumptions with a normal probability plot and a test for equal variances.

b. Test to see if at least two of the grains differ with respect to true average thiamin content. Use an $\alpha = .05$ test based on the P-value method.

9. Derive the fundamental identity SST = SSTr + SSE by squaring both sides of Equation (11.2) and summing over all i and j. *Hint*: For any particular i, $\Sigma_j(x_{ij} - \bar{x}_{i\cdot}) = 0$.

10. In single-factor ANOVA with I treatments and J observations per treatment, let $\mu = (1/I)\sum\mu_{i.}$

a. Express $E(\overline{X}_{..})$ in terms of μ. [*Hint*: $\overline{X}_{..} = (1/I)\sum\overline{X}_{i.}$]

b. Compute $E(\overline{X}_{i.}^2)$. (*Hint*: For any rv Y, $E(Y^2) = V(Y) + [E(Y)]^2$.)

c. Compute $E(\overline{X}_{..}^2)$.

d. Compute $E(\text{SSTr})$ and then show that

$$E(\text{MSTr}) = \sigma^2 + \frac{J}{I-1}\Sigma(\mu_i - \mu)^2$$

e. Using the result of part (d), what is $E(\text{MSTr})$ when H_0 is true? When H_0 is false, how does $E(\text{MSTr})$ compare to σ^2?

11.2 *Multiple Comparisons in ANOVA

When the computed value of the F statistic in single-factor ANOVA is not significant, the analysis is terminated because no differences among the μ_i's have been identified. But when H_0 is rejected, the investigator will usually want to know which of the μ_i's are different from one another. A method for carrying out this further analysis is called a **multiple comparisons procedure**.

Several of the most frequently used such procedures are based on the following central idea. First calculate a confidence interval for each pairwise difference $\mu_i - \mu_j$ with $i < j$. Thus if $I = 4$, the six required CIs would be for $\mu_1 - \mu_2$ (but not also for $\mu_2 - \mu_1$), $\mu_1 - \mu_3$, $\mu_1 - \mu_4$, $\mu_2 - \mu_3$, $\mu_2 - \mu_4$, and $\mu_3 - \mu_4$. Then if the interval for $\mu_1 - \mu_2$ does not include 0, conclude that μ_1 and μ_2 *differ significantly* from one another; if the interval does include 0, the two μ's are judged not significantly different. Following the same line of reasoning for each of the other intervals, we end up being able to judge for each pair of μ's whether or not they differ significantly from one another.

The procedures based on this idea differ in the method used to calculate the various CIs. Here we present a popular method that controls the *simultaneous* confidence level for all $I(I-1)/2$ intervals calculated.

Tukey's Procedure

Tukey's procedure involves the use of another probability distribution.

DEFINITION

Let $Z_1, Z_2, \ldots, Z_m$ be m independent standard normal rv's and W be a chi-squared rv, independent of the Z_i's, with ν df. Then the distribution of

$$Q = \frac{\max|Z_i - Z_j|}{\sqrt{W/\nu}} = \frac{\max(Z_1, \ldots, Z_m) - \min(Z_1, \ldots, Z_m)}{\sqrt{W/\nu}}$$

is called the **studentized range distribution**. The distribution has two parameters, m = the number of Z_i's and ν = denominator df. We denote the critical value that captures upper-tail area α under the density curve of Q by $Q_{\alpha,m,\nu}$. A tabulation of these critical values appears in Appendix Table A.10.

The word "range" reflects the fact that the numerator of Q is indeed the range of the Z_i's. Dividing the range by $\sqrt{W/\nu}$ is the same as dividing each individual Z_i by $\sqrt{W/\nu}$. But $Z_i/\sqrt{W/\nu}$ has a (Student) t distribution (Student was the pseudonym used by the statistician Gossett, who derived the *t* distribution but published his work using the pseudonym "Student" because his employer, the Guinness Brewing Co., would not permit publication under his own name); "studentizing" refers to the division by $\sqrt{W/\nu}$. So Q is actually the range of m variables that have the t distribution (but they are not independent because the denominator is the same for each one).

The identification of the quantities in the definition with single-factor ANOVA is as follows:

$$Z_i = \frac{\overline{X}_{i\cdot} - \mu_i}{\sigma/\sqrt{J}} \qquad m = I \qquad W = \frac{\text{SSE}}{\sigma^2} = \frac{I(J-1)\text{MSE}}{\sigma^2} \qquad \nu = I(J-1)$$

Substituting into Q gives

$$Q = \frac{\max\left|\dfrac{\overline{X}_{i\cdot} - \mu_i}{\sigma/\sqrt{J}} - \dfrac{\overline{X}_{j\cdot} - \mu_j}{\sigma/\sqrt{J}}\right|}{\sqrt{\dfrac{I(J-1)\text{MSE}}{\sigma^2}\Big/[I(J-1)]}} = \frac{\max|\overline{X}_{i\cdot} - \overline{X}_{j\cdot} - (\mu_i - \mu_j)|}{\sqrt{\text{MSE}/J}}$$

In this latter expression for Q, the denominator $\sqrt{\text{MSE}/J}$ is the estimated standard deviation of $\overline{X}_{i\cdot} - \mu_i$. By definition of Q and Q_α, $P(Q > Q_\alpha) = \alpha$, so

$$\begin{aligned}
1 - \alpha &= P\left(\frac{\max|\overline{X}_{i\cdot} - \overline{X}_{j\cdot} - (\mu_i - \mu_j)|}{\sqrt{\text{MSE}/J}} \le Q_{\alpha,I,I(J-1)}\right) \\
&= P\left(\frac{|\overline{X}_{i\cdot} - \overline{X}_{j\cdot} - (\mu_i - \mu_j)|}{\sqrt{\text{MSE}/J}} \le Q_{\alpha,I,I(J-1)} \text{ for all } i, j\right) \\
&= P(-Q_\alpha\sqrt{\text{MSE}/J} \le \overline{X}_{i\cdot} - \overline{X}_{j\cdot} - (\mu_i - \mu_j) \le Q_\alpha\sqrt{\text{MSE}/J} \text{ for all } i, j) \\
&= P(\overline{X}_{i\cdot} - \overline{X}_{j\cdot} - Q_\alpha\sqrt{\text{MSE}/J} \le \mu_i - \mu_j \le \overline{X}_{i\cdot} - \overline{X}_{j\cdot} + Q_\alpha\sqrt{\text{MSE}/J} \\
&\qquad \text{for all } i, j)
\end{aligned}$$

(whew!). Replacing $\overline{X}_{i\cdot}$, $\overline{X}_{j\cdot}$, and MSE by the values calculated from the data gives the following result.

PROPOSITION

For each $i < j$, form the interval

$$\bar{x}_{i\cdot} - \bar{x}_{j\cdot} \pm Q_{\alpha,I,I(J-1)}\sqrt{\text{MSE}/J} \qquad (11.4)$$

There are $\binom{I}{2} = I(I-1)/2$ such intervals: one for $\mu_1 - \mu_2$, another for $\mu_1 - \mu_3, \ldots,$ and the last for $\mu_{I-1} - \mu_I$. Then the *simultaneous* confidence level that *every* interval includes the corresponding value of $\mu_i - \mu_j$ is $100(1 - \alpha)\%$. Notice that the second subscript on Q_α is I, whereas the second subscript on F_α used in the F test is $I - 1$.

We will say more about the interpretation of "simultaneous" shortly. Each interval that doesn't include 0 yields the conclusion that the corresponding values of μ_i and μ_j are different—we say that μ_i and μ_j "differ significantly" from one another. For purposes of deciding which μ_i's differ significantly from which others, that is, identifying the intervals that don't include 0, much of the arithmetic associated with calculating the CI's can be avoided. The following box gives details and describes how differences can be displayed using an "underscoring pattern."

TUKEY'S PROCEDURE FOR IDENTIFYING SIGNIFICANTLY DIFFERENT μ_i's

Select α, extract $Q_{\alpha,I,I(J-1)}$ from Appendix Table A.10, and calculate the quantity $w = Q_{\alpha,I,I(J-1)} \cdot \sqrt{\text{MSE}/J}$. Then list the sample means in increasing order and underline those pairs that differ by less than w. Any pair of sample means not underscored by the same line corresponds to a pair of population or treatment means that are judged significantly different. The quantity w is sometimes referred to as Tukey's honestly significant difference (HSD).

Suppose, for example, that $I = 5$ and that

$$\bar{x}_{2\cdot} < \bar{x}_{5\cdot} < \bar{x}_{4\cdot} < \bar{x}_{1\cdot} < \bar{x}_{3\cdot}$$

Then

1. Consider first the smallest mean $\bar{x}_{2\cdot}$. If $\bar{x}_{5\cdot} - \bar{x}_{2\cdot} \ge w$, proceed to step 2. However, if $\bar{x}_{5\cdot} - \bar{x}_{2\cdot} < w$, connect these first two means with a line segment. Then if possible extend this line segment even further to the right to the largest $\bar{x}_{i\cdot}$ that differs from $\bar{x}_{2\cdot}$ by less than w (so the line may connect two, three, or even more means).
2. Now move to $\bar{x}_{5\cdot}$, and again extend a line segment to the largest $\bar{x}_{i\cdot}$ to its right that differs from $\bar{x}_{5\cdot}$ by less than w (it may not be possible to draw this line, or alternatively it may underscore just two means, or three, or even all four remaining means).
3. Continue by moving to $\bar{x}_{4\cdot}$ and repeating, and then finally move to $\bar{x}_{1\cdot}$.

To summarize, starting from each mean in the ordered list, a line segment is extended as far to the right as possible as long as the difference between the means is smaller than w. It is easily verified that a particular interval of the form (11.4) will contain 0 if and only if the corresponding pair of sample means is underscored by the same line segment.

Example 11.4 An experiment was carried out to compare five different brands of automobile oil filters with respect to their ability to capture foreign material. Let μ_i denote the true average amount of material captured by brand i filters ($i = 1, \ldots, 5$) under controlled conditions. A sample of nine filters of each brand was used, resulting in the following sample mean amounts: $\bar{x}_{1\cdot} = 14.5$, $\bar{x}_{2\cdot} = 13.8$, $\bar{x}_{3\cdot} = 13.3$, $\bar{x}_{4\cdot} = 14.3$, and $\bar{x}_{5\cdot} = 13.1$. Table 11.3 is the ANOVA table summarizing the first part of the analysis.

Table 11.3 ANOVA table for Example 11.4

Source of Variation	df	Sum of Squares	Mean Square	f
Treatments (brands)	4	13.32	3.33	37.84
Error	40	3.53	.088	
Total	44	16.85		

Since $F_{.05,4,40} = 2.61$, H_0 is rejected (decisively) at level .05. We now use Tukey's procedure to look for significant differences among the μ_i's. From Appendix Table A.10, $Q_{.05,5,40} = 4.04$ (the second subscript on Q is I and not $I - 1$ as in F), so $w = 4.04\sqrt{.088/9} = .4$. After we arrange the five sample means in increasing order, the two smallest can be connected by a line segment because they differ by less than .4. However, this segment cannot be extended further to the right since $13.8 - 13.1 = .7 \geq .4$. Moving one mean to the right, the pair $\bar{x}_{3\cdot}$ and $\bar{x}_{2\cdot}$ cannot be underscored because these means differ by more than .4. Again moving to the right, the next mean, 13.8, cannot be connected to any further to the right, and finally the last two means can be underscored with the same line segment.

$\bar{x}_{5\cdot}$	$\bar{x}_{3\cdot}$	$\bar{x}_{2\cdot}$	$\bar{x}_{4\cdot}$	$\bar{x}_{1\cdot}$
13.1	13.3	13.8	14.3	14.5
______	______		______	______

Thus brands 1 and 4 are not significantly different from one another, but are significantly higher than the other three brands in their true average amounts captured. Brand 2 is significantly better than 3 and 5 but worse than 1 and 4, and brands 3 and 5 do not differ significantly.

If $\bar{x}_{2\cdot} = 14.15$ rather than 13.8 with the same computed w, then the configuration of underscored means would be

$\bar{x}_{5\cdot}$	$\bar{x}_{3\cdot}$	$\bar{x}_{2\cdot}$	$\bar{x}_{4\cdot}$	$\bar{x}_{1\cdot}$
13.1	13.3	14.15	14.3	14.5
______	______	______	______	______

■

Example 11.5 A biologist wished to study the effects of ethanol on sleep time. A sample of 20 rats, matched for age and other characteristics, was selected, and each rat was given an oral injection having a particular concentration of ethanol per kg of body weight. The rapid eye movement (REM) sleep time for each rat was then recorded for a 24-hour period, with the following results:

Treatment (ethanol)			**REM Time**			$x_{i\cdot}$	$\bar{x}_{i\cdot}$
0 (control)	88.6	73.2	91.4	68.0	75.2	396.4	79.28
1 g/kg	63.0	53.9	69.2	50.1	71.5	307.7	61.54
2 g/kg	44.9	59.5	40.2	56.3	38.7	239.6	47.92
4 g/kg	31.0	39.6	45.3	25.2	22.7	163.8	32.76
						$x_{\cdot\cdot} = 1107.5$	$\bar{x}_{\cdot\cdot} = 55.375$

Does the data indicate that the true average REM sleep time depends on the concentration of ethanol? (This example is based on an experiment reported in "Relationship of Ethanol Blood Level to REM and Non-REM Sleep Time and Distribution in the Rat," *Life Sci.*, 1978: 839–846.)

The $\bar{x}_{i\cdot}$'s differ rather substantially from one another, but there is also a great deal of variability within each sample, so to answer the question precisely we must carry out the ANOVA. With $\sum\sum x_{ij}^2 = 68{,}697.6$ and correction factor $x_{..}^2/(IJ) = (1107.5)^2/20 = 61{,}327.8$, the computing formulas yield

$$\text{SST} = 68{,}697.6 - 61{,}327.8 = 7369.8$$

$$\text{SSTr} = \frac{1}{5}[(396.40)^2 + (307.70)^2 + (239.60)^2 + (163.80)^2] - 61{,}327.8$$

$$= 67{,}210.2 - 61{,}327.8 = 5882.4$$

and

$$\text{SSE} = 7369.8 - 5882.4 = 1487.4$$

Table 11.4 is a SAS ANOVA table. The last column gives the P-value, which is .0001. Actually, the P-value is .0000083, but SAS does not output anything lower than .0001. It does not output .0000 because this could be misinterpreted to say that the P-value is 0. Using a significance level of .05, we reject the null hypothesis H_0: $\mu_1 = \mu_2 = \mu_3 = \mu_4$, since the printed P-value $= .0001 < .05 = \alpha$. True average REM sleep time does appear to depend on ethanol concentration.

Table 11.4 SAS ANOVA table

Analysis of Variance Procedure

Dependent Variable: TIME

Source	DF	Sum of Squares	Mean Square	F Value	Pr > F
Model	3	5882.35750	1960.78583	21.09	0.0001
Error	16	1487.40000	92.96250		
Corrected Total	19	7369.75750			

There are $I = 4$ treatments and 16 df for error, so $Q_{.05,4,16} = 4.05$ and $w = 4.05\sqrt{93.0/5} = 17.47$. Ordering the means and underscoring yields

$\bar{x}_{4\cdot}$	$\bar{x}_{3\cdot}$	$\bar{x}_{2\cdot}$	$\bar{x}_{1\cdot}$
32.76	47.92	61.54	79.28

The interpretation of this underscoring must be done with care, since we seem to have concluded that treatments 2 and 3 do not differ, 3 and 4 do not differ, yet 2 and 4 do differ. The suggested way of expressing this is to say that although evidence allows us to conclude that treatments 2 and 4 differ from one another, neither has been shown to be significantly different from 3. Treatment 1 has a significantly higher true average REM sleep time than any of the other treatments. This treatment involves 0 ethanol (alcohol) and there is a trend toward less sleep with more ethanol, although not all differences are significant.

Figure 11.4 shows SAS output from the application of Tukey's procedure.

```
              Alpha = 0.05 df = 16 MSE = 92.9625
          Critical Value of Studentized Range = 4.046
            Minimum Significant Difference = 17.446

   Means with the same letter are not significantly different.

   Tukey  Grouping           Mean        N     TREATMENT
                 A         79.280        5     0(control)

                 B         61.540        5     1 gm/kg
                 B
     C           B         47.920        5     2 gm/kg
     C
     C                     32.760        5     4 gm/kg
```

Figure 11.4 Tukey's method using SAS

The Interpretation of α in Tukey's Procedure

We stated previously that the *simultaneous* confidence level is controlled by Tukey's method. So what does "simultaneous" mean here? Consider calculating a 95% CI for a population mean μ based on a sample from that population and then a 95% CI for a population proportion p based on another sample selected independently of the first one. Prior to obtaining data, the probability that the first interval will include μ is .95, and this is also the probability that the second interval will include p. Because the two samples are selected independently of one another, the probability that *both* intervals will include the values of the respective parameters is $(.95)(.95) = (.95)^2 \approx .90$. Thus the *simultaneous* or *joint* confidence level for the two intervals is roughly 90%—if pairs of intervals are calculated over and over again from independent samples, in the long run roughly 90% of the time the first interval will capture μ and the second will include p. Similarly, if three CIs are calculated based on independent samples, the simultaneous confidence level will be $100(.95)^3\% \approx 86\%$. Clearly, as the number of intervals increases, the simultaneous confidence level that all intervals capture their respective parameters will decrease.

Now suppose that we want to maintain the simultaneous confidence level at 95%. Then for two independent samples, the individual confidence level for each would have to be $100\sqrt{.95}\% \approx 97.5\%$. The larger the number of intervals, the higher the individual confidence level would have to be to maintain the 95% simultaneous level.

The tricky thing about the Tukey intervals is that they are not based on independent samples—MSE appears in every one, and various intervals share the same $\bar{x}_{i\cdot}$'s (e.g., in the case $I = 4$, three different intervals all use $\bar{x}_{1\cdot}$). This implies that there is no straightforward probability argument for ascertaining the simultaneous confidence level from the individual confidence levels. Nevertheless, if $Q_{.05}$ is used, the simultaneous confidence level is controlled at 95%, whereas using $Q_{.01}$ gives a simultaneous 99% level. To obtain a 95% simultaneous level, the individual level for each interval must be considerably larger than 95%. Said in a slightly different way, to obtain a 5% *experimentwise* or *family* error rate, the individual or per-comparison error rate for each interval must be considerably smaller than .05. MINITAB asks the user to specify the family error rate (e.g., 5%) and then includes on output the individual error rate (see Exercise 16).

Confidence Intervals for Other Parametric Functions

In some situations, a CI is desired for a function of the μ_i's more complicated than a difference $\mu_i - \mu_j$. Let $\theta = \Sigma c_i\mu_i$, where the c_i's are constants. One such function is $\frac{1}{2}(\mu_1 + \mu_2) - \frac{1}{3}(\mu_3 + \mu_4 + \mu_5)$, which in the context of Example 11.4 measures the difference between the group consisting of the first two brands and that of the last three brands. Because the X_{ij}'s are normally distributed with $E(X_{ij}) = \mu_i$ and $V(X_{ij}) = \sigma^2$, $\hat{\theta} = \sum_i c_i\overline{X}_{i\cdot}$ is normally distributed, unbiased for θ, and

$$V(\hat{\theta}) = V\left(\sum_i c_i\overline{X}_{i\cdot}\right) = \sum_i c_i^2 V(\overline{X}_{i\cdot}) = \frac{\sigma^2}{J}\sum_i c_i^2$$

Estimating σ^2 by MSE and forming $\hat{\sigma}_{\hat{\theta}}$ results in a t variable $(\hat{\theta} - \theta)/\hat{\sigma}_{\hat{\theta}}$, which can be manipulated to obtain the following $100(1 - \alpha)\%$ confidence interval for $\Sigma c_i\mu_i$:

$$\sum c_i\bar{x}_{i\cdot} \pm t_{\alpha/2,I(J-1)}\sqrt{\frac{\text{MSE}\sum c_i^2}{J}} \qquad (11.5)$$

Example 11.6 (Example 11.4 continued)

The parametric function for comparing the first two (store) brands of oil filter with the last three (national) brands is $\theta = \frac{1}{2}(\mu_1 + \mu_2) - \frac{1}{3}(\mu_3 + \mu_4 + \mu_5)$, from which

$$\sum c_i^2 = \left(\frac{1}{2}\right)^2 + \left(\frac{1}{2}\right)^2 + \left(-\frac{1}{3}\right)^2 + \left(-\frac{1}{3}\right)^2 + \left(-\frac{1}{3}\right)^2 = \frac{5}{6}$$

With $\hat{\theta} = \frac{1}{2}(\bar{x}_{1\cdot} + \bar{x}_{2\cdot}) - \frac{1}{3}(\bar{x}_{3\cdot} + \bar{x}_{4\cdot} + \bar{x}_{5\cdot}) = .583$ and MSE $= .088$, a 95% interval is

$$.583 \pm 2.021\sqrt{5(.088)/[(6)(9)]} = .583 \pm .182 = (.401, .765)$$ ■

Notice that in this example the coefficients $c_1, \ldots, c_5$ satisfy $\Sigma c_i = \frac{1}{2} + \frac{1}{2} - \frac{1}{3} - \frac{1}{3} - \frac{1}{3} = 0$. When the coefficients sum to 0, the linear combination $\theta = \Sigma c_i\mu_i$ is called a **contrast** among the means, and the analysis is available in a number of statistical software programs.

Sometimes an experiment is carried out to compare each of several "new" treatments to a control treatment. In such situations, a multiple comparisons technique called Dunnett's method is appropriate.

Exercises Section 11.2 (11–21)

11. An experiment to compare the spreading rates of five different brands of yellow interior latex paint available in a particular area used 4 gallons ($J = 4$) of each paint. The sample average spreading rates (ft²/gal) for the five brands were $\bar{x}_{1\cdot} = 462.0$, $\bar{x}_{2\cdot} = 512.8$, $\bar{x}_{3\cdot} = 437.5$, $\bar{x}_{4\cdot} = 469.3$, and $\bar{x}_{5\cdot} = 532.1$. The computed value of F was found to be significant at level $\alpha = .05$. With MSE $= 272.8$, use Tukey's procedure to investigate significant differences in the true average spreading rates between brands.

12. In Exercise 11, suppose $\bar{x}_{3\cdot} = 427.5$. Now which true average spreading rates differ significantly from one another? Be sure to use the method of underscoring to illustrate your conclusions, and write a paragraph summarizing your results.

13. Repeat Exercise 12 supposing that $\bar{x}_{2\cdot} = 502.8$ in addition to $\bar{x}_{3\cdot} = 427.5$.

14. Use Tukey's procedure on the data in Exercise 3 to identify differences in true average flight times among the four types of mosquitos.

15. Use Tukey's procedure on the data of Exercise 5 to identify differences in true average total Fe among the four types of formations (use MSE = 15.64).

16. Reconsider the axial stiffness data given in Exercise 7. ANOVA output from MINITAB follows:

```
Analysis of Variance for stiffness
Source    DF       SS       MS        F       P
length     4    43993    10998    10.48   0.000
Error     30    31475     1049
Total     34    75468

Level      N      Mean     StDev
 4         7    333.21     36.59
 6         7    368.06     28.57
 8         7    375.13     20.83
10         7    407.36     44.51
12         7    437.17     26.00

Pooled StDev = 32.39

Tukey's pairwise comparisons

    Family error rate = 0.0500
Individual error rate = 0.00693

Critical value = 4.10

Intervals for (column level mean) − (row
level mean)
            4          6          8         10
 6      −85.0
         15.4

 8      −92.1      −57.3
          8.3       43.1

10     −124.3      −89.5      −82.4
        −23.9       10.9       18.0

12     −154.2     −119.3     −112.2     −80.0
        −53.8      −18.9      −11.8      20.4
```

a. Is it plausible that the variances of the five axial stiffness index distributions are identical? Explain.

b. Use the output (without reference to our F table) to test the relevant hypotheses.

c. Use the Tukey intervals given in the output to determine which means differ, and construct the corresponding underscoring pattern.

17. Refer to Exercise 4. Compute a 95% t CI for the contrast $\theta = \frac{1}{2}(\mu_1 + \mu_2) - \mu_3$.

18. Consider the accompanying data on plant growth after the application of different types of growth hormone.

	1	13	17	7	14
	2	21	13	20	17
Hormone	*3*	18	15	20	17
	4	7	11	18	10
	5	6	11	15	8

a. Perform an F test at level $\alpha = .05$.

b. What happens when Tukey's procedure is applied?

19. Consider a single-factor ANOVA experiment in which $I = 3$, $J = 5$, $\bar{x}_{1\cdot} = 10$, $\bar{x}_{2\cdot} = 12$, and $\bar{x}_{3\cdot} = 20$. Find a value of SSE for which $f > F_{.05,2,12}$, so that H_0: $\mu_1 = \mu_2 = \mu_3$ is rejected, yet when Tukey's procedure is applied none of the μ_i's can be said to differ significantly from one another.

20. Refer to Exercise 19 and suppose $\bar{x}_{1\cdot} = 10$, $\bar{x}_{2\cdot} = 15$, and $\bar{x}_{3\cdot} = 20$. Can you now find a value of SSE that produces such a contradiction between the F test and Tukey's procedure?

21. The article "The Effect of Enzyme Inducing Agents on the Survival Times of Rats Exposed to Lethal Levels of Nitrogen Dioxide" (*Toxicol. Appl. Pharmacol.*, 1978: 169–174) reports the following data on survival times for rats exposed to nitrogen dioxide (70 ppm) via different injection regimens. There were $J = 14$ rats in each group.

Regimen	$\bar{x}_{i\cdot}$(min)	s_i
1. Control	166	32
2. 3-Methylcholanthrene	303	53
3. Allylisopropylacetamide	266	54
4. Phenobarbital	212	35
5. Chlorpromazine	202	34
6. *p*-Aminobenzoic acid	184	31

a. Test the null hypothesis that true average survival time does not depend on injection regimen against the alternative that there is some dependence on injection regimen using $\alpha = .01$.

b. Suppose that $100(1 - \alpha)\%$ CIs for k different parametric functions are computed from the same ANOVA data set. Then it is easily verified that the simultaneous confidence level is at least $100(1 - k\alpha)\%$. Compute CIs with simultaneous confidence level at least 98% for the contrasts $\mu_1 - \frac{1}{5}(\mu_2 + \mu_3 + \mu_4 + \mu_5 + \mu_6)$ and $\frac{1}{4}(\mu_2 + \mu_3 + \mu_4 + \mu_5) - \mu_6$.

11.3 *More on Single-Factor ANOVA

In this section, we briefly consider some additional issues relating to single-factor ANOVA. These include an alternative description of the model parameters, β for the F test, the relationship of the test to procedures previously considered, data transformation, a random effects model, and formulas for the case of unequal sample sizes.

An Alternative Description of the ANOVA Model

The assumptions of single-factor ANOVA can be described succinctly by means of the "model equation"

$$X_{ij} = \mu_i + \varepsilon_{ij}$$

where ε_{ij} represents a random deviation from the population or true treatment mean μ_i. The ε_{ij}'s are assumed to be independent, normally distributed rv's (implying that the X_{ij}'s are also) with $E(\varepsilon_{ij}) = 0$ [so that $E(X_{ij}) = \mu_i$] and $V(\varepsilon_{ij}) = \sigma^2$ [from which $V(X_{ij}) = \sigma^2$ for every i and j]. An alternative description of single-factor ANOVA will give added insight and suggest appropriate generalizations to models involving more than one factor. Define a parameter μ by

$$\mu = \frac{1}{I}\sum_{i=1}^{I} \mu_i$$

and the parameters $\alpha_1, \ldots, \alpha_I$ by

$$\alpha_i = \mu_i - \mu \qquad (i = 1, \ldots, I)$$

Then the treatment mean μ_i can be written as $\mu + \alpha_i$, where μ represents the true average overall response in the experiment, and α_i is the effect, measured as a departure from μ, due to the ith treatment. Whereas we initially had I parameters, we now have $I + 1$ $(\mu, \alpha_1, \ldots, \alpha_I)$. However, because $\sum \alpha_i = 0$ (the average departure from the overall mean response is zero), only I of these new parameters are independently determined, so there are as many independent parameters as there were before. In terms of μ and the α_i's, the model becomes

$$X_{ij} = \mu + \alpha_i + \varepsilon_{ij} \qquad (i = 1, \ldots, I, \quad j = 1, \ldots, J)$$

In the next two sections, we will develop analogous models for two-factor ANOVA. The claim that the μ_i's are identical is equivalent to the equality of the α_i's, and because $\sum \alpha_i = 0$, the null hypothesis becomes

$$H_0\colon \quad \alpha_1 = \alpha_2 = \cdots = \alpha_I = 0$$

In Section 11.1, it was stated that MSTr is an unbiased estimator of σ^2 when H_0 is true but otherwise tends to overestimate σ^2. More precisely,

$$E(\text{MSTr}) = \sigma^2 + \frac{J}{I-1}\sum \alpha_i^2$$

When H_0 is true, $\sum \alpha_i^2 = 0$ so $E(\text{MSTr}) = \sigma^2$ (MSE is unbiased whether or not H_0 is true). If $\sum \alpha_i^2$ is used as a measure of the extent to which H_0 is false, then a larger value of $\sum \alpha_i^2$ will result in a greater tendency for MSTr to overestimate σ^2. More generally, formulas for expected mean squares for multifactor models are used to suggest how to form F ratios to test various hypotheses.

Proof of the Formula for *E*(MSTr) For any rv Y, $E(Y^2) = V(Y) + [E(Y)]^2$, so

$$
\begin{aligned}
E(\text{SSTr}) &= E\left(\frac{1}{J}\sum_i X_{i\cdot}^2 - \frac{1}{IJ}X_{\cdot\cdot}^2\right) = \frac{1}{J}\sum_i E(X_{i\cdot}^2) - \frac{1}{IJ}E(X_{\cdot\cdot}^2) \\
&= \frac{1}{J}\sum_i \{V(X_{i\cdot}) + [E(X_{i\cdot})]^2\} - \frac{1}{IJ}\{V(X_{\cdot\cdot}) + [E(X_{\cdot\cdot})]^2\} \\
&= \frac{1}{J}\sum_i \{J\sigma^2 + [J(\mu + \alpha_i)]^2\} - \frac{1}{IJ}[IJ\sigma^2 + (IJ\mu)^2] \\
&= I\sigma^2 + IJ\mu^2 + 2\mu J\sum_i \alpha_i + J\sum_i \alpha_i^2 - \sigma^2 - IJ\mu^2 \\
&= (I-1)\sigma^2 + J\sum_i \alpha_i^2 \qquad \left(\text{since } \sum \alpha_i = 0\right)
\end{aligned}
$$

The result then follows from the relationship $\text{MSTr} = \text{SSTr}/(I-1)$. ■

β for the F Test

Consider a set of parameter values $\alpha_1, \alpha_2, \ldots, \alpha_I$ for which H_0 is not true. The probability of a type II error, β, is the probability that H_0 is not rejected when that set is the set of true values. One might think that β would have to be determined separately for each different configuration of α_i's. Fortunately, since β for the F test depends on the α_i's and σ^2 only through $\sum \alpha_i^2/\sigma^2$, it can be simultaneously evaluated for many different alternatives. For example, $\sum \alpha_i^2 = 4$ for each of the following sets of α_i's for which H_0 is false, so β is identical for all three alternatives:

1. $\alpha_1 = -1, \alpha_2 = -1, \alpha_3 = 1, \alpha_4 = 1$
2. $\alpha_1 = -\sqrt{2}, \alpha_2 = \sqrt{2}, \alpha_3 = 0, \alpha_4 = 0$
3. $\alpha_1 = -\sqrt{3}, \alpha_2 = \sqrt{1/3}, \alpha_3 = \sqrt{1/3}, \alpha_4 = \sqrt{1/3}$

The quantity $J\sum \alpha_i^2/\sigma^2$ is called the **noncentrality parameter** for one-way ANOVA (because when H_0 is false the test statistic has a *noncentral F* distribution with this as one of its parameters), and β is a decreasing function of the value of this parameter. Thus, for fixed values of σ^2 and J, the null hypothesis is more likely to be rejected for alternatives far from H_0 (large $\sum \alpha_i^2$) than for alternatives close to H_0. For a fixed value of $\sum \alpha_i^2$, β decreases as the sample size J on each treatment increases, and it increases as the variance σ^2 increases (since greater underlying variability makes it more difficult to detect any given departure from H_0).

Because hand computation of β and sample size determination for the F test are quite difficult (as in the case of t tests), statisticians have constructed sets of curves from which β can be obtained. Sets of curves for numerator df $\nu_1 = 3$ and $\nu_1 = 4$ are displayed in Figures 11.5 and 11.6, respectively. After the values of σ^2 and the α_i's for which β is desired are specified, these are used to compute the value of ϕ, where $\phi^2 = (J/I)\sum\alpha_i^2/\sigma^2$.

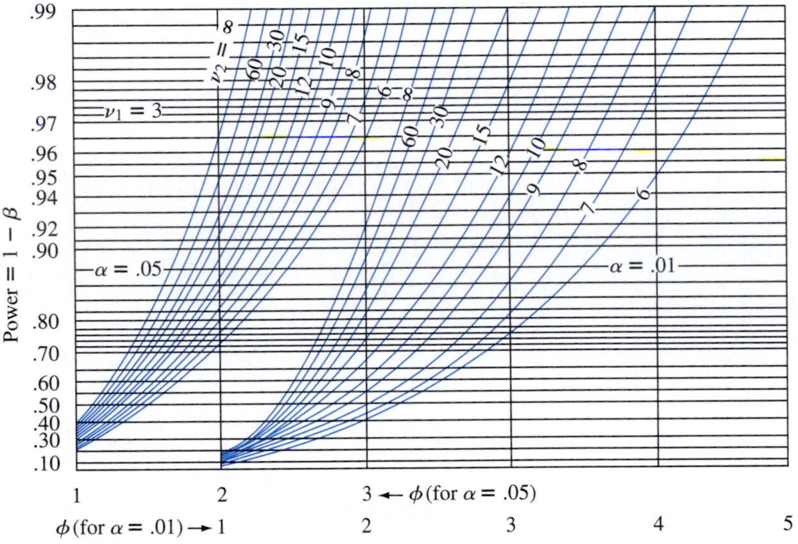

Figure 11.5 Power curves for the ANOVA F test ($\nu_1 = 3$)
(E. S. Pearson and H. O. Hartley, "Charts of the Power Function for Analysis of Variance Tests, Derived from the Non-central F Distribution," *Biometrika*, vol. 38, 1951: 112, by permission of Biometrika Trustees.)

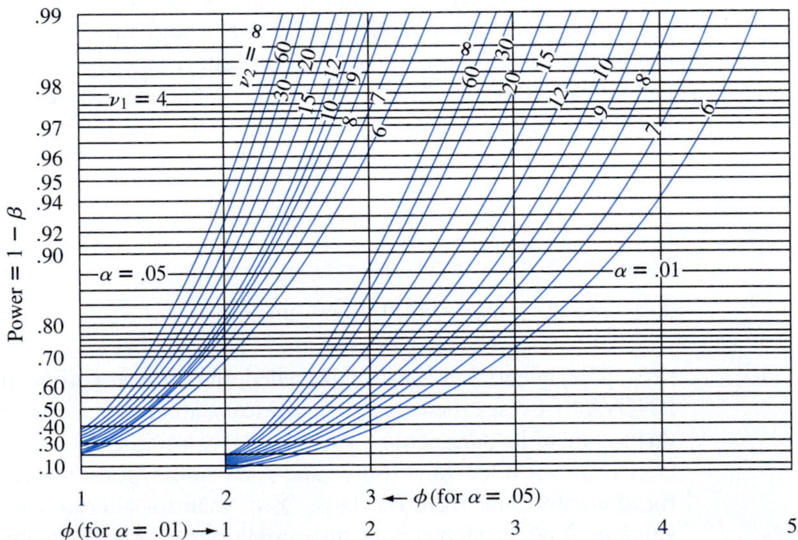

Figure 11.6 Power curves for the ANOVA F test ($\nu_1 = 4$)
(E. S. Pearson and H. O. Hartley, "Charts of the Power Function for Analysis of Variance Tests, Derived from the Non-central F Distribution," *Biometrika*, vol. 38, 1951: 112, by permission of Biometrika Trustees.)

We then enter the appropriate set of curves at the value of ϕ on the horizontal axis, move up to the curve associated with error df ν_2, and move over to the value of power on the vertical axis. Finally, $\beta = 1 -$ power.

Example 11.7 The effects of four different heat treatments on yield point (tons/in^2) of steel ingots are to be investigated. A total of eight ingots will be cast using each treatment. Suppose the true standard deviation of yield point for any of the four treatments is $\sigma = 1$. How likely is it that H_0 will not be rejected at level .05 if three of the treatments have the same expected yield point and the other treatment has an expected yield point that is 1 ton/in^2 greater than the common value of the other three (i.e., the fourth yield is on average 1 standard deviation above those for the first three treatments)?

Suppose that $\mu_1 = \mu_2 = \mu_3$ and $\mu_4 = \mu_1 + 1$, $\mu = (\Sigma\mu_i)/4 = \mu_1 + \frac{1}{4}$. Then $\alpha_1 = \mu_1 - \mu = -\frac{1}{4}$, $\alpha_2 = -\frac{1}{4}$, $\alpha_3 = -\frac{1}{4}$, $\alpha_4 = \frac{3}{4}$ so

$$\phi^2 = \frac{8}{4}\left[\left(\frac{1}{4}\right)^2 + \left(-\frac{1}{4}\right)^2 + \left(-\frac{1}{4}\right)^2 + \left(\frac{3}{4}\right)^2\right] = \frac{3}{2}$$

and $\phi = 1.22$. The degrees of freedom are $\nu_1 = I - 1 = 3$ and $\nu_2 = I(J - 1) = 28$, so interpolating visually between $\nu_2 = 20$ and $\nu_2 = 30$ gives power $\approx .47$ and $\beta \approx .53$. This β is rather large, so we might decide to increase the value of J. How many ingots of each type would be required to yield $\beta \approx .05$ for the alternative under consideration? By trying different values of J, we can verify that $J = 24$ will meet the requirement, but any smaller J will not. ■

As an alternative to the use of power curves, the SAS statistical software package has a function that calculates the cumulative area under a noncentral F curve (inputs F_α, numerator df, denominator df, and ϕ^2), and this area is β. Version 14 of MINITAB does this and also something rather different. The user is asked to specify the maximum difference between μ_i's rather than the individual means. For example, we might wish to calculate the power of the test when $I = 4$, $\mu_1 = 100$, $\mu_2 = 101$, $\mu_3 = 102$, and $\mu_4 = 106$. Then the maximum difference is $106 - 100 = 6$. However, the power depends not only on this maximum difference but on the values of all the μ_i's. In this situation MINITAB calculates the smallest possible value of power subject to $\mu_1 = 100$ and $\mu_4 = 106$, which occurs when the two other μ's are both halfway between 100 and 106. If this power is .85, then we can say that the power is at least .85 and β is at most .15 when the two most extreme μ's are separated by 6 (the common sample size, α, and σ must also be specified). The software will also determine the necessary common sample size if maximum difference and minimum power are specified.

Relationship of the *F* Test to the *t* Test

When the number of populations is just $I = 2$, the ANOVA F is testing H_0: $\mu_1 = \mu_2$ versus H_a: $\mu_1 \neq \mu_2$. In this case, a two-tailed, two-sample t test can also be used. In Section 10.2, we mentioned the pooled t test, which requires equal variances, as an alternative to the two-sample t procedure. With a little algebra, it can be shown that the single-factor ANOVA F test and the two-tailed pooled t test are equivalent; for any given data set, the P-values for the two tests will be identical, so the same conclusion will be reached by either test.

The two-sample t test is more flexible than the F test when $I = 2$ for two reasons. First, it is not based on the assumption that $\sigma_1 = \sigma_2$; second, it can be used to test H_a: $\mu_1 > \mu_2$ (an upper-tailed t test) or H_a: $\mu_1 < \mu_2$ as well as H_a: $\mu_1 \neq \mu_2$. As mentioned at the end of Section 11.1, there is a generalization of the two-sample t test for $I \geq 3$ samples with population variances not necessarily the same.

Single-Factor ANOVA When Sample Sizes Are Unequal

When the sample sizes from each population or treatment are not equal, let $J_1, J_2, \ldots, J_I$ denote the I sample sizes and let $n = \Sigma_i J_i$ denote the total number of observations. The accompanying box gives ANOVA formulas and the test procedure.

$$\text{SST} = \sum_{i=1}^{I}\sum_{j=1}^{J_i}(X_{ij} - \overline{X}_{..})^2 = \sum_{i=1}^{I}\sum_{j=1}^{J_i}X_{ij}^2 - \frac{1}{n}X_{..}^2 \qquad \text{df} = n - 1$$

$$\text{SSTr} = \sum_{i=1}^{I}\sum_{j=1}^{J_i}(\overline{X}_{i.} - \overline{X}_{..})^2 = \sum_{i=1}^{I}\frac{1}{J_i}X_{i.}^2 - \frac{1}{n}X_{..}^2 \qquad \text{df} = I - 1$$

$$\text{SSE} = \sum_{i=1}^{I}\sum_{j=1}^{J_i}(X_{ij} - \overline{X}_{i.})^2 = \text{SST} - \text{SSTr} \qquad \text{df} = \sum(J_i - 1) = n - I$$

Test statistic value:

$$f = \frac{\text{MSTr}}{\text{MSE}} \quad \text{where MSTr} = \frac{\text{SSTr}}{I - 1} \quad \text{and} \quad \text{MSE} = \frac{\text{SSE}}{n - I}$$

Rejection region: $f \geq F_{\alpha,I-1,n-I}$

The correction factor (CF) $X_{..}^2/n$ is subtracted when computing both SST and SSTr. These formulas are derived in the same way (see Exercise 28) as the similar formulas in Section 11.1, except that it is harder here to show that MSTr/MSE has the F distribution under H_0.

Example 11.8 The article "On the Development of a New Approach for the Determination of Yield Strength in Mg-Based Alloys" (*Light Metal Age*, Oct. 1998: 51–53) presented the following data on elastic modulus (GPa) obtained by a new ultrasonic method for specimens of a certain alloy produced using three different casting processes.

Process	Observations								J_i	$x_{i.}$	$\bar{x}_{i.}$
Permanent molding	45.5	45.3	45.4	44.4	44.6	43.9	44.6	44.0	8	357.7	44.71
Die casting	44.2	43.9	44.7	44.2	44.0	43.8	44.6	43.1	8	352.5	44.06
Plaster molding	46.0	45.9	44.8	46.2	45.1	45.5			6	273.5	45.58
									22	983.7	

Let μ_1, μ_2, and μ_3 denote the true average elastic moduli for the three different processes under the given circumstances. The relevant hypotheses are H_0: $\mu_1 = \mu_2 = \mu_3$ versus H_a: at least two of the μ_i's are different. The test statistic is, of course, $F = \text{MSTr}/\text{MSE}$, based on $I - 1 = 2$ numerator df and $n - I = 22 - 3 = 19$ denominator df. Relevant quantities include

$$\sum\sum x_{ij}^2 = 43{,}998.73 \qquad \text{CF} = \frac{983.7^2}{22} = 43{,}984.80$$

$$\text{SST} = 43{,}998.73 - 43{,}984.80 = 13.93$$

$$\text{SSTr} = \frac{357.7^2}{8} + \frac{352.5^2}{8} + \frac{273.5^2}{6} - 43{,}984.80 = 7.93$$

$$\text{SSE} = 13.93 - 7.93 = 6.00$$

The remaining computations are displayed in the accompanying ANOVA table. Since $F_{.001,2,19} = 10.16 < 12.56 = f$, the P-value is smaller than .001. Thus the null hypothesis should be rejected at any reasonable significance level; there is compelling evidence for concluding that true average elastic modulus somehow depends on which casting process is used.

Source of Variation	df	Sum of Squares	Mean Square	f
Treatments	2	7.93	3.965	12.56
Error	19	6.00	.3158	
Total	21	13.93		

Multiple Comparisons When Sample Sizes Are Unequal

There is more controversy among statisticians regarding which multiple comparisons procedure to use when sample sizes are unequal than there is in the case of equal sample sizes. The procedure that we present here is recommended in the excellent book *Beyond ANOVA: Basics of Applied Statistics* (see the chapter bibliography) for use when the I sample sizes $J_1, J_2, \ldots, J_I$ are reasonably close to one another ("mild imbalance"). It modifies Tukey's method by using averages of pairs of $1/J_i$'s in place of $1/J$.

Let

$$w_{ij} = Q_{\alpha,I,n-I} \cdot \sqrt{\frac{\text{MSE}}{2}\left(\frac{1}{J_i} + \frac{1}{J_j}\right)}$$

Then the probability is *approximately* $1 - \alpha$ that

$$\overline{X}_{i\cdot} - \overline{X}_{j\cdot} - w_{ij} \le \mu_i - \mu_j \le \overline{X}_{i\cdot} - \overline{X}_{j\cdot} + w_{ij}$$

for every i and j ($i = 1, \ldots, I$ and $j = 1, \ldots, I$) with $i \ne j$.

The simultaneous confidence level $100(1-\alpha)\%$ is only approximate rather than exact as it is with equal sample sizes. The underscoring method can still be used, but now the w_{ij} factor used to decide whether $\bar{x}_{i.}$ and $\bar{x}_{j.}$ can be connected will depend on J_i and J_j.

Example 11.9 (Example 11.8 continued)

The sample sizes for the elastic modulus data were $J_1 = 8$, $J_2 = 8$, $J_3 = 6$, and $I = 3$, $n - I = 19$, MSE $= .316$. A simultaneous confidence level of approximately 95% requires $Q_{.05,3,19} = 3.59$, from which

$$w_{12} = 3.59\sqrt{\frac{.316}{2}\left(\frac{1}{8} + \frac{1}{8}\right)} = .713 \qquad w_{13} = .771 \qquad w_{23} = .771$$

Since $\bar{x}_{1.} - \bar{x}_{2.} = 44.71 - 44.06 = .65 < w_{12}$, μ_1 and μ_2 are judged not significantly different. The accompanying underscoring scheme shows that μ_1 and μ_3 appear to differ significantly, as do μ_2 and μ_3.

2. Die	1. Permanent	3. Plaster
44.06	44.71	45.58

(44.06 and 44.71 are underscored by a common line.) ■

Data Transformation

The use of ANOVA methods can be invalidated by substantial differences in the variances $\sigma_1^2, \ldots, \sigma_I^2$ (which until now have been assumed equal with common value σ^2). It sometimes happens that $V(X_{ij}) = \sigma_i^2 = g(\mu_i)$, a known function of μ_i (so that when H_0 is false, the variances are not equal). For example, if X_{ij} has a Poisson distribution with parameter λ_i (approximately normal if $\lambda_i \geq 10$), then $\mu_i = \lambda_i$ and $\sigma_i^2 = \lambda_i$, so $g(\mu_i) = \mu_i$ is the known function. In such cases, one can often transform the X_{ij}'s to $h(X_{ij})$'s so that they will have approximately equal variances (while hopefully leaving the transformed variables approximately normal), and then the F test can be used on the transformed observations. The basic idea is that, if $h(\cdot)$ is a smooth function, then we can express it approximately using the first terms of a Taylor series, $h(X_{ij}) \approx h(\mu_i) + h'(\mu_i)(X_{ij} - \mu_i)$. Then $V[h(X_{ij})] \approx V(X_{ij}) \cdot [h'(\mu_i)]^2 = g(\mu_i) \cdot [h'(\mu_i)]^2$. We wish to find the function $h(\cdot)$ for which $g(\mu_i) \cdot [h'(\mu_i)]^2 = c$ (a constant) for every i. Solving this for $h'(\mu_i)$ and integrating gives the following result:

PROPOSITION

If $V(X_{ij}) = g(\mu_i)$, a known function of μ_i, then a transformation $h(X_{ij})$ that "stabilizes the variance" so that $V[h(X_{ij})]$ is approximately the same for each i is given by $h(x) \propto \int [g(x)]^{-1/2}\,dx$.

In the Poisson case, $g(x) = x$, so $h(x)$ should be proportional to $\int x^{-1/2}\,dx = 2x^{1/2}$. Thus Poisson data should be transformed to $h(x_{ij}) = \sqrt{x_{ij}}$ before the analysis.

A Random Effects Model

The single-factor problems considered so far have all been assumed to be examples of a **fixed effects** ANOVA model. By this we mean that the chosen levels of the factor under study are the only ones considered relevant by the experimenter. The single-factor fixed effects model is

$$X_{ij} = \mu + \alpha_i + \varepsilon_{ij} \qquad \sum \alpha_i = 0 \tag{11.6}$$

where the ε_{ij}'s are random and both μ and the α_i's are fixed parameters whose values are unknown.

In some single-factor problems, the particular levels studied by the experimenter are chosen, either by design or through sampling, from a large population of levels. For example, to study the effects on task performance time of using different operators on a particular machine, a sample of five operators might be chosen from a large pool of operators. Similarly, the effect of soil pH on the yield of maize plants might be studied by using soils with four specific pH values chosen from among the many possible pH levels. When the levels used are selected at random from a larger population of possible levels, the factor is said to be random rather than fixed, and the fixed effects model (11.6) is no longer appropriate. An analogous **random effects** model is obtained by replacing the fixed α_i's in (11.6) by random variables. The resulting model description is

---

$$X_{ij} = \mu + A_i + \varepsilon_{ij} \quad \text{with } E(A_i) = E(\varepsilon_{ij}) = 0 \tag{11.7}$$

$$V(\varepsilon_{ij}) = \sigma^2 \qquad V(A_i) = \sigma_A^2$$

with all A_i's and ε_{ij}'s normally distributed and independent of one another.

The condition $E(A_i) = 0$ in (11.7) is similar to the condition $\Sigma\alpha_i = 0$ in (11.6); it states that the expected or average effect of the ith level measured as a departure from μ is zero.

For the random effects model (11.7), the hypothesis of no effects due to different levels is H_0: $\sigma_A^2 = 0$, which says that different levels of the factor contribute nothing to variability of the response. *Although the hypotheses in the single-factor fixed and random effects models are different, they are tested in exactly the same way*, by forming $F = \text{MSTr}/\text{MSE}$ and rejecting H_0 if $f \ge F_{\alpha,I-1,n-I}$. This can be justified intuitively by noting that $E(\text{MSE}) = \sigma^2$ (as for fixed effects), whereas

$$E(\text{MSTr}) = \sigma^2 + \frac{1}{I-1}\left(n - \frac{\sum J_i^2}{n}\right)\sigma_A^2 \tag{11.8}$$

where $J_1, J_2, \ldots, J_I$ are the sample sizes and $n = \Sigma J_i$. The factor in parentheses on the right side of (11.8) is nonnegative, so again $E(\text{MSTr}) = \sigma^2$ if H_0 is true and $E(\text{MSTr}) > \sigma^2$ if H_0 is false.

Example 11.10 The study of nondestructive forces and stresses in materials furnishes important information for efficient design. The article "Zero-Force Travel-Time Parameters for Ultrasonic Head-Waves in Railroad Rail" (*Materials Eval.*, 1985: 854–858) reports on a study of travel time for a certain type of wave that results from longitudinal stress of rails used for railroad track. Three measurements were made on each of six rails randomly selected from a population of rails. The investigators used random effects ANOVA to decide whether some variation in travel time could be attributed to "between-rail variability."

The data is given in the accompanying table (each value, in nanoseconds, resulted from subtracting 36.1 μsec from the original observation) along with the derived ANOVA table. The value of the F ratio is highly significant, so H_0: $\sigma_A^2 = 0$ is rejected in favor of the conclusion that differences between rails are a source of travel-time variability.

Rail	Travel Time			$x_{i.}$
1	55	53	54	162
2	26	37	32	95
3	78	91	85	254
4	92	100	96	288
5	49	51	50	150
6	80	85	83	248
			$x_{..}$	= 1197

Source of Variation	df	Sum of Squares	Mean Square	f
Treatments	5	9310.5	1862.1	115.2
Error	12	194.0	16.17	
Total	17	9504.5		

■

Exercises Section 11.3 (22–34)

22. The following data refers to yield of tomatoes (kg/plot) for four different levels of salinity; salinity level here refers to electrical conductivity (EC), where the chosen levels were EC = 1.6, 3.8, 6.0, and 10.2 nmhos/cm:

1.6	59.5	53.3	56.8	63.1	58.7
3.8	55.2	59.1	52.8	54.5	
6.0	51.7	48.8	53.9	49.0	
10.2	44.6	48.5	41.0	47.3	46.1

Use the F test at level $\alpha = .05$ to test for any differences in true average yield due to the different salinity levels.

23. Apply the modified Tukey's method to the data in Exercise 22 to identify significant differences among the μ_i's.

24. The following partial ANOVA table is taken from the article "Perception of Spatial Incongruity" (*J. Nervous Mental Disease*, 1961: 222) in which the abilities of three different groups to identify a perceptual incongruity were assessed and compared. All individuals in the experiment had been hospitalized to undergo psychiatric treatment. There were 21 individuals in the depressive group, 32 individuals in the functional "other" group, and 21 individuals in the brain-damaged group. Complete the ANOVA table and carry out the F test at level $\alpha = .01$.

Source	df	Sum of Squares	Mean Square	f
Groups			76.09	
Error				
Total		1123.14		

25. Lipids provide much of the dietary energy in the bodies of infants and young children. There is a growing interest in the quality of the dietary lipid supply during infancy as a major determinant of growth, visual and neural development, and long-term health. The article "Essential Fat Requirements of Preterm Infants" (*Amer. J. Clin. Nutrit.*, 2000: 245S–250S) reported the following data on total polyunsaturated fats (%) for infants who were randomized to four different feeding regimens: breast milk, corn-oil-based formula, soy-oil-based formula, or soy-and-marine-oil-based formula:

Regimen	Sample Size	Sample Mean	Sample SD
Breast milk	8	43.0	1.5
CO	13	42.4	1.3
SO	17	43.1	1.2
SMO	14	43.5	1.2

a. What assumptions must be made about the four total polyunsaturated fat distributions before carrying out a single-factor ANOVA to decide whether there are any differences in true average fat content?

b. Carry out the test suggested in part (a). What can be said about the *P*-value?

26. Samples of six different brands of diet/imitation margarine were analyzed to determine the level of physiologically active polyunsaturated fatty acids (PAPFUA, in percentages), resulting in the following data:

Imperial	14.1	13.6	14.4	14.3	
Parkay	12.8	12.5	13.4	13.0	12.3
Blue Bonnet	13.5	13.4	14.1	14.3	
Chiffon	13.2	12.7	12.6	13.9	
Mazola	16.8	17.2	16.4	17.3	18.0
Fleischmann's	18.1	17.2	18.7	18.4	

(The preceding numbers are fictitious, but the sample means agree with data reported in the January 1975 issue of *Consumer Reports*.)

a. Use ANOVA to test for differences among the true average PAPFUA percentages for the different brands.

b. Compute CIs for all $(\mu_i - \mu_j)$'s.

c. Mazola and Fleischmann's are corn-based, whereas the others are soybean-based. Compute a CI for

$$\frac{(\mu_1 + \mu_2 + \mu_3 + \mu_4)}{4} - \frac{(\mu_5 + \mu_6)}{2}$$

[*Hint*: Modify the expression for $V(\hat{\theta})$ that led to (11.5) in the previous section.]

27. Although tea is the world's most widely consumed beverage after water, little is known about its nutritional value. Folacin is the only B vitamin present in any significant amount in tea, and recent advances in assay methods have made accurate determination of folacin content feasible. Consider the accompanying data on folacin content for randomly selected specimens of the four leading brands of green tea.

Brand	**Observations**						
1	7.9	6.2	6.6	8.6	8.9	10.1	9.6
2	5.7	7.5	9.8	6.1	8.4		
3	6.8	7.5	5.0	7.4	5.3	6.1	
4	6.4	7.1	7.9	4.5	5.0	4.0	

(Data is based on "Folacin Content of Tea," *J. Amer. Dietetic Assoc.*, 1983: 627–632.) Does this data suggest that true average folacin content is the same for all brands?

a. Carry out a test using $\alpha = .05$ via the *P*-value method.

b. Assess the plausibility of any assumptions required for your analysis in part (a).

c. Perform a multiple comparisons analysis to identify significant differences among brands.

28. In single-factor ANOVA with sample sizes J_i ($i = 1, \ldots, I$), show that $\text{SSTr} = \Sigma J_i(\overline{X}_{i\cdot} - \overline{X}_{\cdot\cdot})^2 = \Sigma_i J_i \overline{X}_{i\cdot}^2 - n\overline{X}_{\cdot\cdot}^2$, where $n = \Sigma J_i$.

29. When sample sizes are equal ($J_i = J$), the parameters $\alpha_1, \alpha_2, \ldots, \alpha_I$ of the alternative parameterization are restricted by $\Sigma\alpha_i = 0$. For unequal sample sizes, the most natural restriction is $\Sigma J_i\alpha_i = 0$. Use this to show that

$$E(\text{MSTr}) = \sigma^2 + \frac{1}{I-1}\sum J_i\alpha_i^2$$

What is $E(\text{MSTr})$ when H_0 is true? [This expectation is correct if $\Sigma J_i\alpha_i = 0$ is replaced by the restriction $\Sigma\alpha_i = 0$ (or any other single linear restriction on the α_i's used to reduce the model to I independent parameters), but $\Sigma J_i\alpha_i = 0$ simplifies the algebra and yields natural estimates for the model parameters (in particular, $\hat{\alpha}_i = \bar{x}_{i\cdot} - \bar{x}_{\cdot\cdot}$).]

30. Reconsider Example 11.7 involving an investigation of the effects of different heat treatments on the yield point of steel ingots.

a. If $J = 8$ and $\sigma = 1$, what is β for a level .05 F test when $\mu_1 = \mu_2$, $\mu_3 = \mu_1 - 1$, and $\mu_4 = \mu_1 + 1$?

b. For the alternative of part (a), what value of J is necessary to obtain $\beta = .05$?

c. If there are $I = 5$ heat treatments, $J = 10$, and $\sigma = 1$, what is β for the level .05 F test when four of the μ_i's are equal and the fifth differs by 1 from the other four?

31. When sample sizes are not equal, the noncentrality parameter is $\Sigma J_i\alpha_i^2/\sigma^2$ and $\phi^2 = (1/I)\Sigma J_i\alpha_i^2/\sigma^2$. Referring to Exercise 22, what is the power of the test when $\mu_2 = \mu_3$, $\mu_1 = \mu_2 - \sigma$, and $\mu_4 = \mu_2 + \sigma$?

32. In an experiment to compare the quality of four different brands of reel-to-reel recording tape, five 2400-ft reels of each brand (A–D) were selected and the number of flaws in each reel was determined.

A:	10	5	12	14	8
B:	14	12	17	9	8
C:	13	18	10	15	18
D:	17	16	12	22	14

It is believed that the number of flaws has approximately a Poisson distribution for each brand. Analyze the data at level .01 to see whether the expected number of flaws per reel is the same for each brand.

33. Suppose that X_{ij} is a binomial variable with parameters n and p_i (so it is approximately normal when $np_i \geq 10$ and $nq_i \geq 10$). Then since $\mu_i = np_i$, $V(X_{ij}) = \sigma_i^2 = np_i(1 - p_i) = \mu_i(1 - \mu_i/n)$. How should the X_{ij}'s be transformed so as to stabilize the variance? [*Hint*: $g(\mu_i) = \mu_i(1 - \mu_i/n)$.]

34. Simplify $E(\text{MSTr})$ for the random effects model when $J_1 = J_2 = \cdots = J_I = J$.

11.4 *Two-Factor ANOVA with $K_{ij} = 1$

In many experimental situations there are two factors of simultaneous interest. For example, suppose an investigator wishes to study permeability of woven material used to construct automobile air bags (related to the ability to absorb energy). An experiment might be carried out using $I = 4$ temperature levels (10°C, 15°C, 20°C, 25°C) and $J = 3$ levels of fabric denier (420-D, 630-D, 840-D).

When factor A consists of I levels and factor B consists of J levels, there are IJ different combinations (pairs) of levels of the two factors, each called a treatment. With K_{ij} = the number of observations on the treatment consisting of factor A at level i and factor B at level j, we focus in this section on the case $K_{ij} = 1$, so that the data consists of IJ observations. We will first discuss the fixed effects model, in which the only levels of interest for the two factors are those actually represented in the experiment. The case in which one or both factors are random is discussed briefly at the end of the section.

Example 11.11 Is it really as easy to remove marks on fabrics from erasable pens as the word *erasable* might imply? Consider the following data from an experiment to compare three different brands of pens and four different wash treatments with respect to their ability to remove marks on a particular type of fabric (based on "An Assessment of the Effects of Treatment, Time, and Heat on the Removal of Erasable Pen Marks from Cotton and Cotton/Polyester Blend Fabrics," *J. Testing Eval.*, 1991: 394–397). The response variable is a quantitative indicator of overall specimen color change; the lower this value, the more marks were removed.

		Washing Treatment				
		1	**2**	**3**	**4**	**Total**
Brand of Pen	**1**	.97	.48	.48	.46	2.39
	2	.77	.14	.22	.25	1.38
	3	.67	.39	.57	.19	1.82
	Total	2.41	1.01	1.27	.90	5.59

Is there any difference in the true average amount of color change due either to the different brands of pen or to the different washing treatments? ■

As in single-factor ANOVA, double subscripts are used to identify random variables and observed values. Let

X_{ij} = the random variable (rv) denoting the measurement when factor A is held at level i and factor B is held at level j

x_{ij} = the observed value of X_{ij}

The x_{ij}'s are usually presented in a two-way table in which the ith row contains the observed values when factor A is held at level i and the jth column contains the observed values when factor B is held at level j. In the erasable-pen experiment of Example 11.11, the number of levels of factor A is $I = 3$, the number of levels of factor B is $J = 4$, $x_{13} = .48$, $x_{22} = .14$, and so on.

Whereas in single-factor ANOVA we were interested only in row means and the grand mean, here we are interested also in column means. Let

$$\overline{X}_{i\cdot} = \begin{array}{c}\text{the average of measurements obtained}\\ \text{when factor } A \text{ is held at level } i\end{array} = \frac{\sum_{j=1}^{J} X_{ij}}{J}$$

$$\overline{X}_{\cdot j} = \begin{array}{c}\text{the average of measurements obtained}\\ \text{when factor } B \text{ is held at level } j\end{array} = \frac{\sum_{i=1}^{I} X_{ij}}{I}$$

$$\overline{X}_{\cdot\cdot} = \text{the grand mean} = \frac{\sum_{i=1}^{I}\sum_{j=1}^{J} X_{ij}}{IJ}$$

with observed values $\overline{x}_{i\cdot}$, $\overline{x}_{\cdot j}$, and $\overline{x}_{\cdot\cdot}$. Totals rather than averages are denoted by omitting the horizontal bar (so $x_{\cdot j} = \Sigma_i x_{ij}$, etc.). Intuitively, to see whether there is any effect due to the levels of factor A, we should compare the observed $\overline{x}_{i\cdot}$'s with one another, and information about the different levels of factor B should come from the $\overline{x}_{\cdot j}$'s.

The Model

Proceeding by analogy to single-factor ANOVA, one's first inclination in specifying a model is to let μ_{ij} = the true average response when factor A is at level i and factor B at level j, giving IJ mean parameters. Then let

$$X_{ij} = \mu_{ij} + \varepsilon_{ij}$$

where ε_{ij} is the random amount by which the observed value differs from its expectation and the ε_{ij}'s are assumed normal and independent with common variance σ^2. Unfortunately, there is no valid test procedure for this choice of parameters. The reason is that under the alternative hypothesis of interest, the μ_{ij}'s are free to take on any values whatsoever, whereas σ^2 can be any value greater than zero, so that there are $IJ + 1$ freely varying parameters. But there are only IJ observations, so after using each x_{ij} as an estimate of μ_{ij}, there is no way to estimate σ^2.

To rectify this problem of a model having more parameters than observed values, we must specify a model that is realistic yet involves relatively few parameters.

Assume the existence of I parameters $\alpha_1, \alpha_2, \ldots, \alpha_I$ and J parameters $\beta_1, \beta_2, \ldots, \beta_J$ such that

$$X_{ij} = \alpha_i + \beta_j + \varepsilon_{ij} \qquad (i = 1, \ldots, I, \quad j = 1, \ldots, J) \tag{11.9}$$

so that

$$\mu_{ij} = \alpha_i + \beta_j \tag{11.10}$$

Including σ^2, there are now $I + J + 1$ model parameters, so if $I \geq 3$ and $J \geq 3$, there will be fewer parameters than observations [in fact, we will shortly modify (11.10) so that even $I = 2$ and/or $J = 2$ will be accommodated].

The model specified in (11.9) and (11.10) is called an **additive model** because each mean response μ_{ij} is the sum of an effect due to factor A at level i (α_i) and an effect due to factor B at level j (β_j). The difference between mean responses for factor A at level i and level i' when B is held at level j is $\mu_{ij} - \mu_{i'j}$. When the model is additive,

$$\mu_{ij} - \mu_{i'j} = (\alpha_i + \beta_j) - (\alpha_{i'} + \beta_j) = \alpha_i - \alpha_{i'}$$

which is independent of the level j of the second factor. A similar result holds for $\mu_{ij} - \mu_{ij'}$. Thus additivity means that the difference in mean responses for two levels of one of the factors is the same for all levels of the other factor. Figure 11.7(a) shows a set of mean responses that satisfy the condition of additivity (which implies parallel lines), and Figure 11.7(b) shows a nonadditive configuration of mean responses.

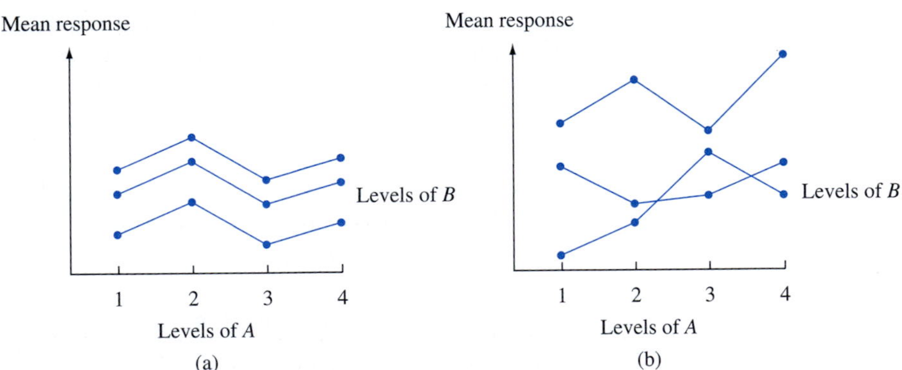

Figure 11.7 Mean responses for two types of model: (a) additive; (b) nonadditive

Example 11.12 (Example 11.11 continued)

When we plot the observed x_{ij}'s in a manner analogous to that of Figure 11.7, we get the result shown in Figure 11.8. Although there is some "crossing over" in the observed x_{ij}'s, the configuration is reasonably representative of what would be expected under additivity with just one observation per treatment.

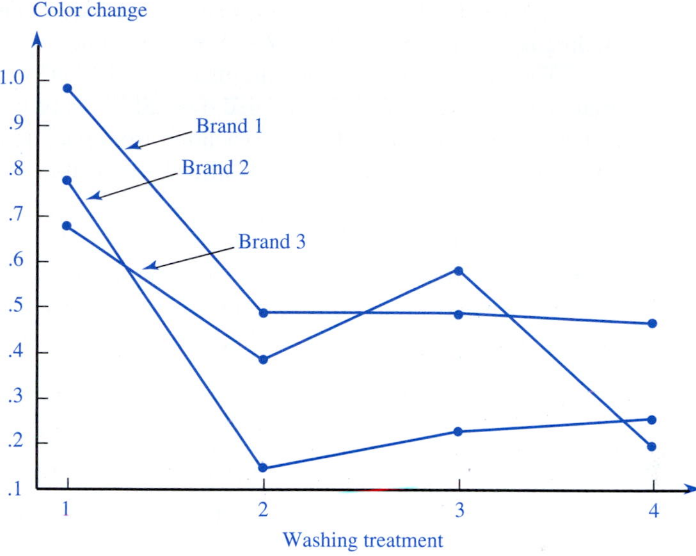

Figure 11.8 Plot of data from Example 11.11 ■

Expression (11.10) is not quite the final model description because the α_i's and β_j's are not uniquely determined. Following are two different configurations of the α_i's and β_j's that yield the same additive μ_{ij}'s.

	$\beta_1 = 1$	$\beta_2 = 4$
$\alpha_1 = 1$	$\mu_{11} = 2$	$\mu_{12} = 5$
$\alpha_2 = 2$	$\mu_{21} = 3$	$\mu_{22} = 6$

	$\beta_1 = 2$	$\beta_2 = 5$
$\alpha_1 = 0$	$\mu_{11} = 2$	$\mu_{12} = 5$
$\alpha_2 = 1$	$\mu_{21} = 3$	$\mu_{22} = 6$

By subtracting any constant c from all α_i's and adding c to all β_j's, other configurations corresponding to the same additive model are obtained. This nonuniqueness is eliminated by use of the following model.

$$X_{ij} = \mu + \alpha_i + \beta_j + \varepsilon_{ij} \tag{11.11}$$

where $\Sigma_{i=1}^{I}\alpha_i = 0$, $\Sigma_{j=1}^{J}\beta_j = 0$, and the ε_{ij}'s are assumed independent, normally distributed, with mean 0 and common variance σ^2.

This is analogous to the alternative choice of parameters for single-factor ANOVA discussed in Section 11.3. It is not difficult to verify that (11.11) is an additive model in which the parameters are uniquely determined (for example, for the μ_{ij}'s mentioned previously, $\mu = 4$, $\alpha_1 = -.5$, $\alpha_2 = .5$, $\beta_1 = -1.5$, and $\beta_2 = 1.5$). Notice that there are only

$I - 1$ independently determined α_i's and $J - 1$ independently determined β_j's, so (including μ) (11.11) specifies $I + J - 1$ mean parameters.

The interpretation of the parameters of (11.11) is straightforward: μ is the true grand mean (mean response averaged over all levels of both factors), α_i is the effect of factor A at level i (measured as a deviation from μ), and β_j is the effect of factor B at level j. Unbiased (and maximum likelihood) estimators for these parameters are

$$\hat{\mu} = \overline{X}_{..} \qquad \hat{\alpha}_i = \overline{X}_{i.} - \overline{X}_{..} \qquad \hat{\beta}_j = \overline{X}_{.j} - \overline{X}_{..}$$

There are two different hypotheses of interest in a two-factor experiment with $K_{ij} = 1$. The first, denoted by H_{0A}, states that the different levels of factor A have no effect on true average response. The second, denoted by H_{0B}, asserts that there is no factor B effect.

$$\begin{aligned} H_{0A}&: \alpha_1 = \alpha_2 = \cdots = \alpha_I = 0 \\ &\qquad \text{versus } H_{aA}\text{: at least one } \alpha_i \neq 0 \\ H_{0B}&: \beta_1 = \beta_2 = \cdots = \beta_J = 0 \\ &\qquad \text{versus } H_{aB}\text{: at least one } \beta_j \neq 0 \end{aligned} \tag{11.12}$$

(No factor A effect implies that all α_i's are equal, so they must all be 0 since they sum to 0, and similarly for the β_j's.)

Test Procedures

The description and analysis now follow closely that for single-factor ANOVA. The relevant sums of squares and their computing forms are given by

$$\begin{aligned} \text{SST} &= \sum_{i=1}^{I}\sum_{j=1}^{J}(X_{ij} - \overline{X}_{..})^2 = \sum_{i=1}^{I}\sum_{j=1}^{J}X_{ij}^2 - \frac{1}{IJ}X_{..}^2 && \text{df} = IJ - 1 \\ \text{SSA} &= \sum_{i=1}^{I}\sum_{j=1}^{J}(\overline{X}_{i.} - \overline{X}_{..})^2 = \frac{1}{J}\sum_{i=1}^{I}X_{i.}^2 - \frac{1}{IJ}X_{..}^2 && \text{df} = I - 1 \\ \text{SSB} &= \sum_{i=1}^{I}\sum_{j=1}^{J}(\overline{X}_{.j} - \overline{X}_{..})^2 = \frac{1}{I}\sum_{j=1}^{J}X_{.j}^2 - \frac{1}{IJ}X_{..}^2 && \text{df} = J - 1 \\ \text{SSE} &= \sum_{i=1}^{I}\sum_{j=1}^{J}(X_{ij} - \overline{X}_{i.} - \overline{X}_{.j} + \overline{X}_{..})^2 && \text{df} = (I-1)(J-1) \end{aligned} \tag{11.13}$$

and the fundamental identity

$$\text{SST} = \text{SSA} + \text{SSB} + \text{SSE} \tag{11.14}$$

allows SSE to be determined by subtraction.

The expression for SSE results from replacing μ, α_i, and β_j in $\Sigma[X_{ij} - (\mu + \alpha_i + \beta_j)]^2$ by their respective estimators. Error df is IJ − number of mean parameters estimated =

$IJ - [1 + (I - 1) + (J - 1)] = (I - 1)(J - 1)$. As in single-factor ANOVA, total variation is split into a part (SSE) that is not explained by either the truth or the falsity of H_{0A} or H_{0B} and two parts that can be explained by possible falsity of the two null hypotheses.

Forming F ratios as in single-factor ANOVA, we can show as in Section 11.1 that if H_{0A} is true, the corresponding F ratio has an F distribution with numerator df $= I - 1$ and denominator df $= (I - 1)(J - 1)$; an analogous result applies when testing H_{0B}.

Hypotheses	Test Statistic Value	Rejection Region
H_{0A} versus H_{aA}	$f_A = \dfrac{\text{MSA}}{\text{MSE}}$	$f_A \geq F_{\alpha,I-1,(I-1)(J-1)}$
H_{0B} versus H_{aB}	$f_B = \dfrac{\text{MSB}}{\text{MSE}}$	$f_B \geq F_{\alpha,J-1,(I-1)(J-1)}$

Example 11.13 (Example 11.12 continued)

The $x_{i.}$'s (row totals) and $x_{.j}$'s (column totals) for the color change data are displayed along the right and bottom margins of the data table in Example 11.11. In addition, $\Sigma\Sigma x_{ij}^2 = 3.2987$ and the correction factor is $x_{..}^2/(IJ) = (5.59)^2/12 = 2.6040$. The sums of squares are then

$$\text{SST} = 3.2987 - 2.6040 = .6947$$

$$\text{SSA} = \frac{1}{4}[(2.39)^2 + (1.38)^2 + (1.82)^2] - 2.6040 = .1282$$

$$\text{SSB} = \frac{1}{3}[(2.41)^2 + (1.01)^2 + (1.27)^2 + (.90)^2] - 2.6040 = .4797$$

$$\text{SSE} = .6947 - (.1282 + .4797) = .0868$$

The accompanying ANOVA table (Table 11.5) summarizes further calculations.

Table 11.5 ANOVA table for Example 11.13

Source of Variation	df	Sum of Squares	Mean Square	f
Factor A (pen brand)	$I - 1 = 2$	SSA = .1282	MSA = .0641	$f_A = 4.43$
Factor B (wash treatment)	$J - 1 = 3$	SSB = .4797	MSB = .1599	$f_B = 11.05$
Error	$(I - 1)(J - 1) = 6$	SSE = .0868	MSE = .01447	
Total	$IJ - 1 = 11$	SST = .6947		

The critical value for testing H_{0A} at level of significance .05 is $F_{.05,2,6} = 5.14$. Since $4.43 < 5.14$, H_{0A} cannot be rejected at significance level .05. Based on this (small) data set, we cannot conclude that true average color change depends on brand of pen. Because $F_{.05,3,6} = 4.76$ and $11.05 \geq 4.76$, H_{0B} is rejected at significance level .05 in favor of the assertion that color change varies with washing treatment. A statistical computer package gives P-values of .066 and .007 for these two tests.

How can plausibility of the normality and constant variance assumptions be investigated graphically? Define the predicted values (also called fitted values)

$\hat{x}_{ij} = \hat{\mu} + \hat{\alpha}_i + \hat{\beta}_j = \bar{x}_{..} + (\bar{x}_{i.} - \bar{x}_{..}) + (\bar{x}_{.j} - \bar{x}_{..}) = \bar{x}_{i.} + \bar{x}_{.j} - \bar{x}_{..}$, and the residuals (the differences between the observations and predicted values) $x_{ij} - \hat{x}_{ij} = x_{ij} - \bar{x}_{i.} - \bar{x}_{.j} + \bar{x}_{..}$. We can check the normality assumption with a normal plot of the residuals, Figure 11.9(a), and we can check the constant variance assumption with a plot of the residuals against the fitted values, Figure 11.9(b).

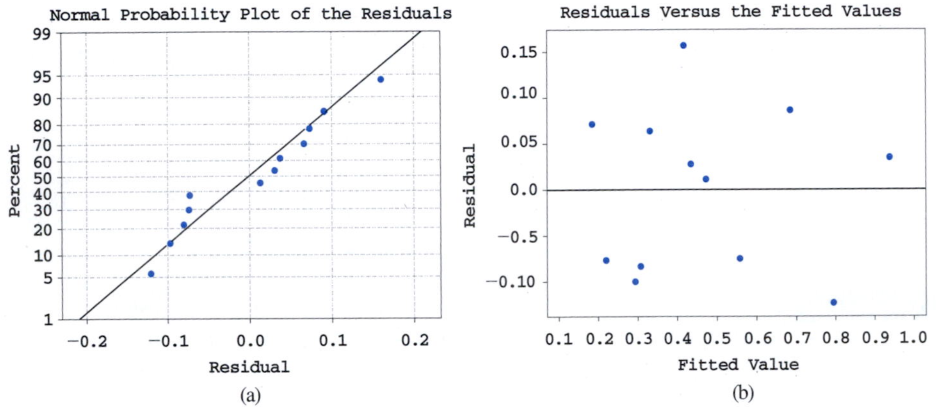

Figure 11.9 Plots from MINITAB for Example 11.13

The normal plot is reasonably straight, so there is no reason to question normality for this data set. On the plot of the residuals against the fitted values, we are looking for differences in vertical spread as we move horizontally across the graph. For example, if there were a narrow range for small fitted values and a wide range for high fitted values, this would suggest that the variance is higher for larger responses (this happens often, and it can sometimes be cured by replacing each observation by its logarithm). No such problem occurs here, so there is no evidence against the constant variance assumption. ■

Expected Mean Squares

The plausibility of using the F tests just described is demonstrated by determining the expected mean squares. After some tedious algebra,

$$E(\text{MSE}) = \sigma^2 \quad \text{(when the model is additive)}$$

$$E(\text{MSA}) = \sigma^2 + \frac{J}{I-1}\sum_{i=1}^{I}\alpha_i^2$$

$$E(\text{MSB}) = \sigma^2 + \frac{I}{J-1}\sum_{j=1}^{J}\beta_j^2$$

When H_{0A} is true, MSA is an unbiased estimator of σ^2, so F is a ratio of two unbiased estimators of σ^2. When H_{0A} is false, MSA tends to overestimate σ^2, so H_{0A} should be rejected when the ratio F_A is too large. Similar comments apply to MSB and H_{0B}.

Multiple Comparisons

When either H_{0A} or H_{0B} has been rejected, Tukey's procedure can be used to identify significant differences between the levels of the factor under investigation. The steps in the analysis are identical to those for a single-factor ANOVA:

1. For comparing levels of factor A, obtain $Q_{\alpha,I,(I-1)(J-1)}$.
   For comparing levels of factor B, obtain $Q_{\alpha,J,(I-1)(J-1)}$.
2. Compute

$$\begin{aligned} w &= Q \cdot (\text{estimated standard deviation of the sample means being compared}) \\ &= \begin{cases} Q_{\alpha,I,(I-1)(J-1)} \cdot \sqrt{\text{MSE}/J} & \text{for factor } A \text{ comparisons} \\ Q_{\alpha,J(I-1)(J-1)} \cdot \sqrt{\text{MSE}/I} & \text{for factor } B \text{ comparisons} \end{cases} \end{aligned}$$

   (because, e.g., the standard deviation of $\overline{X}_{i\cdot}$ is $\sigma/\sqrt{J}$).
3. Arrange the sample means in increasing order, underscore those pairs differing by less than w, and identify pairs not underscored by the same line as corresponding to significantly different levels of the given factor.

Example 11.14 (Example 11.13 continued)

Identification of significant differences among the four washing treatments requires $Q_{.05,4,6} = 4.90$ and $w = 4.90\sqrt{(.01447)/3} = .340$. The four factor B sample means (column averages) are now listed in increasing order, and any pair differing by less than .340 is underscored by a line segment:

$$\begin{array}{cccc} \bar{x}_{4\cdot} & \bar{x}_{2\cdot} & \bar{x}_{3\cdot} & \bar{x}_{1\cdot} \\ .300 & .337 & .423 & .803 \\ \cline{1-3} \end{array}$$

Washing treatment 1 appears to differ significantly from the other three treatments, but no other significant differences are identified. In particular, it is not apparent which among treatments 2, 3, and 4 is best at removing marks. ■

Randomized Block Experiments

In using single-factor ANOVA to test for the presence of effects due to the I different treatments under study, once the IJ subjects or experimental units have been chosen, treatments should be allocated in a completely random fashion. That is, J subjects should be chosen at random for the first treatment, then another sample of J chosen at random from the remaining $IJ - J$ subjects for the second treatment, and so on.

It frequently happens, though, that subjects or experimental units exhibit differences with respect to other characteristics that may affect the observed responses. For example, some patients might be healthier than others. When this is the case, the presence or absence of a significant F value may be due to these differences rather than to the presence or absence of factor effects. This was the reason for introducing a paired experiment in Chapter 10. The generalization of the paired experiment to $I > 2$ is called a **randomized block** experiment. An extraneous factor, "blocks," is constructed by dividing the IJ units into J groups with I units in each group. This grouping or blocking is done in such a way that within each block, the I units are homogeneous with respect to other factors thought to affect the responses. Then within each homogeneous block, the I treatments are randomly assigned to the I units or subjects in the block.

Example 11.15 A consumer product–testing organization wished to compare the annual power consumption for five different brands of dehumidifier. Because power consumption depends on the prevailing humidity level, it was decided to monitor each brand at four different levels ranging from moderate to heavy humidity (thus blocking on humidity level). Within each level, brands were randomly assigned to the five selected locations. The resulting amount of power consumption (annual kWh) appears in Table 11.6.

Table 11.6 Power consumption data for Example 11.15

Treatments (brands)	Blocks (humidity level) 1	2	3	4	$x_{i\cdot}$	$\bar{x}_{i\cdot}$
1	685	792	838	875	3190	797.50
2	722	806	893	953	3374	843.50
3	733	802	880	941	3356	839.00
4	811	888	952	1005	3656	914.00
5	828	920	978	1023	3749	937.25
$x_{\cdot j}$	3779	4208	4541	4797	17,325	

Since $\Sigma\Sigma x_{ij}^2 = 15{,}178{,}901.00$ and $x_{\cdot\cdot}^2/(IJ) = 15{,}007{,}781.25$

$$\text{SST} = 15{,}178{,}901.00 - 15{,}007{,}781.25 = 171{,}119.75$$

$$\text{SSA} = \frac{1}{4}[60{,}244{,}049] - 15{,}007{,}781.25 = 53{,}231.00$$

$$\text{SSB} = \frac{1}{5}[75{,}619{,}995] - 15{,}007{,}781.25 = 116{,}217.75$$

and

$$\text{SSE} = 171{,}119.75 - 53{,}231.00 - 116{,}217.75 = 1671.00$$

The ANOVA calculations are summarized in Table 11.7.

Table 11.7 ANOVA table for Example 11.15

Source of Variation	df	Sum of Squares	Mean Square	f
Treatments (brands)	4	53,231.00	13,307.75	$f_A = 95.57$
Blocks	3	116,217.75	38,739.25	$f_B = 278.20$
Error	12	1671.00	139.25	
Total	19	171,119.75		

Since $F_{.05,4,12} = 3.26$ and $f_A = 95.57 \geq 3.26$, H_0 is rejected in favor of H_a, and we conclude that power consumption does depend on the brand of humidifier. To identify significantly different brands, we use Tukey's procedure. $Q_{.05,5,12} = 4.51$ and $w = 4.51\sqrt{139.25/4} = 26.6$.

$\bar{x}_{1\cdot}$	$\bar{x}_{3\cdot}$	$\bar{x}_{2\cdot}$	$\bar{x}_{4\cdot}$	$\bar{x}_{5\cdot}$
797.50	839.00	843.50	914.00	937.25

The underscoring indicates that the brands can be divided into three groups with respect to power consumption.

Because the block factor is of secondary interest, $F_{.05,3,12}$ is not needed, though the computed value of F_B is clearly highly significant. Figure 11.10 shows SAS output for this data. Notice that in the first part of the ANOVA table, the sums of squares (SS's) for treatments (brands) and blocks (humidity levels) are combined into a single "model" SS.

```
                        Analysis of Variance Procedure
Dependent Variable: POWERUSE

                                   Sum of             Mean
Source                  DF        Squares           Square        F Value        Pr > F

Model                    7     169448.750        24206.964         173.84        0.0001
Error                   12       1671.000          139.250
Corrected Total         19     171119.750

              R-Square          C.V.           Root MSE       POWERUSE Mean

              0.990235      1.362242            11.8004           866.25000

Source              DF          Anova SS        Mean Square        F Value        PR > F

BRAND                4         53231.000          13307.750          95.57        0.0001
HUMIDITY             3        116217.750          38739.250         278.20        0.0001

              Alpha = 0.05 df = 12 MSE = 139.25
         Critical Value of Studentized Range = 4.508
           Minimum Significant Difference = 26.597

Means with the same letter are not significantly different.
     Tukey Grouping          Mean          N          BRAND
                  A       937.250          4              5
                  A
                  A       914.000          4              4
                  B       843.500          4              2
                  B
                  B       839.000          4              3
                  C       797.500          4              1
```

Figure 11.10 SAS output for power consumption data ■

In many experimental situations in which treatments are to be applied to subjects, a single subject can receive all I of the treatments. Blocking is then often done on the subjects themselves to control for variability between subjects; each subject is then said to act as its own control. Social scientists sometimes refer to such experiments as repeated-measures designs. The "units" within a block are then the different "instances" of treatment application. Similarly, blocks are often taken as different time periods, locations, or observers.

In most randomized block experiments in which subjects serve as blocks, the subjects actually participating in the experiment are selected from a large population. The subjects then contribute random rather than fixed effects. This does not affect the procedure for comparing treatments when $K_{ij} = 1$ (one observation per "cell," as in this section), but the procedure is altered if $K_{ij} = K > 1$. We will shortly consider two-factor models in which effects are random.

More on Blocking When $I = 2$, either the F test or the paired differences t test can be used to analyze the data. The resulting conclusion will not depend on which procedure is used, since $T^2 = F$ and $t^2_{\alpha/2,v} = F_{\alpha,1,v}$.

Just as with pairing, blocking entails both a potential gain and a potential loss in precision. If there is a great deal of heterogeneity in experimental units, the value of the variance parameter σ^2 in the one-way model will be large. The effect of blocking is to filter out the variation represented by σ^2 in the two-way model appropriate for a randomized block experiment. Other things being equal, a smaller value of σ^2 results in a test that is more likely to detect departures from H_0 (i.e., a test with greater power).

However, other things are not equal here, since the single-factor F test is based on $I(J - 1)$ degrees of freedom (df) for error, whereas the two-factor F test is based on $(I - 1)(J - 1)$ df for error. Fewer degrees of freedom for error results in a decrease in power, essentially because the denominator estimator of σ^2 is not as precise. This loss in degrees of freedom can be especially serious if the experimenter can afford only a small number of observations. Nevertheless, if it appears that blocking will significantly reduce variability, it is probably worth the loss in degrees of freedom.

Models for Random Effects

In many experiments, the actual levels of a factor used in the experiment, rather than being the only ones of interest to the experimenter, have been selected from a much larger population of possible levels of the factor. In a two-factor situation, when this is the case for both factors, a **random effects model** is appropriate. The case in which the levels of one factor are the only ones of interest and the levels of the other factor are selected from a population of levels leads to a **mixed effects model**. The two-factor random effects model when $K_{ij} = 1$ is

---

$$X_{ij} = \mu + A_i + B_j + \varepsilon_{ij} \qquad (i = 1, \ldots, I, \quad j = 1, \ldots, J)$$

where the A_i's, B_j's, and ε_{ij}'s are all independent, normally distributed rv's with mean 0 and variances σ_A^2, σ_B^2, and σ^2, respectively.

The hypotheses of interest are then H_{0A}: $\sigma_A^2 = 0$ (level of factor A does not contribute to variation in the response) versus H_{aA}: $\sigma_A^2 > 0$ and H_{0B}: $\sigma_B^2 = 0$ versus H_{aB}: $\sigma_B^2 > 0$. Whereas $E(\text{MSE}) = \sigma^2$ as before, the expected mean squares for factors A and B are now

$$E(\text{MSA}) = \sigma^2 + J\sigma_A^2 \qquad E(\text{MSB}) = \sigma^2 + I\sigma_B^2$$

Thus when H_{0A} (H_{0B}) is true, $F_A (F_B)$ is still a ratio of two unbiased estimators of σ^2. It can be shown that a level α test for H_{0A} versus H_{aA} still rejects H_{0A} if $f_A \geq F_{\alpha,I-1,(I-1)(J-1)}$, and, similarly, the same procedure as before is used to decide between H_{0B} and H_{aB}.

For the case in which factor A is fixed and factor B is random, the mixed model is

$$X_{ij} = \mu + \alpha_i + B_j + \varepsilon_{ij} \qquad (i = 1, \ldots, I, \quad j = 1, \ldots, J)$$

where $\Sigma\alpha_i = 0$ and the B_j's and ε_{ij}'s are normally distributed with mean 0 and variances σ_B^2 and σ^2, respectively.

Now the two null hypotheses are

$$H_{0A}\text{: } \alpha_1 = \cdots = \alpha_I = 0 \quad \text{and} \quad H_{0B}\text{: } \sigma_B^2 = 0$$

with expected mean squares

$$E(\text{MSE}) = \sigma^2 \qquad E(\text{MSA}) = \sigma^2 + \frac{J}{I-1}\Sigma\alpha_i^2 \qquad E(\text{MSB}) = \sigma^2 + I\sigma_B^2$$

The test procedures for H_{0A} versus H_{aA} and H_{0B} versus H_{aB} are exactly as before. For example, in the analysis of the color change data in Example 11.11, if the four wash treatments were randomly selected, then because $f_B = 11.05$ and $F_{.05,3,6} = 4.76$, H_{0B}: $\sigma_B^2 = 0$ is rejected in favor of H_{aB}: $\sigma_B^2 > 0$. An estimate of the "variance component" σ_B^2 is then given by $(\text{MSB} - \text{MSE})/I = .0485$.

Summarizing, when $K_{ij} = 1$, although the hypotheses and expected mean squares differ from the case of both effects fixed, the test procedures are identical.

Exercises Section 11.4 (35–48)

35. The number of miles of useful tread wear (in 1000's) was determined for tires of each of five different makes of subcompact car (factor A, with $I = 5$) in combination with each of four different brands of radial tires (factor B, with $J = 4$), resulting in $IJ = 20$ observations. The values SSA = 30.6, SSB = 44.1, and SSE = 59.2 were then computed. Assume that an additive model is appropriate.

a. Test H_0: $\alpha_1 = \alpha_2 = \alpha_3 = \alpha_4 = \alpha_5 = 0$ (no differences in true average tire lifetime due to makes of cars) versus H_a: at least one $\alpha_i \neq 0$ using a level .05 test.

b. H_0: $\beta_1 = \beta_2 = \beta_3 = \beta_4 = 0$ (no differences in true average tire lifetime due to brands of tires) versus H_a: at least one $\beta_j \neq 0$ using a level .05 test.

36. Four different coatings are being considered for corrosion protection of metal pipe. The pipe will be buried in three different types of soil. To investigate whether the amount of corrosion depends either on the coating or on the type of soil, 12 pieces of pipe are selected. Each piece is coated with one of the four coatings and buried in one of the three types of soil for a fixed time, after which the amount of corrosion (depth of maximum pits, in .0001 in.) is determined. The depths are shown in this table:

		Soil Type (B)		
		1	**2**	**3**
Coating (A)	**1**	64	49	50
	2	53	51	48
	3	47	45	50
	4	51	43	52

a. Assuming the validity of the additive model, carry out the ANOVA analysis using an ANOVA table to see whether the amount of corrosion depends on either the type of coating used or the type of soil. Use $\alpha = .05$.

b. Compute $\hat{\mu}$, $\hat{\alpha}_1$, $\hat{\alpha}_2$, $\hat{\alpha}_3$, $\hat{\alpha}_4$, $\hat{\beta}_1$, $\hat{\beta}_2$, and $\hat{\beta}_3$.

	Subject					
Stimulus	1	2	3	4	$x_{i.}$	$\bar{x}_{i.}$
L1	8.0	17.3	52.0	22.0	99.3	24.8
L2	6.9	19.3	63.7	21.6	111.5	27.9
Tone (T)	9.3	18.8	60.0	28.3	116.4	29.1
L1 + L2	9.2	24.9	82.4	44.9	161.4	40.3
L1 + T	12.0	31.7	83.8	37.4	164.9	41.2
L2 + T	9.4	33.6	96.6	40.6	180.2	45.1
$x_{.j}$	54.8	145.6	438.5	194.8	833.7	

37. The data set shown above is from the article "Compounding of Discriminative Stimuli from the Same and Different Sensory Modalities" (*J. Experiment. Anal. Behav.*, 1971: 337–342). Rat response was maintained by fixed interval schedules of reinforcement in the presence of a tone or two separate lights. The lights were of either moderate (L1) or low (L2) intensity. Observations are given as the mean number of responses emitted by each subject during single and compound stimuli presentations over a 4-day period. Carry out an appropriate analysis.

38. In an experiment to see whether the amount of coverage of light-blue interior latex paint depends either on the brand of paint or on the brand of roller used, 1 gallon of each of four brands of paint was applied using each of three brands of roller, resulting in the following data (number of square feet covered).

		Roller Brand		
		1	2	3
Paint Brand	1	454	446	451
	2	446	444	447
	3	439	442	444
	4	444	437	443

a. Construct the ANOVA table. [*Hint*: The computations can be expedited by subtracting 400 (or any other convenient number) from each observation. This does not affect the final results.]

b. State and test hypotheses appropriate for deciding whether paint brand has any effect on coverage. Use $\alpha = .05$.

c. Repeat part (b) for brand of roller.

d. Use Tukey's method to identify significant differences among brands. Is there one brand that seems clearly preferable to the others?

e. Check the normality and constant variance assumptions graphically.

39. In an experiment to assess the effect of the angle of pull on the force required to cause separation in electrical connectors, four different angles (factor A) were used and each of a sample of five connectors (factor B) was pulled once at each angle ("A Mixed Model Factorial Experiment in Testing Electrical Connectors," *Indust. Qual. Control*, 1960: 12–16). The data appears in the accompanying table.

		B				
		1	2	3	4	5
A	0°	45.3	42.2	39.6	36.8	45.8
	2°	44.1	44.1	38.4	38.0	47.2
	4°	42.7	42.7	42.6	42.2	48.9
	6°	43.5	45.8	47.9	37.9	56.4

Does the data suggest that true average separation force is affected by the angle of pull? State and test the appropriate hypotheses at level .01 by first constructing an ANOVA table (SST = 396.13, SSA = 58.16, and SSB = 246.97).

40. A particular county employs three assessors who are responsible for determining the value of residential property in the county. To see whether these assessors differ systematically in their assessments, 5 houses are selected, and each assessor is asked to determine the market value of each house. With factor A denoting assessors ($I = 3$) and factor B denoting houses ($J = 5$), suppose SSA = 11.7, SSB = 113.5, and SSE = 25.6.

a. Test H_0: $\alpha_1 = \alpha_2 = \alpha_3 = 0$ at level .05. (H_0 states that there are no systematic differences among assessors.)

b. Explain why a randomized block experiment with only 5 houses was used rather than a one-way ANOVA experiment involving a total of 15 different houses with each assessor asked to assess 5 different houses (a different group of 5 for each assessor).

41. The article "Rate of Stuttering Adaptation Under Two Electro-Shock Conditions" (*Behav. Res. Therapy*, 1967: 49–54) gives adaptation scores for three different treatments: (1) no shock, (2) shock following each stuttered word, and (3) shock during each moment of stuttering. These treatments were used on each of 18 stutterers.

a. Summary statistics include $x_{1.} = 905, x_{2.} = 913, x_{3.} = 936, x_{..} = 2754, \Sigma_j x_{.j}^2 = 430{,}295$, and $\Sigma\Sigma x_{ij}^2 = 143{,}930$. Construct the ANOVA table and test at level .05 to see whether true average adaptation score depends on the treatment given.

b. Judging from the F ratio for subjects (factor B), do you think that blocking on subjects was effective in this experiment? Explain.

42. The article "The Effects of a Pneumatic Stool and a One-Legged Stool on Lower Limb Joint Load and Muscular Activity During Sitting and Rising" (*Ergonomics*, 1993: 519–535) gives the accompanying data on the effort required of a subject to arise from four different types of stools (Borg scale). Perform an analysis of variance using $\alpha = .05$, and follow this with a multiple comparisons analysis if appropriate.

		Subject									
		1	2	3	4	5	6	7	8	9	$\bar{x}_{i.}$
Type of Stool	1	12	10	7	7	8	9	8	7	9	8.56
	2	15	14	14	11	11	11	12	11	13	12.44
	3	12	13	13	10	8	11	12	8	10	10.78
	4	10	12	9	9	7	10	11	7	8	9.22

43. The strength of concrete used in commercial construction tends to vary from one batch to another. Consequently, small test cylinders of concrete sampled from a batch are "cured" for periods up to about 28 days in temperature- and moisture-controlled environments before strength measurements are made. Concrete is then "bought and sold on the basis of strength test cylinders" (ASTM C 31 Standard Test Method for Making and Curing Concrete Test Specimens in the Field). The accompanying data resulted from an experiment carried out to compare three different curing methods with respect to compressive strength (MPa). Analyze this data.

Batch	Method A	Method B	Method C
1	30.7	33.7	30.5
2	29.1	30.6	32.6
3	30.0	32.2	30.5
4	31.9	34.6	33.5
5	30.5	33.0	32.4
6	26.9	29.3	27.8
7	28.2	28.4	30.7
8	32.4	32.4	33.6
9	26.6	29.5	29.2
10	28.6	29.4	33.2

44. Check the normality and constant variance assumptions graphically for the data of Example 11.15.

45. Suppose that in the experiment described in Exercise 40 the five houses had actually been selected at random from among those of a certain age and size, so that factor B is random rather than fixed. Test H_0: $\sigma_B^2 = 0$ versus H_a: $\sigma_B^2 > 0$ using a level .01 test.

46. a. Show that a constant d can be added to (or subtracted from) each x_{ij} without affecting any of the ANOVA sums of squares.

b. Suppose that each x_{ij} is multiplied by a nonzero constant c. How does this affect the ANOVA sums of squares? How does this affect the values of the F statistics F_A and F_B? What effect does "coding" the data by $y_{ij} = cx_{ij} + d$ have on the conclusions resulting from the ANOVA procedures?

47. Use the fact that $E(X_{ij}) = \mu + \alpha_i + \beta_j$ with $\Sigma\alpha_i = \Sigma\beta_j = 0$ to show that $E(\bar{X}_{i.} - \bar{X}_{..}) = \alpha_i$, so that $\hat{\alpha}_i = \bar{X}_{i.} - \bar{X}_{..}$ is an unbiased estimator for α_i.

48. The power curves of Figures 11.5 and 11.6 can be used to obtain $\beta = P(\text{type II error})$ for the F test in two-factor ANOVA. For fixed values of $\alpha_1, \alpha_2, \ldots, \alpha_I$, the quantity $\phi^2 = (J/I)\Sigma\alpha_i^2/\sigma^2$ is computed. Then the figure corresponding to $\nu_1 = I - 1$ is entered on the horizontal axis at the value ϕ, the power is read on the vertical axis from the curve labeled $\nu_2 = (I-1)(J-1)$, and $\beta = 1 - \text{power}$.

a. For the corrosion experiment described in Exercise 36, find β when $\alpha_1 = 4, \alpha_2 = 0, \alpha_3 = \alpha_4 = -2$, and $\sigma = 4$. Repeat for $\alpha_1 = 6, \alpha_2 = 0, \alpha_3 = \alpha_4 = -3$, and $\sigma = 4$.

b. By symmetry, what is β for the test of H_{0B} versus H_{aB} in Example 11.11 when $\beta_1 = .3, \beta_2 = \beta_3 = \beta_4 = -.1$, and $\sigma = .3$?

11.5 *Two-Factor ANOVA with $K_{ij} > 1$

In Section 11.4, we analyzed data from a two-factor experiment in which there was one observation for each of the IJ combinations of levels of the two factors. To obtain valid test procedures, the μ_{ij}'s were assumed to have an additive structure with $\mu_{ij} = \mu + \alpha_i + \beta_j$, $\Sigma\alpha_i = \Sigma\beta_j = 0$. Additivity means that the difference in true average responses for any two levels of the factors is the same for each level of the other factor. For example, $\mu_{ij} - \mu_{i'j} = (\mu + \alpha_i + \beta_j) - (\mu + \alpha_{i'} + \beta_j) = \alpha_i - \alpha_{i'}$ independent of the level j of the second factor. This is shown in Figure 11.7(a), in which the lines connecting true average responses are parallel.

Figure 11.7(b) depicts a set of true average responses that does not have additive structure. The lines connecting these μ_{ij}'s are not parallel, which means that the difference in true average responses for different levels of one factor does depend on the level of the other factor. When additivity does not hold, we say that there is **interaction** between the different levels of the factors. The assumption of additivity allowed us in Section 11.4 to obtain an estimator of the random error variance σ^2 (MSE) that was unbiased whether or not either null hypothesis of interest was true. When $K_{ij} > 1$ for at least one (i, j) pair, a valid estimator of σ^2 can be obtained without assuming additivity. In specifying the appropriate model and deriving test procedures, we will focus on the case $K_{ij} = K > 1$, so the number of observations per "cell" (for each combination of levels) is constant.

Parameters for the Fixed Effects Model with Interaction

Rather than use the μ_{ij}'s themselves as model parameters, it is usual to use an equivalent set that reveals more clearly the role of interaction. Let

$$\mu = \frac{1}{IJ}\sum_i \sum_j \mu_{ij} \qquad \overline{\mu}_{i.} = \frac{1}{J}\sum_j \mu_{ij.} \qquad \overline{\mu}_{.j} = \frac{1}{I}\sum_i \mu_{ij} \tag{11.15}$$

Thus μ is the expected response averaged over all levels of both factors (the true grand mean), $\overline{\mu}_{i.}$ is the expected response averaged over levels of the second factor when the first factor A is held at level i, and similarly for $\overline{\mu}_{.j}$. Now define

$$\begin{aligned} \alpha_i &= \overline{\mu}_{i.} - \mu = \text{the effect of factor } A \text{ at level } i \\ \beta_j &= \overline{\mu}_{.j} - \mu = \text{the effect of factor } B \text{ at level } j \\ \gamma_{ij} &= \mu_{ij} - (\mu + \alpha_i + \beta_j) \end{aligned} \tag{11.16}$$

from which

$$\mu_{ij} = \mu + \alpha_i + \beta_j + \gamma_{ij} \tag{11.17}$$

The model is additive if and only if all γ_{ij}'s $= 0$. The γ_{ij}'s are referred to as the **interaction parameters**. The α_i's are called the **main effects for factor A**, and the β_j's are the

main effects for factor *B*. Although there are I α_i's, J β_j's, and IJ γ_{ij}'s in addition to μ, the conditions $\Sigma\alpha_i = 0$, $\Sigma\beta_j = 0$, $\Sigma_j\gamma_{ij} = 0$ for any i, and $\Sigma_i\gamma_{ij} = 0$ for any j [all by virtue of (11.15) and (11.16)] imply that only IJ of these new parameters are independently determined: μ, $I - 1$ of the α_i's, $J - 1$ of the β_j's, and $(I - 1)(J - 1)$ of the γ_{ij}'s.

There are now three sets of hypotheses that will be considered:

$$H_{0AB}\colon\ \gamma_{ij} = 0 \text{ for all } i, j \quad \text{versus} \quad H_{aAB}\colon \text{ at least one } \gamma_{ij} \neq 0$$

$$H_{0A}\colon\ \alpha_1 = \cdots = \alpha_I = 0 \quad \text{versus} \quad H_{aA}\colon \text{ at least one } \alpha_i \neq 0$$

$$H_{0B}\colon\ \beta_1 = \cdots = \beta_J = 0 \quad \text{versus} \quad H_{aB}\colon \text{ at least one } \beta_j \neq 0$$

The no-interaction hypothesis H_{0AB} is usually tested first. If H_{0AB} is not rejected, then the other two hypotheses can be tested to see whether the main effects are significant. But once H_{0AB} is rejected, we believe that the effect of factor A at any particular level depends on the level of B (and vice versa). It then does not make sense to test H_{0A} or H_{0B}. In this context a picture similar to that of Figure 11.7(b) is helpful in visualizing the way the factors interact. Here the cell means are used instead of x_{ij}; this type of graph is sometimes called an **interaction plot**.

In case of interaction, it may be appropriate to do a one-way ANOVA to compare levels of A separately for each level of B. For example, suppose factor A involves four kinds of glue, factor B involves three types of material, the response is strength of the glue joint, and the strength rankings of the glues clearly depend on which material is being glued. In this situation with interaction, it makes sense to do three separate one-way ANOVA analyses, one for each material.

Notation, Model, and Analysis

We now use triple subscripts for both random variables and observed values, with X_{ijk} and x_{ijk} referring to the kth observation (replication) when factor A is at level i and factor B is at level j. The model is then

$$X_{ijk} = \mu + \alpha_i + \beta_j + \gamma_{ij} + \varepsilon_{ijk} \qquad (11.18)$$

$$i = 1, \ldots, I, \quad j = 1, \ldots, J, \quad k = 1, \ldots, K$$

where the ε_{ij}'s are independent and normally distributed, each with mean 0 and variance σ^2.

Again a dot in place of a subscript means that we have summed over all values of that subscript, whereas a horizontal bar denotes averaging. Thus $X_{ij.}$ is the total of all K observations made for factor A at level i and factor B at level j [all observations in the (i, j)th cell], and $\overline{X}_{ij.}$ is the average of these K observations.

Example 11.16 Three different varieties of tomato (Harvester, Ife No. 1, and Pusa Early Dwarf) and four different plant densities (10, 20, 30, and 40 thousand plants per hectare) are being considered for planting in a particular region. To see whether either variety or plant density affects yield, each combination of variety and plant density is used in three different plots, resulting in the data on yields in Table 11.8 (based on the article "Effects of Plant Density on Tomato Yields in Western Nigeria," *Experiment. Agric.*, 1976: 43–47).

Table 11.8 Yield data for Example 11.16

	Planting Density													
Variety	**10,000**			**20,000**			**30,000**			**40,000**			$x_{i..}$	$\bar{x}_{i..}$
H	10.5	9.2	7.9	12.8	11.2	13.3	12.1	12.6	14.0	10.8	9.1	12.5	136.0	11.33
Ife	8.1	8.6	10.1	12.7	13.7	11.5	14.4	15.4	13.7	11.3	12.5	14.5	146.5	12.21
P	16.1	15.3	17.5	16.6	19.2	18.5	20.8	18.0	21.0	18.4	18.9	17.2	217.5	18.13
$x_{.j.}$	103.3			129.5			142.0			125.2			500.00	
$\bar{x}_{.j.}$	11.48			14.39			15.78			13.91				13.89

Here, $I = 3$, $J = 4$, and $K = 3$, for a total of $IJK = 36$ observations. ■

To test the hypotheses of interest, we again define sums of squares and present computing formulas:

$$\text{SST} = \sum_i \sum_j \sum_k (X_{ijk} - \bar{X}_{...})^2 = \sum_i \sum_j \sum_k X_{ijk}^2 - \frac{1}{IJK} X_{...}^2 \qquad \text{df} = IJK - 1$$

$$\text{SSE} = \sum_i \sum_j \sum_k (X_{ijk} - \bar{X}_{ij.})^2$$

$$= \sum_i \sum_j \sum_k X_{ijk}^2 - \frac{1}{K} \sum_i \sum_j X_{ij.}^2 \qquad \text{df} = IJ(K - 1)$$

$$\text{SSA} = \sum_i \sum_j \sum_k (\bar{X}_{i..} - \bar{X}_{...})^2 = \frac{1}{JK} \sum_i X_{i..}^2 - \frac{1}{IJK} X_{...}^2 \qquad \text{df} = I - 1$$

$$\text{SSB} = \sum_i \sum_j \sum_k (\bar{X}_{.j.} - \bar{X}_{...})^2 = \frac{1}{IK} \sum_j X_{.j.}^2 - \frac{1}{IJK} X_{...}^2 \qquad \text{df} = J - 1$$

$$\text{SSAB} = \sum_i \sum_j \sum_k (\bar{X}_{ij.} - \bar{X}_{i..} - \bar{X}_{.j.} + \bar{X}_{...})^2 \qquad \text{df} = (I - 1)(J - 1)$$

The fundamental identity

$$\text{SST} = \text{SSA} + \text{SSB} + \text{SSAB} + \text{SSE}$$

implies that **interaction sum of squares** SSAB can be obtained by subtraction.

The computing formulas are all obtained by expanding the squared expressions and summing. The fundamental identity is obtained by squaring and summing an expression similar to Equation (11.2).

Total variation is thus partitioned into four pieces: unexplained (SSE—which would be present whether or not any of the three null hypotheses was true) and three pieces that may be explained by the truth or falsity of the three H_0's. Each of four mean squares is defined by MS = SS/df. The expected mean squares suggest that each set of hypotheses should be tested using the appropriate ratio of mean squares with MSE in the denominator:

$$E(\text{MSE}) = \sigma^2$$

$$E(\text{MSA}) = \sigma^2 + \frac{JK}{I-1}\sum_{i=1}^{I}\alpha_i^2$$

$$E(\text{MSB}) = \sigma^2 + \frac{IK}{J-1}\sum_{j=1}^{J}\beta_j^2$$

$$E(\text{MSAB}) = \sigma^2 + \frac{K}{(I-1)(J-1)}\sum_{i=1}^{I}\sum_{j=1}^{J}\gamma_{ij}^2$$

Each of the three mean square ratios can be shown to have an F distribution when the associated H_0 is true, which yields the following level α test procedures.

Hypotheses	**Test Statistic Value**	**Rejection Region**
H_{0A} versus H_{aA}	$f_A = \dfrac{\text{MSA}}{\text{MSE}}$	$f_A \geq F_{\alpha,I-1,IJ(K-1)}$
H_{0B} versus H_{aB}	$f_B = \dfrac{\text{MSB}}{\text{MSE}}$	$f_B \geq F_{\alpha,J-1,IJ(K-1)}$
H_{0AB} versus H_{aAB}	$f_{AB} = \dfrac{\text{MSAB}}{\text{MSE}}$	$f_{AB} \geq F_{\alpha,(I-1)(J-1),IJ(K-1)}$

As before, the results of the analysis are summarized in an ANOVA table.

Example 11.17 (Example 11.16 continued)

From the given data, $x_{...}^2 = (500)^2 = 250{,}000$

$$\sum_i\sum_j\sum_k x_{ijk}^2 = (10.5)^2 + (9.2)^2 + \cdots + (18.9)^2 + (17.2)^2 = 7404.80$$

$$\sum_i x_{i..}^2 = (136.0)^2 + (146.5)^2 + (217.5)^2 = 87{,}264.50$$

and

$$\sum_j x_{.j.}^2 = 63{,}280.18$$

The cell totals ($x_{ij.}$'s) are

	10,000	20,000	30,000	40,000
H	27.6	37.3	38.7	32.4
Ife	26.8	37.9	43.5	38.3
P	48.9	54.3	59.8	54.5

from which $\Sigma_i \Sigma_j x_{ij.}^2 = (27.6)^2 + \cdots + (54.5)^2 = 22{,}100.28$. Then

$$\text{SST} = 7404.80 - \frac{1}{36}(250{,}000) = 7404.80 - 6944.44 = 460.36$$

$$\text{SSA} = \frac{1}{12}(87{,}264.50) - 6944.44 = 327.60$$

$$\text{SSB} = \frac{1}{9}(63{,}280.18) - 6944.44 = 86.69$$

$$\text{SSE} = 7404.80 - \frac{1}{3}(22{,}100.28) = 38.04$$

and

$$\text{SSAB} = 460.36 - 327.60 - 86.69 - 38.04 = 8.03$$

Table 11.9 summarizes the computations.

Table 11.9 ANOVA table for Example 11.17

Source of Variation	df	Sum of Squares	Mean Square	f
Varieties	2	327.60	163.8	$f_A = 103.02$
Density	3	86.69	28.9	$f_B = 18.18$
Interaction	6	8.03	1.34	$f_{AB} = .84$
Error	24	38.04	1.59	
Total	35	460.36		

Since $F_{.01,6,24} = 3.67$ and $f_{AB} = .84$ is not ≥ 3.67, H_{0AB} cannot be rejected at level .01, so we conclude that the interaction effects are not significant. Now the presence or absence of main effects can be investigated. Since $F_{.01,2,24} = 5.61$ and $f_A = 103.02 \geq 5.61$, H_{0A} is rejected at level .01 in favor of the conclusion that different varieties do affect the true average yields. Similarly, $f_B = 18.18 \geq 4.72 = F_{.01,3,24}$, so we conclude that true average yield also depends on plant density.

Figure 11.11 shows the interaction plot. Notice the nearly parallel lines for the three tomato varieties, in agreement with the F test showing no significant interaction. The yield for Pusa Early Dwarf appears to be significantly above the yields for the other two varieties, and this is in accord with the highly significant F for varieties. Furthermore, all three varieties show the same pattern in which yield increases as the density goes up, but decreases beyond 30,000 per hectare. This suggests that planting more seed will increase the yield, but eventually overcrowding causes the yield to drop.

In this example one of the two factors is quantitative, and this is naturally the factor used for the horizontal axis in the interaction plot. In case both of the factors are quantitative, the choice for the horizontal axis would be arbitrary, but a case can be made for two plots to try it both ways. Indeed, MINITAB has an option to allow both plots to be included in the same graph.

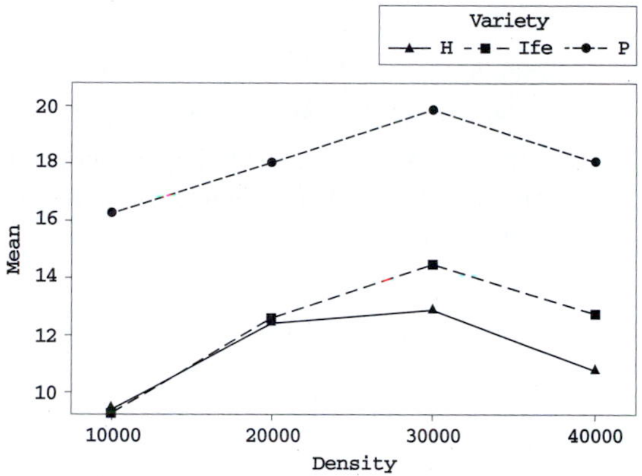

Figure 11.11 Interaction plot from MINITAB for the tomato yield data

To verify plausibility of the normality and constant variance assumptions we can construct plots similar to those of Section 11.4. Define the predicted values (fitted values) to be the cell means, $\hat{x}_{ijk} = \bar{x}_{ij.}$, so the residuals, the differences between the observations and predicted values, are $x_{ijk} - \bar{x}_{ij.}$. The normal plot of the residuals is Figure 11.12(a), and the plot of the residuals against the fitted values is Figure 11.12(b). The normal plot is sufficiently straight that there should be no concern about the normality assumption. The plot of residuals against predicted values has a fairly uniform vertical spread, so there is no cause for concern about the constant variance assumption.

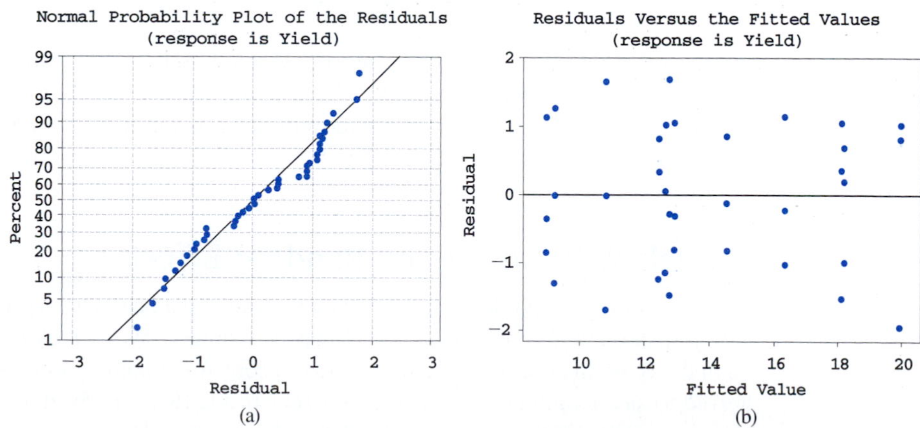

Figure 11.12 Plots from MINITAB to verify assumptions for Example 11.17 ■

Multiple Comparisons

When the no-interaction hypothesis H_{0AB} is not rejected and at least one of the two main-effect null hypotheses is rejected, Tukey's method can be used to identify significant differences in levels. To identify differences among the α_i's when H_{0A} is rejected:

1. Obtain $Q_{\alpha,I,IJ(K-1)}$, where the second subscript I identifies the number of levels being compared and the third subscript refers to the number of degrees of freedom for error.
2. Compute $w = Q\sqrt{\text{MSE}/(JK)}$, where JK is the number of observations averaged to obtain each of the $\bar{x}_{i..}$'s compared in step 3.
3. Order the $\bar{x}_{i..}$'s from smallest to largest and, as before, underscore all pairs that differ by less than w. Pairs not underscored correspond to significantly different levels of factor A.

To identify different levels of factor B when H_{0B} is rejected, replace the second subscript in Q by J, replace JK by IK in w, and replace $\bar{x}_{i..}$ by $\bar{x}_{.j.}$.

Example 11.18 (Example 11.17 continued)

For factor A (varieties), $I = 3$, so with $\alpha = .01$ and $IJ(K-1) = 24$, $Q_{.01,3,24} = 4.55$. Then $w = 4.55\sqrt{1.59/12} = 1.66$, so ordering and underscoring gives

$\bar{x}_{1..}$	$\bar{x}_{2..}$	$\bar{x}_{3..}$
11.33	12.21	18.13

The Harvester and Ife varieties do not appear to differ significantly from one another in effect on true average yield, but both differ from the Pusa variety.

For factor B (density), $J = 4$ so $Q_{.01,4,24} = 4.91$ and $w = 4.91\sqrt{1.59/9} = 2.06$.

$\bar{x}_{.1.}$	$\bar{x}_{.4.}$	$\bar{x}_{.2.}$	$\bar{x}_{.3.}$
11.48	13.91	14.39	15.78

Thus with experimentwise error rate .01, which is quite conservative, only the lowest density appears to differ significantly from all others. Even with $\alpha = .05$ (so that $w = 1.64$), densities 2 and 3 cannot be judged significantly different from one another in their effect on yield. ■

Models with Mixed and Random Effects

In some situations, the levels of either factor may have been chosen from a large population of possible levels, so that the effects contributed by the factor are random rather than fixed. As in Section 11.4, if both factors contribute random effects, the model is referred to as a random effects model, whereas if one factor is fixed and the other is random, a mixed effects model results. We will now consider the analysis for a mixed effects model in which factor A (rows) is the fixed factor and factor B (columns) is the random factor. When either factor is random, interaction effects will also be random.

The case in which both factors are random is dealt with in Exercise 57. The mixed effects model is

$$X_{ijk} = \mu + \alpha_i + B_j + G_{ij} + \varepsilon_{ijk}$$
$$i = 1, \ldots, I, \quad j = 1, \ldots J, \quad k = 1, \ldots, K$$

Here μ and α_i's are constants with $\sum \alpha_i = 0$, and the B_j's, G_{ij}'s, and ε_{ijk}'s are independent, normally distributed random variables with expected value 0 and variances σ_B^2, σ_G^2, and σ^2, respectively.*

H_{0A}: $\alpha_1 = \alpha_2 = \cdots = \alpha_I = 0$ versus H_{aA}: at least one $\alpha_i \neq 0$

H_{0B}: $\sigma_B^2 = 0$ versus H_{aB}: $\sigma_B^2 > 0$

H_{0G}: $\sigma_G^2 = 0$ versus H_{aG}: $\sigma_G^2 > 0$

It is customary to test H_{0A} and H_{0B} only if the no-interaction hypothesis H_{0G} cannot be rejected.

The relevant sums of squares and mean squares needed for the test procedures are defined and computed exactly as in the fixed effects case. The expected mean squares for the mixed model are

$$E(\text{MSE}) = \sigma^2$$
$$E(\text{MSA}) = \sigma^2 + K\sigma_G^2 + \frac{JK}{I-1}\sum \alpha_i^2$$
$$E(\text{MSB}) = \sigma^2 + K\sigma_G^2 + IK\sigma_B^2$$

and

$$E(\text{MSAB}) = \sigma^2 + K\sigma_G^2$$

Thus, to test the no-interaction hypothesis, the ratio $f_{AB} = \text{MSAB}/\text{MSE}$ is again appropriate, with H_{0G} rejected if $f_{AB} \geq F_{\alpha,(I-1)(J-1),IJ(K-1)}$. However, for testing H_{0A} versus H_{aA}, the expected mean squares suggest that although the numerator of the F ratio should still be MSA, the denominator should be MSAB rather than MSE. MSAB is also the denominator of the F ratio for testing H_{0B}.

For testing H_{0A} versus H_{aA} (factors A fixed, B random), the test statistic value is $f_A = \text{MSA}/\text{MSAB}$, and the rejection region is $f_A \geq F_{\alpha,I-1,(I-1)(J-1)}$. The test of H_{0B} versus H_{aB} utilizes $f_B = \text{MSB}/\text{MSAB}$, and the rejection region is $f_B \geq F_{\alpha,J-1,(I-1)(J-1)}$.

*This is referred to as an "unrestricted" model. An alternative "restricted" model requires that $\sum_i G_{ij} = 0$ for each j (so the G_{ij}'s are no longer independent). Expected mean squares and F ratios appropriate for testing certain hypotheses depend on the choice of model. MINITAB's default option gives output for the unrestricted model.

Example 11.19 A process engineer has identified two potential causes of electric motor vibration, the material used for the motor casing (factor A) and the supply source of bearings used in the motor (factor B). The accompanying data on the amount of vibration (microns) resulted from an experiment in which motors with casings made of steel, aluminum, and plastic were constructed using bearings supplied by five randomly selected sources.

	Supply Source				
Material	**1**	**2**	**3**	**4**	**5**
Steel	13.1 13.2	16.3 15.8	13.7 14.3	15.7 15.8	13.5 12.5
Aluminum	15.0 14.8	15.7 16.4	13.9 14.3	13.7 14.2	13.4 13.8
Plastic	14.0 14.3	17.2 16.7	12.4 12.3	14.4 13.9	13.2 13.1

Only the three casing materials used in the experiment are under consideration for use in production, so factor A is fixed. However, the five supply sources were randomly selected from a much larger population, so factor B is random. The relevant null hypotheses are

$$H_{0A}\colon \alpha_1 = \alpha_2 = \alpha_3 = 0 \qquad H_{0B}\colon \sigma_B^2 = 0 \qquad H_{0AB}\colon \sigma_G^2 = 0$$

MINITAB output appears in Figure 11.13.

```
Factor          Type      Levels      Values
casmater       fixed           3           1      2      3
source        random           5           1      2      3      4      5

Source                 DF         SS          MS          F          P
casmater                2     0.7047      0.3523       0.24      0.790
source                  4    36.6747      9.1687       6.32      0.013
casmater*source         8    11.6053      1.4507      13.03      0.000
Error                  15     1.6700      0.1113
Total                  29    50.6547

Source                 Variance   Error   Expected Mean Square for Each Term
                       component  term    (using unrestricted model)
1 casmater                          3     (4) +2(3) +Q[1]
2 source                 1.2863     3     (4) +2(3) +6(2)
3 casmater*source        0.6697     4     (4) +2(3)
4 Error                  0.1113           (4)
```

Figure 11.13 Output from MINITAB's balanced ANOVA option for the data of Example 11.19

The printed 0.000 P-value for interaction means that it is less than .0005 (the actual value is .000018). To interpret the significant interaction we use the interaction plot, Figure 11.14, which has both versions, one with source on the x-axis and one with material on the x-axis. Interaction is evident, because the best material (the one with the least vibration) depends strongly on source. For source 1 the best material is steel, for source 3 the best material is plastic, and for source 4 the best material is aluminum. Because of this interaction, we ordinarily would not interpret the main effects, but one cannot help noticing that there is strong dependence of vibration on source. Source 2 is bad for all three materials and source 3 is pretty good for all three materials. When one-way ANOVA analyses are done to compare the five sources for each of the three materials, all

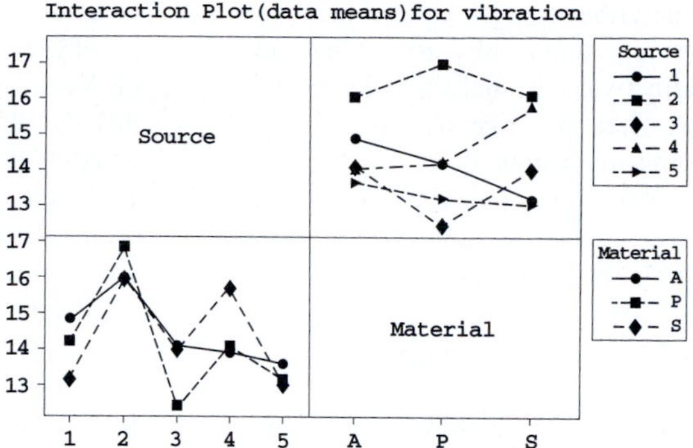

Figure 11.14 MINITAB interaction plot for the data of Example 11.19

three show highly significant differences. This is consistent with the P-value of 0.013 for source in Figure 11.13. We can conclude that, although the interaction causes the best material to depend on the source, the source also makes a difference of its own. ■

When at least two of the K_{ij}'s are unequal, the ANOVA computations are much more complex than for the case $K_{ij} = K$, and there are no nice formulas for the appropriate test statistics. One of the chapter references can be consulted for more information.

Exercises Section 11.5 (49–57)

49. In an experiment to assess the effects of curing time (factor A) and type of mix (factor B) on the compressive strength of hardened cement cubes, three different curing times were used in combination with four different mixes, with three observations obtained for each of the 12 curing time–mix combinations. The resulting sums of squares were computed to be SSA = 30,763.0, SSB = 34,185.6, SSE = 97,436.8, and SST = 205,966.6.

a. Construct an ANOVA table.

b. Test at level .05 the null hypothesis H_{0AB}: all γ_{ij}'s = 0 (no interaction of factors) against H_{0AB}: at least one $\gamma_{ij} \neq 0$.

c. Test at level .05 the null hypothesis H_{0A}: $\alpha_1 = \alpha_2 = \alpha_3 = 0$ (factor A main effects are absent) against H_{aA}: at least one $\alpha_i \neq 0$.

d. Test H_{0B}: $\beta_1 = \beta_2 = \beta_3 = \beta_4 = 0$ versus H_{aB}: at least one $\beta_j \neq 0$ using a level .05 test.

e. The values of the $\bar{x}_{i..}$'s were $\bar{x}_{1..} = 4010.88$, $\bar{x}_{2..} = 4029.10$, and $\bar{x}_{3..} = 3960.02$. Use Tukey's procedure to investigate significant differences among the three curing times.

50. The article "Towards Improving the Properties of Plaster Moulds and Castings" (*J. Engrg. Manuf.*,

Sand Addition (%)	Carbon Fiber Addition (%)	Casting Hardness	Wet-Mold Strength
0	0	61.0	34.0
0	0	63.0	16.0
15	0	67.0	36.0
15	0	69.0	19.0
30	0	65.0	28.0
30	0	74.0	17.0
0	.25	69.0	49.0
0	.25	69.0	48.0
15	.25	69.0	43.0
15	.25	74.0	29.0
30	.25	74.0	31.0
30	.25	72.0	24.0
0	.50	67.0	55.0
0	.50	69.0	60.0
15	.50	69.0	45.0
15	.50	74.0	43.0
30	.50	74.0	22.0
30	.50	74.0	48.0

1991: 265–269) describes several ANOVAs carried out to study how the amount of carbon fiber and sand additions affect various characteristics of the molding process. Here we give data on casting hardness and on wet-mold strength.

a. An ANOVA for wet-mold strength gives SSSand = 705, SSFiber = 1278, SSE = 843, and SST = 3105. Test for the presence of any effects using $\alpha = .05$.

b. Carry out an ANOVA on the casting hardness observations using $\alpha = .05$.

c. Make an interaction plot with sand percentage on the horizontal axis, and discuss the results of part (b) in terms of what the plot shows.

51. The accompanying data resulted from an experiment to investigate whether yield from a certain chemical process depended either on the formulation of a particular input or on mixer speed.

		Speed 60	Speed 70	Speed 80
Formulation	1	189.7	185.1	189.0
		188.6	179.4	193.0
		190.1	177.3	191.1
	2	165.1	161.7	163.3
		165.9	159.8	166.6
		167.6	161.6	170.3

A statistical computer package gave SS(Form) = 2253.44, SS(Speed) = 230.81, SS(Form*Speed) = 18.58, and SSE = 71.87.

a. Does there appear to be interaction between the factors?

b. Does yield appear to depend on either formulation or speed?

c. Calculate estimates of the main effects.

d. Verify that the residuals are 0.23, −0.87, 0.63, 4.50, −1.20, −3.30, −2.03, 1.97, 0.07, −1.10, −0.30, 1.40, 0.67, −1.23, 0.57, −3.43, −0.13, 3.57.

e. Construct a normal plot from the residuals given in part (d). Do the ϵ_{ijk}'s appear to be normally distributed?

f. Plot the residuals against the predicted values (cell means) to see if the population variance appears reasonably constant.

52. In an experiment to investigate the effect of "cement factor" (number of sacks of cement per cubic yard) on flexural strength of the resulting concrete ("Studies of Flexural Strength of Concrete. Part 3: Effects of Variation in Testing Procedure," *Proceedings ASTM*, 1957: 1127–1139), $I = 3$ different factor values were used, $J = 5$ different batches of cement were selected, and $K = 2$ beams were cast from each cement factor/batch combination. Summary values include $\Sigma\Sigma\Sigma x_{ijk}^2 = 12{,}280{,}103$, $\Sigma\Sigma x_{ij.}^2 = 24{,}529{,}699$, $\Sigma x_{i..}^2 = 122{,}380{,}901$, $\Sigma x_{.j.}^2 = 73{,}427{,}483$, and $x_{...} = 19{,}143$.

a. Construct the ANOVA table.

b. Assuming a mixed model with cement factor (A) fixed and batches (B) random, test the three pairs of hypotheses of interest at level .05.

53. A study was carried out to compare the writing lifetimes of four premium brands of pens. It was thought that the writing surface might affect lifetime, so three different surfaces were randomly selected. A writing machine was used to ensure that conditions were otherwise homogeneous (e.g., constant pressure and a fixed angle). The accompanying table shows the two lifetimes (min) obtained for each brand–surface combination. In addition, $\Sigma\Sigma\Sigma x_{ijk}^2 = 11{,}499{,}492$, and $\Sigma\Sigma x_{ij.}^2 = 22{,}982{,}552$.

Brand of Pen	Writing Surface 1	Writing Surface 2	Writing Surface 3	$x_{i..}$
1	709, 659	713, 726	660, 645	4112
2	668, 685	722, 740	692, 720	4227
3	659, 685	666, 684	678, 750	4122
4	698, 650	704, 666	686, 733	4137
$x_{.j.}$	5413	5621	5564	16,598

Carry out an appropriate ANOVA, and state your conclusions.

54. The accompanying data was obtained in an experiment to investigate whether compressive strength of concrete cylinders depends on the type of capping material used or variability in different batches ("The Effect of Type of Capping Material on the Compressive Strength of Concrete Cylinders," *Proceedings ASTM*, 1958: 1166–1186). Each number is a cell total ($x_{ij.}$) based on $K = 3$ observations.

Capping Material	Batch 1	2	3	4	5
1	1847	1942	1935	1891	1795
2	1779	1850	1795	1785	1626
3	1806	1892	1889	1891	1756

In addition, $\Sigma\Sigma\Sigma x_{ijk}^2 = 16{,}815{,}853$ and $\Sigma\Sigma\ x_{ij.}^2 = 50{,}443{,}409$. Obtain the ANOVA table and then test at level .01 the hypotheses H_{0G} versus H_{aG}, H_{0A} versus H_{aA}, and H_{0B} versus H_{aB}, assuming that capping is a fixed effect and batches is a random effect.

55. **a.** Show that $E(\overline{X}_{i..} - \overline{X}_{...}) = \alpha_i$, so that $\overline{X}_{i..} = \overline{X}_{...}$ is an unbiased estimator for α_i (in the fixed effects model).

b. With $\hat{\gamma}_{ij} = \overline{X}_{ij.} - \overline{X}_{i..} - \overline{X}_{.j.} + \overline{X}_{...}$, show that $\hat{\gamma}_{ij}$ is an unbiased estimator for γ_{ij} (in the fixed effects model).

56. Show how a $100(1 - \alpha)\%$ t CI for $\alpha_i - \alpha_i'$ can be obtained. Then compute a 95% interval for $\alpha_2 - \alpha_3$ using the data from Example 11.16. [*Hint:* With $\theta = \alpha_2 - \alpha_3$, the result of Exercise 55a indicates how to obtain $\hat{\theta}$. Then compute $V(\hat{\theta})$ and $\sigma_{\hat{\theta}}$ and obtain an estimate of $\sigma_{\hat{\theta}}$ by using $\sqrt{\text{MSE}}$ to estimate σ (which identifies the appropriate number of df).]

57. When both factors are random in a two-way ANOVA experiment with K replications per combination of factor levels, the expected mean squares are $E(\text{MSE}) = \sigma^2$, $E(\text{MSA}) = \sigma^2 + K\sigma_G^2 + JK\sigma_A^2$, $E(\text{MSB}) = \sigma^2 + K\sigma_G^2 + IK\sigma_B^2$, and $E(\text{MSAB}) = \sigma^2 + K\sigma_G^2$.

a. What F ratio is appropriate for testing H_{0G}: $\sigma_G^2 = 0$ versus H_{aG}: $\sigma_G^2 > 0$?

b. Answer part (a) for testing H_{0A}: $\sigma_A^2 = 0$ versus H_{aA}: $\sigma_A^2 > 0$ and H_{0B}: $\sigma_B^2 = 0$ versus H_{aB}: $\sigma_B^2 > 0$.

Supplementary Exercises (58–70)

58. An experiment was carried out to compare flow rates for four different types of nozzle.

a. Sample sizes were 5, 6, 7, and 6, respectively, and calculations gave $f = 3.68$. State and test the relevant hypotheses using $\alpha = .01$.

b. Analysis of the data using a statistical computer package yielded P-value = .029. At level .01, what would you conclude, and why?

59. The article "Computer-Assisted Instruction Augmented with Planned Teacher/Student Contacts" (*J. Experiment. Ed.*, Winter 1980–1981: 120–126) compared five different methods for teaching descriptive statistics. The five methods were traditional lecture and discussion (L/D), programmed textbook instruction (R), programmed text with lectures (R/L), computer instruction (C), and computer instruction with lectures (C/L). Forty-five students were randomly assigned, 9 to each method. After completing the course, the students took a 1-hour exam. In addition, a 10-minute retention test was administered 6 weeks later. Summary quantities are given.

Method	Exam $\bar{x}_{i.}$	Exam s_i	Retention Test $\bar{x}_{i.}$	Retention Test s_i
L/D	29.3	4.99	30.20	3.82
R	28.0	5.33	28.80	5.26
R/L	30.2	3.33	26.20	4.66
C	32.4	2.94	31.10	4.91
C/L	34.2	2.74	30.20	3.53

The grand mean for the exam was 30.82, and the grand mean for the retention test was 29.30.

a. Does the data suggest that there is a difference among the five teaching methods with respect to true mean exam score? Use $\alpha = .05$.

b. Using a .05 significance level, test the null hypothesis of no difference among the true mean retention test scores for the five different teaching methods.

60. Numerous factors contribute to the smooth running of an electric motor ("Increasing Market Share

Through Improved Product and Process Design: An Experimental Approach," *Qual. Engrg.*, 1991: 361–369). In particular, it is desirable to keep motor noise and vibration to a minimum. To study the effect that the brand of bearing has on motor vibration, five different motor bearing brands were examined by installing each type of bearing on different random samples of six motors. The amount of motor vibration (measured in microns) was recorded when each of the 30 motors was running. The data for this study follows. State and test the relevant hypotheses at significance level .05, and then carry out a multiple comparisons analysis if appropriate.

							Mean
Brand 1	13.1	15.0	14.0	14.4	14.0	11.6	13.68
Brand 2	16.3	15.7	17.2	14.9	14.4	17.2	15.95
Brand 3	13.7	13.9	12.4	13.8	14.9	13.3	13.67
Brand 4	15.7	13.7	14.4	16.0	13.9	14.7	14.73
Brand 5	13.5	13.4	13.2	12.7	13.4	12.3	13.08

61. An article in the British scientific journal *Nature* ("Sucrose Induction of Hepatic Hyperplasia in the Rat," August 25, 1972: 461) reports on an experiment in which each of five groups consisting of six rats was put on a diet with a different carbohydrate. At the conclusion of the experiment, the DNA content of the liver of each rat was determined (mg/g liver), with the following results:

Carbohydrate	$\bar{x}_{i.}$
Starch	2.58
Sucrose	2.63
Fructose	2.13
Glucose	2.41
Maltose	2.49

a. Assuming also that $\Sigma\Sigma x_{ij}^2 = 183.4$, is the true average DNA content affected by the type of carbohydrate in the diet? Construct an ANOVA table and use a .05 level of significance.

b. Construct a t CI for the contrast

$$\theta = \mu_1 - (\mu_2 + \mu_3 + \mu_4 + \mu_5)/4$$

which measures the difference between the average DNA content for the starch diet and the combined average for the four other diets. Does the resulting interval include zero?

c. What is β for the test when true average DNA content is identical for three of the diets and falls below this common value by 1 standard deviation (σ) for the other two diets?

62. Four laboratories (1–4) are randomly selected from a large population, and each is asked to make three determinations of the percentage of methyl alcohol in specimens of a compound taken from a single batch. Based on the accompanying data, are differences among laboratories a source of variation in the percentage of methyl alcohol? State and test the relevant hypotheses using significance level .05.

1:	85.06	85.25	84.87
2:	84.99	84.28	84.88
3:	84.48	84.72	85.10
4:	84.10	84.55	84.05

63. The critical flicker frequency (cff) is the highest frequency (in cycles/sec) at which a person can detect the flicker in a flickering light source. At frequencies above the cff, the light source appears to be continuous even though it is actually flickering. An investigation carried out to see whether true average cff depends on iris color yielded the following data (based on the article "The Effects of Iris Color on Critical Flicker Frequency," *J. Gen. Psych.*, 1973: 91–95):

	Iris Color		
	1. Brown	**2. Green**	**3. Blue**
	26.8	26.4	25.7
	27.9	24.2	27.2
	23.7	28.0	29.9
	25.0	26.9	28.5
	26.3	29.1	29.4
	24.8		28.3
	25.7		
	24.5		
J_i	8	5	6
$x_{i.}$	204.7	134.6	169.0
$\bar{x}_{i.}$	25.59	26.92	28.17
$n = 19$	$x_{..} = 508.3$		

a. State and test the relevant hypotheses at significance level .05 by using the F table to obtain an upper and/or lower bound on the P-value. (*Hint*: $\Sigma\Sigma x_{ij}^2 = 13{,}659.67$ and CF $= 13{,}598.36$.)

b. Investigate differences between iris colors with respect to mean cff.

64. Recall from Section 11.2 that if $c_1, c_2, \ldots, c_I$ are numbers satisfying $\sum c_i = 0$, then $\sum c_i\mu_i = c_1\mu_1 + \cdots + c_I\mu_I$ is called a *contrast* in the μ_i's. Notice that with $c_1 = 1, c_2 = -1, c_3 = \ldots = c_I = 0$, $\sum c_i\mu_i = \mu_1 - \mu_2$, which implies that every pairwise difference between μ_i's is a contrast (so is, e.g., $\mu_1 - .5\mu_2 - .5\mu_3$). A method attributed to Scheffé gives simultaneous CIs with simultaneous confidence level $100(1-\alpha)\%$ for *all* possible contrasts (an infinite number of them!). The interval for $\sum c_i\mu_i$ is

$$\sum c_i\bar{x}_{i.} \pm \left(\sum \frac{c_i^2}{J_i}\right)^{1/2} \cdot [(I-1)\cdot \text{MSE}\cdot F_{\alpha,I-1,n-I}]^{1/2}$$

Using the critical flicker frequency data of Exercise 63, calculate the Scheffé intervals for the contrasts $\mu_1 - \mu_2$, $\mu_1 - \mu_3$, $\mu_2 - \mu_3$, and $.5\mu_1 + .5\mu_2 - \mu_3$ (the last contrast compares blue to the average of brown and green). Which contrasts appear to differ significantly from 0, and why?

65. Four types of mortars—ordinary cement mortar (OCM), polymer impregnated mortar (PIM), resin mortar (RM), and polymer cement mortar (PCM)—were subjected to a compression test to measure strength (MPa). Three strength observations for each mortar type are given in the article "Polymer Mortar Composite Matrices for Maintenance-Free Highly Durable Ferrocement" (*J. Ferrocement*, 1984: 337–345) and are reproduced here. Construct an ANOVA table. Using a .05 significance level, determine whether the data suggests that the true mean strength is not the same for all four mortar types. If you determine that the true mean strengths are not all equal, use Tukey's method to identify the significant differences.

OCM	32.15	35.53	34.20
PIM	126.32	126.80	134.79
RM	117.91	115.02	114.58
PCM	29.09	30.87	29.80

66. Suppose the x_{ij}'s are "coded" by $y_{ij} = cx_{ij} + d$. How does the value of the F statistic computed from the y_{ij}'s compare to the value computed from the x_{ij}'s? Justify your assertion.

67. In Example 11.10, subtract $\bar{x}_{i.}$ from each observation in the ith sample ($i = 1, \ldots, 6$) to obtain a set of 18 residuals. Then construct a normal probability plot and comment on the plausibility of the normality assumption.

68. The results of a study on the effectiveness of line drying on the smoothness of fabric were summarized in the article "Line-Dried vs. Machine-Dried Fabrics: Comparison of Appearance, Hand, and Consumer Acceptance" (*Home Econ. Res. J.*, 1984: 27–35). Smoothness scores were given for nine different types of fabric and five different drying methods: (1) machine dry, (2) line dry, (3) line dry followed by 15-min tumble, (4) line dry with softener, and (5) line dry with air movement. Regarding the different types of fabric as blocks, construct an ANOVA table.

a. Using a .05 significance level, test to see whether there is a difference in the true mean smoothness score for the drying methods.

b. Make a plot like Figure 11.8 with fabric on the horizontal axis. Discuss the result of part (a) in terms of the plot.

c. Did the two methods involving the dryer yield significantly smoother fabric compared to the other three?

	Drying Method				
Fabric	**1**	**2**	**3**	**4**	**5**
Crepe	3.3	2.5	2.8	2.5	1.9
Double knit	3.6	2.0	3.6	2.4	2.3
Twill	4.2	3.4	3.8	3.1	3.1
Twill mix	3.4	2.4	2.9	1.6	1.7
Terry	3.8	1.3	2.8	2.0	1.6
Broadcloth	2.2	1.5	2.7	1.5	1.9
Sheeting	3.5	2.1	2.8	2.1	2.2
Corduroy	3.6	1.3	2.8	1.7	1.8
Denim	2.6	1.4	2.4	1.3	1.6

69. The water absorption of two types of mortar used to repair damaged cement was discussed in the article "Polymer Mortar Composite Matrices for Maintenance-Free, Highly Durable Ferrocement" (*J. Ferrocement*, 1984: 337–345). Specimens of ordinary cement mortar (OCM) and polymer cement mortar (PCM) were submerged for varying lengths of time (5, 9, 24, or 48 hours), and water absorption (% by weight) was recorded. With mortar type as factor A (with two levels) and submersion period as factor B (with four levels), three observations were made for each factor level combination. Data included in the article was used to compute the sums of squares, which were SSA = 322.667, SSB = 35.623, SSAB = 8.557, and SST = 372.113. Use this information to construct an ANOVA table. Test the appropriate hypotheses at a .05 significance level.

70. Four plots were available for an experiment to compare clover accumulation for four different sowing rates ("Performance of Overdrilled Red Clover with Different Sowing Rates and Initial Grazing Managements," *New Zeal. J. Experiment. Agric.*, 1984: 71–81). Since the four plots had been grazed differently prior to the experiment and it was thought that this might affect clover accumulation, a randomized block experiment was used with all four sowing rates tried on a section of each plot. Use the given data to test the null hypothesis of no difference in true mean clover accumulation (kg DM/ha) for the different sowing rates.

a. Test to see if the different sowing rates make a difference in true mean clover accumulation.

b. Make appropriate plots to go with your analysis in (a): Make a plot like the one in Figure 11.8, make a normal plot of the residuals, and plot the residuals against the predicted values. Explain why, based on the plots, the assumptions do not appear to be satisfied for this data set.

c. Repeat part (a) replacing the observations with their natural logarithms.

d. Repeat the plots of (b) for the analysis in (c). Do the logged observations appear to satisfy the assumptions better?

e. Summarize your conclusions for this experiment. Does mean clover accumulation increase with increasing sowing rate?

	Sowing Rate (kg/ha)			
Plot	**3.6**	**6.6**	**10.2**	**13.5**
1	1155	2255	3505	4632
2	123	406	564	416
3	68	416	662	379
4	62	75	362	564

Bibliography

Miller, Rupert, *Beyond ANOVA: The Basics of Applied Statistics*, Wiley, New York, 1986. An excellent source of information about assumption checking and alternative methods of analysis.

Montgomery, Douglas, *Design and Analysis of Experiments* (5th ed.), Wiley, New York, 2001. An up-to-date presentation of ANOVA models and methodology.

Neter, John, William Wasserman, and Michael Kutner, *Applied Linear Statistical Models* (4th ed.), Irwin, Homewood, IL, 1996. The second half of this book contains a well-presented survey of ANOVA; the level is comparable to that of the present text, but the discussion is more comprehensive, making the book an excellent reference.

Ott, R. Lyman, and Michael Longnecker, *An Introduction to Statistical Methods and Data Analysis* (5th ed.), Duxbury Press, Belmont, CA, 2001. Includes several chapters on ANOVA methodology that can profitably be read by students desiring a nonmathematical exposition; there is a good chapter on various multiple comparison methods.

CHAPTER TWELVE

Regression and Correlation

Introduction

The general objective of a regression analysis is to determine the relationship between two (or more) variables so that we can gain information about one of them through knowing values of the other(s). Much of mathematics is devoted to studying variables that are *deterministically* related. Saying that x and y are related in this manner means that once we are told the value of x, the value of y is completely specified. For example, suppose we decide to rent a van for a day and that the rental cost is \$25.00 plus \$.30 per mile driven. If we let $x =$ the number of miles driven and $y =$ the rental charge, then $y = 25 + .3x$. If we drive the van 100 miles ($x = 100$), then $y = 25 + .3(100) = 55$. As another example, if the initial velocity of a particle is v_0 and it undergoes constant acceleration a, then distance traveled $= y = v_0x + \frac{1}{2}ax^2$, where $x =$ time.

There are many variables x and y that would appear to be related to one another, but not in a deterministic fashion. A familiar example to many students is given by variables $x =$ high school grade point average (GPA) and $y =$ college GPA. The value of y cannot be determined just from knowledge of x, and two different students could have the same x value but have very different y values. Yet there is a tendency for those students who have high (low) high school GPAs also to have high (low) college GPAs. Knowledge of a student's high school GPA should be quite helpful in enabling us to predict how that person will do in college.

Other examples of variables related in a nondeterministic fashion include $x =$ age of a child and $y =$ size of that child's vocabulary, $x =$ size of an engine in cubic centimeters and $y =$ fuel efficiency for an automobile equipped with that engine, and x = applied tensile force and y = amount of elongation in a metal strip.

Regression analysis is the part of statistics that deals with investigation of the relationship between two or more variables related in a nondeterministic fashion. In this chapter, we generalize a deterministic linear relation to obtain a linear probabilistic model for relating two variables x and y. We then develop procedures for making inferences based on data obtained from the model, and obtain a quantitative measure (the correlation coefficient) of the extent to which the two variables are related. Techniques for assessing the adequacy of any particular regression model are then considered. We next introduce multiple regression analysis as a way of relating y to two or more variables—for example, relating fuel efficiency of an automobile to weight, engine size, number of cylinders, and transmission type. The last section of the chapter shows how matrix algebra techniques can be used to facilitate a concise and elegant development of regression procedures.

12.1 The Simple Linear and Logistic Regression Models

The key idea in developing a probabilistic relationship between a **dependent** or **response** variable y and an **independent**, **explanatory**, or **predictor** variable x is to realize that once the value of x has been fixed, there is still uncertainty in what the resulting y value will be. That is, for a fixed value of x, we now think of the dependent variable as being random. This random variable will be denoted by Y and its observed value by y. For example, suppose an investigator plans a study to relate y = yearly energy usage of an industrial building (1000's of BTUs) to x = the shell area of the building (ft^2). If one of the buildings selected for the study has a shell area of 25,000 ft^2, the resulting energy usage might be 2,215,000 or it might be 2,348,000 or any one of a number of other possibilities. Since we don't know a priori what the value of energy usage will be (because usage is determined partly by factors other than shell area), usage is regarded as a random variable Y.

We now relate the independent and dependent variables by an **additive model equation**:

$$\begin{aligned} Y &= \text{some particular deterministic function of } x + \text{a random deviation} \\ &= f(x) + \varepsilon \end{aligned} \tag{12.1}$$

The symbol ε represents a random deviation or random "error" (random variable) which is assumed to have mean value 0. This rv incorporates all variation in the dependent variable due to factors other than x. Figure 12.1 shows the graph of a particular $f(x)$. Without the random deviation ε, whenever x is fixed prior to making an observation on the dependent variable, the resulting (x, y) point would fall exactly on the graph. That is, y

would be entirely determined by x. The role of the random deviation ε is to allow a nondeterministic relationship. Now if the value of ε is positive, the resulting (x, y) point falls above the graph of $f(x)$, whereas when ε is negative, the resulting point falls below the graph. The assumption that ε has mean value 0 implies that we *expect* the point (x, y) to fall right on the graph, but we virtually never see what we expect—the observed point will almost always deviate upward or downward from the graph.

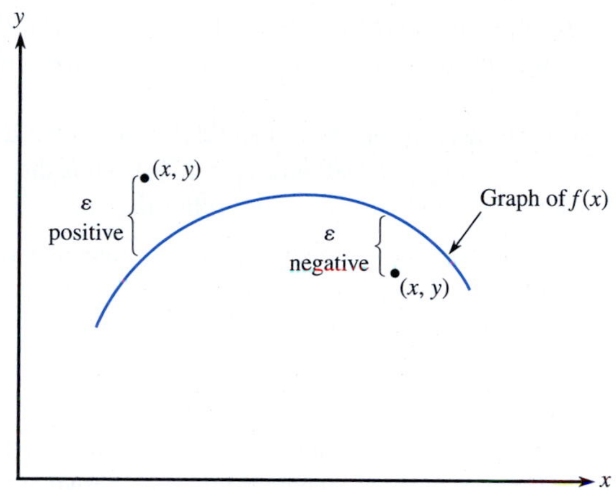

Figure 12.1 Observations resulting from the model Equation (12.1)

How should the deterministic part of the model equation be selected? Occasionally some sort of theoretical argument will suggest an appropriate choice of $f(x)$. However, in practice the specification of $f(x)$ is almost always made by obtaining sample data consisting of n (x, y) pairs. A picture of the resulting observations (x_1, y_1), (x_2, y_2), . . . , (x_n, y_n), called a **scatter plot**, is then constructed. In this scatter plot each (x_i, y_i) is represented as a point in a two-dimensional coordinate system. The pattern of points in the plot should suggest an appropriate $f(x)$.

Example 12.1 Visual and musculoskeletal problems associated with the use of video display terminals (VDTs) have become rather common in recent years. Some researchers have focused on vertical gaze direction as a source of eye strain and irritation. This direction is known to be closely related to ocular surface area (OSA), so a method of measuring OSA is needed. The accompanying representative data on y = OSA (cm^2) and x = width of the palpebral fissure (i.e., the horizontal width of the eye opening, in cm) is from the article "Analysis of Ocular Surface Area for Comfortable VDT Workstation Layout" (*Ergonomics*, 1996: 877–884). The order in which observations were obtained was not given, so for convenience they are listed in increasing order of x values.

i	1	2	3	4	5	6	7	8	9	10	11	12	13	14	15
x_i	.40	.42	.48	.51	.57	.60	.70	.75	.75	.78	.84	.95	.99	1.03	1.12
y_i	1.02	1.21	.88	.98	1.52	1.83	1.50	1.80	1.74	1.63	2.00	2.80	2.48	2.47	3.05

i	16	17	18	19	20	21	22	23	24	25	26	27	28	29	30
x_i	1.15	1.20	1.25	1.25	1.28	1.30	1.34	1.37	1.40	1.43	1.46	1.49	1.55	1.58	1.60
y_i	3.18	3.76	3.68	3.82	3.21	4.27	3.12	3.99	3.75	4.10	4.18	3.77	4.34	4.21	4.92

Thus $(x_1, y_1) = (.40, 1.02)$, $(x_5, y_5) = (.57, 1.52)$, and so on. A MINITAB scatter plot is shown in Figure 12.2; we used an option that produced a dotplot of both the x values and y values individually along the right and top margins of the plot, which makes it easier to visualize the distributions of the individual variables (histograms or boxplots are alternative options). Here are some things to notice about the data and plot:

- Several observations have identical x values yet different y values (e.g., $x_8 = x_9 = .75$, but $y_8 = 1.80$ and $y_9 = 1.74$). Thus the value of y is *not* determined solely by x but also by various other factors.
- There is a strong tendency for y to increase as x increases. That is, larger values of OSA tend to be associated with larger values of fissure width—a positive relationship between the variables.
- It appears that the value of y could be predicted from x by finding a line that is reasonably close to the points in the plot (the authors of the cited article superimposed such a line on their plot). In other words, there is evidence of a substantial (though not perfect) linear relationship between the two variables.

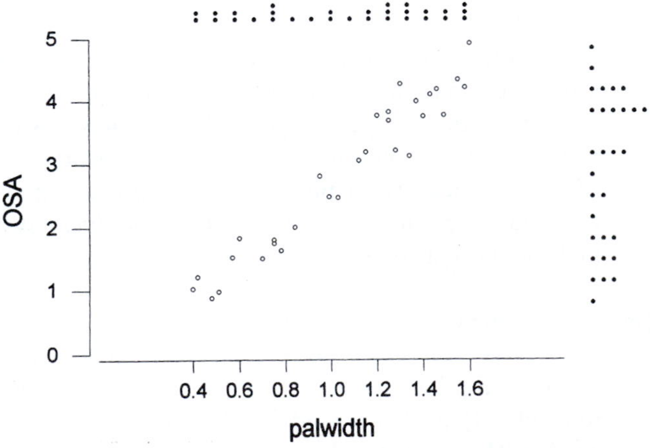

Figure 12.2 Scatter plot from MINITAB for the data from Example 12.1, along with dotplots of x and y values ■

The horizontal and vertical axes in the scatter plot of Figure 12.2 intersect at the point (0, 0). In many data sets, the values of x or y or the values of both variables differ considerably from zero relative to the range(s) of the values. For example, a study of how air conditioner efficiency is related to maximum daily outdoor temperature might involve observations for temperatures ranging from 80°F to 100°F. When this is the case, a more informative plot would show the appropriately labeled axes intersecting at some point other than (0, 0).

Example 12.2 Forest growth and decline phenomena throughout the world have attracted considerable public and scientific interest. The article "Relationships Among Crown Condition, Growth, and Stand Nutrition in Seven Northern Vermont Sugarbushes" (*Canad. J. Forest Res.*, 1995: 386–397) included a scatter plot of y = mean crown dieback (%), one indicator of growth retardation, and x = soil pH (higher pH corresponds to more acidic soil), from which the following observations were taken:

x	3.3	3.4	3.4	3.5	3.6	3.6	3.7	3.7	3.8	3.8
y	7.3	10.8	13.1	10.4	5.8	9.3	12.4	14.9	11.2	8.0

x	3.9	4.0	4.1	4.2	4.3	4.4	4.5	5.0	5.1
y	6.6	10.0	9.2	12.4	2.3	4.3	3.0	1.6	1.0

Figure 12.3 shows two MINITAB scatter plots of this data. In Figure 12.3(a), MINITAB selected the scale for both axes. We obtained Figure 12.3(b) by specifying minimum and maximum values for x and y so that the axes would intersect roughly at the point (0, 0). The second plot is more crowded than the first one; such crowding can make it more difficult to ascertain the general nature of any relationship. For example, it can be more difficult to spot curvature in a crowded plot.

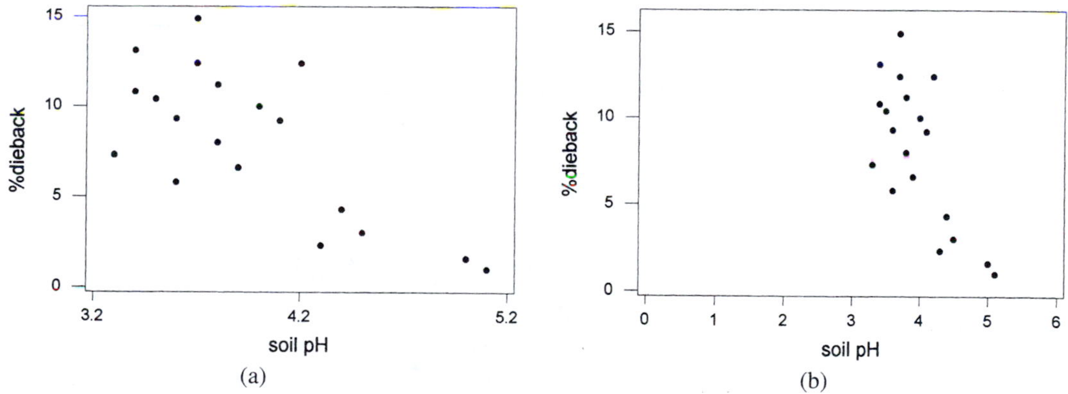

Figure 12.3 MINITAB scatter plots of data in Example 12.2

Large values of percentage dieback tend to be associated with low soil pH, a negative or inverse relationship. Furthermore, the two variables appear to be at least approximately linearly related, although the points would be spread out about any straight line drawn through the plot. ■

A Linear Probabilistic Model

For a deterministic linear relationship $y = \beta_0 + \beta_1 x$, the slope coefficient β_1 is the guaranteed increase in y when x increases by 1 unit and the intercept coefficient β_0 is the value of y when $x = 0$. A graph of $y = \beta_0 + \beta_1 x$ is of course a straight line. The slope gives the amount by which the line rises or falls when we move 1 unit to the right, and the intercept is the height at which the line crosses the vertical axis. For example, the line $y = 100 - 5x$ specifies an increase of -5 (i.e., a decrease of 5) for each 1-unit increase in x, and the vertical intercept of the line is 100. When a scatter plot of bivariate data consisting of n (x, y) pairs shows a reasonably substantial linear pattern, it is natural to specify $f(x)$ in the model Equation (12.1) to be a linear function. Rather than assuming that the dependent variable itself is a linear function of x, the model assumes that the *expected* value of Y is a linear function of x. For any fixed x value, the observed value of Y will deviate by a random amount from its expected value.

THE SIMPLE LINEAR REGRESSION MODEL

There are parameters β_0, β_1, and σ^2 such that for any fixed value of the independent variable x, the dependent variable is related to x through the model equation

$$Y = \beta_0 + \beta_1 x + \varepsilon$$

The random deviation (random variable) ε is assumed to be normally distributed with mean value 0 and variance σ^2, and this mean value and variance are the same regardless of the fixed x value. The observed pairs $(x_1, y_1), (x_2, y_2), \ldots, (x_n, y_n)$ are regarded as having been generated independently of one another from the model equation (fix $x = x_1$ and observe $Y_1 = \beta_0 + \beta_1 x_1 + \varepsilon_1$, then fix $x = x_2$ and observe $Y_2 = \beta_0 + \beta_1 x_2 + \varepsilon_2$, and so on; assuming that the ε's are independent of one another implies that the Y's are also).

Figure 12.4 gives an illustration of data resulting from the simple linear regression model.

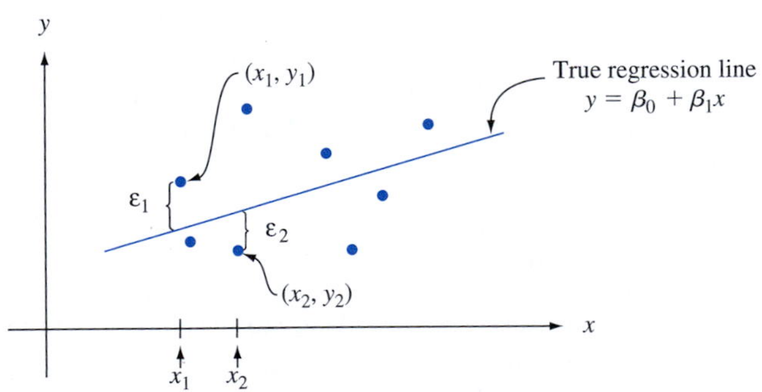

Figure 12.4 Points corresponding to observations from the simple linear regression model

The first two model parameters β_0 and β_1 are the coefficients of the **population** or **true regression line** $\beta_0 + \beta_1 x$. The slope parameter β_1 is now interpreted as the *expected* or *true average* increase in Y associated with a 1-unit increase in x. The variance parameter σ^2 (or equivalently the standard deviation σ) controls the inherent amount of variability in the data. When σ^2 is very close to 0, virtually all of the (x_i, y_i) pairs in the sample should correspond to points quite close to the population regression line. But if σ^2 greatly exceeds 0, a number of points in the scatter plot should fall far from the line. So the larger the value of σ, the greater will be the tendency for observed points to deviate from the population line by substantial amounts. Roughly speaking, the magnitude of σ is the size of a "typical" deviation from the population line.

The following notation will help clarify implications of the model relationship. Let x^* denote a particular value of the independent variable x, and

$$\mu_{Y \cdot x^*} = \text{the expected (i.e., mean) value of } Y \text{ when } x = x^*$$

$$\sigma^2_{Y \cdot x^*} = \text{the variance of } Y \text{ when } x = x^*$$

Alternative notation for these quantities is $E(Y|x^*)$ and $V(Y|x^*)$. For example, if $x =$ applied stress (kg/mm^2) and $y =$ time to fracture (hr), then $\mu_{Y \cdot 20}$ denotes the expected time to fracture when applied stress is 20 kg/mm^2. If we conceptualize an entire population of (x, y) pairs resulting from applying stress to specimens, then $\mu_{Y \cdot 20}$ is the average of all values of the dependent variable for which $x = 20$. The variance $\sigma^2_{Y \cdot 20}$ describes the spread in the distribution of all y values for which applied stress is 20.

Now consider replacing x in the model equation by the fixed value x^*. Then the only randomness on the right-hand side is from the random deviation ε. Recalling that the mean value of a numerical constant is the numerical constant and the variance of a constant is zero, we have that

$$\mu_{Y \cdot x^*} = E(\beta_0 + \beta_1 x^* + \varepsilon) = \beta_0 + \beta_1 x^* + E(\varepsilon) = \beta_0 + \beta_1 x^*$$
$$\sigma^2_{Y \cdot x^*} = V(\beta_0 + \beta_1 x^* + \varepsilon) = V(\beta_0 + \beta_1 x^*) + V(\varepsilon) = 0 + \sigma^2 = \sigma^2$$

The first sequence of equalities says that the mean value of Y when $x = x^*$ is the height of the population regression line above the value x^*. That is, *the population regression line is the line of mean Y values*—the *mean* Y value is a linear function of the independent variable. The second sequence of equalities tells us that the amount of variability in the distribution of Y is the same at any particular x value as it is at any other x value—this is the property of homogeneous variation about the population regression line. If the independent variable is age of a preschool child and the dependent variable is the child's vocabulary size, data suggests that the mean vocabulary size increases linearly with age. However, there is more variability in vocabulary size for 4-year-old children than for 2-year-old children, so there is not constant variation in Y about the population line and the simple linear regression model is therefore not appropriate. The constant variance property implies that points should spread out about the population regression line to the same extent throughout the range of x values in the sample, rather than fanning out more as x increases or as x decreases.

Additionally, the sum of a constant and a normally distributed variable is itself normally distributed, and the addition of the constant affects only the mean value and not the variance. So for any fixed value x^*, $Y (= \beta_0 + \beta_1 x^* + \varepsilon)$ has a normal distribution. The foregoing properties are summarized in Figure 12.5.

Example 12.3 Suppose the relationship between applied stress x and time-to-failure y is described by the simple linear regression model with true regression line $y = 65 - 1.2x$ and $\sigma = 8$. Then on average there is a 1.2-hr decrease in time to rupture associated with an increase of 1 kg/mm^2 in applied stress. For any fixed value of x^* of stress, time to rupture is normally distributed with mean value $65 - 1.2x^*$ and standard deviation 8. Roughly speaking, in

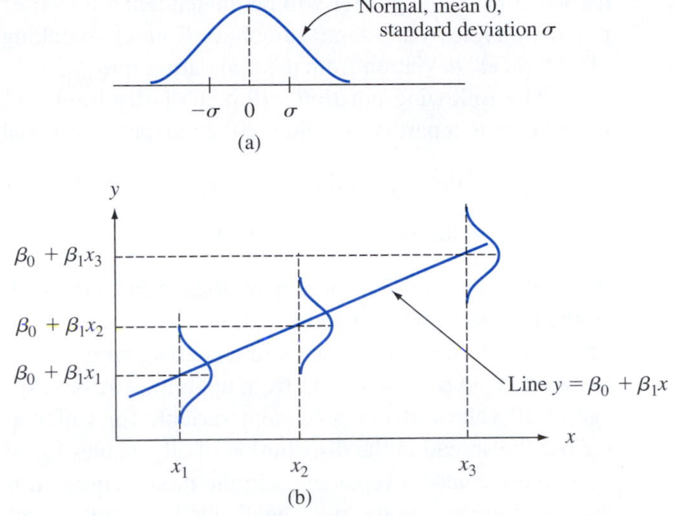

Figure 12.5 (a) Distribution of ε; (b) distribution of Y for different values of x

the population consisting of all (x, y) points, the magnitude of a typical deviation from the true regression line is about 8. For $x = 20$, Y has mean value $\mu_{Y \cdot 20} = 65 - 1.2(20) = 41$, so

$$P(Y > 50 \text{ when } x = 20) = P\left(Z > \frac{50 - 41}{8}\right) = 1 - \Phi(1.13) = .1292$$

When applied stress is 25, $\mu_{Y \cdot 25} = 35$, so the probability that time-to-failure exceeds 50 is

$$P(Y > 50 \text{ when } x = 25) = P\left(Z > \frac{50 - 35}{8}\right) = 1 - \Phi(1.88) = .0301$$

These probabilities are illustrated as the shaded areas in Figure 12.6.

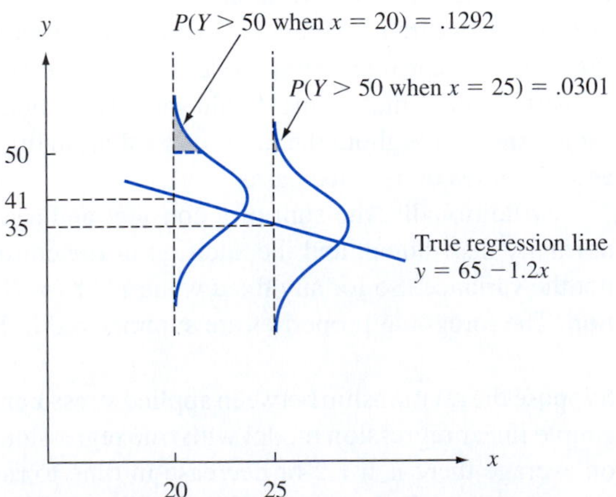

Figure 12.6 Probabilities based on the simple linear regression model

Suppose that Y_1 denotes an observation on time-to-failure made with $x = 25$ and Y_2 denotes an independent observation made with $x = 24$. Then $Y_1 - Y_2$ is normally distributed with mean value $E(Y_1 - Y_2) = \beta_1 = -1.2$, variance $V(Y_1 - Y_2) = \sigma^2 + \sigma^2 = 128$, and standard deviation $\sqrt{128} = 11.314$. The probability that Y_1 exceeds Y_2 is

$$P(Y_1 - Y_2 > 0) = P\left[Z > \frac{0 - (-1.2)}{11.314}\right] = P(Z > .11) = .4562$$

That is, even though we expected Y to decrease when x increases by 1 unit, the probability is fairly high (but less than .5) that the observed Y at $x + 1$ will be larger than the observed Y at x. ■

The Logistic Regression Model

The simple linear regression model is appropriate for relating a quantitative response variable y to a quantitative predictor x. Suppose that y is a dichotomous variable with possible values 1 and 0 corresponding to success and failure. Let $p = P(S) = P(Y = 1)$. Frequently, the value of p will depend on the value of some quantitative variable x. For example, the probability that a car needs warranty service of a certain kind might well depend on the car's mileage, or the probability of avoiding an infection of a certain type might depend on the dosage in an inoculation. Instead of using just the symbol p for the success probability, we now use $p(x)$ to emphasize the dependence of this probability on the value of x. The simple linear regression equation $Y = \beta_0 + \beta_1 x + \varepsilon$ is no longer appropriate, for taking the mean value on each side of the equation gives

$$\mu_{Y \cdot x} = 1 \cdot p(x) + 0 \cdot [1 - p(x)] = p(x) = \beta_0 + \beta_1 x$$

Whereas $p(x)$ is a probability and therefore must be between 0 and 1, $\beta_0 + \beta_1 x$ need not be in this range.

Instead of letting the mean value of y be a linear function of x, we now consider a model in which some function of the mean value of y is a linear function of x. In other words, we allow $p(x)$ to be a function of $\beta_0 + \beta_1 x$ rather than $\beta_0 + \beta_1 x$ itself. A function that has been found quite useful in many applications is the **logit function**

$$p(x) = \frac{e^{\beta_0 + \beta_1 x}}{1 + e^{\beta_0 + \beta_1 x}}$$

Figure 12.7 shows a graph of $p(x)$ for particular values of β_0 and β_1 with $\beta_1 > 0$. As x increases, the probability of success increases. For β_1 negative, the success probability would be a decreasing function of x.

Logistic regression means assuming that $p(x)$ is related to x by the logit function. Straightforward algebra shows that

$$\frac{p(x)}{1 - p(x)} = e^{\beta_0 + \beta_1 x}$$

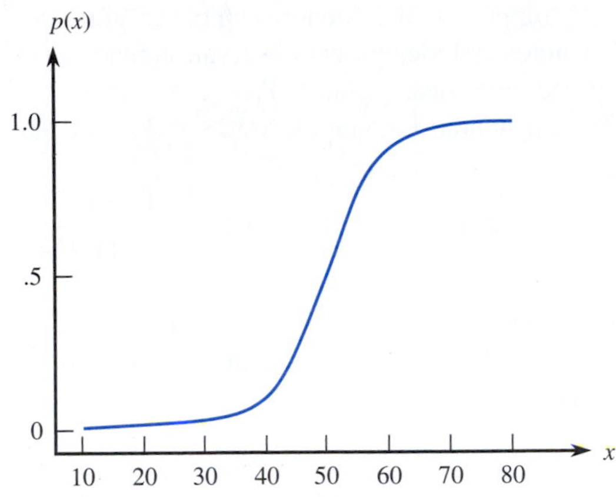

Figure 12.7 A graph of a logit function

The expression on the left-hand side is called the *odds ratio*. If, for example $p(60) = \frac{3}{4}$, then $p(60)/[1 - p(60)] = \frac{3}{4}/(1 - \frac{3}{4}) = 3$ and when $x = 60$ a success is three times as likely as a failure. This is described by saying that the odds are 3 to 1 because the success probability is 3 times the failure probability. Taking the natural log of the expression for the odds ratio shows that the log odds is a linear function of the predictor:

$$\ln\left[\frac{p(x)}{1 - p(x)}\right] = \beta_0 + \beta_1 x$$

In particular, the slope parameter β_1 is the change in the log odds associated with a 1-unit increase in x. This implies that the odds ratio itself changes by the multiplicative factor e^{β_1} when x increases by 1 unit.

Example 12.4 It seems reasonable that the size of a cancerous tumor should be related to the likelihood that the cancer will spread (metastasize) to another site. The article "Molecular Detection of p16 Promoter Methylation in the Serum of Patients with Esophageal Squamous Cell Carcinoma" (*Cancer Res.*, 2001: 3135–3138) investigated the spread of esophageal cancer to the lymph nodes. With x = size of a tumor (cm) and $Y = 1$ if the cancer does spread, consider the logistic regression model with $\beta_1 = .5$ and $\beta_0 = -2$ (values suggested by data in the article). Then

$$p(x) = \frac{e^{-2+.5x}}{1 + e^{-2+.5x}}$$

from which $p(2) = .27$ and $p(8) = .88$ (tumor sizes for patients in the study ranged from 1.7 cm to 9.0 cm). Because $e^{-2+.5(6.77)} \approx 4$, the odds for a 6.77-cm tumor are 4, so that it is four times as likely as not that a tumor of this size will spread to the lymph nodes. ■

Exercises Section 12.1 (1–12)

1. The efficiency ratio for a steel specimen immersed in a phosphating tank is the weight of the phosphate coating divided by the metal loss (both in mg/ft^2). The article "Statistical Process Control of a Phosphate Coating Line" (*Wire J. Internat.*, May 1997: 78–81) gave the accompanying data on tank temperature (x) and efficiency ratio (y).

Temp.	170	172	173	174	174	175	176
Ratio	.84	1.31	1.42	1.03	1.07	1.08	1.04

Temp.	177	180	180	180	180	180	181
Ratio	1.80	1.45	1.60	1.61	2.13	2.15	.84

Temp.	181	182	182	182	182	184	184
Ratio	1.43	.90	1.81	1.94	2.68	1.49	2.52

Temp.	185	186	188
Ratio	3.00	1.87	3.08

a. Construct stem-and-leaf displays of both temperature and efficiency ratio, and comment on interesting features.

b. Is the value of efficiency ratio completely and uniquely determined by tank temperature? Explain your reasoning.

c. Construct a scatter plot of the data. Does it appear that efficiency ratio could be very well predicted by the value of temperature? Explain your reasoning.

2. The article "Exhaust Emissions from Four-Stroke Lawn Mower Engines" (*J. Air Water Manag. Assoc.*, 1997: 945–952) reported data from a study in which both a baseline gasoline mixture and a reformulated gasoline were used. Consider the following observations on age (yr) and NO_x emissions (g/kWh):

Engine	1	2	3	4	5
Age	0	0	2	11	7
Baseline	1.72	4.38	4.06	1.26	5.31
Reformulated	1.88	5.93	5.54	2.67	6.53

Engine	6	7	8	9	10
Age	16	9	0	12	4
Baseline	.57	3.37	3.44	.74	1.24
Reformulated	.74	4.94	4.89	.69	1.42

Construct scatter plots of NO_x emissions versus age. What appears to be the nature of the relationship between these two variables? (*Note*: The authors of the cited article commented on the relationship.)

3. Bivariate data often arises from the use of two different techniques to measure the same quantity. As an example, the accompanying observations on x = hydrogen concentration (ppm) using a gas chromatography method and y = concentration using a new sensor method were read from a graph in the article "A New Method to Measure the Diffusible Hydrogen Content in Steel Weldments Using a Polymer Electrolyte-Based Hydrogen Sensor" (*Welding Res.*, July 1997: 251s–256s).

x	47	62	65	70	70	78	95	100	114	118
y	38	62	53	67	84	79	93	106	117	116

x	124	127	140	140	140	150	152	164	198	221
y	127	114	134	139	142	170	149	154	200	215

Construct a scatter plot. Does there appear to be a very strong relationship between the two types of concentration measurements? Do the two methods appear to be measuring roughly the same quantity? Explain your reasoning.

4. A study to assess the capability of subsurface flow wetland systems to remove biochemical oxygen demand (BOD) and various other chemical constituents resulted in the accompanying data on x = BOD mass loading (kg/ha/d) and y = BOD mass removal (kg/ha/d) ("Subsurface Flow Wetlands—A Performance Evaluation," *Water Environ. Res.*, 1995: 244–247).

x	3	8	10	11	13	16	27	30	35	37	38	44	103	142
y	4	7	8	8	10	11	16	26	21	9	31	30	75	90

a. Construct boxplots of both mass loading and mass removal, and comment on any interesting features.

b. Construct a scatter plot of the data, and comment on any interesting features.

5. The article "Objective Measurement of the Stretchability of Mozzarella Cheese" (*J. Texture Stud.*, 1992: 185–194) reported on an experiment to investigate how the behavior of mozzarella cheese varied with temperature. Consider the accompanying data on x = temperature and y = elongation (%) at failure

Figure 12.8 and the foregoing discussion suggest that our estimate of $y = \beta_0 + \beta_1 x$ should be a line that provides in some sense a best fit to the observed data points. This is what motivates the principle of least squares, which can be traced back to the mathematicians Gauss and Legendre around the year 1800. According to this principle, a line provides a good fit to the data if the vertical distances (deviations) from the observed points to the line are small (see Figure 12.9). The measure of the goodness of fit is the sum of the squares of these deviations. The best-fit line is then the one having the smallest possible sum of squared deviations.

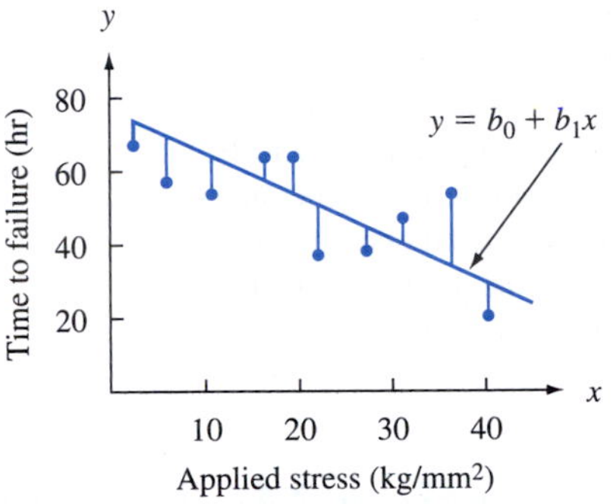

Figure 12.9 Deviations of observed data from line $y = b_0 + b_1 x$

PRINCIPLE OF LEAST SQUARES

The vertical deviation of the point (x_i, y_i) from the line $y = b_0 + b_1 x$ is

$$\text{height of point} - \text{height of line} = y_i - (b_0 + b_1 x_i)$$

The sum of squared vertical deviations from the points $(x_1, y_1), \ldots, (x_n, y_n)$ to the line is then

$$f(b_0, b_1) = \sum_{i=1}^{n} [y_i - (b_0 + b_1 x_i)]^2$$

The point estimates of β_0 and β_1, denoted by $\hat{\beta}_0$ and $\hat{\beta}_1$ and called the **least squares estimates**, are those values that minimize $f(b_0, b_1)$. That is, $\hat{\beta}_0$ and $\hat{\beta}_1$ are such that $f(\hat{\beta}_0, \hat{\beta}_1) \leq f(b_0, b_1)$ for any b_0 and b_1. The **estimated regression line** or **least squares line** is then the line whose equation is $y = \hat{\beta}_0 + \hat{\beta}_1 x$.

The minimizing values of b_0 and b_1 are found by taking partial derivatives of $f(b_0, b_1)$ with respect to both b_0 and b_1, equating them both to zero [analogously to $f'(b) = 0$ in univariate calculus], and solving the equations

$$\frac{\partial f(b_0, b_1)}{\partial b_0} = \sum 2(y_i - b_0 - b_1 x_i)(-1) = 0$$

$$\frac{\partial f(b_0, b_1)}{\partial b_1} = \sum 2(y_i - b_0 - b_1 x_i)(-x_i) = 0$$

Cancellation of the factor 2 and rearrangement gives the following system of equations, called the **normal equations**:

$$\begin{aligned} nb_0 + (\textstyle\sum x_i)b_1 &= \textstyle\sum y_i \\ (\textstyle\sum x_i)b_0 + (\textstyle\sum x_i^2)b_1 &= \textstyle\sum x_i y_i \end{aligned}$$

The normal equations are linear in the two unknowns b_0 and b_1. Provided that at least two of the x_i's are different, the least squares estimates are the unique solution to this system.

The least squares estimate of the slope coefficient β_1 of the true regression line is

$$b_1 = \hat{\beta}_1 = \frac{\sum(x_i - \bar{x})(y_i - \bar{y})}{\sum(x_i - \bar{x})^2} = \frac{S_{xy}}{S_{xx}} \tag{12.2}$$

Computing formulas for the numerator and denominator of b_1 are

$$S_{xy} = \sum x_i y_i - \frac{(\sum x_i)(\sum y_i)}{n} \qquad S_{xx} = \sum x_i^2 - \frac{(\sum x_i)^2}{n}$$

(the S_{xx} formula was derived in Chapter 1 in connection with the sample variance, and the derivation of the S_{xy} formula is similar).

The least squares estimate of the intercept b_0 of the true regression line is

$$b_0 = \hat{\beta}_0 = \frac{\sum y_i - \hat{\beta}_1 \sum x_i}{n} = \bar{y} - \hat{\beta}_1 \bar{x} \tag{12.3}$$

Because of the normality assumption, $\hat{\beta}_0$ and $\hat{\beta}_1$ are also the maximum likelihood estimates (see Exercise 23).

The computational formulas for S_{xy} and S_{xx} require only the summary statistics $\sum x_i$, $\sum y_i$, $\sum x_i^2$, $\sum x_i y_i$ ($\sum y_i^2$ will be needed shortly); the x and y deviations are then not needed. In computing $\hat{\beta}_0$, use extra digits in $\hat{\beta}_1$ because, if $\bar{x}$ is large in magnitude, rounding may affect the final answer. We emphasize that *before $\hat{\beta}_1$ and $\hat{\beta}_0$ are computed, a scatter plot should be examined to see whether a linear probabilistic model is plausible*. If the points do not tend to cluster about a straight line with roughly the same degree of spread for all x, other models should be investigated. In practice, plots and regression calculations are usually done by using a statistical computer package.

Example 12.5 Global warming is a major issue, and CO_2 emissions are an important part of the discussion. What is the effect of increased CO_2 levels on the environment? In particular, what is the effect of these higher levels on the growth of plants and trees? The article "Effects of Atmospheric CO_2 Enrichment on Biomass Accumulation and Distribution in Eldarica Pine Trees" (*J. Experiment. Botany*, 1994: 345–349) describes the results of growing pine trees with increasing levels of CO_2 in the air. There were two trees at each of four levels of CO_2 concentration, and the mass of each tree was measured after

11 months of the experiment. Here are the observations with x = atmospheric concentration of CO_2 (μL/L, or ppm) and y = tree mass (kg), along with x^2, xy, and y^2. The mass measurements were read from a graph in the article.

Obs	x	y	x^2	xy	y^2
1	408	1.1	166,464	448.8	1.21
2	408	1.3	166,464	530.4	1.69
3	554	1.6	306,916	886.4	2.56
4	554	2.5	306,916	1385.0	6.25
5	680	3.0	462,400	2040.0	9.00
6	680	4.3	462,400	2924.0	18.49
7	812	4.2	659,344	3410.4	17.64
8	812	4.7	659,344	3816.4	22.09
Sum	4908	22.7	3,190,248	15,441.4	78.93

Thus $\bar{x} = 4908/8 = 613.5$, $\bar{y} = 22.7/8 = 2.838$, and

$$\hat{\beta}_1 = \frac{S_{xy}}{S_{xx}} = \frac{15{,}441.4 - (4908)(22.7)/8}{3{,}190{,}248 - (4908)^2/8}$$

$$= \frac{1514.95}{179{,}190} = .00845443 \approx .00845$$

$$\hat{\beta}_0 = 2.838 - (.00845443)(613.5) = -2.349$$

We estimate that the expected change in tree mass associated with a 1-part-per-million increase in CO_2 concentration is .00845. The equation of the estimated regression line (least squares line) is then $y = -2.35 + .00845x$. Figure 12.10, generated by the statistical computer package S-Plus, shows that the least squares line provides an excellent summary of the relationship between the two variables.

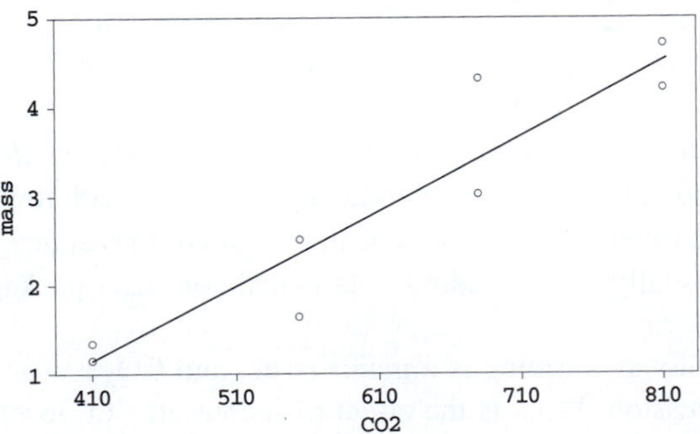

Figure 12.10 A scatter plot of the data in Example 12.5 with the least squares line superimposed, from S-Plus ■

The estimated regression line can immediately be used for two different purposes. For a fixed x value x^*, $\hat{\beta}_0 + \hat{\beta}_1 x^*$ (the height of the line above x^*) gives either (1) a point estimate of the expected value of Y when $x = x^*$ or (2) a point prediction of the Y value that will result from a single new observation made at $x = x^*$.

The least squares line should not be used to make a prediction for an x value much beyond the range of the data, such as $x = 250$ or $x = 1000$ in Example 12.5. The **danger of extrapolation** is that the fitted relationship (a line here) may not be valid for such x values. (In the foregoing example, $x = 250$ gives $\hat{y} = -.235$, a patently ridiculous value of mass, but extrapolation will not always result in such inconsistencies.)

Example 12.6 Refer to the tree-mass–CO_2 data in the previous example. With a little extrapolation, a point estimate for true average mass for all specimens with CO_2 concentration 365 is

$$\hat{\mu}_{Y\cdot 365} = \hat{\beta}_0 + \hat{\beta}_1(365) = -2.35 + .00845(365) = .73$$

With a little more extrapolation, a point estimate for true average mass for all specimens with CO_2 concentration 315 is

$$\hat{\mu}_{Y\cdot 315} = \hat{\beta}_0 + \hat{\beta}_1(315) = -2.35 + .00845(315) = .31$$

The values 315 and 365 are chosen based on actual values: The average world atmospheric CO_2 concentration rose from 315 to 365 parts per million between 1960 and 2000. Even if the prediction equation is somewhat inaccurate when extrapolated to the left, it is clear that changes in carbon dioxide are making a big difference in the growth of trees. Notice that in Figure 12.10 the tree mass increases by a factor of more than 4 while the CO_2 concentration increases by just a factor of 2. ■

Estimating σ^2 and σ

The parameter σ^2 determines the amount of variability inherent in the regression model. A large value of σ^2 will lead to observed (x_i, y_i)'s that are quite spread out about the true regression line, whereas when σ^2 is small the observed points will tend to fall very close to the true line (see Figure 12.11). An estimate of σ^2 will be used in confidence interval (CI) formulas and hypothesis-testing procedures presented in the next two sections. Because the equation of the true line is unknown, the estimate is based on the extent to which the sample observations deviate from the estimated line. Many large deviations (residuals) suggest a large value of σ^2, whereas if all deviations are small in magnitude it indicates that σ^2 is small.

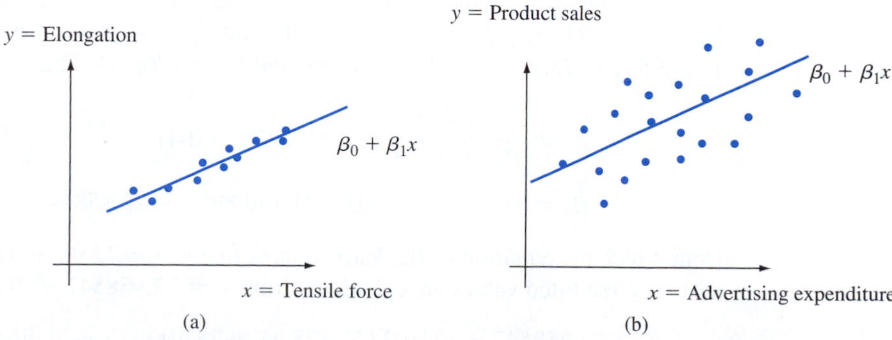

Figure 12.11 Typical sample for σ^2: (a) small; (b) large

DEFINITION The **fitted** (or **predicted**) **values** $\hat{y}_1, \hat{y}_2, \ldots, \hat{y}_n$ are obtained by successively substituting the x values $x_1, \ldots, x_n$ into the equation of the estimated regression line: $\hat{y}_1 = \hat{\beta}_0 + \hat{\beta}_1 x_1, \hat{y}_2 = \hat{\beta}_0 + \hat{\beta}_1 x_2, \ldots, \hat{y}_n = \hat{\beta}_0 + \hat{\beta}_1 x_n$. The **residuals** are the vertical deviations $y_1 - \hat{y}_1, y_2 - \hat{y}_2, \ldots, y_n - \hat{y}_n$ from the estimated line.

In words, the predicted value $\hat{y}_i$ is the value of y that we would predict or expect when using the estimated regression line with $x = x_i$; $\hat{y}_i$ is the height of the estimated regression line above the value x_i for which the ith observation was made. The residual $y_i - \hat{y}_i$ is the difference between the observed y_i and the predicted $\hat{y}_i$. If the residuals are all small in magnitude, then much of the variability in observed y values appears to be due to the linear relationship between x and y, whereas many large residuals suggest quite a bit of inherent variability in y relative to the amount due to the linear relation. Assuming that the line in Figure 12.9 is the least squares line, the residuals are identified by the vertical line segments from the observed points to the line. When the estimated regression line is obtained via the principle of least squares, the sum of the residuals should in theory be zero (an immediate consequence of the first normal equation; see Exercise 24). In practice, the sum may deviate a bit from zero due to rounding.

Example 12.7 Japan's high population density has resulted in a multitude of resource usage problems. One especially serious difficulty concerns waste removal. The article "Innovative Sludge Handling Through Pelletization Thickening" (*Water Res.*, 1999: 3245–3252) reported the development of a new compression machine for processing sewage sludge. An important part of the investigation involved relating the moisture content of compressed pellets (y, in %) to the machine's filtration rate (x, in kg-DS/m/hr). The following data was read from a graph in the paper:

x	125.3	98.2	201.4	147.3	145.9	124.7	112.2	120.2	161.2	178.9
y	77.9	76.8	81.5	79.8	78.2	78.3	77.5	77.0	80.1	80.2

x	159.5	145.8	75.1	151.4	144.2	125.0	198.8	132.5	159.6	110.7
y	79.9	79.0	76.7	78.2	79.5	78.1	81.5	77.0	79.0	78.6

Relevant summary quantities (*summary statistics*) are $\Sigma x_i = 2817.9$, $\Sigma y_i = 1574.8$, $\Sigma x_i^2 = 415{,}949.85$, $\Sigma x_i y_i = 222{,}657.88$, and $\Sigma y_i^2 = 124{,}039.58$, from which $\bar{x} = 140.895$, $\bar{y} = 78.74$, $S_{xx} = 18{,}921.8295$, and $S_{xy} = 776.434$. Thus

$$\hat{\beta}_1 = \frac{776.434}{18{,}921.8295} = .04103377 \approx .041$$

$$\hat{\beta}_0 = 78.74 - (0.4103377)(140.895) = 72.958547 \approx 72.96$$

from which the equation of the least squares line is $\hat{y} = 72.96 + .041x$. For numerical accuracy, the fitted values are calculated from $\hat{y}_i = 72.958547 + .04103377x_i$:

$$\hat{y}_1 = 72.958547 + .04103377(125.3) \approx 78.100 \qquad y_1 - \hat{y}_1 \approx -.200, \text{ etc.}$$

A positive residual corresponds to a point in the scatter plot that lies above the graph of the least squares line, whereas a negative residual results from a point lying below the line. All predicted values (fits) and residuals appear in the accompanying table.

Obs	Filtrate	Moistcon	Fit	Residual
1	125.3	77.9	78.100	−0.200
2	98.2	76.8	76.988	−0.188
3	201.4	81.5	81.223	0.277
4	147.3	79.8	79.003	0.797
5	145.9	78.2	78.945	−0.745
6	124.7	78.3	78.075	0.225
7	112.2	77.5	77.563	−0.063
8	120.2	77.0	77.891	−0.891
9	161.2	80.1	79.573	0.527
10	178.9	80.2	80.299	−0.099
11	159.5	79.9	79.503	0.397
12	145.8	79.0	78.941	0.059
13	75.1	76.7	76.040	0.660
14	151.4	78.2	79.171	−0.971
15	144.2	79.5	78.876	−0.624
16	125.0	78.1	78.088	0.012
17	198.8	81.5	81.116	0.384
18	132.5	77.0	78.396	−1.396
19	159.6	79.0	79.508	−0.508
20	110.7	78.6	77.501	1.099

■

In much the same way that the deviations from the mean in a one-sample situation were combined to obtain the estimate $s^2 = \Sigma(x_i - \bar{x})^2/(n-1)$, the estimate of σ^2 in regression analysis is based on squaring and summing the residuals. We will continue to use the symbol s^2 for this estimated variance, so don't confuse it with our previous s^2.

DEFINITION The **error sum of squares** (equivalently, residual sum of squares), denoted by SSE, is

$$\text{SSE} = \sum (y_i - \hat{y}_i)^2 = \sum [y_i - (\hat{\beta}_0 + \hat{\beta}_1 x_i)]^2$$

and the least squares estimate of σ^2 is

$$\hat{\sigma}^2 = s^2 = \frac{\text{SSE}}{n-2} = \frac{\sum (y_i - \hat{y}_i)^2}{n-2}$$

The divisor $n - 2$ in s^2 is the number of degrees of freedom (df) associated with the estimate (or, equivalently, with the error sum of squares). This is because to obtain s^2, the two parameters β_0 and β_1 must first be estimated, which results in a loss of 2 df (just as μ had to be estimated in one-sample problems, resulting in an estimated variance based on $n - 1$ df). Replacing each y_i in the formula for s^2 by the rv Y_i gives the estimator S^2. It can be shown that S^2 is an unbiased estimator for σ^2 (though the estimator S is biased for σ). The mle of σ^2 has divisor n rather than $n - 2$, so it is biased.

Example 12.8 (Example 12.7 continued)

The residuals for the filtration rate–moisture content data were calculated previously. The corresponding error sum of squares is

$$\text{SSE} = (-.200)^2 + (-.188)^2 + \cdots + (1.099)^2 = 7.968$$

The estimate of σ^2 is then $\hat{\sigma}^2 = s^2 = 7.968/(20 - 2) = .4427$, and the estimated standard deviation is $\hat{\sigma} = s = \sqrt{.4427} = .665$. Roughly speaking, .665 is the magnitude of a typical deviation from the estimated regression line. ■

Computation of SSE from the defining formula involves much tedious arithmetic because both the predicted values and residuals must first be calculated. Use of the following computational formula does not require these quantities.

$$\text{SSE} = \sum y_i^2 - \hat{\beta}_0 \sum y_i - \hat{\beta}_1 \sum x_i y_i$$

This expression results from substituting $\hat{y}_i = \hat{\beta}_0 + \hat{\beta}_1 x_i$ into $\Sigma(y_i - \hat{y}_i)^2$, squaring the summand, carrying the sum through to the resulting three terms, and simplifying (see Exercise 24). This computational formula is especially sensitive to the effects of rounding in $\hat{\beta}_0$ and $\hat{\beta}_1$, so use as many digits as your calculator will provide.

Example 12.9

The article "Promising Quantitative Nondestructive Evaluation Techniques for Composite Materials" (*Materials Eval.*, 1985: 561–565) reports on a study to investigate how the propagation of an ultrasonic stress wave through a substance depends on the properties of the substance. The accompanying data on fracture strength (x, as a percentage of ultimate tensile strength) and attenuation (y, in neper/cm, the decrease in amplitude of the stress wave) in fiberglass-reinforced polyester composites was read from a graph that appeared in the article. The simple linear regression model is suggested by the substantial linear pattern in the scatter plot.

x	12	30	36	40	45	57	62	67	71	78	93	94	100	105
y	3.3	3.2	3.4	3.0	2.8	2.9	2.7	2.6	2.5	2.6	2.2	2.0	2.3	2.1

The necessary summary quantities are $n = 14$, $\Sigma x_i = 890$, $\Sigma x_i^2 = 67{,}182$, $\Sigma y_i = 37.6$, $\Sigma y_i^2 = 103.54$, and $\Sigma x_i y_i = 2234.30$, from which $S_{xx} = 10{,}603.4285714$, $S_{xy} = -155.98571429$, $\hat{\beta}_1 = -.0147109$, and $\hat{\beta}_0 = 3.6209072$. The computational formula for SSE gives

$$\begin{aligned} \text{SSE} &= 103.54 - (3.6209072)(37.6) - (-.0147109)(2234.30) \\ &= .2624532 \end{aligned}$$

so $s^2 = .2624532/12 = .0218711$ and $s = .1479$. When $\hat{\beta}_0$ and $\hat{\beta}_1$ are rounded to three decimal places in the computational formula for SSE, the result is

$$\text{SSE} = 103.54 - (3.621)(37.6) - (-.015)(2234.30) = .905$$

which is more than three times the correct value. ■

The Coefficient of Determination

Figure 12.12 shows three different scatter plots of bivariate data. In all three plots, the heights of the different points vary substantially, indicating that there is much variability in observed y values. The points in the first plot all fall exactly on a straight line. In this case, all (100%) of the sample variation in y can be attributed to the fact that x and y are linearly related in combination with variation in x. The points in Figure 12.12(b) do not fall exactly on a line, but compared to overall y variability, the deviations from the least squares line are small. It is reasonable to conclude in this case that much of the observed y variation can be attributed to the approximate linear relationship between the variables postulated by the simple linear regression model. When the scatter plot looks like that of Figure 12.12(c), there is substantial variation about the least squares line relative to overall y variation, so the simple linear regression model fails to explain variation in y by relating y to x.

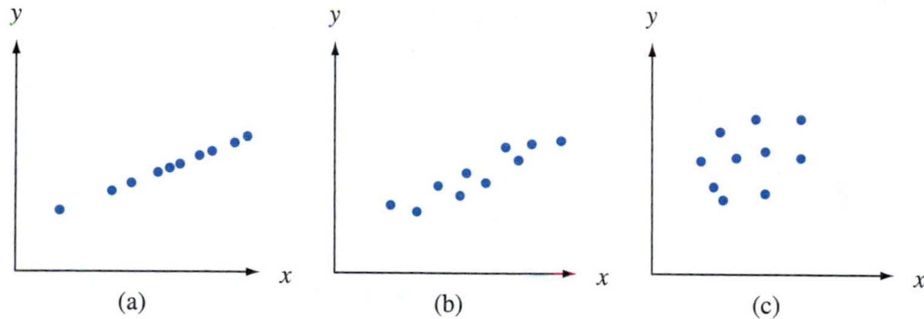

Figure 12.12 Explaining y variation: (a) all variation explained; (b) most variation explained; (c) little variation explained

The error sum of squares SSE can be interpreted as a measure of how much variation in y is left unexplained by the model—that is, how much cannot be attributed to a linear relationship. In Figure 12.12(a), SSE = 0, and there is no unexplained variation, whereas unexplained variation is small for the data of Figure 12.12(b) and much larger in Figure 12.12(c). A quantitative measure of the total amount of variation in observed y values is given by the **total sum of squares**

$$\text{SST} = S_{yy} = \sum (y_i - \bar{y})^2 = \sum y_i^2 - \left(\sum y_i\right)^2/n$$

Total sum of squares is the sum of squared deviations about the sample mean of the observed y values. Thus the same number $\bar{y}$ is subtracted from each y_i in SST, whereas SSE involves subtracting each different predicted value $\hat{y}_i$ from the corresponding observed y_i. Just as SSE is the sum of squared deviations about the least squares line $y = \hat{\beta}_0 + \hat{\beta}_1 x$, SST is the sum of squared deviations about the horizontal line at height $\bar{y}$ (since then vertical deviations are $y_i - \bar{y}$), as pictured in Figure 12.13. Furthermore, because the sum of squared deviations about the least squares line is smaller than the sum of squared deviations about *any* other line, SSE < SST unless the horizontal line itself is the least squares line. The ratio SSE/SST is the proportion of total variation that cannot be explained by the simple linear regression model, and 1 − SSE/SST (a number between 0 and 1) is the proportion of observed y variation explained by the model.

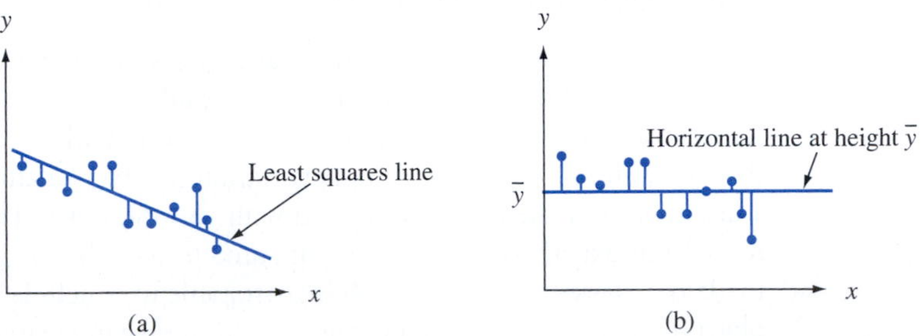

Figure 12.13 Sums of squares illustrated: (a) SSE = sum of squared deviations about the least squares line; (b) SST = sum of squared deviations about the horizontal line

---

DEFINITION

The **coefficient of determination**, denoted by r^2, is given by

$$r^2 = 1 - \frac{\text{SSE}}{\text{SST}}$$

It is interpreted as the proportion of observed y variation that can be explained by the simple linear regression model (attributed to an approximate linear relationship between y and x).

In equivalent words, r^2 is the proportion by which the error sum of squares is reduced by the regression line compared to the horizontal line. For example, if SST = 20 and SSE = 2, then $r^2 = 1 - \frac{2}{20} = .90$, so the regression reduces the error sum of squares by .90 = 90%.

The higher the value of r^2, the more successful is the simple linear regression model in explaining y variation. When regression analysis is done by a statistical computer package, either r^2 or $100r^2$ (the percentage of variation explained by the model) is a prominent part of the output. If r^2 is small, an analyst may want to search for an alternative model (either a nonlinear model or a multiple regression model that involves more than a single independent variable) that can more effectively explain y variation.

Example 12.10
(Example 12.5 continued)

The scatter plot of the CO_2 concentration data in Figure 12.10 indicates a fairly high r^2 value. With

$$\hat{\beta}_0 = -2.349293 \qquad \hat{\beta}_1 = .00845443 \qquad \sum y_i = 22.7$$
$$\sum x_i y_i = 15{,}441.4 \qquad \sum y_i^2 = 78.93$$

we have

$$\text{SST} = 78.93 - \frac{22.7^2}{8} = 14.519$$

$$\text{SSE} = 78.93 - (-2.349293)(22.7) - (.00845443)(15{,}441.4) = 1.711$$

The coefficient of determination is then

$$r^2 = 1 - \frac{1.711}{14.519} = 1 - .118 = .882$$

That is, 88.2% of the observed variation in mass is attributable to (can be explained by) the approximate linear relationship between mass and CO_2 concentration, a fairly impressive result. By the way, although it is common to have r^2 values of .88 or more in engineering, the physical sciences, and the biological sciences, r^2 is likely to be much smaller in social sciences such as psychology and sociology. An r^2 as big as .5 would be unusual in predicting one test score from another. In particular, when third-grade verbal IQ score is used to predict third-grade written IQ score for the 33 students of Example 1.2, r^2 is only .28.

Figure 12.14 shows partial MINITAB output for the CO_2 concentration data of Examples 12.5 and 12.10; the package will also provide the predicted values and residuals upon request, as well as other information. The formats used by other packages differ slightly from that of MINITAB, but the information content is very similar. Quantities such as the standard deviations, t ratios, and the details of the ANOVA table are discussed in Section 12.3.

```
The regression equation is
Kg = -2.35 + 0.00845 CO2

Predictor          Coef            SE Coef        T          P

Constant        -2.3493←β̂0        0.7966      -2.95      0.026
CO2              0.008454←β̂1      0.001261     6.70      0.001

S = 0.533964    R-Sq = 88.2%←100r²    R-Sq(adj) = 86.3%

Analysis of Variance

Source            DF       SS            MS         F        P
Regression         1    12.808        12.808     44.92    0.001
Residual Error     6     1.711←SSE     0.285
Total              7    14.519←SST
```

Figure 12.14 MINITAB output for the regression of Examples 12.5 and 12.10 ■

For regression there is an analysis of variance identity like the fundamental identity, Equation (11.1) in Section 11.1. Add and subtract $\hat{y}_i$ in the total sum of squares:

$$\text{SST} = \sum (y_i - \bar{y})^2 = \sum [(y_i - \hat{y}_i) + (\hat{y}_i - \bar{y})]^2 = \sum (y_i - \hat{y}_i)^2 + \sum (\hat{y}_i - \bar{y})^2$$

Notice that the middle (cross-product) term is missing on the right, but see Exercise 24 for the justification. Of the two sums on the right, the first is $\text{SSE} = \Sigma(y_i - \hat{y}_i)^2$ and the second is something new, the **regression sum of squares,** $\text{SSR} = \Sigma(\hat{y}_i - \bar{y})^2$. Interpret the regression sum of squares as the amount of total variation that *is* explained by the model. The analysis of variance identity for regression is

$$\text{SST} = \text{SSE} + \text{SSR} \tag{12.4}$$

The coefficient of determination in Example 12.10 can now be written in a slightly different way:

$$r^2 = 1 - \frac{\text{SSE}}{\text{SST}} = \frac{\text{SST} - \text{SSE}}{\text{SST}} = \frac{\text{SSR}}{\text{SST}}$$

the ratio of explained variation to total variation. The ANOVA table in Figure 12.14 shows that SSR = 12.808, from which $r^2 = 12.808/14.519 = .882$.

Terminology and Scope of Regression Analysis

The term *regression analysis* was first used by Francis Galton in the late nineteenth century in connection with his work on the relationship between father's height x and son's height y. After collecting a number of pairs (x_i, y_i), Galton used the principle of least squares to obtain the equation of the estimated regression line with the objective of using it to predict son's height from father's height. In using the derived line, Galton found that if a father was above average in height, the son would also be expected to be above average in height, *but not by as much as the father was.* Similarly, the son of a shorter-than-average father would also be expected to be shorter than average, but not by as much as the father. Thus the predicted height of a son was "pulled back in" toward the mean; because *regression* can be defined as moving backward, Galton adopted the terminology *regression line.* This phenomenon of being pulled back in toward the mean has been observed in many other situations (e.g., batting averages from year to year in baseball) and is called the **regression effect** or **regression to the mean**. See also Section 5.3 for a discussion of this topic in the context of the bivariate normal distribution.

Because of the regression effect, care must be exercised in experiments that involve selecting individuals based on below-average scores. For example, if students are selected because of below-average performance on a test, and they are then given special instruction, then the regression effect predicts improvement even if the instruction is useless. A similar warning applies in studies of underperforming businesses or hospital patients.

Our discussion thus far has presumed that the independent variable is under the control of the investigator, so that only the dependent variable Y is random. This was not, however, the case with Galton's experiment; fathers' heights were not preselected, but instead both X and Y were random. Methods and conclusions of regression analysis can be applied both when the values of the independent variable are fixed in advance and when they are random, but because the derivations and interpretations are more straightforward in the former case, we will continue to work explicitly with it. For more commentary, see the excellent book by John Neter et al. listed in the chapter bibliography.

Exercises Section 12.2 (13–30)

13. Exercise 4 gave data on x = BOD mass loading and y = BOD mass removal. Values of relevant summary quantities are

$n = 14 \quad \Sigma x_i = 517$
$\Sigma y_i = 346 \quad \Sigma x_i^2 = 39{,}095$
$\Sigma y_i^2 = 17{,}454 \quad \Sigma x_i y_i = 25{,}825$

a. Obtain the equation of the least squares line.
b. Predict the value of BOD mass removal for a single observation made when BOD mass loading is 35, and calculate the value of the corresponding residual.
c. Calculate SSE and then a point estimate of σ.
d. What proportion of observed variation in removal can be explained by the approximate linear relationship between the two variables?
e. The last two x values, 103 and 142, are much larger than the others. How are the equation of the least squares line and the value of r^2 affected by deletion of the two corresponding observations from the sample? Adjust the given values of the summary quantities, and use the fact that the new value of SSE is 311.79.

14. The accompanying data on x = current density (mA/cm^2) and y = rate of deposition (μm/min) appeared in the article "Plating of 60/40 Tin/Lead Solder for Head Termination Metallurgy" (*Plating and Surface Finishing*, Jan. 1997: 38–40). Do you agree with the claim by the article's author that "a linear relationship was obtained from the tin–lead rate of deposition as a function of current density"? Explain your reasoning.

x	20	40	60	80
y	.24	1.20	1.71	2.22

15. Refer to the data given in Exercise 1 on tank temperature and efficiency ratio.
a. Determine the equation of the estimated regression line.
b. Calculate a point estimate for true average efficiency ratio when tank temperature is 182.
c. Calculate the values of the residuals from the least squares line for the four observations for which temperature is 182. Why do they not all have the same sign?
d. What proportion of the observed variation in efficiency ratio can be attributed to the simple linear regression relationship between the two variables?

16. As an alternative to the use of father's height to predict son's height, Galton also used the midparent height, the average of the father's and mother's heights. Here are the heights of nine female students along with their midparent heights in inches:

Midparent	66.0	65.5	71.5	68.0	70.0	65.5	67.0
Daughter	64.0	63.0	69.0	69.0	69.0	65.0	63.0
Midparent	70.5	69.5	64.5	67.5			
Daughter	68.5	69.0	64.0	67.0			

a. Make a scatter plot of daughter's height against the midparent height and comment on the strength of the relationship.
b. Is the daughter's height completely and uniquely determined by the midparent height? Explain.
c. Use the accompanying MINITAB output to obtain the equation of the least squares line for predicting daughter height from midparent height, and then predict the height of a daughter whose midparent height is 70 inches. Would you feel comfortable using the least squares line to predict daughter height when midparent height is 74 inches? Explain.

```
Predictor      Coef    SE Coef      T        P
Constant       1.65      13.36    0.12    0.904
midparent    0.9555     0.1971    4.85    0.001

S = 1.45061  R-Sq = 72.3% R-Sq(adj) = 69.2%

Analysis of Variance

Source       DF      SS      MS       F       P
Regression    1  49.471  49.471   23.51   0.001
Residual      9  18.938   2.104
 Error
Total        10  68.409
```

d. What are the values of SSE, SST, and the coefficient of determination? How well does the midparent height account for the variation in daughter height?
e. Notice that for most of the families, the midparent height exceeds the daughter height. Is this what is meant by regression to the mean? Explain.

17. The article "Characterization of Highway Runoff in Austin, Texas, Area" (*J. Environ. Engrg.*, 1998: 131–137) gave a scatter plot, along with the least squares line, of x = rainfall volume (m^3) and y = runoff volume (m^3) for a particular location. The accompanying values were read from the plot.

x	5	12	14	17	23	30	40	47
y	4	10	13	15	15	25	27	46

x	55	67	72	81	96	112	127
y	38	46	53	70	82	99	100

a. Does a scatter plot of the data support the use of the simple linear regression model?
b. Calculate point estimates of the slope and intercept of the population regression line.
c. Calculate a point estimate of the true average runoff volume when rainfall volume is 50.
d. Calculate a point estimate of the standard deviation σ.
e. What proportion of the observed variation in runoff volume can be attributed to the simple linear regression relationship between runoff and rainfall?

18. A regression of y = calcium content (g/L) on x = dissolved material (mg/cm^2) was reported in the article "Use of Fly Ash or Silica Fume to Increase the Resistance of Concrete to Feed Acids" (*Mag. Concrete Res.*, 1997: 337–344). The equation of the estimated regression line was $y = 3.678 + .144x$, with $r^2 = .860$, based on $n = 23$.
a. Interpret the estimated slope .144 and the coefficient of determination .860.
b. Calculate a point estimate of the true average calcium content when the amount of dissolved material is 50 mg/cm^2.
c. The value of total sum of squares was SST = 320.398. Calculate an estimate of the error standard deviation σ in the simple linear regression model.

19. The following data is representative of that reported in the article "An Experimental Correlation of Oxides of Nitrogen Emissions from Power Boilers Based on Field Data" (*J. Engrg. Power*, July 1973: 165–170), with x = burner area liberation rate (MBtu/hr-ft^2) and y = NO_x emission rate (ppm):

x	100	125	125	150	150	200	200
y	150	140	180	210	190	320	280

x	250	250	300	300	350	400	400
y	400	430	440	390	600	610	670

a. Assuming that the simple linear regression model is valid, obtain the least squares estimate of the true regression line.
b. What is the estimate of expected NO_x emission rate when burner area liberation rate equals 225?
c. Estimate the amount by which you expect NO_x emission rate to change when burner area liberation rate is decreased by 50.
d. Would you use the estimated regression line to predict emission rate for a liberation rate of 500? Why or why not?
e. Does the simple linear regression model appear to do an effective job of explaining variation in emission rate? Justify your assertion.

20. A number of studies have shown lichens (certain plants composed of an alga and a fungus) to be excellent bioindicators of air pollution. The article "The Epiphytic Lichen *Hypogymnia physodes* as a Biomonitor of Atmospheric Nitrogen and Sulphur Deposition in Norway" (*Environ. Monitoring Assessment*, 1993: 27–47) gives the following data (read from a graph) on x = NO_3^- wet deposition (g N/m^2) and y = lichen N (% dry weight):

x	.05	.10	.11	.12	.31	.37	.42
y	.48	.55	.48	.50	.58	.52	1.02

x	.58	.68	.68	.73	.85	.92
y	.86	.86	1.00	.88	1.04	1.70

The author used simple linear regression to analyze the data. Use the accompanying MINITAB output to answer the following questions:
a. What are the least squares estimates of β_0 and β_1?
b. Predict lichen N for an NO_3^- deposition value of .5.
c. What is the estimate of σ?
d. What is the value of total variation, and how much of it can be explained by the model relationship?

```
The regression equation is lichen
N = 0.365 + 0.967 no3 depo

Predictor      Coef     Stdev   t-ratio      P
Constant    0.36510   0.09904      3.69  0.004
no3 depo     0.9668    0.1829      5.29  0.000

S = 0.1932  R-sq = 71.7%  R-sq (adj) = 69.2%

Analysis of Variance

SOURCE        DF       SS       MS      F      P
Regression     1   1.0427   1.0427  27.94  0.000
Error         11   0.4106   0.0373
Total         12   1.4533
```

21. The article "Effects of Bike Lanes on Driver and Bicyclist Behavior" (*ASCE Transportation Engrg. J.*, 1977: 243–256) reports the results of a regression analysis with x = available travel space in feet (a convenient measure of roadway width, defined as the distance between a cyclist and the roadway center line) and separation distance y between a bike and a passing car (determined by photography). The data, for ten streets with bike lanes, follows:

x	12.8	12.9	12.9	13.6	14.5
y	5.5	6.2	6.3	7.0	7.8

x	14.6	15.1	17.5	19.5	20.8
y	8.3	7.1	10.0	10.8	11.0

a. Verify that $\Sigma x_i = 154.20$, $\Sigma y_i = 80$, $\Sigma x_i^2 = 2452.18$, $\Sigma x_i y_i = 1282.74$, and $\Sigma y_i^2 = 675.16$.

b. Derive the equation of the estimated regression line.

c. What separation distance would you predict for another street that has 15.0 as its available travel space value?

d. What would be the estimate of expected separation distance for all streets having available travel space value 15.0?

22. The accompanying data was read from a graph that appeared in the article "Reactions on Painted Steel Under the Influence of Sodium Chloride, and Combinations Thereof" (*Indust. Engrg. Chem. Products Res. Dev.*, 1985: 375–378). The independent variable is SO_2 deposition rate (mg/m^2/d) and the dependent variable is steel weight loss (g/m^2).

x	14	18	40	43	45	112
y	280	350	470	500	560	1200

a. Construct a scatter plot. Does the simple linear regression model appear to be reasonable in this situation?

b. Calculate the equation of the estimated regression line.

c. What percentage of observed variation in steel weight loss can be attributed to the model relationship in combination with variation in deposition rate?

d. Because the largest x value in the sample greatly exceeds the others, this observation may have been very influential in determining the equation of the estimated line. Delete this observation and recalculate the equation. Does the new equation appear to differ substantially from the original one? (You might consider predicted values.)

23. Show that the mle's of β_0 and β_1 are indeed the least squares estimates. *Hint*: The pdf of Y_i is normal with mean $\mu_i = \beta_0 + \beta_1 x_i$ and variance σ^2; the likelihood is the product of the n pdf's.

24. Denote the residuals by $e_1, \ldots, e_n$ ($e_i = y_i - \hat{y}_i$).

a. Show that $\Sigma e_i = 0$ and $\Sigma x_i e_i = 0$. *Hint*: Examine the two normal equations.

b. Show that $\hat{y}_i - \bar{y} = \hat{\beta}_1(x_i - \bar{x})$.

c. Use (a) and (b) to derive the analysis of variance identity for regression, Equation (12.4), by showing that the cross-product term is 0.

d. Use (b) and Equation (12.4) to verify the computational formula for SSE.

25. A regression analysis is carried out with y = temperature, expressed in °C. How do the resulting values of $\hat{\beta}_0$ and $\hat{\beta}_1$ relate to those obtained if y is reexpressed in °F? Justify your assertion. *Hint*: new $y_i = y_i' = 1.8y_i + 32$.

26. Show that b_1 and b_0 of Expressions (12.2) and (12.3) satisfy the normal equations.

27. Show that the "point of averages" $(\bar{x}, \bar{y})$ lies on the estimated regression line.

28. Suppose an investigator has data on the amount of shelf space x devoted to display of a particular product and sales revenue y for that product. The investigator may wish to fit a model for which the true regression line passes through (0, 0). The appropriate model is $Y = \beta_1 x + \varepsilon$. Assume that $(x_1, y_1), \ldots, (x_n, y_n)$ are observed pairs generated from this model, and derive the least squares estimator of β_1. (*Hint*: Write the sum of squared deviations as a function of b_1, a trial value, and use calculus to find the minimizing value of b_1.)

29. **a.** Consider the data in Exercise 20. Suppose that instead of the least squares line passing through the points $(x_1, y_1), \ldots, (x_n, y_n)$, we wish the least squares line passing through $(x_1 - \bar{x}, y_1), \ldots, (x_n - \bar{x}, y_n)$. Construct a scatter plot of the (x_i, y_i) points and then of the $(x_i - \bar{x}, y_i)$ points. Use the plots to explain intuitively how the two least squares lines are related to one another.

b. Suppose that instead of the model $Y_i = \beta_0 + \beta_1 x_i + \varepsilon_i (i = 1, \ldots, n)$, we wish to fit a model of the form $Y_i = \beta_0^* + \beta_1^*(x_i - \bar{x}) + \varepsilon_i \, (i = 1, \ldots, n)$. What are the least squares estimators of β_0^* and β_1^*, and how do they relate to $\hat{\beta}_0$ and $\hat{\beta}_1$?

30. Consider the following three data sets, in which the variables of interest are x = commuting distance and y = commuting time. Based on a scatter plot and the values of s and r^2, in which situation would simple linear regression be most (least) effective, and why?

	1		2		3	
	x	y	x	y	x	y
	15	42	5	16	5	8
	16	35	10	32	10	16
	17	45	15	44	15	22
	18	42	20	45	20	23
	19	49	25	63	25	31
	20	46	50	115	50	60
S_{xx}	17.50		1270.8333		1270.8333	
S_{xy}	29.50		2722.5		1431.6667	
$\hat{\beta}_1$	1.685714		2.142295		1.126557	
$\hat{\beta}_0$	13.666672		7.868852		3.196729	
SST	114.83		5897.5		1627.33	
SSE	65.10		65.10		14.48	

12.3 Inferences About the Regression Coefficient β_1

In virtually all of our inferential work thus far, the notion of sampling variability has been pervasive. In particular, properties of sampling distributions of various statistics have been the basis for developing confidence interval formulas and hypothesis-testing methods. The key idea here is that the value of virtually any quantity calculated from sample data—the value of virtually any statistic—is going to vary from one sample to another.

Example 12.11 Reconsider the data on x = burner area liberation rate and y = NO_x emission rate from Exercise 19 in the previous section. There are 14 observations, made at the x values 100, 125, 125, 150, 150, 200, 200, 250, 250, 300, 300, 350, 400, and 400, respectively. Suppose that the slope and intercept of the true regression line are $\beta_1 = 1.70$ and $\beta_0 = -50$, with $\sigma = 35$ (consistent with the values $\hat{\beta}_1 = 1.7114$, $\hat{\beta}_0 = -45.55$, $s = 36.75$). We proceeded to generate a sample of random deviations $\tilde{\varepsilon}_1, \ldots, \tilde{\varepsilon}_{14}$ from a normal distribution with mean 0 and standard deviation 35, and then added $\tilde{\varepsilon}_i$ to $\beta_0 + \beta_1 x_i$ to obtain 14 corresponding y values. Regression calculations were then carried out to obtain the estimated slope, intercept, and standard deviation. This process was repeated a total of 20 times, resulting in the values given in Table 12.1.

Table 12.1 Simulation results for Example 12.11

$\hat{\beta}_1$	$\hat{\beta}_0$	s	$\hat{\beta}_1$	$\hat{\beta}_0$	s
1. 1.7559	−60.62	43.23	11. 1.7843	−67.36	41.80
2. 1.6400	−49.40	30.69	12. 1.5822	−28.64	32.46
3. 1.4699	−4.80	36.26	13. 1.8194	−83.99	40.80
4. 1.6944	−41.95	22.89	14. 1.6469	−32.03	28.11
5. 1.4497	5.80	36.84	15. 1.7712	−52.66	33.04
6. 1.7309	−70.01	39.56	16. 1.7004	−58.06	43.44
7. 1.8890	−95.01	42.37	17. 1.6103	−27.89	25.60
8. 1.6471	−40.30	43.71	18. 1.6396	−24.89	40.78
9. 1.7216	−42.68	23.68	19. 1.7857	−77.31	32.38
10. 1.7058	−63.31	31.58	20. 1.6342	−17.00	30.93

There is clearly variation in values of the estimated slope and estimated intercept, as well as the estimated standard deviation. The equation of the least squares line thus varies from one sample to the next. Figure 12.15 shows a dotplot of the estimated slopes as well as graphs of the true regression line and the 20 sample regression lines. ■

The slope β_1 of the population regression line is the true average change in the dependent variable y associated with a 1-unit increase in the independent variable x. The slope of the least squares line, $\hat{\beta}_1$, gives a point estimate of β_1. In the same way that a confidence interval for μ and procedures for testing hypotheses about μ were based on properties of the sampling distribution of $\overline{X}$, further inferences about β_1 are based on thinking of $\hat{\beta}_1$ as a statistic and investigating its sampling distribution.

The values of the x_i's are assumed to be chosen before the experiment is performed, so only the Y_i's are random. The estimators (statistics, and thus random variables) for β_0 and β_1 are obtained by replacing y_i by Y_i in (12.2) and (12.3):

$$\hat{\beta}_1 = \frac{\sum(x_i - \overline{x})(Y_i - \overline{Y})}{\sum(x_i - \overline{x})^2} \qquad \hat{\beta}_0 = \frac{\sum Y_i - \hat{\beta}_1 \sum x_i}{n}$$

Similarly, the estimator for σ^2 results from replacing each y_i in the formula for s^2 by the rv Y_i:

$$\hat{\sigma}^2 = S^2 = \frac{\sum Y_i^2 - \hat{\beta}_0 \sum Y_i - \hat{\beta}_1 \sum x_i Y_i}{n - 2}$$

The denominator of $\hat{\beta}_1$, $S_{xx} = \Sigma(x_i - \overline{x})^2$, depends only on the x_i's and not on the Y_i's, so it is a constant. Then because $\Sigma(x_i - \overline{x})\overline{Y} = \overline{Y}\Sigma(x_i - \overline{x}) = \overline{Y} \cdot 0 = 0$, the slope estimator can be written as

$$\hat{\beta}_1 = \frac{\sum(x_i - \overline{x})Y_i}{S_{xx}} = \sum c_i Y_i \qquad \text{where } c_i = \frac{x_i - \overline{x}}{S_{xx}}$$

That is, $\hat{\beta}_1$ is a linear function of the independent rv's $Y_1, Y_2, \ldots, Y_n$, each of which is normally distributed. Invoking properties (discussed in Section 6.3) of a linear function of random variables leads to the following results (Exercise 40).

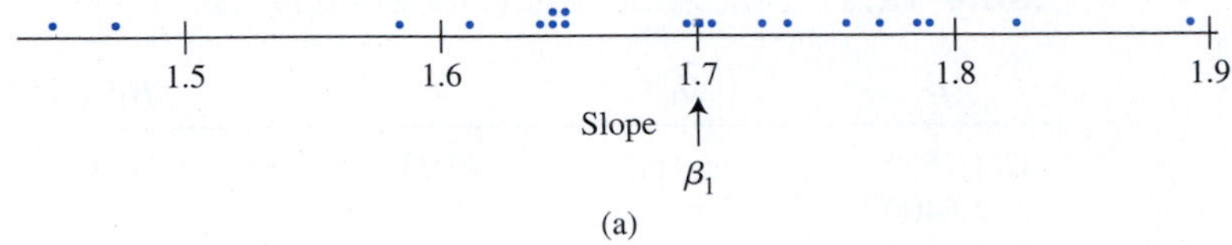

(a)

(b)

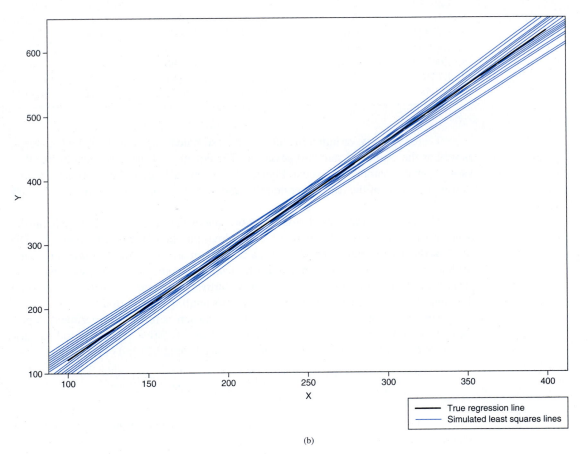

Figure 12.15 Simulation results from Example 12.11: (a) dotplot of estimated slopes; (b) graphs of the true regression line and 20 least squares lines (from S-Plus)

---

1. The mean value of $\hat{\beta}_1$ is $E(\hat{\beta}_1) = \mu_{\hat{\beta}_1} = \beta_1$, so $\hat{\beta}_1$ is an unbiased estimator of β_1 (the distribution of $\hat{\beta}_1$ is always centered at the value of β_1).
2. The variance and standard deviation of $\hat{\beta}_1$ are

$$V(\hat{\beta}_1) = \sigma^2_{\hat{\beta}_1} = \frac{\sigma^2}{S_{xx}} \qquad \sigma_{\hat{\beta}_1} = \frac{\sigma}{\sqrt{S_{xx}}} \tag{12.5}$$

where $S_{xx} = \Sigma(x_i - \bar{x})^2 = \Sigma x_i^2 - (\Sigma x_i)^2/n$. Replacing σ by its estimate s gives an estimate for $\sigma_{\hat{\beta}_1}$ (the estimated standard deviation, i.e., estimated standard error, of $\hat{\beta}_1$):

$$s_{\hat{\beta}_1} = \frac{s}{\sqrt{S_{xx}}}$$

(This estimate can also be denoted by $\hat{\sigma}_{\hat{\beta}_1}$.)

3. The estimator $\hat{\beta}_1$ has a normal distribution (because it is a linear function of independent normal rv's).

According to (12.5), the variance of $\hat{\beta}_1$ equals the variance σ^2 of the random error term—or, equivalently, of any Y_i—divided by $\Sigma(x_i - \bar{x})^2$. Because $\Sigma(x_i - \bar{x})^2$ is a measure of how spread out the x_i's are about $\bar{x}$, we conclude that making observations at x_i values that are quite spread out results in a more precise estimator of the slope parameter (smaller variance of $\hat{\beta}_1$), whereas values of x_i all close to one another imply a highly variable estimator. Of course, if the x_i's are spread out too far, a linear model may not be appropriate throughout the range of observation.

Many inferential procedures discussed previously were based on standardizing an estimator by first subtracting its mean value and then dividing by its estimated standard deviation. In particular, test procedures and a CI for the mean μ of a normal population utilized the fact that the standardized variable $(\bar{X} - \mu)/(S/\sqrt{n})$—that is, $(\bar{X} - \mu)/S_{\hat{\mu}}$—had a t distribution with $n - 1$ df. A similar result here provides the key to further inferences concerning β_1.

THEOREM The assumptions of the simple linear regression model imply that the standardized variable

$$T = \frac{\hat{\beta}_1 - \beta_1}{S/\sqrt{S_{xx}}} = \frac{\hat{\beta}_1 - \beta_1}{S_{\hat{\beta}_1}}$$

has a t distribution with $n - 2$ df.

The T ratio can be written as

$$T = \frac{\hat{\beta}_1 - \beta_1}{S/\sqrt{S_{xx}}} = \frac{\dfrac{\hat{\beta}_1 - \beta_1}{\sigma/\sqrt{S_{xx}}}}{\sqrt{\dfrac{(n-2)S^2/\sigma^2}{n-2}}}$$

The theorem is a consequence of the following facts: $(\hat{\beta}_1 - \beta_1)/(\sigma/\sqrt{S_{xx}}) \sim N(0, 1)$, $(n-2)S^2/\sigma^2 \sim \chi^2_{n-2}$, and $\hat{\beta}_1$ is independent of S^2. That is, T is a standard normal rv divided by the square root of an independent chi-squared rv over its df, so T has the specified t distribution.

A Confidence Interval for β_1

As in the derivation of previous CIs, we begin with a probability statement:

$$P\left(-t_{\alpha/2,n-2} < \frac{\hat{\beta}_1 - \beta_1}{S_{\hat{\beta}_1}} < t_{\alpha/2,n-2}\right) = 1 - \alpha$$

Manipulation of the inequalities inside the parentheses to isolate β_1 and substitution of estimates in place of the estimators gives the CI formula.

A $100(1 - \alpha)\%$ **CI for the slope β_1** of the true regression line is

$$\hat{\beta}_1 \pm t_{\alpha/2,n-2} \cdot s_{\hat{\beta}_1}$$

This interval has the same general form as did many of our previous intervals. It is centered at the point estimate of the parameter, and the amount it extends out to each side of the estimate depends on the desired confidence level (through the t critical value) and on the amount of variability in the estimator $\hat{\beta}_1$ (through $s_{\hat{\beta}_1}$, which will tend to be small when there is little variability in the distribution of $\hat{\beta}_1$ and large otherwise).

Example 12.12 Is it possible to predict graduation rates from freshman test scores? Based on the average SAT score of entering freshmen at a university, can we predict the percentage of those freshmen who will get a degree there within six years? We use a random sample of 20 universities from the 248 national universities listed in the 2005 edition of *America's Best Colleges*, published by *U.S. News & World Report*.

Obs	Rank	University	Grad Rate	SAT	Private or State
1	2	Princeton	98	1465.00	P
2	13	Brown	96	1395.00	P
3	15	Johns Hopkins	88	1380.00	P
4	69	Pittsburgh	65	1215.00	S
5	77	SUNY-Binghamton	80	1235.00	P
6	94	Kansas	58	1011.10	S
7	102	Dayton	76	1055.54	P
8	107	Illinois Inst Tech	67	1166.65	P
9	125	Arkansas	48	1055.54	S
10	139	Florida Inst Tech	54	1155.00	P
11	147	New Mexico Inst Mining	42	1099.99	S
12	158	Temple	54	1080.00	S
13	172	Montana	45	944.43	S
14	174	New Mexico	42	899.99	S
15	178	South Dakota	51	944.43	S
16	183	Virginia Commonwealth	42	1060.00	S
17	186	Widener	70	1005.00	P
18	187	Alabama A&M	38	722.21	S
19	243	Toledo	44	877.77	S
20	245	Wayne State	31	833.32	S

The SAT scores were actually given in the form of first and third quartiles, so the average of those two numbers is used here. Notice that some of the SAT scores are not integers. Those values were computed from ACT scores using the NCAA formula SAT $= -55.556 + 44.444 \cdot$ ACT, which is equivalent to saying that there is a linear relationship with 17 on the ACT corresponding to 700 on the SAT, and 26 on the ACT corresponding to 1100 on the SAT.

The scatter plot of the data in Figure 12.16 suggests the appropriateness of the linear regression model; graduation rate increases approximately linearly with SAT.

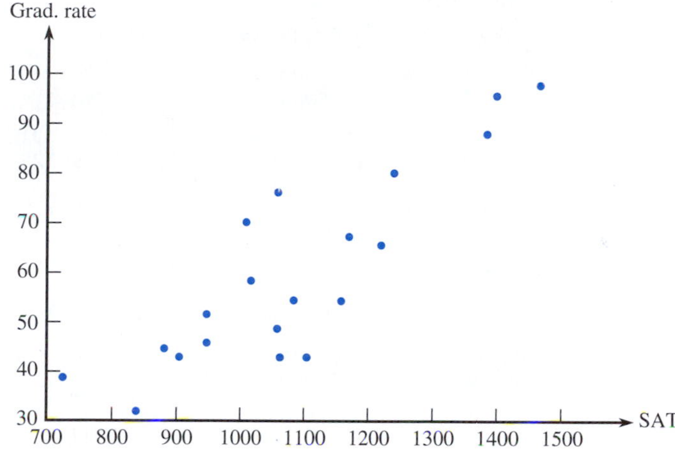

Figure 12.16 Scatter plot of the data from Example 12.12

The values of the summary statistics required for calculation of the least squares estimates are

$$\sum x_i = 21{,}600.97 \qquad \sum y_i = 1189 \qquad \sum x_i^2 = 24{,}034{,}220.545$$

$$\sum x_i y_i = 1{,}346{,}524.53 \qquad \sum y_i^2 = 78{,}113$$

from which $S_{xy} = 62{,}346.86$, $S_{xx} = 704{,}125.298$, $\hat{\beta}_1 = .08854513$, $\hat{\beta}_0 = -36.1830309$, SST $= 7426.95$, SSE $= 1906.439$, and $r^2 = 1 - 1906.439/7426.95 = .7433$. Roughly 74% of the observed variation in graduation rate can be attributed to the simple linear regression model relationship between graduation rate and SAT. Error df is $20 - 2 = 18$, giving $s^2 = 1906.439/18 = 105.9$ and $s = 10.29$.

The estimated standard deviation of $\hat{\beta}_1$ is

$$s_{\hat{\beta}_1} = \frac{s}{\sqrt{S_{xx}}} = \frac{10.29}{\sqrt{704{,}125}} = .01226$$

The t critical value for a confidence level of 95% is $t_{.025,18} = 2.101$. The confidence interval is

$$.0885 \pm (2.101)(.01226) = .0885 \pm .0258 = (.063, .114)$$

With a high degree of confidence, we estimate that an average increase in percentage graduation rate of between .063 and .114 is associated with a 1-point increase in SAT. Multiplying by 100 gives the change in graduation percentage corresponding to a

100-point increase in SAT, 8.85 ± 2.58, between 6.3 and 11.4. This shows that a substantial increase in graduation rate accompanies an increase of 100 SAT points. Is this a causal relationship, so a university president can count on an increased graduation rate if the admissions process becomes more selective in terms of entrance exam scores? One can imagine contrary scenarios, such as that more serious students attend more prestigious colleges, with higher entrance requirements and higher graduation rates, and that

```
The REG Procedure
Model: Linear_Regression_Model
Dependent Variable: GradRate

                              Analysis of Variance

                                  Sum of            Mean
Source                 DF        Squares          Square      F Value      Pr > F

Model                   1     5520.51091       5520.51091       52.12      <.0001
Error                  18     1906.43909        105.91328
Corrected Total        19     7426.95000

Root MSE             10.29142     R-Square      0.7433
Dependent Mean       59.45000     Adj R-Sq      0.7290
Coeff Var            17.31105

                              Parameter Estimates

               Parameter    Standard
Variable   DF   Estimate       Error    t Value  Pr > |t|    95% Confidence  Limits

Intercept   1  -36.18303    13.44467      -2.69    0.0149    -64.42924   -7.93682
SAT         1    0.08855     0.01226       7.22    <.0001      0.06278    0.11431

The REG Procedure
Model: Linear_Regression_Model
Dependent Variable: GradRate

                 Output Statistics

         Dep Var     Predicted
Obs     GradRate         Value        Residual

  1      98.0000       93.5356          4.4644
  2      96.0000       87.3374          8.6626
  3      88.0000       86.0092          1.9908
  4      65.0000       71.3993         -6.3993
  5      80.0000       73.1702          6.8298
  6      58.0000       53.3449          4.6551
  7      76.0000       57.2799         18.7201
  8      67.0000       67.1181         -0.1181
  9      48.0000       57.2799         -9.2799
 10      54.0000       66.0866        -12.0866
 11      42.0000       61.2157        -19.2157
 12      54.0000       59.4457         -5.4457
 13      45.0000       47.4416         -2.4416
 14      42.0000       43.5067         -1.5067
 15      51.0000       47.4416          3.5584
 16      42.0000       57.6748        -15.6748
 17      70.0000       52.8048         17.1952
 18      38.0000       27.7651         10.2349
 19      44.0000       41.5392          2.4608
 20      31.0000       37.6034         -6.6034
```

Figure 12.17 SAS output for the data of Example 12.12

prestige would not be affected by an increase in entrance requirements. However, it seems more likely that prestige would benefit from higher test scores, so this scenario is not a very good argument against causality. In any case, at least one university president claimed that increasing test scores resulted in a higher graduation rate.

Looking at the SAS output of Figure 12.17, we find the value of $s_{\hat{\beta}_1}$ under Parameter Estimates as the second number in the Standard Error column. All of the widely used statistical packages include this estimated standard error in output. There is also an estimated standard error for the statistic $\hat{\beta}_0$. Confidence intervals for β_1 and β_0 appear on the output. For all of the statistics, compare the values on the SAS output with the values that we calculated.

The output shows the values of graduation rate, predicted values, and residuals. Matching the rows in Figure 12.17 with the corresponding rows in the original listing of the data, it is possible to see that the residuals for the private universities are mostly positive. However, it is much easier to see this in Figure 12.18, where the private universities are labeled "P" and the public universities are labeled "S." Of the eight private universities, six are above their predictions (positive residual) and one is barely below. Private universities mostly seem to achieve a higher graduation rate for a given entrance exam score (for more on this issue, see the rest of the story in Sections 12.6 and 12.7). It is interesting to speculate about why this might occur. Is there a more nurturing atmosphere with more individual attention at private schools? On the other hand, private universities might attract students who are more likely to graduate regardless of the campus atmosphere.

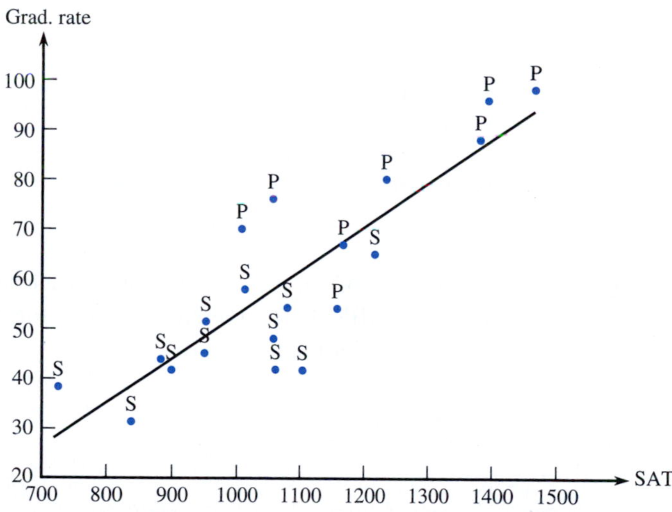

Figure 12.18 Comparing private and state universities ■

Hypothesis-Testing Procedures

As before, the null hypothesis in a test about β_1 will be an equality statement. The null value (value of β_1 claimed true by the null hypothesis) will be denoted by β_{10} (read "beta one nought," *not* "beta ten"). The test statistic results from replacing β_1 in the standardized variable T by the null value β_{10}—that is, from standardizing the estimator of β_1 under the assumption that H_0 is true. The test statistic thus has a t distribution with $n - 2$ df when H_0 is true, so the type I error probability is controlled at the desired level α by using an appropriate t critical value.

The most commonly encountered pair of hypotheses about β_1 is H_0: $\beta_1 = 0$ versus H_a: $\beta_1 \neq 0$. When this null hypothesis is true, $\mu_{Y \cdot x} = \beta_0$ independent of x, so knowledge of x gives no information about the value of the dependent variable. A test of these two hypotheses is often referred to as the *model utility test* in simple linear regression. Unless n is quite small, H_0 will be rejected and the utility of the model confirmed precisely when r^2 is reasonably large. The simple linear regression model should not be used for further inferences (estimates of mean value or predictions of future values) unless the model utility test results in rejection of H_0 for a suitably small α.

Null hypothesis: H_0: $\beta_1 = \beta_{10}$

Test statistic value: $t = \dfrac{\hat{\beta}_1 - \beta_{10}}{s_{\hat{\beta}_1}}$

Alternative Hypothesis	**Rejection Region for Level α Test**
H_a: $\beta_1 > \beta_{10}$	$t \geq t_{\alpha,n-2}$
H_a: $\beta_1 < \beta_{10}$	$t \leq -t_{\alpha,n-2}$
H_a: $\beta_1 \neq \beta_{10}$	either $t \geq t_{\alpha/2,n-2}$ or $t \leq -t_{\alpha/2,n-2}$

A P-value based on $n - 2$ df can be calculated just as was done previously for t tests in Chapters 9 and 10.

The **model utility test** is the test of H_0: $\beta_1 = 0$ versus H_a: $\beta_1 \neq 0$, in which case the test statistic value is the ***t* ratio** $t = \hat{\beta}_1 / s_{\hat{\beta}_1}$.

Example 12.13 Let's carry out the model utility test at significance level $\alpha = .05$ for the data of Example 12.12. We use the MINITAB regression output in Figure 12.19, which can be compared with the SAS output of Figure 12.17.

```
The regression equation is
Grad Rate = −36.2 + 0.0885 SAT

Predictor      Coef      SE Coef          T          P
Constant     −36.18        13.44      −2.69      0.015
SAT         0.08855      0.01226←s_β̂1      7.22←t = β̂1/s_β̂1     0.000←P-value for model utility test

S = 10.2914     R-Sq = 74.3%     R-Sq(adj) = 72.9%

Analysis of Variance

Source          DF      SS      MS      F      P
Regression       1  5520.5  5520.5  52.12  0.000
Residual Error  18  1906.4   105.9
Total           19  7427.0
```

Figure 12.19 MINITAB output for Example 12.13

The parameter of interest is β_1, the expected change in graduation rate associated with an increase of 1 in SAT score. The hypothesis H_0: $\beta_1 = 0$ will be rejected in favor of H_a: $\beta_1 \neq 0$ if the t ratio $t = \hat{\beta}_1/s_{\hat{\beta}_1}$ satisfies either $t \geq t_{\alpha/2,n-2} = t_{.025,18} = 2.101$ or $t \leq -2.101$.

From Figure 12.19, $\hat{\beta}_1 = .08855$, $s_{\hat{\beta}_1} = .01226$, and

$$t = \frac{.08855}{.01226} = 7.22 \quad \text{(also on output)}$$

Clearly, $7.22 \geq 2.101$, so H_0 is resoundingly rejected. Alternatively, the P-value is twice the area captured under the 18 df t curve to the right of 7.22. MINITAB gives P-value = .000, so H_0 should be rejected at any reasonable α. This confirmation of the utility of the simple linear regression model gives us license to calculate various estimates and predictions as described in Section 12.4.

Notice that, in contrast, SAS in Figure 12.17 gives a P-value of $<.0001$. This is better than the MINITAB P-value of .000 because the MINITAB value could be incorrectly read as 0. Of course the actual value is positive, approximately .0000010. When rounded to three decimals this gives the value .000 printed by MINITAB.

Given the confidence interval of Example 12.12, the result of the hypothesis test should be no surprise. It should be clear, in the two-tailed test for H_0: $\beta_1 = 0$ at level α, that H_0 is rejected if and only if the $100(1 - \alpha)\%$ confidence interval fails to include 0. In the present instance, the 95% confidence interval did not include 0, so we should have known that the two-tailed test at level .05 would reject H_0: $\beta_1 = 0$. ■

Regression and ANOVA

The splitting of the total sum of squares $\sum(y_i - \bar{y})^2$ into a part consisting of SSE, which measures unexplained variation, and a part consisting of SSR, which measures variation explained by the linear relationship, is strongly reminiscent of one-way ANOVA. In fact, the null hypothesis H_0: $\beta_1 = 0$ can be tested against H_a: $\beta_1 \neq 0$ by constructing an ANOVA table (Table 12.2) and rejecting H_0 if $f \geq F_{\alpha,1,n-2}$.

Table 12.2 ANOVA table for simple linear regression

Source of Variation	df	Sum of Squares	Mean Square	f
Regression	1	SSR	SSR	$\frac{\text{SSR}}{\text{SSE}/(n-2)}$
Error	$n - 2$	SSE	$s^2 = \frac{\text{SSE}}{n-2}$	
Total	$n - 1$	SST		

The F test gives exactly the same result as the model utility t test because $t^2 = f$ and $t^2_{\alpha/2,n-2} = F_{\alpha,1,n-2}$. Virtually all computer packages that have regression options include such an ANOVA table in the output. For example, Figure 12.17 shows SAS output for the university data of Example 12.12. The ANOVA table at the top of the output has $f = 52.12$ with a P-value of $<.0001$ (the actual value is about .0000010) for the model utility test.

The table of parameter estimates gives $t = 7.22$, again with $P = <.0001$ (the actual value is about .0000010), and $t^2 = (7.22)^2 = 52.12 = f$.

Fitting the Logistic Regression Model

Recall from Section 12.1 that in the logistic regression model, the dependent variable Y is 1 if the observation is a success and 0 otherwise. The probability of success is related to a quantitative predictor x by the logit function $p(x) = e^{\beta_0+\beta_1 x}/(1 + e^{\beta_0+\beta_1 x})$. Fitting the model to sample data requires that the parameters β_0 and β_1 be estimated. The standard way of doing this is by the method of maximum likelihood. Suppose, for example, that $n = 5$ and that the observations made at x_2, x_4, and x_5 are successes whereas the other two observations are failures. Then the likelihood function is

$$[1 - p(x_1)][p(x_2)][1 - p(x_3)][p(x_4)][p(x_5)]$$
$$= \left[\frac{1}{1 + e^{\beta_0+\beta_1 x_1}}\right]\left[\frac{e^{\beta_0+\beta_1 x_2}}{1 + e^{\beta_0+\beta_1 x_2}}\right]\left[\frac{1}{1 + e^{\beta_0+\beta_1 x_3}}\right]\left[\frac{e^{\beta_0+\beta_1 x_4}}{1 + e^{\beta_0+\beta_1 x_4}}\right]\left[\frac{e^{\beta_0+\beta_1 x_5}}{1 + e^{\beta_0+\beta_1 x_5}}\right]$$

Unfortunately it is not at all straightforward to maximize this likelihood, and there are no nice formulas for the mle's $\hat{\beta}_0$ and $\hat{\beta}_1$. The maximization process must be carried out using iterative numerical methods. The details are involved, but fortunately the most popular statistical software packages will do this on request and provide quantitative and graphical indications of how well the model fits.

In particular, the mle $\hat{\beta}_1$ is provided along with its estimated standard deviation $s_{\hat{\beta}_1}$. For large n, the estimator has approximately a normal distribution and the standardized variable $(\hat{\beta}_1 - \beta_1)/S_{\hat{\beta}_1}$ has approximately a standard normal distribution. This allows for calculation of a confidence interval for β_1 as well as for testing H_0: $\beta_1 = 0$, according to which the value of x has no impact on the likelihood of success. Some software packages report the value of the chi-squared statistic z^2 rather than z itself, along with the corresponding P-value for a two-tailed test.

Example 12.14 Here is data on launch temperature and the incidence of failure for O-rings in 24 space shuttle launches prior to the *Challenger* disaster of January 1986.

Temperature	Failure	Temperature	Failure	Temperature	Failure
53	Y	68	N	75	N
56	Y	69	N	75	Y
57	Y	70	N	76	N
63	N	70	Y	76	N
66	N	70	Y	78	N
67	N	70	Y	79	N
67	N	72	N	80	N
67	N	73	N	81	N

Figure 12.20 shows JMP output for a logistic regression analysis. We have chosen to let p denote the probability of failure. Failures tended to occur at lower temperatures and successes at higher temperatures, so the graph of $\hat{p}$ decreases as temperature increases. The estimate of β_1 is $\hat{\beta}_1 = -.1713$, and the estimated standard deviation of $\hat{\beta}_1$ is $s_{\hat{\beta}_1} = .08344$. The value of z for testing $H_0\colon \beta_1 = 0$, which asserts that temperature does not affect the likelihood of O-ring failure, is $\hat{\beta}_1 / s_{\hat{\beta}_1} = -.1713/.08344 = -2.05$. The P-value is .0404 (twice the area under the z curve to the left of -2.05). JMP reports the value of a chi-squared statistic, which is just z^2, and the chi-squared P-value differs from that for z only because of rounding. For each 1-degree increase in temperature, the odds of failure decrease by a factor of $e^{\hat{\beta}_1} = e^{-.1713} \approx .84$. The launch temperature for the *Challenger* mission was only 31°F. Because this value is much smaller than any temperature in our sample, it is dangerous to extrapolate the estimated relationship. Nevertheless, it appears that for a temperature this small, O-ring failure is almost a sure thing. The logistic regression gives the estimated probability at $x = 31$

$$p(31) = \frac{e^{\beta_0 + \beta_1 31}}{1 + e^{\beta_0 + \beta_1 31}} = \frac{e^{10.8753 - .17132(31)}}{1 + e^{10.8753 - .17132(31)}} = .996182$$

and the odds associated with this probability are $.996182/(1 - .996182) = .996182/.003818 \approx 261$. Thus, if the logistic regression can be extrapolated down to 31, the probability of failure is .996182, the probability of success is .003818, and the predicted odds are 261 to 1 against success. Too bad this calculation was not done before launch!

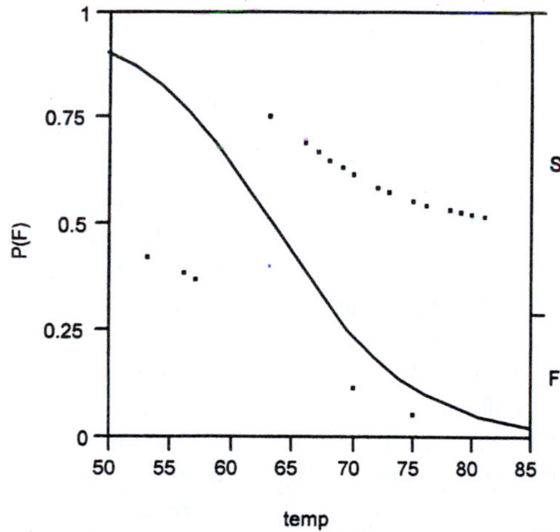

Parameter Estimates

Term	Estimate	Std Error	ChiSquare	Prob>ChiSq
Intercept	10.8753321	5.7031291	3.64	0.0565
temp	-0.1713202	0.0834419	4.22	0.0401

Figure 12.20 Logistic regression output from JMP

Exercises Section 12.3 (31–44)

31. Reconsider the situation described in Example 12.5, in which x = CO_2 concentration and y = mass of 11-month-old pine trees. Suppose the simple linear regression model is valid for x between 450 and 750, and that $\beta_1 = .008$ and $\sigma = .5$. Consider an experiment in which $n = 7$, and the x values at which observations are made are $x_1 = 450$, $x_2 = 500$, $x_3 = 550$, $x_4 = 600$, $x_5 = 650$, $x_6 = 700$, and $x_7 = 750$.

a. Calculate $\sigma_{\hat{\beta}_1}$, the standard deviation of $\hat{\beta}_1$.

b. What is the probability that the estimated slope based on such observations will be between .006 and .010?

c. Suppose it is also possible to make a single observation at each of the $n = 11$ values 525, 540, 555, 570, . . . , 675. If a major objective is to estimate β_1 as accurately as possible, would the experiment with $n = 11$ be preferable to the one with $n = 7$?

32. Exercise 17 of Section 12.2 gave data on x = rainfall volume and y = runoff volume (both in m^3). Use the accompanying MINITAB output to decide whether there is a useful linear relationship between rainfall and runoff, and then calculate a confidence interval for the true average change in runoff volume associated with a 1-m^3 increase in rainfall volume.

```
The regression equation is runoff =
-1.13 + 0.827 rainfall

Predictor      Coef     Stdev  t-ratio      P
Constant     -1.128     2.368    -0.48  0.642
rainfall    0.82697   0.03652    22.64  0.000

s = 5.240 R-sq = 97.5%  R-sq(adj) = 97.3%
```

33. Exercise 16 of Section 12.2 included MINITAB output for a regression of daughter's height on the midparent height.

a. Use the output to calculate a confidence interval with a confidence level of 95% for the slope β_1 of the population regression line, and interpret the resulting interval.

b. Suppose it had previously been believed that when midparent height increased by 1 inch, the associated true average change in the daughter's height would be at least 1 inch. Does the sample data contradict this belief? State and test the relevant hypotheses.

34. Refer to the MINITAB output of Exercise 20, in which x = NO_3^- wet deposition and y = lichen N (%).

a. Carry out the model utility test at level .01, using the rejection region approach.

b. Repeat part (a) using the P-value approach.

c. Suppose it had previously been believed that when NO_3^- wet deposition increases by .1 g N/m^2, the associated change in expected lichen N is at least .15%. Carry out a test of hypotheses at level .01 to decide whether the data contradicts this prior belief.

35. How does lateral acceleration—side forces experienced in turns that are largely under driver control—affect nausea as perceived by bus passengers? The article "Motion Sickness in Public Road Transport: The Effect of Driver, Route, and Vehicle" (*Ergonomics*, 1999: 1646–1664) reported data on x = motion sickness dose (calculated in accordance with a British standard for evaluating similar motion at sea) and y = reported nausea (%). Relevant summary quantities are

$$n = 17,\ \sum x_i = 222.1,\ \sum y_i = 193,$$
$$\sum x_i^2 = 3056.69,\ \sum x_i y_i = 2759.6,$$
$$\sum y_i^2 = 2975$$

Values of dose in the sample ranged from 6.0 to 17.6.

a. Assuming that the simple linear regression model is valid for relating these two variables (this is supported by the raw data), calculate and interpret an estimate of the slope parameter that conveys information about the precision and reliability of estimation.

b. Does it appear that there is a useful linear relationship between these two variables? Answer the question by employing the P-value approach.

c. Would it be sensible to use the simple linear regression model as a basis for predicting % nausea when dose = 5.0? Explain your reasoning.

d. When MINITAB was used to fit the simple linear regression model to the raw data, the observation (6.0, 2.50) was flagged as possibly having a substantial impact on the fit. Eliminate this observation from the sample and recalculate the estimate of part (a). Based on this, does the observation appear to be exerting an undue influence?

36. Mist (airborne droplets or aerosols) is generated when metal-removing fluids are used in machining operations to cool and lubricate the tool and workpiece. Mist generation is a concern to OSHA, which has recently lowered substantially the workplace standard. The article "Variables Affecting Mist Generation from Metal Removal Fluids" (*Lubricat. Engrg.*, 2002: 10–17) gave the accompanying data on x = fluid flow velocity for a 5% soluble oil (cm/sec) and y = the extent of mist droplets having diameters smaller than 10 μm (mg/m^3):

x	89	177	189	354	362	442	965
y	.40	.60	.48	.66	.61	.69	.99

a. The investigators performed a simple linear regression analysis to relate the two variables. Does a scatter plot of the data support this strategy?

b. What proportion of observed variation in mist can be attributed to the simple linear regression relationship between velocity and mist?

c. The investigators were particularly interested in the impact on mist of increasing velocity from 100 to 1000 (a factor of 10 corresponding to the difference between the smallest and largest x values in the sample). When x increases in this way, is there substantial evidence that the true average increase in y is less than .6?

d. Estimate the true average change in mist associated with a 1 cm/sec increase in velocity, and do so in a way that conveys information about precision and reliability.

37. Refer to the data on x = liberation rate and y = NO_x emission rate given in Exercise 19.

a. Does the simple linear regression model specify a useful relationship between the two rates? Use the appropriate test procedure to obtain information about the P-value and then reach a conclusion at significance level .01.

b. Compute a 95% CI for the expected change in emission rate associated with a 10 MBtu/hr-ft^2 increase in liberation rate.

38. Carry out the model utility test using the ANOVA approach for the filtration rate–moisture content data of Example 12.7. Verify that it gives a result equivalent to that of the t test.

39. Use the rules of expected value to show that $\hat{\beta}_0$ is an unbiased estimator for β_0 (assuming that $\hat{\beta}_1$ is unbiased for β_1).

40. a. Verify that $E(\hat{\beta}_1) = \beta_1$ by using the rules of expected value from Chapter 6.

b. Use the rules of variance from Chapter 6 to verify the expression for $V(\hat{\beta}_1)$ given in this section.

41. Verify that if each x_i is multiplied by a positive constant c and each y_i is multiplied by another positive constant d, the t statistic for testing H_0: $\beta_1 = 0$ versus H_a: $\beta_1 \neq 0$ is unchanged in value (the value of $\hat{\beta}_1$ will change, which shows that the magnitude of $\hat{\beta}_1$ is not by itself indicative of model utility).

42. The probability of a type II error for the t test for H_0: $\beta_1 = \beta_{10}$ can be computed in the same manner as it was computed for the t tests of Chapter 9. If the alternative value of β_1 is denoted by β_1', the value of

$$d = \frac{|\beta_{10} - \beta_1'|}{\sigma\sqrt{\dfrac{n-1}{\sum x_i^2 - (\sum x_i)^2/n}}}$$

is first calculated, then the appropriate set of curves in Appendix Table A.17 is entered on the horizontal axis at the value of d, and β is read from the curve for $n - 2$ df. An article in the *J. Public Health Engrg.* reports the results of a regression analysis based on $n = 15$ observations in which x = filter application temperature (°C) and y = % efficiency of BOD removal. Here BOD stands for biochemical oxygen demand, and it is a measure of organic matter in sewage. Calculated quantities include $\sum x_i = 402$, $\sum x_i^2 = 11{,}098$, $s = 3.725$, and $\hat{\beta}_1 = 1.7035$. Consider testing at level .01 H_0: $\beta_1 = 1$, which states that the expected increase in % BOD removal is 1 when filter application temperature increases by 1°C, against the alternative H_a: $\beta_1 > 1$. Determine P(type II error) when $\beta_1' = 2$, $\sigma = 4$.

43. Kyphosis, or severe forward flexion of the spine, may persist despite corrective spinal surgery. A study carried out to determine risk factors for kyphosis reported the following ages (months) for 40 subjects at the time of the operation; the first 18 subjects did have kyphosis and the remaining 22 did not.

Kyphosis	12	15	42	52	59	73
	82	91	96	105	114	120
	121	128	130	139	139	157
No kyphosis	1	1	2	8	11	18
	22	31	37	61	72	81
	97	112	118	127	131	140
	151	159	177	206		

Use the accompanying MINITAB logistic regression output to decide whether age appears to have a significant impact on the presence of kyphosis.

Predictor	Coef	StDev	z	P	Odds Ratio	95% Lower	CI Upper
Constant	−0.5727	0.6024	−0.95	0.342			
age	0.004296	0.005849	0.73	0.463	1.00	0.99	1.02

44. The following data resulted from a study commissioned by a large management consulting company to investigate the relationship between amount of job experience (months) for a junior consultant and the likelihood of the consultant being able to perform a certain complex task.

Success 8 13 14 18 20 21 21 22 25 26 28 29 30 32

Failure 4 5 6 6 7 9 10 11 11 13 15 18 19 20 23 27

Interpret the accompanying MINITAB logistic regression output, and sketch a graph of the estimated probability of task performance as a function of experience.

Predictor	Coef	StDev	z	P	Odds Ratio	95% Lower	CI Upper
Constant	−3.211	1.235	−2.60	0.009			
age	0.17772	0.06573	2.70	0.007	1.19	1.05	1.36

12.4 Inferences Concerning $\mu_{Y \cdot x^*}$ and the Prediction of Future Y Values

Let x^* denote a specified value of the independent variable x. Once the estimates $\hat{\beta}_0$ and $\hat{\beta}_1$ have been calculated, $\hat{\beta}_0 + \hat{\beta}_1 x^*$ can be regarded either as a point estimate of $\mu_{Y \cdot x^*}$ (the expected or true average value of Y when $x = x^*$) or as a prediction of the Y value that will result from a single observation made when $x = x^*$. The point estimate or prediction by itself gives no information concerning how precisely $\mu_{Y \cdot x^*}$ has been estimated or Y has been predicted. This can be remedied by developing a CI for $\mu_{Y \cdot x^*}$ and a prediction interval (PI) for a single Y value.

Before we obtain sample data, both $\hat{\beta}_0$ and $\hat{\beta}_1$ are subject to sampling variability—that is, they are both statistics whose values will vary from sample to sample. This variability was shown in Example 12.11 at the beginning of Section 12.3. Suppose, for example, that $\beta_0 = 50$ and $\beta_1 = 2$. Then a first sample of (x, y) pairs might give $\hat{\beta}_0 = 52.35$, $\hat{\beta}_1 = 1.895$, a second sample might result in $\hat{\beta}_0 = 46.52$, $\hat{\beta}_1 = 2.056$, and so on. It follows that $\hat{Y} = \hat{\beta}_0 + \hat{\beta}_1 x^*$ itself varies in value from sample to sample, so it is a statistic. If the intercept and slope of the population line are the aforementioned values 50 and 2, respectively, and $x^* = 10$, then this statistic is trying to estimate the value $50 + 2(10) = 70$. The estimate from a first sample might be $52.35 + 1.895(10) = 71.30$, from a second sample might be $46.52 + 2.056(10) = 67.08$, and so on. In the same way that a confidence interval for β_1 was based on properties of the sampling distribution of $\hat{\beta}_1$, a confidence interval for a mean y value in regression is based on properties of the sampling distribution of the statistic $\hat{\beta}_0 + \hat{\beta}_1 x^*$.

Substitution of the expressions for $\hat{\beta}_0$ and $\hat{\beta}_1$ into $\hat{\beta}_0 + \hat{\beta}_1 x^*$ followed by some algebraic manipulation leads to the representation of $\hat{\beta}_0 + \hat{\beta}_1 x^*$ as a linear function of the Y_i's:

$$\hat{\beta}_0 + \hat{\beta}_1 x^* = \sum_{i=1}^{n} \left[\frac{1}{n} + \frac{(x^* - \bar{x})(x_i - \bar{x})}{\sum (x_j - \bar{x})^2} \right] Y_i = \sum_{i=1}^{n} d_i Y_i$$

The coefficients $d_1, d_2, \ldots, d_n$ in this linear function involve the x_i's and x^*, all of which are fixed. Application of the rules of Section 6.3 to this linear function gives the following properties. (Exercise 55 requests a derivation of Property 2).

Let $\hat{Y} = \hat{\beta}_0 + \hat{\beta}_1 x^*$, where x^* is some fixed value of x. Then

1. The mean value of $\hat{Y}$ is

$$E(\hat{Y}) = E(\hat{\beta}_0 + \hat{\beta}_1 x^*) = \mu_{\hat{\beta}_0 + \hat{\beta}_1 x^*} = \beta_0 + \beta_1 x^*$$

Thus $\hat{\beta}_0 + \hat{\beta}_1 x^*$ is an unbiased estimator for $\beta_0 + \beta_1 x^*$ (i.e., for $\mu_{Y \cdot x^*}$).

2. The variance of $\hat{Y}$ is

$$V(\hat{Y}) = \sigma_{\hat{Y}}^2 = \sigma^2 \left[\frac{1}{n} + \frac{(x^* - \bar{x})^2}{\sum x_i^2 - (\sum x_i)^2/n} \right] = \sigma^2 \left[\frac{1}{n} + \frac{(x^* - \bar{x})^2}{S_{xx}} \right]$$

and the standard deviation $\sigma_{\hat{Y}}$ is the square root of this expression. The estimated standard deviation of $\hat{\beta}_0 + \hat{\beta}_1 x^*$, denoted by $s_{\hat{Y}}$ or $s_{\hat{\beta}_0 + \hat{\beta}_1 x^*}$, results from replacing σ by its estimate s:

$$s_{\hat{Y}} = s_{\hat{\beta}_0 + \hat{\beta}_1 x^*} = s \sqrt{\frac{1}{n} + \frac{(x^* - \bar{x})^2}{S_{xx}}}$$

3. $\hat{Y}$ has a normal distribution (because the Y_i's are normally distributed and independent).

The variance of $\hat{\beta}_0 + \hat{\beta}_1 x^*$ is smallest when $x^* = \bar{x}$ and increases as x^* moves away from $\bar{x}$ in either direction. Thus the estimator of $\mu_{Y \cdot x^*}$ is more precise when x^* is near the center of the x_i's than when it is far from the x values at which observations have been made. This implies that both the CI and PI are narrower for an x^* near $\bar{x}$ than for an x^* far from $\bar{x}$. Most statistical computer packages provide both $\hat{\beta}_0 + \hat{\beta}_1 x^*$ and $s_{\hat{\beta}_0 + \hat{\beta}_1 x^*}$ for any specified x^* upon request.

Inferences Concerning $\mu_{Y \cdot x^*}$

Just as inferential procedures for β_1 were based on the t variable obtained by standardizing β_1, a t variable obtained by standardizing $\hat{\beta}_0 + \hat{\beta}_1 x^*$ leads to a CI and test procedures here.

THEOREM The variable

$$T = \frac{\hat{\beta}_0 + \hat{\beta}_1 x^* - (\beta_0 + \beta_1 x^*)}{S_{\hat{\beta}_0 + \hat{\beta}_1 x^*}} = \frac{\hat{Y} - (\beta_0 + \beta_1 x^*)}{S_{\hat{Y}}} \quad (12.6)$$

has a t distribution with $n - 2$ df.

As for β_1 in the previous section, a probability statement involving this standardized variable can be manipulated to yield a confidence interval for $\mu_{Y \cdot x^*}$.

A $100(1 - \alpha)\%$ **CI for $\mu_{Y \cdot x^*}$**, the expected value of Y when $x = x^*$, is

$$\hat{\beta}_0 + \hat{\beta}_1 x^* \pm t_{\alpha/2, n-2} \cdot s_{\hat{\beta}_0 + \hat{\beta}_1 x^*} = \hat{y} \pm t_{\alpha/2, n-2} \cdot s_{\hat{Y}} \quad (12.7)$$

This CI is centered at the point estimate for $\mu_{Y \cdot x^*}$ and extends out to each side by an amount that depends on the confidence level and on the extent of variability in the estimator on which the point estimate is based.

Example 12.15 Recall the university data of Example 12.12, where the dependent variable was graduation rate and the predictor was the average SAT for entering freshmen. Results from Example 12.12 include $\sum x_i = 21{,}600.97$, $S_{xx} = 704{,}125.298$, $\hat{\beta}_1 = .088545$, $\hat{\beta}_0 = -36.18$, $s = 10.29$, and therefore $\bar{x} = 21{,}600.97/20 = 1080$. Let's now calculate a confidence interval, using a 95% confidence level, for the mean graduation rate for all universities having an average freshman SAT of 1200—that is, a confidence interval for $\beta_0 + \beta_1(1200)$. The interval is centered at

$$\hat{y} = \hat{\beta}_0 + \hat{\beta}_1(1200) = -36.18 + .0885(1200) = 70.07$$

The estimated standard deviation of the statistic $\hat{Y}$ is

$$s_{\hat{Y}} = s\sqrt{\frac{1}{n} + \frac{(x^* - \bar{x})^2}{S_{xx}}} = 10.29\sqrt{\frac{1}{20} + \frac{(1200 - 1080)^2}{704{,}125}} = 2.731$$

The 18 df t critical value for a 95% confidence level is 2.101, from which we determine the desired interval to be

$$70.07 \pm (2.101)(2.731) = 70.07 \pm 5.74 = (64.33, 75.81)$$

The narrowness of this interval suggests that we have reasonably precise information about the mean value being estimated. Remember that if we recalculated this interval for sample after sample, in the long run about 95% of the calculated intervals would include $\beta_0 + \beta_1(1200)$. We can only hope that this mean value lies in the single interval that we have calculated.

Figure 12.21 shows MINITAB output resulting from a request to calculate confidence intervals for the mean graduation rate when the SAT is 1100 and 1200. Because this optional output was requested, the confidence intervals (Figure 12.21) were appended to the bottom of the regression output given in Figure 12.19. Note that the first interval is narrower than the second, because 1100 is much closer to $\bar{x}$ than is 1200. Figure 12.22 shows curves corresponding to the confidence limits for each different x value. Notice how the curves get farther and farther apart as x moves away from $\bar{x}$. The output labeled PI in Figure 12.21 and the curves labeled PI in Figure 12.22 refer to prediction intervals, to be discussed shortly.

```
Predicted Values for New Observations
New
Obs     Fit   SE Fit        95% CI              95% PI
  1   61.22     2.31   (56.35, 66.08)   (39.06, 83.38)
  2   70.07     2.73   (64.33, 75.81)   (47.70, 92.44)
```

Figure 12.21 MINITAB regression output for the data of Example 12.15

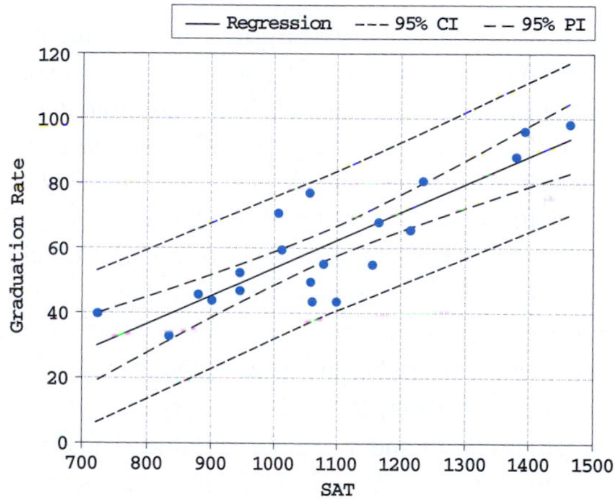

Figure 12.22 MINITAB scatter plot with confidence intervals and prediction intervals for the data of Example 12.15 ■

In some situations, a CI is desired not just for a single x value but for two or more x values. Suppose an investigator wishes a CI both for $\mu_{Y \cdot v}$ and for $\mu_{Y \cdot w}$ where v and w are two different values of the independent variable. It is tempting to compute the interval (12.7) first for $x = v$ and then for $x = w$. Suppose we use $\alpha = .05$ in each computation to get two 95% intervals. Then if the variables involved in computing the two intervals were independent of one another, the joint confidence coefficient would be $(.95) \cdot (.95) \approx .90$.

Unfortunately, the intervals are not independent because the same $\hat{\beta}_0$, $\hat{\beta}_1$, and S are used in each. We therefore cannot assert that the joint confidence level for the two intervals is exactly 90%. However, Exercise 78 of Chapter 8 derives the Bonferroni inequality showing that, if the $100(1 - \alpha)\%$ CI (12.7) is computed both for $x = v$ and for

$x = w$ to obtain joint CIs for $\mu_{Y \cdot v}$ and $\mu_{Y \cdot w}$, then *the joint confidence level on the resulting pair of intervals is at least* $100(1 - 2\alpha)\%$. In particular, using $\alpha = .05$ results in a joint confidence level of *at least* 90%, whereas using $\alpha = .01$ results in at least 98% confidence. For example, in Example 12.15 a 95% CI for $\mu_{Y \cdot 1100}$ was (56.35, 66.08) and a 95% CI for $\mu_{Y \cdot 1200}$ was (64.33, 75.81). The simultaneous or joint confidence level for the two statements $56.35 < \mu_{Y \cdot 1100} < 66.08$ and $64.33 < \mu_{Y \cdot 1200} < 75.81$ is at least 90%.

The joint CIs are referred to as *Bonferroni intervals.* The method is easily generalized to yield joint intervals for k different $\mu_{Y \cdot x}$'s. *Using the interval* (12.7) *separately first for* $x = x_1^*$, *then for* $x = x_2^*, \ldots,$ *and finally for* $x = x_k^*$ *yields a set of k CIs for which the joint or simultaneous confidence level is guaranteed to be at least* $100(1 - k\alpha)\%$.

Tests of hypotheses about $\beta_0 + \beta_1 x^*$ are based on the test statistic T obtained by replacing $\beta_0 + \beta_1 x^*$ in the numerator of (12.6) by the null value μ_0. For example, the assertion H_0: $\beta_0 + \beta_1(1200) = 75$ in Example 12.15 says that when the average SAT is 1200, expected (i.e., true average) graduation rate is 75%. The test statistic value is then $t = [\hat{\beta}_0 + \hat{\beta}_1(1200) - 75]/s_{\hat{\beta}_0 + \hat{\beta}_1(1200)}$, and the test is upper-, lower-, or two-tailed according to the inequality in H_a.

A Prediction Interval for a Future Value of Y

Analogous to the CI (12.7) for $\mu_{Y \cdot x^*}$, one frequently wishes to obtain an interval of plausible values for the value of Y associated with some future observation when the independent variable has value x^*. In the example in which vocabulary size y is related to the age x of a child, for $x = 6$ years (12.7) would provide a CI for the true average vocabulary size of *all* 6-year-old children. Alternatively, we might wish an interval of plausible values for the vocabulary size of a particular 6-year-old child.

A CI refers to a parameter, or population characteristic, whose value is fixed but unknown to us. In contrast, a future value of Y is not a parameter but instead a random variable; for this reason we refer to an interval of plausible values for a future Y as a **prediction interval** rather than a confidence interval. For the confidence interval we use the error of estimation, $\beta_0 + \beta_1 x^* - (\hat{\beta}_0 + \hat{\beta}_1 x^*)$, a difference between a fixed (but unknown) quantity and a random variable. The error of prediction is $Y - (\hat{\beta}_0 + \hat{\beta}_1 x^*) = \beta_0 + \beta_1 x^* + \varepsilon - (\hat{\beta}_0 + \hat{\beta}_1 x^*)$, a difference between two random variables. With the additional random ε term, there is more uncertainty in prediction than in estimation, so a PI will be wider than a CI. Because the future value Y is independent of the observed Y_i's,

$$\begin{aligned} V[Y - (\hat{\beta}_0 + \hat{\beta}_1 x^*)] &= \text{variance of prediction error} \\ &= V(Y) + V(\hat{\beta}_0 + \hat{\beta}_1 x^*) \\ &= \sigma^2 + \sigma^2 \left[\frac{1}{n} + \frac{(x^* - \bar{x})^2}{S_{xx}} \right] \\ &= \sigma^2 \left[1 + \frac{1}{n} + \frac{(x^* - \bar{x})^2}{S_{xx}} \right] \end{aligned}$$

Furthermore, because $E(Y) = \beta_0 + \beta_1 x^*$ and $E(\hat{\beta}_0 + \hat{\beta}_1 x^*) = \beta_0 + \beta_1 x^*$, the expected value of the prediction error is $E[Y - (\hat{\beta}_0 + \hat{\beta}_1 x^*)] = 0$. It can then be shown that the standardized variable

$$T = \frac{Y - (\hat{\beta}_0 + \hat{\beta}_1 x^*)}{S\sqrt{1 + \dfrac{1}{n} + \dfrac{(x^* - \bar{x})^2}{S_{xx}}}}$$

has a t distribution with $n - 2$ df. Substituting this T into the probability statement $P(-t_{\alpha/2,n-2} < T < t_{\alpha/2,n-2}) = 1 - \alpha$ and manipulating to isolate Y between the two inequalities yields the following interval.

A $100(1 - \alpha)\%$ **PI for a future Y observation to be made when $x = x^*$ is**

$$\hat{\beta}_0 + \hat{\beta}_1 x^* \pm t_{\alpha/2,n-2} \cdot s\sqrt{1 + \frac{1}{n} + \frac{(x^* - \bar{x})^2}{S_{xx}}} \qquad (12.8)$$

$$= \hat{\beta}_0 + \hat{\beta}_1 x^* \pm t_{\alpha/2,n-2} \cdot \sqrt{s^2 + s^2_{\hat{\beta}_0 + \hat{\beta}_1 x^*}}$$

$$= \hat{y} \pm t_{\alpha/2,n-2} \cdot \sqrt{s^2 + s^2_{\hat{Y}}}$$

The interpretation of the prediction level $100(1 - \alpha)\%$ is identical to that of previous confidence levels—if (12.8) is used repeatedly, in the long run the resulting intervals will actually contain the observed y values $100(1 - \alpha)\%$ of the time. Notice that the 1 underneath the initial square root symbol makes the PI (12.8) wider than the CI (12.7), though the intervals are both centered at $\hat{\beta}_0 + \hat{\beta}_1 x^*$. Also, as $n \to \infty$ the width of the CI approaches 0, whereas the width of the PI approaches $2z_{\alpha/2}\sigma$ (because even with perfect knowledge of β_0 and β_1, there will still be uncertainty in prediction).

Example 12.16 Let's return to the university data of Example 12.15 and calculate a 95% prediction interval for a graduation rate that would result from selecting a single university whose average SAT is 1200. Relevant quantities from that example are

$$\hat{y} = 70.07 \qquad s_{\hat{Y}} = 2.731 \qquad s = 10.29$$

For a prediction level of 95% based on $n - 2 = 18$ df, the t critical value is 2.101, exactly what we previously used for a 95% confidence level. The prediction interval is then

$$70.07 \pm (2.101)\sqrt{10.29^2 + 2.731^2} = 70.07 \pm (2.101)(10.646)$$
$$= 70.07 \pm 22.37 = (47.70, 92.44)$$

Plausible values for a single observation on graduation rate when SAT is 1200 are (at the 95% prediction level) between 47.70% and 92.44%. The 95% confidence interval for graduation rate when SAT is 1200 was (64.33, 75.81). The prediction interval is much wider than this because of the extra 10.29^2 under the square root. Figure 12.22, the MINITAB output for Example 12.15, shows this interval as well as the confidence interval. ■

The Bonferroni technique can be employed as in the case of confidence intervals. If a $100(1 - \alpha)\%$ PI is calculated for each of k different values of x, the simultaneous or joint prediction level for all k intervals is at least $100(1 - k\alpha)\%$.

Exercises Section 12.4 (45–55)

45. Recall Examples 12.5 and 12.6 of Section 12.2, where the simple linear regression model was applied to 8 observations on x = CO_2 concentration and y = mass (kg) of pine trees at age 11 months. Further calculations give $s = .534$ and $\hat{y} = 2.723$, $s_{\hat{Y}} = .190$ when $x = 600$ and $\hat{y} = 3.992$, $s_{\hat{Y}} = .256$ when $x = 750$.

a. Explain why $s_{\hat{Y}}$ is larger when $x = 750$ than when $x = 600$.

b. Calculate a confidence interval with a confidence level of 95% for the true average mass of all trees grown with a CO_2 concentration of 600 ppm.

c. Calculate a prediction interval with a prediction level of 95% for the mass of a tree grown with a CO_2 concentration of 600 ppm.

d. If a 95% CI is calculated for the true average mass when CO_2 concentration is 750, what will be the simultaneous confidence level for both this interval and the interval calculated in part (b)?

46. Reconsider the filtration rate–moisture content data introduced in Example 12.7 (see also Example 12.8).

a. Compute a 90% CI for $\beta_0 + 125\beta_1$, true average moisture content when the filtration rate is 125.

b. Predict the value of moisture content for a single experimental run in which the filtration rate is 125 using a 90% prediction level. How does this interval compare to the interval of part (a)? Why is this the case?

c. How would the intervals of parts (a) and (b) compare to a CI and PI when filtration rate is 115? Answer without actually calculating these new intervals.

d. Interpret the hypotheses H_0: $\beta_0 + 125\beta_1 = 80$ and H_0: $\beta_0 + 125\beta_1 < 80$, and then carry out a test at significance level .01.

47. The article "The Incorporation of Uranium and Silver by Hydrothermally Synthesized Galena" (*Econ. Geol.*, 1964: 1003–1024) reports on the determination of silver content of galena crystals grown in a closed hydrothermal system over a range of temperature. With x = crystallization temperature in °C and y = Ag_2S in mol%, the data follows:

x	398	292	352	575	568	450	550
y	.15	.05	.23	.43	.23	.40	.44

x	408	484	350	503	600	600
y	.44	.45	.09	.59	.63	.60

from which $\sum x_i = 6130$, $\sum x_i^2 = 3{,}022{,}050$, $\sum y_i = 4.73$, $\sum y_i^2 = 2.1785$, $\sum x_i y_i = 2418.74$, $\hat{\beta}_1 = .00143$, $\hat{\beta}_0 = -.311$, and $s = .131$.

a. Estimate true average silver content when temperature is 500°C using a 95% confidence interval.

b. How would the width of a 95% CI for true average silver content when temperature is 400°C compare to the width of the interval in part (a)? Answer without computing this new interval.

c. Calculate a 95% CI for the true average change in silver content associated with a 1°C increase in temperature.

d. Suppose it had previously been believed that when crystallization temperature was 400°C, true average silver content would be .25. Carry out a test at significance level .05 to decide whether the sample data contradicts this prior belief.

48. The simple linear regression model provides a very good fit to the data on rainfall and runoff volume given in Exercise 17 of Section 12.2. The equation of the least squares line is $\hat{y} = -1.128 + .82697x$, $r^2 = .975$, and $s = 5.24$.

a. Use the fact that $s_{\hat{Y}} = 1.44$ when rainfall volume is 40 m^3 to predict runoff in a way that conveys information about reliability and precision. Does the resulting interval suggest that precise information about the value of runoff for this future observation is available? Explain your reasoning.

b. Calculate a PI for runoff when rainfall is 50 using the same prediction level as in part (a). What can be said about the simultaneous prediction level for the two intervals you have calculated?

49. You are told that a 95% CI for expected lead content when traffic flow is 15, based on a sample of $n = 10$ observations, is (462.1, 597.7). Calculate a CI with confidence level 99% for expected lead content when traffic flow is 15.

50. Refer to Exercise 21 in which x = available travel space in feet and y = separation distance in feet between a bicycle and a passing car.

a. MINITAB gives $s_{\hat{\beta}_0+\hat{\beta}_1(15)} = .186$ and $s_{\hat{\beta}_0+\hat{\beta}_1(20)} = .360$. Explain why one is much larger than the other.

b. Calculate a 95% CI for expected separation distance when available travel space is 15 feet. (Use $s_{\hat{\beta}_0+\hat{\beta}_1(15)} = .186$.)

c. Calculate a 95% PI for a single instance of separation distance when available travel space is 20 feet. (Use $s_{\hat{\beta}_0+\hat{\beta}_1(20)} = .360$.)

51. Plasma etching is essential to the fine-line pattern transfer in current semiconductor processes. The article "Ion Beam-Assisted Etching of Aluminum with Chlorine" (*J. Electrochem. Soc.*, 1985: 2010–2012) gives the accompanying data (read from a graph) on chlorine flow (x, in SCCM) through a nozzle used in the etching mechanism and etch rate (y, in 100 A/min).

x	1.5	1.5	2.0	2.5	2.5	3.0	3.5	3.5	4.0
y	23.0	24.5	25.0	30.0	33.5	40.0	40.5	47.0	49.0

The summary statistics are $\sum x_i = 24.0$, $\sum y_i = 312.5$, $\sum x_i^2 = 70.50$, $\sum x_i y_i = 902.25$, $\sum y_i^2 = 11{,}626.75$, $\hat{\beta}_0 = 6.448718$, $\hat{\beta}_1 = 10.602564$.

a. Does the simple linear regression model specify a useful relationship between chlorine flow and etch rate?

b. Estimate the true average change in etch rate associated with a 1-SCCM increase in flow rate using a 95% confidence interval, and interpret the interval.

c. Calculate a 95% CI for $\mu_{Y \cdot 3.0}$, the true average etch rate when flow = 3.0. Has this average been precisely estimated?

d. Calculate a 95% PI for a single future observation on etch rate to be made when flow = 3.0. Is the prediction likely to be accurate?

e. Would the 95% CI and PI when flow = 2.5 be wider or narrower than the corresponding intervals of parts (c) and (d)? Answer without actually computing the intervals.

f. Would you recommend calculating a 95% PI for a flow of 6.0? Explain.

g. Calculate simultaneous CI's for true average etch rate when chlorine flow is 2.0, 2.5, and 3.0, respectively. Your simultaneous confidence level should be at least 97%.

52. Consider the following four intervals based on the data of Exercise 20 (Section 12.2):

a. A 95% CI for lichen nitrogen when NO_3^- is .5

b. A 95% PI for lichen nitrogen when NO_3^- is .5

c. A 95% CI for lichen nitrogen when NO_3^- is .8

d. A 95% PI for lichen nitrogen when NO_3^- is .8

Without computing any of these intervals, what can be said about their widths relative to one another?

53. The decline of water supplies in certain areas of the United States has created the need for increased understanding of relationships between economic factors such as crop yield and hydrologic and soil factors. The article "Variability of Soil Water Properties and Crop Yield in a Sloped Watershed" (*Water Resources Bull.*, 1988: 281–288) gives data on grain sorghum yield (y, in g/m-row) and distance upslope (x, in m) on a sloping watershed. Selected observations are given in the accompanying table.

x	0	10	20	30	45	50	70
y	500	590	410	470	450	480	510
x	80	100	120	140	160	170	190
y	450	360	400	300	410	280	350

a. Construct a scatter plot. Does the simple linear regression model appear to be plausible?

b. Carry out a test of model utility.

c. Estimate true average yield when distance upslope is 75 by giving an interval of plausible values.

54. Infestation of crops by insects has long been of great concern to farmers and agricultural scientists. The article "Cotton Square Damage by the Plant Bug, *Lygus hesperus*, and Abscission Rates" (*J. Econ. Entomology*, 1988: 1328–1337) reports data on x = age of a cotton plant (days) and y = % damaged squares. Consider the accompanying $n = 12$ observations (read from a scatter plot in the article).

x	9	12	12	15	18	18
y	11	12	23	30	29	52
x	21	21	27	30	30	33
y	41	65	60	72	84	93

a. Why is the relationship between x and y not deterministic?

b. Does a scatter plot suggest that the simple linear regression model will describe the relationship between the two variables?

c. The summary statistics are $\sum x_i = 246$, $\sum x_i^2 = 5742$, $\sum y_i = 572$, $\sum y_i^2 = 35{,}634$ and $\sum x_i y_i = 14{,}022$. Determine the equation of the least squares line.

d. Predict the percentage of damaged squares when the age is 20 days by giving an interval of plausible values.

55. Verify that $V(\hat{\beta}_0 + \hat{\beta}_1 x)$ is indeed given by the expression in the text. [*Hint:* $V(\sum d_i Y_i) = \sum d_i^2 \cdot V(Y_i)$.]

12.5 Correlation

In many situations the objective in studying the joint behavior of two variables is to see whether they are related, rather than to use one to predict the value of the other. In this section, we first develop the sample correlation coefficient r as a measure of how strongly related two variables x and y are in a sample and then relate r to the correlation coefficient ρ defined in Chapter 5.

The Sample Correlation Coefficient r

Given n pairs of observations $(x_1, y_1), (x_2, y_2), \ldots, (x_n, y_n)$, it is natural to speak of x and y having a positive relationship if large x's are paired with large y's and small x's with small y's. Similarly, if large x's are paired with small y's and small x's with large y's, then a negative relationship between the variables is implied. Consider the quantity

$$S_{xy} = \sum_{i=1}^{n}(x_i - \bar{x})(y_i - \bar{y}) = \sum_{i=1}^{n} x_i y_i - \frac{\left(\sum_{i=1}^{n} x_i\right)\left(\sum_{i=1}^{n} y_i\right)}{n}$$

Then if the relationship is strongly positive, an x_i above the mean $\bar{x}$ will tend to be paired with a y_i above the mean $\bar{y}$, so that $(x_i - \bar{x})(y_i - \bar{y}) > 0$, and this product will also be positive whenever both x_i and y_i are below their respective means. Thus a positive relationship implies that S_{xy} will be positive. An analogous argument shows that when the relationship is negative, S_{xy} will be negative, since most of the products $(x_i - \bar{x})(y_i - \bar{y})$ will be negative. This is illustrated in Figure 12.23.

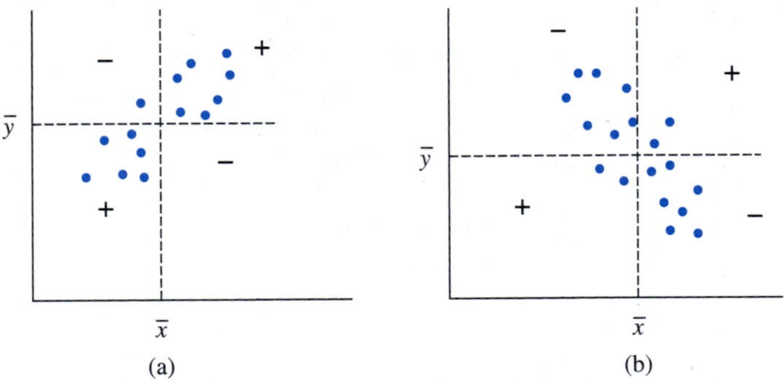

Figure 12.23 (a) Scatter plot with S_{xy} positive; (b) scatter plot with S_{xy} negative [+ means $(x_i - \bar{x})(y_i - \bar{y}) > 0$, and − means $(x_i - \bar{x})(y_i - \bar{y}) < 0$]

Although S_{xy} seems a plausible measure of the strength of a relationship, we do not yet have any idea of how positive or negative it can be. Unfortunately, S_{xy} has a serious defect: By changing the unit of measurement for either x or y, S_{xy} can be made either arbitrarily large in magnitude or arbitrarily close to zero. For example, if $S_{xy} = 25$ when x is measured in meters, then $S_{xy} = 25{,}000$ when x is measured in millimeters and .025 when x is expressed in kilometers. A reasonable condition to impose on any measure of how strongly x and y are related is that the calculated measure should not depend on the particular unit used to measure them. This condition is achieved by modifying S_{xy} to obtain the sample correlation coefficient.

DEFINITION

The **sample correlation coefficient** for the n pairs $(x_1, y_1), \ldots, (x_n, y_n)$ is

$$r = \frac{S_{xy}}{\sqrt{\sum (x_i - \bar{x})^2}\sqrt{\sum (y_i - \bar{y})^2}} = \frac{S_{xy}}{\sqrt{S_{xx}}\sqrt{S_{yy}}} \qquad (12.9)$$

Example 12.17 An accurate assessment of soil productivity is critical to rational land-use planning. Unfortunately, as the author of the article "Productivity Ratings Based on Soil Series" (*Prof. Geographer*, 1980: 158–163) argues, an acceptable soil productivity index is not so easy to come by. One difficulty is that productivity is determined partly by which crop is planted, and the relationship between yield of two different crops planted in the same soil may not be very strong. To illustrate, the article presents the accompanying data on corn yield x and peanut yield y (mT/ha) for eight different types of soil.

x	2.4	3.4	4.6	3.7	2.2	3.3	4.0	2.1
y	1.33	2.12	1.80	1.65	2.00	1.76	2.11	1.63

With $\sum x_i = 25.7$, $\sum y_i = 14.40$, $\sum x_i^2 = 88.31$, $\sum x_i y_i = 46.856$, and $\sum y_i^2 = 26.4324$,

$$S_{xx} = 88.31 - \frac{(25.7)^2}{8} = 88.31 - 82.56 = 5.75$$

$$S_{yy} = 26.4324 - \frac{(14.40)^2}{8} = .5124$$

$$S_{xy} = 46.856 - \frac{(25.7)(14.40)}{8} = .5960$$

from which

$$r = \frac{.5960}{\sqrt{5.75}\sqrt{.5124}} = .347$$

■

Properties of r

The most important properties of r are as follows:

1. The value of r does not depend on which of the two variables under study is labeled x and which is labeled y.
2. The value of r is independent of the units in which x and y are measured.
3. $-1 \le r \le 1$
4. $r = 1$ if and only if (iff) all (x_i, y_i) pairs lie on a straight line with positive slope, and $r = -1$ iff all (x_i, y_i) pairs lie on a straight line with negative slope.

5. The square of the sample correlation coefficient gives the value of the coefficient of determination that would result from fitting the simple linear regression model—in symbols, $(r)^2 = r^2$.

Property 1 should be evident. Exercise 66 asks you to verify Property 2. To derive Property 5, recall the regression analysis of variance identity (12.4) [SST = SSE + SSR = SSE + $\sum(\hat{y}_i - \bar{y})^2$]. It is easily shown [Exercise 24(b)] that $\hat{y}_i - \bar{y} = \hat{\beta}_1(x_i - \bar{x})$, and therefore

$$\sum(\hat{y}_i - \bar{y})^2 = \hat{\beta}_1^2 \sum(x_i - \bar{x})^2 = \left[\frac{\sum(x_i - \bar{x})(y_i - \bar{y})}{\sum(x_i - \bar{x})^2}\right]^2 \sum(x_i - \bar{x})^2$$

$$= \frac{\left[\sum(x_i - \bar{x})(y_i - \bar{y})\right]^2}{\sum(x_i - \bar{x})^2 \sum(y_i - \bar{y})^2} \sum(y_i - \bar{y})^2 = (r)^2\text{SST}$$

Here $(r)^2$ is the square of the correlation coefficient. Substituting this result into the identity ANOVA gives SST = SSE + $(r)^2$ SST, so $(r)^2$ = (SST − SSE)/SST, completing the derivation of Property 5.

Properties 3 and 4 follow from Property 5. Because $(r)^2$ = (SST − SSE)/ SST, and the numerator cannot be bigger than the denominator, Property 3 follows immediately. Furthermore, because the ratio can be 1 if and only if SSE = 0, we conclude that $r^2 = 1$ if and only if all the points fall on a straight line. If the correlation is positive this will be a line with positive slope, and if the correlation is negative it will be a line with negative slope, so we have verified Property 4.

Property 1 stands in marked contrast to what happens in regression analysis, where virtually all quantities of interest (the estimated slope, estimated y-intercept, s^2, etc.) depend on which of the two variables is treated as the dependent variable. However, Property 5 shows that the proportion of variation in the dependent variable explained by fitting the simple linear regression model does not depend on which variable plays this role.

Property 2 is equivalent to saying that r is unchanged if each x_i is replaced by cx_i and if each y_i is replaced by dy_i (a change in the scale of measurement), as well as if each x_i is replaced by $x_i - a$ and y_i by $y_i - b$ (which changes the location of zero on the measurement axis). This implies, for example, that r is the same whether temperature is measured in °F or °C.

Property 3 tells us that the maximum value of r, corresponding to the largest possible degree of positive relationship, is $r = 1$, whereas the most negative relationship is identified with $r = -1$. According to Property 4, the largest positive and largest negative correlations are achieved only when all points lie along a straight line. Any other configuration of points, even if the configuration suggests a deterministic relationship between variables, will yield an r value less than 1 in absolute magnitude. Thus *r measures the degree of linear relationship* among variables. A value of r near 0 is not evidence of the lack of a strong relationship, but only the absence of a linear relation, so that such a value of r must be interpreted with caution. Figure 12.24 illustrates several configurations of points associated with different values of r.

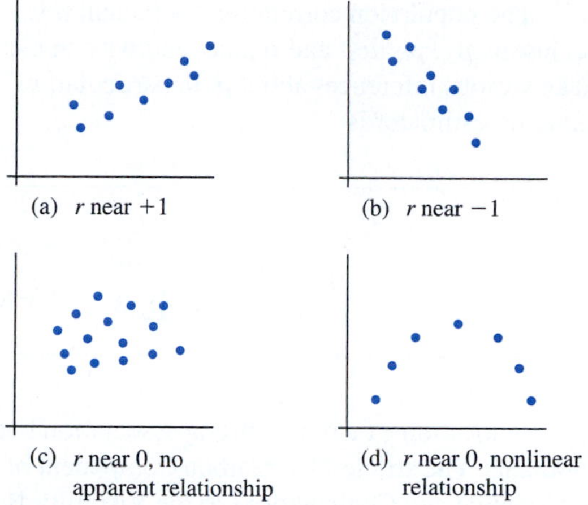

Figure 12.24 Scatter plots for different values of r

A frequently asked question is, "When can it be said that there is a strong correlation between the variables, and when is the correlation weak?" A reasonable rule of thumb is to say that the correlation is weak if $0 \le |r| \le .5$, strong if $.8 \le |r| \le 1$, and moderate otherwise. It may surprise you that $r = .5$ is considered weak, but $r^2 = .25$ implies that in a regression of y on x, only 25% of observed y variation would be explained by the model. In Example 12.17, the correlation between corn yield and peanut yield would be described as weak.

The Population Correlation Coefficient ρ and Inferences About Correlation

The correlation coefficient r is a measure of how strongly related x and y are in the observed sample. We can think of the pairs (x_i, y_i) as having been drawn from a bivariate population of pairs, with (X_i, Y_i) having joint probability distribution $f(x, y)$. In Chapter 5, we defined the correlation coefficient $\rho(X, Y)$ by

$$\rho = \rho(X, Y) = \frac{\text{Cov}(X, Y)}{\sigma_X \cdot \sigma_Y}$$

where

$$\text{Cov}(X, Y) = \begin{cases} \sum_x \sum_y (x - \mu_X)(y - \mu_Y)f(x, y) & (X, Y) \text{ discrete} \\ \int_{-\infty}^{\infty}\int_{-\infty}^{\infty} (x - \mu_X)(y - \mu_Y)f(x, y)\,dx\,dy & (X, Y) \text{ continuous} \end{cases}$$

If we think of $f(x, y)$ as describing the distribution of pairs of values within the entire population, ρ becomes a measure of how strongly related x and y are in that population. Properties of ρ analogous to those for r were given in Chapter 5.

The population correlation coefficient ρ is a parameter or population characteristic, just as μ_X, μ_Y, σ_X, and σ_Y are, and we can use the sample correlation coefficient to make various inferences about ρ. In particular, r is a point estimate for ρ, and the corresponding estimator is

$$\hat{\rho} = R = \frac{\sum(X_i - \overline{X})(Y_i - \overline{Y})}{\sqrt{\sum(X_i - \overline{X})^2}\sqrt{\sum(Y_i - \overline{Y})^2}}$$

Example 12.18 In some locations, there is a strong association between concentrations of two different pollutants. The article "The Carbon Component of the Los Angeles Aerosol: Source Apportionment and Contributions to the Visibility Budget" (*J. Air Pollution Control Fed.*, 1984: 643–650) reports the accompanying data on ozone concentration x (ppm) and secondary carbon concentration y (μg/m^3).

x	.066	.088	.120	.050	.162	.186	.057	.100
y	4.6	11.6	9.5	6.3	13.8	15.4	2.5	11.8

x	.112	.055	.154	.074	.111	.140	.071	.110
y	8.0	7.0	20.6	16.6	9.2	17.9	2.8	13.0

The summary quantities are $n = 16$, $\sum x_i = 1.656$, $\sum y_i = 170.6$, $\sum x_i^2 = .196912$, $\sum x_i y_i = 20.0397$, and $\sum y_i^2 = 2253.56$, from which

$$r = \frac{20.0397 - (1.656)(170.6)/16}{\sqrt{.196912 - (1.656)^2/16}\sqrt{2253.56 - (170.6)^2/16}}$$

$$= \frac{2.3826}{(.1597)(20.8456)} = .716$$

The point estimate of the population correlation coefficient ρ between ozone concentration and secondary carbon concentration is $\hat{\rho} = r = .716$. ■

The small-sample intervals and test procedures presented in Chapters 8–10 were based on an assumption of population normality. To test hypotheses about ρ, we make an analogous assumption about the distribution of pairs of (x, y) values in the population. We are now assuming that *both* X and Y are random, with joint distribution given by the bivariate normal pdf introduced in Section 5.3.

If $X = x$, recall that the (conditional) distribution of Y is normal with mean $\mu_{Y \cdot x} = \mu_2 + (\rho\sigma_2/\sigma_1)(x - \mu_1)$ and variance $(1 - \rho^2)\sigma_2^2$. This is exactly the model used in simple linear regression with $\beta_0 = \mu_2 - \rho\mu_1\sigma_2/\sigma_1$, $\beta_1 = \rho\sigma_2/\sigma_1$, and $\sigma^2 = (1 - \rho^2)\sigma_2^2$ independent of x. The implication is that *if the observed pairs (x_i, y_i) are actually drawn from a bivariate normal distribution, then the simple linear regression model is an appropriate way of studying the behavior of Y for fixed x.* If $\rho = 0$, then $\mu_{Y \cdot x} = \mu_2$

independent of x; in fact, when $\rho = 0$ the joint probability density function $f(x, y)$ can be factored into a part involving x only and a part involving y only, which implies that X and Y are independent variables.

Example 12.19 As discussed in Section 5.3, contours of the bivariate normal distribution are elliptical, and this suggests that a scatter plot of observed (x, y) pairs from such a joint distribution should have a roughly elliptical shape. The scatter plot in Figure 12.25 of y = visceral fat (cm^2) by the CT method versus x = visceral fat (cm^2) by the US method for a sample of $n = 100$ obese women appeared in the paper "Methods of Estimation of Visceral Fat: Advantages of Ultrasonography" (*Obesity Res.*, 2003: 1488–1494). Computerized tomography is considered the most accurate technique for body fat measurement, but is costly, time-consuming, and involves exposure to ionizing radiation; the US method is noninvasive and less expensive.

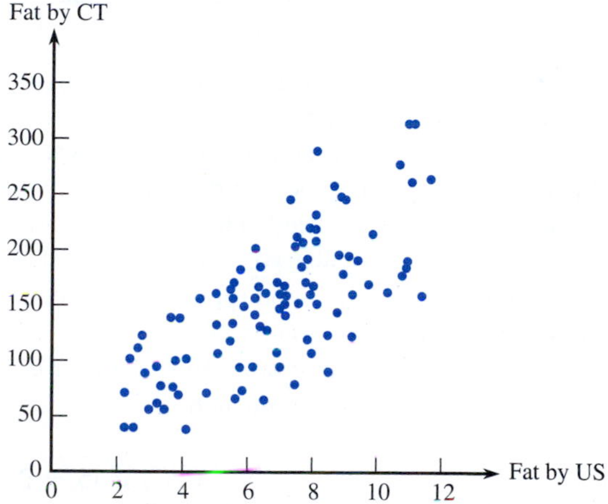

Figure 12.25 Scatter plot of data from Example 12.19

The pattern in the scatter plot seems consistent with an assumption of bivariate normality. Here $r = .71$, which is not all that impressive ($r^2 = .50$), but the investigators reported that a test of H_0: $\rho = 0$ (to be introduced shortly) gives P-value $< .001$. Of course we would want values from the two methods to be very highly correlated before regarding one as an adequate substitute for the other. ■

Assuming that the pairs are drawn from a bivariate normal distribution allows us to test hypotheses about ρ and to construct a CI. There is no completely satisfactory way to check the plausibility of the bivariate normality assumption. A partial check involves constructing two separate normal probability plots, one for the sample x_i's and another for the sample y_i's, since bivariate normality implies that the marginal distributions of both X and Y are normal. If either plot deviates substantially from a straight-line pattern, the following inferential procedures should not be used when the sample size n is small. Also, as discussed in Example 12.19, the scatter plot should show a roughly elliptical shape.

TESTING FOR THE ABSENCE OF CORRELATION

When H_0: $\rho = 0$ is true, the test statistic

$$T = \frac{R\sqrt{n-2}}{\sqrt{1-R^2}}$$

has a t distribution with $n - 2$ df (see Exercise 65).

Alternative Hypothesis	Rejection Region for Level α Test
H_a: $\rho > 0$	$t \geq t_{\alpha,n-2}$
H_a: $\rho < 0$	$t \leq -t_{\alpha,n-2}$
H_a: $\rho \neq 0$	either $t \geq t_{\alpha/2,n-2}$ or $t \leq -t_{\alpha/2,n-2}$

A P-value based on $n - 2$ df can be calculated as described previously.

Example 12.20 Neurotoxic effects of manganese are well known and are usually caused by high occupational exposure over long periods of time. In the fields of occupational hygiene and environmental hygiene, the relationship between lipid peroxidation, which is responsible for deterioration of foods and damage to live tissue, and occupational exposure had not been previously reported. The article "Lipid Peroxidation in Workers Exposed to Manganese" (*Scand. J. Work Environ. Health*, 1996: 381–386) gave data on x = manganese concentration in blood (ppb) and y = concentration (μmol/L) of malondialdehyde, which is a stable product of lipid peroxidation, both for a sample of 22 workers exposed to manganese and for a control sample of 45 individuals. The value of r for the control sample was .29, from which

$$t = \frac{(.29)\sqrt{45-2}}{\sqrt{1-(.29)^2}} \approx 2.0$$

The corresponding P-value for a two-tailed t test based on 43 df is roughly .052 (the cited article reported only P-value $> .05$). We would not want to reject the assertion that $\rho = 0$ at either significance level .01 or .05. For the sample of exposed workers, $r = .83$ and $t = 6.7$, which is clear evidence of a positive relationship in the entire population of exposed workers from which the sample was selected. Although in general correlation does not necessarily imply causation, it is plausible here that higher levels of manganese cause higher levels of peroxidation. ■

Because ρ measures the extent to which there is a linear relationship between the two variables in the population, the null hypothesis H_0: $\rho = 0$ states that there is no such population relationship. In Section 12.3, we used the t ratio $\hat{\beta}_1/s_{\hat{\beta}_1}$ to test for a linear relationship between the two variables in the context of regression analysis. It turns out that the two test procedures are completely equivalent because $r\sqrt{n-2}/\sqrt{1-r^2} = \hat{\beta}_1/s_{\hat{\beta}_1}$ (Exercise 65). When interest lies only in assessing the strength of any linear relationship rather than in fitting a model and using it to estimate or predict, the test statistic formula just presented requires fewer computations than does the t ratio.

Other Inferences Concerning ρ

The procedure for testing H_0: $\rho = \rho_0$ when $\rho_0 \neq 0$ is not equivalent to any procedure from regression analysis. The test statistic is based on a transformation of R called the Fisher transformation.

PROPOSITION When $(X_1, Y_1), \ldots, (X_n, Y_n)$ is a sample from a bivariate normal distribution, the rv

$$V = \frac{1}{2}\ln\left(\frac{1+R}{1-R}\right) \tag{12.10}$$

has approximately a normal distribution with mean and variance

$$\mu_V = \frac{1}{2}\ln\left(\frac{1+\rho}{1-\rho}\right) \qquad \sigma_V^2 = \frac{1}{n-3}$$

The rationale for the transformation is to obtain a function of R that has a variance independent of ρ; this would not be the case with R itself. Also, the approximation will not be valid if n is quite small.

The test statistic for testing H_0: $\rho = \rho_0$ is

$$Z = \frac{V - \frac{1}{2}\ln[(1+\rho_0)/(1-\rho_0)]}{1/\sqrt{n-3}}$$

Alternative Hypothesis	**Rejection Region for Level α Test**
H_a: $\rho > \rho_0$	$z \geq z_\alpha$
H_a: $\rho < \rho_0$	$z \leq -z_\alpha$
H_a: $\rho \neq \rho_0$	either $z \geq z_{\alpha/2}$ or $z \leq -z_{\alpha/2}$

A P-value can be calculated in the same manner as for previous z tests.

Example 12.21 As far back as Leonardo da Vinci, it was known that height and wingspan (measured fingertip to fingertip between outstretched hands) are closely related. For these measurements (in inches) from 16 students in a statistics class notice how close the two values are.

Student	1	2	3	4	5	6	7	8
Height	63.0	63.0	65.0	64.0	68.0	69.0	71.0	68.0
Wingspan	62.0	62.0	64.0	64.5	67.0	69.0	70.0	72.0

Student	9	10	11	12	13	14	15	16
Height	68.0	72.0	73.0	73.5	70.0	70.0	72.0	74.0
Wingspan	70.0	72.0	73.0	75.0	71.0	70.0	76.0	76.5

The scatter plot in Figure 12.26 shows an approximately linear shape, and the point cloud is roughly elliptical. Also, the normal plots for the individual variables are roughly linear, so the bivariate normal distribution can reasonably be assumed.

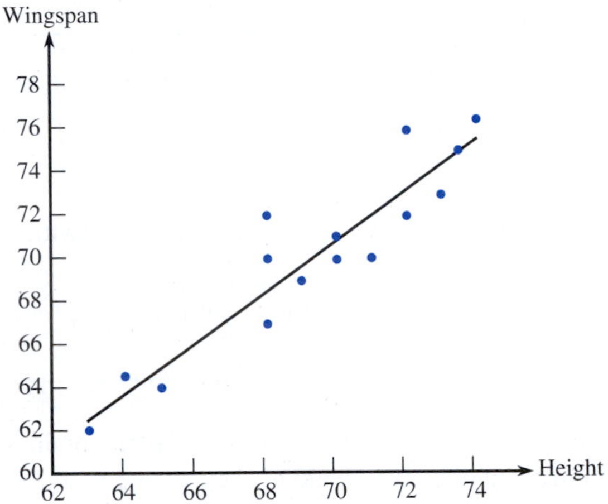

Figure 12.26 Scatter plot of data from Example 12.21

The correlation is computed to be .9422. Can we conclude that wingspan and height are highly correlated, in the sense that $\rho > .8$? To carry out a test of H_0: $\rho = .8$ versus H_a: $\rho > .8$, we Fisher transform .9422 and .8:

$$\frac{1}{2}\ln\left(\frac{1 + .9422}{1 - .9422}\right) = 1.757 \qquad \frac{1}{2}\ln\left(\frac{1 + .8}{1 - .8}\right) = 1.099$$

The calculation is easily done on a calculator with hyperbolic functions, because the inverse hyperbolic tangent is equivalent to the Fisher transformation. That is, $\tanh^{-1}(.9422) = 1.757$ and $\tanh^{-1}(.8) = 1.099$. Compute the test statistic value $z = (1.757 - 1.099)/(1/\sqrt{16 - 3}) = 2.37$. Since $2.37 > 1.645$, at level .05 we can reject H_0: $\rho = .8$ in favor of H_a: $\rho > .8$. Indeed, because $2.37 > 2.33$, it is also true that we can reject H_0 in this one-tailed test at the .01 level, and conclude that wingspan is highly correlated with height. ■

To obtain a CI for ρ, we first derive an interval for $\mu_V = \frac{1}{2}\ln[(1 + \rho)/(1 - \rho)]$. Standardizing V, writing a probability statement, and manipulating the resulting inequalities yields

$$\left(v - \frac{z_{\alpha/2}}{\sqrt{n - 3}}, v + \frac{z_{\alpha/2}}{\sqrt{n - 3}}\right) \tag{12.11}$$

as a $100(1 - \alpha)\%$ interval for μ_V, where $v = \frac{1}{2}\ln[(1 + r)/(1 - r)]$. This interval can then be manipulated to yield a CI for ρ.

A $100(1-\alpha)\%$ confidence interval for ρ is

$$\left(\frac{e^{2c_1}-1}{e^{2c_1}+1}, \frac{e^{2c_2}-1}{e^{2c_2}+1}\right)$$

where c_1 and c_2 are the left and right endpoints, respectively, of the interval (12.11).

Example 12.22

(Example 12.21 continued)

The sample correlation coefficient between wingspan and height was $r = .9422$, giving $v = 1.757$. With $n = 16$, a 95% confidence interval for μ_V is $1.757 \pm 1.96/\sqrt{16-3} = (1.213, 2.301) = (c_1, c_2)$.

The 95% interval for ρ is

$$\left[\frac{e^{2(1.213)}-1}{e^{2(1.213)}+1}, \frac{e^{2(2.301)}-1}{e^{2(2.301)}+1}\right] = (.838, .980)$$

This calculation is easy on a calculator with hyperbolic functions because all that is needed is the inverse of the Fisher transformation, which is the hyperbolic tangent. We get (tanh(1.213), tanh(2.301)) = (.838, .980). Notice that this interval excludes .8, and that our hypothesis test in Example 12.21 would have rejected H_0: $\rho = .8$ in favor of H_a: $\rho > .8$ at the .025 level. ■

Absent the assumption of bivariate normality, a bootstrap procedure can be used to obtain a CI for ρ or test hypotheses.

In Chapter 5, we cautioned that a large value of the correlation coefficient (near 1 or -1) implies only association and not causation. This applies to both ρ and r. It is easy to find strong but weird correlations in which neither variable is causally related to the other. For example, since Prohibition ended in the 1930s, beer consumption and church attendance have correlated very highly. Of course, the reason is that both variables have increased in accord with population growth.

Exercises Section 12.5 (56–67)

56. The article "Behavioural Effects of Mobile Telephone Use During Simulated Driving" (*Ergonomics*, 1995: 2536–2562) reported that for a sample of 20 experimental subjects, the sample correlation coefficient for x = age and y = time since the subject had acquired a driving license (yr) was .97. Why do you think the value of r is so close to 1? (The article's authors gave an explanation.)

57. The Turbine Oil Oxidation Test (TOST) and the Rotating Bomb Oxidation Test (RBOT) are two different procedures for evaluating the oxidation stability of steam turbine oils. The article "Dependence of Oxidation Stability of Steam Turbine Oil on Base Oil Composition" (*J. Soc. Tribologists and Lubricat. Engrs.*, Oct. 1997: 19–24) reported the accompanying observations on x = TOST time (hr) and y = RBOT time (min) for 12 oil specimens.

TOST	4200	3600	3750	3675	4050	2770
RBOT	370	340	375	310	350	200
TOST	4870	4500	3450	2700	3750	3300
RBOT	400	375	285	225	345	285

a. Calculate and interpret the value of the sample correlation coefficient (as did the article's authors).

b. How would the value of r be affected if we had let x = RBOT time and y = TOST time?

c. How would the value of r be affected if RBOT time were expressed in hours?

d. Construct a scatter plot and normal probability plots and comment.

e. Carry out a test of hypotheses to decide whether RBOT time and TOST time are linearly related.

58. Toughness and fibrousness of asparagus are major determinants of quality. This was the focus of a study reported in "Post-Harvest Glyphosphate Application Reduces Toughening, Fiber Content, and Lignification of Stored Asparagus Spears" (*J. Amer. Soc. Horticult. Sci.*, 1988: 569–572). The article reported the accompanying data (read from a graph) on x = shear force (kg) and y = percent fiber dry weight.

x	46	48	55	57	60	72	81	85	94
y	2.18	2.10	2.13	2.28	2.34	2.53	2.28	2.62	2.63

x	109	121	132	137	148	149	184	185	187
y	2.50	2.66	2.79	2.80	3.01	2.98	3.34	3.49	3.26

$$n = 18, \Sigma x_i = 1950, \Sigma x_i^2 = 251{,}970,$$
$$\Sigma y_i = 47.92, \Sigma y_i^2 = 130.6074, \Sigma x_i y_i = 5530.92$$

a. Calculate the value of the sample correlation coefficient. Based on this value and a scatter plot, how would you describe the nature of the relationship between the two variables?

b. If a first specimen has a larger value of shear force than does a second specimen, what tends to be true of percent fiber dry weight for the two specimens?

c. If shear force is expressed in pounds, what happens to the value of r? Why?

d. If the simple linear regression model were fit to this data, what proportion of observed variation in percent fiber dry weight could be explained by the model relationship?

e. Carry out a test at significance level .01 to decide whether there is a positive linear association between the two variables.

59. The authors of the paper "Objective Effects of a Six Months' Endurance and Strength Training Program in Outpatients with Congestive Heart Failure" (*Med. Sci. Sports Exercise*, 1999: 1102–1107) presented a correlation analysis to investigate the relationship between maximal lactate level x and muscular endurance y. The accompanying data was read from a plot in the paper.

x	400	750	770	800	850	1025	1200
y	3.80	4.00	4.90	5.20	4.00	3.50	6.30

x	1250	1300	1400	1475	1480	1505	2200
y	6.88	7.55	4.95	7.80	4.45	6.60	8.90

$$S_{xx} = 36.9839, S_{yy} = 2{,}628{,}930.357, S_{xy} = 7377.704$$

A scatter plot shows a linear pattern.

a. Test to see whether there is a positive correlation between maximal lactate level and muscular endurance in the population from which this data was selected.

b. If a regression analysis were to be carried out to predict endurance from lactate level, what proportion of observed variation in endurance could be attributed to the approximate linear relationship? Answer the analogous question if regression is used to predict lactate level from endurance—and answer both questions without doing any regression calculations.

60. Hydrogen content is conjectured to be an important factor in porosity of aluminum alloy castings. The article "The Reduced Pressure Test as a Measuring Tool in the Evaluation of Porosity/Hydrogen Content in A1–7 Wt Pct Si-10 Vol Pct SiC(p) Metal Matrix Composite" (*Metallurgical Trans.*, 1993: 1857–1868) gives the accompanying data on x = content and y = gas porosity for one particular measurement technique.

x	.18	.20	.21	.21	.21	.22	.23
y	.46	.70	.41	.45	.55	.44	.24

x	.23	.24	.24	.25	.28	.30	.37
y	.47	.22	.80	.88	.70	.72	.75

MINITAB gives the following output in response to a CORRELATION command:

```
Correlation of Hydrcon and
Porosity = 0.449
```

a. Test at level .05 to see whether the population correlation coefficient differs from 0.

b. If a simple linear regression analysis had been carried out, what percentage of observed variation in porosity could be attributed to the model relationship?

61. Physical properties of six flame-retardant fabric samples were investigated in the article "Sensory and Physical Properties of Inherently Flame-Retardant Fabrics" (*Textile Res.*, 1984: 61–68). Use the accompanying data and a .05 significance level to determine whether there is a significant correlation between stiffness x (mg-cm) and thickness y (mm). Is the result of the test surprising in light of the value of r?

x	7.98	24.52	12.47	6.92	24.11	35.71
y	.28	.65	.32	.27	.81	.57

62. The article "Increases in Steroid Binding Globulins Induced by Tamoxifen in Patients with Carcinoma of the Breast" (*J. Endocrinology*, 1978: 219–226) reports data on the effects of the drug tamoxifen on change in the level of cortisol-binding globulin (CBG) of patients during treatment. With age $= x$ and $\Delta\text{CBG} = y$, summary values are $n = 26$, $\sum x_i = 1613$, $\sum(x_i - \bar{x})^2 = 3756.96$, $\sum y_i = 281.9$, $\sum(y_i - \bar{y})^2 = 465.34$, and $\sum x_i y_i = 16{,}731$.

a. Compute a 90% CI for the true correlation coefficient ρ.

b. Test H_0: $\rho = -.5$ versus H_a: $\rho < -.5$ at level .05.

c. In a regression analysis of y on x, what proportion of variation in change of cortisol-binding globulin level could be explained by variation in patient age within the sample?

d. If you decide to perform a regression analysis with age as the dependent variable, what proportion of variation in age is explainable by variation in ΔCBG?

63. A sample of $n = 500$ (x, y) pairs was collected and a test of H_0: $\rho = 0$ versus H_a: $\rho \neq 0$ was carried out. The resulting P-value was computed to be .00032.

a. What conclusion would be appropriate at level of significance .001?

b. Does this small P-value indicate a very strong relationship between x and y (a value of ρ that differs considerably from 0)? Explain.

c. Now suppose a sample of $n = 10{,}000$ (x, y) pairs resulted in $r = .022$. Test H_0: $\rho = 0$ versus H_a: $\rho \neq 0$ at level .05. Is the result statistically significant? Comment on the practical significance of your analysis.

64. Let x be number of hours per week of studying and y be grade point average. Suppose we have one sample of (x, y) pairs for females and another for males. Then we might like to test H_0: $\rho_1 - \rho_2 = 0$ against the alternative that the two population correlation coefficients are different.

a. Use the results of the proposition concerning the transformed variable $V = .5\ln[(1 + R)/(1 - R)]$ to propose an appropriate test statistic and rejection region (let R_1 and R_2 denote the two sample correlation coefficients).

b. The paper "Relational Bonds and Customer's Trust and Commitment: A Study on the Moderating Effects of Web Site Usage" (*Service Indust. J.*, 2003: 103–124) reported that $n_1 = 261$, $r_1 = .59$, $n_2 = 557$, $r_2 = .50$, where the first sample consisted of corporate Website users and the second of nonusers; here r is the correlation between an assessment of the strength of economic bonds and performance. Carry out the test for this data (as did the authors of the cited paper).

65. Verify that the t ratio for testing H_0: $\beta_1 = 0$ in Section 12.3 is identical to t for testing H_0: $\rho = 0$.

66. Verify Property 2 of the correlation coefficient: The value of r is independent of the units in which x and y are measured, that is, if $x_i' = ax_i + c$ and $y_i' = by_i + d$, then r for the (x_i', y_i') pairs is the same as r for the (x_i, y_i) pairs, if $a > 0$ and $b > 0$

67. Consider a time series—that is, a sequence of observations $X_1, X_2, \ldots$ on some response variable (e.g., concentration of a pollutant) over time—with observed values $x_1, x_2, \ldots, x_n$ over n time periods. Then the *lag 1 autocorrelation coefficient* is defined as

$$r_1 = \frac{\sum_{i=1}^{n-1}(x_i - \bar{x})(x_{i+1} - \bar{x})}{\sum_{i=1}^{n}(x_i - \bar{x})^2}$$

Autocorrelation coefficients $r_2, r_3, \ldots$ for lags 2, 3, . . . are defined analogously.

a. Calculate the values of r_1, r_2, and r_3 for the temperature data from Exercise 79 of Chapter 1.

b. Consider the $n - 1$ pairs $(x_1, x_2), (x_2, x_3), \ldots, (x_{n-1}, x_n)$. What is the difference between the formula for the sample correlation coefficient r applied to these pairs and the formula for r_1? What if n, the length of the series, is large? What about r_2 compared to r for the $n - 2$ pairs $(x_1, x_3), (x_2, x_4), \ldots, (x_{n-2}, x_n)$?

c. Analogous to the population correlation coefficient ρ, let ρ_i $(i = 1, 2, 3, \ldots)$ denote the theoretical or long-run autocorrelation coefficients at the various lags. If all these ρ's are zero, there is no (linear) relationship between observations in the series at *any* lag. In this case, if n is large, each R_i has approximately a normal distribution with mean 0 and standard deviation $1/\sqrt{n}$ and different R_i's are almost independent. Thus H_0: $\rho = 0$ can be rejected at a significance level of approximately .05 if either $r_i \geq 2/\sqrt{n}$ or $r_i \leq -2/\sqrt{n}$. If $n = 100$ and $r_1 = .16$, $r_2 = -.09$, $r_3 = -.15$, is there evidence of theoretical autocorrelation at any of the first three lags?

d. If you are testing the null hypothesis in (c) for more than one lag, why might you want to increase the cutoff constant 2 in the rejection region? *Hint*: What about the probability of committing at least one type I error?

12.6 *Aptness of the Model and Model Checking

A plot of the observed pairs (x_i, y_i) is a necessary first step in deciding on the form of a mathematical relationship between x and y. It is possible to fit many functions other than a linear one ($y = b_0 + b_1x$) to the data, using either the principle of least squares or another fitting method. Once a function of the chosen form has been fitted, it is important to check the fit of the model to see whether it is in fact appropriate. One way to study the fit is to superimpose a graph of the best-fit function on the scatter plot of the data. However, any tilt or curvature of the best-fit function may obscure some aspects of the fit that should be investigated. Furthermore, the scale on the vertical axis may make it difficult to assess the extent to which observed values deviate from the best-fit functions.

Residuals and Standardized Residuals

A more effective approach to assessment of model adequacy is to compute the fitted or predicted values $\hat{y}_i$ and the residuals $e_i = y_i - \hat{y}_i$ and then plot various functions of these computed quantities. We then examine the plots either to confirm our choice of model or for indications that the model is not appropriate. Suppose the simple linear regression model is correct, and let $y = \hat{\beta}_0 + \hat{\beta}_1 x$ be the equation of the estimated regression line. Then the ith residual is $e_i = y_i - (\hat{\beta}_0 + \hat{\beta}_1 x_i)$. To derive properties of the residuals, let $e_i = Y_i - \hat{Y}_i$ represent the ith residual as a random variable (rv) (before observations are actually made). Then

$$E(Y_i - \hat{Y}_i) = E(Y_i) - E(\hat{\beta}_0 + \hat{\beta}_1 x_i) = \beta_0 + \beta_1 x_i - (\beta_0 + \beta_1 x_i) = 0 \qquad (12.12)$$

Because $\hat{Y}_i$ $(= \hat{\beta}_0 + \hat{\beta}_1 x_i)$ is a linear function of the Y_j's, so is $Y_i - \hat{Y}_i$ (the coefficients depend on the x_j's). Thus the normality of the Y_j's implies that each residual is normally distributed. It can also be shown (Exercise 76) that

$$V(Y_i - \hat{Y}_i) = \sigma^2 \cdot \left[1 - \frac{1}{n} - \frac{(x_i - \bar{x})^2}{S_{xx}}\right] \qquad (12.13)$$

Replacing σ^2 by s^2 and taking the square root of Equation (12.13) gives the estimated standard deviation of a residual.

Let's now standardize each residual by subtracting the mean value (zero) and then dividing by the estimated standard deviation.

The **standardized residuals** are given by

$$e_i^* = \frac{y_i - \hat{y}_i}{s\sqrt{1 - \frac{1}{n} - \frac{(x_i - \bar{x})^2}{S_{xx}}}} \qquad i = 1, \ldots, n \qquad (12.14)$$

Notice that the variances of the residuals differ from one another. If n is reasonably large, though, the bracketed term in (12.13) will be approximately 1, so some sources use e_i/s as the standardized residual. Computation of the e_i^*'s can be tedious, but the most widely used statistical computer packages automatically provide these values and (upon request) can construct various plots involving them.

Example 12.23 Example 12.12 presented data on x = average SAT score for entering freshmen and y = six-year percentage graduation rate. Here we reproduce the data along with the fitted values and their estimated standard deviations, residuals and their estimated standard deviations, and standardized residuals. The estimated regression line is $y = -36.18 + .08855x$, and $r^2 = .729$. Notice that estimated standard deviations of the residuals (in the s_e column) differ somewhat, so $e^* \neq e/s$. The standard deviations of the residuals are higher near $\bar{x}$, in contrast to the standard deviations of the predicted values, which are lower near $\bar{x}$.

x	y	$\hat{y}$	$s_{\hat{Y}}$	e	s_e	e^*
722.21	38	27.7651	4.9554	10.2349	9.020	1.135
833.32	31	37.6034	3.8016	−6.6034	9.564	−0.690
877.77	44	41.5392	3.3838	2.4608	9.719	0.253
899.99	42	43.5067	3.1894	−1.5067	9.785	−0.154
944.43	45	47.4416	2.8394	−2.4416	9.892	−0.247
944.43	51	47.4416	2.8394	3.5584	9.892	0.360
1005.00	70	52.8048	2.4785	17.1952	9.989	1.721
1011.10	58	53.3449	2.4517	4.6551	9.995	0.466
1055.54	48	57.2799	2.3208	−9.2799	10.026	−0.926
1055.54	76	57.2799	2.3208	18.7201	10.026	1.867
1060.00	42	57.6748	2.3143	−15.6748	10.028	−1.563
1080.00	54	59.4457	2.3012	−5.4457	10.031	−0.543
1099.99	42	61.2157	2.3142	−19.2157	10.028	−1.916
1155.00	54	66.0866	2.4780	−12.0866	9.989	−1.210
1166.65	67	67.1181	2.5345	−0.1181	9.974	−0.012
1215.00	65	71.3993	2.8346	−6.3993	9.893	−0.647
1235.00	80	73.1702	2.9845	6.8298	9.849	0.693
1380.00	88	86.0092	4.3392	1.9908	9.332	0.213
1395.00	96	87.3374	4.4963	8.6626	9.257	0.936
1465.00	98	93.5356	5.2522	4.4644	8.850	0.504

■

Diagnostic Plots

The basic plots that many statisticians recommend for an assessment of model validity and usefulness are the following:

1. y_i on the vertical axis versus x_i on the horizontal axis
2. y_i on the vertical axis versus $\hat{y}_i$ on the horizontal axis
3. e_i^* (or e_i) on the vertical axis versus x_i on the horizontal axis
4. e_i^* (or e_i) on the vertical axis versus $\hat{y}_i$ on the horizontal axis
5. A normal probability plot of the standardized residuals (or residuals)

Plots 3 and 4 are called **residual plots** against the independent variable and fitted (predicted) values, respectively.

If Plot 2 yields points close to the 45° line [slope +1 through (0, 0)], then the estimated regression function gives accurate predictions of the values actually observed. Thus Plot 2 provides a visual assessment of model effectiveness in making predictions. Provided that the model is correct, neither residual plot should exhibit distinct patterns. The residuals should be randomly distributed about 0 according to a normal distribution, so all but a very few standardized residuals should lie between −2 and +2 (i.e., all but a few residuals within 2 standard deviations of their expected value 0). The plot of standardized residuals versus $\hat{y}$ is really a combination of the two other plots, showing implicitly both how residuals vary with x and how fitted values compare with observed values. This latter plot is the single one most often recommended for multiple regression analysis. Plot 5 allows the analyst to assess the plausibility of the assumption that ε has a normal distribution.

Example 12.24 (Example 12.23 continued)

Figure 12.27 presents the five plots just recommended along with a sixth plot. The plot of y versus $\hat{y}$ confirms the impression given by r^2 that x is fairly effective in predicting y. The residual plots show no unusual pattern or discrepant values. The normal probability plot of the standardized residuals is quite straight. In summary, the first five plots leave us with no qualms about either the appropriateness of a simple linear relationship or the fit to the given data.

Notice that plotting against x yields the same shape as a plot against the predicted values. Is this surprising? The predicted value is a linear function of x, so the plots will have the same appearance. Given that the plots look the same, why include both? This is preparation for the next section, where more than one predictor is allowed, and plotting against x is not the same as plotting against the predicted values.

The sixth plot in Figure 12.27 is in accord with what was found graphically in Example 12.12. In that example, Figure 12.18 showed that private universities might tend to have better graduation rates than state universities. For another graphical view of this, we show in the last plot of Figure 12.27 the standardized residuals plotted against a variable that is 0 for state universities and 1 for private universities. In this graph the private universities do seem to have an advantage, but we will need to wait until the next section for a hypothesis test, which requires including this new variable as a second predictor in the model.

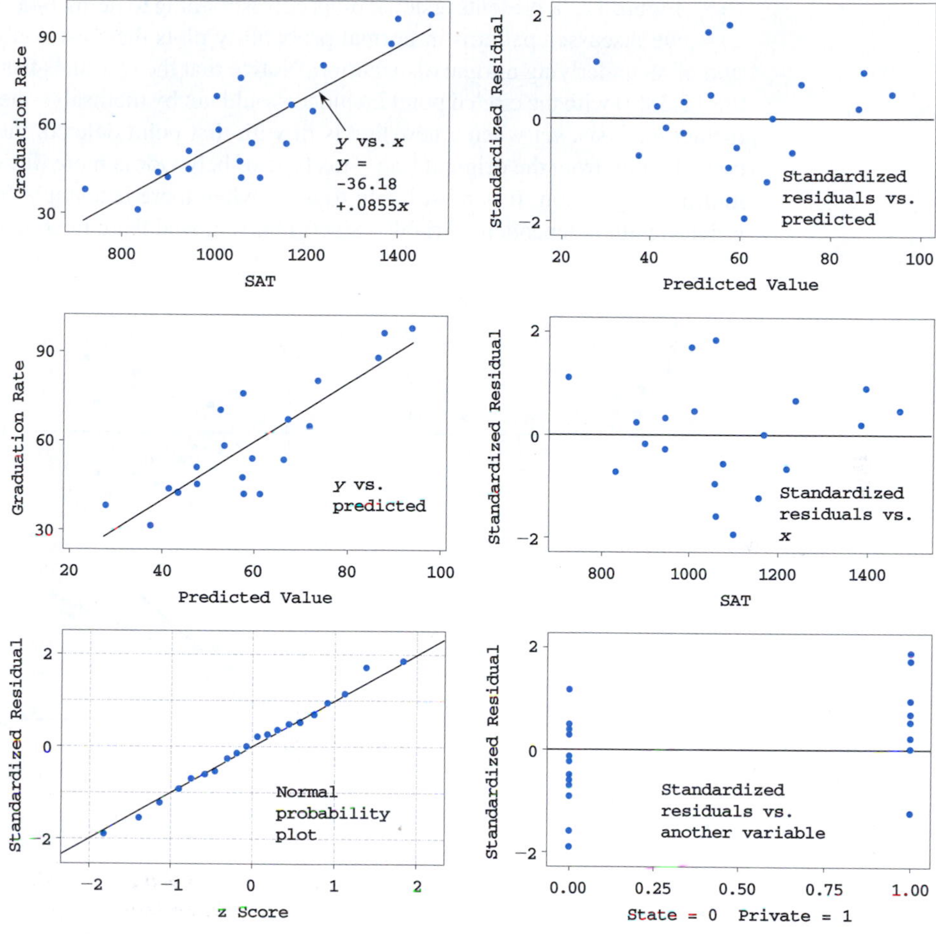

Figure 12.27 Plots from MINITAB for the data from Example 12.24

Difficulties and Remedies

Although we hope that our analysis will yield plots like the first five of Figure 12.27, quite frequently the plots will suggest one or more of the following difficulties:

1. A nonlinear probabilistic relationship between x and y is appropriate.
2. The variance of ε (and of Y) is not a constant σ^2 but depends on x.
3. The selected model fits the data well except for a very few discrepant or outlying data values, which may have greatly influenced the choice of the best-fit function.
4. The error term ε does not have a normal distribution (this is related to item 3).
5. When the subscript i indicates the time order of the observations, the ε_i's exhibit dependence over time.
6. One or more relevant independent variables have been omitted from the model.

Figure 12.28 presents residual plots corresponding to items 1–3, 5, and 6. In Chapter 4, we discussed patterns in normal probability plots that cast doubt on the assumption of an underlying normal distribution. Notice that the residuals from the data in Figure 12.28(d) with the circled point included would not by themselves necessarily suggest further analysis, yet when a new line is fit with that point deleted, the new line differs considerably from the original line. This type of behavior is more difficult to identify in multiple regression. It is most likely to arise when there is a single (or very few) data point(s) with independent variable value(s) far removed from the remainder of the data.

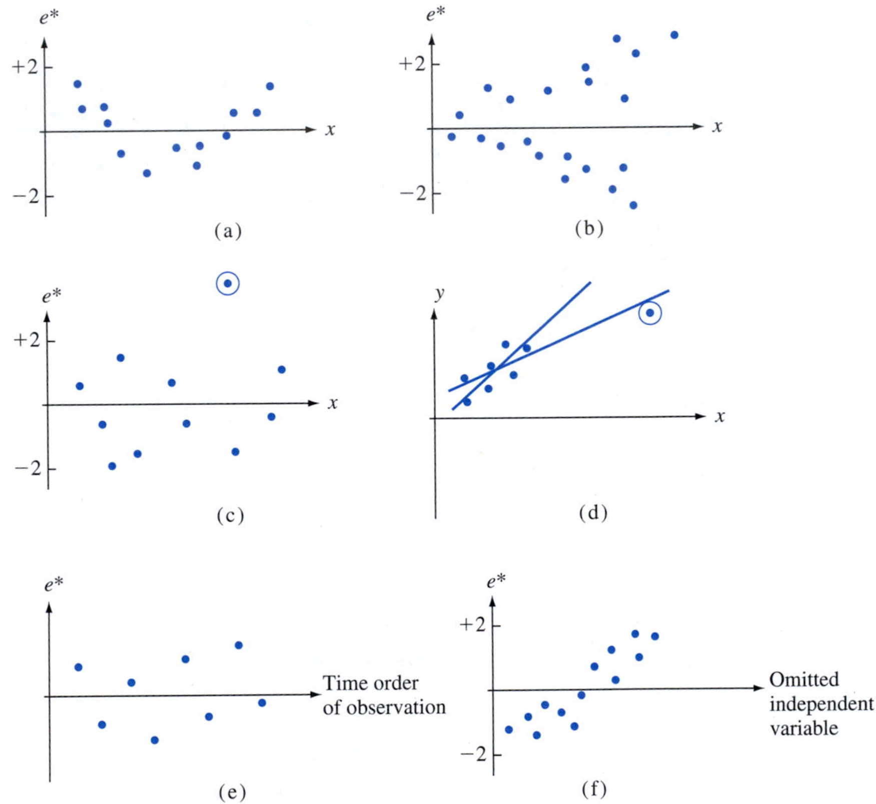

Figure 12.28 Plots that indicate abnormality in data: (a) nonlinear relationship; (b) nonconstant variance; (c) discrepant observation; (d) observation with large influence; (e) dependence in errors; (f) variable omitted

We now indicate briefly what remedies are available for the types of difficulties. For a more comprehensive discussion, one or more of the references on regression analysis should be consulted. If the residual plot looks something like that of Figure 12.28(a), exhibiting a curved pattern, then a nonlinear function of x may be fit.

The residual plot of Figure 12.28(b) suggests that, although a straight-line relationship may be reasonable, the assumption that $V(Y_i) = \sigma^2$ for each i is of doubtful validity. When the error term ε satisfies the independence and constant variance assumptions (normality is not needed) for the simple linear regression model of Section 12.1, it can be shown that among all unbiased estimators of β_0 and β_1, the ordinary least squares estimators have minimum variance. These estimators give equal weight to each (x_i, Y_i). If the variance of Y increases with x, then Y_i's for large x_i should be given less weight than those with small x_i. This suggests that β_0 and β_1 should be estimated by minimizing

$$f_w(b_0, b_1) = \sum w_i[y_i - (b_0 + b_1x_i)]^2 \tag{12.15}$$

where the w_i's are weights that decrease with increasing x_i. Minimization of Expression (12.15) yields **weighted least squares** estimates. For example, if the standard deviation of Y is proportional to x (for $x > 0$)—that is, $V(Y) = kx^2$—then it can be shown that the weights $w_i = 1/x_i^2$ yield minimum variance estimators of β_0 and β_1. The books by John Neter et al. and by S. Chatterjee et al. contain more detail (see the chapter bibliography). Weighted least squares is used quite frequently by econometricians (economists who use statistical methods) to estimate parameters.

When plots or other evidence suggest that the data set contains outliers or points having large influence on the resulting fit, one possible approach is to omit these outlying points and recompute the estimated regression equation. This would certainly be correct if it were found that the outliers resulted from errors in recording data values or experimental errors. If no assignable cause can be found for the outliers, it is still desirable to report the estimated equation both with and without outliers. Yet another approach is to retain possible outliers but to use an estimation principle that puts relatively less weight on outlying values than does the principle of least squares. One such principle is MAD (minimize absolute deviations), which selects $\hat{\beta}_0$ and $\hat{\beta}_1$ to minimize $\sum | y_i - (b_0 + b_1x_i)|$. Unlike the estimates of least squares, there are no nice formulas for the MAD estimates; their values must be found by using an iterative computational procedure. Such procedures are also used when it is suspected that the ε_i's have a distribution that is not normal but instead has "heavy tails" (making it much more likely than for the normal distribution that discrepant values will enter the sample); robust regression procedures are those that produce reliable estimates for a wide variety of underlying error distributions. Least squares estimators are not robust in the same way that the sample mean $\overline{X}$ is not a robust estimator for μ.

When a plot suggests time dependence in the error terms, an appropriate analysis may involve a transformation of the y's or else a model explicitly including a time variable. Lastly, a plot such as that of Figure 12.28(f), which shows a pattern in the residuals when plotted against an omitted variable, suggests considering a model that includes the omitted variable. We have already seen an illustration of this in Example 12.24. ■

Exercises Section 12.6 (68–77)

68. Suppose the variables x = commuting distance and y = commuting time are related according to the simple linear regression model with $\sigma = 10$.

a. If $n = 5$ observations are made at the x values $x_1 = 5$, $x_2 = 10$, $x_3 = 15$, $x_4 = 20$, and $x_5 = 25$, calculate the standard deviations of the five corresponding residuals.

b. Repeat part (a) for $x_1 = 5$, $x_2 = 10$, $x_3 = 15$, $x_4 = 20$, and $x_5 = 50$.

c. What do the results of parts (a) and (b) imply about the deviation of the estimated line from the observation made at the largest sampled x value?

69. The x values and standardized residuals for the chlorine flow/etch rate data of Exercise 51 (Section 12.4) are displayed in the accompanying table. Construct a standardized residual plot and comment on its appearance.

x	1.50	1.50	2.00	2.50	2.50
e^*	.31	1.02	−1.15	−1.23	.23

x	3.00	3.50	3.50	4.00
e^*	.73	−1.36	1.53	.07

Also construct a normal probability plot and comment.

70. Example 12.7 presented the residuals from a simple linear regression of moisture content y on filtration rate x.

a. Plot the residuals against x. Does the resulting plot suggest that a straight-line regression function is a reasonable choice of model? Explain your reasoning.

b. Using $s = .665$, compute the values of the standardized residuals. Is $e_i^* \approx e_i/s$ for $i = 1, \ldots, n$, or are the e_i^*'s not close to being proportional to the e_i's?

c. Plot the standardized residuals against x. Does the plot differ significantly in general appearance from the plot of part (a)?

71. Wear resistance of certain nuclear reactor components made of Zircaloy-2 is partly determined by properties of the oxide layer. The following data appears in an article that proposed a new nondestructive testing method to monitor thickness of the layer ("Monitoring of Oxide Layer Thickness on Zircaloy-2 by the Eddy Current Test Method," *J. Testing Eval.*, 1987: 333–336). The variables are x = oxide-layer thickness (μm) and y = eddy-current response (arbitrary units).

x	0	7	17	114	133
y	20.3	19.8	19.5	15.9	15.1

x	142	190	218	237	285
y	14.7	11.9	11.5	8.3	6.6

a. The authors summarized the relationship by giving the equation of the least squares line as $y = 20.6 - .047x$. Calculate and plot the residuals against x and then comment on the appropriateness of the simple linear regression model.

b. Use $s = .7921$ to calculate the standardized residuals from a simple linear regression. Construct a standardized residual plot and comment.

72. As the air temperature drops, river water becomes supercooled and ice crystals form. Such ice can significantly affect the hydraulics of a river. The article "Laboratory Study of Anchor Ice Growth" (*J. Cold Regions Engrg.*, 2001: 60–66) described an experiment in which ice thickness (mm) was studied as a function of elapsed time (hr) under specified conditions. The following data was read from a graph in the article: $n = 33$; $x = .17, .33, .50, .67, \ldots, 5.50$; $y = .50, 1.25, 1.50, 2.75, 3.50, 4.75, 5.75, 5.60, 7.00, 8.00, 8.25, 9.50, 10.50, 11.00, 10.75, 12.50, 12.25, 13.25, 15.50, 15.00, 15.25, 16.25, 17.25, 18.00, 18.25, 18.15, 20.25, 19.50, 20.00, 20.50, 20.60, 20.50, 19.80$.

a. The r^2 value resulting from a least squares fit is .977. Given the high r^2, does it seem appropriate to assume an approximate linear relationship?

b. The residuals, listed in the same order as the x values, are

−1.03	−0.92	−1.35	−0.78	−0.68	−0.11	0.21
−0.59	0.13	0.45	0.06	0.62	0.94	0.80
−0.14	0.93	0.04	0.36	1.92	0.78	0.35
0.67	1.02	1.09	0.66	−0.09	1.33	−0.10
−0.24	−0.43	−1.01	−1.75	−3.14		

Plot the residuals against x, and reconsider the question in (a). What does the plot suggest?

73. The accompanying data on x = true density (kg/mm^3) and y = moisture content (% d.b.) was read from a plot in the article "Physical Properties of Cumin Seed" (*J. Agric. Engrg. Res.*, 1996: 93–98).

x	7.0	9.3	13.2	16.3	19.1	22.0
y	1046	1065	1094	1117	1130	1135

The equation of the least squares line is $y = 1008.14 + 6.19268x$ (this differs very slightly from the equation given in the article); $s = 7.265$ and $r^2 = .968$.

a. Carry out a test of model utility and comment.

b. Compute the values of the residuals and plot the residuals against x. Does the plot suggest that a linear regression function is inappropriate?

c. Compute the values of the standardized residuals and plot them against x. Are there any unusually large (positive or negative) standardized residuals? Does this plot give the same message as the plot of part (b) regarding the appropriateness of a linear regression function?

74. Continuous recording of heart rate can be used to obtain information about the level of exercise intensity or physical strain during sports participation, work, or other daily activities. The article "The Relationship Between Heart Rate and Oxygen Uptake During Non–Steady State Exercise" (*Ergonomics*, 2000: 1578–1592) reported on a study to investigate using heart rate response (x, as a percentage of the maximum rate) to predict oxygen uptake (y, as a percentage of maximum uptake) during exercise. The accompanying data was read from a graph in the paper.

HR	43.5	44.0	44.0	44.5	44.0	45.0	48.0	49.0
VO_2	22.0	21.0	22.0	21.5	25.5	24.5	30.0	28.0

HR	49.5	51.0	54.5	57.5	57.7	61.0	63.0	72.0
VO_2	32.0	29.0	38.5	30.5	57.0	40.0	58.0	72.0

Use a statistical software package to perform a simple linear regression analysis. Considering the list of potential difficulties in this section, see which of them apply to this data set.

75. Consider the following four (x, y) data sets; the first three have the same x values, so these values are listed only once (Frank Anscombe, "Graphs in Statistical Analysis," *Amer. Statist.*, 1973: 17–21):

1–3	**1**	**2**	**3**	**4**	**4**
x	y	y	y	x	y
10.0	8.04	9.14	7.46	8.0	6.58
8.0	6.95	8.14	6.77	8.0	5.76
13.0	7.58	8.74	12.74	8.0	7.71
9.0	8.81	8.77	7.11	8.0	8.84
11.0	8.33	9.26	7.81	8.0	8.47
14.0	9.96	8.10	8.84	8.0	7.04
6.0	7.24	6.13	6.08	8.0	5.25
4.0	4.26	3.10	5.39	19.0	12.50
12.0	10.84	9.13	8.15	8.0	5.56
7.0	4.82	7.26	6.42	8.0	7.91
5.0	5.68	4.74	5.73	8.0	6.89

For each of these four data sets, the values of the summary statistics $\sum x_i$, $\sum x_i^2$, $\sum y_i$, $\sum y_i^2$, and $\sum x_i y_i$ are virtually identical, so all quantities computed from these five will be essentially identical for the four sets—the least squares line ($y = 3 + .5x$), SSE, s^2, r^2, t intervals, t statistics, and so on. The summary statistics provide no way of distinguishing among the four data sets. Based on a scatter plot and a residual plot for each set, comment on the appropriateness or inappropriateness of fitting a straight-line model; include in your comments any specific suggestions for how a "straight-line analysis" might be modified or qualified.

76. **a.** Express the ith residual $Y_i - \hat{Y}_i$ (where $\hat{Y}_i = \hat{\beta}_0 + \hat{\beta}_1 x_i$) in the form $\sum c_j Y_j$, a linear function of the Y_j's. Then use rules of variance to verify that $V(Y_i - \hat{Y}_i)$ is given by Expression (12.13).

b. As x_i moves farther away from $\bar{x}$, what happens to $V(\hat{Y}_i)$ and to $V(Y_i - \hat{Y}_i)$?

77. If there is at least one x value at which more than one observation has been made, there is a formal test procedure for testing

H_0: $\mu_{Y \cdot x} = \beta_0 + \beta_1 x$ for some values β_0, β_1 (the true regression function is linear)

versus

H_a: H_0 is not true (the true regression function is not linear)

Suppose observations are made at $x_1, x_2, \ldots, x_c$. Let $Y_{11}, Y_{12}, \ldots, Y_{1n_1}$ denote the n_1 observations when $x = x_1$; . . . ; $Y_{c1}, Y_{c2}, \ldots, Y_{cn_c}$ denote the n_c observations when $x = x_c$. With $n = \sum n_i$ (the total number of observations), SSE has $n - 2$ df. We break SSE into two pieces, SSPE (pure error) and SSLF (lack of fit), as follows:

$$\text{SSPE} = \sum_i \sum_j (Y_{ij} - \overline{Y}_{i\cdot})^2$$
$$= \sum\sum Y_{ij}^2 - \sum n_i (\overline{Y}_{i\cdot})^2$$
$$\text{SSLF} = \text{SSE} - \text{SSPE}$$

The n_i observations at x_i contribute $n_i - 1$ df to SSPE, so the number of degrees of freedom for SSPE is $\sum_i (n_i - 1) = n - c$ and the degrees of freedom for SSLF is $n - 2 - (n - c) = c - 2$. Let MSPE = SSPE/$(n - c)$ and MSLF = SSLF/$(c - 2)$. Then it can be shown that whereas $E(\text{MSPE}) = \sigma^2$ whether or not H_0 is true, $E(\text{MSLF}) = \sigma^2$ if H_0 is true and $E(\text{MSLF}) > \sigma^2$ if H_0 is false.

Test statistic: $F = \dfrac{\text{MSLF}}{\text{MSPE}}$

Rejection region: $f \geq F_{\alpha, c-2, n-c}$

The following data comes from the article "Changes in Growth Hormone Status Related to Body Weight of Growing Cattle" (*Growth*, 1977: 241–247), with x = body weight and y = metabolic clearance rate/body weight.

x	110	110	110	230	230	230	360
y	235	198	173	174	149	124	115

x	360	360	360	505	505	505	505
y	130	102	95	122	112	98	96

(So $c = 4$, $n_1 = n_2 = 3$, $n_3 = n_4 = 4$.)

a. Test H_0 versus H_a at level .05 using the lack-of-fit test just described.

b. Does a scatter plot of the data suggest that the relationship between x and y is linear? How does this compare with the result of part (a)? (A nonlinear regression function was used in the article.)

12.7 *Multiple Regression Analysis

In multiple regression, the objective is to build a probabilistic model that relates a dependent variable y to more than one independent or predictor variable. Let k represent the number of predictor variables ($k \geq 2$) and denote these predictors by $x_1, x_2, \ldots, x_k$. For example, in attempting to predict the selling price of a house, we might have $k = 3$ with x_1 = size (ft^2), x_2 = age (years), and x_3 = number of rooms.

DEFINITION

The **general additive multiple regression model equation** is

$$Y = \beta_0 + \beta_1 x_1 + \beta_2 x_2 + \cdots + \beta_k x_k + \varepsilon \qquad (12.16)$$

where $E(\varepsilon) = 0$ and $V(\varepsilon) = \sigma^2$. In addition, for purposes of testing hypotheses and calculating CIs or PIs, it is assumed that ε is normally distributed. It is assumed that the ε's associated with various observations, and thus the Y_i's themselves, are independent of one another.

Let $x_1^*, x_2^*, \ldots, x_k^*$ be particular values of $x_1, \ldots, x_k$. Then (12.16) implies that

$$\mu_{Y \cdot x_1^*, \ldots, x_k^*} = \beta_0 + \beta_1 x_1^* + \cdots + \beta_k x_k^* \qquad (12.17)$$

Thus, just as $\beta_0 + \beta_1 x$ describes the mean Y value as a function of x in simple linear regression, the **true** (or **population**) **regression function** $\beta_0 + \beta_1 x_1 + \cdots + \beta_k x_k$ gives the expected value of Y as a function of $x_1, \ldots, x_k$. The β_i's are the **true** (or **population**) **regression coefficients**. The regression coefficient β_1 is interpreted as the expected change in Y associated with a 1-unit increase in x_1 *while* $x_2, \ldots, x_k$ *are held fixed.* Analogous interpretations hold for $\beta_2, \ldots, \beta_k$.

Estimating Parameters

The data in simple linear regression consists of n pairs $(x_1, y_1), \ldots, (x_n, y_n)$. Suppose that a multiple regression model contains two predictor variables, x_1 and x_2. Then each observation will consist of three numbers (a triple): a value of x_1, a value of x_2, and a value of y. More generally, with k independent or predictor variables, each observation will consist of $k + 1$ numbers (a "$k + 1$ tuple"). The values of the predictors in the individual observations are denoted using double-subscripting:

$$x_{ij} = \text{the value of the } j\text{th predictor } x_j \text{ in the } i\text{th observation} \\ (i = 1, \ldots, n; j = 1, \ldots, k)$$

Thus the first subscript is the observation number and the second subscript is the predictor number. For example, x_{83} is the value of the 3rd predictor in the 8th observation (to avoid confusion, a comma can be inserted between the two subscripts, e.g., $x_{12,3}$). The first observation in our data set is then $(x_{11}, x_{12}, \ldots, x_{1k}, y_1)$, the second is $(x_{21}, x_{22}, \ldots, x_{2k}, y_2)$, and so on.

Consider candidates $b_0, b_1, \ldots, b_k$ for estimates of the β_i's and the corresponding candidate regression function $b_0 + b_1x_1 + \cdots + b_kx_k$. Substituting the predictor values for any individual observation into this candidate function gives a prediction for the y value that would be observed, and subtracting this prediction from the actual observed y value gives the prediction error. The **principle of least squares** says we should square these prediction errors, sum, and then take as the least squares estimates $\hat{\beta}_0, \hat{\beta}_1, \ldots, \hat{\beta}_k$, the values of the b_i's that minimize the sum of squared prediction errors. To carry out this program, form the criterion function (sum of squared prediction errors)

$$g(b_0, b_1, \ldots, b_k) = \sum_{i=1}^{n} [y_i - (b_0 + b_1x_{i1} + \cdots + b_kx_{ik})]^2$$

then take the partial derivative of $g(\cdot)$ with respect to each b_i ($i = 0, 1, \ldots, k$), and equate these $k + 1$ partial derivatives to 0. The result is a system of $k + 1$ equations, the **normal equations**, in the $k + 1$ unknowns (the b_i's). Very importantly, the fact that the criterion function is quadratic implies that the normal equations are linear in the unknowns.

$$\begin{aligned} nb_0 + (\textstyle\sum x_{i1})b_1 + (\sum x_{i2})b_2 + \cdots + (\sum x_{ik})b_k &= \textstyle\sum y_i \\ (\textstyle\sum x_{i1})b_0 + (\sum x_{i1}^2)b_1 + (\sum x_{i1}x_{i2})b_2 + \cdots + (\sum x_{i1}x_{ik})b_k &= \textstyle\sum x_{i1}y_i \\ &\vdots \\ (\textstyle\sum x_{ik})b_0 + (\sum x_{i1}x_{ik})b_1 + \cdots + (\sum x_{i,k-1}x_{ik})b_{k-1} + (\sum x_{ik}^2)b_k &= \textstyle\sum x_{ik}y_i \end{aligned}$$

We will assume that the system has a unique solution, the least squares estimates $\hat{\beta}_0, \hat{\beta}_1, \hat{\beta}_2, \ldots, \hat{\beta}_k$. The next section uses matrix algebra to deal with the system of equations and develop inferential procedures for multiple regression. For the moment, though, we shall take advantage of the fact that all of the commonly used statistical software packages are programmed to solve the equations and provide the results needed for inference.

Sometimes interest in the individual regression coefficients is the main reason for doing the regression. The article "Autoregressive Modeling of Baseball Performance and Salary Data," *Proceedings Statisti. Graph. Sect., Amer. Statisti. Assoc.*, 1988: 132–137, describes a multiple regression of runs scored as a function of singles, doubles, triples, home runs, and walks (combined with hit-by-pitcher). The estimated regression equation is

$$\text{runs} = -2.49 + .47 \text{ singles} + .76 \text{ doubles} + 1.14 \text{ triples} \\ + 1.54 \text{ home runs} + .39 \text{ walks}$$

This is very similar to the popular slugging percentage statistic, which gives weight 1 to singles, 2 to doubles, 3 to triples, and 4 to home runs. However, the slugging percentage gives no weight to walks, whereas the regression puts weight .39 on walks, more than 80% of the weight it assigns to singles. The importance of walks is well known among statisticians who follow baseball, and some statistically savvy people in major league baseball management are now emphasizing walks in choosing players.

Example 12.25 The article "Factors Affecting Achievement in the First Course in Calculus" (*J. Experiment. Ed.*, 1984: 136–140) discussed the ability of several variables to predict y = freshman calculus grade (on a scale of 0–100). The variables included x_1 = an algebra placement test given in the first week of class, x_2 = ACT math score, x_3 = ACT natural science score, and x_4 = high school percentile rank. Here are the scores for the first five and the last five of the 80 students (the data set is available on the data disk):

Observation	Algebra	ACTM	ACTNS	HS Rank	Grade
1	21	27	23	68	62
2	16	29	32	99	75
3	22	30	32	98	95
4	25	34	28	90	78
5	22	29	23	99	95
⋮	⋮	⋮	⋮	⋮	⋮
76	22	29	26	88	85
77	17	29	33	92	75
78	26	27	29	95	88
79	26	28	30	99	95
80	21	28	30	99	85

The JMP statistical computer package gave the following least squares estimates:

$$\hat{\beta}_0 = 36.12 \qquad \hat{\beta}_1 = .9610 \qquad \hat{\beta}_2 = .2718 \qquad \hat{\beta}_3 = .2161 \qquad \hat{\beta}_4 = .1353$$

Thus we estimate that .9610 is the average increase in final grade associated with a 1-point increase in the algebra placement score when the other three predictors are held fixed. Alternatively, a 10-point increase in the algebra pretest score, with the other scores held fixed, corresponds to a 9.6-point increase in the final grade, an increase of approximately one letter grade if A = 90s, B = 80s, etc. The other estimated coefficients are interpreted in a similar manner.

The estimated regression equation is

$$y = 36.12 + .9610x_1 + .2718x_2 + .2161x_3 + .1353x_4$$

A point prediction of final grade for a single student with an algebra test score of 25, ACTM score of 28, ACTNS score of 26, and a high school percentile rank of 90 is

$$\hat{y} = 36.12 + .9610(25) + .2718(28) + .2161(26) + .1353(90) = 85.55$$

a middle B. This is also a point estimate of the mean for the population of all students with an algebra test score of 25, ACTM score of 28, ACTNS score of 26, and a high school percentile rank of 90. ■

$\hat{\sigma}^2$ and the Coefficient of Multiple Determination

Substituting the values of the predictors from the successive observations into the equation for an estimated regression function gives the **predicted** or **fitted values** $\hat{y}_1, \hat{y}_2, \ldots, \hat{y}_n$. For example, since the values of the four predictors for the last observation in Example 12.25 are 21, 28, 30, and 99, respectively, the corresponding predicted value is $\hat{y}_{80} = 83.79$. The **residuals** are the differences $y_1 - \hat{y}_1, \ldots, y_n - \hat{y}_n$. In simple linear regression, they were the vertical deviations from the least squares line, but in general there is no geometric interpretation in multiple regression (the exception is the case $k = 2$, where the estimated regression function specifies a plane in three dimensions and the residuals are the vertical deviations from the plane). The last residual in Example 12.25 is $85 - 83.79 = 1.21$. The closer the residuals are to 0, the better our estimated equation is doing in predicting the y values actually observed.

The residuals are sometimes important not just for judging the quality of a regression. Several enterprising students developed a multiple regression model using age, size in square feet, and other factors to predict the price of four-unit apartment buildings. They found that one building had a strongly negative residual, meaning that the price was much lower than predicted. As it turned out, the reason was that the owner had "cash-flow" problems, and needed to sell quickly, so the students got an unusually good deal.

As in simple linear regression, the estimate of the variance parameter σ^2 is based on the sum of squared residuals (or sum of squared errors) $\text{SSE} = \Sigma(y_i - \hat{y}_i)^2$. Previously we divided SSE by $n - 2$ to obtain the estimate. The explanation was that the two estimates $\hat{\beta}_0$ and $\hat{\beta}_1$ had to be calculated, entailing a loss of 2 degrees of freedom. For each parameter there is a normal equation that can be expressed as a constraint on the residuals, with a loss of 1 df. In multiple regression with k predictors, $k + 1$ df are lost in estimating the β_i's (don't forget the constant term β_0). Here are the normal equations rewritten as constraints on the residuals:

$$\sum [y_i - (b_0 + x_{i1}b_1 + x_{i2}b_2 + \cdots + x_{ik}b_k)] = 0$$
$$\sum x_{i1}[y_i - (b_0 + x_{i1}b_1 + x_{i2}b_2 + \cdots + x_{ik}b_k)] = 0$$
$$\vdots$$
$$\sum x_{ik}[y_i - (b_0 + x_{i1}b_1 + x_{i2}b_2 + \cdots + x_{ik}b_k)] = 0$$

The first equation says that the sum of the residuals is 0, the second equation says that the first predictor times the residual sums to 0, and so on. These $k + 1$ constraints allow any $k + 1$ residuals to be determined from the others. This implies that SSE is based on $n - (k + 1)$ df, and this is the divisor in the estimate of σ^2:

$$\hat{\sigma}^2 = s^2 = \frac{\text{SSE}}{n - (k + 1)} = \text{MSE} \qquad \hat{\sigma} = s = \sqrt{s^2}$$

SSE can once again be regarded as a measure of unexplained variation in the data—the extent to which observed variation in y cannot be attributed to the model relationship. The total sum of squares SST, defined as $\Sigma(y_i - \bar{y})^2$ as in simple linear regression, is a measure of total variation in the observed y values. Taking the ratio of these sums of squares and subtracting from 1 gives the **coefficient of multiple determination**

$$R^2 = 1 - \frac{\text{SSE}}{\text{SST}}$$

Sometimes called just the **coefficient of determination** or the **squared multiple correlation**, R^2 is interpreted as the proportion of observed variation that can be attributed to (explained by) the model relationship. Thinking of SST as the error sum of squares using just the constant model (with β_0 as the only term in the model) having $\bar{y}$ as the predictor, R^2 is the proportion by which the model reduces the error sum of squares. For example, if SST $= 20$ and SSE $= 5$, then the model reduces the error sum of squares by 75%, so $R^2 = .75$. The closer R^2 is to 1, the greater the proportion of observed variation that can be explained by the fitted model.

Unfortunately, there is a potential problem with R^2: Its value can be inflated by including predictors in the model that are relatively unimportant or even frivolous. For example, suppose we plan to obtain a sample of 20 recently sold houses in order to relate sale price to various characteristics of a house. Natural predictors include interior size, lot size, age, number of bedrooms, and distance to the nearest school. Suppose we also include in the model the diameter of the doorknob on the door of the master bedroom, the height of the toilet bowl in the master bath, and so on until we have 19 predictors. Then unless we are extremely unlucky in our choice of predictors, the value of R^2 will be 1 (because 20 coefficients are estimated from 20 observations)! Rather than seeking a model that has the highest possible R^2 value, which can be achieved just by "packing" our model with predictors, what is desired is a relatively simple model based on just a few important predictors whose R^2 value is high.

It is therefore desirable to adjust R^2 to take account of the fact that its value may be quite high just because many predictors were used relative to the amount of data. The **adjusted coefficient of multiple determination** is defined by

$$R_a^2 = 1 - \frac{\text{MSE}}{\text{MST}} = 1 - \frac{\text{SSE}/[n - (k + 1)]}{\text{SST}/(n - 1)} = 1 - \frac{n - 1}{n - (k + 1)} \frac{\text{SSE}}{\text{SST}}$$

The ratio multiplying SSE/SST in adjusted R^2 exceeds 1 (the denominator is smaller than the numerator), so adjusted R^2 is smaller than R^2 itself, and in fact will be much smaller when k is large relative to n. A value of R_a^2 much smaller than R^2 is a warning flag that the chosen model has too many predictors relative to the amount of data.

Example 12.26 Continuing with the previous example in which a model with four predictors was fit to the calculus data consisting of 80 observations, the JMP software package gave SSE $=$ 7346.05 and SST $=$ 10,332.20, from which $s = 9.90$, $R^2 = .289$, and $R_a^2 = .251$. The estimated standard deviation s is very close to 10, which corresponds to one letter grade on the usual A $=$ 90s, B $=$ 80s, . . . , scale. About 29% of observed variation in grade can be attributed to the chosen model. The difference between R^2 and R_a^2 is not very dramatic, a reflection of the fact that $k = 4$ is much smaller than $n = 80$. ■

A Model Utility Test

In multiple regression, is there a single indicator that can be used to judge whether a particular model will be useful? The value of R^2 certainly communicates a preliminary message, but this value is sometimes deceptive because it can be greatly inflated by using a large number of predictors (large k) relative to the sample size n (this is the rationale behind adjusting R^2).

The model utility test in simple linear regression involved the null hypothesis H_0: $\beta_1 = 0$, according to which there is no useful relation between y and the single predictor x. Here we consider the assertion that $\beta_1 = 0, \beta_2 = 0, \ldots, \beta_k = 0$, which says that there is no useful relationship between y and *any* of the k predictors. If at least one of these β's is not 0, the corresponding predictor(s) is (are) useful. The test is based on a statistic that has a particular F distribution when H_0 is true.

Null hypothesis: H_0: $\beta_1 = \beta_2 = \cdots = \beta_k = 0$

Alternative hypothesis: H_a: at least one $\beta_i \neq 0 \qquad (i = 1, \ldots, k)$

Test statistic value:

$$f = \frac{R^2/k}{(1 - R^2)/[n - (k + 1)]}$$

$$= \frac{\text{SSR}/k}{\text{SSE}/[n - (k + 1)]} = \frac{\text{MSR}}{\text{MSE}} \qquad (12.18)$$

where SSR = regression sum of squares = SST − SSE

Rejection region for a level α test: $f \geq F_{\alpha,k,n-(k+1)}$

See the next section for an explanation of why the ratio MSR/MSE has an F distribution under the null hypothesis.

Except for a constant multiple, the test statistic here is $R^2/(1 - R^2)$, the ratio of explained to unexplained variation. If the proportion of explained variation is high relative to unexplained, we would naturally want to reject H_0 and confirm the utility of the model. However, if k is large relative to n, the factor $[(n - (k + 1))/k]$ will decrease f considerably.

Example 12.27 Returning to the calculus data of Example 12.25, a model with $k = 4$ predictors was fitted, so the relevant hypotheses are

$$H_0: \beta_1 = \beta_2 = \beta_3 = \beta_4 = 0$$

$$H_a: \text{at least one of these four } \beta\text{'s is not } 0$$

Figure 12.29 shows output from the JMP statistical package. The value of the model utility F ratio is

$$f = \frac{R^2/k}{(1 - R^2)/[n - (k + 1)]} = \frac{.289/4}{.711/(80 - 5)} = 7.62$$

This value also appears in the F Ratio column of the ANOVA table in Figure 12.29. The ANOVA table in the JMP output shows that P-value $< .0001$, and in fact the P-value for $f = 7.62$ is approximately .000033. This is a highly significant result. The null hypothesis should be rejected at any reasonable significance level. We conclude that there *is* a useful linear relationship between y and *at least one* of the four predictors in the model.

This does not mean that all four predictors are useful; we will say more about this subsequently.

Summary of Fit

RSquare	0.289014
RSquare Adj	0.251095
Root Mean Square Error	9.896834
Mean of Response	80.15
Observations (or Sum Wgts)	80

Analysis of Variance

Source	DF	Sum of Squares	Mean Square	F Ratio
Model	4	2986.150	746.538	7.6218
Error	75	7346.050	97.947	Prob > F
C. Total	79	10332.200		<.0001

Parameter Estimates

Term	Estimate	Std Error	t Ratio	Prob>|t|	Lower 95%	Upper 95%
Intercept	36.121531	10.7519	3.36	0.0012	14.702651	57.540411
Alg Place	0.960992	0.26404	3.64	0.0005	0.4349971	1.4869868
ACTM	0.2718147	0.453505	0.60	0.5507	−0.631614	1.1752438
ACTNS	0.2161047	0.313215	0.69	0.4924	−0.407851	0.8400606
HS Rank	0.1353158	0.103642	1.31	0.1957	−0.07115	0.3417815

Figure 12.29 Multiple regression output from JMP for the data of Example 12.27 ■

Inferences in Multiple Regression

Before testing hypotheses, constructing CIs, and making predictions, one should first examine diagnostic plots to see whether the model needs modification or whether there are outliers in the data. The recommended plots are (standardized) residuals versus each independent variable, residuals versus $\hat{y}$, y versus $\hat{y}$, and a normal probability plot of the standardized residuals. Potential problems are suggested by the same patterns discussed in Section 12.6. Of particular importance is the identification of observations that have a large influence on the fit.

Because each $\hat{\beta}_i$ is a linear function of the y_i's, the standard deviation of each $\hat{\beta}_i$ is the product of σ and a function of the x_{ij}'s, so an estimate $s_{\hat{\beta}_i}$ is obtained by substituting s for σ. A formula for $s_{\hat{\beta}_i}$ is given in the next section, and the result is part of the output from all standard regression computer packages. Inferences concerning a single $\hat{\beta}_i$ are based on the standardized variable

$$T = \frac{\hat{\beta}_i - \beta_i}{S_{\hat{\beta}_i}}$$

which, assuming the model is correct, has a t distribution with $n - (k + 1)$ df.

The point estimate of $\hat{\mu}_{Y \cdot x_1^*, \ldots, x_k^*}$, the expected value of Y when $x_1 = x_1^*, \ldots, x_k = x_k^*$, is $\hat{\mu}_{Y \cdot x_1^*, \ldots, x_k^*} = \hat{\beta}_0 + \hat{\beta}_1 x_1^* + \cdots + \hat{\beta}_k x_k^*$. The estimated standard deviation of the corresponding estimator is a complicated expression involving the sample x_{ij}'s, but a simple matrix formula is given in the next section. The better statistical computer packages will calculate it on request. Inferences about $\mu_{Y \cdot x_1^*, \ldots, x_k^*}$ are based on standardizing its estimator to obtain a t variable having $n - (k + 1)$ df.

---

1. A $100(1-\alpha)\%$ CI for β_i, the coefficient of x_i in the regression function, is

$$\hat{\beta}_i \pm t_{\alpha/2,n-(k+1)} \cdot s_{\hat{\beta}_i}$$

2. A test for H_0: $\beta_i = \beta_{i0}$ uses the t statistic value $t = (\hat{\beta}_i - \beta_{i0})/s_{\hat{\beta}_i}$ based on $n-(k+1)$ df. The test is upper-, lower-, or two-tailed according to whether H_a contains the inequality $>$, $<$, or $\neq$.
3. A $100(1-\alpha)\%$ CI for $\mu_{Y\cdot x_1^*,\ldots,x_k^*}$ is

$$\hat{\mu}_{Y\cdot x_1^*,\ldots,x_k^*} \pm t_{\alpha/2,n-(k+1)} \cdot (\text{estimated SD of } \hat{\mu}_{Y\cdot x_1^*,\ldots,x_k^*}) = \hat{y} \pm t_{\alpha/2,n-(k+1)} \cdot s_{\hat{Y}}$$

where $\hat{Y}$ is the statistic $\hat{\beta}_0 + \hat{\beta}_1 x_1^* + \cdots + \hat{\beta}_k x_k^*$ and $\hat{y}$ is the calculated value of $\hat{Y}$.
4. A $100(1-\alpha)\%$ PI for a future y value is

$$\hat{\mu}_{Y\cdot x_1^*,\ldots,x_k^*} \pm t_{\alpha/2,n-(k+1)} \cdot \left[s^2 + (\text{estimated SD of } \hat{\mu}_{Y\cdot x_1^*,\ldots,x_k^*})^2\right]^{1/2} =$$

$$\hat{y} \pm t_{\alpha/2,n-(k+1)} \cdot \sqrt{s^2 + s_{\hat{Y}}^2}$$

Simultaneous intervals for which the simultaneous confidence or prediction level is controlled can be obtained by applying the Bonferroni technique.

Example 12.28 (Example 12.27 continued)

The JMP output for the calculus data includes 95% confidence intervals for the coefficients. Let's verify the interval for β_1, the coefficient for algebra placement score:

$$\hat{\beta}_1 \pm t_{.025,80-5} s_{\hat{\beta}_1} = .961 \pm 1.992(.264) = .961 \pm .526 = (.435, 1.487)$$

which agrees with the interval in Figure 12.29. Thus if ACTM score, ACTNS score, and percentile rank are fixed, we estimate that on average an increase between .435 and 1.487 in grade is associated with a one-point increase in algebra score.

We found in Example 12.25 that, if a student has an algebra test score of 25, ACTM score of 28, ACTNS score of 26, and high school percentile rank of 90, then the predicted value is 85.55. The estimated standard deviation for this predicted value can be obtained from JMP, with the result $s_{\hat{Y}} = 1.882$, so a 95% confidence interval for the expected grade is

$$\begin{aligned}\hat{\mu}_{Y\cdot 25,28,26,90} \pm t_{.025,80-5} s_{\hat{Y}} &= 85.55 \pm 1.992(1.882)\\ &= 85.55 \pm 3.75 = (81.8, 89.3)\end{aligned}$$

which can also be obtained from JMP. This interval is for the mean score of all students with the predictor values 25, 28, 26, and 90. Regarding scores in the 80's as B's, we can say with 95% confidence that the expected grade is a B. Now consider the estimated standard deviation for the error in predicting the final grade of a single student with the predictor values 25, 28, 26, and 90. This is

$$\sqrt{s^2 + s_{\hat{Y}}^2} = \sqrt{9.897^2 + 1.882^2} = 10.074$$

Therefore, a 95% prediction interval for the final grade of a single student with predictor scores 25, 28, 26, and 90 is

$$\hat{\mu}_{Y\cdot 25,28,26,90} \pm t_{.025,80-5}10.074 = 85.55 \pm 1.992(10.074)$$
$$= 85.55 \pm 20.07 = (65.5, 105.6)$$

Of course, this PI is much wider than the corresponding CI. Although we are highly confident that the expected score is a B, the score for a single student could be as low as a D or as high as an A. Notice that the upper end of the interval exceeds the maximum score of 100, so it would be appropriate to truncate the interval to (65.5, 100). ■

Frequently, the hypothesis of interest has the form H_0: $\beta_i = 0$ for a particular i. For example, after fitting the four-predictor model in Example 12.25, the investigator might wish to test H_0: $\beta_2 = 0$. According to H_0, as long as the predictors x_1, x_3, and x_4 remain in the model, x_2 contains no useful information about y. The test statistic value is the ***t* ratio** $\hat{\beta}_i/s_{\hat{\beta}_i}$. Many statistical computer packages report the t ratio and corresponding P-value for each predictor included in the model. For example, Figure 12.29 shows that as long as algebra pretest score, ACT natural science, and high school percentile rank are retained in the model, the predictor x_2 = ACT math score can be deleted. The P-value for the test is .55, much too large to reject the null hypothesis.

It is interesting to look at the correlations between the predictors and the response variable in Example 12.25. Here are the correlations and the corresponding P-values (in parentheses):

	alg plc	ACTmath	ACTns	rank
calc grade	0.491	0.353	0.259	0.324
	(0.000)	(0.0013)	(0.020)	(0.003)

Do these values seem inconsistent with the multiple regression results? There is a highly significant correlation between calculus grade and ACT math score, but in the multiple regression the ACT math score is redundant, not needed in the model. The idea is that ACT math score also has highly significant correlations with the other predictors, so much of its predictive ability is retained in the model when this variable is deleted. In order to be a statistically significant predictor in the multiple regression model, a variable must provide additional predictive ability beyond what is offered by the other predictors.

The r and R^2 values for the calculus data are disappointing. Given the importance placed on predictors such as ACT scores and high school rank in college admissions and NCAA eligibility, we might expect that these scores would give better predictions.

Assessing Model Adequacy

The standardized residuals in multiple regression result from dividing each residual by its estimated standard deviation; a matrix formula for the standard deviation is given in the next section. We recommend a normal probability plot of the standardized residuals as a basis for validating the normality assumption. Plots of the standardized residuals versus each predictor and versus $\hat{y}$ should show no discernible pattern. The book by Neter et al. in the chapter bibliography discusses other diagnostic plots.

Example 12.29 Figure 12.30 from JMP shows a histogram and normal probability plot of the standardized residuals for the calculus data discussed in the preceding examples. The plot is sufficiently straight that there is no reason to doubt the assumption of normally distributed errors.

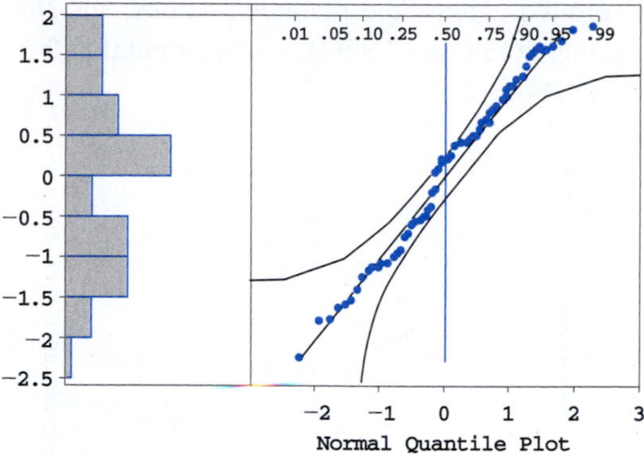

Figure 12.30 A normal probability plot and histogram of the standardized residuals for the calculus data

Figure 12.31 shows plots of the standardized residuals versus the predictors for the calculus data. There is not much evidence of a pattern in plots (b), (c), and (d), other than

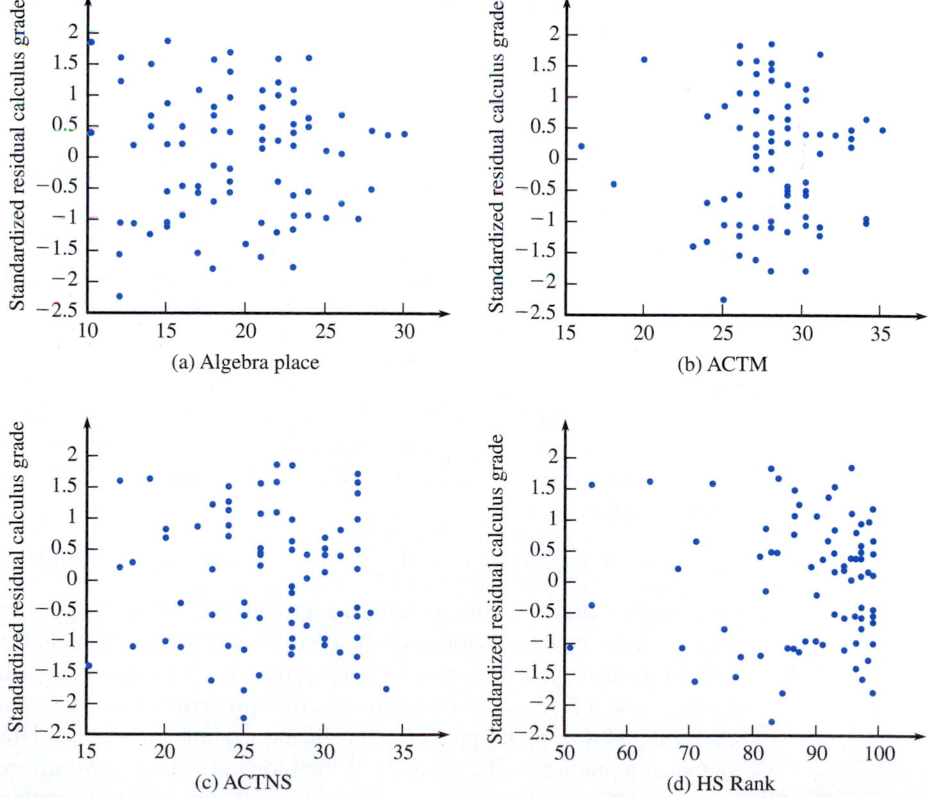

Figure 12.31 Standardized residuals versus predictors for the calculus data

randomness. However, the first plot does show some indication that the variance might be lower at the high end.

The graphs in Figure 12.32 show the calculus grade and the standardized residuals plotted against the predicted values, and these also show narrowing on the right. Looking at Figure 12.32(a), it is apparent that this would have to occur, because no score can be above 100.

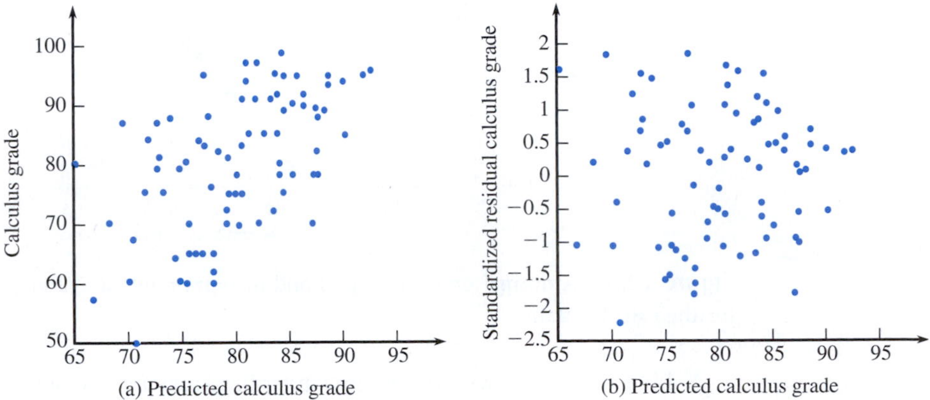

Figure 12.32 Diagnostic plots for the calculus data: (a) y versus $\hat{y}$, (b) standardized residual versus $\hat{y}$

Multiple Regression Models

We now consider various ways of creating predictors to specify informative models.

Polynomial Regression Let's return for a moment to the case of bivariate data consisting of n (x,y) pairs. Suppose that a scatter plot shows a parabolic rather than linear shape. Then it is natural to specify a **quadratic regression model**:

$$Y = \beta_0 + \beta_1 x + \beta_2 x^2 + \varepsilon$$

The corresponding population regression function $\beta_0 + \beta_1 x + \beta_2 x^2$ is quadratic rather than linear, and gives the mean or expected value of Y for any particular x.

So what does this have to do with multiple regression? Let's rewrite the quadratic model equation as follows:

$$Y = \beta_0 + \beta_1 x_1 + \beta_2 x_2 + \varepsilon \qquad \text{where} \quad x_1 = x \quad \text{and} \quad x_2 = x^2$$

Now this looks exactly like a multiple regression equation with two predictors. You may object on the grounds that one of the predictors is a mathematical function of the other one. Appeal denied! It is not only legitimate for a predictor in a multiple regression model to be a function of one or more other predictors but often desirable in the sense that a model with such a predictor may be judged much more useful than the model without such a predictor. The message at the moment is that *quadratic regression is a special case of multiple regression.* Thus any software package capable of carrying out a multiple regression analysis can fit the quadratic regression model. The same is true of cubic regression and even higher order polynomial models, though in practice very rarely are such higher order predictors needed.

The interpretation of β_i given previously for the general multiple regression model is not legitimate in quadratic regression. This is because $x_2 = x^2$, so the value of x_2 cannot be increased while $x_1 = x$ is held fixed. More generally, the interpretation of regression coefficients requires extra care when some predictor variables are mathematical functions of others.

Models with Interaction Suppose that an industrial chemist is interested in the relationship between product yield (y) from a certain reaction and two independent variables, x_1 = reaction temperature and x_2 = pressure at which the reaction is carried out. The chemist initially proposes the relationship

$$y = 1200 + 15x_1 - 35x_2 + \varepsilon$$

for temperature values between 80 and 100 in combination with pressure values ranging from 50 to 70. The population regression function $1200 + 15x_1 - 35x_2$ gives the mean y value for any particular values of the predictors. Consider this mean y value for three different particular temperature values:

$$x_1 = 90: \quad \text{mean } y \text{ value} = 1200 + 15(90) - 35x_2 = 2550 - 35x_2$$

$$x_1 = 95: \quad \text{mean } y \text{ value} = 2625 - 35x_2$$

$$x_1 = 100: \quad \text{mean } y \text{ value} = 2700 - 35x_2$$

Graphs of these three mean y value functions are shown in Figure 12.33(a). Each graph is a straight line, and the three lines are parallel, each with a slope of -35. Thus irrespective of the fixed value of temperature, the average change in yield associated with a 1-unit increase in pressure is -35.

When pressure x_2 increases, the decline in average yield should be more rapid for a high temperature than for a low temperature, so the chemist has reason to doubt the

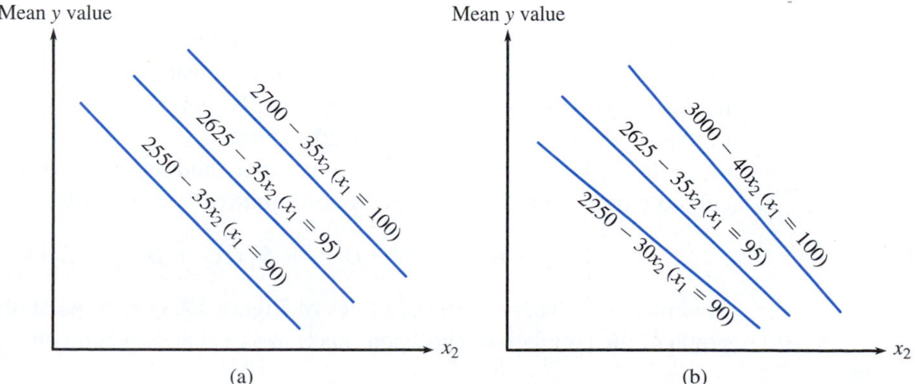

Figure 12.33 Graphs of the mean y value for two different models: (a) $1200 + 15x_1 - 35x_2$; (b) $-4500 + 75x_1 + 60x_2 - x_1x_2$

appropriateness of the proposed model. Rather than the lines being parallel, the line for a temperature of 100 should be steeper than the line for a temperature of 95, and that line in turn should be steeper than the line for $x_1 = 90$. A model that has this property includes, in addition to predictors x_1 and x_2, a third predictor variable, $x_3 = x_1x_2$. One such model is

$$y = -4500 + 75x_1 + 60x_2 - x_1x_2 + \varepsilon$$

for which the population regression function is $-4500 + 75x_1 + 60x_2 - x_1x_2$. This gives

$$\begin{aligned}
(\text{mean } y \text{ value when temperature is 100}) &= -4500 + (75)(100) + 60x_2 - 100x_2 \\
&= 3000 - 40x_2 \\
(\text{mean value when temperature is 95}) &= 2625 - 35x_2 \\
(\text{mean value when temperature is 90}) &= 2250 - 30x_2
\end{aligned}$$

These are graphed in Figure 12.33(b), where it is clear that the three slopes are different. Now each different value of x_1 yields a line with a different slope, so the average change in yield associated with a 1-unit increase in x_2 depends on the value of x_1. When this is the case, the two variables are said to *interact*.

DEFINITION

If the change in the mean y value associated with a 1-unit increase in one independent variable depends on the value of a second independent variable, there is **interaction** between these two variables. Denoting the two independent variables by x_1 and x_2, we can model this interaction by including as an additional predictor $x_3 = x_1x_2$, the product of the two independent variables.

The general equation for a multiple regression model based on two independent variables x_1 and x_2 and also including an interaction predictor is

$$y = \beta_0 + \beta_1x_1 + \beta_2x_2 + \beta_3x_3 + \varepsilon \qquad \text{where } x_3 = x_1x_2$$

When x_1 and x_2 do interact, this model will usually give a much better fit to resulting data than would the no-interaction model. Failure to consider a model with interaction too often leads an investigator to conclude incorrectly that the relationship between y and a set of independent variables is not very substantial.

In applied work, quadratic predictors x_1^2 and x_2^2 are often included to model a curved relationship. This leads to the **full quadratic** or **complete second-order model**

$$y = \beta_0 + \beta_1x_1 + \beta_2x_2 + \beta_3x_1x_2 + \beta_4x_1^2 + \beta_5x_2^2 + \varepsilon$$

This model replaces the straight lines of Figure 12.33 with parabolas (each one is the graph of the population regression function as x_2 varies when x_1 has a particular value).

Example 12.30 Investigators carried out a study to see how various characteristics of concrete are influenced by x_1 = % limestone powder and x_2 = water–cement ratio, resulting in the accompanying data ("Durability of Concrete with Addition of Limestone Powder," *Mag. Concrete Res.*, 1996: 131–137).

x_1	x_2	x_1x_2	28-Day Comp Str. (MPa)	Adsorbability (%)
21	.65	13.65	33.55	8.42
21	.55	11.55	47.55	6.26
7	.65	4.55	35.00	6.74
7	.55	3.85	35.90	6.59
28	.60	16.80	40.90	7.28
0	.60	0.00	39.10	6.90
14	.70	9.80	31.55	10.80
14	.50	7.00	48.00	5.63
14	.60	8.40	42.30	7.43
			$\bar{y} = 39.317$, SST = 278.52	$\bar{y} = 7.339$, SST = 18.356

Consider first compressive strength as the dependent variable y. Fitting the first-order model results in

$$y = 84.82 + .1643x_1 - 79.67x_2 \qquad \text{SSE} = 72.25 \text{ (df} = 6)$$
$$R^2 = .741 \qquad R_a^2 = .654$$

whereas including an interaction predictor gives

$$y = 6.22 + 5.779x_1 + 51.33x_2 - 9.357x_1x_2$$
$$\text{SSE} = 29.35 \text{ (df} = 5) \qquad R^2 = .895 \qquad R_a^2 = .831$$

Based on this latter fit, a prediction for compressive strength when % limestone = 14 and water–cement ratio = .60 is

$$\hat{y} = 6.22 + 5.779(14) + 51.33(.60) - 9.357(8.4) = 39.32$$

Fitting the full quadratic relationship results in virtually no change in the R^2 value. However, when the dependent variable is adsorbability, the following results are obtained: $R^2 = .747$ when just two predictors are used, .802 when the interaction predictor is added, and .889 when the five predictors for the full quadratic relationship are used. ■

Models with Predictors for Categorical Variables Thus far we have explicitly considered the inclusion of only quantitative (numerical) predictor variables in a multiple regression model. Using simple numerical coding, qualitative (categorical) variables, such as type of college (private or state) or type of wood (pine, oak, or walnut), can also be incorporated into a model. Let's first focus on the case of a dichotomous variable, one with just two possible categories—male or female, U.S. or foreign manufacture, and so on. With any such variable, we associate a **dummy** or **indicator variable** $\boldsymbol{x}$ whose possible values 0 and 1 indicate which category is relevant for any particular observation.

Example 12.31 Recall the graduation rate data introduced in Example 12.12 and plotted in Example 12.24. There it appeared that private universities might do better for a given SAT score. To test this we will use a model with y = graduation rate, x_2 = average freshman SAT score, and x_1 = a variable defined to indicate private or public status. Define

$$x_1 = \begin{cases} 1 & \text{if the university is private} \\ 0 & \text{if the university is public} \end{cases}$$

and consider the multiple regression model

$$Y = \beta_0 + \beta_1 x_1 + \beta_2 x_2 + \varepsilon$$

The mean graduation rate depends on whether the university is public or private:

$$\text{mean graduation rate} = \beta_0 + \beta_2 x_2 \qquad \text{when } x_1 = 0 \quad \text{(public)}$$

$$\text{mean graduation rate} = \beta_0 + \beta_1 + \beta_2 x_2 \qquad \text{when } x_1 = 1 \quad \text{(private)}$$

Thus there are two parallel lines with vertical separation β_1 as shown in Figure 12.34(a). The coefficient β_1 is the difference in mean graduation rates between private and public universities with SAT held fixed. If $\beta_1 > 0$, then on average, for a given SAT, private universities will have a higher graduation rate.

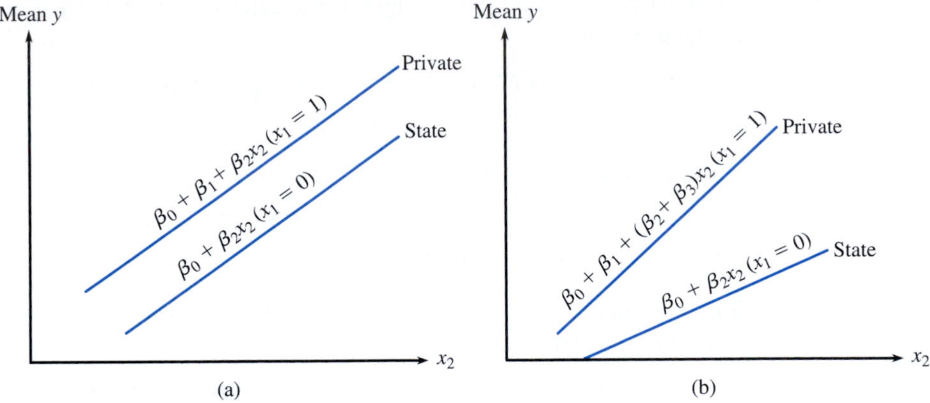

Figure 12.34 Regression functions for models with one dummy variable (x_1) and one quantitative variable (x_2): (a) no interaction; (b) interaction

A second possibility is a model with a product (interaction) term:

$$Y = \beta_0 + \beta_1 x_1 + \beta_2 x_2 + \beta_3 x_1 x_2 + \varepsilon$$

Now the mean graduation rates for the two types of university are

$$\text{mean graduation rate} = \beta_0 + \beta_2 x_2 \qquad \text{when } x_1 = 0 \quad \text{(public)}$$

$$\text{mean graduation rate} = \beta_0 + \beta_1 + (\beta_2 + \beta_3)x_2 \qquad \text{when } x_1 = 1 \quad \text{(private)}$$

Thus we have two lines where β_1 is the difference in intercepts and β_3 is the difference in slopes, as shown in Figure 12.34(b). Unless $\beta_3 = 0$, the lines will not be parallel and there will be interaction, which means that the separation between public and private universities depends on SAT.

The usual procedure is to test the interaction hypothesis H_0: $\beta_3 = 0$ versus H_a: $\beta_3 \neq 0$ first. If we do not reject H_0 (no interaction) then we can use the parallel model to see if there is a separation (β_1) between lines. Of course, it does not make sense to estimate the difference between lines if the difference depends on x_2, which is the case when there is interaction.

Figure 12.35 shows SAS output for these two tests. The coefficient for interaction has a *P*-value of 0.4047, so there is no reason to reject the null hypothesis H_0: $\beta_3 = 0$. Since we do not reject the hypothesis of no interaction, let's look at the results for the difference β_1 in the model with two parallel lines. The variable Priv1_St0 is x_1, the dummy variable with value 1 for private and 0 for state universities. The *P*-value for its coefficient is .0034, so we can reject the hypothesis that it is 0 at the .01 level. The value of the coefficient is 16.93, which means that a private university is estimated to have a graduation rate about 17 percentage points higher than a state university with the same freshman SAT. This is huge, especially in comparison with the coefficient for SAT, which is .05929. Dividing .05929 into $\hat{\beta}_1 = 16.93$ gives 286, which means that it takes 286 SAT points to make up the difference between private and public universities. To put it another way, a private university with average freshman SAT of 1000 is estimated to have the same graduation rate as a state university with SAT of 1286. Notice that the 95% confidence interval for β_1 is quite wide, (6.44, 27.42), but even 6.44 percentage points would be a substantial difference in graduation rates.

Test Interaction

```
                       Analysis of Variance
                              Sum of
Source              DF       Squares     Mean Square      F Value     Pr > F
Model                3    6343.01499      2114.33833        31.21     <.0001
Error               16    1083.93501        67.74594
Corrected Total     19    7426.95000

Root MSE             8.23079     R-Square     0.8541
Dependent Mean      59.45000     Adj R-Sq     0.8267
Coeff Var           13.84490

                        Parameter Estimates
                        Parameter     Standard
Variable        DF       Estimate        Error      t Value     Pr > |t|
Intercept        1       -0.52145     18.16644        -0.03       0.9775
SAT              1        0.04822      0.01840         2.62       0.0186
Priv1_St0        1       -7.86223     29.39747        -0.27       0.7925
Inter            1        0.02240      0.02617         0.86       0.4047
```

Test Private versus State

```
                       Analysis of Variance
                             Sum of
Source              DF      Squares     Mean Square      F Value     Pr > F
Model                2   6293.39873      3146.69936        47.19     <.0001
Error               17   1133.55127        66.67949
Corrected Total     19   7426.95000

Root MSE             8.16575     R-Square     0.8474
Dependent Mean      59.45000     Adj R-Sq     0.8294
Coeff Var           13.73549

                        Parameter Estimates
                  Parameter  Standard                             95%
Variable      DF   Estimate     Error  t Value  Pr > |t|   Confidence   Limits
Intercept      1  -11.35960  12.92137    -0.88    0.3916   -38.62131  15.90210
SAT            1    0.05929   0.01298     4.57    0.0003     0.03190   0.08668
Priv1_St0      1   16.92772   4.97206     3.40    0.0034     6.43759  27.41785
```

Figure 12.35 SAS Output for interaction model and parallel model ■

You might think that the way to handle a three-category situation is to define a single numerical variable with coded values such as 0, 1, and 2 corresponding to the three categories. This is incorrect, because it imposes an ordering on the categories that is not necessarily implied by the problem context. The correct way to incorporate three categories is to define *two* different dummy variables. Suppose, for example, that y is a score on a posttest taken after instruction, x_1 is the score on an ability pretest taken before instruction, and that there are three methods of instruction in a mathematics unit: (1) with symbols, (2) without symbols, (3) a mixture with and without symbols. Then let

$$x_2 = \begin{cases} 1 & \text{instruction method 1} \\ 0 & \text{otherwise} \end{cases} \qquad x_3 = \begin{cases} 1 & \text{instruction method 2} \\ 0 & \text{otherwise} \end{cases}$$

For an individual taught with method 1, $x_2 = 1$ and $x_3 = 0$, whereas for an individual taught with method 2, $x_2 = 0$ and $x_3 = 1$. For an individual taught with method 3, $x_2 = x_3 = 0$, and it is not possible that $x_2 = x_3 = 1$ because an individual cannot be taught simultaneously by both methods 1 and 2. The no-interaction model would have only the predictors x_1, x_2, and x_3. The following interaction model allows the mean change in posttest score associated with a 1-unit increase in pretest score to depend on the method of instruction:

$$Y = \beta_0 + \beta_1 x_1 + \beta_2 x_2 + \beta_3 x_3 + \beta_4 x_1 x_2 + \beta_5 x_1 x_3 + \varepsilon$$

Construction of a picture like Figure 12.34 with a graph for each of the three possible (x_2, x_3) pairs gives three nonparallel lines (unless $\beta_4 = \beta_5 = 0$). How would we interpret statistically significant interaction? Suppose that it occurs to the extent that the lines for methods 1 and 2 cross. In particular, if the line for method 1 is higher on the right and lower on the left, it means that symbols work well for high-ability students but not as well for low-ability students.

More generally, incorporating a categorical variable with c possible categories into a multiple regression model requires the use of $c - 1$ indicator variables (e.g., five methods of instruction would necessitate using four indicator variables). Thus even one categorical variable can add many predictors to a model.

Indicator variables can be used for categorical variables without any other variables in the model. For example, consider Example 11.2, which compared three different compounds in their ability to prevent fabric soiling. Using a regression with two dummy variables gives the following regression ANOVA table, just like the one in Example 11.2:

Source	DF	SS	MS	F	P
Regression	2	0.06085	0.03043	0.99	0.401
Residual Error	12	0.37008	0.03084		
Total	14	0.43093			

Analysis that involves both quantitative and categorical predictors, as in Example 12.31, is called **analysis of covariance**, and the quantitative variable is called a **covariate**. Sometimes more than one covariate is used.

Other Models The logistic regression model introduced in Section 12.1 can be extended to incorporate more than one predictor. Various nonlinear models are also used frequently in applied work. An example is the multiple exponential model

$$Y = e^{\beta_0 + \beta_1 x_1 + \cdots + \beta_k x_k} \cdot \varepsilon$$

Taking logs on both sides shows that $\ln(Y) = \beta_0 + \beta_1 x_1 + \cdots + \beta_k x_k + \varepsilon'$, where $\varepsilon' = \ln(\varepsilon)$. This is the usual multiple regression model with $\ln(Y)$ as the response variable.

Exercises Section 12.7 (78–88)

78. Cardiorespiratory fitness is widely recognized as a major component of overall physical well-being. Direct measurement of maximal oxygen uptake (VO_2max) is the single best measure of such fitness, but direct measurement is time-consuming and expensive. It is therefore desirable to have a prediction equation for VO_2max in terms of easily obtained quantities. Consider the variables

$y = VO_2\text{max (L/min)}$ $\quad x_1 = \text{weight (kg)}$

$x_2 = \text{age (yr)}$

$x_3 = \text{time necessary to walk 1 mile (min)}$

$x_4 = \text{heart rate at the end of the walk (beats/min)}$

Here is one possible model, for male students, consistent with the information given in the article "Validation of the Rockport Fitness Walking Test in College Males and Females" (*Res. Q. Exercise Sport*, 1994: 152–158):

$$Y = 5.0 + .01x_1 - .05x_2 - .13x_3 - .01x_4 + \varepsilon$$
$$\sigma = .4$$

a. Interpret β_1 and β_3.

b. What is the expected value of VO_2max when weight is 76 kg, age is 20 yr, walk time is 12 min, and heart rate is 140 beats/min?

c. What is the probability that VO_2max will be between 1.00 and 2.60 for a single observation made when the values of the predictors are as stated in part (b)?

79. Let y = sales at a fast-food outlet (\$1000's), x_1 = number of competing outlets within a 1-mile radius, x_2 = population within a 1-mile radius (1000's of people), and x_3 be an indicator variable that equals 1 if the outlet has a drive-up window and 0 otherwise. Suppose that the true regression model is

$$Y = 10.00 - 1.2x_1 + 6.8x_2 + 15.3x_3 + \varepsilon$$

a. What is the mean value of sales when the number of competing outlets is 2, there are 8000 people within a 1-mile radius, and the outlet has a drive-up window?

b. What is the mean value of sales for an outlet without a drive-up window that has three competing outlets and 5000 people within a 1-mile radius?

c. Interpret β_3.

80. The article "Analysis of the Modeling Methodologies for Predicting the Strength of Air-Jet Spun Yarns" (*Textile Res. J.*, 1997: 39–44) reported on a study carried out to relate yarn tenacity (y, in g/tex) to yarn count (x_1, in tex), percentage polyester (x_2), first nozzle pressure (x_3, in kg/cm^2), and second nozzle pressure (x_4, in kg/cm^2). The estimate of the constant term in the corresponding multiple regression equation was 6.121. The estimated coefficients for the four predictors were −.082, .113, .256, and −.219, respectively, and the coefficient of multiple determination was .946. Assume that $n = 25$.

a. State and test the appropriate hypotheses to decide whether the fitted model specifies a useful linear relationship between the dependent variable and at least one of the four model predictors.

b. Calculate the value of adjusted R^2 and comment.

c. Calculate a 99% confidence interval for true mean yarn tenacity when yarn count is 16.5, yarn contains 50% polyester, first nozzle pressure is 3, and second nozzle pressure is 5 if the estimated standard deviation of predicted tenacity under these circumstances is .350.

81. The ability of ecologists to identify regions of greatest species richness could have an impact on the preservation of genetic diversity, a major objective of the World Conservation Strategy. The article "Prediction of Rarities from Habitat Variables: Coastal Plain Plants on Nova Scotian Lakeshores" (*Ecology*, 1992: 1852–1859) used a sample of n = 37 lakes to obtain the estimated regression equation

$$y = 3.89 + .033x_1 + .024x_2 + .023x_3 - .0080x_4 - .13x_5 - .72x_6$$

where y = species richness, x_1 = watershed area, x_2 = shore width, x_3 = poor drainage (%), x_4 = water color (total color units), x_5 = sand (%), and x_6 = alkalinity. The coefficient of multiple determination was reported as $R^2 = .83$. Carry out a test of model utility.

82. An investigation of a die casting process resulted in the accompanying data on x_1 = furnace temperature, x_2 = die close time, and y = temperature difference on the die surface ("A Multiple-Objective Decision-Making Approach for Assessing Simultaneous Improvement in Die Life and Casting Quality in a Die Casting Process," *Qual. Engrg.*, 1994: 371–383).

x_1	1250	1300	1350	1250	1300
x_2	6	7	6	7	6
y	80	95	101	85	92

x_1	1250	1300	1350	1350
x_2	8	8	7	8
y	87	96	106	108

MINITAB output from fitting the multiple regression model with predictors x_1 and x_2 is given here.

```
The regression equation is
tempdiff = -200 + 0.210 furntemp
               + 3.00 clostime

Predictor        Coef       Stdev   t-ratio       p
Constant      -199.56       11.64    -17.14   0.000
furntemp     0.210000    0.008642     24.30   0.000
clostime       3.0000      0.4321      6.94   0.000

s = 1.058   R-sq = 99.1%     R-sq(adj) = 98.8%

Analysis of Variance

SOURCE       DF       SS       MS        F       p
Regression    2   715.50   357.75   319.31   0.000
Error         6     6.72     1.12
Total         8   722.22
```

a. Carry out the model utility test.

b. Calculate and interpret a 95% confidence interval for β_2, the population regression coefficient of x_2.

c. When $x_1 = 1300$ and $x_2 = 7$, the estimated standard deviation of $\hat{y}$ is $s_{\hat{y}} = .353$. Calculate a 95% confidence interval for true average temperature difference when furnace temperature is 1300 and die close time is 7.

d. Calculate a 95% prediction interval for the temperature difference resulting from a single experimental run with a furnace temperature of 1300 and a die close time of 7.

e. Use appropriate diagnostic plots to see if there is any reason to question the regression model assumptions.

83. An experiment carried out to study the effect of the mole contents of cobalt (x_1) and the calcination temperature (x_2) on the surface area of an iron–cobalt hydroxide catalyst (y) resulted in the accompanying data ("Structural Changes and Surface Properties of $Co_xFe_{3-x}O_4$ Spinels," *J. Chem. Tech. Biotech.*, 1994: 161–170).

x_1	.6	.6	.6	.6	.6	1.0	1.0
x_2	200	250	400	500	600	200	250
y	90.6	82.7	58.7	43.2	25.0	127.1	112.3

x_1	1.0	1.0	1.0	2.6	2.6	2.6	2.6
x_2	400	500	600	200	250	400	500
y	19.6	17.8	9.1	53.1	52.0	43.4	42.4

x_1	2.6	2.8	2.8	2.8	2.8	2.8
x_2	600	200	250	400	500	600
y	31.6	40.9	37.9	27.5	27.3	19.0

A request to the SAS package to fit $\beta_0 + \beta_1 x_1 + \beta_2 x_2 + \beta_3 x_3$, where $x_3 = x_1 x_2$ (an interaction predictor) yielded the accompanying output.

a. Predict the value of surface area when cobalt content is 2.6 and temperature is 250, and calculate the value of the corresponding residual.

b. Since $\hat{\beta}_1 = -46.0$, is it legitimate to conclude that if cobalt content increases by 1 unit while the values of the other predictors remain fixed, surface area can be expected to decrease by roughly 46 units? Explain your reasoning.

c. Does there appear to be a useful relationship between y and the predictors?

d. Given that mole contents and calcination temperature remain in the model, does the interaction predictor x_3 provide useful information about y? State and test the appropriate hypotheses using a significance level of .01.

e. The estimated standard deviation of $\hat{Y}$ when mole contents is 2.0 and calcination temperature is 500 is $s_{\hat{Y}} = 4.69$. Calculate a 95% confidence interval for the mean value of surface area under these circumstances.

f. Based on appropriate diagnostic plots, is there any reason to question the regression model assumptions?

SAS output for Exercise 83

```
Dependent Variable: SURFAREA
                              Analysis of Variance

  Source        DF      Sum of Squares        Mean Square         F Value       Prob>F
   Model         3         15223.52829         5074.50943          18.924       0.0001
   Error        16          4290.53971          268.15873
C Total         19         19514.06800

Root MSE               16.37555            R-square           0.7801
Dep Mean               48.06000            Adj R-sq           0.7389
C.V.                   34.07314

                              Parameter Estimates

                        Parameter          Standard         T for H0:          Prob
Variable        DF       Estimate             Error       Parameter = 0        > |T|
INTERCEP         1     185.485740       21.19747682           8.750          0.0001
COBCON           1     -45.969466       10.61201173          -4.332          0.0005
TEMP             1      -0.301503        0.05074421          -5.942          0.0001
CONTEMP          1       0.088801        0.02540388           3.496          0.0030
```

84. A regression analysis carried out to relate y = repair time for a water filtration system (hr) to x_1 = elapsed time since the previous service (months) and x_2 = type of repair (1 if electrical and 0 if mechanical) yielded the following model based on $n = 12$ observations: $y = .950 + .400x_1 + 1.250x_2$. In addition, SST = 12.72, SSE = 2.09, and $s_{\hat{\beta}_2} = .312$.

- **a.** Does there appear to be a useful linear relationship between repair time and the two model predictors? Carry out a test of the appropriate hypotheses using a significance level of .05.
- **b.** Given that elapsed time since the last service remains in the model, does type of repair provide useful information about repair time? State and test the appropriate hypotheses using a significance level of .01.
- **c.** Calculate and interpret a 95% CI for β_2.
- **d.** The estimated standard deviation of a prediction for repair time when elapsed time is 6 months and the repair is electrical is .192. Predict repair time under these circumstances by calculating a 99% prediction interval. Does the interval suggest that the estimated model will give an accurate prediction? Why or why not?

85. The article "The Undrained Strength of Some Thawed Permafrost Soils" (*Canad. Geotech. J.*, 1979: 420–427) contains the following data on undrained shear strength of sandy soil (y, in kPa), depth (x_1, in m), and water content (x_2, in %).

Obs	y	x_1	x_2	$\hat{y}$	$y - \hat{y}$	e^*
1	14.7	8.9	31.5	23.35	−8.65	−1.50
2	48.0	36.6	27.0	46.38	1.62	.54
3	25.6	36.8	25.9	27.13	−1.53	−.53
4	10.0	6.1	39.1	10.99	−.99	−.17
5	16.0	6.9	39.2	14.10	1.90	.33
6	16.8	6.9	38.3	16.54	.26	.04
7	20.7	7.3	33.9	23.34	−2.64	−.42
8	38.8	8.4	33.8	25.43	13.37	2.17
9	16.9	6.5	27.9	15.63	1.27	.23
10	27.0	8.0	33.1	24.29	2.71	.44
11	16.0	4.5	26.3	15.36	.64	.20
12	24.9	9.9	37.8	29.61	−4.71	−.91
13	7.3	2.9	34.6	15.38	−8.08	−1.53
14	12.8	2.0	36.4	7.96	4.84	1.02

The predicted values and residuals were computed by fitting a full quadratic model, which resulted in the estimated regression function

$$y = -151.36 - 16.22x_1 + 13.48x_2 + .094x_1^2 - .253x_2^2 + .492x_1x_2$$

- **a.** Do plots of e^* versus x_1, e^* versus x_2, and e^* versus $\hat{y}$ suggest that the full quadratic model should be modified? Explain your answer.
- **b.** The value of R^2 for the full quadratic model is .759. Test at level .05 the null hypothesis stating that there is no linear relationship between the dependent variable and any of the five predictors.

c. Each of the null hypotheses H_0: $\beta_i = 0$ versus H_a: $\beta_i \neq 0$, $i = 1, 2, 3, 4, 5$, is not rejected at the 5% level. Does this make sense in view of the result in (b)? Explain.

d. It is shown in Section 12.8 that $V(Y) = \sigma^2 = V(\hat{Y}) + V(Y - \hat{Y})$. The estimate of σ is $\hat{\sigma} = s = 6.99$ (from the full quadratic model). First obtain the estimated standard deviation of $Y - \hat{Y}$, and then estimate the standard deviation of $\hat{Y}$ (i.e., $\hat{\beta}_0 + \hat{\beta}_1 x_1 + \hat{\beta}_2 x_2 + \hat{\beta}_3 x_1^2 + \hat{\beta}_4 x_2^2 + \hat{\beta}_5 x_1 x_2$) when $x_1 = 8.0$ and $x_2 = 33.1$. Finally, compute a 95% CI for mean strength. [*Hint:* What is $(y - \hat{y})/e^*$?]

e. Sometimes an investigator wishes to decide whether a group of m predictors ($m > 1$) can simultaneously be eliminated from the model. The null hypothesis says that all β's associated with these m predictors are 0, which is interpreted to mean that as long as the other $k - m$ predictors are retained in the model, the m predictors under consideration collectively provide no useful information about y. The test is carried out by first fitting the "full" model with all k predictors to obtain SSE(full) and then fitting the "reduced" model consisting just of the $k - m$ predictors *not* being considered for deletion to obtain SSE(red). The test statistic is

$$F = \frac{[\text{SSE(red)} - \text{SSE(full)}]/m}{\text{SSE(full)}/[n - (k + 1)]}$$

The test is upper-tailed and based on m numerator df and $n - (k + 1)$ denominator df. Fitting the first-order model with just the predictors x_1 and x_2 results in SSE $= 894.95$. State and test at significance level .05 the null hypothesis that none of the three second-order predictors (one interaction and two quadratic predictors) provides useful information about y provided that the two first-order predictors are retained in the model.

86. The following data on y = glucose concentration (g/L) and x = fermentation time (days) for a particular blend of malt liquor was read from a scatter plot in the article "Improving Fermentation Productivity with Reverse Osmosis" (*Food Tech.*, 1984: 92–96):

x	1	2	3	4	5	6	7	8
y	74	54	52	51	52	53	58	71

a. Verify that a scatter plot of the data is consistent with the choice of a quadratic regression model.

b. The estimated quadratic regression equation is $y = 84.482 - 15.875x + 1.7679x^2$. Predict the value of glucose concentration for a fermentation time of 6 days, and compute the corresponding residual.

c. Using SSE $= 61.77$, what proportion of observed variation can be attributed to the quadratic regression relationship?

d. The $n = 8$ standardized residuals based on the quadratic model are 1.91, −1.95, −.25, .58, .90, .04, −.66, and .20. Construct a plot of the standardized residuals versus x and a normal probability plot. Do the plots exhibit any troublesome features?

e. The estimated standard deviation of $\hat{\mu}_{Y\cdot 6}$—that is, $\hat{\beta}_0 + \hat{\beta}_1(6) + \hat{\beta}_2(36)$—is 1.69. Compute a 95% CI for $\mu_{Y\cdot 6}$.

f. Compute a 95% PI for a glucose concentration observation made after 6 days of fermentation time.

87. Utilization of sucrose as a carbon source for the production of chemicals is uneconomical. Beet molasses is a readily available and low-priced substitute.

Obs	Linoleic	Kerosene	Antiox	Betacaro
1	30.00	30.00	10.00	0.7000
2	30.00	30.00	10.00	0.6300
3	30.00	30.00	18.41	0.0130
4	40.00	40.00	5.00	0.0490
5	30.00	30.00	10.00	0.7000
6	13.18	30.00	10.00	0.1000
7	20.00	40.00	5.00	0.0400
8	20.00	40.00	15.00	0.0065
9	40.00	20.00	5.00	0.2020
10	30.00	30.00	10.00	0.6300
11	30.00	30.00	1.59	0.0400
12	40.00	20.00	15.00	0.1320
13	40.00	40.00	15.00	0.1500
14	30.00	30.00	10.00	0.7000
15	30.00	46.82	10.00	0.3460
16	30.00	30.00	10.00	0.6300
17	30.00	13.18	10.00	0.3970
18	20.00	20.00	5.00	0.2690
19	20.00	20.00	15.00	0.0054
20	46.82	30.00	10.00	0.0640

The article "Optimization of the Production of β-Carotene from Molasses by *Blakeslea trispora*" (*J. Chem. Tech. Biotech.*, 2002: 933–943) carried out a multiple regression analysis to relate the dependent variable y = amount of β-carotene (g/dm^3) to the three predictors: amount of linoleic acid, amount of kerosene, and amount of antioxidant (all g/dm^3).

a. Fitting the complete second-order model in the three predictors resulted in $R^2 = .987$ and adjusted $R^2 = .974$, whereas fitting the first-order model gave $R^2 = .016$. What would you conclude about the two models?

b. For $x_1 = x_2 = 30$, $x_3 = 10$, a statistical software package reported that $\hat{y} = .66573$, $s_{\hat{Y}} = 01785$ based on the complete second-order model. Predict the amount of β-carotene that would result from a single experimental run with the designated values of the independent variables, and do so in a way that conveys information about precision and reliability.

88. Snowpacks contain a wide spectrum of pollutants that may represent environmental hazards. The article "Atmospheric PAH Deposition: Deposition Velocities and Washout Ratios"(*J. Environ. Engrg.*, 2002: 186–195) focused on the deposition of polyaromatic hydrocarbons. The authors proposed a multiple regression model for relating deposition over a specified time period (y, in μg/m^2) to two rather complicated predictors x_1 (μg-sec/m^3) and x_2 (μg/m^2) defined in terms of PAH air concentrations for various species, total time, and total amount of precipitation. Here is data on the species fluoranthene and corresponding MINITAB output:

obs	x1	x2	flth
1	92017	.0026900	278.78
2	51830	.0030000	124.53
3	17236	.0000196	22.65
4	15776	.0000360	28.68
5	33462	.0004960	32.66
6	243500	.0038900	604.70
7	67793	.0011200	27.69
8	23471	.0006400	14.18
9	13948	.0004850	20.64
10	8824	.0003660	20.60
11	7699	.0002290	16.61
12	15791	.0014100	15.08
13	10239	.0004100	18.05
14	43835	.0000960	99.71
15	49793	.0000896	58.97
16	40656	.0026000	172.58
17	50774	.0009530	44.25

```
The regression equation is

flth = -33.5 + 0.00205 x1 + 29836 x2

Predictor        Coef      SE Coef       T       P
Constant       -33.46        14.90   -2.25   0.041
x1          0.0020548    0.0002945    6.98   0.000
x2              29836        13654    2.19   0.046

S = 44.28   R-Sq = 92.3%   R-Sq(adj) = 91.2%

Analysis of Variance

Source          DF       SS       MS       F       P
Regression       2   330989   165495   84.39   0.000
Residual Error  14    27454     1961
Total           16   358443
```

Formulate questions and perform appropriate analyses.

Construct the appropriate residual plots, including plots against the predictors. Based on these plots, justify adding a quadratic term, and fit the model with this additional term. Is this term statistically significant, and does it help the appearance of the diagnostic plots?

12.8 *Regression with Matrices

In Section 12.7 we used the following additive model equation to relate a dependent variable y to independent variables $x_1, \ldots, x_k$:

$$y = \beta_0 + \beta_1 x_1 + \beta_2 x_2 + \cdots + \beta_k x_k + \varepsilon$$

where ε is normally distributed with mean 0 and variance σ^2, and the various ε's are independent of one another. Simple linear regression is the special case in which $k = 1$.

Suppose that we have n observations, each consisting of a y value and values of the k predictors (so each observation consists of $k + 1$ numbers). Then we can write

$$\begin{bmatrix} y_1 \\ \vdots \\ y_n \end{bmatrix} = \begin{bmatrix} \beta_0 + \beta_1 x_{11} + \beta_2 x_{12} + \cdots + \beta_k x_{1k} + \varepsilon_1 \\ \vdots \\ \beta_0 + \beta_1 x_{n1} + \beta_2 x_{n2} + \cdots + \beta_k x_{nk} + \varepsilon_n \end{bmatrix}$$

For example, if there are $n = 6$ cars, where y is horsepower, x_1 is engine size (liters), and x_2 indicates fuel type (regular or premium), then we are trying to predict horsepower as a linear function of the $k = 2$ predictors engine size and fuel type. The equations can be written much more compactly using vectors and matrices. To do this, form a column vector of observations on y, a column vector of regression coefficients, and a vector of random deviations:

$$\mathbf{y} = \begin{bmatrix} y_1 \\ \vdots \\ y_n \end{bmatrix} \qquad \boldsymbol{\beta} = \begin{bmatrix} \beta_0 \\ \beta_1 \\ \vdots \\ \beta_k \end{bmatrix} \qquad \boldsymbol{\varepsilon} = \begin{bmatrix} \varepsilon_1 \\ \vdots \\ \varepsilon_n \end{bmatrix}$$

Also form an $n \times (k + 1)$ matrix in which the first column consists of 1's (corresponding to the constant term in the model), the second column consists of the values of the first predictor x_1 (i.e., of $x_{11}, x_{21}, \ldots, x_{n1}$, the third column has the values of x_2, and so on.

$$\mathbf{X} = \begin{bmatrix} 1 & x_{11} & \cdots & x_{1k} \\ \vdots & \vdots & & \vdots \\ 1 & x_{n1} & \cdots & x_{nk} \end{bmatrix}$$

The $\mathbf{X}$ matrix has a row for each observation, consisting of 1 and then the values of the k predictors. The equations relating the observed y's to the x_i's can then be expressed very concisely as

$$\begin{bmatrix} y_1 \\ \vdots \\ y_n \end{bmatrix} = \mathbf{y} = \mathbf{X}\boldsymbol{\beta} + \boldsymbol{\varepsilon} = \begin{bmatrix} 1 & x_{11} & \cdots & x_{1k} \\ \vdots & \vdots & & \vdots \\ 1 & x_{n1} & \cdots & x_{nk} \end{bmatrix} \begin{bmatrix} \beta_0 \\ \beta_1 \\ \vdots \\ \beta_k \end{bmatrix} + \begin{bmatrix} \varepsilon_1 \\ \vdots \\ \varepsilon_n \end{bmatrix}$$

We now estimate $\beta_0, \beta_1, \beta_2, \ldots, \beta_k$ using the principle of least squares: Find $b_0, b_1, b_2, \ldots, b_k$ to minimize

$$\sum_{i=1}^{n} [y_i - (b_0 + b_1 x_{i1} + b_2 x_{i2} + \cdots + b_k x_{ik})]^2 = (\mathbf{y} - \mathbf{X}\mathbf{b})'(\mathbf{y} - \mathbf{X}\mathbf{b}) = \|\mathbf{y} - \mathbf{X}\mathbf{b}\|^2$$

where $\mathbf{b}$ is the column vector with entries $b_0, b_1, \ldots, b_k$ and $\|\mathbf{u}\|$ is the length of $\mathbf{u}$.

The following normal equations result from equating to zero the partial derivative with respect to each coefficient.

$$\begin{aligned} b_0 \sum_{i=1}^{n} 1 + b_1 \sum_{i=1}^{n} x_{i1} + \cdots + b_k \sum_{i=1}^{n} x_{ik} &= \sum_{i=1}^{n} y_i \\ b_0 \sum_{i=1}^{n} x_{i1} + b_1 \sum_{i=1}^{n} x_{i1}x_{i1} + \cdots + b_k \sum_{i=1}^{n} x_{i1}x_{ik} &= \sum_{i=1}^{n} x_{i1}y_i \\ &\vdots \\ b_0 \sum_{i=1}^{n} x_{ik} + b_1 \sum_{i=1}^{n} x_{ik}x_{i1} + \cdots + b_k \sum_{i=1}^{n} x_{ik}x_{ik} &= \sum_{i=1}^{n} x_{ik}y_i \end{aligned}$$

In matrix form this is

$$\begin{bmatrix} \sum_{i=1}^{n} 1 & \sum_{i=1}^{n} x_{i1} & \cdots & \sum_{i=1}^{n} x_{ik} \\ \sum_{i=1}^{n} x_{i1} & \sum_{i=1}^{n} x_{i1}x_{i1} & \cdots & \sum_{i=1}^{n} x_{i1}x_{ik} \\ \vdots & \vdots & & \vdots \\ \sum_{i=1}^{n} x_{ik} & \sum_{i=1}^{n} x_{ik}x_{i1} & \cdots & \sum_{i=1}^{n} x_{ik}x_{ik} \end{bmatrix} \begin{bmatrix} b_0 \\ b_1 \\ \vdots \\ b_k \end{bmatrix} = \begin{bmatrix} \sum_{i=1}^{n} y_i \\ \sum_{i=1}^{n} x_{i1}y_i \\ \vdots \\ \sum_{i=1}^{n} x_{ik}y \end{bmatrix}$$

The matrix on the left is just $\boldsymbol{X}'\boldsymbol{X}$ and the matrix on the right is $\boldsymbol{X}'\boldsymbol{y}$, where $\boldsymbol{X}'$ indicates $\boldsymbol{X}$-transpose, so the normal equations become $\boldsymbol{X}'\boldsymbol{X}\boldsymbol{b} = \boldsymbol{X}'\boldsymbol{y}$. We will assume throughout this section that $\boldsymbol{X}'\boldsymbol{X}$ has an inverse. The vector of estimated coefficients is then $\hat{\boldsymbol{\beta}} = \boldsymbol{b} = [\boldsymbol{X}'\boldsymbol{X}]^{-1}\boldsymbol{X}'\boldsymbol{y}$.

Example 12.32 Using data from six cars, we try to predict horsepower (hp) using engine size (liters) and fuel type. Here is the data set:

Make	hp	Liters	Fuel
Dodge Neon	132	2.0	Regular
Ford Focus SVT	170	2.0	Premium
BMW	184	2.5	Premium
Subaru	165	2.5	Regular
Saturn	182	3.0	Regular
Jaguar	231	3.0	Premium

The hp column will be used for $\boldsymbol{y}$, and engine size values are placed in the second column of $\boldsymbol{X}$, but numbers must be used instead of words in the third column. We use 0 for "regular" and 1 for "premium." Any two numbers could be used instead of 0 and 1, but this choice is convenient in terms of the interpretation of the coefficients. This gives

$$\boldsymbol{X} = \begin{bmatrix} 1 & 2.0 & 0 \\ 1 & 2.0 & 1 \\ 1 & 2.5 & 1 \\ 1 & 2.5 & 0 \\ 1 & 3.0 & 0 \\ 1 & 3.0 & 1 \end{bmatrix} \quad \boldsymbol{y} = \begin{bmatrix} 132 \\ 170 \\ 184 \\ 165 \\ 182 \\ 231 \end{bmatrix} \quad \boldsymbol{X}'\boldsymbol{X} = \begin{bmatrix} 6 & 15 & 3 \\ 15 & 38.5 & 7.5 \\ 3 & 7.5 & 3 \end{bmatrix} \quad \boldsymbol{X}'\boldsymbol{y} = \begin{bmatrix} 1064 \\ 2715.5 \\ 585 \end{bmatrix}$$

Therefore,

$$\hat{\boldsymbol{\beta}} = [\boldsymbol{X}'\boldsymbol{X}]^{-1}\boldsymbol{X}'\boldsymbol{y} = \begin{bmatrix} \frac{79}{12} & -\frac{5}{2} & -\frac{1}{3} \\ -\frac{5}{2} & 1 & 0 \\ -\frac{1}{3} & 0 & \frac{2}{3} \end{bmatrix} \begin{bmatrix} 1064 \\ 2715.5 \\ 585 \end{bmatrix} = \begin{bmatrix} 20.92 \\ 55.50 \\ 35.33 \end{bmatrix}$$

The coefficient 55.5 for engine size means that, if the fuel type is held constant, then we estimate that horsepower will increase on average by 55.5 when the engine size increases by 1 liter. Similarly, the coefficient 35.33 for fuel means that, if the engine size is held constant, then we estimate that horsepower will increase on average by 35.33 when the fuel type increases by 1. However, increasing fuel type by 1 unit means switching from regular fuel to premium fuel, so the difference in horsepower corresponding to the difference in fuels is 35.33. Notice that this is the difference between the average for the three premium-fuel cars and the average for the three regular-fuel cars. ■

The estimated regression coefficients can be used to obtain the predicted values. Recall that $\hat{y}_i = \hat{\beta}_0 + \hat{\beta}_1 x_{i1} + \hat{\beta}_2 x_{i2} + \cdots + \hat{\beta}_k x_{ik}$. The expression for $\hat{y}_i$ is the product of the ith row of $\boldsymbol{X}$ and the $\boldsymbol{\beta}$ vector. The vector of predicted values is then

$$\begin{bmatrix} \hat{y}_1 \\ \vdots \\ \hat{y}_n \end{bmatrix} = \hat{\boldsymbol{y}} = \boldsymbol{X}\hat{\boldsymbol{\beta}} = \boldsymbol{X}[\boldsymbol{X}'\boldsymbol{X}]^{-1}\boldsymbol{X}'\boldsymbol{y}$$

Because **y-hat** is the product of $\boldsymbol{H} = \boldsymbol{X}[\boldsymbol{X}'\boldsymbol{X}]^{-1}\boldsymbol{X}'$ and $\boldsymbol{y}$, the matrix $\boldsymbol{H}$ is called the **hat matrix**. A residual is $\boldsymbol{y}_i - \hat{\boldsymbol{y}}_i$, so the vector of n residuals is

$$\boldsymbol{y} - \hat{\boldsymbol{y}} = \boldsymbol{y} - \boldsymbol{H}\boldsymbol{y} = (\boldsymbol{I} - \boldsymbol{H})\boldsymbol{y}$$

The error sum of squares SSE is the sum of the n squared residuals,

$$\text{SSE} = (\boldsymbol{y} - \hat{\boldsymbol{y}})'(\boldsymbol{y} - \hat{\boldsymbol{y}}) = \|\boldsymbol{y} - \hat{\boldsymbol{y}}\|^2$$

An unbiased estimator of σ^2 is the mean square error $\text{MSE} = s^2 = \text{SSE}/[n - (k + 1)]$. Notice that the estimated variance is the average [with $n - (k + 1)$ in place of n] squared residual. The divisor $n - (k + 1)$ is used because SSE is proportional to a chi-squared rv with $n - (k + 1)$ degrees of freedom under the assumptions given at the beginning of this section, including the assumption that $\boldsymbol{X}'\boldsymbol{X}$ be invertible.

We can rewrite the normal equations in the form

$$0 = \boldsymbol{X}'\boldsymbol{y} - \boldsymbol{X}'\boldsymbol{X}\hat{\boldsymbol{\beta}} = \boldsymbol{X}'(\boldsymbol{y} - \boldsymbol{X}\hat{\boldsymbol{\beta}}) = \boldsymbol{X}'(\boldsymbol{y} - \hat{\boldsymbol{y}}) \tag{12.19}$$

Because the transpose of $\boldsymbol{X}$ times the residual vector is zero, each of the columns of $\boldsymbol{X}$, including the column of 1's, is perpendicular to the residual vector $\boldsymbol{y} - \hat{\boldsymbol{y}}$. In particular, because the dot product of the column of 1's with the residual vector is zero, the sum of the residuals is zero. There are $k + 1$ columns of $\boldsymbol{X}$, and the dot product of each column with the residual vector is zero, so there are $k + 1$ conditions satisfied by the residual vector. This helps to explain intuitively why there are only $n - (k + 1)$ degrees of freedom for SSE.

Letting $\bar{\boldsymbol{y}}$ be the vector with n identical components $\bar{y}$, the total sum of squares SST is the sum of the squared deviations from $\bar{y}$, $\text{SST} = \|\boldsymbol{y} - \bar{\boldsymbol{y}}\|^2$. Similarly, the regression sum of squares SSR is defined to be the sum of the squared deviations of the predicted values from $\bar{y}$, $\text{SSR} = \|\hat{\boldsymbol{y}} - \bar{\boldsymbol{y}}\|^2$. As before, the ANOVA relationship is

$$\text{SST} = \text{SSE} + \text{SSR} \tag{12.20}$$

This can be obtained by subtracting and adding $\hat{\boldsymbol{y}}$:

$$\begin{aligned}\text{SST} &= \|\boldsymbol{y} - \bar{\boldsymbol{y}}\|^2 = [(\boldsymbol{y} - \hat{\boldsymbol{y}}) + (\hat{\boldsymbol{y}} - \bar{\boldsymbol{y}})]'[(\boldsymbol{y} - \hat{\boldsymbol{y}}) + (\hat{\boldsymbol{y}} - \bar{\boldsymbol{y}})]\\ &= \|\boldsymbol{y} - \hat{\boldsymbol{y}}^2\| + \|\hat{\boldsymbol{y}} - \bar{\boldsymbol{y}}\|^2 = \text{SSE} + \text{SSR}\end{aligned}$$

The cross terms in the matrix product are zero because of Equation (12.19) (see Exercise 100).

Recall that the null hypothesis in the model utility test is H_0: $\beta_1 = \cdots = \beta_k = 0$, in which case the model consists of just β_0. That is, under H_0 the observations all have the same mean $\mu = \beta_0$. For a normal random sample with mean μ and standard deviation σ, a proposition in Section 6.4 shows that SST/σ^2 has the chi-squared distribution with $n - 1$ degrees of freedom. Dividing Equation (12.20) by σ^2 gives

$$\frac{\text{SST}}{\sigma^2} = \frac{\text{SSE}}{\sigma^2} + \frac{\text{SSR}}{\sigma^2}$$

It can be shown that SSE and SSR are independent of one another. We know $\text{SST}/\sigma^2 \sim \chi^2_{n-1}$ under the null hypothesis and $\text{SSE}/\sigma^2 \sim \chi^2_{n-k-1}$. Then, by a proposition in Section 6.4, SSR/σ^2 is distributed as chi-squared with df $[n - 1] - [n - (k + 1)] = k$. Recall from Section 6.4 that the F distribution is the ratio of two independent chi-squared variables that have been divided by their degrees of freedom. Applying this to SSR/σ^2 and SSE/σ^2 leads to the F ratio

$$\frac{\dfrac{\text{SSR}}{\sigma^2 k}}{\dfrac{\text{SSE}}{\sigma^2[n - (k + 1)]}} = \frac{\dfrac{\text{SSR}}{k}}{\dfrac{\text{SSE}}{n - (k + 1)}} = \frac{\text{MSR}}{\text{MSE}} \sim F_{k,n-(k+1)} \qquad (12.21)$$

Here $\text{MSR} = \text{SSR}/k$ and MSE was previously defined as $\text{SSE}/[n - (k + 1)]$. The F ratio MSR/MSE is a standard part of regression output for statistical computer packages. It tests the null hypothesis H_0: $\beta_1 = \cdots = \beta_k = 0$, the hypothesis of a constant mean model. This is the model utility test, and it tests the hypothesis that the explanatory variables are useless for predicting y. Rejection of H_0 occurs for large values of the F ratio. This should be intuitively reasonable, because if the prediction quality is good, then SSE should be small and SSR should be large, and therefore the F ratio should be large. The dividing line between large and small is set using the upper tail of the F distribution. In particular, H_0 is typically rejected if the F ratio exceeds $F_{.05,k,n-(k+1)}$.

Another measure of the relationship between y and the predictors is the R^2 statistic, the coefficient of multiple determination, which is the fraction SSR/SST:

$$R^2 = \frac{\text{SSR}}{\text{SST}} = \frac{\text{SST} - \text{SSE}}{\text{SST}} = 1 - \frac{\text{SSE}}{\text{SST}} \qquad (12.22)$$

By the analysis of variance, Equation (12.20), this is always between 0 and 1. The R^2 statistic is also called the squared multiple correlation. For example, suppose $\text{SST} = 200$, $\text{SSR} = 120$, and therefore $\text{SSE} = 80$. Then $R^2 = 1 - (\text{SSE}/\text{SST}) = 1 - 80/200 = .60$, so the error sum of squares is 60% less than the total sum of squares. This is sometimes interpreted by saying that the regression explains 60% of the variability of y, which means that the regression has reduced the error sum of squares by 60% from what it would be (SST) with just a constant model and no predictors.

The F ratio and R^2 are equivalent statistics in the sense that one can be obtained from the other. For example, dividing numerator and denominator through by SST in Equation (12.21) and using Equation (12.22), we find that the F ratio is [see Equation (12.18)]

$$F = \frac{R^2/k}{(1 - R^2)/[n - (k + 1)]}$$

In the special case of just one predictor, $k = 1$, $F = (n - 2)R^2/(1 - R^2)$, and the multiple correlation is just the absolute value of the ordinary correlation coefficient. This F is the square of the statistic $T = \sqrt{n-2}R/\sqrt{1 - R^2}$ given in Section 12.5.

Example 12.33 (Example 12.32 continued) The predicted values and residuals are easily obtained:

$$\hat{\mathbf{y}} = \mathbf{X}\hat{\boldsymbol{\beta}} = \begin{bmatrix} 1 & 2 & 0 \\ 1 & 2 & 1 \\ 1 & 2.5 & 1 \\ 1 & 2.5 & 0 \\ 1 & 3 & 0 \\ 1 & 3 & 1 \end{bmatrix} \begin{bmatrix} 20.92 \\ 55.50 \\ 35.33 \end{bmatrix} = \begin{bmatrix} 131.92 \\ 167.25 \\ 195.00 \\ 159.67 \\ 187.42 \\ 222.75 \end{bmatrix}$$

$$\mathbf{y} - \hat{\mathbf{y}} = \begin{bmatrix} 132 \\ 170 \\ 184 \\ 165 \\ 182 \\ 231 \end{bmatrix} - \begin{bmatrix} 131.92 \\ 167.25 \\ 195.00 \\ 159.67 \\ 187.42 \\ 222.75 \end{bmatrix} = \begin{bmatrix} .08 \\ 2.75 \\ -11.00 \\ 5.33 \\ -5.42 \\ 8.25 \end{bmatrix}$$

Therefore, the error sum of squares is $\text{SSE} = \|\mathbf{y} - \hat{\mathbf{y}}\|^2 = .08^2 + \cdots + (-8.25)^2 = 254.42$ and $\text{MSE} = s^2 = \text{SSE}/[n - (k + 1)] = 254.42/[6 - (2 + 1)] = 84.81$. The square root of this yields the estimated standard deviation $s = 9.209$, which is a form of average for the magnitude of the residuals. However, notice that only one of the six residuals exceeds s in magnitude. The total sum of squares is given by $\text{SST} = \|\mathbf{y} - \bar{\mathbf{y}}\|^2 = \Sigma(y_i - 177.33)^2 = 5207.33$. The regression sum of squares can be obtained by subtraction using the analysis of variance, $\text{SSR} = \text{SST} - \text{SSE} = 5207.33 - 254.42 = 4952.92$. The sums of squares and the computation of the F test and R^2 are often done through an analysis of variance table, as copied in Figure 12.36 from SAS output.

Analysis of Variance

Source	DF	Sum of Squares	Mean Square	F Value	Pr > F
Model	2	4952.91667	2476.45833	29.20	0.0108
Error	3	254.41667	84.80556		
Corrected Total	5	5207.33333			

Figure 12.36 Analysis of variance table from SAS

The regression sum of squares is called the model sum of squares here. The mean square is the sum of squares divided by the degrees of freedom, and the F value is the ratio of mean squares. Because the P-value is less than .05, we reject the null hypothesis (that both the liters and fuel population coefficients are 0) at the .05 level. The coefficient of multiple determination is $R^2 = \text{SSR}/\text{SST} = 4952.92/5207.33 = .9511$. We say that the two predictors account for 95% of the variance of horsepower because the error sum of squares is reduced by 95% compared to the total sum of squares. ■

Covariance Matrices

In order to develop hypothesis tests and confidence intervals for the regression coefficients, the standard deviations of the estimated coefficients are needed. These can be obtained from a certain *covariance matrix*, a matrix with the variances on the diagonal and the covariances in the off-diagonal elements. If $\boldsymbol{U}$ is a column vector of random variables $U_1, \ldots, U_n$ with means $\mu_1 = E(U_1), \ldots, \mu_n = E(U_n)$, let $\boldsymbol{\mu}$ be the vector of these n means and define

$$\text{Cov}(\boldsymbol{U}) = \begin{bmatrix} \text{Cov}(U_1,U_1) & \cdots & \text{Cov}(U_1,U_n) \\ \vdots & \ddots & \vdots \\ \text{Cov}(U_n,U_1) & \cdots & \text{Cov}(U_n,U_n) \end{bmatrix} \tag{12.23}$$

$$= \begin{bmatrix} E[(U_1 - \mu_1)(U_1 - \mu_1)] & \cdots & E[(U_1 - \mu_1)(U_n - \mu_n)] \\ \vdots & \ddots & \vdots \\ E[(U_n - \mu_n)(U_1 - \mu_1)] & \cdots & E[(U_n - \mu_n)(U_n - \mu_n)] \end{bmatrix}$$

$$= E\left\{ \begin{bmatrix} U_1 - \mu_1 \\ \vdots \\ U_n - \mu_n \end{bmatrix} [U_1 - \mu_1, \ldots, U_n - \mu_n] \right\} = E\{[\boldsymbol{U} - \boldsymbol{\mu}][\boldsymbol{U} - \boldsymbol{\mu}]'\}$$

When $n = 1$ this reduces to just the ordinary variance. The key to finding the needed covariance matrix is this proposition:

PROPOSITION

If $\boldsymbol{A}$ is a matrix whose entries are constants and $\boldsymbol{V} = \boldsymbol{AU}$, then $\text{Cov}(\boldsymbol{V}) = \boldsymbol{A}\,\text{Cov}(\boldsymbol{U})\boldsymbol{A}'$.

Proof By the linearity of the expectation operator, $E(\boldsymbol{V}) = E(\boldsymbol{AU}) = \boldsymbol{A}E(\boldsymbol{U})$. Then

$$\begin{aligned} \text{Cov}(\boldsymbol{V}) &= E\{[\boldsymbol{AU} - E(\boldsymbol{AU})][\boldsymbol{AU} - E(\boldsymbol{AU})]'\} \\ &= E(\boldsymbol{A}[\boldsymbol{U} - E(\boldsymbol{U})]\{\boldsymbol{A}[\boldsymbol{U} - E(\boldsymbol{U})]\}') = E\{\boldsymbol{A}[\boldsymbol{U} - E(\boldsymbol{U})][\boldsymbol{U} - E(\boldsymbol{U})]'\boldsymbol{A}'\} \\ &= \boldsymbol{A}E\{[\boldsymbol{U} - E(\boldsymbol{U})][\boldsymbol{U} - E(\boldsymbol{U})]'\}\boldsymbol{A}' = \boldsymbol{A}\,\text{Cov}(\boldsymbol{U})\boldsymbol{A}' \end{aligned}$$ ■

Let's apply the proposition to find the covariance matrix of $\hat{\boldsymbol{\beta}}$. Because $\hat{\boldsymbol{\beta}} = [\boldsymbol{X}'\boldsymbol{X}]^{-1}\boldsymbol{X}'\boldsymbol{Y}$, we use $\boldsymbol{A} = [\boldsymbol{X}'\boldsymbol{X}]^{-1}\boldsymbol{X}'$ and $\boldsymbol{U} = \boldsymbol{Y}$. The transpose of $\boldsymbol{A}$ is $\boldsymbol{A}' = \{[\boldsymbol{X}'\boldsymbol{X}]^{-1}\boldsymbol{X}'\}' = \boldsymbol{X}[\boldsymbol{X}'\,\boldsymbol{X}]^{-1}$. The covariance matrix of $\boldsymbol{Y}$ is just the variance σ^2 times the n-dimensional identity matrix, that is, $\sigma^2\boldsymbol{I}$, because the observations are independent and all have the same variance σ^2. Then the proposition says

$$\text{Cov}(\hat{\boldsymbol{\beta}}) = \boldsymbol{A}\,\text{Cov}(\boldsymbol{Y})\boldsymbol{A}' = [\boldsymbol{X}'\boldsymbol{X}]^{-1}\boldsymbol{X}'[\sigma^2 I]\,\boldsymbol{X}[\boldsymbol{X}'\boldsymbol{X}]^{-1} = \sigma^2[\boldsymbol{X}'\boldsymbol{X}]^{-1} \tag{12.24}$$

We also need to find the expected value of $\hat{\boldsymbol{\beta}}$:

$$E(\hat{\boldsymbol{\beta}}) = E([\boldsymbol{X}'\boldsymbol{X}]^{-1}\boldsymbol{X}'\boldsymbol{Y}) = [\boldsymbol{X}'\boldsymbol{X}]^{-1}\boldsymbol{X}'E(\boldsymbol{Y})$$
$$= [\boldsymbol{X}'\boldsymbol{X}]^{-1}\boldsymbol{X}'E(\boldsymbol{X}\boldsymbol{\beta} + \boldsymbol{\varepsilon}) = [\boldsymbol{X}'\boldsymbol{X}]^{-1}\boldsymbol{X}'\boldsymbol{X}\boldsymbol{\beta} = \boldsymbol{\beta}$$

That is, $\hat{\boldsymbol{\beta}}$ is an unbiased estimator of β (for each i, $\hat{\beta}_i$ is unbiased for estimating β_i).

Write the inverse matrix as $[\boldsymbol{X}'\boldsymbol{X}]^{-1} = C = [c_{ij}]$. In particular, let $c_{00}, c_{11}, \ldots, c_{kk}$ be the diagonal elements of this inverse matrix. Then $V(\hat{\beta}_j) = \sigma^2 c_{jj}$. Also, $\hat{\beta}_j$ is a linear combination of $Y_1, \ldots, Y_n$, which are independent normal, so $(\hat{\beta}_j - \beta_j)/\sigma\sqrt{c_{jj}} \sim N(0, 1)$. It follows that (this requires the independence of S and the estimated regression coefficients, which we will not prove) $(\hat{\beta}_j - \beta_j)/(S\sqrt{c_{jj}}) \sim t_{n-(k+1)}$. This leads to the confidence interval and hypothesis test for coefficients of Section 12.7.

The 95% confidence interval for β_j is

$$\hat{\beta}_j \pm t_{.025,n-(k+1)} s\sqrt{c_{jj}} \tag{12.25}$$

We can test the hypothesis H_0: $\beta_j = \beta_{j0}$ using the t ratio

$$T = \frac{\hat{\beta}_j - \beta_{j0}}{S\sqrt{c_{jj}}} \sim t_{n-(k+1)}$$

Statistical software packages usually provide output for testing H_0: $\beta_j = 0$ against the two-sided alternative H_a: $\beta_j \neq 0$. In particular, we would reject H_0 in favor of H_a at the 5% level if $|t|$ exceeds $t_{.025,n-(k+1)}$. Usually, with computer output there is no need to use statistical tables for hypothesis tests because P-values for these tests are included.

Example 12.34 (Example 12.33 continued)

For the engine horsepower scenario we found that $s = 9.209$, $\hat{\beta}_0 = 20.92$, $\hat{\beta}_1 = 55.5$, $\hat{\beta}_2 = 35.33$, and $[\boldsymbol{X}'\boldsymbol{X}]^{-1}$ has elements $c_{00} = \frac{79}{12}$, $c_{11} = 1$, $c_{22} = \frac{2}{3}$. Therefore, we get these 95% confidence intervals:

$$\hat{\beta}_1 \pm t_{.025,6-(2+1)} s\sqrt{c_{11}} = 55.5 \pm 3.182(9.209)\sqrt{1} = 55.50 \pm 29.30 = (26.2, 84.8)$$
$$\hat{\beta}_2 \pm t_{.025,6-(2+1)} s\sqrt{c_{22}} = 35.33 \pm 3.182(9.209)\sqrt{\tfrac{2}{3}} = 35.33 \pm 23.93 = (11.4, 59.3)$$

We can also do the individual t tests for the coefficients:

$$\frac{\hat{\beta}_1 - 0}{s\sqrt{c_{11}}} = \frac{55.5 - 0}{9.209\sqrt{1}} = 6.03 \qquad \text{two-tailed } P\text{-value} = .009$$
$$\frac{\hat{\beta}_2 - 0}{s\sqrt{c_{22}}} = \frac{35.33 - 0}{9.209\sqrt{\tfrac{2}{3}}} = 4.70 \qquad \text{two-tailed } P\text{-value} = .018$$

Both of these exceed $t_{.025,6-2-1} = 3.182$ in absolute value (and their P-values are less than .05), so for both of them we reject at the 5% level the null hypothesis that the coefficient is 0, in favor of the two-sided alternative. These conclusions are consistent with the fact that the corresponding confidence intervals do not include zero. Also, recall that the F test rejected at the 5% level the null hypothesis that both coefficients are zero. As our intuition suggests, horsepower increases with engine size and horsepower is higher when the engine requires premium fuel. ■

The Hat Matrix

The foregoing proposition can be used to find estimated standard deviations for predicted values and residuals. Recall that the vector of predicted values can be obtained by multiplying the hat matrix $\boldsymbol{H}$ times the $\boldsymbol{Y}$ vector, $\boldsymbol{HY} = \hat{\boldsymbol{Y}}$. First, in order to apply the proposition, let's obtain the transpose of $\boldsymbol{H}$. With the help of the rules $(\boldsymbol{AB})' = \boldsymbol{B}'\boldsymbol{A}'$ and $(\boldsymbol{A}^{-1})' = (\boldsymbol{A}')^{-1}$, we find that $\boldsymbol{H}$ is symmetric:

$$\boldsymbol{H}' = \{\boldsymbol{X}[\boldsymbol{X}'\boldsymbol{X}]^{-1}\boldsymbol{X}'\}' = (\boldsymbol{X}')'\{[\boldsymbol{X}'\boldsymbol{X}]^{-1}\}'\boldsymbol{X}' = \boldsymbol{X}\{[\boldsymbol{X}'\boldsymbol{X}]'\}^{-1}\boldsymbol{X}' = \boldsymbol{X}[\boldsymbol{X}'\boldsymbol{X}]^{-1}\boldsymbol{X}' = \boldsymbol{H}$$

Therefore,

$$\text{Cov}(\hat{\boldsymbol{Y}}) = \boldsymbol{H}\,\text{Cov}(\hat{\boldsymbol{Y}})\boldsymbol{H}' = \boldsymbol{X}[\boldsymbol{X}'\boldsymbol{X}]^{-1}\boldsymbol{X}'[\sigma^2\boldsymbol{I}]\boldsymbol{X}[\boldsymbol{X}'\boldsymbol{X}]^{-1}\boldsymbol{X}' \qquad (12.26)$$

$$= \sigma^2\boldsymbol{X}[\boldsymbol{X}'\boldsymbol{X}]^{-1}\boldsymbol{X}' = \sigma^2\boldsymbol{H}$$

A similar calculation shows that the covariance matrix of the residuals is

$$\text{Cov}(\boldsymbol{Y} - \hat{\boldsymbol{Y}}) = \sigma^2(\boldsymbol{I} - \boldsymbol{H}) \qquad (12.27)$$

Of course, the true variance $\boldsymbol{\sigma}\sigma^2$ is generally unknown, so the estimate s^2 = MSE is used instead.

Example 12.35 (Example 12.34 continued) If residuals and predicted values are requested from SAS, then the output includes the information in Figure 12.37.

Obs	Dep Var hp	Predicted Value	Std Error Mean Predict	Residual	Std Error Residual	Student Residual
1	132.0000	131.9167	7.0335	0.0833	5.944	0.0140
2	170.0000	167.2500	7.0335	2.7500	5.944	0.463
3	184.0000	195.0000	5.3168	−11.0000	7.519	−1.463
4	165.0000	159.6667	5.3168	5.3333	7.519	0.709
5	182.0000	187.4167	7.0335	−5.4167	5.944	−0.911
6	231.0000	222.7500	7.0335	8.2500	5.944	1.388

Figure 12.37 Predicted values and residuals from SAS

The column labeled "Std Error Mean Predict" has the estimated standard deviations for the predicted values and it contains the square roots of the $s^2\boldsymbol{H}$ matrix diagonal elements. The column labeled "Std Error Residual" has the estimated standard deviations for the residuals, and it contains the square roots of the $s^2(\boldsymbol{I} - \boldsymbol{H})$ diagonal elements. The column labeled "Student Residual" is what we defined as the standardized residual in Section 12.6. It is the ratio of the previous two columns. ■

The hat matrix is also important as a measure of the influence of individual observations. Because $\hat{\boldsymbol{y}} = \boldsymbol{Hy}$, $\hat{y}_i = h_{i1}y_1 + h_{i2}y_2 + \cdots + h_{in}y_n$, and therefore $\partial\hat{y}_i/\partial y_i = h_{ii}$. That is, the partial derivative of $\hat{y}_i$ with respect to y_i is the ith diagonal element of the hat matrix. In other words, the ith diagonal element of $\boldsymbol{H}$ measures the influence of the ith observation on its predicted value. The diagonal elements of $\boldsymbol{H}$ are sometimes called the **leverages** to indicate their influence over the regression. An observation with very high leverage will tend to pull the regression toward it, and its residual will tend to be small. Of course, $\boldsymbol{H}$ depends only on the values of the predictors, so the leverage measures only one aspect of influence. If the influence of an observation is defined in terms of the effect on the predicted values when the observation is omitted, then an influential observation is one that has both large leverage and a large (in absolute value) residual.

Example 12.36 Table 12.3 contains data on height, foot length, and wingspan (distance between fingertips when arms are outstretched) for a sample of 16 students. The last column has the leverages for the regression of wingspan on height and foot length.

Table 12.3 Height, foot length, and wingspan data

Obs	Height	Foot	Wingspan	Leverage
1	63.0	9.0	62.0	0.239860
2	63.0	9.0	62.0	0.239860
3	65.0	9.0	64.0	0.228236
4	64.0	9.5	64.5	0.223625
5	68.0	9.5	67.0	0.196418
6	69.0	10.0	69.0	0.083676
7	71.0	10.0	70.0	0.262182
8	68.0	10.0	72.0	0.067207
9	68.0	10.5	70.0	0.187088
10	72.0	10.5	72.0	0.151959
11	73.0	11.0	73.0	0.143279
12	73.5	11.0	75.0	0.168719
13	70.0	11.0	71.0	0.245380
14	70.0	11.0	70.0	0.245380
15	72.0	11.0	76.0	0.128790
16	74.0	11.2	76.5	0.188340

In Figure 12.38 we show the plot of height against foot length, along with the leverage for each point. Notice that the points at the extreme right and left of the plot tend to have high leverage, and the points near the center tend to have low leverage.

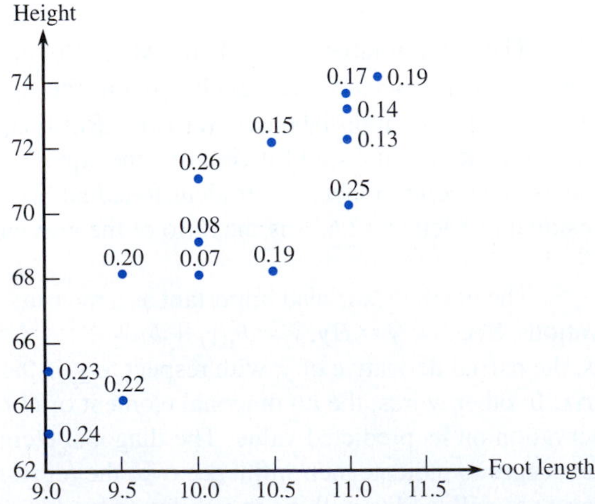

Figure 12.38 Plot of height and foot length showing leverage

However, it is interesting that the point with highest leverage is not at the extremes of height or foot length. This is student number 7, with a 10-inch foot and height of 70 inches, and the high leverage comes from the height being extreme relative to foot length. Indeed, when there are several predictors, high leverage often occurs when values of one predictor are extreme relative to the values of other predictors. For example, if height and weight are predictors, then an extremely overweight or underweight subject would likely have high leverage.

Figure 12.39 shows some useful output from MINITAB, including the model utility test, the regression coefficients, and the correlations among the variables. The correlation table shows all three correlations among the three variables along with their P-values. Clearly, the three variables are very strongly related. However, when wingspan is regressed on height and foot length, the P-value for foot length is greater than .05, so we can consider eliminating foot length from the regression equation. Does it make sense for foot length to be very strongly related to wingspan, as measured by correlation, but for the foot length term to be not statistically significant in the regression equation? The difference is that the regression test is asking whether foot length is needed in addition to height. Because the two predictors are themselves highly correlated, foot length is redundant in the sense that it offers little prediction ability beyond what is contributed by height.

```
Analysis of Variance

Source            DF        SS        MS        F        P
Regression         2    294.79    147.40    67.33    0.000
Residual Error    13     28.46      2.19
Total             15    323.25

Predictor       Coef    SE Coef         T        P
Constant      -6.085      8.018     -0.76    0.461
height        0.8060     0.2305      3.50    0.004
foot           1.973      1.044      1.89    0.081

S = 1.47956    R-Sq = 91.2%    R-Sq(adj) = 89.8%

Correlations: height, foot, wingspan
            height       foot
foot         0.892
             0.000

wingspan     0.942      0.911
             0.000      0.000
```

Figure 12.39 Regression output for height, foot length, wingspan

Exercises Section 12.8 (89–102)

89. Fit the model $y = \beta_0 + \beta_1 x_1 + \beta_2 x_2 + \varepsilon$ to the data.

x_1	−1	−1	1	1
x_2	−1	1	−1	1
y	1	1	0	4

a. Determine $\mathbf{X}$ and $\mathbf{y}$ and express the normal equations in terms of matrices.

b. Determine the $\hat{\boldsymbol{\beta}}$ vector, which contains the estimates for the three coefficients in the model.

c. Determine $\hat{y}$, the predictions for the four observations, and also the four residuals. Find SSE by summing the four squared residuals. Use this to get the estimated variance MSE.

d. Use the MSE and c_{11} to get a 95% confidence interval for β_1.

e. Carry out a t test for the hypothesis H_0: $\beta_1 = 0$ against a two-tailed alternative, and interpret the result.

f. Form the analysis of variance table and carry out the F test for the hypothesis H_0: $\beta_1 = \beta_2 = 0$. Find R^2 and interpret.

90. Consider the model $y = \beta_0 + \beta_1 x_1 + \varepsilon$ for the data

x_1	−.5	−.5	−.5	−.5	.5	.5	.5	.5
y	1	2	2	3	8	9	7	8

a. Determine the $\mathbf{X}$ and $\mathbf{y}$ matrices and express the normal equations in terms of matrices.

b. Determine the $\hat{\boldsymbol{\beta}}$ vector, which contains the estimates for the two coefficients in the model.

c. Determine $\hat{\mathbf{y}}$, the predictions for the eight observations, and also obtain the eight residuals.

d. Find SSE by summing the eight squared residuals. Use this to get the estimated variance MSE.

e. Use the MSE and c_{11} to get a 95% confidence interval for β_1.

f. Carry out a t test for the hypothesis H_0: $\beta_1 = 0$ against a two-tailed alternative.

g. Carry out the F test for the hypothesis H_0: $\beta_1 = 0$. How is this related to part (f)?

91. Suppose that the model consists of just $y = \beta_0 + \varepsilon$, so $k = 0$. Estimate β_0 from $[\mathbf{X}'\mathbf{X}]^{-1}\mathbf{X}'\mathbf{y}$. Find simple expressions for s and c_{00}, and use them along with Equation (12.25) to express simply the 95% confidence interval for β_0. Your result should be equivalent to the one-sample t confidence interval in Section 8.3.

92. Suppose we have $(x_1, y_1), \ldots, (x_n, y_n)$. Let $k = 1$ and $x_{i1} = x_i - \bar{x}$, $i = 1, \ldots, n$, so our model is $y_i = \beta_0 + \beta_1(x_i - \bar{x}) + \varepsilon$, $i = 1, \ldots, n$.

a. Obtain $\hat{\beta}_0$ and $\hat{\beta}_1$ from $[\mathbf{X}'\mathbf{X}]^{-1}\mathbf{X}'\mathbf{y}$.

b. Find c_{00} and c_{11} and use them to simplify the confidence intervals [Equation (12.25)] for β_0 and β_1.

c. In terms of computing $[\mathbf{X}'\mathbf{X}]^{-1}$, why is it better to have $x_{i1} = x_i - \bar{x}$ rather than $x_{i1} = x_i$?

93. Suppose that we have $y_1, \ldots, y_m \sim N(\mu_1, \sigma^2)$, $y_{m+1}, \ldots, y_{m+n} \sim N(\mu_2, \sigma^2)$, and all $m + n$ observations are independent. These are the assumptions of the pooled t procedure in Section 10.2. Let $k = 1$, $x_{11} = .5, \ldots, x_{m1} = .5, x_{m+1,1} = -.5, \ldots, x_{m+n,1} = -.5$. For convenience in inverting $\mathbf{X}'\mathbf{X}$ assume $m = n$.

a. Obtain $\hat{\beta}_0$ and $\hat{\beta}_1$ from $[\mathbf{X}'\mathbf{X}]^{-1}\mathbf{X}'\mathbf{y}$.

b. Find simple expressions for $\hat{y}$, SSE, s, c_{11}.

c. Use parts (a) and (b) to find a simple expression for the 95% confidence interval [Equation (12.25)] for β_1. Letting $\bar{y}_1$ be the mean of the first m observations and $\bar{y}_2$ be the mean of the next n observations, your result should be

$$\hat{\beta}_1 \pm t_{.025,m+n-2}\, s\sqrt{\frac{1}{m} + \frac{1}{n}} = \bar{y}_1 - \bar{y}_2 \pm t_{.025,m+n-2}$$

$$\cdot \sqrt{\frac{\sum_{i=1}^{m}(y_i - \bar{y}_1)^2 + \sum_{i=m+1}^{m+n}(y_i - \bar{y}_2)^2}{m + n - 2}}\sqrt{\frac{1}{m} + \frac{1}{n}}$$

which is the pooled variance confidence interval discussed in Section 9.2.

d. Let $m = 3$ and $n = 3$, with $y_1 = 117$, $y_2 = 119$, $y_3 = 127$, $y_4 = 129$, $y_5 = 138$, $y_6 = 139$. These are the prices in thousands for three houses in Brookwood and then three houses in Pleasant Hills. Apply parts (a), (b), and (c) to this data set.

94. The constant term is not always needed in the regression equation. In general, if the dependent variable should be 0 when the independent variables are 0, then the constant term is not needed. Then it may be preferable to omit β_0 and use the model $y = \beta_1 x_1 + \beta_2 x_2 + \cdots + \beta_k x_k + \varepsilon$. Here we focus on the special case $k = 1$.

a. Differentiate the appropriate sum of squares to derive the one normal equation for estimating β_1.

b. Express your normal equation in matrix terms, $\mathbf{X}'\mathbf{X}\boldsymbol{\beta} = \mathbf{X}'\mathbf{y}$, where $\mathbf{X}$ consists of one column with the values of the predictor variable.

c. Apply the result from part (b) to the data of Example 12.32, using hp for $\mathbf{y}$ and liters for $\mathbf{X}$.

d. Explain why deletion of the constant term might be appropriate for the data set in part (c).

e. By fitting a regression model with a constant term added to the model of part (c), test the hypothesis that the constant is not needed.

95. Assuming that the analysis of variance table is available, show how the last three columns of Figure 12.37 (the columns related to residuals) can be obtained from the previous columns.

96. Given that the residuals are $\mathbf{y} - \hat{\mathbf{y}} = (\mathbf{I} - \mathbf{H})\mathbf{y}$, show that $\text{Cov}(\mathbf{Y} - \hat{\mathbf{Y}}) = (\mathbf{I} - \mathbf{H})\sigma^2$.

97. Use Equation (12.26) and Equation (12.27) to show that each of the leverages is between 0 and 1, and therefore the variances of the predicted values and residuals are between 0 and σ^2.

98. Consider the special case $y = \beta_0 + \beta_1 x + \varepsilon$, so $k = 1$ and $\boldsymbol{X}$ consists of a column of 1's and a column of the values $x_1, \ldots, x_n$ of x.

a. Write the normal equations in matrix form, and solve by inverting $\boldsymbol{X}'\boldsymbol{X}$. *Hint*: If $ad \neq bc$, then

$$\begin{bmatrix} a & b \\ c & d \end{bmatrix}^{-1} = \frac{1}{ad - bc}\begin{bmatrix} d & -b \\ -c & a \end{bmatrix}$$

Check your answers against those in Section 12.2.

b. Use the inverse of $\boldsymbol{X}'\boldsymbol{X}$ to obtain expressions for the variances of the coefficients, and check your answers against the results given in Sections 12.3 and 12.4 ($\hat{\beta}_0$ is the predicted value corresponding to $x^* = 0$).

c. Compare the predictions from this model with the predictions from the model of Exercise 92. Comparing other aspects of the two models, discuss similarities and differences. Mention, in particular, the hat matrix, the predicted values, and the residuals.

99. Continue Exercise 92.

a. Find the elements of the hat matrix and use them to obtain the variance of the predicted values. Noting the result of Exercise 98(c), compare your result with the expression for $V(\hat{y})$ given in Section 12.4.

b. Using the diagonal elements of the hat matrix, obtain the variance of the residuals and compare with the expression given in Section 12.6.

c. Compare the variances of predicted values for an x that is close to $\bar{x}$ and an x that is far from $\bar{x}$.

d. Compare the variances of residuals for an x that is close to $\bar{x}$ and an x that is far from $\bar{x}$.

e. Give intuitive explanations for the results of parts (c) and (d).

100. Carry out the details of the derivation for the analysis of variance, Equation (12.20).

101. The measurements here are similar to those in Example 12.36, except that here the students did the measuring at home, and the results suffered in accuracy.

Wingspan	Foot	Height
74	13.0	75
56	8.5	66
65	10.0	69
66	9.5	66
62	9.0	54
69	11.0	72
75	12.0	75
66	9.0	63
66	9.0	66
63	8.5	63

a. Regress wingspan on the other two variables. Carry out the test of model utility and the tests for the two individual regression coefficients of the predictors.

b. Obtain the diagonal elements of the hat matrix (leverages). Identify the point with the highest leverage. What is unusual about the point? Given the instructor's assertion that there were no students in the class less than 5 ft tall, would you say that there was an error? Give another reason that this student's measurements seem wrong.

c. For the other points with high leverages, what distinguishes them from the points with ordinary leverage values?

d. Examining the residuals, find another student whose data might be wrong.

e. Discuss the elimination of questionable points in order to obtain valid regression results.

102. Here is a method for obtaining the variance of the residuals in simple (one predictor) linear regression, as given by Equation (12.13).

a. We have shown in Equations (12.26) and (12.27) that $\text{Cov}(\hat{\boldsymbol{Y}}) = \sigma^2\boldsymbol{H}$ and $\text{Cov}(\boldsymbol{Y} - \hat{\boldsymbol{Y}}) = \sigma^2(\boldsymbol{I} - \boldsymbol{H})$. Now show that $V(Y_i - \hat{Y}_i) = \sigma^2 - V(\hat{Y}_i)$.

b. Use part (a) and $V(\hat{Y}_i)$ from Section 12.4 to show Equation (12.13) for simple linear regression

$$V(Y_i - \hat{Y}_i) = \sigma^2 \cdot \left[1 - \frac{1}{n} - \frac{(x_i - \bar{x})^2}{S_{xx}}\right]$$

Supplementary Exercises (103–119)

103. The presence of hard alloy carbides in high chromium white iron alloys results in excellent abrasion resistance, making them suitable for materials handling in the mining and materials processing industries. The accompanying data on x = retained austenite content (%) and y = abrasive wear loss (mm^3) in pin wear tests with garnet as the abrasive was read from a plot in the article "Microstructure-Property Relationships in High Chromium White Iron Alloys" (*Internat. Materials Rev.*, 1996: 59–82).

x	4.6	17.0	17.4	18.0	18.5	22.4	26.5	30.0	34.0
y	.66	.92	1.45	1.03	.70	.73	1.20	.80	.91

x	38.8	48.2	63.5	65.8	73.9	77.2	79.8	84.0
y	1.19	1.15	1.12	1.37	1.45	1.50	1.36	1.29

SAS output for Exercise 103

```
Dependent Variable: ABRLOSS

                          Analysis of Variance

Source          DF     Sum of Squares     Mean Square      F Value     Prob > F
Model            1            0.63690         0.63690       15.444       0.0013
Error           15            0.61860         0.04124
C Total         16            1.25551

           Root MSE      0.20308      R-square      0.5073
           Dep Mean      1.10765      Adj R-sq      0.4744
           C.V.         18.33410

                              Parameter Estimates

                       Parameter         Standard        T for H0:
Variable        DF      Estimate            Error     Parameter = 0    Prob > |T|

INTERCEP         1      0.787218       0.09525879             8.264        0.0001
AUSTCONT         1      0.007570       0.00192626             3.930        0.0013
```

a. What proportion of observed variation in wear loss can be attributed to the simple linear regression model relationship?

b. What is the value of the sample correlation coefficient?

c. Test the utility of the simple linear regression model using $\alpha = .01$.

d. Estimate the true average wear loss when content is 50% and do so in a way that conveys information about reliability and precision.

e. What value of wear loss would you predict when content is 30%, and what is the value of the corresponding residual?

104. An investigation was carried out to study the relationship between speed (ft/sec) and stride rate (number of steps taken/sec) among female marathon runners. Resulting summary quantities included $n = 11$, $\Sigma(\text{speed}) = 205.4$, $\Sigma(\text{speed})^2 = 3880.08$, $\Sigma(\text{rate}) = 35.16$, $\Sigma(\text{rate})^2 = 112.681$, and $\Sigma(\text{speed})(\text{rate}) = 660.130$.

a. Calculate the equation of the least squares line that you would use to predict stride rate from speed.

b. Calculate the equation of the least squares line that you would use to predict speed from stride rate.

c. Calculate the coefficient of determination for the regression of stride rate on speed of part (a) and for the regression of speed on stride rate of part (b). How are these related?

d. How is the product of the two slope estimates related to the value calculated in (c)?

105. In Section 12.4, we presented a formula for the variance $V(\hat{\beta}_0 + \hat{\beta}_1 x^*)$ and a CI for $\beta_0 + \beta_1 x^*$. Taking $x^* = 0$ gives $\sigma^2_{\hat{\beta}_0}$ and a CI for β_0. Use the data of Example 12.12 to calculate the estimated standard deviation of $\hat{\beta}_0$ and a 95% CI for the y-intercept of the true regression line.

106. Show that $\text{SSE} = S_{yy} - \hat{\beta}_1 S_{xy}$, which gives an alternative computational formula for SSE.

107. Suppose that x and y are positive variables and that a sample of n pairs results in $r \approx 1$. If the sample correlation coefficient is computed for the (x, y^2) pairs, will the resulting value also be approximately 1? Explain.

108. Let s_x and s_y denote the sample standard deviations of the observed x's and y's, respectively [so $s_x^2 = \Sigma(x_i - \bar{x})^2/(n - 1)$ and similarly for s_y^2].

a. Show that an alternative expression for the estimated regression line $y = \hat{\beta}_0 + \hat{\beta}_1 x$ is

$$y = \bar{y} + r \cdot \frac{s_y}{s_x}(x - \bar{x})$$

b. This expression for the regression line can be interpreted as follows. Suppose $r = .5$. What then is the predicted y for an x that lies 1 SD (s_x units) above the mean of the x_i's? If r were 1, the prediction would be for y to lie 1 SD above its mean $\bar{y}$, but since $r = .5$, we predict a y that is only .5 SD ($.5s_y$ unit) above $\bar{y}$. Using the data in Exercise 62 for a patient whose age is 1 SD below the average age in the sample, by how many standard deviations is the patient's predicted ΔCBG above or below the average ΔCBG for the sample?

109. In biofiltration of wastewater, air discharged from a treatment facility is passed through a damp porous membrane that causes contaminants to dissolve in water and be transformed into harmless products. The accompanying data on x = inlet temperature (°C) and y = removal efficiency (%) was the basis for a scatter plot that appeared in the article "Treatment of Mixed Hydrogen Sulfide and Organic Vapors in a Rock Medium Biofilter"(*Water Environ. Res.*, 2001: 426–435).

Obs	Temp	Removal %	Obs	Temp	Removal %
1	7.68	98.09	17	8.55	98.27
2	6.51	98.25	18	7.57	98.00
3	6.43	97.82	19	6.94	98.09
4	5.48	97.82	20	8.32	98.25
5	6.57	97.82	21	10.50	98.41
6	10.22	97.93	22	16.02	98.51
7	15.69	98.38	23	17.83	98.71
8	16.77	98.89	24	17.03	98.79
9	17.13	98.96	25	16.18	98.87
10	17.63	98.90	26	16.26	98.76
11	16.72	98.68	27	14.44	98.58
12	15.45	98.69	28	12.78	98.73
13	12.06	98.51	29	12.25	98.45
14	11.44	98.09	30	11.69	98.37
15	10.17	98.25	31	11.34	98.36
16	9.64	98.36	32	10.97	98.45

Calculated summary quantities are $\Sigma x_i = 384.26$, $\Sigma y_i = 3149.04$, $\Sigma x_i^2 = 5099.2412$, $\Sigma x_i y_i = 37{,}850.7762$, and $\Sigma y_i^2 = 309{,}892.6548$.

a. Does a scatter plot of the data suggest appropriateness of the simple linear regression model?

b. Fit the simple linear regression model, obtain a point prediction of removal efficiency when temperature = 10.50, and calculate the value of the corresponding residual.

c. Roughly what is the size of a typical deviation of points in the scatter plot from the least squares line?

d. What proportion of observed variation in removal efficiency can be attributed to the model relationship?

e. Estimate the slope coefficient in a way that conveys information about reliability and precision, and interpret your estimate.

f. Personal communication with the authors of the article revealed that one additional observation was not included in their scatter plot: (6.53, 96.55). What impact does this additional observation have on the equation of the least squares line and the values of s and r^2?

110. Normal hatchery processes in aquaculture inevitably produce stress in fish, which may negatively impact growth, reproduction, flesh quality, and susceptibility to disease. Such stress manifests itself in elevated and sustained corticosteroid levels. The article "Evaluation of Simple Instruments for the Measurement of Blood Glucose and Lactate, and Plasma Protein as Stress Indicators in Fish"(*J. World Aquaculture Soc.*, 1999: 276–284) described an experiment in which fish were subjected to a stress protocol and then removed and tested at various times after the protocol had been applied. The accompanying data on x = time (min) and y = blood glucose level (mmol/L) was read from a plot.

x	2	2	5	7	12	13	17	18	23	24	26	28
y	4.0	3.6	3.7	4.0	3.8	4.0	5.1	3.9	4.4	4.3	4.3	4.4

x	29	30	34	36	40	41	44	56	56	57	60	60
y	5.8	4.3	5.5	5.6	5.1	5.7	6.1	5.1	5.9	6.8	4.9	5.7

Use the methods developed in this chapter to analyze the data, and write a brief report summarizing your conclusions (assume that the investigators are particularly interested in glucose level 30 min after stress).

111. The article "Evaluating the BOD POD for Assessing Body Fat in Collegiate Football Players" (*Medi. Sci. Sports Exercise*, 1999: 1350–1356) reports on a new air displacement device for measuring body fat. The customary procedure utilizes the hydrostatic weighing device, which measures the percentage of body fat by means of water displacement. Here is representative data read from a graph in the paper.

BOD	2.5	4.0	4.1	6.2	7.1	7.0	8.3	9.2	9.3	12.0	12.2
HW	8.0	6.2	9.2	6.4	8.6	12.2	7.2	12.0	14.9	12.1	15.3

BOD	12.6	14.2	14.4	15.1	15.2	16.3	17.1	17.9	17.9
HW	14.8	14.3	16.3	17.9	19.5	17.5	14.3	18.3	16.2

a. Use various methods to decide whether it is plausible that the two techniques measure on average the same amount of fat.

b. Use the data to develop a way of predicting an HW measurement from a BOD POD measurement, and investigate the effectiveness of such predictions.

112. Reconsider the situation of Exercise 103, in which x = retained austenite content using a garnet abrasive and y = abrasive wear loss were related via the simple linear regression model $Y = \beta_0 + \beta_1 x + \varepsilon$. Suppose that for a second type of abrasive, these variables are also related via the simple linear regression model $Y = \gamma_0 + \gamma_1 x + \varepsilon$ and that $V(\varepsilon) = \sigma^2$ for both types of abrasive. If the data set consists of n_1 observations on the first abrasive and n_2 on the second and if SSE_1 and SSE_2 denote the two error sums of squares, then a pooled estimate of σ^2 is $\hat{\sigma}^2 = (\text{SSE}_1 + \text{SSE}_2)/(n_1 + n_2 - 4)$. Let SS_{x1} and SS_{x2} denote $\Sigma(x_i - \bar{x})^2$ for the data on the first and second abrasives, respectively. A test of H_0: $\beta_1 - \gamma_1 = 0$ (equal slopes) is based on the statistic

$$T = \frac{\hat{\beta}_1 - \hat{\gamma}_1}{\hat{\sigma}\sqrt{\dfrac{1}{\text{SS}_{x1}} + \dfrac{1}{\text{SS}_{x2}}}}$$

When H_0 is true, T has a t distribution with $n_1 + n_2 - 4$ df. Suppose the 15 observations using the alternative abrasive give $\text{SS}_{x2} = 7152.5578$, $\hat{\gamma}_1 = .006845$, and $\text{SSE}_2 = .51350$. Using this along with the data of Exercise 103, carry out a test at level .05 to see whether expected change in wear loss associated with a 1% increase in austenite content is identical for the two types of abrasive.

113. Show that the ANOVA version of the model utility test discussed in Section 12.3 (with test statistic F = MSR/MSE) is in fact a likelihood ratio test for H_0: $\beta_1 = 0$ versus H_a: $\beta_1 \neq 0$. *Hint*: We have already pointed out that the least squares estimates of β_0 and β_1 are the mle's. What is the mle of β_0 when H_0 is true? Now determine the mle of σ^2 both in Ω (when β_1 is not necessarily 0) and in Ω_0 (when H_0 is true).

114. Show that the t ratio version of the model utility test is equivalent to the ANOVA F statistic version of the test. (Equivalent here means that rejecting H_0: $\beta_1 = 0$ when either $t \geq t_{\alpha/2,n-2}$ or $t \leq -t_{\alpha/2,n-2}$ is the same as rejecting H_0 when $f \geq F_{\alpha,1,n-2}$.)

115. When a scatter plot of bivariate data shows a pattern resembling an exponentially increasing or decreasing curve, the following *multiplicative* exponential model is often used: $Y = \alpha e^{\beta x} \cdot \varepsilon$.

a. What does this multiplicative model imply about the relationship between $Y' = \ln(Y)$ and x? *Hint*: Take logs on both sides of the model equation and let $\beta_0 = \ln(\alpha)$, $\beta_1 = \beta$, $\varepsilon' = \ln(\varepsilon)$, and suppose that ε has a lognormal distribution.

b. The accompanying data resulted from an investigation of how ethylene content of lettuce seeds (y, in nL/g dry wt) varied with exposure time (x, in min) to an ethylene absorbent ("Ethylene Synthesis in Lettuce Seeds: Its Physiological Significance," *Plant Physiol.*, 1972: 719–722).

x	2	20	20	30	40	50	60	70	80	90	100
y	408	274	196	137	90	78	51	40	30	22	15

Fit the simple linear regression model to this data, and check model adequacy using the residuals.

c. Is a scatter plot of the data consistent with the exponential regression model? Fit this model by first carrying out a simple linear regression analysis using $\ln(y)$ as the dependent variable and x as the independent variable. How good a fit is the simple linear regression model to the "transformed" data (the $[x, \ln(y)]$ pairs)? What are point estimates of the parameters α and β?

d. Obtain a 95% prediction interval for ethylene content when exposure time is 50 min. *Hint*: First obtain a PI for $\ln(y)$ based on the simple linear regression carried out in (c).

116. No tortilla chip aficionado likes soggy chips, so it is important to identify characteristics of the production process that produce chips with an appealing texture. The following data on x = frying time (sec) and y = moisture content (%) appeared in the article "Thermal and Physical Properties of Tortilla Chips as a Function of Frying Time" (*J. Food Process. Preserv.*, 1995: 175–189).

x	5	10	15	20	25	30	45	60
y	16.3	9.7	8.1	4.2	3.4	2.9	1.9	1.3

a. Construct a scatter plot of the data and comment.

b. Construct a scatter plot of the $(\ln(x), \ln(y))$ pairs (i.e., transform both x and y by logs) and comment.

c. Consider the *multiplicative* power model $Y = \alpha x^{\beta}\varepsilon$. What does this model imply about the relationship between $y' = \ln(y)$ and $x' = \ln(x)$ (assuming that ε has a lognormal distribution)?

d. Obtain a prediction interval for moisture content when frying time is 25 seconds. *Hint*: First carry out a simple linear regression of y' on x' and calculate an appropriate prediction interval.

117. The article "Determination of Biological Maturity and Effect of Harvesting and Drying Conditions on Milling Quality of Paddy" (*J. Agric. Engrg. Res.*, 1975: 353–361) reported the following data on date of harvesting (x, the number of days after flowering) and yield of paddy, a grain farmed in India (y, in kg/ha).

x	16	18	20	22	24	26	28	30
y	2508	2518	3304	3423	3057	3190	3500	3883

x	32	34	36	38	40	42	44	46
y	3823	3646	3708	3333	3517	3241	3103	2776

a. Construct a scatter plot of the data. What model is suggested by the plot?

b. Use a statistical software package to fit the model suggested in (a) and test its utility.

c. Use the software package to obtain a prediction interval for yield when the crop is harvested 25 days after flowering, and also a confidence interval for expected yield in situations where the crop is harvested 25 days after flowering. How do these two intervals compare to one another? Is this result consistent with what you learned in simple linear regression? Explain.

d. Use the software package to obtain a PI and CI when $x = 40$. How do these intervals compare to the corresponding intervals obtained in (c)? Is this result consistent with what you learned in simple linear regression? Explain.

e. Carry out a test of hypotheses to decide whether the quadratic predictor in the model fit in (b) provides useful information about yield (presuming that the linear predictor remains in the model).

118. The article "Validation of the Rockport Fitness Walking Test in College Males and Females" (*Res. Q. Exercise Sport*, 1994: 152–158) recommended the following estimated regression equation for relating y = VO_2max (L/min, a measure of cardiorespiratory fitness) to the predictors x_1 = gender (female = 0, male = 1), x_2 = weight (lb), x_3 = 1-mile walk time (min), and x_4 = heart rate at the end of the walk (beats/min):

$$y = 3.5959 + .65661x_1 + .0096x_2 - .0996x_3 - .0080x_4$$

a. How would you interpret the estimated coefficient $\hat{\beta}_3 = -0996$?

b. How would you interpret the estimated coefficient $\hat{\beta}_1 = .6566$?

c. Suppose that an observation made on a male whose weight was 170 lb, walk time was 11 min, and heart rate was 140 beats/min resulted in VO_2max = 3.15. What would you have predicted for VO_2max in this situation, and what is the value of the corresponding residual?

d. Using SSE = 30.1033 and SST = 102.3922, what proportion of observed variation in VO_2max can be attributed to the model relationship?

e. Assuming a sample size of $n = 20$, carry out a test of hypotheses to decide whether the chosen model specifies a useful relationship between VO_2max and at least one of the predictors.

119. A sample of $n = 20$ companies was selected, and the values of y = stock price and $k = 15$ predictor variables (such as quarterly dividend, previous year's earnings, and debt ratio) were determined. When the multiple regression model using these 15 predictors was fit to the data, $R^2 = .90$ resulted.

a. Does the model appear to specify a useful relationship between y and the predictor variables? Carry out a test using significance level .05. [*Hint:* The F critical value for 15 numerator and 4 denominator df is 5.86.]

b. Based on the result of part (a), does a high R^2 value by itself imply that a model is useful? Under what circumstances might you be suspicious of a model with a high R^2 value?

c. With n and k as given previously, how large would R^2 have to be for the model to be judged useful at the .05 level of significance?

Bibliography

Chatterjee, Samprit, Ali Hadi, and Bertram Price, *Regression Analysis by Example* (3rd ed.), Wiley, New York, 2000. A brief but informative discussion of selected topics.

Daniel, Cuthbert, and Fred Wood, *Fitting Equations to Data* (2nd ed.), Wiley, New York, 1980. Contains many insights and methods that evolved from the authors' extensive consulting experience.

Draper, Norman, and Harry Smith, *Applied Regression Analysis* (3rd ed.), Wiley, New York, 1999. A comprehensive and authoritative book on regression.

Hoaglin, David, and Roy Welsch, "The Hat Matrix in Regression and ANOVA," *American Statistician*, 1978: 17–23. Describes methods for detecting influential observations in a regression data set.

Neter, John, Michael Kutner, Christopher Nachtsheim, and William Wasserman, *Applied Linear Statistical Models* (4th ed.), Irwin, Homewood, IL, 1996. The first 15 chapters constitute an extremely readable and informative survey of regression analysis.

CHAPTER THIRTEEN

Goodness-of-Fit Tests and Categorical Data Analysis

Introduction

In the simplest type of situation considered in this chapter, each observation in a sample is classified as belonging to one of a finite number of categories (e.g., blood type could be one of the four categories O, A, B, or AB). With p_i denoting the probability that any particular observation belongs in category i (or the proportion of the population belonging to category i), we wish to test a null hypothesis that completely specifies the values of all the p_i's (such as H_0: $p_1 = .45$, $p_2 = .35$, $p_3 = .15$, $p_4 = .05$, when there are four categories). The test statistic will be a measure of the discrepancy between the observed numbers in the categories and the expected numbers when H_0 is true. Because a decision will be reached by comparing the computed value of the test statistic to a critical value of the chi-squared distribution, the procedure is called a chi-squared goodness-of-fit test.

Sometimes the null hypothesis specifies that the p_i's depend on some smaller number of parameters without specifying the values of these parameters. For example, with three categories the null hypothesis might state that $p_1 = \theta^2$, $p_2 = 2\theta(1 - \theta)$, and $p_3 = (1 - \theta)^2$. For a chi-squared test to be performed, the values of any unspecified parameters must be estimated from the sample data. These problems are discussed in Section 13.2. The methods are then applied to test a null hypothesis that the sample comes from a particular family of distributions, such as the Poisson family (with λ estimated from the sample) or the normal family (with μ and σ estimated).

Chi-squared tests for two different situations are presented in Section 13.3. In the first, the null hypothesis states that the p_i's are the same for several different populations. The second type of situation involves taking a sample from a single population and classifying each individual with respect to two different categorical factors (such as religious preference and political party registration). The null hypothesis in this situation is that the two factors are independent within the population.

13.1 Goodness-of-Fit Tests When Category Probabilities Are Completely Specified

A binomial experiment consists of a sequence of independent trials in which each trial can result in one of two possible outcomes, S (for success) and F (for failure). The probability of success, denoted by p, is assumed to be constant from trial to trial, and the number n of trials is fixed at the outset of the experiment. In Chapter 9, we presented a large-sample z test for testing H_0: $p = p_0$. Notice that this null hypothesis specifies both $P(S)$ and $P(F)$, since if $P(S) = p_0$, then $P(F) = 1 - p_0$. Denoting $P(F)$ by q and $1 - p_0$ by q_0, the null hypothesis can alternatively be written as H_0: $p = p_0$, $q = q_0$. The z test is two-tailed when the alternative of interest is $p \neq p_0$.

A **multinomial experiment** generalizes a binomial experiment by allowing each trial to result in one of k possible outcomes, where $k > 2$. For example, suppose a store accepts three different types of credit cards. A multinomial experiment would result from observing the type of credit card used—type 1, type 2, or type 3—by each of the next n customers who pay with a credit card. In general, we will refer to the k possible outcomes on any given trial as categories, and p_i will denote the probability that a trial results in category i. If the experiment consists of selecting n individuals or objects from a population and categorizing each one, then p_i is the proportion of the population falling in the ith category (such an experiment will be approximately multinomial provided that n is much smaller than the population size).

The null hypothesis of interest will specify the value of each p_i. For example, in the case $k = 3$, we might have H_0: $p_1 = .5$, $p_2 = .3$, $p_3 = .2$. The alternative hypothesis will state that H_0 is not true—that is, that at least one of the p_i's has a value different from that asserted by H_0 (in which case at least two must be different, since they sum to 1). The symbol p_{i0} will represent the value of p_i claimed by the null hypothesis. In the example just given, $p_{10} = .5$, $p_{20} = .3$, and $p_{30} = .2$.

Before the multinomial experiment is performed, the number of trials that will result in category i ($i = 1, 2, \ldots,$ or k) is a random variable—just as the number of successes and the number of failures in a binomial experiment are random variables. This random variable will be denoted by N_i and its observed value by n_i. Since each trial results in exactly one of the k categories, $\sum N_i = n$, and the same is true of the n_i's. As an example, an experiment with $n = 100$ and $k = 3$ might yield $N_1 = 46$, $N_2 = 35$, and $N_3 = 19$.

The expected number of successes and expected number of failures in a binomial experiment are np and nq, respectively. When H_0: $p = p_0$, $q = q_0$ is true, the expected numbers of successes and failures are np_0 and nq_0, respectively. Similarly, in a multinomial experiment the expected number of trials resulting in category i is $E(N_i) =$

np_i $(i = 1, \ldots, k)$. When H_0: $p_1 = p_{10}, \ldots, p_k = p_{k0}$ is true, these expected values become $E(N_1) = np_{10}$, $E(N_2) = np_{20}, \ldots, E(N_k) = np_{k0}$. For the case $k = 3$, H_0: $p_1 = .5$, $p_2 = .3$, $p_3 = .2$, and $n = 100$, $E(N_1) = 100(.5) = 50$, $E(N_2) = 30$, and $E(N_3) = 20$ when H_0 is true. The n_i's are often displayed in a tabular format consisting of a row of k cells, one for each category, as illustrated in Table 13.1. The expected values when H_0 is true are displayed just below the observed values. The N_i's and n_i's are usually referred to as *observed cell counts* (or *observed cell frequencies*), and $np_{10}, np_{20}, \ldots, np_{k0}$ are the corresponding *expected cell counts* under H_0.

Table 13.1 Observed and expected cell counts

Category:	$i = 1$	$i = 2$	...	$i = k$	Row Total
Observed	n_1	n_2	...	n_k	n
Expected	np_{10}	np_{20}	...	np_{k0}	n

The n_i's should all be reasonably close to the corresponding np_{i0}'s when H_0 is true. On the other hand, several of the observed counts should differ substantially from these expected counts when the actual values of the p_i's differ markedly from what the null hypothesis asserts. The test procedure involves assessing the discrepancy between the n_i's and the np_{i0}'s, with H_0 being rejected when the discrepancy is sufficiently large. It is natural to base a measure of discrepancy on the squared deviations $(n_1 - np_{10})^2$, $(n_2 - np_{20})^2, \ldots, (n_k - np_{k0})^2$. An obvious way to combine these into an overall measure is to add them together to obtain $\Sigma(n_i - np_{i0})^2$. However, suppose $np_{10} = 100$ and $np_{20} = 10$. Then if $n_1 = 95$ and $n_2 = 5$, the two categories contribute the same squared deviations to the proposed measure. Yet n_1 is only 5% less than what would be expected when H_0 is true, whereas n_2 is 50% less. To take relative magnitudes of the deviations into account, we will divide each squared deviation by the corresponding expected count and then combine.

Before giving a more detailed description, we must discuss the *chi-squared distribution*. This distribution was first introduced in Section 6.4 and was used in Chapter 8 to obtain a confidence interval for the variance σ^2 of a normal population. The chi-squared distribution has a single parameter, called the number of degrees of freedom (df) of the distribution, with possible values 1, 2, 3, ... Analogous to the critical value $t_{\alpha,\nu}$ for the t distribution, $\chi^2_{\alpha,\nu}$ is the value such that α of the area under the χ^2 curve with ν df lies to the right of $\chi^2_{\alpha,\nu}$ (see Figure 13.1). Selected values of $\chi^2_{\alpha,\nu}$ are given in Appendix Table A.7.

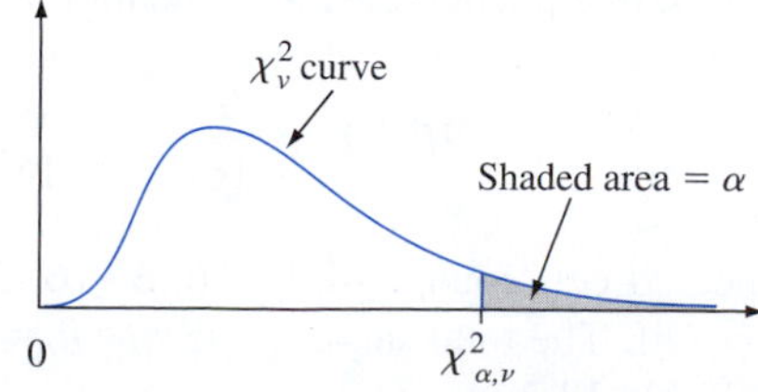

Figure 13.1 A critical value for a chi-squared distribution

THEOREM Provided that $np_i \geq 5$ for every i ($i = 1, 2, \ldots, k$), the variable

$$\chi^2 = \sum_{i=1}^{k} \frac{(N_i - np_i)^2}{np_i} = \sum_{\text{all cells}} \frac{(\text{observed} - \text{expected})^2}{\text{expected}}$$

has approximately a chi-squared distribution with $k - 1$ df.

The fact that df $= k - 1$ is a consequence of the restriction $\sum N_i = n$. Although there are k observed cell counts, once any $k - 1$ are known, the remaining one is uniquely determined. That is, there are only $k - 1$ "freely determined" cell counts, and thus $k - 1$ df.

If np_{i0} is substituted for np_i in χ^2, the resulting test statistic has a chi-squared distribution when H_0 is true. Rejection of H_0 is appropriate when $\chi^2 \geq c$ (because large discrepancies between observed and expected counts lead to a large value of χ^2), and the choice $c = \chi^2_{\alpha,k-1}$ yields a test with significance level α.

Null hypothesis: H_0: $p_1 = p_{10}, p_2 = p_{20}, \ldots, p_k = p_{k0}$

Alternative hypothesis: H_a: at least one p_i does not equal p_{i0}

Test statistic value: $$\chi^2 = \sum_{\text{all cells}} \frac{(\text{observed} - \text{expected})^2}{\text{expected}} = \sum_{i=1}^{k} \frac{(n_i - np_{i0})^2}{np_{i0}}$$

Rejection region: $\chi^2 \geq \chi^2_{\alpha,k-1}$

Example 13.1 If we focus on two different characteristics of an organism, each controlled by a single gene, and cross a pure strain having genotype AABB with a pure strain having genotype aabb (capital letters denoting dominant alleles and small letters recessive alleles), the resulting genotype will be AaBb. If these first-generation organisms are then crossed among themselves (a dihybrid cross), there will be four phenotypes depending on whether a dominant allele of either type is present. Mendel's laws of inheritance imply that these four phenotypes should have probabilities $\frac{9}{16}, \frac{3}{16}, \frac{3}{16}$, and $\frac{1}{16}$ of arising in any given dihybrid cross.

The article "Linkage Studies of the Tomato" (*Trans. Royal Canad. Institut.*, 1931: 1–19) reports the following data on phenotypes from a dihybrid cross of tall cut-leaf tomatoes with dwarf potato-leaf tomatoes. There are $k = 4$ categories corresponding to the four possible phenotypes, with the null hypothesis being

$$H_0\text{: } p_1 = \frac{9}{16}, p_2 = \frac{3}{16}, p_3 = \frac{3}{16}, p_4 = \frac{1}{16}$$

The expected cell counts are $9n/16$, $3n/16$, $3n/16$, and $n/16$, and the test is based on $k - 1 = 3$ df. The total sample size was $n = 1611$. Observed and expected counts are given in Table 13.2.

Table 13.2 Observed and expected cell counts for Example 13.1

	$i = 1$ Tall, Cut-Leaf	$i = 2$ Tall, Potato-Leaf	$i = 3$ Dwarf, Cut-Leaf	$i = 4$ Dwarf, Potato-Leaf
n_i	926	288	293	104
np_{i0}	906.2	302.1	302.1	100.7

The contribution to χ^2 from the first cell is

$$\frac{(n_1 - np_{10})^2}{np_{10}} = \frac{(926 - 906.2)^2}{906.2} = .433$$

Cells 2, 3, and 4 contribute .658, .274, and .108, respectively, so $\chi^2 = .433 + .658 + .274 + .108 = 1.473$. A test with significance level .10 requires $\chi^2_{.10,3}$, the number in the 3 df row and .10 column of Appendix Table A.7. This critical value is 6.251. Since 1.473 is not at least 6.251, H_0 cannot be rejected even at this rather large level of significance. The data is quite consistent with Mendel's laws. ■

Although we have developed the chi-squared test for situations in which $k > 2$, it can also be used when $k = 2$. The null hypothesis in this case can be stated as H_0: $p_1 = p_{10}$, since the relations $p_2 = 1 - p_1$ and $p_{20} = 1 - p_{10}$ make the inclusion of $p_2 = p_{20}$ in H_0 redundant. The alternative hypothesis is H_a: $p_1 \neq p_{10}$. These hypotheses can also be tested using a two-tailed z test with test statistic

$$Z = \frac{(N_1/n) - p_{10}}{\sqrt{\frac{p_{10}(1 - p_{10})}{n}}} = \frac{\hat{p}_1 - p_{10}}{\sqrt{\frac{p_{10}p_{20}}{n}}}$$

Surprisingly, the two test procedures are completely equivalent. This is because it can be shown that $Z^2 = \chi^2$ and $(z_{\alpha/2})^2 = \chi^2_{1,\alpha}$, so that $\chi^2 \geq \chi^2_{1,\alpha}$ if and only if (iff) $|Z| \geq z_{\alpha/2}$.* If the alternative hypothesis is either H_a: $p_1 > p_{10}$ or H_a: $p_1 < p_{10}$, the chi-squared test cannot be used. One must then revert to an upper- or lower-tailed z test.

As is the case with all test procedures, one must be careful not to confuse statistical significance with practical significance. A computed χ^2 that exceeds $\chi^2_{\alpha,k-1}$ may be a result of a very large sample size rather than any practical differences between the hypothesized p_{i0}'s and true p_i's. Thus if $p_{10} = p_{20} = p_{30} = \frac{1}{3}$, but the true p_i's have values .330, .340, and .330, a large value of χ^2 is sure to arise with a sufficiently large n. Before rejecting H_0, the $\hat{p}_i$'s should be examined to see whether they suggest a model different from that of H_0 from a practical point of view.

*The fact that $(z_{\alpha/2})^2 = \chi^2_{1,\alpha}$ is a consequence of the relationship between the standard normal distribution and the chi-squared distribution with 1 df; if $Z \sim N(0, 1)$, then Z^2 has a chi-squared distribution with $\nu = 1$. See the first proposition in Section 6.4.

P-Values for Chi-Squared Tests

The chi-squared tests in this chapter are all upper-tailed, so we focus on this case. Just as the *P*-value for an upper-tailed *t* test is the area under the t_ν curve to the right of the calculated *t*, the *P*-value for an upper-tailed chi-squared test is the area under the χ^2_ν curve to the right of the calculated χ^2. Appendix Table A.7 provides limited *P*-value information because only five upper-tail critical values are tabulated for each different ν. We have therefore included another appendix table, analogous to Table A.8, that facilitates making more precise *P*-value statements.

The fact that *t* curves were all centered at zero allowed us to tabulate *t*-curve tail areas in a relatively compact way, with the left margin giving values ranging from 0.0 to 4.0 on the horizontal *t* scale and various columns displaying corresponding upper-tail areas for various df's. The rightward movement of chi-squared curves as df increases necessitates a somewhat different type of tabulation. The left margin of Appendix Table A.11 displays various upper-tail areas: .100, .095, .090, . . . , .005, and .001. Each column of the table is for a different value of df, and the entries are values on the horizontal chi-squared axis that capture these corresponding tail areas. For example, moving down to tail area .085 and across to the 4 df column, we see that the area to the right of 8.18 under the 4 df chi-squared curve is .085 (see Figure 13.2).

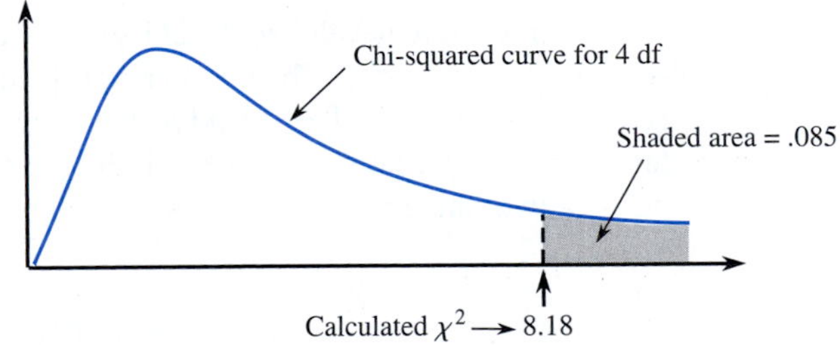

Figure 13.2 A *P*-value for an upper-tailed chi-squared test

To capture this same upper-tail area under the 10 df curve, we must go out to 16.54. In the 2 df column, the top row shows that if the calculated value of the chi-squared variable is smaller than 4.60, the captured tail area (the *P*-value) exceeds .10. Similarly, the bottom row in this column indicates that if the calculated value exceeds 13.81, the tail area is smaller than .001 (*P*-value $< .001$).

χ^2 When the p_i's Are Functions of Other Parameters

Frequently the p_i's are hypothesized to depend on a smaller number of parameters $\theta_1, \ldots, \theta_m$ $(m < k)$. Then a specific hypothesis involving the θ_i's yields specific p_{i0}'s, which are then used in the χ^2 test.

Example 13.2 In a well-known genetics article ("The Progeny in Generations F_{12} to F_{17} of a Cross Between a Yellow-Wrinkled and a Green-Round Seeded Pea," *J. Genetics*, 1923: 255–331), the early statistician G. U. Yule analyzed data resulting from crossing garden peas. The dominant alleles in the experiment were Y = yellow color and R = round shape, resulting in the double dominant YR. Yule examined 269 four-seed pods resulting from a dihybrid cross and counted the number of YR seeds in each pod. Letting X denote the number of YR's in a randomly selected pod, possible X values are 0, 1, 2, 3, 4, which we identify with cells 1, 2, 3, 4, and 5 of a rectangular table (so, for example, a pod with $X = 4$ yields an observed count in cell 5).

The hypothesis that the Mendelian laws are operative and that genotypes of individual seeds within a pod are independent of one another implies that X has a binomial distribution with $n = 4$ and $\theta = \frac{9}{16}$. We thus wish to test H_0: $p_1 = p_{10}, \ldots, p_5 = p_{50}$, where

$$\begin{aligned} p_{i0} &= P(i - 1 \text{ YR's among 4 seeds when } H_0 \text{ is true}) \\ &= \binom{4}{i-1}\theta^{i-1}(1-\theta)^{4-(i-1)} \qquad i = 1, 2, 3, 4, 5;\ \theta = \frac{9}{16} \end{aligned}$$

Yule's data and the computations are in Table 13.3 with expected cell counts $np_{i0} = 269p_{i0}$.

Table 13.3 Observed and expected cell counts for Example 13.2

Cell *i*: **YR peas/pod:**	**1** **0**	**2** **1**	**3** **2**	**4** **3**	**5** **4**
Observed	16	45	100	82	26
Expected	9.86	50.68	97.75	83.78	26.93
$\dfrac{(\text{observed} - \text{expected})^2}{\text{expected}}$	3.823	.637	.052	.038	.032

Thus $\chi^2 = 3.823 + \cdots + .032 = 4.582$. Since $\chi^2_{.01,k-1} = \chi^2_{.01,4} = 13.277$, H_0 is not rejected at level .01. Appendix Table A.11 shows that because $4.582 < 7.77$, the P-value for the test exceeds .10. H_0 should not be rejected at any reasonable significance level. ■

χ^2 When the Underlying Distribution Is Continuous

We have so far assumed that the k categories are naturally defined in the context of the experiment under consideration. The χ^2 test can also be used to test whether a sample comes from a specific underlying continuous distribution. Let X denote the variable being sampled and suppose the hypothesized pdf of X is $f_0(x)$. As in the construction of a frequency distribution in Chapter 1, subdivide the measurement scale of X into k intervals $[a_0, a_1), [a_1, a_2), \ldots, [a_{k-1}, a_k)$, where the interval $[a_{i-1}, a_i)$ includes the value a_{i-1} but not a_i. The cell probabilities specified by H_0 are then

$$p_{i0} = P(a_{i-1} \le X < a_i) = \int_{a_{i-1}}^{a_i} f_0(x)\,dx$$

The cells should be chosen so that $np_{i0} \ge 5$ for $i = 1, \ldots, k$. Often they are selected so that the np_{i0}'s are equal.

Example 13.3 To see whether the time of onset of labor among expectant mothers is uniformly distributed throughout a 24-hour day, we can divide a day into k periods, each of length $24/k$. The null hypothesis states that $f(x)$ is the uniform pdf on the interval [0, 24], so that $p_{i0} = 1/k$. The article "The Hour of Birth" (*British J. Preventive and Social Med.*, 1953: 43–59) reports on 1186 onset times, which were categorized into $k = 24$ 1-hour intervals beginning at midnight, resulting in cell counts of 52, 73, 89, 88, 68, 47, 58, 47, 48, 53, 47, 34, 21, 31, 40, 24, 37, 31, 47, 34, 36, 44, 78, and 59. Each expected cell count is $1186 \cdot \frac{1}{24} = 49.42$, and the resulting value of χ^2 is 162.77. Since $\chi^2_{.01,23} = 41.637$, the computed value is highly significant, and the null hypothesis is resoundingly rejected. Generally speaking, it appears that labor is much more likely to commence very late at night than during normal waking hours. ■

For testing whether a sample comes from a specific normal distribution, the fundamental parameters are $\theta_1 = \mu$ and $\theta_2 = \sigma$, and each p_{i0} will be a function of these parameters.

Example 13.4 The developers of a new standardized exam want it to satisfy the following criteria: (1) actual time taken to complete the test is normally distributed, (2) $\mu = 100$ min, and (3) exactly 90% of all students will finish within a 2-hour period. In the pilot testing of the standardized test, 120 students are given the test, and their completion times are recorded. For a chi-squared test of normally distributed completion time it is decided that $k = 8$ intervals should be used. The criteria imply that the 90th percentile of the completion time distribution is $\mu + 1.28\sigma = 2$ hours $= 120$ minutes. Since $\mu = 100$, this implies that $\sigma = 15.63$.

The eight intervals that divide the standard normal scale into eight equally likely segments are [0, .32), [.32, .675), [.675, 1.15), [1.15, ∞), and their four counterparts on the other side of 0. For $\mu = 100$ and $\sigma = 15.63$, these intervals become [100, 105), [105, 110.55), [110.55, 117.97), and [117.97, ∞). Thus $p_{i0} = \frac{1}{8} = .125$ ($i = 1, \ldots, 8$), so each expected cell count is $np_{i0} = 120(.125) = 15$. The observed cell counts were 21, 17, 12, 16, 10, 15, 19, and 10, resulting in a χ^2 of 7.73. Since $\chi^2_{.10,7} = 12.017$ and 7.73 is not ≥ 12.017, there is no evidence for concluding that the criteria have not been met. ■

Exercises Section 13.1 (1–11)

1. What conclusion would be appropriate for an upper-tailed chi-squared test in each of the following situations?

a. $\alpha = .05$, df $= 4$, $\chi^2 = 12.25$
b. $\alpha = .01$, df $= 3$, $\chi^2 = 8.54$
c. $\alpha = .10$, df $= 2$, $\chi^2 = 4.36$
d. $\alpha = .01$, $k = 6$, $\chi^2 = 10.20$

2. Say as much as you can about the P-value for an upper-tailed chi-squared test in each of the following situations:

a. $\chi^2 = 7.5$, df $= 2$ **b.** $\chi^2 = 13.0$, df $= 6$
c. $\chi^2 = 18.0$, df $= 9$ **d.** $\chi^2 = 21.3$, $k = 5$
e. $\chi^2 = 5.0$, $k = 4$

3. A statistics department at a large university maintains a tutoring center for students in its introductory service courses. The center has been staffed with the expectation that 40% of its clients would be from the business statistics course, 30% from engineering statistics, 20% from the statistics course for social science students, and the other 10% from the course for agriculture students. A random sample of $n = 120$ clients revealed 52, 38, 21, and 9 from the four courses. Does this data suggest that the percentages on which staffing was based are not correct? State and test the relevant hypotheses using $\alpha = .05$.

4. It is hypothesized that when homing pigeons are disoriented in a certain manner, they will exhibit no preference for any direction of flight after takeoff (so that the direction X should be uniformly distributed on the interval from 0° to 360°). To test this, 120 pigeons are disoriented, let loose, and the direction of flight of each is recorded; the resulting data follows. Use the chi-squared test at level .10 to see whether the data supports the hypothesis.

Direction	0–<45°	45–<90°	90–<135°
Frequency	12	16	17
Direction	135–<180°	180–<225°	225–<270°
Frequency	15	13	20
Direction	270–<315°	315–<360°	
Frequency	17	10	

5. An information retrieval system has ten storage locations. Information has been stored with the expectation that the long-run proportion of requests for location i is given by $p_i = (5.5 - |i - 5.5|)/30$. A sample of 200 retrieval requests gave the following frequencies for locations 1–10, respectively: 4, 15, 23, 25, 38, 31, 32, 14, 10, and 8. Use a chi-squared test at significance level .10 to decide whether the data is consistent with the a priori proportions (use the P-value approach).

6. Sorghum is an important cereal crop whose quality and appearance could be affected by the presence of pigments in the pericarp (the walls of the plant ovary). The article "A Genetic and Biochemical Study on Pericarp Pigments in a Cross Between Two Cultivars of Grain Sorghum, Sorghum Bicolor" (*Heredity*, 1976: 413–416) reports on an experiment that involved an initial cross between CK60 sorghum (an American variety with white seeds) and Abu Taima (an Ethiopian variety with yellow seeds) to produce plants with red seeds and then a self-cross of the red-seeded plants. According to genetic theory, this F_2 cross should produce plants with red, yellow, or white seeds in the ratio 9:3:4. The data from the experiment follows; does the data confirm or contradict the genetic theory? Test at level .05 using the P-value approach.

Seed Color	Red	Yellow	White
Observed Frequency	195	73	100

7. Criminologists have long debated whether there is a relationship between weather conditions and the incidence of violent crime. The author of the article "Is There a Season for Homicide?" (*Criminology*, 1988: 287–296) classified 1361 homicides according to season, resulting in the accompanying data. Test the null hypothesis of equal proportions using $\alpha = .01$ by using the chi-squared table to say as much as possible about the P-value.

Winter	Spring	Summer	Fall
328	334	372	327

8. The article "Psychiatric and Alcoholic Admissions Do Not Occur Disproportionately Close to Patients' Birthdays" (*Psych. Rep.*, 1992: 944–946) focuses on the existence of any relationship between date of patient admission for treatment of alcoholism and patient's birthday. Assuming a 365-day year (i.e., excluding leap year), in the absence of any relation, a patient's admission date is equally likely to be any one of the 365 possible days. The investigators established four different admission categories: (1) within 7 days of birthday, (2) between 8 and 30 days, inclusive, from the birthday, (3) between 31 and 90 days, inclusive, from the birthday, and (4) more than 90 days from the birthday. A sample of 200 patients gave observed frequencies of 11, 24, 69, and 96 for categories 1, 2, 3, and 4, respectively. State and test the relevant hypotheses using a significance level of .01.

9. The response time of a computer system to a request for a certain type of information is hypothesized to have an exponential distribution with parameter $\lambda = 1$ [so if X = response time, the pdf of X under H_0 is $f_0(x) = e^{-x}$ for $x \geq 0$].

a. If you had observed $X_1, X_2, \ldots, X_n$ and wanted to use the chi-squared test with five class intervals having equal probability under H_0, what would be the resulting class intervals?

b. Carry out the chi-squared test using the following data resulting from a random sample of 40 response times:

.10	.99	1.14	1.26	3.24	.12	.26	.80
.79	1.16	1.76	.41	.59	.27	2.22	.66
.71	2.21	.68	.43	.11	.46	.69	.38
.91	.55	.81	2.51	2.77	.16	1.11	.02
2.13	.19	1.21	1.13	2.93	2.14	.34	.44

10. a. Show that another expression for the chi-squared statistic is

$$\chi^2 = \sum_{i=1}^{k} \frac{N_i^2}{np_{i0}} - n$$

Why is it more efficient to compute χ^2 using this formula?

b. When the null hypothesis is H_0: $p_1 = p_2 = \cdots = p_k = 1/k$ (i.e., $p_{i0} = 1/k$ for all i), how does the formula of part (a) simplify? Use the simplified expression to calculate χ^2 for the pigeon/direction data in Exercise 4.

11. a. Having obtained a random sample from a population, you wish to use a chi-squared test to decide whether the population distribution is standard normal. If you base the test on six class intervals having equal probability under H_0, what should be the class intervals?

b. If you wish to use a chi-squared test to test H_0: the population distribution is normal with $\mu = .5$, $\sigma = .002$ and the test is to be based on six equiprobable (under H_0) class intervals, what should be these intervals?

c. Use the chi-squared test with the intervals of part (b) to decide, based on the following 45 bolt diameters, whether bolt diameter is a normally distributed variable with $\mu = .5$ in., $\sigma = .002$ in.

.4974	.4976	.4991	.5014	.5008	.4993
.4994	.5010	.4997	.4993	.5013	.5000
.5017	.4984	.4967	.5028	.4975	.5013
.4972	.5047	.5069	.4977	.4961	.4987
.4990	.4974	.5008	.5000	.4967	.4977
.4992	.5007	.4975	.4998	.5000	.5008
.5021	.4959	.5015	.5012	.5056	.4991
.5006	.4987	.4968			

13.2 *Goodness-of-Fit Tests for Composite Hypotheses

In the previous section, we presented a goodness-of-fit test based on a χ^2 statistic for deciding between H_0: $p_1 = p_{10}, \ldots, p_k = p_{k0}$ and the alternative H_a stating that H_0 is not true. The null hypothesis was a *simple* hypothesis in the sense that each p_{i0} was a specified number, so that the expected cell counts when H_0 was true were uniquely determined numbers.

In many situations, there are k naturally occurring categories, but H_0 states only that the p_i's are functions of other parameters $\theta_1, \ldots, \theta_m$ without specifying the values of these θ's. For example, a population may be in equilibrium with respect to proportions of the three genotypes AA, Aa, and aa. With p_1, p_2, and p_3 denoting these proportions (probabilities), one may wish to test

$$H_0\colon\ p_1 = \theta^2, p_2 = 2\theta(1 - \theta), p_3 = (1 - \theta)^2 \qquad (13.1)$$

where θ represents the proportion of gene A in the population. This hypothesis is **composite** because knowing that H_0 is true does not uniquely determine the cell probabilities and expected cell counts but only their general form. To carry out a χ^2 test, the unknown θ_i's must first be estimated.

Similarly, we may be interested in testing to see whether a sample came from a particular family of distributions without specifying any particular member of the family. To use the χ^2 test to see whether the distribution is Poisson, for example, the parameter λ must be estimated. In addition, because there are actually an infinite number of possible values of a Poisson variable, these values must be grouped so that there are a finite number of cells. If H_0 states that the underlying distribution is normal, use of a χ^2 test must be preceded by a choice of cells and estimation of μ and σ.

χ^2 When Parameters Are Estimated

As before, k will denote the number of categories or cells and p_i will denote the probability of an observation falling in the ith cell. The null hypothesis now states that each p_i is a function of a small number of parameters $\theta_1, \ldots, \theta_m$ with the θ_i's otherwise unspecified:

$$H_0\text{: } p_1 = \pi_1(\boldsymbol{\theta}), \ldots, p_k = \pi_k(\boldsymbol{\theta}) \qquad \text{where } \boldsymbol{\theta} = (\theta_1, \ldots, \theta_m) \tag{13.2}$$

H_a: the hypothesis H_0 is not true

For example, for H_0 of (13.1), $m = 1$ (there is only one θ), $\pi_1(\theta) = \theta^2$, $\pi_2(\theta) = 2\theta(1 - \theta)$, and $\pi_3(\theta) = (1 - \theta)^2$.

In the case $k = 2$, there is really only a single rv, N_1 (since $N_1 + N_2 = n$), which has a binomial distribution. The joint probability that $N_1 = n_1$ and $N_2 = n_2$ is then

$$P(N_1 = n_1, N_2 = n_2) = \binom{n}{n_1} p_1^{n_1} \cdot p_2^{n_2} \propto p_1^{n_1} \cdot p_2^{n_2}$$

where $p_1 + p_2 = 1$ and $n_1 + n_2 = n$. For general k, the joint distribution of $N_1, \ldots, N_k$ is the multinomial distribution (Section 5.1) with

$$P(N_1 = n_1, \ldots, N_k = n_k) \propto p_1^{n_1} \cdot p_2^{n_2} \cdot \cdots \cdot p_k^{n_k} \tag{13.3}$$

When H_0 is true, (13.3) becomes

$$P(N_1 = n_1, \ldots, N_k = n_k) \propto [\pi_1(\theta)]^{n_1} \cdot \cdots \cdot [\pi_k(\theta)]^{n_k} \tag{13.4}$$

To apply a chi-squared test, $\theta = (\theta_1, \ldots, \theta_m)$ must be estimated.

METHOD OF ESTIMATION

Let $n_1, n_2, \ldots, n_k$ denote the observed values of $N_1, \ldots, N_k$. Then $\hat{\theta}_1, \ldots, \hat{\theta}_m$ are those values of the θ_i's that maximize (13.4), that is, the maximum likelihood estimators (see Section 7.2).

Example 13.5 In humans there is a blood group, the MN group, that is composed of individuals having one of the three blood types M, MN, and N. Type is determined by two alleles, and there is no dominance, so the three possible genotypes give rise to three phenotypes. A population consisting of individuals in the MN group is in equilibrium if

$$\begin{aligned} P(\text{M}) &= p_1 = \theta^2 \\ P(\text{MN}) &= p_2 = 2\theta(1 - \theta) \\ P(\text{N}) &= p_3 = (1 - \theta)^2 \end{aligned}$$

for some θ. Suppose a sample from such a population yielded the results shown in Table 13.4.

Table 13.4 Observed counts for Example 13.5

Type:	**M**	**MN**	**M**	
Observed	125	225	150	$n = 500$

Then

$$[\pi_1(\theta)]^{n_1}[\pi_2(\theta)]^{n_2}[\pi_3(\theta)]^{n_3} = [(\theta^2)]^{n_1}[2\theta(1-\theta)]^{n_2}[(1-\theta)^2]^{n_3}$$
$$= 2^{n_2} \cdot \theta^{2n_1+n_2} \cdot (1-\theta)^{n_2+2n_3}$$

Maximizing this with respect to θ (or, equivalently, maximizing the natural logarithm of this quantity, which is easier to differentiate) yields

$$\hat{\theta} = \frac{2n_1 + n_2}{[(2n_1 + n_2) + (n_2 + 2n_3)]} = \frac{2n_1 + n_2}{2n}$$

With $n_1 = 125$ and $n_2 = 225$, $\hat{\theta} = 475/1000 = .475$. ■

Once $\boldsymbol{\theta} = (\theta_1, \ldots, \theta_m)$ has been estimated by $\hat{\boldsymbol{\theta}} = (\hat{\theta}_1, \ldots, \hat{\theta}_m)$, the estimated expected cell counts are the $n\pi_i(\hat{\boldsymbol{\theta}})$'s. These are now used in place of the np_{i0}'s of Section 13.1 to specify a χ^2 statistic.

THEOREM Under general "regularity" conditions on $\theta_1, \ldots, \theta_m$ and the $\pi_i(\boldsymbol{\theta})$'s, if $\theta_1, \ldots, \theta_m$ are estimated by the method of maximum likelihood as described previously and n is large,

$$\chi^2 = \sum_{\text{all cells}} \frac{(\text{observed} - \text{estimated expected})^2}{\text{estimated expected}} = \sum_{i=1}^{k} \frac{[N_i - n\pi_i(\hat{\boldsymbol{\theta}})]^2}{n\pi_i(\hat{\boldsymbol{\theta}})}$$

has approximately a chi-squared distribution with $k - 1 - m$ df when H_0 of (13.2) is true. An approximately level α test of H_0 versus H_a is then to reject H_0 if $\chi^2 \geq \chi^2_{\alpha,k-1-m}$. In practice, the test can be used if $n\pi_i(\hat{\theta}) \geq 5$ for every i.

Notice that *the number of degrees of freedom is reduced by the number of θ_i's estimated.*

Example 13.6 (Example 13.5 continued)

With $\hat{\theta} = .475$ and $n = 500$, the estimated expected cell counts are $n\pi_1(\hat{\theta}) = 500(\hat{\theta})^2 = 112.81$, $n\pi_2(\hat{\theta}) = (500)(2)(.475)(1 - .475) = 249.38$, and $n\pi_3(\hat{\theta}) = 500 - 112.81 - 249.38 = 137.81$. Then

$$\chi^2 = \frac{(125 - 112.81)^2}{112.81} + \frac{(225 - 249.38)^2}{249.38} + \frac{(150 - 137.81)^2}{137.81} = 4.78$$

Since $\chi^2_{.05,k-1-m} = \chi^2_{.05,3-1-1} = \chi^2_{.05,1} = 3.843$ and $4.78 \geq 3.843$, H_0 is rejected. Appendix Table A.11 shows that P-value $\approx .029$. ■

Example 13.7 Consider a series of games between two teams, I and II, that terminates as soon as one team has won four games (with no possibility of a tie). A simple probability model for such a series assumes that outcomes of successive games are independent and that the probability of team I winning any particular game is a constant θ. We arbitrarily designate I the better team, so that $\theta \geq .5$. Any particular series can then terminate after 4, 5, 6, or 7 games. Let $\pi_1(\theta)$, $\pi_2(\theta)$, $\pi_3(\theta)$, $\pi_4(\theta)$ denote the probability of termination in 4, 5, 6, and 7 games, respectively. Then

$$\begin{aligned}
\pi_1(\theta) &= P(\text{I wins in 4 games}) + P(\text{II wins in 4 games}) \\
&= \theta^4 + (1-\theta)^4 \\
\pi_2(\theta) &= P(\text{I wins 3 of the first 4 and the fifth}) \\
&\quad + P(\text{I loses 3 of the first 4 and the fifth}) \\
&= \binom{4}{3}\theta^3(1-\theta)\cdot\theta + \binom{4}{1}\theta(1-\theta)^3\cdot(1-\theta) \\
&= 4\theta(1-\theta)[\theta^3 + (1-\theta)^3] \\
\pi_3(\theta) &= 10\theta^2(1-\theta)^2[\theta^2 + (1-\theta)^2] \\
\pi_4(\theta) &= 20\theta^3(1-\theta)^3
\end{aligned}$$

The article "Seven-Game Series in Sports" by Groeneveld and Meeden (*Math. Mag.*, 1975: 187–192) tested the fit of this model to results of National Hockey League playoffs during the period 1943–1967 (when league membership was stable). The data appears in Table 13.5.

Table 13.5 Observed and expected counts for the simple model

Cell:	**1**	**2**	**3**	**4**	
Number of games played:	**4**	**5**	**6**	**7**	
Observed Frequency	15	26	24	18	$n = 83$
Estimated Expected Frequency	16.351	24.153	23.240	19.256	

The estimated expected cell counts are $83\pi_i(\hat{\theta})$, where $\hat{\theta}$ is the value of θ that maximizes

$$\begin{aligned}
&\{\theta^4 + (1-\theta)^4\}^{15}\cdot\{4\theta(1-\theta)[\theta^3 + (1-\theta)^3]\}^{26} \\
&\quad\cdot\{10\theta^2(1-\theta)^2[\theta^2 + (1-\theta)^2]\}^{24}\cdot\{20\theta^3(1-\theta)^3\}^{18}
\end{aligned} \tag{13.5}$$

Standard calculus methods fail to yield a nice formula for the maximizing value $\hat{\theta}$, so it must be computed using numerical methods. The result is $\hat{\theta} = .654$, from which $\pi_i(\hat{\theta})$ and the estimated expected cell counts are computed. The computed value of χ^2 is .360, and (since $k - 1 - m = 4 - 1 - 1 = 2$) $\chi^2_{.10,2} = 4.605$. There is thus no reason to reject the simple model as applied to NHL playoff series.

The cited article also considered World Series data for the period 1903–1973. For the simple model, $\chi^2 = 5.97$, so the model does not seem appropriate. The suggested reason for this is that for the simple model

$$P(\text{series lasts six games} \mid \text{series lasts at least six games}) \geq .5 \tag{13.6}$$

whereas of the 38 series that actually lasted at least six games, only 13 lasted exactly six. The following alternative model is then introduced:

$$\pi_1(\theta_1, \theta_2) = \theta_1^4 + (1 - \theta_1)^4$$
$$\pi_2(\theta_1, \theta_2) = 4\theta_1(1 - \theta_1)[\theta_1^3 + (1 - \theta_1)^3]$$
$$\pi_3(\theta_1, \theta_2) = 10\theta_1^2(1 - \theta_1)^2\theta_2$$
$$\pi_4(\theta_1, \theta_2) = 10\theta_1^2(1 - \theta_1)^2(1 - \theta_2)$$

The first two π_i's are identical to the simple model, whereas θ_2 is the conditional probability of (13.6) (which can now be any number between 0 and 1). The values of $\hat{\theta}_1$ and $\hat{\theta}_2$ that maximize the expression analogous to Expression (13.5) are determined numerically as $\hat{\theta}_1 = .614$, $\hat{\theta}_2 = .342$. A summary appears in Table 13.6, and $\chi^2 = .384$. Since two parameters are estimated, df $= k - 1 - m = 1$ with $\chi^2_{.10,1} = 2.706$, indicating a good fit of the data to this new model.

Table 13.6 Observed and expected counts for the more complex model

Number of games played:	**4**	**5**	**6**	**7**
Observed Frequency	12	16	13	25
Estimated Expected Frequency	10.85	18.08	12.68	24.39

■

One of the regularity conditions on the θ_i's in the theorem is that they be functionally independent of one another. That is, no single θ_i can be determined from the values of other θ_i's, so that m is the number of functionally independent parameters estimated. A general rule of thumb for degrees of freedom in a chi-squared test is the following.

$$\chi^2 \text{ df} = \begin{pmatrix}\text{number of freely} \\ \text{determined cell counts}\end{pmatrix} - \begin{pmatrix}\text{number of independent} \\ \text{parameters estimated}\end{pmatrix}$$

This rule will be used in connection with several different chi-squared tests in the next section.

Goodness of Fit for Discrete Distributions

Many experiments involve observing a random sample $X_1, X_2, \ldots, X_n$ from some discrete distribution. One may then wish to investigate whether the underlying distribution is a member of a particular family, such as the Poisson or negative binomial family. In the case of both a Poisson and a negative binomial distribution, the set of possible values is infinite, so the values must be grouped into k subsets before a chi-squared test can be used. The groupings should be done so that the expected frequency in each cell (group) is at least 5. The last cell will then correspond to X values of $c, c + 1, c + 2, \ldots$ for some value c.

This grouping can considerably complicate the computation of the $\hat{\theta}_i$'s and estimated expected cell counts. This is because the theorem requires that the $\hat{\theta}_i$'s be obtained from the cell counts $N_1, \ldots, N_k$ rather than the sample values $X_1, \ldots, X_n$.

Example 13.8 Table 13.7 presents count data on the number of *Larrea divaricata* plants found in each of 48 sampling quadrats, as reported in the article "Some Sampling Characteristics of Plants and Arthropods of the Arizona Desert" (*Ecology*, 1962: 567–571).

Table 13.7 Observed counts for Example 13.8

Cell:	**1**	**2**	**3**	**4**	**5**
Number of plants:	**0**	**1**	**2**	**3**	**≥4**
Frequency	9	9	10	14	6

The author fit a Poisson distribution to the data. Let λ denote the Poisson parameter and suppose for the moment that the six counts in cell 5 were actually 4, 4, 5, 5, 6, 6. Then denoting sample values by $x_1, \ldots, x_{48}$, nine of the x_i's were 0, nine were 1, and so on. The likelihood of the observed sample is

$$\frac{e^{-\lambda}\lambda^{x_1}}{x_1!} \cdot \cdots \cdot \frac{e^{-\lambda}\lambda^{x_{48}}}{x_{48}!} = \frac{e^{-48\lambda}\lambda^{\Sigma x_i}}{x_1! \cdot \cdots \cdot x_{48}!} = \frac{e^{-48\lambda}\lambda^{101}}{x_1! \cdot \cdots \cdot x_{48}!}$$

The value of λ for which this is maximized is $\hat{\lambda} = \Sigma x_i/n = 101/48 = 2.10$ (the value reported in the article).

However, the $\hat{\lambda}$ required for χ^2 is obtained by maximizing Expression (13.4) rather than the likelihood of the full sample. The cell probabilities are

$$\pi_i(\lambda) = \frac{e^{-\lambda}\lambda^{i-1}}{(i-1)!} \qquad i = 1, 2, 3, 4$$

$$\pi_5(\lambda) = 1 - \sum_{i=0}^{3} \frac{e^{-\lambda}\lambda^i}{i!}$$

so the right-hand side of (13.4) becomes

$$\left[\frac{e^{-\lambda}\lambda^0}{0!}\right]^9 \left[\frac{e^{-\lambda}\lambda^1}{1!}\right]^9 \left[\frac{e^{-\lambda}\lambda^2}{2!}\right]^{10} \left[\frac{e^{-\lambda}\lambda^3}{3!}\right]^{14} \left[1 - \sum_{i=0}^{3} \frac{e^{-\lambda}\lambda^i}{i!}\right]^6 \qquad (13.7)$$

There is no nice formula for $\hat{\lambda}$, the maximizing value of λ in this latter expression, so it must be obtained numerically. ■

Because the parameter estimates are usually much more difficult to compute from the grouped data than from the full sample, they are often computed using this latter method. When these "full" estimators are used in the chi-squared statistic, the distribution of the statistic is altered and a level α test is no longer specified by the critical value $\chi^2_{\alpha,k-1-m}$.

THEOREM Let $\hat{\theta}_1, \ldots, \hat{\theta}_m$ be the maximum likelihood estimators of $\theta_1, \ldots, \theta_m$ based on the full sample $X_1, \ldots, X_n$, and let χ^2 denote the statistic based on these estimators. Then the critical value c_α that specifies a level α upper-tailed test satisfies

$$\chi^2_{\alpha,k-1-m} \le c_\alpha \le \chi^2_{\alpha,k-1} \tag{13.8}$$

The test procedure implied by this theorem is the following:

If $\chi^2 \ge \chi^2_{\alpha,k-1}$, reject H_0.
If $\chi^2 \le \chi^2_{\alpha,k-1-m}$, do not reject H_0. (13.9)
If $\chi^2_{\alpha,k-1-m} < \chi^2 < \chi^2_{\alpha,k-1}$, withhold judgment.

Example 13.9 (Example 13.8 continued)

Using $\hat{\lambda} = 2.10$, the estimated expected cell counts are computed from $n\pi_i(\hat{\lambda})$, where $n = 48$. For example,

$$n\pi_1(\hat{\lambda}) = 48 \cdot \frac{e^{-2.1}(2.1)^0}{0!} = (48)(e^{-2.1}) = 5.88$$

Similarly, $n\pi_2(\hat{\lambda}) = 12.34$, $n\pi_3(\hat{\lambda}) = 12.96$, $n\pi_4(\hat{\lambda}) = 9.07$, and $n\pi_5(\hat{\lambda}) = 48 - 5.88 - \cdots - 9.07 = 7.75$. Then

$$\chi^2 = \frac{(9 - 5.88)^2}{5.88} + \cdots + \frac{(6 - 7.75)^2}{7.75} = 6.31$$

Since $m = 1$ and $k = 5$, at level .05 we need $\chi^2_{.05,3} = 7.815$ and $\chi^2_{.05,4} = 9.488$. Because $6.31 \le 7.815$, we do not reject H_0; at the 5% level, the Poisson distribution provides a reasonable fit to the data. Notice that $\chi^2_{.10,3} = 6.251$ and $\chi^2_{.10,4} = 7.779$, so at level .10 we would have to withhold judgment on whether the Poisson distribution was appropriate.

For comparison we can with a little additional effort maximize Expression (13.7). Use of a graphing calculator gives $\hat{\lambda} = 2.047$. Because this differs very little from 2.10, there is little change in the results. Using 2.047, we get the estimated expected cell counts 6.197, 12.687, 12.985, 8.860, and 7.271, and the resulting value of χ^2 is 6.230. Comparing this with $\chi^2_{.05,3} = 7.815$, we do not reject the Poisson null hypothesis at the .05 level. Because 6.230 does not quite exceed $\chi^2_{.10,3} = 6.251$, we also do not reject the null hypothesis at the 10% level. ■

Sometimes even the maximum likelihood estimates based on the full sample are quite difficult to compute. This is the case, for example, for the two-parameter (generalized) negative binomial distribution. In such situations, method-of-moments estimates are often used and the resulting χ^2 compared to $\chi^2_{\alpha,k-1-m}$, though it is not known to what extent the use of moments estimators affects the true critical value.

Goodness of Fit for Continuous Distributions

The chi-squared test can also be used to test whether the sample comes from a specified family of continuous distributions, such as the exponential family or the normal family. The choice of cells (class intervals) is even more arbitrary in the continuous case than in the discrete case. To ensure that the chi-squared test is valid, the cells should be chosen independently of the sample observations. Once the cells are chosen, it is almost always quite difficult to estimate unspecified parameters (such as μ and σ in the normal case) from the observed cell counts, so instead mle's based on the full sample are computed. The critical value c_α again satisfies (13.8), and the test procedure is given by (13.9).

Example 13.10 The Institute of Nutrition of Central America and Panama (INCAP) has carried out extensive dietary studies and research projects in Central America. In one study reported in the November 1964 issue of the *American Journal of Clinical Nutrition* ("The Blood Viscosity of Various Socioeconomic Groups in Guatemala"), serum total cholesterol measurements for a sample of 49 low-income rural Indians were reported as follows (in mg/L):

204	108	140	152	158	129	175	146	157	174	192	194	144
152	135	223	145	231	115	131	129	142	114	173	226	155
166	220	180	172	143	148	171	143	124	158	144	108	189
136	136	197	131	95	139	181	165	142	162			

Is it plausible that serum cholesterol level is normally distributed for this population? Suppose that prior to sampling, it was believed that plausible values for μ and σ were 150 and 30, respectively. The seven equiprobable class intervals for the standard normal distribution are $(-\infty, -1.07)$, $(-1.07, -.57)$, $(-.57, -.18)$, $(-.18, .18)$, $(.18, .57)$, $(.57, 1.07)$, and $(1.07, \infty)$, with each endpoint also giving the distance in standard deviations from the mean for any other normal distribution. For $\mu = 150$ and $\sigma = 30$, these intervals become $(-\infty, 117.9)$, $(117.9, 132.9)$, $(132.9, 144.6)$, $(144.6, 155.4)$, $(155.4, 167.1)$, $(167.1, 182.1)$, and $(182.1, \infty)$.

To obtain the estimated cell probabilities $\pi_1(\hat{\mu}, \hat{\sigma}), \ldots, \pi_7(\hat{\mu}, \hat{\sigma})$, we first need the mle's $\hat{\mu}$ and $\hat{\sigma}$. In Chapter 7, the mle of σ was shown to be $[\Sigma(x_i - \bar{x})^2/n]^{1/2}$ (rather than s), so with $s = 31.75$,

$$\hat{\mu} = \bar{x} = 157.02 \qquad \hat{\sigma} = \left[\frac{\sum (x_i - \bar{x})^2}{n}\right]^{1/2} = \left[\frac{(n-1)s^2}{n}\right]^{1/2} = 31.42$$

Each $\pi_i(\hat{\mu}, \hat{\sigma})$ is then the probability that a normal rv X with mean 157.02 and standard deviation 31.42 falls in the ith class interval. For example,

$$\pi_2(\hat{\mu}, \hat{\sigma}) = P(117.9 \le X \le 132.9) = P(-1.25 \le Z \le -.77) = .1150$$

so $n\pi_2(\hat{\mu}, \hat{\sigma}) = 49(.1150) = 5.64$. Observed and estimated expected cell counts are shown in Table 13.8.

The computed χ^2 is 4.60. With $k = 7$ cells and $m = 2$ parameters estimated, $\chi^2_{.05,k-1} = \chi^2_{.05,6} = 12.592$ and $\chi^2_{.05,k-1-m} = \chi^2_{.05,4} = 9.488$. Since $4.60 \le 9.488$, a normal distribution provides quite a good fit to the data.

Table 13.8 Observed and expected counts for Example 13.10

Cell:	$(-\infty, 117.9)$	$(117.9, 132.9)$	$(132.9, 144.6)$	$(144.6, 155.4)$
Observed	5	5	11	6
Estimated Expected	5.17	5.64	6.08	6.64

Cell:	$(155.4, 167.1)$	$(167.1, 182.1)$	$(182.1, \infty)$
Observed:	6	7	9
Estimated Expected	7.12	7.97	10.38

■

Example 13.11 The article "Some Studies on Tuft Weight Distribution in the Opening Room" (*Textile Res. J.*, 1976: 567–573) reports the accompanying data on the distribution of output tuft weight X (mg) of cotton fibers for the input weight $x_0 = 70$.

Interval:	0–8	8–16	16–24	24–32	32–40	40–48	48–56	56–64	64–70
Observed Frequency	20	8	7	1	2	1	0	1	0
Expected Frequency	18.0	9.9	5.5	3.0	1.8	.9	.5	.3	.1

The authors postulated a truncated exponential distribution:

$$H_0\colon f(x) = \frac{\lambda e^{-\lambda x}}{1 - e^{-\lambda x_0}} \qquad 0 \le x \le x_0$$

The mean of this distribution is

$$\mu = \int_0^{x_0} x f(x)\, dx = \frac{1}{\lambda} - \frac{x_0 e^{-\lambda x_0}}{1 - e^{-\lambda x_0}}$$

The parameter λ was estimated by replacing μ by $\bar{x} = 13.086$ and solving the resulting equation to obtain $\hat{\lambda} = .0742$ (so $\hat{\lambda}$ is a method-of-moments estimate and not an mle). Then with $\hat{\lambda}$ replacing λ in $f(x)$, the estimated expected cell frequencies as displayed previously are computed as

$$40\pi_i(\hat{\lambda}) = 40P(a_{i-1} \le X < a_i) = 40\int_{a_{i-1}}^{a_i} f(x)\, dx = \frac{40(e^{-\hat{\lambda}a_{i-1}} - e^{-\hat{\lambda}a_i})}{1 - e^{-\hat{\lambda}x_0}}$$

where $[a_{i-1}, a_i)$ is the ith class interval. To obtain expected cell counts of at least 5, the last six cells are combined to yield observed counts 20, 8, 7, 5 and expected counts of

18.0, 9.9, 5.5, 6.6. The computed value of chi-squared is then $\chi^2 = 1.34$. Because $\chi^2_{.05,2} = 5.992$, H_0 is not rejected, so the truncated exponential model provides a good fit. ■

A Special Test for Normality

Probability plots were introduced in Section 4.6 as an informal method for assessing the plausibility of any specified population distribution as the one from which the given sample was selected. The straighter the probability plot, the more plausible is the distribution on which the plot is based. A normal probability plot is used for checking whether *any* member of the normal distribution family is plausible. Let's denote the sample x_i's when ordered from smallest to largest by $x_{(1)}, x_{(2)}, \ldots, x_{(n)}$. Then the strategy suggested for checking normality was to plot the n points $(x_{(i)}, y_i)$, where $y_i = \Phi^{-1}[(i - .5)/n)]$.

A quantitative measure of the extent to which points cluster about a straight line is the sample correlation coefficient r introduced in Chapter 12. Consider calculating r for the n pairs $(x_{(1)}, y_1), \ldots, (x_{(n)}, y_n)$. The y_i's here are not observed values in a random sample from a y population, so properties of this r are quite different from those described in Section 12.5. However, it is true that the more r deviates from 1, the less the probability plot resembles a straight line (remember that a probability plot must slope upward). This idea can be extended to yield a formal test procedure: Reject the hypothesis of population normality if $r \le c_\alpha$, where c_α is a critical value chosen to yield the desired significance level α. That is, the critical value is chosen so that when the population distribution is actually normal, the probability of obtaining an r value that is at most c_α (and thus incorrectly rejecting H_0) is the desired α. The developers of the MINITAB statistical computer package give critical values for $\alpha = .10$, .05, and .01 in combination with different sample sizes. Because no theory exists for the distribution of r for a normal plot, the critical values are determined by computer simulation. These critical values are based on a slightly different definition of the y_i's than that given previously. The new values give slightly better approximations to the expected values of the ordered normal observations.

MINITAB will also construct a normal probability plot based on these y_i's. The plot will be almost identical in appearance to that based on the previous y_i's. When there are several tied $x_{(i)}$'s, MINITAB computes r by using the average of the corresponding y_i's as the second number in each pair.

Let $y_i = \Phi^{-1}[(i - .375)/(n + .25)]$ and compute the sample correlation coefficient r for the n pairs $(x_{(1)}, y_1), \ldots, (x_{(n)}, y_n)$. The Ryan–Joiner test of

H_0: the population distribution is normal

versus

H_a: the population distribution is not normal

consists of rejecting H_0 when $r \le c_\alpha$. Critical values c_α are given in Appendix Table A.12 for various significance levels α and sample sizes n.

Example 13.12 The following sample of $n = 20$ observations on dielectric breakdown voltage of a piece of epoxy resin first appeared in Example 4.35.

y_i	−1.871	−1.404	−1.127	−.917	−.742	−.587	−.446	−.313	−.186	−.062
$x_{(i)}$	24.46	25.61	26.25	26.42	26.66	27.15	27.31	27.54	27.74	27.94

y_i	.062	.186	.313	.446	.587	.742	.917	1.127	1.404	1.871
$x_{(i)}$	27.98	28.04	28.28	28.49	28.50	28.87	29.11	29.13	29.50	30.88

We asked MINITAB to carry out the Ryan–Joiner test, and the result appears in Figure 13.3. The test statistic value is $r = .9881$, and Appendix Table A.12 gives .9600 as the critical value that captures lower-tail area .10 under the r sampling distribution curve when $n = 20$ and the underlying distribution is actually normal. Since $.9881 > .9600$, the null hypothesis of normality cannot be rejected even for a significance level as large as .10.

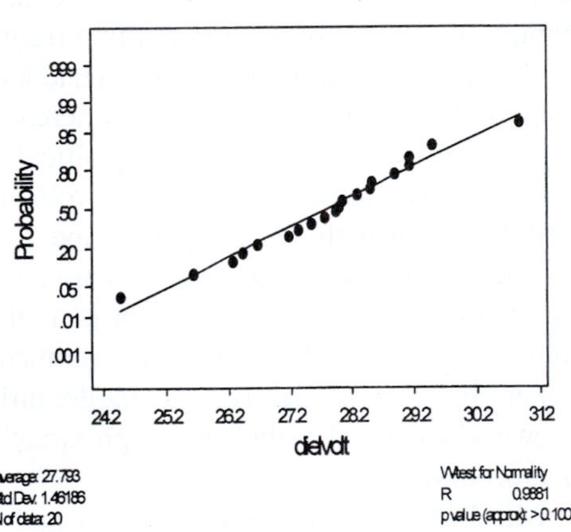

Figure 13.3 MINITAB output from the Ryan–Joiner test for the data of Example 13.12 ■

Exercises Section 13.2 (12–22)

12. Consider a large population of families in which each family has exactly three children. If the genders of the three children in any family are independent of one another, the number of male children in a randomly selected family will have a binomial distribution based on three trials.

a. Suppose a random sample of 160 families yields the following results. Test the relevant hypotheses by proceeding as in Example 13.5.

Number of Male Children	0	1	2	3
Frequency	14	66	64	16

b. Suppose a random sample of families in a nonhuman population resulted in observed frequencies of 15, 20, 12, and 3, respectively. Would the chi-squared test be based on the same number of degrees of freedom as the test in part (a)? Explain.

13. A study of sterility in the fruit fly ("Hybrid Dysgenesis in *Drosophila melanogaster:* The Biology of Female and Male Sterility," *Genetics*, 1979: 161–174) reports the following data on the number of ovaries developed for each female fly in a sample of size 1388. One model for unilateral sterility states that each ovary develops with some probability p independently of the other ovary. Test the fit of this model using χ^2.

x = Number of Ovaries Developed	0	1	2
Observed Count	1212	118	58

14. The article "Feeding Ecology of the Red-Eyed Vireo and Associated Foliage-Gleaning Birds" (*Ecological Monographs*, 1971: 129–152) presents the accompanying data on the variable X = the number of hops before the first flight and preceded by a flight. The author then proposed and fit a geometric probability distribution [$p(x) = P(X = x) = p^{x-1} \cdot q$ for $x = 1, 2, \ldots$, where $q = 1 - p$] to the data. The total sample size was $n = 130$.

x	1	2	3	4	5	6	7	8	9	10	11	12
Number of Times x Observed	48	31	20	9	6	5	4	2	1	1	2	1

a. The likelihood is $(p^{x_1-1} \cdot q) \cdot \cdots \cdot (p^{x_n-1} \cdot q) = p^{\Sigma x_i - n} \cdot q^n$. Show that the mle of p is given by $\hat{p} = (\Sigma x_i - n)/\Sigma x_i$, and compute $\hat{p}$ for the given data.

b. Estimate the expected cell counts using $\hat{p}$ of part (a) [expected cell counts $= n \cdot (\hat{p})^{x-1} \cdot \hat{q}$ for $x = 1, 2, \ldots$], and test the fit of the model using a χ^2 test by combining the counts for $x = 7, 8, \ldots$, and 12 into one cell ($x \geq 7$).

15. A certain type of flashlight is sold with the four batteries included. A random sample of 150 flashlights is obtained, and the number of defective batteries in each is determined, resulting in the following data:

Number Defective	0	1	2	3	4
Frequency	26	51	47	16	10

Let X be the number of defective batteries in a randomly selected flashlight. Test the null hypothesis that the distribution of X is Bin(4, θ). That is, with $p_i = P(i \text{ defectives})$, test

$$H_0\text{: } p_i = \binom{4}{i}\theta^i(1 - \theta)^{4-i} \qquad i = 0, 1, 2, 3, 4$$

[*Hint*: To obtain the mle of θ, write the likelihood (the function to be maximized) as $\theta^u(1 - \theta)^v$, where the exponents u and v are linear functions of the cell counts. Then take the natural log, differentiate with respect to θ, equate the result to 0, and solve for $\hat{\theta}$.]

16. In a genetics experiment, investigators looked at 300 chromosomes of a particular type and counted the number of sister-chromatid exchanges on each ("On the Nature of Sister-Chromatid Exchanges in 5-Bromodeoxyuridine-Substituted Chromosomes," *Genetics*, 1979: 1251–1264). A Poisson model was hypothesized for the distribution of the number of exchanges. Test the fit of a Poisson distribution to the data by first estimating λ and then combining the counts for $x = 8$ and $x = 9$ into one cell.

x = Number of Exchanges	0	1	2	3	4	5	6	7	8	9
Observed Counts	6	24	42	59	62	44	41	14	6	2

17. An article in *Annals of Mathematical Statistics* reports the following data on the number of borers in each of 120 groups of borers. Does the Poisson pmf provide a plausible model for the distribution of the number of borers in a group? (*Hint*: Add the frequencies for 7, 8, . . . , 12 to establish a single category "≥7.")

Number of Borers	0	1	2	3	4	5	6	7	8	9	10	11	12
Frequency	24	16	16	18	15	9	6	5	3	4	3	0	1

18. The article "A Probabilistic Analysis of Dissolved Oxygen–Biochemical Oxygen Demand Relationship in Streams" (*J. Water Resources Control Fed.*, 1969: 73–90) reports data on the rate of oxygenation in streams at 20°C in a certain region. The sample mean and standard deviation were computed as $\bar{x} = .173$ and $s = .066$, respectively. Based on the accompanying frequency distribution, can it be concluded that oxygenation rate is a normally distributed variable? Use the chi-squared test with $\alpha = .05$.

Rate (per day)	Frequency
Below .100	12
.100–below .150	20
.150–below .200	23
.200–below .250	15
.250 or more	13

19. Each headlight on an automobile undergoing an annual vehicle inspection can be focused either too high (H), too low (L), or properly (N). Checking the two headlights simultaneously (and not distinguishing between left and right) results in the six possible outcomes HH, LL, NN, HL, HN, and LN. If the probabilities (population proportions) for the single headlight focus direction are $P(H) = \theta_1$, $P(L) = \theta_2$, and $P(N) = 1 - \theta_1 - \theta_2$ and the two headlights are focused independently of one another, the probabilities of the six outcomes for a randomly selected car are the following:

$$p_1 = \theta_1^2 \quad p_2 = \theta_2^2 \quad p_3 = (1 - \theta_1 - \theta_2)^2$$
$$p_4 = 2\theta_1\theta_2 \quad p_5 = 2\theta_1(1 - \theta_1 - \theta_2)$$
$$p_6 = 2\theta_2(1 - \theta_1 - \theta_2)$$

Use the accompanying data to test the null hypothesis

$$H_0\text{: } p_1 = \pi_1(\theta_1, \theta_2), \ldots, p_6 = \pi_6(\theta_1, \theta_2)$$

where the $\pi_i(\theta_1, \theta_2)$'s are given previously.

Outcome	*HH*	*LL*	*NN*	*HL*	*HN*	*LN*
Frequency	49	26	14	20	53	38

(*Hint*: Write the likelihood as a function of θ_1 and θ_2, take the natural log, then compute $\partial/\partial\theta_1$ and $\partial/\partial\theta_2$, equate them to 0, and solve for $\hat{\theta}_1, \hat{\theta}_2$.)

20. The article "Compatibility of Outer and Fusible Interlining Fabrics in Tailored Garments (*Textile Res. J.*, 1997: 137–142) gave the following observations on bending rigidity (μN · m) for medium-quality fabric specimens, from which the accompanying MINITAB output was obtained:

24.6	12.7	14.4	30.6	16.1	9.5	31.5
17.2	46.9	68.3	30.8	116.7	39.5	73.8
80.6	20.3	25.8	30.9	39.2	36.8	46.6
15.6	32.3					

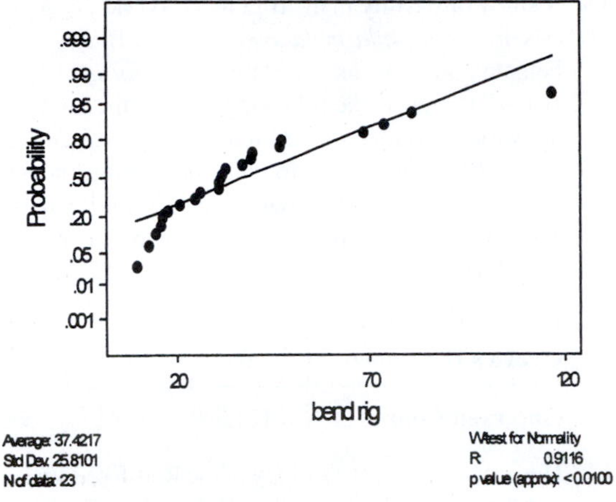

Would you use a one-sample t confidence interval to estimate true average bending rigidity? Explain your reasoning.

21. The article from which the data in Exercise 20 was obtained also gave the accompanying data on the composite mass/outer fabric mass ratio for high-quality fabric specimens.

1.15	1.40	1.34	1.29	1.36	1.26	1.22
1.40	1.29	1.41	1.32	1.34	1.26	1.36
1.36	1.30	1.28	1.45	1.29	1.28	1.38
1.55	1.46	1.32				

MINITAB gave $r = .9852$ as the value of the Ryan–Joiner test statistic and reported that P-value $>.10$. Would you use the one-sample t test to test hypotheses about the value of the true average ratio? Why or why not?

22. The article "Nonbloated Burned Clay Aggregate Concrete" (*J. Materials*, 1972: 555–563) reports the following data on 7-day flexural strength of nonbloated burned clay aggregate concrete samples (psi):

257	327	317	300	340	340	343
374	377	386	383	393	407	407
434	427	440	407	450	440	456
460	456	476	480	490	497	526
546	700					

Test at level .10 to decide whether flexural strength is a normally distributed variable.

13.3 Two-Way Contingency Tables

In the previous two sections, we discussed inferential problems in which the count data was displayed in a rectangular table of cells. Each table consisted of one row and a specified number of columns, where the columns corresponded to categories into which the population had been divided. We now study problems in which the data also consists of counts or frequencies, but the data table will now have I rows ($I \geq 2$) and J columns, so IJ cells. There are two commonly encountered situations in which such data arises:

1. There are I populations of interest, each corresponding to a different row of the table, and each population is divided into the same J categories. A sample is taken from the ith population ($i = 1, \ldots, I$), and the counts are entered in the cells in the ith row of the table. For example, customers of each of $I = 3$ department store chains might have available the same $J = 5$ payment categories: cash, check, store credit card, Visa, and MasterCard.

2. There is a single population of interest, with each individual in the population categorized with respect to two different factors. There are I categories associated with the first factor, and J categories associated with the second factor. A single sample is taken, and the number of individuals belonging in both category i of factor 1 and category j of factor 2 is entered in the cell in row i, column j ($i = 1, \ldots, I$; $j = 1, \ldots, J$). As an example, customers making a purchase might be classified according to department in which the purchase was made, with $I = 6$ departments, and according to method of payment, with $J = 5$ as in (1) above.

Let n_{ij} denote the number of individuals in the sample(s) falling in the (i, j)th cell (row i, column j) of the table—that is, the (i, j)th cell count. The table displaying the n_{ij}'s is called a **two-way contingency table**; a prototype is shown in Table 13.9.

Table 13.9 A two-way contingency table

	1	**2**	**...**	***j***	**...**	***J***
1	n_{11}	n_{12}	$\cdots$	n_{1j}	$\cdots$	n_{1J}
2	n_{21}					$\vdots$
$\vdots$	$\vdots$					
i	n_{i1}	$\cdots$		n_{ij}	$\cdots$	
$\vdots$	$\vdots$					
I	n_{I1}	$\cdots$				n_{IJ}

In situations of type 1, we want to investigate whether the proportions in the different categories are the same for all populations. The null hypothesis states that the populations are *homogeneous* with respect to these categories. In type 2 situations, we investigate whether the categories of the two factors occur independently of one another in the population.

Testing for Homogeneity

We assume that each individual in every one of the I populations belongs in exactly one of J categories. A sample of n_i individuals is taken from the ith population; let $n = \sum n_i$ and

$$n_{ij} = \text{the number of individuals in the } i\text{th sample who fall into category } j$$

$$n_{\cdot j} = \sum_{i=1}^{I} n_{ij} = \begin{array}{l}\text{the total number of individuals among} \\ \text{the } n \text{ sampled who fall into category } j\end{array}$$

The n_{ij}'s are recorded in a two-way contingency table with I rows and J columns. The sum of the n_{ij}'s in the ith row is n_i, whereas the sum of entries in the jth column is $n_{\cdot j}$.

Let

$$p_{ij} = \begin{array}{l}\text{the proportion of the individuals in} \\ \text{population } i \text{ who fall into category } j\end{array}$$

Thus, for population 1, the J proportions are $p_{11}, p_{12}, \ldots, p_{1J}$ (which sum to 1) and similarly for the other populations. The **null hypothesis of homogeneity** states that the proportion of individuals in category j is the same for each population and that this is true for every category; that is, for every j, $p_{1j} = p_{2j} = \cdots = p_{Ij}$.

When H_0 is true, we can use $p_1, p_2, \cdots, p_J$ to denote the population proportions in the J different categories; these proportions are common to all I populations. The expected number of individuals in the ith sample who fall in the jth category when H_0 is true is then $E(N_{ij}) = n_i \cdot p_j$. To estimate $E(N_{ij})$, we must first estimate p_j, the proportion in category j. Among the total sample of n individuals, $N_{\cdot j}$ fall into category j, so we use $\hat{p}_j = N_{\cdot j}/n$ as the estimator (this can be shown to be the maximum likelihood estimator of p_j). Substitution of the estimate $\hat{p}_j$ for p_j in $n_i p_j$ yields a simple formula for estimated expected counts under H_0:

$$\hat{e}_{ij} = \text{estimated expected count in cell } (i, j) = n_i \cdot \frac{n_{\cdot j}}{n} \tag{13.10}$$

$$= \frac{(i\text{th row total})(j\text{th column total})}{n}$$

The test statistic also has the same form as in previous problem situations. The number of degrees of freedom comes from the general rule of thumb. In each row of Table 13.9 there are $J - 1$ freely determined cell counts (each sample size n_i is fixed), so there are a total of $I(J - 1)$ freely determined cells. Parameters $p_1, \ldots, p_J$ are estimated, but because $\sum p_i = 1$, only $J - 1$ of these are independent. Thus df $= I(J - 1) - (J - 1) = (J - 1)(I - 1)$.

Null hypothesis: H_0: $p_{1j} = p_{2j} = \cdots = p_{Ij} \quad j = 1, 2, \ldots, J$

Alternative hypothesis: H_a: H_0 is not true

Test statistic value:

$$\chi^2 = \sum_{\text{all cells}} \frac{(\text{observed} - \text{estimated expected})^2}{\text{estimated expected}} = \sum_{i=1}^{I}\sum_{j=1}^{J} \frac{(n_{ij} - \hat{e}_{ij})^2}{\hat{e}_{ij}}$$

Rejection region: $\chi^2 \geq \chi^2_{\alpha,(I-1)(J-1)}$

P-value information can be obtained as described in Section 13.1. The test can safely be applied as long as $\hat{e}_{ij} \geq 5$ for all cells.

Example 13.13 A company packages a particular product in cans of three different sizes, each one using a different production line. Most cans conform to specifications, but a quality control engineer has identified the following reasons for nonconformance: (1) blemish on can; (2) crack in can; (3) improper pull tab location; (4) pull tab missing; (5) other. A sample of nonconforming units is selected from each of the three lines, and each unit is categorized according to reason for nonconformity, resulting in the following contingency table data:

		Reason for Nonconformity					
		Blemish	**Crack**	**Location**	**Missing**	**Other**	**Sample Size**
Production Line	**1**	34	65	17	21	13	150
	2	23	52	25	19	6	125
	3	32	28	16	14	10	100
	Total	89	145	58	54	29	375

Does the data suggest that the proportions falling in the various nonconformance categories are not the same for the three lines? The parameters of interest are the various proportions, and the relevant hypotheses are

H_0: the production lines are homogeneous with respect to the five nonconformance categories; that is, $p_{1j} = p_{2j} = p_{3j}$ for $j = 1, \ldots, 5$

H_a: the production lines are not homogeneous with respect to the categories

The estimated expected frequencies (assuming homogeneity) must now be calculated. Consider the first nonconformance category for the first production line. When the lines are homogeneous,

estimated expected number among the 150 selected units that are blemished

$$= \frac{(\text{first row total})(\text{first column total})}{\text{total of sample sizes}} = \frac{(150)(89)}{375} = 35.60$$

The contribution of the cell in the upper-left corner to χ^2 is then

$$\frac{(\text{observed} - \text{estimated expected})^2}{\text{estimated expected}} = \frac{(34 - 35.60)^2}{35.60} = .072$$

The other contributions are calculated in a similar manner. Figure 13.4 shows MINITAB output for the chi-squared test. The observed count is the top number in each cell, and directly below it is the estimated expected count. The contribution of each cell to χ^2 appears below the counts, and the test statistic value is $\chi^2 = 14.159$. All estimated expected counts are at least 5, so combining categories is unnecessary. The test is based on $(3 - 1)(5 - 1) = 8$ df. Appendix Table A.11 shows that the values that capture upper-tail areas of .08 and .075 under the 8 df curve are 14.06 and 14.26, respectively. Thus the P-value is between .075 and .08; MINITAB gives P-value $= .079$. The null hypothesis of homogeneity should not be rejected at the usual significance levels of .05 or .01, but it would be rejected for the higher α of .10.

```
Expected counts are printed below observed counts

          blem     crack       loc   missing     other    Total
1           34        65        17        21        13      150
         35.60     58.00     23.20     21.60     11.60

2           23        52        25        19         6      125
         29.67     48.33     19.33     18.00      9.67

3           32        28        16        14        10      100

         23.73     38.67     15.47     14.40      7.73

Total       89       145        58        54        29      375

Chisq = 0.072 + 0.845 + 1.657 + 0.017 + 0.169 + 1.498 + 0.278 +
        1.661 + 0.056 + 1.391 + 2.879 + 2.943 + 0.018 + 0.011 +
        0.664 = 14.159

df = 8, p = 0.079
```

Figure 13.4 MINITAB output for the chi-squared test of Example 13.13 ■

Testing for Independence

We focus now on the relationship between two different factors in a single population. The number of categories of the first factor will be denoted by I and the number of categories of the second factor by J. Each individual in the population is assumed to belong in exactly one of the I categories associated with the first factor and exactly one of the J categories associated with the second factor. For example, the population of interest might consist of all individuals who regularly watch the national news on television, with the first factor being preferred network (ABC, CBS, NBC, PBS, CNN, or FOX, so $I = 6$) and the second factor political philosophy (liberal, moderate, conservative, giving $J = 3$).

For a sample of n individuals taken from the population, let n_{ij} denote the number among the n who fall both in category i of the first factor and category j of the second factor. The n_{ij}'s can be displayed in a two-way contingency table with I rows and

J columns. In the case of homogeneity for I populations, the row totals were fixed in advance, and only the J column totals were random. Now only the total sample size is fixed, and both the $n_{i\cdot}$'s and $n_{\cdot j}$'s are observed values of random variables. To state the hypotheses of interest, let

$$\begin{aligned} p_{ij} &= \text{the proportion of individuals in the population who belong in} \\ &\quad\ \text{category } i \text{ of factor 1 and category } j \text{ of factor 2} \\ &= P(\text{a randomly selected individual falls in both category } i \text{ of factor 1} \\ &\qquad \text{and category } j \text{ of factor 2}) \end{aligned}$$

Then

$$p_{i\cdot} = \sum_j p_{ij} = P(\text{a randomly selected individual falls in category } i \text{ of factor 1})$$

$$p_{\cdot j} = \sum_i p_{ij} = P(\text{a randomly selected individual falls in category } j \text{ of factor 2})$$

Recall that two events A and B are independent if $P(A \cap B) = P(A) \cdot P(B)$. *The null hypothesis here says that an individual's category with respect to factor* 1 *is independent of the category with respect to factor* 2. In symbols, this becomes $p_{ij} = p_{i\cdot} \cdot p_{\cdot j}$ for every pair (i, j).

The expected count in cell (i, j) is $n \cdot p_{ij}$, so when the null hypothesis is true, $E(N_{ij}) = n \cdot p_{i\cdot} \cdot p_{\cdot j}$. To obtain a chi-squared statistic, we must therefore estimate the $p_{i\cdot}$'s $(i = 1, \ldots, I)$ and $p_{\cdot j}$'s $(j = 1, \ldots, J)$. The (maximum likelihood) estimates are

$$\hat{p}_{i\cdot} = \frac{n_{i\cdot}}{n} = \text{sample proportion for category } i \text{ of factor 1}$$

and

$$\hat{p}_{\cdot j} = \frac{n_{\cdot j}}{n} = \text{sample proportion for category } j \text{ of factor 2}$$

This gives estimated expected cell counts identical to those in the case of homogeneity.

$$\begin{aligned} \hat{e}_{ij} &= n \cdot \hat{p}_{i\cdot} \cdot \hat{p}_{\cdot j} = n \cdot \frac{n_{i\cdot}}{n} \cdot \frac{n_{\cdot j}}{n} = \frac{n_{i\cdot} \cdot n_{\cdot j}}{n} \\ &= \frac{(i\text{th row total})(j\text{th column total})}{n} \end{aligned}$$

The test statistic is also identical to that used in testing for homogeneity, as is the number of degrees of freedom. This is because the number of freely determined cell counts is $IJ - 1$, since only the total n is fixed in advance. There are I estimated $p_{i\cdot}$'s, but only $I - 1$ are independently estimated since $\sum p_{i\cdot} = 1$, and similarly $J - 1$ $p_{\cdot j}$'s are independently estimated, so $I + J - 2$ parameters are independently estimated. The rule of thumb now yields df $= IJ - 1 - (I + J - 2) = IJ - I - J + 1 = (I - 1) \cdot (J - 1)$.

Null hypothesis: H_0: $p_{ij} = p_{i\cdot} \cdot p_{\cdot j} \quad i = 1, \ldots, I; j = 1, \ldots, J$

Alternative hypothesis: H_a: H_0 is not true

Test statistic value:

$$\chi^2 = \sum_{\text{all cells}} \frac{(\text{observed} - \text{estimated expected})^2}{\text{estimated expected}} = \sum_{i=1}^{I} \sum_{j=1}^{J} \frac{(n_{ij} - \hat{e}_{ij})^2}{\hat{e}_{ij}}$$

Rejection region: $\chi^2 \geq \chi^2_{\alpha,(I-1)(J-1)}$

Again, P-value information can be obtained as described in Section 13.1. The test can safely be applied as long as $\hat{e}_{ij} \geq 5$ for all cells.

Example 13.14 A study of the relationship between facility conditions at gasoline stations and aggressiveness in the pricing of gasoline ("An Analysis of Price Aggressiveness in Gasoline Marketing," *J. Marketing Res.*, 1970: 36–42) reports the accompanying data based on a sample of $n = 441$ stations. At level .01, does the data suggest that facility conditions and pricing policy are independent of one another? Observed and estimated expected counts are given in Table 13.10.

Table 13.10 Observed and estimated expected counts for Example 13.14

		Observed Pricing Policy				Expected Pricing Policy			
		Aggressive	Neutral	Nonaggressive	$n_{i\cdot}$				
	Substandard	24	15	17	56	17.02	22.10	16.89	56
Condition	Standard	52	73	80	205	62.29	80.88	61.83	205
	Modern	58	86	36	180	54.69	71.02	54.29	180
	$n_{\cdot j}$	134	174	133	441	134	174	133	441

Thus

$$\chi^2 = \frac{(24 - 17.02)^2}{17.02} + \cdots + \frac{(36 - 54.29)^2}{54.29} = 22.47$$

and because $\chi^2_{.01,4} = 13.277$, the hypothesis of independence is rejected.

We conclude that knowledge of a station's pricing policy does give information about the condition of facilities at the station. In particular, stations with an aggressive pricing policy appear more likely to have substandard facilities than stations with a neutral or nonaggressive policy. ■

Ordinal Factors and Logistic Regression

Sometimes a factor has ordinal categories, meaning that there is a natural ordering. For example, there is a natural ordering to freshman, sophomore, junior, senior. In such situations we can use a method that often has greater power to detect relationships. Consider the case in which the first factor is ordinal and the other has two categories. Denote by X the level of the first (ordinal) factor, the rows, which will be the predictor in the model. Then Y designates the column, either 1 or 2 and will be the dependent variable in the model. It is convenient for purposes of logistic regression to label column 1 as $Y = 0$ (failure) and column 2 as $Y = 1$ (success), corresponding to the usual notation for binomial trials. In terms of logistic regression, $p(x)$ is the probability of success given that $X = x$:

$$p(x) = P(Y = 1\,|\,X = x) = P(j = 2\,|\,i = x) = \frac{p_{x2}}{p_{x1} + p_{x2}}$$

Then the logistic model of Chapter 12 says that

$$e^{\beta_0 + \beta_1 x} = \frac{p(x)}{1 - p(x)} = \frac{p_{x2}}{p_{x1}}$$

In terms of the odds of success in a row (estimated by the ratio of the two counts), the model says that the odds change proportionally (by the fixed multiple e^{β_1}) from row to row. For example, suppose a test is given in grades 1, 2, 3, and 4 with successes and failures as follows:

Grade	Failed	Passed	Estimated Odds
1	45	45	1
2	30	60	2
3	18	72	4
4	10	80	8

Here the model fits perfectly, with odds ratio $e^{\beta_1} = 2$, so $\beta_1 = \ln(2)$ and $\beta_0 = -\ln(2)$. In general, it should be clear that β_1 is the natural log of the odds ratio between successive rows. When a table with I rows and 2 columns has roughly a common odds ratio from row to row, then the logistic model should be a good fit if the rows are labeled with consecutive integers.

We focus on the slope β_1 because the relationship between the two factors hinges on this parameter. The hypothesis of no relationship is equivalent to H_0: $\beta_1 = 0$, which is usually tested against a two-tailed alternative.

Example 13.15 The relationship between TV watching and physical fitness was considered in the article "Television Viewing and Physical Fitness in Adults" (*Res. Q. Exercise Sport*, 1990: 315–320). Subjects were asked about their television-viewing habits and were classified as physically fit if they scored in the excellent or very good category on a step test. Table 13.11 shows the results in the form of a 4×2 table. The TV column gives time per day (hr) spent watching.

The rows need to be given specific numeric values for computational purposes, and it is convenient to just make these 1, 2, 3, 4, because consecutive integers correspond to the assumption of a common odds ratio from row to row. The columns may need to be labeled as 0 and 1 for input to a program. The logistic regression results from MINITAB are shown in Figure 13.5, where the estimated coefficient $\hat{\beta}_1$ for TV is given

Table 13.11 TV versus fitness results

TV Time	Unfit	Fit
0	147	35
1–2	629	101
3–4	222	28
5+	34	4

as $-.29$ and the odds ratio is given as $.75 = e^{-.29}$. This means that, for each increase in TV watching category, the odds of being fit decline to about $\frac{3}{4}$ of the previous value. There is a loss of 25% for each increment in TV. The output shows two tests for β_1, a z based on the ratio of the coefficient to its estimated standard error, and G, which is based on a likelihood ratio test and gives the chi-squared approximation for the difference of log likelihoods. The two tests usually give very similar results, with G being approximately the square of z. In this case they agree that the P-value is around .02, which means that we should reject at the .05 level the hypothesis that $\beta_1 = 0$ and conclude that there is a relationship between TV watching and fitness. Of course, the existence of a relationship does not imply anything about one causing the other. By the way, a chi-squared test yields $\chi^2 = 6.161$ with 3 df, $P = .104$, so with this test we would not conclude that there is a relationship, even at the 10% level. There is an advantage in using logistic regression for this kind of data.

```
Logistic Regression Table

                                                          Odds     95%      CI
Predictor        Coef      SE Coef       Z         P     Ratio    Lower   Upper
Constant     -1.21316     0.267486    -4.54    0.000
TV           -0.290693    0.125588    -2.31    0.021      0.75     0.58    0.96

Log-Likelihood = -483.205
Test that all slopes are zero: G = 5.501, DF = 1, P-Value = 0.019
```

Figure 13.5 Logistic regression for TV versus fitness ■

Suppose there are two ordinal factors, each with more than two levels. This too can be handled with logistic regression, but it requires a procedure called ordinal logistic regression that allows an ordinal dependent variable. When one factor is ordinal and the other is not, the analysis can be done with multinomial (also called nominal or polytomous) logistic regression, which allows a non-ordinal dependent variable.

Models and methods for analyzing data in which each individual is categorized with respect to three or more factors (multidimensional contingency tables) are discussed in several of the references in the chapter bibliography.

Exercises Section 13.3 (23–35)

23. Reconsider the Cubs data of Exercise 56 in Chapter 10. Form a 2 × 2 table for the data and use a χ^2 statistic to test the hypothesis of equal population proportions. The χ^2 statistic should be the square of the z statistic in Exercise 56 of Chapter 10. How are the P-values related?

24. The accompanying data refers to leaf marks found on white clover samples selected from both long-grass areas and short-grass areas ("The Biology of the Leaf Mark Polymorphism in *Trifolium repens* L.," *Heredity*, 1976: 306–325). Use a χ^2 test to decide whether the true proportions of different marks are identical for the two types of regions.

	Type of Mark L	LL	Y + YL	O	Others	Sample Size
Long-Grass Areas	409	11	22	7	277	726
Short-Grass Areas	512	4	14	11	220	761

25. The following data resulted from an experiment to study the effects of leaf removal on the ability of fruit of a certain type to mature ("Fruit Set, Herbivory, Fruit Reproduction, and the Fruiting Strategy of *Catalpa speciosa*," *Ecology*, 1980: 57–64).

Treatment	Number of Fruits Matured	Number of Fruits Aborted
Control	141	206
Two leaves removed	28	69
Four leaves removed	25	73
Six leaves removed	24	78
Eight leaves removed	20	82

Does the data suggest that the chance of a fruit maturing is affected by the number of leaves removed? State and test the appropriate hypotheses at level .01.

26. The article "Human Lateralization from Head to Foot: Sex-Related Factors" (*Science*, 1978: 1291–1292) reports for both a sample of right-handed men and a sample of right-handed women the number of individuals whose feet were the same size, had a bigger left than right foot (a difference of half a shoe size or more), or had a bigger right than left foot.

	L > R	L = R	L < R	Sample Size
Men	2	10	28	40
Women	55	18	14	87

Does the data indicate that gender has a strong effect on the development of foot asymmetry? State the appropriate null and alternative hypotheses, compute the value of χ^2, and obtain information about the P-value.

27. The article "Susceptibility of Mice to Audiogenic Seizure Is Increased by Handling Their Dams During Gestation" (*Science*, 1976: 427–428) reports on research into the effect of different injection treatments on the frequencies of audiogenic seizures.

Treatment	No Response	Wild Running	Clonic Seizure	Tonic Seizure
Thienylalanine	21	7	24	44
Solvent	15	14	20	54
Sham	23	10	23	48
Unhandled	47	13	28	32

Does the data suggest that the true percentages in the different response categories depend on the nature of the injection treatment? State and test the appropriate hypotheses using $\alpha = .005$.

28. The accompanying data on sex combinations of two recombinants resulting from six different male genotypes appears in the article "A New Method for Distinguishing Between Meiotic and Premeiotic

Recombinational Events in *Drosophila melanogaster*" (*Genetics*, 1979: 543–554). Does the data support the hypothesis that the frequency distribution among the three sex combinations is homogeneous with respect to the different genotypes? Define the parameters of interest, state the appropriate H_0 and H_a, and perform the analysis.

		Sex Combination		
		M/M	**M/F**	**F/F**
Male Genotype	**1**	35	80	39
	2	41	84	45
	3	33	87	31
	4	8	26	8
	5	5	11	6
	6	30	65	20

29. Each individual in a random sample of high school and college students was cross-classified with respect to both political views and marijuana usage, resulting in the data displayed in the accompanying two-way table ("Attitudes About Marijuana and Political Views," *Psych. Rep.*, 1973: 1051–1054). Does the data support the hypothesis that political views and marijuana usage level are independent within the population? Test the appropriate hypotheses using level of significance .01.

		Usage Level		
		Never	**Rarely**	**Frequently**
Political Views	**Liberal**	479	173	119
	Conservative	214	47	15
	Other	172	45	85

30. Show that the chi-squared statistic for the test of independence can be written in the form

$$\chi^2 = \sum_{i=1}^{I}\sum_{j=1}^{J}\left(\frac{N_{ij}^2}{\hat{E}_{ij}}\right) - n$$

Why is this formula more efficient computationally than the defining formula for χ^2?

31. Suppose that in Exercise 29 each student had been categorized with respect to political views, marijuana usage, and religious preference, with the categories of this latter factor being Protestant, Catholic, and other. The data could be displayed in three different two-way tables, one corresponding to each category of the third factor. With $p_{ijk} = P$(political category i, marijuana category j, and religious category k), the null hypothesis of independence of all three factors states that $p_{ijk} = p_{i..}p_{.j.}p_{..k}$. Let n_{ijk} denote the observed frequency in cell (i, j, k). Show how to estimate the expected cell counts assuming that H_0 is true ($\hat{e}_{ijk} = n\hat{p}_{ijk}$, so the $\hat{p}_{ijk}$'s must be determined). Then use the general rule of thumb to determine the number of degrees of freedom for the chi-squared statistic.

32. Suppose that in a particular state consisting of four distinct regions, a random sample of n_k voters is obtained from the kth region for $k = 1, 2, 3, 4$. Each voter is then classified according to which candidate (1, 2, or 3) he or she prefers and according to voter registration (1 = Dem., 2 = Rep., 3 = Indep.). Let p_{ijk} denote the proportion of voters in region k who belong in candidate category i and registration category j. The null hypothesis of homogeneous regions is H_0: $p_{ij1} = p_{ij2} = p_{ij3} = p_{ij4}$ for all i, j (i.e., the proportion within each candidate/registration combination is the same for all four regions). Assuming that H_0 is true, determine $\hat{p}_{ijk}$ and $\hat{e}_{ijk}$ as functions of the observed n_{ijk}'s, and use the general rule of thumb to obtain the number of degrees of freedom for the chi-squared test.

33. Consider the accompanying 2 × 3 table displaying the sample proportions that fell in the various combinations of categories (e.g., 13% of those in the sample were in the first category of both factors).

	1	**2**	**3**
1	.13	.19	.28
2	.07	.11	.22

a. Suppose the sample consisted of $n = 100$ people. Use the chi-squared test for independence with significance level .10.

b. Repeat part (a) assuming that the sample size was $n = 1000$.

c. What is the smallest sample size n for which these observed proportions would result in rejection of the independence hypothesis?

34. Use logistic regression to test the relationship between leaf removal and fruit growth in Exercise 25. Compare the *P*-value with what was found in Exercise 25. (Remember that $\chi_1^2 = z^2$.) Explain why you expected the logistic regression to give a smaller *P*-value.

35. A random sample of 100 faculty at a university gives the results shown below for professorial rank versus gender.
   a. Test for a relationship at the 5% level using a chi-squared statistic.
   b. Test for a relationship at the 5% level using logistic regression.
   c. Compare the *P*-values in parts (a) and (b). Is this in accord with your expectations? Explain.
   d. Interpret your results. Assuming that today's assistant professors are tomorrow's associate professors and professors, do you see implications for the future?

Rank	Male	Female
Professor	25	9
Assoc Prof	20	8
Asst Prof	18	20

Supplementary Exercises (36–47)

36. The article "Birth Order and Political Success" (*Psych. Rep.*, 1971: 1239–1242) reports that among 31 randomly selected candidates for political office who came from families with four children, 12 were firstborn, 11 were middleborn, and 8 were lastborn. Use this data to test the null hypothesis that a political candidate from such a family is equally likely to be in any one of the four ordinal positions.

37. The results of an experiment to assess the effect of crude oil on fish parasites are described in the article "Effects of Crude Oils on the Gastrointestinal Parasites of Two Species of Marine Fish" (*J. Wildlife Diseases*, 1983: 253–258). Three treatments (corresponding to populations in the procedure described) were compared: (1) no contamination, (2) contamination by 1-year-old weathered oil, and (3) contamination by new oil. For each treatment condition, a sample of fish was taken, and then each fish was classified as either parasitized or not parasitized. Data compatible with that in the article is given. Does the data indicate that the three treatments differ with respect to the true proportion of parasitized and nonparasitized fish? Test using $\alpha = .01$.

Treatment	Parasitized	Nonparasitized
Control	30	3
Old oil	16	8
New oil	16	16

38. Qualifications of male and female head and assistant college athletic coaches were compared in the article "Sex Bias and the Validity of Believed Differences Between Male and Female Interscholastic Athletic Coaches" (*Res. Q. Exercise Sport*, 1990: 259–267). Each person in random samples of 2225 male coaches and 1141 female coaches was classified according to number of years of coaching experience to obtain the accompanying two-way table. Is there enough evidence to conclude that the proportions falling into the experience categories are different for men and women? Use $\alpha = .01$.

	Years of Experience				
Gender	**1–3**	**4–6**	**7–9**	**10–12**	**13+**
Male	202	369	482	361	811
Female	230	251	238	164	258

39. The authors of the article "Predicting Professional Sports Game Outcomes from Intermediate Game Scores" (*Chance*, 1992: 18–22) used a chi-squared test to determine whether there was any merit to the idea that basketball games are not settled until the last quarter, whereas baseball games are over by the seventh inning. They also considered football and hockey. Data was collected for 189 basketball games, 92 baseball games, 80 hockey games, and 93 football games. The games analyzed were sampled

randomly from all games played during the 1990 season for baseball and football and for the 1990–1991 season for basketball and hockey. For each game, the late-game leader was determined, and then it was noted whether the late-game leader actually ended up winning the game. The resulting data is summarized in the accompanying table.

Sport	Late-Game Leader Wins	Late-Game Leader Loses
Basketball	150	39
Baseball	86	6
Hockey	65	15
Football	72	21

The authors state, "*Late-game leader* is defined as the team that is ahead after three quarters in basketball and football, two periods in hockey, and seven innings in baseball. The chi-square value on three degrees of freedom is 10.52 ($P < .015$)."

a. State the relevant hypotheses and reach a conclusion using $\alpha = .05$.

b. Do you think that your conclusion in part (a) can be attributed to a single sport being an anomaly?

40. The accompanying two-way frequency table appears in the article "Marijuana Use in College" (*Youth and Society*, 1979: 323–334). Each of 445 college students was classified according to both frequency of marijuana use and parental use of alcohol and psychoactive drugs. Does the data suggest that parental usage and student usage are independent in the population from which the sample was drawn? Use the P-value method to reach a conclusion.

		Standard Level of Marijuana use		
		Never	Occasional	Regular
Parental Use of Alcohol and Drugs	Neither	141	54	40
	One	68	44	51
	Both	17	11	19

41. In a study of 2989 cancer deaths, the location of death (home, acute-care hospital, or chronic-care facility) and age at death were recorded, resulting in the given two-way frequency table ("Where Cancer Patients Die," *Public Health Rep.*, 1983: 173). Using a .01 significance level, test the null hypothesis that age at death and location of death are independent.

	Location		
Age	Home	Acute-Care	Chronic-Care
15–54	94	418	23
55–64	116	524	34
65–74	156	581	109
Over 74	138	558	238

42. In a study to investigate the extent to which individuals are aware of industrial odors in a certain region ("Annoyance and Health Reactions to Odor from Refineries and Other Industries in Carson, California," *Environ. Res.*, 1978: 119–132), a sample of individuals was obtained from each of three different areas near industrial facilities. Each individual was asked whether he or she noticed odors (1) every day, (2) at least once/week, (3) at least once/month, (4) less often than once/month, or (5) not at all, resulting in the data and output from SPSS on page 74. State and test the appropriate hypotheses.

43. Many shoppers have expressed unhappiness because grocery stores have stopped putting prices on individual grocery items. The article "The Impact of Item Price Removal on Grocery Shopping Behavior" (*J. Marketing*, 1980: 73–93) reports on a study in which each shopper in a sample was classified by age and by whether he or she felt the need for item pricing. Based on the accompanying data, does the need for item pricing appear to be independent of age?

	Age				
	<30	30–39	40–49	50–59	≥ 60
Number in Sample	150	141	82	63	49
Number Who Want Item Pricing	127	118	77	61	41

Crosstabulation: AREA By CATEGORY

CATEGORY → AREA (Count / Exp Val / Row Pct / Col Pct)	1.00	2.00	3.00	4.00	5.00	Row Total
1.00	20 12.7 20.6% 52.6%	28 24.7 28.9% 37.8%	23 18.0 23.7% 42.6%	14 16.0 14.4% 29.2%	12 25.7 12.4% 15.6%	97 33.3%
2.00	14 12.4 14.7% 36.8%	34 24.2 35.8% 45.9%	21 17.6 22.1% 38.9%	14 15.7 14.7% 29.2%	12 25.1 12.6% 15.6%	95 32.6%
3.00	4 12.9 4.0% 10.5%	12 25.2 12.1% 16.2%	10 18.4 10.1% 18.5%	20 16.3 20.2% 41.7%	53 26.2 53.5% 68.8%	99 34.0%
Column Total	38 13.1%	74 25.4%	54 18.6%	48 16.5%	77 26.5%	291 100.0%

Chi-Square	D.F.	Significance	Min E.F.	Cells with E.F. < 5
70.64156	8	.0000	12.405	None

44. Let p_1 denote the proportion of successes in a particular population. The test statistic value in Chapter 9 for testing H_0: $p_1 = p_{10}$ was $z = (\hat{p}_1 - p_{10})/\sqrt{p_{10}p_{20}/n}$, where $p_{20} = 1 - p_{10}$. Show that for the case $k = 2$, the chi-squared statistic value of Section 13.1 satisfies $\chi^2 = z^2$. [*Hint*: First show that $(n_1 - n\,p_{10})^2 = (n_2 - n\,p_{20})^2$.]

45. The NCAA basketball tournament begins with 64 teams that are apportioned into four regional tournaments, each involving 16 teams. The 16 teams in each region are then ranked (seeded) from 1 to 16. During the 12-year period from 1991 to 2002, the top-ranked team won its regional tournament 22 times, the second-ranked team won 10 times, the third-ranked team won 5 times, and the remaining 11 regional tournaments were won by teams ranked lower than 3. Let P_{ij} denote the probability that the team ranked i in its region is victorious in its game against the team ranked j. Once the P_{ij}'s are available, it is possible to compute the probability that any particular seed wins its regional tournament (a complicated calculation because the number of outcomes in the sample space is quite large). The paper "Probability Models for the NCAA Regional Basketball Tournaments"(*Amer. Statist.*, 1991: 35–38) proposed several different models for the P_{ij}'s.

a. One model postulated $P_{ij} = .5 - \lambda(i - j)$ with $\lambda = \frac{1}{32}$ (from which $P_{16,1} = \frac{1}{32}$, $P_{16,2} = \frac{2}{32}$, etc.). Based on this, P(seed #1 wins) = .27477, P(seed #2 wins) = .20834, and P(seed #3 wins) = .15429. Does this model appear to provide a good fit to the data?

b. A more sophisticated model has $P_{ij} = .5 + .2813625(z_i - z_j)$, where the z's are measures of relative strengths related to standard normal percentiles [percentiles for successive highly seeded teams are closer together than is the case for teams seeded lower, and .2813625 ensures that the range of probabilities is the same as for the model in part (a)]. The resulting probabilities of seeds 1, 2, or 3 winning their regional tournaments are .45883, .18813, and .11032, respectively. Assess the fit of this model.

46. Have you ever wondered whether soccer players suffer adverse effects from hitting "headers"? The authors of the article "No Evidence of Impaired Neurocognitive Performance in Collegiate Soccer Players" (*Amer. J. Sports Med.* 2002: 157–162) investigated this issue from several perspectives.

a. The paper reported that 45 of the 91 soccer players in their sample had suffered at least one concussion, 28 of 96 nonsoccer athletes had suffered at least one concussion, and only 8 of 53 student controls had suffered at least one concussion. Analyze this data and draw appropriate conclusions.

b. For the soccer players, the sample correlation coefficient calculated from the values of x = soccer exposure (total number of competitive seasons played prior to enrollment in the study) and y = score on an immediate memory recall test was $r = -.220$. Interpret this result.

c. Here is summary information on scores on a controlled oral word-association test for the soccer and nonsoccer athletes:

$n_1 = 26, \bar{x}_1 = 37.50, s_1 = 9.13, n_2 = 56,$
$\bar{x}_2 = 39.63, s_2 = 10.19$

Analyze this data and draw appropriate conclusions.

d. Considering the number of prior nonsoccer concussions, the values of mean ± SD for the three groups were .30 ± .67, .49 ± .87, and .19 ± .48. Analyze this data and draw appropriate conclusions.

47. Do the successive digits in the decimal expansion of π behave as though they were selected from a random number table (or came from a computer's random number generator)?

a. Let p_0 denote the long-run proportion of digits in the expansion that equal 0, and define $p_1, \ldots, p_9$ analogously. What hypotheses about these proportions should be tested, and what is df for the chi-squared test?

b. H_0 of part (a) would not be rejected for the nonrandom sequence 012 . . . 901 . . . 901 . . . . Consider nonoverlapping groups of two digits, and let p_{ij} denote the long-run proportion of groups for which the first digit is i and the second digit is j. What hypotheses about these proportions should be tested, and what is df for the chi-squared test?

c. Consider nonoverlapping groups of 5 digits. Could a chi-squared test of appropriate hypotheses about the p_{ijklm}'s be based on the first 100,000 digits? Explain.

d. The paper "Are the Digits of π an Independent and Identically Distributed Sequence?" (*Amer. Statist.*, 2000: 12–16) considered the first 1,254,540 digits of π, and reported the following P-values for group sizes of 1, . . . , 5: .572, .078, .529, .691, .298. What would you conclude?

Bibliography

Agresti, Alan, *An Introduction to Categorical Data Analysis*, Wiley, New York, 1996. An excellent treatment of various aspects of categorical data analysis by one of the most prominent researchers in this area.

Everitt, B. S., *The Analysis of Contingency Tables* (2nd ed.), Halsted Press, New York, 1992. A compact but informative survey of methods for analyzing categorical data, exposited with a minimum of mathematics.

Mosteller, Frederick, and Richard Rourke, *Sturdy Statistics*, Addison-Wesley, Reading, MA, 1973. Contains several very readable chapters on the varied uses of chi-square.

CHAPTER FOURTEEN

Alternative Approaches to Inference

Introduction

In this final chapter we consider some inferential methods that are different in important ways from those considered earlier. Recall that many of the confidence intervals and test procedures developed in Chapters 9–12 were based on some sort of a normality assumption. As long as such an assumption is at least approximately satisfied, the actual confidence and significance levels will be at least approximately equal to the "nominal" levels, those prescribed by the experimenter through the choice of particular *t* or *F* critical values. However, if there is a substantial violation of the normality assumption, the actual levels may differ considerably from the nominal levels (e.g., the use of $t_{.025}$ in a confidence interval formula may actually result in a confidence level of only 88% rather than the nominal 95%). In the first three sections of this chapter, we develop *distribution-free* or *nonparametric* procedures which are valid for a wide variety of underlying distributions rather than being tied to normality. We have actually already introduced several such methods: the bootstrap intervals and permutation tests are valid without restrictive assumptions on the underlying distribution(s).

Section 14.4 introduces the Bayesian approach to inference. The standard *frequentist* view of inference is that the parameter of interest, θ, has a fixed but unknown value. Bayesians, however, regard θ as a random variable having a *prior* probability distribution which incorporates whatever is known about its value. Then to learn more about θ, a sample from the *conditional* distribution $f(x \mid \theta)$ is obtained, and Bayes' theorem is used to produce the *posterior* distribution of

θ given the data $x_1, \ldots, x_n$. All Bayesian methods are based on this posterior distribution.

All inferential methods heretofore considered are based on specifying a sample size n prior to collecting data and then obtaining that many observations. Some statistical methods, though, are *sequential* in nature: Observations are selected one by one and the investigator has the option of terminating the experiment after just a few observations or else continuing to collect data for a much longer time. This is particularly important in some medical contexts, for example when a new treatment for a disease is being compared to the current favorite. It may be quite obvious after just a small number of patients have been treated that the new treatment improves significantly on the other one—or that the new treatment is much poorer than the current favorite. Alternatively, a great deal of data may be needed before it is obvious which of the two treatments is the better one. We close out our exposition by focusing on several very simple sequential problems and results.

14.1 *The Wilcoxon Signed-Rank Test

A research chemist replicated a particular experiment a total of 10 times and obtained the following values of reaction temperature, ordered from smallest to largest:

$$-.57 \quad -.19 \quad -.05 \quad .76 \quad 1.30 \quad 2.02 \quad 2.17 \quad 2.46 \quad 2.68 \quad 3.02$$

The distribution of reaction temperature is of course continuous. Suppose the investigator is willing to assume that this distribution is symmetric, so that the pdf satisfies $f(\tilde{\mu} + t) = f(\tilde{\mu} - t)$ for any $t > 0$, where $\tilde{\mu}$ is the median of the distribution (and also the mean μ provided that the mean exists). This condition on $f(x)$ simply says that the height of the density curve above a value any particular distance to the right of the median is the same as the height that same distance to the left of the median. The assumption of symmetry may at first thought seem quite bold, but remember that we have frequently assumed a normal distribution. Since a normal distribution is symmetric, the assumption of symmetry without any additional distributional specification is actually a weaker assumption than normality.

Let's now consider testing the null hypothesis that $\tilde{\mu} = 0$. This amounts to saying that a temperature of any particular magnitude, say 1.50, is no more likely to be positive (+1.50) than to be negative (−1.50). A glance at the data casts doubt on this hypothesis; for example, the sample median is 1.66, which is far larger in magnitude than any of the three negative observations.

Figure 14.1 shows graphs of two symmetric pdf's, one for which H_0 is true and the other for which the median of the distribution considerably exceeds 0. In the first case we expect the magnitudes of the negative observations in the sample to be comparable to those of the positive sample observations. However, in the second case observations of large absolute magnitude will tend to be positive rather than negative.

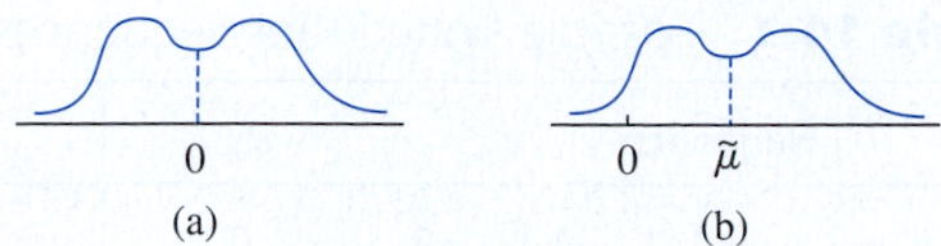

Figure 14.1 Distributions for which (a) $\tilde{\mu} = 0$; (b) $\tilde{\mu} \gg 0$

For the sample of ten reaction temperatures, let's for the moment disregard the signs of the observations and rank the absolute magnitudes from 1 to 10, with the smallest getting rank 1, the second smallest rank 2, and so on. Then apply the sign of each observation to the corresponding rank (so some signed ranks will be negative, e.g. -3, whereas others will be positive, e.g. 8). The test statistic will be S_+ = the sum of the positively signed ranks.

Absolute Magnitude	.05	.19	.57	.76	1.30	2.02	2.17	2.46	2.68	3.02
Rank	1	2	3	4	5	6	7	8	9	10
Signed Rank	-1	-2	-3	4	5	6	7	8	9	10

$$s_+ = 4 + 5 + 6 + 7 + 8 + 9 + 10 = 49$$

When the median of the distribution is much greater than 0, most of the observations with large absolute magnitudes should be positive, resulting in positively signed ranks and a large value of s_+. On the other hand, if the median is 0, magnitudes of positively signed observations should be intermingled with those of negatively signed observations, in which case s_+ will not be very large. Thus we should reject H_0: $\tilde{\mu} = 0$ when s_+ is "quite large"— the rejection region should have the form $s_+ \geq c$.

The critical value c should be chosen so that the test has a desired significance level (type I error probability), such as .05 or .01. This necessitates finding the distribution of the test statistic S_+ when the null hypothesis is true. Let's consider $n = 5$, in which case there are $2^5 = 32$ ways of applying signs to the five ranks 1, 2, 3, 4, and 5 (each rank could have a $-$ sign or a $+$ sign). The key point is that when H_0 is true, *any collection of five signed ranks has the same chance as does any other collection.* That is, the smallest observation in absolute magnitude is equally likely to be positive or negative, the same is true of the second smallest observation in absolute magnitude, and so on. Thus the collection $-1, 2, 3, -4, 5$ of signed ranks is just as likely as the collection $1, 2, 3, 4, -5$, and just as likely as any one of the other 30 possibilities.

Table 14.1 lists the 32 possible signed-rank sequences when $n = 5$ along with the value s_+ for each sequence. This immediately gives the "null distribution" of S_+ displayed in Table 14.2. For example, Table 14.1 shows that three of the 32 possible sequences have $s_+ = 8$, so $P(S_+ = 8 \text{ when } H_0 \text{ is true}) = \frac{1}{32} + \frac{1}{32} + \frac{1}{32} = \frac{3}{32}$. Notice that the null distribution is symmetric about 7.5 [more generally, symmetrically distributed over the possible values $0, 1, 2, \ldots, n(n + 1)/2$]. This symmetry is important in relating the rejection region of lower-tailed and two-tailed tests to that of an upper-tailed test.

Table 14.1 Possible signed-rank sequences for $n = 5$

Sequence					s_+	Sequence					s_+
−1	−2	−3	−4	−5	0	−1	−2	−3	+4	−5	4
+1	−2	−3	−4	−5	1	+1	−2	−3	+4	−5	5
−1	+2	−3	−4	−5	2	−1	+2	−3	+4	−5	6
−1	−2	+3	−4	−5	3	−1	−2	+3	+4	−5	7
+1	+2	−3	−4	−5	3	+1	+2	−3	+4	−5	7
+1	−2	+3	−4	−5	4	+1	−2	+3	+4	−5	8
−1	+2	+3	−4	−5	5	−1	+2	+3	+4	−5	9
+1	+2	+3	−4	−5	6	+1	+2	+3	+4	−5	10
−1	−2	−3	−4	+5	5	−1	−2	−3	+4	+5	9
+1	−2	−3	−4	+5	6	+1	−2	−3	+4	+5	10
−1	+2	−3	−4	+5	7	−1	+2	−3	+4	+5	11
−1	−2	+3	−4	+5	8	−1	−2	+3	+4	+5	12
+1	+2	−3	−4	+5	8	+1	+2	−3	+4	+5	12
+1	−2	+3	−4	+5	9	+1	−2	+3	+4	+5	13
−1	+2	+3	−4	+5	10	−1	+2	+3	+4	+5	14
+1	+2	+3	−4	+5	11	+1	+2	+3	+4	+5	15

Table 14.2 Null distribution of S_+ when $n = 5$

s_+	0	1	2	3	4	5	6	7
$p(s_+)$	$\frac{1}{32}$	$\frac{1}{32}$	$\frac{1}{32}$	$\frac{2}{32}$	$\frac{2}{32}$	$\frac{3}{32}$	$\frac{3}{32}$	$\frac{3}{32}$
s_+	8	9	10	11	12	13	14	15
$p(s_+)$	$\frac{3}{32}$	$\frac{3}{32}$	$\frac{3}{32}$	$\frac{2}{32}$	$\frac{2}{32}$	$\frac{1}{32}$	$\frac{1}{32}$	$\frac{1}{32}$

For $n = 10$ there are $2^{10} = 1024$ possible signed-rank sequences, so a listing would involve much effort. Each sequence, though, would have probability $\frac{1}{1024}$ when H_0 is true, from which the distribution of S_+ when H_0 is true can be easily obtained.

We are now in a position to determine a rejection region for testing H_0: $\tilde{\mu} = 0$ versus H_a: $\tilde{\mu} > 0$ that has a suitably small significance level α. Consider the rejection region $R = \{s_+: s_+ \geq 13\} = \{13, 14, 15\}$. Then

$$\begin{aligned}\alpha &= P(\text{reject } H_0 \text{ when } H_0 \text{ is true}) \\ &= P(S_+ = 13, 14, \text{ or } 15 \text{ when } H_0 \text{ is true}) \\ &= \tfrac{1}{32} + \tfrac{1}{32} + \tfrac{1}{32} = \tfrac{3}{32} \\ &= .094\end{aligned}$$

so that $R = \{13, 14, 15\}$ specifies a test with approximate level .1. For the rejection region $\{14, 15\}$, $\alpha = 2/32 = .063$. For the sample $x_1 = .58$, $x_2 = 2.50$, $x_3 = -.21$, $x_4 = 1.23$, $x_5 = .97$, the signed rank sequence is $-1, +2, +3, +4, +5$, so $s_+ = 14$ and at level .063 H_0 would be rejected.

A General Description of the Wilcoxon Signed-Rank Test

Because the underlying distribution is assumed symmetric, $\mu = \tilde{\mu}$, so we will state the hypotheses of interest in terms of μ rather than $\tilde{\mu}$.*

ASSUMPTION $X_1, X_2, \ldots, X_n$ is a random sample from a continuous and symmetric probability distribution with mean (and median) μ.

When the hypothesized value of μ is μ_0, the absolute differences $|x_1 - \mu_0|, \ldots, |x_n - \mu_0|$ must be ranked from smallest to largest.

Null hypothesis: H_0: $\mu = \mu_0$

Test statistic value: s_+ = the sum of the ranks associated with positive $(x_i - \mu_0)$'s

Alternative Hypothesis	**Rejection Region for Level α Test**
H_a: $\mu > \mu_0$	$s_+ \geq c_1$
H_a: $\mu < \mu_0$	$s_+ \leq c_2$ [where $c_2 = n(n + 1)/2 - c_1$]
H_a: $\mu \neq \mu_0$	either $s_+ \geq c$ or $s_+ \leq n(n + 1)/2 - c$

where the critical values c_1 and c obtained from Appendix Table A.13 satisfy $P(S_+ \geq c_1) \approx \alpha$ and $P(S_+ \geq c) \approx \alpha/2$ when H_0 is true.

Example 14.1 A producer of breakfast cereals wants to verify that a filler machine is operating correctly. The machine is supposed to fill one-pound boxes with 460 grams, on the average. This is a little above the 453.6 grams needed for one pound. When the contents are weighed, it is found that 15 boxes yield the following measurements:

454.4	470.8	447.5	453.2	462.6	445.0	455.9	458.2
461.6	457.3	452.0	464.3	459.2	453.5	465.8	

It is believed that deviations of any magnitude from 460 g are just as likely to be positive as negative (in accord with the symmetry assumption) but the distribution may not be normal. Therefore, the Wilcoxon signed-rank test will be used to see if the filler machine is calibrated correctly.

* If the tails of the distribution are "too heavy," as was the case with the Cauchy distribution of Chapter 7, then μ will not exist. In such cases, the Wilcoxon test will still be valid for tests concerning $\tilde{\mu}$.

The hypotheses are H_0: $\mu = 460$ versus H_a: $\mu \neq 460$, where μ is the true average weight. Subtracting 460 from each measurement gives

-5.6	10.8	-12.5	-6.8	2.6	-15.0	-4.1	-1.8	1.6	-2.7
-8.0	4.3	$-.8$	-6.5	5.8					

The ranks are obtained by ordering these from smallest to largest without regard to sign.

Absolute Magnitude	.8	1.6	1.8	2.6	2.7	4.1	4.3	5.6	5.8	6.5	6.8	8.0	10.8	12.5	15.0
Rank	1	2	3	4	5	6	7	8	9	10	11	12	13	14	15
Sign	−	+	−	+	−	−	+	−	+	−	−	−	+	−	−

Thus $s_+ = 2 + 4 + 7 + 9 + 13 = 35$. From Appendix Table A.13, $P(S_+ \geq 95) = P(S_+ \leq 25) = .024$ when H_0 is true, so the two-tailed test with approximate level .05 rejects H_0 when either $s_+ \geq 95$ or ≤ 25 [the exact α is $2(.024) = .048$]. Since $s_+ = 35$ is not in the rejection region, it cannot be concluded at level .05 that μ differs from 460. Even at level .094 (approximately .1), H_0 cannot be rejected, since $P(S_+ \leq 30) = P(S_+ \geq 90) = .047$ implies that s_+ values between 30 and 90 are not significant at that level. The P-value of the data is thus greater than .1. ■

Although a theoretical implication of the continuity of the underlying distribution is that ties will not occur, in practice they often do because of the discreteness of measuring instruments. If there are several data values with the same absolute magnitude, then they would be assigned the average of the ranks they would receive if they differed very slightly from one another. For example, if in Example 14.1 $x_8 = 458.2$ is changed to 458.4, then two different values of $(x_i - 460)$ would have absolute magnitude 1.6. The ranks to be averaged would be 2 and 3, so each would be assigned rank 2.5.

Paired Observations

When the data consisted of pairs $(X_1, Y_1), \ldots, (X_n, Y_n)$ and the differences $D_1 = X_1 - Y_1, \ldots, D_n = X_n - Y_n$ were normally distributed, in Chapter 10 we used a paired t test for hypotheses about the expected difference μ_D. If normality is not assumed, hypotheses about μ_D can be tested by using the Wilcoxon signed-rank test on the D_i's provided that the distribution of the differences is continuous and symmetric. *If X_i and Y_i both have continuous distributions that differ only with respect to their means* (so the Y distribution is the X distribution shifted by $\mu_1 - \mu_2 = \mu_D$), *then D_i will have a continuous symmetric distribution* (it is *not* necessary for the X and Y distributions to be symmetric individually). The null hypothesis is H_0: $\mu_D = \Delta_0$, and the test statistic S_+ is the sum of the ranks associated with the positive $(D_i - \Delta_0)$'s.

Example 14.2 About 100 years ago an experiment was done to see if drugs could help people with severe insomnia ("The Action of Optical Isomers, II: Hyoscines," *J. Physiol.*, 1905: 501–510). There were 10 patients who had trouble sleeping, and each patient tried several medications. Here we compare just the control (no medication) and levohyoscine. Does the drug offer an improvement in average sleep time? The relevant hypotheses are H_0: $\mu_D = 0$ versus H_a: $\mu_D < 0$. Here are the sleep times, differences, and signed ranks.

Patient	1	2	3	4	5	6	7	8	9	10
Control	0.6	1.1	2.5	2.8	2.9	3.0	3.2	4.7	5.5	6.2
Drug	2.5	5.7	8.0	4.4	6.3	3.8	7.6	5.8	5.6	6.1
Difference	−1.9	−4.6	−5.5	−1.6	−3.4	−.8	−4.4	−1.1	−.1	.1
Signed rank	−6	−9	−10	−5	−7	−3	−8	−4	−1.5	1.5

Notice that there is a tie for the lowest rank, so the two lowest ranks are split between observations 9 and 10, and each receives rank 1.5. Appendix Table A.13 shows that for a test with significance level approximately .05, the null hypothesis should be rejected if $s_+ \leq (10)(11)/2 - 44 = 11$. The test statistic value is 1.5, which falls in the rejection region. We therefore reject H_0 at significance level .05 in favor of the conclusion that the drug gives greater mean sleep time. The accompanying MINITAB output shows the test statistic value and also the corresponding P-value, which is $P(S_+ \leq 1.5$ when H_0 is true).

```
Test of median = 0.000000 versus median < 0.000000

                   N
                  for      Wilcoxon                   Estimated
           N     Test     Statistic          P           Median
diff      10       10           1.5      0.005           -2.250
```
■

Efficiency of the Wilcoxon Signed-Rank Test

When the underlying distribution being sampled is normal, either the t test or the signed-rank test can be used to test a hypothesis about μ. The t test is the best test in such a situation because among all level α tests it is the one having minimum β. Since it is generally agreed that there are many experimental situations in which normality can be reasonably assumed, as well as some in which it should not be, two questions must be addressed in an attempt to compare the two tests:

1. When the underlying distribution is normal (the "home ground" of the t test), how much is lost by using the signed-rank test?
2. When the underlying distribution is not normal, can a significant improvement be achieved by using the signed-rank test?

If the Wilcoxon test does not suffer much with respect to the t test on the "home ground" of the latter, and performs significantly better than the t test for a large number of other distributions, then there will be a strong case for using the Wilcoxon test.

Unfortunately, there are no simple answers to the two questions. Upon reflection, it is not surprising that the t test can perform poorly when the underlying distribution has "heavy tails" (i.e., when observed values lying far from μ are relatively more likely than they are when the distribution is normal). This is because the behavior of the t test depends on the sample mean and variance, which are both unstable in the presence of heavy tails. The difficulty in producing answers to the two questions is that β for the Wilcoxon test is very difficult to obtain and study for *any* underlying distribution, and the same can be said for the t test when the distribution is not normal. Even if β were easily obtained, any measure of efficiency would clearly depend on which underlying distribution was postulated. A number of different efficiency measures have been proposed by statisticians; one that many statisticians regard as credible is called **asymptotic relative efficiency** (ARE). The ARE of one test with respect to another is essentially the limiting ratio of sample sizes necessary to obtain identical error probabilities for the two tests. Thus if the ARE of one test with respect to a second equals .5, then when sample sizes are large, twice as large a sample size will be required of the first test to perform as well as the second test. Although the ARE does not characterize test performance for small sample sizes, the following results can be shown to hold:

1. When the underlying distribution is normal, the ARE of the Wilcoxon test with respect to the t test is approximately .95.
2. For any distribution, the ARE will be at least .86 and for many distributions will be much greater than 1.

We can summarize these results by saying that, in large-sample problems, the Wilcoxon test is never very much less efficient than the t test and may be much more efficient if the underlying distribution is far from normal. Though the issue is far from resolved in the case of sample sizes obtained in most practical problems, studies have shown that the Wilcoxon test performs reasonably and is thus a viable alternative to the t test.

Exercises Section 14.1 (1–8)

1. Reconsider the situation described in Exercise 32 of Section 9.2, and use the Wilcoxon test with $\alpha = .05$ to test the relevant hypotheses.

2. Use the Wilcoxon test to analyze the data given in Example 9.9.

3. The accompanying data is a subset of the data reported in the article "Synovial Fluid pH, Lactate, Oxygen and Carbon Dioxide Partial Pressure in Various Joint Diseases" (*Arthritis and Rheumatism*, 1971: 476–477). The observations are pH values of synovial fluid (which lubricates joints and tendons) taken from the knees of individuals suffering from arthritis. Assuming that true average pH for nonarthritic individuals is 7.39, test at level .05 to see whether the data indicates a difference between average pH values for arthritic and nonarthritic individuals.

7.02	7.35	7.34	7.17	7.28	7.77	7.09
7.22	7.45	6.95	7.40	7.10	7.32	7.14

4. A random sample of 15 automobile mechanics certified to work on a certain type of car was selected, and the time (in minutes) necessary for each one to diagnose a particular problem was determined, resulting in the following data:

30.6	30.1	15.6	26.7	27.1	25.4	35.0	30.8
31.9	53.2	12.5	23.2	8.8	24.9	30.2	

Use the Wilcoxon test at significance level .10 to decide whether the data suggests that true average diagnostic time is less than 30 minutes.

5. Both a gravimetric and a spectrophotometric method are under consideration for determining phosphate content of a particular material. Twelve samples of the material are obtained, each is split in half, and a determination is made on each half using one of the two methods, resulting in the following data:

Sample	1	2	3	4
Gravimetric	54.7	58.5	66.8	46.1
Spectrophotometric	55.0	55.7	62.9	45.5

Sample	5	6	7	8
Gravimetric	52.3	74.3	92.5	40.2
Spectrophotometric	51.1	75.4	89.6	38.4

Sample	9	10	11	12
Gravimetric	87.3	74.8	63.2	68.5
Spectrophotometric	86.8	72.5	62.3	66.0

Use the Wilcoxon test to decide whether one technique gives on average a different value than the other technique for this type of material.

6. The signed-rank statistic can be represented as $S_+ = W_1 + W_2 + \cdots + W_n$, where $W_i = i$ if the sign of the $x_i - \mu_0$ with the ith largest absolute magnitude is positive (in which case i is included in S_+) and $W_i = 0$ if this value is negative ($i = 1, 2, 3, \ldots, n$). Furthermore, when H_0 is true, the W_i's are independent and $P(W = i) = P(W = 0) = .5$.

a. Use these facts to obtain the mean and variance of S_+ when H_0 is true. *Hint*: The sum of the first n positive integers is $n(n + 1)/2$, and the sum of the squares of the first n positive integers is $n(n + 1)(2n + 1)/6$.

b. The W_i's are not identically distributed (e.g., possible values of W_2 are 2 and 0 whereas possible values of W_5 are 5 and 0), so our Central Limit Theorem for identically distributed and independent variables cannot be used here when n is large. However, a more general CLT can be used to assert that when H_0 is true and $n > 20$, S_+ has approximately a normal distribution with mean and variance obtained in (a). Use this to propose a large-sample standardized signed-rank test statistic and then an appropriate rejection region with level α for each of the three commonly encountered alternative hypotheses. *Note*: When there are ties in the absolute magnitudes, it is still correct to standardize S_+ by subtracting the mean from (a), but there is a correction for the variance which can be found in books on nonparametric statistics.

c. A particular type of steel beam has been designed to have a compressive strength (lb/in^2) of at least 50,000. An experimenter obtained a random sample of 25 beams and determined the strength of each one, resulting in the following data (expressed as deviations from 50,000):

−10	−27	36	−55	73	−77	−81
90	−95	−99	113	−127	−129	136
−150	−155	−159	165	−178	−183	−192
−199	−212	−217	−229			

Carry out a test using a significance level of approximately .01 to see if there is strong evidence that the design condition has been violated.

7. The accompanying 25 observations on fracture toughness of base plate of 18% nickel maraging steel were reported in the article "Fracture Testing of Weldments" (*ASTM Special Publ. No. 381*, 1965: 328–356). Suppose a company will agree to purchase this steel for a particular application only if it can be strongly demonstrated from experimental evidence that true average toughness exceeds 75. Assuming that the fracture toughness distribution is symmetric, state and test the appropriate hypotheses at level .05, and compute a *P*-value. [*Hint*: Use Exercise 6(b).]

69.5	71.9	72.6	73.1	73.3	73.5	74.1	74.2	75.3
75.5	75.7	75.8	76.1	76.2	76.2	76.9	77.0	77.9
78.1	79.6	79.7	80.1	82.2	83.7	93.7		

8. Suppose that observations $X_1, X_2, \ldots, X_n$ are made on a process at times $1, 2, \ldots, n$. On the basis of this data, we wish to test

H_0: the X_i's constitute an independent and identically distributed sequence

versus

H_a: X_{i+1} tends to be larger than X_i for $i = 1, \ldots, n$ (an increasing trend)

Suppose the X_i's are ranked from 1 to n. Then when H_a is true, larger ranks tend to occur later in the sequence, whereas if H_0 is true, large and small ranks tend to be mixed together. Let R_i be the rank of X_i and consider the test statistic $D = \sum_{i=1}^{n} (R_i - i)^2$. Then small values of D give support to H_a (e.g., the smallest value is 0 for $R_1 = 1, R_2 = 2, \ldots, R_n = n$), so H_0 should be rejected in favor of H_a if $d \le c$. When H_0 is true, any sequence of ranks has probability $1/n!$. Use this to find c for which the test has a level as close to .10 as possible in the case $n = 4$. [*Hint*: List the 4! rank sequences, compute d for each one, and then obtain the null distribution of D. See the Lehmann book (in the chapter bibliography), p. 290, for more information.]

14.2 *The Wilcoxon Rank-Sum Test

When at least one of the sample sizes in a two-sample problem is small, the t test requires the assumption of normality (at least approximately). There are situations, though, in which an investigator would want to use a test that is valid even if the underlying distributions are quite nonnormal. We now describe such a test, called the **Wilcoxon rank-sum test**. An alternative name for the procedure is the Mann–Whitney test, though the Mann–Whitney test statistic is sometimes expressed in a slightly different form from that of the Wilcoxon test. The Wilcoxon test procedure is distribution-free because it will have the desired level of significance for a very large class of underlying distributions.

ASSUMPTIONS $X_1, \ldots, X_m$ and $Y_1, \ldots, Y_n$ are two independent random samples from continuous distributions with means μ_1 and μ_2, respectively. The X and Y distributions have the same shape and spread, the only possible difference between the two being in the values of μ_1 and μ_2.

When H_0: $\mu_1 - \mu_2 = \Delta_0$ is true, the X distribution is shifted by the amount Δ_0 to the right of the Y distribution; whereas when H_0 is false, the shift is by an amount other than Δ_0.

Development of the Test When $m = 3$, $n = 4$

Let's first test H_0: $\mu_1 - \mu_2 = 0$. If μ_1 is actually much larger than μ_2, then most of the observed x's will fall to the right of the observed y's. However, if H_0 is true, then the observed values from the two samples should be intermingled. The test statistic will provide a quantification of how much intermingling there is in the two samples.

Consider the case $m = 3$, $n = 4$. Then if all three observed x's were to the right of all four observed y's, this would provide strong evidence for rejecting H_0 in favor of H_a: $\mu_1 - \mu_2 \ne 0$, and a similar conclusion is appropriate if all three x's fall below all four of the y's. Suppose we pool the x's and y's into a combined sample of size $m + n = 7$ and rank these observations from smallest to largest, with the smallest receiving rank 1 and the largest, rank 7. If either most of the largest ranks or most of the smallest ranks were associated with X observations, we would begin to doubt H_0. This suggests the test statistic

$$W = \text{the sum of the ranks in the combined sample associated with } X \text{ observations} \tag{14.1}$$

For the values of m and n under consideration, the smallest possible value of W is $w = 1 + 2 + 3 = 6$ (if all three x's are smaller than all four y's), and the largest possible value is $w = 5 + 6 + 7 = 18$ (if all three x's are larger than all four y's).

As an example, suppose $x_1 = -3.10$, $x_2 = 1.67$, $x_3 = 2.01$, $y_1 = 5.27$, $y_2 = 1.89$, $y_3 = 3.86$, and $y_4 = .19$. Then the pooled ordered sample is -3.10, .19, 1.67, 1.89, 2.01, 3.86, and 5.27. The X ranks for this sample are 1 (for -3.10), 3 (for 1.67), and 5 (for 2.01), so the computed value of W is $w = 1 + 3 + 5 = 9$.

The test procedure based on the statistic (14.1) is to reject H_0 if the computed value w is "too extreme"—that is, $\geq c$ for an upper-tailed test, $\leq c$ for a lower-tailed test, and either $\geq c_1$ or $\leq c_2$ for a two-tailed test. The critical constant(s) c (c_1, c_2) should be chosen so that the test has the desired level of significance α. To see how this should be done, recall that when H_0 is true, all seven observations come from the same population. This means that under H_0, any possible triple of ranks associated with the three x's—such as (1, 4, 5), (3, 5, 6), or (5, 6, 7)—has the same probability as any other possible rank triple. Since there are $\binom{7}{3} = 35$ possible rank triples, under H_0 each rank triple has probability $\frac{1}{35}$. From a list of all 35 rank triples and the w value associated with each, the probability distribution of W can immediately be determined. For example, there are four rank triples that have w value 11—(1, 3, 7), (1, 4, 6), (2, 3, 6), and (2, 4, 5)—so $P(W = 11) = \frac{4}{35}$. The summary of the listing and computations appears in Table 14.3.

Table 14.3 Probability distribution of W ($m = 3$, $n = 4$) when H_0 is true

w	6	7	8	9	10	11	12	13	14	15	16	17	18
$P(W = w)$	$\frac{1}{35}$	$\frac{1}{35}$	$\frac{2}{35}$	$\frac{3}{35}$	$\frac{4}{35}$	$\frac{4}{35}$	$\frac{5}{35}$	$\frac{4}{35}$	$\frac{4}{35}$	$\frac{3}{35}$	$\frac{2}{35}$	$\frac{1}{35}$	$\frac{1}{35}$

The distribution of Table 14.3 is symmetric about the value $w = (6 + 18)/2 = 12$, which is the middle value in the ordered list of possible W values. This is because the two rank triples (r, s, t) (with $r < s < t$) and $(8 - t, 8 - s, 8 - r)$ have values of w symmetric about 12, so for each triple with w value below 12, there is a triple with w value above 12 by the same amount.

If the alternative hypothesis is H_a: $\mu_1 - \mu_2 > 0$, then H_0 should be rejected in favor of H_a for large W values. Choosing as the rejection region the set of W values {17, 18}, $\alpha = P(\text{type I error}) = P(\text{reject } H_0 \text{ when } H_0 \text{ is true}) = P(W = 17 \text{ or } 18 \text{ when } H_0 \text{ is true}) = \frac{1}{35} + \frac{1}{35} = \frac{2}{35} = .057$; the region {17, 18} therefore specifies a test with level of significance approximately .05. Similarly, the region {6, 7}, which is appropriate for H_a: $\mu_1 - \mu_2 < 0$, has $\alpha = .057 \approx .05$. The region {6, 7, 17, 18}, which is appropriate for the two-sided alternative, has $\alpha = \frac{4}{35} = .114$. The W value for the data given several paragraphs previously was $w = 9$, which is rather close to the middle value 12, so H_0 would not be rejected at any reasonable level α for any one of the three H_a's.

General Description of the Rank-Sum Test

The null hypothesis H_0: $\mu_1 - \mu_2 = \Delta_0$ is handled by subtracting Δ_0 from each X_i and using the $(X_i - \Delta_0)$'s as the X_i's were previously used. Recalling that for any positive integer K, the sum of the first K integers is $K(K + 1)/2$, the smallest possible value of the statistic W is $m(m + 1)/2$, which occurs when the $(X_i - \Delta_0)$'s are all to the left of the Y sample. The largest possible value of W occurs when the $(X_i - \Delta_0)$'s lie entirely to the right of the Y's; in this case, $W = (n + 1) + \cdots + (m + n) =$ (sum of first $m + n$ integers) $-$ (sum of first n integers), which gives $m(m + 2n + 1)/2$. As with the special case $m = 3, n = 4$, the distribution of W is symmetric about the value that is halfway between the smallest and largest values; this middle value is $m(m + n + 1)/2$. Because of this symmetry, probabilities involving lower-tail critical values can be obtained from corresponding upper-tail values.

Null hypothesis: $H_0\text{: } \mu_1 - \mu_2 = \Delta_0$

Test statistic value: $w = \sum_{i=1}^{m} r_i$ where r_i = rank of $(x_i - \Delta_0)$ in the combined sample of $m + n(x - \Delta_0)$'s and y's

Alternative Hypothesis	**Rejection Region**
$H_a\text{: } \mu_1 - \mu_2 > \Delta_0$	$w \geq c_1$
$H_a\text{: } \mu_1 - \mu_2 < \Delta_0$	$w \leq m(m + n + 1) - c_1$
$H_a\text{: } \mu_1 - \mu_2 \neq \Delta_0$	either $w \geq c$ or $w \leq m(m + n + 1) - c$

where $P(W \geq c_1$ when H_0 is true$) \approx \alpha$, $P(W \geq c$ when H_0 is true$) \approx \alpha/2$.

Because W has a discrete probability distribution, there will not always exist a critical value corresponding exactly to one of the usual levels of significance. Appendix Table A.14 gives upper-tail critical values for probabilities closest to .05, .025, .01, and .005, from which level .05 or .01 one- and two-tailed tests can be obtained. The table gives information only for $m = 3, 4, \ldots, 8$ and $n = m, m + 1, \ldots, 8$ (i.e., $3 \leq m \leq n \leq 8$). For values of m and n that exceed 8, a normal approximation can be used (Exercise 14). To use the table for small m and n, though, *the X and Y samples should be labeled so that* $m \leq n$.

Example 14.3 The urinary fluoride concentration (parts per million) was measured both for a sample of livestock grazing in an area previously exposed to fluoride pollution and for a similar sample grazing in an unpolluted region:

Polluted	21.3	18.7	23.0	17.1	16.8	20.9	19.7
Unpolluted	14.2	18.3	17.2	18.4	20.0		

Does the data indicate strongly that the true average fluoride concentration for livestock grazing in the polluted region is larger than for the unpolluted region? Use the Wilcoxon rank-sum test at level $\alpha = .01$.

The sample sizes here are 7 and 5. To obtain $m \leq n$, label the unpolluted observations as the x's ($x_1 = 14.2, \ldots, x_5 = 20.0$) and the polluted observations as the y's. Thus μ_1 is the true average fluoride concentration without pollution, and μ_2 is the true average concentration with pollution. The alternative hypothesis is H_a: $\mu_1 - \mu_2 < 0$ (pollution causes an increase in concentration), so a lower-tailed test is appropriate. From Appendix Table A.14 with $m = 5$ and $n = 7$, $P(W \geq 47 \text{ when } H_0 \text{ is true}) \approx .01$. The critical value for the lower-tailed test is therefore $m(m + n + 1) - 47 = 5(13) - 47 = 18$; H_0 will now be rejected if $w \leq 18$. The pooled ordered sample follows; the computed W is $w = r_1 + r_2 + \cdots + r_5$ (where r_i is the rank of x_i) $= 1 + 5 + 4 + 6 + 9 = 25$. Since 25 is not ≤ 18, H_0 is not rejected at (approximately) level .01.

x	y	y	x	x	x	y	y	x	y	y	y
14.2	16.8	17.1	17.2	18.3	18.4	18.7	19.7	20.0	20.9	21.3	23.0
1	2	3	4	5	6	7	8	9	10	11	12

■

Ties are handled as suggested for the signed-rank test in the previous section.

Efficiency of the Wilcoxon Rank-Sum Test

When the distributions being sampled are both normal with $\sigma_1 = \sigma_2$, and therefore have the same shapes and spreads, either the pooled t test or the Wilcoxon test can be used (the two-sample t test assumes normality but not equal variances, so assumptions underlying its use are more restrictive in one sense and less in another than those for Wilcoxon's test). In this situation, the pooled t test is best among all possible tests in the sense of minimizing β for any fixed α. However, an investigator can never be absolutely certain that underlying assumptions are satisfied. It is therefore relevant to ask (1) how much is lost by using Wilcoxon's test rather than the pooled t test when the distributions are normal with equal variances and (2) how W compares to T in nonnormal situations.

The notion of test efficiency was discussed in the previous section in connection with the one-sample t test and Wilcoxon signed-rank test. The results for the two-sample tests are the same as those for the one-sample tests. When normality and equal variances both hold, the rank-sum test is approximately 95% as efficient as the pooled t test in large samples. That is, the t test will give the same error probabilities as the Wilcoxon test using slightly smaller sample sizes. On the other hand, the Wilcoxon test will always be at least 86% as efficient as the pooled t test and may be much more efficient if the underlying distributions are very nonnormal, especially with heavy tails. The comparison of the Wilcoxon test with the two-sample (unpooled) t test is less clear-cut. The t test is not known to be the best test in any sense, so it seems safe to conclude that as long as the population distributions have similar shapes and spreads, the behavior of the Wilcoxon test should compare quite favorably to the two-sample t test.

Lastly, we note that β calculations for the Wilcoxon test are quite difficult. This is because the distribution of W when H_0 is false depends not only on $\mu_1 - \mu_2$ but also on the shapes of the two distributions. For most underlying distributions, the nonnull distribution of W is virtually intractable. This is why statisticians have developed large-sample (asymptotic relative) efficiency as a means of comparing tests. With the capabilities of modern-day computer software, another approach to calculation of β is to carry out a simulation experiment.

Exercises Section 14.2 (9–16)

9. In an experiment to compare the bond strength of two different adhesives, each adhesive was used in five bondings of two surfaces, and the force necessary to separate the surfaces was determined for each bonding. For adhesive 1, the resulting values were 229, 286, 245, 299, and 250, whereas the adhesive 2 observations were 213, 179, 163, 247, and 225. Let μ_i denote the true average bond strength of adhesive type i. Use the Wilcoxon rank-sum test at level .05 to test H_0: $\mu_1 = \mu_2$ versus H_a: $\mu_1 > \mu_2$.

10. The article "A Study of Wood Stove Particulate Emissions" (*J. Air Pollution Control Assoc.*, 1979: 724–728) reports the following data on burn time (hours) for samples of oak and pine. Test at level .05 to see whether there is any difference in true average burn time for the two types of wood.

Oak	1.72	.67	1.55	1.56	1.42	1.23	1.77	.48
Pine	.98	1.40	1.33	1.52	.73	1.20		

11. A modification has been made to the process for producing a certain type of "time-zero" film (film that begins to develop as soon as a picture is taken). Because the modification involves extra cost, it will be incorporated only if sample data strongly indicates that the modification has decreased true average developing time by more than 1 second. Assuming that the developing-time distributions differ only with respect to location if at all, use the Wilcoxon rank-sum test at level .05 on the accompanying data to test the appropriate hypotheses.

Original Process	8.6	5.1	4.5	5.4	6.3	6.6	5.7	8.5
Modified Process	5.5	4.0	3.8	6.0	5.8	4.9	7.0	5.7

12. The article "Measuring the Exposure of Infants to Tobacco Smoke" (*New Engl. J. Med.*, 1984: 1075–1078) reports on a study in which various measurements were taken both from a random sample of infants who had been exposed to household smoke and from a sample of unexposed infants. The accompanying data consists of observations on urinary concentration of cotinine, a major metabolite of nicotine (the values constitute a subset of the original data and were read from a plot that appeared in the article). Does the data suggest that true average cotinine level is higher in exposed infants than in unexposed infants by more than 25? Carry out a test at significance level .05.

Unexposed	8	11	12	14	20	43	111	
Exposed	35	56	83	92	128	150	176	208

13. Reconsider the situation described in Exercise 100 of Chapter 10 and the accompanying MINITAB output (the Greek letter eta is used to denote a median).

```
Mann-Whitney Confidence Interval and Test
good          N = 8          Median = 0.540
poor          N = 8          Median = 2.400
Point estimate for ETA1-ETA2 is     -1.155
95.9 Percent CI for ETA1-ETA2 is(-3.160,
-0.409) W = 41.0
Test of ETA1 = ETA2 vs ETA1 < ETA2 is
significant at 0.0027
```

a. Verify that the value of MINITAB's test statistic is correct.

b. Carry out an appropriate test of hypotheses using a significance level of .01.

14. The Wilcoxon rank-sum statistic can be represented as $W = R_1 + R_2 + \cdots + R_m$, where R_i is the rank of $X_i - \Delta_0$ among all $m + n$ such differences. When H_0 is true, each R_i is equally likely to be one of the first $m + n$ positive integers; that is, R_i has a discrete uniform distribution on the values $1, 2, 3, \ldots, m + n$.

a. Determine the mean value of each R_i when H_0 is true and then show that the mean value of W is $m(m + n + 1)/2$. *Hint*: Use the hint given in Exercise 6(a).

b. The variance of each R_i is easily determined. However, the R_i's are not independent random variables because, for example, if $m = n = 10$ and we are told that $R_1 = 5$, then R_2 must be one of the *other* 19 integers between 1 and 20. However, if a and b are any two distinct positive integers between 1 and $m + n$ inclusive, $P(R_i = a \text{ and } R_j = b) = 1/[(m + n)(m + n - 1)]$ since two integers are being sampled without replacement from among $1, 2, \ldots, m + n$. Use this fact to show that $\text{Cov}(R_i, R_j) = -(m + n + 1)/12$ and then show that the variance of W is $mn(m + n + 1)/12$.

c. A central limit theorem for a sum of *non-independent* variables can be used to show that when $m > 8$ and $n > 8$, W has approximately a normal distribution with mean and variance given by the results of (a) and (b). Use this to propose a large-sample standardized rank-sum test statistic and then describe the rejection region that has approximate significance level α for testing H_0 against each of the three commonly encountered alternative hypotheses. *Note*: When there are ties in the observed values, a correction for the variance derived in (b) should be used in standardizing W; please consult a book on nonparametric statistics for the result.

15. The accompanying data resulted from an experiment to compare the effects of vitamin C in orange juice and in synthetic ascorbic acid on the length of odontoblasts in guinea pigs over a 6-week period ("The Growth of the Odontoblasts of the Incisor Tooth as a Criterion of the Vitamin C Intake of the Guinea Pig," *J. Nutrit.*, 1947: 491–504). Use the Wilcoxon rank-sum test at level .01 to decide whether true average length differs for the two types of vitamin C intake. Compute also an approximate P-value. [*Hint*: See Exercise 14.]

Orange Juice	8.2	9.4	9.6	9.7	10.0	14.5
	15.2	16.1	17.6	21.5		
Ascorbic Acid	4.2	5.2	5.8	6.4	7.0	7.3
	10.1	11.2	11.3	11.5		

16. Test the hypotheses suggested in Exercise 15 using the following data:

Orange Juice	8.2	9.5	9.5	9.7	10.0	14.5
	15.2	16.1	17.6	21.5		
Ascorbic Acid	4.2	5.2	5.8	6.4	7.0	7.3
	9.5	10.0	11.5	11.5		

[*Hint*: See Exercise 14.]

14.3 *Distribution-Free Confidence Intervals

The method we have used so far to construct a confidence interval (CI) can be described as follows: Start with a random variable (Z, T, χ^2, F, or the like) that depends on the parameter of interest and a probability statement involving the variable, manipulate the inequalities of the statement to isolate the parameter between random endpoints, and finally substitute computed values for random variables. Another general method for obtaining CIs takes advantage of a relationship between test procedures and CIs. A $100(1-\alpha)\%$ CI for a parameter θ can be obtained from a level α test for H_0: $\theta = \theta_0$ versus H_a: $\theta \neq \theta_0$. This method will be used to derive intervals associated with the Wilcoxon signed-rank test and the Wilcoxon rank-sum test.

Before using the method to derive new intervals, reconsider the t test and the t interval. Suppose a random sample of $n = 25$ observations from a normal population yields summary statistics $\bar{x} = 100$, $s = 20$. Then a 90% CI for μ is

$$\left(\bar{x} - t_{.05,24}\cdot\frac{s}{\sqrt{25}},\ \ \bar{x} + t_{.05,24}\cdot\frac{s}{\sqrt{25}}\right) = (93.16, 106.84) \qquad (14.2)$$

Suppose that instead of a CI, we had wished to test a hypothesis about μ. For H_0: $\mu = \mu_0$ versus H_a: $\mu \neq \mu_0$, the t test at level .10 specifies that H_0 should be rejected if t is either ≥ 1.711 or ≤ -1.711, where

$$t = \frac{\bar{x} - \mu_0}{s/\sqrt{25}} = \frac{100 - \mu_0}{20/\sqrt{25}} = \frac{100 - \mu_0}{4} \qquad (14.3)$$

Consider now the null value $\mu_0 = 95$. Then $t = 1.25$, so H_0 is not rejected. Similarly, if $\mu_0 = 104$, then $t = -1$, so again H_0 is not rejected. However, if $\mu_0 = 90$, then $t = 2.5$, so H_0 is rejected, and if $\mu_0 = 108$, then $t = -2$, so H_0 is again rejected. By considering other values of μ_0 and the decision resulting from each one, the following general fact emerges: *Every number inside the interval* (14.2) *specifies a value of* μ_0 *for which t of* (14.3) *leads to nonrejection of* H_0, *whereas every number outside interval* (14.2) *corresponds to a t for which* H_0 *is rejected*. That is, for the fixed values of n, $\bar{x}$, and s, the set of all μ_0 values for which testing H_0: $\mu = \mu_0$ versus H_a: $\mu \neq \mu_0$ results in nonrejection of H_0 is precisely the interval (14.2).

PROPOSITION Suppose we have a level α test procedure for testing H_0: $\theta = \theta_0$ versus H_a: $\theta \neq \theta_0$. For fixed sample values, let A denote the set of all values θ_0 for which H_0 is not rejected. Then A is a $100(1 - \alpha)\%$ CI for θ.

There are actually pathological examples in which the set A defined in the proposition is not an interval of θ values, but instead the complement of an interval or something even stranger. To be more precise, we should really replace the notion of a CI with that of a confidence set. In the cases of interest here, the set A does turn out to be an interval.

The Wilcoxon Signed-Rank Interval

To test H_0: $\mu = \mu_0$ versus H_a: $\mu \neq \mu_0$ using the Wilcoxon signed-rank test, where μ is the mean of a continuous symmetric distribution, the absolute values $|x_1 - \mu_0|, \ldots, |x_n - \mu_0|$ are ordered from smallest to largest, with the smallest receiving rank 1 and the largest, rank n. Each rank is then given the sign of its associated $x_i - \mu_0$, and the test statistic is the sum of the positively signed ranks. The two-tailed test rejects H_0 if s_+ is either $\geq c$ or $\leq n(n + 1)/2 - c$, where c is obtained from Appendix Table A.13 once the desired level of significance α is specified. For fixed $x_1, \ldots, x_n$, the $100(1 - \alpha)\%$ signed-rank interval will consist of all μ_0 for which H_0: $\mu = \mu_0$ is not rejected at level α. To identify this interval, it is convenient to express the test statistic S_+ in another form.

$$S_+ = \text{the number of pairwise averages } (X_i + X_j)/2 \text{ with } i \leq j \text{ that are } \geq \mu_0 \tag{14.4}$$

That is, if we average each x_j in the list with each x_i to its left, including $(x_j + x_j)/2$ (which is just x_j), and count the number of these averages that are $\geq \mu_0$, s_+ results. In moving from left to right in the list of sample values, we are simply averaging every pair of observations in the sample [again including $(x_j + x_j)/2$] exactly once, so the order in which the observations are listed before averaging is not important. The equivalence of the two methods for computing s_+ is not difficult to verify. The number of pairwise averages is $\binom{n}{2} + n$ (the first term due to averaging of different observations and the second due to averaging each x_i with itself), which equals $n(n + 1)/2$. If either too many or too few of these pairwise averages are $\geq \mu_0$, H_0 is rejected.

Example 14.4 The following observations are values of cerebral metabolic rate for rhesus monkeys: $x_1 = 4.51$, $x_2 = 4.59$, $x_3 = 4.90$, $x_4 = 4.93$, $x_5 = 6.80$, $x_6 = 5.08$, $x_7 = 5.67$. The 28 pairwise averages are, in increasing order,

4.51	4.55	4.59	4.705	4.72	4.745	4.76	4.795	4.835	4.90
4.915	4.93	4.99	5.005	5.08	5.09	5.13	5.285	5.30	5.375
5.655	5.67	5.695	5.85	5.865	5.94	6.235	6.80		

The first few and the last few of these are pictured on a measurement axis in Figure 14.2.

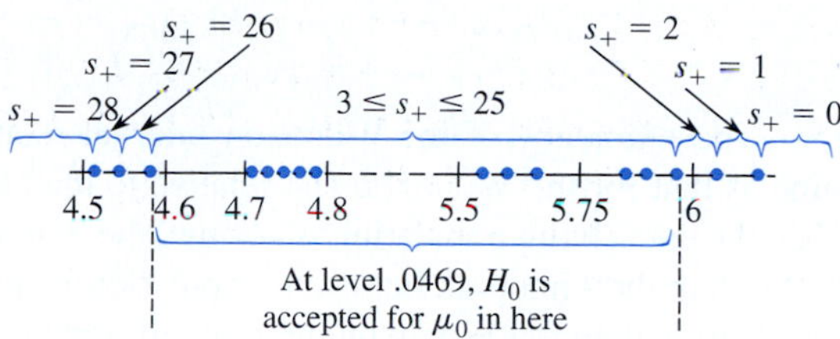

Figure 14.2 Plot of the data for Example 14.4

Because of the discreteness of the distribution of S_+, $\alpha = .05$ cannot be obtained exactly. The rejection region $\{0, 1, 2, 26, 27, 28\}$ has $\alpha = .046$, which is as close as possible to .05, so the level is approximately .05. Thus if the number of pairwise averages $\geq \mu_0$ is between 3 and 25, inclusive, H_0 is not rejected. From Figure 14.2 the (approximate) 95% CI for μ is (4.59, 5.94). ■

In general, once the pairwise averages are ordered from smallest to largest, the endpoints of the Wilcoxon interval are two of the "extreme" averages. To express this precisely, let the smallest pairwise average be denoted by $\overline{x}_{(1)}$, the next smallest by $\overline{x}_{(2)}, \ldots,$ and the largest by $\overline{x}_{(n(n+1)/2)}$.

PROPOSITION If the level α Wilcoxon signed-rank test for H_0: $\mu = \mu_0$ versus H_a: $\mu \neq \mu_0$ is to reject H_0 if either $s_+ \geq c$ or $s_+ \leq n(n+1)/2 - c$, then a $100(1-\alpha)\%$ CI for μ is

$$(\overline{x}_{(n(n+1)/2-c+1)}, \overline{x}_{(c)}) \qquad (14.5)$$

In words, the interval extends from the dth smallest pairwise average to the dth largest average, where $d = n(n+1)/2 - c + 1$. Appendix Table A.15 gives the values of c that correspond to the usual confidence levels for $n = 5, 6, \ldots, 25$.

Example 14.5 (Example 14.4 continued) For $n = 7$, an 89.1% interval (approximately 90%) is obtained by using $c = 24$ (since the rejection region $\{0, 1, 2, 3, 4, 24, 25, 26, 27, 28\}$ has $\alpha = .109$). The interval is $(\overline{x}_{(28-24+1)}, \overline{x}_{(24)}) = (\overline{x}_{(5)}, \overline{x}_{(24)}) = (4.72, 5.85)$, which extends from the fifth smallest to the fifth largest pairwise average. ■

The derivation of the interval depended on having a single sample from a continuous symmetric distribution with mean (median) μ. When the data is paired, the interval constructed from the differences $d_1, d_2, \ldots, d_n$ is a CI for the mean (median) difference μ_D. In this case, the symmetry of X and Y distributions need not be assumed; as long as the X and Y distributions have the same shape, the $X - Y$ distribution will be symmetric, so only continuity is required.

For $n > 20$, the large-sample approximation (Exercise 6) to the Wilcoxon test based on standardizing S_+ gives an approximation to c in (14.5). The result [for a $100(1 - \alpha)\%$ interval] is

$$c \approx \frac{n(n + 1)}{4} + z_{\alpha/2}\sqrt{\frac{n(n + 1)(2n + 1)}{24}}$$

The efficiency of the Wilcoxon interval relative to the t interval is roughly the same as that for the Wilcoxon test relative to the t test. In particular, for large samples when the underlying population is normal, the Wilcoxon interval will tend to be slightly longer than the t interval, but if the population is quite nonnormal (symmetric but with heavy tails), then the Wilcoxon interval will tend to be much shorter than the t interval. And as we emphasized earlier in our discussion of bootstrapping, in the presence of nonnormality the actual confidence level of the t interval may differ considerably from the nominal (e.g. 95%) level.

The Wilcoxon Rank-Sum Interval

The Wilcoxon rank-sum test for testing H_0: $\mu_1 - \mu_2 = \Delta_0$ is carried out by first combining the $(X_i - \Delta_0)$'s and Y_j's into one sample of size $m + n$ and ranking them from smallest (rank 1) to largest (rank $m + n$). The test statistic W is then the sum of the ranks of the $(X_i - \Delta_0)$'s. For the two-sided alternative, H_0 is rejected if w is either too small or too large.

To obtain the associated CI for fixed x_i's and y_j's, we must determine the set of all Δ_0 values for which H_0 is not rejected. This is easiest to do if we first express the test statistic in a slightly different form. The smallest possible value of W is $m(m + 1)/2$, corresponding to every $(X_i - \Delta_0)$ less than every Y_j, and there are mn differences of the form $(X_i - \Delta_0) - Y_j$. A bit of manipulation gives

$$\begin{aligned} W &= \left[\text{number of } (X_i - Y_j - \Delta_0)\text{'s} \geq 0\right] + \frac{m(m + 1)}{2} \qquad (14.6) \\ &= \left[\text{number of } (X_i - Y_j)\text{'s} \geq \Delta_0\right] + \frac{m(m + 1)}{2} \end{aligned}$$

Thus rejecting H_0 if the number of $(x_i - y_j)$'s $\geq \Delta_0$ is either too small or too large is equivalent to rejecting H_0 for small or large w.

Expression (14.6) suggests that we compute $x_i - y_j$ for each i and j and order these mn differences from smallest to largest. Then if the null value Δ_0 is neither smaller than most of the differences nor larger than most, H_0: $\mu_1 - \mu_2 = \Delta_0$ is not rejected. Varying Δ_0 now shows that a CI for $\mu_1 - \mu_2$ will have as its lower endpoint one of the ordered $(x_i - y_i)$'s, and similarly for the upper endpoint.

PROPOSITION

Let $x_1, \ldots, x_m$ and $y_1, \ldots, y_n$ be the observed values in two independent samples from continuous distributions that differ only in location (and not in shape). With $d_{ij} = x_i - y_j$ and the ordered differences denoted by $d_{ij(1)}, d_{ij(2)}, \ldots, d_{ij(mn)}$, the general form of a $100(1-\alpha)\%$ CI for $\mu_1 - \mu_2$ is

$$(d_{ij(mn-c+1)}, d_{ij(c)}) \tag{14.7}$$

where c is the critical constant for the two-tailed level α Wilcoxon rank-sum test.

Notice that the form of the Wilcoxon rank-sum interval (14.7) is very similar to the Wilcoxon signed-rank interval (14.5); (14.5) uses pairwise averages from a single sample, whereas (14.7) uses pairwise differences from two samples. Appendix Table A.16 gives values of c for selected values of m and n.

Example 14.6 The article "Some Mechanical Properties of Impregnated Bark Board" (*Forest Products J.*, 1977: 31–38) reports the following data on maximum crushing strength (psi) for a sample of epoxy-impregnated bark board and for a sample of bark board impregnated with another polymer:

Epoxy (x's)	10,860	11,120	11,340	12,130	14,380	13,070
Other (y's)	4,590	4,850	6,510	5,640	6,390	

Obtain a 95% CI for the true average difference in crushing strength between the epoxy-impregnated board and the other type of board.

From Appendix Table A.16, since the smaller sample size is 5 and the larger sample size is 6, $c = 26$ for a confidence level of approximately 95%. The d_{ij}'s appear in Table 14.4. The five smallest d_{ij}'s $[d_{ij(1)}, \ldots, d_{ij(5)}]$ are 4350, 4470, 4610, 4730, and 4830; and the five largest d_{ij}'s are (in descending order) 9790, 9530, 8740, 8480, and 8220. Thus the CI is $(d_{ij(5)}, d_{ij(26)}) = (4830, 8220)$.

Table 14.4 Differences (d_{ij}) for the rank-sum interval in Example 14.6

		y_j: 4590	4850	5640	6390	6510
x_i	**10,860**	6270	6010	5220	4470	4350
	11,120	6530	6270	5480	4730	4610
	11,340	6750	6490	5700	4950	4830
	12,130	7540	7280	6490	5740	5620
	13,070	8480	8220	7430	6680	6560
	14,380	9790	9530	8740	7990	7870

■

When m and n are both large, the Wilcoxon test statistic has approximately a normal distribution (Exercise 14). This can be used to derive a large-sample approximation for the value c in interval (14.7). The result is

$$c \approx \frac{mn}{2} + z_{\alpha/2}\sqrt{\frac{mn(m+n+1)}{12}} \tag{14.8}$$

As with the signed-rank interval, the rank-sum interval (14.7) is quite efficient with respect to the t interval; in large samples, (14.7) will tend to be only a bit longer than the t interval when the underlying populations are normal and may be considerably shorter than the t interval if the underlying populations have heavier tails than do normal populations. And once again, the actual confidence level for the t interval may be quite different from the nominal level in the presence of substantial nonnormality.

Exercises Section 14.3 (17–22)

17. The article "The Lead Content and Acidity of Christchurch Precipitation" (*New Zeal. J. Sci.*, 1980: 311–312) reports the accompanying data on lead concentration (μg/L) in samples gathered during eight different summer rainfalls: 17.0, 21.4, 30.6, 5.0, 12.2, 11.8, 17.3, and 18.8. Assuming that the lead-content distribution is symmetric, use the Wilcoxon signed-rank interval to obtain a 95% CI for μ.

18. Compute the 99% signed-rank interval for true average pH μ (assuming symmetry) using the data in Exercise 3. [*Hint*: Try to compute only those pairwise averages having relatively small or large values (rather than all 105 averages).]

19. Compute a CI for μ_D of Example 14.2 using the data given there; your confidence level should be roughly 95%.

20. The following observations are amounts of hydrocarbon emissions resulting from road wear of bias-belted tires under a 522-kg load inflated at 228 kPa and driven at 64 km/hr for 6 hours ("Characterization of Tire Emissions Using an Indoor Test Facility," *Rubber Chem. Tech.*, 1978: 7–25): .045, .117, .062, and .072. What confidence levels are achievable for this sample size using the signed-rank interval? Select an appropriate confidence level and compute the interval.

21. Compute the 90% rank-sum CI for $\mu_1 - \mu_2$ using the data in Exercise 9.

22. Compute a 99% CI for $\mu_1 - \mu_2$ using the data in Exercise 10.

14.4 *Bayesian Methods

Consider making an inference about some parameter θ. The "frequentist" or "classical" approach, which we have followed until now in this book, is to regard the value of θ as fixed but unknown, observe data from a joint pmf or pdf $f(x_1, \ldots, x_n; \theta)$, and use the observations to draw appropriate conclusions. The Bayesian or "subjective" paradigm is different. Again the value of θ is unknown, but Bayesians say that all available information about it—intuition, data from past experiments, expert opinions, and so on—can be incorporated into a *prior distribution*, usually a prior pdf $g(\theta)$ since there will typically be a continuum of possible values of the parameter rather than just a discrete set. If there is substantial knowledge about θ, the prior will be quite peaked and highly concentrated about some central value, whereas a lack of information manifests itself in a relatively flat "uninformative" prior. These possibilities are illustrated in Figure 14.3.

In essence we are now thinking of the actual value of θ as the observed value of a random variable Θ, though unfortunately we ourselves don't get to observe the value. The (prior) distribution of this random variable is $g(\theta)$. Now, just as in the frequentist

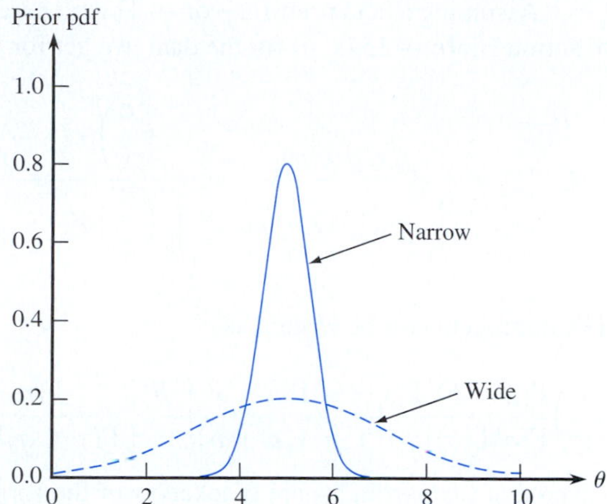

Figure 14.3 A narrow, concentrated prior and a wider, less informative prior

scenario, an experiment is performed to obtain data. The joint pmf or pdf of the data given the value of θ is $p(x_1, \ldots, x_n | \theta)$ or $f(x_1, \ldots, x_n | \theta)$. We use a vertical line segment here rather than the earlier semicolon to emphasize that we are conditioning on the value of a random variable.

At this point, an appropriate version of Bayes' theorem is used to obtain the *posterior* distribution $h(\theta | x_1, \ldots, x_n)$ of the parameter. In the Bayesian world, this posterior distribution contains all current information about θ. In particular, the mean of this posterior distribution gives a point estimate of the parameter. An interval $[a, b]$ having posterior probability .95 gives a 95% *credibility* interval, the Bayesian analog of a 95% confidence interval (but the interpretation is different). After presenting the necessary version of Bayes' theorem, we illustrate the Bayesian approach with two examples.

Bayes' theorem here needs to be a bit more general than in Section 2.4 to allow for the possibility of continuous distributions. This version gives the posterior distribution $h(\theta | x_1, x_2, \ldots, x_n)$ as a product of the prior pdf times the conditional pdf, with a denominator to ensure that the total posterior probability is 1:

$$h(\theta | x_1, x_2, \ldots, x_n) = \frac{f(x_1, x_2, \ldots, x_n | \theta) g(\theta)}{\int_{-\infty}^{\infty} f(x_1, x_2, \ldots, x_n | \theta) g(\theta)\, d\theta}$$

Example 14.7 Suppose we want to make an inference about a population proportion p. Since the value of this parameter must be between 0 and 1, and the family of standard beta distributions is concentrated on the interval [0, 1], a particular beta distribution is a natural choice for a prior on p. In particular, consider data from a survey of 1574 American adults reported by the National Science Foundation in May 2002. Of those responding, 803 (51%) incorrectly said that antibiotics kill viruses. In accord with the discussion in Section 3.5, the data can be considered either a random sample of size 1574 from the Bernoulli distribution (binomial with number of trials = 1) or a single observation from the binomial distribution with $n = 1574$. We use the latter approach here, but Exercise 23 involves showing that the Bernoulli approach is equivalent.

Assuming a beta prior for p on [0,1] with parameters a and b and the binomial distribution Bin(n = 1574, p) for the data, we get for the posterior distribution,

$$h(p\,|\,x) = \frac{f(x\,|\,p)g(p)}{\int_{-\infty}^{\infty} f(x\,|\,p)g(p)\,dp} = \frac{\binom{n}{x}p^x(1-p)^{n-x}\frac{\Gamma(a+b)}{\Gamma(a)\Gamma(b)}p^{a-1}(1-p)^{b-1}}{\int_0^1 \binom{n}{x}p^x(1-p)^{n-x}\frac{\Gamma(a+b)}{\Gamma(a)\Gamma(b)}p^{a-1}(1-p)^{b-1}\,dp}$$

The numerator can be written as

$$\binom{n}{x}\frac{\Gamma(a+b)}{\Gamma(a)\Gamma(b)}\,\frac{\Gamma(x+a)\Gamma(n-x+b)}{\Gamma(n+a+b)}\left[\frac{\Gamma(n+a+b)}{\Gamma(x+a)\Gamma(n-x+b)}p^{x+a-1}(1-p)^{n-x+b-1}\right]$$

Given that the part in square brackets is of the form of a beta pdf on [0, 1], its integral over this interval is 1. The part in front of the square brackets is shared by the numerator and denominator, and will therefore cancel. Thus

$$h(p\,|\,x) = \frac{\Gamma(n+a+b)}{\Gamma(x+a)\Gamma(n-x+b)}p^{x+a-1}(1-p)^{n-x+b-1}$$

That is, the posterior distribution of p is itself a beta distribution with parameters $x + a$ and $n - x + b$.

If we were using the traditional non-Bayesian frequentist approach to statistics, and we wanted to give an estimate of p for this example, we would give the usual estimate from Section 8.2, x/n = 803/1574 = .51. The usual Bayesian estimate is the posterior mean, the expected value of p given the data. Recalling that the mean of the beta distribution on [0, 1] is $\alpha/(\alpha + \beta)$, we obtain

$$E(p\,|\,x) = (x + a)/(n + a + b) = (803 + a)/(1574 + a + b)$$

for the posterior mean.

Suppose that $a = b = 1$, so the beta prior distribution reduces to the uniform distribution on [0, 1]. Then $E(p\,|\,x) = (803 + 1)/(1574 + 2) = .51$, and in this case the Bayesian and frequentist results are essentially the same. It should be apparent that, if a and b are small compared to n, then the prior distribution will not matter much. Indeed, if a and b are close to 0 and positive, then $E(p\,|\,x) \approx x/n$. We should hesitate to set a and b equal to 0, because this would make the beta prior pdf not integrable, but it does nevertheless give a reasonable posterior distribution if x and $n - x$ are positive. When a prior distribution is not integrable it is said to be **improper**.

In Bayesian inference, is there an interval corresponding to the confidence interval for p given in Section 8.2? We have the posterior distribution for p, so we can take the central 95% of this distribution and call it a 95% credibility interval, as mentioned at the beginning of this section. In the case with beta prior and $a = 1$, $b = 1$, we have a beta posterior with $\alpha = 804$, $\beta = 772$. Using the inverse cumulative beta distribution function from MINITAB (or almost any major statistical package) evaluated at .025 and .975, we obtain the interval (.4855, .5348). For comparison the 95% confidence interval from Equation (8.10) of Section 8.2 is (.4855, .5348). The intervals are not exactly the same, although they do agree to four decimals. The simpler formula, Equation (8.11), gives the answer (.4855, .5349), which is very close because of the large sample size.

It is interesting that, although the frequentist and Bayesian intervals agree to four decimals, they have very different interpretations. For the Bayesian interval we can say that the probability is .95 that p is in the interval, given the data. However, this is not correct for the frequentist interval, because p is not random and the endpoints are not random after they have been specified, and therefore no probability statement is appropriate. Here the .95 applies to the aggregate of confidence intervals, of which in the long run 95% should include the true p.

The confidence intervals and credibility interval all include .5, so they allow the possibility that $p = .5$. Another way to view this possibility in Bayesian terms is to see whether the posterior distribution is consistent with $p = .5$. We actually consider the related hypothesis $p \le .5$. Using $a = 1$ and $b = 1$ again, we find from MINITAB that the beta distribution with $\alpha = 804$ and $\beta = 772$ has probability .2100 of being less than or equal to .5. The corresponding one-tailed frequentist P-value is the probability, assuming $p = .5$, of at least 803 successes in 1574 trials, which is .2173. Both the Bayesian and frequentist values are much greater than .05, and there is no reason to reject .5 as a possible value for p.

To clarify the relationship between $E(p\,|\,x)$ and x/n, we can write $E(p\,|\,x)$ as a weighted average of the prior mean $a/(a + b)$ and x/n.

$$E(p\,|\,x) = \frac{a+b}{n+a+b}\,\frac{a}{a+b} + \frac{n}{n+a+b}\,\frac{x}{n}$$

The weights can be interpreted in terms of the sum of the two parameters of the beta distribution, which is often called the ***concentration parameter***. The weights are proportional to the concentration parameter $a + b$ of the prior distribution and the number n of observations. The weight of the prior depends on the size of $a + b$ in relation to n, and the concentration parameter of the posterior distribution is the total $a + b + n$.

It is also useful to interpret the posterior pdf in terms of the concentration parameter. Because the first parameter is the sum $x + a$ and the second is the sum $(n - x) + b$, the effect of a is to add to the number of successes and the effect of b is to add to the number of failures. In particular, setting a to 1 and b to 1 resulted in a posterior with the equivalent of 803 + 1 successes and (1574 − 803) + 1 failures, for a total of 1574 + 2 observations. From this viewpoint, the total observations are the $a + b$ provided by the prior plus the n provided by the data, and this addition also gives the concentration parameter of the posterior in terms of the concentration parameter of the prior.

How should we specify the prior distribution? The beta distribution is convenient, because it is easy with this specification to find the posterior distribution, but what about a and b? Suppose we have asked 10 adults about the effect of antibiotics on viruses, and it is reasonable to assume that the 10 are a random sample. If 6 of the 10 say that antibiotics kill viruses, then we set $a = 6$ and $b = 10 - 6 = 4$. That is, we have a beta-distributed prior with parameters 6 and 4. Then the posterior distribution is beta with parameters $803 + 6 = 809$ and $(1574 - 803) + 4 = 775$. The posterior is the same as if we had started with $a = 0$ and $b = 0$ and observed 809 who said that antibiotics kill viruses and 775 who said no. In other words, observations can be incorporated into the prior and count just as if they were part of the NSF survey. ■

Life in the Bayesian world is sometimes more complicated. Perhaps the prior observations are not of a quality equivalent to that of the survey, but we would still like to use them to form a prior distribution. If we regard them as being only half as good, then we could use the same proportions but cut the a and b in half, using 3 and 2 instead of 6 and 4. There is certainly a subjective element to this, and it suggests why some statisticians are hesitant about using Bayesian methods. When everyone can agree about the prior distribution, there is little controversy about the Bayesian procedure, but when the prior is very much a matter of opinion people tend to disagree about its value.

Example 14.8 Assume a random sample $X_1, X_2, \ldots, X_n$ from the normal distribution with known variance, and assume a normal prior distribution for μ. In particular, consider the IQ scores of 18 first-grade boys

113 108 140 113 115 146 136 107 108 119 132 127 118
108 103 103 122 111

from the private speech data introduced in Example 1.2. Because the IQ has a standard deviation of 15 nationwide, we assume $\sigma = 15$ is valid here. For the prior distribution it is reasonable to use a mean of $\mu_0 = 110$, a ballpark figure for previous years in this school. It is harder to prescribe a standard deviation for the prior, but we will use $\sigma_0 = 7.5$. This is the standard deviation for the average of four independent observations if the individual standard deviation is 15. As a result, the effect on the posterior mean will turn out to be the same as if there were four additional observations with average 110.

To compute the posterior distribution of the mean μ, we use Bayes' theorem:

$$h(\mu \mid x_1, x_2, \ldots, x_n) = \frac{f(x_1, x_2, \ldots, x_n \mid \mu)g(\mu)}{\displaystyle\int_{-\infty}^{\infty} f(x_1, x_2, \ldots, x_n \mid \mu)g(\mu)\, d\mu}$$

The numerator is

$$f(x_1, x_2, \ldots, x_n \mid \mu)g(\mu) = \frac{1}{\sqrt{2\pi}\sigma} e^{-.5(x_1-\mu)^2/\sigma^2} \cdot \ldots \cdot \frac{1}{\sqrt{2\pi}\sigma} e^{-.5(x_n-\mu)^2/\sigma^2}$$

$$\times \frac{1}{\sqrt{2\pi}\sigma_0} e^{-.5(\mu-\mu_0)^2/\sigma_0^2}$$

$$= \frac{1}{(2\pi)^{(n+1)/2}\sigma^n\sigma_0} e^{-.5[(x_1-\mu)^2/\sigma^2 + \cdots + (x_n-\mu)^2/\sigma^2 + (\mu-\mu_0)^2/\sigma_0^2]}$$

The trick here is to complete the square in the exponent, which yields

$$(-.5/\sigma_1^2)(\mu - \mu_1)^2 + C$$

where C does not involve μ and

$$\sigma_1^2 = \frac{1}{\dfrac{n}{\sigma^2} + \dfrac{1}{\sigma_0^2}} \qquad \mu_1 = \frac{\dfrac{\sum x_i}{\sigma^2} + \dfrac{\mu_0}{\sigma_0^2}}{\dfrac{n}{\sigma^2} + \dfrac{1}{\sigma_0^2}} = \frac{\dfrac{n\bar{x}}{\sigma^2} + \dfrac{\mu_0}{\sigma_0^2}}{\dfrac{n}{\sigma^2} + \dfrac{1}{\sigma_0^2}}$$

The posterior is then

$$h(\mu \mid x_1, x_2, \ldots, x_n) = \frac{\dfrac{\sigma_1}{(2\pi)^{n/2}\sigma^n\sigma_0} \cdot \dfrac{1}{(2\pi)^{.5}\sigma_1} e^{(-.5/\sigma_1^2)(\mu-\mu_1)^2} e^C}{\dfrac{\sigma_1}{(2\pi)^{n/2}\sigma^n\sigma_0} e^C \displaystyle\int_{-\infty}^{\infty} \frac{1}{(2\pi)^{.5}\sigma_1} e^{(-.5/\sigma_1^2)(\mu-\mu_1)^2}\, d\mu}$$

The integral is 1 because it is the area under a normal pdf, and the part in front of the integral cancels out, leaving a posterior distribution that is normal with mean μ_1 and standard deviation σ_1.

Notice that the posterior mean μ_1 is a weighted average of the prior mean μ_0 and the data mean $\bar{x}$, with weights that are the reciprocals of the prior variance and the variance of $\bar{x}$. It makes sense to define the **precision** as the reciprocal of the variance because a lower variance implies a more precise measurement, and the weights then are the corresponding precisions. Furthermore, the posterior variance is the reciprocal of the sum of the reciprocals of the two variances, but this can be described much more simply by saying that the posterior precision is the sum of the prior precision plus the precision of $\bar{x}$.

Numerically, we have

$$\frac{1}{\sigma_1^2} = \frac{1}{\sigma^2/n} + \frac{1}{\sigma_0^2} = \frac{1}{15^2/18} + \frac{1}{7.5^2} = .09778 = \frac{1}{10.227} = \frac{1}{3.198^2}$$

$$\mu_1 = \frac{\dfrac{n\bar{x}}{\sigma^2} + \dfrac{\mu_0}{\sigma_0^2}}{\dfrac{n}{\sigma^2} + \dfrac{1}{\sigma_0^2}} = \frac{\dfrac{18(118.28)}{15^2} + \dfrac{110}{7.5^2}}{\dfrac{18}{15^2} + \dfrac{1}{7.5^2}} = 116.77$$

The posterior distribution is normal with mean $\mu_1 = 116.77$ and standard deviation $\sigma_1 = 3.198$. The mean μ_1 is a weighted average of $\bar{x} = 118.28$ and $\mu_0 = 110$, so μ_1 is necessarily between them. As n becomes large the weight given to μ_0 declines, and μ_1 will be closer to $\bar{x}$.

Knowing the mean and standard deviation, we can use the normal distribution to find an interval with 95% probability for μ. This 95% credibility interval is (110.502, 123.038). For comparison the 95% confidence interval using $\bar{x} = 118.28$ and $\sigma = 15$ is $\bar{x} \pm 1.96\,\sigma/\sqrt{n} = (111.35, 125.21)$. Notice that this interval must be wider. Because the precisions add to give the posterior precision, the posterior precision is greater than the prior precision and it is greater than the data precision. Therefore, it is guaranteed that the posterior standard deviation σ_1 will be less than σ_0 and less than the data standard deviation $\sigma/\sqrt{n}$.

Both the credibility interval and the confidence interval exclude 110, so we can be fairly sure that μ exceeds 110. Another way of looking at this is to calculate the posterior probability of μ being less than or equal to 110. Using $\mu_1 = 116.77$ and $\sigma_1 = 3.198$, we obtain the probability .0171, so this too supports the idea that μ exceeds 110.

How should we go about choosing μ_0 and σ_0 for the prior distribution? Suppose we have four prior observations for which the mean is 110. The standard deviation of the mean is $15/\sqrt{4}$. We therefore choose $\mu_0 = 110$ and $\sigma_0 = 7.5$, the same values used for this example. If the four values are combined with the 18 values from the data set, then the mean of all 22 is $116.77 = \mu_1$ and the standard deviation is $15/\sqrt{22} = 3.198 = \sigma_1$. The 95% confidence interval for the mean, based on the average of all 22 observations, is the same as the Bayesian 95% credibility interval. This says that if you have some preliminary data values that are just as good as the regular data values that will be obtained, then base the prior distribution on the preliminary data. The posterior mean and its standard deviation will be the same as if the preliminary data were combined with the regular data, and the 95% credibility interval will be the same as the 95% confidence interval.

It should be emphasized that, even if the confidence interval is the same as the credibility interval, they have different interpretations. To interpret the Bayesian credibility interval, we can say that the probability is .95 that μ is in the interval (110.502, 123.038). However, for the frequentist confidence interval such a probability statement does not make sense because μ and the endpoints of the interval are all constants after the interval has been calculated. Instead we have the more complicated interpretation that, in repeated realizations of the confidence interval, 95% of the intervals will include the true μ in the long run.

What should be done if there are no prior observations and there are no strong opinions about the prior mean μ_0? In this case the prior standard deviation σ_0 can be taken as some large number much bigger than σ, such as $\sigma_0 = 1000$ in our example. The result is that the prior will have essentially no effect, and the posterior distribution will be based on the data, $\mu_1 = \bar{x} = 118.28$ and $\sigma_1 = \sigma = 15$. The 95% credibility interval will be the same as the 95% confidence interval based on the 18 observations, (111.35, 125.21), but of course the interpretation is different. ■

In both examples it turned out that the posterior distribution belongs to the same family as the prior distribution. When this happens we say that this distribution is **conjugate** to the data distribution. Exercises 31 and 32 offer additional examples of conjugate distributions.

Exercises Section 14.4 (23–32)

23. For the data of Example 14.7 assume a beta prior distribution and assume that the 1574 observations are a random sample from the Bernoulli distribution. Use Bayes' theorem to derive the posterior distribution, and compare your answer with the result of Example 14.7.

24. Here are the IQ scores for the 15 first-grade girls from the study mentioned in Example 14.8.

102	96	106	118	108	122	115	113
109	113	82	110	121	110	99	

Assume the same prior distribution used in Example 14.8, and assume that the data is a random sample from a normal distribution with mean μ and $\sigma = 15$.

a. Find the posterior distribution of μ.

b. Find a 95% credibility interval for μ.

c. Add four observations with average 110 to the data and find a 95% confidence interval for μ using the 19 observations. Compare with the result of (b).

d. Change the prior so the prior precision is very small but positive, and then recompute (a) and (b).

e. Find a 95% confidence interval for μ using the 15 observations and compare with the credibility interval of (d).

25. Laplace's rule of succession says that if there have been n Bernoulli trials and they have all been successes, then the probability of a success on the next trial is $(n + 1)/(n + 2)$. For the derivation Laplace used a beta prior with $a = 1$ and $b = 1$ for binomial data, as in Example 14.7.

a. Show that, if $a = 1$ and $b = 1$ and there are n successes in n trials, then the posterior mean of p is $(n + 1)/(n + 2)$.

b. Explain (a) in terms of total successes and failures; that is, explain the result in terms of two prior trials plus n later trials.

c. Laplace applied his rule of succession to compute the probability that the sun will rise tomorrow using 5000 years, or n = 1,826,214 days of history in which the sun rose every day. Is Laplace's method equivalent to including two prior days when the sun rose once and failed to rise once? Criticize the answer in terms of total successes and failures.

26. For the scenario of Example 14.8 assume the same normal prior distribution but assume that the data set is just one observation $\bar{x} = 118.28$ with standard deviation $\sigma/\sqrt{n} = 15/\sqrt{18} = 3.5355$. Use Bayes' theorem to derive the posterior distribution, and compare your answer with the result of Example 14.8.

27. Let X have the beta distribution on [0, 1] with parameters $\alpha = \nu_1/2$ and $\beta = \nu_2/2$, where $\nu_1/2$ and $\nu_2/2$ are positive integers. Define $Y = (X/\alpha)/[(1 - X)/\beta]$. Show that Y has the F distribution with degrees of freedom ν_1, ν_2.

28. In a study by Erich Brandt of 70 restaurant bills, 40 of the 70 were paid using cash. We assume a random sample and estimate the posterior distribution of the binomial parameter p, the population proportion paying cash.

a. Use a beta prior distribution with $a = 2$ and $b = 2$.

b. Use a beta prior distribution with $a = 1$ and $b = 1$.

c. Use a beta prior distribution with a and b very small and positive.

d. Calculate a 95% credibility interval for p using (c). Is your interval compatible with $p = .5$?

e. Calculate a 95% confidence interval for p using Equation (8.10) of Section 8.2, and compare with the result of (d).

f. Calculate a 95% confidence interval for p using Equation (8.11) of Section 8.2, and compare with the results of (d) and (e).

g. Compare the interpretations of the credibility interval and the confidence intervals.

h. Based on the prior in (c), test the hypothesis $p \le .5$ using the posterior distribution to find $P(p \le .5)$.

29. Exercise 27 gives an alternative way of finding beta probabilities when software for the beta distribution is unavailable.

a. Use Exercise 27 together with the F table to obtain a 90% credibility interval for Exercise 28(c). [*Hint*: To find c such that .05 is the probability that F is to the left of c, reverse the degrees of freedom and take the reciprocal of the value for α =.05.]

b. Repeat (a) using software for the beta distribution and compare with the result of (a).

30. If α and β are large, then the beta distribution can be approximated by the normal distribution using the beta mean and variance given in Section 4.5. This is useful in case beta distribution software is unavailable. Use the approximation to compute the credibility interval in Example 14.7.

31. Assume a random sample $X_1, X_2, \ldots, X_n$ from the Poisson distribution with mean λ. If the prior distribution for λ has a gamma distribution with parameters α and β, show that the posterior distribution is also gamma distributed. What are its parameters?

32. Consider a random sample $X_1, X_2, \ldots, X_n$ from the normal distribution with mean 0 and precision τ (use τ as a parameter instead of $\sigma^2 = 1/\tau$). Assume a gamma-distributed prior for τ and show that the posterior distribution of τ is also gamma. What are its parameters?

14.5 *Sequential Methods

Statistical methods usually assume a fixed sample size, and this makes sense for most problems. However, there are some situations in which observations need to be evaluated in sequence as they become available. Suppose, for example, we need to test a new drug for coronary artery disease. The drug is to be compared to a placebo, a pill that has no chemical effect on the body. It is important not to continue the trial beyond the point where it is evident that the drug is significantly better or worse than the placebo. In particular, if the drug is clearly effective, then it would be unfair to continue giving the placebo to any more patients.

Assuming that we want to monitor the results as the observations become available, why not perform a sequence of tests? Suppose we do a test at level .05 and then do another test at level .05 when there are more observations? If the null hypothesis is true, then there is a .05 probability that the first test incorrectly rejects the null hypothesis. Then when the second test is done, it gives an additional opportunity to reject the null hypothesis incorrectly, and therefore the two tests cause the total probability of incorrectly rejecting the null hypothesis to exceed .05. More tests will make the situation even worse, with type I error probability exceeding what is intended.

Valid sequential procedures were developed for testing the quality of materials as they were produced for use in World War II, primarily by Abraham Wald. Assume independent observations each with the same distribution, $X_1, X_2, \ldots, X_n$. The basic idea is to compute the ratio $\lambda = L_1/L_0$, where L_1 is the likelihood (i.e., joint pmf or pdf) under the alternative hypothesis and L_0 is the likelihood under the null hypothesis. Sometimes we will include a subscript to indicate the number of observations, $\lambda_n = L_{1,n}/L_{0,n}$. If λ is very large, then L_1 is much bigger than L_0, and this suggests that H_a explains the data much better than H_0, so we should reject H_0 in favor of H_a. On the other hand, if λ is very small, then similar reasoning suggests that we accept H_0 (in sequential analysis it is traditional to say "accept H_0" rather than "do not reject H_0"). The hard part is deciding where to draw the line with regard to "very large" and "very small."

Recall that a *simple* hypothesis is one for which the pmf or pdf is completely specified, for example H_0: $p = .5$ when X has a Bernoulli distribution with parameter p. Wald gave a theorem about determining cutoff values for rejecting or accepting a simple null hypothesis H_0 in a sequential test against a simple alternative hypothesis H_a. The cutoff values A and B are set ahead of time, and after each observation the likelihood ratio λ is evaluated and compared with them. If λ exceeds the upper value B, then we reject the null hypothesis. If λ is below the lower value A, then we accept the null hypothesis. Otherwise, more observations are required. The cutoff values are computed in terms of just the type I error probability α and the type II error probability β, and the test is called the sequential probability ratio test (SPRT).

WALD THEOREM

Let α and β be the desired type I and type II error probabilities which satisfy $0 < \alpha < 1 - \beta < 1$. Choose the lower cutoff value to be $A = \beta/(1 - \alpha)$ and the upper cutoff value to be $B = (1 - \beta)/\alpha$. Then for the sequential test the approximate probability of type I error is α and the approximate probability of type II error is β. The test terminates with probability 1; that is, there is zero chance that testing will go on forever.

Proof The proof that the test terminates for sure can be carried out using the logarithm of the likelihood, which can be written as a sum of independent, identically distributed rv's, but we will omit the details.

When the sequential test stops at the nth observation and chooses H_a we have

$$\lambda_n = \frac{L_{1,n}}{L_{0,n}} \approx B = \frac{1-\beta}{\alpha}$$

so that

$$L_{0,n} \approx \frac{\alpha L_{1,n}}{1-\beta}$$

This is approximate because the likelihood ratio exceeds B by a small amount when B is crossed. To get the true type I error probability α^*, we sum $L_{0,n}$ in the discrete case (or integrate in the continuous case) over the region R_n of rejection (crossing B) at observation n, and then sum over all n:

$$\alpha^* = \sum_{n=1}^{\infty}\sum_{R_n} L_{0,n} \approx \sum_{n=1}^{\infty}\sum_{R_n} \frac{\alpha L_{1,n}}{1-\beta} = \frac{\alpha}{1-\beta}\sum_{n=1}^{\infty}\sum_{R_n} L_{1,n}$$

Because the test eventually either accepts or rejects the null hypothesis, the double sum on the right is $1 - \beta^*$, where β^* is the true type II error probability. Thus

$$\alpha^* \approx \frac{\alpha(1-\beta^*)}{1-\beta}$$

from which

$$\frac{\alpha^*}{1-\beta^*} \approx \frac{\alpha}{1-\beta} \qquad (14.9)$$

Similarly, if the sequential test stops at the nth observation and chooses H_0 we have $\lambda_n = L_{1,n}/L_{0,n} \approx A = \beta/(1-\alpha)$; that is, $L_{1,n} \approx \beta L_{0,n}/(1-\alpha)$. Now sum $L_{1,n}$ in the discrete case (or integrate in the continuous case) over the region Q_n of accepting H_0 (crossing A) at observation n, and then sum over all n:

$$\beta^* = \sum_{n=1}^{\infty}\sum_{Q_n} L_{1,n} \approx \sum_{n=1}^{\infty}\sum_{Q_n} \frac{\beta L_{0,n}}{1-\alpha} = \frac{\beta}{1-\alpha}\sum_{n=1}^{\infty}\sum_{Q_n} L_{0,n} = \frac{\beta(1-\alpha^*)}{1-\alpha}$$

This shows

$$\frac{\beta^*}{1-\alpha^*} \approx \frac{\beta}{1-\alpha} \qquad (14.10)$$

Solving the two simultaneous Equations (14.9) and (14.10) gives $\alpha \approx \alpha^*$ and $\beta \approx \beta^*$, so the actual and desired error probabilities are approximately the same. ■

Sequential Testing for the Bernoulli Parameter

Let's apply the SPRT to a series of independent Bernoulli trials, under the assumption that the probability of success is either p_0 or p_1, where $p_0 < p_1$. The relevant hypotheses are then H_0: $p = p_0$ and H_a: $p = p_1$. The series of trials is recorded in the form $X_1, X_2, \ldots,$

$X_n, \ldots$, where for each observation 1 represents a success and 0 represents a failure. If k is the number of successes after n observations then the likelihood function is $L = p^k(1 - p)^{n-k}$. The likelihood ratio is

$$\lambda_n = \frac{L_{1,n}}{L_{0,n}} = \frac{p_1^k(1 - p_1)^{n-k}}{p_0^k(1 - p_0)^{n-k}} = \left(\frac{p_1}{p_0}\right)^k \left(\frac{1 - p_1}{1 - p_0}\right)^{n-k}$$

The test continues as long as

$$\frac{\beta}{1 - \alpha} = A < \lambda_n < B = \frac{1 - \beta}{\alpha}$$

In terms of natural logarithms this says

$$\ln\left(\frac{\beta}{1 - \alpha}\right) < \ln\left(\frac{p_1}{p_0}\right)^{\mathrm{k}} + \ln\left(\frac{1 - p_1}{1 - p_0}\right)^{\mathrm{n-k}} < \ln\left(\frac{1 - \beta}{\alpha}\right)$$

Simplifying,

$$\ln\left(\frac{\beta}{1 - \alpha}\right) < k \ln\left(\frac{p_1}{p_0}\right) + (n - k)\ln\left(\frac{1 - p_1}{1 - p_0}\right) < \ln\left(\frac{1 - \beta}{\alpha}\right)$$

Isolating k in the middle gives

$$\frac{\ln[\beta/(1 - \alpha)] + n \ln[(1 - p_0)/(1 - p_1)]}{\ln(p_1/p_0) + \ln[(1 - p_0)/(1 - p_1)]} < k \tag{14.11}$$

$$< \frac{\ln[(1 - \beta)/\alpha] + n \ln[(1 - p_0)/(1 - p_1)]}{\ln(p_1/p_0) + \ln[(1 - p_0)/(1 - p_1)]}$$

This shows that data collection will continue as long as k is between two linear functions of n. If these two linear functions are plotted against n, the result will be two parallel lines. If k is also plotted, then the study continues as long as k is between the lines, but the study terminates and decides in favor of H_a if the upper line is crossed, or the study terminates and decides in favor of H_0 if the lower line is crossed.

Example 14.9 The drug isocarboxazid was studied for relief of angina caused by coronary heart disease ("Sequential Trial of Isocarboxazid in Angina Pectoris," *British Med. J.*, Feb. 24, 1962: 513–515). Each patient took both the drug and a placebo, a chemically inert pill, in random order. One treatment was given for four weeks, followed by three weeks with no treatment, followed by four weeks on the other treatment. To avoid bias the trial was double-blind, meaning that the treatments were not known by the patients or the doctors while judging the outcome. After each treatment the patient was asked about improvement. A success was recorded if the drug did better than the placebo, a failure was recorded if the patient preferred the placebo, and ties were ignored.

The null hypothesis was H_0: $p = .5$ and the alternative hypothesis was H_a: $p = .75$. The type I error probability was set at $\alpha = .025$ and the type II error probability was set at $\beta = .05$. Then the lower and upper bounds for the likelihood ratio are

$$A = \frac{\beta}{1 - \alpha} = \frac{.05}{1 - .025} = .0513$$

$$B = \frac{1-\beta}{\alpha} = \frac{1-.05}{.025} = 38$$

This indicates that one likelihood must be much bigger than the other in order for a decision to be reached. However, rather than calculate the likelihood ratio, it is simpler to calculate the linear boundaries of Equation (14.11):

$$.631n - 2.704 < k < .631n + 3.311$$

The pattern of successes and failures was recorded as

F, S, S, S, S, S, F, S, S, S, S, S, F, S, S, S, F, F, S, S, S, S, S

Figure 14.4 plots k along with the boundary lines, and it shows the upper boundary line being crossed when $n = 23$. Therefore, we conclude that the drug is more effective than the placebo in relieving pain from heart disease.

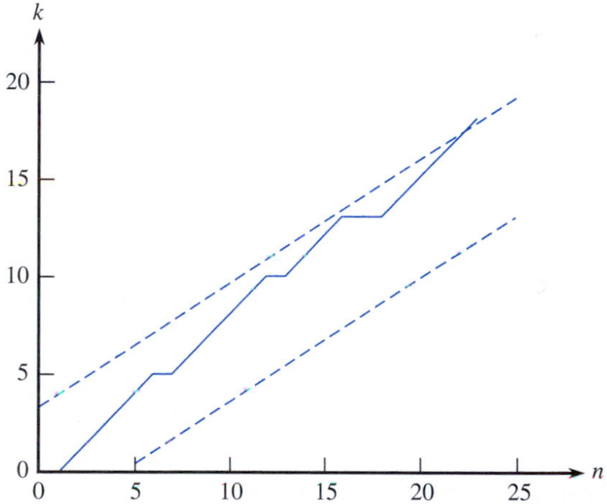

Figure 14.4 Plotting the successes of the isocarboxazid trial

At the end of the trial there were 18 successes, 5 failures, 7 ties in which both treatments seemed effective, and 25 ties in which both treatments seemed ineffective. For only a third of the patients (18 of 55) was the drug an improvement, so a patient taking isocarboxazid would not have good odds that it would outperform an inert pill. Although there is no question about the statistical significance of the trial, there might be some question about the practical significance for angina patients. ■

The Expected Sample Size

Wald's theorem guarantees that the experiment will end eventually, but how many observations will be required? We need at least one more observation as long as

$$\frac{\beta}{1-\alpha} = A < \lambda_n < B = \frac{1-\beta}{\alpha}$$

Here λ_n is the product of n values of the pdf ratio:

$$\lambda_n = \left[\frac{f_1(X_1)}{f_0(X_1)}\right]\left[\frac{f_1(X_2)}{f_0(X_2)}\right]\cdots\left[\frac{f_1(X_n)}{f_0(X_n)}\right]$$

Taking logarithms and letting $Z_i = \ln[f_1(X_i)/f_0(X_i)]$, the experiment continues as long as

$$\ln\left(\frac{\beta}{1-\alpha}\right) < Z_1 + \cdots + Z_n < \ln\left(\frac{1-\beta}{\alpha}\right)$$

Recall from Section 6.3 that for a fixed number n of random variables $Z_1, \ldots, Z_n$ having the same mean value μ, $E(\sum Z_i) = n\mu$. Here, though, the number of observations prior to termination of the SPRT is a random variable, which we denote by N. Wald showed that the expected value of a sum of a *random* number of identically distributed and independent random variables is the product of the expected number of observations and the expected value of an individual variable:

$$E\left[\sum_{i=1}^{N} Z_i\right] = E(N)\cdot\mu \tag{14.12}$$

In our situation, $\mu = E\{\ln[f_1(X_i)/f_0(X_i)]\}$.

Equation (14.12) can be solved for $E(N)$ if $E[\ln(\lambda_N)] = E(\sum Z_i)$ can be determined, so we now consider $E[\ln(\lambda_N)]$. If H_0 is in fact true, then the log likelihood ratio has probability α of crossing the upper boundary $\ln[(1-\beta)/\alpha]$, which causes incorrect rejection of H_0, and it has probability $1-\alpha$ of crossing the lower boundary $\ln[\beta/(1-\alpha)]$, which causes correct acceptance of H_0. Thus

$$H_0 \text{ true} \Rightarrow E[\ln(\lambda_N)] \approx \alpha\ln\left(\frac{1-\beta}{\alpha}\right) + (1-\alpha)\ln\left(\frac{\beta}{1-\alpha}\right) \tag{14.13}$$

Analogous reasoning gives

$$H_a \text{ true} \Rightarrow E[\ln(\lambda_N)] \approx (1-\beta)\ln\left(\frac{1-\beta}{\alpha}\right) + \beta\ln\left(\frac{\beta}{1-\alpha}\right) \tag{14.14}$$

Thus we have the following result.

PROPOSITION When H_0 is true,

$$E(N) \approx \frac{\alpha\ln[(1-\beta)/\alpha] + (1-\alpha)\ln[\beta/(1-\alpha)]}{E\{\ln[f_1(X_i)/f_0(X_i)]\}} \tag{14.15}$$

and when H_a is true,

$$E(N) \approx \frac{(1-\beta)\ln[(1-\beta)/\alpha] + \beta\ln[\beta/(1-\alpha)]}{E\{\ln[f_1(X_i)/f_0(X_i)]\}} \tag{14.16}$$

These are approximations because α and β are approximations to the true type I and type II error probabilities. Also, because of overshoot when the boundary is crossed, A and B are not the exact values of λ_n.

In Equations (14.15) and (14.16) the denominator depends on the hypothesis. For example, consider Bernoulli trials. Then

$$
\begin{aligned}
E\left[\ln\left(\frac{f_1(X_i)}{f_0(X_i)}\right)\right] &= E\left[\ln\left(\frac{p_1^{X_i}(1-p_1)^{1-X_i}}{p_0^{X_i}(1-p_0)^{1-X_i}}\right)\right] \\
&= E\left[X_i \ln\left(\frac{p_1}{p_0}\right)\right] + E\left[(1-X_i)\ln\left(\frac{(1-p_1)}{(1-p_0)}\right)\right] \\
&= p\ln\left(\frac{p_1}{p_0}\right) + (1-p)\ln\left(\frac{(1-p_1)}{(1-p_0)}\right)
\end{aligned}
$$

If H_0 is true, then Equation (14.15) becomes

$$
E(N) = \frac{\alpha\ln[(1-\beta)/\alpha] + (1-\alpha)\ln[\beta/(1-\alpha)]}{p_0\ln(p_1/p_0) + (1-p_0)\ln[(1-p_1)/(1-p_0)]} \tag{14.17}
$$

and when H_a is true Equation (14.16) becomes

$$
E(N) = \frac{(1-\beta)\ln[(1-\beta)/\alpha] + \beta\ln[\beta/(1-\alpha)]}{p_1\ln(p_1/p_0) + (1-p_1)\ln[(1-p_1)/(1-p_0)]} \tag{14.18}
$$

Example 14.10 (Example 14.9 continued)

Let's determine the expected number of observations needed to terminate the SPRT in the case of Bernoulli (S/F) observations. With $p_0 = .5$, $p_1 = .75$, $\alpha = .025$, $\beta = .05$, Equation (14.17) for the expected number of observations under H_0 becomes

$$
E(N) = \frac{.025\ln[(1-.05)/.025] + (1-.025)\ln[.05/(1-.025)]}{.5\ln(.75/.5) + (1-.5)\ln[(1-.75)/(1-.5)]} = 19.5
$$

Under H_a, Equation (14.18) for the expected number of observations becomes

$$
E(N) = \frac{(1-.05)\ln[(1-.05)/.025] + .05\ln[.05/(1-.025)]}{.75\ln(.75/.5) + (1-.75)\ln[(1-.75)/(1-.5)]} = 25.3
$$

Recall that the experiment actually used $n = 23$ observations, although many more were needed if the count included the ties that were neither successes nor failures for the drug.

It is interesting to compare these results with the number of observations needed in a traditional fixed-sample-size experiment as described in Section 9.3. Applying the formula given there for the sample size needed to achieve specified α and β,

$$
\begin{aligned}
n &= \left[\frac{z_\alpha\sqrt{p_0(1-p_0)} + z_\beta\sqrt{p_1(1-p_1)}}{p_1 - p_0}\right]^2 \\
&= \left[\frac{1.96\sqrt{.5(1-.5)} + 1.645\sqrt{.75(1-.75)}}{.75 - .5}\right]^2 = 46
\end{aligned}
$$

The value 46 exceeds the expected values for the number needed in the SPRT. ■

The following result provides a strong rationale for using the SPRT to test simple hypotheses.

PROPOSITION If either H_0 or H_a is true, then $E(N)$ for the SPRT is smaller than the n required by the corresponding fixed-sample-size experiment.

It should be emphasized that the proposition considers only the expected sample size. Even though the expected number for the sequential trial is less, the actual number is random, and can occasionally be greater than the number required for the fixed-sample-size test.

There are some disadvantages to the SPRT approach. For Bernoulli data, it is necessary to specify the null and alternative values of p precisely, and this is not always easy. In general, Wald's method requires a simple null hypothesis and a simple alternative hypothesis, which will not always be appropriate.

Although it is possible to deal with normal data using the SPRT (Exercise 33), it requires that the variance be known in order to have simple hypotheses. There are more complicated sequential methods that do not require known variance, but they are harder to apply. Furthermore, the major statistical packages do not include SPRT procedures, so one is stuck with homemade software or hand calculations.

At the end of an experiment it is natural to want to form a confidence interval and make other inferences on the parameter of interest. However, this is not appropriate when the SPRT is used. Suppose that a trial has just ended by rejecting the null hypothesis as in Example 14.9. This means that the trial has ended at a high value for the number of successes, and it suggests that the estimate for p might be biased on the high side. Indeed, this is the case, and it is very difficult to get an unbiased estimator of p at the end of a sequential trial.

At the beginning of this section, we criticized the idea of just applying a statistical test repeatedly as more observations are obtained. It is true that repeated testing at the .05 level will result in a true type I error probability of more than .05 for the whole trial. However, by doing the repeated testing at appropriate values below .05 and limiting the number of tests, it is possible to do a legitimate experiment at the .05 level. In particular, *group sequential analysis* generally involves grouping the observations so there are two to ten groups. When the observations have all been obtained for a group, the observations of this and previous groups are combined and tested at an appropriate level below .05, so the overall level for the whole experiment is .05. This method has the advantage of being sequential, monitoring the results of a drug trial as the data is obtained and allowing for the discontinuation of the trial in case the results are clear. That is, the group sequential method has the main advantage of the SPRT, which allows for a short experiment in case the drug is very effective (or very ineffective). However, the group sequential methods are more versatile, with a greater variety of statistical methods than can be derived by Wald's sequential likelihood approach. There are now programs for use with S-Plus, including the S+SeqTrial module and also user-written software, which facilitate the design and analysis of group sequential trials.

Exercises Section 14.5 (33–36)

33. a. Apply the Wald procedure to derive a stopping rule for normal data with known $\sigma = 15$. Let H_0: $\mu = 100$, H_a: $\mu = 110$, $\alpha = .05$, $\beta = .05$. Express the continuance of sampling in terms of the sum of the first n observations. This should show that sampling continues as long as the sum is between two linear functions of n.

b. Apply your procedure to the following data (see Example 1.2), IQ scores from a class of 33 students:

118	108	110	136	106	102	132	115	111
113	113	140	82	118	110	109	96	99
113	115	127	107	119	108	122	103	122
108	103	146	113	121	108			

(*Hint*: The sums can be obtained from software, for example, the partial sums function in MINITAB.) What is your conclusion and how many observations are required?

c. Find the expected number of observations required under H_0 and under H_a. In terms of the specifications for this problem, what causes these two numbers to be the same?

d. Find the number required for a fixed sample to achieve the same α and β, and compare with the answer to part (c).

34. For the treatment of anxiety and tension, Sainsbury and Lucas ("Sequential Methods Applied to the Study of Prochlorperazine," *British Med. J.*, 1959: 737–740) conducted an experiment to see if prochlorperazine was effective. The design was similar to the one used in Example 14.9, with both the drug and a placebo being taken by each patient in random order double-blind; that is, neither the patient nor the physician knew which treatment was being given while the treatments were being evaluated. Each subject was asked which treatment was preferred, so the result is a Bernoulli rv. For this experiment $p_0 = .5$, $p_1 = .8$, $\alpha = .05$, $\beta = .05$.

a. Determine the linear boundaries such that the experiment continues if the number of successes remains within the lines.

b. Given the data

$S, S, F, F, F, F, S, S, F, S, S, F, S, F, F$

determine the result of the experiment.

c. Find the expected number of observations needed under H_0 and under H_a.

d. Find the number of trials needed if the experiment were conducted with a fixed sample. Compare with the answer to part (c) and the number actually used.

35. Consider the Bernoulli case and Equation (14.11) for continued sampling. Instead of the number of successes k, we can use the difference $y = k - (n - k) = 2k - n$ between the number of successes and the number of failures.

a. Convert Equation (14.11) to express the bounds for continued sampling in terms of y.

b. Use part (a) to express the bounds for continued sampling in Example 14.9 in terms of y.

c. Sketch the lines of part (b) and also show y for the 23 observations. That is, redo Figure 14.4 in terms of y. Describe what happens to y as each new observation is included.

36. Assume observations $X_1, X_2, \ldots, X_n, \ldots$ from the Poisson distribution with mean μ.

a. Apply the SPRT to test H_0: $\mu = \mu_0$ against H_a: $\mu = \mu_1$, where $\mu_0 < \mu_1$. Obtain linear bounds similar to Equation (14.11), except that Σx_i takes the place of k.

b. The following observations are monthly accident totals in a factory workplace.

11 7 9 5 10 11 16 9 7 3 5 5

Apply part (a) with $\mu_0 = 5$, $\mu_1 = 10$, $\alpha = .01$, and $\beta = .05$ to do the test. (*Hint*: The sums and the linear boundaries can be obtained from software. For example, the partial sums function in MINITAB can be used for the sum.)

Supplementary Exercises (37–46)

37. The article "Effects of a Rice-Rich Versus Potato-Rich Diet on Glucose, Lipoprotein, and Cholesterol Metabolism in Noninsulin-Dependent Diabetics" (*Amer. J. Clin. Nutrit.*, 1984: 598–606) gives the accompanying data on cholesterol-synthesis rate for eight diabetic subjects. Subjects were fed a standardized diet with potato or rice as the major carbohydrate source. Participants received both diets for specified periods of time, with cholesterol-synthesis rate (mmol/day) measured at the end of each dietary period. The analysis presented in this article used a distribution-free test. Use such a test with significance level .05 to determine whether the true mean cholesterol-synthesis rate differs significantly for the two sources of carbohydrates.

Subject	1	2	3	4	5	6	7	8
Potato	1.88	2.60	1.38	4.41	1.87	2.89	3.96	2.31
Rice	1.70	3.84	1.13	4.97	.86	1.93	3.36	2.15

38. The study reported in "Gait Patterns During Free Choice Ladder Ascents" (*Hum. Movement Sci.*, 1983: 187–195) was motivated by publicity concerning the increased accident rate for individuals climbing ladders. A number of different gait patterns were used by subjects climbing a portable straight ladder according to specified instructions. The ascent times for seven subjects who used a lateral gait and six subjects who used a four-beat diagonal gait are given.

Lateral	.86	1.31	1.64	1.51	1.53	1.39	1.09
Diagonal	1.27	1.82	1.66	.85	1.45	1.24	

a. Carry out a test using $\alpha = .05$ to see whether the data suggests any difference in the true average ascent times for the two gaits.

b. Compute a 95% CI for the difference between the true average gait times.

39. The **sign test** is a very simple procedure for testing hypotheses about a population median assuming only that the underlying distribution is continuous. To illustrate, consider the following sample of 20 observations on component lifetime (hr):

1.7	3.3	5.1	6.9	12.6	14.4	16.4
24.6	26.0	26.5	32.1	37.4	40.1	40.5
41.5	72.4	80.1	86.4	87.5	100.2	

We wish to test H_0: $\tilde{\mu} = 25.0$ versus H_a: $\tilde{\mu} > 25.0$. The test statistic is Y = the number of observations that exceed 25.

a. Consider rejecting H_0 if $Y \geq 15$. What is the value of α (the probability of a type I error) for this test? *Hint*: Think of a "success" as a lifetime that exceeds 25.0. Then Y is the number of successes in the sample. What kind of a distribution does Y have when $\tilde{\mu} = 25.0$?

b. What rejection region of the form $Y \geq c$ specifies a test with a significance level as close to .05 as possible? Use this region to carry out the test for the given data.

Note: The test statistic is the number of differences $X_i - 25.0$ that have positive signs, hence the name *sign test*.

40. Refer to Exercise 39, and consider a confidence interval associated with the sign test, the **sign interval**. The relevant hypotheses are now H_0: $\tilde{\mu} = \tilde{\mu}_0$ versus H_a: $\tilde{\mu} \neq \tilde{\mu}_0$. Let's use the following rejection region: either $Y \geq 15$ or $Y \leq 5$.

a. What is the significance level for this test?

b. The confidence interval will consist of all values $\tilde{\mu}_0$ for which H_0 is not rejected. Determine the CI for the given data, and state the confidence level.

41. The single-factor ANOVA model considered in Chapter 11 assumed the observations in the ith sample were selected from a normal distribution with mean μ_i and variance σ^2, that is, $X_{ij} = \mu_i + \varepsilon_{ij}$ where the ε's are normal with mean 0 and variance σ^2. The normality assumption implies that the F test is not distribution-free. We now assume that the ε's all come from the same continuous, but not necessarily normal, distribution, and develop a distribution-free test of the null hypothesis that all I μ_i's are identical. Let $N = \sum J_i$, the total number of observations in the data set (there are J_i observations in the ith sample). Rank these N observations from 1 (the smallest) to N, and let $\overline{R}_i$ be the average of the ranks for the observations in the ith sample. When H_0 is true, we expect the rank of any particular observation and therefore also $\overline{R}_i$ to be $(N + 1)/2$. The data argues against H_0 when some of the $\overline{R}_i$'s differ considerably from $(N + 1)/2$. The *Kruskal–Wallis* test statistic is

$$K = \frac{12}{N(N+1)} \sum J_i \left(\overline{R}_i - \frac{N+1}{2} \right)^2$$

When H_0 is true and either (1) $I = 3$, all $J_i \geq 6$ or (2) $I > 3$, all $J_i \geq 5$, the test statistic has approximately a chi-squared distribution with $I - 1$ df.

The accompanying observations on axial stiffness index resulted from a study of metal-plate connected trusses in which five different plate lengths—4 in., 6 in., 8 in., 10 in., and 12 in.—were used ("Modeling Joints Made with Light-Gauge

Metal Connector Plates," *Forest Products J.*, 1979: 39–44).

$i = 1$ (4″):	309.2	309.7	311.0	316.8
	326.5	349.8	409.5	
$i = 2$ (6″):	331.0	347.2	348.9	361.0
	381.7	402.1	404.5	
$i = 3$ (8″):	351.0	357.1	366.2	367.3
	382.0	392.4	409.9	
$i = 4$ (10″):	346.7	362.6	384.2	410.6
	433.1	452.9	461.4	
$i = 5$ (12″):	407.4	410.7	419.9	441.2
	441.8	465.8	473.4	

Use the *K–W* test to decide at significance level .01 whether the true average axial stiffness index depends somehow on plate length.

42. The article "Production of Gaseous Nitrogen in Human Steady-State Conditions" (*J. Appl. Physiol.*, 1972: 155–159) reports the following observations on the amount of nitrogen expired (in liters) under four dietary regimens: (1) fasting, (2) 23% protein, (3) 32% protein, and (4) 67% protein. Use the Kruskal–Wallis test (Exercise 41) at level .05 to test equality of the corresponding μ_i's.

1.	4.079	4.859	3.540	5.047	3.298
	4.679	2.870	4.648	3.847	
2.	4.368	5.668	3.752	5.848	3.802
	4.844	3.578	5.393	4.374	
3.	4.169	5.709	4.416	5.666	4.123
	5.059	4.403	4.496	4.688	
4.	4.928	5.608	4.940	5.291	4.674
	5.038	4.905	5.208	4.806	

43. The model for the data from a randomized block experiment for comparing I treatments was $X_{ij} = \mu + \alpha_i + \beta_j + \varepsilon_{ij}$, where the α's are treatment effects, the β's are block effects, and the ε's were assumed normal with mean 0 and variance σ^2. We now replace normality by the assumption that the ε's have the same continuous distribution. A distribution-free test of the null hypothesis of no treatment effects, called *Friedman's test*, involves first ranking the observations in each block separately from 1 to I. The rank average $\overline{R}_i$ is then calculated for each of the I treatments. If H_0 is true, the expected value of each rank average is $(I + 1)/2$. The test statistic is

$$F_r = \frac{12J}{I(I+1)} \sum \left(\overline{R}_i - \frac{I+1}{2} \right)^2$$

For even moderate values of J, the test statistic has approximately a chi-squared distribution with $I - 1$ df when H_0 is true.

The article "Physiological Effects During Hypnotically Requested Emotions" (*Psychosomatic Med.*, 1963: 334–343) reports the following data (x_{ij}) on skin potential in millivolts when the emotions of fear, happiness, depression, and calmness were requested from each of eight subjects.

	Blocks (Subjects)			
	1	**2**	**3**	**4**
Fear	23.1	57.6	10.5	23.6
Happiness	22.7	53.2	9.7	19.6
Depression	22.5	53.7	10.8	21.1
Calmness	22.6	53.1	8.3	21.6
	5	**6**	**7**	**8**
Fear	11.9	54.6	21.0	20.3
Happiness	13.8	47.1	13.6	23.6
Depression	13.7	39.2	13.7	16.3
Calmness	13.3	37.0	14.8	14.8

Use Friedman's test to decide whether emotion has an effect on skin potential.

44. In an experiment to study the way in which different anesthetics affect plasma epinephrine concentration, ten dogs were selected and concentration was measured while they were under the influence of the anesthetics isoflurane, halothane, and cyclopropane ("Sympathoadrenal and Hemodynamic Effects of Isoflurane, Halothane, and Cyclopropane in Dogs," *Anesthesiology*, 1974: 465–470). Test at level .05 to see whether there is an anesthetic effect on concentration. *Hint*: See Exercise 43.

	Dog				
	1	**2**	**3**	**4**	**5**
Isoflurane	.28	.51	1.00	.39	.29
Halothane	.30	.39	.63	.38	.21
Cyclopropane	1.07	1.35	.69	.28	1.24
	6	**7**	**8**	**9**	**10**
Isoflurane	.36	.32	.69	.17	.33
Halothane	.88	.39	.51	.32	.42
Cyclopropane	1.53	.49	.56	1.02	.30

45. Suppose we wish to test

H_0: the X and Y distributions are identical

versus

H_a: the X distribution is less spread out than the Y distribution

The accompanying figure pictures X and Y distributions for which H_a is true. The Wilcoxon rank-sum test is not appropriate in this situation because when H_a is true as pictured, the Y's will tend to be at the extreme ends of the combined sample (resulting in small and large Y ranks), so the sum of X ranks will result in a W value that is neither large nor small.

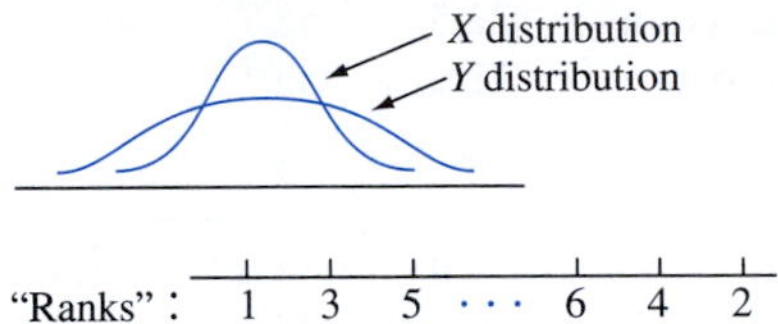

Consider modifying the procedure for assigning ranks as follows: After the combined sample of $m + n$ observations is ordered, the smallest observation is given rank 1, the largest observation is given rank 2, the second smallest is given rank 3, the second largest is given rank 4, and so on. Then if H_a is true as pictured, the X values will tend to be in the middle of the sample and thus receive large ranks. Let W' denote the sum of the X ranks and consider rejecting H_0 in favor of H_a when $w' \geq c$. When H_0 is true, every possible set of X ranks has the same probability, so W' *has the same distribution as does* W *when* H_0 *is true*. Thus c can be chosen from Appendix Table A.14 to yield a level α test. The accompanying data refers to medial muscle thickness for arterioles from the lungs of children who died from sudden infant death syndrome (x's) and a control group of children (y's). Carry out the test of H_0 versus H_a at level .05.

SIDS	4.0	4.4	4.8	4.9	
Control	3.7	4.1	4.3	5.1	5.6

Consult the Lehmann book (in the chapter bibliography) for more information on this test, called the *Siegel–Tukey test*.

46. The ranking procedure described in Exercise 45 is somewhat asymmetric, because the smallest observation receives rank 1 whereas the largest receives rank 2, and so on. Suppose both the smallest and the largest receive rank 1, the second smallest and second largest receive rank 2, and so on, and let W'' be the sum of the X ranks. The null distribution of W'' is not identical to the null distribution of W, so different tables are needed. Consider the case $m = 3$, $n = 4$. List all 35 possible orderings of the three X values among the seven observations (e.g., 1, 3, 7 or 4, 5, 6), assign ranks in the manner described, compute the value of W'' for each possibility, and then tabulate the null distribution of W''. For the test that rejects if $w'' \geq c$, what value of c prescribes approximately a level .10 test? This is the *Ansari–Bradley test*; for additional information, see the book by Hollander and Wolfe in the chapter bibliography.

Bibliography

Berry, Donald A., *Statistics: A Bayesian Perspective*, Brooks/Cole—Thomson Learning, Belmont, CA, 1996. An elementary introduction to Bayesian ideas and methodology.

Gelman, Andrew, John B. Carlin, Hal S. Stern, and Donald B. Rubin, *Bayesian Data Analysis* (2nd ed.), Chapman and Hall, London, 2003. An up-to-date survey of theoretical, practical, and computational issues in Bayesian inference.

Hollander, Myles, and Douglas Wolfe, *Nonparametric Statistical Methods* (2nd ed.), Wiley, New York, 1999. A very good reference on distribution-free methods with an excellent collection of tables.

Lehmann, Erich, *Nonparametrics: Statistical Methods Based on Ranks*, Holden-Day, San Francisco, 1975. An excellent discussion of the most important distribution-free methods, presented with a great deal of insightful commentary.

Siegmund, David, *Sequential Analysis: Tests and Confidence Intervals*, Springer-Verlag, Berlin, 1986. An authoritative and sophisticated treatment of sequential analysis.

Appendix Tables

Table A.1 Cumulative Binomial Probabilities

$$B(x; n, p) = \sum_{y=0}^{x} b(y; n, p)$$

a. $n = 5$

	p														
x	**0.01**	**0.05**	**0.10**	**0.20**	**0.25**	**0.30**	**0.40**	**0.50**	**0.60**	**0.70**	**0.75**	**0.80**	**0.90**	**0.95**	**0.99**
0	.951	.774	.590	.328	.237	.168	.078	.031	.010	.002	.001	.000	.000	.000	.000
1	.999	.977	.919	.737	.633	.528	.337	.188	.087	.031	.016	.007	.000	.000	.000
2	1.000	.999	.991	.942	.896	.837	.683	.500	.317	.163	.104	.058	.009	.001	.000
3	1.000	1.000	1.000	.993	.984	.969	.913	.812	.663	.472	.367	.263	.081	.023	.001
4	1.000	1.000	1.000	1.000	.999	.998	.990	.969	.922	.832	.763	.672	.410	.226	.049

b. $n = 10$

	p														
x	**0.01**	**0.05**	**0.10**	**0.20**	**0.25**	**0.30**	**0.40**	**0.50**	**0.60**	**0.70**	**0.75**	**0.80**	**0.90**	**0.95**	**0.99**
0	.904	.599	.349	.107	.056	.028	.006	.001	.000	.000	.000	.000	.000	.000	.000
1	.996	.914	.736	.376	.244	.149	.046	.011	.002	.000	.000	.000	.000	.000	.000
2	1.000	.988	.930	.678	.526	.383	.167	.055	.012	.002	.000	.000	.000	.000	.000
3	1.000	.999	.987	.879	.776	.650	.382	.172	.055	.011	.004	.001	.000	.000	.000
4	1.000	1.000	.998	.967	.922	.850	.633	.377	.166	.047	.020	.006	.000	.000	.000
5	1.000	1.000	1.000	.994	.980	.953	.834	.623	.367	.150	.078	.033	.002	.000	.000
6	1.000	1.000	1.000	.999	.996	.989	.945	.828	.618	.350	.224	.121	.013	.001	.000
7	1.000	1.000	1.000	1.000	1.000	.998	.988	.945	.833	.617	.474	.322	.070	.012	.000
8	1.000	1.000	1.000	1.000	1.000	1.000	.998	.989	.954	.851	.756	.624	.264	.086	.004
9	1.000	1.000	1.000	1.000	1.000	1.000	1.000	.999	.994	.972	.944	.893	.651	.401	.096

c. $n = 15$

	p														
x	**0.01**	**0.05**	**0.10**	**0.20**	**0.25**	**0.30**	**0.40**	**0.50**	**0.60**	**0.70**	**0.75**	**0.80**	**0.90**	**0.95**	**0.99**
0	.860	.463	.206	.035	.013	.005	.000	.000	.000	.000	.000	.000	.000	.000	.000
1	.990	.829	.549	.167	.080	.035	.005	.000	.000	.000	.000	.000	.000	.000	.000
2	1.000	.964	.816	.398	.236	.127	.027	.004	.000	.000	.000	.000	.000	.000	.000
3	1.000	.995	.944	.648	.461	.297	.091	.018	.002	.000	.000	.000	.000	.000	.000
4	1.000	.999	.987	.836	.686	.515	.217	.059	.009	.001	.000	.000	.000	.000	.000
5	1.000	1.000	.998	.939	.852	.722	.402	.151	.034	.004	.001	.000	.000	.000	.000
6	1.000	1.000	1.000	.982	.943	.869	.610	.304	.095	.015	.004	.001	.000	.000	.000
7	1.000	1.000	1.000	.996	.983	.950	.787	.500	.213	.050	.017	.004	.000	.000	.000
8	1.000	1.000	1.000	.999	.996	.985	.905	.696	.390	.131	.057	.018	.000	.000	.000
9	1.000	1.000	1.000	1.000	.999	.996	.966	.849	.597	.278	.148	.061	.002	.000	.000
10	1.000	1.000	1.000	1.000	1.000	.999	.991	.941	.783	.485	.314	.164	.013	.001	.000
11	1.000	1.000	1.000	1.000	1.000	1.000	.998	.982	.909	.703	.539	.352	.056	.005	.000
12	1.000	1.000	1.000	1.000	1.000	1.000	1.000	.996	.973	.873	.764	.602	.184	.036	.000
13	1.000	1.000	1.000	1.000	1.000	1.000	1.000	1.000	.995	.965	.920	.833	.451	.171	.010
14	1.000	1.000	1.000	1.000	1.000	1.000	1.000	1.000	1.000	.995	.987	.965	.794	.537	.140

(continued)

Table A.1 Cumulative Binomial Probabilities (*cont.*) $B(x; n, p) = \sum_{y=0}^{x} b(y; n, p)$

d. $n = 20$

								p							
x	**0.01**	**0.05**	**0.10**	**0.20**	**0.25**	**0.30**	**0.40**	**0.50**	**0.60**	**0.70**	**0.75**	**0.80**	**0.90**	**0.95**	**0.99**
0	.818	.358	.122	.012	.003	.001	.000	.000	.000	.000	.000	.000	.000	.000	.000
1	.983	.736	.392	.069	.024	.008	.001	.000	.000	.000	.000	.000	.000	.000	.000
2	.999	.925	.677	.206	.091	.035	.004	.000	.000	.000	.000	.000	.000	.000	.000
3	1.000	.984	.867	.411	.225	.107	.016	.001	.000	.000	.000	.000	.000	.000	.000
4	1.000	.997	.957	.630	.415	.238	.051	.006	.000	.000	.000	.000	.000	.000	.000
5	1.000	1.000	.989	.804	.617	.416	.126	.021	.002	.000	.000	.000	.000	.000	.000
6	1.000	1.000	.998	.913	.786	.608	.250	.058	.006	.000	.000	.000	.000	.000	.000
7	1.000	1.000	1.000	.968	.898	.772	.416	.132	.021	.001	.000	.000	.000	.000	.000
8	1.000	1.000	1.000	.990	.959	.887	.596	.252	.057	.005	.001	.000	.000	.000	.000
9	1.000	1.000	1.000	.997	.986	.952	.755	.412	.128	.017	.004	.001	.000	.000	.000
10	1.000	1.000	1.000	.999	.996	.983	.872	.588	.245	.048	.014	.003	.000	.000	.000
11	1.000	1.000	1.000	1.000	.999	.995	.943	.748	.404	.113	.041	.010	.000	.000	.000
12	1.000	1.000	1.000	1.000	1.000	.999	.979	.868	.584	.228	.102	.032	.000	.000	.000
13	1.000	1.000	1.000	1.000	1.000	1.000	.994	.942	.750	.392	.214	.087	.002	.000	.000
14	1.000	1.000	1.000	1.000	1.000	1.000	.998	.979	.874	.584	.383	.196	.011	.000	.000
15	1.000	1.000	1.000	1.000	1.000	1.000	1.000	.994	.949	.762	.585	.370	.043	.003	.000
16	1.000	1.000	1.000	1.000	1.000	1.000	1.000	.999	.984	.893	.775	.589	.133	.016	.000
17	1.000	1.000	1.000	1.000	1.000	1.000	1.000	1.000	.996	.965	.909	.794	.323	.075	.001
18	1.000	1.000	1.000	1.000	1.000	1.000	1.000	1.000	.999	.992	.976	.931	.608	.264	.017
19	1.000	1.000	1.000	1.000	1.000	1.000	1.000	1.000	1.000	.999	.997	.988	.878	.642	.182

(*continued*)

Table A.1 Cumulative Binomial Probabilities (*cont.*)

$$B(x; n, p) = \sum_{y=0}^{x} b(y; n, p)$$

e. $n = 25$

		p														
		0.01	**0.05**	**0.10**	**0.20**	**0.25**	**0.30**	**0.40**	**0.50**	**0.60**	**0.70**	**0.75**	**0.80**	**0.90**	**0.95**	**0.99**
	0	.778	.277	.072	.004	.001	.000	.000	.000	.000	.000	.000	.000	.000	.000	.000
	1	.974	.642	.271	.027	.007	.002	.000	.000	.000	.000	.000	.000	.000	.000	.000
	2	.998	.873	.537	.098	.032	.009	.000	.000	.000	.000	.000	.000	.000	.000	.000
	3	1.000	.966	.764	.234	.096	.033	.002	.000	.000	.000	.000	.000	.000	.000	.000
	4	1.000	.993	.902	.421	.214	.090	.009	.000	.000	.000	.000	.000	.000	.000	.000
	5	1.000	.999	.967	.617	.378	.193	.029	.002	.000	.000	.000	.000	.000	.000	.000
	6	1.000	1.000	.991	.780	.561	.341	.074	.007	.000	.000	.000	.000	.000	.000	.000
	7	1.000	1.000	.998	.891	.727	.512	.154	.022	.001	.000	.000	.000	.000	.000	.000
	8	1.000	1.000	1.000	.953	.851	.677	.274	.054	.004	.000	.000	.000	.000	.000	.000
	9	1.000	1.000	1.000	.983	.929	.811	.425	.115	.013	.000	.000	.000	.000	.000	.000
	10	1.000	1.000	1.000	.994	.970	.902	.586	.212	.034	.002	.000	.000	.000	.000	.000
	11	1.000	1.000	1.000	.998	.980	.956	.732	.345	.078	.006	.001	.000	.000	.000	.000
x	12	1.000	1.000	1.000	1.000	.997	.983	.846	.500	.154	.017	.003	.000	.000	.000	.000
	13	1.000	1.000	1.000	1.000	.999	.994	.922	.655	.268	.044	.020	.002	.000	.000	.000
	14	1.000	1.000	1.000	1.000	1.000	.998	.966	.788	.414	.098	.030	.006	.000	.000	.000
	15	1.000	1.000	1.000	1.000	1.000	1.000	.987	.885	.575	.189	.071	.017	.000	.000	.000
	16	1.000	1.000	1.000	1.000	1.000	1.000	.996	.946	.726	.323	.149	.047	.000	.000	.000
	17	1.000	1.000	1.000	1.000	1.000	1.000	.999	.978	.846	.488	.273	.109	.002	.000	.000
	18	1.000	1.000	1.000	1.000	1.000	1.000	1.000	.993	.926	.659	.439	.220	.009	.000	.000
	19	1.000	1.000	1.000	1.000	1.000	1.000	1.000	.998	.971	.807	.622	.383	.033	.001	.000
	20	1.000	1.000	1.000	1.000	1.000	1.000	1.000	1.000	.991	.910	.786	.579	.098	.007	.000
	21	1.000	1.000	1.000	1.000	1.000	1.000	1.000	1.000	.998	.967	.904	.766	.236	.034	.000
	22	1.000	1.000	1.000	1.000	1.000	1.000	1.000	1.000	1.000	.991	.968	.902	.463	.127	.002
	23	1.000	1.000	1.000	1.000	1.000	1.000	1.000	1.000	1.000	.998	.993	.973	.729	.358	.026
	24	1.000	1.000	1.000	1.000	1.000	1.000	1.000	1.000	1.000	1.000	.999	.996	.928	.723	.222

Table A.2 Cumulative Poisson Probabilities

$$F(x; \lambda) = \sum_{y=0}^{x} \frac{e^{-\lambda}\lambda^y}{y!}$$

		λ									
		.1	**.2**	**.3**	**.4**	**.5**	**.6**	**.7**	**.8**	**.9**	**1.0**
	0	.905	.819	.741	.670	.607	.549	.497	.449	.407	.368
	1	.995	.982	.963	.938	.910	.878	.844	.809	.772	.736
	2	1.000	.999	.996	.992	.986	.977	.966	.953	.937	.920
x	3		1.000	1.000	.999	.998	.997	.994	.991	.987	.981
	4				1.000	1.000	1.000	.999	.999	.998	.996
	5							1.000	1.000	1.000	.999
	6										1.000

(continued)

Table A.2 Cumulative Poisson Probabilities (*cont.*)

$$F(x; \lambda) = \sum_{y=0}^{x} \frac{e^{-\lambda}\lambda^y}{y!}$$

x	λ = 2.0	3.0	4.0	5.0	6.0	7.0	8.0	9.0	10.0	15.0	20.0
0	.135	.050	.018	.007	.002	.001	.000	.000	.000	.000	.000
1	.406	.199	.092	.040	.017	.007	.003	.001	.000	.000	.000
2	.677	.423	.238	.125	.062	.030	.014	.006	.003	.000	.000
3	.857	.647	.433	.265	.151	.082	.042	.021	.010	.000	.000
4	.947	.815	.629	.440	.285	.173	.100	.055	.029	.001	.000
5	.983	.916	.785	.616	.446	.301	.191	.116	.067	.003	.000
6	.995	.966	.889	.762	.606	.450	.313	.207	.130	.008	.000
7	.999	.988	.949	.867	.744	.599	.453	.324	.220	.018	.001
8	1.000	.996	.979	.932	.847	.729	.593	.456	.333	.037	.002
9		.999	.992	.968	.916	.830	.717	.587	.458	.070	.005
10		1.000	.997	.986	.957	.901	.816	.706	.583	.118	.011
11			.999	.995	.980	.947	.888	.803	.697	.185	.021
12			1.000	.998	.991	.973	.936	.876	.792	.268	.039
13				.999	.996	.987	.966	.926	.864	.363	.066
14				1.000	.999	.994	.983	.959	.917	.466	.105
15					.999	.998	.992	.978	.951	.568	.157
16					1.000	.999	.996	.989	.973	.664	.221
17						1.000	.998	.995	.986	.749	.297
18							.999	.998	.993	.819	.381
19							1.000	.999	.997	.875	.470
20								1.000	.998	.917	.559
21									.999	.947	.644
22									1.000	.967	.721
23										.981	.787
24										.989	.843
25										.994	.888
26										.997	.922
27										.998	.948
28										.999	.966
29										1.000	.978
30											.987
31											.992
32											.995
33											.997
34											.999
35											.999
36											1.000

Table A.3 Standard Normal Curve Areas

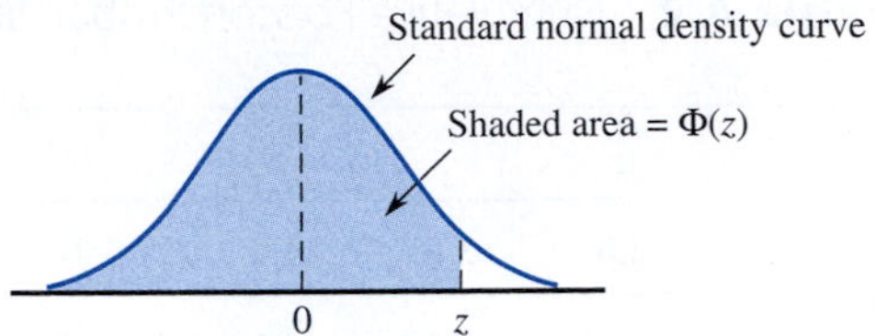

z	.00	.01	.02	.03	.04	.05	.06	.07	.08	.09
−3.4	.0003	.0003	.0003	.0003	.0003	.0003	.0003	.0003	.0003	.0002
−3.3	.0005	.0005	.0005	.0004	.0004	.0004	.0004	.0004	.0004	.0003
−3.2	.0007	.0007	.0006	.0006	.0006	.0006	.0006	.0005	.0005	.0005
−3.1	.0010	.0009	.0009	.0009	.0008	.0008	.0008	.0008	.0007	.0007
−3.0	.0013	.0013	.0013	.0012	.0012	.0011	.0011	.0011	.0010	.0010
−2.9	.0019	.0018	.0017	.0017	.0016	.0016	.0015	.0015	.0014	.0014
−2.8	.0026	.0025	.0024	.0023	.0023	.0022	.0021	.0021	.0020	.0019
−2.7	.0035	.0034	.0033	.0032	.0031	.0030	.0029	.0028	.0027	.0026
−2.6	.0047	.0045	.0044	.0043	.0041	.0040	.0039	.0038	.0037	.0036
−2.5	.0062	.0060	.0059	.0057	.0055	.0054	.0052	.0051	.0049	.0038
−2.4	.0082	.0080	.0078	.0075	.0073	.0071	.0069	.0068	.0066	.0064
−2.3	.0107	.0104	.0102	.0099	.0096	.0094	.0091	.0089	.0087	.0084
−2.2	.0139	.0136	.0132	.0129	.0125	.0122	.0119	.0116	.0113	.0110
−2.1	.0179	.0174	.0170	.0166	.0162	.0158	.0154	.0150	.0146	.0143
−2.0	.0228	.0222	.0217	.0212	.0207	.0202	.0197	.0192	.0188	.0183
−1.9	.0287	.0281	.0274	.0268	.0262	.0256	.0250	.0244	.0239	.0233
−1.8	.0359	.0352	.0344	.0336	.0329	.0322	.0314	.0307	.0301	.0294
−1.7	.0446	.0436	.0427	.0418	.0409	.0401	.0392	.0384	.0375	.0367
−1.6	.0548	.0537	.0526	.0516	.0505	.0495	.0485	.0475	.0465	.0455
−1.5	.0668	.0655	.0643	.0630	.0618	.0606	.0594	.0582	.0571	.0559
−1.4	.0808	.0793	.0778	.0764	.0749	.0735	.0722	.0708	.0694	.0681
−1.3	.0968	.0951	.0934	.0918	.0901	.0885	.0869	.0853	.0838	.0823
−1.2	.1151	.1131	.1112	.1093	.1075	.1056	.1038	.1020	.1003	.0985
−1.1	.1357	.1335	.1314	.1292	.1271	.1251	.1230	.1210	.1190	.1170
−1.0	.1587	.1562	.1539	.1515	.1492	.1469	.1446	.1423	.1401	.1379
−0.9	.1841	.1814	.1788	.1762	.1736	.1711	.1685	.1660	.1635	.1611
−0.8	.2119	.2090	.2061	.2033	.2005	.1977	.1949	.1922	.1894	.1867
−0.7	.2420	.2389	.2358	.2327	.2296	.2266	.2236	.2206	.2177	.2148
−0.6	.2743	.2709	.2676	.2643	.2611	.2578	.2546	.2514	.2483	.2451
−0.5	.3085	.3050	.3015	.2981	.2946	.2912	.2877	.2843	.2810	.2776
−0.4	.3446	.3409	.3372	.3336	.3300	.3264	.3228	.3192	.3156	.3121
−0.3	.3821	.3783	.3745	.3707	.3669	.3632	.3594	.3557	.3520	.3482
−0.2	.4207	.4168	.4129	.4090	.4052	.4013	.3974	.3936	.3897	.3859
−0.1	.4602	.4562	.4522	.4483	.4443	.4404	.4364	.4325	.4286	.4247
−0.0	.5000	.4960	.4920	.4880	.4840	.4801	.4761	.4721	.4681	.4641

(continued)

Table A.3 Standard Normal Curve Areas (*cont.*) $\Phi(z) = P(Z \leq z)$

z	.00	.01	.02	.03	.04	.05	.06	.07	.08	.09
0.0	.5000	.5040	.5080	.5120	.5160	.5199	.5239	.5279	.5319	.5359
0.1	.5398	.5438	.5478	.5517	.5557	.5596	.5636	.5675	.5714	.5753
0.2	.5793	.5832	.5871	.5910	.5948	.5987	.6026	.6064	.6103	.6141
0.3	.6179	.6217	.6255	.6293	.6331	.6368	.6406	.6443	.6480	.6517
0.4	.6554	.6591	.6628	.6664	.6700	.6736	.6772	.6808	.6844	.6879
0.5	.6915	.6950	.6985	.7019	.7054	.7088	.7123	.7157	.7190	.7224
0.6	.7257	.7291	.7324	.7357	.7389	.7422	.7454	.7486	.7517	.7549
0.7	.7580	.7611	.7642	.7673	.7704	.7734	.7764	.7794	.7823	.7852
0.8	.7881	.7910	.7939	.7967	.7995	.8023	.8051	.8078	.8106	.8133
0.9	.8159	.8186	.8212	.8238	.8264	.8289	.8315	.8340	.8365	.8389
1.0	.8413	.8438	.8461	.8485	.8508	.8531	.8554	.8577	.8599	.8621
1.1	.8643	.8665	.8686	.8708	.8729	.8749	.8770	.8790	.8810	.8830
1.2	.8849	.8869	.8888	.8907	.8925	.8944	.8962	.8980	.8997	.9015
1.3	.9032	.9049	.9066	.9082	.9099	.9115	.9131	.9147	.9162	.9177
1.4	.9192	.9207	.9222	.9236	.9251	.9265	.9278	.9292	.9306	.9319
1.5	.9332	.9345	.9357	.9370	.9382	.9394	.9406	.9418	.9429	.9441
1.6	.9452	.9463	.9474	.9484	.9495	.9505	.9515	.9525	.9535	.9545
1.7	.9554	.9564	.9573	.9582	.9591	.9599	.9608	.9616	.9625	.9633
1.8	.9641	.9649	.9656	.9664	.9671	.9678	.9686	.9693	.9699	.9706
1.9	.9713	.9719	.9726	.9732	.9738	.9744	.9750	.9756	.9761	.9767
2.0	.9772	.9778	.9783	.9788	.9793	.9798	.9803	.9808	.9812	.9817
2.1	.9821	.9826	.9830	.9834	.9838	.9842	.9846	.9850	.9854	.9857
2.2	.9861	.9864	.9868	.9871	.9875	.9878	.9881	.9884	.9887	.9890
2.3	.9893	.9896	.9898	.9901	.9904	.9906	.9909	.9911	.9913	.9916
2.4	.9918	.9920	.9922	.9925	.9927	.9929	.9931	.9932	.9934	.9936
2.5	.9938	.9940	.9941	.9943	.9945	.9946	.9948	.9949	.9951	.9952
2.6	.9953	.9955	.9956	.9957	.9959	.9960	.9961	.9962	.9963	.9964
2.7	.9965	.9966	.9967	.9968	.9969	.9970	.9971	.9972	.9973	.9974
2.8	.9974	.9975	.9976	.9977	.9977	.9978	.9979	.9979	.9980	.9981
2.9	.9981	.9982	.9982	.9983	.9984	.9984	.9985	.9985	.9986	.9986
3.0	.9987	.9987	.9987	.9988	.9988	.9989	.9989	.9989	.9990	.9990
3.1	.9990	.9991	.9991	.9991	.9992	.9992	.9992	.9992	.9993	.9993
3.2	.9993	.9993	.9994	.9994	.9994	.9994	.9994	.9995	.9995	.9995
3.3	.9995	.9995	.9995	.9996	.9996	.9996	.9996	.9996	.9996	.9997
3.4	.9997	.9997	.9997	.9997	.9997	.9997	.9997	.9997	.9997	.9998

Table A.4 The Incomplete Gamma Function

$$F(x; \alpha) = \int_0^x \frac{1}{\Gamma(\alpha)} y^{\alpha-1} e^{-y} dy$$

x \ α	1	2	3	4	5	6	7	8	9	10
1	.632	.264	.080	.019	.004	.001	.000	.000	.000	.000
2	.865	.594	.323	.143	.053	.017	.005	.001	.000	.000
3	.950	.801	.577	.353	.185	.084	.034	.012	.004	.001
4	.982	.908	.762	.567	.371	.215	.111	.051	.021	.008
5	.993	.960	.875	.735	.560	.384	.238	.133	.068	.032
6	.998	.983	.938	.849	.715	.554	.394	.256	.153	.084
7	.999	.993	.970	.918	.827	.699	.550	.401	.271	.170
8	1.000	.997	.986	.958	.900	.809	.687	.547	.407	.283
9		.999	.994	.979	.945	.884	.793	.676	.544	.413
10		1.000	.997	.990	.971	.933	.870	.780	.667	.542
11			.999	.995	.985	.962	.921	.857	.768	.659
12			1.000	.998	.992	.980	.954	.911	.845	.758
13				.999	.996	.989	.974	.946	.900	.834
14				1.000	.998	.994	.986	.968	.938	.891
15					.999	.997	.992	.982	.963	.930

Table A.5 Critical Values for t Distributions

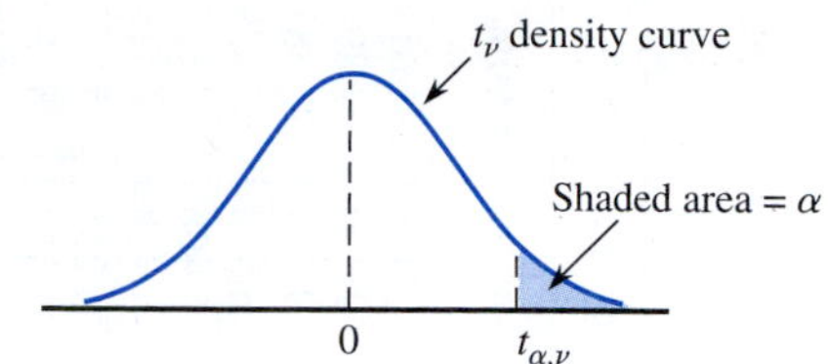

	α						
ν	.10	.05	.025	.01	.005	.001	.0005
1	3.078	6.314	12.706	31.821	63.657	318.31	636.62
2	1.886	2.920	4.303	6.965	9.925	22.326	31.598
3	1.638	2.353	3.182	4.541	5.841	10.213	12.924
4	1.533	2.132	2.776	3.747	4.604	7.173	8.610
5	1.476	2.015	2.571	3.365	4.032	5.893	6.869
6	1.440	1.943	2.447	3.143	3.707	5.208	5.959
7	1.415	1.895	2.365	2.998	3.499	4.785	5.408
8	1.397	1.860	2.306	2.896	3.355	4.501	5.041
9	1.383	1.833	2.262	2.821	3.250	4.297	4.781
10	1.372	1.812	2.228	2.764	3.169	4.144	4.587
11	1.363	1.796	2.201	2.718	3.106	4.025	4.437
12	1.356	1.782	2.179	2.681	3.055	3.930	4.318
13	1.350	1.771	2.160	2.650	3.012	3.852	4.221
14	1.345	1.761	2.145	2.624	2.977	3.787	4.140
15	1.341	1.753	2.131	2.602	2.947	3.733	4.073
16	1.337	1.746	2.120	2.583	2.921	3.686	4.015
17	1.333	1.740	2.110	2.567	2.898	3.646	3.965
18	1.330	1.734	2.101	2.552	2.878	3.610	3.922
19	1.328	1.729	2.093	2.539	2.861	3.579	3.883
20	1.325	1.725	2.086	2.528	2.845	3.552	3.850
21	1.323	1.721	2.080	2.518	2.831	3.527	3.819
22	1.321	1.717	2.074	2.508	2.819	3.505	3.792
23	1.319	1.714	2.069	2.500	2.807	3.485	3.767
24	1.318	1.711	2.064	2.492	2.797	3.467	3.745
25	1.316	1.708	2.060	2.485	2.787	3.450	3.725
26	1.315	1.706	2.056	2.479	2.779	3.435	3.707
27	1.314	1.703	2.052	2.473	2.771	3.421	3.690
28	1.313	1.701	2.048	2.467	2.763	3.408	3.674
29	1.311	1.699	2.045	2.462	2.756	3.396	3.659
30	1.310	1.697	2.042	2.457	2.750	3.385	3.646
32	1.309	1.694	2.037	2.449	2.738	3.365	3.622
34	1.307	1.691	2.032	2.441	2.728	3.348	3.601
36	1.306	1.688	2.028	2.434	2.719	3.333	3.582
38	1.304	1.686	2.024	2.429	2.712	3.319	3.566
40	1.303	1.684	2.021	2.423	2.704	3.307	3.551
50	1.299	1.676	2.009	2.403	2.678	3.262	3.496
60	1.296	1.671	2.000	2.390	2.660	3.232	3.460
120	1.289	1.658	1.980	2.358	2.617	3.160	3.373
∞	1.282	1.645	1.960	2.326	2.576	3.090	3.291

Table A.6 Tolerance Critical Values for Normal Population Distributions

	Two-Sided Intervals						One-Sided Intervals					
Confidence Level	95%			99%			95%			99%		
% of Population Captured	≥90%	≥95%	≥99%	≥90%	≥95%	≥99%	≥90%	≥95%	≥99%	≥90%	≥95%	≥99%
Sample Size n 2	32.019	37.674	48.430	160.193	188.491	242.300	20.581	26.260	37.094	103.029	131.426	185.617
3	8.380	9.916	12.861	18.930	22.401	29.055	6.156	7.656	10.553	13.995	17.370	23.896
4	5.369	6.370	8.299	9.398	11.150	14.527	4.162	5.144	7.042	7.380	9.083	12.387
5	4.275	5.079	6.634	6.612	7.855	10.260	3.407	4.203	5.741	5.362	6.578	8.939
6	3.712	4.414	5.775	5.337	6.345	8.301	3.006	3.708	5.062	4.411	5.406	7.335
7	3.369	4.007	5.248	4.613	5.488	7.187	2.756	3.400	4.642	3.859	4.728	6.412
8	3.136	3.732	4.891	4.147	4.936	6.468	2.582	3.187	4.354	3.497	4.285	5.812
9	2.967	3.532	4.631	3.822	4.550	5.966	2.454	3.031	4.143	3.241	3.972	5.389
10	2.839	3.379	4.433	3.582	4.265	5.594	2.355	2.911	3.981	3.048	3.738	5.074
11	2.737	3.259	4.277	3.397	4.045	5.308	2.275	2.815	3.852	2.898	3.556	4.829
12	2.655	3.162	4.150	3.250	3.870	5.079	2.210	2.736	3.747	2.777	3.410	4.633
13	2.587	3.081	4.044	3.130	3.727	4.893	2.155	2.671	3.659	2.677	3.290	4.472
14	2.529	3.012	3.955	3.029	3.608	4.737	2.109	2.615	3.585	2.593	3.189	4.337
15	2.480	2.954	3.878	2.945	3.507	4.605	2.068	2.566	3.520	2.522	3.102	4.222
16	2.437	2.903	3.812	2.872	3.421	4.492	2.033	2.524	3.464	2.460	3.028	4.123
17	2.400	2.858	3.754	2.808	3.345	4.393	2.002	2.486	3.414	2.405	2.963	4.037
18	2.366	2.819	3.702	2.753	3.279	4.307	1.974	2.453	3.370	2.357	2.905	3.960
19	2.337	2.784	3.656	2.703	3.221	4.230	1.949	2.423	3.331	2.314	2.854	3.892
20	2.310	2.752	3.615	2.659	3.168	4.161	1.926	2.396	3.295	2.276	2.808	3.832
25	2.208	2.631	3.457	2.494	2.972	3.904	1.838	2.292	3.158	2.129	2.633	3.601
30	2.140	2.549	3.350	2.385	2.841	3.733	1.777	2.220	3.064	2.030	2.516	3.447
35	2.090	2.490	3.272	2.306	2.748	3.611	1.732	2.167	2.995	1.957	2.430	3.334
40	2.052	2.445	3.213	2.247	2.677	3.518	1.697	2.126	2.941	1.902	2.364	3.249
45	2.021	2.408	3.165	2.200	2.621	3.444	1.669	2.092	2.898	1.857	2.312	3.180
50	1.996	2.379	3.126	2.162	2.576	3.385	1.646	2.065	2.863	1.821	2.269	3.125
60	1.958	2.333	3.066	2.103	2.506	3.293	1.609	2.022	2.807	1.764	2.202	3.038
70	1.929	2.299	3.021	2.060	2.454	3.225	1.581	1.990	2.765	1.722	2.153	2.974
80	1.907	2.272	2.986	2.026	2.414	3.173	1.559	1.965	2.733	1.688	2.114	2.924
90	1.889	2.251	2.958	1.999	2.382	3.130	1.542	1.944	2.706	1.661	2.082	2.883
100	1.874	2.233	2.934	1.977	2.355	3.096	1.527	1.927	2.684	1.639	2.056	2.850
150	1.825	2.175	2.859	1.905	2.270	2.983	1.478	1.870	2.611	1.566	1.971	2.741
200	1.798	2.143	2.816	1.865	2.222	2.921	1.450	1.837	2.570	1.524	1.923	2.679
250	1.780	2.121	2.788	1.839	2.191	2.880	1.431	1.815	2.542	1.496	1.891	2.638
300	1.767	2.106	2.767	1.820	2.169	2.850	1.417	1.800	2.522	1.476	1.868	2.608
∞	1.645	1.960	2.576	1.645	1.960	2.576	1.282	1.645	2.326	1.282	1.645	2.326

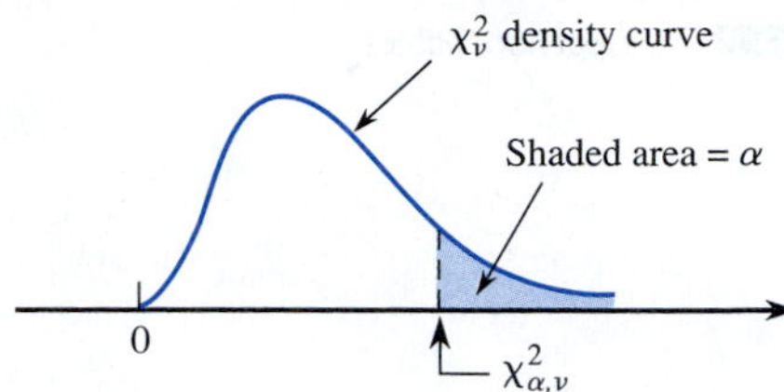

Table A.7 Critical Values for Chi-Squared Distributions

					α					
ν	.995	.99	.975	.95	.90	.10	.05	.025	.01	.005
1	0.000	0.000	0.001	0.004	0.016	2.706	3.843	5.025	6.637	7.882
2	0.010	0.020	0.051	0.103	0.211	4.605	5.992	7.378	9.210	10.597
3	0.072	0.115	0.216	0.352	0.584	6.251	7.815	9.348	11.344	12.837
4	0.207	0.297	0.484	0.711	1.064	7.779	9.488	11.143	13.277	14.860
5	0.412	0.554	0.831	1.145	1.610	9.236	11.070	12.832	15.085	16.748
6	0.676	0.872	1.237	1.635	2.204	10.645	12.592	14.440	16.812	18.548
7	0.989	1.239	1.690	2.167	2.833	12.017	14.067	16.012	18.474	20.276
8	1.344	1.646	2.180	2.733	3.490	13.362	15.507	17.534	20.090	21.954
9	1.735	2.088	2.700	3.325	4.168	14.684	16.919	19.022	21.665	23.587
10	2.156	2.558	3.247	3.940	4.865	15.987	18.307	20.483	23.209	25.188
11	2.603	3.053	3.816	4.575	5.578	17.275	19.675	21.920	24.724	26.755
12	3.074	3.571	4.404	5.226	6.304	18.549	21.026	23.337	26.217	28.300
13	3.565	4.107	5.009	5.892	7.041	19.812	22.362	24.735	27.687	29.817
14	4.075	4.660	5.629	6.571	7.790	21.064	23.685	26.119	29.141	31.319
15	4.600	5.229	6.262	7.261	8.547	22.307	24.996	27.488	30.577	32.799
16	5.142	5.812	6.908	7.962	9.312	23.542	26.296	28.845	32.000	34.267
17	5.697	6.407	7.564	8.682	10.085	24.769	27.587	30.190	33.408	35.716
18	6.265	7.015	8.231	9.390	10.865	25.989	28.869	31.526	34.805	37.156
19	6.843	7.632	8.906	10.117	11.651	27.203	30.143	32.852	36.190	38.580
20	7.434	8.260	9.591	10.851	12.443	28.412	31.410	34.170	37.566	39.997
21	8.033	8.897	10.283	11.591	13.240	29.615	32.670	35.478	38.930	41.399
22	8.643	9.542	10.982	12.338	14.042	30.813	33.924	36.781	40.289	42.796
23	9.260	10.195	11.688	13.090	14.848	32.007	35.172	38.075	41.637	44.179
24	9.886	10.856	12.401	13.848	15.659	33.196	36.415	39.364	42.980	45.558
25	10.519	11.523	13.120	14.611	16.473	34.381	37.652	40.646	44.313	46.925
26	11.160	12.198	13.844	15.379	17.292	35.563	38.885	41.923	45.642	48.290
27	11.807	12.878	14.573	16.151	18.114	36.741	40.113	43.194	46.962	49.642
28	12.461	13.565	15.308	16.928	18.939	37.916	41.337	44.461	48.278	50.993
29	13.120	14.256	16.147	17.708	19.768	39.087	42.557	45.772	49.586	52.333
30	13.787	14.954	16.791	18.493	20.599	40.256	43.773	46.979	50.892	53.672
31	14.457	15.655	17.538	19.280	21.433	41.422	44.985	48.231	52.190	55.000
32	15.134	16.362	18.291	20.072	22.271	42.585	46.194	49.480	53.486	56.328
33	15.814	17.073	19.046	20.866	23.110	43.745	47.400	50.724	54.774	57.646
34	16.501	17.789	19.806	21.664	23.952	44.903	48.602	51.966	56.061	58.964
35	17.191	18.508	20.569	22.465	24.796	46.059	49.802	53.203	57.340	60.272
36	17.887	19.233	21.336	23.269	25.643	47.212	50.998	54.437	58.619	61.581
37	18.584	19.960	22.105	24.075	26.492	48.363	52.192	55.667	59.891	62.880
38	19.289	20.691	22.878	24.884	27.343	49.513	53.384	56.896	61.162	64.181
39	19.994	21.425	23.654	25.695	28.196	50.660	54.572	58.119	62.426	65.473
40	20.706	22.164	24.433	26.509	29.050	51.805	55.758	59.342	63.691	66.766

For $\nu > 40$, $\chi^2_{\alpha,\nu} \approx \nu\left(1 - \frac{2}{9\nu} + z_\alpha\sqrt{\frac{2}{9\nu}}\right)^3$

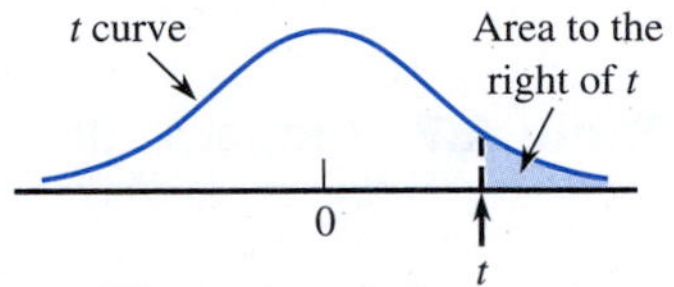

Table A.8 *t* Curve Tail Areas

t \ ν	1	2	3	4	5	6	7	8	9	10	11	12	13	14	15	16	17	18
0.0	.500	.500	.500	.500	.500	.500	.500	.500	.500	.500	.500	.500	.500	.500	.500	.500	.500	.500
0.1	.468	.465	.463	.463	.462.	.462	.462	.461	.461	.461	.461	.461	.461	.461	.461	.461	.461	.461
0.2	.437	.430	.427	.426	.425	.424	.424	.423	.423	.423	.423	.422	.422	.422	.422	.422	.422	.422
0.3	.407	.396	.392	.390	.388	.387	.386	.386	.386	.385	.385	.385	.384	.384	.384	.384	.384	.384
0.4	.379	.364	.358	.355	.353	.352	.351	.350	.349	.349	.348	.348	.348	.347	.347	.347	.347	.347
0.5	.352	.333	.326	.322	.319	.317	.316	.315	.315	.314	.313	.313	.313	.312	.312	.312	.312	.312
0.6	.328	.305	.295	.290	.287	.285	.284	.283	.282	.281	.280	.280	.279	.279	.279	.278	.278	.278
0.7	.306	.278	.267	.261	.258	.255	.253	.252	.251	.250	.249	.249	.248	.247	.247	.247	.247	.246
0.8	.285	.254	.241	.234	.230	.227	.225	.223	.222	.221	.220	.220	.219	.218	.218	.218	.217	.217
0.9	.267	.232	.217	.210	.205	.201	.199	.197	.196	.195	.194	.193	.192	.191	.191	.191	.190	.190
1.0	.250	.211	.196	.187	.182	.178	.175	.173	.172	.170	.169	.169	.168	.167	.167	.166	.166	.165
1.1	.235	.193	.176	.167	.162	.157	.154	.152	.150	.149	.147	.146	.146	.144	.144	.144	.143	.143
1.2	.221	.177	.158	.148	.142	.138	.135	.132	.130	.129	.128	.127	.126	.124	.124	.124	.123	.123
1.3	.209	.162	.142	.132	.125	.121	.117	.115	.113	.111	.110	.109	.108	.107	.107	.106	.105	.105
1.4	.197	.148	.128	.117	.110	.106	.102	.100	.098	.096	.095	.093	.092	.091	.091	.090	.090	.089
1.5	.187	.136	.115	.104	.097	.092	.089	.086	.084	.082	.081	.080	.079	.077	.077	.077	.076	.075
1.6	.178	.125	.104	.092	.085	.080	.077	.074	.072	.070	.069	.068	.067	.065	.065	.065	.064	.064
1.7	.169	.116	.094	.082	.075	.070	.065	.064	.062	.060	.059	.057	.056	.055	.055	.054	.054	.053
1.8	.161	.107	.085	.073	.066	.061	.057	.055	.053	.051	.050	.049	.048	.046	.046	.045	.045	.044
1.9	.154	.099	.077	.065	.058	.053	.050	.047	.045	.043	.042	.041	.040	.038	.038	.038	.037	.037
2.0	.148	.092	.070	.058	.051	.046	.043	.040	.038	.037	.035	.034	.033	.032	.032	.031	.031	.030
2.1	.141	.085	.063	.052	.045	.040	.037	.034	.033	.031	.030	.029	.028	.027	.027	.026	.025	.025
2.2	.136	.079	.058	.046	.040	.035	.032	.029	.028	.026	.025	.024	.023	.022	.022	.021	.021	.021
2.3	.131	.074	.052	.041	.035	.031	.027	.025	.023	.022	.021	.020	.019	.018	.018	.018	.017	.017
2.4	.126	.069	.048	.037	.031	.027	.024	.022	.020	.019	.018	.017	.016	.015	.015	.014	.014	.014
2.5	.121	.065	.044	.033	.027	.023	.020	.018	.017	.016	.015	.014	.013	.012	.012	.012	.011	.011
2.6	.117	.061	.040	.030	.024	.020	.018	.016	.014	.013	.012	.012	.011	.010	.010	.010	.009	.009
2.7	.113	.057	.037	.027	.021	.018	.015	.014	.012	.011	.010	.010	.009	.008	.008	.008	.008	.007
2.8	.109	.054	.034	.024	.019	.016	.013	.012	.010	.009	.009	.008	.008	.007	.007	.006	.006	.006
2.9	.106	.051	.031	.022	.017	.014	.011	.010	.009	.008	.007	.007	.006	.005	.005	.005	.005	.005
3.0	.102	.048	.029	.020	.015	.012	.010	.009	.007	.007	.006	.006	.005	.004	.004	.004	.004	.004
3.1	.099	.045	.027	.018	.013	.011	.009	.007	.006	.006	.005	.005	.004	.004	.004	.003	.003	.003
3.2	.096	.043	.025	.016	.012	.009	.008	.006	.005	.005	.004	.004	.003	.003	.003	.003	.003	.002
3.3	.094	.040	.023	.015	.011	.008	.007	.005	.005	.004	.004	.003	.003	.002	.002	.002	.002	.002
3.4	.091	.038	.021	.014	.010	.007	.006	.005	.004	.003	.003	.003	.002	.002	.002	.002	.002	.002
3.5	.089	.036	.020	.012	.009	.006	.005	.004	.003	.003	.002	.002	.002	.002	.002	.001	.001	.001
3.6	.086	.035	.018	.011	.008	.006	.004	.004	.003	.002	.002	.002	.002	.001	.001	.001	.001	.001
3.7	.084	.033	.017	.010	.007	.005	.004	.003	.002	.002	.002	.002	.001	.001	.001	.001	.001	.001
3.8	.082	.031	.016	.010	.006	.004	.003	.003	.002	.002	.001	.001	.001	.001	.001	.001	.001	.001
3.9	.080	.030	.015	.009	.006	.004	.003	.002	.002	.001	.001	.001	.001	.001	.001	.001	.001	.001
4.0	.078	.029	.014	.008	.005	.004	.003	.002	.002	.001	.001	.001	.001	.001	.001	.001	.000	.000

(continued)

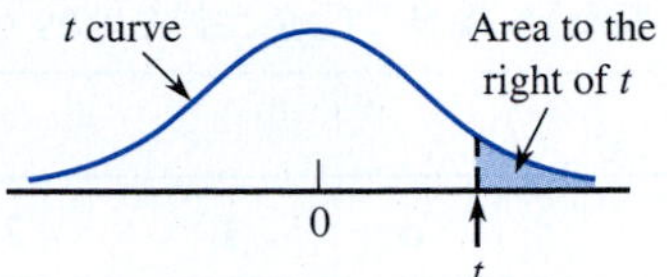

Table A.8 *t* Curve Tail Areas (*cont.*)

$t \backslash \nu$	19	20	21	22	23	24	25	26	27	28	29	30	35	40	60	120	$\infty (= z)$
0.0	.500	.500	.500	.500	.500	.500	.500	.500	.500	.500	.500	.500	.500	.500	.500	.500	.500
0.1	.461	.461	.461	.461	.461	.461	.461	.461	.461	.461	.461	.461	.460	.460	.460	.460	.460
0.2	.422	.422	.422	.422	.422	.422	.422	.422	.421	.421	.421	.421	.421	.421	.421	.421	.421
0.3	.384	.384	.384	.383	.383	.383	.383	.383	.383	.383	.383	.383	.383	.383	.383	.382	.382
0.4	.347	.347	.347	.347	.346	.346	.346	.346	.346	.346	.346	.346	.346	.346	.345	.345	.345
0.5	.311	.311	.311	.311	.311	.311	.311	.311	.311	.310	.310	.310	.310	.310	.309	.309	.309
0.6	.278	.278	.278	.277	.277	.277	.277	.277	.277	.277	.277	.277	.276	.276	.275	.275	.274
0.7	.246	.246	.246	.246	.245	.245	.245	.245	.245	.245	.245	.245	.244	.244	.243	.243	.242
0.8	.217	.217	.216	.216	.216	.216	.216	.215	.215	.215	.215	.215	.215	.214	.213	.213	.212
0.9	.190	.189	.189	.189	.189	.189	.188	.188	.188	.188	.188	.188	.187	.187	.186	.185	.184
1.0	.165	.165	.164	.164	.164	.164	.163	.163	.163	.163	.163	.163	.162	.162	.161	.160	.159
1.1	.143	.142	.142	.142	.141	.141	.141	.141	.141	.140	.140	.140	.139	.139	.138	.137	.136
1.2	.122	.122	.122	.121	.121	.121	.121	.120	.120	.120	.120	.120	.119	.119	.117	.116	.115
1.3	.105	.104	.104	.104	.103	.103	.103	.103	.102	.102	.102	.102	.101	.101	.099	.098	.097
1.4	.089	.089	.088	.088	.087	.087	.087	.087	.086	.086	.086	.086	.085	.085	.083	.082	.081
1.5	.075	.075	.074	.074	.074	.073	.073	.073	.073	.072	.072	.072	.071	.071	.069	.068	.067
1.6	.063	.063	.062	.062	.062	.061	.061	.061	.061	.060	.060	.060	.059	.059	.057	.056	.055
1.7	.053	.052	.052	.052	.051	.051	.051	.051	.050	.050	.050	.050	.049	.048	.047	.046	.045
1.8	.044	.043	.043	.043	.042	.042	.042	.042	.042	.041	.041	.041	.040	.040	.038	.037	.036
1.9	.036	.036	.036	.035	.035	.035	.035	.034	.034	.034	.034	.034	.033	.032	.031	.030	.029
2.0	.030	.030	.029	.029	.029	.028	.028	.028	.028	.028	.027	.027	.027	.026	.025	.024	.023
2.1	.025	.024	.024	.024	.023	.023	.023	.023	.023	.022	.022	.022	.022	.021	.020	.019	.018
2.2	.020	.020	.020	.019	.019	.019	.019	.018	.018	.018	.018	.018	.017	.017	.016	.015	.014
2.3	.016	.016	.016	.016	.015	.015	.015	.015	.015	.015	.014	.014	.014	.013	.012	.012	.011
2.4	.013	.013	.013	.013	.012	.012	.012	.012	.012	.012	.012	.011	.011	.011	.010	.009	.008
2.5	.011	.011	.010	.010	.010	.010	.010	.010	.009	.009	.009	.009	.009	.008	.008	.007	.006
2.6	.009	.009	.008	.008	.008	.008	.008	.008	.007	.007	.007	.007	.007	.007	.006	.005	.005
2.7	.007	.007	.007	.007	.006	.006	.006	.006	.006	.006	.006	.006	.005	.005	.004	.004	.003
2.8	.006	.006	.005	.005	.005	.005	.005	.005	.005	.005	.005	.004	.004	.004	.003	.003	.003
2.9	.005	.004	.004	.004	.004	.004	.004	.004	.004	.004	.004	.003	.003	.003	.003	.002	.002
3.0	.004	.004	.003	.003	.003	.003	.003	.003	.003	.003	.003	.003	.002	.002	.002	.002	.001
3.1	.003	.003	.003	.003	.003	.002	.002	.002	.002	.002	.002	.002	.002	.002	.001	.001	.001
3.2	.002	.002	.002	.002	.002	.002	.002	.002	.002	.002	.002	.002	.001	.001	.001	.001	.001
3.3	.002	.002	.002	.002	.002	.001	.001	.001	.001	.001	.001	.001	.001	.001	.001	.001	.000
3.4	.002	.001	.001	.001	.001	.001	.001	.001	.001	.001	.001	.001	.001	.001	.001	.000	.000
3.5	.001	.001	.001	.001	.001	.001	.001	.001	.001	.001	.001	.001	.001	.001	.000	.000	.000
3.6	.001	.001	.001	.001	.001	.001	.001	.001	.001	.001	.001	.001	.000	.000	.000	.000	.000
3.7	.001	.001	.001	.001	.001	.001	.001	.001	.000	.000	.000	.000	.000	.000	.000	.000	.000
3.8	.001	.001	.001	.000	.000	.000	.000	.000	.000	.000	.000	.000	.000	.000	.000	.000	.000
3.9	.000	.000	.000	.000	.000	.000	.000	.000	.000	.000	.000	.000	.000	.000	.000	.000	.000
4.0	.000	.000	.000	.000	.000	.000	.000	.000	.000	.000	.000	.000	.000	.000	.000	.000	.000

Table A.9 Critical Values for *F* Distributions

ν_2 = denominator df	α	ν_1 = numerator df								
		1	**2**	**3**	**4**	**5**	**6**	**7**	**8**	**9**
1	.100	39.86	49.50	53.59	55.83	57.24	58.20	58.91	59.44	59.86
	.050	161.45	199.50	215.71	224.58	230.16	233.99	236.77	238.88	240.54
	.010	4052.2	4999.5	5403.4	5624.6	5763.6	5859.0	5928.4	5981.1	6022.5
	.001	405284	500000	540379	562500	576405	585937	592873	598144	602284
2	.100	8.53	9.00	9.16	9.24	9.29	9.33	9.35	9.37	9.38
	.050	18.51	19.00	19.16	19.25	19.30	19.33	19.35	19.37	19.38
	.010	98.50	99.00	99.17	99.25	99.30	99.33	99.36	99.37	99.39
	.001	998.50	999.00	999.17	999.25	999.30	999.33	999.36	999.37	999.39
3	.100	5.54	5.46	5.39	5.34	5.31	5.28	5.27	5.25	5.24
	.050	10.13	9.55	9.28	9.12	9.01	8.94	8.89	8.85	8.81
	.010	34.12	30.82	29.46	28.71	28.24	27.91	27.67	27.49	27.35
	.001	167.03	148.50	141.11	137.10	134.58	132.85	131.58	130.62	129.86
4	.100	4.54	4.32	4.19	4.11	4.05	4.01	3.98	3.95	3.94
	.050	7.71	6.94	6.59	6.39	6.26	6.16	6.09	6.04	6.00
	.010	21.20	18.00	16.69	15.98	15.52	15.21	14.98	14.80	14.66
	.001	74.14	61.25	56.18	53.44	51.71	50.53	49.66	49.00	48.47
5	.100	4.06	3.78	3.62	3.52	3.45	3.40	3.37	3.34	3.32
	.050	6.61	5.79	5.41	5.19	5.05	4.95	4.88	4.82	4.77
	.010	16.26	13.27	12.06	11.39	10.97	10.67	10.46	10.29	10.16
	.001	47.18	37.12	33.20	31.09	29.75	28.83	28.16	27.65	27.24
6	.100	3.78	3.46	3.29	3.18	3.11	3.05	3.01	2.98	2.96
	.050	5.99	5.14	4.76	4.53	4.39	4.28	4.21	4.15	4.10
	.010	13.75	10.92	9.78	9.15	8.75	8.47	8.26	8.10	7.98
	.001	35.51	27.00	23.70	21.92	20.80	20.03	19.46	19.03	18.69
7	.100	3.59	3.26	3.07	2.96	2.88	2.83	2.78	2.75	2.72
	.050	5.59	4.74	4.35	4.12	3.97	3.87	3.79	3.73	3.68
	.010	12.25	9.55	8.45	7.85	7.46	7.19	6.99	6.84	6.72
	.001	29.25	21.69	18.77	17.20	16.21	15.52	15.02	14.63	14.33
8	.100	3.46	3.11	2.92	2.81	2.73	2.67	2.62	2.59	2.56
	.050	5.32	4.46	4.07	3.84	3.69	3.58	3.50	3.44	3.39
	.010	11.26	8.65	7.59	7.01	6.63	6.37	6.18	6.03	5.91
	.001	25.41	18.49	15.83	14.39	13.48	12.86	12.40	12.05	11.77
9	.100	3.36	3.01	2.81	2.69	2.61	2.55	2.51	2.47	2.44
	.050	5.12	4.26	3.86	3.63	3.48	3.37	3.29	3.23	3.18
	.010	10.56	8.02	6.99	6.42	6.06	5.80	5.61	5.47	5.35
	.001	22.86	16.39	13.90	12.56	11.71	11.13	10.70	10.37	10.11
10	.100	3.29	2.92	2.73	2.61	2.52	2.46	2.41	2.38	2.35
	.050	4.96	4.10	3.71	3.48	3.33	3.22	3.14	3.07	3.02
	.010	10.04	7.56	6.55	5.99	5.64	5.39	5.20	5.06	4.94
	.001	21.04	14.91	12.55	11.28	10.48	9.93	9.52	9.20	8.96
11	.100	3.23	2.86	2.66	2.54	2.45	2.39	2.34	2.30	2.27
	.050	4.84	3.98	3.59	3.36	3.20	3.09	3.01	2.95	2.90
	.010	9.65	7.21	6.22	5.67	5.32	5.07	4.89	4.74	4.63
	.001	19.69	13.81	11.56	10.35	9.58	9.05	8.66	8.35	8.12
12	.100	3.18	2.81	2.61	2.48	2.39	2.33	2.28	2.24	2.21
	.050	4.75	3.89	3.49	3.26	3.11	3.00	2.91	2.85	2.80
	.010	9.33	6.93	5.95	5.41	5.06	4.82	4.64	4.50	4.39
	.001	18.64	12.97	10.80	9.63	8.89	8.38	8.00	7.71	7.48

(continued)

Table A.9 Critical Values for *F* Distributions (*cont.*)

ν_1 = numerator df

10	12	15	20	25	30	40	50	60	120	1000
60.19	60.71	61.22	61.74	62.05	62.26	62.53	62.69	62.79	63.06	63.30
241.88	243.91	245.95	248.01	249.26	250.10	251.14	251.77	252.20	253.25	254.19
6055.8	6106.3	6157.3	6208.7	6239.8	6260.6	6286.8	6302.5	6313.0	6339.4	6362.7
605621	610668	615764	620908	624017	626099	628712	630285	631337	633972	636301
9.39	9.41	9.42	9.44	9.45	9.46	9.47	9.47	9.47	9.48	9.49
19.40	19.41	19.43	19.45	19.46	19.46	19.47	19.48	19.48	19.49	19.49
99.40	99.42	99.43	99.45	99.46	99.47	99.47	99.48	99.48	99.49	99.50
999.40	999.42	999.43	999.45	999.46	999.47	999.47	999.48	999.48	999.49	999.50
5.23	5.22	5.20	5.18	5.17	5.17	5.16	5.15	5.15	5.14	5.13
8.79	8.74	8.70	8.66	8.63	8.62	8.59	8.58	8.57	8.55	8.53
27.23	27.05	26.87	26.69	26.58	26.50	26.41	26.35	26.32	26.22	26.14
129.25	128.32	127.37	126.42	125.84	125.45	124.96	124.66	124.47	123.97	123.53
3.92	3.90	3.87	3.84	3.83	3.82	3.80	3.80	3.79	3.78	3.76
5.96	5.91	5.86	5.80	5.77	5.75	5.72	5.70	5.69	5.66	5.63
14.55	14.37	14.20	14.02	13.91	13.84	13.75	13.69	13.65	13.56	13.47
48.05	47.41	46.76	46.10	45.70	45.43	45.09	44.88	44.75	44.40	44.09
3.30	3.27	3.24	3.21	3.19	3.17	3.16	3.15	3.14	3.12	3.11
4.74	4.68	4.62	4.56	4.52	4.50	4.46	4.44	4.43	4.40	4.37
10.05	9.89	9.72	9.55	9.45	9.38	9.29	9.24	9.20	9.11	9.03
26.92	26.42	25.91	25.39	25.08	24.87	24.60	24.44	24.33	24.06	23.82
2.94	2.90	2.87	2.84	2.81	2.80	2.78	2.77	2.76	2.74	2.72
4.06	4.00	3.94	3.87	3.83	3.81	3.77	3.75	3.74	3.70	3.67
7.87	7.72	7.56	7.40	7.30	7.23	7.14	7.09	7.06	6.97	6.89
18.41	17.99	17.56	17.12	16.85	16.67	16.44	16.31	16.21	15.98	15.77
2.70	2.67	2.63	2.59	2.57	2.56	2.54	2.52	2.51	2.49	2.47
3.64	3.57	3.51	3.44	3.40	3.38	3.34	3.32	3.30	3.27	3.23
6.62	6.47	6.31	6.16	6.06	5.99	5.91	5.86	5.82	5.74	5.66
14.08	13.71	13.32	12.93	12.69	12.53	12.33	12.20	12.12	11.91	11.72
2.54	2.50	2.46	2.42	2.40	2.38	2.36	2.35	2.34	2.32	2.30
3.35	3.28	3.22	3.15	3.11	3.08	3.04	3.02	3.01	2.97	2.93
5.81	5.67	5.52	5.36	5.26	5.20	5.12	5.07	5.03	4.95	4.87
11.54	11.19	10.84	10.48	10.26	10.11	9.92	9.80	9.73	9.53	9.36
2.42	2.38	2.34	2.30	2.27	2.25	2.23	2.22	2.21	2.18	2.16
3.14	3.07	3.01	2.94	2.89	2.86	2.83	2.80	2.79	2.75	2.71
5.26	5.11	4.96	4.81	4.71	4.65	4.57	4.52	4.48	4.40	4.32
9.89	9.57	9.24	8.90	8.69	8.55	8.37	8.26	8.19	8.00	7.84
2.32	2.28	2.24	2.20	2.17	2.16	2.13	2.12	2.11	2.08	2.06
2.98	2.91	2.85	2.77	2.73	2.70	2.66	2.64	2.62	2.58	2.54
4.85	4.71	4.56	4.41	4.31	4.25	4.17	4.12	4.08	4.00	3.92
8.75	8.45	8.13	7.80	7.60	7.47	7.30	7.19	7.12	6.94	6.78
2.25	2.21	2.17	2.12	2.10	2.08	2.05	2.04	2.03	2.00	1.98
2.85	2.79	2.72	2.65	2.60	2.57	2.53	2.51	2.49	2.45	2.41
4.54	4.40	4.25	4.10	4.01	3.94	3.86	3.81	3.78	3.69	3.61
7.92	7.63	7.32	7.01	6.81	6.68	6.52	6.42	6.35	6.18	6.02
2.19	2.15	2.10	2.06	2.03	2.01	1.99	1.97	1.96	1.93	1.91
2.75	2.69	2.62	2.54	2.50	2.47	2.43	2.40	2.38	2.34	2.30
4.30	4.16	4.01	3.86	3.76	3.70	3.62	3.57	3.54	3.45	3.37
7.29	7.00	6.71	6.40	6.22	6.09	5.93	5.83	5.76	5.59	5.44

(*continued*)

Table A.9 Critical Values for *F* Distributions (*cont.*)

ν_2 = denominator df	α	ν_1 = numerator df: 1	2	3	4	5	6	7	8	9
13	**.100**	3.14	2.76	2.56	2.43	2.35	2.28	2.23	2.20	2.16
	.050	4.67	3.81	3.41	3.18	3.03	2.92	2.83	2.77	2.71
	.010	9.07	6.70	5.74	5.21	4.86	4.62	4.44	4.30	4.19
	.001	17.82	12.31	10.21	9.07	8.35	7.86	7.49	7.21	6.98
14	**.100**	3.10	2.73	2.52	2.39	2.31	2.24	2.19	2.15	2.12
	.050	4.60	3.74	3.34	3.11	2.96	2.85	2.76	2.70	2.65
	.010	8.86	6.51	5.56	5.04	4.69	4.46	4.28	4.14	4.03
	.001	17.14	11.78	9.73	8.62	7.92	7.44	7.08	6.80	6.58
15	**.100**	3.07	2.70	2.49	2.36	2.27	2.21	2.16	2.12	2.09
	.050	4.54	3.68	3.29	3.06	2.90	2.79	2.71	2.64	2.59
	.010	8.68	6.36	5.42	4.89	4.56	4.32	4.14	4.00	3.89
	.001	16.59	11.34	9.34	8.25	7.57	7.09	6.74	6.47	6.26
16	**.100**	3.05	2.67	2.46	2.33	2.24	2.18	2.13	2.09	2.06
	.050	4.49	3.63	3.24	3.01	2.85	2.74	2.66	2.59	2.54
	.010	8.53	6.23	5.29	4.77	4.44	4.20	4.03	3.89	3.78
	.001	16.12	10.97	9.01	7.94	7.27	6.80	6.46	6.19	5.98
17	**.100**	3.03	2.64	2.44	2.31	2.22	2.15	2.10	2.06	2.03
	.050	4.45	3.59	3.20	2.96	2.81	2.70	2.61	2.55	2.49
	.010	8.40	6.11	5.19	4.67	4.34	4.10	3.93	3.79	3.68
	.001	15.72	10.66	8.73	7.68	7.02	6.56	6.22	5.96	5.75
18	**.100**	3.01	2.62	2.42	2.29	2.20	2.13	2.08	2.04	2.00
	.050	4.41	3.55	3.16	2.93	2.77	2.66	2.58	2.51	2.46
	.010	8.29	6.01	5.09	4.58	4.25	4.01	3.84	3.71	3.60
	.001	15.38	10.39	8.49	7.46	6.81	6.35	6.02	5.76	5.56
19	**.100**	2.99	2.61	2.40	2.27	2.18	2.11	2.06	2.02	1.98
	.050	4.38	3.52	3.13	2.90	2.74	2.63	2.54	2.48	2.42
	.010	8.18	5.93	5.01	4.50	4.17	3.94	3.77	3.63	3.52
	.001	15.08	10.16	8.28	7.27	6.62	6.18	5.85	5.59	5.39
20	**.100**	2.97	2.59	2.38	2.25	2.16	2.09	2.04	2.00	1.96
	.050	4.35	3.49	3.10	2.87	2.71	2.60	2.51	2.45	2.39
	.010	8.10	5.85	4.94	4.43	4.10	3.87	3.70	3.56	3.46
	.001	14.82	9.95	8.10	7.10	6.46	6.02	5.69	5.44	5.24
21	**.100**	2.96	2.57	2.36	2.23	2.14	2.08	2.02	1.98	1.95
	.050	4.32	3.47	3.07	2.84	2.68	2.57	2.49	2.42	2.37
	.010	8.02	5.78	4.87	4.37	4.04	3.81	3.64	3.51	3.40
	.001	14.59	9.77	7.94	6.95	6.32	5.88	5.56	5.31	5.11
22	**.100**	2.95	2.56	2.35	2.22	2.13	2.06	2.01	1.97	1.93
	.050	4.30	3.44	3.05	2.82	2.66	2.55	2.46	2.40	2.34
	.010	7.95	5.72	4.82	4.31	3.99	3.76	3.59	3.45	3.35
	.001	14.38	9.61	7.80	6.81	6.19	5.76	5.44	5.19	4.99
23	**.100**	2.94	2.55	2.34	2.21	2.11	2.05	1.99	1.95	1.92
	.050	4.28	3.42	3.03	2.80	2.64	2.53	2.44	2.37	2.32
	.010	7.88	5.66	4.76	4.26	3.94	3.71	3.54	3.41	3.30
	.001	14.20	9.47	7.67	6.70	6.08	5.65	5.33	5.09	4.89
24	**.100**	2.93	2.54	2.33	2.19	2.10	2.04	1.98	1.94	1.91
	.050	4.26	3.40	3.01	2.78	2.62	2.51	2.42	2.36	2.30
	.010	7.82	5.61	4.72	4.22	3.90	3.67	3.50	3.36	3.26
	.001	14.03	9.34	7.55	6.59	5.98	5.55	5.23	4.99	4.80

(*continued*)

Table A.9 Critical Values for *F* Distributions (*cont.*)

ν_1 = numerator df

10	12	15	20	25	30	40	50	60	120	1000
2.14	2.10	2.05	2.01	1.98	1.96	1.93	1.92	1.90	1.88	1.85
2.67	2.60	2.53	2.46	2.41	2.38	2.34	2.31	2.30	2.25	2.21
4.10	3.96	3.82	3.66	3.57	3.51	3.43	3.38	3.34	3.25	3.18
6.80	6.52	6.23	5.93	5.75	5.63	5.47	5.37	5.30	5.14	4.99
2.10	2.05	2.01	1.96	1.93	1.91	1.89	1.87	1.86	1.83	1.80
2.60	2.53	2.46	2.39	2.34	2.31	2.27	2.24	2.22	2.18	2.14
3.94	3.80	3.66	3.51	3.41	3.35	3.27	3.22	3.18	3.09	3.02
6.40	6.13	5.85	5.56	5.38	5.25	5.10	5.00	4.94	4.77	4.62
2.06	2.02	1.97	1.92	1.89	1.87	1.85	1.83	1.82	1.79	1.76
2.54	2.48	2.40	2.33	2.28	2.25	2.20	2.18	2.16	2.11	2.07
3.80	3.67	3.52	3.37	3.28	3.21	3.13	3.08	3.05	2.96	2.88
6.08	5.81	5.54	5.25	5.07	4.95	4.80	4.70	4.64	4.47	4.33
2.03	1.99	1.94	1.89	1.86	1.84	1.81	1.79	1.78	1.75	1.72
2.49	2.42	2.35	2.28	2.23	2.19	2.15	2.12	2.11	2.06	2.02
3.69	3.55	3.41	3.26	3.16	3.10	3.02	2.97	2.93	2.84	2.76
5.81	5.55	5.27	4.99	4.82	4.70	4.54	4.45	4.39	4.23	4.08
2.00	1.96	1.91	1.86	1.83	1.81	1.78	1.76	1.75	1.72	1.69
2.45	2.38	2.31	2.23	2.18	2.15	2.10	2.08	2.06	2.01	1.97
3.59	3.46	3.31	3.16	3.07	3.00	2.92	2.87	2.83	2.75	2.66
5.58	5.32	5.05	4.78	4.60	4.48	4.33	4.24	4.18	4.02	3.87
1.98	1.93	1.89	1.84	1.80	1.78	1.75	1.74	1.72	1.69	1.66
2.41	2.34	2.27	2.19	2.14	2.11	2.06	2.04	2.02	1.97	1.92
3.51	3.37	3.23	3.08	2.98	2.92	2.84	2.78	2.75	2.66	2.58
5.39	5.13	4.87	4.59	4.42	4.30	4.15	4.06	4.00	3.84	3.69
1.96	1.91	1.86	1.81	1.78	1.76	1.73	1.71	1.70	1.67	1.64
2.38	2.31	2.23	2.16	2.11	2.07	2.03	2.00	1.98	1.93	1.88
3.43	3.30	3.15	3.00	2.91	2.84	2.76	2.71	2.67	2.58	2.50
5.22	4.97	4.70	4.43	4.26	4.14	3.99	3.90	3.84	3.68	3.53
1.94	1.89	1.84	1.79	1.76	1.74	1.71	1.69	1.68	1.64	1.61
2.35	2.28	2.20	2.12	2.07	2.04	1.99	1.97	1.95	1.90	1.85
3.37	3.23	3.09	2.94	2.84	2.78	2.69	2.64	2.61	2.52	2.43
5.08	4.82	4.56	4.29	4.12	4.00	3.86	3.77	3.70	3.54	3.40
1.92	1.87	1.83	1.78	1.74	1.72	1.69	1.67	1.66	1.62	1.59
2.32	2.25	2.18	2.10	2.05	2.01	1.96	1.94	1.92	1.87	1.82
3.31	3.17	3.03	2.88	2.79	2.72	2.64	2.58	2.55	2.46	2.37
4.95	4.70	4.44	4.17	4.00	3.88	3.74	3.64	3.58	3.42	3.28
1.90	1.86	1.81	1.76	1.73	1.70	1.67	1.65	1.64	1.60	1.57
2.30	2.23	2.15	2.07	2.02	1.98	1.94	1.91	1.89	1.84	1.79
3.26	3.12	2.98	2.83	2.73	2.67	2.58	2.53	2.50	2.40	2.32
4.83	4.58	4.33	4.06	3.89	3.78	3.63	3.54	3.48	3.32	3.17
1.89	1.84	1.80	1.74	1.71	1.69	1.66	1.64	1.62	1.59	1.55
2.27	2.20	2.13	2.05	2.00	1.96	1.91	1.88	1.86	1.81	1.76
3.21	3.07	2.93	2.78	2.69	2.62	2.54	2.48	2.45	2.35	2.27
4.73	4.48	4.23	3.96	3.79	3.68	3.53	3.44	3.38	3.22	3.08
1.88	1.83	1.78	1.73	1.70	1.67	1.64	1.62	1.61	1.57	1.54
2.25	2.18	2.11	2.03	1.97	1.94	1.89	1.86	1.84	1.79	1.74
3.17	3.03	2.89	2.74	2.64	2.58	2.49	2.44	2.40	2.31	2.22
4.64	4.39	4.14	3.87	3.71	3.59	3.45	3.36	3.29	3.14	2.99

(*continued*)

Table A.9 Critical Values for F Distributions (*cont.*)

		ν_1 = numerator df								
	α	1	2	3	4	5	6	7	8	9
25	.100	2.92	2.53	2.32	2.18	2.09	2.02	1.97	1.93	1.89
	.050	4.24	3.39	2.99	2.76	2.60	2.49	2.40	2.34	2.28
	.010	7.77	5.57	4.68	4.18	3.85	3.63	3.46	3.32	3.22
	.001	13.88	9.22	7.45	6.49	5.89	5.46	5.15	4.91	4.71
26	.100	2.91	2.52	2.31	2.17	2.08	2.01	1.96	1.92	1.88
	.050	4.23	3.37	2.98	2.74	2.59	2.47	2.39	2.32	2.27
	.010	7.72	5.53	4.64	4.14	3.82	3.59	3.42	3.29	3.18
	.001	13.74	9.12	7.36	6.41	5.80	5.38	5.07	4.83	4.64
27	.100	2.90	2.51	2.30	2.17	2.07	2.00	1.95	1.91	1.87
	.050	4.21	3.35	2.96	2.73	2.57	2.46	2.37	2.31	2.25
	.010	7.68	5.49	4.60	4.11	3.78	3.56	3.39	3.26	3.15
	.001	13.61	9.02	7.27	6.33	5.73	5.31	5.00	4.76	4.57
28	.100	2.89	2.50	2.29	2.16	2.06	2.00	1.94	1.90	1.87
	.050	4.20	3.34	2.95	2.71	2.56	2.45	2.36	2.29	2.24
	.010	7.64	5.45	4.57	4.07	3.75	3.53	3.36	3.23	3.12
	.001	13.50	8.93	7.19	6.25	5.66	5.24	4.93	4.69	4.50
29	.100	2.89	2.50	2.28	2.15	2.06	1.99	1.93	1.89	1.86
	.050	4.18	3.33	2.93	2.70	2.55	2.43	2.35	2.28	2.22
	.010	7.60	5.42	4.54	4.04	3.73	3.50	3.33	3.20	3.09
	.001	13.39	8.85	7.12	6.19	5.59	5.18	4.87	4.64	4.45
30	.100	2.88	2.49	2.28	2.14	2.05	1.98	1.93	1.88	1.85
	.050	4.17	3.32	2.92	2.69	2.53	2.42	2.33	2.27	2.21
	.010	7.56	5.39	4.51	4.02	3.70	3.47	3.30	3.17	3.07
	.001	13.29	8.77	7.05	6.12	5.53	5.12	4.82	4.58	4.39
40	.100	2.84	2.44	2.23	2.09	2.00	1.93	1.87	1.83	1.79
	.050	4.08	3.23	2.84	2.61	2.45	2.34	2.25	2.18	2.12
	.010	7.31	5.18	4.31	3.83	3.51	3.29	3.12	2.99	2.89
	.001	12.61	8.25	6.59	5.70	5.13	4.73	4.44	4.21	4.02
50	.100	2.81	2.41	2.20	2.06	1.97	1.90	1.84	1.80	1.76
	.050	4.03	3.18	2.79	2.56	2.40	2.29	2.20	2.13	2.07
	.010	7.17	5.06	4.20	3.72	3.41	3.19	3.02	2.89	2.78
	.001	12.22	7.96	6.34	5.46	4.90	4.51	4.22	4.00	3.82
60	.100	2.79	2.39	2.18	2.04	1.95	1.87	1.82	1.77	1.74
	.050	4.00	3.15	2.76	2.53	2.37	2.25	2.17	2.10	2.04
	.010	7.08	4.98	4.13	3.65	3.34	3.12	2.95	2.82	2.72
	.001	11.97	7.77	6.17	5.31	4.76	4.37	4.09	3.86	3.69
100	.100	2.76	2.36	2.14	2.00	1.91	1.83	1.78	1.73	1.69
	.050	3.94	3.09	2.70	2.46	2.31	2.19	2.10	2.03	1.97
	.010	6.90	4.82	3.98	3.51	3.21	2.99	2.82	2.69	2.59
	.001	11.50	7.41	5.86	5.02	4.48	4.11	3.83	3.61	3.44
200	.100	2.73	2.33	2.11	1.97	1.88	1.80	1.75	1.70	1.66
	.050	3.89	3.04	2.65	2.42	2.26	2.14	2.06	1.98	1.93
	.010	6.76	4.71	3.88	3.41	3.11	2.89	2.73	2.60	2.50
	.001	11.15	7.15	5.63	4.81	4.29	3.92	3.65	3.43	3.26
1000	.100	2.71	2.31	2.09	1.95	1.85	1.78	1.72	1.68	1.64
	.050	3.85	3.00	2.61	2.38	2.22	2.11	2.02	1.95	1.89
	.010	6.66	4.63	3.80	3.34	3.04	2.82	2.66	2.53	2.43
	.001	10.89	6.96	5.46	4.65	4.14	3.78	3.51	3.30	3.13

ν_2 = denominator df

(*continued*)

Table A.9 Critical Values for *F* Distributions (*cont.*)

ν_1 = numerator df										
10	**12**	**15**	**20**	**25**	**30**	**40**	**50**	**60**	**120**	**1000**
1.87	1.82	1.77	1.72	1.68	1.66	1.63	1.61	1.59	1.56	1.52
2.24	2.16	2.09	2.01	1.96	1.92	1.87	1.84	1.82	1.77	1.72
3.13	2.99	2.85	2.70	2.60	2.54	2.45	2.40	2.36	2.27	2.18
4.56	4.31	4.06	3.79	3.63	3.52	3.37	3.28	3.22	3.06	2.91
1.86	1.81	1.76	1.71	1.67	1.65	1.61	1.59	1.58	1.54	1.51
2.22	2.15	2.07	1.99	1.94	1.90	1.85	1.82	1.80	1.75	1.70
3.09	2.96	2.81	2.66	2.57	2.50	2.42	2.36	2.33	2.23	2.14
4.48	4.24	3.99	3.72	3.56	3.44	3.30	3.21	3.15	2.99	2.84
1.85	1.80	1.75	1.70	1.66	1.64	1.60	1.58	1.57	1.53	1.50
2.20	2.13	2.06	1.97	1.92	1.88	1.84	1.81	1.79	1.73	1.68
3.06	2.93	2.78	2.63	2.54	2.47	2.38	2.33	2.29	2.20	2.11
4.41	4.17	3.92	3.66	3.49	3.38	3.23	3.14	3.08	2.92	2.78
1.84	1.79	1.74	1.69	1.65	1.63	1.59	1.57	1.56	1.52	1.48
2.19	2.12	2.04	1.96	1.91	1.87	1.82	1.79	1.77	1.71	1.66
3.03	2.90	2.75	2.60	2.51	2.44	2.35	2.30	2.26	2.17	2.08
4.35	4.11	3.86	3.60	3.43	3.32	3.18	3.09	3.02	2.86	2.72
1.83	1.78	1.73	1.68	1.64	1.62	1.58	1.56	1.55	1.51	1.47
2.18	2.10	2.03	1.94	1.89	1.85	1.81	1.77	1.75	1.70	1.65
3.00	2.87	2.73	2.57	2.48	2.41	2.33	2.27	2.23	2.14	2.05
4.29	4.05	3.80	3.54	3.38	3.27	3.12	3.03	2.97	2.81	2.66
1.82	1.77	1.72	1.67	1.63	1.61	1.57	1.55	1.54	1.50	1.46
2.16	2.09	2.01	1.93	1.88	1.84	1.79	1.76	1.74	1.68	1.63
2.98	2.84	2.70	2.55	2.45	2.39	2.30	2.25	2.21	2.11	2.02
4.24	4.00	3.75	3.49	3.33	3.22	3.07	2.98	2.92	2.76	2.61
1.76	1.71	1.66	1.61	1.57	1.54	1.51	1.48	1.47	1.42	1.38
2.08	2.00	1.92	1.84	1.78	1.74	1.69	1.66	1.64	1.58	1.52
2.80	2.66	2.52	2.37	2.27	2.20	2.11	2.06	2.02	1.92	1.82
3.87	3.64	3.40	3.14	2.98	2.87	2.73	2.64	2.57	2.41	2.25
1.73	1.68	1.63	1.57	1.53	1.50	1.46	1.44	1.42	1.38	1.33
2.03	1.95	1.87	1.78	1.73	1.69	1.63	1.60	1.58	1.51	1.45
2.70	2.56	2.42	2.27	2.17	2.10	2.01	1.95	1.91	1.80	1.70
3.67	3.44	3.20	2.95	2.79	2.68	2.53	2.44	2.38	2.21	2.05
1.71	1.66	1.60	1.54	1.50	1.48	1.44	1.41	1.40	1.35	1.30
1.99	1.92	1.84	1.75	1.69	1.65	1.59	1.56	1.53	1.47	1.40
2.63	2.50	2.35	2.20	2.10	2.03	1.94	1.88	1.84	1.73	1.62
3.54	3.32	3.08	2.83	2.67	2.55	2.41	2.32	2.25	2.08	1.92
1.66	1.61	1.56	1.49	1.45	1.42	1.38	1.35	1.34	1.28	1.22
1.93	1.85	1.77	1.68	1.62	1.57	1.52	1.48	1.45	1.38	1.30
2.50	2.37	2.22	2.07	1.97	1.89	1.80	1.74	1.69	1.57	1.45
3.30	3.07	2.84	2.59	2.43	2.32	2.17	2.08	2.01	1.83	1.64
1.63	1.58	1.52	1.46	1.41	1.38	1.34	1.31	1.29	1.23	1.16
1.88	1.80	1.72	1.62	1.56	1.52	1.46	1.41	1.39	1.30	1.21
2.41	2.27	2.13	1.97	1.87	1.79	1.69	1.63	1.58	1.45	1.30
3.12	2.90	2.67	2.42	2.26	2.15	2.00	1.90	1.83	1.64	1.43
1.61	1.55	1.49	1.43	1.38	1.35	1.30	1.27	1.25	1.18	1.08
1.84	1.76	1.68	1.58	1.52	1.47	1.41	1.36	1.33	1.24	1.11
2.34	2.20	2.06	1.90	1.79	1.72	1.61	1.54	1.50	1.35	1.16
2.99	2.77	2.54	2.30	2.14	2.02	1.87	1.77	1.69	1.49	1.22

Table A.10 Critical Values for Studentized Range Distributions

		m										
ν	α	**2**	**3**	**4**	**5**	**6**	**7**	**8**	**9**	**10**	**11**	**12**
5	.05	3.64	4.60	5.22	5.67	6.03	6.33	6.58	6.80	6.99	7.17	7.32
	.01	5.70	6.98	7.80	8.42	8.91	9.32	9.67	9.97	10.24	10.48	10.70
6	.05	3.46	4.34	4.90	5.30	5.63	5.90	6.12	6.32	6.49	6.65	6.79
	.01	5.24	6.33	7.03	7.56	7.97	8.32	8.61	8.87	9.10	9.30	9.48
7	.05	3.34	4.16	4.68	5.06	5.36	5.61	5.82	6.00	6.16	6.30	6.43
	.01	4.95	5.92	6.54	7.01	7.37	7.68	7.94	8.17	8.37	8.55	8.71
8	.05	3.26	4.04	4.53	4.89	5.17	5.40	5.60	5.77	5.92	6.05	6.18
	.01	4.75	5.64	6.20	6.62	6.96	7.24	7.47	7.68	7.86	8.03	8.18
9	.05	3.20	3.95	4.41	4.76	5.02	5.24	5.43	5.59	5.74	5.87	5.98
	.01	4.60	5.43	5.96	6.35	6.66	6.91	7.13	7.33	7.49	7.65	7.78
10	.05	3.15	3.88	4.33	4.65	4.91	5.12	5.30	5.46	5.60	5.72	5.83
	.01	4.48	5.27	5.77	6.14	6.43	6.67	6.87	7.05	7.21	7.36	7.49
11	.05	3.11	3.82	4.26	4.57	4.82	5.03	5.20	5.35	5.49	5.61	5.71
	.01	4.39	5.15	5.62	5.97	6.25	6.48	6.67	6.84	6.99	7.13	7.25
12	.05	3.08	3.77	4.20	4.51	4.75	4.95	5.12	5.27	5.39	5.51	5.61
	.01	4.32	5.05	5.50	5.84	6.10	6.32	6.51	6.67	6.81	6.94	7.06
13	.05	3.06	3.73	4.15	4.45	4.69	4.88	5.05	5.19	5.32	5.43	5.53
	.01	4.26	4.96	5.40	5.73	5.98	6.19	6.37	6.53	6.67	6.79	6.90
14	.05	3.03	3.70	4.11	4.41	4.64	4.83	4.99	5.13	5.25	5.36	5.46
	.01	4.21	4.89	5.32	5.63	5.88	6.08	6.26	6.41	6.54	6.66	6.77
15	.05	3.01	3.67	4.08	4.37	4.59	4.78	4.94	5.08	5.20	5.31	5.40
	.01	4.17	4.84	5.25	5.56	5.80	5.99	6.16	6.31	6.44	6.55	6.66
16	.05	3.00	3.65	4.05	4.33	4.56	4.74	4.90	5.03	5.15	5.26	5.35
	.01	4.13	4.79	5.19	5.49	5.72	5.92	6.08	6.22	6.35	6.46	6.56
17	.05	2.98	3.63	4.02	4.30	4.52	4.70	4.86	4.99	5.11	5.21	5.31
	.01	4.10	4.74	5.14	5.43	5.66	5.85	6.01	6.15	6.27	6.38	6.48
18	.05	2.97	3.61	4.00	4.28	4.49	4.67	4.82	4.96	5.07	5.17	5.27
	.01	4.07	4.70	5.09	5.38	5.60	5.79	5.94	6.08	6.20	6.31	6.41
19	.05	2.96	3.59	3.98	4.25	4.47	4.65	4.79	4.92	5.04	5.14	5.23
	.01	4.05	4.67	5.05	5.33	5.55	5.73	5.89	6.02	6.14	6.25	6.34
20	.05	2.95	3.58	3.96	4.23	4.45	4.62	4.77	4.90	5.01	5.11	5.20
	.01	4.02	4.64	5.02	5.29	5.51	5.69	5.84	5.97	6.09	6.19	6.28
24	.05	2.92	3.53	3.90	4.17	4.37	4.54	4.68	4.81	4.92	5.01	5.10
	.01	3.96	4.55	4.91	5.17	5.37	5.54	5.69	5.81	5.92	6.02	6.11
30	.05	2.89	3.49	3.85	4.10	4.30	4.46	4.60	4.72	4.82	4.92	5.00
	.01	3.89	4.45	4.80	5.05	5.24	5.40	5.54	5.65	5.76	5.85	5.93
40	.05	2.86	3.44	3.79	4.04	4.23	4.39	4.52	4.63	4.73	4.82	4.90
	.01	3.82	4.37	4.70	4.93	5.11	5.26	5.39	5.50	5.60	5.69	5.76
60	.05	2.83	3.40	3.74	3.98	4.16	4.31	4.44	4.55	4.65	4.73	4.81
	.01	3.76	4.28	4.59	4.82	4.99	5.13	5.25	5.36	5.45	5.53	5.60
120	.05	2.80	3.36	3.68	3.92	4.10	4.24	4.36	4.47	4.56	4.64	4.71
	.01	3.70	4.20	4.50	4.71	4.87	5.01	5.12	5.21	5.30	5.37	5.44
∞	.05	2.77	3.31	3.63	3.86	4.03	4.17	4.29	4.39	4.47	4.55	4.62
	.01	3.64	4.12	4.40	4.60	4.76	4.88	4.99	5.08	5.16	5.23	5.29

Table A.11 Chi-Squared Curve Tail Areas

Upper-Tail Area	$\nu = 1$	$\nu = 2$	$\nu = 3$	$\nu = 4$	$\nu = 5$
> .100	< 2.70	< 4.60	< 6.25	< 7.77	< 9.23
.100	2.70	4.60	6.25	7.77	9.23
.095	2.78	4.70	6.36	7.90	9.37
.090	2.87	4.81	6.49	8.04	9.52
.085	2.96	4.93	6.62	8.18	9.67
.080	3.06	5.05	6.75	8.33	9.83
.075	3.17	5.18	6.90	8.49	10.00
.070	3.28	5.31	7.06	8.66	10.19
.065	3.40	5.46	7.22	8.84	10.38
.060	3.53	5.62	7.40	9.04	10.59
.055	3.68	5.80	7.60	9.25	10.82
.050	3.84	5.99	7.81	9.48	11.07
.045	4.01	6.20	8.04	9.74	11.34
.040	4.21	6.43	8.31	10.02	11.64
.035	4.44	6.70	8.60	10.34	11.98
.030	4.70	7.01	8.94	10.71	12.37
.025	5.02	7.37	9.34	11.14	12.83
.020	5.41	7.82	9.83	11.66	13.38
.015	5.91	8.39	10.46	12.33	14.09
.010	6.63	9.21	11.34	13.27	15.08
.005	7.87	10.59	12.83	14.86	16.74
.001	10.82	13.81	16.26	18.46	20.51
< .001	> 10.82	> 13.81	> 16.26	> 18.46	> 20.51

Upper-Tail Area	$\nu = 6$	$\nu = 7$	$\nu = 8$	$\nu = 9$	$\nu = 10$
> .100	< 10.64	< 12.01	< 13.36	< 14.68	< 15.98
.100	10.64	12.01	13.36	14.68	15.98
.095	10.79	12.17	13.52	14.85	16.16
.090	10.94	12.33	13.69	15.03	16.35
.085	11.11	12.50	13.87	15.22	16.54
.080	11.28	12.69	14.06	15.42	16.75
.075	11.46	12.88	14.26	15.63	16.97
.070	11.65	13.08	14.48	15.85	17.20
.065	11.86	13.30	14.71	16.09	17.44
.060	12.08	13.53	14.95	16.34	17.71
.055	12.33	13.79	15.22	16.62	17.99
.050	12.59	14.06	15.50	16.91	18.30
.045	12.87	14.36	15.82	17.24	18.64
.040	13.19	14.70	16.17	17.60	19.02
.035	13.55	15.07	16.56	18.01	19.44
.030	13.96	15.50	17.01	18.47	19.92
.025	14.44	16.01	17.53	19.02	20.48
.020	15.03	16.62	18.16	19.67	21.16
.015	15.77	17.39	18.97	20.51	22.02
.010	16.81	18.47	20.09	21.66	23.20
.005	18.54	20.27	21.95	23.58	25.18
.001	22.45	24.32	26.12	27.87	29.58
< .001	> 22.45	> 24.32	> 26.12	> 27.87	> 29.58

(*continued*)

Table A.11 Chi-Squared Curve Tail Areas (*cont.*)

Upper-Tail Area	$\nu = 11$	$\nu = 12$	$\nu = 13$	$\nu = 14$	$\nu = 15$
> .100	< 17.27	< 18.54	< 19.81	< 21.06	< 22.30
.100	17.27	18.54	19.81	21.06	22.30
.095	17.45	18.74	20.00	21.26	22.51
.090	17.65	18.93	20.21	21.47	22.73
.085	17.85	19.14	20.42	21.69	22.95
.080	18.06	19.36	20.65	21.93	23.19
.075	18.29	19.60	20.89	22.17	23.45
.070	18.53	19.84	21.15	22.44	23.72
.065	18.78	20.11	21.42	22.71	24.00
.060	19.06	20.39	21.71	23.01	24.31
.055	19.35	20.69	22.02	23.33	24.63
.050	19.67	21.02	22.36	23.68	24.99
.045	20.02	21.38	22.73	24.06	25.38
.040	20.41	21.78	23.14	24.48	25.81
.035	20.84	22.23	23.60	24.95	26.29
.030	21.34	22.74	24.12	25.49	26.84
.025	21.92	23.33	24.73	26.11	27.48
.020	22.61	24.05	25.47	26.87	28.25
.015	23.50	24.96	26.40	27.82	29.23
.010	24.72	26.21	27.68	29.14	30.57
.005	26.75	28.29	29.81	31.31	32.80
.001	31.26	32.90	34.52	36.12	37.69
< .001	> 31.26	> 32.90	> 34.52	> 36.12	> 37.69

Upper-Tail Area	$\nu = 16$	$\nu = 17$	$\nu = 18$	$\nu = 19$	$\nu = 20$
> .100	< 23.54	< 24.77	< 25.98	< 27.20	< 28.41
.100	23.54	24.76	25.98	27.20	28.41
.095	23.75	24.98	26.21	27.43	28.64
.090	23.97	25.21	26.44	27.66	28.88
.085	24.21	25.45	26.68	27.91	29.14
.080	24.45	25.70	26.94	28.18	29.40
.075	24.71	25.97	27.21	28.45	29.69
.070	24.99	26.25	27.50	28.75	29.99
.065	25.28	26.55	27.81	29.06	30.30
.060	25.59	26.87	28.13	29.39	30.64
.055	25.93	27.21	28.48	29.75	31.01
.050	26.29	27.58	28.86	30.14	31.41
.045	26.69	27.99	29.28	30.56	31.84
.040	27.13	28.44	29.74	31.03	32.32
.035	27.62	28.94	30.25	31.56	32.85
.030	28.19	29.52	30.84	32.15	33.46
.025	28.84	30.19	31.52	32.85	34.16
.020	29.63	30.99	32.34	33.68	35.01
.015	30.62	32.01	33.38	34.74	36.09
.010	32.00	33.40	34.80	36.19	37.56
.005	34.26	35.71	37.15	38.58	39.99
.001	39.25	40.78	42.31	43.81	45.31
< .001	> 39.25	> 40.78	> 42.31	> 43.81	> 45.31

Table A.12 Critical Values for the Ryan–Joiner Test of Normality

		α		
		.10	**.05**	**.01**
	5	.9033	.8804	.8320
	10	.9347	.9180	.8804
	15	.9506	.9383	.9110
	20	.9600	.9503	.9290
	25	.9662	.9582	.9408
n	**30**	.9707	.9639	.9490
	40	.9767	.9715	.9597
	50	.9807	.9764	.9664
	60	.9835	.9799	.9710
	75	.9865	.9835	.9757

Table A.13 Critical Values for the Wilcoxon Signed-Rank Test

$P_0(S_+ \geq c_1) = P(S_+ \geq c_1 \text{ when } H_0 \text{ is true})$

n	c_1	$P_0(S_+ \geq c_1)$
3	6	.125
4	9	.125
	10	.062
5	13	.094
	14	.062
	15	.031
6	17	.109
	19	.047
	20	.031
	21	.016
7	22	.109
	24	.055
	26	.023
	28	.008
8	28	.098
	30	.055
	32	.027
	34	.012
	35	.008
	36	.004
9	34	.102
	37	.049
	39	.027
	42	.010
	44	.004
10	41	.097
	44	.053
	47	.024
	50	.010
	52	.005
11	48	.103
	52	.051
	55	.027
	59	.009
	61	.005
12	56	.102
	60	.055
	61	.046
	64	.026
	68	.010
	71	.005
13	64	.108
	65	.095
	69	.055
	70	.047
	74	.024
	78	.011
	79	.009
	81	.005
14	73	.108
	74	.097
	79	.052
	84	.025
	89	.010
	92	.005
15	83	.104
	84	.094
	89	.053
	90	.047
	95	.024
	100	.011
	101	.009
	104	.005
16	93	.106
	94	.096
	100	.052
	106	.025
	112	.011
	113	.009
	116	.005
17	104	.103
	105	.095
	112	.049
	118	.025
	125	.010
	129	.005
18	116	.098
	124	.049
	131	.024
	138	.010
	143	.005
19	128	.098
	136	.052
	137	.048
	144	.025
	152	.010
	157	.005
20	140	.101
	150	.049
	158	.024
	167	.010
	172	.005

Table A.14 Critical Values for the Wilcoxon Rank-Sum Test

$P_0(W \geq c) = P(W \geq c \text{ when } H_0 \text{ is true})$

m	n	c	$P_0(W \geq c)$	m	n	c	$P_0(W \geq c)$
3	3	15	.05			40	.004
	4	17	.057		6	40	.041
		18	.029			41	.026
	5	20	.036			43	.009
		21	.018			44	.004
	6	22	.048		7	43	.053
		23	.024			45	.024
		24	.012			47	.009
	7	24	.058			48	.005
		26	.017		8	47	.047
		27	.008			49	.023
	8	27	.042			51	.009
		28	.024			52	.005
		29	.012	6	6	50	.047
		30	.006			52	.021
4	4	24	.057			54	.008
		25	.029			55	.004
		26	.014		7	54	.051
	5	27	.056			56	.026
		28	.032			58	.011
		29	.016			60	.004
		30	.008		8	58	.054
	6	30	.057			61	.021
		32	.019			63	.01
		33	.010			65	.004
		34	.005	7	7	66	.049
	7	33	.055			68	.027
		35	.021			71	.009
		36	.012			72	.006
		37	.006		8	71	.047
	8	36	.055			73	.027
		38	.024			76	.01
		40	.008			78	.005
		41	.004	8	8	84	.052
5	5	36	.048			87	.025
		37	.028			90	.01
		39	.008			92	.005

Table A.15 Critical Values for the Wilcoxon Signed-Rank Interval $(\bar{x}_{(n(n+1)/2-c+1)}, \bar{x}_{(c)})$

n	Confidence Level (%)	c	n	Confidence Level (%)	c	n	Confidence Level (%)	c
5	93.8	15	13	99.0	81	20	99.1	173
	87.5	14		95.2	74		95.2	158
6	96.9	21		90.6	70		90.3	150
	93.7	20	14	99.1	93	21	99.0	188
	90.6	19		95.1	84		95.0	172
7	98.4	28		89.6	79		89.7	163
	95.3	26	15	99.0	104	22	99.0	204
	89.1	24		95.2	95		95.0	187
8	99.2	36		90.5	90		90.2	178
	94.5	32	16	99.1	117	23	99.0	221
	89.1	30		94.9	106		95.2	203
9	99.2	44		89.5	100		90.2	193
	94.5	39	17	99.1	130	24	99.0	239
	90.2	37		94.9	118		95.1	219
10	99.0	52		90.2	112		89.9	208
	95.1	47	18	99.0	143	25	99.0	257
	89.5	44		95.2	131		95.2	236
11	99.0	61		90.1	124		89.9	224
	94.6	55	19	99.1	158			
	89.8	52		95.1	144			
12	99.1	71		90.4	137			
	94.8	64						
	90.8	61						

Table A.16 Critical Values for the Wilcoxon Rank-Sum Interval $(d_{ij(mn-c+1)}, d_{ij(c)})$

	Smaller Sample Size							
	5		6		7		8	
Larger Sample Size	Confidence Level (%)	c	Confidence Level (%)	c	Confidence Level (%)	c	Confidence Level (%)	c
5	99.2	25						
	94.4	22						
	90.5	21						
6	99.1	29	99.1	34				
	94.8	26	95.9	31				
	91.8	25	90.7	29				
7	99.0	33	99.2	39	98.9	44		
	95.2	30	94.9	35	94.7	40		
	89.4	28	89.9	33	90.3	38		
8	98.9	37	99.2	44	99.1	50	99.0	56
	95.5	34	95.7	40	94.6	45	95.0	51
	90.7	32	89.2	37	90.6	43	89.5	48
9	98.8	41	99.2	49	99.2	56	98.9	62
	95.8	38	95.0	44	94.5	50	95.4	57
	88.8	35	91.2	42	90.9	48	90.7	54
10	99.2	46	98.9	53	99.0	61	99.1	69
	94.5	41	94.4	48	94.5	55	94.5	62
	90.1	39	90.7	46	89.1	52	89.9	59
11	99.1	50	99.0	58	98.9	66	99.1	75
	94.8	45	95.2	53	95.6	61	94.9	68
	91.0	43	90.2	50	89.6	57	90.9	65
12	99.1	54	99.0	63	99.0	72	99.0	81
	95.2	49	94.7	57	95.5	66	95.3	74
	89.6	46	89.8	54	90.0	62	90.2	70

	Smaller Sample Size							
	9		10		11		12	
Larger Sample Size	Confidence Level (%)	c	Confidence Level (%)	c	Confidence Level (%)	c	Confidence Level (%)	c
9	98.9	69						
	95.0	63						
	90.6	60						
10	99.0	76	99.1	84				
	94.7	69	94.8	76				
	90.5	66	89.5	72				
11	99.0	83	99.0	91	98.9	99		
	95.4	76	94.9	83	95.3	91		
	90.5	72	90.1	79	89.9	86		
12	99.1	90	99.1	99	99.1	108	99.0	116
	95.1	82	95.0	90	94.9	98	94.8	106
	90.5	78	90.7	86	89.6	93	89.9	101

Table A.17 β Curves for t Tests

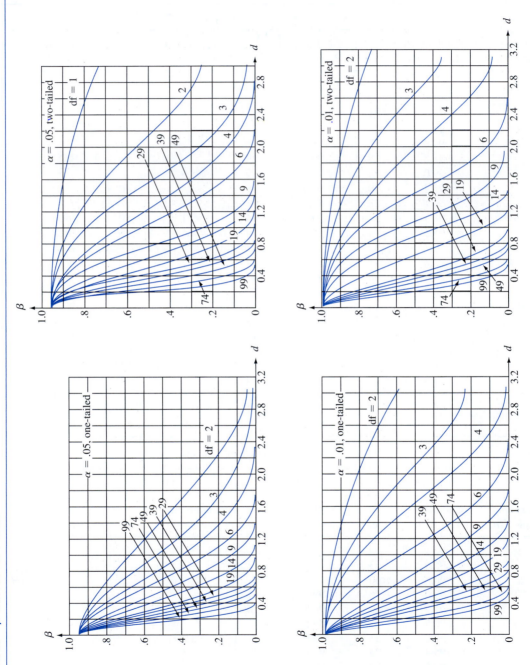

$\alpha = .05$, one-tailed

$\alpha = .05$, two-tailed

$\alpha = .01$, one-tailed

$\alpha = .01$, two-tailed

Answers to Odd-Numbered Exercises

Chapter 1

1. a. Houston Chronicle, Des Moines Register, Chicago Tribune, Washington Post
b. Capital One, Campbell Soup, Merrill Lynch, Prudential
c. Bill Jasper, Kay Reinke, Helen Ford, David Menendez
d. 1.78, 2.44, 3.5, 3.04

3. a. In a sample of 100 DVD players, what are the chances that more than 20 need service while under warranty? What are the chances that none need service while still under warranty?
b. What proportion of all DVD players of this brand and model will need service within the warranty period?

5. a. No, the relevant conceptual population is all scores of all students who participate in the SI in conjunction with this particular statistics course.
b. The advantage of randomly allocating students to the two groups is that the two groups should then be fairly comparable before the study. If the two groups perform differently in the class, we might attribute this to the treatments (SI and control). If it were left to students to choose, stronger or more dedicated students might gravitate toward SI, confounding the results.
c. If all students were put in the treatment group there would be no results with which to compare the treatments.

7. One could generate a simple random sample of all single-family homes in the city, or a stratified random sample by taking a simple random sample from each of the 10 district neighborhoods. From each of the homes in the sample values of relevant variables would be collected. This would be an enumerative study because there exists a finite, identifiable population of objects from which to sample.

9. a. There could be several explanations for the variability of the measurements. Among them could be measuring errors (due to mechanical or technical changes across measurements), recording error, differences in weather conditions at time of measurements, etc.
b. This study involves a conceptual population. There is no sampling frame.

11.

6L	034	
6H	667899	
7L	00122244	
7H		Stem: Tens
8L	001111122344	Leaf: Ones
8H	5557899	
9L	03	
9H	58	

This display brings out the gap in the data: There are no scores in the high 70s.

13. a.

2	23'	Stem unit: 1.0
3	2344567789	Leaf unit: .10
4	01356889	
5	00001114455666789	
6	0000122223344456667789999	
7	00012233455555668	
8	02233448	
9	012233335666788	
10	2344455688	
11	2335999	
12	37	
13	8	
14	36	
15	0035	
16		
17		
18	9	

b. A representative value could be the median, 7.0.
c. The data appear to be highly concentrated, except for a few values on the positive side.
d. No, there is skewness to the right, or positive skewness.
e. The value 18.9 appears to be an outlier, being more than two stem units from the previous value.

15. a.

Number Nonconforming	Frequency	Relative Frequency (Freq/60)
0	7	0.117
1	12	0.200
2	13	0.217
3	14	0.233
4	6	0.100
5	3	0.050
6	3	0.050
7	1	0.017
8	1	0.017
		1.001

(doesn't add exactly to 1 because relative frequencies have been rounded)

b. .917, .867, 1 − .867 = .133
c. The center of the histogram is somewhere around 2 or 3 and it shows that there is some positive skewness in the data.

17. a. .99, or 99%; .71, or 71% **b.** .64, or 64%; .44, or 44%
c. The histogram is *almost* symmetric and unimodal; however, it has a few relative maxima (i.e., modes) and has a very slight positive skew.

19. a. The number of subdivisions having no cul-de-sacs is $\frac{17}{47}$ = .362, or 36.2%. The proportion having at least one cul-de-sac is $\frac{30}{47}$ = .638, or 63.8%.

y	Count	Percent
0	17	36.17
1	22	46.81
2	6	12.77
3	1	2.13
5	1	2.13
$n = 47$		

.362, .638

b.

z	Count	Percent
0	13	27.66
1	11	23.40
2	3	6.38
3	7	14.89
4	5	10.64
5	3	6.38
6	3	6.38
8	2	4.26
$n = 47$		

.894, .830

21. a.

Class	Freq.	Rel. Freq.
0–<100	21	0.21
100–<200	32	0.32
200–<300	26	0.26
300–<400	12	0.12
400–<500	4	0.04
500–<600	3	0.03
600–<700	1	0.01
700–<800	0	0.00
800–<900	1	0.01
	100	1.00

The histogram is skewed right, with a majority of observations between 0 and 300 cycles. The class holding the most observations is between 100 and 200 cycles.

b.

Class	Freq.	Rel. Freq.	Density
0–<50	8	0.08	.0016
50–<100	13	0.13	.0026
100–<150	11	0.11	.0022
150–<200	21	0.21	.0042
200–<300	26	0.26	.0026
300–<400	12	0.12	.0012
400–<500	4	0.04	.0004
500–<600	3	0.03	.0003
600–<900	2	0.02	.00007
	100	1.00	

c. .79

23.

Class	Freq.	Class	Freq.
10–<20	8	1.1–<1.2	2
20–<30	14	1.2–<1.3	6
30–<40	8	1.3–<1.4	7
40–<50	4	1.4–<1.5	9
50–<60	3	1.5–<1.6	6
60–<70	2	1.6–<1.7	4
70–<80	1	1.7–<1.8	5
	40	1.8–<1.9	1
			40

The original distribution is positively skewed. The transformation creates a much more symmetric, mound-shaped histogram.

25. a.

Class	Freq.	Rel. Freq.
0–<50	9	0.18
50–<100	19	0.38
100–<150	11	0.22
150–<200	4	0.08
200–<250	2	0.04
250–<300	2	0.04
300–<350	1	0.02
350–<400	1	0.02
400–<600	1	0.02
	50	1.00

The distribution is skewed to the right, or positively skewed. There is a gap in the histogram, and what appears to be an outlier.

b.

Class	Freq.	Rel. Freq.
2.25–<2.75	2	0.04
2.75–<3.25	2	0.04
3.25–<3.75	3	0.06
3.75–<4.25	8	0.16
4.25–<4.75	18	0.36
4.75–<5.25	10	0.20
5.25–<5.75	4	0.08
5.75–<6.25	3	0.06

The distribution of the natural logs of the original data is much more symmetric than the original.

c. .56, .14

29. a. The frequency distribution is:

Class	Rel. Freq.	Class	Rel. Freq.
0–<150	.193	900–<1050	.019
150–<300	.183	1050–<1200	.029
300–<450	.251	1200–<1350	.005
450–<600	.148	1350–<1500	.004
600–<750	.097	1500–<1650	.001
750–<900	.066	1650–<1800	.002
		1800–<1950	.002

The relative frequency distribution is almost unimodal and exhibits a large positive skew. The typical middle value is somewhere between 400 and 450, although the skewnesss makes it difficult to pinpoint more exactly than this.

b. .775, .014 **c.** .211

31. a. 5.24

b. The median, 2, is much lower because of positive skewness.

c. Trimming the largest and smallest observations yields the 5.9% trimmed mean, 4.4, which is between the mean and median.

33. a. A stem- and leaf display:

Stem	Leaf	
32	55	Stem: ones
33	49	Leaf: tenths
34		
35	6699	
36	34469	
37	03345	
38	9	
39	2347	
40	23	
41		
42	4	

The display is reasonably symmetric, so the mean and median will be close.

b. 370.7, 369.50

c. The largest value (currently 424) could be increased by any amount without changing the median. It can be decreased to any value at least 370 without changing the median.

d. 6.18 min; 6.16 min

35. a. 125

b. If 127.6 is reported as 130, then the median is 130, a substantial change. When there is rounding or grouping, the median can be highly sensitive to a small change.

37. $\tilde{x} = 92$, $\bar{x}_{tr(25)} = 95.07$, $\bar{x}_{tr(10)} = 102.23$, $\bar{x} = 119.3$

Positive skewness causes the mean to be larger than the median. Trimming moves the mean closer to the median.

39. a. $\bar{y} = \bar{x} + c$, $\tilde{y} = \tilde{x} + c$ **b.** $\bar{y} = c\bar{x}$, $\tilde{y} = c\tilde{x}$

41. a. 25.8 **b.** 49.31 **c.** 7.02 **d.** 49.31

43. a. 2887.6, 2888 **b.** 7060.3

45. 24.36

47. $1,961,160

49. −3.5; 1.3, 1.9, 2.0, 2.3, −2.5

51. a. 1, 6, 5

b. The box plot shows positive skewness. The two longest runs are extreme outliers.

c. outlier: greater than 13.5 or less than −6.5
extreme outlier: greater than 21 or less than −14

d. The largest observation could be decreased to 6 without affecting f_s.

53. The box plot shows some positive skewness. The largest value is an extreme outlier.

55. The two distributions are centered in about the same place, but one machine is much more variable than the other. The more precise machine produced one outlier, but this part would not be an outlier if judged by the distribution of the other machine.

57. The test welds have higher burst strengths but their burst strengths are more variable. There are two outliers in the production data.

61. The three flow rates yield similar uniformities, but the values for the 160 flow rate are a little higher.

63. a. 9.59, 59.41. The standard deviations are large, so it is certainly not true that repeated measurements are identical.

b. .396, .323. In terms of the coefficient of variation, the HC emissions are more variable.

65. 10.65

67. a. $\bar{y} = a\bar{x} + b$, $s_y^2 = a^2 s_x^2$ **b.** 100.78, .572

69.

	$\bar{x}$	$\tilde{x}$	s	min	max
PTSD	32.92	37	9.93	10	46
Healthy	52.23	51	14.86	23	72

Healthy individuals have much higher binding measures on average than PTSD individuals. The healthy distribution is reasonably symmetric, whereas the PTSD distribution is negatively skewed.

71. a. Mode = .93. It occurs four times in the data set.

b. The modal category is the one in which the most observations occur.

73. The measures that are sensitive to outliers are the mean and the midrange. The mean is sensitive because all values are used in computing it. The midrange is the most sensitive because it uses only the most extreme values in its computation.

The median, the trimmed mean, and the midfourth are less sensitive to outliers. The median is the most resistant to outliers because it uses only the middle value (or values) in its computation. The midfourth is also quite resistant because it uses the fourths. The resistance of the trimmed mean increases with the trimming percentage.

75. a. $s_y^2 = s_x^2$ and $s_y = s_x$ **b.** $s_z^2 = 1$ and $s_z = 1$

77. b. .552, .102 **c.** 30 **d.** 19

79. a. There may be a tendency to a repeating pattern.

b. The value .1 gives a much smoother series.

c. The smoothed value depends on all previous values of the time series, but the coefficient decreases with k.

d. As t gets large, the coefficient $(1 - \alpha)^{t-1}$ decreases to zero, so there is decreasing sensitivity to the initial value.

Chapter 2

1. a. $A \cap B'$ **b.** $A \cup B$ **c.** $(A \cap B') \cup (B \cap A')$

3. d. $S = \{1324, 1342, 1423, 1432, 2314, 2341, 2413, 2431, 3124, 3142, 4123, 4132, 3214, 3241, 4213, 4231\}$
e. $A = \{1324, 1342, 1423, 1432\}$
f. $B = \{2314, 2341, 2413, 2431, 3214, 3241, 4213, 4231\}$
g. $A \cup B = \{1324, 1342, 1423, 1432, 2314, 2341, 2413, 2431, 3214, 3241, 4213, 4231\}$
$A \cap B = \varnothing$
$A' = \{2314, 2341, 2413, 2431, 3124, 3142, 4123, 4132, 3214, 3241, 4213, 4231\}$

5. a. $A = \{$SSF, SFS, FSS$\}$
b. $B = \{$SSS, SSF, SFS, FSS$\}$
c. $C = \{$SSS, SSF, SFS$\}$
d. $C' = \{$SFF, FSS, FSF, FFS, FFF$\}$
$A \cup C = \{$SSS, SSF, SFS, FSS$\}$
$A \cap C = \{$SSF, SFS$\}$
$B \cup C = \{$SSS, SSF, SFS, FSS$\}$
$B \cap C = \{$SSS, SSF, SFS$\}$

7. a. {111, 112, 113, 121, 122, 123, 131, 132, 133, 211, 212, 213, 221, 222, 223, 231, 232, 233, 311, 312, 313, 321, 322, 323, 331, 332, 333}
b. {111, 222, 333}
c. {123, 132, 213, 231, 312, 321}
d. {111, 113, 131, 133, 311, 313, 331, 333}

9. a. S = {BBBAAAA, BBABAAA, BBAABAA, BBAAABA, BBAAAAB, BABBAAA, BABABAA, BABAABA, BABAAAB, BAABBAA, BAABABA, BAABAAB, BAAABBA, BAAABAB, BAAAABB, ABBBAAA, ABBABAA, ABBAABA, ABBAAAB, ABABBAA, ABABABA, ABABAAB, ABAABBA, ABAABAB, ABAAABB, AABBBAA, AABBABA, AABBAAB, AABABBA, AABABAB, AABAABB, AAABBBA, AAABBAB, AAABABB, AAAABBB}
b. {AAAABBB, AAABABB, AAABBAB, AABAABB, AABABAB}

13. a. .07 **b.** .30 **c.** .57

15. a. They are awarded at least one of the first two projects, .36.
b. They are awarded neither of the first two projects, .64.
c. They are awarded at least one of the projects, .53.
d. They are awarded none of the projects, .47.
e. They are awarded only the third project, .17.
f. Either they fail to get the first two or they are awarded the third, .75.

17. a. .572 **b.** .879

19. a. SAS and SPSS are not the only packages.
b. .7 **c.** .8 **d.** .2

21. a. .8841 **b.** .0435

23. a. .10 **b.** .18, .19 **c.** .41
d. .59 **e.** .31 **f.** .69

25. a. $\frac{1}{15}$ **b.** $\frac{6}{15}$ **c.** $\frac{14}{15}$ **d.** $\frac{8}{15}$

27. a. .98 **b.** .02 **c.** .03 **d.** .24

29. a. $\frac{1}{9}$ **b.** $\frac{8}{9}$ **c.** $\frac{2}{9}$

31. a. 20 **b.** 60 **c.** 10

33. a. 243 **b.** 3645, 10

35. a. 53,130 **b.** 1190 **c.** .0224 **d.** .0235

37. .2

39. .0456

41. a. .0839 **b.** .24975 **c.** .1998

43. a. $\frac{1}{15}$ **b.** $\frac{1}{3}$ **c.** $\frac{2}{3}$

45. a. .447, .5, .2
b. $P(A \mid C) = .4$, the fraction of ethnic group C that has blood type A
$P(C \mid A) = .447$, the fraction of those with blood group A that are of ethnic group C
c. .211

47. a. Of those with a Visa card, .5 is the fraction who also have a MasterCard.
b. Of those with a Visa card, .5 is the fraction who do not have a MasterCard.
c. Of those with MasterCard, .625 is the fraction who also have a Visa card.
d. Of those with MasterCard, .375 is the fraction who do not have a Visa card.
e. Of those with at least one of the two cards, .769 is the fraction who have a Visa card.

49. .217, .178

51. .436, .582

53. .0833

59. a. .067 **b.** .509

61. .287

63. .0588, .673, .99958

65. .466, .288, .247

67. a. Because of independence, the conditional probability is the same as the unconditional probability, .3.
b. .82 **c.** .146

71. .349, .651, $(1-p)^n$, $1-(1-p)^n$

73. .99999969, .226

75. .9981

77. Yes, no

79. a. $2p - p^2$ **b.** $1-(1-p)^n$ **c.** $(1-p)^3$
d. $.9 + .1(1-p)^3$ **e.** .0137

81. .8588, .9896

83. $2\pi(1-\pi)$

85. a. $\frac{1}{3}$, .444 **b.** .15 **c.** .291

87. .45, .32

89. a. $\frac{1}{120}$ **b.** $\frac{1}{5}$ **c.** $\frac{1}{5}$

91. .9046

93. a. .904 **b.** .766

95. .008

97. .362, .348, .290

99. a. $P(G \mid R_1 < R_2 < R_3) = \frac{2}{3}$, so classify as granite if $R_1 < R_2 < R_3$.
b. $P(G \mid R_1 < R_3 < R_2) = .294$, so classify as basalt if $R_1 < R_3 < R_2$.
$P(G \mid R_3 < R_1 < R_2) = \frac{1}{15}$, so classify as basalt if $R_3 < R_1 < R_2$.
c. .175
d. $p > \frac{14}{17}$

101. a. $\frac{1}{24}$ **b.** $\frac{3}{8}$

103. $s = 1$

107. a. $P(B_0 \mid \text{survive}) = b_0/[1 - (b_1 + b_2)cd]$
$P(B_1 \mid \text{survive}) = b_1(1 - cd)/[1 - (b_1 + b_2)cd]$
$P(B_2 \mid \text{survive}) = b_2(1 - cd)/[1 - (b_1 + b_2)cd]$

b. .712, .058, .231

Chapter 3

1.

$\mathcal{S}$:	*FFF*	*SFF*	*FSF*	*FFS*	*FSS*	*SFS*	*SSF*	*SSS*
X:	0	1	1	1	2	2	2	3

3. M = the absolute value of the difference between the outcomes with possible values 0, 1, 2, 3, 4, 5 or 6; $W = 1$ if the sum of the two resulting numbers is even and $W = 0$ otherwise, a Bernoulli random variable.

5. No, X can be a Bernoulli random variable where a success is an outcome in B, with B a particular subset of the sample space.

7. a. Possible values are 0, 1, 2, . . . , 12; discrete
b. With N = # on the list, values are 0, 1, 2, . . . , N; discrete
c. Possible values are 1, 2, 3, 4, . . . ; discrete
d. $\{x: 0 < x < \infty\}$ if we assume that a rattlesnake can be arbitrarily short or long; not discrete
e. With c = amount earned per book sold, possible values are 0, c, $2c$, $3c$, . . . , $10{,}000c$; discrete
f. $\{y: 0 \le y \le 14\}$ since pH must always be between 0 and 14, not discrete
g. With m and M denoting the minimum and maximum possible tension, respectively, possible values are $\{x: m \le x \le M\}$; not discrete
h. Possible values are 3, 6, 9, 12, 15, . . . —i.e., 3(1), 3(2), 3(3), 3(4), . . . giving a first element, etc.; discrete

9. a. X is a discrete random variable with possible values $\{2, 4, 6, 8, \ldots\}$
b. X is a discrete random variable with possible values $\{2, 3, 4, 5, \ldots\}$

11. a. $p(4) = .45$, $p(6) = .40$, $p(8) = .15$, $p(x) = 0$ otherwise
c. .55, .15

13. a. .70 **b.** .45 **c.** .55 **d.** .71 **e.** .65
f. .45

15. a. (1, 2) (1, 3) (1, 4) (1, 5) (2, 3) (2, 4) (2, 5) (3, 4) (3, 5) (4, 5)
b. $p(0) = .3$, $p(1) = .6$, $p(2) = .1$, $p(x) = 0$ otherwise
c. $F(0) = .30$, $F(1) = .90$, $F(2) = 1$. The cdf is

$$F(x) = \begin{cases} 0 & x < 0 \\ .30 & 0 \le x < 1 \\ .90 & 1 \le x < 2 \\ 1 & 2 \le x \end{cases}$$

17. a. .81 **b.** .162
c. The fifth battery must be an A, and one of the first four must also be an A, so $p(5) = P(AUUUA$ or $UAUUA$ or $UUAUA$ or $UUUAA) = .00324$
d. $P(Y = y) = (y - 1)(.1)^{y-2}(.9)^2$, $y = 2,3,4,5, \ldots$

19. $p(1) = p(2) = p(3) = p(4) = \frac{1}{4}$

21. $F(x) = 0$, $x < 0$; .10, $0 \le x < 1$; .25, $1 \le x < 2$; .45, $2 \le x < 3$; .70, $3 \le x < 4$; .90, $4 \le x < 5$; .96, $5 \le x < 6$; 1.00, $6 \le x$

23. a. $p(1) = .30, p(3) = .10, p(4) = .05, p(6) = .15, p(12) = .40$
b. .30, .60

25. a. $p(x) = (\frac{1}{3})(\frac{2}{3})^{x-1}$, $x = 1, 2, 3, \ldots$
b. $p(y) = (\frac{1}{3})(\frac{2}{3})^{y-2}$, $y = 2, 3, 4, \ldots$
c. $p(0) = \frac{1}{6}$, $p(z) = (\frac{25}{54})(\frac{4}{9})^{z-1}$, $z = 1, 2, 3, 4, \ldots$

29. a. .60 **b.** \$110

31. a. 16.38, 272.298, 3.9936 **b.** 401
c. 2496 **d.** 13.66

33. Yes, because $\Sigma(1/x^2)$ is finite.

35. \$700

37. $E[h(X)] = .408 > 1/3.5 = .286$, so you expect to win more if you gamble.

39. $V(-X) = V(X)$

41. a. 32.5 **b.** 7.5
c. $V(X) = E[X(X - 1)] + E(X) - [E(X)]^2$

43. a. $\frac{1}{4}, \frac{1}{9}, \frac{1}{16}, \frac{1}{25}, \frac{1}{100}$
b. $\mu = 2.64$, $\sigma = 1.54$, $P(|X - \mu| \ge 2\sigma) = .04 < .25$, $P(|X - \mu| \ge 3\sigma) = 0 < \frac{1}{9}$
The actual probability can be far below the Chebyshev bound, so the bound is conservative.
c. $\frac{1}{9}$, equal to the Chebyshev bound
d. $P(-1) = .02$, $P(0) = .96$, $P(1) = .02$

45. $M_X(t) = .5e^t/(1 - .5e^t)$, $E(X) = 2$, $V(X) = 2$

47. $p_Y(y) = .75(.25)^{y-1}$, $y = 1, 2, 3, \ldots$

49. $E(X) = 5$, $V(X) = 4$

51. $M_Y(t) = e^{t^2/2}$, $E(X) = 0$, $V(X) = 1$

53. $E(X) = 0$, $V(X) = 2$

59. a. .850 **b.** .200 **c.** .200 **d.** .701
e. .851 **f.** .000 **g.** .570

61. a. .354 **b.** .114 **c.** .919

63. a. .403 **b.** .787 **c.** .773

65. .1478

67. .4068, assuming independence

69. a. .0173 **b.** .8106, .4246 **c.** .0056, .9022, .5858

71. For $p = .9$ the probability is higher for B (.9963 versus .99 for A).
For $p = .5$ the probability is higher for A (.75 versus .6875 for B).

73. c. The tabulation for $p > .5$ is not needed.

75. a. 20, 16 (binomial, $n = 100$, $p = .2$)
b. 70, 21

77. When $p = .5$, the true probability for $k = 2$ is .0414, compared to the bound of .25.
When $p = .5$, the true probability for $k = 3$ is .0026, compared to the bound of .1111.
When $p = .75$, the true probability for $k = 2$ is .0652, compared to the bound of .25.
When $p = .75$, the true probability for $k = 3$ is .0039, compared to the bound of .1111.

79. $M_{n-X}(t) = [p + (1 - p)e^t]^n$, $E(n - X) = n(1 - p)$, $V(n - X) = np(1 - p)$
Intuitively, the means of X and $n - X$ should add to n and their variances should be the same.

81. **a.** .114 **b.** .879 **c.** .121
d. Use the binomial distribution with $n = 15$ and $= .1$.

83. **a.** $h(x; 15, 10, 20)$ **b.** .0325 **c.** .6966

85. **a.** $h(x; 10, 10, 20)$ **b.** .0325
c. $h(x; n, n, 2n)$, $E(X) = n/2$, $V(X) = n^2/[4(2n - 1)]$

87. **a.** $nb(x; 2, .5) = (x + 1).5^{x+2}$, $x = 0, 1, 2, 3, \ldots$
b. $\frac{3}{16}$ **c.** $\frac{11}{16}$ **d.** 2, 4

89. $nb(x; 6, .5)$, $E(X) = 6 = 3(2)$

93. **a.** .932 **b.** .065 **c.** .068 **d.** .491
e. .251

95. **a.** .011 **b.** .441 **c.** .554, .459 **d.** .944

97. **a.** .491 **b.** .133

99. **a.** .122, .808, .283 **b.** 12, 3.464
c. .530, .011

101. **a.** .099 **b.** .135 **c.** 2

103. **a.** 4 **b.** .215 **c.** 1.15 years

105. **a.** .221 **b.** 6,800,000 **c.** $p(x; 1608.5)$

111. **b.** 3.114, .405, .636

113. **a.** $b(x; 15, .75)$ **b.** .6865 **c.** .313
d. $\frac{45}{4}, \frac{45}{16}$ **e.** .309

115. .9914

117. **a.** $p(x; 2.5)$ **b.** .067 **c.** .109

119. 1.813, 3.05

121. $p(2) = p^2$, $p(3) = (1 - p)p^2$, $p(4) = (1 - p)p^2$, $p(x) = [1 - p(2) - \cdots - p(x - 3)](1 - p)p^2$, $x = 5, 6, 7, \ldots$. Alternatively, $p(x) = (1 - p)p(x - 1) + p(1 - p) \cdot p(x - 2)$, $x = 5, 6, 7, \ldots$.
.99950841

123. **a.** 0029 **b.** .0767, .9702

125. **a.** .135 **b.** .00144 **c.** $\sum_{x=0}^{\infty}[p(x; 2)]^5$

127. 3.590

129. **a.** No **b.** .0273

131. **b.** $.6p(x; \lambda) + .4p(x; \mu)$ **c.** $(\lambda + \mu)/2$
d. $(\lambda + \mu)/2 + (\lambda - \mu)^2/4$

133. $P(X = 0) = \sum_{i=1}^{10}(p_i + p_{i+1} + p_{i-1})p_i$, where $p_k = 0$ if $k < 1$ or $k > 10$.
$P(X = j) = \sum_{i=1}^{10}(p_{i+j+1} + p_{i-j-1})p_i$, $j = 1, \ldots, 9$, where $p_k = 0$ if $k < 1$ or $k > 10$.

137. $X \sim b(x; 25, p)$, $E[h(X)] = 500p + 750$, $\sigma_{h(X)} = 100\sqrt{p(1 - p)}$
Independence and constant probability might not be valid because of the effect that customers can have on each other. Also, store employees might affect customer decisions.

139.

x	0	1	2	3	4
$p(x)$	.07776	.10368	.19008	.20736	.17280

x	5	6	7	8
$p(x)$	.13824	.06912	.03072	.01024

Chapter 4

1. **a.** .25 **b.** .5 **c.** $\frac{7}{16}$

3. **b.** .5 **c.** $\frac{11}{16}$ **d.** .6328

5. **a.** $\frac{3}{8}$ **b.** $\frac{1}{8}$ **c.** .2969 **d.** .5781

7. **a.** $f(x) = \frac{1}{10}$ for $25 \le x \le 35$; $f(x) = 0$ otherwise
b. .2 **c.** 4 **d.** .2

9. **a.** .5618 **b.** .4382, .4382 **c.** .0709

11. **a.** $\frac{1}{4}$ **b.** $\frac{3}{16}$ **c.** $\frac{15}{16}$ **d.** $\sqrt{2}$
e. $f(x) = x/2$ for $0 \le x < 2$, and $f(x) = 0$ otherwise

13. **a.** 3 **b.** 0 for $x \le 1$, $1 - 1/x^3$ for $x > 1$
c. $\frac{1}{8}$, .088

15. **a.** $F(x) = 0$ for $x \le 0$, $F(x) = x^3/8$ for $0 < x < 2$, $F(x) = 1$ for $x \ge 2$
b. $\frac{1}{64}$ **c.** .0137, .0137 **d.** 1.817

17. **b.** 90th percentile of Y = 1.8(90th percentile of X) + 32
c. (100p)th percentile of $Y = a$[(100p)th percentile of X] + b

19. **a.** 1.5, .866 **b.** .9245

21. **a.** .8182, .1113 **b.** .044

23. **a.** $A + (B - A)p$
b. $(A + B)/2$, $(B - A)^2/12$, $(B - A)/\sqrt{12}$
c. $(B^{n+1} - A^{n+1})/[(n + 1)(B - A)]$

25. 314.79

27. 248, 3.6

29. $1/(1 - t/4)$, $\frac{1}{4}$, $\frac{1}{16}$

35. $E(X) = 7.167$, $V(X) = 44.44$

37. $M(t) = .15/(.15 - t)$, $E(X) = 6.667$, $V(X) = 44.44$
This distribution is shifted left by .5, so the mean differs by .5 but the variance is the same.

39. **a.** .4850 **b.** .3413 **c.** .4938 **d.** .9876
e. .9147 **f.** .9599 **g.** .9104 **h.** .0791
i. .0668 **j.** .9876

41. **a.** 1.34 **b.** −1.34 **c.** .674 **d.** −.674
e. −1.555

43. **a.** .9772 **b.** .5 **c.** .9104 **d.** .8413
e. .2417 **f.** .6826

45. **a.** .7977 **b.** .0004
c. The top 5% are the values above .3987.

47. The second machine

49. **a.** .2525, .00008 **b.** 39.96

51. .0510

53. **a.** .8664 **b.** .0124 **c.** .2718

55. **a.** .794 **b.** 5.88 **c.** 7.94 **d.** .265

57. No, because of symmetry

59. **a.** Approximate, .0391; binomial, .0437
b. Approximate, .99993; binomial, .99976

61. **a.** .7287 **b.** .8643, .8159

63. **a.** Approximate, .9933; binomial, .9905
b. Approximate, .9874; binomial, .9837
c. Approximate, .8051; binomial, .8066

67. a. .15872 **b.** .0013495 **c.** .999936655
Actual: .15866 .0013499 .999936658
d. .00000028669
.00000028665

69. a. 120 **b.** 1.329 **c.** .371 **d.** .735 **e.** 0

71. a. 5, 4 **b.** .715 **c.** .411

73. a. 1 **b.** 1 **c.** .982 **d.** .129

75. a. .449, .699, .148 **b.** .050, .018

77. a. $\cap A_i$
b. Exponential with $\lambda = .05$
c. Exponential with parameter $n\lambda$

83. a. .8257, .8257, .0636 **b.** .6637 **c.** 172.73

87. a. .9296 **b.** .2975 **c.** 98.18

89. a. 68.03, 122.09 **b.** .3196 **c.** .7257, skewness

91. a. 149.157, 223.595 **b.** .957 **c.** .0416
d. 148.41 **e.** 9.57 **f.** 125.90

93. $\alpha = \beta$

95. b. $\Gamma(\alpha + \beta)\,\Gamma(m + \beta)/[\Gamma(\alpha + \beta + m)\,\Gamma(\beta)]$,
$\beta/(\alpha + \beta)$

97. Yes, since the pattern in the plot is quite linear

99. Yes

101. Yes

103. Form a new variable, the logarithms of the rainfall values, and then construct a normal plot for the new variable. Because of the linearity of this plot, normality is plausible.

105. The normal plot has a nonlinear pattern showing positive skewness.

107. The plot deviates from linearity, especially at the low end, where the smallest three observations are too small relative to the others. The plot works for any λ because λ is a scale parameter.

109. $f_Y(y) = 2/y^3, y > 1$

111. $f_Y(y) = ye^{-y^2/2}, y > 0$

113. $f_Y(y) = \frac{1}{16}, 0 < y < 16$

115. $f_Y(y) = 1/[\pi(1 + y^2)]$

117. $Y = X^2/16$

119. $f_Y(y) = 1/[2\sqrt{y}], 0 < y < 1$

121. $f_Y(y) = 1/[4\sqrt{y}],\ 0 < y < 1,\ f_Y(y) = 1/[8\sqrt{y}],$
$1 < y < 9$

125. $p_Y(y) = (1 - p)^{y-1}p, y = 1, 2, 3, \ldots$

127. a. .4 **b.** .6
c. $F(x) = x/25,\ 0 \le x \le 25;\ F(x) = 0,\ x < 0;$
$F(x) = 1, x > 25$
d. 12.5, 7.22

129. b. $F(x) = 1 - 16/(x + 4)^2, x \ge 0; F(x) = 0, x < 0$
c. .247 **d.** 4 **e.** 16.67

131. a. .6563 **b.** 41.55 **c.** .3179

133. a. .00025, normal approximation; .000859, binomial
b. .0888, normal approximation; .0963, binomial

135. a. $F(x) = 1.5(1 - 1/x),\ 1 \le x \le 3;\ F(x) = 0,\ x < 1;$
$F(x) = 1, x > 3$
b. .9, .4 **c.** 1.6479 **d.** .5333 **e.** .2662

137. a. 1.075, 1.075 **b.** .0614, .3331 **c.** 2.476

139. a. c/λ **b.** $c(.5\lambda - a)/(\lambda - a)$

141. b. $F(x) = .5e^{.2x}, x \le 0; F(x) = 1 - .5e^{-.2x}, x > 0$
c. .5, .6648, .2555, .6703

143. a. $k = (\alpha - 1)5^{\alpha-1}$
b. $F(x) = 0, x \le 5; F(x) = 1 - (5/x)^{\alpha-1}, x > 5$
c. $5(\alpha - 1)/(\alpha - 2)$

145. b. .4602, .3636 **c.** .5950 **d.** 140.178

147. a. Weibull **b.** .5422

149. a. λ **b.** $\alpha x^{\alpha-1}/\beta^{\alpha}$, ↑ for $\alpha > 1$, ↓ for $\alpha < 1$
c. $F(x) = 1 - e^{-\alpha[x - x^2/(2\beta)]},\ 0 \le x \le \beta;\ F(x) = 0,$
$x < 0; F(x) = 1 - e^{-\alpha\beta/2}, x > \beta$
$f(x) = \alpha(1 - x/\beta)e^{-\alpha[x - x^2/(2\beta)]}, 0 \le x \le \beta; f(x) = 0, x < 0,$
$f(x) = 0, x > \beta$
This gives total probability less than 1, so some probability is located at infinity (for items that last forever).

151. a. $E[g(X)] \approx g(\mu), V[g(X)] \approx [g'(\mu)]^2 \cdot V(X)$
b. $\mu_R \approx v/20, \sigma_R \approx v/800$

155. $F(q^*) = .818$

Chapter 5

1. a. .20 **b.** .42 **c.** The probability of at least one hose being in use at each pump is .70.

d.

x	0	1	2	y	0	1	2
$p_X(x)$	.16	.34	.50	$p_Y(y)$	.24	.38	.38

$P(X \le 1) = .50$
e. Dependent, $.30 = P(X = 2 \text{ and } Y = 2) \ne P(X = 2) \cdot P(Y = 2) = (.50)(.38)$

3. a. .15 **b.** .40 **c.** $.22 = P(A) = P(|X_1 - X_2| \ge 2)$
d. .17, .46

5. a. .54 **b.** .00018

7. a. .030 **b.** .120 **c.** .10, .30 **d.** .38
e. Yes, $p(x, y) = p_x(x) \cdot p_y(y)$

9. a. .3/380,000 **b.** .3024 **c.** .3593
d. $10Kx^2 + .05, 20 \le x \le 30$ **e.** no

11. a. $p(x, y) = (e^{-\lambda}\lambda^x/x!)\,(e^{-\theta}\theta^y/y!)$ for $x = 0, 1, 2, \ldots;$
$y = 0, 1, 2, \ldots$
b. $e^{-\lambda-\theta}(1 + \lambda + \theta)$
c. $e^{-\lambda-\theta}(\lambda + \theta)^m/m!$, Poisson with parameter $\lambda + \theta$

13. a. $e^{-x-y}, x \ge 0, y \ge 0$ **b.** .3996 **c.** .5940
d. .3298

15. a. $F(y) = 1 - 2e^{-2\lambda y} + e^{-3\lambda y}$ for $y \ge 0$, $F(y) = 0$ for $y < 0$;
$f(y) = 4\lambda e^{-2\lambda y} - 3\lambda e^{-3\lambda y}$ for $y \ge 0$, $f(y) = 0$ for $y < 0$
b. $2/(3\lambda)$

17. a. .25 **b.** $1/\pi$ **c.** $2/\pi$
d. $f_X(x) = 2\sqrt{R^2 - x^2}/(\pi R^2)$ for $-R \le x \le R, f_Y(y) =$
$2\sqrt{R^2 - y^2}/(\pi R^2)$ for $-R \le y \le R$, no

19. .15

21. L^2

23. $\frac{1}{4}$ hour

25. $-\frac{2}{3}$

27. a. $-.1058$ **b.** $-.0128$

37. a. $f_X(x) = 2x, 0 < x < 1, f_X(x) = 0$ elsewhere
b. $f_{Y|X}(y|x) = 1/x, 0 < y < x < 1$
c. .6
d. No, the domain is not a rectangle
e. $E(Y|X = x) = x/2$, a linear function of x
f. $V(Y|X = x) = x^2/12$

39. a. $f_X(x) = 2e^{-2x}, 0 < x < \infty, f_X(x) = 0, x \le 0$
b. $f_{Y|X}(y|x) = e^{-y+x}, 0 < x < y < \infty$
c. $P(Y > 2|x = 1) = 1/e$
d. No, the domain is not rectangular
e. $E(Y|X = x) = x + 1$, a linear function of x
f. $V(Y|X = x) = 1$

41. a. $E(Y|X = x) = x/2$, a linear function of x; $V(Y|X = x) = x^2/12$
b. $f(x, y) = 1/x, 0 < y < x < 1$
c. $f_Y(y) = -\ln(y), 0 < y < 1$

43. a. $p_{Y|X}(0|1) = 4/17, p_{Y|X}(1|1) = 10/17, p_{Y|X}(2|1) = 3/17$
b. $p_{Y|X}(0|2) = .12, p_{Y|X}(1|2) = .28, p_{Y|X}(2|2) = .60$
c. .40
d. $p_{X|Y}(0|2) = 1/19, p_{X|Y}(1|2) = 3/19, p_{X|Y}(2|2) = 15/19$

45. a. $E(Y|X = x) = x^2/2$ **b.** $V(Y|X = x) = x^4/12$
c. $f_Y(y) = y^{-.5} - 1, 0 < y < 1$

47. a. $p(1, 1) = p(2, 2) = p(3, 3) = 1/9, p(2, 1) = p(3, 1) = p(3, 2) = 2/9$
b. $p_X(1) = 1/9, p_X(2) = 3/9, p_X(3) = 5/9$
c. $p_{Y|X}(1|1) = 1, p_{Y|X}(1|2) = 2/3, p_{Y|X}(2|2) = 1/3, p_{Y|X}(1|3) = .4, p_{Y|X}(2|3) = .4, p_{Y|X}(3|3) = .2$
d. $E(Y|X = 1) = 1, E(Y|X = 2) = 4/3, E(Y|X = 3) = 1.8$, no
e. $V(Y|X = 1) = 0, V(Y|X = 2) = 2/9, V(Y|X = 3) = .56$

49. a. $p_{X|Y}(1|1) = .2, p_{X|Y}(2|1) = .4, p_{X|Y}(3|1) = .4, p_{X|Y}(2|2) = 1/3, p_{X|Y}(3|2) = 2/3, p_{X|Y}(3|3) = 1$
b. $E(X|Y = 1) = 2.2, E(X|Y = 2) = 8/3, E(X|Y = 3) = 3$, no
c. $V(X|Y = 1) = .56, V(X|Y = 2) = 2/9, V(X|Y = 3) = 0$

51. a. $2x - 10$ **b.** 9 **c.** 3 **d.** .0228

53. a. $p_X(x) = .1, x = 0, 1, 2, \ldots, 9$; $p_{Y|X}(y|x) = \frac{1}{9}, y = 0, 1, 2, \ldots, 9, y \ne x$; $p_{X,Y}(x, y) = 1/90, x, y = 0, 1, 2, \ldots, 9, y \ne x$
b. $E(Y|X = x) = 5 - x/9, x = 0, 1, 2, \ldots, 9$, a linear function of x

55. a. $.6x, .24x$ **b.** 60 **c.** 60

57. a. .1410 **b.** .1165 With positive correlation, the deviations from their means of X and Y are likely to have the same sign.

59. a. If $U = X_1 + X_2$, $f_U(u) = u^2, 0 < u < 1$, $f_U(u) = 2u - u^2, 1 < u < 2, f_U(u) = 0$, elsewhere
b. If $V = X_2 - X_1$, $f_V(v) = 2 - 2v, 0 < v < 1$, $f_V(v) = 0$, elsewhere

61. $4y_3[\ln(y_3)]^2, 0 < y_3 < 1$

65. a. $g_5(y) = 5y^4/10^5, \frac{25}{3}$ **b.** $\frac{20}{3}$ **c.** 5 **d.** 1.409

67. $g_{Y_5|Y_1}(y_5|4) = \frac{2}{3}[(y_5 - 4)/6]^3, 4 < y_5 < 10$; 8.8

69. $1/(n + 1), 2/(n + 1), 3/(n + 1), \ldots, n/(n + 1)$

71. $\dfrac{\Gamma(n + 1)\Gamma(i + 1/\theta)}{\Gamma(i)\Gamma(n + 1 + 1/\theta)}, \dfrac{\Gamma(n + 1)\Gamma(i + 2/\theta)}{\Gamma(i)\Gamma(n + 1 + 2/\theta)} - \left[\dfrac{\Gamma(n + 1)\Gamma(i + 1/\theta)}{\Gamma(i)\Gamma(n + 1 + 1/\theta)}\right]^2$

73. $g_{3,8}(y_3, y_8) = (37{,}800/5^{10})y_3^2(y_8 - y_3)^4(5 - y_8)^2$, $0 < y_3 < y_8 < 5$

75. $g_{i,j}(y_i, y_j) = \dfrac{n!}{(i - 1)!(j - i - 1)!(n - j)!} \cdot F(y_i)^{i-1}[F(y_j) - F(y_i)]^{j-i-1}[1 - F(y_j)]^{n-j} f(y_i)f(y_j)$, $-\infty < y_i < y_j < \infty$

77. a. $f_{W_2}(w_2) = n(n - 1)\int_{-\infty}^{\infty}[F(w_1 + w_2) - F(w_1)]^{n-2} f(w_1)f(w_1 + w_2)dw_1$
b. $f_{W_2}(w_2) = n(n - 1)w_2^{n-2}(1 - w_2), 0 < w_2 < 1$

79. a. 3/81,250
b. $f_X(x) = \begin{cases} \int_{20-x}^{30-x} kxy\,dy = k(250x - 10x^2) & 0 \le x \le 20 \\ \int_{0}^{30-x} kxy\,dy = k(450x - 30x^2 + \frac{1}{2}x^3) & 20 \le x \le 30 \end{cases}$

$f_Y(y) = f_X(y)$; dependent
c. .3548 **d.** 25.969 **e.** $-32.19, -.894$
f. 7.651

81. $\frac{7}{6}$

85. c. If $p(0) = .3, p(1) = .5, p(2) = .2$, then 1 is the smaller of the two roots, so extinction is certain in this case with $\mu < 1$. If $p(0) = .2, p(1) = .5, p(2) = .3$, then $\frac{2}{3}$ is the smaller of the two roots, so extinction is not certain with $\mu > 1$.

87. a. $P[(X, Y) \in A] = F(b, d) - F(b, c) - F(a, d) + F(a, b)$
b. $P[(X, Y) \in A] = F(10, 6) - F(10, 1) - F(4, 6) + F(4, 1)$
$P[(X, Y) \in A] = F(b, d) - F(b, c - 1) - F(a - 1, d) + F(a - 1, b - 1)$
c. At each (x^*, y^*), $F(x^*, y^*)$ is the sum of the probabilities at points (x, y) such that $x \le x^*$ and $y \le y^*$. The table of $F(x, y)$ values is

		x	
		100	**250**
y	200	.50	1
	100	.30	.50
	0	.20	.25

d. $F(x, y) = .6x^2y + .4xy^3, 0 \le x \le 1; 0 \le y \le 1$; $F(x, y) = 0, x \le 0$; $F(x, y) = 0, y \le 0$;

$F(x, y) = .6x^2 + .4x,\ 0 \le x \le 1,\ y > 1;$
$F(x, y) = .6y + .4y^3,\ x > 1,\ 0 \le y \le 1;$
$F(x, y) = 1, x > 1, y > 1$
$P(.25 \le x \le .75, .25 \le y \le .75) = .23125$

e. $F(x, y) = 6x^2y^2, x + y \le 1, 0 \le x \le 1; 0 \le y \le 1, x \ge 0, y \ge 0$
$F(x, y) = 3x^4 - 8x^3 + 6x^2 + 3y^4 - 8y^3 + 6y^2 - 1, x + y > 1, x \le 1, y \le 1$
$F(x, y) = 0, x \le 0; F(x, y) = 0, y \le 0;$
$F(x, y) = 3x^4 - 8x^3 + 6x^2, 0 \le x \le 1, y > 1$
$F(x, y) = 3y^4 - 8y^3 + 6y^2, 0 \le y \le 1, x > 1$
$F(x, y) = 1, x > 1, y > 1$

89. **a.** $2x, x$ **b.** 40 **c.** .100

91. $M_W(t) = 2/[(1 - 1000t)(2 - 1000t)]$, 1500

Chapter 6

1. **a.**

$\bar{x}$	25	32.5	40	45	52.5	65
$p(\bar{x})$	.04	.20	.25	.12	.30	.09

$E(\bar{X}) = 44.5 = \mu$

b.

s^2	0	112.5	312.5	800
$p(s^2)$	.38	.20	.30	.12

$E(S^2) = 212.25 = \sigma^2$

3.

x/n	0	.1	.2	.3	.4
$p(x/n)$	0.0000	0.0000	0.0001	0.0008	0.0055

.5	.6	.7	.8	.9	1.0
0.0264	0.0881	0.2013	0.3020	0.2684	0.1074

5. **a.**

$\bar{x}$	1	1.5	2	2.5	3	3.5	4
$p(\bar{x})$	.16	.24	.25	.20	.10	.04	.01

b. $P(\bar{X} \le 2.5) = .85$

c.

r	0	1	2	3
$p(r)$	.30	.40	.22	.08

d. .24

7.

$\bar{x}$	$p(\bar{x})$	$\bar{x}$	$p(\bar{x})$	$\bar{x}$	$p(\bar{x})$
0.0	0.000045	1.4	0.090079	2.8	0.052077
0.2	0.000454	1.6	0.112599	3.0	0.034718
0.4	0.002270	1.8	0.125110	3.2	0.021699
0.6	0.007567	2.0	0.125110	3.4	0.012764
0.8	0.018917	2.2	0.113736	3.6	0.007091
1.0	0.037833	2.4	0.094780	3.8	0.003732
1.2	0.063055	2.6	0.072908	4.0	0.001866

11. **a.** 12, .01 **b.** 12, .005
c. With less variability, the second sample is more closely concentrated near 12.

13. **a.** .9876 **b.** .6915

15. **a.** .8366 **b.** No

17. 43.29

19. **a.** .9802, .4802 **b.** 32

21. **a.** .9839 **b.** .8932

27. **a.** 87,850, 19,100,116
b. In case of dependence, the mean calculation is still valid, but not the variance calculation.

29. **a.** .2871 **b.** .3695

31. .0317; Because each piece is played by the same musicians, there could easily be some dependence. If they perform the first piece slowly, then they might perform the second piece slowly, too.

33. **a.** 45 **b.** 68.33 **c.** −1, 13.67
d. −5, 68.33

35. **a.** 50, 10.308 **b.** .0076 **c.** 50
d. 111.56 **e.** 131.25

37. **a.** .9615 **b.** .0617

39. **a.** .5, $n(n + 1)/4$ **b.** .25, $n(n + 1)(2n + 1)/24$

41. 10:52.74

43. .48

45. **b.** $M_Y(t) = 1/[1 - t^2/(2n)]^n$

47. Because χ^2_ν is the sum of ν independent random variables, each distributed as χ^2_1, the Central Limit Theorem applies.

53. **a.** 3.2 **b.** 10.04, the square of the answer to (a)

57. **a.** $\nu_2/(\nu_2 - 2), \nu_2 > 2$
b. $2\nu_2^2(\nu_1 + \nu_2 - 2)/[\nu_1(\nu_2 - 2)^2(\nu_2 - 4)], \nu_2 > 4$

61. **a.** 4.32

65. **a.** The approximate value, .0228, is smaller because of skewness in the chi-squared distribution.
b. This approximation gives the answer .03237, agreeing with the software answer to this number of decimals.

67. No, the sum of the percentiles is not the same as the percentile of the sum, except that they are the same for the 50th percentile. For all other percentiles, the percentile of the sum is closer to the 50th percentile than is the sum of the percentiles.

69. **a.** 2360, 73.70 **b.** .9713

71. .9685

73. .9093 Independence is questionable because consumption one day might be related to consumption the next day.

75. .8340

77. **a.** $\rho = \sigma^2_W/(\sigma^2_W + \sigma^2_E)$ **b.** $\rho = .9999$

79. 26, 1.64

81. If Z_1 and Z_2 are independent standard normal observations, then let $X = 5Z_1 + 100$, $Y = 2(.5Z_1 + (\sqrt{3}/2)Z_2) + 50$.

Chapter 7

1. **a.** 113.73, $\bar{X}$ **b.** 113, $\tilde{X}$
c. 12.74, S, an estimator for the population standard deviation
d. The sample proportion of students exceeding 100 in IQ is $\frac{30}{33} = .91$.
e. .112, $S/\bar{X}$

3. a. 1.3481, $\overline{X}$ **b.** 1.3481, $\overline{X}$ **c.** 1.78, $\overline{X} + 1.282\, S$
d. .67 **e.** .0846

5. 1,703,000; 1,599,730; 1,601,438

7. a. 120.6 **b.** 1,206,000, 10,000 $\overline{X}$ **c.** .8
d. 120, $\tilde{X}$

9. a. $\overline{X}$, 2.113 **b.** $\sqrt{\lambda/n}$, .119

11. b. $\sqrt{p_1(1-p_1)/n_1 + p_2(1-p_2)/n_2}$
c. In part (b) replace p_1 with X_1/n_1 and replace p_2 with X_2/n_2.
d. −.245 **e.** .0411

15. a. $\hat{\theta} = \Sigma X_i^2/(2n)$ **b.** 74.505

17. b. $\frac{4}{9}$

19. a. $\hat{p} = 2\hat{\lambda} - .30 = .20$
c. $\hat{p} = (100\hat{\lambda} - 9)/70$

21. a. .15 **b.** Yes **c.** .4437

23. a. $\hat{\theta} = (2\bar{x} - 1)/(1 - \bar{x}) = 3$
b. $\hat{\theta} = [-n/\Sigma \ln(x_i)] - 1 = 3.12$

25. $\hat{p} = r/(r + x) = .15$ This is the number of successes over the number of trials, the same as the result in Exercise 21. It is not the same as the estimate of Exercise 17.

27. $\Phi[(113.4 - \bar{x})/\hat{\sigma}] = .54$
Here $\hat{\sigma} = s\sqrt{(n-1)/n}$ is the mle of σ.

29. a. $\hat{\theta} = \Sigma X_i^2/(2n) = 74.505$, the same as in Exercise 15

b. $\sqrt{2\hat{\theta}\ln(2)} = 10.16$

31. $\hat{\lambda} = -\ln(\hat{p})/24 = .0120$

33. No, statistician A does not have more information.

35. $\prod_{i=1}^{n} x_i,\ \Sigma_{i=1}^{n} x_i$

37. $I[.5\max(x_1, x_2, \ldots, x_n) \le \theta \le \min(x_1, x_2, \ldots, x_n)]$

39. a. $2X(n - X)/[n(n-1)]$

41. a. $\overline{X}$ **b.** $\Phi[(\overline{X} - c)/\sqrt{1 - 1/n}]$

43. a. $V(\tilde{\theta}) = \theta^2/[n(n+2)]$ **b.** θ^2/n
c. The variance in (a) is below the bound of (b), but the theorem does not apply because the domain is a function of the parameter.

45. a. $\bar{x}$ **b.** $N(\mu, \sigma^2/n)$
c. Yes, the variance is equal to the Cramér–Rao bound.
d. The answer in (b) shows that the asymptotic distribution of the theorem is actually exact here.

47. a. $2/\sigma^2$
b. The answer in (a) is different from the answer, $1/(2\sigma^4)$, to 46(a), so the information does depend on the parameterization.

49. $\hat{\lambda} = 6/(6t_6 - t_1 - \cdots - t_5) = 6/(x_1 + 2x_2 + \cdots + 6x_6) = .0436$, where $x_1 = t_1, x_2 = t_2 - t_1, \ldots, x_6 = t_6 - t_5$

51. $K = (n-1)/(n+1)$

53. 1.275, $s = 1.462$

55. b. No, $E(\hat{\sigma}^2) = \sigma^2/2$, so $2\hat{\sigma}^2$ is unbiased.

59. .416, .448

61. $\delta(X) = (-1)^X$, $\delta(200) = 1$, $\delta(199) = -1$

63. b. $\hat{\beta} = \Sigma x_i y_i / \Sigma x_i^2 = 30.040$, the estimated time per item; $\hat{\sigma}^2 = (1/n)\,\Sigma(y_i - \hat{\beta}x_i)^2 = 16.912$; $25\,\hat{\beta} = 751$

Chapter 8

1. a. 99.5% **b.** 85% **c.** 2.97 **d.** 1.15

3. a. A narrower interval has a lower probability.
b. No, μ is not random.
c. No, the interval refers to μ, not individual observations.
d. No, a probability of .95 does not guarantee 95 successes in 100 trials.

5. a. (4.52, 5.18) **b.** (4.12, 5.00) **c.** 55 **d.** 94

7. Increase n by a factor of 4. Decrease the width by a factor of 5.

9. a. $(\bar{x} - 1.645\sigma/\sqrt{n}, \infty)$; $(4.57, \infty)$
b. $(\bar{x} - z_\alpha \cdot \sigma/\sqrt{n}, \infty)$
c. $(-\infty, \bar{x} + z_\alpha \cdot \sigma/\sqrt{n})$; $(-\infty, 59.7)$

11. 950; .8724 (normal approximation), .8731 (binomial)

13. a. (.99, 1.07) **b.** 158

15. a. 80% **b.** 98% **c.** 75%

17. .06, which is positive, suggesting that the population mean change is positive

19. (.513, .615)

21. .218

23. a. (.439, .814) **b.** 664

25. a. 381 **b.** 339

29. a. 1.341 **b.** 1.753 **c.** 1.708 **d.** 1.684
e. 2.704

31. a. 2.228 **b.** 2.131 **c.** 2.947 **d.** 4.604
e. 2.492 **f.** 2.715

33. b. (38.081, 38.439) **c.** (100.55, 101.19), yes

35. a. Assuming normality, a 95% lower confidence bound is 8.11. When the bound is calculated from repeated independent samples, roughly 95% of such bounds should be below the population mean.
b. A 95% lower prediction bound is 7.03. When the bound is calculated from repeated independent samples, roughly 95% of such bounds should be below the value of an independent observation.

37. a. 378.85 **b.** 413.09 **c.** (333.88, 407.50)

39. a. 95% prediction interval: (.0498, .0772)
b. 99% tolerance interval with $k = .95$: (.0442, .0828)

41. a. (169.36, 179.37)
b. (134.30, 214.43), which includes 152
c. The second interval is much wider, because it allows for the variability of a single observation.
d. The normal probability plot gives no reason to doubt normality. This is especially important for part (b), but the

large sample size implies that normality is not so critical for (a).

45. a. 18.307 **b.** 3.940 **c.** .95 **d.** .10

47. b. (1.65, 5.60)

49. a. (7.91, 12.00)
b. Because of an outlier, normality is questionable for this data set.
c. In MINITAB, put the data in C1 and execute the following macro 1000 times:

```
Let k3 = N(c1)
sample k3 c1 c3;
replace.
let k1 = mean(c3)
stack k1 c5 c5
end
```

51. a. (26.61, 32.94)
b. Because of outliers, the weight gains do not seem normally distributed.
c. In MINITAB, see Exercise 49(c).

53. a. (38.46, 38.84)
b. Although the normal probability plot is not perfectly straight, there is not enough deviation to reject normality.
c. In MINITAB, see Exercise 49(c).

55. a. (169.13, 205.43)
b. Because of an outlier, normality is questionable for this data set.
c. In MINITAB, see Exercise 49(c).

57. a. In MINITAB, put the data in C1 and execute the following macro 1000 times:

```
Let k3 = N(c1)
sample k3 c1 c3;
replace.
let k1 = stdev(c3)
stack k1 c5 c5
end
```

b. Assuming normality, a 95% confidence interval for s is (3.541, 6.578), but the interval is inappropriate because the normality assumption is clearly not satisfied.

59. a. (.198, .230) **b.** .048
c. A 90% prediction interval is (.149, .279).

61. 246

63. a. A 95% confidence interval for the mean is (.163, .174). Yes, this interval is below the interval for 59(a).
b. (.089, .326)

65. (0.1263, 0.3018)

67. a. Yes **b.** (196.88, 222.62)
c. (139.63, 279.87)

69. c. $V(\hat{\beta}) = \sigma^2/\sum x_i^2$, $\sigma_{\hat{\beta}} = \sigma/\sqrt{\sum x_i^2}$
d. Put the x_i's far from 0 to minimize $\sigma_{\hat{\beta}}$.
e. $\hat{\beta} \pm t_{\alpha/2,n-1}s/\sqrt{\sum x_i^2}$, (29.93, 30.15)

73. a. .00985 **b.** .0578

75. a. $(\bar{x} - (s/\sqrt{n})t_{.025,n-1,\delta}, \bar{x} - (s/\sqrt{n})t_{.975,n-1,\delta})$
b. (3.01, 4.46)

77. a. $1/2^n$ **b.** $n/2^n$ **c.** $(n+1)/2^n$, $1 - (n+1)/2^{n-1}$, (29.9, 39.3) with confidence level 97.85%

79. a. $P(A_1 \cap A_2) = .95^2$ **b.** $P(A_1 \cap A_2) \geq .90$
c. $P(A_1 \cap A_2) \geq 1 - \alpha_1 - \alpha_2$; $P(A_1 \cap \cdots \cap A_k) \geq 1 - \alpha_1 - \alpha_2 - \cdots - \alpha_k$

Chapter 9

1. a. Yes **b.** No **c.** No **d.** Yes
e. No **f.** Yes

5. H_0: $\sigma = .05$ vs. H_a: $\sigma < .05$. Type I error: Conclude that the standard deviation is less than .05 mm when it is really equal to .05 mm. Type II error: Conclude that the standard deviation is .05 mm when it is really less than .05.

7. A type I error here involves saying that the plant is not in compliance when in fact it is. A type II error occurs when we conclude that the plant is in compliance when in fact it isn't. A government regulator might regard the type II error as being more serious.

9. a. R_1
b. A type I error involves saying that the two companies are not equally favored when they are. A type II error involves saying that the two companies are equally favored when they are not.
c. Binomial, $n = 25$, $p = .5$; .0433
d. .3, .4881; .4, .8452; .6, .8452; .7, .4881
e. If only 6 favor the first company, then reject the null hypothesis and conclude that the first company is not preferred.

11. a. H_0: $\mu = 10$ vs. H_a: $\mu \neq 10$ **b.** .0099
c. .5319, .0076 **d.** $c = 2.58$ **e.** $c = 1.96$
f. $\bar{x} = 10.02$, so do not reject H_0
g. Recalibrate if $z \leq -2.58$ or $z \geq 2.58$

13. b. .00043, .0000075, less than .01

15. a. .0301 **b.** .0030 **c.** .0040

17. a. Because $z = 2.56 \geq 2.33$, reject H_0.
b. .84 **c.** 142 **d.** .0052

19. a. Because $z = -2.27 > -2.58$, do not reject H_0.
b. .22 **c.** 22

21. Test H_0: $\mu = .5$ vs. H_a: $\mu \neq .5$.
a. Do not reject H_0 because $t_{.025,12} = 2.179 > |1.6|$.
b. Do not reject H_0 because $t_{.025,12} = 2.179 > |-1.6|$.
c. Do not reject H_0 because $t_{.005,24} = 2.797 > |-2.6|$.
d. Reject H_0 because $t_{.005,24} = 2.797 < |-3.9|$.

23. Because $t = 2.24 \geq 1.708 = t_{.05,25}$, reject H_0: $\mu = 360$. Yes, this suggests contradiction of prior belief.

25. Because $|z| = 3.37 \geq 1.96$, reject the null hypothesis. It appears that average IQ in this population exceeds the national average.

27. a. No, $t = -.02$ **b.** .58
c. $n = 20$ total observations

29. a. Because $t = .50 < 1.89 = t_{.05,7}$ do not reject H_0.
b. .73

31. Because $t = -1.24 > -1.40 = -t_{.10,8}$, we do not have evidence to question the prior belief.

35. a. The distribution is fairly symmetric, without outliers.
b. Because $t = 4.25 > 3.50 = t_{.005,7}$, there is strong evidence to say that the amount poured differs from the

industry standard, and indeed bartenders tend to exceed the standard.

c. Yes, the test in (b) depends on normality, and a normal probability plot gives no reason to doubt the assumption.
d. For a confidence level of 95%, an interval for capturing at least 95% of the amounts is (1.03, 2.60).
e. .643, .185, .016

37. a. Do not reject H_0: $p = .10$ in favor of H_a: $p > .10$ because 16 or more blistered plates would be required for rejection at the .05 level. Because the null hypothesis is not rejected, there could be a type II error.
b. The probability of 15 or fewer blistered plates is .57 if $p = .15$. If $n = 200$, reject when 28 or more are blistered, and the probability of 27 or less is .32 when $p = .15$.
c. 362

39. a. Do not reject H_0: $p = .02$ in favor of H_a: $p < .02$ because 12 or fewer misshelved or unlocatable would be required to reject at the .05 level. There is no strong evidence suggesting that the inventory be postponed.
b. If $p = .01$, the probability of 13 or more misshelved or unlocatable is .21.
c. If $p = .05$, the probability of 12 or fewer misshelved or unlocatable is .00000000006.

41. a. At the .01 level, reject H_0 if the number of qualifiers is 12 or less or if the number is 39 or more. With 40 qualifiers, reject H_0.
b. With 10% qualifiers in the population, the probability of having between 13 and 38 qualifiers in the sample is .039.

43. Using $n = 25$, the probability of 5 or more leaky faucets is .0980 if $p = .10$, and the probability of 4 or fewer leaky faucets is .0905 if $p = .3$. Thus, the rejection region is 5 or more, $\alpha = .0980$, and $\beta = .0905$.

45. a. Reject **b.** Reject **c.** Do not reject
d. Reject **e.** Do not reject

47. a. .0778 **b.** .1841 **c.** .0250 **d.** .0066
e. .5438

49. a. $P = .0403$ **b.** $P = .0176$ **c.** $P = .1304$
d. $P = .6532$ **e.** $P = .0021$ **f.** $P = .000022$

51. Based on the given data, there is no reason to believe that pregnant women differ from others in terms of average serum receptor concentration.

53. a. Because the P-value is .21, no modification is indicated.
b. .9933

55. Because $t = -1.759$ and the P-value $= .089$, which is less than .10, reject H_0:$\mu = 3.0$ against a two-tailed alternative at the 10% level. However, the P-value exceeds .05, so do not reject H_0 at the 5% level. There is just a weak indication that the percentage is not equal to 3% (lower than 3%).

57. a. Test H_0: $\mu = 10$ vs. H_a: $\mu < 10$.
b. Because the P-value is $.017 < .05$, reject H_0, suggesting that the pens do not meet specifications.
c. Because the P-value is $.045 > .01$, do not reject H_0, suggesting there is no reason to say the lifetime is inadequate.
d. Because the P-value is .0011, reject H_0. There is good evidence showing that the pens do not meet specifications.

59. Reject H_0 for $\alpha = .10$, not .05 or .01.

61. a. .98, .85, .43, .004, .0000002
b. .40, .11, .0062, .0000003
c. Because H_0 will be rejected with high probability, even with only slight departure from H_0, it is not very useful to do a .01 level test.

63. b. 36.614 **c.** Yes

65. a. $\sum x_i \ge c$ **b.** Yes

67. Yes, the test is UMP for the alternative H_a: $\theta > .5$ because the tests for H_0: $\theta = .5$ vs. H_a: $\theta = p_0$ all have the same form for $p_0 > .5$.

69. b. .05
c. .04345, .05826; Because $.04345 < .05$, the test is not unbiased.
d. .05114; not most powerful

71. b. The value of the test statistic is 3.041, so the P-value is .081, compared to .089 for Exercise 55.

73. A sample size of 32 should suffice.

75. a. Test H_0: $\mu = 2150$ vs. H_a: $\mu > 2150$.
b. $t = (\bar{x} - 2150)/(s/\sqrt{n})$ **c.** 1.33 **d.** .101
e. Do not reject H_0 at the .05 level.

77. Because $t = .77$ and the P-value is .23, there is no evidence suggesting that coal increases the mean heat flux.

79. Conclude that activation time is too slow at the .05 level, but not at the .01 level.

81. A normal probability plot gives no reason to doubt the normality assumption. Because the sample mean is 9.815, giving $t = 4.75$ and an upper-tail P-value of .00007, reject the null hypothesis at any reasonable level. The true average flame time is too high.

83. Assuming normality, calculate $t = 1.70$, which gives a two-tailed P-value of .102. Do not reject the null hypothesis H_0: $\mu = 1.75$.

85. The P-value for a lower-tail test is .0014, so it is reasonable to reject the idea that $p = .75$ and conclude that fewer than 75% of mechanics can identify the problem.

87. Because $t = 6.43$, giving an upper-tail P-value of .0000002, and conclude that the population mean time exceeds 15 minutes.

89. Because the P-value is $.013 > .01$, do not reject the null hypothesis at the .01 level.

91. a. For the test of H_0: $\mu = \mu_0$ vs. H_a: $\mu > \mu_0$ at level α, reject H_0 if $2\sum x_i/\mu_0 \ge \chi^2_{\alpha,2n}$.
For the test of H_0: $\mu = \mu_0$ vs. H_a: $\mu < \mu_0$ at level α, reject H_0 if $2\sum x_i/\mu_0 \le \chi^2_{1-\alpha,2n}$.
For the test of H_0: $\mu = \mu_0$ vs. H_a: $\mu \ne \mu_0$ at level α, reject H_0 if $2\sum x_i/\mu_0 \ge \chi^2_{\alpha/2,2n}$ or if $2\sum x_i/\mu_0 \le \chi^2_{1-\alpha/2,2n}$.
b. Because $\sum x_i = 737$, the value of the test statistic is $2\sum x_i/\mu_0 = 19.65$, which gives a P-value of .52. There is no reason to reject the null hypothesis.

93. a. Yes

Chapter 10

1. a. $-.4$; it doesn't **b.** .0724, .269
c. Although the CLT implies that the distribution will be approximately normal when the sample sizes are each 100, the distribution will not necessarily be normal when the sample sizes are each 10.

3. Do not reject H_0 because $z = 1.76 < 2.33$.

5. a. H_a says that the average calorie output for sufferers is more than 1 cal/cm^2/min below that for nonsufferers. Reject H_0 in favor of H_a because $z = -2.90 \le -2.33$.
b. .0019 **c.** .819 **d.** 66

7. Yes, because $z = 1.83 \ge 1.645$.

9. a. $\bar{x} - \bar{y} = 6.2$
b. $z = 1.14$, two-tailed P-value $= .25$, so do not reject the null hypothesis that the population means are equal.
c. No, the values are positive and the standard deviation exceeds the mean.
d. 95% CI: (10.0, 29.8)

11. A 95% CI is (.99, 2.41).

13. 22. No.

15. b. It increases.

17. Because $z = 1.36$, there is no reason to reject the hypothesis of equal population means ($P = .17$).

19. Because $z = .59$, there is no reason to conclude that the population mean is higher for the no-involvement group ($P = .28$).

21. Because $t = -3.35 \le -3.30 = t_{.001,42}$, yes, there is evidence that experts do hit harder.

23. b. No **c.** Because $|t| = |-.38| < 2.23 = t_{.025,10}$, no, there is no evidence of a difference.

25. Because the one-tailed P-value is $.0004 < .01$, conclude at the .01 level that the difference is as stated. This could result in a type I error.

27. Yes, because $t = 2.08$ with P-value $= .046$.

29. b. (127.6, 202.0) **c.** 131.8

31. Because $t = 1.82$ with P-value $.046 < .05$, conclude at the .05 level that the difference exceeds 1.

33. a. $(\bar{x} - \bar{y}) \pm t_{\alpha/2,m+n-2} \cdot s_p\sqrt{\dfrac{1}{m} + \dfrac{1}{n}}$ **b.** (−.24, 3.64)
c. (−.34, 3.74), which is wider because of the loss of a degree of freedom

35. a. The slender distribution appears to have a lower mean and lower variance.
b. With $t = 1.88$ and a P-value of .097, there is no significant difference at the .05 level.

37. With $t = 2.19$ and a two-tailed P-value of .031, there is a significant difference at the .05 level but not the .01 level.

39. The second fabric has an outlying difference. With the outlier, we have $t = 1.73$, P-value .064, so there is no significant difference at the .01 level. Without the outlier, we have $t = 2.08$, P-value .041, so there is still no significant difference at the .01 level.

41. b. The 95% confidence interval for the difference of means is (−3.17, −.43), which has only negative values. This suggests that, on the average, there is more sleep with a drug.

43. With $t = 1.87$ and a P-value of .049, the difference is (barely) significantly greater than 5 at the .05 level.

45. a. No **b.** −49.1 **c.** 49.1

47.

	1	2	3	4
x	10	20	30	40
y	11	21	31	41

49. a. Because $|z| = |-4.84| \ge 1.96$, conclude that there is a difference. Rural residents are more favorable to the increase.
b. .9967

51. A 95% CI is (.016, .171)

53. Because $z = 4.27$ with P-value .000010, conclude that the radiation is beneficial.

55. a. H_0: $p_3 = p_2$, H_a: $p_3 > p_2$ **b.** $(X_3 - X_2)/n$
c. $Z = (X_3 - X_2)/\sqrt{X_2 + X_3}$
d. With $z = 2.67$, $P = .004$, reject H_0 at the .01 level.

57. 769

59. Because $z = 3.14$ with $P = .002$, reject H_0 at the .01 level. Conclude that lefties are more accident-prone.

61. a. .0175 **b.** .1642 **c.** .0200 **d.** .0448
e. .0035

63. No, because $f = 1.814 < 6.72 = F_{.01,9,7}$.

65. Because $f = 1.2219$ with $P = .505$, there is no reason to question the equality of population variances.

67. 8.10

69. a. (.158, .735)
b. Here is a macro that can be executed 1000 times in MINITAB:

```
# start with X in C1, Y in C2
let k3=N(c1)
let k4=N(c2)
sample k3 c1 c3;
replace.
sample k4 c2 c4;
replace.
let k1=mean(c3) −mean(c4)
stack k1 c5 c5
end
```

71. a. Here is a macro that can be executed 1000 times in MINITAB:

```
# start with X in C1, Y in C2
let k3=N(c1)
let k4=N(c2)
sample k3 c1 c3;
replace.
sample k4 c2 c4;
replace.
let k2=medi(c3) −medi(c4)
stack k2 c6 c6
end
```

73. a. (.593, 1.246)
b. Here is a macro that can be executed 1000 times in MINITAB:

```
# start with X in C1, Y in C2
let k3=N(c1)
let k4=N(c2)
sample k3 c1 c3;
```

```
replace.
sample k4 c2 c4;
replace.
let k5=stdev(c3)/stdev(c4)
stack k5 c12 c12
end
```

75. **a.** Because $t = -2.62$ with a P-value of .018, conclude that the population means differ. At the 5% level, blueberries are significantly better.
b. Here is a macro that can be executed repeatedly in MINITAB:

```
# start with data in C1, group var
in C2
let k3=N(c1)
Sample k3 c1 c3.
unstack c3 c4 c5;
subs c2.
let k9=mean(c4) −mean(c5)
stack k9 c6 c6
end
```

77. **a.** Because $f = 4.46$ with a two-tailed P-value of .122, there is no evidence of unequal population variances.
b. Here is a macro that can be executed repeatedly in MINITAB:

```
let k1 = n(C1)
Sample K1 c1 c3.
unstack c3 c4 c5;
subs c2.
let k6 = stdev(c4)/stdev(c5)
stack k6 c6 c6
end
```

79. **a.** A MINITAB macro is given in Answer 75(b).

81. **a.** $(-11.85, -6.40)$
b. See Exercise 57(a) in Chapter 8.

85. Significant difference at $\alpha = .05$, .01, and .001.

89. **b.** No, given that the 95% CI includes 0, the two-tailed test at the .05 level does not reject equality of means.

91. $(-299.2, 1517.8)$

93. (1020.2, 1339.9) Because 0 is not in the CI, we would reject equality of means at the .01 level.

95. Because $t = 2.61$ and the one-tailed P-value is .007, the difference is significant at the .05 level using either a one-tailed or a two-tailed test.

97. **a.** Because $t = 3.04$ and the two-tailed P-value is .008, the difference is significant at the .05 level.
b. No, the mean of the concentration distribution depends on both the mean and standard deviation of the log concentration distribution.

99. Because $t = 7.50$ and the one-tailed P-value is .0000001, the difference is highly significant, assuming normality.

101. The two-sample t is inappropriate for paired data. The paired t gives a mean difference .3, $t = 2.67$, and the two-tailed P-value is .045, so the means are significantly different at the .05 level. We are concluding tentatively that the label understates the alcohol percentage.

103. Because the paired $t = 3.88$ and the two-tailed P-value is .008, the difference is significant at the .05 and .01 levels, but not at the .001 level.

105. Because the $z = 2.63$ and the two-tailed P-value is .009, there is a significant difference at the .01 level, suggesting better survival at the higher temperature.

107. .902, .826, .029, .00000003

109. Because $z = 4.25$ and the one-tailed P-value is .00001, the difference is highly significant and companies do discriminate.

111. With $Z = (\bar{X} - \bar{Y})/\sqrt{\bar{X}/n + \bar{Y}/m}$, the result is $z = -5.33$, two-tailed P-value $= .0000001$, so one should conclude that there is a significant difference in parameters.

113. **b.** (1) not bioequivalent (2) not bioequivalent (3) bioequivalent

Chapter 11

1. **a.** Reject H_0: $\mu_1 = \mu_2 = \mu_3 = \mu_4 = \mu_5$ in favor of H_1: $\mu_1, \mu_2, \mu_3, \mu_4, \mu_5$ not all the same, because $f = 5.57 \geq 2.69 = F_{.05,4,30}$.
b. Using Table A.9, $.001 < P\text{-value} < .01$. (The P-value is .0018.)

3. Because $f = 6.43 \geq 2.95 = F_{.05,3,28}$, there are significant differences among the means.

5. Because $f = 10.85 \geq 4.38 = F_{.01,3,36}$, there are significant differences among the means.

Source	df	SS	MS	f	P
Formation	3	509.1	169.7	10.85	0.000
Error	36	563.1	15.6		
Total	39	1072.3			

7. **a.** The Levene test gives $f = 1.47$, P-value .236, so there is no reason to doubt equal variances.
b. Because $f = 10.48 \geq 4.02 = F_{.01,4,30}$, there are significant differences among the means.

Source	df	SS	MS	f	P
Plate length	4	43993	10998	10.48	0.000
Error	30	31475	1049		
Total	34	75468			

11. $w = 36.09$ 3 1 4   2 5

Splitting the paints into two groups, {3, 1, 4}, {2, 5}, there are no significant differences within groups but the paints in the first group differ significantly from those in the second group.

13.

3	1	4	2	5
427.5	462.0	469.3	502.8	532.1

15. $w = 5.92$; At the 1% level the only significant differences are between formation 4 and the first two formations.

2	1	3	4
24.69	26.08	29.95	33.84

17. (−.029, .379)

19. 426

21. **a.** Because $f = 22.60 \geq 3.26 = F_{.01,5,78}$, there are significant differences among the means.
b. (−99.1, −35.7), (29.4, 99.1)

23. The nonsignificant differences are indicated by the underscores.

10	6	3	1
45.5	50.85	55.40	58.28

25. **a.** Assume normality and equal variances.
b. Because $f = 1.71 < 2.20 = F_{.10,3,48}$, P-value = .18, there are no significant differences among the means.

27. **a.** Because $f = 3.75$, P-value = .028, there are significant differences among the means.
b. Because the normal plot looks fairly straight and the P-value for the Levene test is .68, there is no reason to doubt the assumptions of normality and constant variance.
c. The only significant pairwise difference is between brands 1 and 4:

4	3	2	1
5.82	6.35	7.50	8.27

31. .63

33. $\arcsin(\sqrt{x/n})$

35. **a.** Because $f = 1.55 < 3.26 = F_{.05,4,12}$, there are no significant differences among the means.
b. Because $f = 2.98 < 3.49 = F_{.05,3,12}$, there are no significant differences among the means.

37. With $f = 5.49 \geq 4.56 = F_{.01,5,15}$, there are significant differences among the stimulus means. Although not all differences are significant in the multiple comparisons analysis, the means for combined stimuli were higher. Differences among the subject means are not very important here. The normal plot of residuals shows no reason to doubt normality. However, the plot of residuals against the fitted values shows some dependence of the variance on the mean. If logged response is used in place of response, the plots look good and the F test result is similar but stronger. Furthermore, the logged response gives more significant differences in the multiple comparisons analysis. Means:

L1	L2	T	L1+L2	L1+T	L2+T
24.825	27.875	29.1	40.35	41.22	45.05

39. With $f = 2.56 < 2.61 = F_{.10,3,12}$, there are no significant differences among the angle means.

41. **a.** With $f = 1.04 < 3.28 = F_{.05,2,34}$, there are no significant differences among the treatment means.

Source	df	SS	MS	f
Treatment	2	28.78	14.39	1.04
Block	17	2977.67	175.16	12.68
Error	34	469.56	13.81	
Total	53	3476		

b. The very significant f for blocks, which shows that blocks differ strongly, implies that blocking was successful.

43. With $f = 8.69 \geq 6.01 = F_{.01,2,18}$, there are significant differences among the three treatment means.

The normal plot of residuals shows no reason to doubt normality, and the plot of residuals against the fitted values shows no reason to doubt constant variance. There is no significant difference between treatments B and C, but treatment A differs (it is lower) significantly from the others at the .01 level. Means:

A 29.49 B 31.31 C 31.40

45. Because $f = 8.87 \geq 7.01 = F_{.01,4,8}$, reject the hypothesis that the variance for B is 0.

49. **a.**

Source	df	SS	MS	f
A	2	30763	15381.5	3.79
B	3	34185.6	11395.2	2.81
Interaction	6	43581.2	7263.5	1.79
Error	24	97436.8	4059.9	
Total	35	3476		

b. Because $1.79 < 2.04 = F_{.10,6,24}$, there is no significant interaction.
c. Because $3.79 \geq 3.40 = F_{.05,2,24}$, there is a significant difference among the A means at the .05 level.
d. Because $2.81 < 3.01 = F_{.05,6,24}$, there is no significant difference among the B means at the .05 level.
e. Using $w = 64.93$,

3	1	2
3960.2	4010.88	4029.10

51. **a.** With $f = 1.55 < 2.81 = F_{.10,2,12}$, there is no significant interaction at the .10 level.
b. With $f = 376.27 \geq 18.64 = F_{.001,2,12}$, there is a significant difference between the formulation means at the .001 level.

With $f = 19.27 \geq 12.97 = F_{.001,1,12}$, there is a significant difference among the speed means at the .001 level.
c. Main effects: Formulation: (1) 11.19, (2) −11.19
Speed: (60) 1.99, (70) −5.03, (80) 3.04

53. Here is the ANOVA table:

Source	df	SS	MS	f	P
Pen	3	1387.5	462.50	0.68	0.583
Surface	2	2888.1	1444.04	2.11	0.164
Interaction	6	8100.3	1350.04	1.97	0.149
Error	12	8216.0	684.67		
Total	23	20591.8			

With $f = 1.97 < 2.33 = F_{.10,6,12}$, there is no significant interaction at the .10 level.

With $f = .68 < 2.61 = F_{.10,3,12}$, there is no significant difference among the pen means at the .10 level.

With $f = 2.11 < 2.81 = F_{.10,2,12}$, there is no significant difference among the surface means at the .10 level.

57. **a.** F = MSAB/MSE
b. A: F = MSA/MSAB B: F = MSB/MSAB

59. **a.** Because $f = 3.43 \geq 2.61 = F_{.05,4,40}$, there is a significant difference among the exam means at the .05 level.
b. Because $f = 1.65 < 2.61 = F_{.05,4,40}$, there is no significant difference among the retention means at the .05 level.

61. **a.**

Source	df	SS	MS	f
Diet	4	.929	.232	2.15
Error	25	2.690	.108	
Total	29	3.619		

Because $f = 2.15 < 2.76 = F_{.05,4,25}$, there is no significant difference among the diet means at the .05 level.
b. (−.59, .92); Yes, the interval includes 0.
c. .53

63. **a.** Test H_0: $\mu_1 = \mu_2 = \mu_3$ versus H_a: the three means are not all the same. With $f = 4.80$ and $F_{.05,2,16} = 3.63 < 4.80 < 6.23 = F_{.01,2,16}$, it follows that $.01 < P\text{-value} < .05$ (more precisely, $P = .023$). Reject H_0 in favor of H_a at the 5% level but not at the 1% level.
b. Only the first and third means differ significantly at the 5% level.

1	2	3
25.59	26.92	28.17

65. Because $f = 1123 \geq 4.07 = F_{.05,3,8}$, there are significant differences among the means at the .05 level. For Tukey multiple comparisons, $w = 7.12$:

PCM	OCM	RM	PIM
29.92	33.96	125.84	129.30

The means split into two groups of two. The means within each group do not differ significantly, but the means in the top group differ strongly from the means in the bottom group.

67. The normal plot is reasonably straight, so there is no reason to doubt the normality assumption.

69.

Source	df	SS	MS	f
A	1	322.667	322.667	980.5
B	3	35.623	11.874	36.1
AB	3	8.557	2.852	8.7
Error	16	5.266	.329	
Total	23	372.113		

With $f = 8.7 \geq 3.24 = F_{.05,3,16}$, there is significant interaction at the .05 level. In the presence of significant interaction, main effects are not very useful.

Chapter 12

1. **a.** Temperature:

```
17 | 0
17 | 23
17 | 445
17 | 67
17 |             Stem: hundreds and tens
18 | 0000011     Leaf: ones
18 | 2222
18 | 445
18 | 6
18 | 8
```

The distribution is fairly symmetric and bell-shaped with a center around 180.

Ratio:

```
0 | 889
1 | 0000
1 | 3
1 | 4444
1 | 66
1 | 8889     Stem: ones
2 | 11       Leaf: tenths
2 |
2 | 5
2 | 6
2 |
3 | 00
```

The distribution is concentrated between 1 and 2, with some positive skewness.
b. No, x does not determine y: For a given x there may be more than one y.
c. No, there is a wide range of y values for a given x; for example, when temperature is 18.2 the ratio ranges from .9 to 2.68.

3. Yes. Yes.

5. **b.** Yes
c. The relationship of y to x is roughly quadratic.

7. **a.** 5050 psi **b.** 1.3 psi **c.** 130 psi **d.** −130 psi

9. **a.** .095 m^3/min **b.** −.475 m^3/min
c. .83 m^3/min, 1.305 m^3/min **d.** .4207, .3446
e. .0036

11. **a.** −.01 hr, −.10 hr **b.** 3.0 hr, 2.5 hr
c. .3653 **d.** .4624

13. **a.** $y = .63 + .652x$ **b.** 23.46, −2.46
c. 392, 5.72 **d.** 95.6% **e.** $y = 2.29 + .564x$, $r^2 = .688$

15. **a.** $y = -15.2 + .0942x$ **b.** 1.906
c. −1.006, −0.096, 0.034, 0.774 **d.** 45.1%

17. **a.** Yes **b.** Slope, .827; intercept, −1.13
c. 40.22 **d.** 5.24 **e.** 97.5%

19. **a.** $y = -45.6 + 1.71x$ **b.** 339.5 **c.** -85.5
d. No, because 500 is well beyond the range of the data.
e. Yes, the predictor accounts for 96.1% of the variation in y.

21. **b.** $y = -2.18 + .660\,x$ **c.** 7.72 **d.** 7.72

25. $\hat{\beta}_0' = 1.8\hat{\beta}_0 + 32, \quad \hat{\beta}_1' = 1.8\hat{\beta}_1$

29. **a.** Subtracting $\bar{x}$ from each x_i shifts the plot $\bar{x}$ units to the left. The slope is left unchanged, but the new y intercept is $\bar{y}$, the height of the old line at $x = \bar{x}$.
b. $\hat{\beta}_0^* = \bar{Y} = \hat{\beta}_0 + \hat{\beta}_1\bar{x}$ and $\hat{\beta}_1^* = \hat{\beta}_1$

31. **a.** .00189 **b.** .7101
c. No, because here $\sum(x_i - \bar{x})^2$ is 24,750, smaller than the value 70,000 in part (a), so $V(\hat{\beta}_1) = \sigma^2/\sum(x_i - \bar{x})^2$ is higher here.

33. **a.** (.51, 1.40)
b. To test H_0: $\beta_1 = 1$ versus H_a: $\beta_1 < 1$, we compute $t = -.2258 > -1.38 = -t_{.10,9}$, so there is no reason to reject the null hypothesis, even at the 10% level. There is no conflict between the data and the assertion that the slope is at least 1.

35. **a.** $\hat{\beta}_1 = 1.536$, and a 95% CI is (.632, 2.440).
b. Yes, for the test of H_0: $\beta_1 = 0$ versus H_a: $\beta_1 \neq 0$, we find $t = 3.62$, with P-value .0025. At the .01 level, conclude that there is a useful linear relationship.
c. Because 5 is beyond the range of the data, predicting at a dose of 5 might involve too much extrapolation.
d. $\hat{\beta}_1 = 1.683$, and a 95% CI is (.531, 2.835). Eliminating the point causes only moderate change, so the point is not extremely influential.

37. **a.** Yes, for the test of H_0: $\beta_1 = 0$ versus H_a: $\beta_1 \neq 0$, we find $t = 17.11$, with P-value less than 10^{-9}. At the .01 level conclude that there is a useful linear relationship.
b. (14.94, 19.29)

43. No, $z = .73$ and the P-value is .46, so there is no evidence for a significant impact of age on kyphosis.

45. **a.** $s_{\hat{Y}}$ increases as the distance of x from $\bar{x}$ increases.
b. (2.26, 3.19) **c.** (1.34, 4.11) **d.** 90%

47. **a.** (.322, .488)
b. The width would be greater at 400 because $s_{\hat{Y}}$ increases as the distance of x from $\bar{x}$ increases.
c. (.00064, .00222)
d. For a two-tailed test, $|t| = |.256| < 2.201 = t_{.025,11}$, so there is no reason to challenge the belief.

49. (431.2, 628.6)

51. **a.** Yes, for the test of H_0: $\beta_1 = 0$ versus H_a: $\beta_1 \neq 0$, we find $t = 10.62$, with P-value .000014. At the .001 level conclude that there is a useful linear relationship.
b. (8.24, 12.96) With 95% confidence, when the flow rate is increased by 1 SCCM, the associated expected change in etch rate is in the interval.
c. (36.10, 40.41) This is fairly precise.
d. (31.86, 44.65) This is much less precise than the interval in (c).
e. Because 2.5 is closer to the mean, the intervals will be narrower.
f. Because 6 is outside the range of the data, it is unknown whether the regression will apply there.
g. Use a 99% CI at each value: (23.88, 31.43), (29.93, 35.98), (35.07, 41.45).

53. **a.** Yes
b. For the test of H_0: $\beta_1 = 0$ versus H_a: $\beta_1 \neq 0$, we find $t = -4.39$, with P-value $< .001$. At the .001 level conclude that there is a useful linear relationship.
c. A 95% CI is (403.6, 468.2)

57. **a.** $r = .923$, so x and y are strongly correlated.
b. Unaffected
c. Unaffected
d. The normal plots seem consistent with normality, but the scatter plot shows a slight curvature.
e. For the test of H_0: $\rho = 0$ versus H_a: $\rho \neq 0$, we find $t = 7.59$, with P-value .00002. At the .001 level conclude that there is a useful linear relationship.

59. **a.** For the test of H_0: $\rho = 0$ versus H_a: $\rho > 0$, we find $r = .760$, $t = 4.05$, with P-value $< .001$. At the .001 level conclude that there is a positive correlation.
b. Because $r^2 = .578$ we say that the regression accounts for 57.8% of the variation in endurance. This also applies to prediction of lactate level from endurance.

61. For the test of H_0: $\rho = 0$ versus H_a: $\rho \neq 0$, we find $r = .773$, $t = 2.44$, with P-value .072. At the .05 level conclude that there is not a significant correlation. With such a small sample size, a high r is needed for significance.

63. **a.** Reject the null hypothesis in favor of the alternative.
b. No, with a large sample size a small r can be significant.
c. Because $t = 2.200 \geq 1.96 = t_{.025,9998}$ the correlation is statistically (but not necessarily practically) significant at the .05 level.

67. **a.** .184, $-.238$, $-.426$
b. The mean that is subtracted is not the mean $\bar{x}_{1,n-1}$ of $x_1, x_2, \ldots, x_{n-1}$, or the mean $\bar{x}_{2,n}$ of $x_2, x_3, \ldots, x_n$. Also, the denominator of r_1 is not $\sqrt{\sum_1^{n-1}(x_i - \bar{x}_{1,n-1})^2}\sqrt{\sum_2^n(x_i - \bar{x}_{2,n})^2}$. However, if n is large then r_1 is approximately the same as the correlation. A similar relationship applies to r_2.
c. No
d. After performing one test at the .05 level, doing more tests raises the probability of at least one type I error to more than .05.

69. The plot shows no reasons for concern about using the simple linear regression model.

71. **a.** The simple linear regression model may not be a perfect fit because the plot shows some curvature.
b. The plot of standardized residuals is very similar to the residual plot. The normal probability plot gives no reason to doubt normality.

73. **a.** For the test of H_0: $\beta_1 = 0$ versus H_a: $\beta_1 \neq 0$, we find $t = 10.97$, with P-value .0004. At the .001 level conclude that there is a useful linear relationship.
b. The residual plot shows curvature, so the linear relationship of part (a) is questionable.
c. There are no extreme standardized residuals, and the plot of standardized residuals is similar to the plot of ordinary residuals.

75. The first data set seems appropriate for a straight-line model. The second data set shows a quadratic relationship, so the straight-line relationship is inappropriate. The third data set is linear except for an outlier, and removal of the outlier will allow a line to be fitted. The fourth data set has only two values of x, so there is no way to tell if the relationship is linear.

77. a. To test for lack of fit, we find $f = 3.30$, with 3 numerator df and 10 denominator df, so the P-value is .079. At the .05 level we cannot conclude that the relationship is poor.
b. The scatter plot shows that the relationship is not linear, in spite of (a). In this case, the plot is more sensitive than the test.

79. a. 77.3
b. 40.4
c. The coefficient β_3 is the difference in sales caused by the window, all other things being equal.

81. We find $f = 24.4 \geq 5.52 = F_{.001,6,30}$, so there is a significant relationship at the .001 level.

83. a. 48.31, 3.69
b. No, because the interaction term will change.
c. Yes, $f = 18.92$, P-value $< .0001$.
d. Yes, $t = 3.496$, P-value $= .003 < .01$
e. (21.6, 41.6)
f. There appear to be no problems with normality or curvature, but the variance may depend on x_1.

85. a. No
b. With $f = 5.03 \geq 3.69 = F_{.05,5,8}$, there is a significant relationship at the .05 level.
c. Yes, the individual hypotheses deal with the issue of whether an individual predictor can be deleted, not the effectiveness of the whole model.
d. 6.2, 3.3, (16.7, 31.9)
e. With $f = 3.44 < 4.07 = F_{.05,3,8}$, there is no reason to reject the null hypothesis, so the quadratic terms can be deleted.

87. a. The quadratic terms are important in providing a good fit to the data.
b. A 95% PI is (.560, .771).

89. a. $X = \begin{bmatrix} 1 & -1 & -1 \\ 1 & -1 & 1 \\ 1 & 1 & -1 \\ 1 & 1 & 1 \end{bmatrix}$ $\quad y = \begin{bmatrix} 1 \\ 1 \\ 0 \\ 4 \end{bmatrix}$,

$\begin{bmatrix} 4 & 0 & 0 \\ 0 & 4 & 0 \\ 0 & 0 & 4 \end{bmatrix} \hat{\beta} = \begin{bmatrix} 6 \\ 2 \\ 4 \end{bmatrix}$ $\quad$ **b.** $\hat{\beta} = \begin{bmatrix} 1.5 \\ .5 \\ 1 \end{bmatrix}$

c. $\hat{y} = \begin{bmatrix} 0 \\ 2 \\ 1 \\ 3 \end{bmatrix}$ $\quad y - \hat{y} = \begin{bmatrix} 1 \\ -1 \\ -1 \\ 1 \end{bmatrix}$

SSE = 4, MSE = 4
d. (−12.2, 13.2)
e. For the test of H_0: $\beta_1 = 0$ versus H_a: $\beta_1 \neq 0$, we find $|t| = .5 < t_{.025,1} = 12.7$, so do not reject H_0 at the .05 level. The x_1 term does not play a significant role.

f.

Source	df	SS	MS	f
Regression	2	5	2.5	0.625
Error	1	4	4.0	
Total	3	9		

With $f = .625 < 199.5 = F_{.05,2,1}$, there is no significant relationship at the .05 level. $R^2 = .56$.

91. $\hat{\beta}_0 = \bar{y}$, $\quad s = \sqrt{\Sigma(y_i - \bar{y})^2/(n-1)}$, $\quad c_{00} = 1/n$,
$\bar{y} \pm t_{.025,n-1} s/\sqrt{n}$

93. a. $\hat{\beta}_0 = \dfrac{1}{m+n}\Sigma_1^{m+n} y_i = \bar{y}$,

$\hat{\beta}_1 = \dfrac{1}{m}\Sigma_1^m y_i - \dfrac{1}{n}\Sigma_{m+1}^{m+n} y_i = \bar{y}_1 - \bar{y}_2$

b. $\hat{y}_i = \bar{y}_1$, $i = 1, \ldots, m$;
$\hat{y}_i = \bar{y}_2$, $i = m+1, \ldots, m+n$
$\text{SSE} = \Sigma_1^m (y_i - \bar{y}_1)^2 + \Sigma_{m+1}^{m+n}(y_i - \bar{y}_2)^2$, $\quad s = \sqrt{\text{SSE}/(m+n-2)}$, $\quad c_{11} = 4/(m+n)$

d. $\hat{\beta}_0 = 128.17$, $\hat{\beta}_1 = 14.33$, $\quad \hat{y}_i = 121$, $i = 1, 2, 3$;
$\hat{y}_i = 135.33$, $i = 4, 5, 6$
SSE = 116.67, $s = 5.4006$, $\quad c_{11} = \frac{2}{3}$
95% CI for β_1 (2.09, 26.58)

95. Residual = Dep Var − Predicted Value
Std Error Residual = [MSE − (Std Error Predict)^2]$^{.5}$
Student Residual = Residual/Std Error Residual

99. a. $h_{ij} = 1/n + (x_i - \bar{x})(x_j - \bar{x})/\Sigma(x_k - \bar{x})^2$
$V(\hat{Y}_i) = \sigma^2[1/n + (x_i - \bar{x})^2/\Sigma(x_k - \bar{x})^2]$
b. $V(Y_i - \hat{Y}_i) = \sigma^2[1 - 1/n - (x_i - \bar{x})^2/\Sigma(x_k - \bar{x})^2]$
c. The variance of predicted values is greater for an x that is farther from $\bar{x}$.
d. The variance of residuals is lower for an x that is farther from $\bar{x}$.
e. It is intuitive that the variance of prediction should be higher with increasing distance. However, points that are farther away tend to draw the line toward them, so the residual naturally has lower variance.

101. a. With $f = 12.04 \geq 9.55 = F_{.01,2,7}$, there is a significant relationship at the .01 level.

To test H_0: $\beta_1 = 0$ versus H_a: $\beta_1 \neq 0$, we find $|t| = 2.96 \geq t_{.025,7} = 2.36$, so reject H_0 at the .05 level. The foot term is needed.

To test H_0: $\beta_2 = 0$ versus H_a: $\beta_2 \neq 0$, we find $|t| = 0.02 < t_{.025,7} = 2.36$, so do not reject H_0 at the .05 level. The height term is not needed.
b. The highest leverage is .88 for the fifth point. The height for this student is given as 54 inches, too low to be correct for this group of students. Also this value differs by 8 inches from the wingspan, an extreme difference.
c. Point 1 has leverage .55, and this student has height 75, foot length 13, both quite high. Point 2 has leverage .31, and this student has height 66 and foot length 8.5, at the low end. Point 7 has leverage .31 and this student has both height and foot length at the high end.

d. Point 2 has the most extreme residual. This student has a height of 66 inches and a wingspan of 56 inches, a difference of 10 inches, so the extremely low wingspan is probably wrong.
e. For this data set it would make sense to eliminate points 2 and 5 because they seem to be wrong. However, outliers are not always mistakes and one needs to be careful about eliminating them.

103. **a.** 50.73% **b.** .7122
c. To test H_0: $\beta_1 = 0$ versus H_a: $\beta_1 \neq 0$, we have $t = 3.93$, with P-value .0013. At the .01 level conclude that there is a useful linear relationship.
d. (1.056, 1.275)
e. $\hat{y} = 1.014$ $y - \hat{y} = -.214$

105. -36.18, $(-64.43, -7.94)$

107. No, if the relationship of y to x is linear, then the relationship of y^2 to x is quadratic.

109. **a.** Yes **b.** $\hat{y} = 98.293$ $y - \hat{y} = .117$
c. $s = .155$ **d.** $R^2 = .794$
e. 95% CI for β_1: (.0613, .0901)
f. The new observation is an outlier, and has a major impact: The equation of the line changes from $y = 97.50 + .0757x$ to $y = 97.28 + .1603x$; s changes from .155 to .291; R^2 changes from .794 to .616.

111. **a.** The paired t procedure gives $t = 3.54$ with a two-tailed P-value of .002, so at the .01 level we reject the hypothesis of equal means.
b. The regression line is $y = 4.79 + .743\,x$, and the test of H_0: $\beta_1 = 0$ versus H_a: $\beta_1 \neq 0$ gives $t = 7.41$ with a P-value of <.000001, so there is a significant relationship. However, prediction is not perfect, with $R^2 = .753$, so one variable accounts for only 75% of the variability in the other.

115. **a.** Linear
b. After fitting a line to the data, the residuals show a lot of curvature.
c. Yes. The residuals from the logged model show some departure from linearity, but the fit is good in terms of $R^2 = .988$. We find $\hat{\alpha} = 411.98$, $\hat{\beta} = -.03333$.
d. (58.15, 104.18)

117. **a.** The plot suggests a quadratic model.
b. With $f = 25.08$ and a P-value of <.0001, there is a significant relationship at the .0001 level.
c. CI: (3282.3, 3581.3), PI: (2966.6, 3897.0). Of course, the PI is wider, as in simple linear regression, because it needs to include the variability of a new observation in addition to the variability of the mean.
d. CI: (3257.6, 3565.6), PI: (2945.0, 3878.2). These are slightly wider than the intervals in (c), which is appropriate, given that 25 is slightly closer to the mean and the vertex.
e. With $t = -6.73$ and a two-tailed P-value of <.0001, the quadratic term is significant at the .0001 level, so this term is definitely needed.

119. **a.** With $f = 2.4 < 5.86 = F$ with $\alpha = .05$ and degrees of freedom of 15 and 4, there is no significant relationship at the .05 level.
b. No, especially when k is large compared to n
c. .9565

Chapter 13

1. **a.** Reject H_0. **b.** Do not reject H_0.
c. Do not reject H_0. **d.** Do not reject H_0.

3. Do not reject H_0, because $\chi^2 = 1.57 < 7.815 = \chi^2_{.05,3}$.

5. Because $\chi^2 = 6.61$ with P-value .68, do not reject H_0.

7. Because $\chi^2 = 4.03$ with P-value >.10, do not reject H_0.

9. **a.** [0, .223), [.223, .510), [.510, .916), [.916, 1.609), [1.609, ∞)
b. Because $\chi^2 = 1.25$ with P-value >.10, do not reject H_0.

11. **a.** $(-\infty, -.967)$, $[-.967, -.431)$, $[-.431, 0)$, [0, .431), [.431, .967), [.967, ∞)
b. $(-\infty$, .49806), [.49806, .49914), [.49914, .50), [.50, .50086), [.50086, .50194), [.50194, ∞)
c. Because $\chi^2 = 5.53$ with P-value >.10, do not reject H_0.

13. Using $\hat{p} = .0843$, find $\chi^2 = 280.3$ with P-value < .001, so reject the independence model.

15. The likelihood is proportional to $\theta^{233}(1-\theta)^{367}$, from which $\hat{\theta} = .3883$. This gives estimated probabilities .1400, .3555, .3385, .1433, .0227 and expected counts 21.00, 53.32, 50.78, 21.49, 3.41. Because 3.41 < 5, combine the last two categories, giving $\chi^2 = 1.62$ with P-value > .10. Do not reject the binomial model.

17. $\hat{\lambda} = 3.167$ which gives $\chi^2 = 103.9$ with P-value < .001, so reject the assumption of a Poisson model.

19. $\hat{\theta}_1 = .4275$, $\hat{\theta}_2 = .2750$ which gives $\chi^2 = 29.3$ with P-value < .001, so reject the model.

21. Yes, the test gives no reason to reject the null hypothesis of a normal distribution.

23. The P-values are both .243.

25. Let p_{i1} = the probability that a fruit given treatment i matures and p_{i2} = the probability that a fruit given treatment i aborts, so H_0: $p_{i1} = p_{i2}$ for $i = 1, 2, 3, 4, 5$. We find $\chi^2 = 24.82$ with P-value < .001, so reject the null hypothesis and conclude that maturation is affected by leaf removal.

27. If p_{ij} denotes the probability of a type j response when treatment i is applied, then H_0: $p_{1j} = p_{2j} = p_{3j} = p_{4j}$ for $j = 1, 2, 3, 4$. With $\chi^2 = 27.66 \geq 23.587 = \chi^2_{.005,9}$, reject H_0 at the .005 level. The treatment does affect the response.

29. With $\chi^2 = 64.65 \geq 18.47 = \chi^2_{.001,4}$, reject H_0 at the .001 level. Political views are related to marijuana usage. In particular, liberals are more likely to be users.

31. Compute the expected counts by $\hat{e}_{ijk} = n\hat{p}_{ijk} = n\hat{p}_{i..}\hat{p}_{.j.}\hat{p}_{..k} = n\dfrac{n_{i..}}{n}\dfrac{n_{.j.}}{n}\dfrac{n_{..k}}{n}$. For the χ^2 statistic, df = 20.

33. **a.** With $\chi^2 = .681 < 4.605 = \chi^2_{.10,2}$, do not reject independence at the .10 level.
b. With $\chi^2 = 6.81 \geq 4.605 = \chi^2_{.10,2}$, reject independence at the .10 level.
c. 677

35. **a.** With $\chi^2 = 6.45$ and P-value .040, reject independence at the .05 level.
b. With $z = -2.29$ and P-value .022, reject independence at the .05 level.

c. Because the logistic regression takes into account the order in the professorial ranks, it should be more sensitive, so it should give a lower P-value.
d. There are few female professors but many female assistant professors, and the assistant professors will be the professors of the future.

37. With $\chi^2 = 13.005 \geq 9.21 = \chi^2_{.01,2}$, reject the null hypothesis of no effect at the .01 level. Oil does make a difference (more parasites).

39. a. H_0: The population proportion of Late Game Leader Wins is the same for all four sports; H_a: The proportion of Late Game Leader Wins is not the same for all four sports. With $\chi^2 = 10.518 \geq 7.815 = \chi^2_{.05,3}$, reject the null hypothesis at the .05 level. Sports differ in terms of coming from behind late in the game.
b. Yes (baseball)

41. With $\chi^2 = 197.6 \geq 16.8 = \chi^2_{.01,6}$, reject the null hypothesis at the .01 level. The aged are more likely to die in a chronic-care facility.

43. With $\chi^2 = .763 < 7.78 = \chi^2_{.10,4}$, do not reject the hypothesis of independence at the .10 level. There is no evidence that age influences the need for item pricing.

45. a. No, $\chi^2 = 9.02 \geq 7.815 = \chi^2_{.05,3}$.
b. With $\chi^2 = .157 < 6.251 = \chi^2_{.10,3}$, there is no reason to say the model does not fit.

47. a. H_0: $p_0 = p_1 = \cdots = p_9 = .10$ versus H_a: at least one $p_i \neq .10$, with df = 9.
b. H_0: $p_{ij} = .01$ for i and $j = 0, 1, 2, \ldots, 9$ versus H_a: at least one $p_{ij} \neq .01$, with df = 99.
c. No, there must be more observations than cells to do a valid chi-squared test.
d. The results give no reason to reject randomness.

Chapter 14

1. For a two-tailed test of H_0: $\mu = 100$ at level .05, we find that $s_+ = 27$, and because $14 < s_+ < 64$, we do not reject H_0.

3. For a two-tailed test of H_0: $\mu = 7.39$ at level .05, we find that $s_+ = 18$, and because s_+ does not satisfy $21 < s_+ < 84$, we reject H_0.

5. We form the difference and perform a two-tailed test of H_0: $\mu = 0$ at level .05. This gives $s_+ = 72$ and because it does not satisfy $14 < s_+ < 64$, we reject H_0 at the .05 level.

7. Because $s_+ = 162.5$ with P-value .044, reject H_0: $\mu = 75$ in favor of H_a: $\mu > 75$ at the .05 level.

9. With $w = 38$, reject H_0 at the .05 level because the critical region is $\{w \geq 36\}$.

11. Test H_0: $\mu_1 - \mu_2 = 1$ versus H_a: $\mu_1 - \mu_2 > 1$. After subtracting 1 from the original process measurements, we get $w = 65$. Do not reject H_0 because $w < 84$.

13. b. Test H_0: $\mu_1 - \mu_2 = 0$ versus H_a: $\mu_1 - \mu_2 < 0$. With a P-value of .002 we reject H_0 at the .01 level.

15. With $w = 135$, $z = 2.223$, and the approximate P-value is .026, so we would not reject the null hypothesis at the .01 level.

17. (11.15, 23.80)

19. (−.585, .025)

21. (16, 87)

29. a. (.4736, .6669) **b.** (.4736, .6669)

33. a. $105n - 22.5 \ln(19) < \Sigma x_i < 105n - 22.5 \ln(19)$
b. Reject H_0 when $n = 7$.
c. 11.9, 11.9, $\alpha = \beta$
d. 25

35. a.

$$\frac{2\ln[\beta/(1-\alpha)] + n\ln\{p_0(1-p_0)/[p_1(1-p_1)]\}}{\ln(p_1/p_0) + \ln[(1-p_0)/(1-p_1)]} < y < \frac{2\ln[(1-\beta)/\alpha] + \ln\{p_0(1-p_0)/[p_1(1-p_1)]\}}{\ln(p_1/p_0 + \ln[(1-p_0)/(1-p_1)]}$$

b. $.262n - 5.408 < y < .262n + 6.622$
c. y goes up or down by 1

37. For a two-tailed test at level .05, we find that $s_+ = 24$ and because $4 < s_+ < 32$, we do not reject the hypothesis of equal means.

39. a. $\alpha = .0207$; Bin(20, .5)
b. $c = 14$; Because $y = 12$, do not reject H_0.

41. With $k = 20.12 \geq 13.28 = \chi^2_{.01,4}$, reject the null hypothesis of equal means at the 1% level. Axial strength does seem to (as an increasing function) depend on plate length.

43. Because $f_r = 6.45 < 7.815 = \chi^2_{.05,3}$, do not reject the null hypothesis of equal emotion means at the 5% level.

45. Because $w' = 26 < 27$, do not reject the null hypothesis at the 5% level.

Index

T